The Atmosphere and Climate of Mars

Humanity has long been fascinated by the planet Mars. Was its climate ever conducive to life? What is the atmosphere like today, and why did it change so dramatically over time? Twelve spacecraft have successfully flown to Mars since the Viking mission of the 1970s and early 1980s. These orbiters, landers, and rovers have generated vast amounts of data that now span a Martian decade (~18 years). This new volume brings together the many new ideas about the atmosphere and climate system that have emerged, including the complex interplay of the volatile and dust cycles, the atmosphere–surface interactions that connect them over time, and the diversity of the planet's environment and its complex history. Including tutorials and explanations of complicated ideas, students, researchers, and non-specialists alike are able to use this resource to gain a thorough and up-to-date understanding of this most Earth-like of planetary neighbors.

ROBERT M. HABERLE is a senior scientist in the Space Science and Astrobiology Division at NASA Ames Research Center. His main research interests center around the atmosphere and climate of Mars: past, present, and future. He has been involved in multiple NASA missions to Mars including Pathfinder, Mars Global Surveyor, and the Mars Science Laboratory, and he has promoted and developed landed network mission concepts for atmospheric science.

R. TODD CLANCY is a senior scientist with the Space Science Institute of Boulder, Colorado, and his research has focused on observational studies of atmospheres of the Earth, Venus, and Mars.

FRANÇOIS FORGET is a CNRS senior scientist in Paris, where he studies the past and present climate of Mars. He has been heavily involved in the ESA missions Mars Express and Exomars 2016, and is a member of the NASA Mars Reconnaissance Orbiter (MRO) and Insight science teams.

MICHAEL D. SMITH is a senior scientist in the Planetary Systems Laboratory of NASA's Goddard Space Flight Center. His research interests include the meteorology and dynamics of planetary atmospheres, radiative transfer, and remote sensing techniques, and he has been an active participant for more than 20 years on the science teams of eight past, current, and future spacecraft missions to Mars.

RICHARD W. ZUREK is chief scientist for the Mars Program Office at the Jet Propulsion Laboratory (JPL), California Institute of Technology. He also serves as the project scientist for the Mars Reconnaissance Orbiter (MRO), and is involved in the development and implementation of new missions to Mars.

Cambridge Planetary Science

Series Editors:

Fran Bagenal, David Jewitt, Carl Murray, Jim Bell, Ralph Lorenz, Francis Nimmo, Sara Russell

Books in the series:

† Reissued as a paperback

THE ATMOSPHERE AND CLIMATE OF MARS

ROBERT M. HABERLE

NASA Ames Research Center, Moffett Field, California, USA

R. TODD CLANCY

Space Science Institute, Boulder, Colorado, USA

FRANÇOIS FORGET

Laboratoire de Météorologie Dynamique, CNRS, Université Pierre et Marie Curie, Paris, France

MICHAEL D. SMITH

NASA Goddard Space Flight Center, Greenbelt, Maryland, USA

RICHARD W. ZUREK

Jet Propulsion Laboratory, California Institute of Technology, Pasadena, California, USA

CAMBRIDGE
UNIVERSITY PRESS

CAMBRIDGE
UNIVERSITY PRESS

University Printing House, Cambridge CB2 8BS, United Kingdom

One Liberty Plaza, 20th Floor, New York, NY 10006, USA

477 Williamstown Road, Port Melbourne, VIC 3207, Australia

314–321, 3rd Floor, Plot 3, Splendor Forum, Jasola District Centre, New Delhi – 110025, India

79 Anson Road, #06-04/06, Singapore 079906

Cambridge University Press is part of the University of Cambridge.

It furthers the University's mission by disseminating knowledge in the pursuit of
education, learning, and research at the highest international levels of excellence.

www.cambridge.org
Information on this title: www.cambridge.org/9781107016187
DOI: 10.1017/9781139060172

First published 2017
Reprinted 2019

Printed in the United Kingdom by Print on Demand, World Wide

A catalogue record for this publication is available from the British Library.

Library of Congress Cataloging-in-Publication Data
Names: Haberle, Robert M. | Clancy, R. Todd. | Forget, François. |
Smith, Michael D., 1966– | Zurek, Richard W., 1947–
Title: The atmosphere and climate of Mars / [edited by] Robert M. Haberle,
NASA-Ames Research Center, R. Todd Clancy, Space Science Institute,
Boulder, Colorado, François Forget, Laboratoire de Météorologie
Dynamique, Paris, Michael D. Smith, NASA-Goddard Space Flight Center,
Richard W. Zurek, Jet Propulsion Laboratory, California.
Description: Cambridge: Cambridge University Press, 2017. |
Series: Cambridge planetary science; 18 |
Includes bibliographical references and index.
Identifiers: LCCN 2017001528 | ISBN 9781107016187 (hardback: alk. paper)
Subjects: LCSH: Mars (Planet) – Atmosphere. | Mars (Planet) – Climate. |
Planets – Atmospheres.
Classification: LCC QB643.A86 A86 2017 | DDC 551.50999/23–dc23
LC record available at https://lccn.loc.gov/2017001528

ISBN 978-1-107-01618-7 Hardback

DEDICATION

This book is dedicated to the memory of Conway B. Leovy (1933–2011), a true planetary scientist and believer in comparative study of any planet or moon known to have an atmosphere: Earth, Mars, Venus, Titan, and the Solar System's gas giants.

He was fortunate to finish his academic training as space probes were first being launched to the planets, and he was an intellectual leader on many Mars flight science teams, starting with the imaging teams of the Mariners 6 and 7 flyby missions, the Mariner 9 Orbiter, the Viking Lander Meteorology Teams, and the Mars Reconnaissance Orbiter Mars Climate Sounder investigation. His goal was to apply the rapidly developing theory of dynamical meteorology, including advances in numerical simulation, and the growing sophistication of atmospheric radiation models and input databases, to interpret the new data being returned from space. While at the Rand Corporation, Conway pioneered – with Yale Mintz at UCLA – the application of general circulation models to atmospheres other than Earth, but he did not stop there during a long tenure as a professor of atmospheric science and geophysics at the University of Washington, Seattle. His successful research on ozone depletion in the Earth's stratosphere, the effects of vast dust storms on the Mars atmosphere, and the momentum balance on slowly rotating Venus and Titan were just some of the activities for which he was recognized as a Fellow of the American Meteorological Society and as recipient of the Gerard P. Kuiper Prize of the AAS Division for Planetary Sciences.

Expert in many disciplines, Conway was an inspiration to his colleagues and to the many students he mentored (several of whom are authors contributing to this book). He was an avid environmentalist, a dedicated family man, and possessed a kindness, generosity, and modesty not normally found in such intellectual giants. We miss his presence still, but his inquisitive spirit and exceptional integrity continue to inspire us. If here, he would ask us what we had learned and what interesting questions remain. For Mars, this book is our answer.

CONTENTS

A color plate section can be found between pages 402 and 403

CONTRIBUTORS

STANISLAV BARABASH
Swedish Institute of Space Physics, Kiruna, Sweden

JEFFREY R. BARNES
College of Earth, Ocean, and Atmospheric Sciences, Oregon State University, Corvallis, Oregon, USA

JENNIFER BENSON
Jet Propulsion Laboratory, California Institute of Technology, Pasadena, California, USA

STEPHEN W. BOUGHER
Department of Atmospheric, Oceanic, and Space Sciences, University of Michigan, Ann Arbor, Michigan, USA

DAVID A. BRAIN
Laboratory for Atmospheric and Space Physics, University of Colorado, Boulder, Colorado, USA

SHANE BYRNE
Lunar and Planetary Laboratory, University of Arizona, Tucson, Arizona, USA

BRUCE A. CANTOR
Malin Space Science Systems, San Diego, California, USA

MICHAEL H. CARR
United States Geological Survey, Menlo Park, California, USA

DAVID C. CATLING
Department of Earth and Space Sciences, University of Washington, Seattle, Washington, USA

PHILIP R. CHRISTENSEN
School of Earth and Space Exploration, Arizona State University, Tempe, Arizona, USA

R. TODD CLANCY
Space Science Institute, Boulder, Colorado, USA

ANTHONY COLAPRETE
NASA Ames Research Center, Moffett Field, California, USA

FRANK DAERDEN
Royal Belgian Institute for Space Aeronomy, Brussels, Belgium

GREGORY T. DELORY
Space Sciences Laboratory, and Center for Integrative Planetary Sciences, University of California, Berkeley, California, USA

FIRDEVS DURU
Department of Physics and Astronomy, University of Iowa, Iowa City, Iowa, USA

THÉRÈSE ENCRENAZ
LESIA, CNRS, Observatoire de Paris, Meudon, France

ANNA FEDOROVA
Space Research Institute, Russian Academy of Sciences, Moscow, Russia

FRANÇOIS FORGET
Laboratoire de Météorologie Dynamique, CNRS, Université Pierre et Marie Curie, Paris, France

THIERRY FOUCHET
LESIA, CNRS, Observatoire de Paris, Meudon, France

JANE L. FOX
Department of Physics, Wright State University, Dayton, Ohio, USA

BORIS GALPERIN
College of Marine Science, University of South Florida, St Petersburg, Florida, USA

FRANCISCO GONZÁLEZ-GALINDO
Instituto de Astrofísica de Andalucía, IAA/CSIC, Granada, Spain

ROBERT M. HABERLE
NASA Ames Research Center, Moffett Field, California, USA

JAMES W. HEAD
Department of Earth, Environmental and Planetary Sciences, Brown University, Providence, Rhode Island, USA

BRUCE M. JAKOSKY
Laboratory for Atmospheric and Space Physics, University of Colorado, Boulder, Colorado, USA

PHILIP B. JAMES
Space Science Institute, Boulder, Colorado, USA

MELINDA A. KAHRE
NASA Ames Research Center, Moffett Field, California, USA

ARMIN KLEINBÖHL
Jet Propulsion Laboratory, California Institute of Technology, Pasadena, California, USA

VLADIMIR KRASNOPOLSKY
Institute for Astrophysics and Computational Sciences, Catholic University of America, Washington, DC, USA

YVES LANGEVIN
Institut d'Astrophysique Spatiale, CNRS, Université Paris-
Sud, Orsay, France

SØREN E. LARSEN
Risø National Laboratory, Technical University of Denmark,
Roskilde, Denmark

FRANCK LEFÈVRE
Laboratoire Atmosphères, Milieux, Observations Spatiales,
CNRS, Université Pierre et Marie Curie, Paris, France

MARK T. LEMMON
Texas A&M University, College Station, Texas, USA

STEPHEN R. LEWIS
Department of Physical Sciences, The Open University,
United Kingdom

MIGUEL LOPÉZ-VALVERDE
Instituto de Astrofísica de Andalucía, IAA/CSIC,
Granada, Spain

ANNI MÄÄTTÄNEN
Laboratoire Atmosphères, Milieux, Observations Spatiales,
CNRS, Paris, France

JEAN-BAPTISTE MADELEINE
Laboratoire de Météorologie Dynamique, IPSL, Université
Pierre et Marie Curie, Paris, France

MICHAEL T. MELLON
Southwest Research Institute, Boulder, Colorado, USA

TIMOTHY I. MICHAELS
SETI Institute, Mt. View, California, USA

MICHAEL A. MISCHNA
Jet Propulsion Laboratory, California Institute of Technology,
Pasadena, California, USA

RONAN MODOLO
Laboratoire Atmosphères, Milieux, Observations Spatiales,
CNRS, Guyancourt, France

FRANCK MONTMESSIN
Laboratoire Atmosphères, Milieux, Observations Spatiales,
CNRS, Guyancourt, France

JAMES R. MURPHY
Department of Astronomy, New Mexico State University, Las
Cruces, New Mexico, USA

CLAIRE E. NEWMAN
Ashima Research, Pasadena, California, USA

ARAKEL PETROSYAN
Space Research Institute, Russian Academy of Sciences,
Moscow, Russia

THOMAS H. PRETTYMAN
Planetary Science Institute, Tucson, Arizona, USA

SCOT C. R. RAFKIN
Southwest Research Institute, Boulder, Colorado, USA

PETER L. READ
Atmospheric, Oceanic and Planetary Physics Department,
University of Oxford, United Kingdom

NILTON RENNÓ
Department of Atmospheric, Oceanic, and Space Sciences,
University of Michigan, Ann Arbor, Michigan, USA

HANNU SAVIJÄRVI
Department of Physics, University of Helsinki, Helsinki,
Finland

NORBERT SCHÖRGHOFER
Institute for Astronomy, University of Hawaii, Honolulu,
Hawaii, USA

TERO SIILI
Earth Observation Division, Finnish Meteorological Institute,
Helsinki, Finland

CYRIL SIMON-WEDLUND
Department of Radio Science and Engineering, School of
Electrical Engineering, Aalto University, Espoo, Finland

MICHAEL D. SMITH
NASA Goddard Space Flight Center, Greenbelt,
Maryland, USA

AYMERIC SPIGA
Laboratoire de Météorologie Dynamique, CNRS, Université
Pierre et Marie Curie, Paris, France

TIMOTHY N. TITUS
Astrogeology Science Center, United States Geological
Survey, Flagstaff, Arizona, USA

ANTHONY TOIGO
Applied Physics Laboratory, Johns Hopkins University,
Laurel, Maryland, USA

LUIS VÁZQUEZ
Departamento de Matemática Aplicada, Facultad de
Informática, Universidad Complutense de Madrid,
Madrid, Spain

R. JOHN WILSON
NASA Ames Research Center, Moffett Field, California, USA.

PAUL WITHERS
Astronomy Department, Boston University, Boston,
Massachusetts, USA

MICHAEL J. WOLFF
Space Science Institute, Boulder, Colorado, USA

KEVIN J. ZAHNLE
NASA Ames Research Center, Moffett Field, California, USA

RICHARD W. ZUREK
Jet Propulsion Laboratory, California Institute of Technology,
Pasadena, California, USA

GENERAL ACKNOWLEDGMENTS

The editors gratefully acknowledge support for this project from NASA's Planetary Science Division, the Jet Propulsion Laboratory/California Institute of Technology, and the Centre National de la Recherche Scientifique.

Introduction

ROBERT M. HABERLE, R. TODD CLANCY, FRANÇOIS FORGET, MICHAEL D. SMITH, RICHARD W. ZUREK

In 1978, a workshop report entitled *Dynamics of Earth and Planetary Atmospheres* (NASA-JPL, 1978) made the point that, while study of the climate and meteorology of the Earth could proceed without studying other planets, comparative planetology could help identify and evaluate the many physical processes that interact to produce a planet's climate and contribute to its change. Planetary atmospheres other than that of Earth present an opportunity to test theories and general understanding of these processes and their interactions when observed in different environments with diverse forcing functions and boundary conditions. Mars has always played a prominent role in such comparative study with the Earth because the two planets share the fundamental properties of rapid rotation, a relatively thin atmosphere largely heated by radiative and convective exchange with the surface, and a seasonal progression of their climate. Mars also exhibits crucial differences that can test our theory and numerical modeling capabilities in meaningful ways. In fact, the application of state-of-the-art Earth general circulation models to simulate the Martian atmosphere started almost as soon as those were available for Earth.

Of course, Mars is a worthy object of study for other reasons. There is considerable morphologic and compositional evidence of a more Earth-like environment that had liquid water on its surface very early in Mars history. This was a time when life appears to have started on Earth and when life as we know it may have been most likely to originate on Mars. For more than a century, radical planetary climate change and the possible origin and evolution of life have been major themes in the exploration of Mars. Unlike the Earth, where plate tectonics has destroyed most of the rock record from that early age, rocks of equivalent age are still present on the surface of Mars today. Evidence for more recent climate change also exists, while the processes and volatile inventories of the current climate give us clues about the past. And of course, in a more practical vein, the inevitable journey of humans to Mars, the most hospitable of our planetary neighbors, requires a detailed knowledge of its current resources and environment. These too are clearly connected to the climate system. Therefore, Mars is a fascinating planet to study and its climate system is the link to some of its most fundamental mysteries.

Study of the Martian climate system requires the acquisition of data to define it, and the use of numerical models to interpret it. In 1992 the book *Mars* was published (Kieffer et al., 1992). Weighing in at 1302 pages of text (a third devoted to the current atmosphere and past climate change), the book was an ambitious attempt to cover all that was known, or suspected, of the Red Planet. The book summarized the data and supporting modeling from the first two waves of spacecraft (Mariner 4 through Mariner 9 and then Viking) and ground-based exploration of Mars and the interpretations of those data. Major themes in the 1992 book had to do with the seasonal cycles of carbon dioxide, water vapor, and dust, and their inter-annual variation. The relative contributions of surface and subsurface (regolith) reservoirs to these cycles and the relative contributions to the transport of mass, water, and dust by the various components of the general circulation – the Hadley-type circulations, atmospheric waves, condensation flow – were discussed extensively.

The book *Mars* was prepared in anticipation of a third wave of spacecraft exploration that was to begin with Mars Observer in 1992, ending a long hiatus that dated back to the last signals in 1981 from Viking Lander 1, the longest operating element of the Viking mission. A quarter of a century later, that third wave of exploration is well underway. It has returned enormous amounts of data about Mars, addressing fundamental questions in a wide range of scientific disciplines.

This book, *The Atmosphere and Climate of Mars*, reports the tremendous progress that has been made based on the data and analyses that this ongoing third wave of exploration has yielded thus far. However, as seen in the chapters in the current work, the framework in *Mars* of seasonal cycles and the nature of surface sources and sinks and of atmospheric transport between them remain useful constructs. In that sense, as well as building on the data of past exploration, *The Atmosphere and Climate of Mars* is a continuation of the atmospheric chapters in the 1992 book.

At the end of several chapters in *Mars*, the authors identified priorities for making progress. Two major needs were frequently cited. The first was to develop more sophisticated models of transport in the atmosphere and exchange with the surface and subsurface. These models would have better horizontal resolution to characterize atmospheric waves and topographic effects, many more vertical levels to cover the deep dust-driven circulations, and better physical parameterization of surface exchanges of mass, energy and volatiles. Much progress has been made in these areas. Advances in computing power and the application of many physical parameterizations developed for the study of Earth climate have addressed the first of these needs, although still more is needed for the study of atmosphere and climate on both Earth and Mars. Furthermore, application of these parameterizations to Mars is not always straightforward, and the right data are required to validate that they are being properly used.

The second major need was acquisition of extended, more detailed observations of the atmosphere. High on the list were extended climatological records of the present atmospheric temperature, column dust and water amounts and their vertical profiles, the compositions of surface seasonal ices, and the detailed character of the polar caps. One missing request in *Mars* was characterization of clouds and thin ice hazes, whose frequent presence and potency to affect the circulation through radiation was not then known. Given their formation, water could no longer be ignored as a driver of circulation and shaper of thermal structure, despite the relatively ineffective latent heating/cooling of water due to its small abundance. Despite whatever might have been hoped, Mars is not a simpler climate system – it is just different in key respects.

Post-1992 developments in instrumental capabilities and observing facilities (e.g. the Hubble Space Telescope and ground-based microwave instruments) provided new information during the hiatus of space missions to Mars, and their contribution continues today. The anticipated third wave of spacecraft exploration did not begin as expected, as Mars Observer failed to achieve orbit. In 1997 the long hiatus ended with the successful landing of Pathfinder, which carried, among other things, a modest meteorological payload. Then in 1998 systematic, global observations of the Martian surface and atmosphere began with the Mars Global Surveyor (MGS) Orbiter. Orbital observations were replenished at roughly four-year intervals with three more orbiters – NASA's Mars Odyssey (launched in 2001) and Mars Reconnaissance Orbiter (MRO, 2005), separated by the European Space Agency's (ESA) Mars Express (2003). As of early 2017, all these later orbiters continue to observe Mars, picking up from MGS, for which contact was lost in 2006. In October 2014 two new orbiters joined these missions, the NASA Mars Atmosphere and Volatile Evolution (MAVEN) mission, focused on processes of atmospheric escape driven by atmosphere–solar wind interactions, and the India Space Research Organization's Mars Orbiter Mission (MOM), with diverse measurements of the general Martian environment. And in 2018 two more missions with atmospheric science components, the NASA Insight Lander and the ESA–Russia ExoMars Trace Gas Orbiter, are scheduled to begin operations.

Thanks to the unprecedented longevity of MGS, Odyssey, Mars Express, and MRO, there is now almost two decades (nearly a single Martian decade) of atmospheric temperature and aerosol profiling, column water vapor and aerosol opacity, and daily global weather mosaics. We also have surface data from the meteorological/sounding packages on the MER and Curiosity Rovers, and the Pathfinder and Phoenix Landers. These data, together with detailed databases of global topography, surface properties (e.g. albedo, thermal inertia, roughness), and polar cap monitoring, form an extraordinary climate database for Mars, better than any other planet except Earth itself.

This is not to imply that the knowledge base is the same for Earth and Mars. The study of Mars necessarily tends to focus on the more global aspects of the planet and on the surviving evidence of the cumulative action of past processes. Humankind has robotically operated on the surface, comparable in area to the land area of the Earth, in only a few places (seven locales with immobile platforms or short-range rovers). Even though the data return from nine successful orbiters since 1971 is measured in terabits, a volume returned by each of several spacecraft currently observing Earth, we do not have the synoptic coverage of Earth-observing systems, nor is there the *in situ* data from airborne platforms, rockets, and radiosondes that regularly launch on Earth, nor an extensive surface network of meteorological stations making measurements throughout the day and night. Still, the Mars that has emerged from this latest stage of exploration is a planet that has changed dramatically over time and is still changing even today. Part of its allure is that it retains a physical record of much of that change.

This book provides a detailed look at the data from this latest wave of spacecraft and ground-based exploration of Mars and how it has advanced our understanding of the atmosphere and climate on Mars today. It therefore emphasizes what has been learned since the 1992 *Mars* book up through late 2015, and covers the entire atmosphere from the planetary boundary layer at the surface to the exobase at the top. While the emphasis is on the present climate system, the book also addresses how the climate has changed over time, including obliquity-driven climate change of the geologically recent past, and the ancient climate that likely was substantially different than what is observed today. Since processes control the nature of the atmosphere, this book also addresses the state of our understanding of radiation, transport, photochemistry, surface–atmosphere exchange, escape mechanisms, and circulation dynamics, as well as the status of our attempts to model them.

As with studies of Earth, the more extensive and detailed databases now in hand reveal that Mars – like Earth – is a diverse, complex planet. While many old mysteries have been explained, Mars still challenges us to think outside our "Earth-box" with regard to planetary climate and atmospheres. While Mars has played this role for more than a century, it remains a fascinating world in its own right, even as it teaches us about our own.

The Editors
June 2016

REFERENCES

Kieffer, H., Jakosky, B. M., Snyder, C. W., and Mathews, M. S. (Eds) (1992) *Mars* (Tucson, AZ: University of Arizona Press).

NASA-JPL (1978) *Dynamics of Earth and Planetary Atmospheres: A Brief Assessment of Our Present Understanding, Report of the Planetary Atmospheres Workshop*, July 10–16, 1977, Snowmass, Colorado, Contract NAS 7-100, National Aeronautics and Space Administration, JPL Publication 78-46.

Understanding Mars and Its Atmosphere

RICHARD W. ZUREK

2.1 IN THE BEGINNING

To millennia of naked-eye observers, Mars was just another of the "wanderers" in the night sky, varying in brightness as the months passed and distinguished by its reddish color. The development of good telescopes changed that.

By the late 18th century William Herschel (1784) could confidently say that Mars had an atmosphere. His observation that the recently discovered polar caps on Mars (Cassini, 1666) changed size with season, and the edges of the observed disk were not sharp, pointed to the existence of an atmosphere. That Mars had seasons was evident in Herschel's measurement of the axial tilt of Mars, which was remarkably similar to Earth's. The rotation rate of the planet was also very similar, having been well established by tracking Syrtis Major (Cassini, 1666), one of the darkest features on Mars and the first to have been confidently observed (C. Huygens was the first to draw it, in 1659). Early on, the dark areas were assumed to be seas and their names until recent times (e.g. Mare Cimmerium) reflected that early assumption.

Later observations of hazes (obscurations) and distinct clouds confirmed the atmosphere's presence, although the planet's low albedo suggested that there was less air than on Earth. With the advent of much improved photographic capabilities early in the 20th century, Mars was seen to be larger and fuzzier in blue filters than in red, and it was possible to distinguish reliably "white" clouds from "yellow" ones (e.g. Slipher, 1962; Martin et al., 1992). And yet clouds were sufficiently rare that their presence was worthy of note by observers. Viewing the planet's surface was difficult, not just due to the often great distance between Earth and Mars, but also because the Earth-based observer was looking up through Earth's atmosphere and down through that of Mars. Even so, a fascinating vision of our planetary neighbor was taking shape.

By the early 20th century, particularly in the perspective popularized by Percival Lowell (Lowell, 1895, 1896, 1906, 1908), Mars was an older Earth, its mountains worn down and much of its water lost to space or frozen in its crust (a dichotomy we investigate even today). The spidery network of canals drawn by Lowell appeared artificial and he took it as evidence of a race of intelligent beings struggling against a changing climate. Through global engineering, the Martians in his view were redistributing the precious remnant of the planet's water that melted seasonally at the poles to irrigate what otherwise was a desert planet. No mountains had been reported to bar their path. The atmosphere was there, but like the major deserts of the Earth, rain was rare, with most condensation coming as snow near the poles. Each spring, a wave of darkening (see Lowell, 1906) was reported to sweep down from the poles; this was the water coursing through the channels and canals towards the equator, nourishing and darkening what were then regarded as vast regions of vegetation. Except for the larger scale, this was not unlike the irrigated desert in the American southwest where Lowell had built his observatory to view Mars.

It was suspected that Mars, being a smaller planet, would have a less dense atmosphere than Earth. This thinner atmosphere, together with the planet's greater distance from the Sun, meant that the ground and atmosphere would be colder, but in the Lowellian view, it was warm enough. However, the liquid that remained would have to be carefully husbanded. It all made a kind of sense to the general public, for whom the idea of life on other planets seemed no more radical than Darwin's recent theory of evolution.

Scientifically, Lowell and his ideas were very controversial even in his own time. In a scathing review of Lowell's work, Alfred Russel Wallace (1907), famous as an independent developer of the theory of evolution, declared that Mars would be much too cold and that Mars was "not only uninhabited by intelligent beings … but is absolutely UNINHABITABLE". The canals themselves were much debated. Many observers, particularly in the cadre of professional astronomers, simply did not see them. Even many of those who did (and there had been reports even before Schiaparelli's report on the 1877 opposition had brought them into wider view) saw them as disjointed or irregular – few saw the numerous fine lineae and geometric pattern that argued for their artificiality.

Today we know that the canals, especially those quasi-linear versions pointing to artificial origin, have no physical correspondence on the planet; they were the results of the great difficulty of peering through two shimmering atmospheres trying to see features that would have been at the very limit of detectability even had they existed. However, the existence and nature of the canals and of the dark areas were debated well into the 1960s, long after Lowell's death in 1916 and into the early days of the space age. The "wave of darkening" also seemed to be different things to different observers (see the discussion in Martin et al., 1992). Today we know that it is the wind and its redistribution of bright dust that affects the surface albedo. This can darken vast regions, sometimes the cumulative action of hundreds of dust devils leaving their mark. And the belief that there were no mountains on Mars was just wrong, as Mars has major topography, comparable to the continental highs and oceanic basin lows on the Earth. Its Olympus Mons is the tallest of the known volcanoes in the Solar System, reaching ~16 miles above the surrounding plains. But this was not known until Mariner 9, the first Earth spacecraft to orbit another planet, observed the summits of four major volcanoes towering above a global dust haze in 1971.

To convince his critics, Lowell worked – as a good scientist should – to acquire more data that would support his theories. He sought experts who could apply then state-of-the-art spectroscopic instruments in an attempt to quantify how much water was in the Mars atmosphere. Water vapor absorbs sunlight in specific spectral bands. The difficulty is to separate the absorption of sunlight that is reflected from Mars from that absorbed in the more massive – and wetter – atmosphere of the Earth. Lowell and his co-workers realized that the relative motions of Earth and Mars would Doppler-shift the Mars spectral lines away from the Earth lines. Thus, the time to try to detect water in the Mars atmosphere was not when the planets were closest, lined up with the Sun during opposition, but when they were almost in quadrature. The planets were farther apart then, but the greater relative motion could separate the absorption features of the two planetary atmospheres. This approach is used today in our ground-based search for trace gases such as methane in planetary atmospheres. Unfortunately for Lowell, his measurement attempts were at best inconclusive. Ironically, it would be improved spectroscopic methods that first provided solid evidence that Mars and its atmosphere today were not as Earth-like as they once had seemed.

2.2 1962–1972: A DECADE OF CHANGE WITH THE FIRST WAVE OF SPACECRAFT EXPLORATION

In the mid-1950s, de Vaucouleurs (1954) summarized the estimates at that time of atmospheric pressure on Mars. Based on indirect measurements, such as the polarization of reflected sunlight, the Mars surface pressure was estimated at 85 hPa (mbar), as compared to the Earth's average surface pressure of approximately 1 bar (1000 hPa). This was lower than had been expected by many earlier scientists, but not greatly so. In a remarkable book, the *Exploration of Mars* published in 1956, Werner Von Braun and Willy Ley summarized the current knowledge of Mars and outlined how one might explore the planet with emerging rocketry (Von Braun and Ley, 1956). Their Mars landing craft had extensive wings – not unlike the recent space shuttle – because they were still expecting atmospheric pressures on Mars to be ~10% that of Earth – not the ~1% that we know today.

In 1947 Kuiper analyzed bands of CO_2, recorded in telescopic spectroscopic data, to derive an amount of CO_2 for Mars that was only twice that in the Earth's atmosphere (Kuiper, 1952). Because the absorption bands observed could be pressure-broadened, the amount of derived CO_2 was inversely proportional to the square root of the total ambient pressure, which could include hard-to-detect gases like nitrogen or argon. In the 1960s, Spinrad et al. (1963) did what Lowell had failed to do: detect water vapor in the Mars atmosphere. And Kaplan et al. (1964) derived a CO_2 abundance from a weaker CO_2 absorption band observed by Spinrad et al. (1963) that was nearly pressure-independent. When combined with Kuiper's measurements, Spinrad et al. (1966) derived a total surface pressure of 25±15 hPa and 14±7 pr μm for water vapor (1 pr μm is the equivalent depth of water if all the water vapor in a column were condensed to liquid; a typical value for the Earth's column vapor – excluding liquid water drops – is ~5 pr cm, an amount ~3500 times greater than the Spinrad et al. value for Mars). These landmark results indicated a much thinner atmosphere than had been previously suspected (Owen, 1992). This result was soon to be tested in a very novel way.

The clincher came when the first spacecraft flew by Mars in 1965. During its encounter, Mariner 4 transmitted a radio signal through the Mars atmosphere as the spacecraft disappeared behind the planet as seen from Earth (a radio occultation event). Analysis of the refraction of that radio signal by the atmosphere indicated that the total atmospheric pressure was 4–6 hPa. Not only was the atmosphere thin, it would have to be composed almost entirely (>90%) of carbon dioxide. This newly measured pressure could be significantly below the triple point for water (6.1 hPa), so liquid water was not to be expected on the Martian surface. This seemed consistent with impressions left by the Mariner 4 photographs of a narrow swath of the Martian surface that showed only a heavily cratered, Moon-like surface.

A straightforward one-dimensional (vertical) energy balance calculation by Leighton and Murray (1966) showed that a cold Mars atmosphere composed of CO_2 would have another very un-Earth-like feature: temperatures in the winter polar region would be so cold (~140 K) that CO_2, the major constituent of the atmosphere, would condense out – a lot of CO_2. This implied that the seasonal snow was CO_2 (not water) and the polar caps themselves might well be composed of dry (CO_2) ice, not water ice. Furthermore, the mass of the atmosphere would vary throughout the Mars year, with two maxima and two minima, as the atmospheric mass cycled between the two polar regions in response to the seasonally changing insolation.

By the late 1960s, Sagan and Pollack (1969) concluded that albedo changes – even the seasonal "wave of darkening", which seemed such a robust indicator of vegetation – were more likely due to the emplacement and removal of bright, fine-grained dust. (Scattering of sunlight not solely by gas, but also by dust suspended in the atmosphere would have led to the earlier overestimation of the atmospheric pressure derived indirectly from radiometric and polarization measurements.) The seasonal timing of the dust removal was attributed to the seasonal migration of storms from high to low latitudes, as on Earth. This might be aided by the inferred Martian outflow from the poles, a "sublimation" flow in the spring (reversed as a "condensation" flow in the fall) of a significant fraction of the total CO_2 inventory subliming from (condensing onto) the polar caps.

Images from the Mariner 4, 6, and 7 flybys of Mars had all largely sampled its southern hemisphere, revealing it to be heavily cratered. No canal-like features were seen. This, along with the atmospheric results and the demonstration that the seasonal surface albedo changes were meteorological rather than biological in nature, spelled the end of the Lowellian view of modern Mars as an older Earth-like planet. Interest in this Moon-like Mars plunged, but fortunately development of the next Mars mission was already underway. In the meantime, the meteorologists were making progress.

Seymour Hess (1950) published the first "climatology" of the Mars atmosphere. This paper was the first ever published in the *Journal of Meteorology* (now the *Journal of the Atmospheric Sciences*) that dealt with the atmosphere of another planet. The climatology was based on surface temperature measurements by Coblentz and Lampland (1927) and a mere 18 wind vectors derived from tracking clouds. Obviously, Hess relied on his experience as a terrestrial meteorologist and the theoretical relationships between temperature and winds that had been developed already for weather forecasting on Earth. (Hess would later lead the Viking mission meteorology team.)

Observations in the 1960s indicating that the atmosphere had low mass, was mainly composed of carbon dioxide, and rested on a desert-like surface with little heat capacity, had other implications for understanding the Mars atmosphere. In a series of papers (Goody and Belton, 1967; Gierasch and Goody, 1967, 1968), results from one-dimensional radiative–convective transfer calculations indicated that such a Mars atmosphere should respond quickly to solar and infrared radiation. In an atmosphere with so little water vapor, latent heating would be small, unlike the Earth, so heat transport and exchange would be dominated by radiation and dry convection above a heated surface. Given the near absence of clouds and lacking large amounts of trace gases like ozone and water vapor, the Mars atmosphere would let sunlight pass through nearly unattenuated and it would be absorbed by the surface. With little ability to store the heat (again partly a consequence of no liquid water), the surface would undergo a large daily temperature variation.

The atmosphere would be heated by convective and sensible heat transfer from the surface and by some absorption by CO_2 of the infrared radiation emitted from the surface. Although CO_2 is a potent greenhouse gas, there is not much of it in the thin Mars atmosphere, and so air temperatures would rapidly decrease with height through a planetary boundary layer (a few kilometers deep during the day, perhaps several hundred meters at night) and then stay relatively constant until high in the atmosphere (above ~90 km), where temperatures would increase again due to absorption of ultraviolet solar radiation. Our current, more detailed understanding of radiation in the Mars atmosphere, the factors that control it, and how we compute the resulting forcing for numerical models, is discussed in Chapter 6).

This control of temperature in a radiative–convective environment can be expressed in terms of the exponential folding time it would take for an atmospheric temperature perturbation to dissipate by radiation back to a purely radiatively determined equilibrium. On Earth, this time is several days; on Mars, it is a day or so. That meant that it would be more difficult to transport heat, for example, into the polar regions to restrict CO_2 condensation, and that diurnal variations on Mars should be larger in amplitude than even above the highest deserts of the Earth. To understand this quantitatively, it was necessary to put all this new information together in a four-dimensional simulation of the Mars atmospheric state and circulation. Fortunately, such an experiment was already underway.

Yale Mintz (1961) had predicted, based upon terrestrial experience and meteorological scaling arguments, that Mars should have winter storm systems like those on Earth (i.e. baroclinic systems). In the summertime, however, a single Hadley-like cell would dominate, with a physical overturning of the atmosphere in which preferentially heated air rising above the more strongly heated low latitudes moves poleward, cools radiatively, and sinks in mid-latitudes while adiabatically warming. The rising and sinking branches of these cross-equatorial circulations would alternate hemispheres with the seasons.

Leovy and Mintz (1969) tested these ideas by adapting a then state-of-the-art Earth general circulation model that had been developed at UCLA. Given the limitations of their computers (they used the medical school's computer, which was the fastest available to them), the model was restricted to two levels in the vertical, and a horizontal grid with 7°×9° in latitude and longitude (922 points, including the two poles). The model input parameters had to be chosen despite uncertainties in atmospheric composition (it had been suggested that the inert gas argon could compose up to 40% of the atmosphere), almost no knowledge of surface topography (none was assumed in the model), in details of radiative transfer, and in surface heat capacity (Leovy (1966) derived a set of surface thermal inertias). A novel feature for Mars was that the model atmosphere had to gain or lose mass as dictated by the polar radiation budget, unlike the mass conservation typically assumed for Earth.

Their results confirmed Mintz's earlier theoretical expectations to some extent. Storm systems embedded in the polar jet streams were prominent, with their changes over a few days dominating variability at high latitudes. But even these were not vigorous enough that their poleward heat transport would stop condensation in the cold winter polar night. Radiation loss and latent heating of the condensing CO_2 were still the dominant terms in the polar energy balance. Also prominent in the simulations were large diurnal fluctuations in temperature, wind, and pressure. These were large enough to be the second largest component of variation at lower latitudes, behind the seasonal variations but ahead of the day-to-day changes. On Earth, these fluctuations, driven by each day's cycle of solar heating, are muted by the larger thermal mass of the atmosphere and by the action of liquid water, with its large heat capacity, both in the oceans and in the hydrated land. Finally, a Hadley-like circulation did develop in the model simulations, but its structure was more complex (and limited in latitude) than the analytic theory had predicted.

This was a pioneering experiment in many ways. It was the first numerical simulation of the general circulation of another planetary atmosphere, and it used what was nearly the state-of-the-art tools and methodology that were being used for study of the Earth at that time. This approach of adapting advanced four-dimensional circulation models to Mars almost as soon as they were developed for Earth studies continues through the present time (see Chapters 8 and 9). The main difficulties were the limited computing capability and the paucity of data that could be used to define the boundary conditions, to inform the physical parameterizations, and to validate the results. One of the key omissions of this initial general circulation experiment was the (lack of) heating and cooling by airborne dust.

Observers on Earth historically viewed Mars when it was close at opposition, a period that varied in a synodic 17-year cycle so that different seasons were viewed at different oppositions. Some of these oppositions had better viewing than others because the elliptical orbit of Mars would bring the planet closer to the Sun and Earth when Mars was at its perihelion, which currently occurs towards the end of southern spring on Mars. (This seasonal date varies on timescales of hundreds of thousands of years.) Schiaparelli first gave prominence to the *canali* as a result of observing during the favorable opposition of 1877. In another such opposition in 1956, a major dust event was observed for several days. Such events were regarded as rare, but that perception was about to change.

In 1971 Mariner 9 went into orbit around Mars in the middle of a truly global dust storm that had been raging for more than a month before its arrival and which continued to obscure the surface from its view for several months afterwards. Dust was raised 70 km above the surface, with a thin ice haze detached above it (Anderson and Leovy, 1978), and all of the planet, even the poles, was affected. Middle atmospheric temperatures became much warmer for a time (Hanel et al., 1972).

As the atmosphere cleared, a new Mars was revealed (Hartmann and Raper, 1974) due to the global coverage, higher resolution, and better signal-to-noise ratio of the Mariner 9 cameras. A handful of dark spots visible early above the dust pall were revealed to be the summit calderas of massive volcanoes; channels – not canals, but massive channels – were etched on the planet's surface, with converging valley networks revealed in scattered locales. The polar caps and surrounding terrains were extensively layered, suggesting episodic deposition in a series of ice ages, perhaps triggered by the effects of changes in orbital eccentricity and rotation pole obliquity (Chapter 16). Also revealed was a planetary dichotomy, with heavily cratered highlands in the southern hemisphere (which had been overflown by the earlier Mariner spacecraft) and vast, smooth, low-lying plains in the northern hemisphere.

Now interest in Mars soared, as this was once again a dynamic world, one that may have been more Earth-like in its past, one whose climate had obviously changed, and one perhaps capable of change even today. That global dust storm – which remains the most extensive yet seen on Mars – in particular had a definitive impact on thinking about Earth's climate. It was now a plausible reality that the sky could be darkened over most of the Earth for months by a dust cloud from an asteroid impact or even by dust and smoke from a nuclear conflagration. Mars had caught our attention.

For the atmospheric scientists, whose interest in Mars had not waned, there were two new major features that needed to be taken into account: the role of dust in heating the atmosphere (Gierasch and Goody, 1972); and the effect of the large planetary-scale topography. Both affect the basic temperature structure and the general circulation of the Mars atmosphere.

2.3 VIKING: THE SECOND WAVE OF SPACECRAFT EXPLORATION OF MARS

The highly ambitious Viking mission – two orbiters deploying two landers to the surface of Mars – was focused on the search for life. It was predicated on an assumption that, if life had developed anywhere on the planet, it would be everywhere and could be detected by analysis of any soil sample. The orbiter instruments – multispectral cameras, a thermal infrared radiometer, and a water vapor mapping spectrometer – were flown in the hopes that they could help with site selection in terms of safety (surface properties) and of life detection potential (water sources and/or "hot spots"). Launched in 1975, the orbiters and landers explored Mars from 1976 until contact with the last spacecraft (Viking Lander 1) was lost in November 1982. While it did not detect life, Viking expanded our knowledge of Mars immensely, particularly in terms of surface properties and climate.

Isotopic measurements made during entry of the Viking Landers indicated a loss of the lighter isotopes of nitrogen; this argued for massive loss of an early Mars atmosphere through escape to space. Atmospheric measurements from orbit and by the landed meteorology packages were conducted for more than one Mars year. These gave a much better idea of the annual climatology and its inter-annual variation (Hess et al., 1977). In particular, the annual cycles of atmospheric water vapor revealed a seasonal progression of water to low latitudes from a permanent water ice cap at the north pole, eerily reminiscent of past arguments (but no darkening vegetation!) (Farmer et al., 1977).

Measurements in parts of three Mars years also revealed multiple episodes of very large dust storms during the southern spring and summer, the perihelic seasons when Mars is closest to the Sun (Chapters 3 and 10) and insolation is most intense. The effects of dust heating amplifying the already large daily fluctuations of temperature, pressure, and wind and the role of topography in modulating these variations were prominent and soon the subject of numerical simulations (Chapters 6 and 9). Atmospheric dust was now prescribed in models, and there were two sites of meteorological data against which to compare. The richness of these meteorological fields is exhibited by looking at the surface pressure records recorded by the Viking Landers at just two places on Mars (Figure 2.1).

The first thing in Figure 2.1 that catches one's attention is that ~25% of the Mars atmosphere disappears and reappears twice in a Mars year. This is due to the condensation/sublimation of the CO_2 atmosphere onto the polar caps in fall/spring, noted earlier. The differences in the twice-yearly maxima and minima reflect the very elliptical orbit of Mars, with a longer and cool aphelic northern spring–summer, and a shorter and warm perihelic southern spring–summer. The offset of the two curves reflects the ~1.5 km difference in their elevations. Third, there are different meteorological regimes evident at the two Viking Lander sites, with VL1 at 22.4°N (corresponding latitude on Earth is Hawaii) and VL2 at 48°N (on Earth, close to the U.S.–Canada boundary). The quasi-regular variation on timescales of a few days is apparent at the higher latitude in winter–spring and reflects the baroclinic storm systems that feed on the potential energy inherent in the large

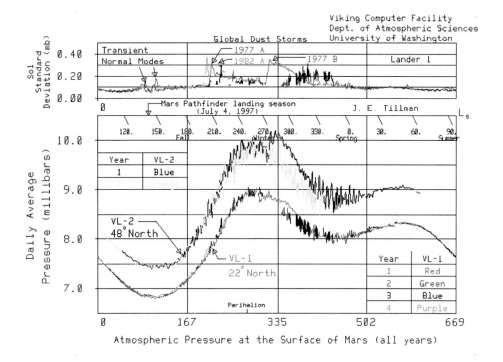

Figure 2.1. The Viking Lander pressure curves measured on the surface at 23°N and 48°N by the Viking Lander 1 and 2 meteorological sensors, respectively. The daily (sol) average and the sol standard deviations are shown. The bottom axis is given in sols, dated from the arrival of Viking Lander 1; the upper axis of the lower panel gives the time of year in L_s, the areocentric longitude of the Sun. The effects of topography, latitude, weather, and even a nearly global dust event are shown (see text). Figure provided courtesy of James Tillman, a veteran of the Viking mission and Viking Lander Meteorology Team. A black and white version of this figure will appear in some formats. For the color version, please refer to the plate section.

latitudinal temperature gradients that develop in those seasons (Chapter 9). Variation at the lower-latitude site is driven by the annual migration of the subsolar point and its associated heating. This drives the rising branch of the overturning circulation and its cross-equatorial transport.

Daily fluctuations of the meteorological fields are shown as the standard deviation of surface pressure within a Martian day in the top panel of Figure 2.1. Note the large amplification of these fluctuations during the major dust storms. These are the thermal atmospheric tides, so-called in analogy to variations of the sea surface on Earth due to the gravity of the Moon and Sun. On both Earth and Mars the atmosphere also responds to these gravitational perturbations by the Sun and moons, but the resulting variations are small compared to the global oscillations driven by the daily heating of the atmosphere. On Earth, this heating is due to ozone and water vapor absorption of sunlight and by convective heating (Chapman and Lindzen, 1970); on Mars, it is due to absorption of sunlight by CO_2 and airborne dust and convective heat exchange with the surface (Lindzen, 1970; Zurek, 1976).

In Figure 2.1 there is an increase in daily mean surface pressure at both Viking sites during the first Mars year (dark curve), though more pronounced at the more northern site, at L_s ~ 280°. (L_s is the areocentric longitude of the Sun measured from vernal equinox, so that L_s = 0°, 90°, 180°, and 270° mark the beginning of northern spring, summer, fall, and winter, respectively; a Mars year is 687 days or 670 sols long (Table 2.1; a sol being a Martian solar day of 24^h37^m).) This increase in surface pressure is due to the second of two planetary-scale dust storms that occurred during the first year of Viking observations. Occurring near the southern summer solstice, when Mars is near perihelion, this dust storm drove a massive overturning of the atmosphere, with air rising in the dustier atmosphere and higher insolation of the southern subtropics. The air crossed the equator and sank in the northern subtropics and middle latitudes, producing a zone of downwelling convergence increasing the column air mass and thus the surface pressure. The greater the heating, the more vertically extended the circulation cell and the further poleward this descending branch can go. On Earth this zone is in the subtropics and low mid-latitudes and accounts for the latitudinal zones of the major deserts on the planet. On Mars this zone can move further poleward as the dust haze that heats the atmosphere increases in opacity, altering the vertical extent of the associated solar heating by dust absorption. Given the transience of dust events on Mars, this expansion of the zone of high surface pressure is temporary on Mars, moving so far north only in those years with major dust events.

After Viking there was a hiatus in the exploration of Mars by spacecraft. While highly successful in expanding our knowledge of Mars, the expense of the mission and the disappointment of not detecting evidence of life, past or present, gave pause to further exploration.

But that did not halt progress. The long-lived Viking mission had left a gold mine of data that would take many years to digest. Furthermore, improved Earth-based observations – both

Table 2.1. *Fundamental metrics for Mars and Earth.*

Metric	Mars	Earth	Ratio
Radius (equatorial) (km)	3396	6378	0.53
Area (10^6 km^2)	144.8	510.1	0.28[a]
Solar day (Mars sol, Earth day, h)	24.66	24.00	1.027
Sidereal day (h)	24.62	23.93	1.029
Rotation frequency, Ω (sidereal, 10^{-5} s^{-1})	7.088	7.292	0.972
Year[c] (Mars sols)	668.6,	355.6,	1.88
Year[c] (Earth days)	687.0	365.25	
Orbital semi-major axis[b] (AU)	1.524	1.000	1.52
Orbit eccentricity[b]	0.0935	0.0167	5.60
Perihelion[b] (AU)	1.38	0.98	1.41
Aphelion[b] (AU)	1.67	1.02	1.64
Obliquity (tilt of rotation axis)[b] (deg)	25.19	23.44	1.075
Gravity (surface, m s^{-2})	3.71	9.80	0.38

[a] This is nearly the same as the land area of the Earth (29%).

[b] These orbital parameters vary much more for Mars than they do for Earth.

[c] By convention, Mars years start with the northern spring equinox when the areocentric solar longitude $L_s = 0° = 360°$. Many scientists analyzing modern spacecraft data follow the Clancy et al. (2000) convention that counts the Mars year of the 1956 great dust storm as Mars Year 1. In this convention, Mars Year 33 started June 18, 2015.

ground-based high-resolution spectrometers and orbital observatories like the Hubble Space Telescope – were still observing Mars (Chapter 3). Their purpose was largely to detect trace gases and to characterize inter-annual variability in dust storm events and in basic temperature structure. In parallel, ever more sophisticated tools were being developed to simulate the observed atmospheric phenomena. Atmospheric general circulation models (GCMs) were now prescribing dust hazes and planetary-scale topography, and they were being run with increased spatial resolution and vertical range; e.g. simulations were done using a six-layer model just prior to the arrival of the Viking spacecraft. Even so, the vertical domains in the models were inadequate to describe the deep Hadley-type circulations that could develop in a dusty atmosphere or the atmospheric tides, with their vertically propagating components. Quasi-analytic models, such as zonally symmetric (two-dimensional in latitude and altitude) (Haberle et al., 1982), and linear atmospheric tidal models (Fourier components in time and longitude), were still used to provide some insight.

For the atmospheric tidal theory, these models indicated that daily changes in meteorological fields were not functions purely of local time, with minima and maxima following the Sun's apparent westward motion. In particular there was a class of eastward-propagating components (Kelvin modes) that could be efficiently excited (as in resonance) by longitudinal variations in topography and column dust

opacity (Zurek, 1976). The presence of such a tidal variation was confirmed via temperature observations by Mariner 9 (Conrath, 1976) and Viking surface pressure data (Zurek and Leovy, 1981). Thus, the structure and temporal variation of these daily fluctuations were more complex than had been first anticipated. Interestingly, the observed prominence of the twice-daily tide during very dusty periods on Mars could be demonstrated to be the same effect as for the Earth's semidiurnal pressure tide being larger than its diurnal counterpart, despite the latter being the bigger component of the daily insolation variation. On both planets, it was heating at higher elevations (in a dust haze on Mars and by stratospheric ozone on Earth) adding up to make the local tide with the larger vertical wavelength (i.e. the semidiurnal tide) the dominant component of surface pressure variation (Chapman and Lindzen, 1970; Zurek, 1980).

Work with these models in the post-Viking period helped illuminate many observations made during the Viking missions. A long hiatus in missions to Mars followed, but there was an effort through various data analysis programs and workshops to exploit the datasets that had been gathered and to utilize increasingly sophisticated models to understand the resulting clues about the present and past climates of Mars. As the prospects for a new (third) wave of Mars exploration loomed on the horizon, the knowledge gained from these efforts was summarized in a massive book (appropriately called *Mars*) covering all the many aspects of Mars, including 420 pages spanning several chapters on the atmosphere of Mars alone. Here are some summary highlights (Kieffer et al., 1992, and references therein):

- The mostly carbon dioxide atmosphere held trace amounts of water vapor and ozone, anticorrelated in their seasonal and spatial distributions by photochemistry, while isotopic signatures of trace nitrogen pointed to massive loss over time of atmospheric gases via escape to space processes.
- The existence of vast channels and valley networks, apparently carved by water, indicated massive water activity on early Mars (3.5–4 Ga).
- The puzzling nature of the polar caps: the low-lying, but still mile-thick, north polar cap, with its layers of water ice exposed during summer, but nearly crater-free surface; the high-altitude, apparently older south polar cap, with a thin layer of carbon dioxide ice persisting throughout the hot southern summer, with the survival of that ice dependent on a remarkably high surface albedo, which appeared to brighten as the seasonal insolation increases.
- Large-scale topography from surface pressure and radio occultation data had outlined the hemispheric dichotomy (heavily cratered, high-altitude southern hemisphere punctuated by the giant impact basins of Hellas and Argyre versus relatively smooth and featureless northern low-lying plains), the massive volcanoes of Olympus Mons, the Tharsis plateau and Elysium, and the Valles Marineris rift systems, the latter extending across the equivalent width of the continental U.S.
- The basic cycles of dust, water, and carbon dioxide:
  - A basic understanding of the recycling of a major fraction of the largely CO_2 atmosphere, controlled by

radiative energy balance and subsurface heat conduction. However, it was surprisingly difficult for the GCMs to get this right, within the observational constraints of polar surface albedo, etc.

- – The episodic nature of the larger dust storms, occurring in some years but not others, but typically in southern spring and summer, a so-called great dust storm season. Local dust storms had been observed in all seasons, but the statistics of their occurrence and nature were poor due to the non-systematic coverage provided by the Viking Orbiters, which were also tasked to provide communication with the landers.
- – An annual, global cycle of water vapor, with the north polar cap being the major source of atmospheric water for the planet, but with major uncertainties as to the role of the regolith as a source or sink, or the extent to which the cycle was closed.
- • Apart from the surface pressure and wind measurements at the two Viking Lander sites, there were few quantitative aspects determined regarding the atmospheric circulation. Basic clues came from observations of atmospheric tracers like ozone, water vapor, and dust and of surface wind streaks. A remarkable warming of the north polar atmosphere during the largest of the 1977 dust storms was observed and attributed to adiabatic warming of the downwelling branch of a Hadley-like circulation extending all the way to high latitudes. However, models continued to be the main means of estimating the general circulation.
- • Basically unknown were the magnetic properties of the planet and details of the upper atmosphere. For example, there were less than two dozen profiles of the ionospheric peak in the upper atmosphere, observed by the Mariner 9 Ultraviolet Spectrometer data (Stewart et al., 1972). Using those data, A. I. Stewart estimated a natural variability (one-sigma) of the upper atmosphere from orbit to orbit of ~30% in atmospheric density, a surprisingly robust number confirmed by later aerobraking Mars orbiters operating above ~100 km (Tolson et al., 2007).

A major discovery not captured in *Mars* during this period came from ground-based microwave spectrometers. In observations by Clancy (also see Chapter 5) the Mars atmosphere appeared to be colder and cloudier than Viking had reported, particularly during northern spring and summer, when Mars was near aphelion in its eccentric orbit and the atmosphere was relatively dust-free. Inter-annual differences in the atmosphere during southern spring and summer were easily ascribed to the episodic occurrence of planetary-scale dust storms. These microwave ground-based observations of Mars during its northern spring–summer suggested a major shift in the modern Mars climate. Subsequent observations and further analysis of Viking infrared thermal mapper data showed that the "aphelion cloud belt", a low-latitude zone of thin water ice clouds, had been present during the Viking era too (Tamppari et al., 2000). While not indicative of a major change in climate, Clancy's discovery showed that a new element (ice clouds) had to be taken into account if we were to advance our understanding of atmospheric structure and circulation.

2.4 THE THIRD WAVE OF SPACECRAFT EXPLORATION OF MARS

Nearly continuous remote sensing from orbit of the Mars atmosphere, a need elegantly articulated in *Mars* (Kieffer et al., 1992), has been a signature feature of the modern program of Mars exploration. However, success was not immediate. An ambitious attempt to restart Mars exploration was stymied first by the loss in 1991 of Mars Observer. A shift to smaller, less expensive, spacecraft resulted in major successes with Mars Pathfinder conducting landed operations for two months in 1997 and, with an eight-year highly productive Mars Global Surveyor mission in 1998. However, further cost-cutting as part of a "faster, better, cheaper" strategy ultimately resulted in the loss of both the Mars Climate Orbiter and the Mars Polar Lander, launched in the 1998–1999 opportunity.

A reinvigorated Mars Exploration Program (Chapter 3) was developed following the loss of those two missions and has resulted in major successes for NASA with the launches in 2001 of the Mars Odyssey (ODY) Orbiter, of two Mars Exploration Rovers (MERs) in 2003 (Opportunity and Spirit), of the Mars Reconnaissance Orbiter (MRO) in 2005, the Phoenix Lander in 2007, and the Mars Science Laboratory (Curiosity) Rover in 2011. The European Space Agency (ESA) also launched in 2003 a highly capable orbiter, Mars Express (MEX), which also carried a small probe, Beagle II, which unfortunately was lost during landing. (Twelve years later, a former member of the operations team detected the craft in an MRO high-resolution camera image, which showed that the craft had successfully landed but only partially deployed its solar panels, blocking the radio antenna.)

Launched in 2007, Phoenix landed at high northern latitudes on Mars, where it operated from May to September 2008 before the harsh northern winter ended the mission. (Images taken the following spring show that the weight of accumulated wintertime CO_2 frost had broken the solar panels.) Contact with the Spirit Rover was lost in 2010 after seven years of exploration in Gusev Crater and its Columbia Hills in what was originally a 90 sol mission. The Opportunity Rover continues to operate 13 years later. (The rate of dust accumulation on the Mars Pathfinder solar panels during its short mission suggested that the MER craft would be starved for solar power after 90 sols. Fortunately, winds have periodically removed dust from the MER panels, renewing solar power generation.) Mars Odyssey, Mars Express, and MRO all continue to explore and return data from Mars orbit, with MRO having returned 300 Tbits (as of March 2017) of often-compressed science data, an amount greater than all other deep-space planetary missions.

In September 2014, two new orbiters, NASA's Mars Atmosphere and Volatile Evolution (MAVEN) mission and the India Space Research Organization's (ISRO) Mars Orbiter Mission (MOM), joined the three working orbiters and two operating rovers (Opportunity and Curiosity). MAVEN's focus during its one Earth year prime mission has been the solar wind interaction with the Mars upper atmosphere and the mechanisms by which volatiles can escape from Mars. Its observations will test models of present escape (Chapters 14 and 15), hopefully adding enough detail of current processes to permit extrapolation back into the ancient regime, when the Sun was ultraviolet-bright, but radiated less total energy (see Chapter 17).

Three meteorological stations have been landed since Viking (Chapter 3). Two had limited lifetimes: Mars Pathfinder operated near the equator for two months in 1997, and Phoenix was limited in 2008 to a summer of observations at a high northern latitude. The third (Curiosity) has operated since 2012. In addition to meteorological measurements, these have provided ground truth for the opacity record by upward-looking observations of the extinction of sunlight at the landing sites (Chapter 10). The Mars Exploration Rovers did not carry meteorological sensors given their tight mass constraints and presumed 90 sol lifetimes, but they have provided many years of overhead atmospheric opacity measurements.

This third wave of exploration of Mars by spacecraft has steadily improved the spatial resolution of our global coverage of Mars. We now have global datasets of the surface which have increased the spatial resolution of visual images of Mars from the 200 m or more per pixel of Viking to more than 99% of the planet covered in a panchromatic band at 6 m/pixel, stereo color images for more than half the planet at resolutions of ~20 m/pixel, and a carefully selected 2.5% of the planet at an unprecedented 30 cm/pixel, a fifth of which is in color. Highly magnetized remnants of the crust, but only in the oldest terrains, have been mapped, indicating that Mars once had a global magnetic field that disappeared early in the planet's history (~4 Ga; Acuna et al., 1999)[1]. The radical reduction in that global field strength possibly led to a much more accelerated loss of atmospheric mass due to the actions of the solar wind, as it could now sweep through the upper atmosphere (Chapters 15 and 17). Characterizing escape processes is the ongoing goal of the MAVEN mission.

Mars topography is now known from laser altimetry to a precision of ~3 m averaged in 1 m spots approximately 100 m apart along the ground track (see Smith et al., 1999). Due to the spacing between ground tracks, this yields a *global* topography with resolutions of ~1 km in latitude and 2 km in longitude at the equator (Smith et al., 2001). Valley networks can be shown to have had streams running downhill and inverted streambeds speak to extensive erosion on Mars some time in its history. The thermal inertia and albedo of the surface are now characterized at 100 m/ pixel (Fergason et al., 2006) permitting, amongst other things, calculations of the ice holding capacity of the near surface.

The actual distribution of near-surface ice (at <1 m depth) has been determined using the orbital observations of subsurface/surface hydrogen in footprints a few hundred meters across (Boynton et al., 2002; Feldman et al., 2002; Mitrofanov et al., 2002). These data show very shallow ice in the middle to high latitudes and adsorbed water and/or hydrated minerals in many locations at lower latitudes. The presence and depth of ice (a few centimeters of overburden at 68°N on the northern plains) was confirmed by the Phoenix Lander digging locally into the Martian ground (Smith et al., 2009; Mellon et al., 2009).

As noted earlier, a signature achievement of the recent exploration program has been acquisition of a multi-year record of atmospheric fields. Daily, global weather maps at ~1 km resolution, together with seasonal maps of column dust opacity and zonally averaged water vapor, now span a Mars decade and reveal a wealth of phenomena: dust storms, weather systems, jet streams (Chapters 5, 9, and 10). Systematic maps of column ozone and carbon monoxide have been added to that record since 2006 (Chapter 13). A greatly expanded database of Mars clouds, including multiple new cloud types such as CO_2 hazes high in the middle atmosphere, together with refined estimates of their particle sizes, has been acquired (Chapter 5). The importance of clouds for radiative, photochemical and dynamical processes is increasingly revealed in the atmospheric observations (see Chapters 4–6 and 9–13).

A second major achievement of modern exploration has been the improved vertical profiling at half-scale height resolution (~5 km) of temperature, dust, and water ice aerosol (McCleese et al., 2010; Chapters 4, 5, 10, and 11). Building on earlier work (Wilson, 2000), these have revealed the longsought global signatures of thermal tidal wave structure in the interior atmosphere (Lee et al., 2009; Chapters 4, 5, 9, and 11). The non-uniformity of dust mixing in the lower atmosphere (Heavens et al., 2011, 2014; Chapters 4 and 10) was long suspected, but is now proven. The potent radiative drive of even thin ice clouds (Kleinböhl et al., 2013; Chapters 5, 6, 9, and 11) was also revealed.

The combination of higher spatial observing and of an extended record of observation has also provided clear evidence that Mars is changing today. Many sand dunes are observed to move (Bridges et al., 2012a,b), seasonal CO_2 slab ice subliming in the spring at high latitudes produces a variety of surface patterns (Hansen et al., 2010, 2013, 2015), and repeat observations have revealed recurring slope lineae (RSL), enigmatic albedo features a few meters wide which darken during the warm seasons, elongate downslope, and then fade away until the following Mars year, when the patterns repeat again (McEwen et al., 2014).

These databases are still being analyzed today and new observations continue to be acquired. In the next section, our current state of knowledge about the Mars atmosphere and climate is previewed, pointing to the more detailed discussions in the following chapters. This overview is divided into three connecting, but distinct, periods of Mars climate evolution: early Mars, with its more Earth-like climate; middle Mars, with its suggested ice ages driven by obliquity and orbital cycle variations; and modern Mars, still changing today.

2.5 MARS ATMOSPHERIC PHENOMENA: WHERE ARE WE NOW?

2.5.1 Early Mars

Billions of years ago, water did flow across its surface in great quantities, imprinting channels and other features on its ancient surface (Chapter 17). Groundwater levels rose and fell, altering the volcanic rock to produce aqueous minerals (e.g. carbonates, clays, sulfates). Exposed at the surface today, those minerals indicate a diversity of ancient surface environments, with different levels of acidity and processes operating at different

---

[1] Throughout the book, we use the abbreviations Ma (mega-annum) and Ga (giga-annum) to denote millions and billions of years of geological age (ago), respectively; and Myr and Gyr to denote time spans of millions and billions of years.

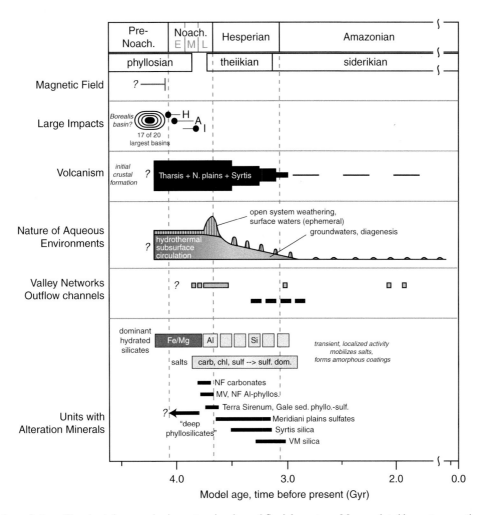

Figure 2.2. Mars through time. The chart shows major impact, volcanic, and fluvial events on Mars as dated by crater counting. The evidence for aqueous environments comes from both morphology (e.g. channels and valley networks) and composition, looking at the locations where surface composition reveals minerals whose formation requires the action of (liquid) water. These two ways of looking at Mars history can each be used to describe coherent epochs of Mars history. The traditional one uses the terms pre-Noachian (prior to 4.1 Ga), Noachian (4.1–3.7 Ga), Hesperian (3.7–3.1 Ga), and Amazonian (3.1 Ga to present). It is the early, apparently wetter, climate at the Noachian–Hesperian boundary that has long intrigued climatologists – and biologists – because of the possibly more Earth-like climate then. After Ehlmann et al. (2011).

temperatures. Ancient rocks dating from a time on Mars comparable to the time that life originated on Earth (~3.5 billion years) still survive on vast stretches of Mars, while they have been largely removed on the Earth through crustal subduction. For the atmosphere, the clues of that ancient past and subsequent evolution are contained in the rocks (see below) and in the trace gases of the atmosphere, including isotopes (e.g. Mahaffy et al., 2013).

The geochemical timeline shown in Figure 2.2 (Ehlmann et al., 2011) has been derived from multiple analyses based on data provided by the post-Viking wave of spacecraft exploration. It suggests an early period of high water activity, though in what form (groundwater, transient lakes and seas, precipitation – or all at different times) is highly debated. Nevertheless, there seems to have been a chemical evolution – perhaps episodic – from alteration of the ancient crust by more neutral pH water (e.g. producing clays) to later action by more acidic water (e.g. producing sulfates) (e.g. Murchie et al., 2009; Ehlmann and Edwards, 2014). The Curiosity

Rover has been working its way toward a set of stratigraphic layers in the lower elevations of Mt. Sharp, the central peak of Gale Crater, that encompass this climate transition. Over the next few years, its analytic laboratories will tell us more about this geologically preserved record of change on early Mars. However, as high-spatial-resolution orbiter coverage of surface composition has expanded, with growing confidence in identification of various mineral types, there appear to have been periods when these different environments (e.g. neutral or high pH) either coexisted or repeated episodically, producing a more complex record. In any case, both orbiter and lander data are consistent with the alteration occurring in shallow seas or lakes and as a result of alteration by groundwater on early Mars. The debate about regional seas or an ocean (e.g. Di Achille and Hynek, 2010) and the role of precipitation continues (Chapter 17).

Ages of the surface material are derived from crater counting: the more heavily cratered a surface is, the older it is. Radiometric dating of Martian meteorites has been done on

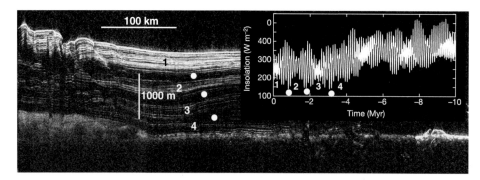

Figure 2.3. (Left) SHARAD radar reflections (Phillips et al., 2008) from depth through more than a kilometer of ice forming the north polar ice cap. Bright lines are radar reflections from layer interfaces, while dark areas have little radar return, presumably from cleaner ice. Note that the layers seem to come in "packets" of bright interfaces separated by darker bands. (Right) The packets are numbered and may correspond to periods of moderate polar irradiation due to obliquity variations (Laskar et al., 2004), when polar ice tends to sublime overall, leaving dustier (more reflective) layers. The darker areas (dots) correspond to periods of minimum polar irradiation, when relatively pure ice may condense overall.

Earth (most are geologically young), but the locations where the meteorites were blasted off the Martian surface are unknown, thus making it difficult to establish their geological context. An intriguing result from Curiosity analyzing drilled material in Gale Crater is that, while the rocks of the Gale Crater locale formed ~3.5 billion years ago (in agreement with crater counts), some of those rocks have been only recently exposed at the surface (~80 million years ago; Farley et al., 2013). This provides clues about the rates and timing of erosion and also points to places where organic material or other bio-signatures could be protected from ultraviolet and cosmic radiation for very long periods of time. In particular, there may have been more massive, probably episodic, erosion in relatively recent geologic times on Mars.

Both the meteoritic record and the *in situ* isotopic measurements made on Mars indicate that there has been massive atmospheric loss. Whether that is related to the early loss of a global magnetic field is still debated. If Mars once had a massive CO_2 atmosphere, which drove an Earth-like hydrologic cycle with precipitation and sustained liquid water on its surface, one path for its loss is the formation of carbonate at the Martian surface (Fanale et al., 1982; Kahn, 1985; Pollack et al., 1987). While carbonate has been detected, the volume of exposed carbonate (equivalent to ≤12 hPa of atmospheric CO_2; Edwards and Ehlmann, 2015) is much less than the hundreds of millibars often invoked in an attempt to simulate an early hydrosphere (Chapter 17).

If the Mars climate was much wetter early in its history, as the surface morphology and composition indicate, where has that water gone? As discussed above, there is evidence both for escape and for buried reservoirs of ice and hydrated materials in the crust (Chapters 15–17).

2.5.2 Middle Mars

The lure of a more Earth-like climate on ancient Mars, a time when life may have started on two planets in our Solar System, has sometimes eclipsed the fact that Mars may have undergone some dramatic climate changes in geologically more recent times; i.e. the middle and late Amazonian periods (Figure 2.2) have attractions of their own. The two polar caps are geologically young, but differ in many key respects, perhaps due to their different elevations. The north polar cap exposed during summer is a mile-thick slab of ice, with a semi-regular pattern of internal layering; it is also relatively young (≤10 million years, based on the flatness of the ground beneath it and the number of craters on its surface). The south polar cap appears much older (a few hundred million years), with its internal layers being much less regular (Phillips et al., 2008, and references therein). Furthermore, the south polar cap appears to contain enough buried CO_2 ice today that, if it were sublimed into the atmosphere, the resulting gas could double the present atmospheric mass (Phillips et al., 2011). At that point, much more of the surface could have pressures above the triple point of water, with more widespread transient liquid water (longer portions of a day and more days in the warm seasons). Such scenarios are now being investigated, with a goal of identifying what and where the physical evidence of such water activity would be.

The layering internal to the polar caps and exposed at their peripheries is due to the varying amounts of dust and ice, with dust being darker visually but more radar-reflective (bright). Whether the cameras and radars are viewing the same physical structures continues to be investigated – both see "packets" of layers suggesting deposition and erosion of layers of volatiles occurring on quasi-periodic timescales (Figure 2.3, Chapters 10, 11, and 16).

The prime source of the water that cycles over Mars each year today is the north polar ice cap. That ice cap is "permanent" in today's climate in the sense that a cap remains at the end of each water ice sublimation season. The amount of water that sublimes is dependent on the solar insolation absorbed by the polar cap and so is dependent on variations over thousands and millions of years of planetary obliquity and orbital eccentricity and phasing (Chapter 16). Figure 2.3 shows the radar-detected internal structure of the north polar cap that, together with the layers seen at the cap edge, suggests ice ages driven by the changes in polar insolation due to changes in the tilt of the Mars rotation axis (obliquity) and to changes in orbital eccentricity and its phasing (Chapters 10, 11, and 16).

Similar Milankovitch cycles are thought to drive ice ages on Earth. These astronomical cycles of insolation are more pronounced for Mars than for Earth. Compared to Earth, Mars is a lumpy planet without oceans and the stabilizing influence of a large Moon; it is also closer to the gas giants of the outer Solar System (especially Jupiter) and thus more subject to their gravitational pull. In particular, while the axial tilt of Earth may vary by a degree or so, that of Mars varies by tens of degrees on timescales of a few hundred thousand to a few million years (Figure 2.3). Further, these variations for Mars are sufficiently chaotic that they cannot be deterministically calculated more than 20 million years into the past for the planetary parameters known today. However, it can be estimated statistically that, while Mars obliquity can range from nearly zero to above 60°, the "average" obliquity is ~40° (Laskar et al., 2004; Chapter 16).

One possible constraint on these ice ages would be characterizing the shallow ice distribution in mid-latitudes at depths of a few meters, greater than probed by the orbital high-energy spectrometers. Today this is being done by detecting new surface albedo changes using moderate-resolution cameras, with their more extensive coverage, and then confirming at higher resolution that the change has the signature of a new impact crater. Moderate-sized meteors (~1 m) can make it through the thin Mars atmosphere and the craters they produce excavate a few meters deep. Some of the over 500 new impact craters detected by MRO in the last 10 years (Byrne et al., 2009; Daubar et al., 2013) have white (icy) bottoms, which eventually go away as the exposed ice sublimes into the atmosphere (Dundas et al., 2014). These have moved the "ice boundary" closer to low latitudes, and the exact location of that boundary would have implications for the ice age scenarios (Chamberlain and Boynton, 2007).

2.5.3 Modern Mars: the Present Atmosphere

For transported aerosols and volatiles, the concept of cycles is a useful framework for understanding where things come from (sources), how they move elsewhere (transport), and where they end up (sinks). As discussed in *Mars* (Kieffer et al., 1992), Viking data enabled quantitative estimation of the principal components of these cycles for dust, water, and carbon dioxide. These cycles are all coupled in at least three ways (Chapters 4–13): (1) the transport circulation (dynamics, including interactions with the surface; Chapters 7–9); (2) the radiation fields that drive photochemistry (Chapter 13) and the circulation, responding to radiatively active gases and aerosols (Chapters 4–6); and (3) microphysics, which directly transforms the physical state of volatiles through sublimation and condensation, e.g. dust acting as ice cloud condensation nuclei, while condensing ice scavenges dust from the atmosphere (Chapters 10 and 11). While the basic outline of these cycles could be filled in using Viking data (as in *Mars*), many features remained uncertain, but fortunately much has been learned from data obtained by the latest wave of spacecraft exploration and its analysis and simulation.

The seasonal cycles of temperature, dust, ice, and water vapor have been observed for almost a full Mars decade and are discussed in detail in Chapters 4, 5, and 10–12. However, many questions remain. Are the annual cycles of water and carbon dioxide closed? For example, is the north polar cap gaining or losing water ice in the present climate? Is the thin

carbon dioxide ice cover of the south polar cap disappearing or re-forming on timescales of a few decades (Thomas et al., 2014, 2016; Chapter 16)? Are the RSL truly the result of melted water ice forming brine flows (McEwen et al., 2014; Ojha et al., 2015)?

The dust cycle is particularly puzzling because of its interannual variability, with the largest storms occurring in some years, but not others. While regional dust events are likely to occur in particular seasonal windows, the processes by which some events grow to planetary scale in some years but not others remain a subject of intense research (Chapters 8–10). A recent hypothesis even implicates a contribution by the Sun–Mars motion around the Solar System's barycenter as a trigger to the largest events (Shirley, 2015; Chapters 8–10).

The thin CO_2 cover capping the south pole "permanent" cap is also a puzzle, as the pits (an arabesque "Swiss cheese" terrain) on the CO_2 cover expand with time (Malin et al., 2001) while other areas brighten, making the long-term effects uncertain (Thomas et al., 2014, 2016). Meanwhile the quantity of CO_2 snow contributing to the seasonal polar cap – as opposed to frost at the surface – has been estimated for the first time and could play a role, given that its emissivity differs from that of frost, to the uneven seasonal sublimation (Hayne et al., 2012, 2014).

Transport couples all these cycles. Winds raise the dust, which absorbs sunlight, heating the atmosphere, and changing the temperatures and winds. Dust is removed by scavenging due to formation of water ice crystals. The circulation redistributes carbon dioxide, whose seasonal sublimation and condensation are driven by the polar radiation balance (Chapters 6 and 12). Modeling atmospheric dynamics is discussed in detail in Chapters 8 and 9. The adequacy of these models can be viewed in the detailed discussions of the planetary boundary layer (Chapter 7), of the various cycles of dust, water, and carbon dioxide (Chapters 10–12), of atmospheric photochemistry (Chapter 13), of the upper atmosphere (Chapters 14 and 15), and of early climate (Chapter 17).

Radiation is the ultimate driver of the atmospheric processes (Chapter 6), while the radiative and energetic interactions between the atmosphere and space are discussed in Chapter 15. Microphysics are naturally discussed in many chapters, but particularly Chapters 5 and 10–12, as it is the microphysical processes that also link various components of the primary volatile and dust cycles.

In the book *Mars* (Kieffer et al., 1992), a common hope for the future was the acquisition of global, long-term datasets and improvements in atmospheric models. Much new data has been acquired (Chapter 3). Despite this bonanza, there are limitations. Coverage and vertical resolution near the surface are poor, in part because of dust opacity. While column abundances of water vapor (and carbon monoxide) have been retrieved, the vertical variation of water vapor is largely unobserved. And it would be useful to have another planetary-scale dust event during which our current orbiters observe the mechanics and environment of dust storm onset in detail; this, of course, depends on Mars. A major deficiency is the lack of wind observations. General limitations are the lack of synoptic and full diurnal coverage, and the lack of surface meteorological networks with simultaneous measurements over extended areas. Surface meteorological packages on Curiosity and a planned

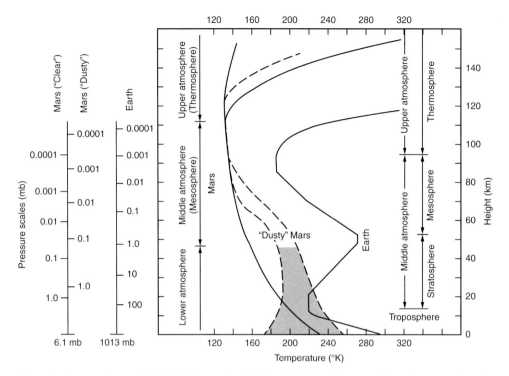

Figure 2.4. The U.S. standard atmosphere, published for the year in which the Viking Landers entered the Mars atmosphere, is shown (solid curve on the right) with the standard nomenclature for terrestrial meteorology. Temperatures derived from the Viking Lander 1 entry (Seiff and Kirk, 1977) serves here as a "standard reference atmosphere" for Mars, with similar nomenclature. Dashed curves with shading reflect the effects on temperature of aerosols during major dust events. Note the different hydrostatic pressure profiles on the left. Figure from Zurek (1992).

2020 Mars rover suffer from lack of vertical profiling above the platforms and from the inadvertent effects of the lander or rover supporting them. Some of these deficiencies may be addressed by a new orbiter, the ESA ExoMars Trace Gas Orbiter (TGO) launched in March 2016. After arrival at Mars in September and following a year of aerobraking, it will acquire new measurements from a ~400 km orbit drifting in local time.

Despite these limitations, as shown in the following chapters, the richness of the acquired data has added much to our understanding of the present cycles of dust, water, and carbon dioxide, and of the weather and climate of the planet.

As also hoped by the authors in *Mars*, atmospheric models have gone far beyond the 5°×7° latitude–longitude, six-level models of the Viking days (Pollack et al., 1976; Chapter 9). Ever greater computing power has enabled 50–100 levels in GCMs, extending model tops to 80 km or into the thermosphere (e.g. Forget et al., 1999; Richardson et al., 2007; Chapters 9 and 14). Annual simulations have been run at resolutions as fine as 1.5°×3° in latitude–longitude and mesoscale models have simulated regions at even higher resolution (Chapter 8). Interactive model runs in which dust is injected, transported around, and then removed from the atmosphere using physical parameterizations, including options for the radiative effects of the dust, are now possible. The effects of ice particle formation on dust and, through the ice aerosol radiative properties, on the general circulation, are now included in many models (Chapters 8–11).

The difficulty is that – as on Earth – the interaction between water and aerosols is complex and challenging, as the models try to bridge the gap between microphysical processes in the atmosphere and surface all the way to the global circulation. It was once

thought (hoped?) that Mars would prove to be a simpler climate system than Earth, given the lack of latent heating by the condensation of significant amounts of water vapor. Measurements, however, have shown that clouds are as important for Mars as for Earth when simulating their climates (Chapters 5, 9, and 11). This is a significant change from the perspective expressed in *Mars*. Furthermore, attempts to understand the ancient climate of Mars (Chapter 17), when conditions may have been more Earth-like, may still require treatments of a full hydrological cycle (thick clouds with precipitation).

2.6 BASIC PARAMETERS AND NOMENCLATURE

2.6.1 Temperature

Figure 2.4 reproduces the "standard" temperature profile and nomenclature for Mars (left) and Earth (curve on right) from the book *Mars* (Kieffer et al., 1992). Although this "standard" Mars profile was based on the Viking entry profile (Seiff and Kirk, 1977), it is remarkably similar to the average of millions of temperature profiles retrieved by the MRO Mars Climate Sounder (Figure 4.2). Chapter 4 discusses in detail temperatures derived from multiple instruments and presents tables and graphs of "standard" profiles for Mars. The nomenclature still applies: The division between the upper and lower atmosphere – i.e. the mesopause or base of the thermosphere – occurs at altitudes ~100 km above the surface of both planets. The similarity of pressure near the mesopause on both planets is somewhat remarkable given the different surface pressures: on Earth, 6.1 hPa occurs in the stratosphere at ~40 km altitude, while it is the surface pressure on Mars (Figure 2.4). On Mars, the lower gravity (resulting in a larger scale height) hydrostatically produces less change of pressure with height.

Above the mesopause, temperatures begin to increase due to the absorption of extreme ultraviolet radiation (Chapter 14). Below this level, the atmosphere on Mars can be divided into two regions, with the middle atmosphere (~50–100 km) marked by a more isothermal region (in reality with superposed waves; see Chapter 4). Near the surface, a planetary boundary layer exists, driven by convective and radiative transfer with the surface; its thickness and stability strongly vary throughout the diurnal cycle (Chapter 7).

The altitude separating the middle and lower regions varies significantly with atmospheric dust opacity, as well as season and geography. It is only during the planetary-scale dust events that Mars has something resembling the Earth's stratosphere, with solar heating of the airborne dust temporarily producing nearly isothermal zones or deep inversions comparable to the effect of ozone heating on Earth. Of course, individual temperature profiles do not vary smoothly with height as shown in Figure 2.4, but have pronounced wave structure, changing with time of day (Chapter 4; e.g. Figure 4.1). Such structures are a more pronounced category of variation on Mars than are their terrestrial counterparts (Chapters 4 and 9).

2.6.2 Clouds

Telescopic detection of distinct clouds was sufficiently rare that it was a noteworthy event. The major attributes recorded were the size, location, and color. The polar hazes were well known, obscuring the skies above the wintertime polar caps. Elsewhere, the two main categories of cloud were distinguished by color: white (presumed to be water-laden) or yellow (presumed to be composed of dust). As the low temperature of the atmosphere became better known, the water in the white clouds was assumed to be water ice (not droplets), and this was confirmed by the Mariner 9 Orbiter (Curran et al., 1973). The origin of the most recurrent of the white clouds (e.g. as in the historical Tharsis clouds or the Olympus "ring" cloud) became better understood once Mariner 9 revealed the topography: these were daily recurrences of orographically forced clouds, appearing prominently on the flanks of the massive Mars volcanoes. Chapter 5 provides a detailed description of cloud morphologies and occurrences.

2.6.3 Dust Storms and Mars Years

Originally, a dust storm was a yellow cloud that was observed to move. Telescopic observers recorded extensive dust storms (i.e. moving clouds that obscured large areas) in 1922 that lasted just four days, with a more prominent and longer-lived storm in 1956. The 1956 event was well documented by many astronomers thanks to an international coordinating effort. It was regarded as truly exceptional (Slipher, 1962) until 1971 when the International Planetary Patrol alerted NASA that a global dust storm awaited its Mariner 9 spacecraft. In recent times, many scientists working with spacecraft observations have followed Clancy et al. (2000) by counting Mars years referenced to designating the Mars year in which this vast storm occurred as Mars Year 1 (MY 1) (1 MY = 1.88 years). Currently, it is MY 33, which began with the Martian northern spring equinox ($L_s = 0°$) on June 18, 2015.

The reference to the largest-scale events as global dust storms (GDS) began with the descriptions and discussions of the possible origin of the 1971 storm that was still raging when Mariner 9 entered Mars orbit (Gierasch, 1974). That truly global event extended from pole to pole. Of two very large events observed by the Viking Orbiters, the first was more hemispheric, while the second affected both hemispheres; neither was as global as the 1971 GDS, and nothing of that scale occurred in the second Viking year. Thus, the GDS nomenclature became great dust storms, associated with obscurations that encircled Mars or covered regions comparable to a planetary radius or more in size, i.e. planetary scale (Zurek, 1982). Furthermore, these new data combined with historic Earth-based observations suggested that the largest dust storms tended to occur in southern spring and summer, the so-called (large or great) "dust storm season". It is important to remember that smaller dust storms occur every Mars year and in every season, although their frequency and place of occurrence vary throughout the year (Chapter 10).

The dust hazes are far more extensive than the dust raising regions themselves, and some researchers have restricted the use of "dust storm" to indicate only those areas where dust is actively being raised into the atmosphere. Both Earth-based telescopic and spacecraft imagers have shown that the planetary-scale events typically have multiple centers of dust raising; this may even be necessary to produce a planetary-scale haze, even though a single local dust storm may be a trigger. To keep up with the added insight provided by daily global coverage that recent orbiters have produced for nearly 17 years, the ensemble of dust raising zones and associated dust hazes obscuring a major fraction of the planet is often referred to as a dust "event". Chapter 10 provides a modern description of the dust cycle and these episodic events.

Table 2.1 shows basic properties of Mars in terms of size, mass, rotation, and orbital parameters, contrasting these with Earth. Clearly, Mars and Earth share some fundamental properties. While smaller than the Earth, the surface area of Mars is nearly the same as the land area of the Earth. The fast rotation of the planets means that much of the meteorological theory developed for Earth should be applicable to Mars, and so we have confidence that the meteorological models developed for Earth can be adapted to Mars.

Table 2.2 (also Chapter 4) shows the basic composition of the atmosphere of Mars, which has been updated using the recent measurements by the analytical laboratory on the Curiosity Rover (Mahaffy et al., 2013). Although largely carbon dioxide, the atmosphere contains several trace gases whose isotopes are useful in understanding the age and exposure of the surface (Farley et al., 2013), the amount of atmosphere loss that may have occurred, and as tracers of transport (e.g. water vapor, carbon monoxide, and argon). Searches for other trace gases (e.g. methane) continue, as they could be signatures of biological and/or geochemical processes.

Table 2.2. *Atmospheric composition and other data.*

	Mars	Earth	Ratio
Composition (vol.%)[a]			
Carbon dioxide CO_2	0.960	0.00036	
Argon Ar	0.0193	0.00934	
Nitrogen N_2	0.0189	0.781	
Oxygen O_2	0.00145	0.209	
Mean molecular weight, m_w (g mol^{-1})	43.6	29.0	
Gas constant, $R = R^*/m_w$ (J K^{-1} kg^{-1})	191	287	
Specific heat, c_p (at 200 K) (J K^{-1} kg^{-1})	735	1000	
$\kappa = R/c_p$	0.259	0.287	
Ratio of specific heats, $\gamma = c_p/c_v$	1.2	1.4	
Surface pressure[b] (at mean radius) (hPa)	6.3	1013	0.62%
Column mass[b] (mean) (kg m^{-2})	170	1030	1.65%
Column water (pr cm water equivalent)[b]	<0.0008	≤8	~0.01%
Planetary equilibrium temperature, T_e (K)	210	256	0.82
Adiabatic temp. lapse rate, $-g/c_p$ (K km^{-1})	−5.0	−9.8	0.47
Radiative time constant (days)	~2	≥20	~10%

[a] Mahaffy et al. (2013).
[b] These quantities are highly variable on Mars, much more than the ~10% variation on Earth.

In Table 2.3 some basic meteorological scaling parameters are also shown. Again, it is the similarities in some, but not all, that makes Mars a high-priority object for comparative study.

2.7 SUMMARY

As noted in the Introduction (Chapter 1), the existence of other planets in our Solar System (and now known to be around other stars, as well) provides natural laboratories that can test our theories and our tools for understanding planetary climates, including our own. Mars plays a special role in comparison with Earth, given its many fundamental similarities – rapid rotation, seasons, and sunlight largely absorbed at its surface – and its key differences – its greater distance from the Sun, a mostly CO_2 atmosphere which seasonally condenses and sublimes, little if any surface water today, and no global magnetic field at present. The similarities mean our terrestrial models should be applicable to Mars, while the differences can truly test our hypotheses of atmospheric behavior, when we have the needed data. This is certainly true for modern Earth and Mars, but an equally intriguing comparison is that of modern Earth and ancient Mars – the more Earth-like Mars with water coursing across its surface, shallow lakes and possibly seas, a thicker

Table 2.3. *Dynamical parameters for Mars and Earth.*

Parameter	Mars	Earth	Units
Scale height (lower atmos.), $T_e R/g$	10.8	7.5	km
Equatorial rotation speed, $a\Omega$	241	465	m s^{-1}
External gravity wave speed, $(gH)^{1/2}$	200	271	m s^{-1}
Lamb's parameter, $4(a\Omega)^2/gH$	5.8	11.8	—
Speed of sound, $(\gamma gH)^{1/2}$	219	321	m s^{-1}
Typical lapse rate (just above boundary layer)	−2.5	−6.5	K km^{-1}
Brunt–Väisälä frequency, N	0.67×10^{-2}	1.12×10^{-2}	s^{-1}
Internal gravity wave speed, NH	72	84	m s^{-1}
Rossby radius of deformation, NH/Ω	1022	1154	km

atmosphere shielded by a global magnetic field. And, finally, that ancient Mars with its rock record still preserved today may tell us about early Earth and early planetary development.

The compelling human interests that today drive exploration in the 21st century are much the same as those that engaged the public early in the 20th, when Lowell and other astronomers were trying to make sense of what they saw. Those interests are expressed today as follows:

- *Life*. Are we alone in this Universe? Did life develop on Mars, perhaps in a more Earth-like early environment? If so, is there evidence of that life preserved on the planet today? Is there life even today on or near its surface? Life on Earth has dramatically affected our atmosphere and climate. What happened on Mars?
- *Climate*. There is evidence that the climate of Mars was different in the past – why? How much is the climate changing today? Can Mars give us clues as to *how* planetary climates change, providing insight into the past and future climates of our own planet?
- *Destination*. Of all the planets, it seemed then, as well as now, that Mars would be most hospitable to explorers from Earth.

These remain compelling themes of Mars exploration.

Today's observations of the present planet are the best means of establishing what its present climate is and how it may have changed. As in the past, atmospheric circulation/climate models remain the best integrators of climate processes as we understand them. They permit us to explore the uncertainties and the possible roles of new physical processes needed to truly understand the Martian climate and its evolution. The data tell us whether the models have the right processes working in the proper way. The current data already tell us that we still have much to learn, even given the impressive returns from our latest (third) wave of space exploration. We still lack key information needed to describe the current Martian atmospheric state and circulation; such data are also needed for validation and improvement of atmospheric models. Some high-priority goals are observations of winds and of detailed boundary layer processes and microphysical processes generally. This is true even after a half-century of space observation, analyses, and model development.

All the current space missions are in extended operations; several (Opportunity, MEX, ODY, MRO) are still returning highly valuable data a decade or more after their launch, while Curiosity, MOM, and MAVEN continue to explore, building on their prime missions. This is a testament to the excellent technical capabilities of the spacecraft and payloads and to the dedicated efforts of the hundreds of women and men who built, launched, and then operated them – and those who continue do so even now.

If we are to have confidence in our understanding of the present and past atmosphere of Mars, we need the data that these aging but capable assets continue to provide, and we also need to fill the many key data gaps with new missions that have the technical capabilities to make the necessary advances in observing (Chapter 18). And we need to continue to improve our models, which continue to be our best means for integrating diverse datasets and physical intuition and for ultimately understanding the impressive climate change that has occurred on Mars.

For now, however, the reader is invited into the further discussions of our present understanding of the atmosphere and climate of Mars, remembering that such understanding is rooted in observational coverage in time and space that is unprecedented – despite its limitations – for any other planet except, of course, our own Earth.

ACKNOWLEDGMENTS

The writing and research of this chapter was carried out at the Jet Propulsion Laboratory, California Institute of Technology, under a contract with the National Aeronautics and Space Administration.

REFERENCES

Acuna, M. H., J. E. P. Connerney, N. F. Ness, et al. (1999), Global distribution of crustal magnetization discovered by the Mars Global Surveyor MAG/ER experiment. *Science* 284: 790–793, doi:10.1126/science.284.5415.790.

Anderson, E. M. and C. B. Leovy (1978), Mariner 9 television limb observations of dust and ice hazes on Mars. *J. Atmos. Sci.* 35: 723–734.

Boynton, W. V., W. C. Feldman, S. W. Squyres, et al. (2002), Distribution of hydrogen in the near surface of Mars: evidence for subsurface ice deposits. *Science* 297: 81–85, doi:10.1126/science.1073722.

Bridges, N., F. Ayoub, J.-P. Avouac, et al. (2012a), Earth-like sand fluxes on Mars. *Nature* 485: 339–342, doi:10.1038/nature11022.

Bridges, N., M. C. Bourke, P. E. Geissler, et al. (2012b), Planet-wide sand motion on Mars. *Geology* 40: 31–34, doi:10.1130/G23273.1.

Byrne, S., C. M. Dundas, M. R. Kennedy, et al. (2009), Distribution of mid-latitude ground ice on Mars from new impact craters. *Science* 325: 1674–1676, doi:10.1126/science.1175307.

Cassini (1666), *J. Savants* 2: 316.

Chamberlain, M. A., and W. V. Boynton (2007), Response of Martian ground ice to orbit-induced climate change. *J. Geophysical Res.–Planets* 112, doi:10.1029/2006JE002801.

Chapman, S., and R. S Lindzen (1970), *Atmospheric Tides* (New York: Gordon and Breach).

Clancy, R. T., B. J. Sandor, M. J. Wolff, et al. (2000), An intercomparison of ground-based millimeter, MGS TES, and Viking atmospheric temperature measurements: seasonal and interannual variability of temperatures and dust loading in the global Mars atmosphere. *J. Geophys. Res.* 105: 9553–9571.

Coblentz, W. W., and C. O. Lampland (1927), Further radiometric measurements and temperature estimates of the planet Mars. *Sci. Paper Natl. Bur. Stds.* 22: 237–276.

Conrath, B. J. (1976), Influence of planetary-scale topography on the diurnal thermal tide during the 1971 Martian dust storm. *J. Atmos. Sci.* 33: 2430–2439.

Curran, R. J., B. J. Conrath, R. A. Hanel, V. G. Kunde, and J. C. Pearl (1973), Mars: Mariner 9 spectroscopic evidence for H_2O ice clouds. *Science* 182: 381–383.

De Vaucouleurs, G. (1954), *Physics of the Planet Mars* (London: Faber and Faber).

Di Achille, G., and B. M. Hynek (2010), Ancient ocean on Mars supported by global distribution of deltas and valleys. *Nature Geoscience* 3: 459–463.

Dundas, C. M., S. Byrne, A. S. McEwen, et al. (2014), HiRISE observations of new impact craters exposing Martian ground ice. *J. Geophys. Res.* 119, doi:10.1002/2013JE004482.

Edwards, C. S., and B. L. Ehlmann (2015), Carbon sequestration on Mars. *Geology* 43, doi:10.1130/G36983.1.

Ehlmann, B. L., and C. S. Edwards (2014), Mineralogy of the Martian surface. *Annual Review of Earth and Planetary Sciences* 42: 291–315, doi:10.1146/annurev-earth-060313-055024.

Ehlmann, B. L., J. F. Mustard, S. L. Murchie, et al. (2011), Subsurface water and clay mineral formation during the early history of Mars. *Nature* 479: 53–60, doi:10.1038/nature10582.

Fanale, F. P., J. R. Salvail, W. B. Banerdt, and R. S. Saunders (1982), Mars: the regolith–atmosphere–cap system and climate change. *Icarus* 50: 381–407.

Farley, K. A., C. Malespin, P. Mahaffy, et al. (2013), In situ radiometric and exposure age dating of the Martian surface. *Science* 343, doi:10.1126/science.1247166.

Farmer, C. B., D. W. Davies, A. L. Holland, D. D. Laporte, and P. E. Doms (1977), Mars: water vapor observations from the Viking Orbiters. *J. Geophys. Res.* 82: 4225–4248.

Feldman, W. C., W. V. Boynton, R. L. Tokar, et al. (2002), Global distribution of neutrons from Mars: results from Mars Odyssey. *Science* 297: 75–78, doi:10.1126/science.1073541.

Fergason, R. L., P. R. Christensen, and H. H. Kieffer (2006), High-resolution thermal inertia derived from the Thermal Emission Imaging System (THEMIS): thermal model and applications. *J. Geophys. Res.* 111, E12004, doi:10.1029/2006JE002735.

Forget, F., F. Hourdin, R. Fournier, et al. (1999), Improved general circulation models of the Martian atmosphere from the surface to above 80 km. *J. Geophys. Res.* 104: 24155–24176.

Gierasch, P. J. (1974), Martian dust storms. *Rev. Geophys. Space Phys.* 12: 730–734.

Gierasch, P. J., and R. M. Goody (1967), An approximate calculation of radiative heating and radiative equilibrium in the Martian atmosphere. *Planet. Space Sci.* 15: 1465–1477.

Gierasch, P. J., and R. M. Goody (1968), A study of the thermal and dynamical structure of the Martian lower atmosphere. *Planet. Space Sci.* 16: 615–646.

Gierasch, P. J., and R. M. Goody (1972), The effect of dust on the temperature of the Mars atmosphere. *J. Atmos. Sci.* 29: 400–402.

Goody, R. M., and M. J. S. Belton (1967), Radiative relaxation times for Mars: a discussion of Martian atmospheric dynamics. *Planet. Space Sci.* 15: 247–256.

Haberle R. M., C. B. Leovy and J. B. Pollack (1982), Some effects of global dust storms on the atmospheric circulation of Mars. *Icarus* 50: 322–367, doi:10.1016/0019-1035(82)90129-4.

Hanel, R. A., B. J. Conrath, W. A. Hovis, et al. (1972), Infrared spectroscopy experiment on the Mariner 9 mission: preliminary results. *Science* 175: 305–308.

Hansen, C. J., N. Thomas, G. Portyankina, et al. (2010), HiRISE observations of gas sublimation-driven activity in Mars southern polar regions: I. Erosion of the surface. *Icarus* 205: 283–295.

Hansen, C. J., S. Byrne, G. Portyankina, et al. (2013), Observations of the northern seasonal polar cap on Mars: I. Spring sublimation activity and processes. *Icarus* 225: 881–897.

Hansen, C. J., S. Diniega, N. Bridges, et al. (2015), Agents of change on Mars' northern dunes: CO_2 ice and wind. *Icarus*, 251: 264–274.

Hartmann, W. K., and O. Raper (1974), *The New Mars: The Discoveries of Mariner 9*, NASA SP-337 (Library of Congress Catalog Card #74-600084).

Hayne, P. O., D. A. Paige, J. T. Schofield, et al. (2012), Carbon dioxide snow clouds on Mars: south polar winter observations by the Mars Climate Sounder. *J. Geophys. Res.* 117, E08014, 10.1029/2011JE004040.

Hayne, P. O., D. A. Paige, N. G. Heavens, et al. (2014), The role of snowfall in forming the seasonal ice caps of Mars: models and constraints from the Mars Climate Sounder. *Icarus* 231: 122–130.

Heavens, N. G., M. I. Richardson, A. Kleinböhl, et al. (2011), The vertical distribution of dust in the Martian atmosphere during northern spring and summer: observations by the Mars Climate Sounder and analysis of zonal average vertical dust profiles. *J. Geophys. Res.*, 116, E4, E04003, doi:10.1029/2010JE003691.

Heavens, N. G., M. S. Johnson, W. A. Abdou, et al. (2014), Seasonal and diurnal variability of detached dust layers in the Mars atmosphere. *J. Geophys. Res. Planets* 119: 1748–1774, doi:10.1002/2014JE004619.

Herschel, W. (1784), On the remarkable appearances of the polar regions of the planet Mars, the inclination of its axis, the position of its poles, and its spheroidical figure; with a few hints relating to its real diameter and atmosphere. *Phil. Trans.* 24: 233–273.

Hess, S. (1950), Some aspects of the meteorology of Mars. *J. Meteorol.* 7: 1–13.

Hess, S. L., R. M. Henry, C. B. Leovy, et al. (1977), Meteorological results from the surface of Mars: Viking 1 and 2. *J. Geophys. Res.* 82, 4559–4574.

Kahn, R. (1985), The evolution of CO_2 on Mars. *Icarus* 62: 175–190.

Kaplan, L. D., G. Munch, and H. Spinrad (1964), An analysis of the spectrum of Mars. *Astrophys. J.* 139: 1–15.

Kieffer, H., B. M. Jakosky, C. W. Snyder, and M. S. Mathews (Eds) (1992), *Mars* (Tucson, AZ: University of Arizona Press).

Kleinböhl, A., R. J. Wilson, D. Kass, J. T. Schofield, and D. J. McCleese (2013), The semidiurnal tide in the middle atmosphere of Mars, *Geophys. Res. Lett.* 40: 1952–1959, doi:10.1002/grl.50497.

Kuiper, G. P. (1952), *The Atmospheres of the Earth and Planets*, rev. ed. (Chicago: University of Chicago Press), 358–361.

Laskar, J., A. C. M. Correia, M. Gastineau, et al. (2004), Long term evolution and chaotic diffusion of the insolation quantities of Mars. *Icarus* 170: 343–364.

Lee, C., W. G. Lawson, M. I. Richardson, et al. (2009), Thermal tides in the Martian middle atmosphere as seen by the Mars Climate Sounder. *J. Geophys. Res.* 114, E03005, doi:10.1029/2008JE003285.

Leighton, R. B., and B. C. Murray (1966), Behavior of carbon dioxide and other volatiles on Mars. *Science* 153: 136–144.

Leovy, C. B. (1966), Note on thermal properties of Mars. *Icarus* 5: 1–6.

Leovy, C. B., and Y. Mintz (1969), Numerical simulation of the atmospheric circulation and climate of Mars. *J. Atmos. Sci.* 26: 1167–1190.

Lindzen, R. S. (1970), The application and applicability of terrestrial atmospheric tidal theory to Venus and Mars. *J. Atmos. Sci.*, 27: 536–549.

Lowell, P. (1895), *Mars* (London: Longmans and Green).

Lowell, P. (1896), *Mars* (Boston: Houghton Mifflin).

Lowell, P. (1906), *Mars and Its Canals* (New York: Macmillan).

Lowell, P. (1908), *Mars as the Abode of Life* (New York: Macmillan).

Mahaffy, P. R., C. R. Webster, S. K. Atreya, et al. (2013), Abundance and isotopic composition of gases in the Martian atmosphere from the Curiosity Rover. *Science* 341: 263–266, doi:10.1126/science.1237966.

Malin, M. C., M. A. Caplinger, and S. D. Davis (2001), Observational evidence for an active surface reservoir of solid carbon dioxide on Mars. *Science* 294: 2146–2148, doi:10.1126/science.1066416.

Martin, L. J., P. B. James, A. Dollfus, K. Iwasaki, and J. Beish (1992), Comparative aspects of the climate of Mars: telescopic observations: visual, photographic, polarimetric. In *Mars*, ed. H. Kieffer et al. (Univ. Arizona Press, Tucson), 34–70.

McCleese, D. G., N. G. Heavens, J. T. Schofield, et al. (2010), The structure and dynamics of the Martian lower and middle atmosphere as observed by the Mars Climate Sounder: 1. Seasonal variations in zonal mean temperature, dust and water ice aerosols. *J. Geophys. Res.* 115, E12016, doi:10.1029/2010JE003677.

McEwen, A. S., C. M. Dundas, S. S. Mattson, et al. (2014), Recurring slope lineae in equatorial regions of Mars. *Nature Geoscience* 7: 53–58.

Mellon, M. T., R. E. Arvidson, H. G. Sizemore, et al. (2009), Ground ice at the Phoenix landing site: stability state and origin. *J. Geophys. Res.* 114, E00E07, doi:10.1029/2009JE003417.

Mintz, Y. (1961), The general circulation of planetary atmospheres. *The Atmospheres of Mars and Venus*, ed. Kellogg and C. Sagan, NAS-NRC Publ. 944: 107–146.

Mitrofanov, I., D. Anfimov, A. Lozyrev, et al. (2002), Maps of subsurface hydrogen from the high energy neutron detector, Mars Odyssey. *Science* 297: 78–81, doi:10.1126/science.1073616.

Murchie, S. L., J. Mustard, B. L. Ehlmann, et al. (2009), A synthesis of Martian aqueous mineralogy after 1 Mars year of observations from the Mars Reconnaissance Orbiter. *J. Geophys. Res.* 114, doi:10.1029/2009JE003342.

Ojha, L., M. B. Wilhelm, S. L. Murchie, et al. (2015), Spectral evidence for hydrated salts in recurring slope lineae on Mars. *Nature Geoscience* 8: 829–832, doi:10:1038/ngeo2546.

Owen, T. (1992) The composition and early history of the atmosphere of Mars. In *Mars*, ed. H. Kieffer et al. (Univ. Arizona Press, Tucson), 818–834.

Phillips, R. J., M. T. Zuber, S. E. Smrekar, et al. (2008), Mars north polar deposits: stratigraphy, age, and geodynamical response. *Science* 320: 1182–1185, doi:10.1126/science.1157546.

Phillips, R. J., B. J. Davis, K. L. Tanaka, et al. (2011), Massive CO_2 ice deposits sequestered in the south polar layered deposits of Mars. *Science* 332: 838–841, doi:10.1126/science.1203091.

Pollack, J. B., C. B. Leovy, Y. H. Mintz, and W. Van Camp (1976), Winds on Mars during the Viking season: predictions based on a general circulation model with topography. *Geophysical Research Letters* 3, doi:10.1029/GL003i008p00479.

Pollack, J. B., J. F. Kasting, S. M. Richardson, and K. Poliakoff (1987), The case for a wet, warm climate on early Mars. *Icarus* 71: 203–224.

Richardson, M. I., A. D. Toigo, and C. E. Newman (2007), PlanetWRF: a general purpose, local to global numerical model for planetary atmospheric and climate dynamics. *J. Geophys. Res.* 112, E09001, doi:10.1029/2006JE002825.

Sagan, C. A., and J. B. Pollack (1969), Windblown dust on Mars. *Nature* 223: 791–794.

Seiff, A., and D. B. Kirk (1977), Structure of the atmosphere of Mars in summer at mid-latitudes. *J. Geophys. Res.* 82: 4364–4378.

Shirley, J. H. (2015), Solar system dynamics and global-scale dust storms on Mars. *Icarus* 252: 128–144, 10.1016/j.icarus.2014.09.038.

Slipher, E. C. (1962), *The Photographic Story of Mars* (Flagstaff: Northland Press).

Smith, D. E., M. T. Zuber, S. C. Solomon, et al. (1999), The global topography of Mars and implications for surface evolution. *Science* 284: 1495–1503.

Smith, D. E., M. T. Zuber, H. V. Frey, et al. (2001), Mars Orbiter Laser Altimeter: experiment summary after the first year of global mapping of Mars. *J. Geophys. Res.: Planets* 106: 23689–23722.

Smith, P. H., L. K. Tamppari, R. E. Arvidson, et al. (2009), H_2O at the Phoenix landing site. *Science* 325: 58–61, doi:10.1126/science.1172339.

Spinrad, H., G., Munch, and L. D. Kaplan (1963), The detection of water vapor on Mars. *Astrophys. J.* 137: 1319–1321.

Spinrad, H., R. A. Schorn, R. Moore, L. P. Giver, and H. J. Smith (1966), High dispersion spectroscopic observations of Mars. I. The CO_2 content and surface pressure. *Astrophys. J.* 146: 331–338.

Stewart, A. I., C. A. Barth, C. W. Hord, and A. L. Lane (1972), Mariner 9 ultraviolet spectrometer experiment: structure of Mars upper atmosphere. *Icarus* 17: 469–474.

Tamppari, L., R. W. Zurek, and D. A. Paige (2000), Viking era water-ice clouds. *J. Geophys. Res.* 105: 4087–4107.

Thomas, P. C., W. Calvin, R. Haberle, et al. (2014), Mass balance of Mars' south polar residual cap from spacecraft imaging. In *Eighth International Conference on Mars*, LPI Contribution 1791: 1085.

Thomas, P. C., W. Calvin, B. Cantor, et al. (2016), Mass balance of Mars' residual south polar cap from CTX images and other data, *Icarus*, in press.

Tolson, R. H., G. M. Keating, R. W. Zurek, et al. (2007), Application of accelerometer data to atmospheric modeling during Mars aerobraking operations. *J. Spacecraft and Rockets* 44(6): (1172–1179).

Von Braun, W., and W. Ley (1956), *The Exploration of Mars* (New York: Viking Press).

Wallace, A. R. (1907), *Is Mars Habitable?* (London: Macmillan).

Wilson, R. J. (2000), Evidence for diurnal period Kelvin waves in the Martian atmosphere from Mars Global Surveyor TES data. *Geophys. Res. Lett.* 27(23): 3889–3892, doi:10.1029/2000GL012028.

Zurek, R. W. (1976), Diurnal tide in the Martian atmosphere. *J. Atmos. Sci.* 33: 321–337.

Zurek, R. W. (1980), Surface pressure response to elevated tidal heating sources: comparison of Earth and Mars. *J. Atmos. Sci.* 37: 1132–1136.

Zurek, R. W. (1982), Martian Dust Storms, An Update. *Icarus* 50: 288–310.

Zurek, R. W. (1992), Comparative aspects of the climate of Mars: an introduction to the current atmosphere. *Mars*, ed. H. Kieffer et al. (University of Arizona Press), Tucson, 799–817.

Zurek, R. W., and C. B. Leovy (1981), Thermal tides in the dusty Martian atmosphere: a verification of theory. *Science* 213 (4506): 437–439, doi:10.1126/science.213.4506.437.

History of Mars Atmosphere Observations

PHILIP B. JAMES, PHILIP R. CHRISTENSEN, R. TODD CLANCY, MARK T. LEMMON, PAUL WITHERS

3.1 INTRODUCTION

William Herschel seems to have been the first astronomer to suggest that Mars has an atmosphere, based upon his observations of brightening that he ascribed to clouds. In the two centuries since Herschel we have come to understand the Martian atmosphere better than any besides our own, due in large part to the advent of space observation of the planet in 1964. The last general review of the history of observations of Mars, including its atmosphere and climate, was contained in the book *Mars* (Kieffer et al., 1992). We shall concentrate on the history of observation of Mars' atmosphere since the publication of that book rather than dwell on earlier developments. After reviewing the contributions of Viking to the study of the Martian atmosphere, recent missions and experiments are divided into three sections: Earth-based spectroscopic observations, remote sensing observations from orbit, and *in situ* observations from landers and rovers. We will not present the history of measurements of the space environment of Mars but refer the reader to Chapters 14 and 15 of this book as well as to the review by Withers (2009).

3.2 PRE-VIKING OBSERVATIONS

Ground-based astronomers made many significant contributions to our knowledge of the Martian atmosphere prior to the first spacecraft observations (Martin et al., 1992). They observed dust and condensate clouds in the Martian atmosphere, determined seasonal polar cycles, and described planet-encircling dust storms (Lowell, 1895; Kuiper, 1957) and mapped their evolution (Martin, 1976). Kuiper (1952) identified CO_2 in spectra of the Martian atmosphere in 1947. Kaplan et al. (1964) determined CO_2 column density and Mars surface pressure through a combined analysis of an optically thin visible (0.87 μm) CO_2 band absorption together with the near-infrared optically thick CO_2 line absorptions observed by Kuiper. Spinrad et al. (1963) first detected Mars atmospheric water from ground-based measurement of weak 0.82 μm band absorption. The distribution of water vapor on Mars has been mapped in time and space by ground-based high-resolution spectroscopy over visible (Barker et al., 1970; Sprague et al., 1996), infrared (Encrenaz et al., 2005), and radio wavelengths (Clancy et al., 1992, at 22 GHz).

 Several spacecraft visited Mars between 1964 and 1975 (Snyder and Moroz, 1992). Radio occultation measurements by Mariner 4 determined that the average surface pressure was approximately 6 mbar (Kliore et al., 1965) and led to the conclusion that the prominent seasonal Martian polar caps were composed of condensed CO_2 and perhaps buffered the atmosphere (Leighton and Murray, 1966). Mariner 9, the first spacecraft to orbit Mars, included a wide-angle camera ideal for cloud studies; the imaging team observed numerous localized dust and condensate clouds in addition to the waning phases of a major planet-encircling dust storm (Leovy et al., 1972). The Mariner 9 Infrared Interferometer/Spectrometer (IRIS) determinations of the atmospheric temperature field (Conrath et al., 1973) provided the first quantitative characterization of the diurnal thermal tide (Pirraglia and Conrath, 1974). These IRIS observations were also the basis for finding that the radii of the dust particles were on the order of a few micrometers and determining the presence of H_2O ice clouds with a mean ice particle radius of 2.0 μm (Curran et al., 1973). The Mariner 9 UVS experiment measured O_3 and discovered its considerable spatial, seasonal variation in the Mars atmosphere (Barth and Dick, 1974). Despite a general lack of success of several Soviet probes launched during this time period, the descent module of the Soviet Mars 6 mission returned the first *in situ* measurements of the vertical structure in Mars' atmosphere in 1974 (Avduyevskiy et al., 1975).

3.3 VIKING AND POST-VIKING YEARS

Much of the basis for our understanding of the Martian atmosphere is built on a foundation of analyses of data acquired by Viking Orbiters and Landers. Even though the orbiters were not designed primarily for atmospheric observations, one Mars year of data were acquired by Viking Orbiter 2 and over two Mars years of data by Viking Orbiter 1. The Viking Landers provided data on the state and composition of the atmosphere that are still primary datasets for composition and surface pressure used in atmospheric studies. Tables 3.1 and 3.2, respectively, compare the characteristics of cameras and infrared experiments most relevant to study of the atmosphere from Viking and subsequent missions.

3.3.1 Viking Orbiters

The Viking mission to Mars consisted of two orbiters and two landers. The identical Viking Imaging Subsystems (VIS) on the two orbiters inherited a basic design from earlier Mariners but

Table 3.1. *Properties of synoptic imaging systems that have been included on orbital missions to Mars.*

	Mariner 9 A	Viking	MOC-WA	MARCI
Type	Slow videcon Framing	Slow videcon Framing	CCD 1D Push-broom	CCD 2D Push-frame
FOV (deg)	11×14	0.97×1.11	140	180
Image size (pixels)	700×832	1056×1182	3456 samples per line	16×1024 framelets
Filters	4 visible, clear; 3 polarization	5 visible, clear	2 visible	5 visible; 2 UV
Eccentricity and periapsis height (km)	$e = 0.601$ $p = 1650$	$e = 0.81$ (varied) $p = 778$	$e = 0.005$ $p = 380$	$e = 0.009$ $p = 255$
Scale (m/pixel)	>460 (at periapsis)	<492 (at apoapsis)	240 at nadir	600–750 at nadir

Table 3.2. *Properties of thermal infrared instruments that have been included on orbital missions to Mars.*

Instrument	Spectral range	Spectral resolution	Spatial resolution	NESR	Mission	Observations
Mariner 6/7 IRS	1.9–14.4 μm	1%			Flyby. July 30 and Aug. 4, 1969	Flyby
Mariner 9 IRIS	5–50 μm	2.4 cm^{-1}	130–1325 km	0.5×10^{-7} W cm^{-2} sr^{-1}/ cm^{-1}	Orbital. Nov. 1971–Oct. 1972	Atmospheric sampling. Variable time of day and resolution
Viking IRTM	7, 9, 11, 15, 20 μm	1–2 μm	8–170 km		Orbital. June 1976–Aug. 1980	Global. Variable time of day and resolution
Viking MAWD	1.38–1.39 μm	1.2 cm^{-1}	3×24 km		Orbital. June 1976–Aug. 1980	Global. Variable time of day and resolution
MGS TES	5.8–50 μm (200–1709 cm^{-1})	6.5, 13 cm^{-1}	3×5 km	2.5–6×10^{-8} W cm^{-2} sr^{-1}/ cm^{-1}	Orbital. Sept. 1997–Nov. 2006	Global. Sun-synchronous; ~2:30 p.m. local time
Odyssey THEMIS	6–16 μm	1 μm	0.1 km	2–3.5×10^{-6} W cm^{-2} sr^{-1} μm^{-1}	Oct. 2001–present	Global. Sun-synchronous; ~5:30 p.m. local time
Mars Express PFS	1.2–45 μm (220–8190 cm^{-1})	1.3 cm^{-1}	≤10 km	1.0×10^{-10} W cm^{-2} sr^{-1}/ cm^{-1}	Dec. 2003–present	Global. Variable time of day and resolution
MER Mini-TES	5–29 μm	9–99 cm^{-1}	20 mrad	1.8–5×10^{-8} W cm^{-2} sr^{-1}/ cm^{-1}	Jan. 2004–present	Upward-looking from surface in two locations
MRO MCS	11.8–42.1 μm	20–100 cm^{-1}	5.0×8.6 km at limb		March 2006–present	Global. Sun-synchronous; ~3:00 p.m. local time

differed in two major respects (Carr et al., 1972). Instead of the wide-angle/narrow-angle combination, a pair of high-resolution optical systems were adopted. And the repetition interval between frames was reduced to only 1/20th of that of Mariner 9 so that sequential images would overlap despite spacecraft motion at periapsis. Together these features allowed the orbiter to cover a region surrounding a putative landing site at high resolution on one orbit from near their initial 1500 km periapses for purposes of site certification. The initial periapsis latitudes were chosen to cover possible landing sites but were later adjusted to around

39°N for Orbiter 1, whose inclination was ~39°, and 55°N for Orbiter 2, which had inclination of ~80°. The resolution of the cameras from near apoapsis was optimum for atmospheric physics, but the result of these orbital constraints was a geographical bias in synoptic imaging. The situation was further complicated by the adjustable walk rate of the longitude of periapsis and by the moving terminator (Snyder, 1977).

A striking example of the effects of orbital constraints is the disparity in dust storm coverage. Viking Orbiter 2 acquired many low-resolution images in the southern hemisphere that, when mosaicked, provided excellent synoptic views of the development of two planet-encircling storms as well as many localized dust storms that occurred in the southern hemisphere during MY 12 (Briggs et al., 1978). On the other hand, the relatively small area encompassed by images in northern latitudes limited the number of dust observations there, perpetuating the false premise that dust storms were mainly a southern hemisphere phenomenon.

The Viking Orbiters made many observations of both condensate and dust clouds that are summarized in a comprehensive paper by Kahn (1984). The non-Sun-synchronous orbits of the Viking spacecraft together with the ability to point the cameras on a scan platform allowed observations of the diurnal behaviors of various atmospheric phenomena that have not been possible on subsequent missions. These included interesting cloud phenomena in the early morning hours near the Tharsis volcanoes: bore waves resulting from cold, downslope flows (Hunt et al., 1981; Kahn and Gierasch, 1982) and a variety of plumes that are not seen at later local times (Hunt et al., 1980).

The two Viking Orbiters, which orbited Mars from June 1976 through August 1980, each carried the Infrared Thermal Mapper (IRTM) and the Mars Atmospheric Water Detector (MAWD) experiments. The IRTM investigated the surface, atmosphere, and polar ices using three infrared channels intended for atmospheric observations: a ~1 μm wide filter centered on the 15 μm CO_2 gas absorption band determined the atmospheric temperature at a pressure of ~0.5 mbar; an 11 μm band observed atmospheric water ice; and a 9 μm band detected silicate dust in the atmosphere (Kieffer et al., 1977). It later was discovered that the 15 μm band had a spectral leak that led to temperatures ~10 K too warm (Wilson and Richardson, 2000). The IRTM had a 5.2 mrad field of view in all bands, giving spatial resolutions of 8–170 km depending on the orbiter's position in its orbit. An important characteristic of the IRTM instruments was the collection of diurnal data from the elliptical Viking orbit (1500 km periapsis, 32 000 km apoapsis) in which the periapsis was allowed to drift in local time throughout the mission. The IRTM investigation successfully mapped atmospheric temperatures, measured the column-integrated dust opacity over two Mars years, and detected and mapped water ice clouds (Kieffer et al., 1977; Martin, 1981; Christensen and Zurek, 1984). IRTM also established the compositions of the residual polar caps: water ice in the north (Kieffer et al., 1976) and CO_2 ice in the south (Kieffer, 1979).

The first detailed mapping of Mars atmospheric water column abundances (typically presented in units of precipitable micrometers (pr μm)) was performed by the Mars Atmospheric Water Detector (MAWD) on-board the two Viking Orbiters (e.g. Farmer et al., 1977; Jakosky and Farmer, 1982). The MAWD experiment was a five-channel solar reflectance grating spectrometer operating in the weak 1.4 μm water vapor bands; MAWD used cooled PbS detectors and had a spectral resolution of 1.2 cm⁻¹ (Farmer et al., 1977). The instantaneous field of view projected onto the surface of the planet at periapsis (1500 km) was 3×24 km; MAWD could be step-scanned across the surface to generate sets of observations to provide regional mapping or localized observations of the diurnal characteristics of selected locations. The observed seasonal dependence of the latitude distribution of the column abundance of vapor was found to be consistent with a model in which the vapor is in equilibrium with the regolith at polar and mid-latitudes, with a permanent reservoir of water ice buried at a depth of 10 cm to 1 m at all latitudes poleward of 40° (Farmer and Doms, 1979). The observed behavior of vapor suggested an annual net transport of the vapor phase from the southern to the northern hemisphere, with deposition of ice of thickness of the order of a few milligrams per square centimeter in the northern polar latitudes (Farmer and Doms, 1979).

3.3.2 Viking Landers

The two Viking Landers are located in Chryse Basin and in Utopia near Mie Crater. Each lander had a camera in which a rotating mirror directed light from a vertical scan to one of 12 diodes, each of which had a different filter or function. The camera physically rotated by small increments to provide the horizontal picture dimension. One of the diodes was used to image the Sun, and these data were used to calculate the opacity of the atmosphere and aerosol properties near the landers (Pollack et al., 1977). In 1982, after the demise of both orbiters, lander images revealed the presence of a major dust storm near $L_s = 200°$.

Viking established the major components of the atmosphere by *in situ* measurements in 1976–1977. N_2 and ^{40}Ar were determined through Viking Lander and descent entry mass spectrometry to contribute 2.7% and 1.6% of the Mars atmosphere by number density (Nier et al., 1976a; Owen and Biemann, 1976). Viking mass spectrometry also detected the noble gas species ^{36,38}Ar, Ne, Kr, and Xe at much smaller concentrations (5 to 0.1 ppmv), as well as upper atmospheric species such as NO and O⁺. Viking mass spectrometry during lander descent and surface operations in 1976–1977 suggested terrestrial ^{12}C/^{13}C and ^{16}O/^{18}O ratios for CO_2, within 5% uncertainties (Nier et al., 1976b). Viking Lander mass spectrometry significantly expanded the range of Mars isotopic observations with measurements of depleted ^{14}N/^{15}N (Nier et al., 1976b), and enhanced isotopic ratios ^{40}Ar/^{36}Ar and ^{129}Xe/^{132}Xe (Owen and Biemann, 1976; Owen et al., 1976, 1977) relative to terrestrial values. Atmospheric composition and isotopic data are discussed in more detail in Chapters 4 and 17 of this book.

Mars atmospheric CO_2 abundances were predicted by Leighton and Murray (1966) to vary seasonally, associated with the retreat and growth of high-latitude CO_2 surface ice as Mars progresses annually about its eccentric orbit. A seasonal variation of ±15% in atmospheric CO_2 abundance (and, hence, surface pressure) was recorded over parts of several Mars years by Viking surface lander pressure sensors (Hess et al., 1980; Tillman, 1988). The Viking Lander pressure data collected over

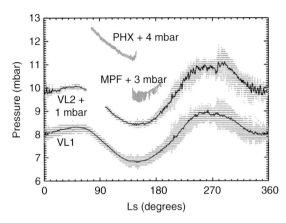

Figure 3.1. Surface pressure measurements (gray dots) from Viking Lander (VL) 1 and 2, Mars Pathfinder (MPF), and Phoenix (PHX) as a function of season. Reproduced from Withers (2012) by permission. Pressure offsets have been applied to several datasets to improve the clarity of this figure. Coarse digitization of the Viking data is apparent. The black lines through the VL1 and VL2 measurements indicate the diurnal mean surface pressure.

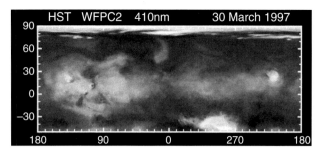

Figure 3.2. This image is a mosaic of three HST images acquired on March 30, 1997 ($L_s = 98°$) using WFPC2 with the 410 nm filter. The belt of clouds centered north of the equator appears shortly before aphelion ($L_s = 71°$) and persists through early summer. The water ice clouds form when water vapor that appears as the north polar cap sublimes is lifted in the ascending branch of the Hadley cell and freezes. The brightest portions of the belt correspond to the Tharsis volcanoes and Elysium Mons, and bright clouds also fill the Hellas Basin.

four Mars years are remarkable for their nearly perfect inter-annual repeatability (Tillman, 1988); inter-annual variation is estimated to be less than 2% (Paige and Wood, 1992). Figure 3.1 shows pressure measurements from the Viking Landers together with those of the subsequent Pathfinder and Phoenix spacecraft. The pressure measurements during the two major dust events that occurred during the Viking mission provided important data for understanding thermal tides on Mars (Zurek and Leovy, 1981).

3.3.3 Phobos 2

The Phobos 2 mission was launched in July 1988 and arrived at Mars in January 1989. Unfortunately, the spacecraft was lost in March 1989 while maneuvering in Martian orbit to encounter Phobos. Although most of the original science objectives were not met, Phobos 2 data have been used in several studies of particles and fields and of the upper Martian atmosphere (Blamont, 1991; Krasnopolsky et al., 1991). Phobos 2 also provided the first solar occultation measurements from Mars orbit, obtaining vertical profile measurements of aerosols and ozone (e.g. Chassifière et al., 1992).

3.3.4 Hubble Space Telescope

Between 1980, when Viking Orbiter 1 failed due to exhaustion of its attitude control gas, and 1998 when Mars Global Surveyor performed limited observations during aerobraking, only the shortened Phobos mission provided observations of Mars by orbiting spacecraft. During the hiatus, the Hubble Space Telescope (HST) conducted several groups of synoptic observations. These were not continuous because the angular separation of Mars and the Sun excluded observations about half of the time, and high demand for the telescope restricted images to roughly every 30° of L_s. However, sequences often included images of Mars separated by 120° of longitude on the same day, thus providing global coverage

(albeit with local times that varied over the disk). Because Mars is a very bright target for HST, exposure times for individual images were very short, allowing images through a number of filters encompassing a broad spectral range from 255 to 1042 nm.

Because of their synoptic quality and spectral resolution, HST images were valuable for the study of clouds, especially after WFPC2 was installed to rectify earlier imaging problems. A band of clouds circling the planet between 10°S and 30°N that has become known as the aphelion cloud belt was discovered in images from the aphelic ($L_s = 60°$) 1995 opposition (James et al., 1996). HST images of the aphelion cloud belt from 1997 are shown in Figure 3.2. The spatial and temporal extent of these clouds and their opacities were subsequently determined from the cumulative HST data from several years (Wolff et al., 1999). These clouds have a significant role in the seasonal climate cycle on Mars (Clancy et al., 1996). Viking data (Christensen, 1998) confirmed that this cloud band was also extant in those years, and it has been a climate fixture during subsequent investigations.

3.4 GROUND-BASED SPECTROSCOPY

Earth-based observations of isotopic ratios and trace components such as O_2, H_2O_2, and CH_4 have retained their importance relative to Mars spacecraft measurements even to the current date because spectroscopic platforms necessary to achieve relevant sensitivities have not been flown on Mars spacecraft so far. Trace constituents in the Mars atmosphere are largely associated with dissociation products of CO_2, water vapor, and N_2 (to a lesser extent). Terrestrial-based detections of photochemical products include O_2, CO, O_3, H_2, D, and H_2O_2. The first detections of such trace components were obtained from ground-based high-resolution spectroscopy at a variety of observing wavelengths. Kaplan et al. (1969) employed high-resolution spectra at 2.4 μm to obtain the first measurements of Mars CO, which was subsequently measured by ground-based microwave spectroscopy at 115 GHz (Kakar et al., 1977). Several years after the possible detection of O_2 by Belton and Hunten (1968), Barker (1972) and Carleton and Traub (1972) employed 0.76 μm spectroscopy to definitively measure the abundance of O_2, which is directly related to CO abundance as a collateral product of CO_2 photolysis. Most recently, the Herschel Observatory measured O_2 with exceptional sensitivity using submillimeter spectroscopy (Hartogh et al., 2010). As with microwave CO and water vapor line absorptions, this submillimeter O_2 line absorption supports vertical profiling associated with its pressure-broadened line width.

Following the initial Mariner 9 observations (Barth et al., 1973), Mars O_3 has been characterized by a wide variety of Hubble Space Telescope (Clancy et al., 1999), ground-based (Espenak et al., 1991; Fast et al., 2006), and Mars spacecraft observations. All of the space-based observations measure ultraviolet absorption in the ozone Hartley band, whereas the unique ground-based infrared (9.7 μm) observations employ very high spectral resolution, heterodyne spectroscopy. Mars O_3 measurements have been interpreted in terms of variable atmospheric water columns (Barth et al., 1973; Perrier et al., 2006), water vapor saturation profiles (Clancy and Nair, 1996), and potential heterogeneous activity on water ice clouds (Lefèvre et al., 2008). The detection of 1.27 μm band emission from electronically excited O_2 (singlet delta) in 1973 (Noxon et al., 1976) is closely related to measurement of Mars O_3, in that O_2 singlet delta primarily results from photodissociation of O_3. Subsequent ground-based (Novak et al., 2002; Krasnopolsky, 2007) measurements of O_2 singlet delta have been interpreted in terms of seasonal and latitudinal distributions of Mars ozone.

The low abundances of CO, O_2, and O_3 returned by the early ground-based and Mariner 9 observations indicated that some form of catalytic chemistry, operating on a much faster timescale than simple three-body recombination ($CO+O+CO_2 \rightarrow 2CO_2$), must be active in the Mars atmosphere. Parkinson and Hunten (1972) and McElroy and Donahue (1972) proposed that odd-hydrogen products of water vapor photolysis (H, HO_2, and OH) play this role, and so enforce the basic stability of the Mars CO_2 atmosphere to photolytic decomposition. Still, no measurement of these odd-hydrogen species exists to date. The Earth-orbiting HST and FUSE observatories have provided ultraviolet spectral measurements of upper atmospheric H_2 (Lyman β; Krasnopolsky and Feldman, 2001) and D (Lyman α; Krasnopolsky et al., 1998) abundances. However, their interpretation is more specific to atmospheric escape of water than to lower atmospheric photochemistry. Two distinct measurements of H_2O_2, a key atmospheric reservoir for HO_2, are more relevant to atmospheric photochemistry. Very high spectral resolution is a primary observational requirement of these ground-based submillimeter (Clancy et al., 2004) and infrared (Encrenaz et al., 2004) H_2O_2 line absorption measurements. These and subsequent infrared H_2O_2 measurements (Encrenaz et al., 2008, 2012) verify the basic applicability of odd-hydrogen catalytic chemistry in the Mars atmosphere (e.g. Lefèvre et al., 2008).

A key exception to the standard photochemical interpretation of Mars atmospheric trace composition is the potential presence of methane in surprising amounts. Absorption spectra from very high-resolution, ground-based spectroscopy at 3.3 μm (Mumma et al., 2009) forms the strongest evidence for Mars atmospheric methane and a substantial spatial, temporal variability in its abundance (<3 to >100 ppbv). This detection has been challenged on the basis of possible terrestrial methane line contaminations (Zahnle et al., 2011) and the fact that the observed variability is difficult to reconcile with potential methane loss mechanisms (Lefèvre and Forget, 2009). More recently, the MSL (Mars Science Laboratory) SAM (Sample Analysis at Mars) experiment indicates lower levels of methane abundance in Gale Crater, with significant time variability (<1 to 7 ppbv) over seasonal timescales (Webster et al., 2015). The still controversial detections of methane have also stimulated interest in other potential non-equilibrium species such as SO_2, for which ground-based upper limits exist (Krasnopolsky, 2006).

The first isotopic ratio measurements for Mars were obtained from ground-based, high-resolution spectroscopic observations by Connes et al. (1969) at near- to thermal IR wavelengths (2–13 μm), providing $^{12}C/^{13}C$ and $^{16}O/^{18}O$ ratio determinations for CO_2 (Young, 1971) and CO (Kaplan et al., 1969). These early analyses indicated terrestrial isotopic values within 15% uncertainties. More recent ground-based measurements of $^{13}C/^{12}C$, $^{16}O/^{18}O$, and $^{17}O/^{18}O$ isotopic ratios from high-resolution IR (4–8 μm) spectra of CO_2 generally support terrestrial isotopic ratios (Krasnopolsky et al., 1996; Encrenaz et al., 2005; Krasnopolsky, 2007), consistent with the initial ground-based and Viking measurements. Key ground-based near-IR (3.7 μm) measurements of water vapor D/H ratios (Owen et al., 1988; Bjoraker et al., 1989) indicated substantial (5×) enhancements of lower atmospheric D, presumably associated with preferential atmospheric escape of H. However, Krasnopolsky et al. (1998) observed HD Lyman alpha scattering with HST to reveal 10× reduced D/H ratios at altitudes above 100 km, suggesting additional, vertical fractionation processes (Cheng et al., 1999; Bertaux and Montmessin, 2001). Strong spatial and seasonal gradients in HDO/H_2O isotopic ratios (Mumma et al., 2003; Novak et al., 2011) also suggest cloud fractionation effects (Montmessin et al., 2005) and surface/subsurface water ice contributions to variations in lower atmospheric water isotope ratios. Since the Viking period, measurements of Mars atmospheric isotopic ratios have been augmented by laboratory analyses of SNC meteorites, which are most relevant to long-term (hundreds of millions of years) trends in isotope ratios (e.g. Bogard and Johnson, 1983; Clayton and Mayeda, 1983, 1988; Gooding et al., 1988; Wright et al., 1990; Jakosky, 1991; Jakosky and Jones, 1997; Farquhar et al., 2007).

3.5 ORBITERS

The decade of the 1990s opened with a major disappointment: the Mars Observer Orbiter, which was intended to rekindle the U.S. Mars exploration program, was lost prior to Mars orbit insertion in 1993. Instruments lost on Mars Observer subsequently were carried to Mars on Mars Global Surveyor (MGS), Mars Odyssey, and Mars Reconnaissance Orbiter (MRO). MGS was the only successful mission launched in the 1990s; the other three orbiters launched in the 1990s – the Russian Mars 96, the Japanese Nozomi, and the U.S. Mars Climate Orbiter – failed to achieve Mars orbit. In a reversal of fortune, all three attempts to orbit spacecraft from 2000 to 2010 were successful: ESA's Mars Express, and NASA's Mars Odyssey and Mars Reconnaissance Orbiter (MRO).

3.5.1 Mars Global Surveyor

The payload of Mars Global Surveyor (MGS), launched in 1996, included three instruments that were part of the payload of the Mars Observer spacecraft that was lost in 1993, including Mars Orbiter Camera (MOC), Thermal Emission Spectrometer (TES), and Mars Orbiter Laser Altimeter (MOLA). MGS achieved its mapping orbit in early 1999 via the process of aerobraking, in which successive passes through the outer portion of the atmosphere are used to gradually adjust the orbit. The aerobraking process was extended for several months due to a damaged solar panel on the spacecraft that necessitated more gentle encounters with the atmosphere. MGS reached a nearly circular, nearly polar, Sun-synchronous (equator crossing at ~2 p.m. local time) orbit with a period of approximately two hours; this orbit was tailor-made for observations of seasonal atmospheric phenomena.

The MOC experiment (Malin et al., 1992) took advantage of this orbit by including two push-broom linear array wide-angle (WA) cameras that had fields of view from horizon to horizon in their nadir-pointed orientation; one WA camera had a blue filter (425 nm), the other red (600 nm). The full resolution of these cameras was about 230 m/pixel at nadir. The wide-angle cameras operated continuously over the illuminated portion of the planet; in order to accommodate the wide-angle strips in the buffer, a variable summing algorithm was routinely applied to produce a constant resolution of about 7.5 km/pixel along the scan lines normal to the spacecraft ground track from nadir to horizon. These summed, MOC WA images from adjacent orbits overlap at the equator, and thus the 12–13 orbits on a given sol provide a complete map of Mars in two colors every day. In addition MOC WA images at full or slightly reduced resolution were acquired routinely as context images for the Narrow Angle (NA) camera and were also targeted at specific locations and times; these images were acquired simultaneously with the global map acquisition.

The MOC wide-angle cameras returned these daily global maps for meteorological investigations nearly every day for slightly more than four Mars years. These maps are optimum for observing seasonal changes and inter-annual variability. But they do not provide very much diurnal information at lower latitudes where the strips extend only 25° to either side of nadir at the equator; thus the images reveal atmospheric phenomena only near the 2 p.m. local time of the equator crossing of the Sun-synchronous orbit. On the other hand, since each of the 12–13 daily orbits is tangent to the 87° latitude circles, good diurnal coverage is available in the polar regions.

Most of the atmospheric studies in visible wavelengths have been based on the global map images and have relied on the utility of the dataset for studies of seasonal dependence. For example, Cantor et al. (2001) used MOC data from the first year of observations to determine the spatial and temporal locations of dust storms; they identified a class of regional storms, referred to by Wang and Ingersoll (2002) as "flushing storms", that move from the northern hemisphere into the southern hemisphere along certain preferred paths. Dust storm observations from the four Mars years observed by MGS are shown in Figure 3.3. There was no planet-encircling dust event during the first year of MRO observations, but Cantor (2006) reported the detailed evolution of a major event that began at autumnal equinox during the second year. Wang and Ingersoll (2002) discussed the seasonal behaviors of synoptic cloud systems,

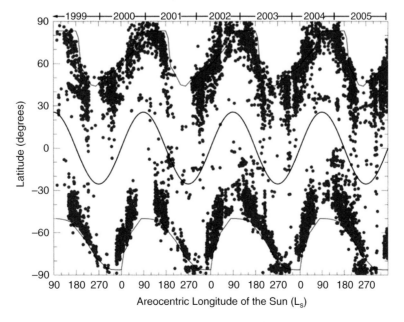

Figure 3.3. Four Mars years of MOC WA images were used to map the distribution of dust storms as a function of latitude and season on the planet. Based on Cantor (2006), with permission. The heavy solid line tracks the subsolar longitude, and finer lines plot the latitudes of the edges of the polar caps. The repetition of the pattern from year to year is apparent. The correlation between dust storms and the edges of the polar caps, indicated by the two thin dark curves, is also evident. This clearly demonstrates the advantages of the MGS/MRO orbits for understanding seasonal behavior. Terrestrial years are noted at the top of the figure.

including the aphelion cloud belt and the polar hoods, using a series of false-color maps constructed from MOC WA images. MOC data were also used to determine the seasonal dependence of the clouds associated with the Tharsis volcanoes and Elysium and their inter-annual variability (Benson et al., 2006), and equatorial mesospheric clouds were discovered in MOC and TES limb observations (Clancy et al., 2007).

The monochromatic, NA camera with resolution of 1.4 m/pixel at nadir has also been used for atmospheric studies. Cantor et al. (2006) used narrow-angle images and full-resolution wide-angle images to study the properties and distribution of Martian dust devils. (Even the largest dust devils are too small to show up in the heavily summed global mapping images.) A total of over 10 000 dust devils were included in the study.

TES was a Fourier-transform interferometric spectrometer (Christensen et al., 1992); it collected data from September 1997 through November 2006. TES collected spectra from 5.8 to 50 μm in a 3$\times$2 array of detectors with a spectral resolution in the center detectors of 12.4–14.8 and 6.2–7.4 cm^{-1}. TES also included broadband thermal (5.1 to 150 μm) and visible/near-IR (0.3 to 2.9 μm) radiometers (Christensen et al., 2001). The TES science objectives were focused on surface composition, so the spatial field of view and spectral resolution were optimized for surface features and materials. As a result, TES had significantly higher spatial resolution (3$\times$5 km) than the Mariner 9 IRIS, but a factor of 4 lower spectral resolution. These three instrument components were calibrated to an absolute accuracy of ~4$\times$10^{-8} W cm^{-2} sr^{-1}/cm^{-1} (0.5 K at 270 K), 1–2%, and ~1–2 K, respectively (Christensen et al., 2001). A significant advantage of the TES over previous instruments was an internal pointing mirror, which provided multiple viewing angles through the Mars atmosphere, allowing surface and atmospheric emission and reflection to be separated and characterized (Smith et al., 2000; Wolff and Clancy, 2003).

TES provided the first systematic mapping of the global thermal structure and water vapor, dust, and ice columns versus Mars season (L_s) and year (Pearl et al., 2001; M. Smith et al., 2001; Smith, 2002, 2008). As such, they constitute a widely employed Mars atmospheric climatology in support of observational and modeling (MGCM) studies for over a decade. They are also key indicators of inter-annual variability, particularly for dust storm activity, in the Mars atmosphere (Smith, 2004). TES data from three Martian years are summarized in Figure 3.4. The life cycles of several regional and planet-encircling dust storms were observed; the storms were found to have major effects on the atmospheric thermal structure, raising temperatures by 15 K throughout several scale heights and leading to an intensification of the Hadley circulation and rapid heating in the opposite hemisphere (M. Smith et al., 2001). The Hadley circulation varied from a nearly symmetrical two-cell structure at equinox to a single, cross-equatorial cell at solstice. Steep temperature gradients in the winter hemispheres produced strong polar vortex winds that reached velocities of 160 m s^{-1}. The occurrence of water ice clouds is closely related to atmospheric temperatures and dust content, with dust storms removing water ice from the atmosphere for extensive periods. Water ice clouds form a distinctive and repeatable band of clouds during aphelion (northern summer) between 10°S and 30°N. Multi-year TES observations showed significant year-to-year variations in both

the dust and water vapor cycles (Smith, 2004). The perihelion season (L_s = 180–360°) is relatively warm, dusty, free of water ice clouds, and shows a relatively high degree of inter-annual variability in dust optical depth and atmospheric temperature. In contrast, the aphelion season (L_s = 0–180°) is relatively cool, cloudy, free of dust, and shows a low degree of inter-annual variability. Water vapor abundance is highest in both hemispheres over the summer polar ice and shows a moderate amount of inter-annual variability at all seasons, due primarily to yearly variable, perihelion-season dust storms. These dust storms increase albedo through deposition of bright dust on the surface causing cooler daytime surface and atmospheric temperatures well after dust optical depth returns to pre-storm values (Smith, 2004).

Emission phase function (EPF) observations are an excellent tool for studying the properties of Martian aerosols. During EPF sequences, a particular location on Mars is observed at a variety of scattering angles as the spacecraft passes overhead; these can be used to determine particle sizes and single-scattering phase functions for aerosols. TES solar-band and infrared (IR) spectral EPF sequences were used to identify two distinct ice cloud types (Clancy et al., 2003): Type 1 ice clouds, having small particle sizes (r_{eff} = 1–2 μm), are most prevalent in the southern hemisphere during aphelion season; and Type 2 ice clouds, having larger particle sizes (r_{eff} = 3–4 μm), appear most prominently in the northern subtropical aphelion cloud belt.

The primary dataset of the Mars Orbiter Laser Altimeter (MOLA) consisted of topographic elevations, which are indispensable for modeling atmospheric physics (Zuber et al., 1992; D. Smith et al., 2001). MOLA supplied its own light source via a pulsed 1.04 μm laser and was not therefore not limited to illuminated regions. MOLA investigators identified clouds in the polar night and determined many of their properties (Neumann et al., 2003). These have been identified as CO_2 ice clouds, suggesting frontal activity and possible CO_2 snow (Colaprete and Toon, 2002).

The vertical structure of the neutral atmosphere has been explored by several experiments, including radio science investigations, aerobraking and entry accelerometers, and UV spectrometers. The MGS Radio Science (RS) investigation conducted thousands of one-way, single-frequency downlink Earth occultation experiments (Hinson et al., 1999, 2000; Tyler et al., 2001). Each occultation produced a vertical profile of atmospheric density, pressure, and temperature from the surface to approximately 40 km. The main strengths of this dataset were excellent sub-kilometer vertical resolution, straightforward absolute calibration, absolute altitude scale (as opposed to pressure levels common to infrared instruments), high accuracy near the surface, and dense sampling of all longitudes due to MGS's Sun-synchronous two hour orbit. The main weaknesses of this dataset, relative to those provided by nadir or limb instruments on orbital platforms, were the limited number of profiles and their limited spatial and local time coverage, imposed by the stringent requirements of occultation geometry. Its main discoveries were detailed characterizations of thermal tides and other dynamical components, as well as topographically forced thermal inversions (Hinson and Wilson, 2004).

The MGS accelerometer investigation (ACC) measured atmospheric density at the spacecraft using the drag relationship between density and measured aerodynamic acceleration

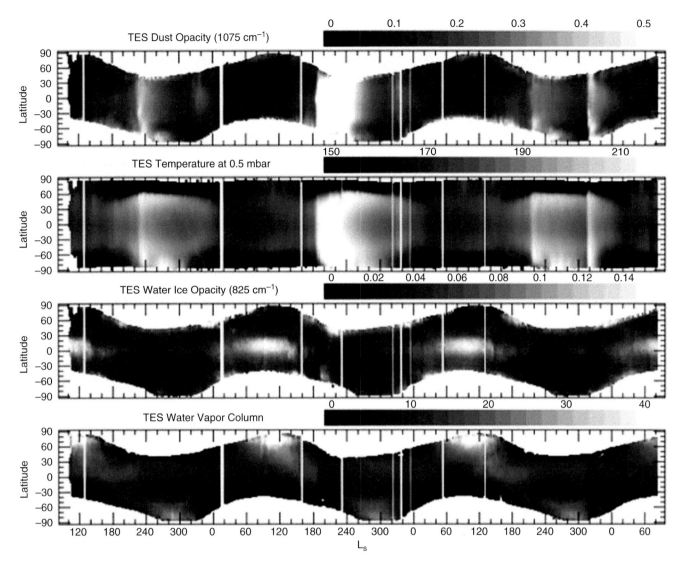

Figure 3.4. Three years of TES observations were used to map the fields of dust opacity, temperature, water ice opacity, and water vapor as a function of season and latitude. The concentration of dust during the season around perihelion is evident. While the inter-annual general repeatability is evident, variability is evidenced by the global dust event that started at equinox in the second Mars year and by the large dust event that occurred at $L_s \sim 310°$ in year 3 but not in the first two years. The figure nicely illustrates the correlation between water vapor near the north pole and the aphelion cloud belt. A black and white version of this figure will appear in some formats. For the color version, please refer to the plate section.

(Keating et al., 1998; Tolson et al., 1999; Withers et al., 2003). A series of density measurements was acquired during the aerobraking phases of the mission whenever spacecraft periapsis was below approximately 160 km, extending to as low as 100 km. Acceleration measurements were averaged over 7 or 39 s to produce two density datasets, the former having better spatial resolution and larger density uncertainties, and the latter having worse spatial resolution and smaller density uncertainties. The main strength of this dataset was its *in situ* sampling of the thermosphere, which is not amenable to remote sensing observations. The main weaknesses of this dataset were its limited spatial and local time coverage, which were determined by the required final orbit, and its lack of associated pressure and temperature measurements, which was caused by the extremely non-vertical spacecraft trajectory. Zonal winds were also inferred (Baird et al., 2007). The major discoveries of ACC

were the large thermospheric density variations due to tides generated at the surface of Mars and the strong thermospheric response to distant regional dust storms.

The MGS radio science investigation also provided measurements of upper atmospheric densities, but for higher periapsis altitudes than those experienced during aerobraking (Tracadas et al., 2001). The atmospheric densities along the spacecraft trajectory, although too small for direct detection by the accelerometer, caused orbital changes that were measured from radio tracking observations. Periapsis densities (170–180 km) were determined for the science phasing orbits (interlude between two aerobraking phases). The main strength of this dataset was its sampling of atmospheric regions that are too tenuous to be studied by most other techniques. The main weaknesses of this dataset were its limited spatial and temporal coverage. Its main discoveries were orbit-to-orbit variabilities on the order of 50% and the strong thermospheric response to distant regional dust storms.

The MGS electron reflectometer (ER) also provided indirect measurements of atmospheric densities around 180 km (Lillis et al., 2005, 2008a,b, 2010; Mitchell et al., 2001). This instrument observed electron flux as a function of energy and pitch angle (angle between electron velocity vector and local magnetic field vector). Downward-traveling electrons, which should be reflected back upwards by the magnetic mirror effect, may be scattered by the atmosphere if they reach sufficiently large atmospheric densities prior to mirroring. Hence asymmetries in the electron pitch angle distribution constrained thermospheric densities. The main strengths of this dataset were its extended spatial and temporal coverage. The main weakness of this dataset is large uncertainties on individual measurements, which require substantial averaging. Its main discoveries were seasonal trends near the exobase and characterization of the dependence of near-exobase densities on atmospheric dust and the solar cycle.

3.5.2 Mars Odyssey

Mars Odyssey arrived in its mapping orbit in February 2002; the orbit is a near-polar, near-circular orbit similar to that of MGS but is Sun-synchronous in the later afternoon, at 4:30–5:00 p.m. (Christensen et al., 2004). The Thermal Emission Imaging System (THEMIS) images in eight ~1 μm wide infrared bands from 6 to 14 μm, which measure ice and dust aerosols, and a single ~1 μm wide band centered on the 15 μm CO_2 absorption band. THEMIS is primarily focused on surface mapping, and has a high spatial resolution of 100 m/pixel at the expense of spectral resolution and continuous coverage. THEMIS, in orbit simultaneously with TES for five years, provided observations at a later time of day than TES and led to significant opportunity to cross-calibrate the atmospheric measurements between the two instruments (Smith et al., 2003). THEMIS has continued to monitor the temperature at approximately the 0.5 mbar level in the atmosphere and has shown a picture of the Martian atmosphere that is driven by year-to-year variations in dust loading. Comparison of water ice optical depths in the aphelion cloud belt measured by THEMIS and TES shows a significantly higher optical depth in the THEMIS late afternoon observations than in the TES early afternoon data, suggesting possible local time variation of clouds (Smith, 2009).

The imaging system included in the THEMIS experiment has fairly high resolution (18 m/pixel) and five spectral wavelengths. Important observations of small clouds in polar troughs and of plume structure in local dust storms have been obtained (Inada et al., 2007). Observations of cloud motions, which can be used to deduce wind speeds, have been observed and studied using Themis Vis observations (McConnochie et al., 2010).

The Neutron Spectrometer component of the Gamma Ray Spectrometer (GRS) on Mars Odyssey has contributed to atmospheric studies by measuring the content of water in the Martian subsurface (Feldman et al., 2004). The Phoenix Lander (see below) verified the large amounts of subsurface ice predicted in the Martian arctic by this experiment, thus confirming the 1979 prediction of Farmer and Doms (1979). Gamma-ray observations also constrained the amount of solid CO_2 in the polar caps

(Feldman et al., 2003); coupled with MOLA observations of the seasonal changes in the ice thickness, the result provided constraints on the density of CO_2 in the caps (Aharonson et al., 2004). The Gamma GRS observations also revealed seasonal enhancements of argon and other non-condensable gases over polar regions during winter (Sprague et al., 2004, 2007).

The Odyssey payload included an aerobraking accelerometer (ACC), whose operations, strengths, and weaknesses were similar to those of the MGS accelerometer (Tolson et al., 2005; Withers, 2006; Crowley and Tolson, 2007), and a radio science investigation that did not include radio occultations. Radio tracking data yielded atmospheric densities at 400 km, as well as estimates of the exospheric scale height and temperature (Mazarico et al., 2007a,b). The main strength of this dataset is its pioneering exploration of exospheric densities at 400 km. The main weaknesses of this dataset are the large uncertainties and the limited spatial and temporal resolution. Its main discovery concerned the sensitivity of densities at 400 km to variations in solar EUV irradiance.

3.5.3 Mars Express

The Mars Express spacecraft has orbited Mars since January 2004. It is in a nearly polar orbit that has periapsis ~250 km and apoapsis 11 500 km. The combination of variable range, lighting, antenna position, etc. make the development of sequences for observations of phenomena at different locations and in different seasons complicated and preclude a general description of coverage here. However, a distinct advantage of the non-Sun-synchronous orbit is extensive local time coverage from Mars Express atmospheric observations, albeit without the global mapping coverage available from Sun-synchronous orbit.

The High Resolution Stereo Camera (HRSC) on Mars Express (Neukum and Jaumann, 2004) is a push-broom instrument with nine charge-coupled device (CCD) line sensors mounted in parallel for simultaneous high-resolution stereo, multicolor, and multi-phase imaging. The nine line sensors have a cross-track field of view of ±6°, ~25 km at periapsis. Although HRSC was designed primarily to study the surface of Mars, its combination of medium resolution, color, and stereo capability make it useful for collecting data relevant to certain atmospheric phenomena. For example, Inada et al. (2008) made use of the fact that the stereo channels detect scattered light at five geometric angles with different path lengths to study the optical depth and composition of hazes observed in Valles Marineris. The HRSC has also imaged early morning fog clouds (Möhlmann et al., 2009) and equatorial high-altitude CO_2 ice clouds (Scholten et al., 2010)

The OMEGA Vis/NIR imaging spectrometer (Bibring et al., 2004) provides coverage of the Martian atmosphere and surface in 352 spectral channels from 0.35 to 5.1 mm (Bibring et al., 2005), with a horizontal resolution ranging from 400 m to 5 km depending on the altitude of the spacecraft. The swath width in pixels is reduced as the range decreases to avoid excessive overlap. Many atmospheric gases have diagnostic features in this wavelength region. The CO_2 absorption band at 2 μm can be used to measure surface pressure, which, when corrected for elevation using MOLA data, can be used to study horizontal pressure gradients, atmospheric oscillations, and topographically induced pressure variations (Forget et al., 2007; Spiga et al., 2007). Seasonal variations in atmospheric CO and H_2O in the Hellas Basin have been addressed using OMEGA data by Encrenaz et al. (2006, 2008). OMEGA is also very well suited to studying the composition and physical state of surface condensates in the polar regions consisting of CO_2, H_2O, and mixtures of these with dust (Langevin et al., 2007). OMEGA near-IR spectra also supported identification of equatorial mesospheric clouds as CO_2 ice, based on 4.26 μm spectral features (Montmessin et al., 2007), and detected polar night upper-level O_2 singlet delta emission (Bertaux et al., 2010)

The Planetary Fourier Spectrometer (PFS) on Mars Express collects data from 8190–220 cm^{-1} (1.2–45 μm) in two separate channels at a uniform spectral resolution of 1.3 cm^{-1} (Formisano et al., 2005). The instrument field of view (FOV) is ~10 km from an altitude of 300 km. The PFS has found the seasonal and latitudinal behavior of the water vapor to be consistent with previous observations, although the absolute water columns are a factor of 1.5 lower than the values obtained by TES (Fouchet et al., 2007). The distribution of water shows local maxima at low latitudes over Tharsis and Arabia. As with previous experiments, an increase in water vapor is observed over the northern seasonal polar cap edge. On a global scale, PFS data show an anticorrelation between water column abundance and surface pressure; this has been suggested to indicate a vertical distribution that is intermediate between control by atmospheric saturation and confinement to a surface layer, and suggesting a possible process of regolith–atmosphere exchange (Fouchet et al., 2007). A report of methane detection by PFS (Formisano et al., 2004) precipitated considerable activity and discussion among ground-based spectroscopists (see ground-based spectroscopy; Section 3.4).

The MEX radio science investigation (MaRS) has conducted over 600 two-way, dual-frequency Earth occultation experiments (Pätzold et al., 2004, 2005; Tellmann et al., 2013). Its data products are similar to those of the MGS radio occultation experiments, although its measurement accuracy is slightly better. The noise in one-way, but not two-way, radio occultation experiments is increased by the noise of the on-board oscillator. Also, the high apoapsis of the eccentric MEX orbit offers a much longer baseline prior to occultations than did the 400 km circular MGS orbit. The main strengths and weaknesses of this dataset were the same as those of the MGS radio occultation dataset, with the exception of the dense longitude sampling provided by MGS's two hour orbit. A major MaRS discovery was measuring spatial variations in the depth of the convective boundary layer (Hinson et al., 2008). MaRS also conducted studies of electron densities within the Mars nightside ionosphere (Withers et al., 2012).

The MEX Spectroscopy for the Investigation of the Characteristics of the Atmosphere (SPICAM) experiment consists of distinct UV and near-IR spectrometers, which can operate in nadir, limb viewing, and Sun/stellar occultation modes (Bertaux et al., 2000, 2005a). The SPICAM UV spectrometer (118–320 nm range, 0.55 nm resolution; Bertaux et al., 2006) has observed over 400 stellar occultations, in which CO_2 absorption over 130–190 nm wavelengths supports vertical profile measurements of CO_2 density, and hence temperatures, for altitudes of 50 to 120 km (Quémerais et al., 2006; Forget et al., 2007). These UV stellar occultation measurements also indicate detached layers of fine (presumably ice) aerosols at altitudes of 100 km (Montmessin et al., 2006). Additional SPICAM UV observations include O_3 column (nadir) measurements (Perrier et al., 2006), unique nighttime ozone profiling in the lower atmosphere (20–60 km; Lebonnois et al., 2006), and detections of aurora (Bertaux et al., 2005b), polar night NO (Bertaux et al., 2005c) and ozone layers (Montmessin and Lefèvre, 2013) in the Mars upper atmosphere. SPICAM also includes a near-IR spectrometer employing an acousto-optical tunable filter in the range of wavelengths from 1.0 to 1.7 μm with a resolution of 0.55 nm at a wavelength of 1.27 μm. The SPICAM IR spectrometer has been used to determine the spatial and seasonal distributions of column water vapor (Fedorova et al., 2006a) and O_2 singlet delta emission (Fedorova et al., 2006b), as well as aerosol particle size determinations (combined UV–IR analysis; Fedorova et al., 2009, 2014). More recently, the development of water vapor profile retrievals from SPICAM solar occultation IR measurements has significantly expanded our knowledge of the vertical distribution of Mars atmospheric water (Maltagliati et al., 2013) and indicated the variable presence of large supersaturation conditions for atmospheric water vapor at altitudes above 25 km (Fedorova et al., 2009; Maltagliati et al., 2011).

3.5.4 Mars Reconnaissance Orbiter

Mars Reconnaissance Orbiter (MRO) reached its Primary Science Orbit in September 2006. The Sun-synchronous orbit is near-circular and near-polar, with an ascending node on the dayside of the planet fixed near 3 p.m. local mean solar time; MRO achieved this orbit via aerobraking (Zurek and Smrekar, 2007). MRO carries several imaging instruments that provide data for atmospheric and polar studies.

The Mars Color Imager (MARCI) is a 180° field-of-view push-frame camera having five color bands in the visible (420, 550, 600, 650, 750 nm) and two in the UV (260, 320 nm) (Bell et al., 2009). The MARCI experiment had originally been included in the Mars Climate Orbiter (MCO) payload, but MCO was lost during Mars orbital insertion in 1999 when it entered the atmosphere at too small an altitude due to a navigational error. MARCI data are typically acquired at a resolution of about 800 m/pixel from nadir, but the UV images are summed in 8×8 blocks. The pole-to-pole swaths on the sunlit side have equator crossings fixed at 3 p.m. local time. The 180° field of view of the optics provides limb-to-limb coverage across the entire ground track even during frequent spacecraft maneuvers. Images acquired on adjacent orbits barely overlap at the equator, but substantial overlap occurs in the polar regions. The capabilities of MARCI exceed those of MOC WA in several respects: UV coverage, spectral resolution, and spatial resolution of the daily maps. Malin et al. (2008) provides a good introduction to the capabilities of MARCI for atmospheric and polar research. A major perihelic dust event was observed by MARCI starting at L_s ~ 260° in MY 28 (Cantor et al., 2008); feedback from the elevated atmospheric dust accelerated the recession of the south polar cap relative to less dusty years (James et al., 2010). UV images have been used to map cloud and ozone columns on Mars with daily global coverage over five Mars years (MY 28–32; Wolff et al., 2011; Clancy et al., 2015), as well as provide the UV absorption properties of Mars dust (Wolff et al., 2010). MARCI cloud image and shadow measurements reveal long cloud trails, most strikingly around Valles Marineris, associated with deep mesoscale vertical updrafts under peak surface heating conditions (Clancy et al., 2009).

The Compact Reconnaissance Imaging Spectrometer for Mars (CRISM) is a hyperspectral imager. In targeted mode the instrument is slewed to remove most along-track motion, and a region of interest is mapped at full spatial and spectral resolution (15–19 m/pixel, 362–3920 nm at 6.55 nm/channel) (Murchie et al., 2007). Ten additional abbreviated, spatially binned images are taken before and after the main image, providing an emission phase function (EPF) of the site for atmospheric study and correction of surface spectra for atmospheric effects. In atmospheric mode, only the EPF is acquired. Wolff et al. (2009) used CRISM EPFs acquired in the vicinity of the Mars Exploration Rovers, which provide ground-truth observations, to study the visible–near-IR wavelength dependence of the dust single-scattering albedo during the 2007 planet-encircling dust event. The seasonal and global distributions of mesospheric (50–80 km) CO_2 and H_2O ice clouds have also been retrieved by Vincendon et al. (2011). In addition to aerosol contributions, CRISM near-IR spectra include significant CO_2, H_2O, O_2 singlet delta, and CO band absorptions. The typical nadir geometry and limited mapping coverage of CRISM observations supports fairly coarse spatial–seasonal coverages for CO, water vapor, and O_2 singlet delta column determinations. These observations indicated unusually low atmospheric water columns in southern summer of MY 28, as well as large CO variations associated with the seasonal CO_2 cycle (M. Smith et al., 2009). Special limb imaging operations with CRISM were initiated in July 2009 and support profiling retrievals for cloud, dust, CO, and water vapor (Smith et al., 2013), dayglow and polar nightglow emissions by O_2 and OH (Clancy et al., 2010, 2012, 2013), and particle sizes for mesospheric CO_2 clouds (Clancy et al., 2014).

MRO has two other imaging systems, both of which have much higher resolution than MARCI. The High Resolution Imaging Science Experiment (HiRISE) is a very high-resolution (0.25–1.3 m/pixel) camera with color and stereo capabilities (McEwen et al., 2007). Although not intended for atmospheric observations, HiRISE observations of polar surfaces are relevant to the small-scale structure of the CO_2 cycle (e.g. Hansen et al., 2010). The Context Camera (CTX) is a monochromatic camera with a pixel scale of 5–6 m/pixel; most of the planet will be mapped at this resolution during the MRO mission, and some regions (e.g. south pole residual cap) will be mapped repetitively (Malin et al., 2007). The resolution of this camera is ideal for studying the inter-annual changes in the unique erosional features in the CO_2 residual south polar cap (Thomas et al., 2009).

The Mars Climate Sounder (MCS) experiment on MRO was the redesigned third incarnation of the Pressure Modulated Infrared Radiometer experiment, originally flown on Mars Observer and Mars Climate Orbiter. It collects IR data in eight spectral bands with band centers from 11.8 to 42.1 µm that vary in width from 20 to 100 cm^{-1} (McCleese et al., 2007). This filter radiometer approach emphasizes high radiometric performance and calibration, high vertical resolution, and continuous limb profiling, and provided the first vertical profiles of water vapor and dust (McCleese et al., 2007). Comparisons between the MCS and the MGS radio science and MGS TES temperatures show very good agreement (Hinson et al., 2004). Kleinböhl et al. (2009) described the algorithm that is used to retrieve atmospheric profiles from MCS limb measurements.

The first systematic observations of the middle atmosphere of Mars (35–80 km) with MCS revealed diurnal thermal variation associated with a diurnal thermal tide (Lee et al., 2009). Polar observations show an intense warming of the middle atmosphere over the south polar region in winter that is at least 10–20 K warmer than predicted, suggesting that the Hadley circulation over the pole may be as much as 50% more vigorous than expected (McCleese et al., 2008). MCS determinations of the vertical distribution of dust, which is important input for atmospheric models, are inconsistent with circulation models during northern spring and summer, suggesting that mesoscale effects of topography on dust lifting may be important (McCleese et al., 2010; Heavens et al., 2011a,b). MCS observations of the vertical structure of the north polar hood clouds, which are present between $L_s = 150°$ and $30°$, indicated that the clouds form in a single layer that extends in altitude from 10 to 40 km above the surface (Benson et al., 2011). More recently, MCS thermal IR profiling has characterized the spatial/seasonal distributions, particle sizes, radiative effects, and surface ice contributions of CO_2 clouds in the polar winter atmosphere (Hayne et al., 2012, 2014).

SHARAD (Shallow Radar) is a sounding radar experiment on MRO. Phillips et al. (2011) have reported the discovery of a buried deposit of CO_2 ice beneath the residual south polar cap with a mass comparable to that of the present atmosphere/cap system. The existence of this reservoir will have significant consequences for the behavior of the climate system as orbital parameters change.

The MRO payload also included an aerobraking accelerometer (ACC), whose operations, strengths, and weaknesses were similar to those of the MGS accelerometer (Tolson et al., 2008), and a radio science investigation. The radio science investigation did not formally include a radio occultation component, but the mission leadership elected to acquire a series of two-way radio occultation measurements as resources permitted (S. W. Asmar, personal communication, 2010). Radio occultation measurements were ongoing, at a typical cadence of once per day after the conclusion of MRO's prime mission, as of 2011. The characteristics, strengths, and weaknesses of this radio occultation dataset are similar to those of its predecessors. Radio tracking to determine atmospheric densities at 250 km was also performed (Mazarico et al., 2008), from which the presence of exospheric variations caused by atmospheric tides was deduced.

3.5.4.1 Mars Atmosphere and Volatile Evolution

The MAVEN mission was competed and selected as the second and final mission of the NASA Mars Scout Program, which was discontinued in 2010 (the Phoenix Lander, described below, was the first mission selected under this program). The MAVEN spacecraft was launched on November 18, 2013 and entered into a highly elliptical Mars orbit (4.5 h period, 75° inclination, 150 km periapsis, 6200 km apoapsis) on September 22, 2014. The science objectives of MAVEN are to conduct Mars upper atmospheric/ionospheric studies centered on atmospheric escape processes that contribute to the long-term evolution of the Mars atmosphere (Jakosky et al., 2015). MAVEN formal science operations began in early November 2014, but were preceded by target-of-opportunity observations during the Mars passage by Comet Siding in mid-October, 2014.

The MAVEN *in situ* and remote sensing instrument complement is designed to provide an integrated description of the time-dependent structure, composition, isotopic ratios, and energetics of the 100–400 km atmospheric region in the context of variable solar forcing (radiation, solar wind, and solar energetic particles). Experiments include a particles and fields instrument package (magnetometer, Langmuir probe, solar inputs), an imaging ultraviolet spectrometer, and a neutral–ion mass spectrometer (Jakosky et al., 2015). This measurement suite is configured to support characterizations of both current and long-term integrated atmospheric loss rates, as key inputs towards our understanding of Mars atmosphere and climate/volatile evolution (see Chapters 15 and 17).

The late timing of MAVEN science operations with respect to publication of this book precludes discussion of specific science results here, although preliminary MAVEN results are addressed in Chapters 15 and 16. At the time of this writing, several MAVEN papers were under review at *Science* and much larger set (~30 papers) of papers were under review for a special issue of *Geophysical Research Letters*.

3.6 LANDERS AND ROVERS

The first attempt to land on Mars in the 1990s was a resounding success. Mars Pathfinder successfully landed in Ares Valles, demonstrating the concept of using inflatable airbags to cushion the spacecraft's impact. This method was later used for the landings of the very successful Mars Exploration Rovers in 2004. The other attempted landing of the 1990s, that of the Mars Polar Lander in 1999, which used the conventional retro rockets for landing, failed. Essentially the same spacecraft was used for the very successful Phoenix Lander in the Martian polar region in 2008. The first attempt by the European Space Agency to land a vehicle on Mars was unsuccessful in late 2003.

3.6.1 Mars Pathfinder

Mars Pathfinder landed in Ares Valles, about 700 km due east of Viking Lander 1, on July 4, 1997, $L_s = 142°$. Data return was limited by the absence of an active orbiter to serve as a relay. The Imager for Mars Pathfinder (IMP) was a stereo camera system having a resolution similar to Viking Lander imaging but a much higher signal-to-noise (S/N) ratio (Smith et al., 1997). Twenty-four filters (12 on each camera) spanning 440–1000 nm provided visible and near-IR coverage; eight of these filters were specifically designed for solar viewing to determine dust opacity and water absorption. The optical depth history determined from the three months of data from these solar filters was reported by Smith and Lemmon (1999), and the properties of the dust were deduced from the sky-imaging data by Tomasko et al. (1999), Markiewicz et al. (1999), and Johnson et al. (2003). The water vapor column was measured on several sols using the weak 935 nm water band (Titov et al., 1999).

The Pathfinder Lander, like the two Viking Landers, carried a Tavis magnetic reluctance diaphragm sensor pressure sensor (Schofield et al., 1997), but calibration uncertainties limit its usefulness in determining change since Viking (Haberle et al., 1999). The surface pressure data showed diurnal, semidiurnal, and higher-order tides (Schofield et al., 1997). Three mast-mounted thermocouple temperature sensors showed that there were stable and turbulent periods of the sol, with up to 20 K rapid variations (Schofield et al., 1997). The Lander's mast also included a hot-wire anemometer that showed diurnally varying slope winds (Schofield et al., 1997) and a windsock (Sullivan et al., 2000).

Dust devils were identified by their pressure, wind, and temperature signatures (Schofield et al., 1997) and in some horizon images (Metzger et al., 1999). The deposition rate of dust on the solar panels was measured (Landis and Jenkins, 2000). The rate of dust deposition was less than the source rate from dust devils, suggesting that the area is a net source for dust (Murphy and Nelli, 2002; Ferri et al., 2003).

An array of magnets of varying strengths was imaged over the 440–1000 nm spectral range of the camera. Magnetic properties of the dust reported by Madsen et al. (1999) suggest that airborne dust is at least weakly magnetic and that the dust is composite, including a strongly magnetic component, likely maghemite.

3.6.2 Mars Exploration Rovers

The two Mars Exploration Rovers landed in January 2004, near $L_s = 330°$ and 340°. Spirit landed first, in Gusev Crater, and operated for 2209 sols to $L_s = 67°$ in the fourth Martian year of the mission. Opportunity landed in Meridiani Planum and is still operating at $L_s = 180°$ of its fifth Martian year. Atmospheric experiments were carried out by the Mini-TES, cameras, and Alpha-Particle X-ray spectrometer (APXS) (Squyres et al., 2003).

The two Mars Exploration Rover Mini-TES instruments collected spectral data from the Martian surface over a spectral range of 5–29 μm (339–1997 cm^{-1}) with a spectral sample interval of 9.99 cm^{-1} (Christensen et al., 2003). Mini-TES has an external, mast-mounted pointing mirror that provides upward-looking atmospheric observations to 30° above the horizon with a spatial resolution of 20 mrad. Mini-TES successfully measured the atmospheric temperature profile in the lower boundary layer with a vertical resolution ranging from less than 100 m near the surface to about 1 km several kilometers above the surface (Smith et al., 2004). Mini-TES observed a very steep, super-adiabatic vertical temperature gradient that is established through the lowest 100 m of the atmosphere by mid-morning. Turbulent convection occurs throughout this lowest layer, with temperature fluctuations of 15–20 K in the

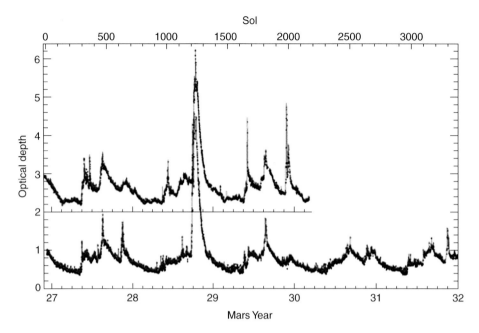

Figure 3.5. Five Mars years of opacity data acquired from the surface of Mars by the Mars Exploration Rovers, Spirit (top), located at −14.57° latitude and 175.48°E longitude, and Opportunity (bottom), at −1.85°, 354.47°. The abscissa scale is in Mars years at the bottom and in Opportunity sols at the top. Minor ticks at the bottom mark every 45° of L_s. The ordinate axis denotes the optical depth at Opportunity; the Spirit optical depth has been displaced by 2.0, so 2.0 must be subtracted from the ordinate scale to derive the optical depth at that location. The major feature in the opacity measurements of both spacecraft is the large, planet-encircling dust event observed in MY 29 just before solstice. Late spring dust can be seen in both datasets in the other years, but the large opacity increases at both sites resulting from the MY 29 is unique in the five-year period. The figure was originally published in *Icarus* (Lemmon et al., 2015) and is reproduced with their permission.

lowest meter above the surface and up to 5 K in the lowest 100 m on timescales of 30–60 s. Turbulent convection continues until late afternoon, at which time the surface becomes cooler than the near-surface atmosphere, convection shuts off, and the near-surface temperature gradient becomes inverted. The inversion layer grows throughout the nighttime hours, reaching a depth of at least 1 km, before rapidly reversing again in the morning (Smith et al., 2006). One of the remarkable achievements of the Mini-TES and TES experiments was the simultaneous observation by both instruments of essentially the same air mass (Wolff et al., 2006). The simultaneous data were used to determine that effective dust particle radii vary from 1.3 μm to 1.8±0.2 μm and that the vertical mixing profile varies from being well mixed to confined in the boundary layer.

Each rover had a mast-mounted Pancam stereo camera with 12 filters over 440–1000 nm, including two for solar imaging (Bell et al., 2003). In addition, the rovers' navigation and hazard avoidance cameras (Navcam and Hazcam) were capable of wider-field sky imaging. Lemmon et al. (in preparation) present the dust optical depth record for the mission (Figure 3.5). Cross-sky imaging found dust aerosol properties consistent with those measured by Viking and Pathfinder (Lemmon et al., 2004). Dust devils were common during southern summer at the Gusev site (Greeley et al., 2006, 2010).

The Rover's Instrument Deployment Device allowed a Microscopic Imager, an Alpha Particle X-Ray Spectrometer, and a Mössbauer Spectrometer to be placed against Martian samples, including dust on magnets carried by the rovers. The

magnets' dust appeared the same as that on Pathfinder, but the Mössbauer data showed the airborne dust to contain magnetite, olivine, pyroxene, and nanophase iron oxides, rather than maghemite, suggestive of a dry past (Goetz et al., 2005; Madsen et al., 2009). The APXS was also used to measure atmospheric argon and constrain transport and mixing in the atmosphere (Arvidson et al., 2011).

3.6.3 Phoenix

The Phoenix Mars Lander (P. Smith et al., 2008, 2009) landed successfully in the Martian arctic at 68.2°N, 234.3°E in May 2008, $L_s = 77°$, and operated through $L_s = 148°$. An important objective was the study of weather in the Martian arctic, an endeavor in which the Surface Stereo Imager (SSI) on Phoenix played an important role. SSI is very similar to the IMP on Pathfinder and to the SSI on the Mars Polar Lander, which was lost in 1999; but it uses larger-format CCDs and has four times the resolution of either of those cameras. As was the case for IMP, six solar filters can be used to determine the atmospheric opacity. Tamppari et al. (2010) reported on a well-organized campaign of coordinated atmospheric observations by Phoenix and MRO. Dust devils were detected in imaging and via their pressure signature (Ellehoj et al., 2010), and the behavior of water ice clouds near the landing site was determined (Moores et al., 2010). Later in the mission, frost (P. Smith et al., 2009; Holstein-Rathlou et al., 2010) and fog (Moores et al., 2011) were seen. Surface frost seen by the SSI was seasonally correlated with the appearance of ice seen near the landing site from orbit (Cull et al., 2010). Spectra of airborne dust captured by magnets were consistent with prior missions, although a component of non-magnetic dust was seen (Drube et al., 2010).

The Light Detection and Ranging (LIDAR) instrument (Whiteway et al., 2008) on Phoenix provided observations of clouds within the Martian boundary layer. The LIDAR observed that dust was well mixed in the bottom 4 km of the atmosphere and peaked near summer solstice. After solstice, LIDAR showed a regular diurnal pattern of cloud formation at around midnight and dissipation before midday (Whiteway et al., 2009). The earliest clouds were thin and occurred at altitudes above 10 km. By $L_s = 113°$, the regular pattern of cloud formation and dissipation had been established within the planetary boundary layer (PBL). This pattern included a surface fog near midnight and a cloud at the top of the PBL around 4 km after 1 a.m. On some sols, fall streaks were seen in the LIDAR signature and suggested fall speeds consistent with 30 μm ice particles. The maximum ice water content of the clouds was estimated to be 1 mg m^{-3} at 6 a.m. with a column ice water content of 6 g m^{-2} (Dickinson et al., 2010). The time of formation of the clouds did not change through the second half of the mission; this is consistent with expected atmospheric cooling if the decrease in water content was confined to the PBL.

The Phoenix Mission measured local pressures using a Barocap sensor from $L_s = 80°$ to 151° (Taylor et al., 2010). The Phoenix data suggest an earlier pressure minimum ($L_s = 143°$) than Viking and Pathfinder ($L_s = 148°$). Haberle and Kahre (2010) corrected pressure data from Viking and Phoenix for elevation and dynamics and concluded that atmospheric CO_2 appeared to have increased between the two missions, consistent with results that suggest that the mass of the south residual cap is decreasing (Malin et al., 2001).

Phoenix was equipped with a robotic arm to dig in the soil to the water ice table (Arvidson et al., 2009). The ice table depth was consistent with vapor diffusive equilibrium with the mean atmospheric water vapor (Mellon et al., 2009). Perchlorate found in the soil (Hecht et al., 2009) may have been created via gas-phase oxidation of chlorine volatiles (Catling et al., 2010). A Thermal and Electrical Conductivity Probe (TECP) measured temperature and relative humidity in the bottom 2 m of the atmosphere on many occasions through the mission and showed substantial adsorption of water into the soil overnight (Zent et al., 2010).

The Thermal Evolved Gas Analyzer instrument includes a magnetic-sector mass spectrometer (Boynton et al., 2009). Measurements of atmospheric CO_2 were used to derive isotopic ratios for $^{13}C/^{12}C$ and $^{18}O/^{16}O$ (Niles et al., 2010). Niles et al. suggest that the isotopic composition indicates a low-temperature water–rock interaction throughout Martian history, including presently active carbonate formation and recent volcanic degassing.

3.6.4 Mars Science Laboratory (MSL)

The Mars Science Laboratory (MSL) was launched on November 26, 2011 and landed in Gale Crater on Mars on August 6, 2012. The Curiosity Rover has explored Gale Crater since that time. Study of the atmospheric environment in Gale Crater to access long-timescale evolution of the atmosphere and to determine the present state of the water and carbon dioxide cycles are major objectives of the mission. Several instruments on Curiosity have contributed and are continuing to contribute to these objectives, including the Sample Analysis at Mars (SAM; Mahaffy et al., 2012) instrument suite, the Chemical Camera instrument suite (ChemCam; Maurice et al., 2012), the Rover Environmental Monitoring Station (REMS; Gómez-Elvira et al., 2012), and science and engineering imaging systems (MastCam; Malin et al., 2005; Navcam; Maki et al., 2012).

The Sample Analysis at Mars (SAM) instrument suite consists of a quadrupole mass spectrometer, a tunable laser spectrometer, and a six-column gas chromatograph that provide complementary information on samples. Precise measurements of the ratio of non-radiogenic argon isotopes provide strong support for a Martian origin of Mars' meteorites and indicate a significant loss of atmosphere since formation (Atreya et al., 2013). The ratio of nitrogen isotopes measured by the SAM mass spectrometer agrees closely with the ratio measured by Viking (Wong et al., 2013); however, the argon and nitrogen abundances measured by MSL (Mahaffy et al., 2013; Franz et al., 2015) differ from those measured by Viking. Due to the inertness of argon and nitrogen, no obvious climate mechanism that could explain such changes has been suggested (Wong et al., 2013). SAM also reports very accurate mass spectrometer measurements for local CO, O_2, as well as Ar and N_2 abundances (Franz et al., 2015).

The SAM tunable laser spectrometer has been employed to obtain measurements of methane concentrations throughout the first Mars year of the MSL mission. Measured background levels of 0.7 ppb are consistent with photochemical models involving solar ultraviolet interaction with organics in meteoric material. However, during one 60 sol period methane concentrations increased significantly to 7 ppb. This suggests an additional local and intermittent source of methane (Webster et al., 2015). Seasonal variability in local O_2 mixing ratios has also reported based on ChemCam passive sky spectroscopy (McConnochie et al., 2014).

REMS measurements of local relative humidity and surface pressure characterize distinct atmospheric behaviors associated with mesoscale dynamics within Gale Crater. REMS humidity measurements indicate drier atmospheric conditions relative to surrounding regions (Savijärvi et al., 2015). REMS measurements of surface pressure variations indicate crater topographic influences on thermal tides (Haberle et al., 2014a) but do not support secular changes in the partitioning of CO_2 between the residual south polar cap and the atmosphere since Viking (Haberle et al., 2014b).

Atmospheric imaging studies have also monitored distinctive cloud and dust behaviors within Gale Crater (Moores et al., 2015a,b). Comparisons of the extinction due to dust within Gale Crater measured by Navcam with the column extinctions measured by MastCam indicate that there is little mixing of dust between the atmosphere and that outside for most of the year and suggest that Gale is a sink for dust (Moore et al., 2016). Mixing does occur during a brief period between $L_s = 270°$ and 290° as predicted by mesocscale models (Tyler and Barnes, 2013). These observations are consistent with REMS measurements of a boundary layer confined below the depth of the crater rim (Moores et al., 2015b).

3.6.5 Descent Profiles

Like aerobraking missions, landed missions to the surface of Mars have carried accelerometers to measure atmospheric densities. Such missions include Pathfinder, Spirit, Opportunity, and Phoenix (Magalhaes et al., 1999; Withers and Smith, 2006; Withers and Catling, 2010). These missions each measured a single profile of atmospheric density between approximately 10 km and 100 km, where the precise vertical range varied with parachute deployment altitude (lower limit) and instrument sensitivity (upper limit). Since these entry trajectories were much closer to vertical than those of aerobraking orbiters, corresponding pressure and temperature profiles were derived from the density profiles. The main strengths of these datasets were their extended vertical range, which linked lower and upper atmospheric regions, excellent (sub-kilometer or better) vertical resolution, which revealed a broad spectrum of waves clearly, and small uncertainties. The main weakness of these datasets was their small number (six total), which makes them hard to interpret within the global context of atmospheric circulation and thermal structure. Their main discoveries have been large temperature oscillations in the middle atmosphere, presumably caused by tides, and connections between upper and lower atmospheric regions.

3.7 LOOKING TO THE FUTURE

There have been great strides in the last several decades towards understanding the atmosphere of Mars, but there are still gaps in the data that will require a new generation of remote sensing instruments to fill. Cancellation of the Trace Gas Orbiter (TGO) mission has postponed observations that would have clarified seasonal and spatial behaviors of trace gas abundances as well as diurnal variations of clouds, especially in equatorial regions. This gap will be remedied by the 2016 ExoMars Trace Gas Orbiter, scheduled to launch in March 2016, which will provide a better understanding of methane and other atmospheric gases that are present in small concentrations in the atmosphere. The Discovery class Insight mission, designed to probe the internal structure of Mars through seismological measurements and conduct surface pressure measurements, will also launch in May 2018. Beyond these 2016 Mars missions, NASA and ESA are developing major Mars surface missions for launch in 2018–2020, even as six currently active Mars missions continue to return new Mars observations (Odyssey, Opportunity, MRO, Curiosity, MAVEN, and MOM). Mars will remain a primary focus for spacecraft investigation, ultimately in support of human exploration.

REFERENCES

Aharonson, O., M. T. Zuber, D. E. Smith, et al. (2004). Depth, distribution, and density of CO_2 deposition on Mars. *J. Geophys. Res.*, 109, E05004, doi:10.1029/2003JE002223.

Arvidson, R. E., R. Bonitz, M. Robinson, et al. (2009). Results from the Mars Phoenix Lander robotic arm experiment, *J. Geophys. Res.*, 114, E00E02, doi:10.1029/2009JE003408.

Arvidson, R. E., J. W. Ashley, J. F. Bell III, et al. (2011). Opportunity Mars Rover mission: overview and selected results from Purgatory ripple to traverses to Endeavour Crater. *J. Geophys. Res.*, 116, E00F15, doi:10.1029/2010JE003746.

Atreya, S. K., M. G. Trainer, H. B. Franz, et al. (2013). Primordial argon isotope fractionation in the atmosphere of Mars, measured by the SAM instrument on Curiosity and implications for atmospheric loss. *Geophys. Res. Letters*, 40, 5605–5609.

Avduyevskiy, V. S., E. L. Akim, V. I. Aleshin, et al. (1975). Martian atmosphere in the landing site of the descent module of Mar-6. NASA transl. into English from *Kosm. Issled. (USSR)*, 13, 1, January–February, 21–32.

Baird, D. T., R. Tolson, S. Bougher, and B. Steers (2007). Zonal wind calculations from Mars Global Surveyor accelerometer and rate data. *J. Spacecraft and Rockets*, 44, 1180–1187, doi: 10.2514/1.28588.

Barker, E. S. (1972). Detection of molecular oxygen in the Martian atmosphere. *Nature*, 238, 447–448.

Barker, E. S., R. A. Schorn, A. Woszczyk, R. G. Tull and S. J. Little (1970). Mars: detection of atmospheric water vapor during the southern hemisphere spring and summer season. *Science*, 170, 1308–1310.

Barth, C. A., and M. L. Dick (1974). Ozone and the polar hood of Mars. *Icarus* 22, 205.

Barth, C. A., C. W. Hord, A. I. Stewart, et al. (1973). Mariner 9 Ultraviolet Spectrometer experiment: seasonal variation of ozone on Mars. *Science* 179, 795–796.

Bell, J. F., III, S. W. Squyres, K. E. Herkenhoff, et al. (2003). The Mars Exploration Rover Athena panoramic camera (Pancam) investigation. *J. Geophys. Res.* 108, 8063, doi:10.1029/2003JE002070.

Bell, J.F., III, and MARCI/CTX et al. (2009). Mars Reconnaissance Orbiter Mars Color Imager (MARCI): instrument description, calibration, and performance. *J. Geophys. Res.*, 114, E08S92, doi:10.1029/2008JE003315.

Belton, M. J. S., and D. M. Hunten (1968). A search for O2 on Mars and Venus: a possible detection of oxygen in the atmosphere of Mars. *Astrophys. J.*, 153, 963–974.

Benson, J. L., P. B. James, B. A. Cantor, R. Remigo (2006). Interannual variability of water ice clouds over major Martian volcanoes observed by MOC. *Icarus* 184, 363–371.

Benson, J., D. Kass, A. Kleinböhl (2011). Mars' north polar hood as observed by the Mars Climate Sounder. *J. Geophys. Res.* 111, E03008, doi:10.1029/2010JE003693.

Bertaux, J.-L., and F. Montmessin (2001). Isotopic fractionation through water vapor condensation: the deuteropause, a cold trap for deuterium in the atmosphere of Mars. *J. Geophys. Res.*, 106, 32879–32884.

Bertaux, J.-L., D. Fonteyn, O. Korablev, et al. (2000). The study of the Martian atmosphere from top to bottom with SPICAM light on Mars Express. *Planet. Space Sci.* 48, 1303–1320.

Bertaux, J.-L., D. Fonteyn, O. Korablev, et al. (2005a). Global structure and composition of the Martian atmosphere with SPICAM on Mars Express. *Adv. Space Res.* 35, 31–36

Bertaux, J.-L., F. Leblanc, O. Witasse, et al. (2005b). Discovery of an aurora on Mars, *Nature* 435, 790–794.

Bertaux, J.-L., F. Leblanc, S. Perrier, et al. (2005c). Nightglow in the upper atmosphere of Mars and implications for atmospheric transport, *Science*, 307, 566–569.

Bertaux, J.-L., O. Korablev, S. Perrier, et al. (2006). SPICAM on Mars Express: observing modes and overview of UV spectrometer data and scientific results. *J. Geophys. Res.*, 111, E10S90, doi:10.1029/2006JE002690.

Bertaux, J.-L., B. Gondet, J. Bibring, F. Montmessin, F. Lefevre (2010). First detection of O_2 recombination nightglow emission at 1.27 μm in the atmosphere of Mars with OMEGA on Mars Express. *Bulletin of AAS*, 42, 10340.

Bibring, J.-P., A. Soufflot, M. Berthé, et al. (2004). OMEGA: Observatoire pour la Minéralogie, l'Eau, les Glaces et l'Activité, in *Mars Express: The Scientific Payload*, edited by A. Wilson, Eur. Space Agency Spec. Publ., ESA-1240, 37–49.

Bibring, J.-P., Y. Langevin, A. Gendrin, et al. (2005). Mars surface diversity as revealed by theOMEGA/Mars Express observations. *Science*, 307, 1576–1581,doi:10.1126/science.1108806.

Bjoraker, G. L., M. J. Mumma, and H. P. Larson (1989). Isotopic abundance ratios for hydrogen and oxygen in the Martian atmosphere. *Bull. Amer. Astron. Soc.*, 21, 991.

Blamont, J. E. (1991). Colloquium on Phobos-Mars mission, Paris, France, Oct. 23–27, 1989. *Proceedings. Planetary and Space Science*, 39.

Bogard, D. D., and P. Johnson (1983). Martian gases in an Antarctic meteorite? *Science*, 221, 651–654.

Boynton, W. V., D. W. Ming, S. P. Kounaves, et al. (2009). Evidence for calcium carbonate at the Mars Phoenix landing site. *Science*, 325, 61.

Briggs, G. A., W. A. Baum, and J. Barnes (1978). Viking Orbiter imaging observations of dust in the Martian atmosphere. *J. Geophys. Res.* 84, 2795–2820.

Cantor, B. A. (2006). MOC observations of the 2001 Mars planet-encircling dust storm. *Icarus*, 186, 60–96.

Cantor, B. A., P. B. James, and M. Caplinger (2001). Martian dust storms: 1999 MOC observations. *J. Geophys. Res.*, 106, 23653–23689.

Cantor, B. A., K. M. Kanak, and K. S. Edgett (2006). Mars Orbiter Camera observations of Martian dust devils and their tracks (September 1997 to January 2006) and evaluation of theoretical vortex models. *J. Geophys. Res.*, 111, E12002, doi:10.1029/2006JE002700.

Cantor, B. A., M. C. Malin, M. J. Wolff, et al. (2008). Observations of the Martian atmosphere by MRO-MARCI, an overview of one Mars year. *Mars Atmosphere: Modeling and ObservationsWorkshop*, Abstract 9075, Williamsburg, VA.

Carleton, N. P., and W. A. Traub (1972). Detection of molecular oxygen on Mars, *Science*, 177, 988–991.

Carr, M. H., W. A. Baum, G. A. Briggs, et al. (1972). Imagine experiment: the Viking Mars Orbiter. *Icarus* 16, 17–33.

Catling, D. C., M. W. Claire, K. J. Zahnle, et al. (2010). Atmospheric origins of perchlorate on Mars and in the Atacama, *J. Geophys. Res.*, 115, E00E11, doi:10.1029/2009JE003425.

Chassefière, E., J. E. Blamont, V. A. Krasnopolsky, et al. (1992). Vertical structure and size distributions of Martian aerosols from solar occultation measurements. *Icarus*, 97, 46–69.

Cheng, B.-M., E. P Chew, C.-P. Liu, et al. (1999). Photo-induced fractionation of water isotopomers in the Martian atmosphere, *Geophys. Res. Lett.*, 26, 3657–3660.

Christensen, P. R. (1998). Variations in Martian surface composition and cloud occurrence determined from thermal infrared spectroscopy: analysis of Viking and Mariner 9 data. *J. Geophys. Res.*, 103, 1733–1746.

Christensen, P. R., and R. W. Zurek (1984). Martian north polar hazes and surface ice: results from the Viking survey/completion mission. *J. Geophys. Res.*, 89, 4587–4596.

Christensen, P. R., D. L. Anderson, S. C. Chase, et al. (1992). Thermal Emission Spectrometer experiment: the Mars Observer Mission, *J. Geophys. Res.*, 97, 7719–7734.

Christensen, P. R., J. L. Bandfield, V. E. Hamilton, et al. (2001). The Mars Global Surveyor Thermal Emission Spectrometer experiment: investigation description and surface science results. *J. Geophys. Res.*, 106, 23823–23871.

Christensen, P. R., G. L. Mehall, S. H. Silverman, et al. (2003). The Miniature Thermal Emission Spectrometer for the Mars Exploration Rovers. *J. Geophys. Res.*, 108, 8064, doi:8010.1029/2003JE002117.

Christensen, P. R., B. M. Jakosky, H. H. Kieffer, et al. (2004). The Thermal Emission Imaging System (THEMIS) for the Mars 2001 Odyssey Mission. *Space Science Reviews*, 110, 85–130.

Clancy, R. T., A. W. Grossman, and D. O. Muhleman (1992). Mapping Mars water vapor with the Very Large Array. *Icarus*, 100, 48–59.

Clancy, R. T., A. W. Grossman, M. J. Wolff, et al. (1996). Water vapor saturation at low altitudes around Mars aphelion: a key to Mars climate? *Icarus*, 122, 36–62.

Clancy, R. T., and H. Nair (1996). Annual (perihelion–aphelion) cycles in the photochemical behavior of the global Mars atmosphere. *J. Geophys. Res.*, 101, 12785–12790.

Clancy, R. T., M. J. Wolff, and P. B. James (1999). Minimal aerosol loading and global increases in atmospheric ozone during the 1996–1997 Martian northern spring season. *Icarus*, 138, 49–63.

Clancy, R.T., M. J. Wolff, and P. R. Christensen (2003). Mars aerosol studies with the MGS TES emission phase function observations: optical depths, particle sizes, and ice cloud types versus latitude and solar longitude. *J. Geophys. Res.*, 108, E95098, doi:10.1029/2003JE002058.

Clancy, R. T., B. J. Sandor, and G. H. Moriarty-Schieven (2004). A measurement of the 362 GHz absorption line of Mars atmospheric HO_2. *Icarus*, 168, 116–12.

Clancy, R. T., M. J. Wolff, B. A. Whitney, B. A. Cantor, and M. D. Smith (2007). Mars equatorial mesospheric clouds: global occurrence and physical properties from Mars Global Surveyor Thermal Emission Spectrometer and Mars Orbiter Camera limb observations. *J. Geophys. Res.*, 112, E04004, doi:10.1029/2006JE002805.

Clancy, R. T., B. J. Sandor, M. J. Wolff et al. (2010). CRISM limb observations of O_2 singlet delta nightglow in the polar winter atmosphere of Mars *Bull. Amer. Astron Soc.*, 42, 1041.

Clancy, R. T., M. J. Wolff, B. A. Cantor, M. C. Malin, and T. I. Michaels (2009). Valles Marineris cloud trails, *J. Geophys. Res.*, 114, E11002, doi:10.1029/2008JE003323.

Clancy, R. T., B. J. Sandor, M. J. Wolff, et al. (2012). Extensive MRO CRISM observations of 1.27 μm O_2 airglow in Mars polar night and their comparison to MRO MCS temperature profiles and LMD GCM simulations. *J. Geophys. Res.*, 117, doi:10.1029/2011JE004018.

Clancy, R. T., B. J. Sandor, A. García-Muñoz, et al. (2013). First detection of Mars atmospheric hydroxyl: CRISM near-IR measurement versus LMD GCM simulation of OH Meinel band emission in the Mars polar winter atmosphere. *Icarus*, 226, 272–281.

Clancy, R. T., M. D. Smith, M. J. Wolff, et al. (2014). CRISM limb observations of Mars mesospheric ice clouds: two new results, *Eighth International Conference on Mars*, July 14–18, Pasadena, CA., LPI Contribution No. 1791, 1006.

Clancy, R. T., M. J. Wolff, F. Lefèvre, et al. (2015). Daily global mapping of Mars ozone column abundances with MARCI UV band imaging. *Icarus*, in review.

Clayton, R. N., and T. K. Mayeda (1983). Oxygen isotopes in eucrites, shergottites, nakhlites, and chassignites. *Earth Planet. Sci. Lett.*, 62, 1–6, 1983.

Clayton, R. N., and T. K. Mayeda (1988). Isotopic composition of carbonate in EETA 79001 and its relation to parent body volatiles. *Geochem. Cosmochim. Acta*, 52, 925–927.

Colaprete, A., and O. B. Toon (2002). Carbon dioxide snowstorms during the polar night on Mars. *J. Geophys. Res.*, 107, E75051, doi:10.1029/2001JE001758.

Connes, J., P. Connes, and J.-P. Maillard, (1969). Atlas des spectres dans le proche infrarouge de Vénus, Mars, Jupiter et Saturne, *Editions du Centre Nationale de la Recherche Scientifique*, 471 p.

Conrath, B., R. Curran, R. Hanel, et al. (1973). Atmospheric and surface properties of Mars obtained by infrared spectroscopy on Mariner 9, *J. Geophys. Res.*, 78, 4267–4278.

Crowley, G., and R. H. Tolson (2007). Mars thermospheric winds from Mars Global Surveyor and Mars Odyssey accelerometers. *Journal of Spacecraft and Rockets* 44, 1188–1194.

Cull, S., R. E. Arvidson, R. V. Morris, et al. (2010). The seasonal ice cycle at the Mars Phoenix landing site: II. Post-landing CRISM and ground observations. *J. Geophys. Res.,* 115, E00E19. doi:10.1029/ 2009JE003410

Curran, R. J., B. J. Conrath, R. A. Hanel, V. G. Kunde, and J. C. Pearl (1973). Mars: Mariner 9 spectroscopic evidence for H_2O ice clouds. *Science,* 182, 381–383.

Dickinson, C., J. A. Whiteway, L. Komguem, J. E. Moores, and M. T. Lemmon (2010). Lidar measurements of clouds in the planetary boundary layer on Mars. *Geophys. Res. Lett.,* 37, L18203. doi:10.1029/2010GL044317.

Drube, L., K. Leer, W. Goetz, et al. (2010). Magnetic and optical properties of airborne dust and settling rates of dust at the Phoenix landing site. *J. Geophys. Res.,* in press. doi:10.1029/ 2009JE003419

Ellehoj, M. D., H. P. Gunnlaugsson, K. M. Bean, et al. (2010). Convective vortices and dust devils at the Phoenix Mars mission landing site. *J. Geophys. Res.,* 115, E00E16. doi:10.1029/ 2009JE003413.

Encrenaz, T., B. Bézard, T. K. Greathouse, et al. (2004). Hydrogen peroxide on Mars: evidence for spatial and seasonal variations. *Icarus,* 170, 424–429.

Encrenaz, T., B. Bézard, T. Owen, et al. (2005). Infrared imaging spectroscopy of Mars: H_2O mapping and determination of CO_2 isotopic ratios, *Icarus,* 179, 43–54.

Encrenaz, T., T. Fouchet, R. Melchiorri, et al. (2006). Seasonal variations of the Martian CO over Hellas as observed by OMEGA/ Mars Express. *Astron. Astrophys.,* 459, 265–270.

Encrenaz, T., T. K. Greathouse, M. J. Richter et al. (2008). Simultaneous mapping of H_2O and H_2O_2 on Mars from infrared high-resolution imaging spectroscopy. *Icarus,* 195, 547–556.

Encrenaz, T., T. K. Greathouse, F. Lefèvre, and S. K. Atreya (2012). Hydrogen peroxide on Mars: observations, interpretation and future plans, *Plan. Space Sci.,* 68, 3–17.

Espenak, F., M. J. Mumma, T. Kostiuk, and D. Zipoy (1991). Ground-based infrared measurements of the global distribution of ozone in the atmosphere of Mars. *Icarus,* 92, 252–262.

Farmer, C. B. and P. E. Doms (1979), Global and seasonal variation of water vapor on Mars and the implications for permafrost, *J. Geophys. Res.,* 84, 2881–2888.

Farmer, C. B., D. W. Davies, A. L. Holland, D. D. LaPorte, and P. E. Doms (1977). Mars: water vapor observations from the Viking Orbiters, *J. Geophys. Res.,* 82, 4225–4248.

Farquhar, J., K. Sang-Tae, and A. Masterson (2007). Implications from sulfur isotopes of the Nakhla meteorite for the origin of sulfate on Mars. *Earth and Planetary Science Letters,* 264, 1–8.

Fast, K., T. Kostiuk, F. Espenak, et al. (2006). Ozone abundance on Mars from infrared heterodyne spectra: I. Acquisition, retrieval, and anticorrelation with water vapor, *Icarus,* 181, 419–431.

Fedorova, A., O. Korablev, J.-L. Bertaux, et al. (2006a). Mars water vapor abundance from SPICAM IR spectrometer: seasonal and geographic distributions. *J. Geophys. Res.* 111, E09S08, doi:10.1029/2006JE002605.

Fedorova, A., O. Korablev, S., Perrier, et al. (2006b). Observation of O_2 1.27 µm dayglow by SPICAM IR: seasonal distribution for the first Martian year of Mars Express. *J. Geophys. Res.,* 111 E09S07, doi:10.1029/2006JE002694.

Fedorova, A. A., O. Korablev, J.-L. Bertaux et al. (2009). Solar infrared occultation observations by SPICAM experiment on Mars-Express: simultaneous measurements of the vertical distributions of H_2O, CO_2 and aerosol. *Icarus,* 200, 96–117.

Fedorova, A. A., F. Montmessin, A. V. Rodin, et al. (2014). Evidence for a bimodal size distribution for the suspended particles on Mars *Icarus,* 231, 239–260.

Feldman, W. C., T. H. Prettyman, W. V. Boynton, et al. (2003). CO_2 frost cap thickness on Mars during northern winter and spring. *J. Geophys. Res.,* 108, E9, 5103, doi:10.1029/2003JE002101.

Feldman, W. C., T. H. Prettyman, S. Maurice, et al. (2004). Global distribution of near-surface hydrogen on Mars, *J. Geophys. Res.,* 109, E09006, doi:10.1029/2003JE002160.

Ferri, F., P. H. Smith, M. T. Lemmon, and N. Rennó (2003). Dust devils as observed by Mars Pathfinder. *J. Geophys. Res.* 108, E12.7–1, doi:10.1029/2000JE001421.

Forget, F., A. Spiga, B. Dolla, et al. (2007). Remote sensing of surface pressure on Mars with the Mars Express/OMEGA spectrometer: 1. Retrieval method, *J. Geophys. Res.,* 112, E08S15, doi:10.1029/2006JE002871.

Formisano, V., S. Atreya, T. Encrenaz, N. Ignatiev, and M. Giuranna (2004). Detection of methane in the atmosphere of Mars. *Science,* 306, 1758–1761.

Formisano, V., F. Angrilli, G. Arnold, et al. (2005). The planetary Fourier spectrometer (PFS) onboard the European Mars Express mission. *Planetary and Space Science,* 53, 10 963–974.

Fouchet, T., E. Lellouch, N. I. Ignatiev, et al. (2007). Martian water vapor: Mars Express PFS/LW observations. *Icarus,* 190, 1 32–49.

Franz, H. B., M. G. Trainer, M. H. Wong, et al. (2015). Reevaluated Martian atmospheric mixing ratios from the mass spectrometer on the Curiosity Rover. *Plan. Space Sci.,* 109–110, 154–158.

Goetz, W., P. Bertelsen, C. S. Binau, et al. (2005). Indication of drier periods on Mars from the chemistry and mineralogy of atmospheric dust. *Nature,* 436, 7047 62–65, doi:10.1038/nature03807.

Gómez-Elvira, J., J. C. Armiens, L. Castañer, et al. (2012). REMS: the environmental sensor suite for the Mars Science Laboratory Rover. *Space Sci. Rev.,* 170, 583–640, doi:10.1007/s11214-012-9921-1.

Gooding, J. L., S. J. Wentworth, and M. E. Zolensky (1988). Calcium carbonate and sulfate of possible extraterrestrial origin in the EETA 79001 meteorite. *Geochim. Cosmochim. Acta,* 52, 909–915.

Greeley, R., P. L. Whelley, R. E. Arvidson, et al. (2006). Active dust devils in Gusev Crater, Mars: observations from the Mars Exploration Rover, Spirit. *J. Geophys. Res.,* 111, E12S09, doi:10.1029/2006JE002743.

Greeley, R., D. Waller, N. Cabrol et al. (2010). Gusev Crater, Mars: observations of three dust devil seasons. *J. Geophys. Res.* 115, E00F02. doi:10.1029/2010JE003608

Haberle, R.M., and M. A. Kahre (2010). Detecting secular climate change on Mars. *Mars* 5, 68–75, 2010; doi:10.1555/ mars.2010.0003.

Haberle, R. M., M. M. Joshi, J. R. Murphy, et al. (1999). General circulation model simulations of the Mars Pathfinder atmospheric structure investigation/meteorology data. *J. Geophys. Res.,* 104, 8957–8974.

Haberle, R. M., J. Gómez-Elvira, M. de la Torre Juárez, et al. (2014a). Preliminary interpretation of the REMS pressure data from the first 100 sols of the MSL mission, *J. Geophys. Res.,* 119, 440–453.

Haberle, R. M., J. Gómez-Elvira, M., de la Torre Juarez, et al. (2014b). Secular climate change on Mars: an update using one Mars year of MSL pressure data. *AGU Fall Meeting* 2014, abstract no. P31B-3947.

Hansen, C. J., N. Thomas, G. Portyankina, et al. (2010). HiRISE observations of gas sublimation-driven activity in Mars' southern polar regions: I. Erosion of the surface. *Icarus,* 205, 263–295.

Hartogh, P., C. Jarchow, E. Lellouch, et al. (2010). Herschel/HIFI observations of Mars: first detection of O_2 at submillimetre wavelengths and upper limits on HCl and H_2O_2. *Astrophys. and Astron.,* 521, L49, doi:10.1051/0004-6361/201015160.

Hayne, P. O., D. A. Paige, J. T. Schofield, et al. (2012). Carbon dioxide snow clouds on Mars: south polar winter observations by the Mars Climate Sounder. *J. Geophys. Res.*, 17, E08014, doi:10.1029/2011JE004040.

Hayne, P. O., D. A. Paige, N. G. Heavens, et al. (2014). The role of snowfall in forming the seasonal ice caps of Mars: models and constraints from the Mars Climate Sounder. *Icarus*, 231, 122–130.

Heavens, N. G., M. I. Richardson, A. Kleinböhl, et al. (2011a). Vertical distribution of dust in the Martian atmosphere during northern spring and summer: high-altitude tropical dust maximum at northern summer solstice. *J. Geophys. Res.*, 116, E01007, doi:10.1029/2010JE003692.

Heavens, N. G., M. I. Richardson, A. Kleinböhl, et al. (2011b). The vertical distribution of dust in the Martian atmosphere during northern spring and summer: observations by the Mars Climate Sounder and analysis of zonal average vertical dust profiles. *J. Geophys. Res.*, 116, E04003, doi:10.1029/2010JE003691.

Hecht, M. H., S. P. Kounaves, R. C. Quinn, et al. (2009). Detection of perchlorate and the soluble chemistry of the Martian soil at the Phoenix Lander site. *Science*, 325, 64, doi:10.1126/science.1172466.

Hess, S. L., J. A. Ryan, J. E. Tillman, R. M. Henry, and C. B. Leovy (1980). The annual cycle of pressure on Mars measured by Viking Landers 1 and 2. *Geophys. Res. Lett.*, 7, 3 197–200. doi:10.1029/GL007i003p00197.

Hinson, D. P., and R. J. Wilson (2004). Temperature inversions, thermal tides, and water ice clouds in the Martian tropics. *J. Geophys. Res.*, 109, E01002. doi:10.1029/2003JE002129.

Hinson, D. P., R. A. Simpson, J. D. Twicken, C. L. Tyler, F. M. Flasar (1999). Initial results from radio occultation measurements with Mars Global Surveyor. *J. Geophys. Res.*, 104, 26997–27012

Hinson, D. P., R. A. Simpson, J. D. Twicken, C. L. Tyler, F. M. Flasar (2000). Erratum: Initial results from radio occultation measurements with Mars Global Surveyor. *J. Geophys. Res.*, 105, 1717–1718.

Hinson, D. P., M. D. Smith, and B. J. Conrath (2004). Comparison of atmospheric temperatures obtained through infrared sounding and radio occultation by Mars Global Surveyor. *J. Geophys. Res.*, 103, E12002, doi: 10.1029/2004JE002344.

Hinson, D. P., M. Pätzold, S. Tellmann, B. Häusler, and G. L. Tyler (2008). The depth of the convective boundary layer on Mars, *Icarus*, 198, 57–66.

Holstein-Rathlou, C., H. P. Gunnlauggson, J. P. Merrison, et al. (2010). Winds at the Phoenix Landing Site. *J. Geophys. Res.*, 115, E00E18, doi:10.1029/2009JE003411.

Hunt, G. E., A. O. Pickersgill, P. B. James, and G. Johnson (1980). Some diurnal properties of clouds over the Martian Volcanoes. *Nature*, 286, 362–364.

Hunt, G. E., A. O. Pickersgill, P. B. James, and N. Evans (1981). Daily and seasonal Viking observations of Martian bore wave systems. *Nature*, 293, 630–633.

Inada, A., M. I. Richardson, T. H. McConnochie, et al. (2007). High-resolution atmospheric observations by the Mars Odyssey Thermal Emission Imaging System. *Icarus*, 192, 378–395.

Inada, A., M. Garcia-Comas, F. Altieri, et al. (2008). Dust haze in Valles Marineris observed by HRSC and OMEGA on board Mars Express. *J. Geophys. Res.*, 113, E02004, doi:10.1029/2007JE002893.

Jakosky, B. M. (1991). Mars volatile evolution: evidence from stable isotopes. *Icarus*, 94, 14–31.

Jakosky, B. M., and C. B. Farmer (1982). The seasonal and global behavior of water vapor in the Martian atmosphere: complete global results of the Viking atmospheric water detector experiment. *J. Geophys. Res.*, 87, 2999–3019.

Jakosky, B. M., and H. Jones (1997). The history of Martian volatiles. *Reviews of Geophysics*, 35, 1–16.

Jakosky, B. M., R. P. Lin, J. M. Grebowsky, et al. (2015). The Mars Atmosphere and Volatile Evolution (MAVEN) Mission. *Space Science Reviews*, doi:10.1007/s11214-015-0139-x.

James, P. B., J. F. Bell III, R. T. Clancy, et al. (1996). Hubble Space Telescope synoptic imaging of Mars: 1995 opposition observations. *J. Geophys. Res.*, 101, 18883–18891.

James, P. B., P. C. Thomas, M. C. Malin (2010). Variability of the south polar cap of Mars in Mars Years 28 and 29. *Icarus*, 208, 82–86.

Johnson, J. R., W. M. Grundy, and M. T. Lemmon (2003). Dust deposition at the Mars Pathfinder landing site: observations and modeling of near-infrared spectra. *Icarus*, 163, 330–346, doi:10.1016/S0019-1035(03)00084-8.

Kahn, R. (1984). The spatial and seasonal distribution of Martian clouds and some meteorological implications. *J. Geophys. Res.*, 89, 6671–6688.

Kahn, R., and P. Gierasch (1982). Long cloud observations on Mars and implications for boundary layer characteristics over slopes. *J. Geophys. Res.*, 87, 867–880.

Kakar, R. K., J. W. Waters, and W. J. Wilson (1977). Mars: microwave detection of carbon monoxide. *Science*, 196, 1090–1091.

Kaplan, L. D., G. Münch, and H. Spinrad (1964). An analysis of the spectrum of Mars. *Astrophys. J.*, 139, 1–15.

Kaplan, L. D., J. Connes, and P. Connes (1969). Carbon monoxide in the Mars atmosphere. *Astrophys. J.*, 157, L187-L192.

Keating, G. M., S. W. Bougher, R. W. Zurek, et al. (1998). The structure of the upper atmosphere of Mars: in situ accelerometer measurements from Mars Global Surveyor. *Science*, 279, 1672–1676.

Kieffer, H. H. (1979). Mars south polar spring and summer temperatures – a residual CO_2 frost. *J. Geophys. Res.*, 84, 8263–8288.

Kieffer, H. H., S. C. Chase, T. Z. Martin, E. D. Miner, and F. D. Palluconi (1976). Martian north pole summer temperatures: dirty water ice. *Science*, 194, 1341–1344.

Kieffer, H. H., T. Z. Martin, A. R. Peterfreund, et al. (1977). Thermal and albedo mapping of Mars during the Viking primary mission. *J. Geophys. Res.*, 82, 4249–4292.

Kieffer, H. H., B. M. Jakosky, C. W. Snyder, and M. S. Matthews (1992). *Mars*. University of Arizona Press, Tucson.

Kleinböhl, A., J. T. Schofield, D. M. Kass, et al. (2009), Mars Climate Sounder limb profile retrieval of atmospheric temperature, pressure, and dust and water ice opacity. *Journal of Geophysical Research*, 114, E10 E10006.

Kliore, A., D. L. Cain, G. S. Levy, et al. (1965). Experiment: results of the First direct measurement of Mars's atmosphere and ionosphere, *Science*, 149, 1243–1248.

Krasnopolsky, V. A. (2006). A sensitive search for SO_2 in the Martian atmosphere: implications for seepage and origin of methane. *Icarus*, 178, 487–492.

Krasnopolsky, V. A. (2007). Long-term spectroscopic observations of Mars using IRTF/CSHELL: mapping of O_2 dayglow, CO, and search for CH_4. *Icarus*, 190, 93–102.

Krasnopolsky, V. A., and P. D. Feldman (2001). Detection of molecular hydrogen in the atmosphere of Mars. *Science*, 294, 1914–1917.

Krasnopolsky, V. A., O. I. Korablev, V. I. Moroz, et al. (1991). Infrared solar occultation sounding of the Martian atmosphere by the Phobos spacecraft. *Icarus*, 94, 32–44.

Krasnopolsky, V. A., M. J. Mumma, G. L. Bjoraker, and D. E. Jennings (1996). Oxygen and carbon isotope ratios in Martian carbon dioxide: measurements and implications for atmospheric evolution. *Icarus*, 124, 553–568.

Krasnopolsky, V. A., M. J. Mumma, and G. R. Gladstone (1998). Detection of atomic deuterium in the upper atmosphere of Mars. *Science*, 280, 1576–1580.

Kuiper, G. P. (1952). *The Atmospheres of the Earth and Planets*, 351–365, Univ. of Chicago Press, Chicago.

Kuiper, G. P. (1957). Visual observations of Mars: 1956. *Ap. J.*, 125, 307–317.

Landis, G. A. and P. P. Jenkins (2000). Measurement of the settling rate of atmospheric dust on Mars by the MAE instrument on Mars Pathfinder. *J. Geophys. Res.*, 105, 1, 1855–1857, doi:10.1029/1999JE001029.

Langevin, Y., J.-P. Bibring, F. Montmessin, et al. (2007). Observations of the south seasonal cap of Mars during recession in 2004–2006 by the OMEGA visible/near-infrared imaging spectrometer on board Mars Express. *J. Geophys. Res.*, 112, E08S12, doi:10.1029/2006JE002841.

Lebonnois, S., E. Quémerais, F. Montmessin, et al. (2006). Vertical distribution of ozone on Mars as measured by SPICAM/Mars Express using stellar occultations. *J. Geophys. Res.*, 111, E09S05, doi:10.1029/2005JE002643.

Lee, C., W. G. Lawson, M. I. Richardson, et al. (2009). Thermal tides in the Martian middle atmosphere as seen by the Mars Climate Sounder. *J. Geophys.Res.*, 114, E03005, doi:10.1029/2008JE003285.

Lefèvre, F., and F. Forget (2009). Observed variations of methane on Mars unexplained by known atmospheric chemistry and physics. *Nature*, 460, 720–723.

Lefèvre, F., J. L. Bertaux, R. T. Clancy, et al. (2008). Heterogeneous chemistry in the atmosphere of Mars. *Nature*, 454, 971–975.

Leighton, R. B. and B. C. Murray (1966). Behavior of carbon dioxide and other volatiles on Mars. *Science*, 153, 136–144.

Lemmon, M. T., M. J. Wolff, M. D. Smith, et al. (2004). Atmospheric Imaging Results from the Mars Exploration Rovers: Spirit and Opportunity. *Science*, 306, 1753–1756, doi:10.1126/science.1104474.

Lemmon, M. T., M. J. Wolff, J. F. Bell III, et al. (2015). Dust aerosol, clouds, and the atmospheric optical depth record over 5 Mars years of the Mars Exploration Rover mission. *Icarus*, 251, 96–111. doi:10.1016/j.icarus.2014.03.029.

Leovy, C., G. Briggs, A. Young, et al. (1972). The Martian atmosphere: Mariner 9 television experiment progress report. *Icarus*, 17, 373–393.

Lillis, R. J., J. H. Engel, D. L. Mitchell, et al. (2005). Probing upper thermospheric neutral densities at Mars using electron reflectometry. *Geophys. Res. Lett.*, 32, L23204, doi:10.1029/2005GL024337.

Lillis, R. J., S. W. Bougher, D. L. Mitchell, et al. (2008a). Continuous monitoring of nightside upper thermospheric mass densities in the Martian southern hemisphere over 4 Martian years using electron reflectometry. *Icarus*, 194, 562–574.

Lillis, R. J., D. L. Mitchell, R. P. Lin, and M. H. Acuña (2008b). Electron reflectometry in the Martian atmosphere. *Icarus*, 194, 544–561.

Lillis, R. J., S. W. Bougher, F. González-Galindo, et al. (2010). Four Martian years of nightside upper thermospheric mass densities derived from electron reflectometry: method extension and comparison with GCM simulations. *J. Geophys. Res.*, 115, E07014, doi:10.1029/2009JE003529.

Lowell, P. (1895). *Mars*. Longmans and Green, London.

Madsen, M. B., S. F. Hviid, H. P. Gunnlaugsson, et al. (1999). The magnetic properties experiments on Mars Pathfinder, *J. Geophys. Res.*, 104, E4, 8761–8779, doi:10.1029/1998JE900006.

Madsen, M. B., W. Goetz, P. Bertelsen, et al. (2009). Overview of the magnetic properties experiments on the Mars Exploration Rovers. *J. Geophys. Res.*, 114, E06S90, doi:10.1029/2008JE003098.

Magalhaes, J., J. T. Schofield, and A. Seiff (1999). Results of the Mars Pathfinder atmospheric structure investigation. *J. Geophys. Res.*, 104, 8943–8956

Mahaffy, P. R., C. R. Webster, M. Cabane, et al. (2012). The Sample Analysis at Mars investigation and instrument suite. *Space Science Reviews*, 170, 1–4, 401–478

Mahaffy, P. R., C. R. Webster, S. K. Atreya et al. (2013). Abundance and isotopic composition of gases in the Martian atmosphere from the Curiosity Rover. *Science*, 341, 263–266.

Maki, J., D. Thiessen, A. Pourangi, et al. (2012). The Mars Science Laboratory engineering cameras. *Space Sci. Rev.*, 170, 77–93.

Malin, M. C., G. E. Danielson, A. P. Ingersoll, et al. (1992). Mars Observer camera. *J. Geophys. Res.*, 97, 7699–7718.

Malin, M. C., M. A. Caplinger, S. D. Davis, et al. (2001). Observational evidence for an active surface reservoir of solid carbon dioxide on Mars. *Science*, 294, 2146–2148.

Malin, M. C., J. F. Bell III, J. Cameron, et al. (2005). The Mast Cameras and Mars Descent Imager (MARDI) for the 2009 Mars Science Laboratory, in *36th Annual Lunar and Planetary Science Conf.*, March 14–18, League City, Texas, Abstract No. 1214.

Malin, M. C., J. F. Bell III, B. A. Cantor, et al. (2007). Context Camera investigation on board the Mars Reconnaissance Orbiter. *Journal of Geophysical Research*, 112, E5, E05S04.

Malin, M. C, W. M. Calvin, B. A. Cantor, et al. (2008). Climate, weather, and north polar observations from the Mars Reconnaissance Orbiter Mars color imager. *Icarus*, 194.

Maltagliati, L., F. Montmessin, A. Fedorova, et al. (2011). Evidence of water vapor in excess of saturation in the atmosphere of Mars. *Science*, 1868–1871.

Maltagliati, L., F. Montmessin, O. Korablev, et al. (2013). Annual survey of water vapor vertical distribution and water-aerosol coupling in the Martian atmosphere observed by SPICAM/MEx solar occultations. *Icarus*, 223, 942–962.

Markiewicz, W. J., R. M. Sablotny, H. U. Keller, et al. (1999). Optical properties of the Martian aerosols as derived from Imager for Mars Pathfinder midday sky brightness data. *J. Geophys. Res.*, 104, 9009–9018.

Martin, L. J. (1976). 1973 dust storm on Mars: maps from hourly photographs. *Icarus*, 29, 363–380.

Martin, L. J., P. B. James, A. Dollfus, et al. (1992). Telescopic observations: visual, photographic, polarimeteric. In *Mars*, University of Arizona Press, Tucson, AZ, 34–70.

Martin, T. Z. (1981). Mean thermal and albedo behavior of the Mars surface and atmosphere over a Martian year, *Icarus*, 45, 427–446.

Maurice, S., R. C. Wiens, M. Saccoccio, et al. (2012). The ChemCam instrument suite on the Mars Science Laboratory (MSL) Rover: science objectives and mast unit description. *Space. Sci. Rev.*, 170, 95–166.

Mazarico, E., M. T. Zuber, F. G. Lemoine, and D. E. Smith (2007a). Martian exospheric density using Mars Odyssey radio tracking data. *J. Geophys. Res.* 112, E05014, doi:10.1029/2006JE002734.

Mazarico, E., M. T. Zuber, F. G. Lemoine, and D. E. Smith (2007b). Atmospheric density during the aerobraking of Mars Odyssey from radio tracking data. *Journal of Spacecraft and Rockets*, 44, 1165–1171.

Mazarico, E., M. T. Zuber, F. G. Lemoine, and D. E. Smith (2008). Observation of atmospheric tides in the Martian exosphere using Mars Reconnaissance Orbiter radio tracking data. *Geophys. Res. Lett.*, 35, L09202, doi:10.1029/2008GL033388.

McCleese, D. J., J. T. Schofield, F. W. Taylor, et al. (2007). Mars Climate Sounder: an investigation of thermal and water vapor structure, dust and condensate distributions in the atmosphere, and energy balance of the polar regions. *Journal of Geophysical Research-Planets*, 112, E5, E05S06.

McCleese, D. J., J. T. Schofield, F. W. Taylor, et al. (2008). Intense polar temperature inversion in the middle atmosphere on Mars. *Nature, Geoscience*, 1, 11, 745–749.

McCleese, D. J., N. G. Heavens, J. T. Schofield, et al. (2010). Structure and dynamics of the Martian lower and middle atmosphere as observed by the Mars Climate Sounder: seasonal variations in zonal mean temperature, dust, and water ice aerosols. *J. Geophys. Res.*, 115, E12016, doi:10.1029/2010JE003677.

McConnochie, T. H., J. F. Bell, D. Savransky, et al. (2010). THEMIS-VIS observations of clouds in the Martian mesosphere: altitudes, wind speeds, and decameter-scale morphology. *Icarus*, 210. 545–565.

McConnochie, T. H., M. D. Smith, S. C. Bender, et al. (2014). The Martian O_2 and H_2O cycles observed with ChemCam passive sky spectroscopy, *American Geophysical Union, Fall Meeting 2014*, abstract #P53D-01.

McElroy, M. B., and T. M. Donahue (1972). Stability of the Martian atmosphere. *Science*, 177, 986–988.

McEwen, A. S., E. M. Eliason, J. W. Bergstrom, et al. (2007). Mars Reconnaissance Orbiter's High Resolution Imaging Science Experiment (HiRISE). *J. Geophys. Res.*, 112, E05S02, doi:10.1029/2005JE002605.

Mellon, M. T., R. E. Arvidson, H. G. Sizemore, et al. (2009). Ground ice at the Phoenix landing site: stability state and origin. *J. Geophys. Res.*, 114, E00E07, doi:10.1029/2009JE003417.

Metzger, S. M., J. R. Johnson, J. R. Carr, T. J. Parker, and M. Lemmon (1999). Dust devil vortices seen by the Mars Pathfinder Camera. *Geophys. Res. L.* 26, 2781–2784, doi:10.1029/1999GL008341.

Mitchell, D. L., R. P. Lin, C. Mazelle, et al. (2001). Probing Mars' crustal magnetic field and ionosphere with the MGS Electron Reflectometer. *J. Geophys. Res.*, 106, 23419–23428.

Möhlmann, D. T. F., M. Niemand, V. Formisano, H. Savijärvi, and P. Wolkenberg (2009). Fog phenomena on Mars. *Plan. Space Sci.*, 57, 1987–1992.

Montmessin, F., and F. Lefèvre (2013). Transport-driven formation of a polar ozone layer on Mars. *Nature Geoscience*, 6, doi:10.1038/NGEO1957.

Montmessin, F., T. Fouchet, and F. Forget (2005). Modeling the annual cycle of HDO in the Martian atmosphere. *J. Geophys. Res.*, 110, E03006, doi:10.1029/2004JE002357.

Montmessin, F., E. Quémerais, J. L. Bertaux, et al. (2006). Stellar occultations at UV wavelengths by the SPICAM instrument: retrieval and analysis of Martian haze profiles. *J. Geophys. Res.*, 111, E09S09, doi:10.1029/2005JE002662.

Montmessin, F., B. Gondet, J.-P. Bibring, et al. (2007). Hyperspectral imaging of convective CO_2 ice clouds in the equatorial mesosphere of Mars. *J. Geophys. Res.*, 112, E11S90, doi:10.1029/2007JE002944.

Moore, C. A., J. E. Moores, M. T. Lemmon, et al. (2016). A full Martian year of line-of-sight extinction within Gale Crater, Mars as acquired by the MSL Navcam through sol 900. *Icarus*, 264, 102–108, doi:10.1016/j.icarus.2015.09.001.

Moores, J. E., M. T. Lemmon, P. H. Smith, L. Komguem, and J. A. Whiteway (2010). Atmospheric dynamics at the Phoenix landing site as seen by the Surface Stereo Imager. *J. Geophys. Res.*, 115, E00E08, doi:10.1029/2009JE003409.

Moores, J. E., L. Komguem, J. A. Whiteway, et al. (2011). Observations of Near-Surface Fog at the Phoenix Landing. *Geophys. Res. L.*, 38, L04203, doi:10.1029/2010GL046315.

Moores, J. E., M. T. Lemmon, S. C. R. Rafkin, et al. (2015a). Atmospheric movies acquired at the Mars Science Laboratory landing site: cloud morphology, frequency and significance to the Gale Crater water cycle and Phoenix mission results. *Adv. Space Res*, 55, 2217–2238.

Moores, J. E., M. T. Lemmon, H. Kahanpää, et al. (2015b). Observational evidence of a suppressed planetary boundary layer in northern Gale Crater: Mars as seen by the Navcam instrument onboard the Mars Science Laboratory Rover. *Icarus*, 249, 129–142.

Mumma, M. J., R. E. Novak, M. A. Disanti, et al. (2003). Seasonal mapping of HDO and H_2O in the Martian atmosphere. *Sixth International Conference on Mars*, Pasadena, Abstract 3186.

Mumma, M. J., G. L. Villanueva, R. E. Novak, et al. (2009). Strong release of methane on Mars in northern summer 2003. *Science*, 323, 1041–1045.

Murchie, S., R. Arvidson, P. Bedini, et al. (2007). Compact Reconnaissance Imaging Spectrometer for Mars (CRISM) on Mars Reconnaissance Orbiter (MRO). *J. Geophys. Res.*, 112, E05S03, doi:10.1029/2006JE002682.

Murphy, J. R. and S. Nelli (2002). Mars Pathfinder convective vortices: frequency of occurrence. *Geophys. Res. Lett.*, 29, 23, 2103, doi:10.1029/2002GL015214.

Neukum, G., and R. Jaumann (2004). HRSC: the High Resolution Stereo Camera of Mars Express. In *Mars Express: the Scientific Payload*. Ed. Andrew Wilson, scientific coordination: Agustin Chicarro. ESA SP-1240, Noordwijk, Netherlands: ESA Publications Division, 17–35.

Neumann, G. A., D. E. Smith, and M. T. Zuber (2003). Two Mars years of clouds detected by the Mars Orbiter Laser Altimeter. *J. Geophys. Res.*, 108, E4, 5023, doi:10.1029/2002JE001849.

Nier, A. O., W. B. Hanson, A. Seitt, et al. (1976a). Composition and structure of the Martian atmosphere: preliminary results from Viking 1. *Science,* 193, 786–788.

Nier, A. O., M. B. McElroy, and Y. L. Yung (1976b). Isotopic composition of the Martian atmosphere. *Science*, 194, 68–70.

Niles, P. B., W. V. Boynton, J. H. Hoffman, et al. (2010). Stable isotope measurements of Martian atmospheric CO_2 at the Phoenix Landing Site. *Science*, 329, 1334–1337. doi:10.1126/science.1192863.

Novak, R. E., M. J. Mumma, M. A. DiSanti, N. D. Russo, and K. Magee-Sauer (2002). Mapping of ozone and water in the atmosphere of Mars near the 1997 aphelion. *Icarus*, 158, 14–23.

Novak, R. E., M. J. Mumma, and G. L. Villanueva (2011). Measurement of the isotopic signatures of water on Mars; implications for studying methane. *Planetary and Space Science*, 59, 163–168.

Noxon, J. F., W. A. Traub, N. P. Carlton, and P. Connes (1976). Detection of O_2 dayglow emission from Mars and the Martian ozone abundance. *Astrophys. J.*, 207, 1025–1035.

Owen, T., and K. Biemann (1976). Composition of the atmosphere at the surface of Mars: detection of argon-36 and preliminary analysis. *Science*, 193, 801–803.

Owen, T., K. Biemann, D. R. Rushneck et al. (1976). The atmosphere of Mars: detection of krypton and xenon. *Science*, 194, 1293–1295.

Owen, T., K. Biemann, D. R. Rushneck, et al. (1977). The composition of the atmosphere at the surface of Mars. *J. Geophys. Res.*, 82, 4635–4639.

Owen, T., J. P. Maillard, C. de Bergh, B. L. Lutz (1988). Deuterium on Mars – the abundance of HDO and the value of D/H. *Science*, 240, 1767–1770.

Paige, D. A., and S. E. Wood (1992). Modeling the Martian seasonal CO_2 cycle 2. Interannual variability. *Icarus*, 99, 15–27.

Parkinson, T. D., and D. M. Hunten (1972). Spectroscopy and aeronomy of O_2 on Mars. *J. Atmos. Sci.*, 29, 1380–1390.

Pätzold, M., F. M. Neubauer, L. Carone, et al. (2004). MaRS: Mars Express Orbiter radio science. In ESA Special Publication 1240, available online at http://sci.esa.int/science-e/www/object/index.cfm?fobjectid = 34885

Pätzold, M., S. Tellmann, B. Häusler, et al. (2005). A sporadic third layer in the ionosphere of Mars. *Science*, 310, 837–839.

Pearl, J. C., M. D. Smith, B. J. Conrath, J. L. Bandfield, and P. R. Christensen (2001). Observations of water-ice clouds by the Mars

Global Surveyor Thermal Emission Spectrometer experiment: the first Martian year. *J. Geophys. Res.*, 12325–12338.

Perrier, S., J. L. Bertaux, F. Lefèvre, et al. (2006). Global distribution of total ozone on Mars from SPICAM/MEX UV measurements. *J. Geophys. Res.*, 111, doi:10.1029/2006JE002681.

Phillips, R. J., B. J. Davis, K. L. Tanaka, et al. (2011). Massive CO_2 ice deposits sequestered in the south polar layered deposits of Mars. *Science*, 332, 838–841, doi:10.1126/science.1203091.

Pirraglia, J. A. and B. J. Conrath (1974). Martian tidal pressure and wind fields obtained from the Mariner 9 Infrared Spectroscopy experiment. *J. Atmos. Sci.*, 31, 318–329.

Pollack, J. B., D. Colburn, R. Kahn, et al. (1977). Properties of aerosols in the Marian atmosphere as inferred from Viking Lander imaging data. *J. Geophys. Res.*, 82, 4479–4496.

Quémerais, E., J.-L. Bertaux, O. Korablev, et al. (2006). Stellar occultations observed by SPICAM on Mars Express. *J. Geophys. Res.*, 111, E09S04, doi:10.1029/2005JE002604.

Savijärvi, H. I., A.-M. Harri, and O. Kemppinen, (2015). Mars Science Laboratory diurnal moisture observations and column simulations. *J. Geophys. Res.*, 120, 1011–1021.

Schofield, J. T., J. R. Barnes, J. R. Crisp, et al. (1997). The Mars Pathfinder Atmospheric Structure Investigation/Meteorology. *Science*, 278, 1752.

Scholten, F., H. Hoffmann, A. Määttänen, et al. (2010). Concatenation of HRSC colour and OMEGA data for the determination and 3D-parameterization of high-altitude CO_2 clouds in the Martian atmosphere. *Plan. Space Sci.*, 58, 1207–1214.

Smith, D. E., M. T. Zuber, H. V. Frey, et al. (2001). Mars Orbiter Laser Altimeter: experiment summary after the first year of global mapping of Mars. *J. Geophys. Res.*, 106, 23689–23722.

Smith, M. D. (2002). The annual cycle of water vapor on Mars as observed by the Thermal Emission Spectrometer. *J. Geophys. Res.*, 197, 5115, doi:5110.1029/2001JE001522.

Smith, M. D. (2004). Interannual variability in TES atmospheric observations of Mars during 1999–2003. *Icarus*, 167, 148–165.

Smith, M. D. (2008). Spacecraft observations of the Martian atmosphere. *Annual Review of Earth and Planetary Sciences*, 36, 191–219, doi:0.1146/annurev.earth.36.031207.124334.

Smith, M. D. (2009). THEMIS observations of Mars aerosol optical depth from 2002–2008. *Icarus*, 202, 444–452.

Smith, M. D., J. L. Bandfield, and P. R. Christensen (2000). Separation of atmospheric and surface spectral features in Mars Global Surveyor Thermal Emission Spectrometer (TES) spectra: models and atmospheric properties. *J. Geophys. Res.*, 105, 9589–9608.

Smith, M. D., J. C. Pearl, B. J. Conrath, and P. R. Christensen (2001). Thermal Emission Spectrometer results: Mars atmospheric thermal structure and aerosol distribution. *J. Geophys. Res.*, 106, 23929–23945.

Smith, M. D., J. L. Bandfield, P. R. Christensen, and M. I. Richardson (2003). Thermal Emission Imaging System (THEMIS) infrared observations of atmospheric dust and water ice cloud optical depth. *J. Geophys. Res.*, 108, doi:10.1029/2003JE002115.

Smith, M. D., M. J. Wolff, M. T. Lemmon, et al. (2004). First atmospheric science results from the Mars Exploration Rovers Mini-TES. *Science*, 306, 1750–1753, doi:10.1126/science.1104257.

Smith, M. D., M. J. Wolff, N. Spanovich, et al. (2006). One Martian year of atmospheric observations using MER Mini-TES. *J. Geophys. Res.*, 111, doi:10.1029/2006JE002770.

Smith, M. D., M. J. Wolff, R. T. Clancy, and S. L. Murchie (2009). CRISM observations of water vapor and carbon monoxide. *J. Geophys. Res.*, 114, E00D03, doi:10.1029/2008JE003288.

Smith, M. D., M. J. Wolff, R. T. Clancy, A. Kleinböhl, and S. L. Murchie (2013). Vertical distribution of dust and water ice aerosols from CRISM limb-geometry observations. *J. Geophys. Res*, 118, 321–334, doi:10.1002/jgre.20047.

Smith, P. H. and M. T. Lemmon (1999). Opacity of the Martian atmosphere measured by the Imager for Mars Pathfinder. *J. Geophys. Res.*, 104, 8975–8985, doi:10.1029/1998JE900017.

Smith, P. H., M. G. Tomasko, D. Britt, et al. (1997). The imager for Mars Pathfinder Experiment. *J. Geophys. Res.*, 102, 4003–4025.

Smith, P. H., L. Tamppari, R. E. Arvidson, et al. (2008). Introduction to special section on the Phoenix Mission: landing site characterization experiments, mission overviews, and expected science. *J. Geophys. Res.*, 113, E00A18, doi:10.1029/2008JE003083.

Smith, P. H., L. K. Tamppari, R. E. Arvidson, et al. (2009). Water at the Phoenix Landing Site. *Science*, 325, 58–61. doi:10.1126/science.1172339.

Snyder, C. W. (1977). The missions of the Viking Orbiters. *J. Geophys. Res.*, 82, 3971–3983.

Snyder, C. W., and V. I. Moroz (1992). Spacecraft exploration of Mars, in *Mars*, eds. Kieffer, H. H., Jakosky, B. M., Snyder, C. W., Matthews, M. S., University of Arizona Press, Tucson, Arizona, USA, 71–119

Spiga, A., F. Forget, B. Dolla, et al. (2007). Remote sensing of surface pressure on Mars with the Mars Express/OMEGA spectrometer: 2. Meteorological maps. *J. Geophys. Res.*, 112, E08S16, doi:10.1029/2006JE002870.

Spinrad, H., G. Münch, and L. D. Kaplan (1963). The detection of water vapor on Mars. *Astrophys. J.*, 137, 1319–1321.

Sprague, A. L., D. M. Hunten, R. E. Hill, B. Rizk, and W. K. Wells (1996). Martian water vapor, 1988–1995. *J. Geophys. Res.*, 101, E10, 23229–23241.

Sprague, A. L., W. V. Boynton, K. E. Kerry, et al. (2004). Mars' south polar Ar enhancement: a tracer for south polar seasonal meridional mixing. *Science*, 306, 1364–1367.

Sprague, A. L., W. V. Boynton, K. E. Kerry, et al. (2007). Mars' atmospheric argon: tracer for understanding Martian atmospheric circulation and dynamics. *J. Geophys. Res.*, 112, E03S02, doi:10.1029/2005JE002597.

Squyres, S. W., R. E. Arvidson, E. T. Baumgartner, et al. (2003). Athena Mars Rover science investigation. *J. Geophys. Res.*, 108, E12, 8062, doi:10.1029/2003JE002121.

Sullivan, R., R. Greeley, M. Kraft, et al. (2000). Results of the imager for Mars Pathfinder windsock experiment. *J. Geophys. Res.*, 105, 24547–24562, doi:10.1029/1999JE001234.

Tamppari, L. K., D. Bass, B. Cantor, et al. (2010). Phoenix and MRO coordinated atmospheric measurements. *J.Geophys. Res.*, 115, E00E17, doi:10.1029/2009JE003415.

Taylor, P. A., H. Kahanpää, W. Weng, et al. (2010). On pressure measurement and seasonal pressure variations during the Phoenix mission. *J. Geophys. Res.*, 115, E00E15, doi:10.1029/2009JE003422.

Tellmann, S., M. Pätzold, B. Häusler, D. P. Hinson, and G. L. Tyler (2013). The structure of Mars lower atmosphere from Mars Express Radio Science (MaRS) occultation measurements. *J. Geophys. Res.*, 118, 306–320, doi:10.1002/jgre.20058.

Thomas, P. C., P. B. James, W. M. Calvin, R. Haberle, and M. C. Malin (2009). Residual south polar cap of Mars: stratigraphy, history, and implications of recent changes. *Icarus*, 203, 352–375.

Tillman, J. E. (1988). Mars global atmospheric oscillations – annually synchronized, transient normal-mode oscillations and the triggering of global dust storms. *J. Geophys. Res.*, 93, 9433–9451.

Titov, D. V., W. J. Markiowicz, N. Thomas, et al. (1999). Measurements of the atmospheric water vapor on Mars by the Imager for Mars Pathfinder. *J. Geophys. Res.*, 104, 9019–9026, doi:10.1029/1998JE900046.

Tolson, R. H., G. J. Cancro, G. M. Keating, et al. (1999). Application of accelerometer data to Mars Global Surveyor aerobraking operations. *Journal of Spacecraft and Rockets*, 36, 323–329.

Tolson, R. H., A. M. Dwyer, P. E. Escalera, et al. (2005). Application of accelerometer data to Mars Odyssey aerobraking and atmospheric modeling. *Journal of Spacecraft and Rockets*, 42, 435–443

Tolson, R. H., E. Bemis, S. Hough, et al. (2008). Atmospheric modeling using accelerometer data during Mars Reconnaissance Orbiter aerobraking operations. *Journal of Spacecraft and Rockets*, 45, 511–518.

Tomasko, M. G., L. R. Doose, M. T. Lemmon, P. H. Smith, and E. Wegryn (1999). Properties of dust in the Martian atmosphere from the Imager for Mars Pathfinder. *J. Geophys. Res.*, 104, 8987–9007, doi:10.1029/1998JE900016.

Tracadas, P. W., M. T. Zuber, D. E. Smith, and F. G. Lemoine (2001). Density structure of the upper thermosphere of Mars from measurements of air drag on the Mars Global Surveyor spacecraft. *J. Geophys. Res.* 106, 23349–23358

Tyler, D., Jr., and J. R. Barnes (2013). Mesoscale modeling of the circulation in the Gale Crater region: an investigation into the complex forcing of convective boundary layer depths. *Mars.*, 8, 58–77.

Tyler, G. L., G. Balmino, D. P. Hinson, et al. (2001). Radio science observations with Mars Global Surveyor: orbit insertion through one Mars year in mapping orbit. *J. Geophys. Res.*, 106, 23327–23348.

Vincendon, M., C. Pilorget, B. Gondet, S. Murchie, and J.-P. Bibring (2011). New near-IR observations of mesospheric CO_2 and H_2O clouds on Mars. *J. Geophys. Res.*, 116, E00J02, doi: 10.1029/2011JE003827.

Wang, H., and A. P. Ingersoll (2002). Martian clouds observed by Mars Global Surveyor Mars Orbiter Camera. *J. Geophys. Res.*, 107, 5078, doi:10.1029/2001JE001815.

Webster, C. R., P. R. Mahaffy, S. K. Atreya et al. (2015). Mars methane detection and variability at Gale Crater. *Science*, 347, 415–417.

Whiteway, J., M. Daly, A. Carswell, et al. (2008). Lidar on the Phoenix mission to Mars. *J. Geophys. Res.*, 113, E00A08, doi:10.1029/2007JE003002.

Whiteway, J. A., L. Komguem, C. Dickinson, et al. (2009). Mars water ice clouds and precipitation. *Science*, 325, 68–70. doi:10.1126/science.1172344.

Wilson, R. J., and M. I. Richardson (2000). The Martian atmosphere during the Viking Mission, I. Infrared measurements of atmospheric temperatures revisited. *Icarus*, 145, 555–579.

Withers, P. (2006). Mars Global Surveyor and Mars Odyssey Accelerometer observations of the Martian upper atmosphere during aerobraking. *Geophys. Res. Lett.*, 33, L02201, doi:10.1029/2005GL024447.

Withers, P. (2009). A review of observed variability in the dayside ionosphere of Mars. *Advances in Space Research*, 44, 3, 277–307.

Withers, P. (2012). Empirical estimates of Martian surface pressure in support of the landing of Mars Science Laboratory. *Space Science Reviews*, 1, 4, 837–860. doi:10.1007/s11214-012-9876-2.

Withers, P., and D. C. Catling (2010). Observations of atmospheric tides on Mars at the season and latitude of the Phoenix atmospheric entry. *Geophys. Res. Lett.*, 37, L24204, doi:10.1029/2010GL045382.

Withers, P., and M. D. Smith (2006). Atmospheric entry profiles from the Mars Exploration Rovers Spirit and Opportunity. *Icarus*, 185, 133–142.

Withers, P., S. W. Bougher, G. M. Keating (2003). The effects of topographically-controlled thermal tides in the Martian upper atmosphere as seen by the MGS accelerometer. *Icarus*, 164, 14–32.

Withers, P., M. O. Fillingim, R. J. Lillis, et al. (2012). Observations of the nightside ionosphere of Mars by the Mars Express Radio Science Experiment (MaRS). *J. Geophys. Res.*, 117, A12307, doi:10.1029/2012JA018185.

Wolff, M. J., and R. T. Clancy (2003). Constraints of the size of Martian aerosols from Thermal Emission Spectrometer observations. *J. Geophys. Res.*, 108, E9, 5097, doi:10.1029/2003JE002057.

Wolff, M. J., P. B. James, J. F. Bell III, R. T. Clancy, and S.W. Lee (1999). Hubble Space Telescope observations of the Martian aphelion cloud belt prior to the Pathfinder mission: seasonal and interannual variations. *J. Geophys. Res.*, 104, 9027–9041.

Wolff, M. J., M. D. Smith, R. T. Clancy, et al. (2006). Constraints on dust aerosols from the Mars Exploration Rovers using MGS overflights and Mini-TES. *J. Geophys. Res.*, 111, E12S17, doi:10.1029/2006JE002786.

Wolff, M. J., M. D. Smith, R. T. Clancy, et al. (2009). Wavelength dependence of dust aerosol single scattering albedo as observed by the Compact Reconnaissance Imaging Spectrometer. *J. Geophys. Res.*, 114, E00D04, doi:10.1029/2009JE003350.

Wolff, M. J., R. T. Clancy, J. D. Goguen, M. C. Malin, and B. A. Cantor (2010). Ultraviolet dust aerosol properties as observed by MARCI. *Icarus*, 208, 143–155.

Wolff, M. J., R. T. Clancy, B. Cantor, and J.-B. Madeleine (2011). Mapping water ice clouds (and ozone) with MRO/MARCI. *The Fourth International Workshop on the Mars Atmosphere: Modelling and Observation*, Paris, France.

Wong, M. H., S. K. Atreya, P. N. Mahaffy, et al. (2013). Isotopes of nitrogen on Mars: atmospheric measurements by Curiosity's mass spectrometer. *Geophys. Res. Lett.*, 40, 6033–6037.

Wright, I. P., M. M. Grady, and C. T. Pillinger (1990). The evolution of atmospheric CO_2 on Mars: the perspective from carbon isotope measurements. *J. Geophys. Res.*, 95, 14, 789–14, 794.

Young, L. D. G. (1971). Interpretation of high resolution spectra of Mars II. Calculations of CO_2 abundance, rotational temperature and surface pressure. *J. Quant. Spectrosc. Radiat. Transfer*, 11, 1075–1086.

Zahnle, K., R. S. Freedman, and D. C. Catling, (2011). Is there methane on Mars? *Icarus*, 212, 493–503.

Zent, A. P., M. H. Hecht, D. R. Cobos, et al. (2010). Initial results from the thermal and electrical conductivity probe (TECP) on Phoenix. *J. Geophys. Res.*, 115, E00E14, doi:10.1029/2009JE003420.

Zuber, M. T., D. E. Smith, S. C. Solomon, et al. (1992). The Mars Observer Laser Altimeter investigation. *J. Geophys. Res.*, 97, 7781–7797.

Zurek, R.W. and C. B. Leovy (1981). Thermal tides in the dusty Martian atmosphere – a verification of theory. *Science*, 213, 437–439.

Zurek, R. W. and S. E. Smrekar (2007). An overview of the Mars Reconnaissance Orbiter (MRO) science mission. *J. Geophys. Res.*, 112, E05S01, doi:10.1029/2006JE002701.

Thermal Structure and Composition

MICHAEL D. SMITH, STEPHEN W. BOUGHER, THÉRÈSE ENCRENAZ, FRANÇOIS FORGET, ARMIN KLEINBÖHL

4.1 INTRODUCTION

Over the past two decades there has been a vast increase in the amount and variety of observations characterizing the thermal structure and composition of the Mars atmosphere. The combination of vigorous ground-based and space-based observational campaigns with the successful operation of numerous spacecraft at Mars including Pathfinder, Mars Global Surveyor (MGS), Mars Odyssey, Mars Express, Mars Exploration Rovers (MER), Mars Reconnaissance Orbiter (MRO), Phoenix, and Mars Science Laboratory (MSL) has built upon the earlier successes of Mariner 9 and Viking to allow the atmosphere of Mars to be characterized in unprecedented detail. While there are still areas that require further observations, with the observational data now in hand it is possible to present a reasonably complete overview of the current Martian climate and atmospheric composition.

Two of the most basic quantities in the characterization of the Martian atmosphere are the atmospheric temperatures (or thermal structure) and the composition of the gases that make up the atmosphere. In general, both of these quantities vary in time and in location, and it is the goal of this chapter to describe and quantify the spatial, diurnal, seasonal, and inter-annual variations. This characterization of the atmospheric state forms the basis for understanding all the physical processes that control the current Martian climate, from the general circulation, to the mechanics of dust storms, the role of clouds, and photochemistry.

Many types of observations have contributed to our knowledge of the thermal structure and composition, and often this has allowed for the validation and cross-calibration between different instruments and techniques. The primary tools for observation have been through thermal infrared sounding, radio and ultraviolet (UV) occultations, and near-infrared spectroscopy. From the Martian surface, a limited number of thermal infrared sounding and meteorological packages that typically include sensors for near-surface temperature and atmospheric pressure have been used. Accelerometer records taken during aerobraking passes, and the entry, descent, and landing of surface landers, have provided additional information.

In this review, we first discuss the thermal structure, with a description of the available observations and the observed spatial and temporal dependence. We next turn our attention to the composition of the gases that make up the Martian atmosphere, including a discussion of isotopic ratios. Water vapor is a special case and is only briefly mentioned in this chapter. A complete discussion of atmospheric water vapor is given in Chapter 11. We conclude this review with a discussion of available Martian climate databases, and we present a set of three reference atmospheres that span the range of typical conditions. Earlier reviews on the thermal structure and composition of the Martian atmosphere can be found in the chapters by Zurek (1992), Zurek et al. (1992), and Owen (1992) in *Mars* (University of Arizona Press, 1992) and in papers by Encrenaz (2001), Liu et al. (2003), and Smith (2008).

4.2 THERMAL STRUCTURE

4.2.1 Overview of Mars Atmosphere Thermal Structure

Figure 4.1 shows the temperature profiles derived from accelerometer observations during the descent through the atmosphere of landed spacecraft (see Section 4.2.2) with labels defining the different regions of the Martian atmosphere. On Earth, the atmosphere is typically divided into the troposphere, stratosphere, mesosphere, and thermosphere based on the vertical gradient of temperatures. The terrestrial troposphere is the lowest layer near the surface, where temperature decreases with height at a rate determined by dry and moist adiabatic processes. Above a height of ~10–15 km on Earth (called the tropopause), temperature increases with height because of solar heating from the ozone layer. Temperature begins to decrease once again with height above about 50 km, forming the Earth's mesosphere, until a height is reached (~90 km) where temperatures increase with height as heat is carried by molecular conduction down from the region where solar extreme UV radiation is absorbed. This highest region is called the thermosphere. On Earth, the transition between the mesosphere and thermosphere is near the homopause (~105 km), which is the level above which individual gases separate diffusively instead of remaining homogeneously mixed.

Vertical regions in the Martian atmosphere can be defined by analogy to those on Earth, although there are necessarily some differences. With no ozone layer or other strong enough absorber, Mars has no stratosphere. Instead, we define a "lower", "middle", and "upper" atmosphere. The lower atmosphere is defined as that region below 50 km (corresponding to a pressure level of ~2 Pa) where temperatures typically decrease with height. The middle atmosphere is defined as the region between 50 and 100 km. Temperatures in the middle atmosphere vary greatly under the influence of tides and waves, and there are relatively fewer observations of temperatures there than in the lower atmosphere. Above 100 km is the upper atmosphere, or thermosphere, where temperatures increase with height from the absorption of solar extreme UV. The Martian homopause is located at about 125 km (Nier and

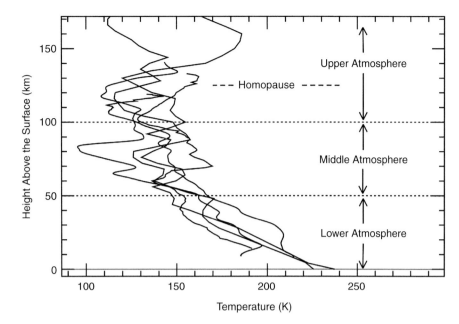

Figure 4.1. An overview of Mars atmosphere thermal structure, defining the "lower", "middle", and "upper" atmosphere. The temperature profiles shown are inferred from accelerometer observations during the descent through the atmosphere of landed spacecraft.

McElroy, 1977). This chapter describes the lower and middle atmosphere. The upper atmosphere is treated in Chapter 14.

4.2.2 Available Observations

4.2.2.1 Thermal Infrared Sounding

The most comprehensive information on the temperature structure of the Martian atmosphere has been obtained by remote sensing instruments observing at thermal infrared wavelengths from orbiting spacecraft. The practice of retrieving planetary atmospheric temperature profiles from thermal infrared spectra taken by a passing or orbiting spacecraft has been well established. From the measured thermal emission of an atmospheric opacity source, such as from a molecular absorption line or band, the atmospheric temperature can be retrieved if the optical depth is known. If the vertical variation of temperature and opacity is not too great, then the observed radiance will be characteristic of the temperature in a layer centered near the level where the absorber optical depth as measured along the line of sight from the observer (spacecraft) reaches unity. When opacity varies with frequency, the height in the atmosphere at which unit optical depth is reached is different for different frequencies. By using this fact, a temperature profile can be retrieved if the vertical distribution of the opacity source is known. For Mars, the strong absorption band of the CO_2 molecule centered at 15 μm (667 cm^{-1}) is very well suited for temperature retrieval. The retrieval of atmospheric temperatures can be performed either by looking nadir through the atmosphere to the surface or by viewing the atmosphere above the limb of the planet at different tangent altitudes. In the limb-viewing geometry, the retrieval is usually performed using a part of the CO_2 band that is optically thin along the line of sight at the considered tangent altitude, so that the major part of the measured radiance originates from the region near the tangent point.

A summary of thermal infrared sounding by Mars space-craft is given in Table 4.1. The first spacecraft to orbit Mars was Mariner 9, arriving in November 1971. Mariner 9 carried the Infrared Interferometer Spectrometer (IRIS) (Hanel et al., 1970, 1972a), a nadir viewing Fourier-transform spectrometer that covered the spectral region between 5 and 50 μm (200 and 2000 cm^{-1}). Temperature profiles from the surface to about 40 km altitude were retrieved from IRIS spectra. Mariner 9 was in an elliptical orbit around Mars, and IRIS observations in the southern hemisphere were primarily in the local morning to early afternoon hours, while observations in the northern hemisphere were primarily in the local late afternoon or at night-time. Because the arrival of Mariner 9 at Mars occurred during a very large global-scale dust storm, the observed atmospheric temperatures were significantly higher than those expected for a clear atmosphere, and most of the initial studies using IRIS retrievals focused on the influence of dust heating on Martian atmospheric temperatures. Even in this early work it was recognized that in the polar regions carbon dioxide ice on the surface was overlaid by a warmer atmosphere, and that the temperature of the middle atmosphere of Mars experienced significant diurnal variations (Hanel et al., 1972b,c). As the Mariner 9 mission progressed, the dust storm gradually dissipated and temperatures decreased (Conrath, 1975). IRIS operated until October 1972, near the end of the Mariner 9 mission.

The next spacecraft with instrumentation for thermal infrared temperature sounding were the two Viking Orbiters that arrived at Mars in 1976. Both orbiters were placed into elliptical orbits. The orbit of Viking Orbiter 1 was Mars-synchronized for most of its five-month primary mission so that periapsis occurred at roughly the same longitude every orbit. Non-synchronous orbits occurred for only a short period during the primary mission and again during the extended mission. Viking Orbiter 2 started in a non-synchronized orbit, then spent a month Mars-synchronized, and was eventually

Table 4.1. *Summary of key parameters of instruments that returned significant thermal atmospheric sounding data from Mars orbit.*

Instrument	Instrument type	Frequency range	Spectral resolution	Dates of operation	Spacecraft
Infrared Interferometer Spectrometer (IRIS)	Fourier-transform spectrometer	200–2000 cm^{-1}	2.4 cm^{-1}	11/1971–10/1972	Mariner 9
Infrared Thermal Mapper (IRTM)	Channel radiometer	420–1640 cm^{-1} (in 5 IR channels, plus 1 vis/NIR channel)	N/A	6/1976–7/ 1980 (VO 1) 8/1976–7/1978 (VO 2)	Viking Orbiters 1 and 2
Thermal Emission Spectrometer (TES)	Fourier-transform spectrometer Visible and infrared bolometers	200–1670 cm^{-1} (FTS)	5 or 10 cm^{-1}	9/1997–10/2006	Mars Global Surveyor
Thermal Emission Imaging System (THEMIS)	Imaging radiometer Vis/NIR imager	650–1590 cm^{-1} (radiometer, at 9 different frequencies)	N/A	2/2002–present	Mars Odyssey
Planetary Fourier Spectrometer (PFS)	Fourier-transform spectrometer	250–8200 cm^{-1}	1.3 cm^{-1}	1/2004–present	Mars Express
Mars Climate Sounder (MCS)	Limb/nadir sounding channel radiometer	220–870 cm^{-1} (in 8 IR channels, plus 1 vis/NIR channel)	N/A	9/2006–present	Mars Reconnaissance Orbiter

brought into a high-inclination non-synchronized orbit to facilitate measurements of the north polar region (Snyder, 1977). Both orbiters carried a thermal emission radiometer called the Infrared Thermal Mapper (IRTM). IRTM had five infrared channels and a broadband channel covering the visible/near-infrared wavelength range (Kieffer et al., 1977). One of the channels covered the 15 μm CO_2 band, providing sensitivity to a broad vertical average of the atmosphere centered about 25 km above the surface, with very little information on the vertical dependence of temperatures. Further analysis of the IRTM 15 μm CO_2 band data by Wilson and Richardson (2000) suggested the channel might have had a spectral leak that allowed surface radiance outside the 15 μm CO_2 band to contribute to the observed signal, but the resulting bias could be removed using surface radiance observations from other IRTM bands. Despite the lack of vertical resolution in the IRTM atmospheric temperature data, its mapping capability with high spatial resolution provided information complementary to the IRIS observations from Mariner 9. Retrievals from IRTM observations were extensively used to study the development of atmospheric temperatures in the south polar region (Kieffer et al., 1977) as well as in the north polar region, and to characterize the diurnal and latitudinal temperature structure in high-dust conditions (Martin and Kieffer, 1979). The IRTM instruments on both Viking Orbiters were operational until close to the end of their respective orbiter missions (Snyder, 1979; Christensen and Zurek, 1984).

After the end of the Viking missions, the next spacecraft to orbit Mars was the Russian Phobos 2 in 1989. Although Phobos 2 carried the Combined Radiometer/Spectrophotometer for Mars (KRFM), an instrument with one channel covering the 15 μm band of CO_2 (Ksanfomality et al., 1989), few observations were taken of Mars and little work has been done to retrieve atmospheric temperatures.

The next orbiting spacecraft to reach Mars was Mars Global Surveyor (MGS) in 1997. During an initial phase of aerobraking and science phasing (September 1997–September 1998), MGS was in a highly elliptical orbit that allowed observations of a large part of the planet at different local times. After the final aerobraking, MGS achieved a Sun-synchronous, near-circular mapping orbit in March 1999. This near-polar orbit with equator crossings at about 2:00 a.m. and 2:00 p.m. local time gave global coverage with good repeatability to study the seasonal variation of temperature. MGS carried the Thermal Emission Spectrometer (TES), a Fourier-transform spectrometer covering the spectral range between 6 and 50 μm (Christensen et al., 2001). Primarily designed to study surface mineralogy, TES took a large majority of observations in the nadir geometry, from which temperature profiles between the surface and ~40 km were retrieved with a vertical resolution of about 10 km (Conrath et al., 2000). A scan mirror gave TES the capability also to observe the limb, which was normally done every 10–20° of latitude along each orbit during the mapping phase of the MGS mission. Retrievals from limb observations extended the vertical range of TES temperature profiles to about 65 km (Smith et al., 2001a). Local time coverage during aerobraking and science phasing allowed studies of the variation of atmospheric temperature related to thermal tides (Banfield et al., 2000), while results from the mapping part of the mission for the first time allowed a detailed investigation of the seasonal temperature variation over the course of the Martian year (Smith et al., 2001b). The TES spectrometer continued its mapping observations for nearly three Mars years until September 2004, allowing extensive studies of inter-annual variations (Liu et al., 2003; Smith, 2004). An engineering instrument on MGS, the Mars Horizon Sensor Assembly, observed the limb in four directions (forward and aft along the orbit path, and to both sides) with a single channel covering the 15 μm CO_2 band. These data have been used to map atmospheric temperatures, and the extra local time coverage provided by the side-looking sensors has enabled isolation of high-frequency tidal components not available from TES (Martin and Murphy, 2003).

While MGS was still operating, Mars Odyssey was the next orbiter to arrive at Mars in October 2001. After a period of aerobraking, a Sun-synchronous orbit was achieved with equator crossings around 5:00 a.m. and 5:00 p.m. local time. Mars Odyssey carries the Thermal Emission Imaging System (THEMIS), capable of taking multispectral images in the thermal infrared at nine wavelengths, and in the visible/near-infrared at five wavelengths (Christensen et al., 2004). One of the thermal infrared bands is centered in the 15 μm CO_2 band, allowing THEMIS to perform mapping of atmospheric temperatures in a fashion similar to IRTM (Smith, 2009). THEMIS began observing in February 2002 and is still operational, providing valuable data for linking TES retrievals with those from later orbiters.

In December 2003, the European Space Agency spacecraft Mars Express arrived at Mars and has been in an elliptical orbit since then. Mars Express carries the Planetary Fourier Spectrometer (PFS), a nadir viewing Fourier-transform spectrometer. Its spectral range includes both the thermal infrared as well as the near-infrared (Formisano et al., 2005). Because of the elliptical orbit of Mars Express, the coverage and spatial resolution of PFS observations is quite irregular in latitude, season, and local time. Analyses of temperature retrievals have concentrated on the thermal structure observed in specific regions such as the Tharsis volcanoes, Hellas, or the southern polar night (Grassi et al., 2005, 2007; Giuranna et al., 2008; Wolkenberg et al., 2010).

The latest orbiting spacecraft at Mars carrying infrared instrumentation is the Mars Reconnaissance Orbiter (MRO), which arrived in March 2006. After a period of aerobraking, MRO was placed into a Sun-synchronous, near-polar mapping orbit with equator crossing times at about 3:00 a.m. and 3:00 p.m. local time. MRO carries the Mars Climate Sounder (MCS), a thermal infrared radiometer designed to take contiguous thermal emission measurements of the Martian surface and atmosphere using limb, nadir, and off-nadir viewing geometries (McCleese et al., 2007). MCS has eight infrared channels and one broadband visible/near-infrared channel. Three channels cover the 15 μm band of CO_2 at different line-of-sight opacities. Each MCS channel consists of a linear array of 21 detectors, which can be pointed vertically at the limb for simultaneous radiance

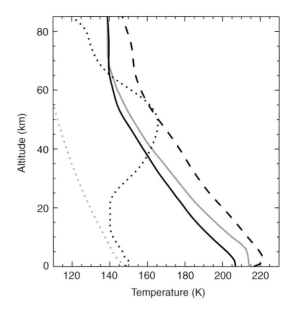

Figure 4.2. Averaged temperature profiles as a function of height as retrieved from the Mars Reconnaissance Orbiter MCS instrument. The solid black line is for the aphelion ($L_s = 0$–180°) season. The dashed line is for the perihelion ($L_s = 180$–360°) season. The black dotted line is for conditions during polar night. The solid gray line shows the COSPAR reference atmosphere, and the gray dotted line shows the CO_2 condensation curve.

measurements at different tangent altitudes. From these measurements, atmospheric temperature profiles can be retrieved from the surface to an altitude of 80–90 km with a vertical resolution of ~5 km (Kleinböhl et al., 2009, 2011). Initial work has focused on the southern polar winter and its dynamically caused temperature maximum in the middle atmosphere, which for the first time can be observed over its full vertical extent (McCleese et al., 2008). By the time of this writing, MCS has been operating in its fourth Martian year and continues to extend the record of Martian atmospheric climatology with unprecedented vertical coverage and resolution (McCleese et al., 2010). Figure 4.2 (discussed further in Section 4.4.1) shows a set of averaged temperature profiles retrieved from MCS observations, demonstrating the large range of atmospheric temperatures observed at Mars.

4.2.2.2 Radio Occultation Observations

Radio occultation experiments work by monitoring the radio signal sent from a spacecraft as it passes behind a planet as viewed from Earth. At both the ingress and egress points, the signal passes through the atmosphere, which both (very slightly) refracts the beam and produces a Doppler shift in the observed frequency. This information can be used to determine the refractive index of the atmosphere as a function of height, which can be converted to temperature given a known atmospheric composition and the assumption of hydrostatic equilibrium.

Radio occultations of Mars have been carried out by Mariner 9 (Kliore et al., 1973), Viking Orbiters (Lindal et al., 1979), Mars Global Surveyor (Hinson et al., 1999), and Mars Express (Pätzold et al., 2009). The resulting temperature profiles cover a

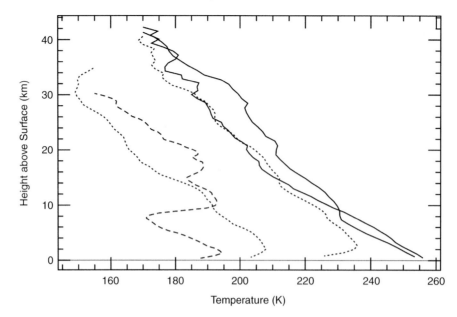

Figure 4.3. Temperature profiles as a function of height as retrieved from Mars Global Surveyor radio occultation observations. Late afternoon temperatures observed during southern mid-latitudes during summer are shown as solid lines. Nighttime temperatures from southern mid-latitudes during summer are shown as dotted lines. A nighttime profile with large waves (dashed line) was observed near the Tharsis volcanoes at $L_s = 150°$.

vertical range from the surface to roughly 45 km with a vertical resolution of about 500 m. The large improvement in vertical resolution over that obtainable using thermal infrared spectra is especially useful for probing near-surface temperatures and for resolving the vertical structure of wave modes. The disadvantage to the radio occultation temperature profiles is their relatively sparse coverage in space and time, which is limited to the times and places where the spacecraft orbital geometry allows an occultation. Radio occultation observations have also been obtained using Mars Reconnaissance Orbiter. Those data are still being analyzed at the time of writing.

Figure 4.3 shows selected temperature profiles retrieved from Mars Global Surveyor radio occultations. In the afternoon (solid line profiles), the retrieved temperature typically decreases with height, and agrees closely with the TES results when convolved to the vertical resolution of the thermal infrared profiles (Hinson et al., 2004). At night (dotted and dashed line profiles), the superior vertical resolution of the radio occultation temperature profiles allows characterization of a near-surface inversion layer, which cannot be resolved in TES profiles. A few radio occultation profiles show large-amplitude waves (dashed line profile), which are thermal tides that have been amplified by the radiative effects of water ice clouds (Hinson and Wilson, 2004) and are also observed by MCS (Wilson, 2011).

4.2.2.3 Stellar Occultation Observations

Stellar occultations, where a spacecraft watches as a star disappears or reappears from behind the limb of Mars, have been used to retrieve atmospheric temperatures between 60 and 130 km using ultraviolet observations made by the Mars Express SPICAM instrument (Bertaux et al., 2006). Because of the strong absorption of radiation by CO_2 below 200 nm, it is possible to estimate the density of the atmosphere and thus retrieve the temperature profile at altitudes where no other remote sensing methods can be used (Forget et al., 2009). To date, the

SPICAM instrument has obtained one Martian year of density and temperature profiles, with 616 profiles retrieved at various latitudes and longitudes (local times), mostly during Mars Year 27 (MY 27). The coverage is largely limited to the nightside, so a detailed analysis of the diurnal cycle with these retrievals is not possible (see Chapter 14).

4.2.2.4 Entry Profiles and Aerobraking

Seven different spacecraft have successfully landed on the surface of Mars. Each carried accelerometers that measured atmospheric drag during descent through the atmosphere. These measurements, along with the known aerodynamic properties of the spacecraft, allow the density, and thus temperature as a function of height, to be inferred, typically covering an altitude range of ~10–100 km above the surface with vertical resolution of 1 km or better. The six profiles from Viking Landers 1 and 2 (Seiff and Kirk, 1977), Pathfinder (Magalhães et al., 1999), Mars Exploration Rovers Spirit and Opportunity (Withers and Smith, 2006), and Phoenix (Withers and Catling, 2010) are shown in Figure 4.4 (see Table 4.2 for the location, seasons, and local times of entry). All of the profiles (except perhaps for Spirit) have large-amplitude oscillations that grow with height, which is characteristic of vertically propagating waves. The very low temperatures inferred from Pathfinder near 80 km are cold enough to allow CO_2 condensation into clouds.

The practice of aerobraking, or dipping a spacecraft's periapsis into the uppermost part of the atmosphere (typically above 120 km altitude) to induce drag to circularize an orbit, produces an indirect measurement of atmospheric density throughout the upper atmosphere. This information can be used to estimate temperatures above 120 km altitude (Keating et al., 1998, 2003; Bougher and Keating, 2006; Tolson et al., 2007). See Chapter 14 for details about each of the MGS, Odyssey and MRO aerobraking sampling campaigns, and the trends and variations of thermospheric densities and inferred temperatures.

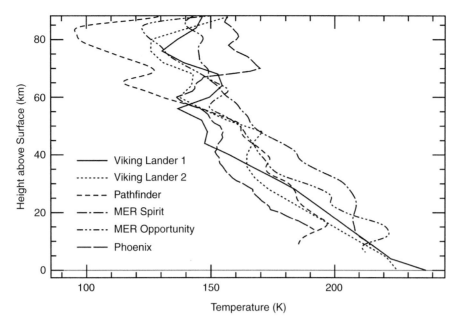

Figure 4.4. Temperature profiles as a function of height above the surface as inferred from accelerometer observations during the descent through the atmosphere of six landed spacecraft. See Table 4.2 for the season, location, and local time of each entry profile.

Table 4.2. *Summary of lander and rover locations, dates of operation, and the seasonal date and local time (hour) of entry, descent, and landing.*

Spacecraft	Location	Dates of operation	Entry season, L_s (deg.)	Entry local time (nearest hour)
Viking Lander 1	22.7°N, 48.2°W	7/1976–11/1982	97	16:00
Viking Lander 2	48.3°N, 226.0°W	9/1976–4/1980	118	10:00
Pathfinder	19.1°N, 33.2°W	7/1997–9/1997	143	03:00
MER Spirit	14.6°S, 184.5°W	1/2004–3/2010	328	14:00
MER Opportunity	2.0°S, 5.5°W	1/2004–present	339	13:00
Phoenix	68.2°N, 125.7°W	5/2008–11/2008	77	16:00
Mars Science Laboratory	4.6°S, 222.6°W	8/2012–present	151	15:00

4.2.2.5 Observations From the Surface of Mars

Observations of atmospheric temperatures from orbit give a very useful large-scale view of the global thermal structure, but even radio occultation observations do not have sufficient vertical resolution to characterize the details of the planetary boundary layer (PBL), the lowest few kilometers of the atmosphere that directly interacts with the surface responding to forcings such as frictional drag and surface heating. PBL temperatures have been measured directly using thermocouples mounted on the Viking Landers (Hess et al., 1977; Sutton et al., 1978), Pathfinder (Schofield et al., 1997), Phoenix (Davy et al., 2010), and Mars Science Laboratory, and have been retrieved from thermal infrared spectra taken by the Mars Exploration Rovers Mini-TES instrument (Smith et al., 2004, 2006; Spanovich et al., 2006).

4.2.3 Observed Thermal Structure

4.2.3.1 Lower Atmosphere Thermal Structure

The seasonal and latitudinal dependence of the lower atmosphere thermal structure is influenced by two types of seasons.

In addition to the familiar summer and winter seasons caused by the obliquity of Mars (25.2°) that are similar to those on Earth, there are also noticeable orbital seasons caused by the eccentricity of the orbit of Mars around the Sun, a type of season not found on Earth. The orbital eccentricity of Mars is 0.0934, which means that Mars is 20% closer to the Sun at perihelion than at aphelion, and receives 40% greater solar insolation at perihelion than at aphelion. At the current epoch, perihelion occurs at $L_s = 251°$, which is close to winter solstice in the northern hemisphere and summer solstice in the southern hemisphere ($L_s = 270°$), so that the obliquity and orbital seasons tend to reinforce each other in the southern hemisphere.

Figure 4.5 shows the seasonal and latitudinal dependence of surface and lower atmospheric temperatures for one typical year, Mars Year 24 (MY 24), $L_s = 145°$ to MY 25, $L_s = 145°$, without global-scale dust storms (Smith et al., 2001b; Liu et al., 2003; Smith, 2004). The thermal structure is controlled by surface temperature, which sets the bottom boundary condition, and by dust and dynamics, which set the vertical dependence. The influence of both the obliquity ("summer" and "winter") and orbital ("perihelion" and "aphelion") seasons is apparent. At high latitudes the obliquity seasons are dominant at all altitudes, with surface and

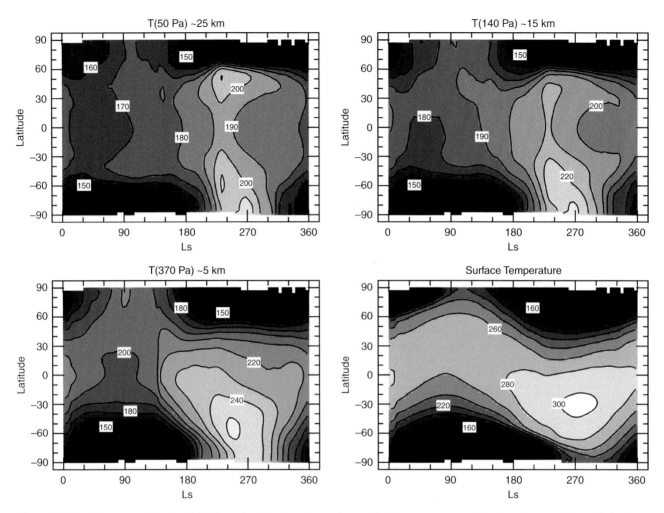

Figure 4.5. Zonally averaged daytime (~2:00 p.m. local time) surface and atmospheric temperature as a function of season (L_s) and latitude as observed between MY 24, $L_s = 145°$ and MY 25, $L_s = 145°$ by Mars Global Surveyor TES. Contours are every 20 K for the surface temperature, and every 10 K for the atmospheric temperatures.

atmospheric temperatures coldest at winter solstice and warmest at summer solstice (at least below 30 km altitude). On the other hand, the orbital seasons are more important at low latitudes, with cool temperatures during the "aphelion season" ($L_s = 0–180°$) and warm temperatures during the "perihelion season" ($L_s = 180–360°$) in both hemispheres. At low latitudes, the influence of the obliquity seasons is most pronounced at the surface and becomes increasingly less important with increasing altitude.

On a local level, a number of properties such as surface albedo, thermal inertia, atmospheric opacity, and the slope and shadowing of the terrain influence surface temperatures, and at the winter poles the condensation temperature of CO_2 ice sets the surface temperature. On a more global scale, the zonally averaged mid-afternoon surface temperatures shown in Figure 4.5 are largely related to the amount of solar insolation received. At any particular time, the maximum surface temperature is found at the subsolar latitude, with a smooth decrease in temperature away from the subsolar latitude toward each pole. The orbital seasons are also apparent with maximum surface temperatures significantly higher (~30–40 K) during the perihelion season than during the aphelion season.

Atmospheric temperatures in the lowest scale height above the surface show similar seasonal and latitudinal trends as surface temperature, although relatively warmer temperatures are maintained to much higher latitudes during the summer season in each hemisphere. Higher above the surface, the temperature difference between aphelion and perihelion is greater and the influence of surface temperature diminishes. At high altitudes, atmospheric temperatures are increasingly controlled by dynamics, tides, and the radiative effects of aerosols. In particular, large dust storms, even those that are not global-scale, can significantly alter the thermal structure of the atmosphere (e.g. Smith et al., 2002). This can be seen in Figure 4.5 in the general heating at $L_s = 225°$, and at $L_s = 270°$ near the south pole, and is discussed in more detail in Section 4.2.3.5. It has also become apparent that the radiative effect of water ice clouds plays an important role in determining both surface and atmospheric temperatures, most importantly by warming the lower atmosphere during the aphelion season at low latitudes when clouds are most prevalent (e.g. Hinson and Wilson, 2004; Wilson et al., 2007; Haberle et al., 2011; Wilson, 2011; Madeleine et al., 2012).

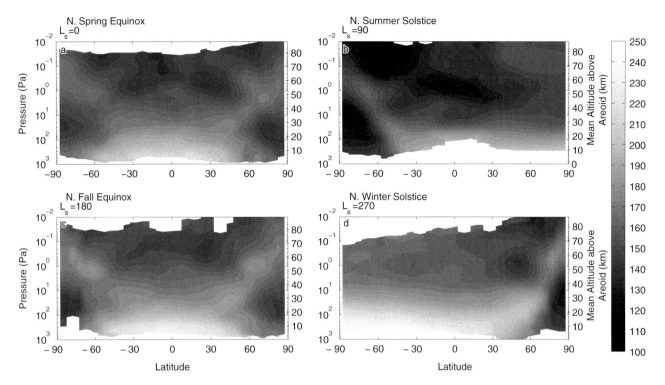

Figure 4.6. Typical zonally averaged daytime (~3:00 p.m. local time) temperatures as a function of latitude and pressure (or height above the surface) as observed by Mars Reconnaissance Orbiter MCS at four seasons throughout the Martian year. Figure after McCleese et al. (2010) and used with permission. A black and white version of this figure will appear in some formats. For the color version, please refer to the plate section.

A more detailed picture of the lower atmospheric thermal structure is presented in the cross-sections shown in Figure 4.6. Temperatures are generally found to depart significantly from radiative equilibrium, indicating strong modifications of the thermal structure by dynamical processes. Under solstice conditions ($L_s = 90°$ and $270°$), maximum solar heating occurs at the summer pole, and near-surface temperatures reach a maximum near there. In the summer hemisphere, the latitudinal temperature gradient is relatively small, and temperatures decrease with height to a minimum value just above the 1 Pa level (~55 km height). Temperatures increase modestly above that level, reaching a maximum in the summer hemisphere at high latitudes at the 0.01–0.1 Pa level (~75 km). Atmospheric temperatures at 75 km altitude during the southern hemisphere summer are about 20 K warmer than during the northern hemisphere summer since southern summer solstice nearly coincides with perihelion.

In the winter polar region, the lack of sunlight leads to very cold temperatures, which can reach the CO_2 frost point close to the pole (Pearl et al., 2001; Kleinböhl et al., 2009). A warmer region overlies this cold region, leading to a temperature inversion. Adiabatic heating due to the downwelling motion of the air causes the higher temperatures over the pole as part of an overturning "Hadley" circulation (Smith et al., 2001a; McCleese et al., 2008, 2010; McDunn et al., 2013). Close to the pole, maximum temperatures are found at around the 0.3 Pa (60–65 km) level. The temperature maximum occurs at lower altitudes at lower latitudes, leading to a characteristic tilt in the polar front from strong adiabatic heating aloft and as cold polar air is advected toward the equator at low altitudes. The latitudinal gradient in temperature is significantly larger and has greater vertical extent in the northern winter polar region than in the southern winter polar region. The temperature in the inversion layer shows maxima around 80°N latitude in the northern and 70°S latitude in the southern polar region. This may be indicative of an additional eddy circulation in the middle atmosphere, modifying the overall Hadley circulation (Heavens et al., 2011).

The thermal structures during the two equinox periods ($L_s = 0°$ and $180°$) are similar to each other and are nearly symmetric about the equator. Warmest atmospheric temperatures are near the surface at the equator. In each hemisphere, temperatures decrease toward the pole at altitudes below ~30 Pa (~30 km). Above that pressure level (at least to the 0.1 Pa level), there is a temperature minimum at the equator and a temperature maximum at middle to high latitudes in each hemisphere. This is indicative of a two-cell circulation, where a Hadley cell forms in each hemisphere (Smith et al., 2001a; McCleese et al., 2010; McDunn et al., 2013).

From the surface, both thermocouple measurements and retrieved temperatures using upward-looking infrared spectra show a consistent diurnal pattern (Figure 4.7). The atmosphere is coolest and stably stratified before dawn. Soon after sunrise, the warming surface heats the atmosphere from the bottom upward. A very steep, super-adiabatic vertical gradient in temperature is established through the lowest 100 m of the atmosphere by mid-morning. Turbulent convection sets in throughout this lowest layer, with temperature fluctuations of 15–20 K

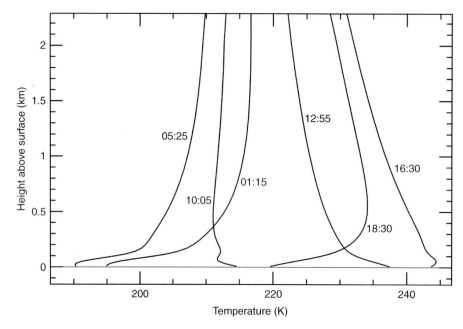

Figure 4.7. Near-surface atmospheric temperatures as a function of height above the surface for different times of day as retrieved from upward-looking observations taken by the Mini-TES instrument on-board the MER Spirit Rover near $L_s = 0°$.

recorded in the lowest meter above the surface and up to 5 K in the lowest 100 m on timescales of 30–60 s. Turbulent convection continues until late afternoon (around 16:30 local time). When the surface becomes cooler than the near-surface atmosphere, convection shuts off, and the near-surface temperature gradient becomes inverted. The inversion layer grows throughout the nighttime hours, reaching a depth of at least 1 km before rapidly reversing again in the morning. See Chapter 7 for a full description of the planetary boundary layer.

4.2.3.2 Middle Atmosphere Thermal Structure

It has long been expected that, on average, Mars atmospheric temperatures decrease with altitude up through the middle atmosphere, above which the absorption of extreme UV radiation creates a warm thermosphere. However, in comparison to the lower atmosphere, very few observations are available above 80 km. Despite this limitation, observations from SPICAM have provided important new information on the vertical and seasonal structure of nightside temperatures in the ~60–125 km (middle atmosphere) range (Forget et al., 2009). Specifically, the SPICAM observations have revealed seasonal variations in: (a) the variation of nightside temperatures with height, (b) the location and magnitude of the nightside mesopause, and (c) the existence of sub-CO_2 frost-point temperatures in the middle atmosphere. The SPICAM observations have also provided constraints on the character of heat balance in the Mars middle atmosphere, although the sparse coverage at high latitudes does not allow polar warming features to be properly addressed in the same way as those clearly observed and characterized by the MCS observations (McCleese et al., 2008; Kleinböhl et al., 2009; McDunn et al., 2013).

The variation of nightside temperatures as a function of height between ~80 and 130 km varies markedly with season (Forget et al., 2009; McDunn et al., 2010) because of the

seasonal expansion of the lower atmosphere (Figure 4.8). For example, southern winter mid-latitudes exhibit a prominent temperature minimum (~104 km) and a large positive lapse rate aloft. In contrast, southern summer mid-latitudes exhibit minimum temperature at higher altitude (above 120 km) with a smaller positive lapse rate aloft. As can be seen in Figure 4.8, the location and magnitude of the minimum temperature varies with season and location (Forget et al., 2009; McDunn et al., 2010), with the minimum colder and lower near aphelion than at perihelion. It is notable (Figure 4.9) that minimum temperatures below the CO_2 frost point are found primarily in the southern winter tropics (Montmessin et al., 2006; Forget et al., 2009).

Two general circulation models (GCMs) show overestimation of the magnitude and pressure level of the southern winter ($L_s = 90$–$120°$) tropical temperature minimum (Forget et al., 2009; McDunn et al., 2010), as shown in Figure 4.10. This discrepancy indicates that either (a) true global winds are slower than simulated, possibly due to seasonally variable gravity wave drag in the Mars atmosphere (e.g. Bougher et al., 2011; Medvedev et al., 2011), or (b) actual CO_2 15 μm cooling is larger than simulated. The latter is likely because of an underestimation of the atomic oxygen abundance in this region that controls the magnitude of CO_2 infrared cooling (e.g. Bougher et al., 1994; Forget et al., 2009; McDunn et al., 2010).

Much work remains to fully characterize the middle atmosphere thermal structure of Mars, including its temporal (diurnal, seasonal, inter-annual) and spatial variability. The recent retrievals from SPICAM observations, which currently cover only MY 27, should be extended into MY 28 once the data are calibrated and the retrievals are made available (Forget et al., 2009). In addition, comparisons with other spatially overlapping *in situ* accelerometer and MCS datasets are just beginning, although the datasets may not be taken during the same Martian year (e.g. Theriot et al., 2006; McDunn et al., 2013).

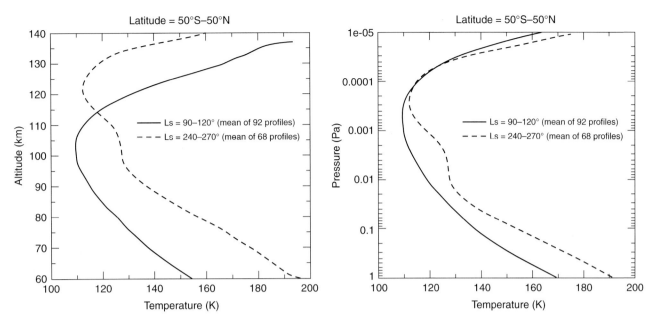

Figure 4.8. A comparison of mean nightside temperature profiles of the middle atmosphere observed around northern summer solstice (L_s = 90–120°, solid line) and winter solstice (L_s = 240–270°, dashed line) retrieved from SPICAM stellar occultations, and shown with altitude or pressure as the vertical coordinate. Each profile is obtained by averaging observations interpolated at the same altitude or pressure level. In altitude coordinates, the seasonal expansion of the lower atmosphere (warmer in southern spring and summer) shifts the temperature profile. Figure from Forget et al. (2009), used with permission.

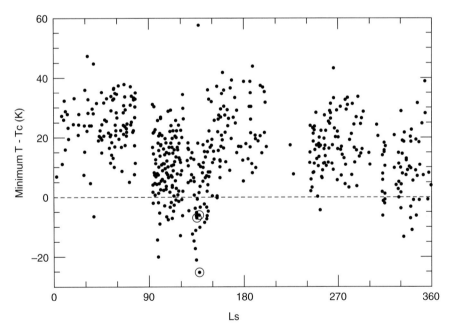

Figure 4.9. The minimum temperature T (related to the CO_2 frost point T_c) encountered in each SPICAM profile plotted as a function of season (L_s). A negative value corresponds to a profile with temperature below the frost point. The mesospheric clouds detected by SPICAM and thought to be high-altitude CO_2 ice clouds are shown by circles. Figure from Forget et al. (2009), used with permission.

4.2.3.3 Global Mean and Extreme Temperatures

Figure 4.11 and Table 4.3 show the globally and annually averaged atmospheric temperature observed by MCS (Kleinböhl et al., 2009, 2011) and TES (Smith, 2004), along with the extreme range of possible temperatures predicted by general circulation models (e.g. Forget et al., 1999, 2006b). The globally and annually averaged temperature profiles from MCS and TES agree to within the combined uncertainty of the two instruments and are shown as a single profile. The maximum temperatures give the upper limits of what could occur under the conditions of an extreme global-scale dust storm. The minimum temperatures, which fall below the CO_2 frost temperature, provide lower limits that could occur in the polar night under conditions of extreme CO_2 depletion or supersaturation (e.g. Colaprete et al., 2008).

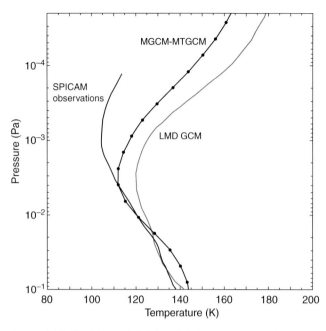

Figure 4.10. SPICAM and GCM modeled temperature profiles for L_s = 90–120° (nightside at southern tropical latitudes, during southern winter). The "SPICAM" curve represents the average of the 39 observations. The "LMD (Laboratoire de Météorologie Dynamique) GCM" curve gives the temperature profile for this same location and seasonal interval from the LMD GCM. The curve with points represents the corresponding average MGCM–MTGCM profile for this location and seasonal interval. Both LMD GCM and MGCM–MTGCM models are described in Chapter 14. Figure adapted from McDunn et al. (2010), and used with permission.

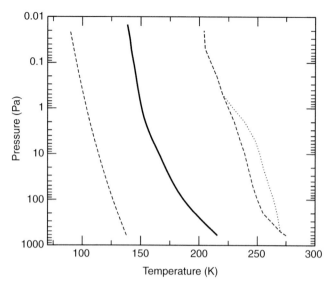

Figure 4.11. The globally and annually averaged (solid line) and minimum and maximum temperatures (dashed and dotted) in the Martian atmosphere as a function of pressure (height). The globally averaged profile was compiled by averaging retrievals from MCS (MY 29, a typical year without global-scale dust storm), TES (below 10 Pa, and from MY 26, also typical) and from the Mars Climate Database (MY 24 scenario, see Section 4.4.3). The extreme curves represent the envelope of all possible temperatures anywhere and anytime on Mars. The minimum temperature here is 10 K below the CO_2 frost point, which implies supersaturation and CO_2 gas depletion. Maximum temperatures for years without planet-encircling dust events (dashed line, right) and years with global-scale dust storms (dotted line) are based on Mars Climate Database GCM predictions.

Table 4.3. *Tabulated values of the globally and annually averaged atmospheric temperatures observed by MCS and TES, and the extreme temperatures predicted by general circulation models (see also Figure 4.11).*

Pressure (Pa)	$T_{average}$ (K)	T_{min} (K)	T_{max} (K)	T_{max} global dust storm (K)
0.02	139.5	90.0	204.0	204.0
0.05	141.6	92.5	205.0	205.0
0.10	143.8	95.0	210.0	210.0
0.20	145.8	97.5	215.0	215.0
0.50	148.4	101.0	220.0	220.0
1.00	150.7	104.0	225.0	229.0
2.00	154.0	106.5	230.0	239.0
5.00	160.0	110.5	235.0	248.0
10.00	165.6	114.0	239.0	252.0
20.00	171.0	117.5	242.0	255.0
50.00	179.0	122.5	246.0	260.0
100.00	187.4	126.0	250.0	264.0
200.00	196.5	130.5	255.0	267.0
500.00	212.0	136.5	270.0	270.0
610.00	215.7	138.0	275.0	275.0

4.2.3.4 Diurnal Variations and Tides

The varying incident insolation over the course of a Martian day causes large differences in surface temperature between day and night. The diurnal variation of surface temperature can be as large as ~100 K in regions of low thermal inertia (e.g. Tharsis, Meridiani) (Smith et al., 2004). Maximum surface temperature is typically reached at local noon or slightly later, while minimum surface temperature occurs in the early morning between about 3:00 a.m. and 5:00 a.m. local time, depending on the thermal inertia of the surface. The temperature structure in the planetary boundary layer (the lowest few kilometers) is primarily driven by the heat flux from the surface, which unlike the Earth is dominated by radiation, with only small sensible heat flux (see Chapter 6 for further details on the energy budget, and Chapter 7 for further details on the planetary boundary layer). During the day, the atmospheric lapse rate in the boundary layer is super-adiabatic, while at night, rapid cooling at the surface establishes a strong temperature inversion. The diurnal variation in thermal forcing also gives rise to global oscillations in atmospheric pressure, temperature, and winds. These oscillations, which are subharmonics of a solar day, are called atmospheric thermal tides. They are to be distinguished from gravitational tides, which are negligible in the

Martian atmosphere (Rondanelli et al., 2006). Sun-synchronous tides propagate at the same speed as the apparent motion of the Sun, hence are called "migrating" tides, and have no longitudinal variation in a local time-oriented reference frame (Forbes, 2004) (see also Chapters 9 and 14).

The first observations of the diurnal variations of Martian atmospheric temperature were provided by the IRIS instrument on Mariner 9, where deviations from the mean temperatures with amplitudes up to ~15 K were observed (Hanel et al., 1972c). These were recognized to be related to the westward-migrating diurnal tide, influenced by the global-scale dust storm that occurred during the Mariner 9 mission (Conrath, 1976). Viking Orbiter IRTM observations also revealed tidal signatures (Martin and Kieffer, 1979; Martin, 1981; Wilson and Richardson, 2000) and oscillations observed in temperature profiles derived from the atmospheric entry of landers, which have also been attributed to the influence of the diurnal tide (Magalhães et al., 1999; Withers and Catling, 2010). Tidal signatures are also evident in the record of surface pressure as measured on the surface by the Viking Landers (Tillman, 1988; Tillman et al., 1993), Pathfinder (Schofield et al., 1997), Phoenix (Taylor et al., 2010), and Mars Science Laboratory (Haberle et al., 2014).

Comprehensive measurements of diurnal temperature variations were made by TES. The most valuable information concerning thermal tides and stationary waves were taken during the aerobraking and science phasing periods of the MGS mission when the orbit drifted with respect to local time, allowing good sampling of the diurnal variation of temperatures. The diurnal thermal tide was found to have an amplitude of about 4 K, defined as the deviation from the mean temperature state (Banfield et al., 2000), with larger amplitudes (exceeding 8 K) during the dustiest conditions. The diurnal tide was also studied using TES data obtained during the MGS mapping period when only two local times were sampled because of the Sun-synchronous orbit. This caused aliasing of higher-order modes into the diurnal response, but Banfield et al. (2003) were still able to identify the diurnal tide in maps of the temperature difference between daytime and nighttime, and showed it to be in reasonable agreement with simulations from a GCM (Wilson and Hamilton, 1996).

New studies of the diurnal tide were enabled by data from MCS. While the Sun-synchronous orbit of MRO leads to the same aliasing in sampling as MGS during its mapping period, the extended vertical range and improved vertical resolution of MCS provided new insight into the tidal structure (Lee et al., 2009). The analysis relies on mean fields of the average diurnal temperature and the deviation from this mean temperature during day and night. Figure 4.12 shows these fields derived from MCS data (Kleinböhl et al., 2009, 2011), which includes temperature retrievals from combined limb and nadir observations for the period L_s = 135–165° of MY 28. The average temperature field shows a temperature structure that is typical for atmospheric conditions close to equinox, with cold temperatures in the upper part of the lower atmosphere of both polar regions, overlaid by layers of warmer temperatures, and cold temperatures in the equatorial middle atmosphere. The temperature deviations show a complex

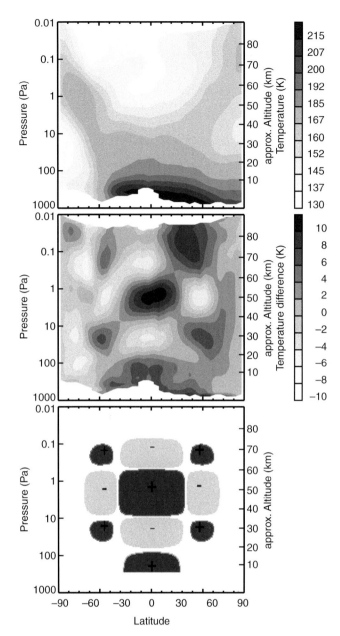

Figure 4.12. Diurnally averaged temperature (top) and the day minus night temperature difference (center) as a function of latitude and pressure for the period L_s = 135–165° in MY 28 derived from MCS data. The bottom panel shows a schematic of the temperature minima and maxima.

pattern of minima and maxima, which is highlighted in the schematic in the bottom panel of Figure 4.12 (Lee et al., 2009). A temperature maximum is found above the equatorial surface. Above the forcing region, the temperature deviations show a nodal pattern consistent with the expectation of a vertically propagating tide with a vertical wavelength of the order of 40 km, or about four pressure scale heights, roughly consistent with classical tidal theory (Zurek, 1976). In the latitudinal domain, the pattern changes sign at around 25–30° latitude, consistent with the behavior expected from the lowest-order westward-migrating Hough mode describing the dominant latitudinal structure of the diurnal tide in

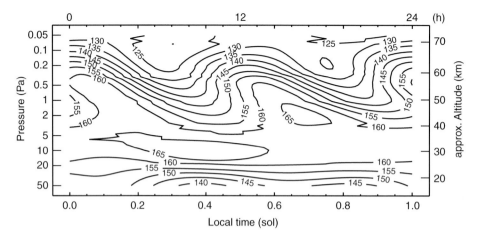

Figure 4.13. Temperature structure versus local time (scale in sols on bottom, scale in "hours" or 1/24 of a sol on top) for 60–65°S at $L_s = 101$–114° of Mars Year 31, derived from fits to zonal averages of MCS data taken at multiple local times.

classical theory (Lindzen, 1970; Zurek, 1976). This tidal structure is well reproduced by GCM modeling (Wilson and Richardson, 2000; Lee et al., 2009).

The observation of semidiurnal tides, tides with a period of half a solar day, also goes back to the Viking Orbiter observations (Martin and Kieffer, 1979; Martin, 1981). Wilson and Richardson (2000) showed that IRTM observations were in agreement with a semidiurnal tide in the tropical middle atmosphere, most prominently during dust storm conditions. While in a clear atmosphere the diurnal tide was dominant in the Martian tropics, the amplitude of the semidiurnal tide increased with increasing dust loading as solar heating became distributed over a larger part of the atmosphere. A semidiurnal tide component was also derived from TES data during aerobraking and science phasing (Banfield et al., 2000). Typical semidiurnal tide amplitudes tended to be on the order of 1 K in the regions and seasons studied. Only in dusty conditions was an amplitude of up to 8 K at an altitude of 3–4 scale heights observed in the tropics during the period $L_s = 255$–270°.

New insights into the character of the semidiurnal tide were obtained by MCS measurements. From the Sun-synchronous orbit of the MRO spacecraft, MCS can use its azimuth actuator to view in different directions away from the orbit track, thereby providing observations at additional local times. At the equator, time differences up to 1.5 hours from the nominal equator crossing times are reached, with increasing local time differences toward the polar regions. Analysis of these measurements reveal that the semidiurnal tide is a significant response of the Martian atmosphere throughout the Martian year (Kleinböhl et al., 2013). A maximum amplitude of ~16 K is found at southern high latitudes at altitudes of 5–7 scale heights during the southern winter season (Figure 4.13). During the northern winter season, comparable semidiurnal amplitudes are observed in the northern high latitudes, and also at low latitudes the semidiurnal amplitude can reach 6–8 K. Modeling studies using Mars general circulation models suggest that the semidiurnal tide can be forced by water ice clouds with a large vertical extent (Kleinböhl et al., 2013). These clouds, which have been observed in the middle atmosphere (Heavens et al., 2010), absorb infrared radiation from the warm Martian surface during daytime and provide an effective

tidal forcing in the middle atmosphere. Considering the radiative effect of water ice clouds in atmospheric models also speeds up the overturning meridional circulation (Madeleine et al., 2012; Kleinböhl et al., 2013), which increases middle atmospheric temperatures at high latitudes in winter, mitigating differences between model results and observations (McCleese et al., 2008) (see also Chapter 9).

The large topography of Mars has a strong influence on the creation of atmospheric tides, as was recognized by Conrath (1976) and Zurek (1976). Zonal variations in topography, as well as thermal inertia and albedo, modulate the tidal forcing through solar heating, and give rise to non-migrating tides. The most prominent mode excited this way is the eastward-propagating diurnal Kelvin wave (DKW), which is caused by the interaction of the westward-migrating diurnal tide with Mars' dominant zonal wavenumber-2 topography, and resonantly amplified in the typical temperature structure of the Martian atmosphere (Zurek, 1976; Wilson and Hamilton, 1996). Tidal modes coupled to topography were observed by TES during the aerobraking and science phasing portion of the MGS mission, and the DKW was identified with an amplitude of 1.5 K in the southern hemisphere (Banfield et al., 2000). The interpretation of observations from a Sun-synchronous orbit (e.g. the mapping portion of the MGS mission, or MRO) is not straightforward because zonal and temporal structures are aliased (Banfield et al., 2003). Non-migrating tides appear as stationary features from Sun-synchronous orbit (Forbes and Hagan, 2000; Wilson, 2000), and model simulations are often required for the identification of modes. Still, such tidal modes, including the DKW, have been identified in TES data from the MGS mapping orbit (Banfield et al., 2003), and shown to be roughly consistent with model results (Wilson and Hamilton, 1996; Wilson, 2000). Also the zonal structure in MCS observations taken during northern high-latitude summer suggests the occurrence of non-migrating tides (Withers and Catling, 2010). At tropical latitudes, eastward-propagating DKWs with zonal wavenumbers 1 through 3 with long vertical wavelengths and amplitudes of 1–3 K have been identified in MCS data (Guzewich et al., 2012). Non-migrating tides can propagate to high altitudes in the atmosphere and cause variations in the

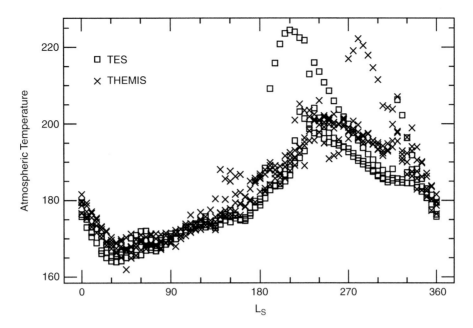

Figure 4.14. Globally averaged values of afternoon atmospheric temperatures at ~25 km above the surface as retrieved from data taken by the Mars Global Surveyor TES and Mars Odyssey THEMIS instruments. The aphelion season ($L_s = 0°$ to $180°$) is relatively cool and shows little variation from Martian year to year, whereas the perihelion season ($L_s = 180°$ to $360°$) is relatively warm and shows large inter-annual variations caused by heating from dust storms. The largest dust storms were observed during MY 25 (TES) and MY 28 (THEMIS).

density structure of the upper atmosphere (Forbes and Hagan, 2000; Wilson, 2000, 2002; Angelats i Coll et al., 2004; Bougher et al., 2004), which has been identified as causes for density variations derived by the MGS accelerometer instrumentation at around 130 km altitude (Forbes et al., 2002; Wilson, 2002; Withers et al., 2003) and dayside ionospheric peak height variations extracted from MGS radio occultation measurements (Bougher et al., 2001, 2004) (see also Chapter 14).

4.2.3.5 Inter-Annual Variations

The continued success and long life of several orbiting spacecraft at Mars has enabled a first look at the magnitude of inter-annual variability in the current climate. Figure 4.14 shows mid-afternoon temperatures averaged between 60°S and 60°N latitude at a pressure level of ~50 Pa (~25 km altitude) as a function of season (L_s) as observed by the Mars Global Surveyor TES and Mars Odyssey THEMIS instruments. The annual cycle described earlier, with a planet-wide warming and cooling of the atmosphere in response to the change in solar insolation between aphelion and perihelion, is seen to be generally repeatable from one Martian year to the next (Liu et al., 2003). The annual minimum temperature occurs near $L_s = 40°$ at a temperature of about 165 K (Smith, 2004). This is somewhat earlier than the seasonal date of aphelion ($L_s = 71°$), and is probably caused by the warming influence of water ice clouds that become more prominent later in the season (e.g. Wilson, 2011). Temperatures at this level are quite repeatable throughout the entire aphelion season, with year-to-year temperature differences generally within the measurement uncertainties. On the other hand, the perihelion season shows significant temperature changes between different Martian years. These differences, which can be 20 K or more on a globally averaged basis, are associated with the intermittent occurrence

of large regional and global-scale dust storms (see Chapter 10 for more information about dust storms). Heating associated with dust storms is caused both by direct solar heating from an increased amount of suspended dust, as well as dynamically by adiabatic heating in the descending branch of an enhanced Hadley circulation (e.g. Haberle et al., 1993; Wilson, 1997). Both mechanisms respond very quickly (within a few days) to changes in dust optical depth (Smith et al., 2002). Large dust storms and their associated atmospheric heating can happen at any seasonal date, but have been observed most frequently at $L_s = 225°$, 270°, and 315°. Excluding the effect of the large dust storms, peak globally averaged temperatures at this level reach about 200 K at about $L_s = 260°$, giving a peak-to-peak annual range of ~35 K.

Figure 4.15 shows the latitudinal and seasonal dependence of nighttime atmospheric temperature at the 50 Pa pressure level (~25 km altitude). By combining observations from the TES, THEMIS, and MCS instruments, we can examine a nearly continuous record of conditions over more than eight Martian years. The heating of the atmosphere in response to regional and global-scale dust storms shows inter-annual variability not only in the seasonal timing (L_s) of events, but also in the magnitude and latitude dependence of the heating. Although dust is often concentrated in the southern hemisphere, the thermal effects are similar in both hemispheres. The atmospheric heating associated with the global-scale dust storm at MY 25, $L_s = 185°$ did not extend poleward of about 60°N latitude, while the global-scale dust storm at MY 28, $L_s = 270°$ led to significant atmospheric heating all the way to the north pole. An intense late-season dust storm at MY 26, $L_s = 315°$ caused a 20 K temperature increase over a much wider range of latitudes than warmings observed during MY 27 and 29 at the same season.

The vertical extent of the thermal effect of the MY 25 global-scale dust storm (Smith et al., 2002) is displayed in Figure 4.16.

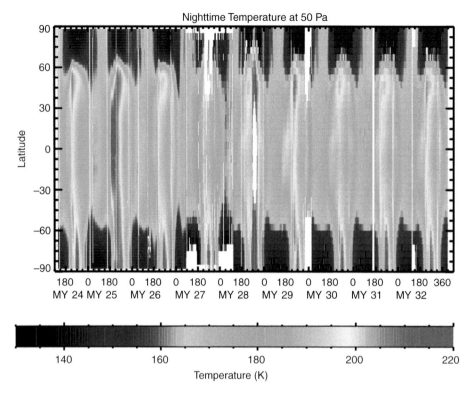

Figure 4.15. Nighttime zonally averaged temperatures at 50 Pa (~25 km above the surface) shown as a function of season (L_s) and latitude covering nine Martian years. Shown is a combination of retrieved temperatures from the Mars Global Surveyor TES (MY 24–27), Mars Odyssey THEMIS (MY 27–present), and Mars Reconnaissance Orbiter MCS (MY 28–present) instruments. A black and white version of this figure will appear in some formats. For the color version, please refer to the plate section.

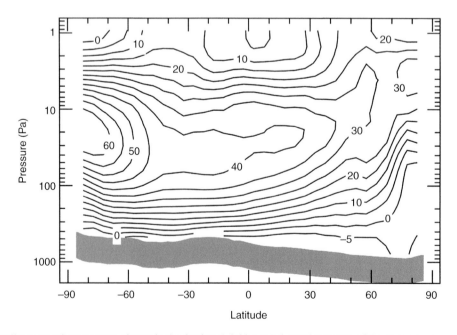

Figure 4.16. The zonally averaged temperature change in the daytime (~2:00 p.m.) thermal structure of the atmosphere caused by planet-encircling dust storm 2001a as observed by Mars Global Surveyor TES. This is the temperature at MY 25, $L_s = 205°$ during the height of the dust storm minus the temperature at the same season the previous Martian year (MY 24, $L_s = 205°$) when there were no large dust storms. The upper boundary of 1 Pa is approximately 55 km above the surface.

Shown is a latitude–height cross-section of mid-afternoon temperatures at MY 25, $L_s = 205°$ during the height of the dust storm compared to the temperature at the same season the previous Martian year (MY 24, $L_s = 205°$) when there were no large dust storms. The temperature difference increases nearly linearly with height, from nearly zero just above the surface to the 20 Pa level (~35 km), and then decreases to generally small values at 1 Pa (~55 km). The largest temperature differences occurred over the south pole, reaching more than 60 K. A 40 K temperature difference was present at the 20 Pa level over the entire southern hemisphere. The latitude gradient of the temperature difference near 60°N latitude below the 10 Pa level shows an intensification of the winter hemisphere front and implies a more vigorous Hadley circulation during the dust storm.

Nighttime atmospheric temperatures show similar trends as the daytime trends described above (Liu et al., 2003; Smith, 2004). Surface temperatures show little inter-annual variability except during the largest dust storms. During the MY 25 global-scale dust storm, daytime surface temperatures were depressed by ~20 K because of diminished solar insolation at the surface. Nighttime surface temperatures were ~20 K higher than seasonal norms, warmed by downward radiation by the optically thick and relatively warm dust in the atmosphere (Wilson and Smith, 2006).

4.3 COMPOSITION

The gas composition of the Martian atmosphere has been studied primarily using ground-based and spacecraft-based spectroscopy and through the analysis of *in situ* samples by the Viking Lander and Mars Science Laboratory mass spectrometers. The large majority of the gas (95%) is in the form of carbon dioxide, with smaller but significant amounts of nitrogen (2%) and argon (2%). Oxygen, carbon monoxide, and water vapor are also present in variable amounts ranging up to about one part per thousand, along with a number of other gases detected at parts per million and parts per billion (Table 4.4). Of these gases, both water vapor and carbon dioxide condense to form thin ice clouds, and the global abundance of carbon dioxide varies by up to 30% on a seasonal basis through condensation and sublimation from seasonal ice caps at each pole. Because of this large seasonal variation in CO_2, the mixing ratio of non-condensable species (e.g. Ar, O_2, CO, etc.) also varies with season. Unless otherwise noted, mixing ratios quoted in this section refer to mean annual values.

In the sections below we give a brief review of each of the most important gases that make up the Martian atmosphere, concluding with a discussion of the isotopic ratios found in the atmosphere. Water vapor is a special case and is discussed in Chapter 11, and further details about the carbon dioxide cycle and the polar caps are discussed in Chapter 12.

4.3.1 Surface Pressure

A measurement of surface pressure is important because it gives a direct indication of the column-integrated mass of the atmosphere. Since the atmosphere of Mars is 95% CO_2, the

Table 4.4. *The abundance of the major gases that make up the Mars atmosphere. The abundance of some species (notably water vapor and ozone) varies greatly with season and location.*

Gaseous species	Average abundance	Reference
CO_2	0.9532	Owen et al. (1977)
N_2	0.027	Owen et al. (1977)
	0.019	Mahaffy et al. (2013)
Ar	0.016	Owen et al. (1977)
	0.019	Mahaffy et al. (2013)
O_2	0.0014	Hartogh et al. (2010)
CO	800 ppm	Smith et al. (2009)
H_2O	15–1500 ppm	Smith (2004)
H_2	15 ppm	Krasnopolsky and Feldman (2001)
Ne	2.5 ppm	Owen et al. (1977)
Kr	0.3 ppm	Owen et al. (1977)
Xe	0.08 ppm	Owen et al. (1977)
O_3	10–350 ppb	Perrier et al. (2006)
H_2O_2	10–40 ppb	Encrenaz et al. (2004)
CH_4	0–40 ppb	Mumma et al. (2009)
	0.7–7 ppb	Webster et al. (2015)

column abundance of CO_2 serves as a reasonable proxy for surface pressure. Surface pressure was first estimated with reasonable accuracy using ground-based spectroscopy of CO_2 lines (Kaplan et al., 1964; Owen and Kuiper, 1964) and radio occultation from Mariner 4 (Kliore et al., 1965). Later, pressure sensors on the Viking Landers (Tillman, 1988; Tillman et al., 1993), Pathfinder (Schofield et al., 1997), Phoenix (Taylor et al., 2010), and Mars Science Laboratory (Haberle, et al., 2014) spacecraft from the surface of Mars provided accurate measurements of surface pressure at the lander sites (see Table 4.2 for the location of each lander and rover), and the retrieval of CO_2 column abundance from orbiting spacecraft (e.g. Forget et al., 2007; Smith et al., 2009) has extended the observations to other locations. Because of their accuracy, frequent sampling, and longevity, the data recorded by the two Viking Landers give the most complete picture to date of the variation of surface pressure at a given location, allowing the study of variations on timescales from hours to inter-annual.

Figure 4.17 shows the daily-averaged surface pressure recorded by the two Viking Landers during the first Martian year of their operation (Tillman, 1988; Tillman et al., 1993). The offset between the two curves is largely caused by the elevation difference (~1.2 km) between the two landing sites. Over the course of a Martian year, surface pressure varies by roughly 30%, decreasing as CO_2 condenses on the seasonal ice cap at the winter pole, and then increasing as CO_2 sublimates from the seasonal ice cap at the summer pole. The timing and differing amplitudes of the two annual minima and maxima are caused by the relative phasing of the seasons with respect to the date of perihelion and aphelion in the eccentric orbit of Mars, with additional strong influence from the albedo and emissivity of the seasonal polar caps (Paige and Wood, 1992). The two

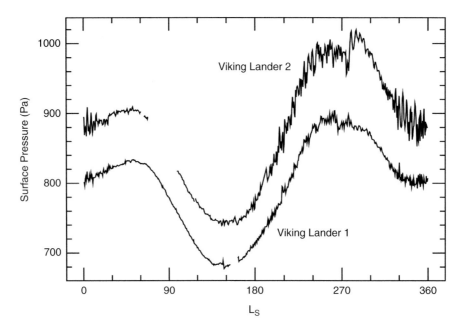

Figure 4.17. Daily averages of surface pressure (Pa) as recorded by the two Viking Lander spacecraft during their first Martian year of their operation.

maxima occur at $L_s = 260°$ and $50°$, and the two minima occur at $L_s = 150°$ and $350°$. A useful and simple analytical expression for approximating the annual surface pressure cycle is given by

$$p_{surf}(L_s) = \langle p_{surf}\rangle \{1 + 0.091 \cos(L_s - 309°) + 0.070 \cos[2(L_s - 248°)]\}$$

where $p_{surf}(L_s)$ is the surface pressure at a particular seasonal date, L_s, and $\langle p_{surf}\rangle$ is the annually averaged value, which is 797 Pa for the Viking Lander 1 site and 871 Pa for the Viking Lander 2 site.

In Figure 4.17, the variations superimposed on the annual cycle are the result of traveling waves, similar to the passage of storm systems on the Earth (e.g. Leovy, 1979). These waves are most prominent during the autumn and winter seasons ($L_s = 180°$ to $360°$ at the northern hemisphere Viking Lander sites). Diurnal and semidiurnal solar thermal tides cause additional variation in surface pressure on timescales of a day or less (Leovy and Zurek, 1979). The amplitude of tides has been observed to increase significantly during large dust storms (Zurek, 1981), and a global-scale dust storm is the cause in the jump in surface pressure at $L_s = 279°$ in the Viking Lander 2 observations shown in Figure 4.17.

Outside the annual cycle and variations caused by waves and tides described above, surface pressure varies spatially in direct response to changes in surface elevation. According to the hydrostatic equation, surface pressure decreases exponentially with increasing surface elevation, with an e-folding length scale known as the "pressure scale height", which has a value in kilometers of approximately 0.05 times the atmospheric temperature in degrees kelvin. The near-surface pressure scale height is typically about 10 km, so that an increase in elevation of 1 km leads to a decrease in surface pressure of about 10%.

The near-infrared CO_2 absorption band centered at 2.0 μm is suitable for the retrieval of CO_2 column abundance, and

is accessible by the Mars Express OMEGA (Forget et al., 2007) and PFS (Grassi et al., 2005) instruments and the Mars Reconnaissance Orbiter CRISM instrument (Smith et al., 2009; Toigo et al., 2013). These orbiters provide spatial sampling of CO_2 column abundance on a global scale, but the temporal sampling at any particular location is not systematic and tends to be at a similar local time.

4.3.2 Nitrogen

Nitrogen lacks suitable absorptions for the spectroscopic determination of abundance, so the only determinations to date have been made by the Viking Lander gas chromatograph mass spectrometer (Owen et al., 1977) and the Mars Science Laboratory Sample Analysis at Mars (SAM) quadrupole mass spectrometer (Mahaffy et al., 2013). The volume mixing ratio of 0.027 determined by Viking is not in agreement with the lower 0.0189 ± 0.0003 value determined by Mars Science Laboratory SAM. Further observations and analysis will be required to reconcile these two observations.

4.3.3 Argon

The first measurements of the abundance of argon in the Martian atmosphere were made by the gas chromatograph spectrometers on the Viking Landers (Biemann et al., 1976; Owen and Biemann, 1976), which provided the volume mixing ratio of 0.016 that served as the standard value until measurements by the SAM quadrupole mass spectrometer on Mars Science Laboratory gave the somewhat higher value of 0.0193 ± 0.0003 (Mahaffy et al., 2013). As a non-condensable gas, the overall abundance of argon in the Martian atmosphere is constant, but its mixing ratio varies with season and location in response to the large seasonal variations of carbon dioxide abundance (Figure 4.17). More recent measurements of argon abundance

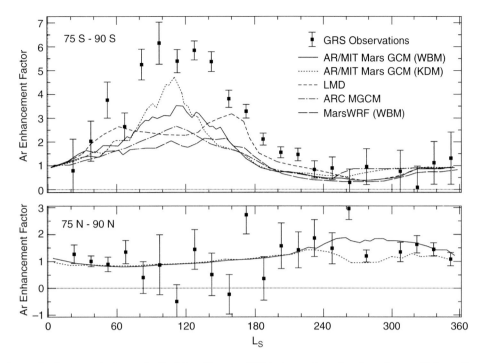

Figure 4.18. The observed enhancement of argon by the Mars Observer GRS instrument as a function of season (L_s) near the north and south poles compared to a number of different GCM models. Enhancement factor is defined as the ratio of argon in the given latitude band to that at the Viking Lander 2 landing site at $L_s = 135°$. The solid and dotted lines show results from the Ashima/MIT Mars GCM using two different radiative transfer schemes (see Lian et al., 2012). The short-dashed line shows results from the Laboratoire de Météorologie Dynamique (LMD) Mars GCM. The dash-dotted line shows results from the NASA Ames GCM. The long-dashed line shows results from the MarsWRF GCM. Figure adapted from Lian et al., 2012, and used with permission.

by Sprague et al. (2004, 2007) using the gamma subsystem of the Gamma Ray Spectrometer (GRS) on-board Mars Odyssey have focused on this seasonal variation in mixing ratio and the potential to use the seasonal and spatial variation of argon mixing ratio as a tracer for understanding the atmospheric circulation and dynamics. Sprague et al. (2007) found a factor of 6 enhancement in argon (defined as the ratio of argon at a given location to that at the Viking Lander 2 landing site at $L_s = 135°$) near the south pole near the onset of southern winter ($L_s = 90°$), but much more limited enhancement near the north pole during northern winter. They also reported statistically significant fluctuations in south polar argon enhancement during the southern winter season that imply transport processes and that are suggestive of standing wave eddies. Most recently, argon has been monitored by the Mars Exploration Rover Alpha Particle X-Ray Spectrometer (APXS) instrument for more than two Martian years using the strength of the 2.957 keV argon line in observations taken of the sky (Economou et al., 2007; Economou, 2008; Arvidson et al., 2011). At the near-equatorial Opportunity landing site, argon shows maximum mixing ratio near $L_s = 180°$, a secondary maximum near $L_s = 0°$, a secondary minimum at $L_s = 90°$, and a broad minimum mixing ratio between $L_s = 240°$ and 300°. Peak-to-peak variation is observed to be about 20% (Economou and Pierrehumbert, 2010), and there is a clear inverse relation to surface pressure (CO_2 abundance) as expected (compare Figure 4.17).

There have been several attempts to reproduce the seasonal variations of argon mixing ratio observed by the Odyssey GRS and MER APXS instruments using general circulation models

(e.g. Nelli et al., 2007; Forget et al., 2008; Guo et al., 2007; Arvidson et al., 2011; Lian et al., 2012). Figure 4.18 shows the results. Although these modeling efforts have been able to reproduce the general trend of the observed seasonal variation, they are unable to match the degree of argon enhancement observed at southern polar latitudes between $L_s = 60°$ and 180°, with maximum enhancement roughly only one-half that observed. This appears to highlight a shortcoming of the representation of transport processes in current GCM models in the challenging environment of the winter polar vortex (Lian et al., 2012).

4.3.4 Oxygen

With a mean mixing ratio above 0.1%, molecular oxygen is the most abundant minor species after nitrogen and argon. Still, its abundance and seasonal variations remain poorly studied because of the weakness of its spectroscopic signatures. Oxygen was first detected using high-resolution spectroscopic observations of Doppler-shifted lines in the 763 nm band, with a column density of 10 cm atm, corresponding to a mixing ratio of 1300 ppm (Barker, 1972; Carleton and Traub, 1972). Another measurement of the same oxygen band led to a slightly weaker value for the oxygen abundance of 8.5 cm atm, or about 1100 ppm (Trauger and Lunine, 1983). From a reanalysis of the Viking Lander data, England and Hrubes (2004) inferred a significantly higher abundance (2500–3300 ppm). A new disk-averaged measurement of 1400±120 ppm has been recently inferred using high-resolution heterodyne spectroscopy at 774

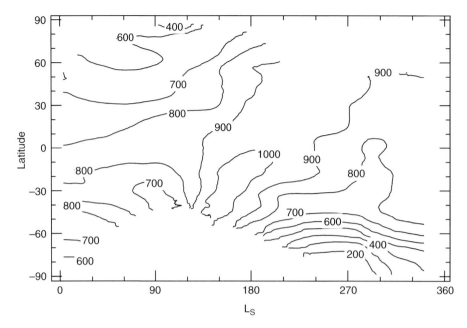

Figure 4.19. The seasonal and latitudinal variation of CO column abundance (ppm) as retrieved from CRISM observations during MY 28–30. Values are zonal averages.

GHz with the HIFI instrument on-board Herschel (Hartogh et al., 2010).

The source of molecular oxygen in the Martian atmosphere is the photolysis of CO_2 and H_2O. The photochemical lifetime of O_2 is about 30 years, much longer than the timescale for global mixing in the Martian atmosphere. Because oxygen is a non-condensable species, its mixing ratio is expected to vary with location and season (Forget et al., 2006a), but these variations have not been monitored so far. Spatially resolved measurements of the O_2 mixing ratio as a function of the Martian season would be useful since oxygen abundance has a direct influence on ozone, a key trace species for Martian photochemistry.

4.3.5 Carbon Monoxide

Carbon monoxide plays a major role in the Martian photochemical cycle of CO_2, and also as a tracer of the middle atmosphere thermal profile and winds. CO is also a non-condensable species for which local and temporal variations are expected over the Martian seasons.

CO was first detected using high-resolution infrared Fourier-transform spectroscopy at 1.6 and 2.35 μm (Kaplan et al., 1969). These disk-averaged observations led to a CO column density of 5.6 cm atm and a total surface pressure of 530 Pa, corresponding to a mixing ratio of 800 ppm. This value was later corroborated by Young and Young (1977), and by the millimeter wavelength observations of CO obtained by Kakar et al. (1977). Good and Schloerb (1981) and Clancy et al. (1983) suggested possible temporal variations of the disk-averaged mixing ratio. Over the following decades, CO was extensively monitored using millimeter and submillimeter heterodyne measurements of several CO and ^{13}CO transitions (Clancy et al., 1990; Lellouch et al., 1991). The spatial resolution of these observations was very limited when single-dish observations were used (Lellouch et al., 1991; Encrenaz et al.,

2001; Cavalié et al., 2008), but has more recently improved since interferometric measurements have become available (Moreno et al., 2009).

Infrared observations of the CO bands, at 4.7 and 2.35 μm, have been performed from the ground (Billebaud et al., 1992, 1998; Krasnopolsky, 2003a) and from spacecraft in orbit around Mars using the OMEGA (Encrenaz et al., 2006) and PFS instruments (Billebaud et al., 2009) on-board Mars Express, and the CRISM instrument on-board MRO (Smith et al., 2009). Krasnopolsky (2003a) first identified latitudinal variations of CO after summer solstice ($L_s = 112°$), with an increase in mixing ratio from 800 ppm at northern latitudes above 23°N up to 1250 ppm at 50°S latitude. Local and seasonal variations of CO mixing ratio were later monitored from spacecraft, and have been found to be in global agreement with GCM simulations for non-condensable minor species (Encrenaz et al., 2006; Forget et al., 2006a; Smith et al., 2009). In particular, the CO abundances retrieved using CRISM observations, which provide a complete view of seasonal and spatial variations (Figure 4.19), showed clear seasonal trends, with minima at the poles at summertime (400 ppm at the north and 200 ppm at the south), in agreement with the models (Smith et al., 2009).

4.3.6 Ozone

Ozone was first detected in the Martian atmosphere from its UV absorption band around 250 nm, first using Mariner 7 data (Barth and Hord, 1971) and then using Mariner 9 spectra (Barth et al., 1972). Strong variations in the O_3 column density (from less than 3 μm atm up to 60 μm atm) were observed, with the highest abundances recorded over the winter poles (Barth et al., 1973; Barth, 1974). Ozone was later monitored from the ground using high-resolution heterodyne spectroscopy in the 9.7 μm band (Espenak et al., 1991; Fast et al., 2006, 2009).

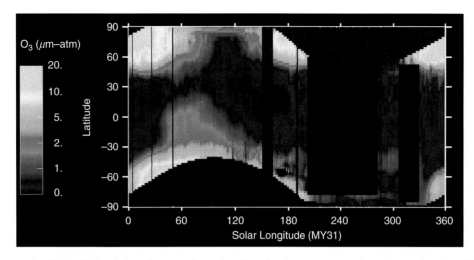

Figure 4.20. The seasonal and latitudinal variation of ozone column abundance in micrometer atmospheres (μm atm) retrieved from MARCI observations during MY 31. Periods between $L_s = 210°$ and $330°$ when high-dust-opacity contaminated ozone retrievals have been excluded. Values are zonal averages. A black and white version of this figure will appear in some formats. For the color version, please refer to the plate section.

A complete monitoring of O_3 as a function of latitude and season has been achieved by the SPICAM instrument on-board Mars Express using the UV band of ozone at 250 nm (Perrier et al., 2006). The general trend showing strong maxima over the winter poles and low abundances at the equinox at all latitudes was confirmed, and observations showed evidence of local variations during northern spring associated with polar vortex oscillations, and for variations associated with topography above Hellas. Daily global images in the UV by the MARCI camera on-board the MRO spacecraft (Clancy et al., 2010) go a step further, providing daily global maps of ozone column abundance over multiple Mars years, revealing its seasonal and spatial variations (Figure 4.20). Vertical profiles of ozone have been retrieved by the SPICAM instrument using the solar occultation mode (Lebonnois et al., 2006). Two ozone layers were identified, one at low altitude (below 30 km), and another one between 30 and 60 km, highly variable with latitude and season.

An indirect measurement of ozone is also obtained from the observation of the oxygen airglow at 1.27 μm, which results from the UV photolysis of O_3. The O_2 emission has been observed by Noxon et al. (1976), Traub et al. (1979), Krasnopolsky (2003b), and Clancy et al. (2012). Using this method only probes the atmospheric ozone above an altitude of 20–25 km since O_2 de-excitation takes place through collisions with CO_2 at lower altitudes. Using the O_2 infrared emission, ozone has been mapped from spacecraft using the OMEGA (Altieri et al., 2009) and PFS instruments (Geminale and Formisano, 2009).

All observations show that ozone abundance is anticorrelated with water vapor abundance, with peak ozone abundance during polar winter at a column density reaching 30 μm atm (8×10^{16} cm^{-2}). The observed behavior of O_3 is consistent with early photochemical models (McElroy and Donahue, 1972; Parkinson and Hunten, 1972) that predict an anticorrelation between ozone and odd oxygen produced by water vapor photolysis (see also Chapter 13 for further discussion).

4.3.7 Hydrogen Peroxide

Hydrogen peroxide is a key trace species in the Martian atmosphere because it is involved in the recycling mechanism that ensures the stability of the Martian atmosphere by making possible the regeneration of CO_2 (Parkinson and Hunten, 1972). It has also been suggested as being responsible for the oxidation of the Martian surface (Oyama and Berdahl, 1977). In particular, H_2O_2 may act as the oxidation agent to form the iron oxides that give the surface its red color. Hydrogen is believed to be formed (together with O_2) from the self-reaction of two HO_2 radicals produced by the photolysis of H_2O. Its photochemical lifetime is very short (a few hours).

Hydrogen peroxide was first detected using submillimeter heterodyne spectroscopy at the James Clerk Maxwell Telescope (JCMT; Clancy et al., 2004). It was also detected and mapped in the thermal infrared using the Texas Echelon Cross Echelle Spectrometer (TEXES) at the Infrared Telescope Facility (Encrenaz et al., 2004; Figure 4.21). The mixing ratio of H_2O_2 varies between less than 10 ppb to about 40 ppb depending on season and location. A discrepancy exists with a measurement obtained by the HIFI instrument on-board the Herschel spacecraft (Hartogh et al., 2010), which gave an upper limit of 2 ppb for $L_s = 77°$. Since H_2O_2 was detected two years earlier with TEXES at a very similar season (15 ppb, $L_s = 80°$), this result might be the signature of inter-annual variations.

4.3.8 Hydrogen

Molecular hydrogen was detected using observations by the Far Ultraviolet Spectroscopic Explorer of four emission bands at 107.17, 109.04, 111.86, and 116.68 nm, giving a column abundance of 1.17 ± 0.13 cm^{-2} above an altitude of 140 km (Krasnopolsky and Feldman, 2001). Molecular hydrogen is a product of the photolysis of water and is formed from the combination of H atoms and HO_2 radicals at altitudes ranging between 20 and 50 km. From a photochemical model of the upper atmosphere that combines observations of deuterium

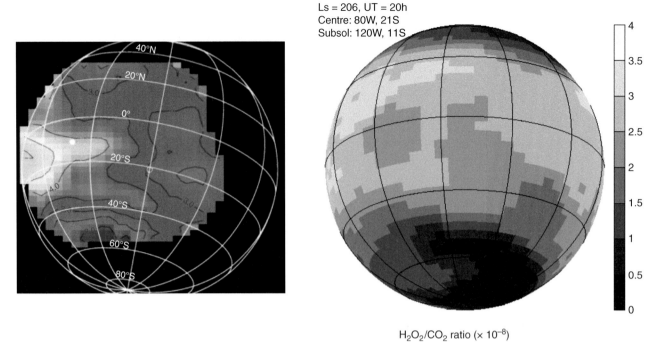

Figure 4.21. (Left) Map of the H_2O_2/CO_2 mixing ratio, derived from the line depth ratio of weak H_2O_2 and CO_2 transitions around 8.0 mm, obtained from TEXES data recorded in June 2003 (MY 26, $L_s = 206°$). The H_2O_2 mixing ratio is indicated in multiples of 10^{-8}; contours are separated by 0.5×10^{-8}. The subsolar point is indicated by a white dot. (Right) Map of the H_2O_2 mixing ratio, in the same units, as modeled by the GCM developed at the Laboratoire de Météorologie Dynamique under the conditions of the observations. Figure is from Encrenaz et al. (2004), and is used with permission. A black and white version of this figure will appear in some formats. For the color version, please refer to the plate section.

(Krasnopolsky et al., 1998) and atomic hydrogen (Anderson and Hord, 1971), Krasnopolsky and Feldman (2001) derive a H_2 mixing ratio of 15±5 ppm in the lower atmosphere. The H_2 and HD mixing ratios inferred from this model favor photochemical fractionation of deuterium between H_2O and H_2, as compared with thermodynamical fractionation. The photochemical lifetime of molecular hydrogen is about 200 years, and its lifetime relative to escape is about 2000 years (Krasnopolsky, 1986).

4.3.9 Nitrogen Monoxide

Molecular nitrogen is dissociated in the Martian upper atmosphere, both by predissociation by solar UV photons and by photoelectron impact dissociation (Barth et al., 1992), producing atomic nitrogen, nitric oxide (NO_2), and nitrogen monoxide (NO). Nitrogen monoxide has been detected in the Martian thermosphere by the Viking Lander entry science mass spectrometer (Nier and McElroy, 1977), with a mixing ratio of about 10^{-4} above an altitude of 120 km. A tentative detection of NO reported from millimeter heterodyne spectroscopy at 251 GHz (Encrenaz et al., 1999) has not been confirmed and was likely the result of an instrumental artifact (Encrenaz, 2001).

4.3.10 Methane

For some time, there has been a lively debate about the possible presence of methane in the Martian atmosphere and its temporal variations. On Earth, most of the atmospheric methane is of biogenic origin. The key issue behind its possible presence on Mars is thus the question of its origin, whether it is abiotic or biogenic. The most recent ground-based observations (Villanueva et al., 2013) and early analysis of atmospheric samples by SAM on-board the Mars Science Laboratory Rover (Webster et al., 2013b) both found no methane. However, more recent analysis of SAM samples using its enrichment process have yielded a background methane mixing ratio of 0.69±0.25 ppb, with episodic instances of elevated abundance of 7.2±2.1 ppb (Webster et al., 2015). The cause of the reported episodic increases in methane abundance is unknown.

The first reported detection of methane on Mars was published by Krasnopolsky et al. (2004), who announced the marginal detection of a few transitions of the CH_4 band near 3.3 μm, using ground-based high-resolution spectroscopy integrated over the Martian disk. The inferred CH_4 mixing ratio was 10±3 ppb. The authors suggested that, in the absence of any identified formation mechanism, this methane was likely to have biogenic origin. A few months later, using the PFS instrument on-board Mars Express, Formisano et al. (2004) announced the detection of CH_4, with a comparable mean mixing ratio, but with fluctuations ranging from 0 to 35 ppb. The authors pointed out that non-biogenic mechanisms, including hydrothermal activity, could not be excluded. Using ground-based high-resolution imaging spectroscopy in the 3 μm region, Mumma et al. (2004, 2009) reported the detection of a strong release of methane on Mars in northern summer 2003, with a mixing ratio as high as 40 ppb, strongly variable with time and location.

The conjunction of several independent measurements, all indicating the existence of a variable methane source, led to a general consensus about the existence of methane on Mars at that time, and many authors speculated about its possible sources, biogenic or abiotic (e.g. Atreya et al., 2007; Lefèvre and Forget, 2009). However, a close examination of the observations shows that most of them are questionable, and the addition of several marginal results does not make a firm detection. Spacecraft observations suffer from low spectral resolution (the entire Q-branch of the CH_4 band is contained within a single spectral channel of PFS) and a limited signal-to-noise ratio, so that large summations are necessary. This is not favorable for detecting a source that reportedly varies with time and location. In addition, the seasonal and spatial variations shown in maps retrieved by different groups do not agree. The ground-based observations obtained by Mumma et al. (2009) have the advantage of very high spectral resolution and signal-to-noise ratio, but they are limited by contamination from the terrestrial atmosphere. The R0 and R1 methane transitions are clearly identified at the expected Doppler-shifted position in the wings of the terrestrial lines (Mumma et al., 2009). However, these wavelengths also coincide with those of terrestrial $^{13}CH_4$ R0 and R1 transitions at the time of the detection when the Martian lines were blue-shifted (Zahnle et al., 2011). To confirm the CH_4 detection by remote sensing in an unambiguous way, one would need to detect the Martian methane when the Martian lines are red-shifted. However, no detection has been reported with this configuration. In addition, no detection has been reported from any of the ground-based observations since 2006 with an upper limit for methane set at 7 ppb (Villanueva et al., 2013). The SAM analysis (Webster et al., 2015) remains the most conclusive detection of methane.

If the time variations in methane abundance described by Mumma et al. (2009) and Webster et al. (2015) are real, then the mechanism behind the variations remains to be understood. The photochemical lifetime of methane according to theory is about 300 years. If methane is destroyed over a timescale of months or years, a new oxidizing mechanism, presently unknown, has to be at work (Lefèvre and Forget, 2009; Zahnle et al., 2011). Therefore, the possible presence of variations in methane abundance is still an open question. Future information will come from further analysis of atmospheric samples from SAM and from the instruments on-board the ExoMars Trace Gas Orbiter.

4.3.11 Upper Limits of Other Trace Species

A number of trace species in the Martian atmosphere have been unsuccessfully searched for using infrared spectroscopy with Mariner 9 (Maguire, 1977), ground-based infrared spectroscopy (Beer et al., 1971; Owen and Sagan, 1972; Villanueva et al., 2013), ground-based millimeter spectroscopy (Encrenaz et al., 1991) and Herschel submillimeter spectroscopy (Hartogh et al., 2010). Table 4.5, updated from Maguire (1977), Owen (1992) and Encrenaz (2001), summarizes the best upper limits presently achieved.

A tentative detection of H_2CO was reported from Phobos solar occultation observations with a mixing ratio of 500 ppb (Korablev et al., 1993). However, subsequent upper limits obtained from ground-based infrared observations showed that

Table 4.5. *The upper limits for the observed abundance of minor atmospheric species. The Maguire (1977) results are from the analysis of Mariner 9 IRIS spectra, while the others result from ground-based spectroscopic observations.*

Gaseous species	Upper limit (ppb)	Reference
C_2H_2	2	Maguire (1977)
C_2H_4	10	Villanueva et al. (2013)
C_2H_6	0.7	Villanueva et al. (2013)
CH_3OH	20	Villanueva et al. (2013)
CH_3Cl	14	Villanueva et al. (2013)
N_2O	80	Villanueva et al. (2013)
NO_2	10	Maguire (1977)
NH_3	5	Maguire (1977)
PH_3	100	Maguire (1977)
SO_2	0.3	Encrenaz et al. (2011)
OCS	70	Encrenaz et al. (1991)
H_2S	20	Encrenaz et al. (1991)
H_2CO	3	Krasnopolsky et al. (1997)
HCl	0.2	Hartogh et al. (2010)
HCN	4.5	Villanueva et al. (2013)
HO_2	200	Villanueva et al. (2013)

the disk-integrated mixing ratio of H_2CO is at least 100 times lower (Krasnopolsky et al., 1997). In the case of HCl, the very low upper limit (0.2 ppb) derived from Herschel observations (Hartogh et al., 2010) confirms that chlorine chemistry is negligible in the atmosphere of Mars.

A special mention should be made for sulfur dioxide, SO_2. After CO_2 and H_2O, SO_2 is the most abundant species outgassed from terrestrial volcanoes. If the Earth and Mars have comparable composition for volcanic outgassed compounds, SO_2 appears to be the best-suited tracer for currently ongoing outgassing activity on Mars. On Earth, the SO_2/CH_4 ratio of outgassed volcanic gases is typically 10^3–10^4. The low upper limits (<0.3 ppb) obtained from high-resolution ground-based spectroscopy (Krasnopolsky, 2005; Encrenaz et al., 2011) suggest that, if methane is occasionally present at the Martian surface at the 10 ppb level as claimed by some infrared observations (Mumma et al., 2009), its origin is not expected to be volcanic.

4.3.12 Isotopic Ratios

Isotopic ratios in volatile species are key parameters as possible diagnostics of the early history of the Martian atmosphere. Carbon and oxygen isotopic ratios were first measured from ground-based infrared observations (Kaplan et al., 1969). Those values were refined, and isotope ratios for nitrogen and noble gases were determined by the Viking Lander (Owen et al., 1977) and Mars Science Laboratory (Mahaffy et al., 2013; Webster et al., 2013a) mass spectrometers. Analysis of trapped gases from SNC meteorites has also provided further useful data (e.g. Carr et al., 1985; Greenwood et al., 2008). The D/H ratio has also been measured from ground-based infrared spectroscopy (Owen et al., 1988). Table 4.6 summarizes the key isotopic ratios discussed below. All isotope ratios are given

Table 4.6. *Key isotope ratios for Mars gases. Results from Krasnopolsky et al. (1997) and Encrenaz et al. (2005) are from ground-based telescopic observations. The Owen et al. (1977) and the Nier and McElroy (1977) results are from the Viking Lander entry mass spectrometer. The Mahaffy et al. (2013) and Atreya et al. (2013) results are from the SAM instrument on the Mars Science Laboratory Rover.*

Isotope ratio	Value with respect to terrestrial value	Reference
$^{13}C/^{12}C$	0.978±0.020	Krasnopolsky et al. (1997)
	1.00±0.11	Encrenaz et al. (2005)
	1.046±0.004	Webster et al. (2013a)
$^{17}O/^{16}O$	1.024±0.005	Webster et al. (2013a)
$^{18}O/^{16}O$	1.018±0.018	Krasnopolsky et al. (1997)
	1.048±0.005	Webster et al. (2013a)
$^{15}N/^{14}N$	1.6±0.2	Nier and McElroy (1977)
$^{38}Ar/^{36}Ar$	1.26±0.03	Atreya et al. (2013)
$^{40}Ar/^{36}Ar$	10.1±1.7	Owen et al. (1977)
	6.4±1.0	Mahaffy et al. (2013)
$^{129}Xe/^{132}Xe$	~2.5	Owen et al. (1977)
D/H	5.5±2.0	Krasnopolsky et al. (1997)
	5.5±1.0	Webster et al. (2013a)

in terms of their value relative to the ratio for terrestrial air. Reviews about the implications of these measurements can be found in Owen (1992) and Bogard et al. (2001). Chapter 15 also discusses volatile escape processes and their implications for Martian atmosphere evolution.

4.3.12.1 Carbon and Oxygen

Using high-resolution Fourier-transform spectra of Mars, Kaplan et al. (1969) first measured $^{13}C/^{12}C$ and $^{18}O/^{16}O$ in CO and CO_2, and found that they were equal to the terrestrial ratios within 15%. Maguire (1977) found a similar result using Mariner 9 IRIS spectra, and Nier and McElroy (1977) used mass spectrometer measurements from the Viking Landers to reduce the uncertainty to 5%. Bjoraker et al. (1989) reported some departures between the Martian and terrestrial values of the $^{17}O/^{16}O$ and $^{18}O/^{16}O$ isotopic ratios measured in H_2O using the Kuiper Airborne Observatory, with a depletion of the heavy isotopes ($^{17}O/^{16}O = 0.95\pm0.01$ and $^{18}O/^{16}O = 0.90\pm0.03$ with respect to the terrestrial values). Later, Krasnopolsky et al. (1996) also reported departures from terrestrial in the $^{18}O/^{17}O$ and $^{13}C/^{12}C$ ratios measured in CO_2 from ground-based infrared spectroscopy, with a relative depletion of the heavier isotopes ($^{18}O/^{17}O = 0.914\pm0.04$ and $^{13}C/^{12}C = 0.94\pm0.15$ relative to the terrestrial values). However, these departures were not confirmed by the analysis of $^{18}O/^{17}O$ and $^{13}C/^{12}C$ in CO_2 obtained by high-resolution spectroscopy measurements in the thermal infrared (Encrenaz et al., 2005), which showed no evidence for a departure between Martian and terrestrial ratios ($^{18}O/^{17}O = 1.03\pm0.09$, $^{13}C/^{12}C = 1.00\pm0.11$ with respect to the terrestrial values).

The most recent and precise measurements by Mars Science Laboratory SAM (Webster et al., 2013a) all show enrichment in the heavier isotope relative to terrestrial values ($^{13}C/^{12}C = 1.046\pm0.004$, $^{17}O/^{16}O = 1.024\pm0.005$, $^{18}O/^{16}O = 1.048\pm0.005$). Enrichment in the heavier carbon and oxygen isotopes is also found in the analysis of Martian meteorites (e.g. Carr et al., 1985; Greenwood et al., 2008). Most carbonates and hydroxyl minerals show an ^{18}O excess ranging from 2% to 4%, while this enrichment in silicate minerals is only 0.5% (Clayton and Mayeda, 1988; Farquhar et al., 1998; Wright et al., 1988; Karlsson et al., 1992; Bogard et al., 2001). The implication seems to be that there is no recycling of H_2O and CO_2 between the volatile reservoirs and the rocky interior. The isotopic composition of volatile species must then be dominated by atmospheric processes such as impacts of cometary material, isotopic fractionation due to atmospheric escape, or photochemical processes (Bogard et al., 2001).

4.3.12.2 Nitrogen and Noble Gases

Prior to the Viking missions, Brinkman (1971) and McElroy (1972) had already suggested that nitrogen might have escaped, leading to enrichment in the heavy isotope ^{15}N. The Viking Lander entry mass spectrometers indeed measured an enrichment of the $^{15}N/^{14}N$ ratio by a factor of 1.6 with respect to the terrestrial value (McElroy et al., 1976; Nier et al., 1976; Nier and McElroy, 1977).

Isotopic ratios of noble gases were measured by the Viking Lander mass spectrometers. Non-radiogenic isotopic ratios of Ar, Kr, and Xe were found to be in agreement with the terrestrial values. In contrast, significant excesses with respect to terrestrial values were measured for $^{40}Ar/^{36}Ar$ (by a factor of 10) and $^{129}Xe/^{132}Xe$ (by a factor of 2.5). The isotopes ^{40}Ar and ^{129}Xe are radiogenic decay isotopes, products of ^{40}K and ^{129}I, respectively. To account for these excesses, it has been proposed that large impacts, occurring early in the planet's history, have led to repeatedly reduce, without fractionation, the amount of the atmosphere. The outgassing of the radiogenic isotopes progressively enriches the remaining atmosphere (Melosh and Vickery, 1989; Turcotte and Schubert, 1988). Jakosky et al. (1994) have studied the effect of atmospheric sputtering as a possible mechanism for atmospheric escape. They conclude that the sputtering mechanism can account for the observed isotopic ratios of N, Ar, and Ne, but the observed Xe fractionation requires hydrodynamical escape, as Xe and Kr are too heavy to escape by sputtering (Bogard et al., 2001).

Most recently, isotope ratios for argon gas have been determined by the SAM quadrupole mass spectrometer on the Mars Science Laboratory Rover. The ratios for both $^{40}Ar/^{36}Ar$ (6.4±1.0; Mahaffy et al., 2013) and $^{38}Ar/^{36}Ar$ (1.26±0.03; Atreya et al., 2013) are both significantly enhanced in the heavier isotope relative to terrestrial values, further strengthening the case for atmospheric escape.

4.3.12.3 Hydrogen

The D/H ratio in Mars could not be measured by Viking, but was later determined using ground-based infrared spectroscopic observations of HDO transitions. Owen et al. (1988) reported a D/H ratio equal to 6±3 times the terrestrial value. Both the ground-based observations by Krasnopolsky et al. (1997) (5.5±2.0) and the Mars Science Laboratory SAM analysis (Webster et al., 2013a) (5.5±1.1) report a comparable result.

Fractionation of deuterium in the Martian atmosphere is expected to occur because of ice/vapor partitioning, especially near the polar caps, where HDO becomes enriched in water ice clouds (Fouchet and Lellouch, 2000). From a GCM simulation of the latitudinal and seasonal variations of D/H, Montmessin et al. (2005) have predicted small global variations (<2%) over the seasons, but large annual changes (by a factor of 2) in the high-latitude regions. According to their study, the global D/H ratio is expected to be more than 15% lower than its concentration in the north permanent ice cap. As a result, the global D/H in H_2O (including the water ice) might be over 6.5 times the terrestrial standard mean ocean water (SMOW) value. Attempts have been made to map D/H as a function of latitude and season using infrared spectroscopic measurements of HDO and H_2O (Mumma et al., 2003; Novak et al., 2007). Significant variations have been measured, with a tendency for the D/H ratio to decrease as water content increases (Fisher, 2007). This trend is in agreement with GCM predictions (Montmessin et al., 2005) and supports the effect of deuterium fractionation by condensation and sublimation expected to produce a minimum D/H over the poles at summer solstice. More measurements will be necessary to fully characterize this effect.

4.4 REFERENCE ATMOSPHERE

In the preceding sections we discussed in detail the thermal structure and composition of the Mars atmosphere as described by the available observations. Here we present a set of "standard Mars atmosphere" profiles of thermal structure that bracket the range of conditions observed at Mars. These can be used when only a simple atmospheric model is required. In reality, the Martian environment is highly variable in space and time. Below we also describe two widely used reference atmosphere databases that have been produced using global circulation model simulations carefully tuned to reproduce observational data. The advantage of such a reference atmosphere model is that all meteorological variables are available for whatever season, location, and local time is desired, and under scenarios with different conditions (e.g. different dust loading). This is especially useful for simulating atmospheric conditions for such applications as modeling the entry, descent, and landing of spacecraft, as well as planning aerobraking operations, and the development of new instruments and new models.

4.4.1 Standard Mars Atmosphere

In Tables 4.7, 4.8, and 4.9 we present three standard atmosphere models for the Mars atmosphere, which are derived from diurnal averages of MCS temperature retrievals (Kleinböhl et al., 2009, 2011). They are also shown in Figure 4.2 in comparison with the COSPAR reference atmosphere (Seiff, 1982). The first model, given in Table 4.7, is representative of the aphelion season, when the atmosphere is relatively clear of dust.

It is the average of temperature retrievals between 45°S and 45°N latitude at $L_s = 0–180°$ during MY 29. The second model, given in Table 4.8, is representative of the perihelion season ($L_s = 180–360°$) when the atmosphere is generally more dusty and warmer. It is composed of retrievals between 45°S and 45°N latitude, primarily from MY 29, a typical year without a global-scale dust storm. The third model, given in Table 4.9, is for conditions near the south pole during winter. It is based on retrievals poleward of 75°S latitude at $L_s = 45–135°$ during MY 29. Temperatures below 20 km altitude closely follow the CO_2 frost point (Figure 4.2).

Tables of mean atmospheric composition were previously given in Tables 4.4 and 4.5. Given the lack of observational data for the vertical variation of gas composition, we do not attempt to provide gas abundance as a function of height. Note that water vapor is a special case. The abundance of water vapor is highly variable in space and time, and there are some limited observations of its vertical distribution. See Chapter 11 for further information on water vapor and the water cycle on Mars. Dust and water ice aerosol abundance and vertical distribution are also highly variable and are discussed in detail in Chapters 5 and 10.

4.4.2 Mars-GRAM

The NASA Mars Global Reference Atmospheric Model (Mars-GRAM) is a GCM-based engineering-level atmospheric model widely used for diverse mission applications. Applications include systems design, performance analysis, and operations planning for aerobraking, entry–descent–landing (EDL), and aerocapture. From the surface to 80 km altitude, Mars-GRAM is based on the NASA Ames Mars General Circulation Model (e.g. Haberle et al., 1993, 1999, 2003). Above 80 km, Mars-GRAM is based on the University of Michigan Mars Thermospheric General Circulation Model (MTGCM) (e.g. Bougher et al., 1999, 2000, 2004) (see Chapter 14). Mars-GRAM has been validated (Justus et al., 2005) against both MGS Radio Science and TES retrievals (Smith, 2004).

In practice, Mars-GRAM is used through an interface in which the standard inputs are geographic position and time. For output, in addition to predictions of the mean density, temperature, pressure, winds, and selected atmospheric constituents, Mars-GRAM includes a perturbation modeling capability commonly used in a Monte Carlo mode to perform high-fidelity engineering end-to-end simulations for EDL planning (Striepe et al., 2002). The perturbations are controlled by input parameters that allow separate scaling of density and wind perturbations, and the adjustment of the wavelengths (spectral range) of the perturbations

There are several options within Mars-GRAM for representing the mean atmosphere. The first option uses the first Martian year of TES observations (MY 24) as a baseline, with user-specified dust optical depth and Mars-GRAM data interpolated from a set of NASA Ames GCM results obtained using selected values of globally uniform dust optical depth. The second option allows the user to use any auxiliary profile of temperature and density versus altitude in addition to the TES MY 24 baseline. In using this option, the values from the auxiliary profile replace data from the original Ames GCM databases.

Table 4.7. *A model reference atmosphere for a "clear" scenario with little dust aerosol. The table shows height above the surface, z (km), atmospheric temperature, T (K), atmospheric pressure, p (Pa), and atmospheric density, ρ (g cm⁻³). Data are based on diurnal averages of observations made by the MCS instrument (Kleinböhl et al., 2009).*

z (km)	T (K)	p (Pa)	ρ (g cm^{-3})	z (km)	T (K)	p (Pa)	ρ (g cm^{-3})
0	206.8	562.000	1.422E–06	43	154.9	4.99047	1.686E–08
1	206.7	511.251	1.294E–06	44	153.9	4.39646	1.495E–08
2	206.3	465.031	1.179E–06	45	152.9	3.86996	1.324E–08
3	205.3	422.854	1.078E–06	46	151.9	3.40366	1.172E–08
4	203.9	384.287	9.860E–07	47	151.0	2.99113	1.036E–08
5	202.4	349.000	9.021E–07	48	150.0	2.62647	9.161E–09
6	200.7	316.711	8.256E–07	49	149.0	2.30425	8.091E–09
7	198.9	287.165	7.554E–07	50	148.0	2.01978	7.140E–09
8	197.2	260.150	6.902E–07	51	147.1	1.76893	6.291E–09
9	195.5	235.475	6.302E–07	52	146.2	1.54798	5.540E–09
10	193.8	212.955	5.749E–07	53	145.4	1.35357	4.870E–09
11	192.2	192.423	5.238E–07	54	144.7	1.18275	4.276E–09
12	190.6	173.723	4.769E–07	55	144.1	1.03287	3.750E–09
13	189.1	156.710	4.336E–07	56	143.6	0.90151	3.285E–09
14	187.6	141.247	3.939E–07	57	143.1	0.78648	2.875E–09
15	186.2	127.207	3.574E–07	58	142.7	0.68584	2.515E–09
16	184.8	114.473	3.241E–07	59	142.3	0.59784	2.198E–09
17	183.5	102.933	2.935E–07	60	141.9	0.52094	1.921E–09
18	182.2	92.487	2.656E–07	61	141.5	0.45375	1.678E–09
19	181.0	83.040	2.400E–07	62	141.1	0.39507	1.465E–09
20	179.9	74.506	2.167E–07	63	140.7	0.34385	1.279E–09
21	178.7	66.803	1.956E–07	64	140.3	0.29914	1.116E–09
22	177.5	59.853	1.764E–07	65	140.0	0.26016	9.722E–10
23	176.3	53.585	1.590E–07	66	139.8	0.22621	8.465E–10
24	175.1	47.938	1.432E–07	67	139.7	0.19665	7.365E–10
25	174.0	42.854	1.289E–07	68	139.6	0.17094	6.406E–10
26	172.8	38.281	1.159E–07	69	139.5	0.14858	5.572E–10
27	171.7	34.171	1.041E–07	70	139.5	0.12913	4.843E–10
28	170.5	30.478	9.352E–08	71	139.6	0.11224	4.206E–10
29	169.3	27.163	8.394E–08	72	139.6	0.09756	3.656E–10
30	168.1	24.188	7.528E–08	73	139.6	0.08480	3.178E–10
31	166.9	21.521	6.746E–08	74	139.7	0.07371	2.761E–10
32	165.8	19.133	6.037E–08	75	139.7	0.06408	2.400E–10
33	164.7	16.996	5.399E–08	76	139.7	0.05570	2.086E–10
34	163.7	15.087	4.822E–08	77	139.7	0.04842	1.813E–10
35	162.6	13.382	4.306E–08	78	139.6	0.04209	1.578E–10
36	161.7	11.861	3.838E–08	79	139.5	0.03659	1.372E–10
37	160.7	10.505	3.420E–08	80	139.4	0.03180	1.193E–10
38	159.7	9.297	3.046E–08	81	139.4	0.02763	1.037E–10
39	158.7	8.222	2.710E–08	82	139.3	0.02401	9.018E–11
40	157.7	7.265	2.410E–08	83	139.2	0.02086	7.842E–11
41	156.8	6.415	2.141E–08	84	139.0	0.01813	6.823E–11
42	155.8	5.660	1.901E–08	85	138.9	0.01575	5.931E–11

Examples of auxiliary profiles include data from TES observations or Mars mesoscale model output at a particular location and time. The final option uses the second and third Martian years of TES observations (MY 25 and 26), with Mars-GRAM data coming from Ames GCM results driven by the observed TES dust optical depth during those years. The user can also adjust the optical depth of the uniformly mixed background dust level, add a seasonal dust optical depth, set the dust particle diameter and density, and provide the starting season, position, duration, intensity, and radius of a dust storm. The most recent version of the model, Mars-GRAM 2010, now includes adjustment factors that are used to alter the input data from the MGCM and MTGCM for the MY 24 (user-controlled dust) case (Justh et al., 2011). The greatest adjustment occurs at large optical depths such as during strong dust storms. The addition of the adjustment factors has led to a better correspondence

Table 4.8. *A model reference atmosphere for a "dusty" scenario for a year without a global-scale dust storm. The table shows height above the surface, z (km), atmospheric temperature, T (K), atmospheric pressure, p (Pa), and atmospheric density, ρ (g cm^{-3}). Data are based on diurnal averages of observations made by the MCS instrument (Kleinböhl et al., 2009).*

z (km)	T (K)	p (Pa)	ρ (g cm^{-3})	z (km)	T (K)	p (Pa)	ρ (g cm^{-3})
0	216.6	562.000	1.357E–06	43	174.1	7.90142	2.374E–08
1	220.3	513.849	1.220E–06	44	172.9	7.05872	2.136E–08
2	221.8	470.319	1.109E–06	45	171.8	6.30115	1.919E–08
3	221.6	430.588	1.017E–06	46	170.6	5.62060	1.724E–08
4	220.8	394.135	9.339E–07	47	169.4	5.00951	1.547E–08
5	219.8	360.637	8.584E–07	48	168.2	4.46121	1.388E–08
6	218.6	329.840	7.894E–07	49	167.0	3.96962	1.244E–08
7	217.4	301.524	7.256E–07	50	165.9	3.52936	1.113E–08
8	216.0	275.491	6.673E–07	51	164.7	3.13535	9.960E–09
9	214.6	251.557	6.133E–07	52	163.6	2.78303	8.900E–09
10	213.3	229.571	5.631E–07	53	162.5	2.46831	7.947E–09
11	212.0	209.390	5.167E–07	54	161.5	2.18748	7.086E–09
12	210.7	190.875	4.740E–07	55	160.6	1.93722	6.311E–09
13	209.4	173.897	4.345E–07	56	159.7	1.71441	5.616E–09
14	208.2	158.341	3.979E–07	57	158.8	1.51619	4.995E–09
15	206.9	144.095	3.644E–07	58	158.0	1.34000	4.437E–09
16	205.6	131.053	3.335E–07	59	157.3	1.18359	3.937E–09
17	204.3	119.120	3.050E–07	60	156.5	1.04482	3.493E–09
18	202.9	108.205	2.790E–07	61	155.9	0.92180	3.093E–09
19	201.5	98.224	2.550E–07	62	155.3	0.81288	2.738E–09
20	200.0	89.102	2.331E–07	63	154.7	0.71647	2.423E–09
21	198.6	80.770	2.128E–07	64	154.2	0.63122	2.142E–09
22	197.1	73.164	1.942E–07	65	153.8	0.55590	1.891E–09
23	195.7	66.226	1.770E–07	66	153.4	0.48941	1.669E–09
24	194.3	59.903	1.613E–07	67	153.1	0.43075	1.472E–09
25	193.0	54.146	1.468E–07	68	152.9	0.37904	1.297E–09
26	191.7	48.909	1.335E–07	69	152.7	0.33348	1.143E–09
27	190.5	44.149	1.212E–07	70	152.5	0.29334	1.006E–09
28	189.4	39.828	1.100E–07	71	152.4	0.25801	8.857E–10
29	188.2	35.907	9.982E–08	72	152.3	0.22691	7.795E–10
30	187.1	32.351	9.046E–08	73	152.1	0.19954	6.864E–10
31	186.0	29.130	8.194E–08	74	151.9	0.17544	6.042E–10
32	184.9	26.213	7.417E–08	75	151.6	0.15421	5.322E–10
33	183.9	23.574	6.707E–08	76	151.3	0.13552	4.686E–10
34	182.9	21.188	6.061E–08	77	150.9	0.11906	4.128E–10
35	181.9	19.033	5.474E–08	78	150.4	0.10456	3.637E–10
36	181.0	17.087	4.939E–08	79	149.9	0.09178	3.203E–10
37	180.0	15.332	4.456E–08	80	149.4	0.08053	2.820E–10
38	179.1	13.749	4.016E–08	81	148.9	0.07063	2.482E–10
39	178.1	12.322	3.620E–08	82	148.3	0.06192	2.184E–10
40	177.2	11.037	3.259E–08	83	147.8	0.05425	1.920E–10
41	176.2	9.880	2.934E–08	84	147.4	0.04752	1.687E–10
42	175.1	8.838	2.641E–08	85	146.9	0.04160	1.482E–10

with TES retrievals, as well as to better agreement with MGS, Odyssey and MRO accelerometer data between 90 and 130 km.

4.4.3 Mars Climate Database

The Mars Climate Database (MCD) is a database of meteorological fields derived from GCM numerical simulations and validated using available observational data (Lewis et al., 1999; Forget et al., 2006b; Millour et al., 2011). The GCM was developed at the Laboratoire de Météorologie Dynamique du CNRS in Paris, France (Forget et al., 1999) in collaboration with the Open University and Oxford University in the UK and the Instituto de Astrofísica de Andalucía in Spain. It includes models of the water cycle (Montmessin et al., 2004), CO_2 cycle (Forget et al., 1998), and photochemistry (Lefèvre et al., 2004, 2008), and has been extended to the upper atmosphere (Angelats

Table 4.9. *A model reference atmosphere for a "polar night" scenario with little dust aerosol. The table shows height above the surface, z (km), atmospheric temperature, T (K), atmospheric pressure, p (Pa), and atmospheric density, ρ (g cm^{-3}). Data are based on diurnal averages of observations made by the MCS instrument (Kleinböhl et al., 2009).*

z (km)	T (K)	p (Pa)	ρ (g cm^{-3})	z (km)	T (K)	p (Pa)	ρ (g cm^{-3})
0	148.2	562.000	1.984E−06	43	163.7	1.87301	5.986E−09
1	149.0	492.663	1.730E−06	44	164.3	1.66235	5.293E−09
2	149.9	432.204	1.508E−06	45	164.8	1.47598	4.686E−09
3	147.9	378.981	1.341E−06	46	165.2	1.31093	4.152E−09
4	147.6	331.973	1.177E−06	47	165.4	1.16458	3.684E−09
5	146.8	290.652	1.036E−06	48	165.6	1.03472	3.269E−09
6	145.8	254.265	9.124E−07	49	165.6	0.91941	2.905E−09
7	144.7	222.219	8.035E−07	50	165.4	0.81688	2.584E−09
8	143.7	194.022	7.064E−07	51	165.0	0.72564	2.301E−09
9	142.7	169.242	6.205E−07	52	164.4	0.64435	2.051E−09
10	141.8	147.492	5.442E−07	53	163.6	0.57188	1.829E−09
11	141.1	128.438	4.762E−07	54	162.5	0.50721	1.633E−09
12	140.6	111.779	4.159E−07	55	161.2	0.44945	1.459E−09
13	140.3	97.242	3.626E−07	56	159.6	0.39784	1.304E−09
14	140.2	84.579	3.156E−07	57	157.8	0.35169	1.166E−09
15	140.2	73.562	2.745E−07	58	155.8	0.31043	1.042E−09
16	140.2	63.979	2.388E−07	59	153.6	0.27355	9.317E−10
17	140.3	55.648	2.075E−07	60	151.4	0.24061	8.315E−10
18	140.3	48.404	1.805E−07	61	149.1	0.21123	7.412E−10
19	140.3	42.103	1.570E−07	62	146.8	0.18506	6.595E−10
20	140.2	36.620	1.367E−07	63	144.6	0.16180	5.854E−10
21	140.2	31.850	1.189E−07	64	142.5	0.14119	5.184E−10
22	140.4	27.704	1.032E−07	65	140.6	0.12296	4.575E−10
23	140.8	24.104	8.957E−08	66	138.8	0.10689	4.029E−10
24	141.6	20.985	7.754E−08	67	137.2	0.09276	3.537E−10
25	142.6	18.286	6.709E−08	68	135.8	0.08037	3.096E−10
26	143.9	15.951	5.799E−08	69	134.6	0.06954	2.703E−10
27	145.3	13.932	5.017E−08	70	133.6	0.06010	2.354E−10
28	146.9	12.186	4.340E−08	71	132.7	0.05189	2.046E−10
29	148.5	10.674	3.761E−08	72	131.9	0.04475	1.775E−10
30	150.2	9.363	3.261E−08	73	131.2	0.03857	1.538E−10
31	151.8	8.225	2.835E−08	74	130.6	0.03321	1.330E−10
32	153.3	7.235	2.469E−08	75	130.0	0.02858	1.150E−10
33	154.7	6.372	2.155E−08	76	129.5	0.02458	9.930E−11
34	156.0	5.618	1.884E−08	77	129.1	0.02113	8.562E−11
35	157.3	4.958	1.649E−08	78	128.7	0.01815	7.379E−11
36	158.3	4.380	1.448E−08	79	128.2	0.01559	6.361E−11
37	159.3	3.872	1.272E−08	80	127.5	0.01338	5.489E−11
38	160.2	3.426	1.119E−08	81	126.6	0.01147	4.739E−11
39	161.0	3.033	9.855E−09	82	125.4	0.00982	4.096E−11
40	161.8	2.686	8.687E−09	83	123.7	0.00839	3.548E−11
41	162.4	2.381	7.671E−09	84	121.9	0.00715	3.070E−11
42	163.1	2.111	6.772E−09	85	120.6	0.00609	2.641E−11

i Coll et al., 2005; González-Galindo et al., 2005, 2009). The Mars Climate Database thus extends from the surface up to ~350 km, and provides mean values and statistics of meteorological variables (atmospheric temperatures, density, pressure, and wind velocity) as well as atmospheric composition and aerosol abundance (including water vapor and ice). The MCD can be accessed either online via an interactive server available at http://www-mars.lmd.jussieu.fr, or from a DVD-ROM version that includes advanced access and post-processing tools (see below).

The seasonal variation of database quantities is given in terms of 12 typical "months", which represent the average over 30° of L_s. The diurnal variation is given in terms of 12 times of day, which represent the average over two Martian "hours". Intermediate seasonal dates and local times are constructed using interpolation between these values.

Three different scenarios are available to account for the level of extreme ultraviolet (EUV) radiation from the Sun, and four different dust scenarios are included to represent variability in Mars atmospheric conditions. The three EUV scenarios are for conditions typical of solar minimum, average, and maximum values. The four dust scenarios include: (1) a baseline case obtained from matching observations by TES during MY 24 and MY 25; (2) a cold scenario corresponding to an extremely clear atmosphere; (3) a warm scenario corresponding to an atmosphere with moderately high dust loading; and (4) a dust storm scenario that represents conditions during the largest global-scale dust storms (see Chapter 10 for more information about dust storms). Day-to-day variability in meteorological values is included in the MCD output as a standard deviation of quantities, and users may reconstruct realistic variability by adding perturbations to mean values in the form of either large-scale perturbations using empirical orthogonal functions derived from the GCM runs, or small-scale perturbations by adding gravity waves with user-defined amplitude and wavelength.

The Mars Climate Database includes a post-processing tool designed to predict surface pressure anywhere and anytime on Mars as accurately as possible (typically within a few percent) by combining the Viking Lander 1 reference pressure record (Sutton et al., 1978) with large-scale meteorological pressure gradients predicted by the GCM and high-resolution (32 pixels per degree) topography obtained by the MGS MOLA instrument (Smith et al., 2001c). In this way, the MCD includes a high-resolution version designed to estimate the three-dimensional atmospheric temperature and density with a ~2 km horizontal resolution.

4.5 SUMMARY AND FUTURE ISSUES

Recent ground-based and spacecraft observations have provided a good overall understanding of the thermal structure and composition of the Mars atmosphere. The current climate pattern shows distinct and generally repeatable patterns in thermal structure and composition as a function of season, latitude, and longitude. These patterns are driven by a combination of orbital and obliquity seasons, and the condensation and sublimation of the seasonal polar ice caps. A rich variety of waves and other phenomena have been observed on shorter timescales, and moderate inter-annual variations are associated with large dust storms, especially during the perihelion season.

Continued monitoring and new types of observations of the Martian atmosphere are needed to more fully characterize the current climate. In particular, better coverage in local time would be helpful to better characterize thermal tides and the diurnal variation of water vapor. Characterization of minor gases (especially methane) and isotope ratios is important for understanding the history of the atmosphere, and better observations of the vertical distribution of gases would provide important constraints for modelers. The planetary boundary layer, with its complicated dynamics and important surface–atmosphere interactions, is not well characterized by current observations.

Fortunately, long-lived spacecraft such as the Mars Reconnaissance Orbiter, Mars Express, and Mars Science Laboratory continue to observe Mars, and several new spacecraft are planned for launch to Mars within the next decade. Observations from the Mars Atmospheric and Volatile Evolution (MAVEN) and the ExoMars Trace Gas Orbiter spacecraft will provide the new information that will build on the existing knowledge presented here to further our understanding of the Martian atmosphere.

ACKNOWLEDGMENTS

We wish to thank T. Owen and T. Fouchet for helpful comments regarding trace gases and isotopes, and R. J. Wilson for helpful comments on the thermal structure and dynamics. We are grateful to N. Heavens and D. Kass for providing Figures 4.6 and 4.15, respectively. D. Banfield and an anonymous referee provided significant and constructive comments. Portions of this work were performed at the Jet Propulsion Laboratory, California Institute of Technology, under contract with the National Aeronautics and Space Administration.

REFERENCES

Altieri, F., Zasova, L., D'Aversa, E., et al. (2009) O$_2$ 1.27 mm emission maps as derived from OMEGA/MEx data, *Icarus*, 204, 499–511.

Angelats i Coll, M., Forget, F., López-Valverde, M. A., et al. (2004) Upper atmosphere of Mars up to 120 km: Mars Global Surveyor accelerometer data analysis with the LMD general circulation model, *J. Geophys. Res.*, 109, E01011, doi:10.1029/2003JE002163.

Angelats i Coll, M., Forget, F., López-Valverde, M. A., and González-Galindo, F. (2005) The first Mars thermospheric general circulation model: the Martian atmosphere from the ground to 240 km, *Geophys. Res. Lett.*, 32, L04201, doi:10.1029/2004GL021368.

Anderson, D. E., and Hord, C. W. (1971) Mariner 6 and 7 ultraviolet spectrometer experiment: analysis of hydrogen Lyman-alpha data, *J. Geophys. Res.*, 76, 6666–6673.

Arvidson, R. E., Ashley, J. W., Bell III, J. F., et al. (2011) Opportunity Mars Rover mission: overview and selected results from Purgatory ripple to traverses to Endeavour Crater, *J. Geophys. Res.*, 116, E00F15, doi:10.1029/2010JE003746.

Atreya, S. K., Mahaffy, P. R., and Wong, A.-S. (2007) Methane and related trace species on Mars: origin, loss, implication for life and habitability, *Planet. Space Sci.*, 55, 358–369.

Atreya, S. K., Trainer, M. G., Franz, H. B., et al. (2013) Primordial argon isotope fractionation in the atmosphere of Mars measured by the SAM instrument on Curiosity and implications for atmospheric loss, *Geophys. Res. Lett.*, 40, 1–5, doi:10.1002/2013GL057763.

Banfield, D., Conrath, B. J., Pearl, J. C., et al. (2000) Thermal tides and stationary waves on Mars as revealed by Mars Global Surveyor Thermal Emission Spectrometer, *J. Geophys. Res.*, 105, 9521–9537.

Banfield, D., Conrath, B. J., Smith, M. D., et al. (2003) Forced waves in the Martian atmosphere from MGS TES nadir data, *Icarus*, 161, 319–345.

Barker, E. S. (1972) Detection of molecular oxygen in the Martian atmosphere, *Nature*, 238, 447–448.

Barth, C. A. (1974) The atmosphere of Mars, *Annu. Rev. Earth Plan. Sci.*, 2, 333–367.

Barth, C. A., and Hord, C. W. (1971) Mariner ultraviolet spectrometer: topography and polar cap, *Science*, 173, 193–201.

Barth, C. A., Hord, C. W., Stewart, A. I., and Lane, A. L. (1972) Mariner 9 ultraviolet spectrometer experiment: initial results, *Science*, 175, 309–312.

Barth, C. A., Hord, C. W., Stewart, A. I., et al. (1973) Mariner 9 ultraviolet spectrometer experiment: seasonal variation of ozone on Mars, *Science*, 179, 795–796.

Barth, C. A., Stewart, A. I. F., Bougher, S. W., et al. (1992) Aeronomy of the current Martian atmosphere, in *Mars* (Kieffer, H. H., Jakosky, B. M., Snyder, C. W., Matthews, M. S., Eds.), University of Arizona Press, Tucson.

Beer, R., Norton, R. H., and Martonchik, J. V. (1971) Astronomical infrared spectroscopy with a Connes-type interferometer: II- Mars, 2500–3500 cm^{-1}, *Icarus*, 15, 1–10.

Bertaux, J.-L., Korablev, O., Perrier, S., et al. (2006) SPICAM on Mars Express: observing modes and overview of UV spectrometer data and scientific results, *J. Geophys. Res.*, 111, E10S90, doi:10.1029/2006JE002690.

Biemann, K., Lafleur, A. L., Owen, T., et al. (1976) The atmosphere of Mars near the surface – isotope ratios and upper limits on noble gases, *Science*, 194, 76–78.

Billebaud, F., Maillard, J.-P., Lellouch, E., and Encrenaz, T. (1992) The spectrum of Mars in the (1–0) band of CO, *Astron. Astrophys.*, 261, 647–657.

Billebaud, F., Rosenqvist, J., Lellouch, E., et al. (1998) Observations of CO in the atmosphere of Mars in the (2–0) vibrational band at 2.35 microns, *Astron. Astrophys.*, 333, 1092–1099.

Billebaud, F., Brillet, J., Lellouch, E., et al. (2009) Observations of CO in the atmosphere of Mars with PFS onboard Mars Express, *Planet. Space Sci.*, 57, 1446–1457.

Bjoraker, G. L., Mumma, M. J., and Larson, H. P. (1989) Isotopic abundance ratios for hydrogen and oxygen in the Martian atmosphere, *Bull. Amer. Astron. Soc.*, 21, 991.

Bogard, D. D., Clayton, R. N., Marti, K., et al. (2001) Martian volatiles: isotopic composition, origin and evolution, *Space Sci. Rev.*, 96, 425–458.

Bougher, S. W., and Keating, G. M. (2006) Mars Reconnaissance Orbiter: aerobraking science analysis, *Bull. Amer. Astron. Soc.*, 38, 605.

Bougher, S. W., Hunten, D. M., and Roble, R. G. (1994) CO_2 Cooling in terrestrial planet thermospheres, *J. Geophys. Res.*, 99, 14609–14622.

Bougher, S. W., Engel, S., Roble, R. G., and Foster, B. (1999) Comparative terrestrial planet thermospheres: 2. Solar cycle variation of global structure and winds at equinox, *J. Geophys. Res.*, 104, 16591–16611, doi:10.1029/1998JE001019.

Bougher, S. W., Engel, S., Roble, R. G., and Foster, B. (2000) Comparative terrestrial planet thermospheres: 3. Solar cycle variation of global structure and winds at solstices, *J. Geophys. Res.*, 105, 17669–17689, doi:10.1029/1999JE001232.

Bougher, S. W., Hinson, D. P., Forbes, J. M., and Engel, S. (2001) MGS Radio Science electron density profiles and implications for the neutral atmosphere, *Geophys. Res. Lett.*, 28, 3091–3094.

Bougher, S. W., Engel, S., Hinson, D. P., and Murphy, J. R. (2004) MGS Radio Science electron density profiles: interannual variability and implications for the neutral atmosphere, *J. Geophys. Res.*, 109, E03010, doi:10.1029/2003JE002154.

Bougher, S. W., McDunn, T., Murphy, J., et al. (2011) Coupling of Mars lower and upper atmosphere revisited: impacts of gravity wave momentum deposition on upper atmosphere structure, in *The Fourth International Workshop on the Mars Atmosphere: Modelling and Observation*, Paris, France.

Brinkman, R. T. (1971) Has nitrogen escaped? *Science*, 174, 944–945.

Carleton, N. P., and Traub, W. A. (1972) Detection of molecular oxygen on Mars, *Science*, 177, 988–992.

Carr, R. H., Grady, M. M., Wright, I. P., and Pillinger, C. T. (1985) Martian meteorite carbon dioxide and weathering products in SNC meteorites, *Nature*, 314, 248–250.

Cavalié, T., Billebaud, F., Encrenaz, T., et al. (2008) Vertical temperature profiles and mesospheric wind retrieval from millimeter observations. Comparison with general circulation predictions, *Astron. Astrophys.*, 89, 795–809.

Christensen, P. R., and Zurek, R. W. (1984) Martian north polar hazes and surface ice: results from the Viking survey/completion mission, *J. Geophys. Res.*, 89, 4587–4596.

Christensen, P. R., Bandfield, J. L., Hamilton, V. E., et al. (2001) Mars Global Surveyor Thermal Emission Spectrometer experiment: investigation description and surface science results, *J. Geophys. Res.*, 106, 23823–23871.

Christensen, P. R., Jakosky, B. M., Kieffer, H. H., et al. (2004) The Thermal Emission Imaging System (THEMIS) for the Mars 2001 Odyssey Mission, *Space Sci. Rev.*, 110, 85–130.

Clancy R. T., Muhleman, D. O., and Jakosky, B. M. (1983) Variability of carbon monoxide in the Mars atmosphere, *Icarus*, 55, 282–301.

Clancy, R. T., Muhlman, D. O., and Berge, G. L. (1990) Global changes in the 0–70 km thermal structure of the Mars atmosphere derived from 1975 to 1989 microwave CO spectra, *J. Geophys. Res.*, 95, 14543–14554.

Clancy, R. T., Sandor, B. J., and Moriarty-Schieven, G. H. (2004) Measurement of the 362 GHz absorption line of Mars atmospheric H_2O_2, *Icarus*, 168, 116–121.

Clancy, R. T., Wolff, M. J., Malin, M. C., and Cantor, B. A. (2010) MARs Color Imager (MARCI) daily global ozone column mapping from the Mars Reconnaissance Orbiter (MRO): a survey of 2006–2010 results, *American Geophysical Union, Fall Meeting 2010*.

Clancy, R. T., Sandor, B. J., Wolff, M. J., et al. (2012) Extensive MRO CRISM observations of 1.27 µm O_2 airglow in Mars polar night and their comparison to MRO MCS temperature profiles and LMD GCM simulations, *J. Geophys. Res.*, 117, E00J10, doi:10.1029/2011JE004018.

Clayton, R. N. and Mayeda, T. K. (1988) Isotopic composition of carbonate in EETA 79001 and its relation to parent body volatiles, *Geochim. Cosmochim. Acta*, 52, 925–927.

Colaprete, A., Barnes, J. R., Haberle, R. M., and Montmessin, F. (2008) CO_2 clouds, CAPE and convection on Mars: observations and general circulation modeling, *Planet. Space Sci.*, 56, 150–180.

Conrath, B. J. (1975) Thermal structure of the Martian atmosphere during the dissipation of the dust storm of 1971, *Icarus*, 24, 36–46.

Conrath, B. J. (1976) Influence of planetary-scale topography on the diurnal thermal tide during the 1971 Martian dust storm, *J. Atmos. Sci.*, 33, 2430–2439.

Conrath, B. J., Pearl, J. C., Smith, M. D., et al. (2000) Mars Global Surveyor Thermal Emission Spectrometer (TES) observations: atmospheric temperatures during aerobraking and science phasing, *J. Geophys. Res.*, 105, 9509–9519.

Davy, R., Davis, J. A., Taylor, P. A., et al. (2010) Initial analysis of air temperature and related data from the Phoenix MET station and their use in estimating turbulent heat fluxes, *J. Geophys. Res.*, 115, E00E13, doi:10.1029/2009JE003444.

Economou, T. E. (2008) Mars atmosphere argon density measurements on MER mission in *The Third International Workshop on the Mars Atmosphere: Modelling and Observations*, Williamsburg, Virginia.

Economou, T. E., and Pierrehumbert, R. T. (2010) Mars atmosphere argon density measurements on MER missions in *41st Lunar and Planetary Sci. Conf.*

Economou, T. E., Pierrehumbert, R., Banfield, D., and Landis, G. A. (2007) Mars atmosphere argon density measurement with the Alpha Particle X-Ray Spectrometer on MER missions in *Seventh International Conference on Mars*, Lunar and Plan. Inst., Houston, Texas.

Encrenaz, T. (2001) The atmosphere of Mars as constrained by remote sensing, *Space Sci. Rev.*, 96, 411–424.

Encrenaz, T., Lellouch, E., Rosenqvist, J., et al. (1991) The atmospheric composition of Mars: ISM and ground-based observational data, *Ann. Geophys.*, 9, 797–803.

Encrenaz, T., Lellouch, E., Paubert, G., and Gulkis, S. (1999) Mars, *IAU Circular* # 7168.

Encrenaz, T., Lellouch, E., Paubert, G., and Gulkis, S. (2001) The water vapor vertical distribution on Mars from millimeter transitions of H_2O and $H_2^{18}O$, *Planet. Space Sci.*, 49, 731–741.

Encrenaz, T., Bézard, B., Greathouse, T. K. et al. (2004) Hydrogen peroxide on Mars: evidence for spatial and seasonal variations, *Icarus*, 170, 424–429.

Encrenaz, T., Bézard, B., Owen, T., et al. (2005) Infrared imaging spectroscopy of Mars: H_2O mapping and determination of CO_2 isotopic ratios, *Icarus*, 179, 43–54.

Encrenaz, T., Fouchet, T., Melchiorri, R., et al. (2006) Seasonal variations of the Martian CO over Hellas as observed by OMEGA/ Mars Express, *Astron. Astrophys.*, 459, 265–270.

Encrenaz, T., Greathouse, T. K., Richter, M. J., et al. (2011) A stringent upper limit to SO_2 in the Martian atmosphere. *Astron. Astrophys.*, 530, A37, doi:10.1051/0004-6361/201116820.

England, C., and Hrubes, J. D. (2004) Molecular oxygen mixing ratio and its seasonal variability in the Martian atmosphere in *Workshop on Oxygen in the Terrestrial Planets*, Santa Fe, New Mexico.

Espenak, F., Mumma, M. J., Kostiuk, T., and Zipoy, D. (1991) Ground-based infrared measurements of the global distribution of ozone in the atmosphere of Mars, *Icarus*, 92, 252–262.

Farquhar, J., Thiemens, M. H., and Jackson, T. (1998) Atmosphere-surface interactions on Mars: delta17O measurements of carbonate from ALH 84001, *Science*, 280, 1580–1582.

Fast, K. E., Kostiuk, T., Hewagama, T., et al. (2006) Ozone abundance on Mars from infrared heterodyne spectra. I. Acquisition, retrieval, and anticorrelation with water vapor, *Icarus*, 181, 419–431.

Fast, K. E., Kostiuk, T., Lefèvre, F., et al. (2009) Comparison of HIPWAC and Mars Express SPICAM observations of ozone on Mars 2006–2008 and variations from 1993 IRHS observations, *Icarus*, 203, 20–27.

Fisher, D. A. (2007) Mars' water isotope (D/H) history in the strata of the north pole cap: inferences about the water cycle, *Icarus*, 187, 430–441.

Forbes, J. M. (2004) Tides in the middle and upper atmospheres of Mars and Venus, *Adv. Space Res.*, 33, 125–131.

Forbes, J. M. and Hagan, M. E. (2000) Diurnal Kelvin wave in the atmosphere of Mars: towards an understanding of "stationary" density structures observed by the MGS accelerometer, *Geophys. Res. Lett.*, 27, 3563–3566.

Forbes, J. M., Bridger, A. C., Hagan, M. E., et al. (2002) Non-migrating tides in the thermosphere of Mars, *J. Geophys. Res.*, 107, 5113, doi:10.1029/2001JE001582.

Forget, F., Hourdin, F., and Talagrand, O. (1998) CO_2 snowfall on Mars: simulation with a general circulation model, *Icarus*, 131, 302–316.

Forget, F., Hourdin, F., Fournier, R., et al. (1999) Improved general circulation models of the Martian atmosphere from the surface to above 80 km, *J. Geophys. Res.*, 104, 24155–24176.

Forget, F., Montabone, L., and Lebonnois, S. (2006a) Modelling the non-condensible gas enrichment in the polar night in *The Second International Workshop on the Mars Atmosphere: Modelling and Observations*, Granada, Spain.

Forget, F., Millour, E., Lebonnois, S., et al. (2006b) The new Mars climate database in *The Second International Workshop on the Mars Atmosphere: Modelling and Observations*, Granada, Spain.

Forget, F., Spiga, A., Dolla, B., et al. (2007) Remote sensing of surface pressure on Mars with the Mars Express/OMEGA spectrometer: 1. Retrieval method, *J. Geophys. Res.*, 112, E08S15, doi:10.1029/2006JE002871.

Forget, F., Millour, E., Montabone, L., and Lefèvre, F. (2008) Non condensable gas enrichment and depletion in the Martian polar regions in *The Third International Workshop on the Mars Atmosphere: Modelling and Observations*, Williamsburg, Virginia.

Forget, F., Montmessin, F., Bertaux, J.-L., et al. (2009) Density and temperatures of the upper Martian atmosphere measured by stellar occultations with Mars Expresss SPICAM, *J. Geophys. Res.*, 114, E01004, doi:10.1029/2008JE003086.

Formisano, V., Atreya, S., Encrenaz, T., et al. (2004) Detection of methane in the atmosphere of Mars, *Science*, 306, 1758–1761.

Formisano, V., Fonti, S., Giuranna, M., et al. (2005) The Planetary Fourier Spectrometer (PFS) onboard the European Mars Express Mission, *Planet. Space Sci.*, 53, 963–974.

Fouchet, T., and Lellouch, E. (2000) Vapor pressure isotope fractionation effects in planetary atmospheres: application to deuterium, *Icarus*, 144, 114–123.

Geminale, A. and Formisano, V. (2009) Study of the oxygen dayglow in Martian atmosphere with the Planetary Fourier Spectrometer on board Mars Express in *European Geophysical Union General Assembly*, Vienna.

Giuranna, M., Grassi, D., Formisano, V., et al. (2008) PFS/MEX observations of the condensing CO_2 south polar cap of Mars, *Icarus*, 197, 386–402.

González-Galindo, F., López-Valverde, M. A., Angelats i Coll, M., and Forget, F. (2005) Extension of a Martian general circulation model to thermospheric altitudes: UV heating and photochemical models, *J. Geophys. Res.*, 110, E09008, doi:10.1029/ 2004JE002312.

González-Galindo F., Forget, F., López-Valverde, M. A., et al. (2009) A Ground-to-Exosphere Martian General Circulation Model. 1. Seasonal, Diurnal and Solar Cycle Variation of Thermospheric Temperatures, *J. Geophys. Res.*, 114, E04001, doi:10.1029/ 2008JE003246.

Good, J. C. and Schloerb, F. P. (1981) Martian CO abundance from the $J = 1 \rightarrow 0$ rotational transition: evidence for temporal variations, *Icarus*, 47, 166–172.

Grassi, D., Fiorenza, C., Zasova, L. V., et al. (2005) The Martian atmosphere above great volcanoes: early planetary Fourier spectrometer observations, *Planet. Space Sci.*, 53, 1017–1034.

Grassi, D., Formisano, V., Forget, F. et al. (2007) The Martian atmosphere in the region of Hellas Basin as observed by the Planetary Fourier Spectrometer (PFS–MEX), *Planet. Space Sci.*, 55, 1346–1357.

Greenwood, J. P., Itoh, S., Sakamoto, N., et al. (2008) Hydrogen isotope evidence for loss of water from Mars through time, *Geophys. Res. Lett.*, 35, L05203, doi:10.1029/2007GL032721.

Guo, X., Richardson, M. I., and Newman, C. E. (2007) Non-condensible gas in a Mars general circulation model in *The Seventh International Conference on Mars*, Pasadena, CA.

Guzewich, S. D., Talaat, E. R., and Waugh, D. W. (2012), Observations of planetary waves and nonmigrating tides by the Mars Climate Sounder, *J. Geophys. Res.*, 117, E03010, doi:10.1029/ 2011JE003924.

Haberle, R. M., Pollack, J. B., Barnes, J. R., et al. (1993) Mars atmospheric dynamics as simulated by the NASA Ames General Circulation Model 1. The zonal-mean circulation, *J. Geophys. Res.*, 98, 3093–3123.

Haberle, R. M., Joshi, M. M., Murphy, J. R., et al. (1999) General circulation model simulations of the Mars Pathfinder atmospheric structure investigation/meteorology data, *J. Geophys. Res.*, 104, 8957–8974.

Haberle, R. M., Murphy, J. R., and Schaeffer, J. (2003) Orbital change experiments with a Mars general circulation model, *Icarus*, 161, 66–89

Haberle, R. M., Montmessin, F., Kahre, M. A., et al. (2011) Radiative effects of water ice clouds on the Martian seasonal water cycle in *The Fourth International Workshop on the Mars Atmosphere: Modelling and Observations*, Paris, France.

Haberle, R. M., Gómez-Elvira, J., de la Torre Juárez, M., et al. (2014) Preliminary interpretation of the REMS pressure data from the first 100 sols of the MSL mission, *J. Geophys. Res.*, 119, 440–453.

Hanel, R. A., Conrath, B. J., Hovis, W. A., et al. (1970) The Infrared Spectroscopy Experiment for Mariner Mars 1971, *Icarus*, 12, 48–62.

Hanel, R. A., Schlachman, B., Breihan, E., et al. (1972a) Mariner 9 Michelson interferometer, *Applied Optics*, 11, 2625–2634.

Hanel, R. A., Conrath, B. J., Hovis, W. A., et al. (1972b) Infrared Spectroscopy Experiment on the Mariner 9 mission: preliminary results, *Science*, 175, 305–308.

Hanel, R. A., Conrath, B. J., Hovis, W. A., et al. (1972c) Investigation of the Martian environment by infrared spectroscopy on Mariner 9, *Icarus*, 17, 423–442.

Hartogh, P., Jarchow, C., Lellouch, E., et al. (2010) Herschel/HIFI observations of Mars: first detection of O_2 at submillimetre wavelengths and upper limits on HCl and H_2O_2, *Astron. Astrophys.* 521, 49.

Heavens, N. G., Benson, J. L., Kass, D. M., et al. (2010) Water ice clouds over the Martian tropics during northern summer, *Geophys. Res. Lett.*, 37, L18202, doi:10.1029/2010GL044610.

Heavens, N. G., McCleese, D. J., Richardson, M. I., et al. (2011) Structure and dynamics of the Martian lower and middle atmosphere as observed by the Mars Climate Sounder: 2. Implications of the thermal structure and aerosol distributions for the mean meridional circulation, *J. Geophys. Res.*, 116, E01010, doi:10.1029/2010JE003713.

Hess, S. L., Henry, R. M., Leovy, C. B., et al. (1977) Meteorological results from the surface of Mars: Viking 1 and 2, *J. Geophys. Res.*, 82, 4559–4574.

Hinson, D. P. and Wilson, R. J. (2004) Temperature inversions, thermal tides, and water ice clouds in the Martian tropics. *J. Geophys. Res.*, 109, E01002, doi:10.1029/2003JE002129.

Hinson, D. P., Simpson, R. A., Twicken, J. D., et al. (1999) Initial results from radio occultation measurements with Mars Global Surveyor, *J. Geophys. Res.*, 104, 26997–27012.

Hinson, D. P., Smith, M. D., and Conrath, B. J. (2004) Comparison of atmospheric temperatures obtained through infrared sounding and radio occultation by Mars Global Surveyor, *J. Geophys. Res.*, 109, E12002, doi:10.1029/2004JE002344.

Jakosky, B. M., Pepin, R. O., Johnson, R. E., and Fox, J. L. (1994) Mars atmospheric loss and isotopic fractionation by solar-wind induced sputtering and photochemical escape, *Icarus*, 111, 271–281.

Justh, H. L., Justus, C. G., and Ramey, H. S. (2011) Mars-GRAM 2010: improving the precision of Mars-GRAM in *The Fourth International Workshop on the Mars Atmosphere: Modelling and Observations*, Paris, France.

Justus C. G., Duvall, A., and Keller, V. W. (2005) Mars aerocapture and validation of Mars- GRAM with TES data in *53rd JANNAF Propulsion Meeting*.

Kakar, R. K., Water, J. W., and Wilson, W. J. (1977) Mars: microwave detection of carbon monoxide, *Science*, 196, 1090–1091.

Kaplan, L. D., Münch, G., and Spinrad, H. (1964) An analysis of the spectrum of Mars, *Astophys. J.*, 139, 1–15.

Kaplan, L. D., Connes, J., and Connes, P. (1969) Carbon monoxide in the Martian atmosphere, *Astrophys. J.*, 157, L187–L192.

Karlsson, H. R., Clayton, R. N., Gibson, Jr., E. K., and Mayeda, T. K. (1992) Water in SNC meteorites: evidence for a Martian hydrosphere, *Science*, 255, 1409–1411.

Keating, G. M., Bougher, S. W., Zurek, R. W., et al. (1998) The structure of the upper atmosphere of Mars: in situ accelerometer measurements from Mars Global Surveyor, *Science*, 279, 1672–1676.

Keating, G. M., Theriot Jr., M., Tolson, R., et al. (2003) Global measurements of the Mars upper atmosphere: in situ accelerometer measurements from Mars Odyssey 2001 and Mars Global Surveyor, in *34th Annual Lunar and Planetary Science Conference*, League City, Texas.

Kieffer, H. H., Martin, T. Z., Peterfreund, A. R., et al. (1977) Thermal and albedo mapping of Mars during the Viking primary mission, *J. Geophys. Res.*, 82, 4249–4292.

Kleinböhl A., Schofield, J. T., Kass, D. M., et al. (2009) Mars Climate Sounder limb profile retrieval of atmospheric temperature, pressure, dust and water ice opacity, *J. Geophys. Res.*, 114, E10006, doi:10.1029/2009JE003358.

Kleinböhl, A., Schofield, J. T., Abdou, W. A., et al. (2011) A single-scattering approximation for infrared radiative transfer in limb geometry in the Martian atmosphere, *J. Quant. Spectrosc. Rad. Transfer*, 112, 1568–1580.

Kleinböhl, A., Wilson, R. J., Kass, D., et al. (2013), The semidiurnal tide in the middle atmosphere of Mars, *Geophys. Res. Lett.*, 40, 1952–1959, doi:10.1002/grl.50497.

Kliore, A. J., Cain, D. L., Levy, G. S., et al. (1965) Occultation experiment: results of the first direct measurement of Mars's atmosphere and ionosphere, *Science*, 149, 1243–1248.

Kliore, A. J., Fjeldbo, G., Seidel, B. L., et al. (1973) S band radio occultation measurements of the atmosphere and topography of Mars with Mariner 9: extended mission coverage of polar and intermediate latitudes, *J. Geophys. Res.*, 78, 4331–4351.

Korablev, O. I., Acherman, M., Krasnopolsky, V. A., et al. (1993) Tentative detection of formaldehyde in the Martian atmosphere, *Planet. Space Sci.*, 41, 441–451.

Krasnopolsky, V. A. (1986) *Photochemistry of the atmospheres of Mars and Venus*. Springer.

Krasnopolsky, V. A. (2003a) Spectroscopic mapping of Mars CO mixing ratio: detection of north–south asymmetry, *J. Geophys. Res.*, 108, 5010, doi:10.1029/2002JE001926.

Krasnopolsky, V. A. (2003b) Mapping of Mars O_2 1.27 mm dayglow at four seasonal points, *Icarus*, 165, 315–325.

Krasnopolsky, V. A. (2005) A sensitive search for SO_2 in the Martian atmosphere: implications for seepage and origin of methane, *Icarus*, 178, 487–492.

Krasnopolsky, V. A. and Feldman, P. D. (2001) Detection of molecular hydrogen in the atmosphere of Mars. *Science*, 294, 1914–1917.

Krasnopolsky, V. A., Mumma, M. J., Bjoraker, G. L., and Jennings, D. E. (1996) Oxygen and carbon isotope ratios in Martian carbon dioxide: measurements and implications for atmospheric evolution, *Icarus*, 124, 553–568.

Krasnopolsky, V. A., Bjoraker, G. L., Mumma, M. J., and Jennings, D. E. (1997) High-resolution spectroscopy of Mars at 3.7 and 8 mm: a sensitive search for H_2O_2, H_2CO, HCl and CH_4, and detection of HDO, *J. Geophys. Res.*, 102, 6525–6534.

Krasnopolsky, V. A., Mumma, M. J., and Gladstone, G. R. (1998) Detection of atomic deuterium in the upper atmosphere of Mars, *Science*, 280, 1576–1580.

Krasnopolsky, V. A., Maillard, J. P., and Owen, T. C. (2004) Detection of methane in the Martian atmosphere: evidence for life?, *Icarus*, 172, 537–547.

Ksanfomality, L. V., Moroz, V. I., Bibring, J.-P., et al. (1989) Spatial variations in thermal and albedo properties of the surface of Phobos, *Nature*, 341, 588–591.

Lebonnois, S., Quémerais, E., Montmessin, F., et al. (2006) Vertical distribution of ozone on Mars as measured by SPICAM/Mars Express using stellar occultations, *J. Geophys. Res.*, 111, E09S05, doi:10.1029/2005JE002643.

Lee, C., Lawson, W. G., Richardson, M. I., et al. (2009) Thermal tides in the Martian middle atmosphere as seen by the Mars Climate Sounder, *J. Geophys. Res.*, 114, E03005, doi:10.1029/2008JE003285.

Lefèvre, F. and Forget, F. (2009) Observed variations of methane on Mars unexplained by known atmospheric chemistry and physics, *Nature*, 460, 720–723.

Lefèvre, F., Lebonnois, S., Montmessin, F., and Forget, F. (2004) Three-dimensional modeling of ozone on Mars, *J. Geophys. Res.*, 109, E07004, doi:10.1029/2004JE002268.

Lefèvre, F., Bertaux, J.-L., Clancy, R. T., et al. (2008) Heterogeneous chemistry on Mars, *Nature*, 454, 971–975.

Lellouch, E., Paubert, G., and Encrenaz, T. (1991) Mapping of CO millimeter-wave lines in Mars atmosphere: the spatial distribution of carbon monoxide on Mars, *Planet. Space Sci.*, 39, 219–224.

Leovy, C. B. (1979) Martian meteorology, *Annu. Rev. Astron. Astrophys.*, 17, 387–413.

Leovy, C. B. and Zurek, R. W. (1979) Thermal tides and Martian dust storms: direct evidence for coupling, *J. Geophys. Res.*, 84, 2956–2968.

Lewis, S. R., Collins, M., Read, P. L., et al. (1999) A climate database for Mars, *J. Geophys. Res.*, 104, 24177–24194.

Lian, Y., Richardson, M. I., Newman, C. E., et al. (2012) The Ashima/MIT Mars GCM and argon in the Martian atmosphere, *Icarus*, 218, 1043–1070.

Lindal, G. F., Hotz, H. B., Sweetnam, D. N., et al. (1979) Viking radio occultation measurements of the atmosphere and topography of Mars – data acquired during 1 Martian year of tracking, *J. Geophys. Res.*, 84, 8443–8456.

Lindzen, R. S. (1970) The application and applicability of terrestrial atmospheric tidal theory to Venus and Mars, *J. Atmos. Sci.*, 27, 536–549.

Liu, J., Richardson, M. I., and Wilson, R. J. (2003) An assessment of the global, seasonal, and interannual spacecraft record of Martian climate in the thermal infrared, *J. Geophys. Res.*, 108, 5089, doi:10.1029/2002JE001921.

Madeleine, J.-B., Forget, F., Millour, E., et al. (2012) The influence of radiatively active water ice clouds on the Martian climate, *Geophys. Res. Lett.*, 39, L23202, doi:10.1029/2012GL053564.

Magalhães, J. A., Schofield, J. T., and Seiff, A. (1999) Results of the Mars Pathfinder atmospheric structure investigation, *J. Geophys. Res.*, 104, 8943–8956.

Maguire, W. C. (1977) Martian isotopic ratios and upper limits for possible minor constituents as derived from Mariner 9 infrared spectrometer data, *Icarus*, 32, 85–97.

Mahaffy, P. R., Webster, C. R., Atreya, S. K., et al. (2013) Abundance and isotopic composition of gases in the Martian atmosphere from the Curiosity Rover, *Science*, 341, 263–266.

Martin, T. Z. (1981) Mean thermal and albedo behavior of the Mars surface and atmosphere over a Martian year, *Icarus*, 45, 427–446.

Martin, T. Z. and Kieffer, H. H. (1979) Thermal infrared properties of the Martian atmosphere: 2. The 15 μm-band measurements, *J. Geophys. Res.*, 84, 2843–2852.

Martin, T. Z. and Murphy, J. R. (2003) Atmospheric wave structure derived from Mars Global Surveyor horizon sensor data, *International Workshop: Mars Atmosphere Modeling and Observations*, Grenada, Spain.

McCleese, D. J., Schofield, J. T., Taylor, F. W., et al. (2007) Mars Climate Sounder: an investigation of thermal and water vapor structure, dust and condensate distributions in the atmosphere, and energy balance of the polar regions, *J. Geophys. Res.*, 112, E05S06, doi:10.1029/2006JE002790.

McCleese D. J., Schofield, J. T., Taylor, F. W., et al. (2008) Intense polar temperature inversion in the middle atmosphere of Mars, *Nature Geosci.*, 1, 745–749, doi:10.1038ngeo332.

McCleese, D. J., Heavens, N. G., Schofield, J. T., et al. (2010) Structure and dynamics of the Martian lower and middle atmosphere as observed by the Mars Climate Sounder: seasonal variations in zonal mean temperature, dust, and water ice aerosol, *J. Geophys. Res.*, 115, E12016, doi:10.1029/2010JE003677.

McDunn, T., Bougher, S. W., Murphy, J., et al. (2010) Simulating the density and thermal structure of the middle atmosphere (~80–130 km) of Mars using the MGCM–MTGCM: a comparison with MEX-SPICAM observations, *Icarus*, 206, 5–17.

McDunn, T., Bougher, S. W., Murphy, J., et al. (2013) Characterization of middle-atmosphere polar warming at Mars, *J. Geophys. Res.*, 118, 161–178, doi:10.1002/jgre.20016.

McElroy, M. B. (1972) Mars: an evolving atmosphere, *Science*, 175, 443–445.

McElroy, M. B. and Donahue, T. M. (1972) Stability of the Martian atmosphere, *Science*, 177, 986–988.

McElroy, M. B., Yung, Y. L., and Nier, A. O. (1976) Isotopic composition of nitrogen: implications for the past history of Mars' atmosphere, *Science*, 194, 70–72.

Medvedev, A., Yigit, E., and Hartogh, P. (2011) Effects of gravity wave drag in the Martian atmosphere: simulations with a GCM in *The Fourth International Workshop on the Mars Atmosphere: Modelling and Observations*, Paris, France.

Melosh, H. J. and Vickery, A. M. (1989) Impact erosion of the primordial atmosphere of Mars, *Nature*, 338, 487–489.

Millour, E., Forget, F., Spiga, A., et al. (2011) An improved Mars climate database in *The Fourth International Workshop on the Mars Atmosphere: Modelling and Observations*, Paris, France.

Montmessin, F., Forget, F., Rannou, P., et al. (2004) The origin and role of water ice clouds in the Martian water cycle as inferred from a General Circulation Model, *J. Geophys. Res.*, 109, E10004, doi:10.1029/2004JE002284.

Montmessin, F., Fouchet, T., and Forget, F. (2005) Modeling the annual cycle of HDO in the Martian atmosphere, *J. Geophys. Res.*, 110, E03006, doi:10.1029/2004JE002357.

Montmessin, F., Bertaux, J.-L., Quémerais, E., et al. (2006) Sub-visible CO_2 ice clouds detected in the mesosphere of Mars, *Icarus*, 183, 403–410.

Moreno, R., Lellouch, E., Forget, F., et al. (2009) Wind measurements in Mars' middle atmosphere: IRAM Plateau de Bure interferometric CO observations, *Icarus*, 201, 549–563.

Mumma, M. J., Novak, R. E., DiSanti, M. A., et al. (2003) Seasonal mapping of HDO and H_2O in the Martian atmosphere in *Sixth International Conference on Mars*, Houston, Texas.

Mumma, M. J., Novak, R. E., DiSanti, M. A., et al. (2004) Detection and mapping of methane and water on Mars, *Bull. Amer. Astron. Soc.*, 36, 1127.

Mumma, M. J., Villanueva, G. L., Novak, R. E., et al. (2009) Strong release of methane on Mars in Northern Summer 2003, *Science*, 323, 1041–1045.

Nelli, S. M., Murphy, J. R., Sprague, A. L., et al. (2007) Dissecting the polar dichotomy of the noncondensable gas enhancement on Mars using the NASA Ames Mars General Circulation Model, *J. Geophys. Res.*, 112, E08S91, doi:10.1029/2006JE002849.

Nier, A. O. and McElroy, M. B. (1977) Composition and structure of Mars' upper atmosphere : results from the neutral mass spectrometers of Viking 1 and 2, *J. Geophys. Res.*, 82, 4341–4349.

Nier, A. O., Hanson, W. B., Seiff, A., et al. (1976) Composition and structure of the Martian atmosphere: preliminary results from Viking 1, *Science*, 193, 786–788.

Novak, R. E., Mumma, M. J., Villanueva, G., et al. (2007) Seasonal mapping of HDO/H₂O in the Martian atmosphere, *Bull. Amer. Astron. Soc.*, 39, 45.

Noxon, J. F., Traub, W. A., Carleton, N. P., and Connes, P. (1976) Detection of O₂ dayglow emission from Mars and the Martian ozone abundance, *Astrophys. J.*, 207, 1025–1035.

Owen, T. (1992) The composition and early history of the atmosphere of Mars, in *Mars* (Kieffer, H. H., Jakosky, B. M., Snyder, C. W., Matthews, M. S., Eds.), University of Arizona Press, Tucson.

Owen, T. and Biemann, K. (1976) Composition of the atmosphere at the surface of Mars – detection of argon-36 and preliminary analysis, *Science*, 193, 801–803.

Owen, T. and Kuiper, G. P. (1964) A determination of the composition and surface pressure of the Martian atmosphere, *Commun. Lunar and Planetary Lab*, 2, 113–132.

Owen, T. and Sagan, S. (1972) Minor constituents in planetary atmospheres: ultraviolet spectroscopy from the Orbiting Astronomical Observatory, *Icarus*, 16, 557–568.

Owen, T., Biemann, K., Rushneck, D. R., et al. (1977) The composition of the atmosphere at the surface of Mars, *J. Geophys. Res.*, 82, 4635–4639.

Owen, T., Maillard, J.-P., de Bergh, C., and Lutz, B. L. (1988) Deuterium on Mars: the abundance of HDO and the value of D/H, *Science*, 240, 1767–1770.

Oyama, V. I. and Berdahl, B. J. (1977) The Viking gas exchange experiment results from Chryse and Utopia surface samples, *J. Geophys. Res.*, 82, 4669–4676.

Paige, D. A. and Wood, S. E. (1992) Modeling the Martian seasonal CO₂ cycle, *Icarus*, 99, 15–27.

Parkinson, T. D. and Hunten, D. M. (1972) Spectroscopy and aeronomy of O₂ on Mars, *J. Atmos. Sci.*, 29, 1380–1390.

Pätzold, M., Tellmann, S., Peter, K., et al. (2009) The structure of the lower Mars ionosphere, *European Planetary Science Congress*, Potsdam, Germany.

Pearl, J. C., Smith, M. D., Conrath, B. J., et al. (2001) Observations of Martian ice clouds by the Mars Global Surveyor Thermal Emission Spectrometer: the first Martian year, *J. Geophys. Res.*, 106, 12325–12338.

Perrier, S., Bertaux, J.-L., Lefèvre, F., et al. (2006) Global distribution of total ozone on Mars from SPICAM/Mars Express UV measurements, *J. Geophys. Res.*, 111, E09S06, doi:10.1029/2006JE002681.

Rondanelli, R., Thayalan, V., Lindzen, R. S., and Zuber, M. T. (2006) Atmospheric contribution to the dissipation of the gravitational tide of Phobos on Mars, *Geophys. Res. Lett.*, 33, L15201, doi:10.1029/2006GL026222.

Schofield, J. T., Barnes, J. R., Crisp, D., et al. (1997) The Mars Pathfinder Atmospheric Structure Investigation/Meteorology, *Science*, 278, 1752–1757.

Seiff, A. (1982) Post-Viking models for the structure of the summer atmosphere of Mars, *Adv. Space Res.*, 2, 3–17.

Seiff, A. and Kirk, D. B. (1977) Structure of the atmosphere of Mars in summer at mid-latitudes, *J. Geophys. Res.*, 82, 4364–4378.

Smith, M. D. (2004) Interannual variability in TES atmospheric observations of Mars during 1999–2003, *Icarus*, 167, 148–165.

Smith, M. D. (2008) Spacecraft observations of the Martian atmosphere, *Annu. Rev. Earth Planet. Sci.*, 36, 191–219.

Smith, M. D. (2009) THEMIS observations of Mars aerosol optical depth from 2002–2008, *Icarus*, 202, 444–452.

Smith, M. D., Pearl, J. C., Conrath, B. J., and Christensen, P. R. (2001a) Thermal Emission Spectrometer results: Mars atmospheric thermal structure and aerosol distribution, *J. Geophys. Res.*, 106, 23929–23945, 2001.

Smith, M. D., Pearl, J. C., Conrath, B. J., and Christensen, P. R. (2001b) One Martian year of atmospheric observations by the Thermal Emission Spectrometer, *Geophys. Res. Let.*, 28, 4263–4266.

Smith, D. E., Zuber, M. T., Frey, H. V., et al. (2001c) Mars Orbiter Laser Altimeter: experiment summary after the first year of global mapping of Mars, *J. Geophys. Res.*, 106, 23689–23722.

Smith, M. D., Conrath, B. J., Pearl, J. C., and Christensen, P. R. (2002) Thermal Emission Spectrometer observations of Martian planet-encircling dust storm 2001A, *Icarus*, 157, 259–263.

Smith, M. D., Wolff, M. J., Lemmon, M. T., et al. (2004) First atmospheric science results from the Mars Exploration Rovers Mini-TES, *Science*, 306, 1750–1753.

Smith, M. D., Wolff, M. J., Spanovich, N., et al. (2006) One Martian year of atmospheric observations using MER Mini-TES, *J. Geophys. Res.*, 111, E12S13, doi:10.1029/2006JE002770.

Smith, M. D., Wolff, M. J., Clancy, R. T., and Murchie, S. L. (2009) Compact Reconnaissance Imaging Spectrometer observations of water vapor and carbon monoxide, *J. Geophys. Res.*, 114, E00D03, doi:10.1029/2008JE003288.

Snyder, C. W. (1977) The missions of the Viking Orbiters, *J. Geophys. Res.*, 82, 3971–3983.

Snyder, C. W. (1979) The extended mission of Viking, *J. Geophys. Res.*, 84, 7917–7933.

Spanovich, N., Smith, M. D., Smith, P. H., et al. (2006) Surface and near-surface atmospheric temperatures from the Mars Exploration Rover landing sites, *Icarus*, 180, 314–320.

Sprague, A. L., Boynton, W. V., Kerry, K. E., et al. (2004) Mars' south polar Ar enhancement: a tracer for south polar seasonal meridional mixing, *Science*, 306, 1364–1367.

Sprague, A. L., Boynton, W. V., Kerry, K. E., et al. (2007) Mars' atmospheric argon: tracer for understanding Martian atmospheric circulation and dynamics, *J. Geophys. Res.*, 112, E03S02, doi:10.1029/2005JE002597.

Striepe, S. A., Way, D. W., Dwyer, A. M., and Balaram, J. (2002) Mars smart lander simulations for entry, descent, and landing in *AIAA Atmospheric Flight Mechanics Conference and Exhibit*.

Sutton, J. L., Leovy, C. B., and Tillman, J. E. (1978) Diurnal variation of the Martian surface layer meteorological parameters during the first 45 sols at two Viking Lander sites, *J. Atmos. Sci.*, 35, 2346–2355.

Taylor, P. A., Kahanpää, H., Weng, W., et al. (2010) On pressure measurement and seasonal pressure variations during the Phoenix mission, *J. Geophys. Res.*, 115, E00E15, doi:10.1029/2009JE003422.

Theriot, M., Keating, G., Blanchard, R., et al. (2006) Inter-annual comparison of temporal and spatial structure in the Martian thermosphere from atmospheric accelerometer measurements of MRO during aerobraking and stellar occultation measurements from SPICAM ultraviolet and infrared atmospheric spectrometer of Mars Express (MEX). *American Astronomical Society*, DPS meeting #38, Abstract #73.02.

Tillman, J. E. (1988) Mars global atmospheric oscillations: annually synchronized, transient normal mode oscillations and the triggering of global dust storms, *J. Geophys. Res.*, 93, 9433–9451.

Tillman, J. E., Johnson, N. C., Guttorp, P., and Percival, D. B. (1993) The Martian annual atmospheric pressure cycle – years without great dust storms, *J. Geophys. Res.*, 104, 8987–9008.

Toigo, A. D., Smith, M. D., Seelos, F. P., and Murchie, S. L. (2013) High spatial and temporal resolution sampling of Martian gas abundances from CRISM spectra, *J. Geophys. Res.*, 118, 89–104, doi:10.1029/2012JE004147.

Tolson, R. H., Keating, G. M., Zurek, R. W., et al. (2007) Application of accelerometer data to atmospheric modeling during Mars aerobraking operations, *J. Spacecraft and Rockets*, 44, 1172–1179.

Traub, W. A., Carleton, N. P., Connes, P., and Noxon, J. F. (1979) The latitude variation of O_2 dayglow and O_3 abundances on Mars, *Astrophys. J.*, 229, 846–850.

Trauger, J. T. and Lunine, J. I. (1983) Spectroscopy of molecular oxygen in the atmospheres of Venus and Mars, *Icarus*, 55, 272–281.

Turcotte, D. L. and Schubert, G. (1988) Tectonic implications of radiogenic gases in planetary atmospheres, *Icarus*, 74, 36–46.

Villanueva, G. L., Mumma, M. J., Novak, R. E., et al. (2013) A sensitive search for organics (CH_4, CH_3OH, H_2CO, C_2H_6, C_2H_2, C_2H_4), hydroperoxyl (HO_2), nitrogen compounds (N_2O, NH_3, HCN) and chlorine species (HCl, CH_3Cl) on Mars using ground-based high-resolution infrared spectroscopy, *Icarus*, 223, 11–27.

Webster, C. R., Mahaffy, P. R., Flesch, G. J., et al. (2013a) Isotope ratios of H, C, and O in CO_2 and H_2O in the Martian atmosphere, *Science*, 341, 260–263.

Webster, C. R., Mahaffy, P. R., Atreya, S. K., et al. (2013b) Low upper limit to methane abundance on Mars, *Science*, doi:10.1126/science.1242902.

Webster, C. R., Mahaffy, P. R., Atreya, S. K., et al. (2015) Mars methane detection and variability at Gale Crater, *Science*, 347, 415–417.

Wilson, R. J. (1997) A general circulation model simulation of the Martian polar warming, *Geophys. Res. Lett.*, 24, 123–127.

Wilson, R. J. (2000) Evidence for diurnal period Kelvin waves in the Martian atmosphere from Mars Global Surveyor TES data, *Geophys. Res. Lett.*, 27, 3889–3892.

Wilson, R. J. (2002) Evidence for non-migrating thermal tides in the Mars upper atmosphere from the Mars Global Surveyor Accelerometer Experiment, *Geophys. Res. Lett.*, 29, 1120, doi:10.1029/2001GL013975.

Wilson, R. J. (2011) Water ice clouds and thermal structure in the Martian tropics as revealed by Mars Climate Sounder in *The Fourth International Workshop on the Mars Atmosphere: Modelling and Observations*, Paris, France.

Wilson, R. J. and Hamilton, K. P. (1996) Comprehensive model simulation of thermal tides in the Martian atmosphere, *J. Atmos. Sci.* 53, 1290–1326.

Wilson, R. J. and Richardson, M. I. (2000) The Martian atmosphere during the Viking mission, I – Infrared measurements of atmospheric temperatures revisited, *Icarus*, 145, 555–579.

Wilson, R. J., and Smith, M. D. (2006) The effects of atmospheric dust on the seasonal variation of Martian surface temperature in *The Second International Workshop on the Mars Atmosphere: Modelling and Observations*, Granada, Spain.

Wilson, R. J., Neumann, G. A., and Smith, M. D. (2007) Diurnal variation and radiative influence of Martian water ice clouds, *Geophys. Res. Lett.*, 34, L02710, doi:10.1029/2006GL027976.

Withers, P. and Catling, D. C. (2010) Observations of atmospheric tides on Mars at the season and latitude of the Phoenix atmospheric entry, *Geophys. Res. Lett.*, 37, L24204, doi:10.1029/2010GL045382.

Withers, P. and Smith, M. D. (2006) Atmospheric entry profiles from the Mars Exploration Rovers Spirit and Opportunity, *Icarus*, 185, 133–142.

Withers, P., Bougher, S. W., and Keating, G. M. (2003) The effects of topography-controlled thermal tides in the Martian upper atmosphere as seen by the MGS accelerometer, *Icarus*, 164, 14–32.

Wolkenberg, P., Formisano, V., Rinaldi, G., and Geminale, A. (2010) The atmospheric temperatures over Olympus Mons on Mars: an atmospheric hot ring, *Icarus*, 207, 110–123.

Wright, I. P., Grady, M. M., and Pillinger, C. T. (1988) Carbon, oxygen and nitrogen isotopic compositions of possible Martian weathering products in EETA 79001, *Geochim. Cosmochim. Acta.*, 52, 917–924.

Young, L. D. G. and Young, A. T. (1977) Interpretation of high-resolution spectra of Mars IV. New calculations of the CO abundance, *Icarus*, 30, 75–79.

Zahnle, K., Freedman, R. S., and Catling, D. C. (2011) Is there methane on Mars? *Icarus*, 212, 493–503.

Zurek, R. W. (1976) Diurnal tide in the Martian atmosphere, *J. Atmos. Sci.*, 33, 321–337.

Zurek, R. W. (1981) Inference of dust opacities for the 1977 Martian great dust storms from Viking Lander 1 pressure data, *Icarus*, 45, 202–215.

Zurek, R. W. (1992) Comparative aspects of the climate of Mars: an introduction to the current atmosphere, in *Mars* (Kieffer, H. H., Jakosky, B. M., Snyder, C. W., Matthews, M. S., Eds.), University of Arizona Press, Tucson.

Zurek, R. W., Barnes, J. R., Haberle, R. M., et al. (1992) Dynamics of the atmosphere of Mars, in *Mars* (Kieffer, H. H., Jakosky, B. M., Snyder, C. W., Matthews, M. S., Eds.), University of Arizona Press, Tucson.

Mars Clouds

R. TODD CLANCY, FRANCK MONTMESSIN, JENNIFER BENSON, FRANK DAERDEN,
ANTHONY COLAPRETE, MICHAEL J. WOLFF

5.1 INTRODUCTION

Early ground-based observations of "blue clearings" and "polar hoods" suggested volatile cloud aerosols distinct from the yellow encircling dust hazes observed to periodically obscure the Mars surface (Dollfus, 1957; Slipher, 1962); for discussion of earlier Mars observations, see Kieffer et al. (1992). Spacecraft imaging observations from the Mariner 6 and 7 flybys in 1969 observed polar and detached limb hazes, but their interpretation in terms of atmospheric water ice clouds was ambiguous (Leovy et al., 1971). The first definitive measurements of Mars water ice cloud composition were obtained from the Mariner 9 Orbiter, employing thermal infrared spectroscopy. Analysis of IRIS (Infrared Interferometer Spectrometer) spectra (Hanel et al., 1972), associated with clouds over Arsia Mons, indicated water ice cloud particle radii of order 1–2 μm (Curran et al., 1973). Mariner 9 imaging observations of water ice clouds were also obtained late in the mission during the declining phase of one of the most intense Mars global dust storms ever recorded. These images captured many of the key characteristics of large-scale cloud behavior in the Mars atmosphere, including Tharsis ridge and Hellas Basin clouds, polar and high-altitude hazes, and waveform clouds (Leovy et al., 1973; Briggs and Leovy, 1974; Anderson and Leovy, 1978).

Our understanding of the full extent of cloud physical forms and their topographic and dynamical contexts was greatly expanded by color imaging conducted from the two Viking Orbiters between 1976 and 1982 (French et al., 1981; Kahn, 1984). The extensive Viking imaging data discriminated dust and ice aerosols, and provided limb views that revealed seasonal and spatial variability of cloud-top altitudes (Jaquin et al., 1986). Atmospheric water vapor column measurements by the Viking MAWD (Mars Atmospheric Water Detector) experiment (Farmer et al., 1977) provided a full seasonal climatology of atmospheric water vapor (Jakosky and Farmer, 1977), which did not obviously correlate with the presence of water ice clouds in the Mars atmosphere (e.g. French et al., 1981). Inaccurate temperature profile characterizations drawn from the Viking Orbiter IRTM (Infrared Thermal Mapper) experiment (Kieffer et al., 1972) led to a mischaracterization of vapor saturation conditions in the Mars atmosphere (Wilson and Richardson, 2000). Consequently, the global nature of the aphelion cloud belt (ACB; Figure 5.1) and key transport, radiative, and photochemical roles played by Mars clouds were belatedly recognized from Earth-based observations (Clancy et al., 1996; Clancy and Nair, 1996) and subsequently recognized in Viking IRTM measurements (Tamppari et al., 2000, 2003). Implied cloud-related processes impact the global and temporal variations of Mars atmospheric temperatures, aerosols, water vapor, photochemistry, meridional/vertical transport, and climate evolution in respects that continue to stimulate new modeling efforts (Michelangeli et al., 1993; Richardson et al., 2002; Hinson and Wilson, 2004; Montmessin et al., 2004, 2007a; Lefèvre et al., 2004, 2008; Michaels et al., 2006; Wilson et al., 2007, 2008; Nelli et al., 2010; Daerden et al., 2010; Haberle et al., 2011; Madeleine et al., 2012a; Wilson, 2011a,b; Rodin et al., 2011; Clancy et al., 2012; Spiga et al., 2012; Madeleine et al., 2014; Yiğit et al., 2015).

The resurgence of Mars spacecraft observations with the Phobos, Pathfinder (MPF), and Mars Global Surveyor (MGS) missions in the late 1990s began a new era of Mars exploration that continues to this day (Mars Odyssey (ODY), Mars Exploration Rovers (MER), Mars Express (MEX), Phoenix (PHX), Mars Reconnaissance Orbiter (MRO), Mars Science Laboratory (MSL), the Mars Atmosphere and Volatile Evolution mission (MAVEN), and Mars Orbiter Mission (MOM)). Mars atmospheric dust and ice aerosol studies have particularly benefited from a variety of spacecraft observations over this period, including imaging, limb sounding, occultation (solar and stellar), lidar, and emission phase function (EPF) measurements from ultraviolet to infrared wavelengths.

MGS TES (Thermal Emission Spectrometer) water vapor, cloud, dust column mapping, and temperature profiling (Christensen et al., 1998) definitively characterized the seasonal and spatial extent of the aphelion and polar hood cloud systems, in terms of vertical saturation conditions and aerosol content. ODY THEMIS (Thermal Emission Imaging System) observations (Christensen et al., 2003) have extended interannual coverage of the aphelion cloud belt (Smith, 2009). MGS MOLA (Mars Orbiter Laser Altimeter) observations (Smith et al., 2001) provided our first (inferred) detections of polar CO_2 ice clouds and CO_2 precipitation, including their distinctive vertical, spatial, and seasonal variations. TES solar-band and infrared EPF observations indicated distinct spatial–seasonal variations in Mars cloud scattering properties and particle sizes. MGS MOC (Mars Orbiter Camera) cloud and dust daily global imaging (Malin et al., 1992) revealed seasonally repeatable formation of cloud forms at polar latitudes and orographic cloud forms over the Mars volcanoes. TES and MOC limb observations, and nadir observations by THEMIS, MEX HRSC (High Resolution Stereo Camera) (Neukum et al., 2004), and MRO CRISM (Compact Reconnaissance Imaging Spectrometer for Mars) (Murchie et al., 2007) have supported detections of equatorial high-altitude clouds (above 60 km), determined to be CO_2 in composition from MEX OMEGA (Observatoire pour la Minéralogie, l'Eau, les Glaces et l'Activité) (Bibring et al.,

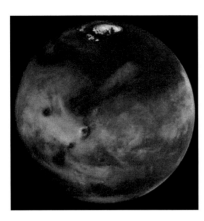

Figure 5.1. Hubble Space Telescope imaging of Mars was obtained during its 1997 opposition (James et al., Space Telescope Science Institute, Press release). The distinctive aphelion cloud belt (ACB), composed of water ice clouds, encompasses the northern subtropics of Mars around the cold aphelion (northern summer) portion of the Mars orbit. A black and white version of this figure will appear in some formats. For the color version, please refer to the plate section.

2004) near-IR spectra (and perhaps viewed from Pathfinder early morning sky imaging in 1998). The MEX SPICAM (Spectroscopy for the Investigation of the Characteristics of the Atmosphere of Mars) experiment (Bertaux et al., 2000) has returned observations of upper altitude hazes, and the global response of middle atmospheric ozone abundances to the orbital variation of the global hygropause level. MRO MARCI (Mars Color Imager) observations (Malin et al., 2008) have identified perihelion cloud trails, and extended daily global imaging of cloud and dust activity from 1998 (MOC) to current times. MRO MCS (Mars Climate Sounder) (McCleese et al., 2007) provides the first dedicated thermal infrared limb sounding dataset for Mars, generating vertical profiles for clouds, dust, and temperatures from the surface up to 80 km with both global and diurnal (3 a.m., 3 p.m.) coverage. The Pathfinder, MER, Phoenix, and MSL Landers imaged cirrus cloud forms drifting across the sky, and Phoenix LIDAR (Whiteway et al., 2008) observations revealed precipitation of large water ice particles under 7–10 km altitudes at high northern latitudes in late summer.

Whereas the above discussion focuses on mission and experimental developments, the main body of this chapter emphasizes the numerous observational and modeling results (with references) pertaining to Mars clouds. This presentation incorporates a set of sections regarding the morphologies, temporal/spatial distributions, processes, and properties of water ice clouds. Mars water ice clouds present a large range of morphologies, including their horizontal aspects and spatial scales. Mars water ice clouds exhibit distinctive global and seasonal (including orbital) variations that reflect variations in atmospheric thermal structure, surface topography, and surface water vapor supply (from the polar cap and perhaps subsurface water ice reservoirs). Mars water ice clouds play fundamental roles in the radiative (Chapter 6), volatile transport (Chapter 11), and photochemical (Chapter 13) characters of the global Mars atmosphere. Their description constitutes the primary focus of this chapter. A separate treatment

of Mars CO_2 ice clouds concludes the chapter, as CO_2 clouds reflect a narrower environmental regime on Mars and less is known about their specific properties and processes. The fundamental importance of CO_2 clouds regards the coupled surface–atmosphere CO_2 cycle of Mars. This remarkable CO_2 global cycle forces key Mars atmospheric behaviors; including large (±15%) variations in Mars surface pressure, polar and high-latitude variations in atmospheric composition associated with non-condensable gases (such as Ar and CO), and distinctive polar (and perhaps equatorial mesospheric) dynamics and transport. Chapter 12 presents these behaviors in considerable detail.

5.2 SPECIFIC CLOUD MORPHOLOGIES AND OCCURRENCES

Mars atmospheric conditions of very low temperatures and pressures lead to the formation of water ice versus liquid cloud particles. Similarly (and in contrast to terrestrial atmospheric conditions), latent heat effects of water cloud formation are minor in terms of convective forcing on Mars, which is dominated by specific meteorological conditions or solar heating by the surface and lifted dust (Chapter 9). Furthermore, Mars water ice clouds are often associated with regional-to-global circulation and saturation conditions, as opposed to local convective overturning, and so typically exhibit cirrus rather than cumulus forms. (Terrestrial cirrus clouds have been distinguished in terms of their ice composition as well as morphological characteristics (e.g. Sassen, 2002). As all Mars clouds are ice, here we distinguish cirrus clouds as the collection of thin, fibrous, sheet, and wavy forms distinct from puffy, towering, cumulus forms.) Notable cumuliform exceptions include cloud streets (Figure 5.2) and frontal clouds associated with strong thermal gradients over the summer polar caps or high surface elevations

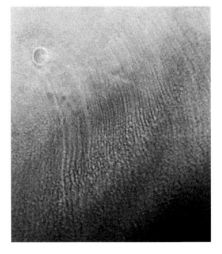

Figure 5.2. Condensate cloud streets progress to the south, off the seasonal north polar edge at 64.5°N, 243.0°W, during the mid-afternoon local time on February 7, 2008 ($L_s = 28.5°$). Revised figure from Cantor et al. 2010 (MARCI band 1 image, P16_007185_0285_MA_00N234W_080207_1.ddd). Provided by Bruce Cantor, Malin Space Science Systems.

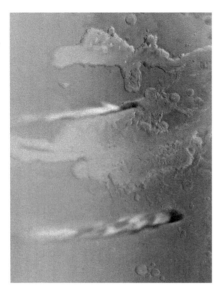

Figure 5.3. Characteristic Valles Marineris locations for perihelion cloud trails on Mars. These fine-particle water ice clouds align with the predominantly zonal circulation in the 40–60 km region of the Mars atmosphere, obtaining horizontal extents of >400 km. Reproduced from Clancy et al. (2009).

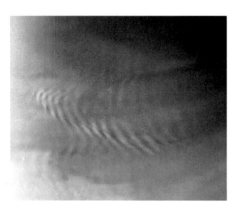

Figure 5.4. Lee-wave clouds over a crater and extending to the east in western Utopia (MARCI image, B05_011670_1956_MA_00N291W_090121_5.ddd). Provided by Bruce Cantor, Malin Space Science Systems.

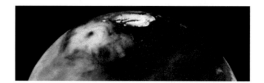

Figure 5.5. Hubble Space Telescope violet imaging obtained during the 1999 opposition of Mars, presents a distinctive spiral storm (Bell et al., Space Telescope Science Institute, News Release Number STScI-1999–22). Water ice clouds extend over ~1000 km scales in morning hours over high latitudes (65°N) in northern spring ($L_s = 63°$).

Figure 5.6. Cirrus CO_2 ice clouds appear at equatorial (0.2°S, 95.9°W) mesospheric altitudes (70–75 km), as imaged by the MEX HRSC (blue band). From Scholten et al. (2010). The east–west extent of this northern early spring ($L_s = 13.4°$), afternoon (local time, LT = 4:18 p.m.) image spans 50 km.

(e.g. Kahn, 1984; Cantor et al., 2010), and perihelion (notably Valles Marineris) cloud trails associated with strong localized updrafts under peak surface heating conditions (Figure 5.3). Convective cloud forms may also locally characterize CO_2 condensation in the polar night (Colaprete et al., 2008). A wide range of morphologies, dimensions, and altitudes exist for Mars clouds (Table 5.1). The spatial and seasonal range of cloud morphologies such as lee wave (Figure 5.4), gravity wave, plume, streak, street (Figure 5.2), and spiral (Figure 5.5) clouds, as well as formless cloud hazes and fogs, were cataloged from Viking (Gierasch et al., 1979; French et al., 1981; Kahn, 1984) and MGS MOC (Cantor et al., 2002; Wang and Ingersoll, 2002) imaging.

The expanded wavelength and spatial sampling capabilities of post-Viking observations reveal additional cloud types, often associated with high altitudes. These include perihelion cloud trails at 40–50 km altitudes (Clancy et al., 2009; Figure 5.3), equatorial mesospheric (60–80 km) clouds of CO_2 composition (Figure 5.6; the basis of many observational studies, see Table 5.1 and Section 5.7), and trough clouds over the polar residual caps (argued to play a key role in the evolution of polar cap morphology (I.B. Smith et al., 2013). Non-equatorial mesospheric haze layers have also been reported, as possibly CO_2 (Montmessin et al., 2006a; Määttänen et al., 2010), as well as water ice composition (e.g. Chassefière et al., 1992; Clancy et al., 2009; Fedorova et al., 2009; McConnochie et al., 2010; Vincendon et al., 2011; also Viking limb observations, Jaquin et al., 1986). Extraordinary views of Mars clouds from the surface have been provided from the Pathfinder, MER, and Phoenix Landers. In the case of MER (Lemmon et al., 2015) and Pathfinder sky images, characteristic cirrus morphologies are quite strikingly displayed (Figure 5.7) with measurable wind entrainment velocities. Phoenix LIDAR observations

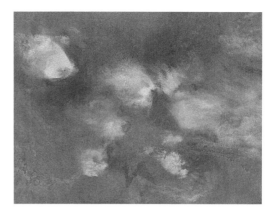

Figure 5.9 Typical afternoon clouds over Tharsis Montes run diagonally from upper center toward lower left; Ascraeus Mons, Pavonis Mons, and Arsia Mons. The simple cylindrical map projected mosaic shows MOC-WA images (M00-01510, M00-01531, M00-01548, M00-01558), from April 10, 1999 during northern summer (L_s = 122.3°). Provided by Bruce Cantor, Malin Space Science Systems.

Figure 5.7. Cirrus clouds are imaged from the Mars surface by the Opportunity Pancam (MERB, sol 269, October 26, 2004, LT = 9:23), with the rim of the Endurance Crater appearing at the bottom of the image. These clouds are part of the ACB, based upon the season (L_s = 106°) and location (2°S, 6°W) of the observation.

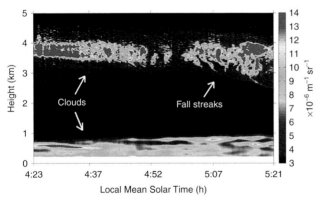

Figure 5.10. Morning (LT = 9:20 a.m.) fog fills the Coprates Chasma (14.2°S, 58°W) within Valles Marineris, as imaged by the MEX/HRSC in Mars northern spring (L_s = 38°, May 25, 2004). Modified from a color figure in Möhlmann et al. (2009).

Figure 5.8. Contour plot of backscatter coefficient derived from the Phoenix LIDAR backscatter signal at wavelength 532 nm in the early morning hours of mission sol 99, at L_s = 122° (Whiteway et al., 2009). A near-surface fog and a cloud layer at an altitude of 4 km appear over the Phoenix Lander at this time. Strikingly, fall streaks indicate gravitational precipitation of ice particles from the top cloud layer. Reproduced from Daerden et al. (2010).

(Whiteway et al., 2009; Dickinson et al., 2010) reveal ground fogs and classic fall streak clouds, in high northern latitude late summer at relatively low altitudes, ~4 km (Figure 5.8).

Upper-level clouds are also apparent with reduced clarity in MSL imaging within Gale Crater (Moores et al., 2015).

Discrete Mars clouds, such as lee waves, are often generated in association with specific topographic features on Mars, including the Tharsis volcanoes (Figure 5.9), crater rims (Figure 5.4), and polar spiral troughs. Wave cloud patterns are also observed in the absence of identifiable topographic forcing, and have been attributed to dynamical instability (e.g. Kahn, 1984) or generic gravity waves. Near-surface cloud hazes or fogs are observed to fill low-elevation regions such as Noctis Labyrinthus, Valles

Table 5.1. *Mars clouds are listed by type, including a range of conditions and properties associated with their occurrence. References are listed in the last column and in the table footnotes.*

Cloud type	Locations	Form	Processes	Season (L_s), LT	Spatial scale (km)	Altitude (km)	Ice	Particle R_{eff} (μm)	Refs
Global structure									
ACB	10°S–30°N	Cirrus and cumulus	Annual (eccentricity), aphelion Hadley circulation, topographic forcing	40–140°, solar tide forced variations	Global, with many cloud types listed below	10–40	H_2O	2–5	1, 2, 5, 15, 16, 21, 22, 24, 26, 27, 30, 31, 32, 42, 45, 47, 54, 58, 61, 68, 70
Polar hood	40–70°N,S	Cirrus, more extensive in the north	Seasonal (obliquity), wind shear and planetary eddy structure	Fall–winter, thermal tides force diurnal variations	Circumpolar haze with streak and wave cloud types	0–50	H_2O	Up to 70 below 5 km, 1–2 above 10–20 km	3, 5, 21, 24, 25, 28, 31, 39, 40, 45, 47, 52, 53, 54, 57, 58, 66
Polar winter CO_2	60–85°N,S	Cirrus and cumulus (?)	Lee and gravity waves, snow (?), moist convection (?)	Winter polar night	NPC topographic waves, 10–40, SPC cloud towers (?)	0–15	CO_2	10 (?)	17, 19, 20, 23, 28, 29, 38, 60, 67
Mid-high latitude									
NP frontal, "comma clouds"	Northern high latitudes, 50–80°N	Cirrus, frontal arcs	Polar wave systems associated with cap thermal gradients	40–110°, spring, late summer	≥500–1000 band arcs	5–30 (?)	H_2O dust	1–2 (mid-latitudes)	43, 46
NP spiral	60–75°N	Cirrus spiral arcs	Polar baroclinic disturbance, anticyclonic	30–180°, repeat ~125°, morning	200–1000	10–30 (?)	H_2O		6, 13, 24, 25, 43, 46
Trough clouds	Spiral troughs in residual caps	Water ice snow?	Katabatic jump	Polar spring–summer	10–25 × 100–300	<1 (?)	H_2O		63
Streaks	ACB, N and S polar hoods	Cirrus E–W linear	Strong winds, vertical shear, low temperatures	40–110° fall–winter (≥40° latitudes)	~200 (ACB) ≥500–1000 (polar hood)	10–30	H_2O	1–2 (mid-latitudes)	10, 12, 13, 24, 27, 30
Streets "actinae"	ACB, NPC boundary	Cumulus 2D spacing	Free convection	100–140° NPC, spring, early fall	1–10 "cells" 100–300 long streets	10–20 (?)	H_2O	2–4 (ACB)	10, 12, 24, 30, 46
Lee waves	50–80°N,S, crater rims	Cirrus	Locally forced gravity waves	Spring, fall–winter	100–500, λ ~ 3–80	10–20	H_2O		3, 7, 10, 12, 24, 43, 46
Ground fog	South basins and canyons, high latitudes	Ice haze	Cold near-surface temperatures	Night, early morning	Fills low regions	0–5	H_2O		5, 10, 12, 28, 40, 44, 53, 55, 56, 61
Volcano regions									
Central disk, with/ without rays	Tharsis volcanoes and Alba Patera	Cirrus, cumulus	Upslope winds	0–180° (ACB) peak in late afternoon	Disk comparable to volcano extent, rays ~ 30×300	15–30, 5 (Alba Patera)	H_2O	2–4 (ACB)	2, 3, 4, 5, 8, 15, 16, 22, 24, 26, 27, 30, 32, 35, 61
Bore waves long clouds	Tharsis volcanoes and saddles	Cirrus	Downslope winds, "hydraulic jump"	70–140° (ACB) early morning	1000	20–25	H_2O		9, 11, 34

Table 5.1. (Cont.)

Cloud type	Locations	Form	Processes	Season (L_s), LT	Spatial scale (km)	Altitude (km)	Ice	Particle R_{eff} (µm)	Refs
Mountain lee waves	Ascraeus, Olympus Mons	Cirrus, two-tailed plume	Orographic displacement of easterly winds	Northern summer (ACB), early morning	200–500	10–25	H_2O		5, 9, 35
Waves	Tharsis plateau	Cirrus	Wind shear instability	0–180° (ACB) morning	$\lambda \sim$ 5–30	5–15	H_2O		9, 10, 11, 12, 13
High altitude									
Trails	Very location-specific (10–35°S)	Cumulus, E–W linear	Forced convection peak surface heating	240–270°, early afternoon	80×500	40–50	H_2O	0.2–0.5 (also ~1.0)	41
Diffuse and layer hazes	50°S–50°N	Cirrus	Low temperatures	150–330°	Regional to global scale	60–110	H_2O, CO_2 (?)	0.2–1, 0.1	14, 32, 49, 52, 54, 57, 64, 65, 69
Equatorial CO_2	10°S–10°N, 300–360°W, 50–130°W	Cirrus, E–W oriented	Global minima in mesospheric temperatures, gravity waves (?)	0–70°, 100–160° (?), solar tide variations (?)	Thick clouds ~5 30, thin clouds ≥100	65–85	CO_2	0.5–2	18, 33, 36, 37, 38, 49, 50, 51, 57, 62, 64, 69

1. Slipher (1962).
2. Peale (1973).
3. Leovy et al. (1973).
4. Curran et al. (1973).
5. Briggs et al. (1977).
6. Gierasch et al. (1979).
7. Pickersgill and Hunt (1979).
8. Hunt et al. (1980).
9. Pickersgill and Hunt (1981).
10. French et al. (1981).
11. Kahn and Gierasch (1982).
12. Kahn (1984).
13. Hunt and James (1985).
14. Jaquin et al. (1986).
15. Clancy et al. (1996).
16. James et al. (1996).
17. Zuber et al. (1998).
18. Clancy and Sandor (1998).
19. Pettengill and Ford (2000).
20. Ivanov and Muhleman (2001).
21. Tamppari et al. (2000).
22. Pearl et al. (2001).
23. Colaprete and Toon (2002).
24. Wang and Ingersoll (2002).
25. Cantor et al. (2002).
26. Benson et al. (2003).
27. Clancy et al. (2003).
28. Neumann et al. (2003).
29. Tobie et al. (2003).
30. Wolff and Clancy (2003).
31. Smith (2004).
32. Hinson and Wilson (2004).
33. Montmessin et al. (2006a).
34. Sta. Maria et al. (2006).
35. Michaels et al. (2006).
36. Clancy et al. (2007).
37. Montmessin et al. (2007b).
38. Colaprete et al. (2008).
39. Tamppari et al. (2008).
40. Whiteway et al. (2009).
41. Clancy et al. (2009).
42. Lee et al. (2009).
43. Wang and Fisher (2009).
44. Möhlmann et al. (2009).
45. Smith (2009).
46. Cantor et al. (2010).
47. Heavens et al. (2010).
48. Benson et al. (2010).
49. McConnochie et al. (2010).
50. Määttänen et al. (2010).
51. Scholten et al. (2010).
52. Clancy et al. (2010).
53. Nelli et al. (2010).
54. McCleese et al. (2010).
55. Dickinson et al. (2010).
56. Moores et al. (2011).
57. Vincendon et al. (2011).
58. Hale et al. (2011).
59. Benson et al. (2011).
60. Hayne et al. (2012).
61. Madeleine et al. (2012b).
62. Spiga et al. (2012).
63. I.B. Smith et al. (2013).
64. Sefton-Nash et al. (2013).
65. Maltagliati et al. (2013).
66. Lemmon (2014).
67. Hayne et al. (2014).
68. Guzewich et al. (2014).
69. Clancy et al. (2014).
70. Fedorova et al. (2014).

Marineris (Figure 5.10), and Hellas Basin, as well as high latitudes in spring and fall seasons (French et al., 1981; Neumann et al., 2003). These near-surface hazes exhibit strong diurnal variations associated with extreme diurnal variation of atmospheric temperature within 1–2 km of the Mars surface (e.g. Nelli et al., 2010; Moores et al., 2011). Cloud hazes at higher altitudes (30–50 km) are observed to cap the vertical distribution of dust, suggestive of key microphysical relationships between dust and ice aerosols (Clancy et al., 1996, 2007; Rodin et al., 1997; Montmessin et al., 2004; Heavens et al., 2011; M.D. Smith et al., 2013).

Significant orbital (eccentricity-driven) and seasonal (obliquity-driven) variations of both topographic clouds and large-scale cloud hazes are expressed in such global-scale phenomena as the ACB (Clancy et al., 1996; Benson et al., 2003; Wang and Ingersoll, 2002; Smith, 2004) and the fall/spring polar hoods of Mars (Wang and Ingersoll, 2002; Tamppari et al., 2008; Cantor et al., 2010; Benson et al., 2010, 2011), respectively. Notable seasonal and spatial correlations are also exhibited for smaller-scale features such as striking spiral cloud systems over the north polar cap (Cantor et al., 2002) and perihelion cloud trails (Clancy et al., 2009). Table 5.1 presents a summary of Mars cloud types, including their seasonal and spatial distributions as determined from existing Mars observations (both imaging and spectroscopic). The reference list associated with this table, which emphasizes observational results, indicates the extensive research effort devoted to Mars cloud studies. The reader is directed to these references for a more thorough treatment of specific cloud forms, occurrences, and origins than can be provided here.

5.3 THE GLOBAL-SCALE DISTRIBUTION OF MARS WATER ICE CLOUDS

We consider global spacecraft datasets in the presentation of the large-scale behaviors of Mars water ice clouds provided by the MGS TES and MOC experiments, the ODY THEMIS experiment, and the MRO MCS and Mars Color Imager (MARCI) (Malin et al., 2008) experiments, operating over the period 1998–2011 (MY 24 to MY 31; where MY (Mars year) 1 begins at Mars $L_s = 0°$ in 1955). This emphasis reflects the unique mapping aspects of the TES, THEMIS, MOC, MARCI, and MCS cloud datasets that allow a fairly uniform sampling in space and time over multiple Mars years. The thermal infrared TES, THEMIS, and MCS datasets provide measurements at two fixed local times (2–5 a.m. versus 2–5 p.m.) and obtain column (TES, THEMIS) or vertical profiling (MCS) cloud measurements. MCS observations provide the first detailed description of the vertical distribution of Mars water ice clouds with full spatial and temporal coverage, although a substantial set of aerosol opacity and particle size profiles have been retrieved from SPICAM solar (Fedorova et al., 2009, 2014; Määttänen et al., 2013) and stellar (Montmessin et al., 2006b) occultation measurements. MOC wide-angle imaging provides daily global maps of dayside clouds, which appear distinctly against the low-albedo Mars surface at blue (400–450 nm) wavelengths. MARCI wide-angle imaging extends MOC imaging observations in a very consistent fashion through current MRO operations, and so effectively constitutes the longest continuous dataset for the Mars atmosphere. The significant contributions from other key spacecraft observations (Viking, Phobos, MER, MEX, Phoenix) are emphasized in the following sections, in which specific properties and processes associated with Mars water ice clouds are discussed.

The global distribution of Mars water ice clouds may be categorized by three broad regimes: the aphelion cloud belt, the polar hoods, and "high-altitude haze". As indicated previously, a range of cloud forms are exhibited by these cloud structures, which occur under globally extended conditions of water vapor saturation associated with global-scale dynamical regimes. The first two categories represent clouds in the lower atmosphere (e.g. below 40 km) of sufficient optical depth ($\tau_{visible} \geq 0.05$) to be detected in nadir sounding/imaging observations. Figures 5.11 and 5.12 demonstrate the appearance of these clouds in global maps of MARCI 320 nm and TES 12 µm retrievals for water ice cloud optical depths, respectively. A low-latitude (aphelion) cloud belt appears around Mars aphelion ($L_s = 71°$) in northern summer, whereas high-latitude clouds appear in the fall–winter–spring seasons in both hemispheres. Only the mid-latitude margins of the latter (polar hood, PH) clouds are apparent in these nadir datasets due to measurement limitations at the cold, un-illuminated high

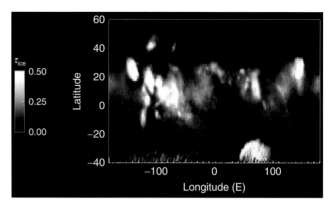

Figure 5.11. Global map of MARCI 320 nm retrieved optical depths accumulated from daily global imaging over $L_s = 120–135°$ in 2008 (MY 29). Distinct clouds over high Mars volcanoes (such as the Tharsis ridge) dominate latitudinal/longitudinal variability in the ACB low-latitude cloud band. Similarly, prominent clouds over Hellas Basin dominate the (daylighted) margins of the SPH. Figure provided by Michael Wolff, Space Science Institute.

latitudes. Both features exhibit strong inter-annual repeatability, as indicated in the multiple MY presentation of the 12 µm cloud map (L_s versus latitude; LT = 2 p.m. for 1999–2004 TES observations, LT = 5 p.m. for 2004–2011 THEMIS observations). Considerable spatial (latitude versus west longitude) variations are present in the aphelion cloud belt, as indicated in the UV cloud map (for $L_s = 120–135°$, LT = 3 p.m.). The third cloud category of "high-altitude hazes" refers to optically thin clouds present at altitudes above 30–40 km. The global and seasonal distributions of these clouds have only recently become apparent, as determined from dedicated MCS limb profiling. Figure 5.13 provides a global description of the Mars water ice cloud distribution (scale bar) versus atmospheric pressure, latitude, and season for a local time of 3 a.m., based upon MCS profile mapping observations in MY 29. The polar hoods and the ACB appear prominently, primarily as lower-altitude phenomena (below 10 Pa, or 30–40 km). Optically thin, high-altitude water ice clouds appear most distinctly at low latitudes around the northern fall equinox ($L_s = 180°$), and extend through late southern summer fall ($L_s = 330°$) at northern winter latitudes (Sefton-Nash et al., 2013; Clancy et al., 2014). The detailed spatial, seasonal, and diurnal characteristics of these Mars cloud regimes are discussed individually below.

5.3.1 Aphelion Cloud Belt

Water ice clouds develop in the tropics during northern spring and summer around Mars aphelion ($L_s = 71°$), forming a low-latitude (10°S–30°N) belt of clouds referred to as the aphelion cloud belt (ACB; Clancy et al., 1996; Smith, 2004), the equatorial cloud belt (James et al., 1996), or the subtropical cloud belt (e.g. McCleese et al., 2010). The ACB plays prominent roles in influencing the seasonal and latitudinal distributions of Mars atmospheric water by restricting cross-equatorial water vapor transport in the aphelion northern summer Hadley circulation (Clancy et al., 1996; Montmessin et al., 2004; see Section 6.4 and Chapter 11). The ACB also imparts orbital variation in Mars photochemistry by modulating production of key catalytic HO_x radicals produced from water vapor photolysis (Clancy

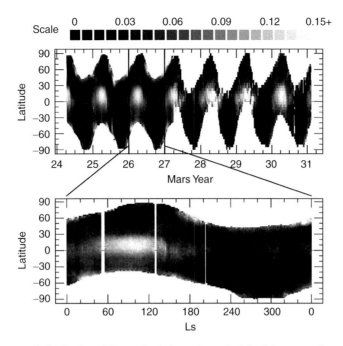

Figure 5.12. The global (zonally averaged) distribution of 12 μm cloud absorption optical depth is presented versus latitude and L_s for multiple Mars years, corresponding to the 1999–2011 period. The period MY 26 is expanded for better viewing of a typical annual behavior. The period MY 24–26 incorporates MGS TES measurements (Smith, 2004), whereas the period MY 27–31 incorporates MO THEMIS measurements (Smith, 2009). The ACB and PH cloud structures apparent in this figure exhibit modest inter-annual variations. Figure provided by Michael Smith, Goddard Space Flight Center. A black and white version of this figure will appear in some formats. For the color version, please refer to the plate section.

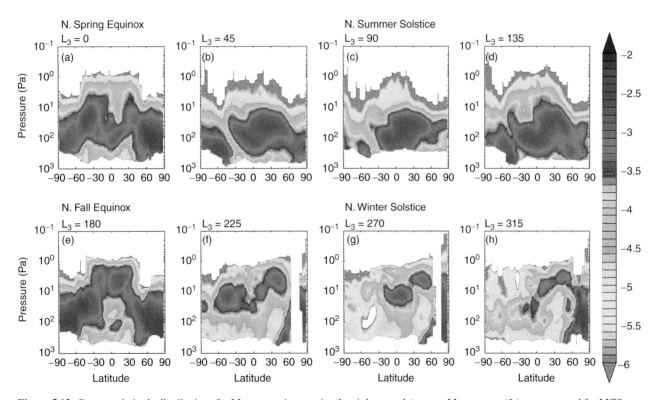

Figure 5.13. Pressure–latitude distributions for Mars water ice opacity (km^{-1}, $\log_{10}$ scale) versus Mars season (L_s) are presented for MCS nighttime (LT = 3 a.m.) retrievals during MY 29. The presented altitude region of MCS retrieval is approximately 0 (~600 Pa) to 60 km (~1 Pa). Figure is reproduced from McCleese et al. (2010).

and Nair, 1996; Lefèvre et al., 2004; Chapter 13). It may also contribute to long-term, orbital variations in polar water ice reservoirs, associated with long-term (Milankovitch cycle) variations in the seasonal timing of perihelion (Clancy et al., 1996; Montmessin et al., 2007a; Chapter 16). These behaviors reflect a strong orbital variation in the altitude of the global hygropause, associated with orbital forcing of Mars global temperatures (+20 K from aphelion to perihelion). Vertical advection of water vapor within the northern solstice Hadley circulation couples with the minimum hygropause altitudes around Mars aphelion to establish a global belt of water ice clouds centered on the northern subtropics (Clancy et al., 1996).

The ACB presents a distinctly non-uniform cloud distribution versus latitude and longitude. Thicker clouds, topographically associated with large volcanoes, also form within the ACB (Clancy et al., 1996; James et al., 1996; Pearl et al., 2001; Wang and Ingersoll, 2002; Benson et al., 2003; Smith, 2004). Data from the Thermal Emission Spectrometer (TES) reveal that the ACB begins to form around $L_s = 0°$, with average column optical depths of 0.08 at 825 cm⁻¹ (12 μm). Peak extended optical depths (0.15–0.2 at 12 μm) and spatial extent (10°S–30°N) develop by $L_s = 80°$ (corresponding visible optical depths are larger by approximate factors of 1 and 4, for ice particle radii of 4 and 1 μm, respectively). Around $L_s = 140°$, the ACB begins to dissipate, and by $L_s = 180°$ has largely disappeared (Smith, 2004). Imaging data from the Mars Orbiter Camera (MOC) also show that the ACB becomes longitudinally continuous by about $L_s = 60°$ and remains so until $L_s \sim 140°$, after which only topographic clouds associated with volcanoes and Valles Marineris remain by late northern summer (Wang and Ingersoll, 2002). A cloudy region associated with Arsia Mons remains throughout the year (Wang and Ingersoll, 2002; Benson et al., 2003). Orographic clouds present typical visible optical depths of ~0.5, although maximum visible optical depths exceed unity for individual clouds (Benson et al., 2003). The optical depths of orographic clouds are generally highest over the volcano flanks and decrease in an extended cloud plume displaced (typically) westward of the volcano summit (e.g. Wang and Ingersoll, 2002).

These nadir, or column, measurements of water ice clouds emphasize spatial and seasonal variations of the bulk atmospheric cloud content. Limb profiling observations allow definition of seasonal and spatial variations in the vertical distribution of clouds, necessary to understand detailed cloud microphysics and radiative forcing. The general altitude range of ACB clouds extends over 15–40 km, with considerable diurnal variation (Glenar et al., 2003; Heavens et al., 2010). Heavens et al. employed MCS data to examine the diurnal and seasonal variability of cloud height and mass mixing ratio (q_{ice}) in the tropics during northern summer. Mass mixing ratios are normalized by the vertical dependence of atmospheric density and so relate more directly to local radiative/thermal forcing of clouds. For comparison, a uniformly mixed 10 pr μm column of water vapor (pr μm = precipitable micrometers, the water column equivalent as ice at 1 g cm⁻³) corresponds to $q_{water} \sim 70$ ppm. Heavens et al. (2010) indicate that altitudes of peak mass mixing ratios ($q_{ice} = 5–10$ ppm) for daytime tropical clouds remain relatively constant (25–30 km, or 20 Pa) throughout northern summer. In contrast, the altitudes of peak nighttime cloud mixing ratios increase from the 20 Pa pressure level in early summer ($L_s = 110–120°$) to 4

Pa (~40 km) by $L_s = 160°$. Peak mass mixing ratios for daytime and nighttime clouds also increase over this period, from 5 ppm in early summer to 10–20 ppm by $L_s = 160°$.

A significant qualification to these derived cloud mass mixing ratios is their dependence on ice particle size, which MCS does not measure. The adoption of a fixed particle size (geometric cross-section weighted radius, $R_{eff} = 1.4$ μm), which is substantially smaller than cloud particle sizes determined for peak opacity regions of the ACB (see Section 5.5.1), can lead to an error by a factor of 2 in derived q_{ice} values. MCS q_{ice} values at lower levels (e.g. <20–30 km) are also affected by decreasing limb transmission. Even so, Heavens et al. (2010) argue that the observed behaviors of derived q_{ice} indicate higher altitudes and lower overall opacities for ACB cloud formation than are currently modeled in MGCM (Mars general circulation model) simulations of northern summer Hadley circulation. This result may conflict with the level of nighttime surface temperature enhancements associated with cloud radiation, which indicates larger cloud opacities (Wilson et al., 2007; Wilson, 2011a).

MCS observations clearly provide confirmation of strong coupling between thermal tide (diurnal) forcing and the ACB cloud vertical distribution, as presented in Lee et al. (2009) and Wilson (2011a). As the saturation pressure of water vapor is a strong function of temperature, water ice clouds will tend to condense near altitudes of temperature minima corresponding to antinodes of the solar tide. This behavior is most strikingly presented at $L_s = 160°$. Furthermore, modeling shows that cloud infrared radiative absorption of diurnally variable surface radiation significantly modulates this diurnal forcing of thermal and ACB vertical dependences (Hinson and Wilson, 2004; Wilson et al., 2008; Haberle et al., 2011; Madeleine et al., 2012a).

5.3.2 Polar Hoods

The north (NPH) and south (SPH) polar hood clouds of Mars are associated with very cold atmospheric temperatures at mid- to high latitudes during fall–winter–spring seasons. These clouds presumably influence the amount of water ice that is incorporated and subsequently released in the (>99% CO_2) seasonal polar ice caps of Mars. Polar hood clouds are temporally and spatially variable, associated with storm systems, streak clouds, lee waves, and hazes (Kahn, 1984; Wang and Ingersoll, 2002; Inada et al., 2007; Wang and Fisher, 2009). NPH clouds can also exhibit regional (e.g. Acidalia Planitia) correlations with topographically oriented winter storm zones (Hollingsworth et al., 1996), leading to stationary wavenumber-2 polar distributions (Wang and Ingersoll, 2002; Tamppari et al., 2008). The NPH has been observed since very early telescopic observations (e.g. Martin et al., 1992), due to its greater latitudinal/seasonal extents and higher cloud opacities relative to the SPH. The SPH also exhibits a distinctive "disappearance" near southern winter solstice (Wang and Ingersoll, 2002; Tamppari et al., 2008; Benson et al., 2010; see below) that further detracts from its visibility relative to the NPH. A number of related factors lead to these north–south asymmetries in polar hood formation, including much lower surface elevations (higher surface pressures) at northern versus southern high latitudes, higher atmospheric water content

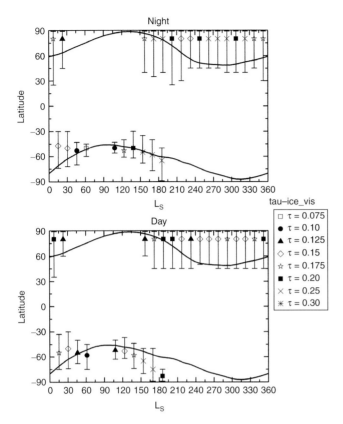

Figure 5.14. MCS limb observations characterize the latitudinal extent (vertical lines) and optical depths (symbols with scale to right, $\tau_{visible}$ scaled from MCS IR measurements according to Kleinböhl et al. (2009)) of the north and south polar hoods during night (~3 a.m., top panel) and day (~3 p.m., bottom panel); local times in MY 29. The solid curve shows the climatological latitude of the seasonal polar cap edges taken from TES observations (Titus, 2005). Compiled from data in Benson et al. (2010, 2011).

at northern versus southern winter mid-latitudes, and stronger meridional (poleward) circulation in the northern versus southern winter hemispheres (e.g. Montmessin et al., 2004).

The limb sounding, IR spectroscopic capabilities of MCS support the most complete global description of polar hood clouds in terms of sensitivity, latitudinal extent, and vertical coverage (with the significant caveat that limb viewing precludes resolution of near-surface and small-scale cloud structures). MCS data show that NPH water ice clouds are present for about 75% of the Martian year, forming around $L_s = 150°$ in late northern summer and dissipating around $L_s = 30°$ the following spring (Benson et al., 2011). The NPH generally extends from the pole to a variable mid-latitude extent (Figure 5.14). This mid-latitude boundary is less variable during daytime (bottom panel, ~3 p.m.), and typically extends to about 45°N latitude. At the northern spring equinox, however, daytime clouds extend further south to 35°N latitude. The nighttime mid-latitude boundary (top panel, ~3 a.m.) is more variable, but on average extends further south than the dayside NPH, at times reaching 25°N latitude. Within the periods $L_s = 150–195°$ and $L_s = 240–330°$, NPH optical depths are significantly higher at night than during the day (~0.25 versus ~0.15, $\tau_{visible}$ as scaled from MCS IR measurements according to Kleinböhl et al. (2009)). Outside of these periods, the

NPH optical depth remains relatively constant (~0.18) until it begins to dissipate by $L_s = 30°$ (Benson et al., 2011).

The seasonal patterns of diurnal variation in NPH cloud optical depths result from two distinct behaviors. Over the $L_s = 150–195°$ period, NPH clouds are located near the diurnal tide antinode at northern mid-latitudes, where daytime temperatures at the latitude and pressure of the clouds are 10–20 K warmer than nighttime temperatures (Lee et al., 2009). Over the $L_s = 240–330°$ period, the diurnal optical depth variation is associated with the 15–25 K diurnal temperature variations of the polar winter vortex. In both cases, cooler nighttime temperatures increase condensation rates and create optically thicker clouds at night than during the day (Benson et al., 2011). MCS observations reveal the vertical structure of the NPH clouds as a single broad layer extending over 10–40 km in altitude during all local times (where the lower 10 km region is not retrieved from MCS limb scans). The vertical dependence is consistent with a relatively constant mass mixing ratio throughout the cloud, with little spatial or seasonal deviation from this character. Northern polar temperature profiles during the seasons of NPH formation are either isothermal or exhibit a broad temperature minimum (25–30 km) at the altitude where ice forms, resulting in a single cloud layer (lower panels of Figure 5.15).

MCS observations indicate that the south polar hood (SPH) is present for about half the Martian year, during southern fall $L_s = 10–70°$ (phase 1) and southern winter $L_s = 100–200°$ (phase 2). A distinctive cloud minimum occurs over $L_s = 70–100°$ (Figure 5.14; Benson et al., 2010). The SPH cloud belt appears as an annulus about the pole for the majority of the time that clouds are present (Figure 5.14; Benson et al., 2010; Brown et al., 2010; Mateshvili et al., 2009). During phase 1, the cloud belt initially extends over a wide latitude range, from 30° to 75°S, with a visible optical depth between 0.075 and 0.15. Towards the middle of phase 1, the cloud belt narrows in latitudinal extent and decreases in optical depth. By the end of this period ($L_s = 70°$), nighttime SPH clouds have dissipated and daytime SPH clouds are limited to lower opacities (0.05–0.10) around 60°S. During phase 2, the SPH forms as a partial band of low-opacity clouds south of the Tharsis region. This band expands over all longitudes by $L_s = 130°$ (daytime) to $L_s = 145°$ (nighttime), with a visible optical depth between 0.125 and 0.25 (Benson et al., 2010). The SPH progresses poleward and disappears by $L_s = 200°$. The vertical structure of SPH clouds throughout their occurrence is characterized by a lower cloud layer centered at ~10 km and an upper layer of diurnally varying altitude (25–35 km). The peak extinction of the upper layer is generally smaller during nighttime versus daytime by about a factor of 10. Both the two-layer structure and the diurnal variation of the upper-level extinction are associated with strong tidal control of condensation altitudes at mid- to high southern latitudes, as demonstrated in MCS temperature profiles (upper panels in Figure 5.15).

5.3.3 High-Altitude Hazes

Regional-scale water ice clouds outside of the polar hood and ACB environments have also been observed at higher altitudes (40–80 km, 20–0.1 Pa), although with considerably lower column optical depths ($\tau_{visible} \leq 0.01$). Jaquin et al. (1986) analyzed Viking limb imaging to determine that the average altitudes of

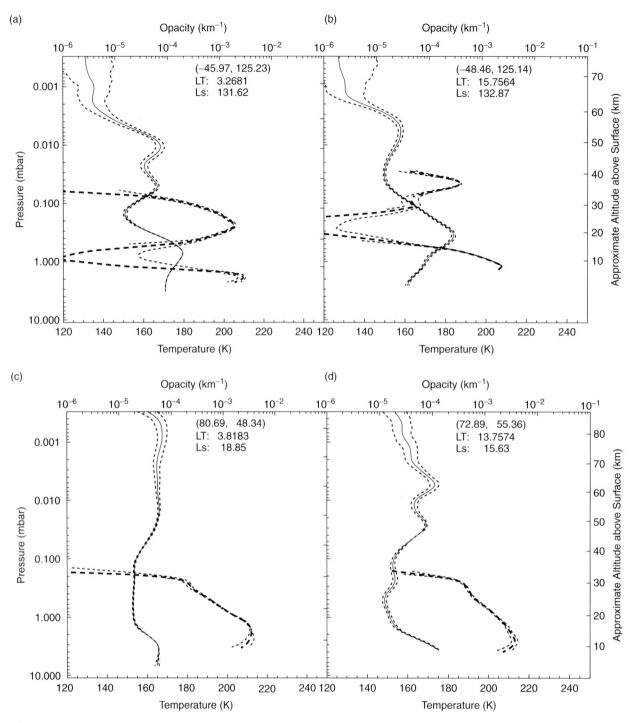

Figure 5.15. MCS profiles of temperature (light solid lines, lower axis), and 12 μm ice extinction (heavy dashed lines, upper axis), where dotted lines indicate the retrieval uncertainties for each profile: (a) SPH nighttime; (b) SPH daytime; (c) NPH nighttime; (d) NPH daytime. Both pressure (mbar) and corresponding altitude (km) scales are provided on the left and right sides, respectively, for each individual profile. The latitude, longitude, and local time are shown in the top right of each panel. Figure is provided by Jennifer Benson, Jet Propulsion Laboratory.

30°S–30°N upper-level haze layers peak near 70 km around Mars perihelion (northern winter). A similar behavior was found in MEX SPICAM stellar occultation measurements (Montmessin et al., 2006b), although the global coverage did not clearly separate L_s versus latitudinal variations. Based upon subsequent MCS observations, it appears likely that northern winter upper-level ice hazes contribute substantially to this

observed seasonal variation. MCS global profiling observations indicate the presence of upper-level water ice hazes during equinoctial seasons at low latitudes and at low to mid-latitudes in the perihelion (northern winter) seasons (McCleese et al., 2010; Figure 5.13). Equinoctial upper-level clouds above 40 km exhibit substantially higher optical depths in the northern fall ($L_s = 180°$) versus southern fall ($L_s = 0°$) seasons, perhaps due to the larger overall atmospheric water vapor abundance during the northern fall season (McCleese et al., 2010). These equinoctial clouds exhibit strong diurnal variation (nighttime maxima), presumably corresponding to the altitude-increasing amplitudes of large thermal tides in the Mars equatorial atmosphere (e.g. Hinson and Wilson, 2004). Winter upper-level ice clouds are primarily confined to the northern hemisphere, with somewhat larger daytime versus nighttime opacities (McCleese et al., 2010). TES visible and thermal IR limb radiance observations conducted in the declining phase of the 2001 global dust storm characterize these clouds as detached limb hazes centered at 70 km altitude, with small particle sizes (<1 μm radii; Clancy et al., 2010). Montmessin et al. (2006b) also determined smaller ice particle sizes for upper-level ice hazes (0.1 to >0.3 μm radii). An attendant feature of these northern winter high-altitude clouds is a very clean (aerosol-free) corridor centered at 30–40°N, extending from the surface to the lower-altitude boundaries of these clouds (McCleese et al., 2010).

More recently, specific mesospheric (50–100 km) analyses with MCS (Sefton-Nash et al., 2013) and CRISM limb observations (Clancy et al., 2014) have indicated extensive mesospheric water ice clouds at low to mid-latitudes throughout the perihelion season (L_s = 150–360°). The Sefton-Nash et al. analysis mapped the distribution of "loop" clouds at 50–90 km altitudes, an observational term describing the appearance of mesospheric discrete clouds (horizontal scales of less than ~100 km) in MCS limb scan sequences. Distinct aphelion and perihelion season latitudinal distributions are mapped for these MCS "loop" clouds, perhaps broadly associated with aphelion low-latitude CO_2 ice versus low to mid-latitude perihelion water ice clouds. CRISM visible to near-IR limb spectra are a much-reduced dataset relative to MCS mapping limb scans, but obtain diagnostic compositional and particle size measurements. Fine water ice clouds (R_{eff} = 0.3–0.8 μm) at low to mid-latitudes predominate during the perihelion season, whereas larger particle size (R_{eff} = 0.5–1.5 μm) CO_2 ice clouds appear at low latitudes in the aphelion season (Clancy et al., 2014). MCS and CRISM limb observations indicate a typical 55–70 km altitude range for mesospheric ice clouds.

Dedicated microphysical modeling efforts have not been pursued for mesospheric water ice hazes to date, although Listowski et al. (2014) recently modeled the microphysics of mesospheric CO_2 clouds (Section 7.2). Rapid ice particle fall rates are expected in this low-pressure environment (e.g. Jaquin et al., 1986), such that active supply of water vapor and condensation nuclei (CN, fine dust particles, meteoritic smoke) are required to support their extended presence (Montmessin et al., 2006b). Vertical, poleward advection of water vapor and dust in the global Hadley circulation is likely important in the maintenance of the northern winter upper-level ice haze in particular. The Listowski et al. (2014) study indicates that lack of an identifiable CN source is a significant issue for mesospheric CO_2 cloud microphysics during the cold, relatively dust-free aphelion season.

5.3.4 Inter-Annual Variability

Given the prominent inter-annual variability of dust storm activity on Mars (e.g. Smith, 2004), one might expect to observe related inter-annual variability of cloud activity. On the other hand, mid- and low-latitude clouds on Mars are most active during the aphelion season, when regional and global dust storms are absent. In fact, the large-scale distribution of Mars water ice clouds is remarkably repeatable from year to year, as evident in Figure 5.12. Focused comparisons of Viking and MGS IR cloud datasets also indicate modest inter-annual variability in global cloud behaviors (Hale et al., 2011). The limited cases of inter-annual variability are generally associated with related increases in atmospheric temperature and dust from regional or planet-encircling dust storms. The observed correlations among cloud, dust, and temperatures are consistent with model predictions that increased dust opacities lead to increased atmospheric temperatures through dust absorption of solar flux, and such higher atmospheric temperatures restrict water vapor saturation conditions necessary for cloud formation (Richardson et al., 2002).

The inter-annual behavior of the ACB is perhaps best studied in this regard, in relation to MGS cloud observations preceding and following the MY 25 planet-encircling dust storm. TES observations show that the amplitude of water ice cloud optical depth and the spatial distribution of clouds in latitude and longitude are nearly identical over three Mars years (MY 24–26). In particular, Smith (2004) determined minimal changes in the ACB spatial, seasonal, or optical depth characteristics between MY 25 (before the MY 25, or 2001, planet-encircling dust storm) and MY 26 (after the dust storm). Smith (2004) listed three reasons to explain this result: (1) the spatial pattern of clouds within the ACB is mostly controlled by topography, forcing a repeatable pattern from year to year due to local circulation; (2) the change in the water condensation level between MY 25 and MY 26 was minimal at the latitude and pressure level of the clouds; and (3) the ACB contains only a small amount of the total atmospheric water content, so observed inter-annual variability in water vapor may not lead to changes in water ice cloud optical depth. However, individual volcano clouds outside the aphelion period were clearly influenced by the MY 25 planet-encircling dust storm. Volcano clouds over Pavonis Mons and Arsia Mons dissipated earlier in MY 25 than in MY 24, presumably a result of increased dust and atmospheric temperatures associated with the dust storm (Benson et al., 2003, 2006). Cloud activity over Arsia Mons was continuous in MY 24, but disappeared completely for about 85° of L_s in MY 25. When clouds reappeared after the dust storm, they exhibited reduced spatial extent relative to the previous year (Benson et al., 2006). Smaller, regional dust storms can also influence cloud formation. For example, Pearl et al. (2001) noted a suppression of cloud activity over Arsia Mons in MY 23, associated with Noachis Terra regional dust lifting.

The NPH also exhibits limited variability that appears related to contemporaneous dust storms, in that its southern boundary extends farther south when regional dust storm activity is not present. Early ground-based observations of Mars indicated reduced NPH cloud extent after major dust storms in MY 1 (1956), 9 (1971), and 10 (1973) (Martin, 1975). Recent spacecraft observations show similar behavior. After a MY 28 (2007) planet-encircling dust storm (e.g. Cantor et al., 2008), NPH clouds observed over L_s = 300–330° were confined further northward than in MY 29, when there was no preceding dust storm. The daytime NPH optical depth was also lower in MY 28, although the nighttime optical depth is similar in both years (Benson et al., 2011). In MY 29 from $L_s = 234–277°$ a planet-encircling dust event in the southern hemisphere increased global atmospheric temperatures (Kass et al., 2009), confining the NPH nighttime clouds to more northerly, cooler regions. Changes in the onset ($L_s = 130–160°$) and longitudinal structure of the NPH, in association with the MY 25 global dust event, are also reported (Tamppari et al., 2008).

5.4 WATER ICE CLOUDS IN THE LOWER BOUNDARY LAYER: PHOENIX LIDAR OBSERVATIONS

Clouds within the planetary boundary layer (PBL, altitudes ≤ 10 km; see Chapter 7) of Mars are not well observed for several reasons. Nadir images clearly detect early morning fogs within low-elevation regions such as Valles Marineris (Figure 5.10), but can determine neither their vertical distribution nor specific cloud properties. Viking (Colburn et al., 1989) and Pathfinder (Savijärvi, 1999) Lander measurements have provided indications for the local formation of surface fog in early morning hours. Limb profiling observations from Mars orbit are generally limited to cloud vertical profiles above ~10 km, due to limited limb path transmission associated with dust and cloud extinction as well as topographic obstruction of limb tangent views (e.g. Kleinböhl et al., 2009). Lidar observations are best suited to the study of low-elevation clouds and have been obtained twice, in very different circumstances, for Mars. The MGS MOLA experiment was designed to measure global topography from orbit (Smith et al., 2001), such that cloud returns were primarily restricted to the special circumstances of polar night CO_2 clouds (e.g. Pettengill and Ford, 2000; see Section 5.5). The first Mars experiment to measure vertical profiles of water ice clouds in the PBL was the Light Detection and Ranging (LIDAR) instrument (Whiteway et al., 2009) on the Phoenix Lander mission (Smith et al., 2008). Phoenix landed in the northern subpolar plains on Mars (68°N, 234°E) on May 25, 2008 ($L_s = 76.5°$). The high-latitude location and late northern summer season of Phoenix LIDAR cloud observations pertain to the NPH, although early in season relative to extensive NPH development.

The LIDAR experiment observed sporadic clouds below 10 km altitude during the northern summer season. From $L_s = 113°$ until the end of the mission at $L_s = 150°$, LIDAR consistently observed a fog layer forming after midnight within the bottom 1 km of the atmosphere, and a second cloud layer forming after 1–2 a.m. LT near the top of the local planetary boundary layer (PBL, at 3–5 km; Whiteway et al., 2009; Dickinson et al., 2010). Both of these cloud features dissipated during daytime hours, extending from early morning to late afternoon hours over the course of the mission. Their optical depths increased over the night (with a maximum typically around 6 a.m.) and with the progressing season of mission operations (Dickinson et al., 2010). Phoenix LIDAR observations provide vertical profiles of cloud extinction opacity at a vertical resolution of 40 m and a time resolution of 20 s (Dickinson et al., 2010). These capabilities provide unprecedented views of the internal structure and temporal evolution of Mars clouds in the northern high-latitude PBL. Most strikingly, fall streaks were found on many occasions (Figure 5.8). These are a common feature in terrestrial cirrus clouds and are indicative of substantial ice particle precipitation.

Analysis with a one-dimensional (1D) radiative transfer model, employing dust profile and near-surface (2 m level; Whiteway et al., 2009) Phoenix air temperature measurements, indicates that the PBL (3–5 km level) clouds formed at temperatures of −65°C, similar to the temperatures at which cirrus clouds form on Earth (Davy et al., 2010). The observed time sequence of the fall streaks at $L_s = 122°$ is consistent with the fall speed of a prolate ellipsoid with a volume-equivalent radius of 35 μm (Fuchs, 1964), which approximates hexagonal columnar crystals of length 150 μm and width 50 μm as typically found in cirrus clouds on Earth (Whiteway et al., 2004; Gallagher et al., 2005). Based upon correlations between optical extinction and ice water content (IWC) for terrestrial cirrus clouds, Whiteway et al. (2009) and Dickinson et al. (2010) derive the total ice water content of the measured clouds from the LIDAR signal. The vertically integrated IWC was typically a few pr μm, to be compared with a total water column over the Phoenix site which decreased from 44 pr μm on $L_s = 120°$ to 22 pr μm on $L_s = 148°$ (Tamppari et al., 2010). Moreover, model temperature calculations indicate that this observed decrease in total water column is almost entirely located in the PBL (Dickinson et al., 2010).

The Phoenix LIDAR measurements reveal a local water cycle that plays a role in the seasonal decrease in total water column. Daytime turbulent mixing distributes water vapor uniformly throughout the PBL, and the diurnal radiative forcing produces clouds at night at the coldest points, i.e. ice fog near the surface and clouds at the top of the residual PBL. The PBL cloud particles grow large enough to generate substantial precipitation. Sublimation of these descending ice particles in the morning hours leads to a redistribution of water vapor in the PBL, as the cycle repeats itself daily. This picture is supported by a detailed microphysical model analysis, as presented in Figure 5.16 (Daerden et al., 2010; see also Pathak et al., 2008; Nelli et at., 2010). Temperatures from radiative transfer modeling were used to drive a detailed ice cloud model (based on Larsen et al., 2004; Daerden et al. 2007). The volume mixing ratio (v.m.r.) of water vapor throughout the residual boundary layer prior to cloud formation was specified as 1300 ppmv, to ensure some supersaturation at the top of the residual PBL at the detected cloud formation time. Water vapor was distributed uniformly with height throughout the PBL according to the observed strong daytime mixing. Ice crystals in the model are formed by heterogeneous nucleation (Määttänen et al., 2005; Vehkamäki et al. 2007) onto the dust particle cores, at supersaturation ratios near 10%. This value for supersaturation is in agreement with recent laboratory studies that confirm 10% supersaturation for temperatures as low as 185 K (Iraci et al., 2010; Phebus et al., 2011), as the clouds discussed here form at temperatures around 210 K.

The Daerden et al. (2010) model simulates cloud patterns that are comparable to the LIDAR observations, particularly with respect to descending cloud layers similar to fall streaks (Figure 5.16b). Effective radii of particles in the cloud layer near the PBL top range from 10 to 20 μm, while in the fall streak they grow up to 50 μm. The integrated cloud ice water content amounts to 2 pr μm above 200 m, with an additional 4 pr μm predicted by the model below 200 m, of which the largest part is deposited as frost on the surface. Based upon SSI imaging of the LIDAR beam scattered by the near-surface

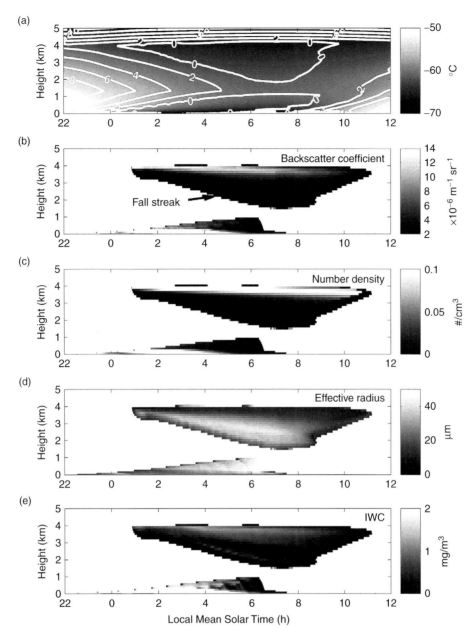

Figure 5.16. (a) Temperatures calculated by a radiative transfer model for the Phoenix location at $L_s = 122°$. Contour lines indicate elevation above frost-point temperature assuming a constant initial water profile of 0.0013 v.m.r. Results from microphysical model calculations driven by these temperatures: (b) backscatter coefficient at 532 nm; (c) ice particle number density; (d) ice particle effective volume-equivalent radius, the length of an ice ellipsoid is ~4.16× this value; and (e) cloud ice water content. Figure is reproduced from Daerden et al. (2010).

fog, Moores et al. (2011) determined that up to 2.5 pr μm or 6% of the total water column is taken up by the surface. In total, formation of the PBL clouds represents a 15% change in the water column. Precipitation of the clouds and their sublimation during late morning solar heating perturbs the water vapor profile. Subsequently, the daytime turbulent mixing and convection gradually return the water vapor profile back to its original shape. In this way, water vapor remains confined to the PBL throughout the late summer season. As the PBL height decreases over the season due to decreasing solar illumination, water will be confined increasingly closer to the surface, and ultimately incorporated in the developing CO₂ seasonal ice cap. In addition to this, poleward transport of water vapor at

higher levels initiates development of overlying NPH clouds, and these ice particles presumably fall to the surface throughout northern winter and fall seasons (e.g. Montmessin et al., 2004).

5.5 PHYSICAL AND RADIATIVE PROPERTIES OF MARS WATER AND CO₂ ICE CLOUDS

The physical properties of Mars clouds are not well characterized, although it is clear that cross-section-weighted particle radii (R_{eff}) for water ice clouds are typically small (1–4 μm) and vary to a substantial degree among cloud types (related in part to their altitudes of formation, see below). Cloud particle size determinations for CO_2 ice clouds have involved combined microphysical modeling/observational studies. Polar night CO_2 cloud studies indicate larger particle sizes ($R_{eff} \sim 30$–200 μm; Tobie et al., 2003; Colaprete et al., 2008; Hayne et al., 2014), and measurement constraints for equatorial mesospheric CO_2 clouds suggest a range of smaller particle radii ($R_{eff} \sim 0.2$–1.5 μm; Clancy et al., 2007, 2014; Montmessin et al., 2007b; McConnochie et al., 2010). Mars cloud observations generally reflect secondary goals of mission experiments, such that their analysis requires complex radiative transfer employing a range of scattering assumptions and approximations. These limitations endure even though cloud particle sizes and their spatial/seasonal variations are critical aspects of cloud microphysics and the effects of clouds on atmospheric processes such as water transport. The radiative properties of Mars clouds can be modeled with more confidence in that the optical indices of water or CO_2 ice clouds are fairly well known (including temperature dependences; see Chapter 6). However, variable cloud particle sizes may still contribute uncertainties in the detailed absorption and scattering properties of clouds.

5.5.1 Cloud Physical Properties

Mars water ice clouds are composed of particles generally 10–100 times smaller than commonly observed for terrestrial cirrus clouds. Such small cloud particle sizes reflect the very low pressure and water vapor content of the Mars atmosphere relative to conditions for the terrestrial upper troposphere. The largest observed values for Mars ice cloud particle sizes, determined for peak ACB opacity regions (3–5 μm; Clancy et al., 2003; Wolff and Clancy, 2003; Madeleine et al., 2012b; Guzewich et al., 2014) and the high-latitude PBL (10–70 μm; Whiteway et al., 2009; Lemmon, 2014), are somewhat comparable to terrestrial sub-visible cirrus clouds present at (or slightly above) the tropical tropopause ($R_{eff} \sim 2$–100 μm; Jensen et al., 2010). However, the predominant range of retrieved particle sizes for Mars clouds is $R_{eff} = 1$–2 μm (Curran et al., 1973; Chassefière et al., 1992; Pearl et al., 2001; Clancy et al., 2003; Wolff and Clancy, 2003). Apart from the distinct fall streak LIDAR (Whiteway et al., 2009) and forward-scattering-lobe imaging (Lemmon, 2014) observations from Phoenix, Mars cloud particle size determinations have relied on observations of wavelength-dependent optical depths that are fairly limited in terms of coverage and accuracy. MGS/TES solar-band and mid-IR spectral EPF observations supported measurements of cloud visible/thermal IR (12 μm) optical depth ratios, which have been used to distinguish ACB cloud (type 2, $R_{eff} = 2$–3 m) versus southern mid-latitude cloud (type 1, $R_{eff} = 1$–2 μm) particle sizes (Clancy et al., 2003). The detailed 12–18 μm (Wolff and Clancy, 2003) and 3–3.5 μm (Madeleine et al., 2012b) spectral shapes of water ice absorption, which provide improved particle size sensitivity to larger cloud particle sizes ($R_{eff} = 3$–5 μm), also indicate larger ice particle sizes in the ACB. Such ACB cloud particle sizes are roughly consistent with MGCM microphysical simulations (Montmessin et al., 2004), although particle sizes as large as 8 μm have been simulated over Mars volcanoes (Michaels et al., 2006). Smaller cloud particle sizes

Figure 5.17. Equidimensional faceted ice particle shapes, suggested to be present in terrestrial upper troposphere cirrus clouds (Yang et al., 2003; Bailey and Hallett, 2009), may best approximate the scattering behavior of Mars water ice clouds (Wolff et al., 2011). Droxtals represent a specific example of such a crystalline shape. Figure is reproduced from Yang et al. (2003).

have been determined for perihelion cloud trials ($R_{eff} = 0.2$–1.0; Clancy et al., 2009) and upper mesospheric hazes of indeterminate composition ($R_{eff} < 0.1$ μm; Montmessin et al., 2006a), based upon ultraviolet-to-visible wavelength-dependent optical depth behaviors. Additional specifics of cloud particle sizes, such as size variance and functional distribution, are largely unconstrained by observations (e.g. Wolff and Clancy, 2003). Modified gamma (Deirmidjian, 1964), standard gamma, and log-normal distributions have been employed in water ice particle size analyses, employing relatively narrow size distributions typical of non-convective clouds on Earth (Mason, 1971; $v_{eff} = 0.1$–0.2 μm, Chassefière et al., 1992; Clancy et al., 2003; Fedorova et al., 2009; Madeleine et al., 2012b). More recently, Fedorova et al. (2014) analyzed SPICAM solar occultation observations with combined UV/near-IR spectral coverage to retrieve bimodal aerosol particle size distributions over 10–50 km altitudes during the aphelion season. Two distinct aerosol sizes ($R_{eff} \sim 1$ and 0.05 μm) are identified, although their compositions (dust versus ice) are not spectroscopically determined.

Cloud particle shapes are also not well characterized. However, the observed single-scattering phase functions of type 1 water ice clouds (Clancy et al., 2003; Wolff et al., 2011; see Chapter 6) have been interpreted in terms of equidimensional faceted shapes such as droxtals, small ice crystals proposed to occur in high-altitude cirrus and polar stratospheric clouds (Yang et al., 2003; Figure 5.17). These shapes appear to correlate with low supersaturation conditions (few percent) as well as cold temperatures in the terrestrial upper troposphere (below −40°C; Bailey and Hallett, 2009). Such shapes minimize strong backscattering enhancements associated with columnar faceted ice crystals, and so better fit the observed scattering angle behavior of Mars clouds (Wolff et al., 2011). In principle, polarization properties of cloud particles also provide evidence of particle shapes. Cloud polarization observations do exist for Mars clouds (e.g. Anderson and Leovy, 1978; Martin et al., 1992), but are not sufficiently diagnostic to draw specific conclusions.

5.5.2 Cloud Radiative Properties

Chapter 6 presents a detailed description of the wavelength-dependent ice optical constants and scattering parameters. Here, we summarize the salient features of atmospheric radiation associated with Mars clouds. The radiative effects of Mars dust aerosols have been considered in much greater detail than those for clouds, due to the typical predominance of atmospheric dust versus ice opacities (particularly during global-to-regional dust storm events). More importantly, Mars dust strongly absorbs solar radiation and so contributes significant direct heating to the Mars atmospheric radiative balance (see Chapters 6 and 10). Mars clouds, CO_2 and water ice, present no such solar absorption over ultraviolet ($\lambda > 200$ nm) to visible wavelengths. Their ability to scatter solar radiation also plays a modest role in that the solar radiation is predominantly scattered toward the surface rather than backward to space. Hence, cloud radiative cooling/heating contributions are primarily limited to the near- and thermal IR regimes, associated with surface and atmospheric thermal IR effects rather than with direct solar coupling. Nevertheless, clouds do impact atmospheric temperatures due to their absorption of thermal IR radiation, forcing distinct diurnal/tidal variations (Wilson et al., 2008). These affects are shown to influence global simulations of atmospheric thermal structure and cloud optical depths (Haberle et al., 2011; Madeleine et al., 2012a; Wilson and Guzewich, 2014), global water vapor transport (Navarro et al., 2014), and upper-level circulation and polar nightglow (Clancy et al., 2012).

The radiative properties of ice clouds also determine their observational characteristics. Water ice exhibits prominent thermal IR absorption bands over 11–16 μm and 35–45 μm, which appear as absorption features in nadir observations (e.g. Smith et al., 2000) and as emission features in limb spectra (Pearl et al., 2001; Clancy et al., 2007; McClesse et al., 2010). The detailed spectral dependence of these features depends on the water ice particle sizes, as indicated above. Water ice also presents near-IR absorption bands in solar reflectance observations (such as from MRO/CRISM and MEX/OMEGA) at 1.5, 2.0, 2.5, and 3.1 μm wavelengths, although use of the 2.0 and 2.5 μm bands is strongly impacted by the presence of gaseous (CO_2, H_2O) band absorptions (Langevin et al., 2007; Vincendon et al., 2011; Madeleine et al., 2012a). The strong 3.1 μm absorption band of Mars atmospheric clouds exhibits useful sensitivity to ice particle size, as indicated above. Mars CO_2 ice clouds present absorption bands centered at 2.7 and 4.25 μm that are typically obscured by strong gaseous band absorptions contributed by the CO_2 atmosphere. Equatorial CO_2 clouds at high altitudes (60–80 km) present measurable scattering in these bands due to much reduced CO_2 gaseous columns at these altitudes (discussed below; Montmessin et al., 2007b). Weaker CO_2 ice bands at 1.43, 1.6, and 2.0 μm are not presented in the near-IR spectra of these clouds. These bands are, however, strongly exhibited in the reflectances of polar CO_2 surface ices, associated with very large (millimeter to centimeter) ice grain sizes (Langevin et al., 2007). At longer wavelengths, CO_2 has also been spectrally distinguished from water ice and dust aerosols in the polar night, associated with MCS thermal infrared bands over 11 to 40 μm (Hayne et al., 2012). However, the much smaller water ice versus CO_2 cloud particle sizes play a substantial role in this spectral separation.

5.6 SIMULATING WATER ICE CLOUDS IN MARS GENERAL CIRCULATION MODELS

The formation of water ice clouds affects the radiative balance (Chapter 6), volatile transport (Chapter 11), and

photochemistry (Chapter 13) of the global Mars atmosphere. Consequently, cloud microphysical simulations in Mars general circulation models (MGCMs) are a significant development in the study of global Mars atmosphere. The formation of clouds in the Martian atmosphere is the end product of a complex chain of processes involving thermodynamics, microphysics, and atmospheric dynamics. However, the essential factor for cloud formation and further evolution is temperature due to its nonlinear control of the equilibrium between the gaseous and the condensed phases of water. Major cloud manifestations on Mars, as detailed in the previous sections, are clearly related to seasons and locations associated with colder climatic conditions. To first order, the seasonal behavior of Martian water ice clouds reflects that of atmospheric temperature variations in the troposphere (in this case, defined as the lowermost 30–50 km atmospheric region exhibiting a significant lapse rate). However, clouds also depend on the availability of condensable material such that atmospheric transport provides significant control on clouds as well. The recent introduction of water ice cloud representations in MGCMs has enabled investigation of Martian cloud formation processes within a consistent and detailed climatic framework (Richardson et al., 2002; Montmessin et al., 2004). As for Earth cloud modeling, however, global models face difficulties associated with the constraints imposed by the specifics of the cloud physics. A central issue concerns the accommodation of the large-scale approach of MGCMs with the restricted spatial and temporal scales relevant to clouds.

Terrestrial models have handled this problem by introducing empirical formulations specifically dedicated to cloud processes, such as parameterization of subgrid-scale cloud cover (e.g. Marchand and Ackerman, 2010). This approach is not practicable for MGCM cloud simulations because required water vapor/cloud observational constraints at sufficiently resolved scales are not available for the Mars atmosphere. However, a major difference between the Mars and Earth cloud systems is the absence on Mars of feedbacks created by the exchanges of latent heat during condensation and sublimation. The trace concentration of water in the Martian atmosphere presents negligible potential for modifying the local environment through absorption or release of latent heat, inducing ≤1 K changes in temperature (Zurek et al., 1992). On the Earth, latent heat release during cloud droplet formation enhances buoyant instabilities and amplifies convective motions, forcing updraft and downdraft motions to occur at spatial scales (kilometers or so) that are not resolved by MGCMs. For this reason, the largest uncertainties in terrestrial climate modeling arise from the parameterization of these subgrid-scale phenomena that shape most of the terrestrial cloud morphologies. As indicated in the Fourth Intergovernmental Panel on Climate Change (IPCC) report (Solomon et al., 2007): *"Nowhere in climate models is there greater potential for error than in the treatment of clouds. This is especially true of low clouds, which cool the climate system, and which the IPCC has admitted are the largest source of uncertainty in global warming projections."* The negligible role of latent heat on Mars is a major simplification of the water ice cloud formation process, which has permitted rapid progress in its study since the first introduction in an MGCM (Richardson et al., 2002).

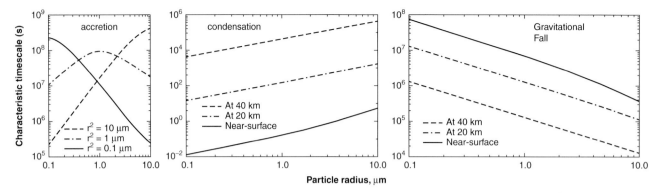

Figure 5.18. Typical timescales of key microphysical processes affecting Martian water ice clouds are presented as a function of the cloud particle radius in micrometers. (Left) The particle accretion (coagulation) timescale is plotted as a function of the secondary accreting particle radius, r_2. (Middle) The particle growth timescale, defined as the time required for a particle to reach twice its initial radius, is plotted for altitudes of 0, 20, and 40 km. (Right) The timescale required for a cloud particle to fall over one atmospheric scale height is plotted for altitudes of 0, 20, and 40 km. Figure is provided by Franck Montmessin, LATMOS/CNRS, Paris.

5.6.1 Cloud Microphysics

Before addressing specific approaches to Mars cloud simulations, it is useful to describe the microphysical processes involved in cloud formation and evolution. At the scale of a Martian cloud particle, four major mechanisms need to be considered; nucleation, condensation/sublimation, gravitational fall (sedimentation), and accretion (coagulation). Accretion does not play a significant role in Mars cloud formation, as demonstrated by Rossow (1978) and Montmessin et al. (2004). The relatively low number density (≤ 10 cm^{-3}) of cloud or dust particles in the Martian atmosphere significantly reduces collision (hence, accretion) rates between particles associated with random Brownian motion or gravitational settling (Figure 5.18, left panel). In addition, accretion probability is reduced by the typically small dispersion of the Mars water ice cloud particle sizes, as indicated by models (Michelangeli et al., 1993; Colaprete et al., 1999; Daerden et al., 2010). Prior to growth by condensation, a water ice crystal must nucleate out of the vapor phase in excess of saturation. On Mars, nucleation is theoretically promoted by the presence of suspended dust, as there is a rather high crystallographic affinity between water ice and dust substrates. Nevertheless, high supersaturation ratios may still be necessary for ice cloud nucleation based upon recent experimental studies by Iraci et al. (2010) and Phebus et al. (2011). A saturation ratio in the range of 1.2 to 2 is in principle sufficient to trigger nucleation events (Määttänen et al., 2005; Iraci et al., 2010). Nucleation is nearly instantaneous and therefore does not induce substantial delay in cloud formation associated with a 20–100% supersaturation threshold. This threshold is equivalent to cooling a 200 K atmosphere by only 1–5 K, a modest temperature change considering the >10 K diurnal variations of Mars atmospheric temperatures.

Theoretical considerations (e.g. Michelangeli et al., 1993) also indicate that condensation is a relatively fast process in the lower Mars atmosphere (Figure 5.18, middle panel), with typical timescales on the order of seconds to minutes, and so remains quite sensitive to changes in the local thermodynamical environment. In such a case, condensation and sublimation

respond quickly to changes in saturation conditions, forcing water vapor partial pressure to constantly achieve equilibrium pressure over ice, regardless of the other processes at work. However, particle growth can be severely delayed when the free molecular path becomes far greater than particle size, a situation that typically occurs at low pressure. This effect should become prominent above 30–40 km altitudes, increasing condensation timescales up to days or weeks (Figure 5.18, middle panel). The recent detection of water vapor in excess of saturation in the tropics above 30 km (Maltagliati et al., 2011, 2013) is the likely consequence of condensation in a low-pressure medium. Cloud particle sedimentation rates present an altitude dependence opposite to condensation (Figure 5.18, right panel), with timescales on the order of weeks to months near the ground versus hours to days at 40 km for a micrometer-sized particle.

These basic considerations allow one to infer important trends for cloud formation in the Martian atmosphere that are of particular interest in the study of their global distributions and effects on global transport of atmospheric water, as discussed in detail in Chapter 11. In terms of cloud particle properties, the above microphysical conditions lead to distinctly larger ice particle sizes in the lower atmosphere. The recent observations made by the Phoenix Lander and described in a previous section illustrate this conclusion. The 20–40 μm radius found for the low-lying summer polar clouds (Whiteway et al., 2009; Daerden et al., 2010) dramatically contrasts with perihelion cloud trail ice particles at 40–50 km altitudes, with radii <1 μm (Clancy et al., 2009). In either case, negligible accretion rates and the dominance of condensation/sublimation over sedimentation in the lower atmosphere generally lead to non-precipitation (to the surface) for Mars water ice clouds, as crystals leaving a cloud zone will quickly sublime and so disappear before reaching the ground. However, very different conditions apply during Mars polar night, where the atmosphere is saturated throughout the troposphere and is thus more favorable for precipitation to the surface (and, hence, formation of the seasonal ice caps; Richardson et al., 2002; Montmessin et al., 2004; Nelli et al., 2010).

5.6.2 Representation of Water Ice Clouds in Mars General Circulation Models

Cloud formation is effectively paced by the rate of change of the local saturation conditions. The latter can vary in two ways, either by local supply/removal of water vapor or by temperature variations. The exponential dependence on temperature for the water vapor pressure in equilibrium over ice generally means that temperature rather than water vapor concentration is the prime driver of cloud formation. Diurnal and seasonal variations of temperature establish the dominant periodicities in the observed activity of Mars clouds (Colburn et al., 1989; Tamppari et al., 2000, 2003; Smith, 2004). It is interesting to note that the wettest region on Mars is the summertime north pole where the atmosphere locally reaches its full water holding capacity, with a water column abundance exceeding 50 pr μm (see Chapter 11). Yet comparably enhanced cloud manifestation has not been observed to occur there. Cloud formation in this season actually occurs at low latitudes where considerably reduced water vapor columns are present (10 pr μm). The key variable, in this case, is vigorous vertical transport of water vapor to higher altitudes (10–25 km), where (vertically) decreasing temperatures lead to saturation conditions and cloud formation.

Initial studies of Mars cloud simulations, employing 1D microphysical models (Michelangeli et al., 1993; Colaprete et al., 1999; Montmessin et al., 2004), emphasized the importance of this vertical transport of water vapor to fuel cloud formation at altitudes (i.e. temperatures) of water vapor saturation. The average temperature profile of Mars exhibits a steady decrease of temperature from the surface up to 40 km (the tropopause), above which the structure becomes dominated by wave activity. In an average sense, clouds can only be fueled by a water vapor supply coming from below their base where the vapor pressure is significantly higher and which therefore maintains a vertical gradient of water vapor with the upper atmospheric layers. Except for the specific case of the wintertime polar regions, water supply is facilitated by convection within the daytime boundary layer and by global-scale advection at higher elevations. One-dimensional models necessitate an *ad hoc* representation of atmospheric water vapor transport associated with vertical mixing. There are substantial differences between the behavior of an advected air mass, such as that occurring in a convective updraft, and the vertically symmetric diffusion approach of standard 1D models that is more appropriate to environments dominated by turbulence. The calculation of self-consistent global thermal and transport conditions, which are in fact modulated by cloud radiative behaviors (Haberle et al., 2011; Madeleine et al., 2012a, 2014), requires the incorporation of cloud microphysics in MGCMs.

However, the demanding computational requirements for highly detailed cloud microphysics currently prevent such treatment within a three-dimensional climate model framework. For instance, the 1D microphysical models of Michelangeli et al. (1993), Daerden et al. (2010), and Burlakov and Rodin (2012) allotted 40–100 size bins to the description of particle size distributions for simulated aerosol species associated with a variety of ice/dust composite structures, an approach not compatible with current capabilities of three-dimensional models.

There is a necessary trade-off between consistent dynamical and microphysical representations when both are to be studied concomitantly. The requirements for resolving the highest zonal wavenumbers of dynamical phenomena (thus requiring high spatial resolution) and the necessity of multi-annual simulations to achieve equilibrated water cycles (of which water ice clouds are a component) leave very few degrees of freedom to simulate clouds rigorously in an MGCM. Furthermore, microphysical processes occur on timescales of the order of seconds or less (e.g. nucleation and condensation), as compared with the MGCM typical dynamical timescales that are usually in the range of tens of minutes. Consequently, simplified approaches have been the common rule in the initial MGCM-led cloud studies, often neglecting nucleation processes and the interactions between cloud and dust particles, as well as cloud feedbacks on the radiative budget (Richardson et al., 2002). Houben et al. (1997) used an even simpler description, mimicking cloud-related processes via an instantaneous transfer of the water vapor in excess of saturation below the condensation level (i.e. assuming that condensation and sedimentation occur instantaneously).

Currently, the prevailing method of MGCM cloud modeling is to reduce cloud properties inside a model grid element (usually larger than hundreds of kilometers in horizontal scale) to a single representative particle responding to the MGCM predicted evolution of atmospheric variables (Richardson et al., 2002; Montmessin et al., 2004; Böttger et al., 2005; Montmessin et al., 2007a; Nelli et al., 2010). The "single-particle approach" can feature either a fixed particle size – such as in Richardson et al. (2002) and Böttger et al. (2005) – or an evolving particle size according to the predicted amount of condensed water equally partitioned among a prescribed, vertically varying number of condensation nuclei, as in Montmessin et al. (2004). The radius of cloud particles determines the gravitational fall speed, thus influencing cloud lifetime, vertical extent, and thickness. A predictive radius thus allows water ice particles to vary in size and properties in response to the thermodynamical conditions. In the study of Mars recent climate changes, the predictive radius approach proved particularly efficient in simulating large (>20 μm) cloud particles, appropriate to the prediction of equatorial precipitation events associated with the formation of piedmont glaciers on the western flanks of Tharsis volcanoes (Forget et al., 2006).

As shown by Richardson et al. (2002) and Montmessin et al. (2004), the influence of cloud properties on the global seasonal water cycle is critical, controlling most of the exchanges of water occurring between the two hemispheres and modulating the cross-equatorial flow of water at aphelion by up to a factor of 3 (Montmessin et al., 2004). The most recent cloud models coupled to MGCMs now describe particle size distribution through their first-order moments (e.g. cloud particle number concentrations, volume, etc.), allowing one to represent cloud particle properties and related processes in more detail (Haberle et al., 2011; Rodin et al., 2011). Among a range of numerical developments in cloud modeling (such as cloud radiative interactions; see Section 6.5), a general trend toward progressive microphysical sophistication proceeds apace (Wilson et al., 2007, 2008; Haberle et al., 2011; Madeleine et al., 2012a).

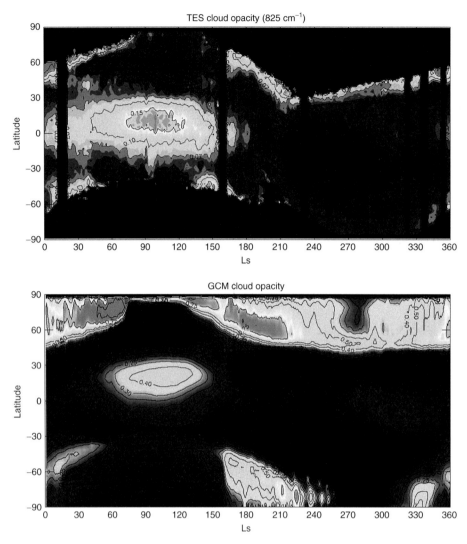

Figure 5.19. A comparison between observed TES and Mars GCM predicted cloud optical depth (12 μm) distributions as a function of season and latitude. Notice that TES observations do not provide optical depth retrievals for un-illuminated high-latitude regions. The generally good agreement between model and data supports the modeled cloud microphysics in the context of the GCM simulated thermal and circulation regimes. Reproduced from Montmessin et al. (2004).

5.6.3 Dynamical Control of Global Cloud Structures on Mars

Mars atmospheric circulation (Chapter 9) shares many similarities with the terrestrial circulation; it is characterized by large Hadley cells redistributing heat from the summer to the winter hemispheres around solstices and by planetary-scale baroclinic waves shaping the weather patterns at the mid-latitudes. Distinct from the Earth, however, the Mars solstitial circulation is dominated by a single, vertically deep overturning cell (Leovy and Mintz, 1969), extending well above the tropopause level (near 40 km). These cells mix atmospheric columns over considerable depths, lofting aerosols up to 70 km during southern spring and summer (Fedorova et al., 2009; Clancy et al., 2010). At mid- to high latitudes, geostrophic flows become highly unstable, giving rise to baroclinic instabilities exhibiting seasonally varying wavenumber patterns (Zurek, 1992). In addition, the presence of large topographical landmarks (volcanic ridges, impact basins) forces the development of planetary-scale stationary waves. Altogether, these dynamical factors force global- and regional-scale manifestations of water ice clouds on Mars through their influences on atmospheric temperature distributions and water vapor transport.

Figure 5.19 presents a global comparison of the seasonal (solar longitude, L_s) Martian water ice cloud evolution as obtained from the MGS/TES infrared spectrometer (Smith, 2004), and as simulated by the Laboratoire de Météorologie Dynamique (LMD) MGCM (Montmessin et al., 2004). The agreement between these observed and model contour maps (versus L_s, latitude) of zonal average cloud optical depths (λ = 12 μm) generally validates the cloud microphysical approximations adopted in such MGCM cloud simulations. Notice, however, that model tropical optical depths exceed the TES values by a factor ranging from 1.5 to >2. A significant fraction of this overestimation reflects radiative warming feedback by tropical clouds on tropospheric temperature at aphelion (Wilson et al., 2008; Madeleine et al., 2012a), which is

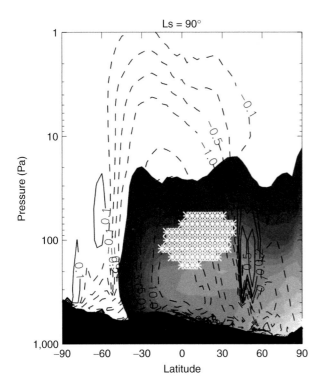

Figure 5.20. Zonally averaged, altitude cross-sections for a GCM simulation of water vapor (shaded gray, from 10 = dark to 1000 ppm = light), cloud concentration (white asterisks), and circulation mass streamfunction (lines). This aphelion season exhibits a predominantly single Hadley overturning cell with counterclockwise orientation (dashed lines), in which water vapor is transported to the ACB formation region centered over the northern subtropics. Reproduced from Montmessin et al. (2004).

not included in this model. Reasonable model agreement is also obtained for water ice cloud particle size properties, based on model comparisons with cloud particle size determinations from TES observations (Clancy et al., 2003; Wolff and Clancy, 2003). Two global Martian cloud structures discussed earlier are prominent in Figure 5.19, the tropical aphelion cloud belt (ACB) and the polar hood (PH) clouds forming in both hemispheres during fall and winter. Again, the PH clouds are only partially displayed in the TES infrared optical depth measurements, which are limited to Sun-lighted (warm surface temperature) conditions (Smith, 2004).

5.6.3.1 The Aphelion Cloud Belt

The origin of the ACB was initially ascribed to vertical lifting of water vapor within the upwelling branch of the northern summer Hadley circulation (Figure 5.20) and the coincidence of this circulation with the cold Mars aphelion ($L_s = 71°$) atmosphere (Clancy et al., 1996). This behavior is now well demonstrated in MGCM cloud simulations (Figures 5.19 and 5.20). Hence, a close coincidence exists between the timing of the ACB and that of the northern summer solstice Hadley cell (which currently occurs near Mars aphelion). A similar cell, but with an opposite orientation and a deeper development, exists during the southern spring and summer. However, the significantly (~20 K) warmer climate of the perihelion ($L_s = 251°$)

season shifts the condensation level above 40 km, where cloud growth rates are smaller and gravitational settling rates are larger. Consequently, much reduced cloud optical depths result for this period (Figures 5.12 and 5.13). Another factor key to the onset and the decay of the ACB is transport of water vapor from the subliming north polar cap, where the major exposed surface reservoir of water ice resides.

A detailed description of the transport mechanisms associated with ACB formation is provided in Chapter 11. To summarize, water vapor meridional export from the subliming north polar cap is locked into a wave-3 stationary configuration dominated by a near-surface shallow channel. This channel tends to concentrate transport of water vapor from the northern high latitudes to the western part of the Tharsis plateau. The latitudinal/vertical dependence of Hadley circulation enforces large-scale convergence of water vapor in the northern tropics with subsequent upwelling and adiabatic cooling that force cloud formation above 10 km altitudes. This mechanism is similar to the Earth inter-tropical convergence zone where deep convective clouds are generated. The ACB exhibits a pronounced zonal structure with focused enhancements of cloud opacity near the volcanoes (Benson et al., 2003). This behavior is associated with the high-latitude transport channels that export water from north polar regions, and with the reinforcement of clouds over the volcano regions of Mars. These topographic obstacles force stronger uplifts and atmospheric cooling compared to the otherwise flat terrain of the equatorial region. According to observations and modeling, the cloud equivalent water content in the ACB equals up to 5 gm m^{-2}. The persistently fainter clouds at other longitudes in the 10°S to 30°N area (e.g. Figures 5.1 and 5.11) are the result of the zonal-mean upward orientation of the circulation. Recent mesoscale modeling dedicated to the volcano clouds of the ACB also reveals a stronger local circulation than predicted by the MGCMs, which further disrupts the zonal structure of circulation and locally increases cloud formation (Michaels et al., 2006). These volcano-induced flows are caused by the synergism of upslope thermal flow and lee volcano waves (Figure 5.21). Such fine circulation aspects lie beyond the coarse resolution capability of the MGCMs, yet they may account for a major portion of the upward transport of species in the Hadley cell. Additional factors, such as the influence of the thermal tides, play complementary roles in shaping the variability of the ACB as addressed subsequently (Section 6.5).

Montmessin et al. (2004) estimate that clouds constitute ~10% of Mars atmospheric water content on a global, annual average. To the degree that cloud particles remain entrained in the atmosphere, they tend to preserve atmospheric water content in cold atmospheric regions of minimal water vapor abundance, and so generally enhance global atmospheric water abundance and transport (as discussed in the following PH discussion in Section 5.6.3.2 and Chapter 11). However, clouds may also impede atmospheric water transport. The ACB, in particular, was proposed to play such a role in modulating the seasonal evolution of Mars' water cycle (Clancy et al., 1996). The model of Montmessin et al. (2004) confirms that southward cross-equatorial advection of water released from northern summer polar ice caps is restricted by cloud particle sedimentation within the northern subtropical ACB. The magnitude of this effect depends on the ACB cloud particle sizes. Larger particle

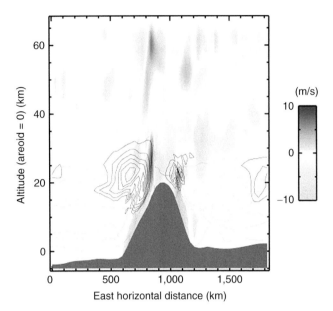

Figure 5.21. Results from a mesoscale simulation of cloud microphysics and transport over a Mars volcano are presented as instantaneous vertical W–E cross-sections through the Olympus Mons summit (vertical velocity shaded in light = down and dark = up), and contours of cloud ice mass mixing ratios (contour lines, in the range $(1–5)\times10^{-5}$) during the afternoon. Reproduced from a color figure in Michaels et al. (2006).

sizes lead to more rapid particle fall rates, confining the atmospheric water content to lower altitudes and so more efficiently restricting southward transport of atmospheric water (vapor plus ice) in the upper-level branch of the Hadley circulation. Montmessin et al. (2004) estimate that likely (modeled and observed) cloud particle sizes for ACB ($R_{eff} \sim$ 3–6 μm) reduce cross-equatorial transport of atmospheric water released from the northern polar ice cap by ~50%. This mechanism may also impact the partitioning of water between the two hemispheres over the past 100 000 years, corresponding to the orbital evolution of the season for Mars aphelion (Montmessin et al., 2007a; Madeleine et al., 2014).

5.6.3.2 The Polar Hoods

Observed seasonal and spatial distributions of polar hood (PH) clouds (e.g. Figures 5.12–5.15) are substantially reproduced by existing MGCM cloud simulations (Figure 5.19). PH clouds are distinguished from ACB clouds in that they form closer to the surface and provide favorable conditions for ice precipitation (Montmessin et al., 2004). Two separate mechanisms contribute to the onset and the maintenance/decay of the PH of both hemispheres, although conditions specific to each hemisphere lead to somewhat asymmetric behaviors. At northern high latitudes, residual humidity associated with north polar cap sublimation gives rise to the low-lying (~5 km) clouds observed by Phoenix towards the end of northern summer (after L_s = 120°, see Section 5.6.3.1). During northern fall, the polar vortex boundary progresses towards the equator. PH clouds accompany this progression, associated with residual water vapor evolved from the summer polar cap. This "wet"

atmospheric background fuels PH cloud formation during late summer and fall (e.g. Wang and Ingersoll, 2002; Montmessin et al., 2004; Smith, 2004; Tamppari et al., 2008; Benson et al., 2010; Nelli et al., 2010). PH latitudinal extension ceases, in conjunction with the polar vortex, after equinox (L_s = 180°) at around 45°N (Benson et al., 2010), when models predict a significant reduction of PH activity (Montmessin et al., 2004). A second stage of winter PH development begins near L_s = 300° with the emerging role of traveling disturbances. This period is associated with a particular phenomenon first described by Houben et al. (1997), a continuous and poleward recycling of volatiles released by the subliming edge of the seasonal polar frost. Water ice in the mid-latitude seasonal ice cap is exposed to increasing levels of sunlight and re-sublimes into the atmosphere near the edge of the polar vortex. At the same time, increased Rossby wave activity is predicted to occur. Eddies disorganize the zonal symmetry of the vortex, creating deep poleward intrusions of the wet air masses that reignite PH formation (Montmessin et al., 2004). This mechanism persists until northern spring, the final stage of the polar vortex retraction phase. Although the same mechanism is predicted to occur at southern high latitudes in corresponding seasons, the significantly warmer climate of late fall and early spring seasons in the south reduces the potential for PH formation, as confirmed by MEX/OMEGA (Langevin et al., 2007) and MRO/MCS (Benson et al., 2010) observations.

Modeled particle sizes (radii ~ 3–6 μm; Montmessin et al., 2004) and the lower altitude (0–30 km) occurrence for global PH clouds (as distinct from the diurnal, lower PBL cloud layer observed by Phoenix LIDAR) correspond to fall lifetimes sufficient to support large-scale (≥1000 km) horizontal transport (Montmessin et al., 2004). For this reason, PH clouds can play a pivotal role in Mars' water cycle regulation, through transport of atmospheric water in the form of ice over cold polar night regions where water vapor is negligible. The net PH-induced effect on Mars' global water cycle consists of a significant reduction of the poleward flux of water during the winter–spring waning stages of the vortex. Since PH clouds can be transported out of the vortex via the cold equatorward phase of the eddies, they partially balance the poleward flux of water vapor and so sustain larger amounts of water vapor away from the poles. Hence, on an annual average, such PH-induced water transport may increase the globally integrated atmospheric inventory of water by a factor of 2 (Montmessin et al., 2004).

5.6.4 Radiative Effects of Clouds and Thermal Tides

Most recently, global and mesoscale studies of Martian clouds have considered the significant interactions between clouds, radiation, and thermal tides. Tidal interactions are especially important in the context of the ACB, imparting large day-to-night modulations of the ACB cloud properties and vertical extent. Tidal perturbations of the thermal structure in the tropics are reflected by observed deep (>10–20 K) temperature inversions that are particularly prominent over Tharsis where various tidal modes interfere constructively (Hinson and Wilson, 2004). Prominent nighttime clouds are modeled to form within an elevated layer above Tharsis. They descend steadily through the night, in response to the downward phase propagation of the thermal tides, to merge eventually before dawn with the ground fog, after which time they begin to dissipate (Hinson and Wilson, 2004). Thermal tides initially induce adiabatic cooling and temperature inversions below which cloud layers form. In turn, the infrared radiative cooling produced by these clouds intensifies the temperature inversion and thus forces a nonlinear coupling between the clouds and the tides. Wilson et al. (2007) indicate that higher-opacity daytime clouds are typically associated with strong circulations localized to volcanoes as in Michaels et al. (2006), whereas nighttime, tidally driven clouds reflect circulation influences on larger scales. Thermal tides associated with Tharsis topography are thus predicted to play a major role in the diurnal variation and vertical structure of tropical water ice clouds. Thermal tides also affect the vertical distributions of polar hood clouds, as indicated in Figure 5.15 (Benson et al., 2010, 2011).

Recent MGCM simulations incorporating cloud radiative effects have also indicated global influences on atmospheric thermal structure and the transport of water vapor and dust. These effects are due to water ice absorption (and emission) of infrared radiation, and associated effects on the vertical distribution of dust, a long-recognized contributor to heating and cooling rates in the Mars atmosphere. Incorporation of cloud IR radiation in MGCM simulations leads to improved model–data comparisons for atmospheric thermal structure within the ACB and PH cloud regimes in particular (Wilson and Guzewich, 2014), but also leads to much higher cloud optical depths than indicated by observations (Haberle et al., 2011; Madeleine et al., 2012b) and poorer agreement between observed and (currently) modeled atmospheric water distributions (Navarro et al., 2014). Cloud radiative effects also impact MGCM meridional, zonal, and polar vortex circulations (e.g. Wilson, 2011b), a factor that appears to play a role in the behavior of polar winter oxygen airglow (Clancy et al., 2012). Finally, Mars water ice clouds strongly impact levels of odd hydrogen (H, OH, HO₂, and H₂O₂), associated with removal of water vapor as the ultimate source of these key catalytic species (Clancy and Nair, 1996; see Chapter 13). MGCM simulations of Mars photochemistry employing cloud microphysics demonstrate the correlation of aphelion and polar hood cloud-forming regions with atmospheric ozone abundances (Lefèvre et al., 2004), as well as suggesting that heterogeneous chemistry may occur on the surfaces of water ice clouds (Lefèvre et al., 2008). Most recently, cloud radiative effects have been integrated into Mars paleoclimate simulations (Madeleine et al., 2014).

5.7 CO₂ ICE CLOUDS IN THE MARS ATMOSPHERE

Every Martian year, nearly 30% of the global atmospheric mass condenses to form the seasonal polar caps. During the polar night, the radiative cooling of the surface and atmosphere is balanced by the release of latent heat from condensing CO₂. The formation of carbon dioxide clouds within the polar night alters the thermodynamic state of the atmosphere through the

distribution of latent heat release as well as scattering/absorption of surface infrared radiation (Chapter 12 provides a thorough development of these processes as well as appropriate references; see also James et al., 1992). MGS radio occultation, TES, and MRO MCS measurements in the polar regions show temperature profiles that frequently follow the saturation curve for CO₂ in the lower portion (below ~30 km) of the atmosphere (Hinson and Wilson, 2002; Colaprete et al., 2003; Hayne et al., 2012, 2014). These observations suggest that atmospheric temperatures in this region are buffered by atmospheric condensation of CO₂. Furthermore, MGS MOLA (Pettengill and Ford, 2000) and MRO MCS (Hayne et al., 2012) aerosol measurements indicate substantial cloud opacities at 0–25 km altitudes, consistent with CO₂ cloud formation.

The precise role of such CO₂ clouds in the growth and recession of the seasonal CO₂ ice caps remains unclear. While direct condensation of CO₂ on the surface is believed to be the dominant process for growth and energetics of the Mars seasonal polar caps, accumulation from precipitating CO₂ snow may contribute substantially over regional scales (Colaprete et al., 2005, 2008; Hayne et al., 2014). Hence, the process of ice deposition via cloud precipitation may lend clues to specific characteristics and distributions of polar frosts, including a distinctly brighter south polar seasonal ice cap (Paige and Ingersoll, 1985), which supports a year-round (residual) south polar CO₂ ice cap (Kieffer, 1979), variable seasonal ice morphologies, and long-term (several to 100 years) variations in the south polar CO₂ residual cap (Thomas et al., 2009). In a very different context, CO₂ ice clouds in the low-latitude, upper atmosphere of Mars have recently been discovered (e.g. Montmessin et al., 2007b). It remains to be seen whether there are climate implications associated with these clouds, as they are short-lived and present negligible atmospheric mass perturbations.

5.7.1 Observations of CO₂ Clouds

Carbon dioxide clouds most commonly form during the polar night, thus making their direct visual observation difficult. However, a number of spectral observations in the infrared, as well as active lidar observations, clearly indicate their presence. The most definitive spectroscopic determination of Mars CO₂ ice aerosols pertains to equatorial mesospheric clouds, associated with MEX/OMEGA imaging spectra of scattered solar flux in diagnostic near-IR CO₂ ice bands (Montmessin et al., 2007b). Observations of CO₂ ice clouds in the polar nights have been historically less direct. Mariner 9 and Viking Orbiter thermal IR observations have shown evidence for apparent surface brightness temperatures below the expected CO₂ saturation temperature in the polar regions, as so-called "cold spots" (Kieffer et al., 1976). Similar observations were obtained from MGS TES (Kieffer et al., 2000) and MRO MCS (Hayne et al., 2012). MCS limb profiles also present pervasive polar night aerosols extending from the near-surface to 20–30 km, which are highly correlated with the cold spots. Furthermore, radiative transfer calculations fit to MCS thermal IR radiance profiles (specifically, 12, 22, and 32 μm wavelengths) are spectrally indicative of CO₂ cloud composition (Hayne et al., 2012).

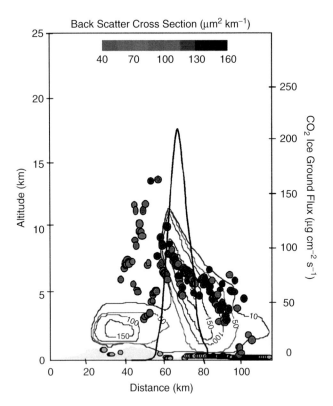

Figure 5.22. MOLA lidar backscatter cross-sections (labeled contours, at 1.067 nm) are simulated in a microphysical model incorporating 5 m s⁻¹ surface flow over the topography as measured by MOLA during pass number 226 (represented by the light shaded region at the surface). MOLA backscatter cross-sections for the same pass are shown as solid shaded circles with gray scale provided at the top of the figure. Also shown (dark, peaked line with right axis) is the calculated mass flux of CO_2 ice particles to the surface for a segment of MOLA pass 226. For the simulation presented here in a 1 hour snow storm, approximately 0.75 g cm⁻² of CO_2 snow may be deposited to the surface. Figure is reproduced from Colaprete and Toon (2002).

The MGS Mars Orbiter Laser Altimeter (MOLA), designed for laser ranging measurements of the Martian surface, detected laser returns (reflections) at altitudes extending from the surface to as high as 20 km, and typically 10–15 km (albeit affected by MOLA measurement ambiguity; Neumann et al., 2003). These lidar reflections are associated with CO_2 cloud scattering (based upon negligible water vapor content and, hence, water ice clouds) in the polar night, and so provide a unique dataset as to the vertical structure of CO_2 clouds and their particle properties. Prominent structures in the observed CO_2 lidar returns appear consistent with sloping fronts of propagating buoyancy waves, extending from the surface to 5–10 km heights (Pettengill and Ford, 2000; Ivanov and Muhleman, 2001). Figure 5.22 presents a set of such MOLA reflections in comparison to a simulation incorporating moist CO_2 convection as forced by flow over a measured topographic obstacle (Colaprete and Toon, 2002).

MOLA lidar observations also indicate distinct differences between the altitudes (densest return at 6–8 km in the north versus 1–2 km in the south) and types of reflections seen at the two winter poles. MOLA had the ability to capture reflections across four time gates, with each gate corresponding to width in the returned laser pulse spread over time. In affect, these gates distinguish vertical gradients in the aerosol scattering. In southern polar winter many more channel-1 echoes, or reflections from a highly extinctive (i.e. vertically narrow) cloud, were observed relative to the northern polar winter. It has been argued that these channel-1 reflections could be the result of either large, flat CO_2 crystals (Neumann et al., 2003) or higher concentrations of much smaller CO_2 ice grains (Colaprete et al., 2003). Colaprete et al. (2003) point to distinctive wave-2 and wave-3 variations in these channel-1 reflections that may be influenced by the large Hellas and Argyre Basins at southern high latitudes.

MGS TES and MRO MCS measurements of surface and atmospheric brightness temperatures below expected CO_2 saturation temperatures have been attributed to significant CO_2 cloud scattering in the lower polar winter atmosphere. MCS limb observations of aerosol extinction over 0–30 km altitudes indicate that optically thick CO_2 clouds persist above 80°S over much of the southern polar winter cap, with less optically thick, variable cloud cover extending down to 60°S. These MCS limb cloud detections correlate with the distribution of anomalously low surface IR brightness ("cold spots"), and are also consistent with vigorous CO_2 snowfall in the lower polar winter atmosphere (<30 km, and perhaps concentrated below 4 km; Hayne et al., 2012). MCS observations appear to confirm the hypothesis that polar winter "cold spots" are radiative anomalies (reduced emissivity) associated with scattering by significant CO_2 snowfall within the atmosphere. However, surface emissivity variations appear also to play a role, and may be associated with granular surface deposits formed by precipitation of CO_2 ice clouds (Hayne et al., 2012).

Zonal asymmetries in the apparent distribution of south polar CO_2 snowfall have been suggested to arise from a persistent wavenumber-1 stationary wave forced by the Hellas and Argyre Basins, which pushes the polar vortex toward the perennial CO_2 cap near 45°W (Colaprete et al., 2005). In the north, where topography is much more subdued, models suggest eastward-propagating planetary waves interact to produce cloudy longitudes 60–120°E and 90–120°W (Kuroda et al., 2013). Cooling rate calculations and snow particle settling models constrained by the MCS observations suggest that CO_2 snowfall may contribute a substantial amount of material to the growing seasonal caps, ~3–20% in south polar mid-winter (Hayne et al., 2014). MCS limb observations further suggest larger particle sizes for northern (~100 μm) versus southern (~10 μm) polar winter CO_2 clouds, as well as larger overall northern cloud opacities (Hayne and Paige, 2009). Such smaller cloud particle sizes in the southern polar winter may correspond to the channel-1 clouds observed by MOLA (Colaprete et al., 2005).

Mars CO_2 clouds have also been observed at mesospheric (50–100 km) altitudes, most prominently over equatorial latitudes. During the atmospheric entry of Mars Pathfinder, atmospheric temperatures consistent with CO_2 saturation conditions were observed at altitudes near 80 km (Schofield et al., 1997). Subsequent sky imaging observations from the Pathfinder Lander revealed occasional pre-dawn, blue-color clouds. Clancy and Sandor (1998) argued that these pre-dawn clouds and the cold mesospheric temperatures were consistent with high-altitude CO_2 clouds. Visible mapping observations of equatorial cloud scattering layers at 60–80 km resulted from MGS TES limb scans and MOC limb imaging (Figure 5.23),

Figure 5.23. This MOC WA blue image of the Mars atmospheric limb was observed on May 5, 2004 (L_s = 28.9°, r17–00363: UT = 00:48:32). Scattering along the viewed limb path reveals high-altitude clouds (~68 km) extending from 5.2°S to 4.8°S (i.e. ~25 km N–S extent), at a longitude of 75°W, and LT of ~1:15 p.m. Although the MOC image does not indicate composition, the location and season of this cloud correspond to CO_2 composition clouds determined from MEX/OMEGA near-IR spectra (Montmessin et al., 2007b). Figure is reproduced from Clancy et al. (2007).

indicating that such mesospheric clouds are primarily equatorial (15°N–15°S) with distinctive seasonal (notably L_s = 0–70°, 100–160°) and longitudinal (30°E to 120°W) dependences as well (Clancy et al., 2007). The observed spatial and temporal distributions of cloud occurrence suggested tidally influenced CO_2 clouds. This relationship is supported by dynamical model predictions of temperature minima at seasons and local times where CO_2 clouds are observed (Montmessin et al., 2007b; Määttänen et al., 2010; González-Galindo et al., 2011), although additional cooling by gravity waves seems to be necessary to reach CO_2 condensation temperatures (Spiga et al., 2012). However, these TES and MOC limb observations do not provide a spectroscopic determination of composition, and the long limb paths preclude definitions of cloud forms or spatially resolved cloud optical depths. The THEMIS instrument on Mars Odyssey obtained nadir, spatially resolved images of these clouds at mesospheric altitudes (based upon parallax among the filter images) at equivalent seasons and locations, with visible optical depth determinations between 0.05 and 0.5 (Inada et al., 2007; McConnochie et al., 2010).

Definitive measurements of CO_2 composition for these mesospheric clouds were provided by the MEX OMEGA near-IR imaging spectrometer. Diagnostic 2.7 μm and 4.24, 4.26 μm "emission" peaks within gaseous CO_2 absorption bands unambiguously identify high-altitude scattering by CO_2 clouds (Montmessin et al., 2007b); the spectral details of these emission peaks can be sensitive to composite particle structures (Isenor et al., 2013). These key OMEGA observations also provide optical depths ($\tau_{visible}$ > 0.2) for spatially resolved clouds and indications of surprisingly large cloud particle sizes (1–2 μm) at mesospheric altitudes. Most recently, CRISM limb spectra over λ = 0.4–4 μm have also spectroscopically identified mesospheric CO_2 cloud compositions and particle sizes (Clancy et al., 2014).

High-spatial-resolution imaging observations from MEX HRSC (Määttänen et al., 2010; Scholten et al., 2010) and MRO

CRISM (Vincendon et al., 2011) generally indicate filamentary cirrus cloud structures aligned with the strong mesospheric circulation (Figure 5.6). As indicated in Määttänen et al. (2010), there is also reasonable agreement among all of the observational datasets in terms of the vertical, seasonal, and spatial distributions exhibited by equatorial mesospheric CO_2 clouds. Mid-latitude and later season (after L_s = 140°) mesospheric clouds observed in several of these datasets may actually correspond to water rather than CO_2 ice clouds (Vincendon et al., 2011), which also appear to predominate over the perihelion season (L_s = 160–330°; Sefton-Nash et al., 2013; Clancy et al., 2014). THEMIS twilight imaging (4–6 p.m., at >90° incidence angles) detected an extensive set of mesospheric (40–70 km) cloud layers over 40–50°N latitudes in northern fall–winter (L_s = 200–300°; McConnochie et al., 2010). A water ice composition for these clouds would be consistent with their association to NPH water ice clouds prevalent at lower altitudes. On the other hand, several mid-latitude occurrences of mesospheric CO_2 ice clouds have been identified (Vincendon et al., 2011).

At higher altitudes (>90 km) and latitudes, MEX SPICAM stellar occultation measurements reveal nighttime aerosol extinction layers above which temperatures are determined to fall below the saturation temperature of CO_2 (Montmessin et al., 2006a). Much smaller visible optical depths (<0.01) and particle sizes (<0.1 μm) are determined for these aerosols, which further distinguishes them from lower-level equatorial mesospheric CO_2 clouds. Although the SPICAM observations do not provide a spectroscopic determination of composition, Montmessin et al. (2006a) interpret the CO_2 saturation temperatures observed above these layers as indicative of a CO_2 aerosol composition. Vincendon et al. (2011) present counterarguments in support of a water ice composition, such that their composition remains unsettled.

5.7.2 CO₂ Cloud Formation Microphysics

The nucleation and growth of CO_2 clouds in the Mars atmosphere represents a special case of cloud formation, one in which the primary atmospheric constituent (CO_2) is condensing to form clouds. Such behavior is unique among the planetary atmospheres of our Solar System, apart from the tenuous atmospheres of Triton and Pluto. In the case of a minor constituent condensing (e.g. water vapor), diffusion of this minor gas through the primary gas can be an important factor in cloud nucleation and growth. On Mars, CO_2 cloud growth is limited only by the release of latent heat and self-diffusion of the primary gas to the growing crystal (Colaprete et al., 2003). The formation of CO_2 clouds is generally believed to occur via heterogeneous nucleation, the formation of a CO_2 ice embryo on the surface of a substrate such as a dust grain (Colaprete et al., 2003; Määttänen et al., 2005; Listowski et al., 2014). Recent MCS aerosol profile measurements in the polar night suggest that fine water ice and dust particles may also serve as nucleation centers (Hayne et al., 2014). Even in the case of heterogeneous nucleation, some degree of supersaturation of the CO_2 gas is required before ice embryo energy barriers are overcome and nucleation occurs. Glandorf et al. (2002) employed laboratory measurements to determine a 35% supersaturation level for CO_2 cloud nucleation. This critical saturation level is consistent with observations of supersaturation levels in the winter polar regions (Hinson and Wilson, 2002; Colaprete et al., 2008). For water and CO_2 ice clouds, critical saturation is strongly influenced by the nuclei size, nuclei properties, and temperature (Colaprete et al., 2003; Määttänen et al., 2005; Iraci et al., 2010; Phebus et al., 2011; Ladino and Abbatt, 2013; Listowski et al., 2013). Consequently, CO_2 cloud particle properties, including the number and size of particles in the cloud, will depend on the local population and composition of condensation nuclei. For example, saturation ratios ≥3 have been modeled for mesospheric CO_2 clouds due to smaller sizes of potential condensation nuclei at low mesospheric pressures, for which sources remain undetermined (Listowski et al., 2014).

The analysis of the characteristics of polar night clouds detected by MGS/MOLA and comparisons with atmospheric model simulations suggest that most of these clouds are made of gravitationally settling ice particles that successively grow and sublimate by crossing cold and warm phases of orographic gravity waves, as generated by local topography in the stable polar night atmosphere (Colaprete et al., 2003; Tobie et al., 2003). When winds are directed up large-scale slopes (such as the polar layered deposits), condensation induced by associated adiabatic cooling may create large-scale clouds that are locally modulated by small-scale gravity waves (Forget et al., 1998; Tobie et al., 2003). It is also possible that the formation and growth of CO_2 clouds within a supersaturated region can provide sufficient local latent heating to produce buoyant convection (Colaprete et al., 2003). In the case of equatorial mesospheric clouds, the total amount of potential convective energy will be small. In the case of polar night CO_2 clouds, such convection might play a significant role in the redistribution of energy and affect the overall temperature structure of the polar night atmosphere (Colaprete et al., 2008). In the extreme polar regions, where meridional heat transport is small, radiative cooling alone is sufficient to promote rapid and persistent CO_2 condensation up to ~30 km (Hayne et al., 2014).

5.7.3 Implications of CO_2 Clouds for Early Mars Climates

Carbon dioxide clouds are expected to play a significant role in the early Martian climate (see Chapter 17). One explanation for the fluvial features seen in the ancient cratered terrain of Mars is that, during their formation, Mars was warm enough for liquid water to be stable at the surface (e.g. Carr, 1986). Climate models show that the greenhouse effects of a thick, 2–5 bar CO_2 atmosphere would have been sufficient to raise the mean surface temperature to above 273 K (Pollack et al., 1987). The formation of CO_2 clouds within these thick atmospheres is extremely likely, leading to additional atmospheric cooling (solar reflection) and heating (IR absorption) contributions (Kasting, 1991). Net radiative effects have been calculated to increase surface temperatures (Forget and Pierrehumbert, 1997), depending on the assumed physical characteristics of the clouds (Colaprete and Toon, 2003; Mischna et al., 2000). Global climate simulations of the early Mars atmospheres have confirmed the warming effect of CO_2 ice clouds, but indicate surface temperature increases less than 15 K due to limited cloud coverage and optical depths (Forget et al., 2013).

5.8 CONCLUSIONS

Mars clouds include both water and CO_2 compositions and exhibit a wide range of properties, conditions, and cloud formation processes. Discrete, regional, and global cloud forms occur, with strong seasonal and orbital dependences, extending from surface fogs to upper mesospheric hazes, and from perihelion peak heating conditions to frigid polar night conditions. Many of these cloud types are unique to the Mars atmosphere. Wave, cumulus, and cirrus forms have been observed from lander and orbiting imagers, whereas spectroscopic cloud observations span the ultraviolet to thermal infrared wavelength range. Lidar observations also exist for polar night CO_2 and high-latitude PBL water ice clouds. Mars water ice cloud particle sizes are much smaller than cirrus clouds in the terrestrial atmosphere, with particle radii in the 1–3 μm range, although smaller and larger cloud particles exist under specific conditions. These smaller particle sizes reflect the microphysics of cloud formation in the low-water-content, atmospheric pressure conditions of the Mars atmosphere. Mars water ice clouds have been simulated in PBL, mesoscale, and global circulation models employing a range of microphysical sophistication. Such models investigate the effects of water ice clouds in solar thermal tides, global transport of water (ice and vapor), photochemistry, and radiative forcing. The Mars atmosphere also presents the unique circumstance of cloud formation from the primary atmospheric constituent, CO_2. Condensation of atmospheric CO_2 in the polar night, at the surface and in the form of precipitating clouds, leads to large seasonal variation in Mars atmospheric pressure. The formation of more extensive CO_2 clouds under higher atmospheric pressure regimes may also have contributed key radiative effects in the early Mars atmosphere.

Despite the existing range of cloud measurement and modeling results, fundamental questions remain regarding processes and effects of clouds in the Mars atmosphere. Remarkably, dedicated water vapor profiling measurements do not exist for the global Mars atmosphere, which considerably restricts diagnostic assessments of cloud microphysics, atmospheric volatile transport, and photochemistry. Extensive global profile measurements of water ice clouds have only recently been obtained, and definitive interpretation of these opacity profiles is impeded by very limited cloud particle size information. Furthermore, diurnal variations of cloud optical depths and vertical distributions are observed and modeled to be significant, whereas diurnal measurement coverage is extremely limited. In summary, we observe a range of unique cloud behaviors and processes in the Mars atmosphere relative to those displayed in the Earth's atmosphere. We know that clouds play critical roles in terrestrial climate forcing and suspect that they played equally important roles in the early Mars climate. A critical component to assessing such early Mars climate issues is a more complete description of cloud influences on the current Mars atmosphere. Such an enhanced cloud description would also significantly improve our understanding of the complex radiative, dynamical, and photochemical environment constituted by today's Mars atmosphere.

REFERENCES

Anderson, E., and Leovy, C. (1978) Mariner 9 television limb observations of dust and ice hazes on Mars, *J. Atmos. Sci.*, 35, 723–734.

Bailey, M. P., and Hallett, J. (2009) A comprehensive habit diagram for atmospheric ice crystals: confirmation from the laboratory, AIRS II, and other field studies, *J. Atmos. Sci.*, 66, 2888–2899.

Benson, J. L., Bonev, B. P., James, P. B., Shan, K. J., Cantor, B. A., and Caplinger, M. A. (2003) The seasonal behavior of water ice clouds in the Tharsis and Valles Marineris regions of Mars: Mars Orbiter Camera Observations, *Icarus*, 165, 34–52.

Benson, J. L., James, P. B., Cantor, B. A., and Remigio, R. (2006) Interannual variability of water ice clouds over major Martian volcanoes observed by MOC, *Icarus*, 184, 365–371.

Benson, J. L., Kass, D. M., Kleinböhl, A., et al. (2010) Mars' south polar hood as observed by the Mars Climate Sounder, *J. Geophys. Res.*, 115, E12015, doi:10.1029/2009JE003554.

Benson, J. L., Kass, D. M., and Kleinböhl, A. (2011) Mars' north polar hood as observed by the Mars Climate Sounder, *J. Geophys. Res.*, 116, E03008, doi:10.1029/2010JE003693.

Bertaux, J.-L., Fonteyn, D., Korablev, O., et al. (2000) The study of the Martian atmosphere from top to bottom with SPICAM light on Mars Express, *Planet. Space Sci.*, 48, 1303–1320.

Bibring, J.-P., Soufflot, A., Berthé, M., et al. (2004) OMEGA: Observatoire pour la Minéralogie, l'Eau, les Glaces et l'Activité, in *Mars Express: The Scientific Payload*, Eur. Space Agency Spec. Publ., ESA-SP 1240, 37–50.

Böttger, H. M., Lewis, S. R., Read, R. L., and Forget, F. (2005) The effects of the Martian regolith on GCM water cycle simulations, *Icarus*, 177, 174–189.

Briggs, G. A., and Leovy, C. B. (1974) Mariner 9 observations of the Mars north polar hood, *Bull. Amer. Met. Soc.*, 55, 278–296.

Briggs, G., Klaasen, K., Thorpe, T., and Wellman, J. (1977) Martian dynamical phenomena during June-November 1976: Viking Orbiter imaging results, *J. Geophys. Res.*, 82, 4121–4149.

Brown, A. J., Calvin, W. M., McGuire, P. C., and Murchie, S. L. (2010) Compact Reconnaissance Imaging Spectrometer for Mars (CRISM) south polar mapping: first Mars year of observations, *J. Geophys. Res.*, 112, E04004, doi: 10.1029/2006JE002805.

Burlakov, A. V., and Rodin, A. V. (2012) A one-dimensional numerical model of H_2O cloud formation in the Martian atmosphere, *Solar System Res.*, 46, No. 1, 18–30.

Cantor, B., Malin, M., and Edgett, K. S. (2002) Multiyear Mars Orbiter Camera (MOC) observations of repeated Martian weather phenomena during the northern summer season, *J. Geophys. Res.*, 107, 10.1029/2001JE001588.

Cantor, B. A., Malin, M. C., Wolff, M. J., et al. (2008) Observations of the Martian atmosphere by MRO-MARCI: an overview of 1 Mars year, in *Proceedings of the Third International Workshop on the Mars Atmosphere: Modeling and Observations*, LPI Contribution No. 1447, 9075.

Cantor, B. A., James, P. B., and Calvin, W. M. (2010) MARCI and MOC observations of the atmosphere and surface cap in the north polar region of Mars, *Icarus*, 208, 61–81.

Carr, M. H. (1986) Mars: a water rich planet?, *Icarus*, 56, 187–216.

Chassefière, E., Blamont, J. E., Krasnopolsky, V. A., et al. (1992) Vertical structure and size distribution of Martian aerosols from solar occultation measurements, *Icarus*, 97, 46–69.

Christensen, P. R., Anderson, D. L., Chase, S. C., et al. (1998) Initial results from the Mars Global Surveyor thermal emission spectrometer experiment, *Science*, 279, 1682–1685.

Christensen, P. R., Jakosky, B. M., Kieffer, H. H., et al. (2003) The Thermal Emission Imaging System (THEMIS) for the Mars 2001 Odyssey mission, *Space Sci. Rev.*, 110, 85–130.

Clancy, R. T., and Nair, H. (1996) Annual (perihelion-aphelion) cycles in the photochemical behavior of the global Mars atmosphere, *J. Geophys. Res.*, 101, 12785–12790.

Clancy, R. T., and Sandor, B. J. (1998) CO_2 ice clouds in the upper atmosphere of Mars, *Geophys. Res. Lett.*, 25, 489–492.

Clancy, R. T., Grossman, A. W., Wolff, M. J., et al. (1996) Water vapor saturation at low altitudes around Mars aphelion: a key to Mars climate? *Icarus*, 122, 36–62.

Clancy, R. T., Wolff, M. J., and Christensen, P. R. (2003) Mars aerosol studies with the MGS TES emission phase function observations: optical depths, particle sizes, and ice cloud types versus latitude and solar longitude, *J. Geophys. Res.*, 108, E9, 1–20, 2003.

Clancy, R. T., Wolff, M. J., Whitney, B. A., Cantor, B. A., and Smith, M. D. (2007) Mars equatorial mesospheric clouds: global occurrence and physical properties from Mars Global Surveyor TES and MOC limb observations, *J. Geophys. Res.*, 112, E04004.

Clancy, R. T., Wolff, M. J., Malin, M. C., Cantor, B. A., and Michaels, T. I. (2009) Valles Marineris cloud trails, *J. Geophys. Res.*, 114, E11, doi:10.1029/2008JE003323.

Clancy, R. T., Wolff, M. J., Whitney, B. A., et al. (2010) Extension of atmospheric dust loading to high altitudes during the 2001 global dust storm: MGS TES limb observations, *Icarus*, 207, 98–109.

Clancy, R. T., Sandor, B. J., Wolff, M. J., et al. (2012) Extensive MRO CRISM observations of 1.27 µm O_2 airglow in Mars polar night and their comparison to MRO MCS temperature profiles and LMDGCM simulations, *J. Geophys. Res.*, 117, E00J10, doi:10.1029/2011JE004018.

Clancy, R. T., Smith, M. D., Wolff, M. J., et al. (2014) CRISM limb observations of Mars mesospheric ice clouds: two new results, *Eighth International Conference on Mars*, July 14–18, Pasadena, CA., LPI Contribution No. 1791, 1006.

Colaprete, A. and Toon, O. B. (2002) Carbon dioxide snow storms during the polar night on Mars, *J. Geophys. Res.*, 107, E7, 5051, 10.1029/2001JE001758.

Colaprete, A., and Toon, O. B. (2003) Carbon dioxide clouds in an early dense Martian atmosphere, *J. Geophys. Res.*, 108, E4, 5025, doi:10.1029/2002JE001967.

Colaprete, A., Toon, O. B., and Magalhães, J. A. (1999) Cloud formation under Mars Pathfinder conditions, *J. Geophys. Res.*, 104, 9043–9053,doi:10.1029/1998JE900018.

Colaprete, A., Haberle, R. M., and Toon, O. B. (2003) Formation of convective carbon dioxide clouds near the south pole of Mars, *J. Geophys. Res.*, 108, E7, 5081, doi:10.1029/2003JE002053.

Colaprete, A., Barnes, J. R., Haberle, R. M., et al. (2005) Albedo of the south pole of Mars determined by topographical forcing of atmosphere dynamics, *Nature*, 435, 184–188.

Colaprete, A., Barnes, J. R., Haberle, R. M., and Montmessin, F. (2008) CO_2 clouds, CAPE and convection on Mars: observations and general circulation modeling, *Planet. Space Sci.*, 56, 150–180.

Colburn, D. S., Pollack, J. B., and Haberle, R. M. (1989) Diurnal variations in optical depth at Mars, *Icarus*, 79, 159–189.

Curran, R. J., Conrath, B. J., Hanel, R. A., Kunde, V. G., and Pearl, J. C. (1973) Mars: Mariner 9 spectroscopic evidence for H_2O ice clouds, *Science*, 182, 381–383.

Daerden, F., Larsen, N., Chabrillat, S., et al. (2007) A 3D-CTM with detailed online PSC microphysics: analysis of the Antarctic winter 2003 by comparison with satellite observations, *Atmos. Chem. Phys.*, 7, 1755–1772.

Daerden, F., Whiteway, J. A., Davy, R., et al. (2010) Simulating observed boundary layer clouds on Mars, *Geophys. Res. Lett.*, 37, L04203, doi:10.1029/2009GL041523.

Davy, R., Davis, J. A., Taylor, P. A., et al. (2010) Initial analysis of air temperature and related data from the Phoenix MET station and their use in estimating turbulent heat fluxes, *J. Geophys. Res.*, 115, E00E13, doi:10.1029/2009JE003444.

Deirmidjian, D. (1964) Scattering and polarization properties of water clouds and hazes in the visible and infrared, *Appl. Opt.*, 3, 187–202.

Dickinson, C., Whiteway, J. A., Komguem, L., Moores, J. E., and Lemmon, M. T. (2010) Lidar measurements of clouds in the planetary boundary layer on Mars, *Geophys. Res. Lett.*, 37, doi:10.1029/2010GL044317.

Dollfus, A. (1957) Etude des planets pour la polarization de leur Lumiere, *Ann. Astrophys. Suppl.*, 4, 3–114.

Farmer, C. B., Davies, D. W., Holland, A. L., Laporte, D. D., and Doms, P. E. (1977) Mars: water vapor observations from the Viking Orbiters, *J. Geophys. Res.*, 82, 4225–4248.

Fedorova, A. A., Korablev, O. I., Bertaux, J.-L., et al. (2009) Solar infrared occultation observations by SPICAM experiment on Mars-Express: simultaneous measurements of the vertical distributions of H_2O, CO_2 and aerosol, *Icarus*, 200, 96–117.

Fedorova, A. A., Montmession, F., Rodin, A. V., et al. (2014) Evidence for a bimodal size distribution for suspended aerosol particles on Mars, *Icarus*, 231, 239–260.

Forget, F., and Pierrehumbert, R. T. (1997) Warming early Mars with carbon dioxide clouds that scatter infrared radiation, *Science*, 278, 1273–1276.

Forget, F., Hourdin, F., and Talagrand, O. (1998) CO_2 snowfall on Mars: simulation with a general circulation model, *Icarus*, 131, 302–316.

Forget, F., Haberle, R. M., Montmessin, F., and Levrard, B. (2006) Formation of glaciers on Mars by atmospheric precipitation at high obliquity, *Science*, 311, 368–371.

Forget, F., Wordsworth, R., Millour, E., et al. (2013) 3D modelling of the early Martian climate under a denser CO_2 atmosphere: temperatures and CO_2 ice clouds, *Icarus*, 222, 81–99.

French, R. G., Gierasch, P. J., Popp, B. D., and Yerdon, R. J. (1981) Global patterns in cloud forms on Mars, *Icarus*, 45, 468–493.

Fuchs, N. A., *The Mechanics of Aerosols*, Pergamon, New York, 1964.

Gallagher, M. W., Connolly, P. J., Whiteway, J., et al. (2005) An overview of the microphysical structure of cirrus clouds observed during EMERALD-1, *Q. J. R. Meteorol. Soc.*, 131, 1143–1169, doi:10.1256/qj.03.138.

Gierasch, P., Thomas, P., French, R., and Veverka, J. (1979) Spiral clouds on Mars: a new atmospheric phenomenon, *Geophys. Res. Lett.*, 6, 405–408.

Glandorf, D. L., Colaprete, A., Tolbert, M. A., and Toon, O. B. (2002) CO2 snow on Mars and early Earth: experimental constraints, *Icarus*, 160, 66–72.

Glenar, D. A., Samuelson, R. E., Pearl, J. C., Bjoraker, G. L., and Blaney, D. (2003) Spectral imaging of Martian water ice clouds and their diurnal behavior during the 1999 aphelion season ($L_s = 130°$), *Icarus*, 161, 297–318.

González-Galindo, F., Määttänen, A., Forget, F., and Spiga, A. (2011) The Martian mesosphere as revealed by CO_2 cloud observations and General Circulation Modeling, *Icarus*, 216, 10–22.

Guzewich, S. D., Smith, M. D., and Wolff, M. J. (2014) Aerosol particle size retrievals from the Compact Reconnaissance Imaging Spectrometer for Mars, *The Fifth International Workshop of the Mars Atmosphere: Modeling and Observation*, Oxford, UK, January 13–16.

Haberle, R. M., Montmessin, F., Kahre, M. A., et al. (2011) Radiative effects of water ice clouds on the Martian seasonal water cycle, *The Fourth International Workshop of the Mars Atmosphere: Modeling and Observation*, Paris, France, February 8–11.

Hale, A. S., Tamppari, L. K., Bass, D. S., and Smith, M. D. (2011) Martian water ice clouds: a view from Mars Global Surveyor Thermal Emission Spectrometer, *J. Geophys. Res.*, 116, E04004, doi:10.1029/2009JE003449.

Hanel, R., Conrath, B., Hovis, W., et al. (1972) Investigation of the Martian environment by infrared spectroscopy on Mariner 9, *Icarus*, 17, 423–442.

Hayne, P. O., and Paige, D. A. (2009) Snow clouds and the carbon dioxide cycle on Mars, *American Geophysical Fall Meeting*, abstract #P53A-1, San Francisco, December 14–18.

Hayne, P. O., Paige, D. A., Schofield, J. T, et al. (2012) Carbon dioxide snow clouds on Mars: south polar winter observations by the Mars Climate Sounder, *J. Geophys. Res.*, 117, E08014, doi:10.1029/2011JE004040.

Hayne, P. O., Paige, D. A., Heavens, N. G., et al. (2014) The role of snowfall in forming the seasonal ice caps of Mars: models and constraints from the Mars Climate Sounder, *Icarus*, 231, 122–130.

Heavens, N. G., Benson, J. L., Kass, D. M., et al. (2010) Water ice clouds over the Martian tropics during northern summer, *Geophys. Res. Lett.*, 37, L18202, doi:10.1029/2010GL044610.

Heavens, N. G., Richardson, M. I., Kleinböhl, A., et al. (2011) Vertical distribution of dust in the Martian atmosphere during northern spring and summer: high-altitude tropical dust maximum at northern summer solstice, *J. Geophys. Res.*, 116, E01007, doi:10.1029/2010JE003692.

Hinson, D. P., and Wilson, R. J. (2002) Transient eddies in the southern hemisphere of Mars, *Geophys. Res. Lett.*, 29(7), doi: 10.1029/2001GL014103.

Hinson, D. P., and Wilson, R. J. (2004) Temperature inversions, thermal tides, and water ice clouds in the Martian tropics, *J. Geophys. Res.*, 109, E01002, doi:10.1029/2003JE002129.

Hollingsworth, J. L., Haberle, R. M., Barnes, J. R., et al. (1996) Orographic control of storm zones on Mars, *Nature*, 380, 413–416.

Houben, H., Haberle, R. M., Young, R. E., and Zent, A. P. (1997) Modeling the Martian seasonal water cycle, *J. Geophys. Res.*, 102, 9069–9084.

Hunt, G. E., and James, P. B. (1985) Martian cloud systems: current knowledge and future observations, *Adv. Space. Res.*, 5, 93–99.

Hunt, G. E., Pickersgill, A. O., James, P. B., and Johnson, G. (1980) Some diurnal properties of clouds over the Martian volcanoes, *Nature*, 286, 362–364.

Inada, A., Richardson, M. I., McConnochie, T. H., et al. (2007) High-resolution atmospheric observations by the Mars Odyssey Thermal Emission Imaging System, *Icarus*, 192, 378–395.

Iraci, L. T., Phebus, B. D., Stone, B. M., and Colaprete, A. (2010) Water ice cloud formation on Mars is more difficult that presumed: laboratory studies of ice nucleation on surrogate materials, *Icarus*, 210, 985–991.

Isenor, M., Escribano, R., Preston, T. C., and Signorell, R. (2013) Predicting the infrared band profiles for CO_2 cloud particles on Mars, *Icarus*, 223, 591–601.

Ivanov, A. B., and Muhleman, D. O. (2001) Cloud reflection observations: results from the Mars Orbiter Laser Altimeter, *Icarus*, 154, 190–206.

Jakosky, B. M., and Farmer, C. B. (2002) Transient eddies in the southern hemisphere of Mars, *Geophys. Res. Lett.*, 29(7), doi: 10.1029/2001GL014103.

James, P. B., Kieffer, H. H., and Paige, D. A. (1992) The seasonal cycle of carbon dioxide on Mars, in *Mars*, eds. H. H. Kieffer et al., University of Arizona Press, Tuscon, AZ.

James, P. B., Bell III, J. F., Clancy, R. T., et al. (1996) Global imaging of Mars by Hubble Space Telescope during the 1995 opposition, *J. Geophys. Res.*, 101, 18883–18890.

Jaquin, F., Gierasch, P., and Kahn, R. (1986) The vertical structure of limb hazes in the Martian atmosphere, *Icarus*, 68, 442–461.

Jensen, E. J., Pfister, L., Bui, T.-P., Lawson, P., and Baumgardner, D. (2010) Ice nucleation and cloud microphysical properties in tropical tropopause layer cirrus, *Atmos. Chem. Phys.*, 10, 1369–1384, doi: 10.5194/acp-10-1369-2010.

Kahn, R. (1984) The spatial and seasonal distribution of Martian clouds and some meteorological implications, *J. Geophys. Res.*, 89, 6671–6688.

Kahn, R., and Gierasch, P. (1982) Observations of Mars and implications for boundary layer characteristics over slopes, *J. Geophys. Res.*, 87, 867–880.

Kass, D. M., Hale, A. S., Schofield, J. T., et al. (2009) MCS Views of Atmospheric Thermal Structure During the 2009 Planet Encircling Dust Event, *Mars Dust Cycle Workshop*, NASA/CP-2010–216477, 15–19.

Kasting, J. F. (1991) CO_2 condensation and the climate of early Mars, *Icarus*, 94, 1–13.

Kieffer, H. H. (1979) Mars south polar spring and summer temperatures: a residual CO_2 frost, *J. Geophys. Res.*, 84, 8263–8288.

Kieffer, H. H., Neugebauer, G., Munch, G., Chase, Jr., S. C., and Miner, E. (1972) Infrared thermal mapping experiment: the Viking Mars Orbiter, *Icarus*, 16, 47–56.

Kieffer, H. H., Chase, S. C., Miner, E. D., et al. (1976) Infrared thermal mapping of the Martian surface and atmosphere: first results, *Science*, 193, 780–786.

Kieffer, H. H., Jakosky, B. M., and Snyder, C. W. (1992) The planet Mars: from antiquity to the present, in *Mars*, eds. H. H. Kieffer et al., University of Arizona Press, Tuscon, AZ, 1–33.

Kieffer, H. H., Titus, T. N., Mullins, K. F., and Christensen, P. R. (2002) Mars south polar spring and summer behavior observed by TES: seasonal cap evolution controlled by frost grain size, *J. Geophys. Res.*, 105, 9653–9699.

Kleinböhl, A., Schofield, J. T., Kass, D. M., et al. (2009) Mars Climate Sounder limb profile retrieval of atmospheric temperature, pressure, and dust and water ice opacity, *J. Geophys. Res.*, 105, 9653–9699.

Kuroda, T., Medvedev, A. S., Kasaba, Y., and Hartogh, P. (2013) Carbon dioxide ice clouds, snowfalls, and baroclinic waves in the northern winter polar atmosphere, *Geophys. Res. Lett.*, 40, 1484–1488, doi:10.1002/grl.50326

Ladino, L. A., and Abbatt, J. P. D. (2013) Laboratory investigation of Martian water ice cloud formation using dust aerosol simulants, *J. Geophys Res.*, 109, 14–25, doi:10.1029/2012JE004238.

Langevin, Y., Bibring, J.-P., Montmessin, F., et al. (2007) Observations of the south seasonal cap of Mars during recession in 2004–2006 by the OMEGA visible/near-infrared imaging spectrometer on board Mars Express, *J. Geophys. Res.*, 112, E08S12, doi:10.1029/2006JE002841.

Larsen, N., Knudsen, B. M., Svendsen, S. H., et al. (2004) Formation of solid particles in synoptic-scale Arctic PSCs in early winter 2002/2003, *Atmos. Chem. and Phys.*, 4, 1–13.

Lee, C., Lawson, W. G., Richardson, M. I., et al. (2009) Thermal tides in the Martian middle atmosphere as seen by the Mars Climate Sounder, *J. Geophys. Res.*, 114, E03005, doi:10.1029/2008JE003285.

Lefèvre, F., Lebonnois, S., Montmessin, F., and Forget, F. (2004) Three-dimensional modeling of ozone on Mars, *J. Geophys Res.*, 109, 10.1029/2004JE002268.

Lefèvre, F., Bertaux, J.-L., Clancy, R. T., et al. (2008) Heterogeneous chemistry in the atmosphere of Mars, *Nature*, 454, 971–975.

Lemmon, M. T. (2014) Large water ice aerosols in Martian north polar clouds, *The Fifth International Workshop of the Mars Atmosphere: Modeling and Observation*, Oxford, UK, January 13–16.

Lemmon, M. T., Wolff, M. J., Bell III, J. F., et al. (2015) Dust aerosol, clouds, and the atmospheric optical depth record over 5 Mars years of the Mars Exploration Rover mission, *Icarus*, 251, 96–111.

Leovy, C., and Mintz, Y. (1969) Numerical simulation of the atmospheric circulation and climate of Mars, *J. Atmos. Sci.*, 26, 1167–1190.

Leovy, C. B., Smith, B. A., Young, A. T., and Leighton, R. B. (1971) Mariner Mars 1969: atmospheric results, *J. Geophys. Res.*, 76, 297–312.

Leovy, C. B., Briggs, G. A., and Smith, B. A. (1973) Mars atmosphere during the Mariner 9 extended mission: television results, *J. Geophys. Res.*, 78, 4252–4266.

Listowski, C., Määttänen, A., Riipinen, Montmessin, F., and Lefèvre, F. (2013) Near-pure condensation in the Martian atmosphere: CO_2 ice crystal growth, *J. Geophys. Res.*, 118, 2153–2171, doi:10.1002/jgre.20149.

Listowski, C., Määttänen, A., Montmessin, F., Spiga, A., and Lefèvre, F. (2014) Modeling the microphysics of CO_2 ice clouds within wave-induced cold pockets in the Martian mesosphere, *Icarus*, 237, 239–261.

Määttänen, A., Vehkamäki, H., Lauri, A., et al. (2005) Nucleation studies in the Martian atmosphere, *J. Geophys. Res.*, 110, E02002, doi:10.1029/2004JE002308.

Määttänen A., Montmessin, F., Gondet B., et al. (2010) Mapping the mesospheric CO_2 clouds on Mars: MEx/OMEGA and MEx/HRSC observations and challenges for atmospheric models, *Icarus*, 209, 452–469.

Määttänen A., Listowski, C., Montmessin, F., et al. (2013) A complete climatology of the aerosol vertical distribution on Mars from MEx/SPICAM UV solar occultations, *Icarus*, 223, 892–941.

Madeleine, J.-B., Forget, F., Millour, E., Navarro, T., and Spiga, A. (2012a) The influence of radiatively active water ice clouds on the Mars climate, *Geophys. Res. Lett.*, 39, L23202, doi:10.1029/2012GL053564.

Madeleine, J.-B., Forget, F., Spica, A., et al. (2012b) Aphelion water-ice cloud mapping and property retrieval using the OMEGA imaging spectrometer onboard Mars Express, *J. Geophys. Res.*, 117, E00J07, doi:10.1029/2011JE003940.

Madeleine, J.-B., Head, J. W., Forget, F., et al. (2014) Recent ice ages on Mars: the role of radiatively active clouds and cloud microphysics, *Geophys. Res. Lett.*, 41, 4873–4879, doi:10.1002/2014GL059861.

Malin, M. C., Danielson, G. E., Ingersoll, A. P., et al. (1992) Mars Observer Camera, *J. Geophys. Res.*, 97, 7699–7718.

Malin, M. C., Calvin, W. M., Cantor, B. A., et al. (2008) Climate, weather, and north polar observations from the Mars Reconnaissance Orbiter Mars Color Imager, *Icarus*, 194, 501–512.

Maltagliati, L., Montmessin, F., Fedorova, A., et al. (2011) Evidence of water vapor in excess of saturation in the atmosphere of Mars, *Science*, 333, 1868–1871.

Maltagliati, L., Montmessin, F., Korablev, O., et al. (2013) Annual survey of water vapor vertical distribution and water-aerosol coupling in the Martian atmosphere observed by SPICAM/MEx solar occultations, *Icarus*, 223, 942–962.

Marchand, R., and Ackerman, T. (2010) An analysis of cloud cover in multiscale modeling framework global climate model simulations using 4 and 1 km horizontal grids, *J. Geophys. Res.*, 115, D16207, doi:10.1029/2009JD013423.

Martin, L. J. (1975) North Polar Hood Observations during Martian Dust Storms, *Icarus*, 26, 341–352.

Martin, L. J., James, P. B., Dollfus, A., Iwasaki, K., and Beish, J. D. (1992) Telescopic observations: visual, photographic, polarimetric, in *Mars*, eds. H. H. Kieffer et al., University of Arizona Press, Tucson, AZ, 34–70.

Mason, B. J. (1971) *The Physics of Clouds*, Clarendon Press, Oxford.

Mateshvili, N., Fussen, D., Vanhellemont, F., et al. (2009) Water ice clouds in the Martian atmosphere: two Martian years of SPICAM nadir UV measurements, *Plan. Space Sci.*, 57, 1022–1031.

McCleese, D. J., Schofield, J. T., Taylor, F. W., et al. (2007) Mars Climate Sounder: an investigation of thermal and water vapor structure, dust and condensate distributions in the atmosphere, and energy balance of the polar regions, *J. Geophys. Res.*, 112, E05S06, doi:10.1029/2006JE002790.

McCleese, D. J., Heavens, N. G., Schofield, J. T., et al. (2010) Structure and dynamics of the Martian lower and middle atmosphere as observed by the Mars Climate Sounder: seasonal variations in zonal mean temperature, dust, and water ice aerosols, *J. Geophys. Res.*, 115, E12016, doi:10.1029/2010JE003677.

McConnochie, T. H., Bell III, J. F., Savransky, D., et al. (2010) THEMIS-VIS observations of clouds in the Martian mesosphere: altitudes, wind speeds, and decameter-scale morphology, *Icarus*, 210, 545–565.

Michaels, T. I., Colaprete, A., and Rafkin, S. C. R. (2006) Significant vertical water transport by mountain-induced circulations on Mars, *Geophys. Res. Lett.*, 33, L16201, doi:10.1029/2006GL026562.

Michelangeli, D. V., Toon, O. B., Haberle, R. B., and Pollack, J. B. (1993) Numerical simulations of the formation and evolution of water ice clouds in the Martian atmosphere, *Icarus*, 100, 261–285.

Mischna, M. A., Kasting, J. F., Pavlov, A., and Freedman, R. (2000) Influence of carbon dioxide clouds on early Martian climate, *Icarus*, 145, 546–554.

Möhlmann, D., Niemand, M., Formisano, V., Savijärvi, H., and Wolkenberg, P. (2009) Fog phenomena on Mars, *Planet. Space Sci.*, 57, Issue 14–15, 1987–1992.

Montmessin, F., Forget, F., Rannou, P., Cabane, M., and Haberle, R. M. (2004) Origin and role of water ice clouds in the Martian water cycle as inferred from a General Circulation Model, *J. Geophys. Res.*, 105, 4109–4121.

Montmessin, F., Bertaux, J.-P., Quémerais, E., et al. (2006a) Subvisible CO_2 clouds detected in the mesosphere of Mars, *Icarus*, 183, 403–410.

Montmessin, F., Quémerais, E., Bertaux, J.-L., et al. (2006b) Stellar occultations at UV wavelengths by the SPICAM instrument: retrievals and analysis of Martian haze profiles, *J. Geophys. Res.*, 111, E09S09, doi:10.1029/2005JE002662.

Montmessin, F., Haberle, R. M., Forget, F., et al. (2007a) On the origin of perennial water ice at the south pole of Mars: a precession-controlled mechanism, *J. Geophys. Res.*, 112, E08S17, doi:10.1029/2007JE002902.

Montmessin, F., Gondet, B., Bibring, J.-P., et al. (2007b) Hyperspectral imaging of convective CO_2 ice clouds in the equatorial mesosphere of Mars, *J. Geophys. Res.*, 112, E11S90, doi:10.1029/2007JE002944.

Moores, J. E., Komguem, L., Whiteway, J. A., et al. (2011) Observations of near-surface fog at the Phoenix Mars landing site, *Geophys. Res. Lett.*, 38, L04203, doi:10.1029/2010GL046315.

Moores, J. E., Lemmon, M. T., Rafkin, S. C. R., et al. (2015) Atmospheric movies acquired at the Mars Science Laboratory landing site: cloud morphology, frequency and significance to the Gale Crater water cycle and Phoenix mission results, *Adv. Space Res.*, 55, 2217–2238.

Murchie, S., Arvidson, R., Bedini, P., et al. (2007) Compact Reconnaissance Imaging Spectrometer for Mars (CRISM) on Mars Reconnaissance Orbiter (MRO), *J. Geophys. Res.*, 112, E05S03.

Navarro, T., Madeleine, J-B., Forget, F., et al. (2014) Global climate modeling of the Martian water cycle with improved microphysics and radiatively active water ice clouds, *J. Geophys. Res.*, 119, 1479–1495, doi:10.1002/2013JE004550.

Nelli, S. M., Rennó, N. O., Murphy, J. R., Feldman, W. C., and Bougher, S. W. (2010) Simulations of atmospheric phenomena at the Phoenix landing site with the Ames general circulation model, *J. Geophys. Res.*, 115, E00E21, doi:10.1029/2010JE003568.

Neukum, G., Jaumann, R., Behnke, T., et al. (2004) HRSC: the High Resolution Stereo Camera of Mars Express, in *Mars Express: The Scientific Payload*, Eur. Space Agency Spec. Publ., ESA-SP1240, 17–35.

Neumann, G. A., Smith, D. E., and Zuber, M. T. (2003) Two years of clouds detected by the Mars Orbiter Laser Altimeter, *J. Geophys. Res.*, 108, E4, 5023, doi:10.1029/2002JE001849.

Paige, D. A., and Ingersoll, A. P. (1985) Annual heat balance of Martian polar caps: Viking observations, *Science*, 228, 1160–1168.

Pathak, J., Michelangeli, D. V., Komguem, L., Whiteway, J., and Tamppari, L. K. (2008) Simulating Martian boundary layer water ice clouds and the lidar measurements for the Phoenix mission, *J. Geophys. Res.*, 113, E00A05, doi:10.1029/2007JE002967.

Peale, S. J. (1973) Water and the Martian W cloud, *Icarus*, 18, 497–501.

Pearl, J. C., Smith, M. D., Conrath, B. J., Bandfield, J. S., and Christensen, P. R. (2001) Observations of Martian ice clouds by the Mars Global Surveyor Thermal Emission Spectrometer: the first Mars year, *J. Geophys. Res.*, 106, E6, 12325–12338.

Pettengill, G. H., and Ford, P. G. (2000) Winter clouds over the north Martian polar cap, *Geophys. Res. Lett.*, 27, 609–612.

Phebus, B. D., Johnson, A. V., Mar, B., et al. (2011) Water ice nucleation characteristics of JSC Mars-1 regolith simulant under simulated Martian atmospheric conditions, *J. Geophys. Res.*, 116, E04009, doi:10.1029/2010JE003699.

Pickersgill, A. O., and Hunt, G. E. (1979) The formation of Martian lee waves generated by a crater, *J. Geophys. Res.*, 84, 8317–8331.

Pickersgill, A. O., and Hunt, G. E. (1981) An examination of the formation of linear lee waves generated by giant Martian volcanoes, *J. Atmos. Sci.*, 38, 40–51.

Pollack, J. B., Kasting, J. F., Richardson, S. M., and Poliakoff, K. (1987) The case for a wet, warm climate on early Mars, *Icarus*, 71, 203–224.

Richardson, M. I., Wilson, R. J., and Rodin, A. V. (2002) Water ice clouds in the Martian atmosphere: general circulation model experiments with a simple cloud scheme, *J. Geophys. Res.*, 107(E9), doi:10.1029/2001JE001804.

Rodin, A. V., Korablev, O. I., and Moroz, V. I. (1997) Vertical distribution of water in near-equatorial troposphere of Mars: water vapor and clouds, *Icarus*, 125, 212–229.

Rodin, A. V., Clancy, R. T., and Wilson, R. J. (1999) Dynamical properties of Mars water ice clouds and their dynamical interactions with atmospheric dust and radiation, *Adv. Space. Res.*, 23, 1577–1585.

Rodin, A. V. Burlakov, A. V., Evdokimova, N. A., Fedorova, A. A., and Wilson, R. J. (2011) GCM simulation of the Mars water cycle with detailed cloud microphysics, EPSC-DPS joint meeting, October 2–7, Nantes, France.

Rossow, W. B. (1978) Cloud microphysics: analysis of the clouds of Earth, Venus, Mars, and Jupiter, *Icarus*, 36, 1–50.

Sassen, K. (2002) Cirrus clouds: a modern perspective, in *Cirrus*, ed. Lynch, D. K., Sassen, K., Starr, D. O'C., and Stephens, G., Oxford University Press, New York.

Savijärvi, H. (1999) A model study of the atmospheric boundary layer in the Mars Pathfinder Lander conditions, *Q. J. R. Meteorol. Soc.*, 125, 483–493.

Schofield, J. T., Barnes, J. R., Crisp, D., et al. (1997) The Mars Pathfinder atmospheric structure investigation/meteorology (ASI/MET) experiment, *Science*, 278, 1752–1757.

Scholten, F., Hoffman, H., Määttänen, A., et al. (2010) Concatenation of HRSC colour and OMEGA data for the determination and 3D-parameterization of high-altitude CO_2 clouds in the Martian atmosphere, *Planet. Space Sci.*, 58, 1207–1214.

Sefton-Nash, E., Teanby, N. A., Montabone, L., et al. (2013), Climatology and first-order composition estimates of mesospheric clouds from Mars Climate Sounder limb spectra, *Icarus*, 222, 342–356.

Smith, D. E., Zuber, M. T., Frey, H. V., et al. (2001) Mars Orbiter Laser Altimeter: experiment summary after the first year of global mapping of Mars, *J. Geophys. Res.*, 106, E10, 23689–23722.

Smith, I. B., Holt, J. W., Spiga, A., Howard, A. D., and Parker, G. (2013) The spiral troughs of Mars as cyclic steps, *J. Geophys. Res.*, 118, 1835–1857, doi:10.1002/jgre.20142.

Smith, M. D. (2004) Interannual variability in TES atmospheric observations of Mars during 1999–2003, *Icarus*, 167, 148–165.

Smith, M. D. (2009) THEMIS observations of Mars aerosol optical depth from 2002–2008, *Icarus*, 202, 444–452.

Smith, M. D., Bandfield, J. L., and Christensen, P. R. (2000) Separation of atmospheric and surface spectral features in Mars Global Surveyor Thermal Emission Spectrometer (TES) spectra, *J. Geopyhs. Res.*, 105, E4, 9589–9607.

Smith, M. D., Wolff, M. J., Clancy, R. T., Kleinböhl, A., and Murchie, S. L. (2013) Vertical distribution of dust and water ice aerosols from CRISM limb-geometry observations, *J. Geophys. Res.*, 118, 321–334, doi:10.1002/jgre.20047.

Smith, P. H., Tamppari, L., Arvidson, R. E., et al. (2008) Introduction to special section on the Phoenix mission: landing site characterization experiments, mission overviews, and expected science, *J. Geophys. Res.*, 113, E00A18, doi:10.1029/2008JE003083.

Slipher, E. C. (1962) *The Photographic Story of Mars*, Sky Publishing, Cambridge, MA.

Solomon, S., Qin, D., Manning, M., et al. (2007) Climate Change 2007: The Physical Science Basis, *Contribution of Working Group I to the Fourth Assessment Report of the Intergovernmental Panel on Climate Change (IPCC)*, Cambridge University Press, Cambridge, UK.

Spiga, A., González-Galindo, F., López-Valverde, M.-Á., and Forget, F. (2012) Gravity waves, cold pockets and CO_2 clouds in the Martian mesosphere, *Geophys. Res. Lett.*, 39, L02201, doi:10.1029/2011GL050343.

Sta. Maria, M. R. V., Rafkin, S. C. R., and Michaels, T. I. (2006) Numerical simulation of atmospheric bore waves on Mars, *Icarus*, 185, 383–394.

Tamppari, L. K., Zurek, R. W., and Paige, D. A. (2000) Viking era water-ice clouds, *J. Geophys. Res.*, 105, E2, 4087–4107.

Tamppari, L. K., Zurek, R. W., and Paige, D. A. (2003) Viking-era diurnal water-ice clouds, *J. Geophys. Res.*, 108, E7, 5073, doi:10.1029/2002JE001911.

Tamppari, L. K., Smith, M. D., Kass, D. S., and Hale, A. S. (2008) Water-ice clouds and dust in the north polar region of Mars using MGS TES data, *Planet. Space Sci.*, 56, 227–245.

Tamppari, L. K., Bass, D., Cantor, B., et al. (2010) Phoenix and MRO coordinated atmospheric measurements, *J. Geophys. Res.*, 115, E00E17, doi:10.1029/2009JE003415.

Thomas, P. C., James, P. B., Calvin, W. M., Haberle, R., and Malin, M. C. (2009) Residual south polar cap of Mars: stratigraphy, history, and implications of recent changes, *Icarus*, 203, 352–375.

Titus, T. (2005) Mars polar cap edges tracked over 3 full Mars years, *36th Lunar and Planetary Science Conference*, abstract no. 1993, League City, Texas, March 14–18.

Tobie, G., Forget, F., and Lott, F. (2003) Numerical simulation of the winter polar wave clouds observed by Mars Global Surveyor Mars Orbiter Laser Altimeter, *Icarus*, 164, 33–49.

Vehkamäki, H., Määttänen, A., Lauri, A., Napari, I., and Kulmala, M. (2007) The heterogeneous Zeldovich factor, *Atmos. Chem. Phys.*, 7, 309–313.

Vincendon, M., Pilorget, C., Gondet, B., Murchie, S., and Bibring, J.-P. (2011) New near-IR observations of mesospheric CO_2 and

H_2O clouds on Mars, *J. Geophys. Res.*, 116, E00J02, doi:10.1029/2011JE003827.

Wang, H., and Fisher, J. A. (2009) North polar frontal clouds and dust storms on Mars during spring and summer, *Icarus*, 204, 103–113.

Wang, H., and Ingersoll, A. P. (2002) Martian clouds observed by Mars Global Surveyor Mars Orbiter Camera, *J. Geophys. Res.*, 107, E105078, doi:10.1029/2001JE001815.

Whiteway, J., Cook, C., Gallagher, M., et al. (2004) Anatomy of cirrus clouds: results from the Emerald airborne campaigns, *Geophys. Res. Lett.*, 31, L24102, doi: 10.1029/2004GL021201.

Whiteway, J., Daly, M., Carswell, A., et al. (2008) Lidar on the Phoenix mission to Mars, *J. Geophys. Res.*, 113, E00A08, doi:10.1029/2007JE003002.

Whiteway, J. A., Komguem, L., Dickinson, C., et al. (2009) Mars water-ice clouds and precipitation, *Science*, 325, 68–70, doi:10.1126/science.1172344.

Wilson, R. J. (2011a) Water ice clouds and thermal structure in the Martian tropics as revealed by Mars Climate Sounder, *The Fourth International Workshop of the Mars Atmosphere: Modeling and Observation*, Paris, France, February 8–11.

Wilson, R. J. (2011b) Dust cycle modeling with the GFDL Mars general circulation model, in *The Fourth International Workshop of the Mars Atmosphere: Modeling and Observation*, Paris, France, February 8–11.

Wilson, R. J., and Guzewich, S. D. (2014) Influence of water ice clouds on nighttime tropical temperature structure as seen by the Mars Climate Sounder, *Geophys. Res. Lett.*, 41, 3375–3381, doi:10.1002/2014GL060086.

Wilson, R. J., and Richardson, M. I. (2000) The Martian atmosphere during the Viking mission, infrared measurements of atmospheric temperatures revisited, *Icarus*, 145, 555–579.

Wilson, R. J., Neumann, G. A., and Smith, M. D. (2007) Diurnal variation and radiative influence of Martian water ice clouds, *Geophys. Res. Lett.*, 34, L02710, doi:10.1029/2006GL027976.

Wilson, R. J., Lewis, S. R., Montabone, L., and Smith, M. D. (2008) Influence of water ice clouds on Martian tropical atmospheric temperatures, *Geophys. Res. Lett.*, 35, doi:10.1029/2007GL032405.

Wolff, M.J. and Clancy, R. T. (2003) Constraints on the size of Martian aerosols from Thermal Emission Spectrometer observations, *J. Geopyhs. Res.*, 108, E9, 1–22.

Wolff, M.J., Clancy, R. T., Cantor, B., and Madeleine, J.-B. (2011) Mapping water ice clouds (and ozone) with MRO/MARCI, *The Fourth International Workshop of the Mars Atmosphere: Modeling and Observation*, Paris, France, February 8–11.

Yang, P., Baum, B. A., Heymsfield, A., et al. (2003) Single-scattering properties of droxtals, *J. Quan. Spec. Rad. Transf.*, 79–80, 1159–1169.

Yiğit, E., Medvedev, A. S., and Hartogh, P. (2015) Gravity waves and high-altitude CO_2 ice cloud formation in the Martian atmosphere, Geophys. Res. Lett., 10.1002/2015GL064275.

Zuber, M. T., Smith, D. E., Solomon, S. C., et al. (1998) Observations of the north polar region of Mars from the Mars Orbiter Laser Altimeter, *Science*, 282, 2053–2060, doi:10.1126/science.282.5396.2053.

Zurek, R. W. (1992) Comparative aspects of the climate of Mars: an introduction to the current atmosphere, in *Mars*, University of Arizona Press, 835–933.

Zurek, R. W. Barnes, J. R. Haberle, R. M. et al. (1992) Dynamics of the atmosphere of Mars, *Mars*, University of Arizona Press, 835–933.

6

Radiative Process: Techniques and Applications

MICHAEL J. WOLFF, MIGUEL LOPÉZ-VALVERDE, JEAN-BAPTISTE MADELEINE, R. JOHN WILSON,
MICHAEL D. SMITH, THIERRY FOUCHET, GREGORY T. DELORY

6.1 INTRODUCTION

At first glance, the atmosphere of Mars may appear to be a fairly simple environment to characterize: the composition is dominated by a single gas; the surface pressure today is typically less than 1% that of the Earth; and the most common type of cloud seems to share many characteristics with that of the wispy, terrestrial cirrus cloud. However, upon closer examination, it turns out that the Martian atmosphere contains a degree a complexity comparable to that of the Earth's atmosphere. This can be clearly illustrated through a consideration of the mean energy budget for Mars as provided in Figure 6.1. The atmospheric absorption and emission drive multiple processes that affect the atmospheric state and structure over both short- and long-period timescales. The net heating rate due to solar absorption by aerosols in a given volume can be considered as the primary energy source for dynamical motion and atmospheric transport, at least outside of "polar night" conditions. The absorption by gases can drive important chemical reactions associated with compositional stability and evolution through photolysis. In addition, emission and scattering by the atmosphere can produce an obscuring signal for those interested in examining only isolated components of the radiation budget, such as the geological information that might be contained in the reflectance or emission spectra of the surface. In other words, the details of the various source and sink terms in Figure 6.1 are directly relevant to many aspects of the Martian atmosphere and surface system.

Evaluation of the Martian energy budget involves using numerical models of the radiation field that utilize observations of Mars across a wide range in wavelengths. The specific methods and assumptions involved in the generation of Figure 6.1 will be discussed in Section 6.6.1. For now, it is sufficiently informative to illustrate the effects of the various terms in the energy budget by showing the transformation of the incident solar radiation to the outgoing spectrum in Figure 6.2. In this latter case, we employ typical atmospheric conditions for the equatorial region on Mars near a solar equinox with a surface temperature of 295 K and a moderate amount of dust opacity (no water ice). In addition to the incident solar flux, we include the downwelling flux at the surface as well as the wavelength ranges of some commonly defined bands and spectral intervals. Table 6.1 connects the wavelength ranges to past and current spacecraft and their instruments. Together, Figures 6.1 and 6.2 capture schematically a path by which one might identify and explore the diverse radiative processes in the Martian atmosphere. Of course, the mechanism by which they are all linked

together, the propagation of electromagnetic radiation, represents the fundamental framework for such an investigation.

The transport of light through a (non-vacuum) medium has been a topic of interest to many disciplines ranging from astrophysics to oceanography. As a consequence, this field, also known as radiative transfer (RT), has produced an extensive collection of literature describing the theory and methodology for its application to a variety of environments. The corpus of this work may be considered to begin with works of Schuster (1905) and Schwarzschild (1906, 1914) in which the so-called radiative transfer equation (RTE) is formulated in terms of absorption, multiple-scattering, and emission processes. Shortly afterwards, Eddington (1916), building upon these efforts, introduced the two-term intensity expansion that bears his name, Eddington's approximation. Yet, it was not until the monograph by Chandrasekhar (1950) that the RTE and its solutions were subjected to a comprehensive, mathematical treatment. In fact, more than one of the solution methods developed by Chandrasekhar remains as a basis for current numerical techniques.

Overall, the occurrence of radiative transfer in the scientific literature of the early to mid-20th century is dominated by application to problems of astrophysical interest. Publications concerned with planetary atmospheres, such as that of the Earth, tended to focus on the transfer of heat, i.e. the purely absorbing case for infrared (IR) radiation (cf. Elsasser, 1942; Strong and Plass, 1950; King, 1952). However, with the techniques pioneered by Chandrasekhar (1950), one observes a marked growth in RT-based studies of the Earth's atmosphere that include the effects of scattering and solar illumination (e.g. Sekera, 1957; Coulson, 1959a,b; Coulson et al., 1960; Ueno, 1960; Sobouti, 1962). Such interests led quickly to the appearance of books dedicated to radiative transfer in a planetary atmosphere, with the work by Goody (1964) being a well-known example. Driven in part by the confluence of digital computers and new observations, the subsequent decade witnessed an explosion in the publication rate of RT-based investigations, exceeding one hundred per year by the early 1970s (e.g. Coulson and Fraser, 1975).

The trajectory of RT studies for the atmosphere of Mars follows an arc surprisingly similar to that for the terrestrial atmosphere. After a brief mention in the closing paragraphs of a paper on IR transfer for carbon dioxide bands on the Earth (King, 1952), Martian RT applications appear to begin with the work of Grandjean and Goody (1955). As with Earth applications, early Mars RT studies often focused on the topics of radiative balance and purely absorbing atmospheres (e.g. Ohring et al.,

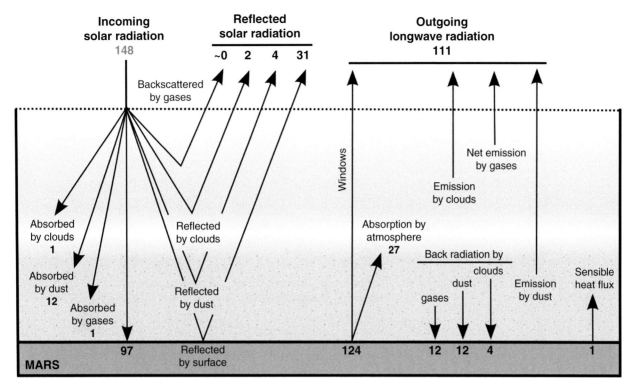

Figure 6.1. Radiation budget of the Martian atmosphere (labeled quantities are absorbed, reflected, or emitted fluxes in W m^{-2}). The main atmospheric constituents that impact the Martian temperature are CO_2 gas, submicrometer to micrometer-size dust particles suspended in the atmosphere, and the micrometer-size water ice particles. The thin Martian atmosphere, which transmits ~80% of both solar and surface IR radiation, is relatively transparent compared to the Earth's atmosphere, which absorbs ~50% of the incoming solar radiation and ~90% of the IR radiation emitted by the surface.

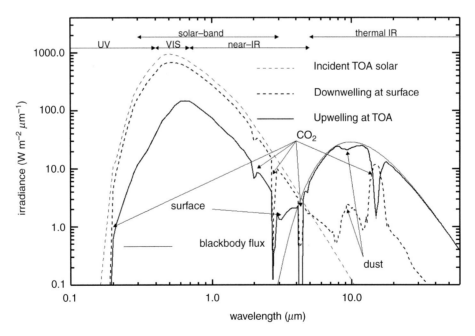

Figure 6.2. The transformation and redistribution of incident solar radiation into solar and thermal IR components. Example incident top-of-atmosphere (TOA) solar, downwelling at the surface, and outgoing TOA irradiance spectra are displayed. A blackbody spectrum for a surface temperature of 295 K is included as a reference. The solar incidence angle is 35° and the atmospheric state is representative of mid-afternoon (2 p.m.) for low-latitude, solar equinox conditions (L_s = 195°). The dust optical depth at 0.67 μm is 0.43 and no water ice is present. We include the common wavelength nomenclature and approximate spectral ranges: VIS = visible, UV = ultraviolet, IR = infrared. Another commonly employed, though somewhat ambiguous, term is *optical*, which we interpret as UV+VIS+NIR. As will be seen in Section 6.4.1, the transition between near-IR (NIR) and thermal IR regimes is somewhat nebulous, and can occur at wavelengths longer than 5 μm for colder temperatures. For our purposes, we chose the dividing line at 5 μm.

Table 6.1 *Spacecraft and instruments with remote sensing datasets of potential interest for studies Martian atmospheric radiative processes.*[a]

Year of arrival	Spacecraft	Instrument	Characteristics
1971	Mariner 9	Ultraviolet Spectrometer (UVS)	Ultraviolet (110–352 nm), spectroscopy
		Visual Imaging System	Visible, images, polarizers
		Infrared Radiometer (IRR)	Thermal IR, bolometer, two channels (10 μm, 22 μm)
		Infrared Interferometer Spectrometer (IRIS)	Thermal IR (6–50 μm), spectroscopy
1971	Mars 3	Photometers	UV–visible
		Radiometers	Visible–IR
		Spectrometer	NIR (2 μm CO_2 band)
1973	Mars 5	Vega, Zufar imagers	Visible
		Photometers	UV, visible, NIR
1977	Viking 1 and 2 (Orbiters)	Visual Imaging System (VIS)	Images, visible
		Mars Atmospheric Water Detector (MAWD)	NIR spectroscopy (1.4 μm H_2O band)
		Infrared Thermal Mapper (IRTM)	Solar-band radiometer, IR radiometers (7–22 μm)
	Viking 1 and 2 (Landers)	Lander Imaging	Images, visible–NIR
1989	Phobos 2	Videospectrometric system (VSK)	Images, visible
		IR radiometer/spectrometer (KRFM)	Visible–NIR and thermal IR spectroscopy
		IR spectrometer (ISM)	NIR
		occultation spectrometer (Auguste)	UV, NIR
1997	Mars Pathfinder (Lander)	Imager for Mars Pathfinder (IMP)	Images, visible–NIR
1997	Mars Global Surveyor (MGS)	Mars Observer Camera (MOC)	Images, visible–NIR
		Thermal Emission Spectrometer (TES)	Solar-band and IR bolometers, spectroscopy (6–50 μm)
2001	Mars Odyssey (MO)	Thermal Emission Imaging System (THEMIS)	Images, visible, thermal IR
2003	Mars Express (MEX)	Spectroscopic Investigation of the Characteristics of the Atmosphere of Mars (SPICAM)	Spectroscopy, passive and occultation, UV, NIR
		High Resolution Stereo Camera (HRSC)	Images, visible–NIR
		Observatoire pour la Minéralogie, l'Eau, les Glaces et l'Activité (OMEGA)	Imaging spectroscopy, visible–NIR
		Planetary Fourier Spectrometer (PFS)	Spectroscopy, visible–NIR, thermal IR
2004	Mars Exploration Rovers (MER)	Panoramic Camera (Pancam)	Images, visible–NIR
		Navigation Camera (Navcam)	Images, unfiltered
		Miniature Thermal Emission Spectrometer (Mini-TES)	Spectroscopy (6–29 μm)
2006	Mars Reconnaissance Orbiter (MRO)	Mars Color Imager (MARCI)	Images, UV–NIR
		High Resolution Imaging Science Experiment (HiRISE)	Images, visible–NIR

Table 6.1 (*Cont.*)

Year of arrival	Spacecraft	Instrument	Characteristics
		Context Imager (CTX)	Images, visible (red)
		Compact Reconnaissance Imaging Spectrometer for Mars (CRISM)	Imaging spectroscopy, visible–NIR
		Mars Climate Sounder (MCS)	Radiometers, solar-band thermal IR (11–42 μm)
2008	Phoenix Lander	Surface Stereo Imager (SSI)	Images, visible–NIR, polarizers
2012	Mars Science Laboratory (MSL, Rover)	Chemistry and Camera (ChemCam)	Spectroscopy, UV–NIR
		Mast Camera (MastCam)	Images, visible–NIR
		Navigation Camera	Images, unfiltered
2014	Mars Atmosphere and Volatile Evolution (MAVEN)	Imaging Ultraviolet Spectrograph (IUVS)	Imaging spectrometer, UV

[a] Table compiled primarily with data from National Space Science Data Center (http://nssdc.gsfc.nasa.gov/nmc/SpacecraftQuery.jsp), with additional input from Moroz (1976), Sagdeev and Zakharov (1989), and Blamont et al. (1991).

1962; Prabhakara and Hogan, 1965; Gierasch and Goody, 1967; Ohring and Mariano, 1968; Leovy and Mintz, 1969), though more sophisticated, non-gray techniques were also employed (e.g. Kaplan et al., 1964; Gray, 1966). Similarly, the number of publications in Mars RT analysis increased appreciably with the availability of new data. However, in this case, the distinctions between the two atmospheres – the smaller Mars atmospheric column and dominance of aerosol effects – provided a detour for Mars RT work. In particular, the Mariner 9 Ultraviolet Spectrometer (UVS) experiment offered a set of novel data that were amenable to interpretation using only transmission models and the so-called single-scattering approximation (i.e. Ajello and Hord, 1973; Ajello et al., 1973; Barth et al., 1973). These simpler approaches continued through the Viking era due in part to the (relatively) large amount of data returned by the missions and to the low to moderate optical depths associated with the Mars atmosphere (e.g. Thorpe, 1979, 1981; Hunt, 1979). Ultimately, the recognition of the importance of aerosol multiple-scattering effects, for both the interpretation of remote sensing data and the atmospheric dynamical simulations, led to the use of more sophisticated RT techniques (cf. Moriyama, 1975; Toon et al., 1977; Pollack et al., 1979; Egan et al., 1980; Zurek, 1981; Martin, 1986; Clancy and Lee, 1991).

The successful landing of Pathfinder and orbital insertion of Mars Global Surveyor (MGS) marked 1997 as the beginning of a new era for RT-based analyses of the Martian atmosphere. The two missions were followed by Mars Odyssey (2001), Mars Express (MEX, 2003), the Mars Exploration Rovers (MER, 2004), and the Mars Reconnaissance Orbiter (MRO, 2006). This proliferation of global datasets in both the optical and thermal IR regimes has been met with the ever-increasing computational resources and newly developed numerical tools from terrestrial and Martian radiative transfer communities. At the present time, with the public availability of the Mars datasets and a variety numerical radiative transfer tools, one has the ability to perform RT calculations relevant to Mars.

Many excellent textbooks exist which allow one to obtain a basic knowledge of RT and the relevant radiative processes. In our experience, it is difficult to recommend a single book for all RT topics. Among our favorites are Goody and Yung (1989), Thomas and Stamnes (2002), Liou (2002), and Hanel et al. (2003). Although they are developed for use in graduate courses, the material may appear daunting to the uninitiated. Petty (2006) offers a more accessible introduction to radiative transfer and atmospheric processes. Pierrehumbert (2010) provides a similar level, but from a climatology studies point of view. However, what is more difficult to find is a single source for the practical knowledge that one needs to perform useful RT calculations for Martian applications, including an introduction to the techniques used in the literature. It is this fact that provides the primary motivation for this chapter. However, it is neither practicable nor fruitful to provide a Mars-centric mini-version of the extant terrestrial textbooks. Instead, we focus on offering brief overviews of the current practices and emphasize the context needed to connect to Mars-based investigations.

In this chapter, we introduce the basic radiative transfer equation and identify key variables and parameters. We discuss relevant geometries and general solution techniques available for Martian remote sensing and meteorological studies in Section 6.2. Section 6.3 presents current approaches for treating molecular and aerosol opacity. In Section 6.4 we delineate the various terms of the source function – solar, thermal, and the so-called non-local thermodynamic equilibrium (non-LTE or NLTE) – with an emphasis on the prescriptions specific to the Martian atmosphere. Next, in Section 6.5, a short overview of energetic particle radiation on the surface of Mars is presented. Finally, Section 6.6 surveys several applications enabled by the preceding sections, including a discussion of the details behind Figures 6.1 and 6.2. Throughout this work, we include citations to historical (classic) and recent literature. Choices in the former category are made for publications that establish precedence or provide context that we have found useful, but the latter type reflects the goal of pedagogy in that the cited work offers greater detail and development than can be provided here. The two groups are neither mutually exclusive nor comprehensive. Our overarching philosophy can be distilled effectively through use of the vernacular: "Here is (much of the) stuff that we wish that we had found in one place all those years ago." As a result, the first several sections focus primarily on providing a description of radiative transfer methods applied *to Mars* (e.g. equations, algorithms, opacity sources). The general reader wishing to know about what these methods have revealed *about Mars* might simply skip to Section 6.6. However, in fact, these methods have also been used in results presented in other chapters; which is another reason for the bias of the authors in the material below on presenting practical details related to the methods themselves.

6.2 RADIATIVE TRANSFER EQUATION (RTE)

Despite its use for almost 100 years, a rigorous derivation of the full RTE (i.e. including polarization) from first principles did not appear in the literature until the 21st century! In Mischenko (2002), one finds an elegant derivation of the vector RTE that follows directly from Maxwell's equations. The results of this effort do not change the classical RTE as formulated by Chandrasekhar (1950). Rather, Mischenko elucidates the important assumptions involved in the use of the RTE, such as the need for the scatterers to be sufficiently separated from each other that the interaction occurs in the far-field zone of the wave propagation (i.e. independent scatterers). Although, this requirement does not pose any prohibition for atmospheric study, it does underline the problematic nature of using the RTE for scattering problems in dense media, e.g. a regolith or planetary surface. Independent of the mathematical foundations of the RTE, the textbooks mentioned above provide excellent introductions to the equation and a multiplicity of solution methods, particularly that of Liou (2002, chapter 6). Below, we present only the basic aspects of the RTE and associated variable definitions needed to understand the necessary sources and sinks of radiation in the Martian atmosphere, e.g. Figure 6.1. We also neglect the polarized nature of light (see Section 6.3.2 for a brief justification), discussing only the scalar aspects of the RTE. We limit discussion of solution techniques primarily to those that are readily available in the public domain.

6.2.1 A Note on Units and Nomenclature

We use the basic MKS system of units, but deviate slightly with respect to wavelength and employ the convention found in the observational literature. Spacecraft and instrument names and abbreviations are listed in Table 6.1. Using the wavelength ranges illustrated in Figure 6.2, references to wavelengths will be nanometers (nm) in the ultraviolet (UV; 100–300 or 400 nm) and visible (VIS, 400–700 nm) and will typically be in micrometers (μm) for the near-infrared to infrared (NIR–IR). In some cases of pure thermal IR, units of inverse length (cm^{-1}) are chosen to allow for the natural sampling of that regime in terms of energy. Similarly, we refer occasionally to the wavelength

dependence of a quantity using the nomenclature of frequency dependence when that is the more common practice. Combinations of the ranges will generally default to the use of micrometers (μm). With respect to the definition of *optical*, we mean UV+VIS+NIR. The term s*olar band* is a subset of the optical, defined by the response of the solar-band bolometers on-board the Viking, MGS, and MRO Orbiters.

6.2.2 General Scalar Equation

For our purposes, it is sufficient to introduce the equation of radiative transfer in a heuristic manner. Consider a pencil beam of light traversing a medium with no sources. If one considers that the intensity or radiance of the beam will change from I_λ to $I_\lambda + dI_\lambda$ after traversing a distance ds, one may describe the reduction

$$dI_\lambda = -\kappa(s,\lambda)\rho(s)I_\lambda ds = -k(s,\lambda)I_\lambda ds \qquad (6.1)$$

where κ and ρ are the mass extinction coefficient (m^2 kg^{-1}) and the mass density (kg m^{-3}) of the attenuating medium and k is the more familiar extinction coefficient (km^{-1}). One may also represent source terms through an "emission coefficient" j_λ (W m^{-3} sr^{-1} μm^{-1})

$$dI_\lambda = -k_\lambda I_\lambda ds + j_\lambda ds \qquad (6.2)$$

Adopting a typical definition of the source function as the ratio of the emission and extinction coefficients, $S_\lambda = j_\lambda / k_\lambda$, we arrive at the basic (scalar) equation of radiative transfer:

$$\frac{dI_\lambda}{ds} = \left(\hat{\Omega}\cdot\nabla\right)I_\lambda = -k_\lambda I_\lambda + k_\lambda S_\lambda \qquad (6.3)$$

where the differential operator or streaming term (d/ds) along the direction of propagation ($\hat{\Omega}$) is written to include only the steady-state case (i.e. $\partial I / \partial t = 0$). For the case of the Martian atmosphere, we include an incident uniform (solar) beam, scattering and absorption opacity, thermal emission, and the potential non-thermal emission processes. For a given wavelength, one can write the source function as

$$S\left(s,\Omega,-\Omega_0\right) = \frac{a(s)}{4\pi}F_0 P\left(s;\Omega,-\Omega_0\right)\exp\left(-\int k(s')ds'\right)$$
$$+ \frac{a(s)}{4\pi}\iint_{4\pi} I\left(s,\Omega'\right)P\left(\Omega,\Omega'\right)d\Omega' + \left(1-a(s)\right)B\left(T(s)\right)+Q(s)$$

$$(6.4)$$

where $-\Omega_0$ is the direction of the incident beam with irradiance F_0, $a(s)$ is the single-scattering albedo, i.e. the ratio of the scattering and extinction coefficients, P is the single-scattering phase function, $B(T)$ is the Planck function for a temperature T, and Q represents non-thermal volume emission such as "airglow" (often $Q(s) = 0$). Each term has an explicit physical meaning. The first term represents the amount of direct solar irradiance that is scattered into the direction Ω. The second term provides for all other scattering scenarios: photons that have been already been scattered at least once in the atmosphere, reflected from the surface, or been emitted from within or below the atmosphere. Finally, the remaining terms allow for thermal and non-thermal sources, respectively. While the emission by a blackbody and the use of the Planck function are familiar concepts, we will defer additional discussion of the equilibrium and non-equilibrium terms until Section 6.4.

As with any differential equation, one must consider the effects of conditions present at the boundaries of the region of interest. Explicit in the above discussion is the "upper" (i.e. top of atmosphere, TOA) condition imposed by the incident solar irradiance (flux), $F_0 = F_\odot$. For a terrestrial planet problem, the analogous "lower" boundary condition involves the reflection of the downwardly propagating radiation and any surface emission. In general, this leads to an expression for the *upwelling* radiation associated with each of the first three terms in (6.4) (ignoring $Q(s)$):

$$\rho(\Omega, -\Omega_0) F_0 \exp\left(-\int k(s')ds'\right)$$
$$+ \iint_{4\pi} I(\Omega, \Omega') \rho(\Omega, \Omega') d\Omega' + \epsilon(\Omega) B\left(T_{surf}\right) \quad (6.5)$$

where $-\Omega_0 \cdot \hat{n}$ is the angle between the incident beam and the surface normal, ϵ is the emissivity of the surface, and ρ is the so-called bidirectional reflectance distribution function (BRDF) as defined by Thomas and Stamnes (2002, p. 134) and Hapke (1993, p. 262). Specific definitions of the BRDF vary and so can be a source of some confusion. For example, another commonly used quantity in the radiative transfer literature is bidirectional reflectance or reflection function (R), related to the BRDF by $\rho = R/\pi$ (i.e. Liou, 2002, pp. 105, 275). Hapke's notation for bidirectional reflectance ($r(\Omega, -\Omega_0)$) is prevalent in Martian remote sensing, and is related to Liou's expression by $R = \pi r / (-\Omega_0 \cdot \hat{n})$ (Hapke, 1993, p. 262).

Angular moments of the intensity, of which irradiance is an example, are useful in the characterization of the radiation field for various radiative processes. The general form of the moment generating equation is

$$M_i = \frac{1}{4\pi} \iint_{4\pi} I(\Omega) \left[\cos(\Omega \cdot \hat{n})\right]^i d\Omega \quad (6.6)$$

where i is an integer. The zeroth moment, known as mean intensity and often represented by the symbol J, is employed in calculations such as photolysis rates and non-equilibrium emission processes. The first moment defines the concept of flux, with M_1 being known as an Eddington flux, found most often in astrophysical applications (e.g. Mihalas, 1978). The more familiar irradiance or flux, at least in planetary atmosphere applications, is an un-normalized version of M_1:

$$F = 4\pi M_1 = \iint_{4\pi} I(\Omega) \cos(\Omega \cdot \hat{n}) d\Omega \quad (6.7)$$

This quantity is used extensively in the energy balance and budget for dynamical models, explicitly appearing in functions such as the heating rate or even in simplified reformulations of the RTE. Use of higher-order moments is less common in planetary atmosphere applications, though M_2 may be found in equations that describe the effects of radiation pressure (e.g. Mihalas, 1978).

Within the RTE and associated equations, the combination of the explicit dependence of the various parameters on the

position s and of the non-trivial boundary conditions reinforces the general complexity associated with solutions for RT problems. That is to say, the functional forms of the quantities found in the RTE will confound the availability of analytical solutions for all but a few special cases. However, at this point, we have only the general form of the time-independent equation. Practical application of the RTE requires the imposition of a specific geometry, i.e. a coordinate system. But before we take that step, we briefly address the issue of polarization.

6.2.3 Neglect of Polarization

The most general formulation of the RTE includes the polarized nature of light (e.g. Hansen and Travis, 1974). The "vector" form of the RTE is generally eschewed as an unnecessary complication, but the fundamental motivation typically is the desire for computational simplicity and efficiency. However, this approach can introduce non-negligible errors in certain situations such as those found in a "Rayleigh atmosphere" (non- or weakly absorbing, small particles) with moderate to large optical depths. As shown by Mishchenko et al. (1994) and Lacis et al. (1998), errors in intensity on the order of 10% are possible, particularly over bright surfaces. The presence of absorbing particles, such as atmospheric dust, reduces the amplitude of the effect by lowering the effective single-scattering albedo of the atmosphere (e.g. Rozanov and Kokhanovsky, 2006). As a result, the thin, dusty Martian atmosphere (containing randomly oriented particles) is not an environment where errors greater than ~1% are likely to occur with the use of the scalar RTE (see also Hansen, 1971). A possible exception to this conclusion might be with wide-field imaging systems, which can have heightened sensitivities to the polarization state of the incident light, particularly in the UV. Although Wolff et al. (2010) discuss this scenario for the MARCI on-board MRO, a more systematic investigation for a wider range of atmospheric conditions may be warranted.

6.2.4 Plane-Parallel Approximation

By far, the most common geometry for applications of the RTE is that of plane-parallel stratified atmosphere, where each layer is considered to extend to infinity in horizontal extent. This is also known as the plane-parallel approximation, with the geometry illustrated in Figure 6.3. Although the coordinate system is one-dimensional in terms of the basic atmospheric state variables, the anisotropic nature of the incident solar radiation and resulting scattered radiation requires the use of two angles defined with respect to the surface normal (e.g. z-axis). Traditionally, the angles are specified using the angular part of the spherical coordinate system, where θ ($= \Omega \cdot \hat{n}$) is the polar (or zenith) angle and $\phi - \phi_0$ is the angle between the planes defined by $\Omega_0 - \hat{n}$ and $\Omega - \hat{n}$, where $\phi - \phi_0 = 0°$ is forward scattering and $\phi - \phi_0 = 180°$ is backscattering; see Figure 6.4. Although downward-propagating radiation has a formal zenith angle of greater than 90°, a common convention is to refer to it using the angle between the beam and vertical axis. Hence, an incoming beam with a formal polar angle of 150° degrees is described as having an "incidence angle" or θ_0 of 30°, with a directional cosine of $\mu_0 = \cos\theta_0$ (where $\theta_0 < 90°$).

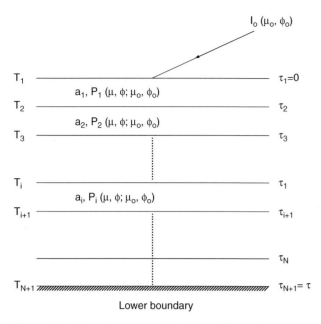

Figure 6.3. Schematic representation of a multilayer plane-parallel geometry. Each layer or level is defined by three radiative properties: single-scattering albedo, single-scattering phase function, and extinction optical depth. In the case of thermal emission, temperature will also be defined.

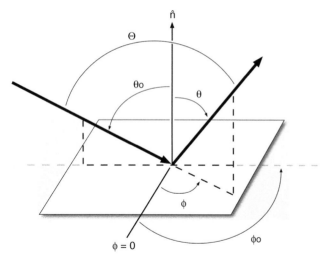

Figure 6.4. Illustration of the polar angle (θ) and azimuth angle (φ) definitions with respect to the surface normal ($\hat{n}$). The scattering angle (Θ) is defined by the plane containing the incident and emergence vectors.

Within the framework described above, the optical depth (τ) is generally defined as the integral of the extinction coefficient from the top of the atmosphere to a height z path length, $\tau(z) = \int_{\infty}^{z} k(z) dz = \int_{\infty}^{s} k(s) \cos\theta ds$. Allowing for the opposite sense of the direction between τ and z (i.e. $d\tau = -dz = -\cos\theta ds = -\mu ds$), (6.3) becomes the more familiar $\mu \, dI / d\tau = I_\lambda - S_\lambda$. One can transform (6.4) in a similar

manner. As such, the general RTE for the plane-parallel case is reduced to the case of a single dimension with three basic radiative properties (τ, a, P) and two atmospheric state variables (e.g. T and pressure); see Figure 6.3. Despite this reduction, the general solution of this equation remains sufficiently complicated that the use of sophisticated numerical techniques represents the standard approach. However, several useful approximations exist that offer significant computational savings over general numerical approaches for RT in the Martian atmosphere.

6.2.4.1 No (Internal) Source Function Approximation

For the case of no internal source function ($S = 0$) and incident radiance I_0 at an angle θ_0 to the local normal, the solution of the resulting ordinary differential equation for downwelling radiance at the lower boundary is the familiar Beer–Lambert–Bouguer law:

$$I = I_0 e^{-\tau/\mu_0} \tag{6.8}$$

(e.g. Thomas and Stamnes, 2002, p. 46). Not surprisingly, this case has rather limited utility for atmospheric studies. Nevertheless, if one allows for a small perturbation (namely $S \neq 0$ but $S \ll I_0$), one finds an application of this solution that is quite powerful. Specifically, one can derive the optical depth along an atmospheric sightline through direct observation of the solar disk. In fact, this special case provides the most robust measurement of the optical depth in the Martian atmosphere (Colburn et al., 1989; Smith and Lemmon, 1999; Lemmon et al., 2004). Similar applications for the terrestrial atmosphere may also be found (Sano et al., 2003). The primary limitations of this approach are found at very low Sun angles and in high-opacity atmospheric states, where the radiance contribution from scattering becomes non-trivial compared to that of the attenuated solar disk.

6.2.4.2 Single-Scattering and Purely Absorbing Approximations

A great deal of computational economy may be achieved if one can ignore the effects of multiple scattering as included through the second term in (6.4). The RTE becomes a simple differential equation with

$$S(s, \Omega, -\Omega_0) = \frac{a(s)}{4\pi} F_0 P(s; \Omega, -\Omega_0) \exp(-\tau/\mu_0) + (1 - a(s)) B(T(s)) \tag{6.9}$$

with a commensurate simplification to the boundary conditions (6.5). The applicability of the approach including only first-order scattering can be determined by analyzing the Neumann series expansion of the general source function, for which (6.9) is the first term. As shown by Thomas and Stamnes (2002, chapter 7), in order to neglect multiple scattering, the scattering optical depth must be small, i.e. $a\tau \ll 1$.

The most obvious application occurs for $a = 0$, which by definition will be valid for all optical depths. This is the purely absorbing or non-scattering atmosphere approximation, $S(s, \Omega, -\Omega_0) = B(T(s)) + Q(s)$. This form of the RTE is commonly employed in the thermal IR regime analyses of

Solar System atmospheres (c.f. Hanel et al., 2003). Specifically, for Mars, the approximation can form the core of operational retrievals for both temperature and aerosol opacity (as it does for TES; Conrath et al., 2000; Smith et al., 2000; Pearl et al., 2001). Despite the fact that $a \neq 0$ for aerosols even in the center of solid-state absorption features, the non-scattering approximation supports TES column-integrated optical depth retrievals with a precision of 10–15% (Wolff and Clancy, 2003). This reflects the dominance of absorption and re-emission processes within the framework of the plane-parallel geometry intrinsic to those data.

Another application of (6.9) regards the non-thermal regime of single scattering of the direct solar beam by aerosols and the surface, referred to as the single-scattering approximation (e.g. chapter 11 in Petty, 2006; chapter 3 in Liou, 2002). The range of validity follows directly from the value of $a\tau$, where one can gain a general sense through the consideration of an atmosphere above non-reflecting surface ($\rho = 0$). For our purposes, we adopt simple atmosphere properties ($a(s) = $ constant, $P\left(s;\Omega,-\Omega_0\right)=1$) and perform an analytical integration of the RTE for a nadir illumination and viewing geometry. Figure 6.5 illustrates the errors associated with such calculations as a function of scattering optical depth for three values of the single-scattering albedo (0.5, 0.8, 0.97), which represent the range of Martian values for the UV through near-IR. A limit of $a\tau < 0.1$ is suggested for errors below the 10–15% level, though large optical depths can be employed for the darker dust due to the diminished importance of scattering. To take into account non-nadir geometries, one can scale the scattering optical depths by $2\mu\mu_0 / (\mu + \mu_0)$. However, even with these limitations, the single-scattering approximation retains utility for a few situations, including the combination of lower optical depths with dark dust (e.g. ultraviolet scattering; cf. Ajello et al., 1973; Ajello and Hord, 1973) and the spectral regions where there is a large amount of molecular opacity (e.g. CO_2 bands; Titov et al., 2000). A recent example reflecting these conditions in the thermal IR is the approximation by

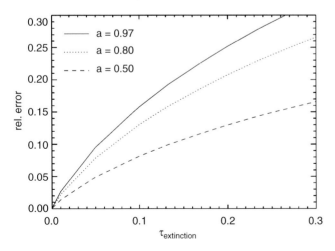

Figure 6.5. Relative error of the single-scattering approximation versus the scattering optical depth ($a\tau$) for the TOA radiance for three values of the single-scattering albedo over a completely absorbing surface. The reference values are calculated for the multiple-scattering case using the DISORT package.

Kleinböhl et al. (2011). They provide a correction to a purely absorbing approach by allowing for the single scattering of surface emission through an analytical, computationally efficient method.

6.2.4.3 Two-Stream Approximations

In terms of explicit citations, two-stream approximations (TSAs) dominate the published literature for both the Earth and Mars. The most prevalent two-stream approaches are based on the Nth approximation ($N = 1$ discrete ordinate) formulation of Chandrasekhar (1950) and the Eddington approximation (Eddington, 1916), where the latter represents the intensity as a two-term Legendre polynomial expansion. Specific TSA applications employ distinct representations of the scattering phase function (e.g. δ-Eddington, Joseph et al., 1976; hemispherical mean, Toon et al., 1989) and "tunable" solution parameters for atmospheric and illumination conditions (e.g. Räisänen, 2002, and references therein). The key advantage of TSAs is the ability to construct analytical expressions for the mean intensity and irradiance/flux (Meador and Weaver, 1980; Toon et al., 1989). Thomas and Stamnes (2002) provide extensive mathematical and numerical developments for TSA applications.

At present, the primary use of TSAs for Martian applications remains in the domain of radiation schemes for mesoscale and general circulation models. Here, the great computational efficiency of an analytical solution is combined with an accuracy for fluxes and heating rates that is better than 2–3% for most plane-parallel illumination conditions. However, errors can reach 10–15% at extreme incidence angles (Wiscombe and Joseph, 1977; Toon et al., 1989; see also discussion in Thomas and Stamnes, 2002, chapter 7). Although improved accuracy can be obtained with the analytical four-stream method ($N = 2$; i.e. Liou, 1974; Cuzzi et al., 1982), as has been done by some in the terrestrial community (e.g. Ayash et al., 2008), this approach has not gained traction in Martian dynamical modeling. One reason for this may be that the error and uncertainties in the input parameters, for example those associated with the highly variable (spatial and temporal) variations in aerosol loading, are likely to be of greater concern than the differences between the two- and four-stream methods.

6.2.4.4 Numerical Solutions

A large number of numerical techniques exist for the solution of the RTE in plane-parallel geometry (e.g. Liou, 2002; Thomas and Stamnes, 2002). The choice of a particular algorithm should be guided by the scope of the situation to be modeled, including computational efficiency, numerical stability, reliability and ease of use, availability of analytical derivatives, ability to isolate various contributions to the source function, etc. For the purposes of this chapter, we focus on a subset that embodies two characteristics that are likely to be of most relevance to the reader: availability as an implementation in a public-domain software package, and past use for Mars applications.

(a) Discrete Ordinate Method

Even a casual inspection of the terrestrial and Martian literature quickly reveals the prevalence of the discrete ordinate method (DOM), originally introduced by Chandrasekhar as the solution of the radiative transfer in the Nth approximation (Chandrasekhar, 1950, chapters II and III). The basic principle of the approach is to replace the source function integral with a quadrature scheme where the ordinates are zeroes of the Legendre polynomial. This discretization leads to a system of coupled differential equations that are amenable to eigenvalue problem techniques. A very popular, public-domain DOM implementation is the DIScrete Ordinate Radiative Transfer (DISORT) package developed by Stamnes and collaborators (Stamnes et al., 1988), which is available via anonymous FTP (http://lllab.phy.stevens.edu/disort/).

The general utility of DISORT can be directly measured by the 1000+ refereed citations to the original DISORT paper of Stamnes et al. (1988), as tabulated by NASA's Astrophysical Data System (http://adsabs.harvard.edu/). Specific applications of DISORT to the Martian atmosphere appear to begin very shortly after the general availability of the code (e.g. Akabane et al., 1990; Clancy and Lee, 1991). Although the Mars-specific usage of DISORT is constant throughout the 1990s (typically using ground-based and Hubble Space Telescope datasets), the datasets produced by MGS and MRO have provided fertile ground for a plethora of DISORT-based studies (e.g. Benson et al., 2003; Clancy et al., 2003; Ignatiev et al., 2005; Wolff et al., 2006; Cantor, 2007; McGuire et al., 2008; Soderblom et al., 2008; Clancy et al., 2009; Wolff et al., 2009).

(b) Spherical Harmonics Method (SHM)

One observes a steep decline in citations between DOM and the second most "popular" plane-parallel RT method, particularly for Martian atmospheric applications. In fact, a single algorithm accounts essentially for the remainder of the Mars-specific references. The technique in question is the public-domain Spherical Harmonics–Discrete Ordinate Method (SHDOM), developed by Evans (1998). Evans combines the efficacy of solving for (and of storing) the source function using the SHM with the more physically intuitive DOM for calculating the radiance. Despite the original development of SHDOM for intrinsically three-dimensional (3D) RT problems (i.e. thick terrestrial clouds), its use by Martian researchers has typically involved the application of the 3D version of the code (SHDOM) to one-dimensional (1D) atmospheric representations (e.g. Fedorova et al., 2004; Korablev et al., 2006). As a result of this trend, which was mirrored to some degree by users for terrestrial application as well, Evans developed and released the purely plane-parallel implementation, SHDOMPP. At present, these two codes are distributed via the Internet: plane-parallel only, SHDOMPP (http://nit.colorado.edu/shdomppda/index.html), and fully 3D (http://nit.colorado.edu/shdom.html).

(c) Two-Stream Codes

A significant number of implementations of this type of algorithm have centered on meteorological models. In fact, two actively maintained Mars-specific codes are the "radiation packages" of the Ames and the PlanetWRF Mars global climate models (MGCMs). The Ames package is based on the Toon et al. (1989)

TSA and is distributed through the Mars Climate Modeling Group portal (http://spacescience.arc.nasa.gov/mars-climate-modeling-group/models.html). A second set of code may be found on the PlanetWRF portal, maintained by Ashima Research (http://planetwrf.com). This code is based upon the TSA formulation of Edwards and Slingo (1996), with further description available in Mischna et al. (2012). Mischna et al. include benchmark comparisons with the Ames code and the popular terrestrial two-stream code TWOSTR that is incorporated into the libRadtran library (Mayer and Kylling, 2005; www.libradtran.org).

6.2.5 Spherical Shell

The potential importance of a curved atmosphere on astrophysical and planetary radiative transfer was recognized early in the era of numerical RT investigations (e.g. Lenoble and Sekera, 1961; Code, 1967; Ueno et al., 1971). These efforts generally employed the spherical shell coordinate system, which restricts the variation of physical properties to the radial direction and so results in a significant simplification of the RTE. However, in the case of an anisotropic source, the angular quantities associated with a local zenith or a tangential coordinate system remain (i.e. θ, θ_0, ϕ, ϕ_0, as in Section 6.2.4). In spherical coordinates, the streaming term for (6.3) becomes (Thomas and Stamnes, 2002, Appendix O):

$$\left(\hat{\Omega}\cdot\nabla\right)=\mu\frac{\partial}{\partial r}+\frac{1-\mu^2}{r}\frac{\partial}{\partial\mu}+\frac{\sqrt{1-\mu^2}\sqrt{1-\mu_0^2}}{r}$$
$$\left[\cos\left(\phi-\phi_0\right)\frac{\partial}{\partial\mu_0}+\frac{\mu_0}{1-\mu_0^2}\sin\left(\phi-\phi_0\right)\frac{\partial}{\partial\left(\phi-\phi_0\right)}\right]$$

(6.10)

where r is the radial distance from the body/system origin ($r = z+R$, where z and R are the height above and the radial distance of the surface in question, respectively).

6.2.5.1 Breakdown of Plane-Parallel Geometry

There is not a simple prescription for determining at which point the plane-parallel results can be deemed suspect, although a survey of the literature suggests that one should be concerned when angles exceed 75–80°. Quantitative insight can be obtained by considering the relationship between vertical and line-of-sight optical depths.

For a spherical atmosphere, the familiar plane-parallel relationship between path length and vertical height remains true only in the differential sense, $ds = dz / \cos\alpha$. Using the geometry illustrated in Figure 6.6, the angle between these two directions (α) can be related to the traditional polar angle θ with spherical trigonometry: $\sin\alpha = [(R + z_0)/(R + z)]\sin\theta$. For $z = 0$ at the surface (compared with the top of the atmosphere as in Section 6.2.4), the optical depth along the path s becomes

$$\tau(s) = \int k(s)ds = \int k(z)/\left[1-\left(\frac{R+z_0}{R+z}\sin\theta\right)^2\right]^{1/2}dz \quad (6.11)$$

This equation may be expressed in terms of the so-called Chapman function, $Ch(X,\theta)$, when the atmospheric opacity

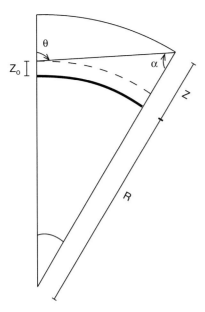

Figure 6.6. The relationship between path length and vertical height for spherical geometry.

can be written as an exponential function with a constant scale length, i.e. $k(z) = k(z_0)e^{-(z-z_0)/H}$, and when one of the limits of the integral is the top of the atmosphere. For purposes of illustrating the limitations of plane-parallel geometry, we need consider only the case for $\theta < 90°$:

$$\tau(s) = k(z_0)H\left(\frac{1}{H}\int_{z_0}^{\infty}\exp\frac{\left[-\dfrac{(z-z_0)}{H}\right]}{\left[1-\left(\dfrac{R+z_0}{R+z}\sin\theta\right)^2\right]^{1/2}}dz\right) \quad (6.12)$$

$$= \tau_0 Ch(X,\theta)$$

where X is the dimensionless parameter $(z_0 + R)/H$ and $\tau_0 = k(z_0)H$ is the column normal optical depth. The recovery of the plane-parallel geometry follows from the asymptotic behavior of the denominator, i.e. $(R+z_0)/(R+z) \to 1$. For the case of $\theta \geq 90°$ (limb geometry), one employs a linear combination of Chapman functions that accounts for the full path through the atmosphere (where the mid-point occurs at the tangent point, $\theta = 90°$) and the "subtraction" of the portion not traversed: $Ch(X,\theta) = 2Ch(X,90°) - Ch(X,180° - \theta)$. For a mathematical discussion of $Ch(X,\theta)$ and its evaluation, as well as an historical context, we refer the reader to Huestis (2001, and references therein). However, this formulation does not include the effects of atmospheric refraction. While very small for Mars, one can find a discussion of the amplitude of this effect in recent terrestrial work (e.g. Rapp-Arrarás and Domingo-Santos, 2011).

For a specific demonstration of the geometrical error introduced by a plane-parallel representation of a spherical atmosphere, we consider the simplified case of a uniformly mixed opacity source in an exponential atmosphere with a constant scale length (10 km) and a surface radius of relevance to Mars (3397 km). In Figure 6.7, we plot the fractional error in the line-of-sight optical depth in a spherical atmosphere using plane-parallel geometry:

$$(\tau_{PP} - \tau_{Ch})/\tau_{Ch} = (\sec\theta - Ch(X,\theta))/Ch(X,\theta)$$

where we calculate $Ch(X,\theta)$ using the numerical technique of Huestis (2001). It is important to keep in mind that $\sec(80°) \sim 5$; so even a 10% error represents an optical depth of 0.5. In terms of RT, there is an error associated with attenuation of the light, as well as with scattering/emission into the beam. Consequently, the limitations of a plane-parallel representation are not likely to scale simply as the error in optical depth.

6.2.5.2 Numerical Solutions

An exploration of the Martian literature reveals three basic responses to the explicit need to treat the spherical nature of the atmosphere: analytical plus approximation, geometric correction to plane-parallel method, and solution of multiple-scattering RTE in curved geometry. Although the number of such Mars publications remains small, the increasing number of Martian datasets that require such treatment merits an overview of currently available approaches and codes.

(a) Analytical Plus Approximation
The analysis of occultation observations fall into this category. In this case, the spherical path through the atmosphere is treated exactly, but the amount of light scattered into the beam is assumed negligible (particularly appropriate when the source of illumination is the Sun). This approach for occultation observations from an orbiting spacecraft has been well developed over the years, with the SPICAM observations providing a large and increasing dataset (i.e. Blamont et al., 1991; Chassefière et al., 1992; Bertaux et al., 2006; Montmessin et al., 2006; Rannou et al., 2006; Lebonnois et al., 2006).

A second algorithm regarding non-occultation measurements also treats the spherical path exactly, but further adopts the single-scattering approximation (as discussed above). In an early application to Mars, Jaquin et al. (1986) applied this approach to study the vertical structure of aerosols. This formulation, similar to the plane-parallel approach of Pang and Ajello (1977) or Thorpe (1979, 1981), offers utility even today because of its computational efficiency, and is employed in the operational retrieval algorithms for the MCS instrument (Kleinböhl et al., 2009, 2011).

(b) Pseudo-Spherical Approximation
The use of a plane-parallel, multiple-scattering solution technique where the path length and incident/emission angles in each atmospheric layer are calculated using spherical geometry is typically referenced as a pseudo-spherical implementation. This can be distinguished from a true spherical treatment by the lack of angular derivatives in the RTE streaming term (6.10). Pseudo-spherical codes developed for terrestrial applications have become relatively common (i.e. SDISORT, Dahlback and Stamnes, 1991; Mayer and Kylling, 2005; VLIDORT, Spurr, 2006). Use for Martian-specific problems has been limited, likely due to easily accessible Mars-specific implementations. One example is that of Wehrbein et al. (1979), where the path length correction compensated for the effects of extreme incidence and emergence angles. For Martian limb observations, such an approach appears to have been even less favored. However, Smith

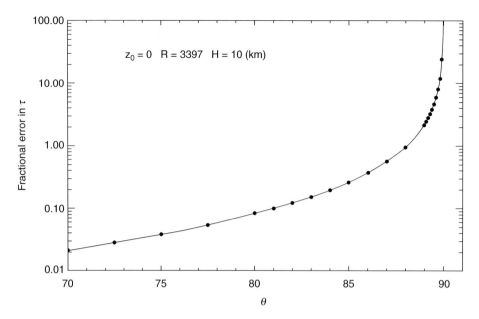

Figure 6.7. The fractional error in the optical depth of a path in a spherical atmosphere calculated using plane-parallel geometry. For the case of an exponential atmosphere with a uniformly mixed opacity source, the relative error is $\left(\sec\theta - Ch(X,\theta)\right)/Ch(X,\theta) \cdot Ch(X,\theta)$ is calculated using the numerical methods of Huestis (2001).

et al. (2013) analyze CRISM limb data employing a pseudo-spherical implementation that is compared against a fully spherical algorithm. Their figure 6 shows a reasonable correspondence between the approximation and a fully spherical treatment. In fact, the agreement would improve to better than 1–2% if they had employed a surface reflectance correction (McLinden and Bourassa, 2010). In spherical geometry, the surface "falls" away and thus provides less reflected light towards the observer than would the infinite plane-parallel representation, i.e. the cosine effect. Currently, there is no publicly available, Mars-specific option, although the more general pseudo-spherical codes of TWOSTR and SDISORT are distributed via the libRadtran package (Mayer and Kylling, 2005; www.libradtran.org).

(c) Spherical Treatment
Full treatment of the RTE with multiple scattering in spherical geometry has often been tackled with the Monte Carlo method (MCM), particularly for Mars (e.g. Wolff et al., 2006; Clancy et al., 2007, 2009, 2010; Vincendon and Langevin, 2010). The relative simplicity of including complicated processes in MCM RT comes at a significant computational cost. On the other hand, the 3D implementation of SHDOM is not necessarily a practical match for spherical shell geometry because of its Cartesian grid. In the realm of terrestrial investigations, such considerations motivated the development of additional algorithms such as the Gauss–Seidel limb scattering (GSLS) and the combined differential–integral methods (Herman et al., 1994, 1995; Rozanov et al., 2005). The accuracy of the various approaches has been compared to MCM benchmarks for Earth atmosphere applications by Loughman et al. (2004) and Pincus and Evans (2009). Similar efforts for Mars and its unique aerosol environment are only just beginning (e.g. Wolff et al., 2012). Although no freely available Mars-specific codes are presently available, it is anticipated that a version of the GSLS will be released in 2017. If one is willing to start with a terrestrial or a general RT code, several options are available,

including SCIATRAN (Rozanov et al., 2005; www.iup.uni-bremen.de/sciatran/), the Atmospheric Radiative Transfer Simulator (ARTS; Eriksson et al., 2011; www.sat.ltu.se/arts/), and Hyperion (Robitaille, 2011; www.hyperion-rt.org).

6.2.6 Three-Dimensional (3D) Radiative Transfer

A plane-parallel (i.e. non-3D) solution of the RTE with an anisotropic source can produce a radiation field that is intrinsically 3D. The one-dimensional aspect of the problem is defined with respect to the variability of the physical medium through which the radiation propagates. The same dimensional restriction applies to RTE solutions for spherical shell geometry as well. Rather, the phrase "three-dimensional radiative transfer" explicitly encompasses the variations in the physical medium associated with the other two dimensions.

Radiative transfer for environments with horizontal variability is often represented through a collection of one-dimensional (1D) calculations. This is known as the independent pixel approximation (IPA; i.e. Marshak et al., 1995, 1999). The practical assumption of the IPA is that the horizontal flow of radiation from "adjacent pixels" is essentially the same as that calculated from the 1D model of the "pixel" in question. Of course, having a cell with a thick water ice cloud (or a local dust storm) but surrounded by cloudless cells would seem to clearly violate the fundamental assumption of the IPA. While this remains a topic of active research for the Earth, the review by Marshak and Davis (2005) suggests that these issues are not particularly relevant for Mars. More specifically, only in the presence of optically thick clouds at the lower end of the mesoscale regime (~10 km, and smaller) will the assumption of the IPA produce appreciable errors in the radiation field.

For Martian studies, 3D effects are likely to have a larger impact for remote sensing studies that attempt to derive atmospheric or aerosol properties. The assumption of constant horizontal properties can either smooth any retrieved distribution or,

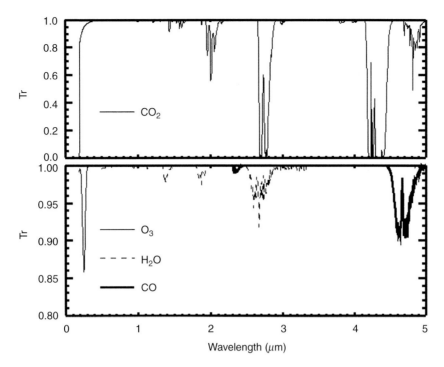

Figure 6.8. Molecular transmittance (*Tr*) for the UV through near-IR for the four primary radiatively active gases. See Section 6.3.2.3 for details on the calculations of *Tr*.

worse, introduce unphysical values. For example, attempting to represent a detached dust layer (see Chapter 10) using spherical shell geometry will introduce a radiative bias at altitudes below the cloud. As illustrated and discussed by Clancy et al. (2007), even if the actual cloud is not present in the sightlines with tangent points below the apparent cloud bottom, the model sightlines will contain the radiative contributions of cloud by definition. Quantitative studies of such effects on retrieved atmospheric properties for Martian datasets are essentially non-existent (at present). Fortunately for the interested reader, three of the codes previously identified have the (3D) capabilities needed to begin such investigations: SHDOM, ARTS, and Hyperion. In terms of more mathematical and numerical detail behind 3D RT, useful starting points include the primers by Davis and Knyazikhin (2005) and Evans and Marshak (2005), as well as the associated code development papers (Evans, 1998; Pincus and Evans, 2009; Eriksson et al., 2011; Robitaille, 2011).

6.3 SOURCES OF ATMOSPHERIC ABSORPTION AND SCATTERING

The radiative processes within an atmosphere are a function of the molecular and physical properties of the radiatively active constituents. For Mars, the situation is simultaneously simpler and more complicated than for the Earth. Among the inventory of gaseous components found in Chapter 4, the strongest gas absorbers are CO_2 and H_2O, but CO and O_3 provide appreciable absorption at some UV and NIR–IR wavelengths. The transmittance associated with the typical abundances of these gases is shown in Figures 6.8 and 6.9. In terms of net absorption or transmittance, only CO_2 provides significant opacity. RT complexity often enters due to the presence of significant

amounts of suspended mineral "dust", as well as the occurrence of non-trivial columns of water (and CO_2) ice at certain locations and seasons. In this section, we outline the prescriptions, data sources, and tools necessary to include the relevant absorption and scattering quantities for Martian RT calculations. The details of the representative molecular transmittance functions will be provided in Section 6.3.2.3. The reader will notice an uneven treatment of the background material for molecular opacities compared with other topics in this chapter. We have found that the coverage of this subject and the background physics is particularly well represented (and self-contained with respect to developing an application for Mars) in several of the textbooks previously cited (e.g. Goody and Yung, 1989; Liou, 2002). We acknowledge that this is our bias, and strongly refer the interested reader to these works.

6.3.1 Atoms and Molecules – Scattering

6.3.1.1 Rayleigh Scattering

The principal scattering process involving the gaseous component of the Martian atmosphere is that originally described by Lord Rayleigh over one hundred years ago (e.g. Strutt, 1871, 1899). Discussion on the theory behind Rayleigh scattering and expressions for the Rayleigh scattering matrix elements may be found in Hansen and Travis (1974) and Liou (2002, pp. 87–97). Here, we define the phase function

$$P_{Rayleigh}(\Theta) = \frac{3}{4}\left(1 + \cos^2\Theta\right), \; \Theta = \cos^{-1}(-\hat{\Omega}_0 \cdot \hat{\Omega}) \qquad (6.13)$$

where the scattering angle Θ is the angle between the incident and emergent rays (Figure 6.4). However, in the case of the scattering cross-section, specific values appropriate for the

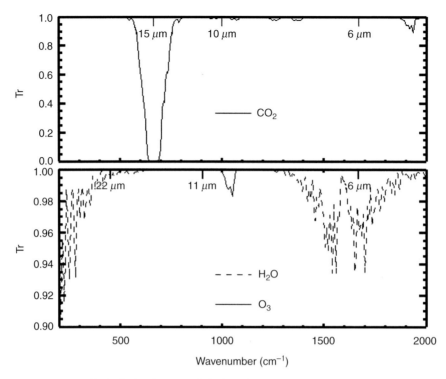

Figure 6.9. Molecular transmittance (*Tr*) in the IR for the two radiatively active gases that have a non-negligible opacity. In order to better illustrate ozone, the ozone abundance in the transmittance here is four times that of Figure 6.8. See Section 6.3.2.3 for details.

carbon dioxide (dominated) atmosphere of Mars are not as readily available.

The Rayleigh scattering cross-section is typically defined per molecule as

$$\sigma\,[\mathrm{cm}^2] = 128\,\pi^5 \alpha_{vol}^2 / \left(3\lambda^4\right)$$
$$= 24\pi^3 / \left(N^2\lambda^4\right)\left(\frac{n^2-1}{n^2+2}\right)^2 F_k \qquad (6.14)$$

where α_{vol} is the molecular volume polarizability (cm^{-3}), λ is the vacuum wavelength (cm), n is the refractive index of the gas with a number density N (cm^{-3}), and F_k is the King factor that takes into account the non-spherical shape of the molecule (see Sneep and Ubachs, 2005, and references within); this convention differs from some authors, where a factor of 3 is shifted from the leading coefficient and included in F_k (e.g. Hansen and Travis, 1974; Goody and Yung, 1989; Liou, 2002). The first expression emphasizes that σ is calculated for a single molecule. For the optical regime, α is essentially independent of wavelength. The connection between the polarizability and refractive index (for density N) is often referred to as the Lorentz–Lorenz equation. The Lorentz–Lorenz equation allows one to apply a measurement of refractive indices for a molecular density of N under terrestrial conditions to the Rayleigh scattering cross-section for the Martian atmosphere.

Ityaksov et al. (2008) present equations for the refractive indices and the King factor for a pure CO_2 atmosphere based upon the work of Bideau-Mehu et al. (1973) and Alms et al. (1975), respectively. Because one of the equations contains an error (a scale factor that is too large by 10^3, as can be demonstrated by going back to the original expression in Bideau-Mehu et al.), we reproduce both here for convenience:

$$\begin{aligned} n(\lambda)-1 = {}&1.1427\times10^3[5799.3/(16.6\times10^9 - \lambda^{-2}) \\ &+120/(7.96\times10^9 - \lambda^{-2}) \\ &+5.33/(5.63\times10^9 - \lambda^{-2}) \\ &+4.32/(4.6\times10^9 - \lambda^{-2}) \\ &+1.22\times10^{-5}/(5.85\times10^6 - \lambda^{-2})] \\ &\text{(for 15°C and 1013hPa)} \end{aligned} \qquad (6.15)$$

$$F_k(\lambda) = 1.14 + 25.3\times10^{-12}/\lambda^2 \qquad (6.16)$$

where the wavelength has the unit of cm. These expressions (with (6.14)) have been verified through laboratory measurements by Sneep and Ubachs (2005) for the visible and by Ityaksov et al. (2008) for the ultraviolet (UV). In addition, for the region of 202–300 nm, Ityaksov et al. fit a simple empirical function to the cross-section behavior by assuming that the wavelength dependence of the refractive indices and the King factor can be modeled by an offset to the traditional λ^{-4} behavior: $\sigma\,[\mathrm{cm}^2] = 1.78\times10^{-46}/\lambda^{4+0.625}$, where λ again has units of cm. Over the specified wavelength range, the agreement is better than a few percent. These two approaches for calculating the Rayleigh scattering cross-section per molecule (for pure CO_2) are shown in Figure 6.10. To illustrate both the validation of the theory and the onset of the ultraviolet CO_2 absorption ($\lesssim 200$ nm), we also include several laboratory measurements for comparison (Sneep and Ubachs, 2005; Ityaksov et al., 2008).

Using the above formulation and the "clear" scenario equation of state from Chapter 4, the normal-incidence Rayleigh scattering optical depth is 0.074 at 250 nm, 0.033 at 300 nm, and 0.009 at 400 nm. Thus, the need for Rayleigh scattering is typically limited to the ultraviolet datasets such as the Mariner 9 UVS (e.g. Ajello et al., 1973; Barth et al., 1973; Curran et al., 1973), Hubble Space Telescope (HST; James et al., 1994;

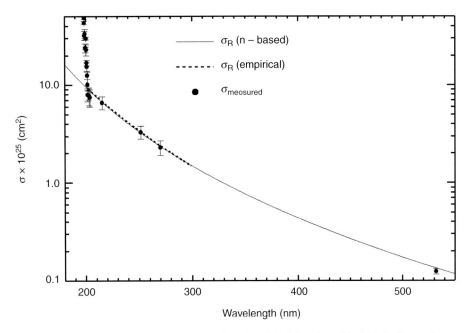

Figure 6.10. The calculated Rayleigh scattering cross-sections from the *n*-based (solid) and empirical (dashed) equations are compared with the laboratory measurements of the total cross-section made by Sneep and Ubachs (2005) and Ityaksov et al. (2008). Notice the onset of the CO_2 absorption near 200 nm, at which point radiative transfer calculations should explicitly include the absorption cross-section.

Clancy et al., 1996a, 1999), the MEX SPICAM (e.g. Bertaux et al., 2006; Perrier et al., 2006; Montmessin et al., 2006) and the MRO MARCI (Malin et al., 2008; Wolff et al., 2010).

6.3.1.2 Raman Scattering

As for Rayleigh scattering, the amplitude of atmospheric Raman scattering depends on the polarizability of the molecules. However, the Raman process is inelastic such that there can be an absorptive component associated with this mechanism. In planetary atmospheres, Raman scattering has been invoked to allow for greater absorption in the ultraviolet and blue, as well as to fill in deep atmospheric absorption and solar lines (cf. Kattawar et al., 1981). This appears to be of practical concern primarily for atmospheres much thicker than that of Mars.

6.3.2 Atoms and Molecules – Absorption

The mathematical treatment of molecular (and atomic) absorption involves quantum mechanics. However, from the perspective of an RT application, one generally employs either direct laboratory measurements of cross-sections (e.g. for electronic transitions, collisionally induced absorption, dimers) or tabulations of spectroscopic parameters (e.g. for predissociation, vibration–rotation, pure rotation transitions) via the extinction coefficient for a specific gaseous species i:

$$k_i(\lambda) = n_i \sigma_i(\lambda) = n_i \sum_j S_{i,j} f_{i,j}(\lambda)$$

(6.17)

where $S_{i,j}$ and $f_{i,j}$ are the line strength and line profile, respectively, for the jth line of the species i (and also functions of temperature and pressure), with the caveat that the third term is not valid for electronic transitions. The choice of using either cross-sections or individual line parameters is usually dictated by the spectral regime of interest. In the UV and into the visible, theoretical

models do not routinely reproduce laboratory measurements of cross-sections across a range of temperatures and pressures with sufficient accuracy (Orphal and Chance, 2003). As such, one is limited to the temperatures and pressures sampled in the laboratory. In contrast, the overall success of theory for vibration–rotation and pure rotation transitions in the near- and thermal IR allows one to use basic formulas and spectroscopic databases to access a wide range of atmospheric conditions.

6.3.2.1 Ultraviolet Cross-Sections

From 100 to 300 nm, the primary non-aerosol opacity sources in the Martian atmosphere are CO_2 and O_3. A nice illustration of the general impact of these two molecules in Martian UV spectra may be found in one of the many papers on stellar occultation measurements by SPICAM (e.g. figure 1 of Montmessin et al., 2006; and in our Figures 6.2 and 6.8). H_2O also absorbs strongly at wavelengths below 180 nm. Although this is less important for atmospheric opacity on Mars, due to the much larger abundance of CO_2, it is important for the photodissociation and loss of water.

Any quantitative treatment of such effects requires laboratory measurements of the cross-sections, such as those tabulated by the Harvard–Smithsonian Atomic and Molecular Physics Group (www.cfa.harvard.edu/amp/ampdata/tstmols.html) and the NASA Jet Propulsion Laboratory (http://jpldataeval.jpl.nasa.gov/). A brief overview of the data available/needed for radiative transfer application is provided below.

(a) CO_2

The compilations of Yoshino et al. (1996) and Parkinson et al. (2003) include cross-section data from 120 to 193 nm for two temperatures, 198 K and 298 K, with 193–200 nm being available for the warmer measurements (www.cfa.harvard.edu/amp/ampdata/cfamols.html#toCO2). Figure 6.11 displays the more relevant case for Martian conditions (198 K) and clearly illustrates

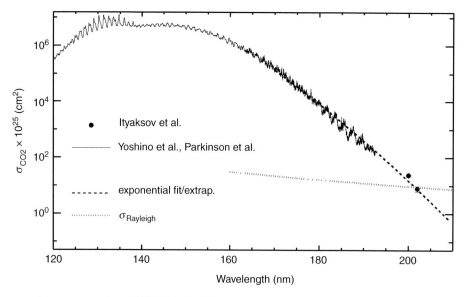

Figure 6.11. CO_2 cross-sections representative of 198 K from Yoshino et al. (1996) and Parkinson et al. (2003), including two points from Ityaksov et al. (2008). A fit to the 155–193 nm region (and the 202 nm point) is used to "fill in" the region between the high-resolution laboratory measurements and the point at which (6.14) is applicable.

the unfortunate gap between 193 nm and the dominance of the scattering cross-section above 204 nm. However, using the 298 K data as a proxy, we find that an exponential function can be used to approximate the mean behavior in this region. We fit a second-order polynomial to the logarithm of the 198 K data with $\lambda > 155$ nm and the 202 nm point of Ityaksov et al. (2008) to produce

$$\sigma\left(cm^2\right) \times \left(1 \times 10^{20}\right)$$
$$= \exp\left(-36.27 + 0.1539\lambda - 0.001234\lambda^2\right) \quad (6.18)$$

where λ is nm. This function and (6.14) are represented as dashed and dotted lines, respectively. The applicability of the cross-section values to other temperatures is discussed by Forget et al. (2009). For warmer temperatures, one can interpolate between the two sets of measurements. For colder temperatures, the situation is more problematic. Forget et al. attempted to address the issue through a sensitivity study, but ultimately employed the 198 K data with the caveat that retrievals using those cross-section will overestimate the CO_2 absorption. The impact of this bias on remote sensing is reported to be at the 5–20% level for column density retrievals from occultation observations (Forget et al., 2009). In terms of radiative processes (e.g. UV heating, photochemistry), the effect of the error will scale roughly with opacity: a cross-section of 10^{-21} cm^2 (~180 nm in Figure 6.11), combined with the densities of the "clear scenario" of Chapter 4, produces an opacity of ~10 km^{-1}, ~1 km^{-1} at 30 km, ~0.1 km^{-1} at 50 km, and ~0.01 km^{-1} at 70 km.

(b) O_3

The most up-to-date set of ozone cross-sections may be found in the compilation by Sander et al. (2011), which is available for download through JPL (http://jpldataeval.jpl.nasa.gov). For Martian applications, only the Hartley band (200–300 nm) is strong enough to be of general interest. Its wavelength-dependent cross-sections values are illustrated in Figure 6.12. Although we do not plot multiple temperatures, it is sufficiently weak that the 218 K data provided may be applied to the typical range

of Martian atmospheric temperatures with errors below a few percent, at most (cf. discussion in section 4A of Sander et al., 2011). Although data tables and interpolation are a common part of radiative transfer tools, the general lack of vibrational structure in the data suggests the plausibility of an analytical function. Combining an exponential and linear function for the Hartley band, we fit the data in the range 210–300 nm:

$$\sigma\left(cm^2\right) \times \left(1 \times 10^{20}\right) = 1132 \exp\left(-z^2/2\right)$$
$$+ 64.50 - 0.2468\lambda, \ z = \frac{\lambda - 254.3}{17.63} \quad (6.19)$$

providing a precision of 1–2% level for 220–280 nm and 3–4% for the 210–220 and 280–295 nm intervals (and should not be used below 205 nm or above 295 nm). To include the 295–340 nm region with a similar level of precision (excluding the band structure near 320–330 nm), one can employ a separate exponential function. Constrained by the 285–340 nm data, the exponential function

$$\sigma\left(cm^2\right) \times \left(1 \times 10^{20}\right)$$
$$= \exp\left(99.86 - 0.7679\lambda + 0.002363\lambda^2 - 2.911 \times 10^{-6}\lambda^3\right)$$
$$(6.20)$$

produces a mean (smoothed) precision of 1–2% over the fitted range (290–340 nm). In Figure 6.12, the resulting empirical fits for both equations are compared to the laboratory data.

(c) Other UV Absorbers

Other UV cross-sections may be of interest in specialized applications such as modeling "ancient Mars" (e.g. SO_2; Córdoba-Jabonero et al., 2003) or self-absorption of non-LTE emission bands (e.g. CO; Stewart, 1972). Because of their relevance for terrestrial studies, laboratory measurements have been made analogously to those discussed above for CO_2 and O_3. The absorption cross-sections may typically be found at the Harvard–Smithsonian and JPL sites.

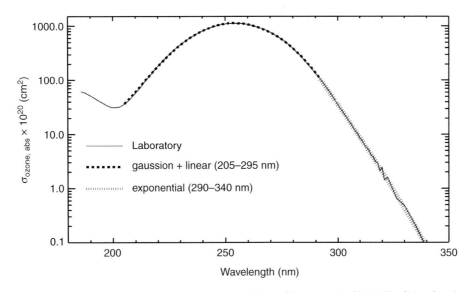

Figure 6.12. Ozone cross-section measurements (at 218 K) from the compilation of Sander et al. (2011). The fitting function for the Hartley band has a general precision of 1–2%, while that for the longer wavelengths has similar precision if one allows for "smoothing" of the structure near 320 nm. See text.

6.3.2.2 Near-Infrared and Infrared Cross-Sections

In contrast to the UV, the use of (6.17) in the NIR and IR regimes has generally been able to rely upon the measurement of basic molecular absorption parameters (e.g. frequency of transition, effects of collisional broadening, etc.). This is driven, in a large part, by the fact that, at these wavelengths, where the vibration–rotation transitions dominate the gas absorption spectrum, the molecular cross-sections vary rapidly with wavelength, and depend much more strongly on changes in pressure and temperature than those associated with electronic transitions. This difference from UV is what led directly to the development of several molecular databases of interest to radiative transfer modeling. Overviews of the general history and impact of such tabulations (with a terrestrial bias) may be found in Rothman et al. (2005) and Rothman (2010). Two of the more popular databases, particularly in terms of application to the Martian atmosphere, are HITRAN (www.cfa.harvard.edu/HITRAN/) and GEISA (http://ara.abct.lmd.polytechnique.fr/index.php?page=geisa-2). The relevant molecular parameters are well described in the documentation available at each site as well as in the associated literature (i.e. HITRAN – Rothman et al., 1987, 1998, 2009; GEISA – Jacquinet-Husson et al., 1998, 2008, 2011). Included with these databases are utility routines that facilitate the application of the molecular parameters to a range of observational conditions, i.e. computation of the total internal partition sums. As such, this leaves only the specification of the line profile function of (6.17) in order to directly calculate molecular opacity.

The frequencies at which the absorption and emission take place are not simply those of the vibration–rotation transitions. Rather, there is frequency distribution, or "line profile", that results from physical effects such as collisions (Lorentzian profile) and thermal motions (Doppler profile). The former tends to dominate at higher pressures and the latter at lower pressures. Wavelength is also a factor for the magnitude of Doppler effects, given the dependence of the line width on frequency. However, the ranges of temperature and pressure in the Martian atmosphere do not typically allow for the use of either case in isolation. Instead, one employs a convolution of the individual line shapes, the so-called Voigt profile. Because a similar situation is present in many terrestrial environments, the relevant mathematical background, including the development of the Lorentzian and Doppler functions themselves, may be found in any of the atmospheric textbooks previously cited (e.g. Goody and Yung, 1989, chapter 3). In terms of application to the Martian atmosphere, the numerical evaluation of the Voigt function may be of more practical interest. Among the plethora of applications in the literature, one finds the oft-cited Humlicek (1982) and Wells (1999). In addition, at the time of publication, we found several public-domain implementations in FORTRAN and C: http://apps.jcns.fz-juelich.de/doku/sc/libcerf, http://cococubed.asu.edu/code_pages/voigt.shtml, www.cs.kent.ac.uk/people/staff/trh/CALGO/680.gz, and http://root.cern.ch/root/html/TMath.html.

However, the far-wing line profiles often differ from the pure Lorentzian or Voigt line shapes. These departures may be attributed to two general factors. The Lorentzian profile is only strictly valid within the context of molecules colliding as solid spheres. This assumption does not take into account weak and distant interactions between molecules. A second problem arises from the fact that molecular energy levels that are "close" to each other cannot be viewed as statistically independent from a quantum mechanical point of view. This line mixing generally populates the lines near the center of a vibration–rotation band, at the expense of those further from the band center. This behavior can be crudely approximated by reducing the opacity in the far-wing regions while increasing the opacity in the line cores. This complication significantly increases the computing cost of the opacity calculation, especially at large pressures. For Mars, this sub-Lorentzian line shape is relevant for the Q-branches within the CO_2 15 μm band, and most significantly for the strong CO_2 ν_3 band at 4.3 μm (e.g. Lellouch et al., 2000). Lellouch et al. demonstrated that the experimental results of Burch et al. (1969) provide a good fit to the Martian spectrum with a prescription that the Lorentzian profile should be multiplied by a factor of 0.5, 0.032, and 0.004 at offsets from the line center of 10, 100, and 200 cm^{-1}, respectively; exponential interpolation was employed for points within that range.

(a) Line-by-Line Technique

With knowledge of line strengths and profile parameters, one can derive the total molecular opacity (for a specific wavelength) by summing the left-hand side of (6.17) over the relevant molecular species i: $k_\lambda = \sum_k k_i(\lambda, P, T)$. This approach forms the basis of the line-by-line (LBL) technique for radiative transfer. The simplicity of LBL is also the general source for its power. The RTE can be solved using molecular opacity that explicitly allows for an essentially arbitrary set of atmosphere conditions and directly includes the effects of monochromatic multiple scattering. A potential drawback is the requirement that the spectral grid be fine enough to resolve all of the wavelength-dependent structure of the gas absorptions bands. For example, to resolve the individual lines in the 15 μm CO_2 band, one must employ several hundred thousand spectral points (assuming a Doppler half-width of 0.0005 cm^{-1}). In general, this computational expense of LBL provides motivation for the development of other techniques; some of which will be briefly discussed below. Nevertheless, (relatively) recent advances in computer hardware have allowed for some applications of the LBL to the Martian atmosphere, though these have tended to focus on remote sensing of smaller spectral regions and weaker molecular bands in the near-IR (e.g. Korablev, 2002; Fedorova et al., 2002, 2009, 2010; Forget et al., 2007). For broader spectral regions and stronger molecular bands, some progress has been made using interpolative methods (i.e. Titov and Haus, 1997), though the errors may be larger than desired (e.g. Ignatiev et al., 2005).

(b) Band Models

A family of techniques called band models were developed early in Mars studies, a time when the observational or computational tools did not exist to identify or quantify the properties of individual lines (transitions) within a band. However, even with the development of such tools, the general intractability of working with actual opacity distributions provided continued support for their use and improvements. As suggested, the word *band* actually refers to a spectral range rather than a particular grouping of absorption lines. The approach involves a frequency average and is formulated in terms of the atmospheric transmittance Tr, which is defined for a given path from a specific point (s) to the "top of the atmosphere" (T):

$$Tr_v(s) = \frac{1}{\Delta v}\int_{\Delta v} Tr_v(s)\,dv = \frac{1}{\Delta v}\int_{\Delta v}\exp\left(-\int_s^T k_v(s)\,ds\right)dv \quad (6.21)$$

Adopting the assumption of a homogeneous atmosphere (i.e. k_v is independent of path and $Tr_v(s) = \exp(-k_v(T-s))$) and expressing k_v in terms of an analytical function, (6.21) can lead to a closed-form solution. The utility and range of validity of any such band model depends on the extent to which the function simulates the actual distribution of line strengths. Along this direction, two general types of models have been developed: regular and random. The former recognizes that some molecular bands can be represented by the periodic repetition of a single line, such as the Q-branch of the 15 μm CO_2 band (Elsasser, 1938). As the name suggests, the latter model exploits the seemingly random line positions of certain bands, which include near-IR water bands (Goody, 1952). Significant

detail, including the historical development of band models, may be found in the excellent review by Goody and Yung (1989, chapter 4). For the purpose of application to the Martian atmosphere, we enumerate three additional points below:

(1) The assumption of a homogeneous atmosphere in the transmittance function requires the use of a scaling approximation to relate the properties of the inhomogeneous path to the use of single values for the atmospheric state variables such as pressure and temperature. The most widely used scaling technique is that of the van de Hulst–Curtis–Godson (HCG) approximation (van de Hulst, 1945; Curtis, 1952; Godson, 1953; see also discussion in Goody and Yung, 1989, chapter 6), where the effective P and T are expectation values along a path weighted by the distribution of the absorbers. Kleinböhl et al. (2009) have proposed a modified HCG scaling for the 15 μm CO_2 band that includes a weighting by absorbers and by pressure for the effective temperature.

(2) Scattering cannot be easily treated in most band models. Due to the dependence of the scattering source function (6.4) on the local radiation field, a breakdown occurs in the necessary conditions of the average transmittance paradigm. This is discussed in more detail in Goody and Yung (1989, chapter 4). However, one can add an absorbing-only opacity source, as was done by Forget et al. (1999).

(3) Although current computational capabilities have led to a de-emphasis on band models, situations may arise for which band transmittance is an appropriate tool for Martian RT. Crisp (1990) developed a quasi-random band model for radiatively active gases relevant to the Martian atmosphere. Hourdin (1992) adapted a terrestrial 15 μm CO_2 band model, which was subsequently used in several global climate models. More recently, motivated by the need for great efficiency, Kleinböhl et al. (2009) employed the band paradigm in the creation of the MCS operational retrieval algorithm.

(c) Correlated-*k* Technique

The method was originally developed by Lacis and Oinas (1991) and refined by Goody et al. (1989). The core of the algorithm involves recasting the frequency integral into one that can be evaluated using a smaller number of quadrature points through a change in variables. Following the notation of Goody et al. (1989), consider a function of the radiation field (L) over the frequency interval Δv in a compositionally homogeneous layer:

$$\bar{L} = \frac{1}{\Delta v}\int_{\Delta v} L(k_v)\,dv = \int_0^\infty L(k) f(k)\,dk = \int_0^1 L(k(g))\,dg \quad (6.22)$$

where $f(k)$ is the distribution function of the absorption coefficient and $g(k)$ is the cumulative distribution function, i.e. $g(k) = \int_0^k f(k')\,dk'$. This approach has considerable merit when dealing with functions that are monotonic with respect to the absorption coefficient, such as radiance or transmittance. The transformation is also the crux of the k-distribution band model (cf. Goody and Yung, 1989). The correlated-k technique arises when the formulation is applied to a non-homogeneous atmosphere. The critical assumption in this step is that the wavelength dependence of the gas absorption coefficients (*k*-distributions) remains sufficiently correlated along the optical path that the last equality in (6.22) is true, at least to the desired precision. While the identity can be demonstrated to be exact for several limiting cases (i.e. weak line, strong line, etc.; cf. Goody et al., 1989), it is not possible for the general case. Perhaps more fundamentally, the approximation is only truly valid for isobaric, isothermal optical paths over which $f(k)$ does not vary in a way that "rescrambles" the monotonic ordering of the opacity. As such, the use of correlated-*k* for new ranges of environmental conditions and spectral intervals should be initially evaluated with LBL calculations.

A significant advantage of the correlated-*k* method is the ability to include scattering processes. Such a capability is a direct result of the general transformation from frequency to *g* space retaining the monochromatic nature of the radiation field, which requires the correlation of the opacity wavelength behavior from layer to layer. In other words, although one cannot associate a single frequency with a given *g* point across all layers in an atmosphere, the radiative transfer equation (6.3) is solved for each *g* using the appropriate $k(g)$ values for each atmospheric layer. The integrals of the needed radiative quantities (e.g. radiance, flux) are then calculated using a discrete version of (6.22), e.g. $\bar{I} = \sum_g w_g I_g$, where w_g is the quadrature weight. In other words, a monochromatic radiative transfer calculation is performed for each term in the series and the resulting weighted sum represents the integrated quantity over that spectral interval. The computational savings of the technique are typically realized by combining the much smaller number of quadrature points required per frequency band/interval (10 or 20 versus thousands or more) with the use of precomputed tables of absorption of coefficients for the needed range of atmospheric conditions. General discussions of the practical information needed for a correlated-*k* implementation may be found in the original papers of Goody et al. (1989) and Lacis and Oinas (1991), as well chapter 4 of Liou (2002). Mischna et al. (2012) give a more specific prescription for the atmosphere of Mars, including the water vapor continuum (i.e. Clough et al., 2005) and collisionally induced opacity (e.g. Gruszka and Borysow, 1997)

Another limitation of correlated-*k* techniques is encountered when more than one gas absorbs in a given spectral range, and the mixing ratio of one (or all) of the gases varies in space or time. Vertical variations can compromise the wavelength-dependent correlation in absorption coefficient, while horizontal variations require spatially varying correlated-*k* coefficients. A standard approach has been to explicitly include the mixing ratio as an additional dimension of the precomputed table, with the so-called single-gas mixture being particular effective (e.g. Goody et al., 1989). At least for the present-day atmosphere of Mars, the situation is much simpler when one considers the small number of strong absorbers (CO_2) and the limited overlap with other species of radiative interest (H_2O and, to a lesser degree, O_3 and CO). In addition, the amount of RAM available in a typical workstation allows one to simply include separate correlated-*k* tables for each of these small number of molecules, i.e. $k_{molecular,g} = k_{CO2,g} + k_{H2O,g}$. Of course, one is still limited by the intrinsic ability of the given $\nu \rightarrow g$ transformation to represent the frequency integral (re-emphasizing the potential need for additional LBL validation).

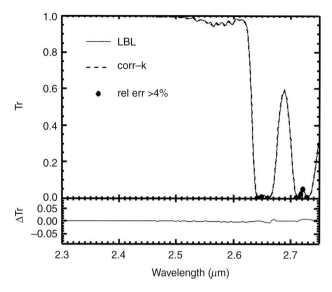

Figure 6.13. Comparison of line-by-line calculations to those for correlated-*k* in a clear atmosphere. The upper panel compares the absolute transmittances while the lower panel shows the difference between the two in absolute units. The filled circles indicate which points have a relative error of larger than 4% in *Tr*, occurring only for very low *Tr* values. When one considers the optical depth, the relative errors are <1%. See text for details of calculations.

The use of the correlated-*k* algorithm for studies of the Martian atmosphere is modestly wide spread (e.g. Smith et al., 1996, 2006, 2009; Conrath et al., 2000; Wolff and Clancy, 2003; Sefton-Nash et al., 2013). It is also a standard part of radiation packages in dynamical models applied to Mars past and present (e.g. Rafkin, 2009; Mischna et al., 2012; Wordsworth et al., 2013). Besides its use in terrestrial applications of relevance to Mars (e.g. modeling of heating by CO_2 and H_2O), the recent work of Mischna et al. (2012) provides additional validation for conditions specific to Mars. While some concern may be justified for its application to remote sensing applications such as small features in high-resolution spectra (Ignatiev et al., 2005), this does not appear to be a general limitation. For example, Figure 6.13 shows the excellent agreement between LBL and correlated-*k* transmittance for H_2O and CO_2 lines near 2.7 μm. The transmittances were calculated using the inhomogeneous atmosphere provided in Table 4.5 ("clear" scenario), with the spectroscopic parameters taken from the HITRAN 2008 database.

(d) Other Techniques

The primary pitfall of the correlated-*k* technique – the lack of a strict correlation of spectroscopic properties across atmospheric levels – can be addressed without having to bear the computation burden of full LBL calculations. West et al. (2010) review a variety of such algorithms, but we wish to highlight two in particular.

(1) West et al. (1990) created a set of spectral mapping techniques that strictly preserve the precise mapping of $\nu \rightarrow g$ through all layers in an atmosphere, essentially allowing arbitrary accuracy with respect to LBL results. This general approach is further refined by Crisp in his development

of the Spectral Mapping Atmospheric Radiative Transfer model, which includes a transformation back to the frequency domain for use with high-resolution spectra (e.g. Crisp et al., 2004; Savijärvi et al., 2005). It is reported (in the references cited) that such spectral mapping methods offer a computational savings of one to three orders over LBL while maintaining an error of less than 1–2%.

(2) Moncet et al. (2008) developed a methodology for rapidly calculated radiances with the accuracy of traditional LBL techniques. The Optimal Spectral Sampling (OSS) technique calculates the radiances within a defined band using a small number of wavelengths whose choice is optimized to produce maximum accuracy. The OSS approach allows for the influence of atmospheric aerosols through a training procedure that should include the full range of atmospheric conditions expected. Thus far, the OSS has been utilized for Martian temperature profile retrieval studies (Eluszkiewicz et al., 2008; Hoffman et al., 2012).

6.3.2.3 Clear-Air Molecular Transmittance Illustrations

Figures 6.8 and 6.9 illustrate the limited nature of molecular opacity effects in the Martian atmosphere, and provide a graphical reference for general band location. The transmittances were calculated using the tabulated UV cross-sections discussed above and the correlated-k algorithm for the NIR and IR (HITRAN 2008 parameters) with spectral resolutions representative of CRISM/OMEGA and TES, respectively. The "clear" scenario model atmosphere from Chapter 4 (Table 4.5) provided the p–T prescription. The molecular abundances were specified using constant mixing ratios with the following values: 0.953 for CO_2; 1.63×10^{-4} for H_2O (10 pr μm); 8.00×10^{-4} for CO; and 6.48×10^{-8} and 2.59×10^{-7} for O_3 (5 μm atm column) in Figures 6.8 and 6.9, respectively.

6.3.3 Aerosols

The importance of aerosol particles for the thermal balance of the Martian atmosphere was first recognized by the inability of workers to reproduce thermal profiles obtained by Mariner 6 and 7 with only molecular opacity (Gierasch and Goody, 1973). Subsequent analyses by Pollack and collaborators strongly reinforced the fundamental role played by aerosols, revealing both positive and negative feedback mechanisms on Martian meteorology (Pollack et al., 1979). While much of the early work concerned only dust particles, the presence of ice aerosols in the atmosphere led one to include their effects in RT calculations. In this section, we outline the important aerosol-specific quantities needed for the solution of the RTE, as well as the location of public-domain tools for their calculation. We also include a discussion of the current state of knowledge of Martian aerosol properties, including basic radiative properties and spatial–temporal distributions.

6.3.3.1 RTE-Related Definitions

Although addressed in the various textbooks cited previously, Hansen and Travis (1974) remains one of the best primers for a general discussion of the aerosol-specific inputs for radiative transfer in a planetary atmosphere. A more recent comprehensive, though somewhat less accessible, treatment may be found in the monograph by Mishchenko et al. (2006). Our goals here are simply to define the variables and equations necessary to include the effects of aerosol in the RTE, and to provide references for readers interested in more detail.

Aerosols enter the RTE through the extinction efficiency coefficient, the single-scattering albedo, and the single-scattering phase function, the k, a, and P (respectively) from (6.3) and (6.4). Although molecular scattering and absorption are calculated with separate algorithms, this is not typically the case for aerosols. Nevertheless, it is useful to consider the scattering and absorbing aspects as separate variables in the following way (with the wavelength and spatial dependences assumed):

$$k = k_{ext} = k_{sca} + k_{abs} \tag{6.23}$$

This equation leads to a simple definition of the RTE single-scattering albedo:

$$a = k_{sca} / k_{ext} = 1 - k_{abs} / k_{ext} \tag{6.24}$$

where the right-hand side of the equality is often employed due to the common practice of computing only the extinction and absorption cross-sections. Practical calculations of k and a (and P) involve averages over an ensemble of non-identical particles. However, it is convenient to start with the single-particle properties for a collection of identical particles and then generalize.

(a) Ensemble of Identical Particles

For a collection of identical particles (with the same orientation), the analog of (6.17) defined in terms of the single particle is

$$k_{ext} = n_0 \sigma_{ext} = n_0 (\sigma_{sca} + \sigma_{abs}), \tag{6.25}$$

where σ (more commonly C in the aerosol literature) again represents the physical cross-section in units of area and n_0 is the number particles per unit volume.

The single-scattering phase function P is a normalized quantity derived from a more basic scattering property called the phase matrix, which is closely related to the so-called scattering and Mueller matrices (e.g. Bohren and Huffman, 1983, chapter 3; Mishchenko et al., 2006, chapter 3). The matrices are 4×4, but for the scalar RTE one only needs to employ the first matrix element. Using the notation of Mishchenko et al. (2006), the phase function for the scalar RTE is related to phase matrix Z by

$$P(\Omega, -\Omega_0) = \frac{4\pi}{\sigma_{sca}} \frac{d\sigma_{sca}}{d\Omega}(\Omega, -\Omega_0) = \frac{4\pi}{\sigma_{sca}} Z_{11}(\Omega, -\Omega_0) \tag{6.26}$$

where the symbol $d\sigma_{sca} / d\Omega$ is the differential scattering cross-section and has units of area (as does Z_{11}). This notation is only schematic; it does not actually indicate a derivative with respect to Ω. The symbol is meant to be a mnemonic device to remember the value of the integral:

$$\sigma_{sca} = \int_{4\pi} \frac{d\sigma_{sca}}{d\Omega} d\Omega \tag{6.27}$$

(e.g. Bohren and Huffman, 1983). The functional dependence of the phase function may be simplified by the assumption of random orientation, in which case the angular arguments are replaced by a single value, the scattering angle Θ (defined in Figure 6.4 and in (6.13)).

Although not explicitly appearing in the RTE, two other definitions are widely used in describing the radiative properties of aerosols: asymmetry parameter and efficiency factor. The former is the first moment of the phase function and represents a measure of the integrated behavior of P:

$$g = \langle \cos\Theta \rangle = \int_{4\pi} P(\Theta)\cos\Theta \, d\Omega \quad (6.28)$$

with $g > 0$ when more light is scattered in the forward direction ($\Theta = 0°$) and $g < 0$ for more in the backward direction ($\Theta = 180°$). The efficiency factor Q is simply defined as the cross-section normalized by the geometrical cross-section G:

$$Q_{ext} = C_{ext}/G, \quad Q_{sca} = C_{sca}/G, \quad Q_{abs} = C_{abs}/G \quad (6.29)$$

The original use of efficiency factors is related to efforts to conceptualize the cross-section in terms of geometrical optics, though this approach does not appear to retain much original meaning in present discourses on scattering theory (Bohren and Huffman, 1983, pp. 72–73). Current use tends to focus on their utility as dimensionless cross-sections.

(b) Ensemble of Non-Identical Particles
Starting at the most basic level, one would include the effects of particle size, shape, and orientation in any ensemble average. If we assume that these properties are statistically independent, the extinction coefficient is

$$k_{ext} = n_0 \int_{4\pi} d\Omega \int_\epsilon d\epsilon \int_r dr \, f_{orientation}(\Omega) f_{shape}(\epsilon) f_{size}(r) \sigma_{ext}(\Omega,\epsilon,r) \quad (6.30)$$

where the f functions represent the normalized probability distributions for each quantity $\left(\int_X f(X)dX = 1\right)$. Under the assumption of randomly oriented particles, one typically folds the average over orientation into the calculation of the single-particle properties themselves. In addition, the general lack of constraint on particle shape for Martian aerosols effectively eliminates the need to include a shape distribution factor. Allowing for the use of the more familiar $n(r)$ notation to represent the distribution of particle sizes, the reduced form of (6.30) combined with (6.25) yields

$$k_{ext} = n_0 \int_{r_{min}}^{r_{max}} f(r) \sigma_{ext}(r)dr = n_0 \langle \sigma_{ext} \rangle$$
$$= n_0 \int_{r_{min}}^{r_{max}} f(r)\left[\sigma_{sca}(r) + \sigma_{ext}(r)\right]dr \quad (6.31)$$
$$= n_0 \left(\langle \sigma_{sca} \rangle + \langle \sigma_{ext} \rangle\right) = k_{sca} + k_{abs}$$

where r is some measure of particle size (to be discussed below) with units of length, the quantity $n_0 f(r)dr$ is the number of particles per unit volume in the size interval from r to $r+dr$, and the brackets indicate the expectation value with respect to the size distribution $f(r)$. This produces the ensemble definition for single-scattering albedo:

$$a = k_{sca}/k_{ext} = \langle \sigma_{sca} \rangle / \langle \sigma_{ext} \rangle \quad (6.32)$$

Because the phase function is a normalized quantity, the generalization of (6.26) requires the use of the ensemble average of each factor:

$$\langle P(\Theta) \rangle = \frac{4\pi}{\langle \sigma_{sca} \rangle} \langle Z_{11}(\Theta) \rangle \neq 4\pi \left\langle \frac{Z_{11}(\Theta)}{\sigma_{sca}} \right\rangle \quad (6.33)$$

The computation of the asymmetry parameter and the efficiency factors simply substitutes the ensemble version of the fundamental variables into (6.28) and (6.29)

$$\langle Q \rangle = \langle C \rangle / \langle G \rangle, \quad \langle g \rangle = \int_{4\pi} \langle P(\Theta) \rangle \cos\Theta \, d\Omega$$

where $\langle G \rangle$ is the size average of the geometrical cross-section, i.e. $\langle G \rangle = \langle \pi r_{eq}^2 \rangle$.

(c) Separate Ensembles
A final type of averaging to consider is that due to separate (i.e. discrete) populations of particles, such as the combination of dust and water ice aerosols. The same principles apply as discussed above, with the integrals replaced by discrete summations. So, our three basic aerosol quantities for the RTE equation expressed in terms of size-averaged values are now

$$k_{ext} = \sum_i k_{ext,i} = \sum_i n_{0,i} \langle \sigma_{ext} \rangle_i, k_{sca} = \sum_i k_{sca,i} = \sum_i n_{0,i} \langle \sigma_{sca} \rangle_i$$
$$a = k_{sca}/k_{ext}$$
$$P(\Theta) = \left(\sum_i \langle P(\Theta) \rangle_i k_{sca,i}\right) / \sum_i k_{sca,i}$$
$$(6.34)$$

where the index i represents the ith population or component.

6.3.3.2 Particle Size Distributions
The introduction of the particle size distribution $n(r)$ reflects the expectation and observation that atmospheric processes will not produce a monodisperse distribution (also known as a monodispersion). The inherent difficulty in determining the shape of this function purely from *ab initio* or empirical means led to the use of mathematically convenient functions, whose choice was informed by theory and measurement. Consequently, in terms of characterization for use in radiative transfer modeling, the more relevant quantities become the moments of the distribution. Building on the physical intuition of cross-section and particle size, Hansen and Travis (1974) proposed mathematical moments that include the additional weighting factor of the geometrical cross-section. Instead of the traditional mean, variance, and skewness, one has instead the effective radius (r_{eff}), effective variance (v_{eff}), and effective skewness (s_{eff}):

$$r_{eff} = \frac{\int_{r_1}^{r_2} r \, \pi r^2 n(r)dr}{\int_{r_1}^{r_2} \pi r^2 n(r)dr} = \frac{1}{\langle G \rangle} \int_{r_1}^{r_2} r \, \pi r^2 n(r)dr \quad (6.35)$$

$$v_{eff} = \frac{1}{\langle G \rangle r_{eff}^2} \int_{r_1}^{r_2} (r - r_{eff})^2 \, \pi r^2 n(r)dr \quad (6.36)$$

$$s_{eff} = \frac{1}{\langle G \rangle r_{eff}^3 v_{eff}^{3/2}} \int_{r_1}^{r_2} (r - r_{eff})^3 \, \pi r^2 n(r)dr \quad (6.37)$$

where $\langle G \rangle$ is the size-averaged geometrical cross-section defined above, and the factor of $1/r_{eff}^2$ introduced in (6.36) is to produce a dimensionless quantity. Their motivation was to minimize the effect of a particular choice of $n(r)$ on the averaged radiative properties. In fact, Hansen and Travis demonstrate that this can typically be accomplished by considering only the first two moments, r_{eff} and v_{eff} (see figures 14 and 15 of Hansen and Travis, 1974). While their deduction is empirical (though rooted in physics), it remains prevalent in the light scattering literature and is occasionally revisited with no apparent change in applicability (e.g. Mishchenko and Travis, 1994).

Despite this de-emphasis on the specific functional form of $n(r)$, one must choose something. In the terrestrial literature, the modified gamma distribution (MGD) has long been employed (e.g. Petty and Huang, 2011). Originally proposed by Deirmendjian (1964) to describe water cloud and hazes, it has the general form

$$n(r) = C_0\, r^a exp\left(-br^\gamma\right) \qquad (6.38)$$

with C_0 being derived from the appropriate normalization condition, i.e.

$$\int n(r)dr = 1 \quad \rightarrow \quad C_0 = \gamma b^{(\alpha+1)/\gamma}/\Gamma[(\alpha+1)/\gamma] \qquad (6.39)$$

where Γ is the well-known gamma function. In addition to the ability of the MGD to reproduce (or at least be consistent with) laboratory and *in situ* measurements, Deirmendjian emphasizes its mathematical connection to potentially observable aspects of the size distribution. While he focused on the mode radius

$$r_c = \left(\frac{\alpha}{b\gamma}\right)^{1/\gamma} \qquad (6.40)$$

r_{eff} and v_{eff} (via (6.35) and (6.36)) can be calculated with a similar convenience and utility as

$$r_{eff} = \frac{\Gamma[(\alpha+4)/\gamma]}{b^{1/\gamma}\Gamma[(\alpha+3)/\gamma]}, \quad v_{eff} = \frac{\Gamma[(\alpha+4)/\gamma]\Gamma[(\alpha+5)/\gamma]}{\Gamma^2[(\alpha+4)/\gamma]} - 1 \qquad (6.41)$$

The MGD was introduced into the realm of Martian studies by Toon et al. (1977), and its use continued in early major studies of Martian aerosols (Pollack et al., 1979; Zurek, 1982; Clancy and Lee, 1991). Whether because of legacy or mathematical convenience, the MGD may be found in the Martian atmospheric literature through the present.

Another function employed to characterize Martian aerosols is that of the gamma distribution, which is the $\gamma = 1$ case of (6.38). However, Hansen and Travis (1974) derived a specific prescription that allows for the parameters to be expressed directly in terms of r_{eff} and v_{eff}:

$$n(r) = \frac{(ab)^{(2b-1)/b}}{\Gamma[(1-2b)/b]} r^{(1-3b)/b}\, exp\left(-r/ab\right) \qquad (6.42)$$

where $r_{eff} = a$, $v_{eff} = b$, and $\int n(r)dr = 1$. Used to represent terrestrial precipitation and cloud sizes (e.g. Petty and Huang, 2011), its application for Mars seems to be driven primarily by the convenience of having only two parameters compared with three for the MGD. Specific examples start to appear in

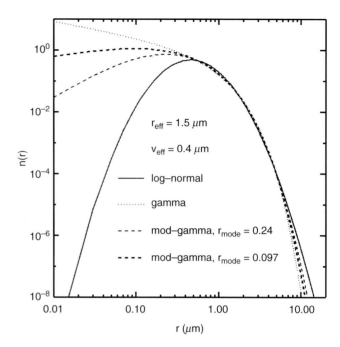

Figure 6.14. Size distribution functions for canonical dust size moments, $r_{eff} = 1.5$ μm and $v_{eff} = 0.4$. The dispersion for smaller particles suggests that the use of only two moments (and two parameter size distributions) may prove problematic for observations at wavelengths more sensitive to small particles, such as the UV.

the mid- to late 1990s (e.g. Chassefière et al., 1995; Tomasko et al., 1999).

A final size distribution to mention here is that of the lognormal distribution (LND). With an origin in laboratory studies of the 1950s, its use is rather sparse even in terrestrial work (cf. Liu and Liu, 1994). However, the adoption of the LND to describe aerosol sizes in several Martian general circulation model (MGCM) studies (e.g. Montmessin et al., 2004; Kahre et al., 2008) suggests the value of its inclusion here. The functional form of the LND may be expressed as

$$n(r) = \frac{1}{\sqrt{2\pi}\sigma_g}\frac{1}{r} exp\left(-\left(\log r - \log r_g\right)/2\sigma_g^2\right) \qquad (6.43)$$

where $\int n(r)dr = 1$. The two distribution parameters can be written in terms of r_{eff} and v_{eff} as

$$r_g = r_{eff}/\left(1+v_{eff}\right)^{5/2}, \qquad \sigma_g^2 = \log\left(1+v_{eff}\right) \qquad (6.44)$$

Despite the general reliance of the two distribution moments, it remains helpful to occasionally visualize a size distribution. Figure 6.14 provides example distributions with $r_{eff} = 1.5$ μm and $v_{eff} = 0.4$. In the case of the modified gamma distribution, the third parameter is specified by the mode radius. The eye is naturally drawn to the dispersion of values for small particle sizes. The studies that demonstrate the need to specify only two size distribution moments are typically those for particles of ~1 μm or greater and from observations in the visible or NIR (e.g. Hansen and Travis, 1974; Mishchenko and Travis, 1994). Given the observations that particles scatter more efficiently where the ratio of size to wavelength is of order unity (see definition and discussion of size parameter in Section 6.3.3.3; also e.g.

Bohren and Huffman, 1983), it is not hard to imagine that the two-moment maxim will be less applicable for shorter wavelength regimes.

6.3.3.3 Calculation of Single-Particle Scattering Properties

At the most fundamental level, the calculation of the particle properties necessary for RT studies involves the solution of Maxwell's equations. More precisely, one employs the equations to relate the characteristics of the scattered electromagnetic wave to that of the incident wave and to the macroscopic properties of the scattering medium. However, the formulation of the general problem and the associated mathematical formalism is beyond the scope of this chapter.

Nevertheless, many texts exist that cover the basic theory and include the foci of practical calculation and applications. The classic book by van de Hulst (1957) remains relevant more than 50 years after publication, providing physical insights and analytical techniques that are beneficial in understanding the role of particle properties in impacting the propagation of light in planetary atmospheres. In terms of a practical guide to the current state of and the basic theoretical framework for modern scattering calculations, the relatively recent monograph of Mishchenko et al. (2002) may represent the best place for the interested reader to begin. In this text, one finds an accessible overview of the theory, a survey of techniques for calculation and measurement, and finally a compendium of examples and applications. In addition, Mishchenko et al. (2002) is available freely and electronically: www.giss.nasa.gov/staff/mmishchenko/books.html.

Here, we concern ourselves here with the details relevant to a discussion of aerosol radiative properties relevant to the Martian atmosphere.

As a starting point, one needs to specify three macroscopic properties for a particle in order to calculate the quantities needed for the solution of the RTE (e.g. k, a, $P(\theta)$):

(1) Complex refractive index m or complex dielectric function ϵ for materials within a particle. One may think of these values as describing the composition of the particle. However, for a complex particle, one may have several compositional constituents, and require multiple sets of m or ϵ. Although the specific use of refractive indices is more common in the scattering literature, the two functions are related through

$$m^2 = \epsilon \quad \text{and} \quad m = n + i\kappa \quad (6.45)$$

where n and κ are often identified individually as the real and imaginary parts of the refractive index. The physical quantities are connected to the volume polarizability of the particle (e.g. Bohren and Huffman, 1983), and thus to the response of the material to the incident wave. Conceptually, one finds that the larger the amplitude of m, the greater the change to the incident light.

(2) Size parameter X, which combines both the particle size and the incident wavelength of light:

$$X = 2\pi r / \lambda \quad (6.46)$$

This dimensionless parameter is the one that appears explicitly in most scattering solutions, characterizing the size of the particle in terms of wavelengths (or number of wave oscillations).

(3) Shape. For many atmospheric studies, the parameter is typically specified as an analytical surface description, such as that for a sphere or a cylinder, with the possible complexity of a concentric layer of identical habit but differing composition. In this case, the shape is interpreted directly as the analytical prescription of the boundary conditions for the solutions to Maxwell's equations. In general, the shape must allow for complex topologies that include irregular particle habits and compositional inhomogeneity.

The choice of a specific algorithm to calculate the desired radiative properties is itself a function of these three macroscopic parameters, though the latter two play a much larger role. In the interest of brevity, we focus on the hierarchy imposed by the size parameter. The various techniques effectively fall into three size parameter regions: $X \ll 1$, the Rayleigh regime; $X \sim 10^0$–10^1, the so-called Lorenz–Mie regime; and $X \gg 1$, the geometrical optical regime. Modern computation capabilities allow for combination of Rayleigh Lorenz–Mie particle size treatments (e.g. Mishchenko et al., 2002, chapter 7). Associating the remaining two size parameter regimes with the additional filter of those techniques that have been or might soon be used in Martian atmospheric studies, several general classes of solutions and associated numerical implementations are listed below.

(a) Separation-of-Variables Method (SVM)

This approach employs a recasting of the solution to Maxwell's equations into that of the vector wave equation. Effectively, this limits the applicability of SVM to spherical and spheroidal particle shapes.

Lorenz, Mie, and others found the SVM solution for spheres independently at the beginning of the 20th century, leading to the nomenclature of Lorenz–Mie for the solution technique as well as with the size parameter range listed above. The numerical implementation has been extensively studied and multiple codes are available publicly, including the appendices of Bohren and Huffman (1983). More practical sources of Lorenz–Mie codes are the SCATTERLIB compilation that has been maintained by Piotr Flatau since 1994 (http://code.google.com/p/scatterlib/) and the SCATTPORT portal maintained by Thomas Wriedt and collaborators (Wriedt and Hellmers, 2008; https://scattport.org). The first computationally tractable extension to layered spheres, e.g. ice-coated dust particles, is found in Toon and Ackerman (1981). As with homogeneous spheres, well-tested and well-described numerical implementations are readily available, with Bohren and Huffman (1983) and SCATTERLIB/SCATTPORT being accessible repositories of source code. With the appropriate use of extended precision arithmetic, the spherical SVM techniques are capable of highly accurate solutions for size $X \sim 10^3$ and beyond (e.g. Mishchenko et al., 2002, chapter 7).

The SVM for homogeneous spheroids was developed in the mid-1970s (Oguchi, 1973; Asano and Yamamoto, 1975). While subsequent developments dramatically improved the numerical issues associated with spheroidal wave functions, the computational cost of the SVM solution remains high, though it provides high accuracy for various spheroid realizations such as extreme axial ratios (i.e. Mishchenko et al., 2002, chapter 6). Available codes may be found in the SCATTPORT compilation, including extensions to multilayer spheroids. Practical numerical considerations keep the range of utility to the Lorenz–Mie regime, $X \lesssim \text{few} \times 10^1$.

(b) *T*-Matrix Method (TMM)

The basis of this technique is the expansion of the incident and scattered electromagnetic fields in terms of spherical vector harmonics. The expansion coefficients of the two waves are related through the so-called transition matrix or *T*-matrix. Developed originally by Waterman (1971), this technique became computationally efficient with the derivation of the analytical formulation for orientational averaging by Mishchenko (1991). Mishchenko and collaborators have produced an extensive body of literature on the importance of non-spherical particle effects through the use of the TMM (reviewed in Mishchenko et al., 2002, chapters 5 and 10). However, it is important to note that their work has focused exclusively on axisymmetric particles, such as circular cylinders and spheroids. More recently, Kahnert (2013a,b) has developed a TMM formulation to handle non-axisymmetric particles such as hexagonal and rectangular prisms. The Mishchenko codes have been developed to be numerically stable and robust, and have been distributed and updated for over 10 years through the NASA GISS website (www.giss.nasa.gov/staff/mmishchenko/t_matrix.html). They are also available via SCATTERLIB and SCATTPORT. The Kahnert formulation has also been included in a public-domain code, TSYM (www.chalmers.se/en/staff/Pages/kahnert.aspx). However, it should be noted that TSYM is described by its author as a "research code", and does not contain some of the "robustness" features of the Mishchenko codes, such as automatic convergence checking.

The quad-precision TMM implementation of Mishchenko is capable of calculating particles on the lower end of the geometric optics regime, i.e. $X \sim 10^2$, but only for modest axial ratios (Mishchenko and Macke, 1999). For axial ratios of ~2–3, the effective limitation is more typically $X \sim 30$–50. The situation is more problematic for large axial ratios, in which case the use of an SVM-based algorithm is indicated.

(c) Finite-Difference Time-Domain Method (FDTDM)

This method involves the direct numerical solution of Maxwell's equations, in particular the curl equations, where the spatial and temporal derivatives are evaluated using finite-difference techniques. Although the original development occurred more than 45 years ago by Yee (1966), the FDTDM did not begin to gain prominence until advances in computer hardware and numerical techniques made accurate calculations tractable. A review of the historical development can be found in Mishchenko et al. (2002, chapter 6), while a succinct mathematical overview exists in Liou (2002, chapter 5). The finite-element basis for FDTDM allows for relatively facile representation of complex shapes and anisotropic compositions, a distinct advantage over the SVM and TMM approaches discussed above. This has led to its application for the scattering properties of terrestrial ice crystals (e.g. Yang and Liou, 1996; Yang et al., 2000, 2003). While there has been little application of FDTDM to Martian aerosols, the growing importance of improved macrophysical properties for water ice particles is likely to change this (see

Chapter 5). Several useful FDTDM codes are available from the SCATTPORT software library. Current implementations appear to be limited to $X \lesssim 15$–20 (Liou, 2002, chapter 5).

(d) Geometrical Optics Approximation (GOA)

The method is eponymous with the largest size parameter regime. Although such size parameters are well beyond those associated with most remote sensing retrievals for Martian dust and water ice aerosols, there is some suggestion that much larger particles may be present in the boundary layer of the north polar region (e.g. Whiteway et al., 2009; Lemmon, 2014). As such, the GOA technique may be relevant for future Martian studies. Both Mishchenko et al. (2002) and Liou (2002) offer an insightful treatment of the theoretical foundations of the GOA, and the interested reader is referred there for more detail. The basic assumption of the technique is that wavelength is so small compared to the particle size that the incident plane wave may be considered as a set of parallel rays whose paths can be traced independently. In this sense, one can consider the GOA to be an asymptotic method. As a result, the practical question of the size parameter at which the approximation is appropriate can only be addressed through comparisons with more accurate methods. For spheres, computations suggest that one must exceed size parameters of 200–300 before use of the GOA would be indicated (e.g. Hansen and Travis, 1974). However, the limit appears to be lower for non-spherical particles, $X \gtrsim 100$–500 (i.e. Macke et al., 1995; Mishchenko and Macke, 1999). Several options are available for GOA computations through links curated by SCATTPORT. Of particular interest should be the collection of tools developed and distributed by Andreas Macke, http://tools.tropos.de (Macke et al., 1995).

6.3.3.4 *The Importance of Non-Spherical Particles*

Mars aerosol particle sizes and atmospheric pressures do not favor particle alignment effects; the disruptive effects of Brownian motion dominate over the aerodynamic tendency to align the short axis of the particle with the local gravitational normal (Jayaweere and Mason, 1965). In such cases, non-spherical aerosol particles in the Martian atmosphere can be considered to be randomly oriented. Nevertheless, real differences exist between the radiative properties of non-spherical versus spherical particles (e.g. Bohren and Huffman, 1983, chapter 13; Hovenier et al., 1986; Mishchenko et al., 2002, chapters 4 and 11).

The extensive use of Lorenz–Mie theory to represent collections of non-spherical particles in both terrestrial and Martian literature has often been a question of computational tractability and expediency. In addition, the limited availability of non-spherical scattering codes complicated the ability of researchers to assess the impact of using Lorenz–Mie algorithms. However, with the advent of publicly available, robust TMM codes in the early 1990s, dozens of papers have appeared on the subject of comparisons between spheres and non-spherical particle scattering. A detailed discussion of this topic, with an emphasis on the rotationally symmetric shapes included in the Mishchenko TMM codes, may be found in Mishchenko et al. (2002, chapter 11). Their results clearly illustrate the sensitivity of the scattering matrix elements to non-sphericity for size parameters beyond $X \sim 1$. Figure 6.15 highlights

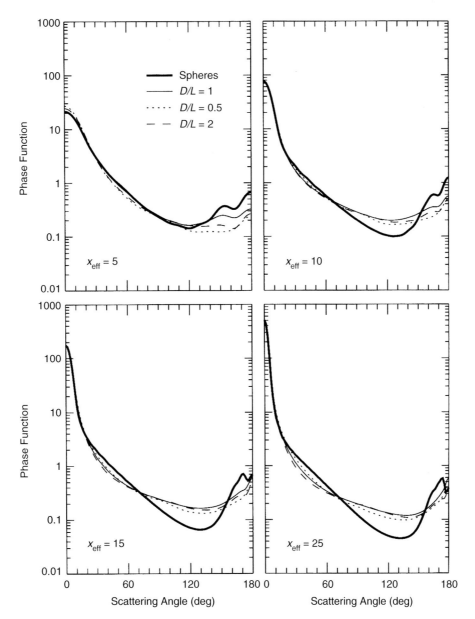

Figure 6.15. Scattering phase functions for (randomly oriented) circular cylinders of moderate diameter-to-length (*D/L*) aspect ratios compared with those for surface-equivalent spheres. The calculations were performed using $v_{eff} = 0.1$ and a refractive index representative of terrestrial aerosols in the visible (m = 1.53+0.008*i*). Reproduced from figure 10.23 in Mishchenko et al. (2002), used courtesy of NASA.

this trend for cylindrical prisms with moderate values of the aspect ratio (diameter-to-length ratio), though departures of the same magnitude occur for other axisymmetric shapes such as spheroids and Chebyshev particles (i.e. Mishchenko et al., 2002). Perhaps less intuitive may be the trends observed for the integrated forward-scattering quantities such as the extinction cross-section and the single-scattering albedo. As shown in Figures 6.16 and 6.17, the largest departures can occur for small size parameters. In addition to underlining the potential biases of the radiative properties calculated using Lorenz–Mie theory, these two figures also provide a caveat for the appropriate choice of the equivalent-sphere convention.

In order to refer to the size of a non-spherical particle by a single value, one may employ the convention of specifying the radius of a sphere with the same volume, the so-called volume-equivalent radius. In the scattering literature, the

phenomenological role played by the particle projected area (through diffraction and interference) has led to the common practice of defining size with respect to a sphere with the same surface area. The calculations shown in Figures 6.15–6.17 use this surface-equivalent definition. However, as discussed by Pollack and Cuzzi (1980), the scattering properties of non-spherical particles for smaller size parameters may often be more similar to those for volume-equivalent spheres. However, differences at larger size parameters increase correspondingly for the volume-equivalent convention.

Despite the tenor of the above discussion, Martian radiative properties are often sufficiently represented by Lorenz–Mie theory, particularly given its computational economy. In particular, for the shapes and sizes typically associated with Martian aerosols, the errors introduced in the radiative properties by the use of spheres in the thermal IR will be negligible with respect to other

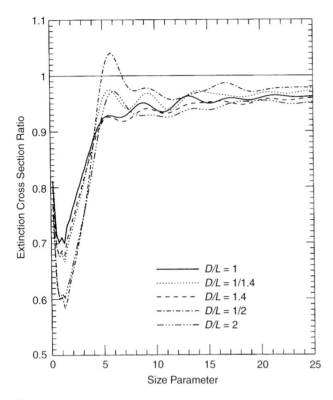

Figure 6.16. The ratio of the extinction cross-section for circular cylinders to that of surface-equivalent spheres as a function of size parameter. Calculations details are the same as those for Figure 6.15. Reproduced from figure 10.17 in Mishchenko et al. (2002), used courtesy of NASA.

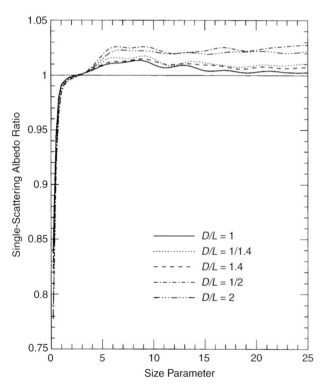

Figure 6.17. As for Figure 6.16, but for the single-scattering albedo. Reproduced from figure 10.20 in Mishchenko et al. (2002), used courtesy of NASA.

parameter uncertainties and measurement errors (cf. figure 1 in Wolff and Clancy, 2003; figure 3 in Wolff et al., 2006).

6.3.3.5 Dust Aerosols

Studies of the radiative properties of Martian atmospheric dust have been pursued since the first spacecraft datasets were analyzed for atmospheric signatures. As a result, there is a significant body of literature that is developed in Chapter 10, as well as the relatively recent review by Smith (2008). Here, we limit ourselves to the references directly relevant to describing (current) options for the three macroscopic parameters, though efforts are made to be as inclusive as possible when only a small number of citations are involved. However, it is important to keep in mind that the observed optical properties depend simultaneously on the particle size, shape, and refractive indices (or composition). As such, even when a study focuses on a particular parameter, the other particle attributes must be specified (ideally in a self-consistent fashion).

(a) Refractive Indices

Complex refractive indices (m) represent a key spectral component in characterizing the dust radiative properties. Part of the initial work analyzing Mariner 9 and Viking observations used the measured values of terrestrial materials that were considered potential analogs of Martian dust, associated with chemically and physically weathered basalts, e.g. montmorillonite in Toon et al. (1977) and palagonite in Clancy et al. (1995). Another line of attack was to derive the refractive indices as part of the overall retrieval process, such as by Pollack et al. (1977, 1979, 1995). To some degree, these efforts were driven by dissatisfaction with the analog values. The culmination of the Pollack et al. series of papers was a set of m that spanned from the UV through the NIR (Ockert-Bell et al., 1997). A study analogous to that of Pollack et al., but with higher-fidelity visible data, was performed Tomasko et al. (1999). For the IR, a similar philosophical approach was begun by Snook (1999) using Mariner 9 IRIS observations of the 1971 global dust event, and continued by Wolff and Clancy (2003) and Wolff et al. (2006) with MGS/TES and MER/Mini-TES data. Finally, the 2007 global dust event offered the opportunity to derive an independent set of refractive indices for the UV through NIR from MRO MARCI and CRISM, as well as MEX OMEGA observations (Määttänen et al., 2009; Wolff et al. 2009, 2010). The results of these efforts in terms of the imaginary and real parts of the refractive index are shown in Figures 6.18 and 6.19, respectively. In the case of the IR regime, the illustrated indices produce better agreement of the models with the data than do the analog materials. However, in the UV–NIR, one must choose between two sets of available refractive index solutions, which themselves embody specific assumptions of particle size and shape.

The early availability of the Ockert-Bell et al. (1997) values and the general importance of the optical regime for both dynamical and remote sensing studies led to their early, widespread adoption by the Mars community (e.g. Forget et al., 1998, 1999; Petrova, 1999; Toigo and Richardson, 2000; Colaprete and Toon, 2000; Montmessin et al., 2002). More recently, Wolff et al. (2009) employed a self-consistent

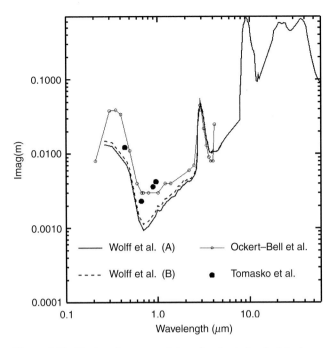

Figure 6.18. The imaginary part of the refractive index for Martian dust aerosol particles derived from observations. The delineation of cases A and B for the Wolff et al. compilation indicates the assumed particle size for the analysis of the 2007 dust storm datasets for UV–NIR: (A) is $r_{eff} = 1.8$ μm and (B) is $r_{eff} = 1.6$ μm. See text for more detail.

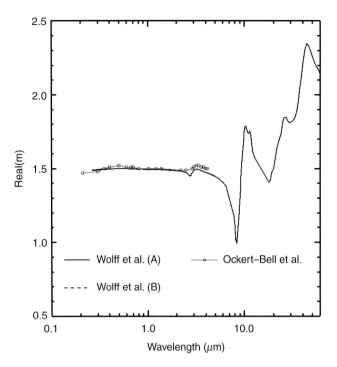

Figure 6.19. As for Figure 6.18, but for the real part of the complex refractive indices. The Tomasko et al. effort assumed a value of Re(m) = 1.5 for all values of Im(m).

model – size, shape, m – to a broader, improved set of observations to obtain refractive indices that disagree with the Ockert-Bell et al. values (notably Figure 6.18). While the discrepancy may be due to an observational bias resulting from low spatial resolution (as suggested by Vincendon et al., 2007), the Wolff et al. indices actually produce radiative properties that are more consistent with those derived empirically from multiple spacecraft than if the Ockert-Bell et al. values are used (i.e. solar-band bolometers on Viking and MGS; Clancy and Lee, 1991; Clancy et al., 2003; see also table 3 of Wolff et al., 2009). Furthermore, the use of the Wolff et al. m values appears to eliminate the need to treat the ratio of visible-to-IR opacity as a free parameter in the dust heating calculations for Martian dynamical models (Madeleine et al., 2011). In any case, regardless of the particular choice for the optical and NIR region, the set of refractive indices shown in Figures 6.18 and 6.19 represent the current state of knowledge for use in the calculation of dust radiative properties.

(b) Shape
Based upon the frequency of appearance in the literature, the most popular modeled shape for Martian dust would be the sphere. Despite the availability of non-spherical particle algorithms such as the TMM code by the mid-1990s, their application to Mars has remained rather limited until recently (e.g. Petrova, 1999; Dlugach et al., 2002; Clancy et al., 2003; Wolff et al., 2009). Certainly a contributing factor is that even the most efficient non-spherical codes are typically much slower than Lorenz–Mie algorithms. However, spheres also possess the advantage of requiring a minimum of needed "information

content" – a sphere is specified by a single parameter, its radius. Until recently, available data did not provide the information content (e.g. coverage in phase angle and in wavelength) needed to address the additional parameters of non-spherical shapes. Nevertheless, one algorithm bears mention: semi-empirical theory (SET; Pollack and Cuzzi, 1980). SET is an empirical algorithm that represents non-spherical (randomly oriented) shapes through a combination of Lorenz–Mie theory and physical/geometrical optics. Different shapes may be accommodated through a set of adjustable parameters that can be fixed through laboratory measurements or derived as part of the retrieval process. However, unlike the use of non-spherical particle techniques discussed previously, the "empirical" aspect of the theory makes it difficult to generalize the resulting values to other wavelengths or particle sizes (e.g. Pollack and Cuzzi, 1980).

Originally employed by Pollack et al. (1977, 1979) with Viking data, SET has been used more recently by Tomasko et al. (1999) to model sky-brightness observations by the Pathfinder IMP camera. The radiometric fidelity and angular resolution of the IMP data allowed Tomasko et al. to derive simultaneously constraints on the particle size, the refractive indices, and the angular scattering distribution for dust aerosols. Despite the mentioned limitations of SET, the resulting single-scattering phase functions have been used as starting points or reference functions for several subsequent efforts to characterize or assess an effective shape for Martian aerosols (Clancy et al., 2003; Wolff et al., 2009; Laan et al., 2009). A comparison of the Tomasko et al. phase function for 960 nm with that of a TMM circular cylinder is shown in Figure 6.20. The particular axial ratio ($D/L = 1$) is chosen because of its use by Wolff

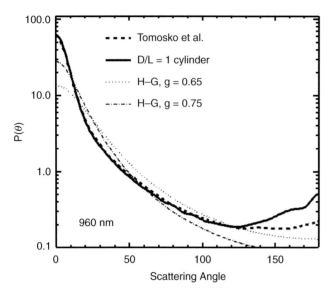

Figure 6.20. The 960 nm scattering phase function derived by Tomasko et al. (1999) is compared to that of the circular cylinder ($D/L = 1$) used by Wolff et al. (2009). Both were calculated using the refractive indices of Tomasko et al. and size distribution moments of specified $r_{eff} = 1.4\ \mu m$ and $v_{eff} = 0.2$. A Henyey–Greenstein function for two values of the asymmetry parameter (g) is also included; each g represents a perturbation of 0.05 about the value reported by Tomasko et al.

et al. (2009) to derive the dust refractive indices. The departure of the TMM phase function in the backscattering direction is related to the interference effects that occur within the regular geometry of a convex analytical shape (e.g. Mishchenko et al., 1997). An improved agreement can be achieved with a slight change in axial ratio ($D/L \sim 1.4$; Clancy et al., 2003), but it comes at the price of increased computational intensity as well as a more limited size parameter range. Within the framework of TMM (and the other methods discussed previously), a more likely solution may involve distributions of axial ratios, as has been done for terrestrial aerosols (Mishchenko et al., 1997), or the explicit use of irregular particle shapes whose habits tend to disrupt the backscatter enhancement (e.g. Laan et al., 2009).

Finally, if one equates the behavior of the phase function with the attribute of particle shape, one may wish to consider the so-called Henyey–Greenstein (HG) phase function,

$$P_{HG}(\Theta) = \frac{1 - g^2}{\left(1 + g^2 - 2g\cos\Theta\right)^{3/2}} \tag{6.47}$$

The HG function offers the convenience of a single-parameter analytical function with the explicit connection to a radiative property that may be computed or prescribed independently (Henyey and Greenstein, 1941). By design, it obeys (6.28), i.e. substitution of P_{HG} gives the expected result: $\cos\Theta = g$. However, the function was designed to reproduce the behavior of submicrometer interstellar dust particles observed in the visible as calculated by Lorenz–Mie theory (Henyey and Greenstein, 1941). Taken at face value, it would seem to have limited applicability to atmospheric aerosols. Nevertheless, the combination of its simplicity and computational efficiency has

led to widespread use in radiative transfer analyses, with the Martian atmosphere being no exception. While such an application to Martian dust in the IR may be appropriate (where the size parameters are similar to those that originally motivated Henyey and Greenstein), this is generally not the case for the visible and NIR. To illustrate the potential limitations or discrepancies associated with the use of P_{HG}, Figure 6.20 includes the HG function for $g = 0.65$ and 0.75. These two g values bracket those associated with the SET and TMM phase functions, which are 0.70 and 0.68, respectively. The situation can be improved by employing a linear combination of HG functions, for example the so-called two-term Henyey–Greenstein function (e.g. Johnson et al., 2006). However, the introduction of additional parameters can quickly dilute its utility beyond that of a "fitting function" and disconnect the resulting parameter values from any clear physical meaning.

(c) Size
Although we have identified the size parameter as the formal macrophysical parameter, particle sizes become the natural variable in practical scattering calculations. As discussed by Smith (2008), studies that employ observations of the total atmospheric column have placed constraints on the average particle sizes in the Martian atmosphere as $r_{eff} = 1.4$–1.7 μm and $v_{eff} = 0.2$–0.5. These ranges of size distribution moments seem to be applicable for a large part of the Martian atmosphere and its spatial–temporal variation. Excursions can occur for "special" locations and times, for example smaller sizes for northern spring locations (e.g. $r_{eff} \approx 1\ \mu m$; Clancy et al., 2003) and larger sizes for high dust loading conditions (e.g. $r_{eff} \approx 2$–2.5 μm; Wolff and Clancy, 2003; Elteto and Toon, 2010a,b). An important caveat for these particle size distributions is that they are column-integrated values. That is to say, they were derived under the assumption of the single size distribution at all altitudes. This approximation for Martian aerosols is driven primarily by the difficulty in extracting size information from plane-parallel viewing geometries, although Wolff et al. (2006) and Vincendon et al. (2007) discuss possible exceptions. Observational constraints on the vertical variation of particle sizes have been limited generally to those obtained from MEX occultation observations: $r_{eff} \sim 0.5$–1.0 μm for altitudes of 15–30 km; $r_{eff} \sim 0.1$–0.3 μm for 35–50 km; and $r_{eff} \sim 0.01$–0.1 μm above 50 km (Montmessin et al., 2006; Rannou et al., 2006; Fedorova et al., 2009). The orbital geometry associated with occultation measurements limits spatial and temporal sampling. Combined analysis of MGS/TES spectral IR and solar-band limb observations indicate substantially larger dust particle sizes (1.5–2.0 μm) at 30–60 km altitudes during peak dust lifting during the 2001 global dust event, falling to 0.5–1.0 μm particle sizes during the clearing phase (Clancy et al., 2010). Extensive efforts to use such synoptic datasets, including MRO/MCS, are currently underway (e.g. Wolff et al., 2012).

Dynamical models also offer insight into the altitude dependence of particle sizes through the connection of gravitational settling and geometric cross-section. The work by Murphy et al. (1990) provided the first detailed constraints on the changes of the dust particle size distribution with height. In terms of the ability to specify a vertical profile of sizes as a

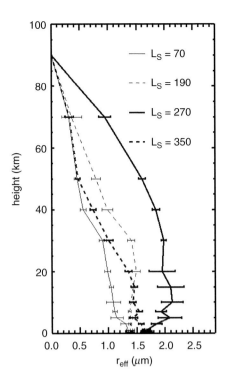

Figure 6.21. Dust size (r_{eff}) vertical distribution from an MGCM simulation for the MER Spirit location. Four mean profiles are determined from a 10 sol period for each the listed seasons (L_s). The error bars are the standard deviation of the particle size at each altitude level for that period. Extracted from the calculations of Kahre et al. (2008), which employ a log-normal size distribution with $v_{eff} = 0.5$.

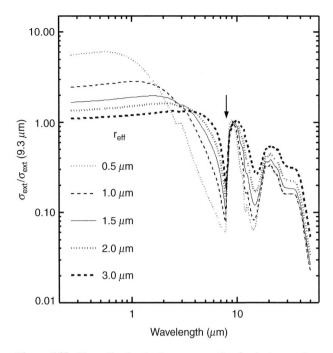

Figure 6.22. Normalized extinction cross-section for dust aerosols computed with case A refractive indices of Wolff et al., circular cylinders with $D/L = 1$, and gamma distribution of surface-equivalent sizes using $v_{eff} = 0.3$. The values are normalized at 9.3 µm, making the ratio the equivalent of the so-called visible-to-infrared ratio. The arrow marks the sharp feature caused by anomalous dispersion in the real part of the refractive indices (Figure 6.19) associated with the 9 µm absorption band.

function of location and season, Kahre et al. (2008) were able to produce results that account for the general column-integrated dust size variations discussed above. Figure 6.21 shows four such profiles that sample the seasonal behavior near the model grid point closest to the MER Spirit site (adapted from Wolff et al., 2009). While the correspondence between the general behaviors of column-integrated r_{eff} values is evident, so is the bias toward larger sizes with respect to the occultation retrievals versus agreement with the TES limb retrievals. Hence, the use of a dust size profile "database" will be a useful tool for Mars atmospheric analysis, particularly for dynamical models and atmospheric retrievals where vertical gradients may be important but self-consistent size information is not available.

(d) Representative Radiative Properties

Because the scattering properties of Martian aerosols can be strongly dependent upon the specific sizes and wavelength ranges of a given application, it is instructive to combine representative macrophysical properties into a set of spectral calculations of the basic radiative properties of extinction, single-scattering albedo, and angular scattering distributions. In order to do so, we specify the refractive indices and particle shape using the "case A" of Figures 6.18 and 6.19 and randomly oriented, equidimensional ($D/L = 1$) cylinders. We perform the integration over particle size distribution using a gamma distribution with $v_{eff} = 0.3$. The results are shown in Figures 6.22–6.24, which we discuss briefly below.

Figure 6.22 plots the extinction cross-section normalized to the center of the 9.3 µm dust (silicate) feature. This convention has the advantage of visualizing the so-called visible-to-infrared ratio, which is often employed as an observational indicator or proxy for particle size (e.g. Zurek, 1982). This type of normalization also suggests the schematic way in which radiative transfer models specify the wavelength dependence of opacity directly from cross-section calculations. Under the assumption of single size distribution over the distance z, (6.31) suggests the following expression:

$$\tau(\lambda) = k_{ext}(\lambda)z = \tau(\lambda_0)\frac{k_{ext}(\lambda)}{k_{ext}(\lambda_0)} = \tau_0 \frac{\sigma_{ext}(\lambda)}{\sigma_{ext}(\lambda_0)} \qquad (6.48)$$

where λ_0 is the reference or normalization wavelength and τ_0 is the optical depth at λ_0. Ultimately, the wavelength and size dependence behavior of τ (or k_{ext}) illustrated in the figure can have significant ramifications for both theoretical and observational Martian atmospheric studies.

The single-scattering albedo spectra in Figure 6.23 capture the complex behavior of light scattered from dust. The absorption "bands" associated with the aerosol composition are easily seen in this radiative property, and marked by the arrows. As discussed in the review by Hamilton et al. (2005), the 3 µm feature is typically associated with the O–H stretch of a hydrated mineral, while the 9 µm and >20 µm absorptions are associated with the Si–O stretch in silicates. The general increase in

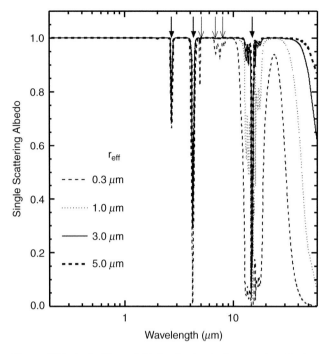

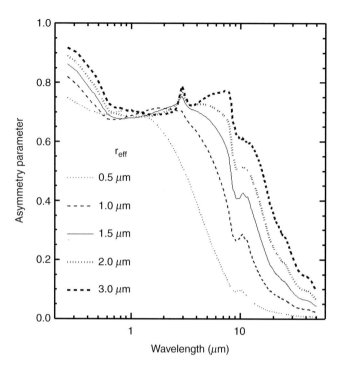

Figure 6.23. As for Figure 6.22, but for the single-scattering albedo. The arrows mark the absorption bands associated with the chemical composition of Martian aerosols. The 3 μm band is identified with the O–H stretch of a hydrated mineral and the 9 μm and >20 μm features with a silicate composition (Hamilton et al., 2005). The general increase in absorption below 0.5 μm is associated with Fe–O electronic charge transfer transitions (e.g. Cloutis et al., 2008).

Figure 6.24. As for Figure 6.22, but for the asymmetry parameter.

6.3.3.6 H₂O Ice Aerosols

The existence of water ice aerosols in the atmosphere of Mars has been long recognized (Curran et al., 1973). However, clouds were considered primarily to be tracers of meteorological activity until Clancy et al. (1996b) established the connection of the Martian water cycle with lower atmospheric temperatures in the aphelion climate cycle. As a result, an emphasis on the macrophysical and radiative properties of water ice particles has occurred only relatively recently. Nevertheless, there is an important and interesting historical record of such studies. As with dust, we refer the reader interested in the more complete list of previous studies to the water ice chapter (Chapter 5) and to the review by Smith (2008). Here, we only summarize the citations and the details necessary to calculate the most relevant water ice aerosol radiative properties, starting with the relevant macrophysical parameters.

absorption below 0.5 μm is associated with Fe–O electronic charge transfer transitions, in particular Fe^{3+}–O and Fe^{2+}–O (e.g. Cloutis et al., 2008). Additionally, the combination of size and refractive index effects produces the dichotomy of smaller particles being more efficient at scattering than larger particles in the UV and optical but much less so in the IR. A final salient point is that the potential impact of vertical gradients in particle size on the absorption of solar radiation is clearly seen. Recalling that the absorption efficiency at a point in the atmosphere is $(1-a)k_{ext}$, the small apparent difference between $a = 0.95$ and $a = 0.90$ actually represents a 100% increase in absorption (heating).

A plot showing the phase function for the range of particle sizes just presented would be fairly complicated (i.e. three-dimensional) and probably not very informative to most readers. Instead, Figure 6.24 shows the asymmetry parameter (6.28) as a proxy for the phase function. The general increase of *g* with decreasing wavelength reflects the growing size of the forward diffraction lobe relative to the side and backscattering directions. However, it also emphasizes the increasing inapplicability of Lorenz–Mie theory or a Henyey–Greenstein function in this size parameter regime. The importance of the phase function shape for remote sensing is relatively straightforward. The dependence of radiative fluxes to changes in *g* is probably less apparent. Nevertheless, because such changes will alter the balance of the upward and downward fluxes, they can become important for atmospheric dynamics through radiative forcing (e.g. Zurek, 1981).

(a) Refractive Indices

Unlike dust particles, the composition of water ice aerosols is less mysterious (ignoring the issue of the composition of any condensation nuclei). That being said, water ice can take one of two crystalline phases for temperatures found within the environment of Mars: cubic (Ic) for temperatures below ~170–180 K, and hexagonal (Ih) for temperatures above that range (e.g. Medcraft et al., 2012). In terms of refractive indices, it is often difficult to tell the difference between the two phases, with experimenters resorting to characterization techniques such as X-ray diffraction or microscopy to be certain. While there is some evidence that the 3 μm water ice band offers the capability to discriminate between Ic and Ih, one commonly finds that the phase is typically assigned by the associated temperature of the ice particles (e.g. Mastrapa et al., 2008; Medcraft et al., 2012).

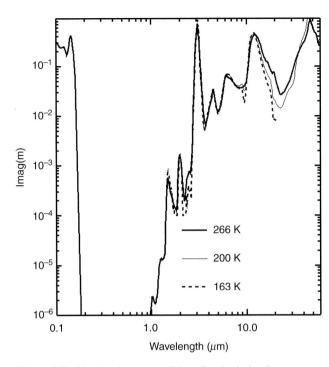

Figure 6.25. The imaginary part of the refractive index for water ice particles for three temperatures: 266 K from Warren and Brandt (2008), 200 K from Iwabuchi and Yang (2011), and 163 K from Toon et al. (1994). As can be seen, the temperature dependence is associated with the various absorption bands. See text for availability of digital version of the data.

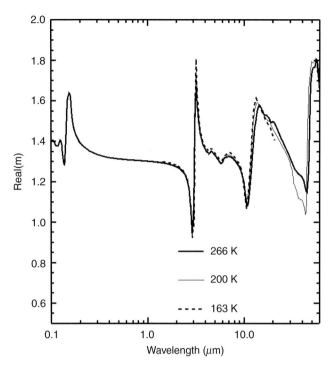

Figure 6.26. As for Figure 6.25, but for the real part of the complex refractive indices.

Warren and Brandt (2008) offer the set of water ice refractive indices with the greatest wavelength coverage. However, it is referenced to a temperature of 266 K. This temperature, and that associated with the predecessor compilation (Warren, 1984), is too warm for many atmospheric applications, even on Earth. This fact stimulated measurements at more relevant temperatures, e.g. 163 K for polar stratospheric clouds (Toon et al., 1994); see also discussions in Grundy and Schmitt (1998), Warren and Brandt (2008), and Medcraft et al. (2012). The disjointed nature of the wavelength and temperature coverage from the resulting datasets is addressed by Iwabuchi and Yang (2011). They provide a compilation of refractive indices for temperatures 160–270 K through a combination of interpolation and extrapolation techniques.

We display the imaginary and real parts of the water ice refractive indices in Figures 25 and 26, respectively. We include a sense of the temperature dependence by showing the Warren and Brandt (2008) compilation at 266 K, the Iwabuchi and Yang (2011) values for 200 K, and the Toon et al. (1994) measurements at 163 K. The Warren and Brandt indices are available at the URL listed in the article (www.atmos.washington.edu/ ice_optical_constants), while those from Iwabuchi and Yang are incorporated into the online version of the article supplementary material. The digital version of the Toon et al. numbers were located in the gray literature (http://gwest.gats-inc.com/ ice_index/Ice_Refractive_Index.html; this site also contains a few other ice datasets that may be of interest to a reader).

(b) Shape

The shape or habit of water ice particles in the Martian atmosphere has not been particularly well constrained by observations. Like dust, the most popular shape for scattering calculations has been that of a sphere. Clancy et al. (2003) derived empirical phase function shapes from analysis of TES solar-band emission phase function (EPF) sequences. This type of data samples a range of scattering angle coverage such that one can separate the phase function behavior of the atmospheric aerosols from that of the surface. By examining a large number of EPFs that were dominated by ice aerosols (as indicated by the TES IR spectra), they constructed several archetype phase function shapes. These are displayed in Figure 6.27 as solid lines and include both the so-called Type 1 ($r_{eff} \sim 1–2$ μm) and Type 2 ($r_{eff} \sim 3–5$ μm) ice clouds (cf. Chapter 5; Clancy et al., 2003); the curve with a shallower "minimum" and a backscatter enhancement is an example of Type 1, while the other two represent Type 2 particles. The effective wavelength of the TES solar band is approximately 700 nm, and we present a Lorenz–Mie phase function calculated with a size distribution appropriate for Type 2 sizes ($r_{eff} = 3.0$ μm, $v_{eff} = 0.1$). The spherical shape is not a good match even allowing for some normalization issues, yet neither are analytical shapes such as circular cylinders or hexagonal prisms (Clancy et al., 2003). In the past decade, terrestrial studies have clearly revealed that, for the temperatures and particle sizes measured in the Martian atmosphere (e.g. $r_{eff} = 1–4$ μm; Wolff and Clancy, 2003), one does not expect a single shape (see Chapter 5; Bailey and Hallett, 2009). Rather, one finds polycrystalline habits. These complex, multi-faceted shapes can often be asymmetric, and thus more complicated to represent in scattering calculations. This led Yang et al. (2003) to propose the use of a droxtal

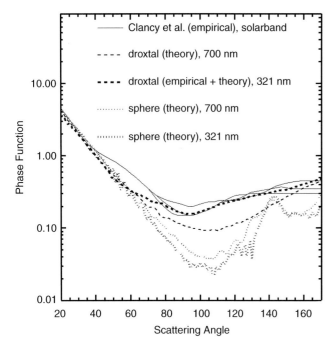

Figure 6.27. Water ice phase function shapes. The empirical phase function shapes derived by Clancy et al. (2003) from TES solar-band analysis of emission phase function sequences are compared with those of analytical calculations for spherical and droxtal particle shapes with $r_{eff} = 3.0$ μm and $v_{eff} = 0.1$ (gamma distribution). The Wolff et al. (2011) phase function used for the MARCI 320 nm band cloud optical depth retrievals, which combines empirical and analytical techniques, is also displayed.

shape, which is a multi-faceted generalization of a hexagonal prism (see Figure 5.17). A numerical calculation for the solar band with an equivalent size distribution of droxtals is also provided in Figure 6.27. This phase function is a much better fit than that for the sphere, and further improvement should be possible given that the employed droxtal parameters ($\theta_1 = 30°$ and $\theta_2 = 70°$; see figure 4 of Yang et al. (2003) for definition) were not optimized. Additional support for the droxtal shape comes from its application to mapping water ice optical depth retrievals using the MARCI wide-angle UV images (Wolff et al., 2011). The wide-angle aspect of the data samples a large range of phase space in a manner analogous to that of an EPF sequence. Mismatches between the model and reality show up as retrieval "bright" or "dark" artifacts that correlate with scattering angle, thus allowing an empirical improvement of the scattering calculations. The resulting $P(\theta)$, currently used by the MARCI team for their cloud mapping retrieval algorithm, is also shown in Figure 6.27, labeled "droxtal(empirical+theory)". The 700 nm curves are relevant potentially to observations in the visible by CRISM or TES (solar band).

(c) Size

As reviewed in Chapter 5, the Type 1 ($r_{eff} \sim 1–2$ μm) clouds are associated with seasonal mid- to high-latitude clouds, while the Type 2 ($r_{eff} \sim 3–5$ μm) particles are typically found in the aphelion cloud belt (Clancy et al., 2003; Wolff and Clancy, 2003; Madeleine et al., 2012a; Guzewich et al., 2014). As with dust particle sizes derived from plane-parallel viewing geometries, these values are

column-integrated. However, the cloud column exhibits a base altitude that is (generally) not at the surface, such that the reported sizes represent more vertically localized averages. The width of the particle size distribution is typically specified using $v_{eff} = 0.1–0.2$, though a significant theoretical or observation justification for these values appears to be somewhat lacking.

Altitude-resolved discrimination of cloud particle sizes remains rather limited. Thin high-altitude hazes (>50 km) have been characterized by UV occultation observations: $r_{eff} < 0.1–0.2$ μm (Montmessin et al., 2006), though in some cases IR occultation data reveal $r_{eff} \sim 0.1–0.3$ μm for some clouds in the 50–60 km range (Fedorova et al., 2009). In both of these cases, the wavelengths employed allow for the discrimination of dust and water ice aerosols. Imaging of cloud shadows by MARCI has provided similar sizes, $r_{eff} \sim 0.2–0.5$ μm for clouds between 40 and 50 km (Clancy et al., 2009). In terms of larger particle sizes, evidence for $r_{eff} \sim 30$ μm has been reported for the boundary layer in the northern polar region (Whiteway et al., 2009; Lemmon, 2014). Finally, the use of the macrophysical and transport schemes in dynamical models to prescribe or constrain particle size distributions represents an area of current research (e.g. Madeleine et al., 2012b). Water ice size profiles analogous to those for dust in Figure 6.21 are anticipated to be available in the near future.

(d) Representative Radiative Properties

Given the ambiguity and numerical complexity surrounding the choice of a non-spherical shape, we calculate representative radiative properties using Lorenz–Mie theory and the 220 K set of refractive indices from Iwabuchi and Yang (2011), shown in Figures 6.25 and 6.26. The size distribution is specified by the gamma distribution with $v_{eff} = 0.1$. We display the resulting normalized extinction cross-sections, single-scattering albedo, and asymmetry parameter in Figures 6.28–6.30. We have chosen a range of r_{eff} (0.3–15 μm) that reflects the dispersion in detected and inferred sizes.

Even allowing for 10–20% errors in the cross-section and albedo calculations associated with using spheres (e.g. Figures 6.16 and 6.17), the extinction and single-scattering albedo functions reveal important trends. The visible-to-IR ratio remains an effective proxy for particle sizes for particles that are not "too big", i.e. $r_{eff} \lesssim 8$ μm. However, the relative depths of the absorption bands offer also a valuable size diagnostic, even for the larger size regime. Unfortunately, certain observational realities do impose limitations, such as large CO_2 features near the 3.1 and 12 μm bands, as well as the need for appreciable optical depth to exploit the relationship between the 1.4, 1.9, and 3.1 μm bands (Wolff and Clancy, 2003; Madeleine et al., 2012a).

The broad trends associated with the asymmetry parameter curves plotted in Figure 6.30 are similar to those for dust aerosols. While the small-scale structure is likely a strong function of the spherical shape, the errors in the general amplitude due to the spherical assumption are likely to be only on the order of 5–10%, as is the case for cylinders versus spheres (Mishchenko et al., 2002, figure 10.21). As with dust aerosols and as illustrated by the ice phase functions in Figure 6.27, the larger values of g suggest the importance of phase function prescription (and the inapplicability of simple empirical functions such as Henyey–Greenstein).

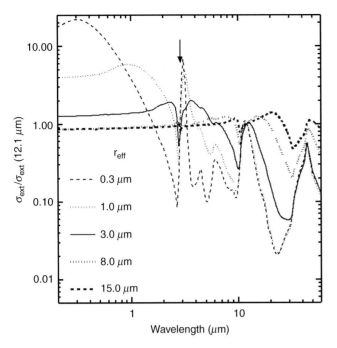

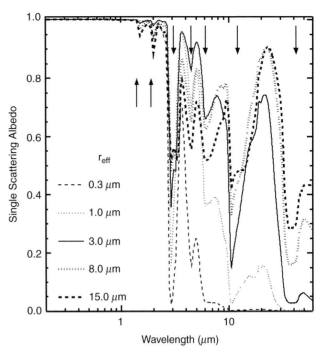

Figure 6.28. The normalized extinction cross-section for water ice aerosols versus wavelength is computed with Iwabuchi and Yang (2011) refractive indices for 220 K, spherical particles, and gamma distribution of $v_{eff} = 0.1$. The values are normalized at 12.1 μm, highlighting the size dependence of the visible-to-infrared ratio. The arrow marks the sharp feature caused by anomalous dispersion in the real part of the refractive indices (Figure 6.26) associated with the 3 μm band.

Figure 6.29. As for Figure 6.28, but for the single-scattering albedo. The arrows mark the location of the several strong water ice absorption bands: 1.4, 1.9, 3.1, 4.5, 6.1, 12, and 42 μm. The relative strengths of the bands can be quite diagnostic of particle size, although it is often difficult to observe several of them because of molecular opacity. In particular, about half of the band areas for the strong 3 μm and 12 μm bands are obscured by very strong CO_2 bands.

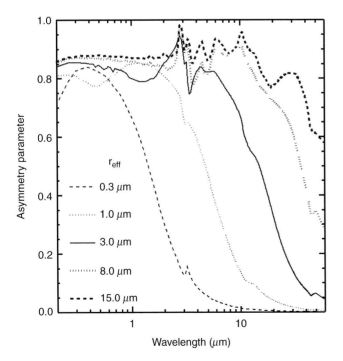

Figure 6.30. As for Figure 6.29, but for the asymmetry parameter.

6.3.3.7 CO_2 Ice Aerosols

The abundance of carbon dioxide in the Martian atmosphere implies the possibility of CO_2 ice clouds, which have been identified in the present Martian atmosphere and have been attributed a potentially important role in the ancient atmosphere. This topic is reviewed in Chapter 5 (see also discussion in Isenor et al., 2013), but we include briefly some framework for the presentation of their radiative properties. At present, CO_2 clouds are found in the limited temporal and spatial domain where conditions permit their formation, namely the mesosphere and the polar night (e.g. Kieffer et al., 2000; Montmessin et al., 2007). However, studies of the early Martian atmosphere indicate that CO_2 clouds may have provided a non-trivial trapping of thermal radiation through downward scattering of the upwelling thermal radiation (Forget and Pierrehumbert, 1997; Pierrehumbert and Erlick, 1998).

(a) Macroscopic Properties

The refractive indices of carbon dioxide ices have been compiled over a fairly wide wavelength range by Warren (1986) and Hansen (1997, 2005). Hansen's results are shown in Figure 6.31. More limited spectral regions have been studied in detail for specific conditions (e.g. Baratta and Palumbo, 1998).

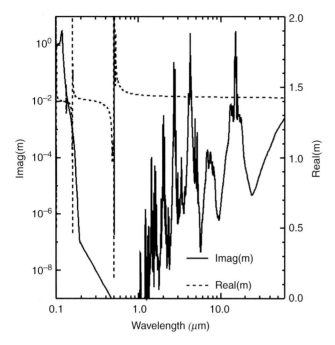

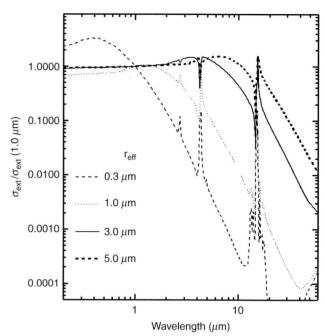

Figure 6.31. Real and imaginary refractive indices for CO_2 ice from the measurements of Hansen (1997, 2005). With the absorption limited to regions where CO_2 gas opacity dominates, scattering issues can dominate the radiative impact of CO_2 particles.

Figure 6.32. Normalized extinction cross-section for CO_2 ice aerosols computed with Hansen (1997, 2005) refractive indices, spherical particles, and gamma distribution of $v_{eff} = 0.1$. The values are normalized at 1.0 μm, reflecting the convention of Määttänen et al. (2010). Unlike dust and water ice, the literature does not seem to contain the repeated use of a particular wavelength. The sharp structure is associated with the strong particle absorption bands at 2.7, 4.3, and 15 μm.

What may be particularly distinctive for CO_2 ice compared to water ice or dust is the essentially non-absorbing nature of most of the range from the UV–IR. As a result, the detection of and the radiative contribution from these particles emphasize their scattering attributes (as opposed to those involving absorption).

Derived observational constraints on particle sizes are similar to those for water ice, $r_{eff} \sim 0.1$–3.0 μm, though much larger sizes (10–100 μm) have been inferred from observation in polar night (Montmessin et al., 2006, 2007; Määttänen et al., 2010; Hayne et al., 2014; see discussion in Chapter 5). The width of the particle size distribution has been set in the range $v_{eff} = 0.1$–0.2, though only poorly constrained at present by observations and theory (e.g. Määttänen et al., 2010). The shapes expected for such airborne particles in the Martian atmosphere in the presence of heterogeneous nucleation are not well constrained by observation, but the cubic crystalline structure suggests octahedral habits while laboratory bulk CO_2 condensation experiments reveal a tendency toward bipyramids (e.g. Foster et al., 1998; Colaprete and Toon, 2003). Although most Martian studies have relied upon Lorenz–Mie theory, these non-spherical habits are tractable with current light scattering algorithms. However, very few researchers have yet to take advantage of such techniques, which also allow for explicit treatment of core–mantle structure and associated properties of the condensation nuclei (Foster et al., 1998; Isenor et al., 2013).

(b) Representative Radiative Properties

As with water ice, representative properties calculated assuming spherical particles can provide useful insights, particularly for the forward scattering/angular integrated quantities. Accordingly, we present such values in Figures 6.32–6.34

using the Hansen (1997, 2005) refractive indices and a gamma distribution with $v_{eff} = 0.1$. The width of the distribution is set for (broadband) comparison with the water ice properties. However, higher-spectral-resolution work in the 4.3 μm band has employed $v_{eff} = 0.2$ (Määttänen et al., 2010).

For comparison with the recently reported CO_2 cloud optical depths by Määttänen et al. (2010), we normalize the extinction cross-sections in Figure 6.32 to 1 μm. The combination of the extinction and single-scattering albedo (Figure 6.33) reveals that, outside of the generally narrow absorption features, the particles are very efficient scatterers. In fact, it is the differences in scattering efficiency associated with particle size variations that allow retrievals of size, i.e. variations of ice scattering with the molecular opacity near the 4.3 μm band (Montmessin et al., 2007; Määttänen et al., 2010). However, based upon these figures, the large particles expected in polar night will have a minimal spectral signature. Meanwhile, as shown in the asymmetry parameter (Figure 6.34), such large particles (wherever they may appear) will be quite forward scattering throughout the IR. This makes it more difficult for such clouds to provide the backscattering greenhouse effect suggested for early Mars by Forget and Pierrehumbert (1997; see also Pierrehumbert and Erlick, 1998).

6.3.3.8 General Spatial/Temporal Distribution of Aerosols – Databases

Discussion about the general distribution of dust and ice aerosols may be found in Chapters 10 and 5, respectively. Here,

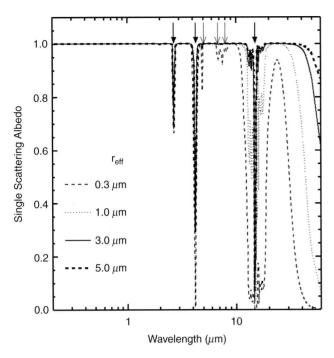

Figure 6.33. As for Figure 6.32, but for the single-scattering albedo. The thick arrows mark the location of the strong CO_2 molecular opacity bands: 2.7, 4.3, and 15 µm. The thin arrows highlight the position of several of the weaker transmittance features from Figure 6.9, which appear for the smaller particle sizes. The shape of the aerosol "absorption features" is diagnostic of particle size, but the presence of significant molecular opacity forces one to rely on the scattering contribution to the spectral shape (i.e. the increase in scattering (Montmessin et al., 2007; Määttänen et al., 2010).

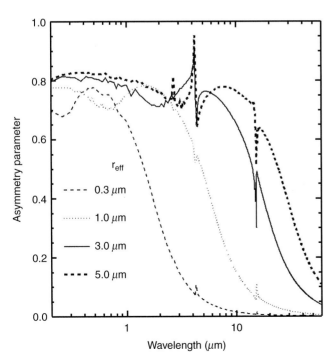

Figure 6.34. As for Figure 6.32, but for the asymmetry parameter.

we focus on the situation where the aerosol content cannot be retrieved from a dataset at hand, but nevertheless needs to be specified for the desired RT modeling. Of course, the complete absence of a dataset, say for dynamical modeling, would also fall under the auspices of this section. A common observational scenario would be the analysis in the near-IR part of the Martian spectrum, where it can be extremely difficult to disentangle surface and atmospheric contributions without special observations like emission phase functions (EPFs). This particular situation has often been the case for the retrieval of the CO_2, H_2O, and CO abundances from the CRISM and OMEGA dataset (Encrenaz et al., 2006, 2008; Melchiorri et al., 2007; Smith et al., 2009; Maltagliati et al., 2011; Toigo et al., 2013), or the water abundance retrieval from the MAWD and SPICAM dataset (Fedorova et al., 2006, 2010).

The optimal course of action would be to employ simultaneous or concurrent aerosol opacity measurements, as was done by Smith et al. (2009) and Toigo et al. (2013), analyzing CRISM data using opacity measurements from THEMIS data (Smith, 2009). The value of temporal connection is further supported by Vincendon et al. (2009), who demonstrated that the seasonal variation of the dust opacity can be much more significant than the geographical variation. More specifically, they retrieved dust opacity from OMEGA data at several locations during Mars years (MY) 27 and 28 (see Clancy et al., 2000, for definition), and demonstrated that such measurements were in good agreement with those of the MER Rovers (Lemmon et al., 2004) when corrected for the elevation differences between the OMEGA scenes and the rover sites. While Vincendon et al. (2009) concluded that dust opacity tends to be zonally homogeneous at mid- to low latitudes, one must allow for significant seasonal or inter-annual variations.

In the absence of concurrent dust observations, the modelers and observers need to rely on data obtained at other epochs (and even locations). The phrase "aerosol climatology" is often associated with this type of data. We adopt it here. We present some details on the basic aerosol climatology datasets currently available. The temporal coverage and the wavelength at which the measurements are made is illustrated in Figure 6.35.

One of the most widely used aerosol climatologies is based upon the TES nadir IR spectra (Smith et al., 2001; Smith, 2004, 2008), which is often used in conjunction with that of THEMIS spectral imaging observations in order to provide a continuous climatology (Smith, 2009). The TES dataset provides a nearly complete coverage for MY 24, $L_s = 104°$ to MY 26, $L_s = 180°$, while the THEMIS climatology starts in MY 25, $L_s = 330°$ and continues through the present (MY 32, at the time of writing). The TES retrieval products are available directly from the Planetary Data System, but have typically been made available upon request to M.D. Smith. In addition, the dust values can also be obtained from the LMD (Laboratoire de Météorologie Dynamique) Mars Climate Database (http://www-mars.lmd. jussieu.fr/) for MY 24, 26, 27, 29, and 30, i.e. Martian years without a global dust storm. In general, the TES/THEMIS aerosol climatology is the most widely accepted aerosol database employed by the Martian community. There are two caveats to using these datasets. The first applies only to TES: the reported optical depths are for absorption only, such that a correction must be applied to generate extinction optical depth (Wolff and

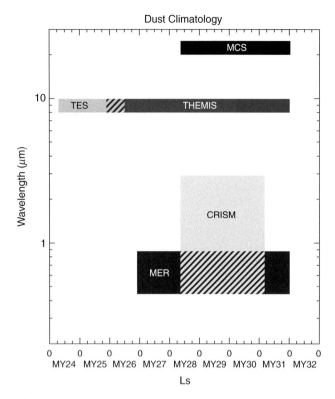

Figure 6.35. Graphical representation of the temporal and wavelength sampling of the available aerosol climatology datasets. Instrument names are defined in Table 6.1. The availability of the CRISM NIR water ice climatology is notional in that one could apply the analyses of Madeleine et al. (2012a), though this has not yet been done.

Clancy, 2003). Secondly, because thermal IR retrievals require a thermal contrast between an absorber (aerosol or gas) and the thermal source (surface), they are relatively insensitive to the abundance of aerosols in the planetary boundary layer (or in polar regions, where the temperature lapse rates are nearly isothermal). The TES and THEMIS retrievals employ specific assumptions about the aerosol vertical distribution. For dust, the distribution is simply that of uniform mixing (i.e. falls off with a vertical scale height equal to the atmospheric scale height). For water ice, the opacity is zero below the water vapor saturation point, and uniformly mixed above.

A more recent addition to thermal IR retrieval datasets is that of the MCS limb aerosol profiles, measuring dust at 22 μm, and water ice at 12 μm. MCS employs a set of 21 detectors aligned parallel to the vertical axis, essentially providing continuous limb observations at a vertical resolution of 5–10 km (Kleinböhl et al., 2009; McCleese et al., 2010, Heavens et al., 2011). Profiles of dust and water ice extinction factors are available through the Planetary Data System. The associated climatology extends from MY 28, $L_s = 133°$ through the present, MY 32. The unique combination of temporal and spatial coverage for vertical profiles (from roughly 20–30 km to 80 km) supports key modeling activities such as thermal tides (Lee et al., 2009), global circulation (Guzewich et al., 2013), non-local aerosol transport (Spiga et al., 2013), and polar cloud distributions (Benson et al., 2010; Hayne et al., 2014).

A final entry to the thermal IR list are the limb observations associated with the TES dataset. McConnochie and Smith (2008) have produced a set of dust and water ice extinction profile retrievals from these data using a pseudo-spherical discrete ordinate algorithm (Smith et al., 2013). The current version of the retrieval uses a temperature profile that explicitly includes the radiative effects of the aerosols. The use of spectral data makes the water ice–dust separation cleaner than is possible with the MCS dataset, but it also suffers from systematic noise issues and has a vertical resolution comparable to the scale height (~10 km). Regardless of specific issues, it offers important atmospheric constraints with synoptic coverage that are not available from other instruments during that time. It is available upon request from T.H. McConnochie.

Moving to the optical regime, a widely cited set of extinction optical depth measurements were derived from the Pancam instrument on-board each of the MER. Due to Pancam's direct solar imaging technique, these retrieved aerosol optical depths are the most accurate of all those discussed here (Lemmon et al., 2004), with a typical absolute uncertainty in optical depth of ±0.04. One caveat is that there is no spectral separation of cloud and dust extinction such that total aerosol optical depth is returned. While limited geographically to the two MER landing site regions – Opportunity in Meridiani Planum, and Spirit in Gusev Crater – the measurements have been applied to other regions through scaling (e.g. Vincendon et al., 2009). The MER optical depth measurements extend from MY 26, $L_s = 328°$ through MY 30, $L_s = 67°$ for Spirit and the present (MY 32, $L_s = 15°$) for Opportunity. They are distributed by the Planetary Data System (http://pds-geosciences.wustl.edu/missions/mer/geo_mer_datasets.htm). Users should note that this dataset represents the total extinction of both dust and water ice (Lemmon et al., 2004).

A less widely used databases, has been derived from the CRISM EPF sequences. Based upon the algorithms presented in Wolff et al. (2009), this method uses the multi-angle coverage to separate the atmospheric and surface contributions to the observed signal. At present, the retrievals have been limited to dust, with the water ice column opacity being set from contemporaneous measurements by MARCI (discussed below). The uncertainty associated with the dust column extinction optical depth retrievals is typically 0.1–0.2 for low to moderate dust loading conditions. Despite the "global coverage" and the more than 20 000 EPF sequences analyzed, issues associated with the interpolation of irregularly sampled, sparse data prevent the simple generation of products similar to those for TES and THEMIS. This CRISM dust climatology covers the period from MY 28, $L_s = 133°$ through MY 31, $L_s = 60°$ and is available upon request from M.J. Wolff; it is also available online (https://gemelli.spacescience.org/twiki/bin/view/RadiativeTransfer/Mars/). Unlike thermal IR observations, CRISMS's reflected sunlight observations are sensitive to the total dust optical depth along the path between the surface and the top of the atmosphere. The retrieved column optical depths are much less sensitive to the details of the assumed aerosol profile than those derived from TES or THEMIS. However, CRISM observations also provide little or no information about the dust vertical profile. The CRISM measurements are more sensitive to the details of the aerosol model: aerosol phase function, single-scattering albedo, etc. These are described in Wolff et al. (2009) with the exception that dust climatology adopts $r_{eff} = 1.5$ μm ($v_{eff} = 0.3$), and the case A refractive indices discussed in this chapter.

There are currently two water ice optical depth aerosol climatologies derived from measurements collected outside of the thermal IR. The first involves water ice columns measured in the UV by MARCI. Using the 321 nm channel, a retrieval algorithm that combines the daily coverage of the MRO orbit with the wide-angle field of view of MARCI has been developed to produce global maps of water ice opacity on a daily basis (Wolff et al., 2014). This use of an imaging dataset is possible due to the relatively high photometric accuracy of this MARCI band (~5%; see Bell et al., 2009; Wolff et al., 2010) and its general insensitivity to dust, at least compared to ice (i.e. $a_{dust} = 0.57$ versus $a_{ice} = 1.0$; see also Figures 6.23 and 6.29). This dataset runs from MY 28, $L_s = 133°$ through the present and is distributed online (https://gemelli.spacescience.org/twiki/bin/view/MarsObservations/MarciObservations/WaterIceClouds). Requests may also be made directly to M.J. Wolff. The limitations of the current version of this climatology are twofold: the algorithm fails over ice-covered surfaces, and it assumes a uniform mixing for water ice starting at the surface.

A second optical regime water ice climatology is derived from the strong 3.1 μm and the weaker 1.4 μm and 1.9 μm water ice bands using OMEGA observations (Madeleine et al., 2012a). The operational and orbital limitations of MEX and OMEGA provide a much smaller database, but it possesses the advantage of including estimates of the average water ice particle size, at least for the thicker clouds. The optical depth uncertainty is estimated to be ~20%. Retrievals are available over the period starting with MY 26, $L_s = 330°$ through the present, and may be requested from J.-B. Madeleine.

6.4 RADIATIVE SOURCES

The RTE source function (6.4) incorporates three distinct sources relevant to planetary atmospheres: external illumination at the top of the atmosphere (F_0); thermal or blackbody emission ($B(T)$) from the atmosphere and surface, which originates from material that is in thermodynamic equilibrium with its environment; and non-thermal equilibrium emission processes (Q). We provide a brief overview of these terms, with an emphasis on their operation in the Martian atmospheric system.

6.4.1 Solar Spectrum

Solar radiation represents an important term in the solution of the RTE for the UV through the NIR (i.e. through 5 μm). In particular, the ability to quantitatively specify the spectral dependence in absolute units is a necessary component for both modeling and observation of the Martian atmosphere in this wavelength range. Fortunately for our purposes, extensive studies of the solar spectral irradiance (SSI) and its properties have been available for many years. In terms of the RTE, the boundary condition imposed by SSI ($F_0 = F_\odot(\lambda)$)

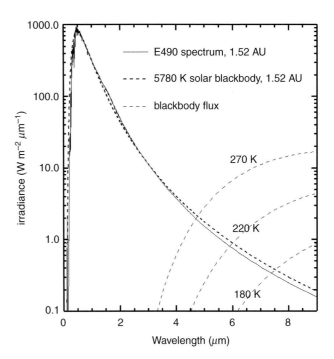

Figure 6.36. Top-of-atmosphere solar irradiance (flux) at a distance of 1.52 AU, compared to the upwelling blackbody irradiance. The solar flux is represented by the E-490 spectrum and by a blackbody at 5780 K (e.g. Lang, 2006). As is illustrated, the crossover between NIR and thermal IR for Mars depends upon the temperature of the scene/region being viewed. See text for details on E-490.

is typically prescribed with a tabulated reference spectrum resampled to the desired model or instrument resolution. An example of one such tabulation may be seen in Figure 6.36 along with a blackbody spectrum for the mean value of the Sun's photosphere (Lang, 2006, p. 23). Figure 6.36 illustrates the spectral regions where a simple solar blackbody radiance poorly fits the SSI, as well as the IR cutoff region for which one needs to include the incident solar flux as a function of the local temperature environment. The transition from solar-dominant to thermal-emission-dominant source functions falls between 4 and 6 μm for most Mars applications, although for CO_2 ice surfaces during twilight hours, the solar contribution may be non-negligible at even longer wavelengths.

Obtaining reference $F_\odot$ across a wide range of wavelengths is complicated by the fact that such spectra are actually composites of observations and models from multiple sources. SSI distributions employed in previous Mars studies, include Wehrli (1985, 1986), Colina et al. (1996), and Berk et al. (1998). However, additional resources are sometimes required. For example, the low resolution of these spectra and the general dearth of observations above 2.5 μm led Fiorenza and Formisano (2005) to produce a solar spectrum through a combination of previous results with high-resolution ground-based data. While their efforts are specific to the capabilities of the Mars Express PFS, the methods and discussion highlight issues that may also be relevant to other Mars datasets. Fortunately, the more recent compilations by Thuillier et al. (2004), Deland and Cebula (2008), Woods et al. (2009), and Fontenla et al. (2011) may address a majority of such concerns. Nevertheless, locating any of the cited spectra in digital format remains less than ideal. We refer readers to the sites maintained by the National Renewable Energy Laboratory (http://rredc.nrel.gov/solar/spectra/), which includes the E490 reference spectrum shown in Figure 6.36; the LASP Interactive Solar Irradiance Datacenter (http://lasp.colorado.edu/lisird/); the Naval Research Laboratory (http://wwwsolar.nrl.navy.mil/solar_spectra.html), which includes the compilations of Thuillier et al. (2004); and a privately maintained archive that includes the high-resolution spectra from Fontenla et al. (2011), as well as extensions and updates in a variety of formats (www.digidyna.com/Results2010/spectra/irradiance/index_spectra_irradiance.html). Another source of solar spectra is the widely used database generated and distributed by Robert Kurucz (http://kurucz.harvard.edu/sun.html), which includes the pioneering high-resolution observations of Brault and Testerman (1972).

The existence of solar variability requires that one consider the time dependence of $F_\odot$ for RT calculations. Terrestrial climate studies characterize SSI variability relative to that of the bolometric irradiance, often called the solar constant ($S_\odot$) or total spectral irradiance (TSI), defined as

$$\text{TSI} = S_\odot = \int_0^\infty F_\odot d\lambda \qquad (6.49)$$

A recent review of TSI measurements and TSI/SSI variability studies may be found in Pagaran et al. (2011) and Ermolli et al. (2013). The current canonical value of TSI is 1361±0.5 W m^{-2}, with the difference between the minimum and the maximum values within a solar cycle being ~1 W m^{-2} (over the past three cycles). Of course, this small percentage in the TSI is not distributed equally across wavelength. The same solar cycle minimum–maximum differences calculated for SSI show ≲0.5% across the visible and NIR, rising to ~1% at 300 nm, to ~4% at 250 nm, and tens of percent below 200 nm (Pagaran et al., 2011; Ermolli et al., 2013). As such, Martian UV studies are most likely to be sensitive to biases introduced by the use of a single reference spectrum. In these cases, corrections to the SSI using spacecraft observations for the phase of the solar cycle may be warranted (Deland and Cebula, 2008; Pagaran et al., 2011).

6.4.2 Thermal (LTE) and Non-LTE Emission

The atmosphere of Mars, like that of any other planet, is not a closed system and therefore is not in strict thermodynamic equilibrium (TE). A working definition of TE would be that one could specify the state of the system with a single parameter, the temperature of the gas. The consequences of this would include the fact that a Maxwellian distribution would describe the gas particles' velocities and kinetic temperature, and the populations of all the energy states (spin, rotational, vibrational, electronic) would follow the Boltzmann distribution for that temperature. Even the radiation field inside the atmosphere would be known, and given by the Planck function. But a planetary atmosphere is not that simple or stable. The combination of intense solar flux at the top of the atmosphere and the diverse set of sources/sinks for atmospheric energy, mass, and momentum can produce changes that may be highly decoupled from the radiation field. Describing the

complete physical state of such a system quickly becomes extremely complicated; so much so that one normally aims at understanding only "some" of its properties and adopts simplifying assumptions such as local thermodynamic equilibrium (LTE).

LTE is a state in which long-distance exchanges of energy and matter do not affect appreciably the local atmosphere, and as a consequence it is approximately valid to describe the system with a single temperature. It is importance to notice that this approximation is derived for the state of the gas, not the radiation field. This is an important difference from strict TE, so it is best illustrated through the example of strongly absorbed solar radiation. In this case, LTE could still prevail if the atmospheric density were sufficiently high such that the molecular collisions adequately diffuse and share the energy between all the molecules, thus maintaining a well-defined local temperature. This single temperature aids in describing at least some of the main atmospheric properties including the source function. LTE is a common assumption for atmospheric radiation, and many of the equations and discussions above have implicitly or explicitly made use of it.

In this section, we review the more general case, known as non-LTE, which occurs wherever the LTE approximation is not valid. In such instances, neither the source function nor the temperature of the different modes of the gas internal energy are known, nor can they be deduced easily (using thermodynamical arguments), from the observation of a single parameter. The classical way to study the gas and its source function under such circumstances is by means of a microscopic statistical model. Such a non-LTE model needs to incorporate the effects of the radiation field, in particular. On the other hand, as presented in previous sections, the usual RTE involves a source function, which in this case remains unknown until the physical state of the gas is determined. Therefore, non-LTE and radiative transfer models are intimately coupled and need to be solved simultaneously.

6.4.2.1 Thermal Versus Non-Thermal Emission Versus Airglow

Non-LTE does not alter the essential problem of radiative transfer in a planetary atmosphere; namely, the means by which the radiation from the Sun is distributed by the atmosphere, transformed into heat, deposited at the surface, and ultimately converted into an emitted radiation field. This component is called planetary or thermal emission, and, as suggested above, can be studied separately from direct solar radiation, i.e. solar-band versus thermal IR. This dichotomy remains valid under both LTE and non-LTE situations, i.e. regardless of the role that each term plays on the physical state of the atmosphere.

There are non-LTE situations where an intense atmospheric emission is observed but is not linked to the "thermal pool" of the atmosphere. This type of atmospheric radiation is normally described by the term non-thermal. Examples of this behavior include fluorescent emissions during daytime following the direct excitation from the solar flux, and airglow emissions following recombination and photodissociation processes in the upper atmosphere.

In the literature, one sees a tendency to distinguish between "airglow" and non-LTE emission, but the boundary is not very clear. Both involve situations where the distribution of the energy level populations is altered by some physical or chemical process, leading to strong emissions – usually enhanced from its LTE case. The term non-LTE is reserved normally for cases where the emission occurs in the IR (including NIR) portion of the spectrum. Conversely, airglow is more closely associated with UV and visible luminescent phenomena, but it is also applied to NIR emissions associated with photochemical reactions (such as for electronically excited O_2 produced from ozone photolysis). Non-LTE theory includes airglow as a particular case, as well as mechanisms that deplete atmospheric emission. The role played by radiative transfer is frequently more important in the NIR and thermal IR part of the spectrum, where strong absorption and emission bands of the main atmospheric species occur, primarily in the range of 1–20 μm. This is the case for terrestrial planet atmospheres, where the strongest ro-vibrational bands of CO_2, H_2O, O_3, CO, and NO can dominate the infrared.

6.4.2.2 Non-LTE Radiative Transfer

Non-LTE phenomena are treated as a distinct topic in some radiative transfer books, which includes Goody and Yung (1989). Other good introductions may be found in Houghton (1986, chapter 5) and in Chamberlain and Hunten (1987, chapter 1). In terms of monographs, López-Puertas and Taylor (2001) may be considered as a basic reference for atmospheric non-LTE topics. They follow the general flow of mathematical derivation and physical reasoning used by Goody and Yung, but they include a set of extended discussions and a thorough review of non-LTE observations in the Earth's upper atmosphere. Here, we draw from that work with the goal of providing context for non-LTE cases in the Martian atmosphere.

The RTE and its solutions studied in previous sections are applicable when the absorbing properties of the gas are known. The RTE is not altered by non-LTE, but its solution is altered by the non-zero nature of the source term, Q, in (6.3). The complication arises from the fact that Q is essentially unknown, given only the input parameters discussed in previous sections. Therefore, non-LTE injects a cautionary note and a complication into RT theory. One must understand the conditions under which the LTE source function is appropriate, or, perhaps more generally, one should define precisely the transition zone over which the breakdown of LTE occurs. The complexity manifests in the computation of the source function for the non-LTE region.

(a) Statistical Equilibrium Equation

Einstein showed that the source function of the radiative transition between two states obeys the Planck law if the associated populations obey the Boltzmann distribution. This provides a correspondence between the source function and the energy level populations of the gas.

In thermal equilibrium, the Boltzmann law defines the distribution of the molecular energy levels, which for two given states is

$$\frac{n_2}{n_1} = \frac{g_2}{g_1} \exp\left(-\frac{E_2 - E_1}{kT}\right) = \frac{g_2}{g_1} \exp\left(-\frac{h\nu_0}{kT}\right) \quad (6.50)$$

where n_i, E_i, and g_i are the populations, energies, and statistical weights of the state i, k is the Boltzmann constant, and T is the temperature of the atmosphere. In the rightmost expression, we are considering a radiative transition between the two states, with ν_0 being the frequency of the associated photon. In thermal equilibrium, all the states obey this law, and thus a common temperature exists for each state. As mentioned previously, one never encounters this ideal situation in an atmosphere. Even in LTE, there is an allowance for a small disequilibrium, normally associated with the role played by the radiation field, but still fulfilling the Boltzmann law. Also, it is possible to find some groups of levels (electronic, or vibrational) that depart from the Boltzmann distribution for the local temperature while other types of levels (kinetic energy, or rotational states) remain in LTE.

In a generic situation, the state populations of the gas follow from a microscopic analysis, and follow the so-called statistical equilibrium equation (SEE), which is an extension of (6.50). The SEE needs to be applied to every energy level and must incorporate all of the relevant microscopic processes (primarily collisional and radiative) that affect the specific state/isotope/species. Assuming stationarity in the state populations, the SEE has the following form for a simple two-level case (see section 3.6 in López-Puertas and Taylor, 2001),

$$\frac{n_2}{n_1} = \frac{B_{12} \bar{J} + k_T' [M]}{A_{12} + B_{21} \bar{J} + k_T [M]} \quad (6.51)$$

Here n_2 and n_1 are the populations of the upper and lower energy states of the transition, normally expressed in cm^{-3}; A_{21}, B_{12}, and B_{21} are the Einstein coefficients for spontaneous emission, absorption, and induced emission between those two states, respectively; $\bar{J}$ is the mean radiance (zeroth moment, (6.6); $4\pi\bar{J}$ is sometimes referred to as the "actinic flux" (e.g. Thomas and Stamnes, 2002)) averaged over the spectral band; and $k_T' [M]$ and $k_T [M]$ refer to the gains and losses in the upper state through collisional processes. In this two-level system, such collisions respond to

$$k_T : n_2 + M \rightarrow n_1 + M + \Delta E \quad (6.52)$$

where M is any atmospheric molecule, whose concentration [M] is expressed in cm^{-3}, k_T is the rate coefficient (per molecule) of the process, normally expressed in cm^3 s^{-1}, so that $k_T[M]$ gives the specific losses (per molecule) of the upper state in s^{-1}, and $k_T' [M] n_1$ the collisional production (per molecule) of the upper state, in cm^3 s^{-1}. This is an example of collisional energy transfer processes called thermal collisions, because the difference in the internal energy of the states ΔE is converted into the kinetic motion of M. During the inverse process with rate coefficient k_T', the excitation energy of the upper level is extracted from the kinetic energy of the gas species M. These collisions are also called V–T, due to the exchange of vibrational and thermal energies.

When the radiative processes are weak or collisions are sufficiently frequent, the SEE reduces to $n_1 / n_2 = k_T' / k_T$, and it can be shown that this is exactly (6.50), solely a function of the local temperature. This results from the thermodynamical argument of detailed balance, as applied to thermal collisions in the special case of strict thermal equilibrium. On the other extreme, when the radiative terms dominate, the level populations will be a function of the ambient radiance. In the intermediate scenario, one can define the breakdown of LTE as the case where both collisional and radiative terms contribute equally to the gains/losses of the population of each state. Since the collisions depend directly on the density of the gas, one expects LTE to prevail in the lower and denser layers of the atmosphere, and non-LTE behavior at high altitudes. However, it is important to note that the breakdown of LTE can occur at different altitudes for each group of radiative transitions, isotopes, and species.

In a real atmosphere, one needs to move beyond the two-level system and incorporate more energy states, isotopes, and molecules, and the possibility of energy exchanges between all of these states via collisions. In the common case of the vibrational states of an atmospheric species like CO_2, the energy transfer occurs through non-thermal collisions, or V–V exchanges, and are given by the rate coefficient, k_{VV}:

$$k_{VV} : n_2 + M(v) \rightarrow n_1 + M(v') + \Delta E_V \qquad (6.53)$$

where the colliding species M may have several vibrational states: v, v'. Again the excess of energy, ΔE_V, may be released as kinetic energy transferred to any of the colliding molecules. The rate coefficients for V–V exchanges are normally much larger than those for thermal collisions, especially if the states n_2 and $M(v')$ are nearly resonant, i.e. when ΔE_V is nearly zero. The inverse process can also take place, and the relation between the direct and inverse rate coefficients is found again by considering the principle of detailed balance under thermodynamic equilibrium:

$$\frac{k'_{VV}}{k_{VV}} = \frac{g_1 g(v)}{g_2 g(v')} \exp\left(-\frac{\Delta E_V}{kT}\right) \qquad (6.54)$$

The SEE in this example becomes

$$\frac{n_2}{n_1} = \frac{B_{12}\bar{J} + k'_T[M] + k'_{VV}[M(v')]}{A_{12} + B_{21}\bar{J} + k_T[M] + k_{VV}[M(v)]} \qquad (6.55)$$

There are other possible non-thermal processes, including the dissociative recombination of ionic species, photolysis of species like ozone (exciting the O_2 states responsible for the 1.27 µm emission), etc. A fairly complete list may be found in López-Puertas and Taylor (2001). It is easy to note that the complexity can increase significantly with an augmentation of the number of included states and the interactions among them. Each state and interaction adds terms to the numerator and denominator of (6.55). Such is the case with the CO_2 vibrational states in the atmosphere of Mars.

(b) Non-LTE Radiative Transfer Equation
In order to include the non-LTE processes discussed above, a source function can be formulated in terms of the Einstein radiation coefficients. For a full derivation, the reader is referred to Goody and Yung (1989) or López-Puertas and Taylor (2001).

For a two-level system, the non-LTE source function (term) becomes

$$S_v = \frac{2h\nu^3}{c}\left[\frac{n_1}{n_2}\frac{g_2}{g_1}\exp\left(\frac{h(\nu - \nu_0)}{kT}\right) - 1\right]^{-1} \qquad (6.56)$$

where c is the speed of light, ν is any frequency of a ro-vibrational transition between those states which is centered at ν_0, and the other symbols are as in (6.50).

If the LTE approximation is introduced (6.50), the source function reduces to the Planck function (as expected). Otherwise, the source function represents the net result of the microscopic processes that affect the two states. And, as mentioned above, it is most common to find incomplete LTE in the atmosphere, where the source function follows the Planck function for some molecular bands and some species, but not for all. Rewriting the source function using the SEE emphasizes this fact:

$$S_{\nu_0} = \frac{\bar{J} + \varepsilon B_{\nu_0}}{1 + \varepsilon}, \qquad \varepsilon = \frac{k_T[M]}{A_{21}}\left[1 - \exp\left(-\frac{h\nu_0}{kT}\right)\right] \qquad (6.57)$$

where ε is defined to be a measure of the relative importance of collisions to radiation. When collisions dominate in LTE, ε is large. Conversely, it approaches zero in the lower densities of the upper atmosphere. The condition $\varepsilon = 1$ is a proper mathematical definition of the "LTE breakdown" layer of a given transition and molecular species. The equation also indicates that, in the absence of strong V–T collisional processes, even thermal radiation can produce non-LTE conditions.

Equation (6.57) contains two unknowns: the source function, B_{ν_0}, and the mean radiance, $\bar{J}$. The RTE also contains these basic variables. Thus, one has essentially a system of two equations (SEE and RTE) and two unknowns, which mathematically represent a solvable system of equations.

(c) Difficulties and Approximate Solutions
The above discussion focused on the two-level problem. Despite its simplicity, it is often applicable to the fundamental ro-vibrational bands of the many atmospheric species with significant transitions in the IR. A common approximation is to assume that the population of the ground state is known. The physical basis of this is that a high fraction of molecules is always in the ground state regardless of the non-LTE population of the excited states. As a result, the source function of the fundamental bands can be used to describe the first excited state of this gas. The extension of this argument to higher states is possible, obtaining a one-to-one correspondence between the number of source functions and the number of excited states. Each one requires a different statistical equilibrium equation (SEE) and each can have a source function very different from the others. However, the number of radiative transfer equations may differ from that of the SEE. A given state may be the upper state of several emission bands, each one requiring a specific RTE solution. As an example, the non-LTE model of López-Valverde and López-Puertas (1994a) for Mars incorporates about 70 vibrational states of four different isotopes of CO_2, with nearly 90 ro-vibrational bands between them.

The situation becomes still more complex when there are important non-thermal processes, like V–V collisional exchanges. This is indeed the case of molecules with several modes of vibration (like CO_2, H_2O, O_3, CH_4, or N_2O), where resonances can make the V–V exchanges very effective. An additional complication can arise from nonlinearities in the collisional processes, requiring iterative solutions. Finally, there can be large uncertainties in many of the collisional rate coefficients. Such difficulties can be added to those discussed for molecular opacity such as the case of overlapping of lines of different bands. Since the lines may have different source functions, many approximate solution methods of the RTE may not be appropriate.

Several formalisms have been developed to tackle these problems, all based on various approximations. Each specific case (emission band, isotope, molecule) requires a specific study to justify the method selected. The list of approximations or a description of the different formalisms is beyond our scope, and the reader is referred to López-Puertas and Taylor (2001). The overriding theme is that approximate solutions are always sought, usually with the purpose of simplifying either the SEE or the RTE (or both).

6.4.2.3 Non-LTE Data and Modeling of the Martian Atmosphere

The first important non-LTE emission observed on Mars was the detection in 1976 of the "Martian laser", a strong emission from the planet at 10 μm (Johnson et al., 1976; Betz et al., 1977; Deming et al., 1983). The observations were carried out from ground-based telescopes and interpreted correctly as a transition between two states of CO_2, the upper one being the first excited state of its asymmetric stretching mode of vibration (001), also responsible for a strong ro-vibrational band at 4.3 μm. The solar absorption in this 4.3 μm fundamental transition was the energy source of the 10 μm emission (Mumma et al., 1981; Deming and Mumma, 1983; Gordiets and Panchenko, 1983).

Previously, Ramanthan and Cess (1974) had employed an approximate analytical formulation of radiative transfer for the fundamental band of CO_2 in 15 μm, whose upper state is the first excited state of the bending mode of vibration (010), responsible for an important cooling of the Earth's mesosphere. They considered more energetic states of the bending mode of vibration, but with the approximation of immediate relaxation to the (010) state. They ignored exchanges with the minor isotopes of CO_2 that were later revealed to be very important.

The model of Gordiets and Panchenko (1983), focusing on the 10 μm laser bands emitted from the 35–110 km altitude range, also used an approximate treatment for the radiative transfer in the 15 μm fundamental band. Their removal-of-radiation approach indicated a breakdown of LTE for the 15 μm fundamental band at about 105 km (in daytime conditions). Their model produced an inversion in the population of the upper and lower states of the 10 μm emission. They incorrectly concluded that the laser emission should be very important for the thermal structure of the Mars mesosphere (e.g. Deming and Mumma, 1983).

The model by Deming and Mumma (1983) was restricted to daytime conditions and made several approximations for the relaxation of the asymmetric stretching mode of vibration. They ignored the first hot band, and limited their radiative transfer to the fundamentals in the 4.3 μm and 15 μm bands. Their approach solved the statistical equilibrium equations of the half-dozen states included with an iterative method, starting with LTE values and assuming further approximations for collisional exchanges. Despite some problematic spectroscopic values for spontaneous emission, they reproduced some observed population inversions, including observed variations in the inversion altitude and strength.

Later, Stepanova and Shved (1985) developed a non-LTE model for the strongest emissions of CO_2 and CO in the IR, though it was still limited to daytime conditions. They incorporated a more realistic radiative transfer treatment for the fundamental bands of CO_2 at 4.3 μm and of CO at 4.7 μm. Among the employed approximations, they treated the bending mode states responsible for the hot bands as being in LTE at all altitudes. However, they provided the first studies of the V–V exchanges between the (001) state of different isotopes.

Another model of interest is that of Bougher and Dickinson (1988) and Bougher et al. (1988). Based on a previous study for Venus (Dickinson and Bougher, 1986), they focused on solar heating by near-IR bands of CO_2 and on the cooling by the 15 μm in the upper Martian atmosphere. The resulting model has been included in various MGCMs for radiative transfer in the upper atmosphere. Their work also included a set of sensitivity studies for several parameters, including rate coefficients. As a result, they provided important insight into the uncertainties inherent in these rates and the resulting impact on the populations and heating rates.

The model that forms the basis of the most recent work on Martian non-LTE was developed in the early 1990s at the Instituto de Astrofísica de Andalucía (IAA; López-Valverde, 1990; López-Valverde and López-Puertas, 1994a,b; López-Puertas and López-Valverde, 1995). While the initial motivation was the opportunity to include non-LTE effects in the thermal profiling of Mars by the ill-fated Mars Observer, the resulting model was more general-purpose: covering daytime and nighttime conditions, CO_2 and CO states, and many more levels than all previous works at the time. The development started with an earlier model for the Earth's mesosphere (López-Puertas et al., 1986). It employed a formalism (modified Curtis matrix) that permits a careful treatment of radiative transfer in the presence of non-thermal processes, including V–V exchanges between isotopes and between CO_2 and CO. The model produces populations for dozens of vibrational states and heating rates that include the numerous state interaction terms. The altitude range ran from the surface to about 200 km, using the most recent spectroscopic data available. A large part of the predictions could not be tested at the time due to a lack of observations. Even today, its users must include indirect validation and comparison with other models.

As an example of results from this model, Figure 6.37 shows the state populations for several important CO_2 vibrational levels (during the daytime), the altitude dependence of the emission, and a comparison to observations. The populations are expressed in terms of vibrational temperatures, which differ for each state and whose departure from the kinetic temperature indicates the breakdown of LTE. During the daytime, all

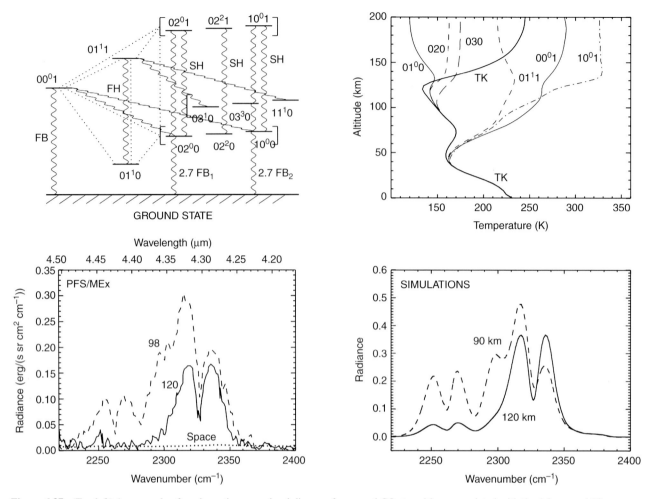

Figure 6.37. (Top left) An example of a schematic energy level diagram for several CO_2 transitions associated with the 4.3 μm and 10 μm NLTE emission. The radiative transitions and collisional exchanges are indicated by solid and dotted lines, respectively. Labels "FB", "FH", and "SH" refer to fundamental, first hot, and second hot bands at 4.3 μm; "2.7 FB" refers to fundamental hot bands at 2.7 μm; and the transitions without labels refer to the 10 μm bands. (Top right) The vibrational temperatures of those states during the daytime, where the model temperature profile is labeled as TK. (Bottom left) Mars limb emission in the 4.3 μm band at 98 and 120 km as observed by MEX/PFS. (Bottom right) NLTE model simulation of this emission at similar tangent altitudes. Figures based on results presented in López-Valverde and López-Puertas (1994b), Formisano et al. (2006), and López-Valverde et al. (2011a).

of the levels manifest enhancements at mesospheric altitudes compared to their LTE values (for an assumed temperature profile, T_K). This excitation is particularly large for the states with quanta in the asymmetric stretching mode (00v), and is driven by the strong absorption of the solar flux by the 2.7 and 4.3 μm bands. The vibrational temperatures increase with altitude and solar flux up to the upper thermosphere, where they become constant. This behavior reveals the absence of collisional quenching due to the low gas densities. At these high altitudes, every photon absorbed is subsequently re-emitted, producing very little heating. At the same time, the excited states of the bending mode vibrational levels are not directly populated by solar absorption. Rather, the primary mechanism is that of V–V relaxation from other states. The possible V–V exchanges represent an intricate network of transitions, as captured in the partial state diagram included in Figure 6.37. The emerging 4.3 μm radiance is the result of a complex combination of contributions, whose amplitude and relative importance vary

with altitude. Consequently, the overall shape of the 4.3 μm emission offers a good test for an NLTE model. In the model calculations included in the figure, the correspondence of the predicted amplitude and shape with that of observed emissions appears quite reasonable.

Naturally, the model also incorporates approximations, both in the treatment of radiative transfer and in the SEE of the state populations. Some of these approximations are based on similar, commonly employed assumptions for terrestrial non-LTE models. For Mars, the following are three of the most important.

(1) The division of the ro-vibrational lines into histograms in order to perform a line-by-line calculation for a given selected line. While the calculation is precise and may be considered a quasi-line-by-line approach, it treats the lines as independent. In other words, it ignores overlapping lines. This limits the validity of the cooling rates to altitudes above about 50 km. However, the state populations

are not much affected by this approximation, particularly since most are in LTE for the Martian troposphere.

(2) Many assumptions are employed to specify rate coefficients that have not been measured in the laboratory. This creates large uncertainties for many collisional processes, particularly those between isotopic species and those involving high-energy states.

(3) LTE is assumed for the rotational distribution of states in every vibrational state included. This approximation may be problematic for the upper thermosphere of Mars, but its impact on most state populations is estimated to be small.

Some of these approximations will be improved upon in future versions of the model. Anticipated changes include more rigorous radiative transfer routines (full line-by-line models), and the extension of the model to rotational non-LTE. Both strategies are now found in similar models of the Earth's atmosphere (Funke et al., 2012).

Comprehensive models like that developed at IAA exhibit two major limitations. The first one is that the inclusion of such a large number of levels and bands requires an iterative solution. This approach, combined with the large number of matrix inversions involved in the radiative coupling solutions, results in a large computational cost. As a consequence, they cannot be employed in GCMs, or in many retrieval algorithms. For such applications, simplified schemes are needed. While such fast algorithms have been developed recently, the additional assumptions and simplifications ultimately reduce the resulting model accuracy.

Another difficulty identified above involves the model validation using observations, particularly those of the upper atmosphere. Fortunately, recent missions like Mars Express are alleviating the lack of relevant data. MEX carries two instruments that permit characterization of the non-LTE emissions by CO_2 and CO, PFS and OMEGA. Although analyses of these data are ongoing, important model validation has already begun (López-Valverde et al., 2005, 2011a; Formisano et al; 2006; Piccialli et al., 2012). Additional tests have been enabled by TES and ground-based observations of the 10 μm emission (Maguire et al., 2002; López-Valverde et al., 2011b).

6.5 ENERGETIC PARTICLES

Mars is exposed to incident energetic charged particle radiation similarly as the Earth, which is mainly composed of galactic cosmic rays (GCRs) and solar energetic particles (SEPs). GCRs are a high-energy (10 MeV–1 GeV/nucleon) radiation source composed primarily of energetic protons, with some helium, heavier elements, and electrons present. GCR flux levels are modulated by up to a factor of 5 over a solar cycle but are otherwise relatively constant. Superimposed on the GCR flux are SEPs, episodic, intense bursts of energetic particles accelerated by coronal mass ejections (CMEs) at the Sun. These events have onset times ranging from minutes to hours, and may last up to several days. Their flux levels are highly variable over orders of magnitude, and typically possess "soft" or "hard" spectra corresponding to peak energies below or above ~150 MeV/nucleon, respectively.

The impact of charged particle radiation at Mars is likely to be very different than that of Earth for two primary reasons. First, Mars lacks any significant, global magnetic field (Acuna et al., 1999), and possesses only a much weaker, remnant field located in isolated, disordered regions of the crust. Second, the atmospheric pressure at the surface of Mars is less than 1% that of Earth, resulting in a lower column depth and a correspondingly lower amount of shielding from incident energetic particle radiation. Hence, compared to Earth, there is the potential for a significantly greater impact of GCRs and SEP events in the ionosphere, atmosphere, and surface of Mars.

Whereas the Earth's magnetosphere shields the ionosphere and upper atmosphere from most SEPs, the main magnetic barriers at Mars consists of a bow shock followed by a magnetic pileup boundary formed by the interaction with the solar wind (Acuna et al., 1998; Bertucci, 2003). Energetic particles such as GCRs and SEPs easily penetrate this region to impact the ionosphere and atmosphere. Magnetic anomalies arising from the localized crustal magnetic fields may provide an effective shield against only the lowest-energy part of the SEP spectrum. In general, soft SEP events will dissipate all of their energy into the atmosphere prior to reaching the surface, whereas more energetic hard events can result in substantial fluxes at the surface. At higher altitudes, SEPs cause atmospheric ionization, sputtering, and heating (Leblanc et al., 2002), which, taken together, could amount to a significant contributor to atmospheric loss processes over time. At lower altitudes and on the surface, SEPs can induce atmospheric chemistry and result in significant radiation exposure for short periods.

The Radiation Assessment Detector (RAD) on the Mars Science Laboratory has been measuring the dose due to GCRs on the surface of Mars since August 2012 (Hassler et al., 2012). RAD has confirmed that the radiation dose rate at the surface is substantially higher, by two to three orders of magnitude, than on Earth given the lower atmospheric and magnetic shielding at Mars. In addition to the solar cycle, a dominant driver for GCR variability on the surface is the atmospheric density as it undergoes diurnal and seasonal variations, further emphasizing its importance as the major contributor to shielding. In the lower ionosphere, GCRs are believed to be a significant contributor to ion-neutral chemistry (Molina-Cuberos et al., 2002). The ability of GCRs to penetrate to the surface and into the regolith implies that it may be a significant contributor to the chemistry of the soil and rocks as well.

The measurement record will be continued through measurements by RAD on MSL in addition to those by the Mars Atmosphere and Volatile Evolution (MAVEN) mission, which began in the fall of 2014. These additions should provide the most comprehensive measurements to date of the complete energetic particle environment at Mars.

6.6 APPLICATIONS

The preceding sections introduce the fundamentals of radiative transfer theory as a basis to perform radiative analyses of the Martian environment. In the following section, we describe several key applications of radiative transfer theory. Firstly, we describe its application to global climate models (GCMs), which use radiative transfer schemes to compute the heating rates associated with the absorption by atmospheric gases and aerosols of short-wave (SW) and long-wave (LW) radiation. A brief overview of atmosphere temperature and aerosol retrievals follows, the latter focusing on aerosol optical depth and macrophysical parameters. Finally, we highlight some specific uses of NLTE algorithms in GCM and remote sensing studies.

6.6.1 Global Climate Models (GCMs)

6.6.1.1 Aerosol Representation in Models

Aerosols are a major source of thermal forcing in the Mars atmosphere such that their representation in regional and global-scale atmospheric models is critical for the successful simulation of atmospheric temperatures and circulation. The circulation, in turn, plays a central role in determining the sources and sinks for dust and ice cloud aerosols as well as their subsequent transport. Consequently, the radiative and dynamical fields are strongly coupled. The latitude–height distribution of dust has a profound influence on the thermal structure of the atmosphere and the intensity of the circulation. In particular, the axisymmetric (meridional overturning) circulation and the thermal tides directly respond to the strength of the radiative forcing (see Chapter 9), as was first appreciated in the calculations by Haberle et al. (1982). These fields undergo a significant seasonal variation as the radiation field migrates with the seasonal march of the subsolar latitude. The dust and ice aerosol fields are correspondingly variable in space and time but with considerably more complexity, such that characterizing their climatology is a major research area. As models have increased in sophistication, closer agreement has been achieved in simulating the observed temperature and aerosol fields returned from spacecraft observations.

There have been a number of approaches in representing aerosols in atmospheric models. Early simulations (e.g. Pollack et al., 1990; Haberle et al., 1993a; Hourdin et al., 1993, 1995; Wilson and Hamilton, 1996) imposed a static dust field that was uniformly distributed in the horizontal direction, and uniformly mixed in height up to a particular altitude, before tapering off as sedimentation dominated eddy mixing (Conrath, 1975). As improved observations of column optical depth became available, simulations were carried out with temporally evolving column optical depths, but still with specified latitude and height structure (Forget et al., 1999; Lewis et al., 1999; Hartogh et al., 2005; Takahashi et al., 2006; Kuroda et al., 2007). The choice of the vertical depth of the dust layer as a function of latitude and season was specified through the Conrath parameter (see Chapter 10) and adjusted by fitting the GCM-simulated zonal-mean temperature response to MGS TES retrievals.

More sophisticated simulations were carried out where the evolving circulation was allowed to account for the spatial and temporal evolution of dust (e.g. Murphy et al., 1995; Wilson, 1997; Newman et al., 2002a,b; Basu et al., 2004, 2006; Kahre et al., 2005, 2006, 2008). Given the balance between sedimentation and transport (resolved scale advection and eddy mixing), the aerosol size distributions play a major role in shaping the evolving latitude–height distribution of dust and clouds

Consequently, atmospheric models have progressed to represent aerosol distributions with a finite number of explicitly represented dust tracers, representing different particle sizes (Murphy et al., 1993, 1995; Richardson and Wilson, 2002; Basu et al., 2004, 2006; Kahre et al., 2006, 2008; Michaels et al., 2006; Wilson et al., 2008a; Greybush et al., 2012). A more recent alternative strategy has been to represent an aerosol distribution by its two moments, the number of aerosol particles N, and a mass mixing ratio. Such a scheme can be efficiently used in atmospheric models as only the moment fields need to be explicitly transported by the model circulation, while physical processes such as sedimentation and cloud microphysics may be calculated following the discretization of the moment-based size distribution into size bins.

Similar issues hold for ice cloud aerosols. Early water cycle modeling assumed that water partitioned itself into vapor and cloud ice particles of a fixed size (e.g. Richardson et al., 2002). Improved results were obtained by allowing a more physically motivated ice particle size evolution guided by microphysics (Montmessin et al., 2004).

More recently, simulations have been carried out so that the dust column opacity is constrained by spacecraft observations, while the simulated circulation is allowed to control the vertical distribution of dust. In one approach (Wilson et al., 2008a; Kahre et al., 2008; Greybush et al., 2012), dust is added (or removed) from the boundary layer as required to allow the simulated dust column to track observations. Another approach (Madeleine et al., 2011) rescales the simulated vertical distribution of dust to keep the simulated column in agreement with the specified value. Dust column opacity scenarios based on spacecraft observations for MY 24–31 have been developed and implemented in the Mars Climate Database (version 5, www-mars.lmd.jussieu.fr/).

6.6.1.2 Simulated Heating Rates

The influence of radiation on atmospheric dynamics can be assessed by examining the net radiative heating due to radiation at all wavelengths. Atmospheric heating is due to the divergence of net upward and downward fluxes, and is conveniently expressed in deg sol^{-1} or in W kg^{-1} as

$$\frac{dT}{dt} = \frac{g}{C_p}\frac{dF_\lambda}{dp} \qquad (6.58)$$

where p is pressure, g is the gravitational acceleration, C_p is the specific heat of dry air, and g/C_p is the dry adiabatic lapse rate. In the absence of dynamics, one can formulate a temperature field that is allowed to evolve to a state of radiative equilibrium so that the IR radiation due to the temperature exactly balances the imposed solar radiation. It is well known that a purely radiative equilibrium state (in the case of an optically thin atmosphere) leads to a strongly unstable layer of atmosphere immediately above the surface, which would be subject to convective mixing by small-scale circulations (e.g. Pierrehumbert, 2010). Sufficient heating from dust aerosols can also stabilize an atmosphere against such instability. Accounting for these processes results in a temperature profile that is in radiative/convective (R/C) equilibrium. In the Martian atmosphere, it is possible for the atmosphere to cool radiatively to temperatures

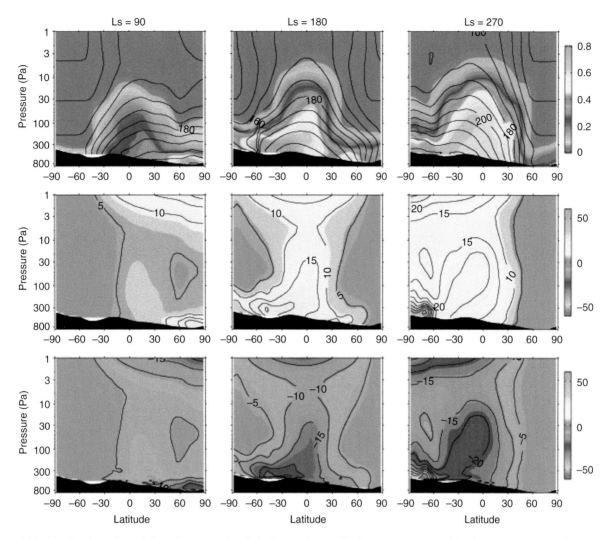

Figure 6.38. Simulated zonally and diurnally averaged radiative/convective equilibrium temperatures and heating rates using specified aerosol distributions in three seasons: $L_s = 90°$ (left), $L_s = 180°$ (center), and $L_s = 270°$ (right). (Top row) Dust is shown as a pressure-normalized visible optical depth (0.67 μm, Pa^{-1}) and the temperature is contoured at 10 K intervals. (Middle row) Solar heating rates (deg sol^{-1}) corresponding to the dust aerosol fields shown in the top row. (Bottom row) Long-wave heating rates. The equatorial column-integrated dust optical depth is ~0.3, 0.6, and 0.7 for the $L_s = 90°$, 180°, and 270° simulations, respectively. A black and white version of this figure will appear in some formats. For the color version, please refer to the plate section.

that would fall below the CO_2 condensation temperature, and so CO_2 condensation may be triggered, with a resulting release of the heat of fusion. While the CO_2 ice cloud formation process is potentially complicated, involving convective mixing and latent heat release (Colaprete et al., 2008), it is appropriate to account for condensation heating in the formulation of an atmospheric state to which large-scale dynamics may respond.

The specification of a temperature state in R/C equilibrium is a strategy often employed in the formulation of simple dynamical models of the atmosphere. The effects of radiative forcing are then parameterized as a relaxation process back to the R/C state, governed by a suitable time scale (α) that may be a function of the temperature and pressure:

$$\frac{dT}{dt} = \alpha(T - T_{R/C})$$

where $\alpha = -(g/C_p)4\sigma T^3$ and σ is the Stefan–Boltzmann constant. R/C temperature fields have been used in a number of

simplified model studies (e.g. Haberle et al., 1982, 1997; Joshi et al., 1995). Radiative damping timescales of 1–4 sols have typically been used in calculations. Haberle et al. (1997) showed that the intensity of the axisymmetric (Hadley) circulation was inversely proportional to the damping time, which illustrates the influence of radiative heating on the atmospheric circulation and the transport of aerosol.

Figure 6.38 shows solar and infrared heating rates from a GCM simulation that has been run without dynamical tendencies, so that the temperature field is in radiative/convective equilibrium. The vertical dust distribution is specified and follows the results of a full simulation that includes dust transport (as represented in Figure 6.39). The SW and LW heating rates are seen essentially to balance each other, as expected, except in the boundary layer (as discussed in Chapter 7) and at winter polar latitudes where convection and CO_2 condensation occur. The solar heating in the lower atmosphere (below 50 km or ~3 Pa) is dominated by the

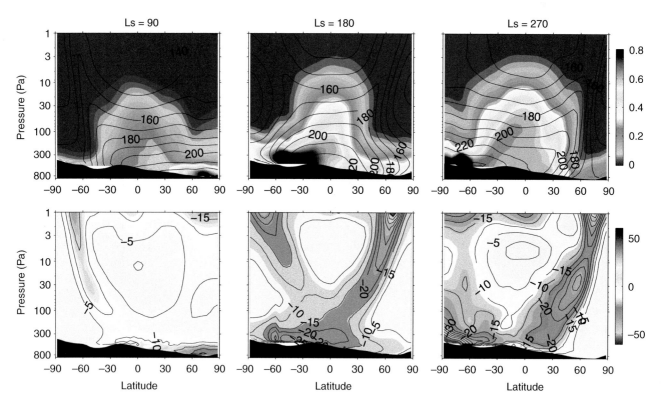

Figure 6.39. Simulated zonally and diurnally average temperature and long-wave heating rates for the specified aerosol distributions shown in Figure 6.38. (Top row) Dust is shown as a pressure-normalized visible optical depth (0.67 μm, Pa^{-1}) and the temperature is contoured at 10 K intervals. (Bottom row) Long-wave heating rates (deg sol^{-1}) corresponding to the dust aerosol fields shown in the top row. A black and white version of this figure will appear in some formats. For the color version, please refer to the plate section.

absorption by dust aerosol. There is very strong LW heating near the surface due to the absorption of upward IR radiation from the surface, as anticipated by Goody and Belton (1967). This depth of penetration is ~40 m for Earth and 1.5 km for Mars. The net radiative heating (SW+LW) is largely balanced with heat transport by the parameterized convective mixing in the simulation, as illustrated in Haberle et al. (1993b). Sensible heat exchange plays a less significant role (see below).

Figure 6.39 shows the complementary situation where the atmospheric dynamical response is allowed. The solar heating is not changed, as it is dependent only on the aerosol field. Both the temperature and LW heating fields have changed substantially in response to the dynamical contribution to energy balance, with the largest changes occurring in the winter hemisphere where dynamically induced adiabatic heating leads to significantly higher temperatures. In this case, the net radiative heating is balanced by dynamical heating. To a large degree, the vertical motion can be approximated as that necessary to yield adiabatic heating/cooling necessary to offset the net radiative heating. Santee and Crisp (1995) diagnosed the circulation from Mariner 9 observations with retrieved temperature and dust by calculating net radiative heating and then reconstructing the zonal-mean circulation necessary to achieve balance. Of course, this balancing is implicit in GCM simulations, and computed by the dynamical core of the model. Therefore, the critical aspect is the degree to which the models accurately simulate observed temperature structure.

6.6.1.3 Water Ice Clouds

The influence of water ice clouds on radiation has only recently been realized to be significant. The first hint of this was provided by the data from the Pathfinder entry (Schofield et al., 1997; Magalhães et al., 1999), which showed a strong thermal inversion between 10 and 16 km. This inversion was attributed to infrared emission by clouds (Colaprete et al., 1999; Haberle et al., 1999). MGS radio occultations (Hinson and Wilson, 2004) and MCS temperature profiles (Wilson and Guzewich, 2014) revealed that similar inversions are common in the tropics during the aphelion season. GCM modeling has shown that inclusion of the radiative effect of clouds plays a critical role in shaping the observed temperature response (Colaprete and Toon, 2000; Hinson and Wilson, 2004; Wilson and Guzewich, 2014). MCS observations of the cloud and temperature fields strongly support the close connection between clouds and temperatures (Lee et al., 2009; Wilson and Guzewich, 2014; Steele et al., 2014). Further characterization of the radiative effect of clouds on zonally averaged tropical temperature has been made by Wilson et al. (2008b), Haberle et al. (2011), and Madeleine et al. (2012b). In particular, Madeleine et al. (2011, 2012b) showed that radiative forcing by dust and ice cloud distributions, as specified by observations, also yields excellent agreement between simulated and observed tropical temperatures. Such correspondence also indicates a high degree of self-consistency in the model details. Wilson et al. (2007) interpreted a seasonal enhancement in the nighttime tropical surface temperatures retrieved by TES as a

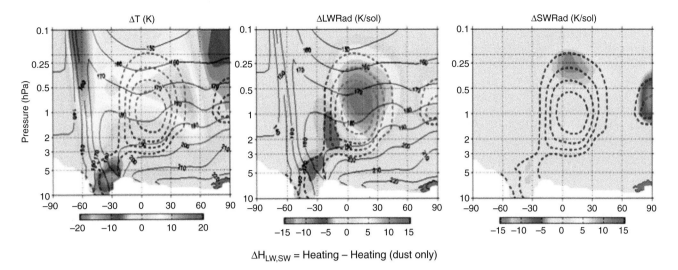

$$\Delta H_{LW,SW} = \text{Heating} - \text{Heating (dust only)}$$

Figure 6.40. (Left) Zonally and diurnally averaged aphelion season ($L_s = 110°$) temperature from a GCM simulation with radiatively active water ice clouds. The temperature field is displayed with contours at 10 K intervals. The (color) shading illustrates the temperature difference against a baseline simulation without water ice clouds. The mass mixing ratio of the simulated cloud (dashed red contours) is shown at intervals of 10 ppm. (Center) The change in LW heating attributable to water ice clouds ($\Delta LW = LW-LW_{dust\ only}$). (Right) As for center panel, but for SW heating. A black and white version of this figure will appear in some formats. For the color version, please refer to the plate section.

consequence of the increase in downwelling thermal emission of water ice clouds. This study emphasized the large diurnal variation in cloud opacity, which is related to the large diurnal variation in atmospheric temperature. The radiative forcing associated with water ice clouds has also been shown to affect the forcing of the semidiurnal tide in the upper (50–80 km) atmosphere (Kleinböhl et al., 2013).

Figure 6.40 shows the increase in zonally and diurnally averaged temperature when radiatively active water ice clouds are included in a GCM simulation. This pattern is similar to that shown in Wilson et al. (2008a), Madeleine et al. (2012b), and Kleinböhl et al. (2013). The change in radiative heating in the diurnal average is dominated by an increase in long-wave heating due to absorption of upwelling radiation from the relatively warm surface. Note that the temperature response is significantly more extensive than the change in heating, which is localized to the region of the tropical water ice cloud. This enhanced temperature response is due to adiabatic heating resulting from the deeper and stronger Hadley circulation associated with cloud radiative heating (Madeleine et al., 2012b; Kahre et al., 2014). It should be emphasized that tropical water ice clouds undergo a significant variation in height as they are modulated by the diurnal tide (Hinson and Wilson, 2004; Madeleine et al., 2012b; Wilson et al., 2014). It is notable that the strong nighttime radiative cooling by low-level clouds is stronger than the cooling of the lower atmosphere shown in Figure 6.40, where the results are diurnally averaged. Inclusion of clouds increases the scattering albedo of the aerosol, resulting in reduced solar absorption, as shown in the right panel of Figure 6.40.

6.6.1.4 Radiation Budget

GCM simulations, such as described above, have become sufficiently realistic to enable an assessment of the global radiation budget of Mars. There have been several attempts to quantify the Martian energy budget, especially in the polar regions (e.g.

Emmanuel, 1968; James and North, 1982; Pollack et al., 1990; Kieffer and Titus, 2001), but this task is made more difficult in non-polar regions by the extensive coverage and large amount of observations required. The following study uses results from the LMD (Laboratoire de Météorologie Dynamique) GCM to quantify the contribution of the different components of the climate system to the global radiation budget. Our goal is to illustrate the processes described in this chapter by reviewing the radiation budget of Mars and discussing some of the differences with a similar budget for the Earth (the numbers of Trenberth et al. (2009) will be used for the case of the Earth). The results are summarized in Figure 6.1, which shows the net annual fluxes associated with the different components of the atmosphere in W m^{-2} for a simulation that incorporates the dust scenario of MY 26 as observed by TES and radiatively active water ice clouds (MCD version 5.0).

The global mean short-wave (SW) flux incident on the top of the Martian atmosphere is given by $F_\odot/4 = 148.5$ W m^{-2}, where the factor of 4 comes from the ratio of the surface area of Mars to the cross-sectional area of the intercepted solar beam. Because of its greater distance from the Sun (~1.52 AU), this is approximately half of that received by the Earth (341 W m^{-2}). Whereas atmospheric absorption of the incident solar radiation is mostly due to molecular opacity on Earth, dust aerosols are the main absorbers on Mars. The limited amount of molecular absorption in the solar band (i.e. Figures 6.2 and 6.8) produces losses of <1%, which contrasts with ~15% for the Earth. Martian dust particles absorb ~8% (12 W m^{-2}) of the incoming solar radiation, and also reflect a small fraction (3%, or 4 W m^{-2}). Water ice aerosols typically absorb less than 1% (1 W m^{-2}) and reflect a little more than 1% (2 W m^{-2}) of the incoming sunlight (see also Figure 6.40). Therefore, only 4% of the incoming solar radiation is simply reflected by the Martian atmosphere. In contrast, about 23% of the incoming solar radiation is reflected by the Earth's atmosphere, mostly by water-based clouds. Of course, the numbers in Figure 6.1 represent an annual radiation budget, and the optical depths of dust and water ice are highly variable in space and time (Smith, 2004; Hinson et al., 2004). Of the 85% of solar radiation that reaches the Mars surface (128 W m^{-2}), ~20% (31 W m^{-2}) is reflected by the surface (the mean albedo of the surface is 0.23), and the total amount of energy that is absorbed by the surface is 65% (97 W m^{-2}) of the TOA incident flux.

The surface emits ~124 W m^{-2} of LW radiation on average. Most of this radiation passes through the atmosphere, where ~20% of the emitted infrared radiation (27 W m^{-2}) is absorbed. The downwelling IR radiation is responsible for a "greenhouse" warming of ~5 K (Haberle, 2013). This contrasts with the case of the Earth, where ~90% of the emitted infrared radiation is absorbed by the atmosphere, mostly by water vapor and CO_2. On Mars, as can be seen in Figure 6.1, CO_2 gas and dust aerosols, in equal amounts, dominate the emission to the surface (12 W m^{-2}). For Mars, most of the molecular absorption occurs in the 15 μm band of CO_2, whereas the LW dust absorption is concentrated in the 9 μm band (see Figures 6.2 and 6.9). Water ice clouds also emit 4 W m^{-2} of infrared radiation to the surface, which highlights the significant role that water ice clouds can play in the Martian environment (Figure 6.40).

Sensible and latent heat fluxes also contribute to the energy budget of Mars, but in a different manner than on Earth. Significant exchanges of latent heat take place in the polar regions, when atmospheric CO_2 condenses out of the atmosphere during winter or when CO_2 ice deposits sublimate during summer. However, the CO_2 cycle is very close to equilibrium and the associated net annual energy budget is zero, which is why this process is not represented in Figure 6.1. Moreover, the exchange of latent heat due to water ice cloud formation is negligible under present-day conditions. In contrast, terrestrial latent heat through water condensation and cloud formation is a major heat source. Sensible heat flux from the surface does provide a net contribution to the Martian energy budget, but remains low at ~1 W m^{-2} on average. This minimal amount is due primarily to the low atmospheric density and thermal inertia on Mars (see Savijärvi, 1999; Spiga et al., 2011). On Earth, this sensible heat flux plays a major role in the surface energy budget and is about 17 W m^{-2} on average. Furthermore, in contrast to Mars, latent heat (evapotranspiration) on Earth accounts for ~80 W m^{-2} at the surface and is redistributed throughout the atmosphere.

Overall, the present-day radiation budget of Mars is largely controlled by dust and water ice aerosols, rather than atmospheric gases. Terrestrial water condensation plays an analogous role to Martian dust aerosols, though the latter is much more seasonally variable and subject to major episodic perturbations. This dependence on aerosol effects, coupled to the relatively short radiative time constant of the Martian atmosphere (~1–2 days; Goody and Belton, 1967), provides a fairly direct explanation for the highly variable thermal structure on Mars.

6.6.1.5 Archetypical Flux Spectra

The spectral distribution of irradiances at the top of the atmosphere (TOA) and at the surface represent useful boundary conditions for atmospheric radiative and surface processes. Not only do these values provide constraints on atmospheric processes of interest, but also they can represent useful input values

to geological or chemical processes such as thermal inertia or aqueous environments (e.g. Zorzano et al., 2009; Kieffer, 2013). Because of the highly variable nature of the Martian radiative environment, flux calculations need to encompass a variety of seasonal conditions. Leveraging the GCM calculations above, we provide spectral irradiances for representative seasonal conditions: equinox (Figure 6.2), southern solstice (Figure 6.41) and northern solstice (Figure 6.42). We have chosen an energetically favorable location (0°N, 0°E) in terms of surface conditions conducive to a putative aqueous environment and with respect to the proximity of the MER Opportunity *in situ* measurements. The flux distributions and associated model inputs are available upon request from M.J. Wolff, and distributed via https://gemelli.spacescience.org/twiki/bin/view/RadiativeTransfer/Mars/.

The three scenarios are described by radiative and thermal conditions taken from the LMD GCM simulations for an "average" aerosol cycle. A local time of 2 p.m. is chosen given the correspondence with the MGS orbit. In addition to the temperature profile and equation of state, the dust vertical profiles, integrated dust column, and the surface temperature are taken directly from model output. The RT calculations are performed using a Mars-specific DISORT implementation and the "case A" dust refractive indices for r_{eff} = 1.5 μm and ν_{eff} = 0.3 (i.e. Wolff et al., 2009). Water ice properties are included using the Type 1 seasonal ice properties with r_{eff} = 2.0 μm and ν_{eff} = 0.1 (Clancy et al., 2003). The molecular opacity for CO_2 is computed using the prescriptions described in Section 6.3.2 for both UV and NIR–thermal IR. Additional details for each of the three models are as follows.

(1) Equinox, or L_s = 195° (Figure 6.2): the column-integrated dust optical depth (at 0.67 μm) is 0.43, water ice opacity (at 12.1 μm) is 0, and the surface temperature is 295 K.
(2) Southern summer solstice, or L_s = 285° (Figure 6.41): τ_{dust}(0.67 μm) = 0.56, τ_{ice}(12.1 μm) = 0.03, T_{surf} = 290 K, water ice cloud base at 25 km with uniform mixing to TOA.
(3) Northern summer solstice, or L_s = 105° (Figure 6.42): τ_{dust}(0.67 μm) = 0.56, τ_{ice}(12.1 μm) = 0.03, T_{surf} = 290 K, water ice cloud base at 10 km with uniform mixing to a height of 25 km.

The CO_2 features are labeled in Figure 6.2. The dust and water ice aerosol features are both labeled for the solstice conditions, and the contributions to emission and absorption clearly identified in the downwelling and TOA flux distributions. For applications involving the UV surface flux, the hard cut-off associated with the short-wave CO_2 absorption is quite striking; the flux drops to zero at 200 nm. Absorption associated with ozone would be negligible for this location. However, only in the polar regions would the Hartley band perturb the spectrum by a non-trivial amount (i.e. Figure 6.8; Clancy et al., 1995, 1999; Perrier et al., 2006).

6.6.2 Remote Sensing and Radiative Transfer

The characterization or retrieval of atmospheric properties is a natural application of radiative transfer methods, particularly given the importance of scattering and aerosols for the Martian atmosphere. The general approach by which RT is applied to such a task is also called the inverse problem. While the mathematical development of this class of algorithms is beyond our scope, a useful place to begin exploring retrieval theory is the

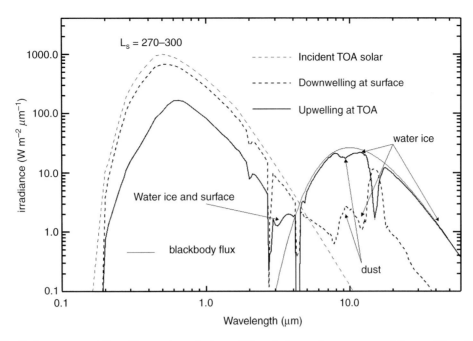

Figure 6.41. Flux distributions representative of 2 p.m. atmospheric conditions during southern summer solstice conditions ($L_s = 285°$) near the equator (0°N, 0°E). A blackbody spectrum for a surface temperature of 290 K is included as a reference. The solar incidence angle is 35°; the dust optical depth (0.67 μm) is 0.56; and the water ice optical depth (12.1 μm) is 0.04.

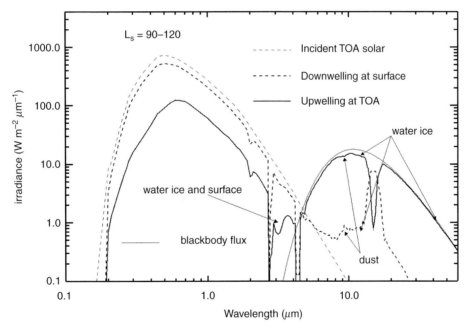

Figure 6.42. Flux distributions representative of 2 p.m. atmospheric conditions during northern summer solstice conditions ($L_s = 105°$) near the equator (0°N, 0°E). A blackbody spectrum for a surface temperature of 268 K is included as a reference. The solar incidence angle is 35°; the dust optical depth (0.67 μm) is 0.20; and the water ice optical depth (12.1 μm) is 0.15.

monograph by Rogers (2000). An advanced topic in inverse theory is that of adjoint radiative transfer theory, which develops a particular form of the RTE that is well suited to the calculation of quantities common employed in inverse theory such as functional derivatives. We refer the interested reader to work by Ustinov (2005, 2007) and Rozanov and Rozanov (2007). For reference, most Martian remote sensing studies have utilized (as we have here) the direct form of the RTE. Below, we provide

a brief description of the RT-based approaches for retrieval of temperature profiles and for various aerosol properties.

6.6.2.1 Retrievals – Temperature Profiles

The practice of retrieving atmospheric temperature as a function of height using thermal IR observations of the CO_2 band centered at 15 μm has been well established by a number of investigations (e.g. Conrath et al., 1973, 2000; Smith et al.,

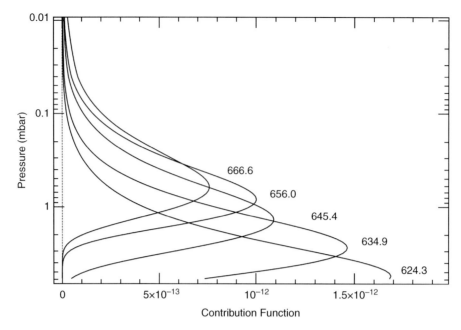

Figure 6.43. Contribution functions for the wavenumbers used in the TES temperature retrievals (e.g. Conrath et al., 2000). These functions were calculated using (6.60) for the nominal TES 10 cm^{-1} resolution. For clarity, we do not include any channels that sample the R-branch.

1996; Grassi et al., 2005; Kleinböhl et al., 2009). In the most basic case, the wavelength dependence of CO_2 absorption is used to sample different vertical levels of the atmosphere. If the vertical variation of temperature is not too great, then the observed radiance will be characteristic of the temperature in a layer centered near the level where CO_2 line-of-sight optical depth reaches unity. Because CO_2 optical depth varies with wavelength, the height in the atmosphere at which unit optical depth is reached also varies with wavelength, and a temperature profile can be retrieved. An alternative approach uses views of the limb at a range of different tangent heights at a single or several wavelengths to sample different atmospheric levels.

In a non-scattering atmosphere approximation, a useful quantity is the weighting function, $W(s,\lambda)$, which describes the local absorption per unit length along the line of sight:

$$W(s,\lambda) = \frac{\partial}{\partial s}\exp(-\tau(s,\lambda)) \qquad (6.59)$$

where s is the distance coordinate along the line of sight, λ is wavelength, and τ is the total optical depth of CO_2 from the observer to a given point, and is a function of both wavelength and location within the atmosphere. When multiplied by the Planck function, it forms the contribution function:

$$C(s,\lambda) = B(T(s),\lambda)W(s,\lambda) \qquad (6.60)$$

which can be considered to be the equivalent of a functional derivative of radiance with respect to temperature (or opacity). From (6.60), we obtain a direct expression for the amount of radiance that each atmospheric level contributes to the observed signal. Ideally, a set of wavelengths (or limb tangent heights) is chosen such that the corresponding contribution functions form a set of narrow, altitude-peaked functions that achieve optimized altitude range and resolution. Figure 6.43 shows the set of contribution functions for MGS/TES observations with nadir geometry

(Conrath et al., 2000). As can be seen directly from the form of the contribution functions, the retrievals are most sensitive to the thermal structure from the surface to the ~0.1 mbar pressure level (~40 km altitude) with a vertical resolution of about one pressure scale height (~10 km). Kleinböhl et al. (2009) give the MRO/MCS contribution functions for limb geometry.

Direct solution of the radiative transfer equation for retrieval of atmospheric temperature is generally an ill-posed problem. The practical effect of this mathematical classification is that many different temperature profiles can be constructed that reproduce the observed radiances to within the uncertainty imposed by instrument error. Therefore, *a priori* constraints are typically employed that explicitly impose a smoothing on the vertical variation of temperature (Conrath, 1972; Conrath et al., 2000), or that implicitly constrain the temperature profile through relaxation from an initial guess (Chahine, 1972; Kleinböhl et al., 2009). It is important to understand that these two approaches can produce differently "looking" profiles from the dataset. For example, the Conrath et al. condition will reduce the amplitude of structure at scales below that of the smoothing function width. Careful consideration of retrieval constraints is necessary when comparing retrieved temperature profiles to those from dynamical models.

Inclusion of scattering by dust and water ice aerosols further complicates the temperature retrieval because contribution functions can no longer be employed without a potentially significant approximation (e.g. single scattering). This may be less problematic for the thermal IR under restricted aerosol loading conditions (Kleinböhl et al., 2011). However, for the general case, nonlinear algorithms must be employed to find best-fit solutions that minimize the difference between observed and computed radiances. A simultaneous or iterative retrieval scheme – such as between atmospheric temperature and aerosol abundance – may also be required, particular for the thermal IR and for strong molecular bands in the NIR (where the opacity can be a strong function of temperature).

6.6.2.2 Retrievals – Aerosols

Retrievals of aerosol properties from observations reflect an even more poorly posed mathematical problem as they are subject to increased uncertainties from radiative transfer input parameters that are not themselves the target of the retrieval. This issue was touched upon in the discussion of aerosol climatology at the end of Section 6.3.3. However, the situation is actually more complicated than finding the appropriate database of optical depths or temperatures profiles, due to issues such as prescribing the surface reflectance, obtaining or calculating a set of self-consistent radiative properties, vertical distribution of aerosols (and their radiative properties), wavelength variation of the various parameters, etc. One is left with the option of simply adopting the best set of input parameters available or leveraging from the synergies of unique observational datasets. The latter choice includes multi-angle observations (Clancy and Lee, 1991; Clancy et al., 2003; Vincendon et al., 2009; Wolff et al., 2009), coordinated observing campaigns (Wolff et al., 2006; Tamppari et al., 2010), and simplified atmospheric conditions to constrain or minimize the number of parameters (such as for high-optical-depth global dust events; Wolff et al., 2009, 2010). The review by Smith (2008) includes an overview of the many previous efforts in the area of aerosol retrievals. Here, we concentrate on the relative precision/accuracy of various RT-based approaches to column-integrated optical depth and macrophysical retrievals.

(a) Column-Integrated Optical Depths

The Pathfinder IMP, Phoenix SSI, and MER Pancam solar extinction measurements represent the most accurate τ dataset available for the Martian atmosphere (Smith and Lemmon, 1999; Lemmon et al., 2004; Tamppari et al., 2010). The Viking VIS observations would certainly not be far behind (Pollack et al., 1977, 1979; Colburn et al., 1989). The strength of these retrievals stems for the simplicity of direct solar imaging: the radiance of the solar disk is so much greater than any atmospheric contribution, with the possible exceptions of low Sun angles and very high dust loading, that the solution to the RTE is the Beer–Lambert–Bouguer law (6.8). As a consequence of no other significant input parameters, the precision of the retrieved values is generally controlled by the radiometric uncertainty of the camera involved.

The emergence phase function (EPF) sequences represent the next level of accuracy for column-integrated optical depths. EPF-based τ retrievals have been carried out for several spacecraft: IRTM (Clancy and Lee, 1991; Clancy et al., 1995), TES (Clancy et al., 2003), OMEGA (Vincendon et al., 2008), and CRISM (Wolff et al., 2009). EPF sequences consist of calibrated reflectance measurements for a single point at multiple angles. This allows simultaneous determinations of the surface albedo and aerosol optical depth by sampling the scattering (or phase) angle over a range of observational geometries for separation of the distinct angular scattering behaviors of the surface and the atmosphere. The aerosol contribution is further distinguished by the change of atmospheric air mass with viewing geometry. Schematically, a down-track region on the surface is targeted and observed repeatedly as the spacecraft approaches and flies by (cf. figure 1 of Clancy and Lee, 1991). An example of a CRISM EPF and a best-fit multiple-scattering radiative transfer model is shown in Figure 6.44, which also displays

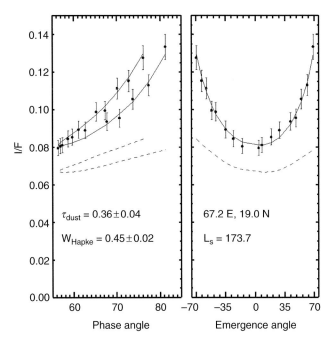

Figure 6.44. Example of a CRISM EPF sequence (points, FRT0000406B) along with the best-fit optical depth retrieval (solid line) and the surface reflectance (dashed line) for 900 nm (Wolff et al., 2009). The *y*-axis is the *I/F* or the observed radiance (*I*) normalized to the TOA solar radiance ($F = F_\odot/\pi$). The two panels are included in order to illustrate variations with respect to the phase angle (left panel, 180°–scattering angle) and the emergence angle (right panel, negative angles are a graphical convention and represent the ingress portion of the EPF). The asymmetry of the surface reflectance compared to that of the total radiance is what allows for the separation of the surface and atmospheric components.

the contribution of the surface reflectance in the absence of an atmosphere (Wolff et al., 2009). The precision of a given EPF retrieval will be dictated by the degree to which the surface can represented by a single surface reflectance function and the atmosphere by a horizontally homogeneous atmosphere.

Other types of multi-angle observations are available, but involve further assumptions or approximations. For example, one can analyze a time series of observations at the same location, where the temporal variation is intended to provide the phase angle variation. The primary complication is that atmospheric state changes between subsequent datasets, i.e. aerosol opacity, temperature structure. This type of sampling, involving the characterization of the atmosphere for each epoch, has been employed with OMEGA data (e.g. Vincendon et al., 2008; Madeleine et al., 2012a). The use of cross-track geometry in a wide-angle imager such as MARCI can also provide the desired phase angle range, but it requires fairly special atmospheric and surface conditions to meet the horizontal homogeneity requirement of the standard EPF algorithm. Using the UV channels of MARCI, Wolff et al. (2010) employed the so-called pseudo-EPF approach during the global dust event of 2007. While the UV surface reflectance is not constant with location, the amplitude of variation is much smaller than in the visible (e.g. figure 1 of Wolff et al., 2010). Water ice clouds exhibit much greater contrast than dust with surface reflectance in the UV such that phase angle coverage is not a prerequisite to map water ice cloud optical depth from nadir UV imaging data (Wolff et al., 2014). Multi-angle sky-imaging observations from the surface of Mars can also mitigate separation with the surface reflectance function, but must then rely heavily upon the constancy of atmospheric properties as a function of azimuth angle (Tomasko et al., 1999; Lemmon et al., 2004).

Observations that include only a single set of photometric angles tend to be most useful for optical depth retrievals when obtained in the thermal IR. For orbital spacecraft, the relatively small contribution of scattering in this wavelength regime allows for optical depth retrievals when the spectral resolution is adequate to resolve an aerosol feature, such as the 9.3 μm silicate dust absorption or the 12 μm water ice band. This spectral leverage directly enables the TES nadir viewing aerosol retrievals, although this requires simultaneous thermal IR measurements of the surface–atmosphere thermal contrast via 15 μm CO_2 line absorption measurements (Smith, 2004, 2008). This behavior can also be approximated with in-band and out-band channels on IR radiometers such as the IRTM, though with poorer precision (Martin, 1986; Martin and Richardson, 1993). For upward-viewing instruments in the IR such as Mini-TES, the reduced importance of the surface contribution allows for retrievals that include multiple scattering (Smith et al., 2006; Wolff et al., 2006). However, such analyses can be compromised by the disparity in the contribution functions between the temperature and aerosol profiles (Wolff et al., 2006).

(b) Macrophysical Retrievals

Access to Martian aerosol properties such as shape, size, and composition (also called "microphysical" properties in the Mars atmospheric studies literature) has been limited by available measurements. Progress has been made primarily by exploiting special atmospheric conditions and sets of synergistic observations that effectively reduce the uncertainty or relative weight of the various radiative transfer input parameters.

Very little can be stated definitively about the particle shape beyond the fact that spheres clearly do not fit available observations, which sample a wide range of geometry. The upward-looking observations of Viking, Pathfinder, and MER offer large variations in all of the photometric angles (i.e. Pollack et al., 1995; Tomasko et al., 1999; Lemmon et al., 2004), providing direct access to the scattering (phase) angle dependence of the aerosol scattering under different illumination and path length conditions. The use of EPFs will sample a more limited range of scattering angles, in general, but are able to include a diversity of other photometric angles via the spatial coverage inherent in an orbital dataset. In the case of orbital wide-angle imaging, the presence of mismatches in radiative transfer-based model–data comparisons that correlate with scattering angle also provides constraints on particle shape, such as has been the experience of the MARCI team for water ice aerosols (Wolff et al., 2011). Undoubtedly, more information is available in existing datasets, but its extraction will require extensive and systematic retrieval efforts that employ multiple-scattering RT algorithms.

Retrieval of composition or refractive indices for dust aerosols has met with more success through special atmospheric circumstances and the combination of observational datasets. Large-scale dust storms on Mars offer a unique opportunity to minimize the relative weight of the surface contribution while maximizing that of the atmosphere. In the UV and optical, MRO observations near the MER sites during the 2007 global dust event provided the dusty atmosphere, as well as an independent, robust measurement of the column-integrated optical depth by Pancam (Wolff et al., 2009, 2010). With dust shape and size prescribed through the use of previous studies (e.g. Tomasko et al., 1999; Clancy et al., 2003; Wolff and Clancy, 2003), the radiative transfer modeling is used to constrain the imaginary refractive index through its radiative transfer proxy, the single-scattering albedo. However, MER- and MRO-based dust absorption analyses reveal inconsistencies with Viking and Pathfinder results (Pollack et al., 1995; Ockert-Bell et al., 1997; Tomasko et al., 1999), which suggests the presence of systematic issues in the analyses of the earlier datasets (cf. Wolff et al., 2009). Additional evidence may be taken from the ability of the MRO and MER retrievals to provide improved agreement between observed and simulated temperature profiles (Madeleine et al., 2011).

A similar use of orbiter "overflights" has also been used to improve thermal IR refractive indices from those derived from Mariner 9 observations of a global dust event (Snook, 1999), employing simultaneous observations by TES (solar band and IR), Mini-TES, and Pancam (Wolff et al., 2006). As a result, the particle size was derived concurrently and the condition of high dust optical depth is not required The most recent set of visible/NIR/thermal IR dust optical constants, based upon these MGS/MER/MRO retrievals, is presented in Figures 6.18 and 6.19.

Particle size retrievals have largely relied on the spectral coverage of observations from the visible to the thermal IR. Thermal IR particle size sensitivity was first applied for dust with the Toon et al. (1977) analysis of Mariner 9 IRIS data, and subsequently extended by Wolff and Clancy (2003), Wolff et al. (2006), and Elteto and Toon (2010a,b) for (Mini-) TES data. Most of the leverage within the thermal IR regime is provided by a comparison of the 9 μm and 20 μm band opacity values, as the gain by using a longer wavelength band is minimal. This can be gleaned through an examination of the extinction cross-section in Figure 6.16, namely the decreasing dispersion between the curves with increasing wavelength; the trends would be even more apparent in plots of the absorption cross-section.

The situation for water ice in the thermal IR is more limited. TES water ice observations are primarily restricted to the 12 μm band (Christensen et al., 2001) such that particle size sensitivity depends upon the shape of the 12 μm band, which itself is degraded by the presence of the 15 μm CO_2 band absorption. Particle size discrimination by TES spectra is generally restricted to differentiating between seasonal ($r_{eff} = 1$–2 μm) and aphelion ($r_{eff} = 3$–5 μm) ice clouds (Wolff and Clancy, 2003). Ongoing work with MCS 20 μm and 42 μm channels indicates the potential for additional spectral leverage towards determination of water ice particle sizes (J.L. Benson, private communication). In addition, the use of the weaker NIR bands, as demonstrated for OMEGA by Madeleine et al. (2012a; see also aerosol database description in Section 6.3.3), supports ice particle size determinations, although the conditions necessary for the retrieval represent only a small fraction of the available data.

Due to the typical range of cloud and dust particle sizes present in the global Mars atmosphere, a source of particle size information comes from combining optical and thermal IR

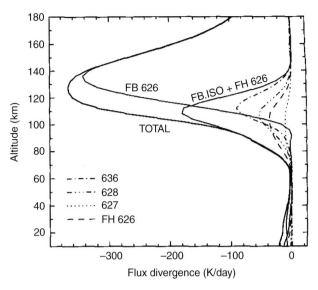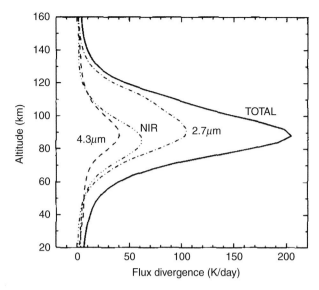

Figure 6.45. (Left) NLTE cooling rates by CO_2 15 μm bands computed for the thermal profile of Kaplan (1988) and the composition model of Rodrigo et al. (1990). Individual contributions from the fundamental bands (FB) of the four most abundant isotopes (626, 636, 628, 627) and the first hot band of 626 are indicated. (Right) Profile of solar heating rates by CO_2 with contributions by all bands in the 4.3 μm and 2.7 μm regions and in the 1–3 μm (NIR). After López-Puertas and López-Valverde (1995), with permission.

observations to exploit the previously discussed visible-to-IR ratio (cf. Section 6.3.3). Even a casual examination of Figures 6.22 and 6.28 reveals a robust size discrimination for dust and ice aerosols, respectively, most appropriate to the 0.5–5 μm particle size regime. This capability is also reasonably insensitive to moderate (and observationally uncorrelated) opacity errors. Originally applied to Viking Lander imaging and IRTM optical depth measurements in a somewhat qualitative fashion (Zurek, 1982; Martin, 1986; Clancy and Lee, 1991; Clancy et al., 1995), the systematic nature of the TES EPF measurements and the leverage of simultaneous IR spectral observations provide for a more quantitative framework to assess the sizes of both dust and water ice aerosols (Clancy et al., 2003). Although the full extent of the TES database has yet to be fully explored, the applicability of the solar band–IR diagnostic capability to limb measurements has also been demonstrated (Clancy et al., 2007, 2010). The power of this approach is also available for use with the growing MCS dataset, particularly given the recent calibration of the solar-band channel by Bandfield et al. (2013).

The ability to derive important constraints on aerosol properties in the Martian atmosphere has been developed and demonstrated in the literature. Extensive datasets exist and continue to accrue that are directly amenable to such radiative-transfer-based analyses. Now, it is basically a question of undertaking the systematic exploration of these accumulating observations by the interested researcher.

6.6.3 Non-Local Thermodynamic Equilibrium

The use of non-LTE can be quite important in several areas, three of which we highlight here. Many emissions in the atmospheres of the terrestrial planets originate under NLTE conditions, as developed in Section 6.4.2. Here we consider the retrieval of atmosphere properties from observations of non-LTE emissions. Specifically, strong NLTE effects, such as

solar fluorescence in the 4.3 μm CO_2 bands observed by PFS and OMEGA, may serve to probe altitudes inaccessible to the standard 15 μm band approach (Section 6.6.2). Atmospheric cooling and heating rates associated with NLTE can also be a fundamental part of the energy budget in the upper atmosphere, as is the case for Mars. An accurate implementation of NLTE radiative heating rates is essential for the computation of realistic thermal structure and dynamical effects at high altitudes. To illustrate the breadth and impact of these applications, we discuss below a few recent studies of the Martian mesosphere and thermosphere.

6.6.3.1 NLTE Parameterizations for GCMs

The radiative balance is usually split into two terms, the thermal cooling and the solar heating. With respect to the former, NLTE models demonstrate the importance of strong 15 μm band (of CO_2) emissions as a key atmospheric energy sink between 70 and 140 km during the daytime (Bougher et al., 1994; González-Galindo et al., 2005). This behavior extends to higher altitudes at night. As for the heating by solar absorption in the various near-IR bands of CO_2, this term dominates the energy balance between 80 and 120 km. Consequently, both terms are essential for thermal and dynamical modeling studies that extend to mesospheric and thermospheric altitudes for Mars. In addition, both processes require a full non-LTE model for accurate calculations; the assumption of LTE emission will introduce large errors at these altitudes.

Figure 6.45 presents 15 μm cooling and the NIR solar heating rates in the upper atmosphere, as computed with a comprehensive NLTE model. The 15 μm cooling profiles include contributions from the four most abundant isotopes of CO_2. The fundamental band of the main isotope, because of its "optically thick" column, can emit effectively only at thermospheric altitudes. As a result, the less opaque isotopic bands represent important cooling agents for the upper mesosphere (90–120 km). As discussed by López-Puertas and López-Valverde (1995), the energy emitted by the minor isotopes is taken from the thermal pool primarily by V–T exchanges of the main isotope, and subsequently transferred by V–V exchanges to these isotopic states. Similarly, the solar heating manifests in two very strong band systems, around 2.7 and 4.3 μm. These are comparable to the weaker overtones and combination bands, labeled as NIR in Figure 6.45. This behavior reveals the competition between radiative transfer and collisional deactivation. On the one hand, at thermospheric altitudes, most of the solar radiation absorbed by the strong bands at 2.7 and 4.3 μm is emitted back to space by those bands, as well as by other optically thin hot and combination bands. Only at mesospheric altitudes (below about 120 km) will molecular collisions successfully compete with radiative relaxation to produce actual heating of the atmosphere. The solar energy ultimately thermalized is only a fraction of the initial absorption, i.e. much smaller than what one would expect from LTE. On the other hand, in the optically thinner spectral regions of the weaker NIR bands, the solar flux can penetrate easily to mesospheric altitudes, where V–T collisions convert this energy very efficiently to heat. This leads to their important contribution to the total solar heating for the 70–100 km range.

A full NLTE model is not a viable option, in terms of computational efficiency, for most dynamical atmospheric models. Therefore, fast parameterizations are required, though they have not been readily available until recently. The lack of such fast methods was a primary limitation in the extension of Martian atmospheric models to high altitudes.

Since the late 1980s, the fast model has been the NCAR Martian Thermospheric GCM (Bougher and Dickinson, 1988), developed originally for Venus (Dickinson, 1972, 1976). More recently, two parameterizations based on the NLTE model represented in Figure 6.45 were proposed: one for the 15 μm cooling (López-Valverde and López-Puertas, 2001; López-Valverde et al., 2008) and another for solar heating (López-Valverde et al., 1998, 2011d). They differ significantly from the NCAR model, not only in updated spectroscopy and collisional rates, but also in the core algorithms. Specifically, the newer models employ a reduction of the full NLTE model to a simplified fast scheme involving only a few CO_2 states and ro-vibrational bands; i.e. they mimic a NLTE mini-model. They were incorporated into the LMD Mars GCM in order to extend this model up to the thermosphere (Forget et al., 1999, 2009). Subsequent updates have been enabled through model–data comparisons involving the SPICAM stellar occultation dataset (López-Valverde and González-Galindo, 2008). This newer class of model may be found in current Martian GCMs and presents two noteworthy advantages: (1) the desired level of approximation can be selected, allowing the user to balance accuracy and computational resources requirements; and (2) the algorithms provide real-time incorporation of the atomic oxygen abundance provided by the GCM for each time step and spatial grid point.

6.6.3.2 Mesospheric Forcing of Thermal Tides

Mesospheric NLTE solar heating provides a contribution to the Martian diurnal thermal tide, and is partly responsible for the distinct temperatures measured by Viking and Pathfinder

in the upper mesosphere (López-Valverde et al., 2000). Using a GCM that includes NLTE heating, González-Galindo et al. (2009a) found a vertically propagating tide with an amplitude of about 20 K, similar to that observed by Viking, SPICAM, and MCS (Seiff and Kirk, 1977; Forget et al., 2009; McCleese et al., 2010). An additional effect of solar heating is an *in situ* forcing that may be relevant to the so-called polar thermospheric warming, first observed by Mars Odyssey (Bougher et al., 2006). According to González-Galindo et al. (2009b), the thermal tides forced by the solar heating appear to be key components of a strong inter-hemispheric transport, which produces the transport from the thermosphere into the polar mesosphere. This transport occurs between 90 and 120 km through strong meridional winds. The peak solar heating manifests at around 110 km, with a maximum tidal amplitude near 120 km. Furthermore, an apparent day-to-night circulation is also added to the summer-to-winter transport (González-Galindo et al., 2009b). In the absence of the daily cycle of heating, the summer-to-winter transport and the thermospheric polar warming are reduced.

6.6.3.3 NLTE Radiative Damping of Gravity Waves

A range of recent observations has renewed interest in an old problem: the potential impact on the upper atmosphere by gravity wave (GW) propagation from the lower atmosphere. The presence of GWs in the Mars atmosphere was recognized before Viking (Briggs and Leovy, 1974). Evidence from radio occultation and accelerometer data indicates that they are prominent at mesospheric altitudes with typical spatial scales of 10 and 200 km in the vertical and the horizontal, respectively (Hinson et al., 1999; Magalhães et al., 1999; Fritts et al., 2006). Clancy and Sandor (1998) speculated that CO_2 ice clouds could form in the mesospheric cold pockets associated with such wave propagation. Recently, MEX and MGS observations have revealed the presence of CO_2 ice clouds in this region (Clancy et al., 2007; Montmessin et al., 2007; González-Galindo et al., 2011). Such cold temperatures were previously difficult to reproduce with GCMs, but recent simulations with a mesoscale model that includes the latest NLTE parameterization of the 15 μm cooling and spatial–temporal scales relevant to GW propagation seem to provide the wave amplitudes required to reduce mesospheric temperatures below the CO_2 condensation value (Spiga et al., 2012). This opens the possibility of developing parameterization schemes for GW propagation and dissipation. An example of this approach may be found in Eckermann et al. (2011). Their model includes a high degree of approximation in its NLTE treatment (e.g. an ultrafast V–V exchange between CO_2 isotopes that likely overestimates the damping rates obtained), but has the virtue of paving the way for the efficient treatment of GW drag in Mars GCMs. A self-consistent parameterization of both NLTE and GW processes remains on the "to-do" list for present MGCMs.

6.6.3.4 Mesospheric Winds From the 10 μm CO_2 Bands

An interesting application of the Martian NLTE CO_2 emission at 10 μm is the derivation of atmospheric wind fields, particularly with the use of high-spectral-resolution ground-based

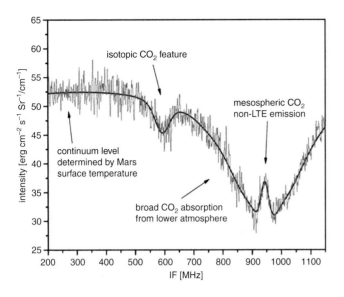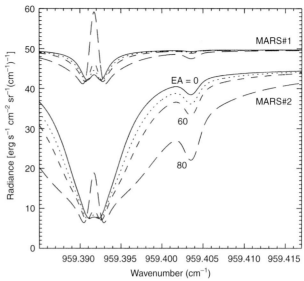

Figure 6.46. Martian atmospheric 10 μm emission observed with ground-based heterodyne spectroscopy (left panel) compared to simulations from an NLTE model (right panel) for emergence angles 0°, 45°, 60°, and 80°, where 0° indicates nadir sounding. Figures from Sonnabend et al. (2006) and López-Valverde et al. (2011b), used with permission.

observations. Figure 6.46 shows one of the best examples of Martian NLTE emission at high spectral resolution. In contrast to Figure 6.37, which shows a spectrum at lower resolution, ground-based heterodyne techniques permit the discrimination of lines and the separation of absorption features in the wings from the line core NLTE emission; as is predicted by NLTE models run at similar resolution (e.g. López-Valverde et al., 2011b). Figure 6.46 clearly indicates that the NLTE emission from the limb is much larger than in nadir geometry, although it remains readily detectable for the nadir. The limb measurements have sufficient spectral resolution to derive mesospheric winds from the observed Doppler shift of the line centers for Mars (Sonnabend et al., 2006), as well as for Venus (Sornig et al., 2008). However, a confounding factor exists as a result of the mixing of the limb- and disk-viewing geometries in the typically large spatial footprint (on the planet) of the instrumental field of view (IFOV). Consequently, the total emission represents a complex combination of Doppler-shift effects and radiative transfer weighting functions. Fortunately, the problem is tractable, but it requires a careful treatment of the NLTE contribution and the convolution of the model with the IFOV function. In a recent realization of this approach López-Valverde et al. (2011c) found that the peak emission of the strongest lines of the 10 μm band originates at ~70 km in the Martian atmosphere, with approximately equal contributions from limb and nadir for typical IFOVs.

6.7 CONCLUSION

Understanding the propagation of electromagnetic radiation through the Martian atmosphere is a fundamental requirement for studies of both the Martian atmosphere and surface. In essence, the impact of radiative transfer theory and the contribution of the processes encountered in the Martian environment cannot be avoided for studies which range from the current and future atmospheric states to the geological interpretation of light reflected and emitted by the surface. This pervasive nature has strongly influenced our emphasis on the more practical details of the radiative transfer equation and the range of parameters that characterize the processes inherent the terms of the equation.

Some have observed that the material above is both too long for a person interested in a specific aspect of the Martian atmosphere and too limited in scope to be of use to students attempting to learn radiative transfer theory. It is also true that there is an uneven treatment in some of the material presented, i.e. amount of space dedicated to aerosol processes and properties as compared to that for molecular opacity. This particular choice was motivated by our experience that the latter is very well described in textbooks and terrestrial atmospheric studies, while the former suffers from a more piecemeal treatment in the literature. Ultimately, as we stated in the introduction, our goal is to provide a collection of material that highlights the important processes and techniques, as well as offering starting points for more comprehensive explorations.

The approach taken in this chapter necessarily underlines past and current work. It might be natural to ask at this point "what's next?". Prognostication remains a difficult business even in the age of easily accessible high-performance computing. Despite this fact, we feel confident in identifying two trends as pushing the boundaries of the state of the art for future Martian radiative transfer studies:

- Limb studies/retrievals. The MGS, MEX, and MRO spacecraft have compiled an extensive dataset. Yet the extant analyses have only begun to explore the information contained within these observations, particularly when one considers that the Mars atmosphere is three-dimensional while many applied techniques are inherently one-dimensional. Studies that improve the efficiency of non-plane-parallel techniques as well as the characterization (and perhaps parameterization) of three-dimensional effects represent areas in need of additional work.

- Fast algorithms for data assimilation studies. Global climate models require efficient radiative transfer calculations for the detailed accounting of radiation (flux) budget. The problem can be considered orders of magnitude more computationally intensive when one includes data assimilation schemes and the forward models for radiance. As a result, the development and validation of very efficient algorithms that include a sufficient level of physical detail (i.e. multiple scattering) offer further directions for research efforts.

ACKNOWLEDGMENTS

We thank Todd Clancy and David Crisp for their thorough and insightful reviews.

REFERENCES

Acuna, M. H., Connerney, J. E., Wasilewski, P., et al. (1998) Magnetic field and plasma observations at Mars: initial results of the Mars Global Surveyor mission, *Science*, 279, 1676.

Acuna, M. H., Connerney, J. E., Ness, N. F., et al. (1999) Global distribution of crustal magnetization discovered by the Mars Global Surveyor MAG/ER experiment, *Science*, 284, 790.

Ajello, J. M. and Hord, C. W. (1973) Mariner 9 ultraviolet spectrometer experiment: morning terminator observations of Mars, *Journal of Atmospheric Sciences*, 30, 1495–1501.

Ajello, J.M., Hord, C. W., Barth, C. A., et al. (1973) Mariner 9 ultraviolet spectrometer experiment: afternoon terminator observations of Mars, *Journal of Geophysical Research*, 78, 4279–4290.

Akabane, T., Iwasaki, K., Saito, Y., and Narumi, Y. (1990) Blue clearing of Syrtis Major at the 1982 opposition, *Journal of Geophysical Research*, 95, 14649–14655.

Alms, G. R., Burnham, A. K., and Flygare, W. H. (1975) Measurement of the dispersion in polarizability anisotropies, *Journal of Chemical Physics*, 63, 3321–3326.

Asano, S., and Yamamoto, G. (1975) Light scattering by a spheroidal particle, *Applied Optics* 14, 29–49.

Ayash, T., Gong, S., and Jia, C. Q. (2008) Implementing the delta-four-stream approximation for solar radiation computations in an atmosphere general circulation model, *Journal of Atmospheric Sciences*, 65, 2448.

Bailey, M. P., and Hallett, J. (2009) A comprehensive habit diagram for atmospheric ice crystals: confirmation from the laboratory, AIRS II, and other field studies, *J. Atmos. Sci.*, 66, 2888–2899.

Bandfield, J. L., Wolff, M. J., Smith, M. D., et al. (2013) Radiometric comparison of Mars Climate Sounder and Thermal Emission spectrometer measurements, *Icarus*, 225, 28–39.

Baratta, G. A. and Palumbo, M. E. (1998) Infrared optical constants of CO and CO_2 thin icy films, *J. Opt. Soc. Am. A*, 15, 3076–3085.

Barth, C. A., Hord, C. W., Stewart, A. I., et al. (1973) Mariner 9 ultraviolet spectrometer experiment: Seasonal variation of ozone on Mars, *Science*, 179, 795–796.

Basu, S., Richardson, M. I., and Wilson, R. J. (2004) Simulation of the Martian dust cycle with the GFDL Mars GCM, *Journal of Geophysical Research (Planets)*, 109, 11006.

Basu, S., Wilson, R. J., Richardson, M. I., and Ingersoll, A. (2006) Simulation of spontaneous and variable global dust storms with the GFDL Mars GCM, *Journal of Geophysical Research (Planets)*, 111, 09004.

Bell, J. F., Wolff, M. J., Malin, M. C., et al. (2009) Mars Reconnaissance Orbiter Mars Color Imager (MARCI): instrument description, calibration, and performance, *Journal of Geophysical Research (Planets)*, 114, E08S92.

Benson, J. L., Bonev, B. P., James, P. B., et al. (2003) The seasonal behavior of water ice clouds in the Tharsis and Valles Marineris regions of Mars: Mars Orbiter Camera observations, *Icarus*, 16, 34–52.

Benson, J. L., Kass, D. M., Kleinböhl, A., et al. (2010) Mars' south polar hood as observed by the Mars Climate Sounder, *Journal of Geophysical Research (Planets)*, 115, 12015.

Berk, A., Bernstein, L. S., Anderson, G. P., et al. (1998) MODTRAN cloud and multiple scattering upgrades with application to AVIRIS, *Remote Sens. Environ.*, 65, 367–375.

Bertaux, J.-L., Korablev, O., Perrier, S., et al. (2006) SPICAM on Mars Express: observing modes and overview of UV spectrometer data and scientific results, *Journal of Geophysical Research (Planets)* 111, E10S90.

Bertucci, C. (2003) Magnetic field draping enhancement at the Martian magnetic pileup boundary from Mars global surveyor observations, *Geophys. Res. Lett.*, 30. 10.1029/2002gl015713.

Betz, A. L., McLaren, R. A.,Johnson, M. A. and Sutton, E. C. (1977) Infrared heterodyne spectroscopy of CO_2 in the atmosphere of Mars, *Icarus*, 30, 650–662.

Bideau-Mehu A., Guern, Y., Abjean, R., and Johannin-Gilles, A. (1973) Interferometric determination of the refractive index of carbon dioxide in the ultraviolet region, *Optics Communications*, 9, 432–434.

Blamont, J. E., Chassefiere, E., Goutail, J. P., et al. (1991) Vertical profiles of dust and ozone in the Martian atmosphere deduced from solar occultation measurements, *Planetary and Space Science*, 39, 175–187.

Bohren C. F. and Huffman, D. R. (1983) *Absorption and Scattering of Light by Small Particles*, John Wiley, USA.

Bougher, S. W. and Dickinson, R. E. (1988) Mars mesosphere and thermosphere. I – Global mean heat budget and thermal structure, *Journal of Geophysical Research*, 93, 7325–7337.

Bougher, S. W., Dickinson, R. E., Roble, R. G. and Ridley, E. C. (1988) Mars thermospheric general circulation model – calculations for the arrival of PHOBOS at Mars, *Geophysical Research Letters*, 15, 1511–1514.

Bougher, S. W., Hunten. D. M., and Roble, R. G., (1994) CO_2 cooling in terrestrial planet thermospheres, *Journal of Geophysical Research*, 99, 14609.

Bougher, S. W., Bell, J. M., Murphy, J. R., et al. (2006) Polar warming in the Mars thermosphere: seasonal variations owing to changing insolation and dust distributions, *Geophysical Research Letters*, 33, 02203.

Brault, J. and Testerman, L. (1972) *Preliminary Edition of Kitt Peak Solar Atlas,* unpublished.

Briggs, G. A., and Leovy, C. B. (1974) Mariner 9 observations of the Mars north polar hood, *Bull. Am. Meteorol. Soc.*, 55, 278–278.

Burch, D. E., Gryvnak, D. A., Patty, R. R., and Bartky, C. E. (1969)Absorption of infrared radiant energy by CO_2 and H_2O. IV. Shapes of collision-broadened CO_2 lines, *J. Opt. Soc. Am.*, 59, 267–280.

Cantor, B. A. (2007) MOC observations of the 2001 Mars planet-encircling dust storm, *Icarus*, 186, 60–96.

Chahine, M. T. (1972) A general relaxation method from inverse solution of the full radiative transfer equation. *J. Atmos. Sci.*, 29, 741–747.

Chamberlain, J. W., and Hunten, D. M. (1987) *Theory of Planetary Atmospheres*, Academic Press, San Diego, USA.

Chandrasekhar, S. (1950) *Radiative Transfer*, Oxford University Press.

Chassefière, E., Blamont, J. E., Krasnopolsky, V. A., et al. (1992) Vertical structure and size distributions of Martian aerosols from solar occultation measurements, *Icarus*, 97, 46–69.

Chassefière, E., Drossart, P., and Korablev, O. (1995) Post-Phobos model for the altitude and size distribution of dust in the low Martian atmosphere, *Journal of Geophysical Research*, 100, 5525–5539.

Christensen, P. R., Bandfield, J. L., Hamilton, V. E., et al. (2001) Mars Global Surveyor Thermal Emission Spectrometer experiment: investigation description and surface science results, *Journal of Geophysical Research*, 106, 23823–23872.

Clancy, R. T. and Lee, S. W. (1991) A new look at dust and clouds in the Mars atmosphere – analysis of emission-phase-function sequences from global Viking IRTM observations, *Icarus*, 93, 135–158.

Clancy, R. T. and Sandor, B. J. (1998) CO_2 ice clouds in the upper atmosphere, *Geophysical Research Letters*, 25, 489–492.

Clancy, R. T., Lee, S. W., Gladstone, G. R., et al. (1995) A new model for Mars atmospheric dust based upon analysis of ultraviolet through infrared observations from Mariner 9, Viking, and PHOBOS, *Journal of Geophysical Research*, 100, 5251–5263.

Clancy, R. T., Wolff, M. J., James, P. B., et al. (1996a), Mars ozone measurements near the 1995 aphelion: Hubble space telescope ultraviolet spectroscopy with the faint object spectrograph, *Journal of Geophysical Research*, 101, 12777–12784.

Clancy, R. T., Grossman, A. W., Wolff, M. J. et al. (1996b) Water vapor saturation at low altitudes around Mars aphelion: a key to Mars climate? *Icarus*, 122, 36–62.

Clancy, R. T., Wolff, M. J., and James, P. B. (1999) Minimal aerosol loading and global increases in atmospheric ozone during the 1996–1997 Martian northern spring season, *Icarus* 138, 49–63.

Clancy, R. T., Sandor, B. J., Wolff, M. J., et al. (2000) An intercomparison of ground-based millimeter, MGS TES, and Viking atmospheric temperature measurements: seasonal and interannual variability of temperatures and dust loading in the global Mars atmosphere, *Journal of Geophysical Research*, 105, 9553–9572.

Clancy, R. T., Wolff, M. J., and Christensen, P. R. (2003) Mars aerosol studies with the MGS TES emission phase function observations: optical depths, particle sizes, and ice cloud types versus latitude and solar longitude, *Journal of Geophysical Research (Planets)*, 108, 5098.

Clancy, R. T., Wolff, M. J., Whitney, B. A., Cantor, B. A. and Smith, M. D. (2007) Mars equatorial mesospheric clouds: global occurrence and physical properties from Mars Global Surveyor Thermal Emission Spectrometer and Mars Orbiter Camera limb observations, *Journal of Geophysical Research (Planets)*, 112, 04004.

Clancy, R. T., Wolff, M. J., Cantor, B. A., Malin, M. C., and Michaels, T. I. (2009) Valles Marineris cloud trails, *Journal of Geophysical Research (Planets)*, 114, 11002.

Clancy, R. T., Wolff, M. J., Whitney, B. A., et al. (2010) Extension of atmospheric dust loading to high altitudes during the 2001 Mars dust storm: MGS TES limb observations, *Icarus*, 207, 98–109.

Clough, S. A., Shephard, M. W., Mlawer, E. J., et al. (2005) Atmospheric radiative transfer modeling: a summary of the AER codes, *Journal of Quantative Spectroscopy and Radiative Transfer*, 91, 233–244.

Cloutis, E. A., McCormack, K. A., Bell, J. F., et al. (2008) Ultraviolet spectral reflectance properties of common planetary minerals, *Icarus* 197, 321–347.

Code, A. D. (1967) Radiative transfer in a spherical Compton scattering atmosphere, *Astrophysical Journal*, 149, 253–263.

Colaprete, A. and Toon, O. B. (2000) The radiative effects of Martian water ice clouds on the local atmospheric temperature profile, *Icarus*, 145, 524–532.

Colaprete, A. and Toon. O. B. (2003) Carbon dioxide clouds in an early dense Martian atmosphere, *Journal of Geophysical Research (Planets)*, 108, 5025.

Colaprete, A., Toon, O. B., and Magalhães, J. A. (1999) Cloud formation under Mars Pathfinder conditions, *Journal of Geophysical Research*, 104, 9043–9054.

Colaprete, A., Barnes, J. R., Haberle, R. M., and Montmessin, F. (2008) CO_2 clouds, CAPE and convection on Mars: observations and general circulation modeling, *Planetary and Space Science*, 56, 150–180.

Colburn, D. S., Pollack, J. B., and Haberle, R. M. (1989) Diurnal variations in optical depth at Mars, *Icarus*, 79, 159–189.

Colina, L., Bohlin, R. C., and Castelli, F. (1996) The 0.12–2.5 micron absolute flux distribution of the Sun for comparison with solar analog stars, *Astronomical Journal*, 112, 307.

Conrath, B. J. (1972) Vertical resolution of temperature profiles obtained from remote radiation measurements, *J. Atmos. Sci.*, 29, 1262–1271.

Conrath, B. J. (1975) Thermal structure of the Martian atmosphere during the dissipation of the dust storm of 1971, *Icarus*, 24, 36–46.

Conrath, B., Curran, R., Hanel, R., et al. (1973) Atmospheric and surface properties of Mars obtained by infrared spectroscopy on Mariner 9, *Journal of Geophysical Research*, 78, 4267–4278.

Conrath B. J., Pearl, J. C., Smith, M. D., Maguire, W. C., Christensen, P. R., et al. (2000) Mars Global Surveyor Thermal Emission Spectrometer (TES) observations: atmospheric temperatures during aerobraking and science phasing, *Journal of Geophysical Research*, 105, 9509–9520.

Córdoba-Jabonero, C., Lara, L. M., Mancho, A. M., Márquez, A., and Rodrigo, R. (2003) Solar ultraviolet transfer in the Martian atmosphere: biological and geological implications, *Planetary and Space Science*, 51, 399–410.

Coulson, K. L. (1959a) Characteristics of the radiation emerging from the top of a Rayleigh atmosphere-I, *Planetary and Space Science*, 1, 265.

Coulson, K. L. (1959b) Characteristics of the radiation emerging from the top of a Rayleigh atmosphere-II Total upward flux and albedo, *Planetary and Space Science*, 1, 277–284.

Coulson, K. L. and Fraser, R. S. (1975) Radiation in the atmosphere, *Reviews of Geophysics and Space Physics*, 13, 732–737.

Coulson, K. L., Dave, J. V., and Sekera, Z. (1960) *Tables Related to Radiation Emerging from a Planetary Atmosphere with Rayleigh Scattering*, University of California Press.

Crisp, D. (1990) Infrared radiative transfer in the dust-free Martian atmosphere, *Journal of Geophysical Research*, 95, 14577–14588.

Crisp, D., Pathare, A., and Ewell, R. C. (2004) The performance of gallium arsenide/germanium solar cells at the Martian surface, *Acta Astronautica*, 54(2), 83–101.

Curran, R. J., Conrath, B. J., Hanel, R. A., Kunde, V. G., and Pearl, J. C. (1973) Mars: Mariner 9 spectroscopic evidence for H_2O ice clouds, *Science*, 182, 381–383.

Curtis, A. R. (1952) Contribution to a discussion of "A statistical model for water vapour absoprtion" by R. M. Goody, *Quart. J. Roy. Meteo Soc.*, 78, 638.

Cuzzi J. N., Ackerman, T. P., and Helmle, L. C. (1982) The delta-four-stream approximation for radiative flux transfer, *Journal of Atmospheric Sciences*, 39, 917–925.

Dahlback, A. and Stamnes, K. (1991) A new spherical model for computing the radiation field available for photolysis and heating at twilight, *Planetary and Space Science*, 39, 671–683.

Deirmendjian, D. (1964) Scattering and polarization properties of water clouds and hazes in the visible and infrared, *Applied Optics*, 3, 187.

Deland, M. T. and Cebula, R. P. (2008) Creation of a composite solar ultraviolet irradiance data set, *Journal of Geophysical Research (Space Physics)*, 113, 11103.

Deming, D. and Mumma, M. J. (1983) Modeling of the 10-micron natural laser emission from the mesospheres of Mars and Venus, *Icarus*, 55, 356–368.

Deming, D., Espenak, F., Jennings, D., et al. (1983) Observations of the 10-micron natural laser emission from the mesospheres of Mars and Venus, *Icarus*, 55, 347–355.

Dickinson, R. E. (1972) Infrared radiative heating and cooling in the Venusian mesosphere. I. Global mean radiative equilibrium, *J. Atmos. Sci.*, 29, 1531–1556.

Dickinson, R. E. (1976) Infrared radiative emission in the venusian mesosphere, *J. Atmos. Sci.*, 33, 290–303.

Dickinson, R. E. and Bougher, S. W. (1986) Venus mesosphere and thermosphere 1. Heat budget and thermal structure, *Journal of Geophysical Research*, 91, 70–80.

Dlugach, Z. M., Mishchenko, M. I., and Morozhenko, A. V. (2002) The effect of the shape of dust aerosol particles in the Martian atmosphere on the particle parameters, *Solar System Research*, 36, 367–373.

Eckermann, S. D., Ma, J., and Zhu, X. (2011) Scale-dependent infrared radiative damping rates on Mars and their role in the deposition of gravity-wave momentum flux, *Icarus*. 211, 429–442.

Eddington, A. S. (1916) On the radiative equilibrium of the stars, *Monthy Notices of the Royal Astronomical Society*, 77, 16–35.

Edwards, J. M., and Slingo, A. (1996) Studies with a flexible new radiation code. I: Choosing a configuration for a large-scale model, *Quart. Journal Royal Meteorol. Soc.*, 122, 689–719.

Egan, W. G., Fischbein, W. L., Smith, L. L., and Hilgeman, T. (1980) High-resolution Martian atmosphere modeling, *Icarus*, 41, 166–174.

Elsasser, W. M. (1938) Mean absorption and equivalent absorption coefficient of a band spectrum, *Phys. Rev.*, 54, 126–129.

Elsasser, W. M. (1942) *Heat Transfer by Infrared Radiation in the Atmosphere*, Cambridge: Harvard University Press.

Elteto, A. and Toon, O. B. (2010a) Retrieval algorithm for atmospheric dust properties from Mars Global Surveyor Thermal Emission Spectrometer data during global dust storm 2001A, *Icarus*, 210, 566–588.

Elteto, A. and Toon, O. B. (2010b) The effects and characteristics of atmospheric dust during Martian global dust storm 2001A, *Icarus*, 210, 589–611.

Eluszkiewicz, J., Moncet, J.-L., Shephard, M. W., et al (2008) Atmospheric and surface retrievals in the Mars polar regions from the Thermal Emission Spectrometer measurements, *Journal of Geophysical Research (Planets)*, 113, 10010.

Emmanuel, C. B. (1968) Radiative equilibrium temperature distribution of the atmosphere of Mars, *Journal of Geophysical Research*, 73, 10, 2156–2202.

Encrenaz, T., Fouchet T., Melchiorri, R., et al. (2006) Seasonal variations of the Martian CO over Hellas as observed by OMEGA/Mars Express, *Astronomy and Astrophysics*, 459, 265–270.

Encrenaz, T., Fouchet T., Melchiorri, R., et al. (2008) A study of the Martian water vapor over Hellas using OMEGA and PFS aboard Mars Express, *Astronomy and Astrophysics*, 484, 547–553.

Eriksson, P., Buehler, S. A., Davis, C. P., Emde, C., and Lemke, O. (2011) ARTS, the Atmospheric Radiative Transfer Simulator, version 2, *Journal of Quantitative Spectroscopy and Radiative Transfer*, 112, 1551–1558.

Ermolli, I., Matthes, K., Dudok de Wit, T., et al. (2013) Recent variability of the solar spectral irradiance and its impact on climate modeling, *Atmospheric Chemistry and Physics*, 13, 3945–3977.

Evans, K. F. (1998) The spherical harmonics discrete ordinate method for three-dimensional atmospherica radiative transfer, *Journal of Atmospheric Science*, 30, 169–179.

Evans, K. and Marshak, A. (2005) Numerical methods, in *3D Radiative Tranfer in Cloudy Atmospheres*, eds. A. Marshak and A. B. Davis, Heidelberg: Springer.

Fedorova, A. A., Lellouch, E., Titov, D. V., de Graauw, T., and Feuchtgruber, H. (2002) Remote sounding of the Martian dust from ISO spectroscopy in the 2.7 μm CO_2 bands, *Planetary and Space Science*, 50, 3–9.

Fedorova, A. A., Rodin, A. V., and Baklanova I. V. (2004) Seasonal cycle of water vapor in the atmosphere of mars as revealed from the MAWD/Viking 1 and 2 experiment, *Solar System Research*, 38, 421–433.

Fedorova, A., Korablev, O., Bertaux, J.-L., et al. (2006) Mars water vapor abundance from SPICAM IR spectrometer: seasonal and geographic distributions, *Journal of Geophysical Research (Planets)*, 111, E09S08.

Fedorova, A., Korablev, O. I., Bertaux, J.-L., et al. (2009) Solar infrared occultation observations by SPICAM experiment on Mars-Express: simultaneous measurements of the vertical distributions of H_2O, CO_2 and aerosol, *Icarus*, 200, 96–117.

Fedorova, A. A., Trokhimovsky, S., Korablev, O., and Montmessin, F. (2010) Viking observation of water vapor on Mars: revision from up-to-date spectroscopy and atmospheric models, *Icarus*, 208, 156–164.

Fiorenza, C. and Formisano, V. (2005) A solar spectrum for PFS data analysis, *Planetary and Space Science*, 53, 1009–1016.

Fontenla, J. M., Harder, J., Livingston, W., Snow, M., and Woods, T. (2011) High-resolution solar spectral irradiance from extreme ultraviolet to far infrared, *Journal of Geophysical Research (Atmospheres)*, 116, 20108.

Forget, F. and Pierrehumbert, R. T. (1998) Warming early Mars with carbon dioxide clouds that scatter infrared radiation, *Science*, 278, 1273.

Forget, F., Hourdin, F., and Talagrand, O. (1998) CO_2 snowfall on Mars: simulation with a general circulation model, *Icarus*, 131, 302–316.

Forget, F., Hourdin, F., Fournier, R., et al. (1999) Improved general circulation models of the Martian atmosphere from the surface to above 80 km, *Journal of Geophysical Research*, 104, 24155–24176.

Forget, F., Spiga, A., Dolla, B., et al (2007) Remote sensing of surface pressure on Mars with the Mars Express/OMEGA spectrometer: 1. Retrieval method, *Journal of Geophysical Research (Planets)*, 112.

Forget, F., Montmessin, F., Bertaux, J.-L., et al. (2009) Density and temperatures of the upper Martian atmosphere measured by stellar occultations with Mars Express SPICAM, *Journal of Geophysical Research (Planets)*, 114, E01004.

Formisano, V., Maturilli, A., Giuranna M., D'Aversa, E., and López-Valverde, M. A. (2006) Observations of non-LTE emission at 4–5 microns with the planetary Fourier spectrometer aboard the Mars Express mission, *Icarus*, 182, 51–67.

Foster, J. L., Chang, A. T. C., Hall, D. K., et al. (1998) Carbon dioxide crystals: an examination of their size, shape, and scattering properties at 37 GHz and comparisons with water ice (snow) measurements, *Journal of Geophysical Research*, 103, 25839–25850.

Fritts, D. C., Wang, L., and Tolson, R. H. (2006) Mean and gravity wave structures and variability in the Mars upper atmosphere inferred from Mars Global Surveyor and Mars Odyssey aerobraking densities, *Journal of Geophysical Research (Space Physics)*, 111, 12304.

Funke, B., López-Puertas, M., García-Comas, M., et al. (2012) GRANADA: a Generic RAdiative traNsfer AnD non-LTE population algorithm, *Journal of Quantitative Spectroscopy and Radiative Transfer*, 113, 1771–1817.

Gierasch, P. and Goody, R. (1967) An approximate calculation of radiative heating and radiative equilibrium in the Martian atmosphere, *Planetary and Space Science*, 15, 1465–1477.

Gierasch, P. J. and Goody, R. M. (1973) A model of a Martian great dust storm, *Journal of Atmospheric Science,* 30, 169–179.

Godson, W. L. (1953) The evaluation of infra-red radiative fluxes due to atmospheric water vapour, *Quart. J. Roy. Meteo. Soc.*, 79, 367.

González-Galindo, F., López-Valverde, M. A., Angelats i Coll, M., and Forget, F. (2005) Extension of a Martian general circulation model to thermospheric altitudes: UV heating and photochemical models, *Journal of Geophysical Research (Planets)*, 110, 09008.

González-Galindo, F., Forget, F., López-Valverde, M. A., and Angelats i Coll, M. (2009a) A ground-to-exosphere Martian general circulation model: 2. Atmosphere during solstice conditions – thermospheric polar warming, *Journal of Geophysical Research (Planets)*, 114, 08004.

González-Galindo, F., Forget, F., López-Valverde, M. A., et al. (2009b) A ground-to-exosphere Martian general circulation model: 1. Seasonal, diurnal, and solar cycle variation of thermospheric temperatures, *Journal of Geophysical Research (Planets)*, 114, 04001.

González-Galindo, F., Määttänen, A., Forget, F., and Spiga, A. (2011) The Martian mesosphere as revealed by CO_2 cloud observations and General Circulation Modeling, *Icarus*, 216, 10–22.

Goody, R. M. (1952) A statistical model for water vapor absorption, *Quart. J. Roy. Meteo. Soc.*, 78, 165–169.

Goody, R. M. (1964) *Atmospheric Radiation: Theoretical Basis*, Oxford University Press.

Goody, R. and Belton, M. J. S. (1967) A discussion of Martian atmospheric dynamics, *Planetary and Space Science*, 15, 247.

Goody, R. M. and Yung, Y. L. (1989) *Atmospheric Radiation: Theoretical Basis*, 2nd edition, Oxford University Press.

Goody, R., West, R., Chen, L., and Crisp, D. (1989) The correlated-k method for radiation calculations in nonhomogeneous atmospheres, *Journal of Quantitative Spectroscopy and Radiative Transfer*, 42, 539–550.

Gordiets, B. F. and Panchenko, V. I. (1983) Nonequilibrium infrared emission and the natural laser effect in the Venus and Mars atmospheres, *Cosmic Research*, 21, 929–939.

Grandjean, J. and Goody, R. M. (1955) The Concentration of Carbon Dioxide in the Atmosphere of Mars, *Astrophysical Journal*, 121, 548.

Grassi, D., Fiorenza, C., Zasova, L. V., et al. (2005) The Martian atmosphere above great volcanoes: early planetary Fourier spectrometer observations, *Planet. Space Sci.*, 53, 1017–1034.

Gray, L. D. (1966) Transmission of the atmosphere of Mars in the region of 2μm, *Icarus*, 5, 390.

Greybush, S. J., Wilson, R. J., Kalnay, E., et al. (2012) Ensemble Kalman filter data assimilation of thermal emission spectrometer (TES) profiles into a Mars global circulation model *Journal of Geophysical Research*, 117, E11008, doi:10.1029/2012JE004097.

Grundy, W. M. and Schmitt, B. (1998) The temperature-dependent near-infrared absorption spectrum of hexagonal H_2O ice, *Journal of Geophysical Research*, 103, 25809–25822.

Gruszka, M. and Borysow, A. (1997) Roto-translational collision-induced absorption of CO_2 for the atmosphere of Venus at frequencies from 0 to 250 cm^{-1}, at temperatures from 200 to 800 K, *Icarus*, 129, 172–177.

Guzewich, S. D., Toigo, A. D., Richardson, M. I., et al. (2013) The impact of a realistic vertical dust distribution on the simulation of the Martian general circulation, *J. Geophys. Res.*, 118, 980–993.

Guzewich, S. D., Smith, M. D., and Wolff, M. J. (2014) Aerosol particle size retrievals from the Compact Reconnaissance Imaging Spectrometer for Mars, *The Fifth International Workshop of the Mars Atmosphere: Modeling and Observation*, Oxford, UK, January 13–16.

Haberle, R. M. (2013) Estimating the power of Mars' greenhouse effect, *Icarus*, 223, 619–620.

Haberle, R. M., Leovy, C. B., and Pollack, J. B. (1982) Some effects of global dust storms on the atmospheric circulation of Mars, *Icarus*, 50, 322–367.

Haberle, R. M., Pollack, J. B., Barnes, J. R., et al. (1993a) Mars atmospheric dynamics as simulated by the NASA AMES general circulation model. I – The zonal-mean circulation, *Journal of Geophysical Research*, 98, 3093–3123.

Haberle, R. M., Houben, H. C., Hertenstein, R., and Herdtle, T. (1993b) A boundary-layer model for Mars: comparison with Viking Lander and entry data, *J. Atmos. Sci.*, 50, 1544–1559.

Haberle, R. M., Houben, H., Barnes, J. R., and Young, R. E. (1997) A simplified three-dimensional model for Martian climate studies, *Journal of Geophysical Research*, 102, 9051–9067.

Haberle, R. M., Joshi, M. M., Murphy, J. R., et al. (1999) General circulation model simulations of the Mars Pathfinder atmospheric structure investigation/meteorology data, *Journal of Geophysical Research*, 104, 8957–8974.

Haberle, R. M., Montmessin, F., Kahre, M. A., et al. (2011) Radiative effects of water ice clouds on the Martian seasonal water cycle in *The Fourth International Workshop on the Mars Atmosphere: Modeling and Observations*, Paris, France.

Hamilton, V. E., McSween, H. Y., and Hapke, B. (2005) Mineralogy of Martian atmospheric dust inferred from thermal infrared spectra of aerosols, *Journal of Geophysical Research (Planets)*, 110, 12006.

Hanel, R. A., Conrath, B. J., Jennings, D. E., and Samuelson, R. E. (2003) *Exploration of the Solar System by Infrared Remote Sensing*, 2nd edition, New York: Cambridge University Press.

Hansen, G. B. (1997) The infrared absorption spectrum of carbon dioxide ice from 1.8 to 333 μm, *Journal of Geophysical Research*, 102, 21569–21588.

Hansen, G. B. (2005) Ultraviolet to near-infrared absorption spectrum of carbon dioxide ice from 0.174 to 1.8 μm, *Journal of Geophysical Research (Planets)*, 110, 11003.

Hansen, J. E. (1971) Multiple scattering of polarized light in planetary atmospheres. II. Sunlight reflected by terrestrial water clouds, *J. Atmos. Sci.*, 28, 1400–1426.

Hansen, J. E. and Travis, L. D. (1974) Light scattering in planetary atmospheres, *Space Sci. Rev.*, 16 (4), 527–610.

Hapke, B. (1993) Theory of Reflectance and Emittance Spectroscopy, *Topics in Remote Sensing*, Cambridge University Press.

Hartogh, P., Medvedev, A. S., Kuroda, T., et al. (2005) Description and climatology of a new general circulation model of the Martian atmosphere, *Journal of Geophysical Research (Planets)*, 110, 11008.

Hassler, D. M., Zeitlin, C., Wimmer-Schweingruber, R. F., et al. (2012) The Radiation Assessment Detector (RAD) Investigation, *Space Sci. Rev.*, 170, 503–558, 10.1007/s11214–012–9913–1.

Hayne, P. O., Paige, D. A., Heavens, N. G., et al. (2014) The role of snowfall in forming the seasonal ice caps of Mars: models and constraints from the Mars Climate Sounder, *Icarus*, 231, 122–130.

Heavens, N. G., Richardson, M. I., Kleinböhl, A., et al. (2011) The vertical distribution of dust in the Martian atmosphere during northern spring and summer: observations by the Mars Climate Sounder and analysis of zonal average vertical dust profiles, *Journal of Geophysical Research (Planets)*, 116, 04003.

Henyey, L. G. and Greenstein, J. L. (1941) Diffuse radiation in the Galaxy, *Astrophysical Journal*, 93, 70–83.

Herman, B. M., Thome, K. J., and Ben-David, A. (1994) Numerical technique for solving the radiative transfer equation for a spherical shell atmosphere, *Applied Optics*, 33, 1760–1770.

Herman, B. M., Flittner, D. E., Caudill, T. R., Thome, K. J., and Ben-David, A. (1995) Comparison of the Gauss-Seidel spherical polarized radiative transfer code with other radiative transfer codes, *Applied Optics*, 34, 4563.

Hinson, D. P. and Wilson, R. J. (2004) Temperature inversions, thermal tides, and water ice clouds in the Martian tropics, *Journal of Geophysical Research (Planets)*, 109, 01002.

Hinson, D. P., Simpson, R. A., Twicken, J. D., et al. (1999) Initial results from radio occultation measurements with Mars Global Surveyor, *Journal of Geophysical Research*, 104, 26997.

Hinson, D. P., Smith, M. D., and Conrath, B. J. (2004) Comparison of atmospheric temperatures obtained through infrared sounding and radio occultation by Mars Global Surveyor, *Journal of Geophysical Research (Planets)*, 109, 12002.

Hoffman, M. J., Eluszkiewicz, J., Weisenstein, D., Uymin, G., and Moncet, J.-L. (2012) Assessment of Mars atmospheric temperature retrievals from the Thermal Emission Spectrometer radiances, *Icarus*, 220, 1031–1039.

Houghton, J. (1986) *Physics of the Atmosphere*, 2nd edition, Cambridge University Press.

Hourdin, F. (1992) A new representation of the absorption by the CO_2 15-microns band for a Martian general circulation model, *Journal of Geophysical Research*, 97, 18319.

Hourdin, F., Le Van, P., Forget, F., and Talagrand, O. (1993) Meteorological variability and the annual surface pressure cycle on Mars, *Journal of Atmospheric Sciences*, 50, 3625–3640.

Hourdin, F., Forget, F., and Talagrand, O. (1995) The sensitivity of the Martian surface pressure and atmospheric mass budget to various parameters: a comparison between numerical simulations and Viking observations, *Journal of Geophysical Research*, 100, 5501–5523.

Hovenier, J. W., van de Hulst, H. C., and van der Mee, C. V. M. (1986) Conditions for the elements of the scattering matrix, *Astronomy and Astrophysics*, 157, 301–310.

Huestis, D. L. (2001) Accurate evaluation of the Chapman function for atmospheric attenuation, *Journal of Quantitative Spectroscopy and Radiative Transfer*, 69, 709–721.

Humlicek, J. (1982) Optimized computation of the Voigt and complex probability functions, *JQSRT*, 27, 437.

Hunt, G. E. (2005) On the opacity of Martian dust storms derived by Viking IRTM spectral measurements, *Journal of Geophysical Research*, 84, 8301–8310.

Ignatiev, N. I., Grassi, D., and Zasova, L. V. (2005) Planetary Fourier spectrometer data analysis: fast radiative transfer models, *Planetary and Space Science*, 53, 1035–1042.

Isenor, M., Escribano, R., Preston, T. C., and Signorell, R. (2013) Predicting the infrared band profiles for CO_2 cloud particles on Mars, *Icarus*, 223, 591–601.

Ityaksov D., Linnartz, H., and Ubachs, W. (2008) Deep-UV absorption and Rayleigh scattering of carbon dioxide, *Chemical Physics Letters*, 462, 31–34.

Iwabuchi, H. and Yang, P. (2011) Temperature dependence of ice optical constants: implications for simulating the single-scattering properties of cold ice clouds, *Journal of Quantitative Spectroscopy and Radiative Transfer*, 112, 2520–2525.

Jacquinet-Husson, N., Scott, N. A., Chédin, A., et al. (1998) The GEISA system in 1996: towards an operational tool for the second generation vertical sounders radiance simulation, *Journal of Quantitative Spectroscopy and Radiative Transfer*, 59, 511–527.

Jacquinet-Husson, N., Scott, N. A., Chédin, A., et al. (2008) The GEISA spectroscopic database: current and future archive for Earth and planetary atmosphere studies, *Journal of Quantitative Spectroscopy and Radiative Transfer*, 109, 1043–1059.

Jacquinet-Husson, N., Crepeau, L., Armante, R., et al. (2011) The 2009 edition of the GEISA spectroscopic database, *Journal of Quantitative Spectroscopy and Radiative Transfer*, 112, 2395–2445.

James, P. B., and North, G. R. (1982) The seasonal CO_2 cycle on Mars: an application of an energy balance climate model, *Journal of Geophysical Research*, 87, B12 2156–2202

James, P. B., Clancy, R. T., Lee, S. W., et al. (1994) Monitoring Mars with the Hubble Space Telescope: 1990–1991 observations, *Icarus*, 109, 79–101.

Jaquin, F., Gierasch, P., and Kahn, R. (1986) The vertical structure of limb hazes in the Martian atmosphere, *Icarus*, 68, 442–461.

Jayaweere, K. and Mason, B. (1965) The behavior of freely falling cylinders and cones in a viscous fluid, *J. Fluid Mec.*, 22, 709–720, 1965.

Johnson, M. A., Betz, A. L., McLaren, R. A., Townes, C. H., and Sutton, E. C. (1976) Nonthermal 10 micron CO_2 emission lines in the atmospheres of Mars and Venus, *Astrophysical Journal*, 208, L145–L148.

Johnson, J. R., Grundy, W. M., Lemmon, M. T., et al. (2006) Spectrophotometric properties of materials observed by Pancam on the Mars Exploration Rovers: 2. Opportunity, *Journal of Geophysical Research (Planets)*, 111, E12S16.

Joseph, J. H., Wiscombe, W. J., and Weinman, J. A. (1976) The delta-Eddington approximation for radiative flux transfer, *Journal of Atmospheric Sciences*, 33, 2452–2459.

Joshi, M. M., Lewis, S. R., Read, P. L., and Catling, D. C. (1995) Western boundary currents in the Martian atmosphere: numerical simulations and observational evidence, *Journal of Geophysical Research*, 100, 5485–5500.

Kahnert, M. (2013a) The T-matrix code Tsym for homogeneous dielectric particles with finite symmetries, *Journal of Quantitative Spectroscopy and Radiative Transfer*, 123, 62–78.

Kahnert, M. (2013b) T-matrix computations for particles with high-order finite symmetries, *Journal of Quantitative Spectroscopy and Radiative Transfer*, 123, 79–91.

Kahre, M. A., Murphy, J. R., Haberle, R. M., et al. (2005) Simulating the Martian dust cycle with a finite surface dust reservoir, *Geophys. Res. Lett.*, 32, L20204.

Kahre, M. A., Murphy, J. R., and Haberle, R. M. (2006) Modeling the Martian dust cycle and surface dust reservoirs with the NASA Ames general circulation model, *Journal of Geophysical Research*, 111, E06008, doi:10.1029/2005JE002588.

Kahre, M. A., Hollingsworth, J. L., Haberle, R. M. (2014) Investigating the effects of water ice cloud radiative forcing on the predicted patterns and strength of dust lifting on Mars. In *AAS/Division for Planetary Sciences Meeting Abstracts*.

Kahre, M. A., Hollingsworth, J. L., Haberle, R. M., and Murphy, J. R. (2008) Investigations of the variability of dust particle sizes in the Martian atmosphere using the NASA Ames General Circulation Model, *Icarus*, 195, 576–597.

Kaplan, D. I. (1988) *Environment of Mars, 1988*. NASA STI/Recon Technical Report N 89.

Kaplan, L. D., Münch, G., and Spinrad, H. (1964) An analysis of the spectrum of Mars, *Astrophysical Journal*, 139, 1.

Kattawar, G. W., Young, A. T., and Humphreys, T. J. (1981) Inelastic scattering in planetary atmospheres. I – The ring effect, without aerosols, *Astrophysical Journal*, 243, 1049–1057.

Kieffer, H. H. (2013) Thermal model for analysis of Mars infrared mapping, *Journal of Geophysical Research (Planets)*, 118, 451–470.

Kieffer, H. H. and Titus, T. N. (2001) TES mapping of Mars' north seasonal cap, *Icarus*, 154, 162–180.

Kieffer, H. H., Titus, T. N., Mullins, K. F., and Christensen, P. R. (2000) Mars south polar spring and summer behavior observed by TES: seasonal cap evolution controlled by frost grain size, *Journal of Geophysical Research*, 105, 9653–9700.

King, J. I. (1952) Transfer Theory for Purely Pressure-broadened Band Spectra, *Astrophysical Journal*, 116, 491–497.

Kleinböhl, A., Schofield, J. T., Kass, D. M., et al. (2009) Mars Climate Sounder limb profile retrieval of atmospheric temperature, pressure, and dust and water ice opacity, *Journal of Geophysical Research (Planets)*, 114, 10006.

Kleinböhl, A., Schofield, J. T., Abdou, W. A., Irwin, P. G. J., and de Kok, R. J. (2011) A single-scattering approximation for infrared radiative transfer in limb geometry in the Martian atmosphere, *Journal of Quantitative Spectroscopy and Radiative Transfer*, 112, 1568–1580.

Kleinböhl, A., Wilson, R. J., Kass, D., Schofield, J. T., and McCleese, D. J. (2013) The semidiurnal tide in the middle atmosphere of Mars, *Geophys. Res. Lett.*, 40, 1952–1959, doi:10.1002/grl.50497.

Korablev, O. I. (2002) Solar occultation measurements of the Martian atmosphere on the Phobos spacecraft: water vapor profile, aerosol parameters, and other results, *Solar System Research*, 36, 12–34.

Korablev, O., Bertaux, J.-L., Fedorova, A., Fonteyn, D., Stepanov, A., et al. (2006) SPICAM IR acousto-optic spectrometer experiment on Mars Express, *Journal of Geophysical Research (Planets)*, 111, E09S03.

Kuroda, T., Medvedev, N. A. S., Hartogh, P., and Takahashi, M. (2007) Seasonal change of the baroclinic wave activity in the northern hemisphere of Mars simulated with a GCM, *Geophys. Res. Lett.*, 34, L09203.

Laan, E. C., Volten, H., Stam, D. M., et al. (2009) Scattering matrices and expansion coefficients of Martian analogue palagonite particles, *Icarus*, 199, 219–230, 2009.

Lacis, A. A. and Oinas, V. (1991) A description of the correlated-k distribution method for modelling nongray gaseous absorption, thermal emission, and multiple scattering in vertically inhomogeneous atmospheres, *Journal of Geophysical Research*, 96, 9027–9064.

Lacis, A. A., Chowdhary, J., Mishchenko, M. I., and Cairns, B. (1998) Modeling errors in diffuse-sky radiation: vector vs. scalar treatment, *Geophysical Research Letters*, 25, 135–138.

Lang, K. R. (2006) *Astrophysical formulae*, volume 1 (3rd edition), Birkhäuser.

Leblanc, F., Luhmann, J. G., Johnson, R. E., and Chassefiere, E. (2002) Some expected impacts of a solar energetic particle event at Mars, *J. Geophys. Res.*, 107, SIA5.

Lebonnois, S., Quémerais, E., Montmessin, F., et al.(2006) Vertical distribution of ozone on Mars as measured by SPICAM/Mars Express using stellar occultations, *Journal of Geophysical Research (Planets)*, 111, E09S05.

Lee, C., Lawson, W. G., Richardson, M. I. et al. (2009) Thermal tides in the Martian middle atmosphere as seen by the Mars Climate Sounder, *Journal of Geophysical Research*, 114, E03005, doi:10.10129/2008JE003285.

Lellouch, E., Encrenaz, T., de Graauw, T., et al. (2000) The 2.4–45 μm spectrum of Mars observed with the infrared space observatory, *Planetary and Space Science*, 48, 1393–1405.

Lemmon, M. T. (2014) Large water ice aerosols in Martian north polar clouds, *The Fifth International Workshop of the Mars Atmosphere: Modeling and Observation*, Oxford, UK, January 13–16.

Lemmon, M. T., Wolff, M. J., Smith, M. D., et al. (2004) Atmospheric imaging results from the Mars Exploration Rovers: Spirit and Opportunity, *Science*, 306, 1753–1756.

Lenoble, J. and Sekera, Z. (1961) Equation of radiative transfer in a planetary spherical atmosphere, *Proceedings of the National Academy of Science*, 47, 372–378, 1961.

Leovy, C. and Mintz, Y. (1969) Numerical simulation of the stmospheric circulation and climate of Mars, *Journal of Atmospheric Sciences*, 26, 1167–1190.

Lewis, S. R., Collins, M., Read, P. L., et al. (1999) A climate database for Mars, *Journal of Geophysical Research*, 104, 24177–24194.

Liou, K. N. (1974) Analytic two-stream and four-stream solutions for radiative transfer, *Journal of Atmospheric Sciences*, 31, 1473.

Liou, K. N. (2002) *An Introduction to Atmospheric Radiation*, 2nd edition, San Diego: Academic Press.

Liu, Y. and Liu, F. (1994) On the description of aerosol particle size distribution, *Atmospheric Research*, 31, 187.

López-Puertas, M. and López-Valverde, M. A. (1995) Radiative energy balance of CO_2 non-LTE infrared emissions in the Martian atmosphere, *Icarus*, 114, 113–129.

López-Puertas, M. and Taylor, F. W. (2001) *Non-LTE Radiative Transfer in the Atmosphere*, World Scientific, Singapore.

López-Puertas, M., Molina, A., Rodrigo, R., and Taylor, F. W. (1986) A non-LTE radiative transfer model for infrared bands in the middle atmosphere. I – Theoretical basis and application to CO_2 15 micron bands, *Journal of Atmospheric and Terrestrial Physics*, 48, 729–748.

López-Valverde, M. A. (1990) Emisiones infrarrojas en la atmósfera de Marte, *Ph.D Thesis*, Granada University.

López-Valverde, M. A. and González-Galindo, F. (2008) Fast computation of CO_2 cooling rates for a Mars GCM, *Third International Workshop on the Mars Atmosphere: Modeling and Observations*, Williamsburg, Virginia, USA.

López-Valverde, M. A. and López-Puertas, M. (1994a) A non-local thermodynamic equilibrium radiative transfer model for infrared emissions in the atmosphere of Mars. 1: Theoretical basis and nighttime populations of vibrational levels, *Journal of Geophysical Research*, 99, 13093–13115.

López-Valverde, M. A. and López-Puertas M. (1994b) A non-local thermodynamic equilibrium radiative transfer model for infrared emission in the atmosphere of Mars. 2: Daytime populations of vibrational levels, *Journal of Geophysical Research*, 99, 13117–13132.

López-Valverde, M. A. and López-Puertas, M. (2001) Atmospheric non-LTE effects and their parameterization for Mars, *ESA Technical Report*.

López-Valverde, M. A., Edwards, D. P., López-Puertas, M., and Roldán, C. (1998) Non-local thermodynamic equilibrium in general circulation models of the Martian atmosphere 1. Effects of the local thermodynamic equilibrium approximation on thermal cooling and solar heating, *Journal of Geophysical Research*, 103, 16799–16812.

López-Valverde, M. A., Haberle, R. M., and López-Puertas, M. (2000) Non-LTE Radiative Mesospheric Study for Mars Pathfinder Entry, *Icarus*, 146, 360–365.

López-Valverde, M. A., López-Puertas, M., López-Moreno, J. J., et al. (2005) Analysis of CO_2 non-LTE emissions at 4.3μm in the

Martian atmosphere as observed by PFS/Mars Express and SWS/ISO, *Planetary and Space Science*, 53, 1079–1087.

López-Valverde, M. A., Gilli, G., García-Comas, M., et al. (2008) The upper atmosphere of Venus observed by Venus Express, *Lecture Notes and Essays in Astrophysics*, 3, 13–32.

López-Valverde, M. A., López-Puertas, M., Funke, B., et al. (2011a) Modeling the atmospheric limb emission of CO_2 at 4.3 μm in the terrestrial planets, *Planetary and Space Science*, 59, 988–998.

López-Valverde, M. A., Sonnabend, G., Sornig, M., and Kroetz, P. (2011b) Modelling the atmospheric CO_2 10 μm non-thermal emission in Mars and Venus at high spectral resolution. *Planetary and Space Science*, 59, 999–1009.

López-Valverde, M. A., Montabone, L., Sonnabend, G., and Sornig, M. (2011c) Mars mesospheric winds: strategy for accurate comparisons between ground based observations and GCM models, in *The Fourth International Workshop on the Mars Atmosphere: Modeling and Observations*, Paris, France.

López-Valverde, M. A., González-Galindo, F., and López-Puertas, M. (2011d) Revisiting the thermal balance of the mesosphere of Mars, in *The Fourth International Workshop on the Mars Atmosphere: Modeling and Observations*, Paris, France.

Loughman, R. P., Griffioen, E., Oikarinen, L., et al. (2004) Comparison of radiative transfer models for limb-viewing scattered sunlight measurements, *Journal of Geophysical Research (Atmospheres)*, 109, 06303.

Määttänen, A., Fouchet, T., Forni, O., et al. (2009) A study of the properties of a local dust storm with Mars Express OMEGA and PFS data, *Icarus*, 201, 504–516.

Määttänen, A., Montmessin, F., Gondet, B., et al. (2010) Mapping the mesospheric CO_2 clouds on Mars: MEx/OMEGA and MEx/HRSC observations and challenges for atmospheric models, *Icarus*, 209, 452–469.

Macke, A., Mishchenko, M. I., Muinonen, K., and Carlson, B. E. (1995) Scattering of light by large nonspherical particles: ray-tracing approximation versus T-matrix method, *Optics Letters*, 20, 1934–1936.

Madeleine, J.-B., Forget, F., Millour, E., Montabone, L., and Wolff, M. J. (2011) Revisiting the radiative impact of dust on Mars using the LMD Global Climate Model, *Journal of Geophysical Research (Planets)*, 116, 11010.

Madeleine, J.-B., Forget, F., Spiga, A., et al. (2012a) Aphelion water-ice cloud mapping and property retrieval using the OMEGA imaging spectrometer onboard Mars Express, *Journal of Geophysical Research (Planets)*, 117, 00J07.

Madeleine, J.-B., Forget, F., Millour, E., Navarro, T., and Spiga, A. (2012b) The influence of radiatively active water ice clouds on the Martian climate, *Geophysical Research Letters*, 39, 23202.

Magalhães, J. A., Schofield, J. T., and Seiff, A. (1999) Results of the Mars Pathfinder atmospheric structure investigation, *Journal of Geophysical Research*, 104, 8943–8956.

Maguire, W. C., Pearl, J. C., Smith, M. D., et al. (2002) Observations of high-altitude CO_2 hot bands in Mars by the orbiting Thermal Emission Spectrometer, *Journal of Geophysical Research (Planets)*, 107, 5063.

Malin, M. C., Calvin, W. M., Cantor, B. A., et al. (2008) Climate, weather, and north polar observations from the Mars Reconnaissance Orbiter Mars Color Imager, *Icarus*, 194, 501–512.

Maltagliati, L., Titov, D. V. Encrenaz, T., et al. (2011) Annual survey of water vapor behavior from the OMEGA mapping spectrometer onboard Mars Express, *Icarus*, 213, 480–495.

Marshak A. and Davis, A. B. (2005) Horizontal fluxes and radiative smoothing, in *3D Radiative Transfer in Cloudy Atmospheres*, eds. A. Marshak, and A. B. Davis, Heidelberg: Springer.

Marshak, A., and Knyazikhin, Y. (2005) A primer in 3D radiative transfer, in *3D Radiative Transfer in Cloudy Atmospheres*, eds. A. Marshak, and A. B. Davis, Heidelberg: Springer.

Marshak, A., Davis, A. Wiscombe, W., and Cahalan, R. (1995) Radiative smoothing in fractal clouds, *Journal of Geophysical Research*, 100, 26247.

Marshak, A., Oreopoulos, L., Davis, A. B., Wiscombe, W. J., and Cahalan, R. F. (1999) Horizontal radiative fluxes in clouds and accuracy of the independent pixel approximation at absorbing wavelengths, *Geophysical Research Letters*, 26, 1585–1588.

Martin, T. Z. (1986) Thermal infrared opacity of the Mars atmosphere, *Icarus*, 66, 2–21.

Martin, T. Z. and Richardson, M. I. (1993) New dust opacity mapping from Viking Infrared Thermal Mapper data, *Journal of Geophysical Research*, 98, 10941.

Mastrapa, R. M., Bernstein, M. P., Sandford, S. A., et al. (2008) Optical constants of amorphous and crystalline H2O-ice in the near infrared from 1.1 to 2.6 μm, *Icarus*, 197, 307–320.

Mayer, B. and Kylling, A. (2005) Technical note: The libRadtran software package for radiative transfer calculations – description and examples of use, *Atmospheric Chemistry and Physics*, 5, 1855–1877.

McCleese, D. J., Heavens, N. G., Schofield, J. T., et al. (2010) Structure and dynamics of the Martian lower and middle atmosphere as observed by the Mars Climate Sounder: seasonal variations in zonal mean temperature, dust, and water ice aerosols, *Journal of Geophysical Research (Planets)*, 115, 12016.

McConnochie, T. H. and Smith, M. D. (2008) Vertically Resolved Aerosol Climatology from Mars Global Surveyor Thermal Emission Spectrometer (MGS-TES) Limb Sounding, *Third International Workshop on the Mars Atmosphere: Modeling and Observations*, Williamsburg, Virginia, USA.

McGuire, P. C., Wolff, M. J., Smith, M. D., et al. (2008) MRO/CRISM retrieval of surface Lambert albedos for multispectral mapping of Mars with DISORT-based radiative transfer modeling: Phase 1 – Using historical climatology for temperatures, aerosol optical depths, and atmospheric pressures, *IEEE Transactions on Geoscience and Remote Sensing*, 46, 4020–4040.

McLinden, C. A. and Bourassa, A. E. (2010) A systematic error in plane-parallel radiative transfer calculations, *Journal of Atmospheric Sciences*, 67, 1695–1699.

Meador W. E and Weaver, W. R. (1980) Two-stream approximations to radiative transfer in planetary atmospheres – a unified description of existing methods and a new improvement, *Journal of Atmospheric Sciences*, 37, 630–643.

Medcraft, C., McNaughton, D., Thompson, C. D., et al. (2012) Size and Temperature Dependence in the Far-IR Spectra of Water Ice Particles, *The Astrophysical Journal*, 758, 17.

Melchiorri, R., Encrenaz, T., Fouchet, T., et al. (2007) Water vapor mapping on Mars using OMEGA/Mars Express, *Planetary and Space Science*, 55, 333–342.

Michaels, T. I., Colaprete, A., and Rafkin, S. C. R. (2006) Significant vertical water transport by mountain-induced circulations on Mars, *Geophysical Research Letters*, 33, 16201.

Mihalas, D. (1978) *Stellar Atmospheres*, 2nd edition. San Francisco: W. H. Freeman.

Mishchenko, M. I. (1991) Light scattering by randomly oriented axially symmetric particles, *Journal of the Optical Society of America A*, 8, 871–882.

Mishchenko, M. I. (2002) Vector radiative transfer equation for arbitrarily shaped and arbitrarily oriented particles: a microphysical derivation from statistical electromagnetics, *Applied Optics*, 41, 7114–7134.

Mishchenko, M. I., Travis, L. D., Kahn, R. A., and West, R. A. (1997) Modeling phase functions for dustlike tropospheric aerosols using a shape mixture of randomly oriented polydisperse spheroids, *Journal of Geophysical Research*, 102, 16831–16847.

Mishchenko, M. I. and Macke, A. (1999) How big should hexagonal ice crystals be to produce halos? *Applied Optics*, 38, 1626–1629.

Mishchenko, M. and Travis, L. D. (1994) Light scattering by polydispersions of randomly oriented spheroids with sizes comparable to wavelengths of observation, *Applied Optics*, 33, 7206–7225.

Mishchenko, M. I., Lacis, A. A., and Travis, L. D. (1994) Errors induced by the neglect of polarization in radiance calculations for Rayleigh-scattering atmospheres, *Journal of Quantitative Spectroscopy and Radiative Transfer*, 51, 491–510.

Mishchenko, M. I., Travis, L. D., and Lacis, A. A. (2002) *Scattering, Absorption, and Emission of Light by Small Particles*, Cambridge University Press.

Mishchenko, M. I., Travis, L. D., and Lacis, A. A. (2006) *Multiple Scattering of Light by Particles*, Cambridge University Press.

Mischna, M. A., Lee, C., and Richardson, M. (2012) Development of a fast, accurate radiative transfer model for the Martian atmosphere, past and present, *Journal of Geophysical Research (Planets)*, 117, 10009.

Molina-Cuberos, G. J., Lichtenegger, H., Schwingenschuh, K., et al. (2002) Ion-neutral chemistry model of the lower ionosphere of Mars, *J. Geophys. Res.*, 107(E5), 5027, 10.1029/2000je001447.

Moncet, J.-L., Uymin, G., Lipton, A. E., and Snell, H. E. (2008) Infrared radiance modeling by Optimal Spectral Sampling, *Journal of Atmospheric Sciences*, 65, 3917.

Montmessin, F., Rannou, P., and Cabane, M. (2002) New insights into Martian dust distribution and water-ice cloud microphysics, *Journal of Geophysical Research (Planets)*, 107, 5037.

Montmessin, F., Forget, F., Rannou, P., Cabane, M., and Haberle, R. M. (2004) Origin and role of water ice clouds in the Martian water cycle as inferred from a general circulation model, *Journal of Geophysical Research (Planets)*, 109, 10004.

Montmessin, F., Quémerais, E., Bertaux, J. L., et al. (2006) Stellar occultations at UV wavelengths by the SPICAM instrument: retrieval and analysis of Martian haze profiles, *Journal of Geophysical Research (Planets)*, 111, E09S09.

Montmessin, F., Gondet, B., Bibring, J.-P., et al. (2007) Hyperspectral imaging of convective CO_2 ice clouds in the equatorial mesosphere of Mars, *Journal of Geophysical Research (Planets)*, 112, 11S90.

Moriyama, S. (1975) Effects of dust on radiation transfer in the Martian atmosphere. II Heating due to absorption of the visible solar radiation and importance of radiative effects of dust on the Martian meteorological phenomena, *Meteorological Society of Japan Journal*, 53, 214–221.

Moroz, V. I. (1976) The atmosphere of Mars, *Space Science Reviews*, 19, 763–843.

Mumma, M. J., Buhl, D., Chin, G., et al. (1981) Discovery of natural gain amplification in the 10-micrometer carbon dioxide laser bands on Mars – a natural laser, *Science*, 212, 45–49.

Murphy, J. R., Haberle, R. M., Toon, O. B., and Pollack, J. B. (1990) Numerical simulations of the decay of Martian global dust storms, *Journal of Geophysical Research*, 95, 14629–14648.

Murphy, J. R., Haberle, R. M., Toon, O. B., and Pollack, J. B. (1993) Martian global dust storms: zonally symmetric numerical simulations including size-dependent particle transport, *Journal of Geophysical Research*, 98, 3197–3220.

Murphy, J. R., Toon, O. B., Haberle, R. M., and Pollack, J. B. (1995) Numerical simulations of the decay of Martian global dust storms, *Journal of Geophysical Research*, 104, 24177–24194.

Newman, C. E., Lewis, S. R., Read, P. L., and Forget, F. (2002a) Modeling the Martian dust cycle. 1: Representations of dust transport processes, *Journal of Geophysical Research*, 107, 5123.

Newman, C. E., Lewis, S. R., Read, P. L., and Forget, F. (2002b) Modeling the Martian dust cycle, 2: Multi-annual radiatively active dust transport simulations, *Journal of Geophysical Research*, 107, 5124.

Ockert-Bell, M. E., Bell III, J. F., Pollack, J. B., McKay, C. P., and Forget, F. (1997) Absorption and scattering properties of the Martian dust in the solar wavelengths, *Journal of Geophysical Research*, 102, 9039–9050.

Oguchi, T. (1973) Attenuation and phase rotation of radio waves due to rain: calculations at 19.3 and 34.8 GHz, *Radio Science*, 8, 31.

Ohring, G. and Mariano, J. (1968) Seasonal and latitudinal variations of the average surface temperature and vertical temperature profile on Mars, *Journal of Atmospheric Sciences*, 25, 673–681.

Ohring, G., Tang, W., and Desanto, G. (1962) Theoretical estimates of the average surface temperature on Mars, *Journal of Atmospheric Sciences*, 19, 444–449.

Orphal J. and Chance, K. (2003) Ultraviolet and visible absorption cross-sections for HITRAN, *Journal of Quantitative Spectroscopy and Radiative Transfer*, 82, 491–504.

Pang, K., and Ajello, J. M. (1977) Complex refractive index of Martian dust – wavelength dependence and composition, *Icarus*, 30, 63–74.

Pagaran, J., Weber, M., Deland, M. T., Floyd, L. E., and Burrows, J. P. (2011) Solar Spectral Irradiance Variations in 240–1600 nm During the Recent Solar Cycles 21–23, *Solar Physics*, 272, 159–188.

Parkinson W. H, Rufus, J., Yoshino, K. (2003) Absolute absorption cross section measurements of CO_2 in the wavelength region 163 200 nm and the temperature dependence, *Chemical Physics*, 290, 251–256.

Pearl J. C, Smith, M. D., Conrath, B. J., Bandfield, J. L., and Christensen, P. R. (2001) Observations of Martian ice clouds by the Mars Global Surveyor Thermal Emission Spectrometer: the first Martian year, *Journal of Geophysical Research*, 106, 12325–12338.

Perrier, S., Bertaux, J.-L., Lefèvre, F., et al. (2006) Global distribution of total ozone on Mars from SPICAM/MEX UV measurements, *Journal of Geophysical Research (Planets)*, 111, E09S06.

Petrova, E. V. (1999) Optical Thickness and Shape of Dust Particles of the Martian Aerosol, *Solar System Research*, 33, 260.

Petty, G. W. (2006) *A First Course in Atmospheric Radiation*, Madison: Sundog Press.

Petty, G. W., and Huang, W. (2011) The modified gamma size distribution applied to inhomogeneous and nonspherical particles: key relationships and conversions, *Journal of Atmospheric Sciences*, 68, 1460–1473.

Piccialli, A., Drossart, P., López-Valverde, M. A., et al. (2012) Characterization of OMEGA/MEx CO_2 non-LTE limb observations on the dayside of Mars, *European Planetary Science Congress*, 504.

Pierrehumbert, R. T. (2010) *Principles of Planetary Climate*, Cambridge University Press.

Pierrehumbert, R. T. and Erlick, C. (1998) On the scattering greenhouse effect of CO_2 ice clouds, *Journal of Atmospheric Sciences*, 55, 1897–1902.

Pincus, R. and Evans, K. F. (2009) Computational cost and accuracy in calculating three-dimensional radiative transfer: results for new implementations of Monte Carlo and SHDOM, *Journal of Atmospheric Sciences*, 66, 3131.

Pollack, J. B. and Cuzzi, J. N. (1980) Scattering by nonspherical particles of size comparable to wavelength – a new semi-empirical theory and its application to tropospheric aerosols, *Journal of Atmospheric Sciences*, 37, 868–881.

Pollack, J. B., Colburn, D., Kahn, R., et al. (1977) Properties of aerosols in the Martian atmosphere, as inferred from Viking Lander imaging data, *Journal of Geophysical Research*, 82, 4479–4496.

Pollack, J. B., Colburn, D. S., Flasar, F. M., et al. (1979) Properties and effects of dust particles suspended in the Martian atmosphere, *Journal of Geophysical Research*, 84, 2929–2945.

Pollack, J. B., Haberle, R. M., Schaeffer, J., and Lee, H. (1990) Simulations of the general circulation of the Martian atmosphere: 1. Polar processes, *Journal of Geophysical Research*, 95, 1447–1473.

Pollack, J. B., Ockert-Bell, M. E., and Shepard, M. K. (1995) Viking Lander image analysis of Martian atmospheric dust, *Journal of Geophysical Research*, 100, 5235–5250.

Prabhakara, C. and Hogan, J. S. (1965) Ozone and carbon dioxide heating in the Martian Atmosphere, *Journal of Atmospheric Sciences*, 22, 97–109.

Rafkin, S. C. R. (2009) A positive radiative-dynamic feedback mechanism for the maintenance and growth of Martian dust storms, *Journal of Geophysical Research (Planets)*, 114, 01009.

Räisänen, P. (2002) Two-stream approximations revisited: a new improvement and tests with GCM data, *Quarterly Journal of the Royal Meteorological Society*, 128, 2397–416.

Ramanthan, V. and Cess, R. D. (1974) Radiative transfer within the mesospheres of Venus and Mars, *Astrophysical Journal*, 188, 407–416.

Rannou, P., Perrier, S., Bertaux, J.-L., et al. (2006) Dust and cloud detection at the Mars limb with UV scattered sunlight with SPICAM, *Journal of Geophysical Research (Planets)*, 111, E09S10.

Rapp-Arrarás, Í. and Domingo-Santos, J. M. (2011), Functional forms for approximating the relative optical air mass, *Journal of Geophysical Research (Atmospheres)*, 116, 24308.

Richardson, M. I. and Wilson, R. J. (2002) A topographically forced asymmetry in the Martian circulation and climate, *Nature*, 416, 298–301.

Richardson, M. I., Wilson, R. J., and Rodin, A. V. (2002) Water ice clouds in the Martian atmosphere: general circulation model experiments with a simple cloud scheme, *Journal of Geophysical Research (Planets)*, 107, 5064.

Robitaille, T. P. (2011) HYPERION: an open-source parallelized three-dimensional dust continuum radiative transfer code, *Astronomy and Astrophysics*, 536, 79.

Rodrigo, R., Garcia-Alvarez, E., Lopez-Gonzalez, M. J., and Lopez-Moreno, J. J. (1990) A nonsteady one-dimensional theoretical model of Mars' neutral atmospheric composition between 30 and 200 km, *Journal of Geophysical Research*, 95, 14795–14810.

Rogers, C. D. (2000) *Inverse Methods for Atmospheric Sounding: Theory and Practice*, World Scientific, Singapore.

Rothman, L. S. (2010) The evolution and impact of the HITRAN molecular spectroscopic database, *Journal of Quantitative Spectroscopy and Radiative Transfer*, 111, 1565–1567.

Rothman, L. S., Gamache, R. R., Goldman, A., et al. (1987) The HITRAN database: 1986 edition, *Applied Optics*, 26, 4058–4097.

Rothman, L. S., Rinsland, C. P., Goldman, A, et al. (1998) The HITRAN Molecular Spectroscopic Database and HAWKS (HITRAN Atmospheric Workstation): 1996 Edition, *Journal of Quantitative Spectroscopy and Radiative Transfer*, 60, 665–710.

Rothman, L. S., Jacquinet-Husson, N., Boulet, C., and Perrin, A. M. (2005) History and future of the molecular spectroscopic databases, *Comptes Rendus Physique*, 6, 897–907.

Rothman, L. S., Gordon, I. E., Barbe, A., et al. (2009) The HITRAN 2008 molecular spectroscopic database, *Journal of Quantitative Spectroscopy and Radiative Transfer*, 110, 533–572.

Rozanov, V. V. and Kokhanovsky, A. A. (2006) The solution of the vector radiative transfer equation using the discrete ordinates technique: selected applications, *Atmospheric Research*, 79, 241–265.

Rozanov, V. V. and Rozanov, A. V. (2007) Generalized form of the direct and adjoint radiative transfer equations, *Journal of Quantitative Spectroscopy and Radiative Transfer*, 104, 155–170.

Rozanov, A., Rozanov, V., Buchwitz, M., et al. (2005) SCIATRAN 2.0 A new radiative transfer model for geophysical applications in the 175 2400 nm spectral region, *Advances in Space Research*, 36, 1015–1019.

Sagdeev, R. Z. and Zakharov, A. V. (1989) Brief history of the Phobos mission, *Nature*, 341, 581–585.

Sander, S. P., Abbatt, J., Barker, J. R., et al. (2011) Chemical Kinetics and Photochemical Data for Use in Atmospheric Studies, Evaluation No. 17, *JPL Publication* 10–6, Jet Propulsion Laboratory, Pasadena (http://jpldataeval.jpl.nasa.gov).

Sano, I., Mukai, S. Yamano, M., et al. (2003) Calibration and validation of retrieved aerosol properties based on AERONET and SKYNET, *Advances in Space Research*, 32, 2159–2164.

Santee, M. L., and Crisp, D. (1995) Diagnostic calculations of the circulation in the Martian atmosphere, *Journal of Geophysical Research*, 100, 5465–5484.

Savijärvi, H. (1999) A model study of the atmospheric boundary layer in the Mars Pathfinder Lander conditions, *Quarterly Journal of the Royal Meteorological Society*, 125, 483–493.

Savijärvi, H., Crisp, D. and Harri, A.-M. (2005) Effects of CO_2 and dust on present-day solar radiation and climate on Mars, *Quarterly Journal of the Royal Meteorological Society*, 131, 2907–2922.

Schofield, J. T., Barnes, J. R., Crisp, D., et al. (1997) The Mars Pathfinder Atmospheric Structure Investigation/Meteorology, *Science*, 278, 1752.

Schuster, A. (1905) Radiation through a foggy atmosphere, *Astrophysical Journal*, 21, 1.

Schwarzschild, K. (1906) Nachrichten von der Königlichen Gesellschaft der Wissenschaften zu Göttingen, *Math.-phys. Klasse*, 195, 41–53.

Schwarzschild, K. (1914) Sitzungsber. Preussichen Akad., *Wiss., Phys.-Math.*, K1, 1183.

Sefton-Nash, E., Teanby, N. A., Montabone, L., et al. (2013) Climatology and first-order composition estimates of mesospheric clouds from Mars Climate Sounder limb spectra, *Icarus*, 222, 342–356.

Seiff, A. and Kirk, D. B. (1977), Structure of the atmosphere of Mars in summer at mid-latitudes, *Journal of Geophysical Research*, 82, 4364–4378.

Sekera, Z. (1957) *Handbuch der Physik. Bd. 48*. Berlin: Springer.

Smith, M. D. (2004) Interannual variability in TES atmospheric observations of Mars during 1999–2003, *Icarus*, 167, 148–165.

Smith, M. D. (2008) Spacecraft observations of the Martian atmosphere, *Annual Review of Earth and Planetary Sciences*, 36, 191–219.

Smith, M. D. (2009) THEMIS observations of Mars aerosol optical depth from 2002–2008, *Icarus*, 202, 444–452.

Smith, M. D., Conrath, B. J., Pearl, J. C., and Ustinov, E. A. (1996) Retrieval of Atmospheric Temperatures in the Martian Planetary Boundary Layer Using Upward-Looking Infrared Spectra, *Icarus*, 124: 586–597.

Smith M. D., Pearl, J. C., Conrath, B. J., and Christensen, P. R. (2000) Mars Global Surveyor Thermal Emission Spectrometer (TES) observations of dust opacity during aerobraking and science phasing, *Journal of Geophysical Research*, 105, 9539–9552.

Smith M. D., Pearl, J. C., Conrath, B. J., and Christensen, P. R. (2001) One Martian year of atmospheric observations by the Thermal

Emission Spectrometer, *Geophysical Research Letters*, 28, 4263–4266.

Smith, M. D., Wolff, M. J., Spanovich, N., et al. (2006) One Martian year of atmospheric observations using MER Mini-TES, *Journal of Geophysical Research (Planets)* 111.

Smith, M. D., Wolff, M. J., Clancy, R. T., and Murchie, S. L. (2009) Compact Reconnaissance Imaging Spectrometer observations of water vapor and carbon monoxide, *Journal of Geophysical Research (Planets)*, 114.

Smith, M. D., Wolff, M. J., Clancy, R. T., Kleinböhl, A., and Murchie, S. L. (2013) Vertical distribution of dust and water ice aerosols from CRISM limb-geometry observations, *Journal of Geophysical Research (Planets)*, 118, 321–334.

Smith, P. H. and Lemmon, M. (1999) Opacity of the Martian atmosphere measured by the Imager for Mars Pathfinder, *Journal of Geophysical Research*, 104, 8975–8986.

Sneep, M. and Ubachs, W. (2005) Direct measurement of the Rayleigh scattering cross section in various gases, *Journal of Quantitative Spectroscopy and Radiative Transfer*, 92, 293–310.

Snook, K. J. (1999) *Optical properties and radiative heating effects of dust suspended in the Mars atmosphere*, Ph.D. thesis, Stanford University, Stanford, California.

Sobouti, Y. (1962) Fluorescent scattering in planetary atmospheres. II. Coupling among transitions, *Astrophysical Journal*, 135, 938.

Soderblom, J. M., Bell III, J. F., Johnson, J. R., Joseph, J., and Wolff, M. J. (2008) Mars Exploration Rover Navigation Camera in-flight calibration, *Journal of Geophysical Research (Planets)*, 113, E03S19.

Sonnabend, G., Sornig, M., Krôtz, P. J., et al. (2006) High spatial resolution mapping of Mars mesospheric zonal winds by infrared heterodyne spectroscopy of CO_2, *Geophysical Research Letters*, 33, 18201.

Sornig, M., Livengood, T., Sonnabend, G., et al. (2008) Venus upper atmosphere winds from ground-based heterodyne spectroscopy of CO_2 at 10 μm wavelength, *Planetary and Space Science*, 56, 1399–1406.

Spiga, A., Forget, F., Madeleine, J.-B., et al. (2011) The impact of Martian mesoscale winds on surface temperature and on the determination of thermal inertia, *Icarus*, 212, 504–519.

Spiga, A., González-Galindo, F., López-Valverde, M.-A., and Forget, F. (2012) Gravity waves, cold pockets and CO_2 clouds in the Martian mesosphere, *Geophysical Research Letters*, 39, 02201.

Spiga, A., Fure, J., Madeleine, J.-P., Määtänen, A., and Forget, F. (2013) Rocket dust storms and detached dust layers in the Martian atmosphere, *J. Geophys. Res.*, 118, 746–767.

Spurr, R. J. D. (2006) VLIDORT: a linearized pseudo-spherical vector discrete ordinate radiative transfer code for forward model and retrieval studies in multilayer multiple scattering media, *Journal of Quantitative Spectroscopy and Radiative Transfer*, 102, 316–342.

Stamnes, K., Tsay, S.-C., Jayaweera, K., and Wiscombe, W. (1988) Numerically stable algorithm for discrete-ordinate-method radiative transfer in multiple scattering and emitting layered media, *Applied Optics*, 27, 2502–2509.

Steele, L. J., Lewis, S. R., and Patel, M. R. (2014) The radiative impact of water ice clouds from a reanalysis of Mars Climate Sounder data, *Geophysical Research Letters*, 41, 4471–4478.

Stepanova, G. I. and Shved, G. M. (1985) Radiation transfer in the 4.3 μm CO_2 band and the 4.7 μm CO band in the atmospheres of Venus and Mars with violation of LTE – populations of vibrational states, *Soviet Astronomy*, 29, 422.

Stewart, A. I. (1972) Mariner 6 and 7 ultraviolet spectrometer experiment: implications of CO_2^+, CO, and O airglow, *Journal of Geophysical Research*, 77, 54–68.

Strong, J. and Plass, G. N. (1950) The effect of pressure broadening of spectral lines on atmospheric temperature, *Astrophysical Journal*, 112, 365–379.

Strutt, J. W. (Lord Rayleigh) (1871) On the light from the sky, its polarization and colour, *Philosophical Magazine*, 41, 447–454.

Strutt, J. W. (Lord Rayleigh) (1899) On the transmission of light through an atmosphere containing small particles in suspension, and on the origin of the blue of the sky. *Philosophical Magazine* 47, 375–384. Reprinted in: *Lord Rayleigh, Scientific papers*, Part IV, New York: Dover Publications, 1964, 397.

Takahashi, Y. O., Fujiwara, H., and Fukunishi, H. (2006) Vertical and latitudinal structure of the migrating diurnal tide in the Martian atmosphere: numerical investigations, *Journal of Geophysical Research (Planets)*, 111, 01003.

Tamppari, L. K., Bass, D., Cantor, B., et al. (2010) Phoenix and MRO coordinated atmospheric measurements, *Journal of Geophysical Research (Planets)*, 115, E00E17.

Thomas, G. E. and Stamnes, K. (2002) *Radiative Transfer in the Atmosphere and Ocean*, Cambridge University Press.

Thorpe, T. E. (1979) A history of Mars atmospheric opacity in the southern hemisphere during the Viking extended mission, *Journal of Geophysical Research,* 84, 6663–6683.

Thorpe, T. E. (1981) Mars atmospheric opacity effects observed in the Northern Hemisphere by Viking Orbiter imaging, *Journal of Geophysical Research*, 86, 11419–11429.

Thuillier, G., Floyd, L., Woods, T. N., et al. (2004) Solar irradiance reference spectra for two solar active levels, *Advances in Space Research*, 34, 256–261, 2004.

Titov, D. V. and Haus, R. (1997) A fast and accurate method of calculation of gaseous transmission functions in planetary atmospheres, *Planetary and Space Science*, 45, 369–377.

Titov D. V., Fedorova, A. A., and Haus, R. (2000) A new method of remote sounding of the Martian aerosols by means of spectroscopy in the 2.7 μm CO_2 band, *Planetary and Space Science*, 48, 67–74.

Toigo, A. D. and Richardson, M. I., et al (2000) Seasonal variation of aerosols in the Martian atmosphere, *Journal of Geophysical Research*, 105, 4109–4122.

Toigo, A. D., Smith, M. D., Seelos, F. P., and Murchie, S. L. (2013) High spatial and temporal resolution sampling of Martian gas abundances from CRISM spectra, *Journal of Geophysical Research (Planets)*, 118, 89–104.

Tomasko, M. G., Doose, L. R., Lemmon, M., et al. (1999) Properties of dust in the Martian atmosphere from the Imager on Mars Pathfinder, *Journal of Geophysical Research*, 104, 8987–9008.

Toon, O. B. and Ackerman, T. P. (1981) Algorithms for the calculation of scattering by stratified spheres, *Applied Optics*, 20, 3657–3660.

Toon, O. B., Pollack, J. B., and Sagan, C. (1977) Physical properties of the particles composing the Martian dust storm of 1971–1972, *Icarus*, 30, 663–696.

Toon, O. B., McKay, C. P., Ackerman, T. P., and Santhanam, K. (1989) Rapid calculation of radiative heating rates and photodissociation rates in inhomogeneous multiple scattering atmospheres, *Journal of Geophysical Research*, 94, 16287–16301.

Toon, O. B., Tolbert, M. A., Koehler, B. G., et al. (1994) Infrared optical constants of H_2O ice, amorphous nitric acid solutions, and nitric acid hydrates, *Journal of Geophysical Research*, 99, 25631.

Trenberth, K. E., Fasullo, J. T., and Kiehl, J. (2009) Earth's global energy budget, *Bull. Amer. Meteor. Soc.*, 90, 311–323.

Ueno, S. (1960) The Probabilistic Method for Problems of Radiative Transfer. X. Diffuse Reflection and Transmission in a Finite Inhomogeneous Atmosphere, *Astrophysical Journal*, 132, 729.

Ueno, S., Kagiwada, H., and Kalaba, R. (1971) Radiative transfer in spherical shell atmospheres with radial symmetry, *Journal of Mathematical Physics*, 12, 1279–1286.

Ustinov, E. A. (2005) Atmospheric weighting functions and surface partial derivatives for remote sensing of scattering planetary atmospheres in thermal spectral region: general adjoint approach, *Journal of Quantitative Spectroscopy and Radiative Transfer*, 92, 351–371.

Ustinov, E. A. (2007) Passive remote sensing of planetary atmospheres and retrievals of atmospheric macro- and microphysical parameters, *Journal of Quantitative Spectroscopy and Radiative Transfer*, 103, 217–230.

van de Hulst, H. C., (1945) Theory of absorption lines in the atmosphere of the Earth, *Ann. Rev. Astrophys.*, 1.

van de Hulst, H. C. (1957) *Light Scattering by Small Particles*, New York: John Wiley.

Vincendon, M. and Langevin, Y. (2010) A spherical Monte-Carlo model of aerosols: validation and first applications to Mars and Titan, *Icarus*, 207, 923–931.

Vincendon, M., Langevin, Y., Poulet, F., Bibring, J.-P., and Gondet, B. (2007) Recovery of surface reflectance spectra and evaluation of the optical depth of aerosols in the near-IR using a Monte Carlo approach: application to the OMEGA observations of high-latitude regions of Mars, *Journal of Geophysical Research (Planets)*, 112, E08S13.

Vincendon, M., Langevin, Y., Poulet, F., et al. (2008) Dust aerosols above the south polar cap of Mars as seen by OMEGA, *Icarus*, 196, 488–505.

Vincendon, M., Langevin, Y., Poulet, F., et al. (2009) Yearly and seasonal variations of low albedo surfaces on Mars in the OMEGA/MEx dataset: constraints on aerosols properties and dust deposits, *Icarus*, 200, 395–405.

Warren, S. G. (1984) Optical constants of ice from the ultraviolet to the microwave, *Applied Optics*, 23, 1206–1225.

Warren, S. G. (1986) Optical constants of carbon dioxide ice, *Appl. Opt.*, 25, 2650–2674.

Warren, S. G. and Brandt, R. E. (2008) Optical constants of ice from the ultraviolet to the microwave: a revised compilation, *Journal of Geophysical Research (Atmospheres)*, 113, 14220.

Waterman, P. C. (1971) Symmetry, unitarity, and geometry in electromagnetic scattering, *Physical Review D*, 3, 825–839.

Wehrbein, W. M., Hord, C. W., and Barth, C. A. (1979) Mariner 9 ultraviolet spectrometer experiment – vertical distribution of ozone on Mars, *Icarus*, 38, 288–299.

Wehrli, C. (1985) *Extraterrestrial Solar Spectrum*, Publication no. 615, Physikalisch-Meteorologisches Observatorium & World Radiation Center (PMO/WRC), Davos Dorf, Switzerland, July.

Wehrli, C. (1986) *Solar Spectral Irradiance, World Climate Research Programme*, Pub. Ser. No. 7, WMO ITD, No. 149, World Radiation Center, Davos-Dorf, Switzerland, 119–126.

Wells, R. J. (1999) Rapid approximation to the Voigt/Faddeeva function and its derivatives, *Journal of Quantitative Spectroscopy and Radiative Transfer*, 62, 29–48.

West, R., Crisp, D., and Chen, L. (1990) Mapping transformations for broadband atmospheric radiation calculations, *Journal of Quantitative Spectroscopy and Radiative Transfer*, 43, 191–199.

West, R., Goody, R., Chen, L., and Crisp, D. (2010) The correlated-k method and related methods for broadband radiation calculations, *Journal of Quantitative Spectroscopy and Radiative Transfer*, 111, 1672–1673.

Whiteway, J. A., Komguem, L., Dickinson, C., et al. (2009) Mars water-ice clouds and precipitation, *Science*, 325, 68.

Wilson, R. J. (1997) A general circulation model simulation of the Martian polar warming, *Geophysical Research Letters*, 24, 123–126.

Wilson, R. J. and Guzewich, S. D. (2014) Influence of water ice clouds on nighttime tropical temperature structure as seen by the Mars Climate Sounder, *Geophys. Res. Lett.*, 41, doi:10.1002/2014GL060082.

Wilson, R. J. and Hamilton, K. (1996) Comprehensive Model Simulation of Thermal Tides in the Martian Atmosphere, *Journal of Atmospheric Sciences*, 53, 1290–1326.

Wilson, R. J., Neumann, G., and Smith, M. D. (2007), The diurnal variation and radiative influence of Martian water ice clouds, *Geophys. Res. Lett.*, 34, L02710.

Wilson, R. J., Haberle, R. M., Noble, J., et al. (2008a) Simulation of the 2001 Planet-encircling Dust Storm with the NASA/NOAA Mars General Circulation Model in *Third International Workshop on the Mars Atmosphere: Modeling and Observations*, Williamsburg, 1447, 9023.

Wilson, R. J., Lewis, S. R., and Montabone, L. (2008b) Influence of water ice clouds on Martian tropical atmospheric temperatures, *Geophys. Res. Lett.*, 35, L07202.

Wilson, R. J., Millour, E., Navarro, T., Forget, F., and Kahre, M. A. (2014) GCM simulations of aphelion season tropical cloud and temperature structure, in *Mars Atmosphere: Modeling and Observations, 5th International Workshop*, Oxford, UK.

Wiscombe, W. J. and Joseph, J. H. (1977) The range of validity of the Eddington approximation, *Icarus* 32, 362–377.

Wolff, M. J. and Clancy, R. T. (2003) Constraints on the size of Martian aerosols from Thermal Emission Spectrometer observations, *Journal of Geophysical Research (Planets)*, 108, 5097.

Wolff, M. J., Smith, M. D., Clancy, R. T., et al. (2006) Constraints on dust aerosols from the Mars Exploration Rovers using MGS overflights and Mini-TES, *Journal of Geophysical Research (Planets)*, 111, E12S17.

Wolff, M. J., Smith, M. D., Clancy, R. T., et al. (2009) Wavelength dependence of dust aerosol single scattering albedo as observed by the Compact Reconnaissance Imaging Spectrometer, *Journal of Geophysical Research (Planets)*, 114, E00D04.

Wolff, M. J., Clancy, R. T., Goguen, J. D., Malin, M. C., and Cantor, B. A. (2010) Ultraviolet dust aerosol properties as observed by MARCI, *Icarus*, 208, 143–155.

Wolff, M. J., Clancy, R. T., Cantor, B., Madeleine, J.-B., and Millour, E. (2011) Mapping water ice clouds (and ozone) with MRO/MARCI, in *The Fourth International Workshop on the Mars Atmosphere: Modelling and Observation*, 8–11 February, Paris, France, 213–216.

Wolff, M. J., Clancy, R. T., Smith, M. D., et al. (2012) Vertical profiles of aerosol particle sizes using MGS/TES and MRO/MCS, Fall Meeting of American Geophysical Union, Abstract, P04.

Wolff, M. J., Clancy, R. T., Cantor, B., and Haberle, R. M. (2014) The MARCI water ice cloud optical depth (public) database, in *The Fifth International Workshop on the Mars Atmosphere: Modeling and Observations*, Oxford, UK, January 13–16.

Woods, T. N., Chamberlin, P. C., Harder, J. W., et al. (2009) Solar irradiance reference spectra (SIRS) for the 2008 whole heliosphere interval (WHI), *Geophysical Research Letters*, 36, 1101.

Wordsworth, R., Forget, F., Millour, E., Head, J. W., et al. (2013) Madeleine and B. Charnay, Global modeling of the early Martian climate under a denser CO_2 atmosphere: water cycle and ice evolution, *Icarus*, 222, 1–19.

Wriedt, T. and Hellmers, J. (2008) New Scattering Information Portal for the light-scattering community, *Journal of Quantitative Spectroscopy and Radiative Transfer*, 109, 1536–1542.

Yang, P. and Liou, K. N. (1996) Finite-difference time domain method for light scattering by small ice crystals in three-dimensional space, *Journal of the Optical Society of America A*, 13, 2072–2085.

Yang, P., Liou, K. N., Mishchenko, M. I., and Gao, B.-C. (2000) Efficient finite-difference time-domain scheme for light scattering by dielectric particles: application to aerosols, *Applied Optics*, 39, 3727–3737.

Yang, P., Baum, B. A., Heymsfield, A., et al. (2003) Single scattering properties of droxtals, *Journal of Quantitative Spectroscopy and Radiative Transfer*, 79, 1159.

Yee, K. (1966) Numerical solution of initial boundary value problems involving Maxwell's equations in isotropic media, *IEEE Transactions on Antennas and Propagation*, 14, 302–307.

Yoshino K., Esmond, J. R., Sun, Y., et al. (1996) Absorption cross section measurements of carbon dioxide in the wavelength region 118.7–175.5 nm and the temperature dependence, *Journal of Quantitative Spectroscopy and Radiative Transfer*, 55, 53–60.

Zorzano, M.-P., Mateo-Martí, E., Prieto-Ballesteros, O., Osuna, S. and Rennó, N. (2009) Stability of liquid saline water on present day Mars, *Geophysical Research Letters*, 36, 20201.

Zurek, R.W. (1981) Inference of dust opacities for the 1977 Martian great dust storms from Viking Lander 1 pressure data, *Icarus*, 45, 202–215.

Zurek, R. W. (1982) Martian great dust storms – an update, *Icarus*, 50, 288–310.

7

The Martian Planetary Boundary Layer

PETER L. READ, BORIS GALPERIN, SØREN E. LARSEN, STEPHEN R. LEWIS, ANNI MÄÄTTÄNEN, ARAKEL PETROSYAN, NILTON RENNÓ, HANNU SAVIJÄRVI, TERO SIILI, AYMERIC SPIGA, ANTHONY TOIGO, LUIS VÁZQUEZ

7.1 INTRODUCTION

The Martian planetary boundary layer (PBL) consists of the layers of the atmosphere closest to the surface, within which interactions between the atmosphere and the surface itself are dominant. In general, this represents the lowest 1–10 km of the atmosphere, within which surface-driven intense convection may take place, with convective plumes and vortices rising to heights in excess of 5–10 km during the day (Thomas and Gierasch, 1985; Haberle et al., 1993b; Larsen et al., 2002; Balme and Greeley, 2006; Hinson et al., 2008). At night, convection is inhibited and radiative cooling produces a stably stratified layer at the surface, and the PBL reduces to a shallow layer forced by mechanical turbulence at the bottom of the stable layer. It is therefore a highly dynamic and variable region of the atmosphere at virtually all locations on Mars, with additional variability induced by interactions with local surface topography.

The PBL is extremely important, both scientifically and operationally. It is the interface between the free atmosphere and the surface and regolith, mediating both short- and long-term exchanges of heat, momentum, dust, water, and a variety of chemical tracers (such as argon and methane) between surface/subsurface and atmospheric reservoirs. It is also the region of the atmosphere through which landed spacecraft need to pass to reach their regions of operations. A clear and quantitative understanding of this part of the atmosphere, and the way in which it interacts with the surface and free atmosphere, should therefore be a vital part of any program to explore and understand the past, present, and future Martian environment. Such an understanding will also enable the anticipation of environmental conditions encountered during spacecraft entries and operations, reliable predictions of which are essential for mission safety and efficient design.

At the present time, our understanding of the Martian PBL and our ability to model it are strongly guided and influenced by studies of its terrestrial counterpart (Larsen et al., 2002). While this may be a valid initial approach, the Martian environment differs from that of the Earth in a number of important aspects. The much lower atmospheric density at the Martian surface may be significant, especially within the thin surface layer, affecting the details of heat, momentum, and mass fluxes. The range of conditions encountered in the Martian PBL may also be substantially more extreme than found typically on Earth, with diurnal contrasts from intensely convective conditions, accompanied by sustained super-adiabatic thermal gradients, to very strongly stably stratified conditions during the night. Such

widespread and extreme variability across the entire planet places extraordinary demands on predictive models to capture accurately and to compute implied vertical transports of heat, momentum, and tracers.

In this chapter, we provide an overview of the Martian PBL, covering the basic theory underlying the physics of atmospheric boundary layers, together with an overview of the body of observations available to characterize the PBL and the methods used to model it. Section 7.2 provides a short introduction to the basic concepts underlying boundary layer theory. Section 7.3 goes on to examine the key similarities and differences between the boundary layers on Mars and the Earth, introducing the critical parameters that determine them, and considers the overall role of the PBL on Mars in the wider global circulation. Section 7.4 provides a detailed discussion of the body of observations that inform our knowledge of the Martian PBL. Section 7.5 surveys the approaches used to capture the behavior of the Martian PBL in models, ranging in sophistication from simple conceptual models, based on bulk formulations of turbulent transports of heat and momentum, to more complex treatments of parameterized Reynolds stress and ultimately to large-eddy simulation (LES) approaches. An overview of the current status of the subject and some of the outstanding issues are provided in Section 7.6, including some discussion of possible options for future measurements.

7.2 ATMOSPHERIC BOUNDARY LAYER PHYSICS

The boundary layer is one of the key subdomains of the troposphere on Earth, where it has a thickness of order 1 km, because it is where the atmosphere interacts with the surface. Within this layer, the meteorological fields, such as wind velocity and temperature, adjust from their values in the free atmosphere towards values and spatial gradients determined by the surface (e.g. see Figure 7.1). For the Earth (though not necessarily for Mars, see Sections 7.3 and 7.5), the heat transfer within this layer is directly related to the solar insolation at the planetary surface and may be carried via both sensible and latent heat fluxes. The momentum transfer in this layer is primarily due to drag forces acting on the free atmospheric motions as a result of the presence of the solid planetary surface. The proximity of the surface initiates mechanical and/or thermal instabilities that enhance the vertical transport and mixing in the boundary layer by the action of the ensuing turbulent motions. The resulting convective plumes and turbulent eddies transport momentum

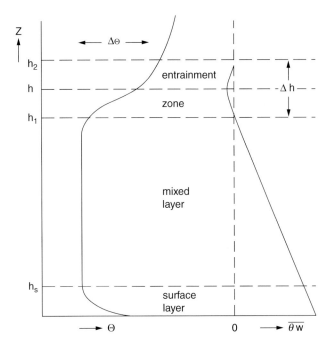

Figure 7.1. Schematic structure of the daytime convective PBL.

and trace species, mixing these quantities within the PBL. The sources of these instabilities are the vertical shear of the horizontal wind and the surface heating, which produce turbulence via shear and convective instabilities, respectively. In this section, we discuss various types of boundary layer and their characteristics, along with the terminology used in boundary layer meteorology.

Generally, atmospheric PBLs can be divided into a number of different subregions, as indicated in Figure 7.1. Closest to the ground is the so-called "surface layer" defined by its proximity to the surface and typically put at $z/h \lesssim 0.1$ (see Figure 7.1); here, z is the vertical coordinate directed upward and h is the boundary layer height. In the surface layer, the vertical fluxes of heat and momentum are almost independent of height and vary by less than 10% of their mean magnitudes (Monin and Yaglom, 1975; Stull, 1988). Above this layer is the bulk of the boundary layer, known as the "mixed layer", where strong mixing may take place up to the top of the boundary layer, which is typically characterized by a jump, $\Delta\Theta$, in potential temperature towards the free atmosphere. This jump is often not discontinuous but forms an entrainment zone, within which mixing from below rapidly decays as air from the free atmosphere is entrained into the underlying flow. This region may also be characterized by a weak reversal in the upward heat flux known as "counter-gradient diffusion" (Deardorff, 1972; Holtslag and Moeng, 1991). Such a local view of heat transport is an oversimplification, however, based on a rather narrow definition of diffusive-like transport and the concept of static stability. A more complete and realistic approach needs to consider the possibility of non-local effects such as "convective overshoot" of buoyant air parcels (see Stull (1991) for further discussion).

One of the most important and useful tools for the study of turbulence is dimensional analysis based upon dimensionless parameters defined in terms of velocity, length, and time scales

that characterize the two basic mechanisms for turbulence production: shear and buoyancy. These dimensionless parameters define flow regimes in PBLs that are dominated by the presence of various turbulent processes – see e.g. Tennekes and Lumley (1972), Monin and Yaglom (1975), Stull (1988), Garratt (1992), and Wyngaard (2010) for detailed reviews.

For simplicity, we only consider a stationary and homogeneous boundary layer. For characterization of its structure and dynamics, we introduce the following parameters: the geostrophic wind G in the free atmosphere ($z > h_2$; see Figure 7.1); the corresponding thermal wind shear, dG/dz; the temperature jump $\Delta\Theta$ at the top of an unstable boundary layer (or entrainment zone; see Figure 7.1); the temperature gradient, $\gamma = d\Theta/dz$, in the free atmosphere above the boundary layer; the Coriolis parameter, $f = 2\Omega \sin \phi$, where Ω is the angular velocity of planetary rotation and ϕ is the latitude; the buoyancy parameter, g/θ_0, with g being the acceleration due to gravity and θ_0 the reference temperature; the surface roughness length, z_0; and the vertical turbulent heat flux at the surface, H_0. All these parameters are related to the phenomena that generate turbulence in the boundary layer.

When turbulence is produced by shear, it is convenient to introduce a velocity scale, u_*, related to the shear stress, τ_{s0}, at the surface, known as the friction velocity:

$$u_* = \sqrt{\frac{\tau_{s0}}{\rho_0}} \tag{7.1}$$

where ρ_0 is the density of the air at the ground (note that quantities with the subscript "0" refer to values at the surface).

When turbulence is produced by buoyancy due to an upward sensible heat flux imposed at the planet's surface, the covariance between vertical velocity and temperature fluctuations, w' and θ', respectively, taken at the surface, determines the surface heat flux, H_0,

$$-(\overline{w'\theta'})_0 = \frac{H_0}{\rho_0 c_p} \tag{7.2}$$

where c_p is the specific heat capacity of the air at constant pressure and the minus sign denotes that the surface is losing energy to the atmosphere. This parameter is often used to scale the intensity of turbulence (Monin and Obukhov, 1954). We also define a length scale known as the Monin–Obukhov length scale, L,

$$L = -u_*^3 \Big/ \left(\kappa \frac{g}{\theta_0} (\overline{w'\theta'})_0 \right) \tag{7.3}$$

where κ is the von Kármán constant, usually taken equal to 0.4. The overbar in (7.2) and (7.3) indicates an ensemble mean value and the prime indicates turbulent fluctuations. Note that L is positive for stable stratification and negative for unstable conditions.

Other appropriate length scales can also be introduced. Recalling that h is the depth of the turbulent layer above the surface and z denotes the height above the surface, we now have three length scales that provide two independent dimensionless combinations: z/h is the location of the point of interest within the boundary layer, and $h/|L|$ represents a stability parameter that is smaller than 1 when shear production dominates and greater than 1 when turbulence is primarily due to buoyant convection.

To characterize the boundary layer dynamics, we also need to introduce time scales that characterize non-stationary processes: (a) T_f, the time scale of external processes that act upon the boundary layer; (b) T_m, the time scale for the development of mean velocity and temperature profiles; (c) $T_t = \ell / U$, the time scale characteristic of the large-scale eddies (here, ℓ and U represent length and velocity scales for the macroscopic structure of turbulence); and (d) T_p, the time scale of turbulence production processes.

Note that u_*, L, and z alone define the scaling structure of the surface layer via the Monin–Obukhov flux-gradient relationships for mean velocity and temperature, which can be written as

$$\Phi_m\left(\frac{z}{L}\right) = \frac{\kappa z}{u_*}\frac{d\overline{u}}{dz} \tag{7.4}$$

$$\Phi_h\left(\frac{z}{L}\right) = Pr_t\frac{\kappa z}{\theta_*}\frac{d\overline{\Theta}}{dz} \tag{7.5}$$

where θ_* is defined by

$$(\overline{w'\theta'})_0 = u_*\theta_* \tag{7.6}$$

and Pr_t is the turbulent Prandtl number in neutrally stratified flows, defined as the ratio of eddy viscosity to eddy thermal diffusivity. These functional expressions relate Φ_m and Φ_h to the eddy viscosity and eddy diffusivity when used in the simplest closure models for the Reynolds-averaged boundary layer equations (see Section 7.5.2). They lead to the logarithmic flux profile relationships for velocity and temperature:

$$\left(\frac{\overline{u}}{u_*}\right) = \frac{1}{\kappa}\ln\left(\frac{z}{z_0}\right) \tag{7.7}$$

$$\left(\frac{\overline{\Theta}}{\theta_*}\right) = \frac{Pr_t}{\kappa}\ln\left(\frac{z}{z_{0T}}\right) \tag{7.8}$$

which govern the surface layer. Here, z_0 is a roughness length that characterizes the inhomogeneity of the surface for the momentum transfer, z_{0T} is an equivalent characteristic for the heat transfer, and $\overline{\Theta} = \Theta - \theta_0$. Equations (7.7) and (7.8) indicate that z_0 and z_{0T} are the heights above the surface at which $\overline{u}$ and $\overline{\Theta}$, respectively, go to zero. Generally, z_0 and z_{0T} are not equal.

On the other hand, molecular exchanges are two orders of magnitude weaker than turbulent diffusion in the first few meters (Martínez et al., 2009a,b), which allows us to neglect molecular diffusion in the surface layer. Moreover, the height of the surface layer and the magnitude of the Coriolis force are found to be of the same order on Mars as on Earth, and consequently Coriolis forces can be neglected in both cases.

We shall treat separately each of the hypotheses required in order to use surface layer similarity theory on Mars. The complexity of the terrain, along with large-scale meteorological phenomena (synoptic perturbations), can greatly alter horizontal homogeneity. However, synoptic perturbations are not usually present at some locations and times of the year

(such as mid-latitude northern summertime); and some sites, such as those for the Mars Pathfinder (hereafter Pathfinder) and Viking missions, although not especially flat, did not present much in the way of sharp topography (Golombek et al., 1997).

Concerning suspended dust, similarity theory takes no account of this peculiarly Martian phenomenon (in that none of the independent variables takes any direct account of dust). Nevertheless, the observed dustiness is usually low (Savijärvi, 1999) in mid-latitude locations during northern summertime, and thus the radiative importance of the dust becomes reduced anyway. Suspended dust may be important, however, for the Martian PBL in complex terrains (Karelsky and Petrosyan, 1995; Karelsky et al., 2007), which may have significant implications for the validity of classical similarity theory.

The portion of the PBL above the surface layer may be characterized by one of three cases: the neutral, convective, or stable boundary layers. For the neutral boundary layer, the heat flux at the surface is close to zero ($(\overline{w'\theta'})_0 \simeq 0$) and buoyancy forces play no significant role, so that u_*, h, and z are the appropriate scaling parameters.

The competition between shear and buoyancy effects is commonly characterized by the dimensionless Richardson number, which is a measure of hydrodynamic stability that depends on the current vertical gradients of temperature and wind speed. It is generically defined as

$$Ri = \frac{N^2}{S^2} \tag{7.9}$$

where N is the Brunt–Väisälä frequency ($N^2 = \beta g\, \partial\Theta / \partial z$, with β the thermal expansion coefficient), measuring the strength of static stability, and S is a measure of the mean vertical shear of the horizontal flow. Small ($\leq O(1)$) or negative values of Ri are often regarded as indicating the likelihood of turbulence generation via instability (through either shear-dominated or buoyancy-dominated processes, though see Galperin et al. (2007) and Section 7.5.2 for further discussion), while larger (positive) values of Ri are associated with intermittent turbulence.

In the convective boundary layer, the convective velocity scale, w_*, is commonly defined as

$$w_* = \left(\frac{g}{\theta_0}(\overline{w'\theta'})_0 h\right)^{1/3} \tag{7.10}$$

(though see also Section 7.5.3.4). The mixed-layer scalings for a convective boundary layer are

$$\frac{\sigma_w}{w_*} \equiv \frac{(\overline{w'^2})^{1/2}}{w_*} = \Phi_w\left(\frac{z}{h}\right) \tag{7.11}$$

$$\frac{\sigma_u}{w_*} \equiv \frac{(\overline{u'^2})^{1/2}}{w_*} = \Phi_u\left(\frac{z}{h}\right) \tag{7.12}$$

and

$$\frac{\sigma_\theta}{\theta_*} \equiv \frac{(\overline{\theta'^2})^{1/2}}{\theta_*} = \Phi_\theta\left(\frac{z}{h}\right) \tag{7.13}$$

where σ_w, σ_u, and $\sigma\theta$ represent the characteristic scales for the standard deviations of vertical velocity w, horizontal velocity u, and temperature θ, respectively; the temperature scale, θ_*, is defined as $\theta_* = (\overline{w'\theta'})_0 / w_*$; and $(\Phi_w, \Phi_u, \Phi_\theta)$ represent dimensionless universal functions of z/h that determine the vertical structure of each quantity.

As for the surface layer, the complexity of the Mars terrain, together with other large-scale phenomena (synoptic perturbations), can greatly alter horizontal homogeneity in the convective mixed layer. Thus, we must be certain to apply this theory only under those circumstances where the flow is sufficiently homogeneous in the horizontal direction.

Along with horizontal homogeneity and stationarity, the other main hypotheses which must be satisfied when applying convective mixed-layer similarity theory are: (i) convection is the dominant heating mechanism throughout the mixed layer; and (ii) surface stress effects become negligible throughout this layer. This is certainly so on Earth under fair weather conditions (André et al., 1978), because the heating rate due to radiation convergence within the boundary layer is negligible, and the winds are light under the aforementioned conditions. On Mars, however, special care must be taken, since this may not always be the case.

There is no uncertainty in neglecting the effects of surface stress throughout the bulk of the Martian convective mixed layer when the winds are calm (especially during low- and mid-latitude northern summertime). A problem may arise with the radiative heating throughout the Martian convective mixed layer, however, which now becomes non-negligible because of the heating due to absorption of the upward long-wave radiative flux (in a mostly CO_2 atmosphere), and also due to the direct absorption of solar radiation by dust (see Section 7.5.3).

Under stable conditions, shear production is the only significant source of turbulence, and so $(\overline{w'\theta'})_0$, τ_s/ρ_0, and z are appropriate scaling parameters, along with the conventional Monin–Obukhov length scale defined by (7.3).

The gradient Richardson number is another important scaling parameter for stable boundary layers, providing an additional measure of static stability; it is defined as

$$Ri_g = \frac{g}{\theta_0}\frac{\partial\overline{\Theta}}{\partial z} \Big/ \left(\frac{\partial\overline{u}}{\partial z}\right)^2 \tag{7.14}$$

This parameter provides an indication of the relative importance of shear production compared with buoyancy effects, with the shear production dominating for numerically small values of Ri_g and buoyancy effects being dominant for numerically large (positive) values of Ri_g.

The Reynolds number, $Re = U\ell / \nu$, is frequently used as a measure of the complexity and degree of turbulence in fluid dynamics. Here, as before, U is a characteristic velocity scale, ν is the kinematic viscosity of the air, and ℓ is a characteristic length scale. When Re exceeds a certain limiting value (typically $\gtrsim 2000$), the flow is likely to become turbulent, and the intensity of turbulence increases with increasing Re.

If viscous dissipation is relatively weak, turbulence kinetic energy (TKE) is freely exchanged between different length scales, systematically cascading energy towards smaller scales over a range of scales known as the inertial subrange. Within this interval, the energy spectrum adopts a self-similar form known as the Kolmogorov law (Tennekes and Lumley, 1972; Pope, 2005),

$$E(k) = \alpha_1 \epsilon^{2/3} k^{-5/3} \tag{7.15}$$

where ϵ is the turbulence energy dissipation rate and α_1 is a dimensionless constant known as the Kolmogorov constant, whose value has been experimentally estimated to be $\alpha_1 = 1.5 \pm 0.15$ (Sreenivasan, 1995). At the largest scales, the limit of the inertial subrange is governed by the scale of shear production, which may be of the order of the domain (or, in the case of the PBL, typically the height of the measurement above the surface). At the smallest scales, the inertial subrange is limited by the scales at which viscous dissipation becomes significant. The latter is characterized by the Kolmogorov scale of viscous dissipation, η, given by

$$\eta = (\nu^3 / \epsilon)^{1/4} \approx (\kappa z \nu^3 / u_*^3)^{1/4} \tag{7.16}$$

where the near-ground approximation of the turbulent dissipation has been used,

$$\epsilon \approx u_*^3 / (\kappa z) \tag{7.17}$$

Taking the scale for surface layer shear production as κz, we express the ratio of the scales of the shear production and the dissipation in terms of the shear production Reynolds number Re_l,

$$\kappa z / \eta = (\kappa Re_l)^{3/4} \tag{7.18}$$

where

$$Re_l = u_* z / \nu \tag{7.19}$$

Close to the surface, the value of ν may determine the conditions for rough or smooth (laminar) flow. These conditions are often formulated in terms of yet another Reynolds number, the roughness Reynolds number,

$$Re_0 = u_* z_0 / \nu \tag{7.20}$$

The transition between rough and smooth flow is generally taken to occur for Re_0 around 2–3 (Brutsaert, 1982).

7.3 COMPARISON AND ROLES OF PLANETARY BOUNDARY LAYERS ON EARTH AND MARS

To a large extent, the description and modeling of the Martian atmospheric PBL are similar to the corresponding features of the Earth's atmospheric boundary layer, with respect both to numerical modeling and to the different scaling laws. Where there are differences, they almost all derive from the differences in density of the two atmospheres. The relatively low density of the Martian atmosphere means that it plays a minor role in the heat budget of the ground.

7.3.1 Energy Budget of PBL and Surface

The sensible heat flux appears negligible in the surface energy budget compared to radiative fluxes (see Figure 7.2). The Martian surface temperature is therefore essentially derived from

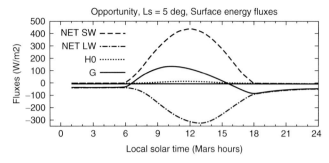

Figure 7.2. Surface energy fluxes in the one-dimensional (1D) simulation of Savijärvi and Kauhanen (2008) of conditions at the Opportunity landing site at $L_s = 5°$, showing the net short-wave flux (dashed line), net long-wave flux (dash-dotted line), sensible heat flux H_0 (dotted line) and net heat flux into the ground G (solid line). Figure adapted from Savijärvi and Kauhanen (2008) with permission.

an equilibrium between the radiative terms and the soil heat conduction (Savijärvi and Kauhanen, 2008). Lacking the moderating influence of the atmosphere, the diurnal variation of the surface temperature tends to be larger than on Earth, except during dust storms (Larsen et al., 2002; Määttänen and Savijärvi, 2004).

While the atmosphere on Mars generally plays a small role for the heat budget of the surface, the opposite is not true. The Martian PBL plays a crucial role in communicating fluxes of heat, momentum, dust, moisture, and chemical species between the solid surface of Mars and the global-scale atmospheric circulation. These fluxes not only inject aerosols and constituents into the atmosphere above the PBL, but they can modify the global atmospheric circulation, either through radiative effects leading to local heating or cooling (such as from absorption by aerosols) or by exchanging momentum and heat between the surface and atmosphere. These last two effects provide a drag on the large-scale flow, by mixing in air with zero mean momentum, and a direct thermal forcing, by mixing heat from the surface.

Heat transfer between the atmosphere and the surface drives the atmosphere into stronger instability at daytime and deeper stability at nighttime than is normally seen on Earth. The stronger daytime convection on Mars is associated with a deep convective boundary layer, extending occasionally to over 10 km altitude above the surface during the daytime (Hinson et al., 2008). This can lead to efficient mixing and vertical transport of quantities such as heat, momentum, dust, moisture, etc. from those parts of the atmosphere in close contact with the surface rapidly to great heights, at which point they may be taken up by the global circulation and distributed around the planet on a timescale of several days. On Mars, the latent heat effects of water are essentially negligible – except perhaps for frost formation and polar clouds (Savijärvi and Määttänen, 2010) – but the ability of dust (and water ice to some extent) to absorb solar and infrared radiation, and so locally to heat the atmosphere, can be thought of as a somewhat analogous effect. This (essentially radiative) heating effect takes place on longer timescales than those on which latent heat can drive convective cumulus clouds to the top of the terrestrial troposphere (~20–40 minutes, assuming a typical updraft wind speed of ~10 m s^{-1} (e.g. see Emanuel, 1994)).

In contrast, nighttime flows normally end up with large height intervals over which the Richardson number is large

enough for turbulence to become intermittent. The small impact of the Martian atmosphere on the surface heat budget is augmented by another aspect of the atmosphere, its lack of significant amounts of water, which on Earth also acts to dampen the diurnal temperature cycle by evaporation and condensation.

On Mars the vertical flux of long-wave net radiation tends to play a larger role in the energy budget of the PBL than on Earth (see also Sections 7.5.1 and 7.5.3). Indeed, close to the ground, it will often dominate over the turbulent heat flux (e.g. see Savijärvi, 1999; Larsen et al., 2002; Savijärvi and Kauhanen, 2008). Model computations by Savijärvi (1999), Savijärvi and Kauhanen (2008), and Spiga et al. (2010) (e.g. see Figure 7.3, where z_i is the depth of the convective boundary layer) indicate that, close to the ground (i.e. the lowest hundred meters or so in the mid-morning), the dominating net radiative forcing of the near-surface air temperature forces the turbulent heat flux to reverse sign relative to normal expectations, i.e. the profiles become super-adiabatic in the daytime sunlit conditions with negative (downward) heat fluxes. Convective processes act to cool the atmosphere during the day rather than warm it, as is the case on Earth. At larger altitudes (more than a few hundred meters above the surface), the turbulent heat flux tends to dominate. Absorption of short- and long-wave radiation by the atmospheric dust loading also influences the heating and cooling of the air, similarly to clouds and dust on Earth.

7.3.2 Boundary Layer Turbulence

The low density of the Martian atmosphere also implies a much higher kinematic viscosity, ν, and heat diffusivity, κ_T, than for Earth, because the dynamic characteristics are divided by the density to derive the kinematic values entering our description of the atmospheric motion. The larger value of ν for the Martian atmospheric PBL in turn influences a number of parameters used to characterize the turbulent conditions of the atmosphere.

The Kolmogorov scale, η, characterizing the smallest scale of turbulence due to the viscous dissipation of small-scale motions, becomes about 20 times larger than in the Earth's PBL. Considering the typical measurement height for Martian wind data, $z = 1$ m, and a u_* value of 0.5 m s^{-1}, we find that η is about 0.3–1 mm on Earth, but nearer 7 mm–2 cm on Mars. Therefore, the characteristic Reynolds number for the flow becomes smaller on Mars and the inertial subrange of the variance spectra for turbulence becomes narrower. Using typical values of atmospheric parameters, we find the ratio of shear production scale to η to be about 50 for Mars but 1250 for Earth. Therefore, we expect the inertial subrange of turbulence in the near-surface atmosphere to be much less important on Mars than on Earth. Similarly, the Reynolds number based on roughness length (7.20) tends to be smaller on Mars than on Earth for similar wind speeds. As described in Section 7.2, the roughness Reynolds number characterizes the interaction very close to the surface, e.g. determining the fluxes between the flow and the surface. This means that the Martian surface flux is characterized as being in a somewhat smoother, more laminar regime than on Earth for similar wind speeds, and z_{0T} and similar scalar parameters will therefore be larger on Mars than on Earth.

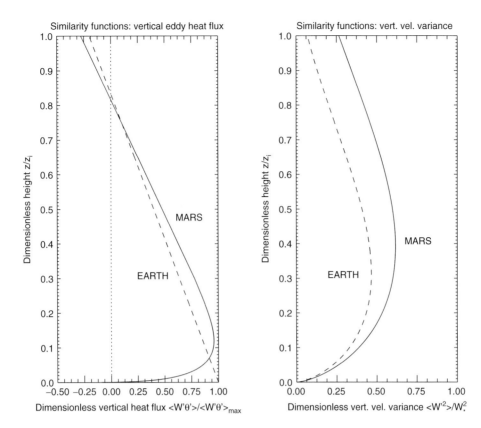

Figure 7.3. Empirical similarity functions in quasi-steady midday conditions, showing the average variation with dimensionless height (normalized by the convective boundary layer height z_i) of dimensionless (a) vertical eddy heat flux and (b) vertical velocity variance on Mars (solid line) and on the Earth (dashed line). Figure adapted from Spiga et al. (2010) with permission.

If one takes into account the above considerations, it seems that one can use the boundary layer formulations from Earth more or less equally to describe the boundary layer on Mars. The overall empirical characteristics are that, close to the surface, we find typical wind speeds on Mars to be much as we know them on Earth. However, we find much stronger diurnal temperature variations on Mars, with more vigorous turbulent temperature and wind fluctuations during the day. These features have been found to be consistent with standard surface boundary layer similarity formulations with respect to the relation between mean profiles, fluxes, and variances (Sutton et al., 1978; Tillman et al., 1994; Larsen et al., 2002; Gunnlaugsson et al., 2008; Davy et al., 2010). See Table 7.1 for estimates of characteristic scales for the unstable boundary layers on Earth and Mars. For neutral to moderately stable conditions, one finds no substantial difference in the description of the Earth and Mars boundary layers, although the relevant parameters for Mars are quite uncertain. For Martian (nighttime) stable conditions, it seems that the surface layer is driven into a more stable state more often than on Earth, in accord with the missing moderation by the turbulent heat flux.

It is uncertain to what extent the importance of the net radiative flux for the temperature field on Mars can be reconciled with current surface layer scaling laws that assume the existence of a layer where the turbulent fluxes are independent of height close to the ground. Also, the equations for the growth of the unstable Martian boundary layer height may be modified by the inclusion of radiative flux terms. The Martian atmospheric surface layer calculated

from the results of models and data analysis cited above appears more consistent with all turbulent fluxes being height-independent, except for the heat flux, with the sum of the net turbulent and radiative heating rates remaining constant with height.

In conclusion, Figure 7.3 and the other LES results (e.g. Spiga et al., 2010) suggest that during daytime the radiative heating is so strong in the surface layer of Mars that the Monin–Obukhov (MO) similarity assumption (based on turbulence being the dominant process) is not well founded as such. However, above this super-adiabatic surface layer, the MO theory appears valid again. Savijärvi (2012a,b) finds the same, validating his 1D model simulations against the Spirit Mini-TES-based soundings of Sorbjan et al. (2009), as did Martínez et al. (2009b) with reference to Mars Pathfinder data under calm and relatively dust-free conditions in northern mid-latitude summer. During nighttime the MO theory appeared as valid as on the Earth, turbulence then being the dominant cooling process near the surface (outside periods of calm winds), while strong longwave cooling dominated at the top of the surface inversion, thus tending to build the stable boundary layer upwards.

7.3.3 Dust Devils and Dust Storms

In direct model simulations, Newman et al. (2002a) showed that boundary layer mixing is essential to lift dust into the body of the atmosphere. If dust is lifted by near-surface wind stress, derived from the large-scale winds, but only injected into the lowest level in a model (5 m above the surface, in the case of

Table 7.1 *Approximate mean daytime unstable surface layer and convective mixed-layer turbulent parameters on Mars and on Earth. Terrestrial values correspond to PBLs formed over flat and homogeneous terrain, and under conditions with no baroclinic disturbances. Note that turbulence characteristics for stable conditions are not included, because for these situations Earth and Mars characteristics are all very similar.*

Parameter	Mars	Earth	Units
Unstable surface layer			
Obukhov length, L	-17	similar	m
Friction velocity, u_*	0.4	0.3	m s^{-1}
Temperature scale, $\lvert T_* \rvert$	1	0.15–0.88	K
Dissipation rate, ϵ_z	0.16	0.01–0.02, $z = 4.32$ m	m^2 s^{-3}
		0.001–0.01, $z = 18$ m	m^2 s^{-3}
Temperature s.d., $\sigma\theta(z)$	3	0.18–0.32, $z = 4$ m	K
Horiz. velocity s.d., $\sigma_u(z)$	2	1.4, $z = 4$ m	m s^{-1}
Vert. velocity s.d., $\sigma_w(z)$	0.5	0.4–0.6, $z = 4$ m	m s^{-1}
		0.38–0.44, $z = 4.32$ m	m s^{-1}
Convective mixed layer			
Boundary layer height, h	6	0.2–2	km
Vertical velocity scale, w_*	4	1–2.5	m s^{-1}
Temperature scale, θ_*	0.1	0.03–0.1	K
Dissipation rate, $\langle \epsilon \rangle$	0.005	0.001–0.005	m^2 s^{-3}
Temperature s.d., $\langle \sigma\theta \rangle$	0.3	0.06–0.2	K
Horiz. velocity s.d., $\langle \sigma_u \rangle$	2.4	0.47–1.13	m s^{-1}
Vert. velocity s.d., $\langle \sigma_w \rangle$	2.4	0.6–1.4	m s^{-1}

Newman et al. (2002a)), then the dust can remain trapped close to the surface and consequently not be transported over large horizontal scales by the model winds. This was particularly true if the dust, or other tracer, in the model is not radiatively active. Feedback from radiative heating can enable the dust to enter the planetary-scale meridional and zonal circulation much more easily by enhancing the local vertical velocity and leads to dust raised in one hemisphere being transported to the other within days as it enters the powerful cross-equatorial Hadley circulation.

Dust devils are one of the most obvious manifestations of boundary layer activity, primarily during summer afternoons, when the atmosphere is relatively clear and the surface is warm. Dust devils are regularly seen on Earth, particularly in desert regions during the summer, but they are even more spectacular on Mars. Dust devils perhaps contribute the largest component of the background dust level during the relatively clear Martian northern hemisphere summer season. Once the amount of dust in the atmosphere becomes larger, the atmosphere becomes relatively warmer and the surface cooler, so reducing the temperature contrast, the intensity of convection, and the height and strength of the convective boundary layer. This negative feedback means that dust devils are unlikely to be responsible for regional or planet-encircling dust storms on Mars, but will instead tend to be inhibited as more dust is lifted (Newman et al., 2002b).

Orbiters and landers have shown that dust devils and dust storms are ubiquitous on Mars (e.g. Thomas and Gierasch, 1985; Greeley and Iversen, 1985; Rennó et al., 2000; Cantor et al., 2001). Together with convective plumes, and turbulence forced by wind shear, these weather phenomena transport dust, water vapor, and other tracer species upwards. Dust is lifted

when the wind speed exceeds a threshold value, and sand particles propelled by drag forces bounce along the surface, ejecting the smaller, harder-to-lift, dust aerosols into the air, in a process known as saltation (Bagnold, 1941). Besides ejecting dust into the air, saltation plays an important role in geological processes such as sediment transport, the formation of sand dunes, and wind erosion (Greeley and Iversen, 1985).

Dust devils form at the bottom of convective plumes. Since their sources of angular momentum are local wind shears, caused either by the convective circulation itself or by larger-scale phenomena, they can rotate clockwise or counterclockwise with equal probability (Rennó et al., 1998). A distinctive feature of intense dust devils is their well-defined dust funnel. Theory indicates that dust is focused around the funnel by a dynamic pressure drop caused by increases in the speeds of the air spiraling towards the vortex (Rennó, 2008). Like waterspouts, tornadoes, and hurricanes, dust devils can be idealized as convective heat engines. They are the smallest and weakest members of this class of weather phenomena (Rennó, 2008). They form when a vortex strong enough to initiate saltation occurs over surfaces composed of loose particles. The intensity of a dust devil depends on the depth of the convective plume and the transfer of heat from the ground into the air (Rennó et al., 1998).

On Mars, dust devils are much bigger and stronger than on Earth. Terrestrial dust devils have typical diameters of less than 10 m and are seldom higher than 500 m (Sinclair, 1973). In contrast, dust devils with diameters between 100 m and 1 km, and heights in excess of 5 km, are observed on Mars (Thomas and Gierasch, 1985; Malin et al., 1999). The dust devils observed in the Pathfinder images have about 700 times the dust content of the local background atmosphere (Metzger et al., 1999).

Measurements by Rennó et al. (2004) indicate that the heat and dust fluxes in terrestrial convective plumes and dust devils can be many orders of magnitude larger than their background values of a few hundred watts per square meter (W m^{-2}) and a few hundred micrograms per square meter per second (μg m^{-2} s^{-1}). There is evidence that, besides dust storms, dust devils play an important role in the Martian dust cycle. For example, the atmospheric dust opacity increased throughout the Pathfinder mission in spite of low wind conditions and the absence of dust storms on the planet. Ferri et al. (2003) showed that the dust flux due to dust devils contributes significantly to the maintenance of dust in the atmosphere of Mars, perhaps even being the primary source of dust into the atmosphere of the Pathfinder landing site at the Ares Vallis region. They might also play an important role in the transport of other tracer species.

On Earth, windblown sand, dust devils, and dust storms produce electric fields ranging from a few hundred volts per meter (V m^{-1}) to 200 kV m^{-1} (Crozier, 1964; Stow, 1969; Schmidt et al., 1998; Rennó and Kok, 2008). Even small terrestrial dust devils can produce electric fields of the order of 10 kV m^{-1} (Rennó et al., 2004), which is of the order of that necessary to produce electric discharges in the thin Martian atmosphere (Melnik and Parrot, 1998). Charge transfer during collisions of sand with dust particles (Rennó et al., 2004), and charge separation by updrafts and turbulent diffusion, produce these large fields (Kok and Rennó, 2008). Numerical simulations by Melnik and Parrot (1998) predict the occurrence of electric discharges in Martian dust storms. Ruf et al. (2009) showed evidence that deep Martian dust storms can produce powerful electric discharges.

Recent studies suggest that electric fields in Martian dust storms produce energetic electrons (Delory et al., 2006), destroy water vapor, and lead to the formation of hydrogen peroxide, a potential sink of methane (Atreya et al., 2006; Kok and Rennó, 2009). Hydrogen peroxide could be responsible for the reactive soil found at the Viking landing sites (Oyama et al., 1977). Moreover, energetic electrons are predicted to directly dissociate methane (Farrell et al., 2006). The large spatial and temporal variability of the methane observed in the Martian atmosphere suggests the existence of as yet unknown strong sources and sinks of this gas (Lefevre and Forget, 2009). Dust electrification and heterogeneous reactions on dust particles might explain this puzzling result.

7.3.4 The Influence of the PBL on Martian Weather and Climate

The planetary boundary layer has a profound effect upon the Martian atmosphere on weather and climate time and spatial scales. Its influence is particularly strong on a planet like Mars with a thin and mostly cloud-free atmosphere, where changes in surface temperature and surface drag have a huge impact compared to the relatively smaller thermal and dynamical inertia of the atmosphere itself. This is illustrated by the small changes in surface temperature with elevation on Mars; temperature contours tend to follow the surface much more so than they do on Earth (Webster, 1977; Nayvelt et al., 1997). This results in a global-scale thermal forcing to the lower atmosphere on Mars, communicated from the surface by conduction over only very

short distances and then, much more strongly, by radiation and convection within the PBL.

The PBL also provides a large-scale friction on the global circulation, by mixing air that has been in direct contact with the surface and that has essentially zero momentum over a range of heights up to at least 10 km. This results in an effective drag to the large-scale winds. Large-scale stationary waves are generated on Mars, linked to the surface topography. Nayvelt et al. (1997) have shown that the near-surface winds associated with stationary waves are sensitive to the parameterized frictional effects of the boundary layer, and a careful choice of parameters is necessary to get good agreement with observed streak directions. Small-scale ($\lesssim$100 km horizontal wavelength) inertia–gravity waves are also readily generated near the surface on Mars, largely through winds blowing over topography, but also from convective and frontal wedge-type forcing in the boundary layer. These waves can have a strong effect on the upper atmospheric winds, where the vertically propagating waves break, typically tending to drag the large-scale winds towards zero (Joshi et al., 1996).

In the lower atmosphere, the varying thickness of the PBL, which tends to be deeper over higher topography (Hinson et al., 2008; Spiga et al., 2010), may itself cause topographic effects (low-level drag over different depths of the atmosphere and generation of stationary waves), which are exaggerated compared to the (already large) actual size of the Martian surface topography – see also the explanation offered for this by Souza et al. (2000).

Finally, the free atmosphere will feed back on the PBL through changes in radiative forcing under very dusty conditions, reducing convection and turbulence. Thus the PBL might itself be seen as a vital component of the Martian dust cycle which can exhibit both positive (at low dust levels) and negative (at high dust levels) feedbacks on the dust cycle and the amount of dust loading in the Martian atmosphere.

7.4 RECENT AND CURRENT MISSIONS AND OBSERVATIONS

Our current knowledge of Mars' atmospheric boundary layer is based on observations made by a large set of spacecraft. The observations can be subdivided into orbital (Section 7.4.3) and surface-based observations, the latter in turn further into *in situ* (Section 7.4.1) and surface-based remotely sensed observations (Section 7.4.2). These observations are of various types, and they characterize not only the atmosphere itself, but also the planetary surface or variables providing indirect diagnostics of the boundary layer, its state and phenomenology.

7.4.1 Surface *In Situ* Observations

Surface *in situ* observations have so far been pointwise time series observations performed by instruments on-board four landers – deployed only in the northern hemisphere and in near-equatorial regions (Figure 7.4). All four landers have provided pressure (p) and temperature (T) time series, whereas only the Viking Landers (VL) have had satisfactory wind (V;

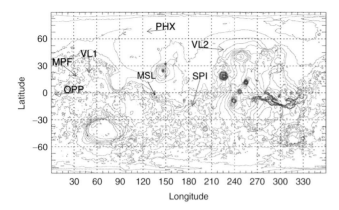

Figure 7.4. Landing sites of missions with relevant PBL observations. A clear majority of the sites are in the northern hemisphere. MOLA topography contours are shown in gray to assist in locating the landing sites.

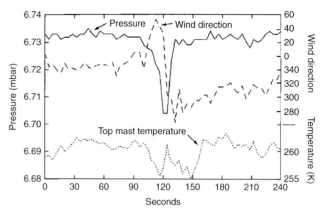

Figure 7.5. Pressure, wind, and temperature changes associated with a dust devil passing through the Pathfinder landing site. The sampling interval was 4 s. Adapted from Schofield et al. (1997) with permission.

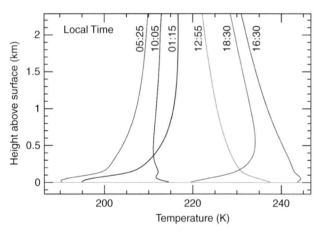

Figure 7.6. Typical temperature profiles retrieved from upward-looking mini-TES observations made over the course of four days by the Spirit lander near $L_s = 5°$, showing the evolution of the daytime near-surface super-adiabatic layer and the nighttime inversion layer. Figure adapted from Smith et al. (2006) with permission.

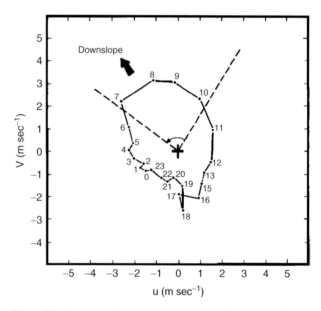

Figure 7.7. Mean wind hodograph for the first 50 sols at the VL2 landing site. Each point represents the vector mean of the northward and eastward components of the wind over the time interval indicated. The sectors indicated by dashed lines are directions suffering from interference from the body of the spacecraft. Figure adapted from Hess et al. (1977) with permission.

speed and direction) sensors. Some quantities have been measured simultaneously at two or three levels above the surface, allowing for estimation of vertical gradients. The datasets exhibit varying temporal coverage, time resolutions, accuracies, and stabilities.

Some measurements have allowed for the characterization of the faster and more short-lived PBL phenomena such as dust devils (see Section 7.3.3). They have been previously observed by orbital (Thomas and Gierasch, 1985) and lander imagers (Metzger et al., 1999; Ferri et al., 2003), but dust devil signatures have also been seen in the landers' high-sampling-rate pressure time series (Ryan and Lucich, 1983; Murphy and Nelli, 2002; Ringrose et al., 2003; Ellehoj et al., 2010; Figure 7.5).

The Mars Exploration Rovers' (MER) payloads include no *in situ* atmospheric sensors (Squyres et al., 2003), but the Mini-TES instruments (Smith et al., 2004; Section 7.4.1) have been noteworthy for being able to make remote sensing measurements that can characterize the microscale region (from

meters to kilometers) around the rovers (Figure 7.6; Section 7.4.2.2). The MERs have operated only in the equatorial region (Figure 7.4).

7.4.1.1 Viking Landers' Meteorological Instrumentation

Viking Landers' payloads included the Viking Meteorology Instrument System (VMIS), which comprised V, T and p sensors at a single level (Chamberlain et al., 1976). The wind speed (V) was measured by an overheat hot-film sensor array, while direction was obtained with a quadrant sensor. Some early observations are shown in Figure 7.7.

Hess et al. (1977) suggested the wind patterns to be predominantly the result of pressure gradients due to a combination of local topographical slopes and coupling with large-scale winds aloft. At the VL1 site both factors were interpreted to be of similar magnitudes, while at the VL2 site the slope is seen to dominate.

The *T* sensors were thermocouple arrays and the *p* sensors were stretched-diaphragm-type reluctance transducers. The *V* and *T* sensors were located approximately 1.6 m above the surface and the *p* sensor was placed inside the lander.

Instrument accuracies were reported as ±15% in wind speed (for speeds over 2 m s⁻¹), ±10% in wind direction, ±1.5 K in temperature, and 0.07 hPa in pressure. The pressure precision was limited by digitization. The *V* and *T* measurements suffered from wind-direction-dependent thermal contamination due to the heat plume emanating from the Radioisotope Thermal Generators (RTGs; Hess et al., 1977; Figure 7.7). The initial observational strategy was to collect data with sample intervals of 4 s or 8 s in 11 min modules spaced 1 h 27 min apart (Hess et al., 1976).

The landers also measured entry profiles – the first ones of their kind for Mars (Seiff and Kirk, 1977). Although sampling essentially single instants (in local time (LT) and season) and locations, the observations provided unprecedented information on the PBL vertical structure. The profiles were interpreted by Seiff and Kirk to show a near-surface convective region and PBL height $h \approx 6$ km at the time of landing.

VL1 landed on July 20, 1976 at 16:13 LT (at areocentric longitude $L_s \approx 96.7°$) and operated until November 13, 1982 ($L_s \approx 226.5°$), operating for 2307 days/2251 sols, more than three Mars years. VL2 landed on September 3, 1976 at 09:49 LT ($L_s \approx 117.2°$) and ended operations on April 11, 1980 ($L_s \approx 91.0°$), operating for 1316 days/1284 sols, almost two Mars years.

7.4.1.2 Mars Pathfinder Meteorological Instrumentation

Pathfinder comprised a larger stationary lander and a small rover. The stationary lander payload included the Atmospheric Structure and Meteorology instrument (ASI/MET) designed for measurement both during entry (especially the parachute-decelerated phase below ≈8 km) and after landing (Seiff et al., 1997). The ASI/MET comprised a *V* sensor, three *T* sensors, and a *p* sensor. Sensor technologies were similar to those used in the VMIS: an overheat *V* sensor (albeit with smaller overheat) located above the surface at $z = 1.1$ m, and thin-wire thermocouple *T* sensors at three levels $z = 0.25$, 0.50, and 1.00 m. Sample *T* observations are shown in Figure 7.8. The diurnal characteristics of the temperatures and the directions of their near-surface vertical gradients stem from the low density of the Martian atmosphere, which causes the atmospheric temperatures to be predominantly driven by those at the surface.

The *p* sensor had two measurement ranges, 0–12 hPa for descent and 6–10 hPa for surface observations. The respective resolutions (at 14 bit) were 0.075 and 0.025 Pa. The sensor had some calibration problems, discussed by Haberle et al. (1999). The *T* sensor design goals were an absolute accuracy of 1 K, relative accuracy of 0.1 K, and resolution of 0.04 K between 160 and 300 K of ambient temperature. The data were sampled at 4 s intervals (synoptic observation sessions) and at 1 s intervals (PBL sessions). Based on testing, the wind speed

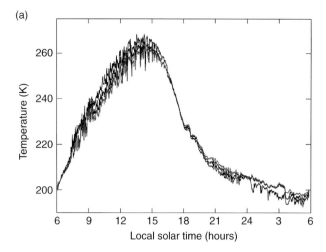

(a)

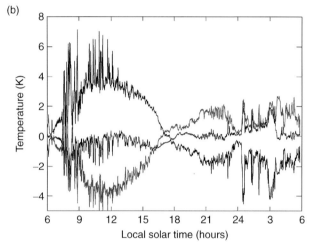

(b)

Figure 7.8. (a) The diurnal variation of atmospheric temperature at the Pathfinder landing site measured by the top (red), middle (black), and bottom (blue) mast thermocouples from 06:00 LT on sol 25 to 06:00 LT on mission sol 26. The *T* sampling interval was 4 s throughout this period, but the plots use 2 min running means for clarity (averaging reduces the amplitude and frequency of the fluctuations present in the raw data). (b) The data plotted as *T* deviations from the mean of all three thermocouples. Adapted from Schofield et al. (1997) with permission. A black and white version of this figure will appear in some formats. For the color version, please refer to the plate section.

accuracy was approximately 1 m s⁻¹ at low wind speeds, worsening to about 4 m s⁻¹ above 20 m s⁻¹. Directional accuracy was of the order of ±10°. Despite the Pathfinder being designed to reduce lander-induced thermal contamination (the MET mast was placed at the end of one petal instead of over the deck), accurate determination of wind speeds has been unsuccessful, although the directions have been extracted satisfactorily.

Pathfinder landed on July 4, 1997 ($L_s \approx 142.4°$) and operated until September 27, 1997 ($L_s \approx 188.0°$), a total of 85 days/83 sols.

7.4.1.3 Phoenix Meteorological Instrumentation

The Phoenix payload included the Meteorological (MET) instrument package to "provide information on the daily and seasonal variations in Mars near-polar weather during Martian late spring and summer" (Taylor et al., 2008). However, Phoenix only measured the atmospheric density profile during the entry and descent phase (Withers and Catling, 2010). The package consisted of T sensors at three vertical levels and a compound p sensor. The T sensors were again thermocouples and were located at $z = 0.25$, 0.50, and 1.00 m, whereas the p sensing system (inside the lander) was based on capacitive silicon diaphragm sensors. Non-MET sensors relevant to observations of the Martian PBL were the imaging system (Section 7.4.2.1), the LIDAR (Section 7.4.2.3) and the Thermal and Electrical Conductivity Probe (TECP; Zent et al., 2009).

The pressure system's accuracy at the beginning of the surface mission was <6 Pa when the temperature in the vicinity of the sensor was <0°C and <11 Pa when $T > 0$°C. Resolution was ≈0.1 Pa and time resolution 2–3 s (Taylor et al., 2010). The T measurement quality requirements were: range 140–280 K, absolute accuracy ±1 K, and resolution 0.5 K (Taylor et al., 2008). The measurements at the lowest level have been shown to be unreliable and highly dependent on wind direction due to thermal contamination caused by the vicinity of the warm spacecraft deck (see also Sections 7.4.1.1 and 7.4.1.2)

Due to resource constraints, the MET package did not include a wind sensor similar to those in the VMIS or ASI/MET packages. Wind measurements were carried out with a simple telltale consisting of a lightweight suspended cylinder being deflected by the wind, with the deflection being estimated utilizing the imaging system. The system has been found to supply reliable data on both mean wind and turbulence, although a simple description of resolution and accuracy is difficult to provide (Gunnlaugsson et al., 2008; Holstein-Rathlou et al., 2010).

Surface–atmosphere fluxes and humidity (first relative humidity observations from the surface) have been estimated based on both MET and TECP (coordinated with LIDAR and MET telltale) observations (Davy et al., 2010; Tamppari et al., 2010; Zent et al., 2010).

Phoenix landed on May 25, 2008 ($L_s \approx 76.3$°) and operated to November 2, 2008 ($L_s \approx 151.1$°), some 161 days/157 sols, almost a quarter of a Martian year).

7.4.1.4 Mars Science Laboratory, First Results

The Mars Science Laboratory (MSL) on-board the large Curiosity Rover landed onto the Gale Crater floor (137.4°E, 4.6°S) in August 2012. MSL's payload includes a comprehensive environmental instrument package, the Rover Environmental Monitoring Station (REMS). The REMS observations most relevant to PBL studies are of wind speed/direction (V), pressure (p), relative humidity (RH), air temperature (T_{air}) and surface temperature (T_s). The pressure sensors are located inside the rover. The other sensors are placed at approximately 1.5 m above the ground, attached to two horizontal booms, attached in turn to the rover's vertical Remote Sensing Mast (RSM) – see Gómez-Elvira et al. (2008, 2012) and Petrosyan et al. (2011) for further details.

The analysis of its meteorological and other data is very much ongoing work at the time of writing. The wind sensor was partially damaged at landing. The MSL air temperatures

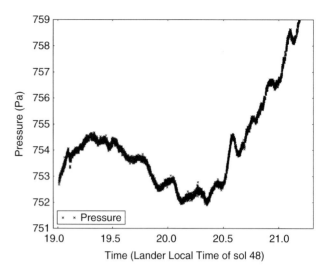

Figure 7.9. MSL REMS-P surface pressure observations during local evening hours 19:00–21:30 LT from sol 48. The observations exhibit rapid oscillations, on top of a strong diurnal–semidiurnal wave, which are likely caused by local PBL circulation phenomena. Figure adapted from Harri et al. (2014b) with permission.

display a similar diurnal cycle to that found at the Pathfinder site (see Figure 7.8), i.e. with large turbulent fluctuations during the day but smaller fluctuations at night. The pressure data imply the influence of strong diurnal thermal tides (Haberle et al., 2014; Harri et al., 2014b) and some novel PBL features. These include rapid evening oscillations and relatively dust-free convective vortices that are structurally analogous to the Pathfinder dust devils (see Figure 7.5).

A plausible interpretation of the evening oscillations (illustrated in Figure 7.9) is that they are due to internal gravity waves excited by emerging drainage flows down the rapidly cooling sides of the crater and the slopes of Mt. Sharp. Mesoscale models indicate the presence of such evening drainage flows (Harri et al., 2014b). They further suggest that the depth of the daytime convective PBL is strongly suppressed over the crater area due to a compensating sinking motion associated with daytime upslope flows (Tyler and Barnes, 2013).

MSL has also made direct humidity observations on Mars (Harri et al., 2014a). The REMS-H device consists of a temperature sensor and three polymeric RH sensors within a dust-protected cage. Figure 7.10 shows the hourly RH values and temperatures at 1.5 m height for MSL sols 15–17 and 80–82. RH evidently reaches values of 45–50% by each sunrise during the first period, whereas the second period was both warmer and drier. REMS-H has now completed its first Martian year of high-quality *in situ* humidity observations, and the analysis is currently underway at the time of writing.

7.4.2 Surface-Based Remote Sensing

The types of surface-based remote sensing instruments deployed to date include imaging systems, Thermal Emission Spectrometers (TES), and LIDARs.

7.4.2.1 Imaging Instruments

All Mars landers have included imaging systems, and most of them have also returned atmospheric imagery from the surface

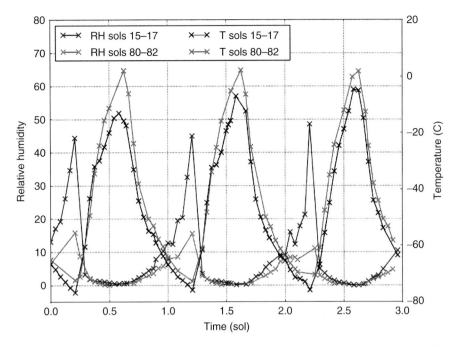

Figure 7.10. MSL REMS-H 1.5 m relative humidity and temperature observations during sols 15–17 ($L_s \sim 158°$) and 80–82 ($L_s \sim 196°$). The first period is cooler and moister; the second is warmer and drier. Figure adapted from Harri et al. (2014a) with permission. A black and white version of this figure will appear in some formats. For the color version, please refer to the plate section.

(e.g. Pollack et al., 1977; Smith et al., 1997; Lemmon et al., 2004; Moores et al., 2010). They have been relevant for studies of atmospheric (and, specifically, PBL) characteristics (such as dust optical thickness, water vapor content, aerosol properties) and phenomena (such as dust devils and condensation clouds). Imaging systems are often limited to the provision of indirect information – such as the occurrence of dust devils, and the presence of surface frost, ground ice, icy soil, frozen brine layers, or liquid brine droplets (Rennó et al., 2009) – on meteorological parameters and phenomena such as pressure, temperature, humidity, and convection. Imaging of dust devils has, however, provided statistics on the dust devils' frequency of occurrence and characteristics (Ferri et al., 2003).

7.4.2.2 Mini-TES Instruments on the MERs

Apart from imaging systems on earlier landers, the MER/ Mini-TES instruments (Christensen et al., 2003) are the first surface-based (infrared) remote sensing instruments on Mars. Atmospheric observations are one of the instruments' three science objectives, the other two pertaining to surface materials. The key qualitative improvement of the Mini-TES over previous atmospheric observations is that they have provided information on the vertical T profile for altitudes $\approx$20–2000 m above the surface (Smith et al., 2004; see Figure 7.6 for examples). By comparison, the other landers' *in situ* T measurements have been carried out only in the lowest meter or two above the surface. In surface-looking mode, the Mini-TES instruments have also observed the ground brightness temperatures (Smith et al., 2004).

The atmospheric temperature retrieval algorithms have been described and errors estimated by Smith et al. (2006). The T errors are estimated to be $\approx$2 K at below $\approx$200 m, increasing to $\approx$5 K at 2 km. The error of aerosol optical depth determination is estimated to be the larger of 0.03 or 10%, and that of the water vapor column abundance $\approx$5 pr µm.

Mini-TES can take a spectrum every 2 s. Each observation in turn is an average over 100–1000 spectra. The profiles are not in the true vertical, since the instrument's line of sight can be at maximum elevation angle of only 30° above the horizontal. Interpretation as a single vertical profile is based on the assumption that the observed atmospheric parameters vary relatively little in the region around the rover. Due to limited rover power resources, the Mini-TES observations typically cover only the morning to afternoon (approximately 10:00–17:00 local time); observations during other times of day are sparse (Smith et al., 2004, 2006).

Coordinated joint observations with MER Mini-TES (Section 7.4.2.2) and orbiting near-infrared instruments (e.g. Mars Global Surveyor/TES, Mars Odyssey/THEMIS, and Mars Express/PFS) have also been made to resolve the full atmospheric profiles and to provide ground truth for validation of the remote sensing observations (Wolkenberg et al., 2009).

The MER-A (Spirit) landed on January 4, 2004 ($L_s \approx 327.6°$) and MER-B (Opportunity) on January 25, 2004 ($L_s \approx 339.0°$). At the time of writing (October 2016) MER-B continues to operate, but no communication from MER-A has been received since March 22, 2010, and active attempts to regain contact have been ended. Both rovers' operational lifetimes have hence exceeded three Martian years.

7.4.2.3 Phoenix LIDAR

The Phoenix payload included the first atmospheric LIDAR deployed onto the Martian surface (Whiteway et al., 2008). The LIDAR has provided information on the vertical and temporal distributions of dust and ice aerosols (and, by proxy, on the PBL height) in the lowest 5–6 km. Observations of water ice clouds at the top of the PBL, of precipitation from the clouds, and of dust have been reported in Whiteway et al. (2009a,b); see also Figure 7.11.

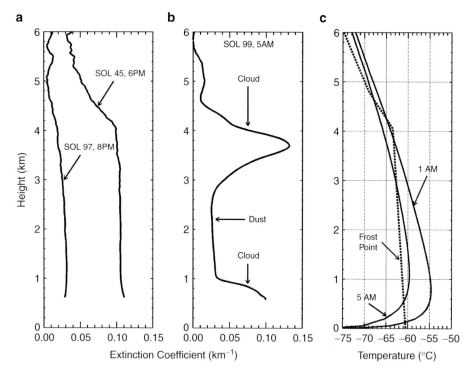

Figure 7.11. Profiles of the optical extinction coefficient at the Phoenix landing site, derived from the LIDAR 532 nm backscatter signal, for (A) sols 45 ($L_s = 97°$) and 97 ($L_s = 121°$), and (B) sol 99 ($L_s = 122°$) (Whiteway et al., 2009a). Each profile is averaged over 1 h and smoothed for a vertical resolution of 40 m. (C) Height profiles of frost-point temperature (dashed line) and atmospheric temperatures at the same site (solid lines), the latter estimated with a numerical PBL simulation model (Davy et al., 2010). Adapted from Whiteway et al. (2009a) with permission.

Figure 7.11(a) shows profiles of optical extinction coefficient in the absence of H_2O clouds, during which dust is the main source of attenuation. These profiles indicate a well-mixed PBL up to an altitude of around 4.5 km, with differences in the mean extinction reflecting slow seasonal changes in dust loading. Figure 7.11(b) shows a profile with H_2O ice clouds present in the form of two layers, located at altitudes around 3.5–4.5 km and 1 km. The identification of these layers as clouds is confirmed from the profiles in Figure 7.11(c) obtained from the model of Davy et al. (2010), which shows temperatures dipping below the H_2O frost point at the approximate altitudes of the clouds. These clouds were sometimes observed to produce intermittent snow-like precipitation over the Phoenix site (Whiteway et al., 2009a).

7.4.3 Orbital Observations

Orbital observations have a somewhat limited capacity to probe the lowest layers of the atmosphere and hence the PBL itself directly, but they are crucial for certain global measurements necessary for PBL studies, such as the properties of the surface. In this section, we list the different observational modes of past and current missions that may be useful for PBL studies.

7.4.3.1 Radio Occultation

The radio occultation (RO) method utilizes three main components: radio transmitters (on one or multiple frequencies) on-board a Mars orbiter, ultra-stable oscillator on-board a Mars orbiter, and receiver(s) either on other orbiter(s) or on Earth.

The receivers observe the radio signal transmitted through Mars' atmosphere, either when the orbiter rises from behind the planet following an occultation or before it sets just prior to the occultation. The primary observable is the phase of the radio signal. From the phase variation, vertical profiles of atmospheric number density, pressure, and temperature can be obtained. Derived quantities relevant for PBL studies include measures of static stability $N(z)$, geopotential height z_Φ, and the PBL height scale h (when defined via the temperature vertical structure). A more detailed description of the RO method can be found in, for example, Tyler et al. (1992).

So far only the Mars orbiter–Earth implementation has been used. Mars orbiter–Mars orbiter implementations (e.g. Kursinski et al., 2004) have been studied and proposed, but not implemented as of yet.

The RO method is generally considered the best orbital remote sensing method for PBL studies. RO observations can probe the atmosphere all the way down to the surface, and (in comparison with other remote sensing methods of the thermal profile) the resolution in the vertical and perpendicular to the line of sight, as well as the overall accuracy, are very good. Weaknesses include poor resolution along the line-of-sight direction (the method inherently averages the atmospheric state over the tangential path the radio signal traverses in the atmosphere) and, in the orbiter–Earth implementation, the coverage is restricted by the spacecraft orbit and the Mars–Earth celestial mechanics.

The retrieved profiles begin either from the surface or from some hundreds of meters above the surface and reach altitudes of tens of kilometers, thus covering easily the extent of the daytime PBL. Vertical and perpendicular-to-line-of-sight resolutions are

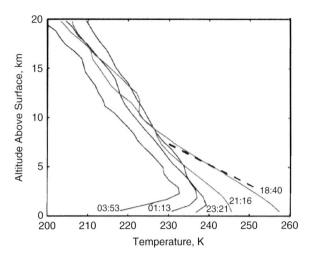

Figure 7.12. Temperature profiles from the southern hemisphere, obtained by Hinson et al. (1999) from MGS RO measurements. Only the lower portion is shown to emphasize structure near the surface. Lines labeled 18:40 and 21:16 denote measurements made before sunset, and those labeled 23:21, 01:13, and 03:53 are measurements after sunset. The profiles show the nighttime development of a temperature inversion within ~3 km of the surface. Profiles were acquired over several months covering the seasonal interval $235° \leq L_s \leq 342°$. The dashed line shows the adiabatic temperature gradient. Figure adapted from Hinson et al. (1999) with permission.

$\approx$500–1000 m (much smaller than typical PBL depths on Mars), whereas along the line of sight the resolution is much coarser, of the order of a few hundred kilometers (Hinson et al., 1999, 2001, 2004; Hinson and Wilson, 2004). The accuracy of the measurements improves with increasing pressure (i.e. decreasing altitude) and is $\approx$0.3–0.4% in the lowest layers, corresponding to 1 K and 2 Pa. In the highest layers of the profiles, the corresponding numbers are 6%, 10 K, and 0.6 Pa.

Hinson et al. (1999) published the first results on the MGS RO measurements and they addressed some results on the observed structure of the radiative–convective PBL. These were also the first remote sensing measurements that could resolve the vertical and temporal variations of the PBL. These observations were limited to the period from evening through nighttime until just before dawn. The stratification below 3 km was slightly subadiabatic in the evening and showed a strong inversion in the pre-dawn hours. They also estimated the daytime convection to reach an altitude of 8–10 km (at around 18:00 local time) in these early summer profiles from southern subtropical and mid-latitudes. An example is shown in Figure 7.12, showing evidence for the development of a radiatively induced temperature inversion in the lowest part of the atmosphere during the night.

Thus, RO profiles with better than 1 km vertical resolution in the lower atmosphere were available during the 10-year-long MGS mission, but the latitude and time coverage were not suitable for PBL convection studies (Hinson et al., 1999). This limitation was addressed with the MEX RO observations (Hinson et al., 2008), which have provided good coverage at latitudes and local times where PBL convection is occurring (low-latitude terrains and around 16:00–17:00 local time). Properties such as the potential temperature down to 1 km above the surface, as well as the PBL depth, can be determined

accurately. The measurements enabled Hinson et al. to identify striking variations in the depth of the convective PBL within the low-latitude Martian regions. The Martian convective PBL appears to extend to higher altitudes over high plateaus (e.g. 8–9 km over Tharsis) than in lower-altitude plains (e.g. 5–6 km over Amazonis) despite similar surface temperatures. Such behavior is related to convection arising from solar heating of the ground, the impact of this heat source on thermal structure being largest where the surface pressure is smallest, at high surface elevations (see also Souza et al., 2000). Modeling studies (Spiga et al., 2010) indicate that this clear correlation of PBL depth with spatial variations in surface elevation and weaker dependence on spatial variations in surface temperature are consequences of the prominent radiative forcing of the Martian PBL (see Section 7.5.3).

7.4.3.2 Other Orbital Measurements

Nadir observations are the most utilized of all the remote sensing modes and they provide important surface and atmospheric data for PBL studies despite the observing geometry often providing coarse or no information on the vertical variations of the observed, non-surface quantities. Observations can be passive or active: radiation emitted from, scattered from, or absorbed in the atmosphere and at the surface is observed with the sensor (e.g. spectrometer or camera) or the reflection of an emission sent from the instrument itself is received at the receiver (lidar, radar).

Many surface parameters important to the atmosphere can be readily retrieved with nadir observations. These include the surface temperature T_s, which itself depends on the thermal inertia (I) and the albedo (α) of the surface material (Kieffer et al., 1976, 1977; Mellon et al., 2000; Christensen et al., 2001; Fergason et al., 2006); T_s controls the flux of heat from surface to the PBL. Equally important is the local topography, mapped with very high resolution using LIDAR (Smith et al., 2001) – see Chapter 2. In addition to the large-scale topography, the surface roughness (z_0) is an important parameter for the development of the vertical profile of the wind in the PBL. Images can be used to map rock abundances at, for example, possible landing sites. This information can be later used to derive z_0 (Heavens et al., 2008; Hébrard et al., 2012).

Imaging instruments can be used to observe clouds, dust devils, and storms (e.g. Cantor et al., 2001, 2002; Drake et al., 2006), and other visible phenomena in the PBL that are the outcomes of atmospheric circulations, which are in turn often affected by the surface properties. These phenomena can reveal information on favored lifting zones, wind speeds, properties of the surface, and the region's applicability as a landing site. In addition, they naturally reveal the state of the atmosphere and the PBL, which needs to be favorable for the formation of such circulations.

Surface pressure (p_s) can also be observed with nadir geometry. For instance, MEX/OMEGA p_s retrievals (Forget et al., 2007; Spiga et al., 2007) have revealed small-scale local atmospheric phenomena (such as waves and pressure changes around craters), possibly related to PBL processes. A more comprehensive review of all of the relevant types of measurement and a list of relevant references is given in Petrosyan et al. (2011).

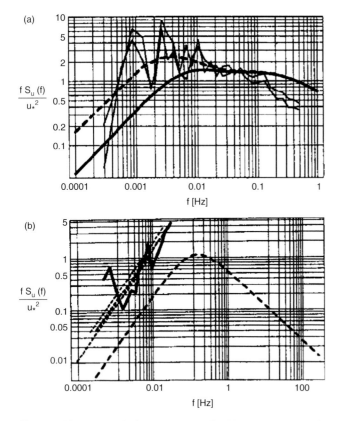

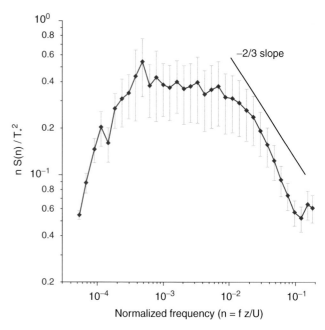

Figure 7.14. Spectra of temperature from the Phoenix Lander (Davy et al., 2010), normalized by T_* and plotted versus the normalized frequency, $n = f z / \overline{u}$, with z being the measurement height, $\overline{u}$ the mean speed, and f the frequency (in Hz). The curve shows a composite daytime (unstable) spectrum, averaged over 32 sols.

Figure 7.13. Spectra of wind speed normalized by u_* and compared with the models. (a) An unstable spectrum from the Pathfinder Lander (Larsen et al., 2002). The low-frequency peak, reflecting the frequency associated with boundary layer height eddies, is clearly seen. At higher frequencies, the effects of the low-pass filter characteristics of the sensors affect the shape of the spectrum. (b) A stable velocity spectrum from the Viking Landers (Tillman et al., 1994). The measured spectrum here is strongly influenced by the relatively low sampling rate and effects of aliasing. The bell shape is the modeled spectrum (see text), the thick line the measured spectrum, and the lines through the data reflect the modeled data spectrum, taking into account the aliasing and the Kolmogorov-scale filtering.

7.4.4　Use of Mars Lander Data to Validate Surface Layer Scaling Laws

Data from the Viking, Pathfinder, and Phoenix missions have been used to validate the Monin–Obukhov (MO) and other PBL turbulence scaling laws – see Tillman et al. (1994) and Larsen et al. (2002) using VL and Pathfinder data, and Gunnlaugsson et al. (2008) and Davy et al. (2010) for Phoenix data. On Earth, such validations normally involve simultaneous measurements of mean profiles of wind, temperature, and humidity, with independent profiles of the corresponding fluxes, and turbulent variances and spectra.

From the Mars landers to date, only datasets that are much less comprehensive are available, basically consisting of wind at one level and temperature at one to three levels, with most of the sensors being able to deliver both mean values and turbulent variances and spectra. Therefore, the MO validation is reduced to comparing the relations between the measured variances and power spectra of temperature and velocity turbulence, with the simultaneous mean wind and temperature profiles, and

evaluating if these relations are in accord with the MO formulations, derived on Earth.

To compare the measured and modeled spectra/variances, the many special measurement aspects of Martian fluctuation measurements relative to normal Earth-based approaches have had to be considered, such as the effects of substantial flow distortions around the landers, sampling problems associated with aliasing (especially for VL), larger Martian Kolmogorov scales than on Earth (VL; see Section 7.3), and unusually low measuring heights. Such issues are compounded by the use of, at best, fairly slowly responding sensors and relatively low data rates (all datasets).

Finally, it should be mentioned that the parameterization of the low-frequency behavior of the velocity fluctuations for unstable conditions requires an estimate of the height of the unstable boundary layer. In practice, this was estimated from the history of the surface heat flux, in a standard way. Examples of spectral presentations are shown in Figures 7.13 and 7.14, illustrating our ability to model especially the wind spectra on Mars (Figure 7.13), using adaptations of the scaling and formulations derived from the Earth's surface layer. Temperature spectra (Figure 7.14) are less well understood, but still show evidence for an onset of the inertial subrange consistent with the expected scaling behavior.

Although the results of the above procedures have provided some validation of the application of the classical MO scaling laws in the Martian PBL, they do not really test the flux–profile relations that constitute the core of the MO similarity approach. They can be regarded as consistency checks of the above type, so long as no inconsistency appears in the results (see the discussion in Section 7.5.2). But, to date, the most direct validation has been performed on the aforementioned spacecraft data.

7.5 MODELING THE MARTIAN PLANETARY BOUNDARY LAYER

The many observational efforts about Mars' atmosphere and ground have been supported and interpreted by a wide range of theoretical and numerical modeling studies, many of which have been based on a hierarchy of atmospheric models. General circulation models (GCMs) produce views of the large-scale synoptic weather and climate (e.g. Leovy and Mintz, 1969; Pollack et al., 1976, 1981; Haberle et al., 1993a; Wilson and Hamilton, 1996; Allison et al., 1999; Forget et al., 1999; Takahashi et al., 2003; Moudden and McConnell, 2005; Segschneider et al., 2005; Takahashi et al., 2006; Richardson et al., 2007), while mesoscale models are increasingly being used to study the fine structure of local weather phenomena and to support various mission objectives, including the selection of safe landing sites and weather forecasts for intended landings (e.g. Rafkin et al., 2001; Toigo and Richardson, 2002; Tyler et al., 2002; Moudden and McConnell, 2005; Wing and Austin, 2006; Kauhanen et al., 2008; Spiga and Forget, 2009). As in the case of the atmospheric and oceanic boundary layers on Earth, models of the PBL on Mars include similarity theories, Reynolds stress models (RSMs), and large-eddy simulations (LESs). All these approaches are intertwined and interconnected, and the degree of their overall consistency can be used as one of the tools for each respective model's validation. Among the similarity theories, Monin–Obukhov theory takes the central stage and has already been briefly reviewed in Sections 7.2, 7.3, and 7.4. The turbulence in the Martian PBL is commonly studied nowadays using LES (see Section 7.5.3) and direct numerical simulations (DNSs), while the local structure of the lower Martian atmosphere has been charted by the help of one- (1D) and two-dimensional (2D) PBL models (e.g. Gierasch and Goody, 1968; Blumsack et al., 1973; Pallman, 1983; Ye et al., 1990; Savijärvi, 1991b; Haberle et al., 1993b; Savijärvi and Siili, 1993; Odaka, 2001). The 1D models still serve as an efficient tool, however, for interpreting the near-surface local conditions at the sites of various landers (e.g. Savijärvi et al., 2004; Savijärvi and Kauhanen, 2008; Savijärvi and Määttänen, 2010).

All of these atmospheric models must provide a representation for the very active turbulence in the PBL of Mars. In practice, the turbulence is usually presented as a diffusive (mixing) process in all prognostic equations, but there is a wide range of approaches that have been adopted to achieve this.

7.5.1 Bulk and Mixing Length Parameterized Models of the Martian PBL

The first Martian atmospheric models, roughly before the 1990s, used either a bulk approach (treating the whole PBL as a single layer) or a mixing length closure, whereby the vertical turbulent diffusion coefficients K_v are given at a few layers through the PBL as the triple product of a squared mixing length ℓ (hence the name), the current wind shear, and a dimensionless semi-empirical stability factor $f(Ri)$ that is a function of a suitably defined Richardson number, based on the Monin–Obukhov similarity theory and calibrated using measurements on the Earth,

$$K_v = \ell^2 \left| \frac{\partial V}{\partial z} \right| f(Ri) \qquad (7.21)$$

The mixing length $\ell(z)$ is a measure of the diameter of the typical turbulent eddies. It is usually assumed to increase linearly with height near the surface and to asymptote higher up towards a constant (the asymptotic length scale (Blackadar, 1957)) which, on Earth, is of the order of one-tenth of the convective boundary layer height h. In this case Ri could be the local bulk Richardson number (see Section 7.2), which provides a measure of hydrodynamic stability dependent on the current vertical gradients of the temperature and wind speed at each height, or an equivalent measure (such as Ri_g; cf. (7.14)).

These diagnostic, first-order closure methods were based on their Earth counterparts. They appear to work quite well on Mars (cf. Section 7.3), thus broadly confirming the planetary validity of the classical similarity theory. They are still adequate if one is mainly interested in the simulation of the basic variables (e.g. grid-scale wind and temperature), and less so in the details of the subgrid-scale turbulence. In a test at York University, where higher-order closures and a mixing-length-type first-order closure were compared in the same 1D model framework at the conditions found by Pathfinder (Weng et al., 2006), the resulting diurnal simulations of the near-surface mean winds and temperatures were quite alike and found to be quite close to the Pathfinder observations. However, the first-order closure methods do not provide a lot of information about the structure of the turbulence itself. Hence the so-called turbulent kinetic energy (TKE) and other higher-order closures, LES, and DNS have recently become popular (see the following subsections).

7.5.1.1 Physics of the Martian PBL as Revealed by First-Order Closure Models

The *in situ* lander observations indicate that the Martian PBL is, at least in the safe, smooth, and flat landing sites explored to date, quite repetitive in its strong diurnal and annual cycles, being super-adiabatic in the daytime sunshine, and sub-adiabatic with a steep surface inversion every night (e.g. Smith et al., 2006). This is partly because clouds in the Martian atmosphere are absent or thin. Clouds do not, therefore, moderate the solar and thermal radiation to the same extent as on Earth, although the ubiquitous suspension of mineral dust tends (in large concentrations) to take over the role of clouds on Mars.

The early first-order closure models were able to reveal that the Martian PBL, and especially its surface layer, is strongly controlled by radiative heating and cooling (i.e. radiative flux divergences): carbon dioxide at the 15 μm band and airborne dust at other infrared (IR) wavelengths readily adopt the daytime heat and nighttime cold of the sand-like ground (regolith) via absorption of thermal radiation emitted by the hot/cold regolith, typical of, for example, the Sahara Desert. The dust also absorbs solar radiation. This adds an extra daytime heating effect, which is small in relatively dust-free conditions but can be strong during heavy dust loading. Thus, the radiation and ground heat conduction schemes should be of relatively high accuracy in all Martian atmospheric models, and especially so for PBL models intended for Mars. Perhaps the largest source of inaccuracy in PBL modeling is our limited knowledge of the local amounts and current radiative properties of the Martian dust. Water vapor can also have a non-negligible effect on both solar and thermal radiative transfer in the Martian PBL (e.g. Savijärvi, 1991a; Määttänen and Savijärvi, 2004). But since the absolute humidity on Mars is small, most models have so far ignored water vapor and other trace gases in their radiation schemes. The CO_2 and water ice-covered winter polar caps form a special, often cloud-topped, arctic PBL.

The small air density, which makes the Martian air react rapidly and strongly to the vertical radiative flux divergences, also makes the turbulent fluxes relatively small in magnitude, so one might be tempted to think that turbulence is ineffective on Mars. But in fact the low density of the air, together with low gravity on Mars, contributes to the vertical turbulent flux divergences actually being quite strong, to the extent that the Martian extrapolar convection and turbulence are more vigorous than those on the Earth (see Section 7.3), thereby mixing the IR-induced daytime heat content and nighttime coldness of the surface layer upward very effectively. The typical depth of the afternoon Martian convective boundary layer is about 7–8 km at low latitudes (cf. 2–3 km on Earth) while the nighttime inversion is about 0.5 km thick (cf. 100–200 m on Earth). The strong daytime turbulence also mixes dust from the ground into the air, feeding, for example, local dust devils and regional and global dust storms. This creates an interesting and poorly known arena for various feedback effects on many scales between the dust, radiation, temperature, and wind distributions.

An interesting feature of the midday super-adiabatic surface layer of Mars (also discussed in Section 7.3.1) is that it appears then to be so strongly heated from below by IR radiation that turbulence is actually forced to cool it down, such that the divergence of the local heat flux reverses direction and becomes positive close to the ground. Therefore, the turbulent heat fluxes most probably attain their morning and midday maxima not at the surface, as on Earth, but a few hundred meters aloft (see Figure 7.3).

7.5.2 TKE Closures for the Martian PBL

The similarity of the physical processes in Martian and terrestrial boundary layers makes it possible to adapt boundary layer parameterizations developed for the Earth's atmosphere and oceans to Martian circulation models. Such a path was taken, for instance, by Forget et al. (1999) and Rafkin et al. (2001), who used a Reynolds stress model (RSM) of the Mellor–Yamada family (Mellor and Yamada, 1982). Traditionally, the Reynolds stress modeling has been a powerful tool in turbulence research, and it is instructive to provide a brief overview of this technique in the context of Martian PBLs.

Generally, RSMs rely upon the Reynolds averaging of the governing equations and a set of closure assumptions relating unknown correlations to the known ones as well as to the mean fields. The foundations of RSMs as well as their extensive use in engineering, environmental, and geophysical sciences have been outlined in numerous review articles and books (e.g. Tennekes and Lumley, 1972; Mellor and Herring, 1973; Monin and Yaglom, 1975; Mellor and Yamada, 1982; Pope, 2005;

Wyngaard, 2010). All RSMs are based upon the Reynolds decomposition of fluctuating flow characteristics, such as the instantaneous velocity, $\tilde{u}_i = U_i + u_i$, where $\overline{\tilde{u}_i} = U_i$ is either the ensemble-, time-, or space-averaged velocity, and u_i its fluctuating counterpart. The subtleties of the averaging have been extensively analyzed (e.g. Monin and Yaglom, 1975; Pope, 2005), and will not be discussed further. We shall refer to either of these averagings as Reynolds averaging and assume that they are used consistently throughout the derivations.

The application of Reynolds averaging to the governing equations in the Boussinesq approximation and utilization of Einstein's summation rule yield (e.g. Mellor and Yamada, 1982; Galperin et al., 1989)

$$\frac{DU_i}{Dt} + \epsilon_{ikl}f_kU_l = \frac{1}{\rho_0}\frac{\partial}{\partial x_l}(-\overline{u_iu_l}) - \frac{1}{\rho_0}\frac{\partial P}{\partial x_i} + g_i\frac{\rho}{\rho_0} \tag{7.22}$$

$$\frac{D\Theta}{Dt} = \frac{\partial}{\partial x_l}(-\overline{u_l\theta}) \tag{7.23}$$

$$\frac{\partial U_i}{\partial x_i} = 0 \tag{7.24}$$

where Θ and θ are the mean and fluctuating potential temperatures, respectively, ρ is the density and ρ_0 a constant reference density, P is the mean pressure, $f_k = 2\Omega(0,\cos\phi,\sin\phi)$ is the Coriolis vector, Ω is the angular velocity of the planet's rotation, and $g_i = (0,0,-g)$ is the acceleration due to gravity directed downwards.

Reynolds averaging of the governing equations for turbulent momentum and heat fluxes, $\overline{u_iu_j}$ and $\overline{u_i\theta}$, respectively, encounter the classical problem of turbulence closure, since those equations involve many unknown correlations. A closed set of equations for these correlations cannot be developed at any correlation order, and so additional "closure" assumptions must be introduced. The most problematic correlations are those involving the pressure because it is a non-local variable (e.g. Mellor and Herring, 1973). Two approaches have been developed to model the pressure–velocity correlation terms, one by Launder et al. (1975) referred to as LRR models, and the other by Mellor and Herring (1973) and Mellor and Yamada (1974) called MY models. Models of the former family are more complicated than their latter counterparts. The LRR approach was initially designed for engineering flows with weak or no shear, while the MY models were more commonly applied to geophysical and environmental flows with strong shear. Later, the LRR models were also adapted to geophysical flows (e.g. Canuto et al., 2001). Models of intermediate complexity have also been proposed (e.g. Kurbatskiy and Kurbatskaya, 2006). The empirical constants in both families of closures have been assumed invariant, although in some models the constants are replaced by flow-dependent variables (e.g. Ristorcelli, 1997). The closures used in current models of the Martian circulation are almost exclusively of the MY family (Forget et al., 1999; Haberle et al., 1999; Rafkin et al., 2001; Moudden and McConnell, 2005) and so only models of this class will be considered hereafter.

Fully prognostic equations for the Reynolds stresses $\overline{u_iu_j}$ and turbulent heat fluxes $\overline{u_i\theta}$ contain the Coriolis terms that significantly complicate the algebra (Mellor and Yamada, 1982; Galperin et al., 1989). In the framework of the MY models, these terms were investigated by Galperin et al. (1989) for the case of stable stratification and by Hassid and Galperin (1994) for the cases of neutral and unstable stratification. It was found for the stable case that the contribution of the Coriolis terms does not exceed about 10% of the total stress, while, for neutral and unstable stratification, the effect of the Coriolis terms can be large. These conclusions could be important for the daytime Martian atmospheric PBL, which is often strongly unstable and very deep. Despite these findings, the explicit Coriolis terms in the Reynolds stress and heat flux equations are usually neglected.

In all applications, RSMs provide expressions for the eddy viscosity, K_M, and eddy diffusivity, K_H, through the equations for the Reynolds stresses and turbulent heat fluxes. In the MY models, these expressions are

$$-(\overline{uw}, \overline{vw}) = K_M \left(\frac{\partial U}{\partial z}, \frac{\partial V}{\partial z} \right) \tag{7.25}$$

$$-\overline{w\theta} = K_H \frac{\partial \Theta}{\partial z} \tag{7.26}$$

where U and V are the components of the mean horizontal velocity, z is the vertical coordinate. Expressions for K_M and K_H are given in the mixing length format as $K_M = q\ell S_M$ and $K_H = q\ell S_H$, where $q^2 = \overline{u_k^2} = 2E_{KT}$, E_{KT} being turbulence kinetic energy (TKE), ℓ is the turbulence macroscale, and S_M and S_H are non-dimensional stability functions.

To compute K_M and K_H, one needs to know q and ℓ. The former is found via its prognostic equation (Mellor and Yamada, 1982),

$$\frac{Dq^2}{Dt} - \frac{\partial}{\partial x_k} \left(q\ell S_q \frac{\partial q^2}{\partial x_k} \right) = -2\overline{u_k u_l} \frac{\partial U_k}{\partial x_l} - 2\beta g_k \overline{u_k \theta} - 2\epsilon \tag{7.27}$$

where β is the thermal expansion coefficient, $\beta = (\partial \rho / \partial \theta)_p / \rho_0$, S_q is the vertical non-dimensional exchange coefficient for q^2 usually taken to be equal to 0.2, and ϵ is the rate of the viscous dissipation given in MY models by

$$\epsilon = \frac{q^3}{B_1 \ell} \tag{7.28}$$

Determination of the turbulence macroscale ℓ presents a formidable problem for turbulence modeling because the macroscale obeys no conservation law. As a result, all existing models for ℓ, either prognostic or diagnostic, are empirical. For neutrally stratified atmospheric PBLs, Blackadar's algebraic equation has been commonly used (Blackadar, 1962),

$$\ell = \ell_B = \frac{\kappa z}{1 + \kappa z / \ell_0} \tag{7.29}$$

where ℓ_0 is a reference length scale. This equation ensures a smooth transition of ℓ from κz in the logarithmic velocity profile near the solid ground to a constant value at the top of the boundary layer. For the Martian atmosphere, Forget et al.

(1999) fix ℓ_0 at 160 m, while Moudden and McConnell (2005) set ℓ_0 to 200 m. Rafkin et al. (2001) use a different definition of ℓ_B, i.e.

$$\ell_B = \frac{\kappa(z + z_0)}{1 + \kappa(z + z_0) / \ell_0} \tag{7.30}$$

where, following Mellor and Yamada (1974),

$$\ell_0 = 0.1 \frac{\int_0^h zq\,dz}{\int_0^h q\,dz} \tag{7.31}$$

and where h is the top of the boundary layer. A similar definition of ℓ_0 was used by Haberle et al. (1993b) but they employed a coefficient of 0.2 instead of 0.1.

In stably or neutrally stratified boundary layers and in flows that combine layers with different senses of stratification, Blackadar's formulation fails and more a sophisticated formulation may be required.

In stably stratified boundary layers, the size of the overturning eddies is conditioned by their kinetic energy exceeding the potential energy of the background. This leads to the length scale limitation first introduced by Deardorff (1976a),

$$\ell \leq \ell_s = 0.53 \frac{q}{N} \tag{7.32}$$

The turbulence macroscale in stably stratified boundary layers can either be clipped to ℓ_s (Galperin et al., 1988; Rafkin et al., 2001) or determined from the equation

$$\ell^{-1} = \ell_B^{-1} + \ell_s^{-1} \tag{7.33}$$

which ensures a smooth transition from the Blackadar formulation to ℓ_s in regions dominated by stable stratification. Some models take into account the dependence of ℓ on the vertical Coriolis parameter f by adding another term to (7.33),

$$\ell^{-1} = \ell_B^{-1} + f / C_f q + \ell_s^{-1} \tag{7.34}$$

where C_f is a constant (see e.g. Zilitinkevich et al., 2007, and references therein).

The gradient Richardson number Ri_g (cf. (7.14)) is an important characteristic parameter for stratified turbulence that determines the strength of stratification with respect to vertical shear. Under strongly stable conditions, a critical value of Ri (denoted as Ri_{cr}) may be attained at which turbulent mixing is often assumed to be fully suppressed. In MY models, Ri_{cr} is less than 0.2, leading to underpredicted mixing in some situations (e.g. Rippeth, 2005). Galperin et al. (2007) discussed the general notion of an Ri_{cr} based upon recent observational and theoretical studies that considered the effects of non-stationarity, internal waves, and flow anisotropization. They concluded that all these factors preclude the full laminarization of turbulence and thus make the concept of an Ri_{cr} devoid of its conventional meaning. Consequently, they suggested that the use of Ri_{cr} as a criterion of turbulence extinction should be avoided.

Turbulence intensity in stably stratified flows can also be judged by another parameter, the buoyancy Reynolds number, $Re_b = \epsilon / v_0 N^2$, v_0 being the molecular viscosity (see e.g. Galperin and Sukoriansky, 2010, and references therein). For

$Re_b = O(1)$, vertical turbulent mixing may become laminarized. However, in horizontal planes, the mixing can still be much larger than in laminar flows. Conclusions similar to those of Galperin et al. (2007) on the absence of a meaningful Ri_{cr} were also reached by Zilitinkevich et al. (2007) based upon the total energy approach, the total energy being the sum of kinetic and potential energies of turbulent fluctuations. A number of RSMs with no Ri_{cr} were subsequently developed (e.g. Canuto et al., 2008; Alexakis, 2009; Kitamura, 2010).

Even though models of the MY family can be, and have been, applied to flows with unstable stratification and strong convection, Forget et al. (1999) express some doubts about the validity of the model in such conditions. Canuto et al. (2005) explored the limitations of models of the MY family in depth and concluded that they effectively force the model to include local interactions only and filter out the non-locality which would otherwise be allowed by the third- and higher-order correlations. To overcome this shortcoming, Canuto et al. (2005) suggest that the third- and, possibly, higher-order correlations should be included in MY-type models.

Specification of the turbulence length scale in flows with multiple regimes has been a persistently difficult problem for modeling. Aside from the algebraic formulations of the kind described earlier, another widely used approach employs prognostic equations for quantities related to ℓ, such as the dissipation rate, ϵ. Models utilizing prognostic equations for E_{KT} and ϵ are often known as K–ϵ models (e.g. Pope, 2005). Mellor and Yamada (1982) consider a prognostic equation for the quantity $q^2 \ell$, which can be constructed by analogy with the equation for q^2,

$$\frac{D(q^2 \ell)}{Dt} - \frac{\partial}{\partial z}\left(q\ell S_q \frac{\partial q^2 \ell}{\partial z} \right) = \ell(E_1 P_s - E_3 P_b)$$

$$- \frac{q^3}{B_1}\left[1 + E_2 \left(\frac{\ell}{\kappa z} \right)^2 \right] \qquad (7.35)$$

where $P_s = K_M S^2$ and $P_b = K_H N^2$ represent the mechanical production and the buoyant destruction of the turbulence energy, respectively, ϵ is the dissipation rate given by (7.28), and the constants E_1, E_2, and E_3 are 1.8, 1.33, and 1.8, respectively. This equation is quite popular in oceanographic modeling and has been employed, for instance, in the Princeton Ocean Model (POM[1]).

The use of turbulence schemes in atmospheric models requires their consistency with the imposed boundary conditions. A conventional method to derive such boundary conditions near the underlying surface is to use a constant-flux layer approximation between the surface and the first grid point, and to employ Monin–Obukhov similarity functions (Φ_q, where $q = m, h$, etc.; cf. Section 7.2) to calculate the required values of the mean profiles. Mellor (1973) showed how Monin–Obukhov similarity functions can be derived directly from the turbulence model. However, in many cases these functions are taken from observations, and so the mismatch between the observed functions and those obtained from the turbulence model may introduce spurious fluxes of momentum, heat, and other quantities.

A new family of RSMs was developed recently, based upon a spectral approach, which is an alternative to the Reynolds stress modeling (Sukoriansky et al., 2005). This approach has been coined a quasi-normal scale elimination (QNSE). Within this theory, internal waves and turbulence are treated as one entity rather than as an *ad hoc* dichotomy. QNSE provides a rigorous procedure of successive coarsening of the system's domain of definition that produces the effective, scale-dependent, vertical viscosity (K_M) and thermal diffusivity (K_H) as well as their horizontal counterparts. Dependent upon the range of eliminated scales, QNSE provides either subgrid-scale parameterization for LESs or an equivalent of a RSM (Sukoriansky et al., 2005). In the latter case, the eddy viscosities and eddy diffusivities become functions of either the local gradient Richardson number, Ri, given by (7.9), or the Froude number, $Fr = \epsilon / NE_{KT}$. One of the important advantages of the QNSE model is the absence of the critical Richardson number, Ri_{cr} (Sukoriansky et al., 2005; Galperin et al., 2007). The QNSE-based expressions for K_M and K_H have been successfully tested in both K–ϵ and K–ℓ applications (e.g. Sukoriansky and Galperin, 2008). Along with the QNSE-based Monin–Obukhov similarity functions for the near-surface layer, these expressions were implemented in recent releases of the Weather Research and Forecasting (WRF[2]) model for Earth applications. A version of the WRF model known as PlanetWRF has been in use recently for global simulations of the Martian atmosphere (Richardson et al., 2007), and other versions are now being used for mesoscale and LES modeling for Mars (Spiga and Forget, 2009; Spiga et al., 2010).

7.5.3 Large-Eddy Simulations of the Martian PBL

7.5.3.1 General Principles

As described in Section 7.4, the exploration of Mars from space has shown the variety and intensity of Martian PBL processes, with daytime convective heat fluxes some three times larger than in the terrestrial environment (Sutton et al., 1978), wide dust devils extending to high altitudes (Thomas and Gierasch, 1985), well-organized convective cloud streets (Malin and Edgett, 2001), super-adiabatic daytime and ultra-stable nighttime near-surface gradients of temperature (Schofield et al., 1997), large turbulent fluctuations of near-surface temperatures (Schofield et al., 1997; Smith et al., 2006), and a mixed-layer depth of the same order of magnitude as the atmospheric scale height (Hinson et al., 2008).

Such a context of observational achievements has motivated the development of a number of dedicated three-dimensional (3D) mesoscale models for Mars (Rafkin et al., 2001; Toigo and Richardson, 2002; Tyler et al., 2002; Richardson et al., 2007; Kauhanen et al., 2008; Spiga and Forget, 2009), so as to resolve Martian circulations at higher resolution than possible with global climate models (GCMs). These efforts have given birth to powerful simulators of the Martian atmospheric circulations at both the mesoscale (hundreds of kilometers to 1 km) and the

---

[1] See www.aos.princeton.edu/WWWPUBLIC/htdocs.pom/.

[2] See www.wrf-model.org/index.php.

microscale (1 km to hundreds of meters). Mesoscale models couple dynamical cores (either hydrostatic or non-hydrostatic), originally developed for terrestrial regional climate modeling, with physical parameterizations of Martian dust, CO_2, and H_2O cycles, developed for Mars GCMs. Of particular interest in PBL studies is the use of mesoscale models for so-called large-eddy simulations (also referred to as microscale modeling or turbulence-resolving simulations). In these simulations, the grid spacing is reduced to a few tens of meters so as to resolve the larger turbulent eddies, responsible for most of the energy transport within the PBL (see e.g. Lilly, 1962). Thus, besides the obvious realistic improvement gained by the 3D computations, LES allows for fewer initial assumptions and parameterizations than single column models.

7.5.3.2 LES Standard Settings

In 3D LES, the numerical integration of the atmospheric fluid dynamic equations is performed through the "dynamical core". In contrast to GCMs, integrations in mesoscale dynamical cores are performed, not over the whole planetary sphere, but in a limited domain over an area of interest. Martian models are adapted from carefully tested dynamical cores developed for the Earth, which integrate the fully compressible, non-hydrostatic Navier–Stokes equations. When fine-scale meteorological motions are resolved, vertical wind accelerations can become comparable to the acceleration due to gravity. Hence hydrostatic balance cannot be assumed in those equations as is typically the case in GCMs.

The dynamical core is coupled with (most often 1D) parameterization schemes, in order to compute physical processes at each grid point specific to the considered planetary environment. Mesoscale models used for LES applications usually come with the full range of physics developed for Martian GCMs, in particular, calculations of the diabatic forcing of atmospheric circulations (radiative transfer, soil thermal diffusion, etc.). As was pointed out for parameterized single-column modeling studies, radiation plays a prominent role in the energy budget of the Martian PBL (Haberle et al., 1993b; Savijärvi, 1999; Davy et al., 2009). Thus, including realistic computations of radiative transfer processes by CO_2 and dust appears necessary in Martian LES, while it may be less crucial in terrestrial LES.

When subgrid-scale dynamical processes are not resolved by the dynamical core, their effects are parameterized in the model's physical schemes. Although PBL mixing through convective plumes is predominantly resolved by LES, turbulent phenomena at smaller scales than typical LES grid spacings (i.e. few tens of meters) are still left unresolved. In LES, handling subgrid-scale PBL mixing requires particular attention (see Section 7.5.2); for Martian applications, the adopted strategy is usually similar to what is done in terrestrial LES (see e.g. Moeng et al., 2007; Basu et al., 2008). Some of the physical parameterizations used in coarser-resolution simulations, however, are not suitable for the very high-resolution LES. In GCMs, in order to ensure numerical stability, and to account for subgrid-scale mixing processes insufficiently handled in the boundary layer parameterization scheme, it is usually necessary to modify and adjust any unstable layer with negative vertical potential temperature gradients (a common near-surface situation during Martian afternoons) into a neutral equivalent (Rafkin, 2003). But the use of convective adjustment is not generally needed in LES since the model explicitly resolves turbulent convective motions: the mixed layer is a direct result of the model's dynamical integrations.

The first model level is located a few meters above the ground (1–4 m), which means surface layer processes also need to be parameterized in LES. Surface layer values for sensible heat flux H_S and friction velocity u_* are passed on to the turbulent diffusion scheme, where they modify momentum and potential temperature at the lowest grid levels (Moeng et al., 2007). Sensible heat flux H_S is evaluated by the bulk aerodynamic formula

$$H_S = \rho c_p \sqrt{C_d}\, u_* \theta_* \qquad (7.36)$$

(cf. Section 7.2), where ρ is atmospheric density and c_p is specific heat capacity. At each grid point and time step, θ_* is effectively the temperature difference between the surface and the first atmospheric layer (at altitude z_1 above the ground). Friction velocity u_* is taken to be the product of $\sqrt{C_d}$ with wind velocity in the first layer (background wind plus resolved turbulent winds).

So far as initial and boundary conditions are concerned, LES are generally idealized numerical experiments rather than "real-case" mesoscale simulations. Periodic boundary conditions are often used to simulate the situation of an infinite flat plane. Surface properties (topography, albedo, thermal inertia) are set as constants over the whole simulation domain, which, typically, is smaller than the resolution element of available observations. Surface static data are extracted from maps derived from recent Martian spacecraft measurements, such as the 64 ppd (pixel-per-degree) MOLA topography (Smith et al., 2001), 8 ppd MGS/TES albedo (Christensen et al., 2001), and 20 ppd TES thermal inertia (Putzig and Mellon, 2007). Large-scale circulations are prescribed through the initial temperature and wind profiles being the same at each grid point. Temperature profiles are usually extracted from large-scale simulations by GCMs (often, but not always, sharing the same physics as the LES model). Usually random (noise) perturbations of ~0.1 K amplitude are added to the initial temperature field so as to break the symmetry of this initial field and to help trigger convective motions. Although the most common kind of simulation encountered in the Martian literature is of "pure free convection", more realistic cases with non-zero background wind velocity have also been performed (e.g. Tyler et al., 2008). Thus far, Martian LES has been conducted with a uniform dust opacity (usually in clear conditions) and surface roughness.

7.5.3.3 First Results

Martian LES has allowed a significant leap forward in our understanding of the PBL dynamics on Mars and allowed the study of the fine-scale structure of the Martian daytime BL, dominated by convective processes (the "convective" BL): mixed-layer growth, polygonal cells, thermal updrafts, and convective vortices (Rafkin et al., 2001; Toigo et al., 2003; Michaels and Rafkin, 2004). Work with 3D LES models for

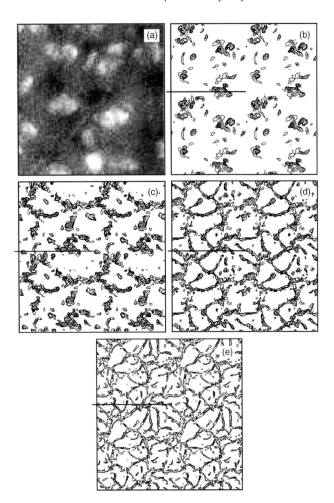

Figure 7.15. (a) An area (48 km×48 km) of likely convective cloud over Syria Planum (portion of Mars Orbiter Camera image M0104901). (b)–(e) Horizontal cross-sections of vertical velocity at 14:41 LT from the LES model of Michaels and Rafkin (2004) for heights of (b) 4342 m, (c) 2092 m, (d) 988 m, and (e) 385 m. Filled areas indicate regions of velocity greater than 2 m s^{-1}. Figure adapted from Michaels and Rafkin (2004) with permission.

Mars quickly followed the first experiments of 2D LES by Odaka et al. (1998): their simulations predicted intense vertical winds (20 m s^{-1}) caused by the convection in the Martian BL, although two-dimensionality is thought to adversely affect the amplitude of the winds due to key aspects of vortex stretching not being permitted.

The first successful attempt to perform 3D LES was reported by Rafkin et al. (2001), who later published a paper entirely dedicated to the topic of Martian LES (Michaels and Rafkin, 2004). Through simulations with horizontal resolutions of 150 and 30 m in situations of moderate background wind (5 m s^{-1}), the authors describe the structure of the convective PBL on Mars. After sunrise, the convection is organized into horizontal linear structures which – between 08:00 and 08:30 LT, under the influence of the horizontal shear – are rapidly turned into open polygonal cells with narrow updrafts at the ridges of the cells and strong subsidence in the middle of each cell (see Figure 7.15). The horizontal structure of the vertical velocity 4 km above the surface (where only the top of the most intense

updrafts remain) is in satisfactory agreement with the organization of convective clouds observed from orbit by MGS/MOC (cf. Figure 7.15a,b). During the afternoon, until the PBL convection collapses around 16:30–17:00, as the convective PBL deepens to reach its maximum altitude (~6 km in the simulations of Michaels and Rafkin (2004)), cells keep on widening while updrafts intensify – as is most likely dictated by the conservation of mass.

The maximal value predicted for the vertical heat flux in the Mars LES by Michaels and Rafkin (2004) is ~1.5 K m s^{-1} and is reached just before noon. It is one order of magnitude larger than in terrestrial deserts (see also Spiga, 2011, and references therein). In addition, contrary to the relative isotropy that is observed on Earth, the contribution of vertical motions to the turbulent kinetic energy seems more prominent than the contribution of the horizontal motions. As was noted in previous studies with single-column models (Haberle et al., 1993b; Savijärvi et al., 2004), convective motions act to cool the near-surface atmosphere ($z \leq 0.1z_i$; cf Figure 7.3) in situations of strong radiative heating (mostly absorption of infrared radiation incoming from the surface by the atmospheric CO_2), instead of warming it, as is the case on Earth. As boundary layer turbulent motions tend to mix heat to counteract the heating gradients, convection transports the radiative heat – plus the contribution of sensible heat flux as on Earth – higher up in the BL.

During the afternoon, when the convection is at its most vigorous, LES shows numerous convective vortices of diameter 100–1000 m in the first 100 m above the surface. However, only a small fraction of these vortices evolve into structures in cyclostrophic equilibrium comparable with the observed dust devils (which have a vertical extent of ~60% of the PBL depth (Cantor et al., 2006) and a depression of ~2 Pa). Toigo et al. (2003) dedicated a paper to an in-depth analysis of these "dust devil-like" vortices by means of LES (see Figure 7.16 for an example of the structures obtained), which discusses some of these issues. The results of Toigo et al. (2003) for convective vortices complements those of Michaels and Rafkin (2004). Toigo et al. (2003) noticed a great similarity with the equivalent terrestrial phenomena (Kanak et al., 2000) and showed that the thermodynamic scaling theory of Rennó et al. (1998) correctly describes the structure of the vortices. In addition, LES enables the assessment of the contribution of each term in the TKE equation. Toigo et al. (2003) showed that the vortices result from an equilibrium between production of TKE by buoyancy and a sink via advection of TKE and dissipation towards the smaller eddies. It is eventually confirmed by the study that convective vortices preferably form at the intersection of the convective cells. A plausible scenario to account for this is the twisting to the vertical of the horizontal vorticity resulting from temperature contrasts in the lowermost levels of the BL. The high sensitivity of the activity of the dust-devil-like vortices to the background wind is also emphasized by Toigo et al. (2003).

7.5.3.4 Recent Efforts and Perspectives

Following the pioneering work described in the previous section, Martian LES studies with various models have confirmed the vigorous nature of the daytime Martian PBL (Richardson

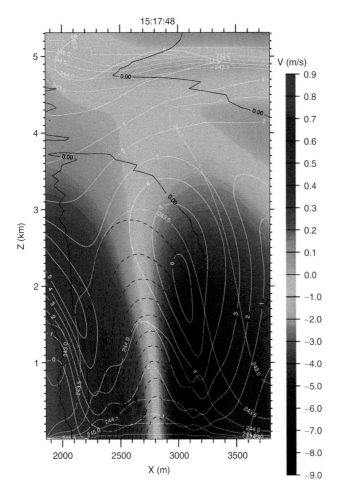

15:17:48

V (m/s)

Figure 7.16. High-resolution simulation ($\Delta x = 10$ m) of the "no wind" simulation dust devil of Toigo et al. (2003). Here is plotted a vertical slice through the center of the dust devil. Background color shows the tangential wind speed. Black contours show the pressure perturbation (in Pa), reaching a maximum difference near the surface of about 1 Pa less than the background. Yellow contours show potential temperature (in K), and the warm core of the dust devil. White contours show upward wind velocity (in m s^{-1}). Upward wind velocity peaks at the walls of the dust devil, and the decrease in upward velocity can be seen in the center of the dust devil core. Figure adapted from Toigo et al. (2003) with permission. A black and white version of this figure will appear in some formats. For the color version, please refer to the plate section.

et al., 2007; Sorbjan, 2007; Tyler et al., 2008; Spiga et al., 2010). These studies have shown that the daytime convective PBL is significantly deeper on Mars than it is on Earth, with typical Martian PBL depths exceeding extreme terrestrial values over desert regions (5 km), while maximum depths are over 10 km at around local time 11:00. Vertical eddy heat fluxes and TKE are respectively of the order 1 K m s^{-1} and 10 m^2 s^{-2}, while updrafts could easily reach local values of 10–15 m s^{-1}. Note that these are not the most extreme values that might be encountered on Mars. As shown in Spiga et al. (2010), in a low-pressure (high-altitude) case where the boundary layer depth is nearly 9 km instead of 5 km, the boundary layer potential temperature is ~50 K warmer, heat flux is more than doubled, and TKE is nearly tripled. By the end of the afternoon, the

activity of convective plumes collapses, but PBL motions are not entirely shut down (see e.g. Spiga and Forget, 2009). The stably stratified free atmosphere above the convective boundary layer is perturbed by the updrafts, which gives rise to internal gravity waves, by a mechanism similar to lee wave generation (Stull, 1976). Due to the propagation of these gravity waves, the upper part of the PBL is still active after the early evening rapid collapse of the well-mixed layer below.

Another consequence of intense boundary layer convection is the presence of convective vortices: Michaels (2006) shows through LES modeling, also taking into account dust lifting, that these structures would indeed account for the formation of dust devils and of subsequent tracks along the ground of darker material, where bright dust is removed by vortices.

To date, LES studies have mostly centered on idealized numerical experiments, which have produced plausible results with respect to the limited observations available. The quantitative validation of LES diagnostics against existing data remains to be done, however. One of the main limiting factors is the paucity of data covering the entire vertical extent of the Martian BL. This limitation was recently addressed with the Mars Express radio occultation experiment (Hinson et al., 2008; see Section 7.4.3.1). Temperature profiles were obtained with good vertical resolution and coverage at latitudes and local times where PBL convection is occurring, permitting an unprecedented estimation of convective PBL depth. In low latitudes, the Martian convective boundary layer appeared to extend to higher altitudes over high plateaus than in lower-altitude plains, despite similar surface temperatures. Surface altimetry strongly influences the regional variability of daytime PBL growth when considering locations at constant latitude and local time. Spiga et al. (2010) show that these dramatic regional variations of convective PBL depth are qualitatively and quantitatively predicted by LES (see Figure 7.17) in spite of their idealized character. High-resolution numerical modeling complements the radio science observations acquired over a considerably larger area than the width of typical convective cells. LES reveals the PBL dynamics associated with the observed regional differences in PBL depth. Intense PBL dynamics is found to underlie the measured depths (up to 9 km), with vertical wind speeds up to 20 m s^{-1}, turbulent heat fluxes up to 2.7 K m s^{-1}, and convective turbulent kinetic energies up to ~26 m^2 s^{-2}.

Through large-eddy simulations, it is also possible to relate the regional variability of PBL depth (Hinson et al., 2008; see Section 7.4) to the aforementioned dominant radiative forcings of the Martian PBL (Spiga et al., 2010; see Sections 7.3 and 7.5.1). Mars appears in striking contrast to terrestrial arid conditions where sensible heat flux dominates (Spiga, 2011). On Earth the afternoon boundary layer generally warms "from below" by sensible heat flux upwelling from the heated surface, whereas on Mars it warms "both from inside and from below", respectively by infrared radiative heating (plus visible absorption by dust) and sensible heat flux. New scaling mixed-layer laws in quasi-steady midday conditions will need to be developed for the Martian case to account for the turbulent heat flux being at its maximum not near the surface but at a few hundred meters above it. Mars confirms that the PBL has to be defined as that part of the atmosphere influenced by the presence of

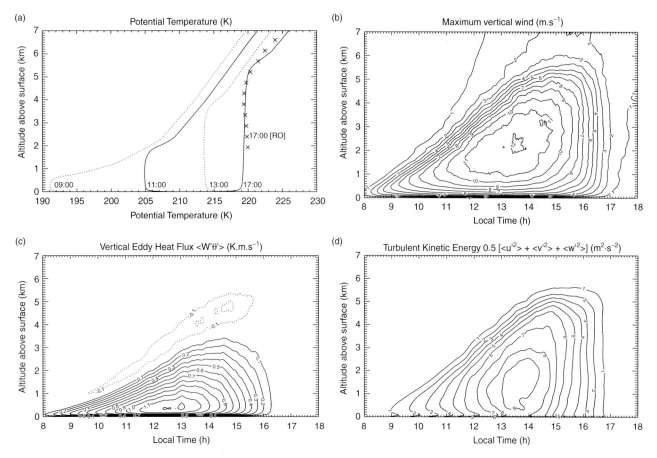

Figure 7.17. Variation of the LES statistics with time and height above ground in case study b (Amazonis Planitia) of Spiga et al. (2010): (a) potential temperature (K) with superimposed radio occultation profile at 17:00 LT, (b) updraft maximum vertical velocity (m s⁻¹), (c) vertical eddy heat flux (K m s⁻¹), and (d) turbulent kinetic energy (m² s⁻²). All displayed quantities are averaged over the simulation domain. Figure adapted from Spiga et al. (2010) with permission.

the surface, and not only by the surface itself. In the Martian environment, the energy that fuels the thermals of typical mean velocity w_* does not originate only from the atmospheric levels immediately adjacent to the surface. Thus, a version of the mixed-layer formulas valid both on Mars and on Earth should substitute the maximum heat flux $\langle w'\theta'\rangle_{max}$ for the surface heat flux $\langle w'\theta'\rangle_0$ (Spiga et al., 2010; see also Sorbjan, 2007). For instance, the relationship

$$W_* = \left[gh\frac{\langle w'\theta'\rangle_{max}}{\langle\Theta\rangle} \right]^{1/3} \tag{7.37}$$

(where h is the PBL depth and Θ is the potential temperature) enables one to calculate the Martian vertical velocity scale W_* consistent with resolved convective motions computed by LES. Alternative definitions for the convective velocity scale have also been made based, for example, on the integral heat flux (Deardorff, 1976b), although these do not seem to differ much from W_* as defined above (Martínez et al., 2011). Values of $W_* = 4$–6.5 m s⁻¹ obtained for Mars account for the vigorous convection compared to Earth (where $W_* < 2$ m s⁻¹), and are in good agreement with similarity estimates based on observations (Martínez et al., 2009a,b). Moreover, scaling by maximum heat flux instead of surface heat flux allows for a rigorous

comparison between Martian and terrestrial convective boundary layers, as shown earlier in Figure 7.3.

Future work in this arena will focus upon the influence of variations of dust opacity, background wind, local topography, and synoptic/mesoscale vertical motions so as to yield more realistic LES results. Although elements of comparison between Mars and the Earth can already be put into a useful perspective, the Martian small-scale variability remains to be explored in greater detail, especially with additional measurements of wind and temperature, in order to validate diagnostics derived from numerical models and to expand the knowledge of small-scale phenomena by new studies in extreme environments. To date, and to the extent of our knowledge, nighttime Martian LES results have not been much discussed in the literature and remain a topic to be explored.

7.5.3.5 LES and Hazards for Mission Operations

Strong turbulent motions in the Martian PBL might cause atmospheric hazards for future Martian landers. In the context of the paucity of turbulent wind measurements in the Martian BL, LES models are interesting and useful tools to address such questions. So far as preparations for missions to Mars are concerned, LES been employed to estimate the atmospheric hazards for entry, descent, and landing at the selected sites of the Mars Exploration Rovers (MER) (Kass et al., 2003; Rafkin and Michaels, 2003; Toigo and Richardson, 2003), Beagle 2 (Rafkin et al., 2004), and Phoenix (Michaels and Rafkin, 2008; Tyler et al., 2008). Rafkin and Michaels (2003) noted that, over Isidis Planitia, where upslope circulations can be encountered, convective updrafts are linearly organized in the horizontal. Furthermore, a return flow layer with moderate subsidence above the upslope circulation suppresses the growth of the afternoon convective BL, which appears only half as deep as over similar plains devoid of this regional circulation feature. This effect is also described by Toigo and Richardson (2003) within Gusev Crater. Tyler et al. (2008) showed through LES modeling in high-latitude regions (Phoenix landing site) that the influence of background wind on the simulated PBL depth is not negligible, but is far less crucial than variations of surface thermophysical properties (namely, albedo and thermal inertia). As also discussed by Michaels and Rafkin (2008), afternoon PBL convection is vigorous, even in those high-latitude regions where surface temperatures are low. Amongst other consequences, this could constitute a significant atmospheric hazard for spacecraft descent and landing. The need for accurate and realistic Martian microscale modeling remains critical for the design of upcoming missions to Mars (e.g. Mars Science Laboratory, ExoMars).

7.6 DISCUSSION AND OUTSTANDING ISSUES

We see from both models and observations available from *in situ* and remote sensing measurements that a great deal of information on the Martian PBL can be gleaned from simple adaptations of our extensive knowledge of the PBL on Earth. Monin–Obukhov similarity theory appears to work quite well for characterizing the surface layer on Mars, provided one takes into account some important differences, e.g. in molecular viscosities and diffusivities of Martian air. The latter implies, amongst other things, a larger Kolmogorov dissipation scale, resulting in a somewhat narrower inertial range within which turbulent motions can evolve before being damped out at small scales. But nevertheless there is enough room for a significant inertial range to develop, allowing for the exchange of energy, momentum, and tracers in a manner consistent with the basic underlying assumptions of the Monin–Obukhov theory.

In the main body of the boundary layer in the transition zone between the PBL and the free atmosphere, it is customary to assume (e.g. in boundary layer parameterizations in global and mesoscale models) that Mars will exhibit flow regimes that are directly equivalent to those encountered on Earth, depending upon whether the basic stratification is stable, neutral or unstable. The limited range of observations available over this height range (10 m–10 km) seem to be consistent with these assumptions, although it is clear that conditions on Mars lead to more extreme forms of convective or ultra-stable stratification than on Earth. The recent development of numerical LES models has enabled at least some aspects of this problem to be investigated

by straightforward computation (see Section 7.5.3), at least under somewhat idealized circumstances. Such investigations have shown that strongly super-adiabatic conditions are likely to be common near the Martian surface during the daytime and are more extreme than encountered even over high-altitude subtropical deserts on the Earth. It is important, therefore, to take this into account when parameterizing low-level convection on Mars in global circulation models using simple convective adjustment schemes. Such schemes may need to be flux-limited in order to provide adequate estimates of the resulting stratification close to the surface. Another important issue, also highlighted in recent LES simulations for Mars, is the likely increased role of direct radiative fluxes within the PBL. Such fluxes are generally assumed to be negligible on Earth, but are almost certainly far from negligible under conditions typically prevailing on Mars, especially during the day. This is currently neglected in most PBL parameterization in global and mesoscale models, as well as in the application of scaling laws such as the Monin–Obukhov scaling. It should receive more attention in future work, but, due to its relative novelty, the discussion cannot be taken further in the present review.

The potentially non-local effects of convective transports are another aspect that many conventional PBL turbulence models do not handle very effectively and are even more of an issue for Mars than for the Earth (Stull, 1991). Such effects are dealt with more effectively in LES, but lower-resolution mesoscale models and GCMs continue to rely on PBL models of the MY family, for which a more sophisticated approach taking into account third-order and greater correlations (Canuto et al., 2005) may be desirable. This is a particularly significant issue for GCMs, which are completely reliant on parameterizations to represent subgrid-scale phenomena such as convection. Such issues may need novel and innovative approaches, of which the parameterized "plume" model of Colaïtis et al. (2013) is a notable recent example. Their approach utilized diagnostics from LES simulations to determine limits on the vertical heat flux in a relatively thin atmosphere such as in the Martian surface layers, enabling a GCM to capture effects such as the sustained formation of super-adiabatic thermal gradients during daytime.

The parameterization of strongly stable conditions in such models has also proved notoriously problematic for both planets for many years. This is mainly because of the common assumption that shear-induced turbulence is fully suppressed if the gradient Richardson number exceeds a critical value of approximately 0.2–0.5. Observational studies on Earth, however, consistently show that, even when Ri_g is much greater than unity, some degree of "turbulent" mixing persists. LES simulations seem to confirm this, with mixing and transport apparently due (at least in part) to the action of gravity waves when the static stability is strong. In this regard, the QNSE approach reviewed in Section 7.5.2 offers a promising approach towards the adaptation of, or even as an alternative to, the conventional Mellor–Yamada family of parameterizations.

One of the traditionally difficult problems for boundary layer meteorology on Earth is the correct replication of all phases of the diurnal cycle. Particularly challenging are the morning and evening transitions when the sense of stratification changes from stable to convective and vice versa.

Despite significant progress made in this area since the early publication by André et al. (1978), difficult problems still hamper faithful representation of both types of transitions. The current state of the art in this area was assessed in the recent paper by Svensson et al. (2011), who compared the performance of 30 different closure models and one LES with observations. The modeled upward sensible heat flux during the morning transition was found to be too weak due to the problems with correct replication of the near-surface air temperatures. The best agreement between the models, LES, and the data was found during the late afternoon. The evening transition was also problematic, however, as the surface cooling leads to the development of an internal, stably stratified boundary layer near the ground while its top borders with an active layer with residual turbulence left over from earlier convective events. This leads to the formation of interleaving layers with different senses of stratification in the vertical and peculiar upside-down boundary layers in which the residual turbulence entrains the developing stable boundary layer from above. Such situations are notoriously difficult to model due to the non-locality, complicated physics, and insufficient resolution. Similar difficulties can be expected to be encountered in modeling of the diurnal cycle in Martian PBLs. In fact, the situation may be exacerbated by the larger contrast of daily and nightly temperatures, larger vertical scales, the prominent role of radiative fluxes, and the effects of dust. On the other hand, processes related to specific humidity and phase transitions (e.g. resulting in radiative fogs) are not expected to be as important on Mars. Clearly, the quality of a turbulence model can be judged by its ability to simulate the diurnal cycle faithfully in diverse environments, and so the Martian atmosphere could be viewed as another test bed for turbulence models. At present, detailed simulations of the Martian diurnal cycle are limited by the availability of suitable data (see e.g. Savijärvi, 2012a,b). This fact underscores the importance of collecting new boundary layer data of sufficient resolution, spatial coverage, and duration in future missions to Mars.

Other important characteristics of strongly stratified flows are the vertical spectra of the horizontal kinetic energy and potential energy. On Earth, the former obeys the universal scaling law, $E_1(k_z) = cN^2k_z^{-3}$, with $c \simeq 0.2$. VanZandt (1982) noticed that the observed vertical spectra of the horizontal velocity in the atmosphere have a tendency to develop a universal distribution for all seasons, meteorological conditions, and geographical locations throughout the atmosphere. This distribution has been referred to as a canonical gravity wave spectrum (Fritts and Alexander, 2003), as its origin has often been attributed to interacting internal gravity waves (see e.g. Dewan, 1979; Dewan and Good, 1986; Fritts and Alexander, 2003; and references therein). Galperin and Sukoriansky (2010) discussed the connection between this spectrum in the free atmosphere and in the PBLs. The vertical spectrum of the potential energy is more difficult to measure but generally it appears to be proportional to $E_1(k_z)$ (see e.g. Cot, 2001). These spectra are not only important atmospheric characteristics but are also critical for understanding of the physics of stably stratified turbulence. It is important that these spectra be measured in the Martian atmosphere in future missions.

In discussing these and other related issues (such as concerning transports of water vapor, dust, and chemical tracers within the PBL), it is important (and sobering!) to keep in mind the relative sparseness of direct observational measurements within the Martian PBL, compared with the rich set of measurements available to validate models and parameterization schemes on Earth. The vast majority of *in situ* measurements on Mars have been obtained from just a few (six, including the MER Rover and Phoenix Lander) spacecraft, at best equipped with meteorological sensors mounted at three vertical levels on measurement booms little longer than 1 m. While these kinds of measurement arrays can (in principle) provide direct measurements of vertical structure, heat, and momentum fluxes and spectra within the surface layer, they leave the rest of the mixed layer and transition zone (forming the bulk of the PBL) virtually unobserved. Even within the set of six successful surface landers, only four were so equipped with proper meteorological instrumentation, and of those, only two (the Viking Landers) had wind sensors that could accurately measure wind velocities sufficiently to measure covariances and momentum fluxes. Practical mission safety requirements have so far limited all available landers to relatively low-lying, flat, and featureless landing sites. While such landscapes may be typical for large tracts of the Martian surface, there are huge areas of mountainous uplands on Mars where we have essentially no *in situ* measurements with which to validate model predictions. Such observational information as exists comes almost entirely from remote sensing instruments in orbit around Mars, which have many limitations in terms of spatial and temporal resolution and coverage. These types of observation, together with just a very few descent profiles measured during the entry, descent, and landing of the extant successful landed spacecraft, also currently constitute our sole sources of direct information on the structure and behavior of the mixed layer and transition zone.

To put this into a terrestrial perspective, such detailed *in situ* coverage is tantamount to attempting to characterize the PBL structure of the entire Earth, based on measurements at (say) just three or four locations in the (non-mountainous parts of the) Sahara Desert and one in northern Canada or Antarctica. Such coverage would be scarcely representative even of the full range of desert landscapes on Earth (or even the Sahara/Sahel region itself). Moreover, only a few of those stations have measured for long enough even to span the entire seasonal cycle over one or a few years. It is hard to imagine how crude our state of knowledge of the Earth's PBL would be, were we to have been limited to this level of coverage.

Given the profound importance of the PBL for both spacecraft operations and scientific modeling of the present and past Martian atmosphere and climate, and the likely strong variability in PBL structure and properties depending on location, season, and time of day, there remains a very clear and compelling need to obtain many more measurements in this critical region of the Martian atmosphere. Even the existing *in situ* measurements have some severe limitations, with relatively few direct measurements of key fluxes of heat and momentum even at the best observed locations. Future instrumental campaigns will need to focus not only on widening the geographical and temporal coverage of measurements in the PBL, but will also need to obtain:

(a) *in situ* measurements with higher temporal resolution (with a sampling frequency in the 1–10 Hz range) in order to enable full characterization of turbulent heat and momentum fluxes, together with full kinetic energy and temperature spectra;

(b) simultaneous measurements of turbulent *and* radiative fluxes close to the surface; and

(c) profiled measurements of the main boundary layer and transition zone, including (ideally) determination of turbulent and radiative fluxes over the lowest 2–5 km of the atmosphere.

Such measurements need to allow for strong temporal variations anticipated in this region of the atmosphere on diurnal, synoptic (2–20 sol), seasonal, and inter-annual timescales at each location in order to characterize fully the complete range of boundary layer processes. Although not hugely ambitious by terrestrial standards, aspirations (a) and (b) are not trivial in their instrumental requirements, requiring some significant upgrades to and redesign of the kind of instrumentation that has already been successfully operated on Mars. For (c), however, a more sophisticated and complex campaign may be necessary, perhaps using a mixture of airborne *in situ* and remote sensing instrumentation.

Such considerations have not tended to command the priority for future spacecraft missions to Mars that one might have expected, in the light of the issues raised above. Moreover, since the Viking, Pathfinder, and Phoenix Landers have already made extensive atmospheric measurements, the notion continues to persist in some quarters that further measurements of this kind have relatively little value. Given (i) the critical importance of wind shear and turbulence, dust, and radiation, etc. from an operational safety perspective – even for exobiologically or geologically focused lander missions – and (ii) the relatively modest cost and resource requirements of the pertinent basic environmental measurements (atmospheric pressure, temperature, and wind), *the omission of these relevant measurements on-board all future Mars landers is difficult to justify*. It is at least somewhat reassuring, therefore, that a basic complement of surface meteorology instrumentation is being considered for many of the currently planned or proposed missions in the near future.

Confirmed missions at the time of writing include NASA's Mars Atmospheric and Volatile Evolution mission (MAVEN; launched in November 2013 and entered orbit in September 2014[3]), the Indian Space Research Organisation's Mars Orbiter Mission (Mangalyaan; also launched in November 2013 and entered Mars orbit in September 2014[4]), and the European Mars Trace Gas Orbiter mission (TGO; which is due for launch in 2016[5]). Of particular interest for the PBL, the TGO will also carry a small lander, the Schiaparelli ExoMars Entry, Descent and Landing Technology Demonstrator Module (or EDM), which will also carry a small science package[6] to the Martian surface. In addition to making atmospheric measurements during descent, the

EDM will carry a surface payload, based on the proposed DREAMS (Dust Characterisation, Risk Assessment, and Environment Analyser on the Martian Surface) package. This consists of a suite of sensors to measure wind speed and direction (MetWind), humidity (DREAMS-H), pressure (DREAMS- P), surface temperature (MarsTem), the transparency of the atmosphere (Solar Irradiance Sensor; SIS), and atmospheric electrification (Atmospheric Radiation and Electricity Sensor; MicroARES), although surface operations are only expected to last up to 8 days.

The 2018 launch window is currently allocated for a rover payload, with emphasis on exobiology. A surface network is a possibility for the 2020 launch window. The latter would be the next logical step scientifically in the field of atmospheric studies. An initial surface network can be expected to have a primarily global and regional (and possibly mission landing and operations assurance) emphasis, but from the PBL study perspective, an adequately instrumented network would permit a significant improvement in the sampling of different environments for surface–atmosphere interactions.

REFERENCES

Alexakis, A. 2009. Stratified shear flow instabilities at large Richardson numbers. *Phys. Fluids*, 21, 054108, doi: 10.1063/1.3147934.

Allison, M., Ross, J. D., and Solomon, N. 1999. Mapping the Martian meteorology. *Fifth International Conference on Mars*, LPI Contribution No. 972, Lunar and Planetary Institute, Tucson, Arizona, 972.

André, J. C., De Moor, G., Lacarrére, P., Therry, G., and du Vachat, R. 1978. Modeling 24-hour evolution of mean and turbulent structures of planetary boundary layer. *J. Atmos. Sci.*, 35, 1861–1883.

Atreya, S. K., Wong, A.-S., Rennó, N. O., et al. 2006. Oxidant enhancement in Martian dust devils and storms: implications for life and habitability. *Astrobiology*, 6, 439–450.

Bagnold, R. A. 1941. *The Physics of Blown Sand and Desert Dunes*. New York: Methuen.

Balme, M., and Greeley, R. 2006. Dust devils on Earth and Mars. *Rev. Geophys.*, 44 (3), RG3003, doi:10.1029/2005RG000188.

Basu, S., Vinuesa, J.-F., and Swift, A. 2008. Dynamic LES Modeling of a Diurnal Cycle. *J. Appl. Met. Clim.*, 47, 1156–1174.

Blackadar, A. K. 1957. Boundary-laycr wind maxima and their significance for the growth of nocturnal inversion. *Bull. Amer. Meteorol. Soc.*, 38, 283–290.

Blackadar, A. K. 1962. The vertical distribution of wind and turbulent exchange in a neutral atmosphere. *J. Geophys. Res.*, 67, 3095–3102.

Blumsack, S. L., Gicrasch, P. J., and Wessel, S. R. 1973. An analytical and numerical study of the Martian planetary boundary layer over slopes. *J. Atmos. Sci.*, 30, 66–80.

Brutsaert, W. H. 1982. Exchange processes at the Earth-atmosphere interface. Plate, E. (ed), *Engineering Meteorology*. Elsevier, 319–369.

Cantor, B. A., James, P. B., Caplinger, M., and Wolff, M. J. 2001. Martian dust storms: 1999 Mars Orbiter Camera observations. *J. Geophys. Res.*, 106 (E10), 23653–23687.

Cantor, B., Malin, M., and Edgett, K. S. 2002. Multiyear Mars Orbiter Camera (MOC) observations of repeated Martian weather phenomena during the northern summer season. *J. Geophys. Res.*, 107 (E3), 3-1–3-8.

---

[3] See www.nasa.gov/mission_pages/maven/main/#.VHWphEv87jJ.

[4] See www.isro.org/mars/updates.aspx.

[5] Unfortunately, during the production phase of this book, the lander failed to land successfully on the Martian surface.

[6] See http://exploration.esa.int/mars/48898-edm-science-payload/.

Cantor, B. A., Kanak, K. M., and Edgett, K. S. 2006. Mars Orbiter Camera observations of Martian dust devils and their tracks (September 1997 to January 2006) and evaluation of theoretical vortex models. *J. Geophys. Res.*, 111 (E12), E12002.

Canuto, V. M., Howard, A., Cheng, Y., and Dubovikov, M. S. 2001. Ocean turbulence. Part I: One-point closure model – momentum and heat vertical diffusivities. *J. Phys. Oceanogr.*, 31, 1413–1426.

Canuto, V. M., Cheng, Y., and Howard, A. M. 2005. What causes divergences in local second-order models? *J. Atmos. Sci.*, 62, 1645–1651.

Canuto, V. M., Cheng, Y., Howard, A. M., and Esau, I. N. 2008. Stably stratified flows: a model with no Ri(cr). *J. Atmos. Sci.*, 65, 2437–2447.

Chamberlain, T. E., Cole, H. L., Dutton, R. G., Greene, G. C., and Tillman, J. E. 1976. Atmospheric measurements on Mars: the Viking meteorology experiment. *Bull. Amer. Meteor. Soc.*, 57, 1094–1104.

Christensen, P. R., Banfield, J. L., Hamilton, V. E., et al. 2001. Mars Global Surveyor Thermal Emission Spectrometer experiment: investigation description and surface science results. *J. Geophys. Res.*, 106 (E10), 23823–23872.

Christensen, P. R., Mehall, G. L., Silverman, S. H., et al. 2003. Miniature Thermal Emission Spectrometer for the Mars Exploration Rovers. *J. Geophys. Res.*, 108 (E12), 23823–23872.

Colaïtis, A., Spiga, A., Hourdin, F., Rio, C., Forget, F., and Millour, E. 2013. A thermal plume model for the Martian convective boundary layer. *J. Geophys. Res.*, 118, 1468–1487.

Cot, C. 2001. Equatorial mesoscale wind and temperature fluctuations in the lower atmosphere. *J. Geophys. Res.*, 106 (D2), 1523–1532.

Crozier, W. D. 1964. The electric field of a New Mexico dust devil. *J. Geophys. Res.*, 69, 5427–5429.

Davy, R., Taylor, P. A., Weng, W., and Li, P.-Y. 2009. A model of dust in the Martian lower atmosphere. *J. Geophys. Res.*, 114 (D4), 4108.

Davy, R., Davis, J. A., Taylor, P. A. 2010. Initial analysis of air temperature and related data from the Phoenix MET station and their use in estimating turbulent heat fluxes. *J. Geophys. Res.*, 115 (E3).

Deardorff, J. W. 1972. Theoretical expression for the countergradient vertical heat flux. *J. Geophys. Res.*, 72, 5900–5904.

Deardorff, J.W. 1976a. Clear and cloud-capped mixed layers – their numerical simulation, structure and growth and parameterization. In *Seminars on the Treatment of the Boundary Layer in Numerical Weather Prediction*, European Center for Medium Range Weather Forecasts, 234–284.

Deardorff, J. W. 1976b. On the entrainment rate of a stratocumulus-topped mixed layer. *Quart. J. R. Meteorol. Soc.*, 102, 563–582.

Delory, G. T., Farrell, W. M., Atreya, S. K. 2006. Oxidant enhancement in Martian dust devils and storms: implications for life and habitability. *Astrobiology*, 6, 451–462.

Dewan, E. M. 1979. Stratospheric wave spectra resembling turbulence. *Science*, 204, 832–835.

Dewan, E. M., and Good, R. E. 1986. Saturation and the "universal" spectrum for vertical profiles of horizontal scalar winds in the atmosphere. *J. Geophys. Res.*, 91, 2742–2748.

Drake, N. B., Tamppari, L. K., Baker, R. D., Cantor, B. A., and Hale, A. S. 2006. Dust devil tracks and wind streaks in the north polar region of Mars: a study of the 2007 Phoenix Mars Lander sites. *Geophys. Res. Lett.*, 33, L19S02.

Ellehoj, M. D., Gunnlaugsson, H. P., Taylor, P. A. 2010. Convective vortices and dust devils at the Phoenix Mars mission landing site. *J. Geophys. Res.*, 115 (E4), E00E16.

Emanuel, K. A. 1994. *Atmospheric Convection*. Oxford and New York: Oxford University Press.

Farrell, W. M., Delory, G. T., and Atreya, S. K. 2006. Martian dust storms as a possible sink of atmospheric methane. *Geophys. Res. Lett.*, 33, L21203.

Fergason, R. L., Christensen, P. R., and Kieffer, H. H. 2006. High-resolution thermal inertia derived from the Thermal Emission Imaging System (THEMIS): thermal model and applications. *J. Geophys. Res.*, 111 (E12), E12004.

Ferri, F., Smith, P. H., Lemmon, M., and Rennó, N. O. 2003. Dust devils as observed by Mars Pathfinder. *J. Geophys. Res.*, 108 (E12), 7–1.

Forget, F., Hourdin, F., Fournier, R. et al. 1999. Improved general circulation models of the Martian atmosphere from the surface to above 80 km. *J. Geophys. Res. Planets*, 104 (E10), 24155–24176.

Forget, F., Spiga, A., Dolla, B., et al. 2007. Remote sensing of surface pressure on Mars with the Mars Express/OMEGA spectrometer: 1. Retrieval method. *J. Geophys. Res.*, 112 (E8), E08S15.

Fritts, D. C., and Alexander, M. J. 2003. Gravity wave dynamics and effects in the middle atmosphere. *Rev. Geophys.*, 41, 1003. doi:10.1029/2001RG000106.

Galperin, B., and Sukoriansky, S. 2010. Geophysical flows with anisotropic turbulence and dispersive waves: flows with stable stratification. *Ocean Dyn.*, 60, 1319–1337.

Galperin, B., Kantha, L. H., Hassid, S., and Rosati, A. R. 1988. A quasi-equilibrium turbulent energy model for geophysical flows. *J. Atmos. Sci.*, 45, 55–62.

Galperin, B., Kantha, L. H., Mellor, G. L., and Rosati, A. R. 1989. Modeling rotating stratified turbulent flows with application to oceanic mixed layers. *J. Phys. Oceanogr.*, 19, 901–916.

Galperin, B., Sukoriansky, S., and Anderson, P. S. 2007. On the critical Richardson number in stably stratified turbulence. *Atmos. Sci. Let.*, 8, 65–69.

Garratt, J. R. 1992. *The atmospheric boundary layer*. Cambridge University Press.

Gierasch, P. J., and Goody, R. M. 1968. A study of the thermal and dynamical structure of the lower Martian atmosphere. *Plan. Space Sci.*, 16, 615–646.

Golombek, M., Cook, R. A., Economou, T., et al. 1997. Overview of the Mars Pathfinder mission and assessment of landing site predictions. *Science*, 278, 1743–1748.

Gómez-Elvira, J., and REMS Team. 2008. Environmental Monitoring Station for Mars Science Laboratory. In *Third International Workshop on The Mars Atmosphere: Modeling and Observations*, November 10–13, 2008, Williamsburg, VA. LPI Contributions, 1447, 9052.

Gómez-Elvira, J., Armiens, C., Castañer, L., et al. 2012. REMS: the environmental sensor suite for the Mars Science Laboratory Rover. *Space Sci. Rev.*, 170, 583–640.

Greeley, R., and Iversen, J. D. 1985. *Wind as a geological process on Earth, Mars, Venus and Titan*. New York: Cambridge University Press.

Gunnlaugsson, H. P., Holstein-Rathlou, C., Merrison, J. P., et al. 2008. Telltale wind indicator for the Mars Phoenix Lander. *J. Geophys. Res.*, 113 (E3), E00A04.

Haberle, R. M., Pollack, J. B., Barnes, J. R., et al. 1993a. Mars atmospheric dynamics as simulated by the NASA Ames general circulation model: 1. The zonal-mean circulation. *J. Geophys. Res.*, 98 (E2), 3093–3123.

Haberle, R. M., Houben, H. C., Hertenstein, R., and Herdtle, T. 1993b. A boundary layer model for Mars: comparison with Viking Lander and entry data. *J. Atmos. Sci.*, 50, 1544–1559.

Haberle, R. M., Joshi, M. M., Murphy, J. R., et al. 1999. General circulation model simulations of the Mars Pathfinder atmospheric structure investigation/meteorology data. *J. Geophys. Res.*, 104 (E4), 8957–8974.

Haberle, R. M., Gómez-Elvira, J., de la Torre Juarez, M., et al. 2014. Preliminary interpretation of the REMS pressure data from the first 100 sols of the MSL mission. *J. Geophys. Res. Planets*, 119, 440–453.

Harri, A.-M., Genzer, M., Kemppinen, O., et al. 2014a. Mars Science Laboratory relative humidity observations – initial results. *J. Geophys. Res. Planets*, 119, 2132–2147.

Harri, A.-M., Genzer, M., Kemppinen, O., et al. 2014b. Pressure observations by the Curiosity Rover – initial results. *J. Geophys. Res. Planets*, 119, 82–92.

Hassid, S., and Galperin, B. 1994. Modeling rotating flows with neutral and unstable stratification. *J. Geophys. Res.*, 99, 12533–12548.

Heavens, N. G., Richardson, M. I., and Toigo, A. D. 2008. Two aerodynamic roughness maps derived from Mars Orbiter Laser Altimeter (MOLA) data and their effects on boundary layer properties in a Mars general circulation model (GCM). *J. Geophys. Res.*, 113 (E2), E02014, doi:10.1029/2007JE002991.

Hébrard, E., Listowski, C., Coll, P., et al. 2012. An aerodynamic roughness length map derived from extended Martian rock abundance data. *J. Geophys. Res.*, 117 (E4), E04008.

Hess, S., Henry, R., Leovy, C., et al. 1976. Preliminary Meteorological Results on Mars from the Viking 1 Lander. *Science*, 193, 788–791.

Hess, S. L., Henry, R. M., Leovy, C. B., Ryan, J. A., and Tillman, J. E. 1977. Meteorological results from the surface of Mars: Viking 1 and 2. *J. Geophys. Res.*, 82, 4559–4574.

Hinson, D. P., and Wilson, R. J. 2004. Temperature inversions, thermal tides, and water ice clouds in the Martian tropics. *J. Geophys. Res.*, 109 (E1), 15.

Hinson, D. P., Simpson, R. A., Twicken, J. D., Tyler, G. L., and Flasar, F. M. 1999. Initial results from radio occultation measurements with Mars Global Surveyor. *J. Geophys. Res.*, 104 (E11), 26997–27012.

Hinson, D. P., Tyler, G. L., Hollingsworth, J. L., and Wilson, R. J. 2001. Radio occultation measurements of forced atmospheric waves on Mars. *J. Geophys. Res.*, 106, 1463–1480.

Hinson, D. P., Smith, M. D., and Conrath, B. J. 2004. Comparison of atmospheric temperatures obtained through infrared sounding and radio occulation by Mars Global Surveyor. *J. Geophys. Res.*, 109 (E12), E12002, doi:10.1029/2004JE002344.

Hinson, D. P., Pätzold, M., Tellmann, S., Häusler, B., and Tyler, G. L. 2008. The depth of the convective boundary layer on Mars. *Icarus*, 198, 57–66.

Holstein-Rathlou, C., Gunnlaugsson, H. P., Merrison, J. P., et al. 2010. Winds at the Phoenix landing site. *J. Geophys. Res.*, 115 (E5), E00E18.

Holtslag, A., and Moeng, C. 1991. Eddy diffusivity and countergradient transport in the convective atmospheric boundary layer. *J. Atmos. Sci.*, 48, 1690–1700.

Joshi, M. M., Lawrence, B. N., and Lewis, S. R. 1996. The effect of spatial variations in unresolved topography on gravity wave drag in the Martian atmosphere. *Geophys. Res. Lett.*, 23, 2927–2930.

Kanak, K. M., Lilly, D. K., and Snow, J. T. 2000. The formation of vertical Vortices in the convective boundary layer. *Quart. J. R. Meteorol. Soc.*, 126, 2789–2810.

Karelsky, K. V., and Petrosyan, A. S. 1995. Numerical simulations of the near surface phenomena on Mars. *Adv. in Space Res.*, 16, 45–48.

Karelsky, K., Petrosyan, A., and Smirnov, I. 2007. A new model for boundary layer flows interacting with particulates in land surface on complex terrain. *Quart. J. Hungarian Meteorol. Service*, 111, 149–159.

Kass, D. M., Schofield, J. T., Michaels, T. I., et al. 2003. Analysis of atmospheric mesoscale models for entry, descent, and landing. *J. Geophys. Res.*, 108 (E12), 8090, doi:10.1029/2003JE002065.

Kauhanen, J., Siili, T., Järvenoja, S., and Savijärvi, H. 2008. The Mars limited area model and simulations of atmospheric circulations for the Phoenix landing area and season of operation. *J. Geophys. Res.*, 113 (E3), E00A14, doi:10.1029/2007JE00301.

Kieffer, H. H., Chase, S. C., Martin, T. Z., Miner, E. D., and Palluconi, F. D. 1976. Martian north pole summer temperatures: dirty water ice. *Science*, 194, 1341–1344.

Kieffer, H. H., Martin, T. Z., Peterfreund, A. R., et al. 1977. Thermal and albedo mapping of Mars during the Viking primary mission. *J. Geophys. Res.*, 82, 4249–4291.

Kitamura, Y. 2010. Modifications to the Mellor–Yamada–Nakanishi–Niino (MYNN) model for the stable stratification case. *J. Met. Soc. Japan*, 88, 857–864.

Kok, J. F., and Rennó, N. O. 2008. Electrostatics in wind-blown sand. *Phys. Rev. Lett.*, 100, 014501.

Kok, J. F., and Rennó, N. O. 2009. Electrification of wind-blown sand on Mars and its implications for atmospheric chemistry. *Geophys. Res. Lett.*, 36, L05202.

Kurbatskiy, A. F., and Kurbatskaya, L. I. 2006. Three-parameter model of turbulence for the atmospheric boundary layer over an urbanized surface. *Izvestiya, Atm. Ocean. Phys.*, 42, 439–455.

Kursinski, E. R., Folkner, W., Zuffada, C., et al. 2004. The Mars Atmospheric Constellation Observatory (MACO) Concept. 393–405 of: Kirchengast, G., Foelsche, U., and Steiner, A. (eds), *Occultations for Probing Atmosphere and Climate*. Springer. Papers from the 1st International Workshop on Occultations for Probing Atmosphere and Climate (OPAC-1).

Larsen, S. E., Jørgensen, H. E., Landberg, L., and Tillman, J. E. 2002. Aspects of the atmospheric surface layers on Mars and Earth. *Boundary-Layer Meteorol.*, 105, 451–470.

Launder, B. E., Reece, G. J., and Rodi, W. 1975. Progress in development of a Reynolds-stress turbulence closure. *J. Fluid Mech.*, 68, 537–566.

Lefevre, F., and Forget, F. 2009. Observed variations of methane on Mars unexplained by known atmospheric chemistry and physics. *Nature*, 460, 720–723.

Lemmon, M., Wolff, M., Smith, M., et al. 2004. Atmospheric imaging results from the Mars Exploration Rovers: Spirit and Opportunity. *Science*, 306, 1753–1756.

Leovy, C. B., and Mintz, Y. 1969. Numerical simulation of the atmospheric circulation and climate of Mars. *J. Atmos. Sci.*, 26, 1167–1190.

Lilly, D. K. 1962. On the numerical simulation of buoyant convection. *Tellus*, 14, 148–172.

Määttänen, A., and Savijärvi, H. 2004. Sensitivity tests with a 1-dimensional boundary layer Mars model. *Boundary-Layer Meteorol.*, 113, 305–320.

Malin, M. C., and Edgett, K. S. 2001. Mars Global Surveyor Mars Orbiter Camera: interplanetary cruise through primary mission. *J. Geophys. Res.*, 106 (E10), 23429–23570.

Malin, M. C., Carr, M. H., Danielson, G. E., et al. 1999. Early views of the Martian surface from the Mars Orbiter Camera of Mars Global Surveyor. *Science*, 279, 1681–1685.

Martínez, G., Valero, F., and Vázquez, L. 2009a. Characterization of the Martian Convective Boundary Layer. *J. Atmos. Sci.*, 66, 2044–2058.

Martínez, G., Valero, F., and Vázquez, L. 2009b. Characterization of the Martian surface layer. *J. Atmos. Sci.*, 66, 187–198.

Martínez, G., Valero, F., and Vázquez, L. 2011. The TKE budget in the convective Martian planetary boundary layer. *Quart. J. R. Meteorol. Soc.*, 137, 2194–2208.

Mellon, M. T., Jakosky, B. M., Kieffer, H. H., and Christensen, P. R. 2000. High-resolution thermal inertia mapping from the Mars Global Surveyor Thermal Emission spectrometer. *Icarus*, 148, 437–455.

Mellor, G. L. 1973. Analytic prediction of properties of stratified planetary surface layers. *J. Atmos. Sci.*, 30, 1061–1069.

Mellor, G. L, and Herring, H. J. 1973. Survey of mean turbulent field closure models. *AIAA J.*, 11, 590–599.

Mellor, G. L., and Yamada, T. 1974. A hierarchy of turbulence closure models for planetary boundary layers. *J. Atmos. Sci.*, 31, 1791–1806.

Mellor, G. L., and Yamada, T. 1982. Development of a turbulence closure model for geophysical fluid problems. *Rev. Geophys. Space Phys.*, 20, 851–875.

Melnik, O., and Parrot, M. 1998. Electrostatic discharge in Martian dust storms. *J. Geophys. Res.*, 103 (A12), 29107–29117.

Metzger, S. M., Carr, J. R., Johnson, J. R., Parker, T. J., and Lemmon, M. 1999. Dust devil vortices seen by the Mars Pathfinder camera. *Geophys. Res. Lett.*, 26, 2781–2784.

Michaels, T. I. 2006. Numerical modeling of Mars dust devils: albedo track generation. *Geophys. Res. Lett.*, 33, L19S08, doi:10.1029/2006GL026268.

Michaels, T. I., and Rafkin, S. C. R. 2004. Large eddy simulation of atmospheric convection on Mars. *Quart. J. R. Meteorol. Soc.*, 130, 1251–1274.

Michaels, T. I., and Rafkin, S. C. R. 2008. Meteorological predictions for candidate 2007 Phoenix Mars Lander sites using the Mars Regional Atmospheric Modeling System (MRAMS). *J. Geophys. Res.*, 113 (E3), E00A07, doi:10.1029/2007JE003013.

Moeng, C. H., Dudhia, J., Klemp, J., and Sullivan, P. 2007. Examining two-way grid nesting for large eddy simulation of the PBL using the WRF model. *Monthly Weather Review*, 135, 2295–2311.

Monin, A. S., and Obukhov, A. M. 1954. Osnovnye zakonomernosti turbulentnogo peremeshivanija v prizemnon sloe atmosfery (Basic laws of turbulent mixing in the atmosphere near the ground). *Trudy Geofiz. Inst. AN SSSR*, 24, 163–187.

Monin, A. S., and Yaglom, A. M. 1975. *Statistical Fluid Mechanics*. MIT Press.

Moores, J. E., Lemmon, M. T., Smith, P. H., Komguem, L., and Whiteway, J. A. 2010. Atmospheric dynamics at the Phoenix landing site as seen by the Surface Stereo Imager. *J. Geophys. Res.*, 115 (E1), E00E08.

Moudden, Y., and McConnell, J. C. 2005. A new model for multiscale modeling of the Martian atmosphere, GM3. *J. Geophys. Res.*, 110 (E4), E00A07, doi:10.1029/2007JE003013.

Murphy, J. R., and Nelli, S. 2002. Mars Pathfinder convective vortices: frequency of occurrence. *J. Geophys. Res.*, 29, 2103, doi:10.1029/2002GL015214.

Nayvelt, L., Gierasch, P. J., and Cook, K. H. 1997. Modeling and observations of Martian stationary waves. *J. Atmos. Sci.*, 54, 986–1013.

Newman, C. E., Lewis, S. R., Read, P. L., and Forget, F. 2002a. Modeling the Martian Dust Cycle, 1. Representations of Dust Transport Processes. *J. Geophys. Res.*, 107 (E12), doi:10.1029/2002JE001910.

Newman, C. E., Lewis, S. R., Read, P. L., and Forget, F. 2002b. Modeling the Martian Dust Cycle, 2. Multiannual Radiatively Active Dust Transport Simulations. *J. Geophys. Res.*, 107 (E12), doi:10.1029/2002JE001920.

Odaka, M. 2001. A numerical simulation of Martian atmospheric convection with a two-dimensional anelastic model: a case of dust-free Mars. *Geophys. Res. Lett.*, 28, 895–898.

Odaka, M., Nakajima, K., Takehiro, S., Ishiwatari, M., and Hayashi, Y. 1998. A numerical study of the Martian atmospheric convection with a two-dimensional anelastic model. *Earth, Planets, and Space*, 50, 431–437.

Oyama, V., Berdahl, B., and Carle, G. 1977. Preliminary findings of Viking gas-exchange experiment and a model for Martian surface chemistry. *Nature*, 265, 110–114.

Pallman, A. J. 1983. The thermal structure of the atmospheric surface layer on Mars as modified by the radiative effect of Aeolian dust. *J. Geophys. Res.*, 88, 5483–5493.

Petrosyan, A., Galperin, B., Larsen, S. E., et al. 2011. The Martian atmospheric boundary layer. *Rev. Geophys.*, 49, RG3005.

Pollack, J. B., Leovy, C. B., Mintz, Y., and Van Camp, W. 1976. Winds on Mars during the Viking season: predictions based on a general circulation model with topography. *Geophys. Res. Lett.*, 3, 479–482.

Pollack, J., Colburn, D., Kahn, R., et al. 1977. Properties of aerosols in the Martian atmosphere, as inferred from Viking Lander imaging data. *J. Geophys. Res.*, 82, 4479–4496.

Pollack, J. B., Leovy, C. B., Greiman, P. W., and Mintz, Y. 1981. A Martian general circulation experiment with large topography. *J. Atmos. Sci.*, 38, 3–29.

Pope, S. B. 2005. *Turbulent Flows*. Cambridge University Press.

Putzig, N. E., and Mellon, M. T. 2007. Apparent thermal inertia and the surface heterogeneity of Mars. *Icarus*, 191, 68–94.

Rafkin, S. C. R. 2003. The effect of convective adjustment on the global circulation of Mars as simulated by a general circulation model. In Albee, A. (ed), *Sixth International Conference on Mars*, 1, 3059.

Rafkin, S. C. R., and Michaels, T. I. 2003. Meteorological predictions for 2003 Mars Exploration Rover high-priority landing sites. *J. Geophys. Res.*, 108 (E12), 8091, doi:10.1029/2002JE002027.

Rafkin, S. C. R., Haberle, R. M., and Michaels, T. I. 2001. The Mars Regional Atmospheric Modeling System: model description and selected simulations. *Icarus*, 151, 228–256.

Rafkin, S. C. R., Michaels, T. I., and Haberle, R. M. 2004. Meteorological predictions for the Beagle 2 mission to Mars. *Geophys. Res. Lett.*, 31, 1703.

Rennó, N. O. 2008. A general theory for convective plumes and vortices. *Tellus A*, 60, 688–699.

Rennó, N. O., and Kok, J. F. 2008. *Electrical activity and dust lifting on Earth, Mars and beyond*. Springer. 419–434.

Rennó, N. O., Burkett, M. L., and Larkin, M. P. 1998. A simple thermodynamic theory for dust devils. *J. Atmos. Sci.*, 55, 3244–3252.

Rennó, N. O., Nash, A. A., Lunine, J., and Murphy, J. 2000. Martian and terrestrial dust devils: test of a scaling theory using Pathfinder data. *J. Geophys. Res.*, 105 (E1), 1859–1865.

Rennó, N. O., Abreu, V., Koch, J., et al. 2004. MATADOR 2002: A field experiment on convective plumes and dust devils. *J. Geophys. Res.*, 109 (E7), E07001, doi:10.1029/2003JE002219.

Rennó, N., Bos, B., Catling, D., et al. 2009. Possible physical and thermodynamical evidence for liquid water at the Phoenix landing site. *J. Geophys. Res.*, 114 (E1), E00E03, doi:10.1029/2009JE003362.

Richardson, M. I., Toigo, A. D., and Newman, C. E. 2007. PlanetWRF: a general purpose, local to global numerical model for planetary atmospheric and climate dynamics. *J. Geophys. Res.*, 112 (E9), E09001, doi:10.1029/2006JE002825.

Ringrose, T. J., Towner, M. C., and Zarnecki, J. C. 2003. Convective vortices on Mars: a reanalysis of Viking Lander 2 meteorological data, sols 1–60. *Icarus*, 163, 78–87.

Rippeth, T. P. 2005. Mixing in seasonally stratified shelf seas: a shifting paradigm. *Phil. Trans. R. Soc. A*, 363, 2837–2854.

Ristorcelli, J. R. 1997. Toward a turbulence constitutive relation for geophysical flows. *Theor. Comput. Fluid Dynamics*, 9, 207–221.

Ruf, C., Rennó, N. O., Kok, J. F., 2009. The emission of non-thermal microwave radiation by a Martian dust storm. *Geophys. Res. Lett.*, 36, L13202.

Ryan, J., and Lucich, R. 1983. Possible Dust Devils, Vortices on Mars. *J. Geophys. Res.*, 88, 11005–11011.

Savijärvi, H. 1991a. A model study of the PBL structure on Mars and the Earth. *Contrib. Atmos. Phys.*, 64, 219–229.

Savijärvi, H. 1991b. Radiative fluxes on a dustfree Mars. *Contrib. Atmos. Phys.*, 64, 103–111.

Savijärvi, H. 1999. A model study of the atmospheric boundary layer in the Mars Pathfinder Lander conditions. *Quart. J. R. Meteorol. Soc.*, 125, 483–493.

Savijärvi, H. 2012a. Mechanisms of the diurnal cycle in the atmospheric boundary layer of Mars. *Quart. J. R. Meteorol. Soc.*, 138, 552–560.

Savijärvi, H. 2012b. The convective boundary layer on Mars: some 1-D simulation results. *Icarus*, 221, 617–623.

Savijärvi, H., and Kauhanen, J. 2008. Surface and boundary layer modelling for the Mars Exploration Rover sites. *Quart. J. R. Meteorol. Soc.*, 134, 635–641.

Savijärvi, H., and Määttänen, A. 2010. Boundary layer simulations for the Mars Phoenix Lander site. *Quart. J. R. Meteorol. Soc.*, 136, 1497–1505.

Savijärvi, H., and Siili, T. 1993. The Martian slope winds and the nocturnal PBL jet. *J. Atmos. Sci.*, 50, 77–88.

Savijärvi, H., Määttänen, A., Kauhanen, J., and Harri, A.-M. 2004. Mars Pathfinder: new data and new model simulations. *Quart. J. R. Meteorol. Soc.*, 130, 669–683.

Schmidt, D. S., Schmidt, R. A., and Dent, J. D. 1998. Electrostatic force on saltating sand. *J. Geophys. Res.*, 103 (D8), 8997–9001.

Schofield, J. T., Barnes, J. R., Crisp, D., et al. 1997. The Mars Pathfinder Atmospheric Structure Investigation/Meteorology (ASI/MET) experiment. *Science*, 278, 1752–1758.

Segschneider, J., Grieger, B., Keller, H. U., et al. 2005. Response of the intermediate complexity Mars Climate Simulator to different obliquity angles. *Planet. Space Sci.*, 53, 659–670.

Seiff, A., and Kirk, D. B. 1977. Structure of the atmosphere of Mars in summer at mid-latitude. *J. Geophys. Res.*, 82, 4364–4388.

Seiff, A., Tillman, J., Murphy, J., et al. 1997. The atmosphere structure and meteorology instrument on the Mars Pathfinder Lander. *J. Geophys. Res.*, 102 (E2), 4045–4056.

Sinclair, P. C. 1973. The lower structure of dust devils. *J. Atmos. Sci.*, 30, 1599–1619.

Smith, D. E., Zuber, M. T., Frey, H. V., et al. 2001. Mars Orbiter Laser Altimeter: experiment summary after the first year of global mapping of Mars. *J. Geophys. Res.*, 106 (E10), 23689–23722.

Smith, M. D., Wolff, M. J., Lemmon, M. T., et al. 2004. First Atmospheric Science Results from the Mars Exploration Rovers Mini-TES. *Science*, 306, 1750–1753.

Smith, M. D., Wolff, M. J., Spanovich, N., et al. 2006. One Martian year of atmospheric observations using MER Mini-TES. *J. Geophys. Res.*, 111 (E12), E12S13.

Smith, P., Tomasko, M., Britt, D., et al. 1997. The Imager for Mars Pathfinder experiment. *J. Geophys. Res.*, 102 (E2), 4003–4025.

Sorbjan, Z. 2007. Statistics of shallow convection on Mars based on large-eddy simulations. Part 1: shearless conditions. *Boundary-Layer Meteorol.*, 123, 121–142.

Sorbjan, Z., Wolff, M., and Smith, M. D. 2009. Thermal structure of the atmospheric boundary layer of Mars based on mini-TES observations. *Quart. J. R. Meteorol. Soc.*, 135, 1776–1787.

Souza, E. P., Rennó, N. O., and Silva Dias, M. A. F. 2000. Convective circulations induced by surface heterogeneities. *J. Atmos. Sci.*, 57, 2915–2922.

Spiga, A. 2011. Elements of comparison between Martian and terrestrial mesoscale meteorological phenomena: katabatic winds and boundary layer convection. *Plan. Space Sci.*, 59, 915–922.

Spiga, A., and Forget, F. 2009. A new model to simulate the Martian mesoscale and microscale atmospheric circulation: validation and first results. *J. Geophys. Res.*, 114 (E2), E02009, doi:10.1029/2008JE003242.

Spiga, A., Forget, F., Dolla, B., et al. 2007. Remote sensing of surface pressure on Mars with the Mars Express/OMEGA spectrometer: 2. Meteorological maps. *J. Geophys. Res.*, 112 (E8), E08S16, doi:10.1029/2006JE002870.

Spiga, A., Forget, F., Lewis, S. R., and Hinson, D. P. 2010. Structure and dynamics of the convective boundary layer on Mars as inferred from large-eddy simulations and remote-sensing measurements. *Quart. J. R. Meteorol. Soc.*, 136, 414–428.

Squyres, S., Arvidson, R., Baumgartner, E., et al. 2003. Athena Mars Rover science investigation. *J. Geophys. Res.*, 108 (E12), 8062, doi:10.1029/2003JE002121.

Sreenivasan, K. 1995. On the universality of the Kolmogorov constant. *Phys. Fluids*, 7, 2778–2784.

Stow, C. D. 1969. Dust and storm electrification. *Weather*, 24, 134–137.

Stull, R. B. 1976. Internal gravity waves generated by penetrative convection. *J. Atmos. Sci.*, 33, 1279–1286.

Stull, R. B. 1988. *An introduction to boundary layer meteorology.* Springer.

Stull, R. B. 1991. Static stability – an update. *Bull. Amer. Meteor. Soc.*, 72, 1521–1529.

Sukoriansky, S., and Galperin, B. 2008. Anisotropic turbulence and internal waves in stably stratified flows (QNSE theory). *Phys. Scr.*, T132, 014036.

Sukoriansky, S., Galperin, B., and Staroselsky, I. 2005. A quasinormal scale elimination model of turbulent flows with stable stratification. *Phys. Fluids*, 17, 085107.

Sutton, J. L., Leovy, C. B., and Tillman, J. E. 1978. Diurnal variations of the Martian surface layer meteorological parameters during the first 45 sols at two Viking Lander sites. *J. Atmos. Sci.*, 35, 2346–2355.

Svensson, G., Holtslag, A. A. M., Kumar, V., et al. 2011. Evaluation of the diurnal cycle in the atmospheric boundary layer over land as represented by a variety of single-column models: the second GABLS experiment. *Boundary Layer Meteorol.*, 140, 177–206.

Takahashi, Y. O., Fujiwara, H., Fukunishi, H., et al. 2003. Topographically induced north–south asymmetry of the meridional circulation in the Martian atmosphere. *J. Geophys. Res.*, 108 (E3), 5018, doi:10.1029/2001JE001638.

Takahashi, Y. O., Fujiwara, H., and Fukunishi, H. 2006. Vertical and latitudinal structure of the migrating diurnal tide in the Martian atmosphere: numerical investigations. *J. Geophys. Res.*, 111 (E1), E01003, doi:10.1029/2005JE002543.

Tamppari, L. K., Bass, D., Cantor, B., et al. 2010. Phoenix and MRO coordinated atmospheric measurements. *J. Geophys. Res.*, 115 (E5), E00E17, doi:10.1029/2009JE003415.

Taylor, P. A, Catling, D. C., Daly, M., et al. 2008. Temperature, pressure, and wind instrumentation in the Phoenix meteorological package. *J. Geophys. Res.*, 113 (E3), E00A10, doi:10.1029/2007JE003015.

Taylor, P. A., Kahanpää, H., Weng, W., et al. 2010. On pressure measurement and seasonal pressure variations during the Phoenix mission. *J. Geophys. Res.*, 115 (E3), E00E15, doi:10.1029/2009JE003422.

Tennekes, H., and Lumley, J. L. 1972. *A First Course in Turbulence.* MIT Press.

Thomas, P., and Gierasch, P. J. 1985. Dust Devils on Mars. *Science*, 230, 175–177.

Tillman, J. E., Landberg, L., and Larsen, S. E. 1994. The boundary layer of Mars: fluxes, stability, turbulent spectra, and growth of the mixed layer. *J. Atmos. Sci.*, 51, 1709–1727.

Toigo, A. D., and Richardson, M. I. 2002. A mesoscale model for the Martian atmosphere. *J. Geophys. Res.*, 107 (E7), 5049, doi:10.1029/ 2000JE001489.

Toigo, A. D., and Richardson, M. I. 2003. Meteorology of proposed Mars Exploration Rover landing sites. *J. Geophys. Res.*, 108 (E12), doi:10.1029/2003JE002064.

Toigo, A. D., Richardson, M. I., Ewald, S. P., and Gierasch, P. J. 2003. Numerical simulation of Martian dust devils. *J. Geophys. Res.*, 108 (E6), 5047, doi:10.1029/2002JE002002.

Tyler, D., and Barnes, J. R. 2013. Mesoscale modeling of the circulation in the Gale Crater region: an investigation into the complex forcing of convective boundary layer depths. *Mars*, 8, 58–77.

Tyler, D., Barnes, J. R., and Haberle, R. M. 2002. Simulation of surface meteorology at the Pathfinder and VL1 sites using a Mars mesoscale model. *J. Geophys. Res.*, 107 (E4), 5018, doi:10.1029/2001JE001618.

Tyler, D., Barnes, J. R., and Skyllingstad, E. D. 2008. Mesoscale and large-eddy simulation model studies of the Martian atmosphere in support of Phoenix. *J. Geophys. Res.*, 113 (E3), E00A12, doi:10.1029/2007JE003012.

Tyler, G. L., Balmino, G., Hinson, D. P., et al. 1992. Radio science investigations with Mars Observer. *J. Geophys. Res.*, 97 (E5), 7759–7780.

VanZandt, T. E. 1982. A universal spectrum of buoyancy waves in the atmosphere. *Geophys. Res. Lett.*, 9, 575–578.

Webster, P. J. 1977. The low latitude circulation of Mars. *Icarus*, 30, 626–649.

Weng, W., Taylor, P. A., and Savijärvi, H. 2006. Modelling the Martian boundary layer. In *2nd International Workshop on Mars Atmosphere Modeling and Observations*. 27 Febrary–3 March, Granada, Spain, 123.

Whiteway, J., Daly, M., Carswell, A., et al. 2008. Lidar on the Phoenix mission to Mars. *J. Geophys. Res.*, 113 (E3), E00A08, doi:10.1029/2007JE003002.

Whiteway, J., Komguem, L., Dickinson, C., et al. 2009a. Phoenix Lidar observations of dust, clouds, and precipitation on Mars. In *Lunar and Planetary Institute Science Conference*, 40, 2202.

Whiteway, J. A., Komguem, L., Dickinson, C., et al. 2009b. Mars water-ice clouds and precipitation. *Science*, 325, 68–70.

Wilson, R. J., and Hamilton, K. P. 1996. Comprehensive model simulation of thermal tides in the Martian atmosphere. *J. Atmos. Sci.*, 53, 1290–1326.

Wing, D. R., and Austin, G. L. 2006. Description of the University of Auckland global Mars mesoscale meteorological model. *Icarus*, 185, 370–382.

Withers, P., and Catling, D. 2010. Observations of atmospheric tides at the season and latitude of the Phoenix atmospheric entry. *Geophys. Res. Lett.*, 37, L24204, doi:10.1029/2010GL045382.

Wolkenberg, P., Grassi, D., Formisano, V., et al. 2009. Simultaneous observations of the Martian atmosphere by Planetary Fourier Spectrometer on Mars Express and Miniature Thermal Emission Spectrometer on Mars Exploration Rover. *J. Geophys. Res.*, 114 (E4), E04012, doi:10.1029/2008JE003216.

Wyngaard, J. C. 2010. *Turbulence in the Atmosphere*. Cambridge University Press.

Ye, Z. J., Segal, M., and Pielke, R. A. 1990. A comparative study of daytime thermally induced upslope flow on Mars and Earth. *J. Atmos. Sci.*, 47, 612–628.

Zent, A. P., Hecht, M. H., Cobos, D. R., et al. 2009. Thermal and Electrical Conductivity Probe (TECP) for Phoenix. *J. Geophys. Res.*, 114 (E3), E00A27, doi:10.1029/2007JE003052.

Zent, A. P., Hecht, M. H., Cobos, D. R., et al. 2010. Initial results from the thermal and electrical conductivity probe (TECP) on Phoenix. *J. Geophys. Res.*, 115 (E3), E00E14, doi:10.1029/2009JE003420.

Zilitinkevich, S. S., Elperin, T., Kleeorin, N., and Rogachevskii, I. 2007. Energy and flux-budget (EFB) turbulence closure model for stably stratified flows. Part I: Steady-state, homogeneous regimes. *Boundary-Layer Meteorol.*, 125, 167–191.

8

Mesoscale Meteorology

SCOT C. R. RAFKIN, AYMERIC SPIGA, TIMOTHY I. MICHAELS

8.1 INTRODUCTION

Mesoscale atmospheric phenomena exist in the temporal–spatial domain between smaller, shorter-lived microscale systems (Chapter 7) and longer-lived, large-scale systems (Chapter 9). In practice, the division between microscale and mesoscale, and mesoscale and large-scale, is fuzzy. While many phenomena that are typically considered mesoscale systems have been observed from imagery and to a lesser extent via remote sensing and *in situ* measurement, numerical models combined with information from terrestrial system analogs are the most effective tools for studying thermal slope flows, gravity waves, bores and density currents, dust storms, and circulations producing water ice and CO_2 ice clouds, all of which have important mesoscale aspects.

Mars mesoscale circulations contribute in important ways to the structure and dynamics of the atmosphere. In many cases, the circulations are much more pronounced and ubiquitous than on Earth. For example, thermal slope flows are thought to dominate the local near-surface meteorology in all but the topographically flattest regions of Mars. In contrast, such flows only dominate on Earth where the large-scale forcing is small and the topographic relief is large. On Mars, the structure and dynamics of the atmosphere are intimately connected to even relatively small variations of topography. As another example, the distribution of dust in the atmosphere is a major driver in determining the thermal structure of Mars. Dust is placed into the atmosphere by, among other phenomena, dust storms that are generally considered mesoscale systems. Thus, mesoscale circulations are a key element of the Mars dust cycle. Vapor transport through mesoscale circulations may also play a similar role in the water cycle. An overview of what constitutes the mesoscale is discussed in the first section of this chapter, with comparisons to Earth where it is instructive to do so. This is followed by an overview of mesoscale numerical models. Then, the variety of mesoscale systems and circulations that exist on Mars are presented from a theoretical, observational, and especially modeling perspective. Water and CO_2 condensate clouds and dust, which are often tied to mesoscale circulations, present an additional level of complexity that mandate a separate and detailed description. Finally, the application of mesoscale models in support of Mars mission design and operations is described.

8.2 THE DEFINITION OF THE MESOSCALE

At first glance, a subjective phenomenological classification of atmospheric phenomena easily eliminates small turbulent eddies, dust devils, and large planetary-scale waves from the mesoscale. Subjective classification also clearly places local katabatic and anabatic circulations, local dust storms, and gravity waves into the mesoscale. Upon further consideration, however, a subjective classification immediately leads to problems and ambiguities. While dust devils on Earth are clearly in the realm of the microscale, dust devils on Mars can be in excess of 0.5 km in diameter (Fisher et al., 2005), which would place them at least close to what might be considered the smaller end of the terrestrial mesoscale spectrum. Immediately, then, the different physical scales of phenomena between Earth and Mars calls into question whether a subjective phenomenological classification appropriate for Earth can be translated directly to Mars; the largest dust devils on Mars (with diameters as much as 1 km and heights of many kilometers) may be more mesoscale than the largest dust devils on Earth, which may be an order of magnitude smaller. Furthermore, Mars dust devils are likely embedded within larger-scale thermal structures that present mesoscale organization (Toigo et al., 2003; Michaels and Rafkin, 2004), much like they are on Earth. But again, the scale of the organization, which is tied at least loosely to the depth of the convective layer, may be quite different; the typical depth of Mars' convective layer is many times larger than Earth's. While the active dust lifting centers of dust storms might be considered mesoscale, such storms can be triggered by larger-scale systems (Wang et al., 2003; Wang, 2007), and the impact of the dust lifted in one region can have a global impact (e.g. Conrath et al., 2000; Medvedev and Hartogh, 2007; Martínez-Alvarado et al., 2009). Circulations perpendicular to katabatic fronts may be only tens of kilometers in scale, but the axis along the front may be hundreds or even a thousand or more kilometers in length (Siili et al., 1999; Kauhanen et al., 2008; Spiga, 2011). To understand the structure and circulations in the cross-front direction demands looking at the system from a mesoscale perspective while the overall length of the front might be considered in the realm of the large scale. The potentially large aspect ratio of these fronts highlights the difficulty in cleanly binning circulations into a single category scale; circulations can have important dynamical aspects at a variety of scales. Similarly, cloud circulations may be mesoscale, but the forcing that produces them may lie within the realm of large-scale circulations. Within the clouds, elements of the circulations may be considered microscale (e.g. turbulent entrainment).

Beyond a subjective classification of phenomenological scale, many objective discriminators may be used to define the mesoscale. From a purely dynamical view, there may be inherent processes or instabilities that constrain atmospheric disturbances to certain scales. Therefore, it may be possible to define one or more dynamically objective classification schemes. For example, theory predicts a narrow, optimum range of scales for baroclinic instability (e.g. Farrell, 1989). In addition to looking for instabilities as a source of classification, the Rossby number and the Rossby radius of deformation (described below) have proven to be useful in establishing a degree of "mesoscale-ness". The required energy cascade from large-scale motions to viscous dissipation must flow through the mesoscale, and it is possible that this process will produce a distinguishable spectral signature. Finally, there is a practical definition, based primarily on historical use and context, as well as distinct classes of atmospheric numerical modeling codes. Each of these different classification possibilities are discussed below.

8.2.1 Rossby Radius of Deformation

Ooyama (1982) proposed using the Rossby radius of deformation, λ_R, to classify dynamical systems,

$$\lambda_R^2 = \frac{(NH)^2}{(\zeta + f)^2} \tag{8.1}$$

where N^2 is the Brunt–Väisälä frequency, H is the scale height, and ζ is the relative vorticity; λ_R is the ratio of a characteristic gravity wave phase speed (NH) to the Coriolis timescale (f) for small values of ζ. Alternatively, the denominator of (8.1) is a measure of the inertial stability, while the numerator is a measure of static stability; one interpretation of λ_R is the ratio of vertical stiffness to horizontal rotational stiffness. Rossby (1938) showed that the length scale with his name emerged in the theoretical consideration of the response of a fluid to a perturbation of horizontal scale L, such that the fluid mass field would adjust primarily to the initial wind field for $L \ll \lambda_R$ and the wind field would adjust to the initial mass field for $L \gg \lambda_R$.

In the latter case, the initial energy of the disturbance remains primarily as potential energy (i.e. the pressure distribution), with only a small fraction converted to kinetic energy in order to produce a dynamically balanced condition where the pressure field is in geostrophic balance with the wind. This condition is more likely for systems with high inertial stability, as measured by the denominator of (8.1), or weak static stability in shallow systems, as measured by the numerator of (8.1). This geostrophic adjustment is achieved through the redistribution of mass by gravity waves and the Coriolis torque on the gravity waves as they propagate away from the initial disturbance. For $L \gg \lambda_R$ gravity wave speed is slow enough so that the Coriolis torque acts over a sufficient time to convert the transient, unbalanced, irrotational gravity wave response into a quasi-two-dimensional rotational flow ($L > H$) in balance with the mass field. Most large-scale and planetary waves fall into this regime.

In the opposite case ($L \ll \lambda_R$), gravity waves triggered by the initial disturbance propagate relatively rapidly compared to the Coriolis timescale, and energy is transported far beyond distance L. At the extreme, the circulations resulting from the disturbance are transient and almost none of the initial energy is transformed into a balanced circulation. Equivalently, the rotational constraints on the system are small compared to the more dominant static stability effects. Circulations with these characteristic properties tend toward three-dimensional ($L < H$), such as free convection and turbulence.

A consistent definition of the mesoscale would be circulations in which L is the same order of magnitude as λ_R. These flows would exhibit a mix of quasi-2D geostrophic circulations and transient, irrotational 3D circulations. The 3D circulations have a non-trivial vertical velocity component, which must ultimately be driven through mass continuity by convergent/divergent horizontal circulations (i.e. irrotational circulations). This classification would be a function of latitude through the Coriolis parameter, f, so that systems of increasingly larger scale could remain as mesoscale at lower latitude. Likewise, the systems would tend toward $H \approx L$. If the rotational energy of a circulation increases with time, the classification of the system will tend toward smaller scales. An important consequence of the latitude dependence is that, under this classification, all systems become mesoscale (or smaller) in the tropics where f becomes vanishingly small. Balanced systems are not dynamically possible in the tropics and all motions will be dominated by transient circulations. Tropical dynamics are inherently mesoscale even if the length scales of phenomena are large.

The scale height H and Coriolis parameter f are approximately the same on Earth and Mars. Thus, according to a Rossby radius definition, the distinction between mesoscale systems on these two planets would be primarily a result of differences in stability and the scale of the system. Stability values can vary significantly on both planets in both space and time, but to first order the stabilities may be assumed approximately the same. Then, ignoring the fluid rotation contribution to the Rossby radius of deformation, the value of λ_R is approximately the same for both planets. Using representative values of N ($\sim 10^{-2}$ s^{-1}) at 45° latitude, and neglecting local vorticity, $\lambda_R \approx 1000$ km. The scale of circulations, however, can be quite different. For example, slope flows on Mars may be a thousand or more kilometers long, while this is extremely rare on Earth. Therefore, what might be taken as a strictly mesoscale system on Earth (e.g. slope flows) may have stronger aspects of large-scale, geostrophic-like dynamics. As systems spin up, and the magnitude of the local vorticity increases, the systems take on increasingly more mesoscale characteristics even if the spatial scale remains the same.

There is clearly an upper limit to the horizontal length scale imposed by the size of the planets themselves. Planetary-scale circulations on Mars simply cannot be as large in absolute size as those on Earth, so that these Martian systems would tend to have more mesoscale characteristics, as minor as those may be. In the case of baroclinic systems, the scale of the disturbances tends to be roughly the same for Earth and Mars, but there are simply fewer of those waves (i.e. a lower wavenumber) on Mars because of the smaller planetary radius. In this case, the degree to which the baroclinic disturbances exhibit mesoscale characteristics would be primarily a function of the stability and the strength of the disturbances as measured through the rotational circulation component.

8.2.2 Richardson Number

An alternative classification based on the Richardson number, $Ri = N^2 / (\partial U / \partial Z)^2$, was described by Emanuel (1986). It has long been recognized that the onset of small-scale turbulence occurs for $Ri \lesssim 0.25$. These microscale circulations have a characteristic depth on the same scale as the depth of the turbulently unstable layer. Often, this is the depth of the convective boundary layer, but elevated layers of turbulence within the free atmosphere are also known. The timescale of these circulations is driven primarily by the Brunt–Väisälä frequency, which is far shorter than the Coriolis timescale; balanced geostrophic circulations are unimportant.

At the other end of the spectrum are circulations with large Ri. These circulations are dominated by the buoyancy restoring force rather than shear vorticity. The relatively high static stability limits the vertical extent of the systems compared to their horizontal scale so that the systems are quasi-2D. For example, in the Charney model of baroclinic instability (Charney, 1947), the horizontal scale, L, is $f(\partial U / \partial Z) / N\beta$, where β is the latitudinal gradient of f and the vertical scale is $H = fL/N$. For typical mid-latitude conditions, L is several thousand kilometers, while H is of order 10 km. The corresponding Ri is ~5.

In between the two extremes are circulations with $Ri \approx 1$. Static stability, which acts to suppress vertical motion, is of roughly equal importance to shear, which tends to drive vertical overturning. The timescale of these circulations, $T = 2\pi L / U$, is long enough that Coriolis torques are important. Circulations with $Ri \approx 1$ tend to be aligned with the shear vector and exhibit characteristics of convective vertical motion. On Earth, convective symmetric instability (e.g. Bennetts and Hoskins, 1979; Emanuel et al., 1987; Clever and Busse, 1992) is probably the most well known, and is often treated as a mesoscale process. On Mars, both observations (Benson et al., 2006) and numerical modeling (Rafkin and Michaels, 2003; Michaels and Rafkin, 2004) have revealed similar shear-aligned organization in the dry convective boundary layer and in convective cloud fields. A consistent definition of the mesoscale can be achieved using Ri as a metric, although the answer will generally not be equivalent to one based on λ_R.

8.2.3 Rossby Number

The Rossby number is constructed from a characteristic length scale L and velocity scale, $Ro = U/fL$. The parameter is a measure of the advective timescale (L/U) to the planetary rotational timescale ($1/f$). Systems with short advective scales will have larger Rossby numbers and will not be as affected by Coriolis torques as their smaller Ro counterparts. Equivalently, Ro is a ratio of the flow acceleration to the Coriolis acceleration. In this sense, there is at least a loose relationship to the Rossby radius of deformation in that both provide a measure of the dynamical

properties of the circulation compared to the rotational component. When systems are relatively fast (i.e. strong winds or short length scales) compared to planetary rotation, they will tend to have less of a geostrophically balanced component and a stronger quasi-geostrophic component. Very large Rossby numbers ($Ro \gg 1$) are not influenced by planetary rotation in any significant ways. Very small Rossby numbers indicate that the air parcels moving through a disturbance will have a long enough residence time to feel appreciable acceleration from the Coriolis force. Systems in this regime are considered large-scale. The mesoscale is in a regime where neither the flow acceleration nor the Coriolis force can be entirely neglected ($Ro \sim 1$).

There is an important branch of theoretical atmospheric dynamics that evaluates the behavior of systems based on the magnitude of the Rossby number. The primitive equations – the system of equations governing momentum, energy, and mass conservation – can be asymptotically expanded in terms of Rossby number (Ro, Ro^2, Ro^3, ...), and the magnitude of each term in those equations can then be evaluated (Dickinson et al., 1968). More exact approximations to the full set of primitive equations are obtained by retaining increasingly higher orders of Ro. For very small Rossby numbers, the geostrophic approximation is obtained by retaining terms of Ro^1. The important aspects of the geostrophic system are that: (1) the Coriolis force exactly balances the pressure gradient force – there is no net acceleration of horizontal wind; and (2) the system is hydrostatic and non-divergent. Without horizontal convergence, there is no vertical motion, and this is dynamically consistent with the hydrostaticity. A purely geostrophic system is dynamically two-dimensional. Oddly enough, however, in the thermodynamic term, vertical motion does appear in the vertical advection term, which means that some amount of ageostrophic motion is necessary for this approximate system of equations, but only in relation to the movement of heat. Vertical motion must be driven by convergence of ageostrophic wind components since the geostrophic wind is non-divergent. Further, the entire system is inconsistent in that there is no equation to predict or diagnose the ageostrophic components of the wind or the vertical velocity resulting from the convergence of those winds.

Even though the very small Rossby number circulations are 2D, vertical shear of the horizontal wind can be considered. Mathematically, this shear is related to the horizontal gradient of temperature, and the relationship can be used to obtain the geostrophic wind (subject to a known wind value for a boundary condition) from an observed thermal field, such as those obtained by the MGS Thermal Emission Spectrometer or the MRO Mars Climate Sounder. The derived values, however, are subject to the dynamical constraints of the geostrophic relationship, and errors in those calculations are then due to ageostrophic effects. Or stated another way, the error can be thought of as being related to the importance of dynamically mesoscale processes. On Earth, estimating the wind using the geostrophic approximation generally leads to an error of order 10%, but it can be substantially larger. Since there are effectively no wind measurements on Mars, the error associated with the technique on Mars is unknown, although it has been assumed to be of the same order of magnitude as on Earth. Many of the circulations in Mars' atmosphere can be of different magnitude than those on Earth, however. For example, the thermal tide is a large-scale divergent feature with low-level (ageostrophic) winds converging into the solar-heated columns of air and divergent flow aloft to produce a net decrease in surface pressure. The geostrophic assumption ignores this direct dynamical effect. On Earth, this is less of an issue because the magnitude of the thermal tide is much smaller. In principle, one might be able to better estimate the error on Mars using output from a climate model; the actual calculated wind could be compared to that derived by the geostrophic balance relationships. However, most climate models employ some level of small-Rossby-number approximations (e.g. hydrostaticity) and are inherently missing the higher-order terms that could be important.

The next higher-order approximation after geostrophy produces the quasi-geostrophic approximation. Here, acceleration of the geostrophic wind is permitted, but only by two processes: (1) advection by the geostrophic wind itself, and (2) Coriolis accelerations on the ageostrophic wind. The system, however, remains hydrostatic. In the continuity equation, vertical motion becomes explicit, and must be balanced by the divergence of the ageostrophic wind. As before, there is no predictive equation for any of the ageostrophic components of the wind. The vertical wind may be solved for diagnostically, however. This then allows for the diagnosis of the ageostrophic divergence, but not the actual ageostrophic wind components. The quasi-geostrophic system is probably the most widely used system of equations to study synoptic-scale (e.g. large-scale) processes on Earth, including the dynamics of Rossby waves and the evolution of baroclinic disturbances.

The hypogeostrophic system of equations (e.g. McWilliams and Gent, 1980) is obtained by retaining all terms up to the second order in Ro. Here, the first-order ageostrophic dynamics now have a prognostic equation. The complete set of second-order Ro equations are quite complex. In practice, the system is simplified through various approximations to the semigeostrophic system (Hoskins, 1975). Semigeostrophic theory neglects geostrophic advection of the ageostrophic wind. However, this simplification can lead to inaccuracies in systems where H is comparable to L, and comparison between solutions using hypogeostrophic and semigeostrophic equations shows noticeable differences even in baroclinic systems.

Clearly, as Ro increases, flows become more ageostrophic. It is also worth noting that geostrophic systems tend to have small aspect ratios $A = L/H$. For example, in the baroclinic systems where $Ro = Ri^{-1/2}$, $A = N/f \ll 1$. On the other hand, for circulations with Ro near unity (e.g. fully isotropic turbulence), Ri is small and $A \approx 1$. In between these two extremes is $Ro \leq 1$, and the aspect ratio suggests circulations that are quasi-3D, but not necessarily isotropic.

A consistent definition of the mesoscale circulations is that for which $Ro \sim 1$ and for which ageostrophic motions as well as balanced rotational flows (e.g. geostrophy) are crucial. In this sense, the mesoscale is an extremely dynamically challenging regime, because neither can background rotation be neglected, as is often done in turbulent microscale studies, nor can high-order ageostrophic motions be neglected, as is often done in large-scale studies.

The hypogeostrophic and semigeostrophic systems of equations are especially valuable for identifying the impact of mesoscale circulations on large-scale dynamical systems, such as baroclinic eddies and frontal systems. While the higher-order systems of equations have been applied widely in dynamical studies of Earth phenomena, this is not yet the case for Mars. A detailed theoretical treatment of the dynamics of Mars' atmosphere using higher-order systems of equations is overdue. Because vertical winds associated with some of the circulations of Mars (e.g. thermal tides and some thermal topographic flows) may be stronger than on Earth, mesoscale dynamical aspects may be more important in situations that have traditionally been considered to fall almost entirely within the large-scale regime.

8.2.4 Energy Cascade

Energy generated at larger scales, for example the energy within the thermal tide or baroclinic storm systems, must on average cascade downward through the mesoscale to smaller scales where the energy is viscously dissipated. The kinetic energy spectrum of the free atmosphere (above the planetary boundary layer) on Earth displays a k^{-2} power law of kinetic energy density for wavenumbers corresponding to scales of greater than a few hundred kilometers (e.g. Vinnichenko, 1970; Lilly and Petersen, 1983). This is consistent with the theory of nearly two-dimensional turbulence, and is in concert with the Ri and Ro classifications for large-scale dynamics. At smaller scales, the kinetic energy spectrum follows a $k^{-5/3}$ slope. The mesoscale may be considered to fall into the transition region between the two regimes. Observations for Earth, however, show an occasional spectral gap where the notional mesoscale energy should be found (Panofsky and van der Hoven, 1955; Vinnichenko, 1970). Apparently, mesoscale phenomena are intermittent, or too spatially variable to produce a continuous, well-defined signal in the atmospheric power spectrum.

There are no datasets of consequence of winds in the free atmosphere of Mars, and surface datasets are extremely limited. Only Viking provides time series data for longer than a few typical synoptic-scale periods. This paucity of data makes it difficult to assess the kinetic energy spectra for the Martian atmosphere even locally, and caution should be exercised when extending the conclusions derived from Viking data to other locations on Mars. As noted by Vinnichenko (1970), the spectrum differences between the free atmosphere and planetary boundary layer on Earth can be dramatically different. For example, surface energy at the synoptic scale is only ~6% of what it is in the free atmosphere. Within the atmospheric surface layer of Mars, observations from the Viking Landers (Tillman et al., 1994; Larsen et al., 2002) and the Mars Pathfinder (Schofield et al., 1997) all seem to suggest that the same basic scaling laws valid for Earth also hold for Mars, although the fundamental scales may be different. This further suggests that, like Earth, the mesoscale on Mars lies in the transition between 3D microscale turbulence and 2D large-scale motions. A consistent definition for the mesoscale would then be those circulations that fill the spectral kinetic energy gap or provide a transition between the k^{-2} and $k^{-5/3}$ regimes.

8.2.5 A Practical Definition

The mesoscale has often been defined, not through subjective classification or dynamical considerations, but by practical considerations. Numerical models of the atmosphere have been indispensable tools in synthesizing sparse observations of Mars into a coherent picture of climate and weather. For many decades these models were exclusively global, general circulation models (GCMs) with horizontal grid spacing of 100 km or greater (or equivalent in the case of spectral models). By default, circulations that could not be resolved by these models were termed mesoscale or microscale. The subgrid-scale circulations were either parameterized or ignored. Within the last decade, terrestrial models with horizontal resolutions and dynamical cores designed to simulate the unresolved circulations of the GCMs have been adapted to Mars. These models have been historically called mesoscale models, and the circulations that they simulate are considered, for practical reasons, mesoscale. Nevertheless, it is of critical importance to appreciate that mesoscale models can also simulate what might be considered both large-scale circulations and microscale circulations under various classification systems. As computing power increases, the horizontal resolution of GCMs is increasing, as is the sophistication of dynamical cores, and the distinction between GCMs and mesoscale models is rapidly blurring. For example, a few GCMs have non-hydrostatic dynamical cores, which permits them to better model mesoscale-like circulations with strong vertical accelerations The association of mesoscale phenomena with what has been historically mesoscale models will likely endure well after the inevitable merger of what was once two distinct numerical modeling enterprises.

8.3 MESOSCALE MODELS

Much of what is known about the mesoscale on Mars is derived from mesoscale model simulations, because there are very limited data that adequately cover the mesoscale temporal–spatial range. Occasionally, clouds and dust make otherwise invisible circulations visible and provide at least indirect information on mesoscale activity.

The first comprehensive mesoscale studies on Earth addressed meteorological phenomena at regional scales mostly through two-dimensional idealized modeling, e.g. tropical hurricanes (Anthes, 1971) and mountain meteorology (Mahrer and Pielke, 1977). Those idealized numerical tools formed the basis for modeling studies of mesoscale atmospheric disturbances in the Martian atmosphere, with a particular emphasis on slope winds (Ye et al., 1990; Savijärvi and Siili, 1993) and orographic waves (Pickersgill and Hunt, 1979; Tobie et al., 2003).

In the 1980s and beyond, a more generalized approach was adopted in terrestrial studies that allowed the same code to be applied to many different scenarios rather than to only one or two specific situations (e.g. only hurricanes or only mountain-induced circulations). Efforts to build platforms capable of reproducing the mesoscale variability in any region of the world at various horizontal scales were pursued (Pielke et al., 1992; Dudhia, 1993; Skamarock and Klemp, 2008). Those efforts yield the framework for versatile three-dimensional mesoscale models presently used in operational weather prediction and meteorological research.

The progress in terrestrial mesoscale modeling, along with the arrival of new atmospheric observations on Mars, some of which related to phenomena left unresolved by global circulation models (GCMs), motivated the development of dedicated three-dimensional mesoscale models for the Martian atmosphere (Rafkin et al., 2001; Toigo et al., 2002; Tyler et al., 2002; Siili et al., 2006; Wing and Austin, 2006; Spiga and Forget, 2009). Recently, GCMs with mesoscale grids have been developed (e.g. Richardson et al., 2007). The decade beginning in 2000 represented a boom phase for Mars mesoscale model development. Now that many models are available, a new phase of model intercomparison and validation against data is needed. Mesoscale model validation is particularly challenging in the absence of adequate observational data. Neither previous or current *in situ* surface data, nor remotely sensed orbital information are adequate. The Mars Exploration Program Assessment Group (MEPAG, 2010) has laid out a set of necessary investigations that include full energy budget studies, including turbulent eddy fluxes, but these have largely been ignored.

Meteorological models aimed at resolving mesoscale phenomena share some of the same structure and design as the GCMs described in Chapter 9. They comprise two main modules. (1) The dynamical core integrates primitive equations for the atmospheric fluid, i.e. Navier–Stokes equations projected on the rotating frame using spherical coordinates. (2) The physical parameterizations represent key diabatic forcing in the primitive equations, i.e. radiative transfer, surface–atmosphere heat and mass exchanges, and latent heat release, as well as unresolved dynamical phenomena, i.e. friction, boundary layer mixing, and wave breaking. Despite this common framework, there are crucial differences between GCMs and mesoscale models. As previously mentioned, the distinction and differences between GCMs and mesoscale models is beginning to blur, but, generally speaking, the differences in dynamical cores and physical parameterizations described below apply to the majority of model codes at the time of writing.

8.3.1 Dynamical Cores

A major difference between most GCMs and mesoscale dynamical cores is the assumption of hydrostatic equilibrium. The hydrostatic equilibrium assumption requires gravitational acceleration to balance exactly the vertical acceleration due to the pressure gradient force, $\partial p / \partial z = -\rho g$. An alternative formulation for hydrostaticity is obtained through integrating this equation between two atmospheric levels z_1 and z_2, $p_2 - p_1 = -\int_{z_1}^{z_2} \rho g \, dz$, where pressure then naturally arises as an equivalent for atmospheric mass.

Hydrostatic equilibrium is only strictly true for a static atmosphere, devoid of any horizontal pressure gradients and vertical acceleration, which is not the case for a real atmosphere (as can be inferred from, for example, geostrophic equilibrium). Notwithstanding this, hydrostatic equilibrium stands as an acceptable approximation as long as acceleration of vertical wind is negligible compared to gravitational acceleration. To illustrate this, Janjic et al. (2001) proposed a simple equation based on the distinction between "true" pressure (P_s), as measured by a barometer at the surface, and hydrostatic pressure (p_s), which satisfies hydrostatic equilibrium:

$$P_s = p_s + \int_0^1 \varepsilon p \, d\sigma' \tag{8.2}$$

where $\varepsilon = (1/g)(dw/dt)$ and σ' denotes a convenient vertical coordinate. As long as vertical accelerations in the atmosphere are low or inhibited, e.g. when atmospheric stability is particularly high, the hydrostatic equilibrium is valid.

Given the low vertical velocities generally involved in large-scale atmospheric phenomena, hydrostatic equilibrium can be assumed in primitive equations used for GCM computations. While hydrostaticity may still be appropriate for many regional-scale phenomena, mesoscale atmospheric dynamics possibly involve strong vertical acceleration (convective updrafts, gravity waves, dust storms) causing the atmospheric state to depart more significantly from hydrostatic equilibrium. Hence, the equation for vertical motions is usually implemented in a more complete form in mesoscale models. The importance of non-hydrostatic effects can be further exemplified by mesoscale gravity waves as expressed through the dispersion relation between wave frequency ω and spatial wavenumber (k,l,m):

$$\omega^2 = f^2 + N^2 \left/ \frac{k^2 + l^2}{m^2} \right. \tag{8.3}$$

under hydrostatic assumption, but

$$\omega^2 = f^2 + N^2 \left/ \frac{k^2 + l^2}{k^2 + l^2 + m^2} \right. \tag{8.4}$$

with all non-hydrostatic contributions (Fritts and Alexander, 2003). In other words, hydrostatic integration leaves part of the gravity wave spectrum unresolved: mesoscale studies devoted to resolving gravity wave phenomena could benefit from non-hydrostatic integrations.

There is another important difference between GCM and mesoscale dynamical cores, which is specific to Mars-like conditions. An intercomparison of Martian mesoscale models carried out in 2003 revealed that non-hydrostatic models had significantly overestimated the diurnal surface pressure cycle compared to hydrostatic models (Tyler and Barnes, 2005). The origin of the problem was the diabatic heating terms in the pressure tendency equation being neglected in non-hydrostatic dynamical cores (Dudhia, 1993) used for Martian applications. This approximation yields negligible differences on Earth, but not on Mars, where the thermal tide has large diabatic forcing. A solution is to include the fully compressible equations in the model. The implementation of this compressibility is model-dependent and not all mesoscale models necessarily include this term. In the case of Spiga and Forget (2009), the pressure tendency equation is replaced by the equivalent, though much simpler, geopotential equation in which the diabatic heating is included. The early studies of Rafkin et al. (2001) still relied upon an Earth-based terrestrial core that neglected compressional effects. This neglect was later found to produce spurious mass sources under the strong radiative forcing in Mars atmosphere. A correction was implemented following Nicholls and Pielke (1994). A

subsequent rewrite of the core allowed for an even more complete treatment.

Different model cores solve the dynamical equations on different numerical grids. A common vertical coordinate for GCMs and some mesoscale models (Toigo et al., 2002; Tyler et al., 2002; Spiga and Forget, 2009) is the σ_p vertical coordinate, which are pressure surfaces normalized by the surface pressure (or some variant of this quantity). Thus, the coordinate system follows the variations in surface pressure in both space and time, and the physical spacing between adjacent vertical levels will move up or down in space and time (assuming the temperature changes), although the pressure difference (i.e. hydrostatic mass) between levels remains quasi-constant. Another possibility is the use of σ_z, which is height normalized by the surface elevation (or some variant of this quantity). This coordinate system follows the topography, and spacing between levels is fixed in time. However, the pressure (i.e. mass) between adjacent levels will vary in space and time. The numerical representation of the dynamics will be different in different cores, and the behavior of the solution can depend strongly on the formulation of the core dynamics on the coordinate system. For example, spurious mass sources in an early version of the MRAMS model were directly linked to the σ_z coordinate system (which allows for changes in mass between vertical levels) and the lack of a compression term. Models with a σ_p coordinate system, in which the mass is more directly related to the vertical coordinate, are less likely to be affected by this problem. Nonetheless, there are distinct advantages and disadvantages to both coordinate systems, which are reviewed in Pielke (2002).

Horizontal grid structures also vary between the models. Generally, the prognostic variables carried in the model are not all located at the same numerical point. Rather, the variables are staggered in space and may include both horizontal and vertical staggering (Haltiner and Williams, 1980). Like vertical coordinate systems, there are advantages and disadvantages to different grid staggering schemes (Arakawa, 1966).

It should be noted that, once specific problems related to the choice of horizontal grid and vertical discretization are solved, different dynamical cores with the same dynamical assumptions (e.g. hydrostatics versus non-hydrostatics) should lead to only small disagreements. The remaining differences in predictions obtained by mesoscale models are then more likely to be related to boundary conditions and physical parameterizations, which are further discussed below. The true origin of disagreements between Mars mesoscale models has not yet been widely investigated.

8.3.2 Physical Parameterizations

Many mesoscale models rely on the same physical parameterizations of dust, CO_2, and H_2O cycles developed for Martian GCMs: the MRAMS model (Rafkin et al., 2001) and the OSU MM5 model (Tyler et al., 2002) contain physics based upon the NASA Ames GCM radiative transfer model (Toon et al., 1989); the Cornell MM5 (Toigo et al., 2002) is based on physics of the GFDL (Geophysical Fluid Dynamics Laboratory) model described in Wilson and Hamilton (1996); and the LMD (Laboratoire de Météorologie Dynamique) mesoscale model

(Spiga and Forget, 2009) is interfaced with the complete physical packages designed for the LMD GCM (Forget et al., 1999). Other parameterizations have been developed independently, are modified versions of those in the parent terrestrial model (e.g. turbulent diffusion in MRAMS), or are adapted from other stand-alone models (e.g. the microphysics code in MRAMS based on the Cloud and Radiation Model for Atmospheres (CARMA)).

Of key importance for the Martian climate, the spatial and temporal variations of dust opacity have until recently been prescribed in mesoscale models similarly to GCMs and derived from 1999–2001 Thermal Emission Spectrometer measurements (Smith et al., 2001) thought to be representative of Martian atmospheric conditions outside of planet-encircling dust storm events (Montabone et al., 2006). Mesoscale studies with active dust cycles are now becoming more common (Rafkin, 2009, 2012; Spiga, 2011).

The main differences in physical parameterizations between GCMs and mesoscale models are of two distinct kinds:

(1) Some physical parameterizations inherited from GCMs for unresolved dynamical processes are unsuitable for mesoscale modeling simply because these dynamical processes are already explicitly resolved by mesoscale modeling. For instance, gravity waves are not resolved in GCMs but these could induce a drag on the large-scale circulation when they break. This impact on the large-scale circulation through momentum transport is parameterized in GCMs by taking into account topographical wave sources enclosed within a GCM grid box (Lott and Miller, 1997; Miller et al., 1989). In mesoscale simulations, the topographic field is described with horizontal resolutions from tens of kilometers to hundreds of meters, hence the gravity wave drag scheme can be switched off. Note that the boundary between what is resolved and what is parameterized is often uncertain, especially in intermediate scales between mesoscale and microscale phenomena, known as the "gray zone" or "Terra Incognita" (Wyngaard, 2004).

(2) New physical parameterizations must be added specifically for mesoscale applications. For instance, mesoscale models resolve steeper topographical contrasts than GCMs, hence the need to account for sloping terrains in the radiation scheme (Rafkin et al., 2002) and more accurate numerics to decompose horizontal and vertical gradients in processes such as turbulent diffusion. It was shown that terrestrial parameterizations, where the solar irradiance reaching an inclined surface is deduced from the value in the horizontal case, can be modified or generalized to Mars-like dusty atmospheres and easily included in mesoscale models (Spiga and Forget, 2008).

Mars mesoscale modeling has also highlighted that, when GCMs and mesoscale models share a given similar physical parameterization, the latter are sometimes more suitable than the former to detect limitations to this parameterization. This is, for example, the case for convective adjustment, which modifies any unstable layer with negative potential temperature gradients (a usual near-surface situation during Martian afternoons) into a neutral equivalent (Hourdin et al., 1993). It is necessary for the sake of numerical stability and physical consistency to include such parameterization in GCMs and mesoscale models because non-physical circulations can develop in an attempt to convectively overturn the unstable layer (see Chapter 7 for further details). Notwithstanding this, as pointed out by Rafkin (2003a), the use of such an artificial convective adjustment scheme might be questionable in Martian atmospheric models, especially when combined with an active turbulent eddy diffusion scheme. Convective adjustment leads to a significant underestimation of both near-surface temperatures and winds in the afternoon: mesoscale modeling carried out in Chryse Planitia by Spiga and Forget (2009) showed that, when convective adjustment is removed, the modeled diurnal variations of near-surface temperature and winds are more consistent with observed variations by Viking and Pathfinder. Yet switching off convective adjustment yields new problems: in the afternoon, near-surface temperatures can be overestimated and spurious vertical motions might appear in the boundary layer. Turbulent mixing parameterizations, such as that of Mellor and Yamada (1974), parameterize the mixing by eddies. An alternative to turbulent mixing parameterizations are thermal models (Siebesma and Cuijpers, 1995) that some groups are now incorporating into GCMs and mesoscale models (Colaitis et al., 2013).

8.3.3 Initial and Boundary Conditions

Unlike GCMs, the limited-area models require both an initial and time-dependent boundary condition for the duration of the simulation. Depending on the latitude of interest for the mesoscale study, different map projections can be used. Polar stereographic projections are most useful in polar regions (Tyler and Barnes, 2005); this projection ensures that mesoscale simulations are devoid of any pole singularity, a usual drawback of grid-point GCMs that requires the use of additional filtering. (Spectral GCMs do not require polar filtering, but may require spectral filters.) Many models allow for a stereographic projection with a pole point different than the actual geographic pole. This capability allows for grids to be centered and focused on areas of interest that may be far from the geographic pole (e.g. locations near the equator).

Tyler et al. (2002) have argued for the use of superhemispheric grids (grids centered on one pole that wrap over the equator and into the other hemisphere) so that the mesoscale model (after initialization) becomes almost entirely responsible for the circulations in the fully covered hemisphere and into the tropics where large-amplitude waves (e.g. the tide and Kelvin waves) propagate longitudinally. The superhemispheric grid eliminates spurious wave reflections at boundaries from these dominant waves because there is no boundary. The drawback to this approach is that a large mother domain is required. It also does not eliminate issues with meridional circulations that must still enter or exit through a grid boundary, but this is no different than a non-superhemispheric grid. However, the boundaries of the mother domain will have much smaller grid spacing than at the pole point, and this will generally introduce a large mismatch in scales between the mesoscale model and the GCM data at those boundaries. The mismatch can lead to spurious waves because, for example, short-wavelength waves produce substantial perturbations compared to the comparatively smooth boundary conditions. However, the boundaries are generally far enough away from the area of interest so that impacts may be minimal. Ensuring a sufficient distance from the boundaries to the area of interest for locations that are tropical requires a very large mother domain that exacerbates the mismatch. The small grid spacing at the boundaries can also drive the need for small dynamical time steps in the model. Finally, for tropical regions of interest, the superhemispheric grid can be problematic, because it must extend far into the hemisphere opposite to the polar projection point. In contrast, a non-superhemispheric grid centered at an area of interest will not avoid boundary condition issues, but will also not have the large mismatch in scales at the edges. Tyler et al. (2002) argue that, since mesoscale circulations on Mars can cover a great extent, attempting to isolate a region for mesoscale studies in any other way will lead to problematic conflicting circulations between the mesoscale model and the GCM-derived boundary conditions. They conclude that a superhemispheric grid is the best approach. This conclusion should be at least partially model-dependent, since the amount of spurious reflections will depend on how boundary conditions are implemented within the mesoscale models. The modeler must weigh the advantages and disadvantages of each approach, but there is general agreement that superhemispheric grids are most advantageous for polar modeling studies.

Another approach to resolve mesoscale phenomena is to run GCMs at higher resolution than usual or use adaptable grid zooming capabilities (Moudden and McConnell, 2005; Richardson et al., 2007; Spiga and Lewis, 2010). This approach is presently not prominent in GCM studies, owing to computational and numerical limitations associated with grids finer than several tens of kilometers. This is likely to change in the near future as computational power increases.

Horizontal boundary conditions must provide winds, temperature, pressure (or two of the three thermodynamic state variables), and advected tracers. One exception is idealized simulations that usually require the use of periodic, symmetric or open boundary conditions rather than specified values (e.g. Rafkin et al., 2001). In "real-case" simulations, the specified boundary conditions and the atmospheric starting state are derived from previously performed GCM simulations that have reached climate equilibrium (typically ~1 year for dry simulations and up to ~10 years for an active water cycle). A relaxation zone of a few grid points in width is implemented at the boundaries of the mesoscale domain to enable both the influence of the large-scale fields on the limited area and the development of the specific mesoscale circulation inside the domain. In Martian conditions, GCM-derived boundary conditions are typically updated every one or two Martian hours. This update frequency is sufficient to resolve the important and dominant large-scale waves such as the diurnal and semidiurnal thermal tide (Tyler et al., 2002). The use of a typical terrestrial update frequency (~6 hours) would alias and poorly resolve these waves.

Another key element in boundary conditions yielding optimal downscaling in the limited-area domain is the consistency in physical parameterizations between the bounding GCM and the associated mesoscale model (Spiga and Forget, 2009). Dimitrijevic and Laprise (2005) showed that using a mesoscale domain constrained on its boundary by GCM results yields unbiased results when the boundary forcing involves a minimum of ~8–10 GCM grid points (possibly lower in situations of complex topography, as shown in Antic et al. (2006)). Hence, the single-domain approach might only be suitable for mesoscale simulations of horizontal resolution of $\Delta x \sim 10$ km (or perhaps coarser). This result also has potential implications for the superhemispheric grids where grid spacing at the edge can approach ~10 km. To reach finer horizontal resolutions of a few kilometers in mesoscale simulations, nested domains must be employed, as first introduced in numerical studies of terrestrial fronts (Harrison and Elsberry, 1972). Thus far, nested simulations are used in most existing mesoscale models for the Martian atmosphere (Rafkin et al., 2001; Tyler et al., 2002; Spiga and Forget, 2009).

Vertical boundaries (bottom and top of the model) are treated in mesoscale modeling rather similarly to GCMs. Two notable differences can be mentioned for the top boundary. Firstly, absorbing ("sponge") layers at the model top have to be stronger in mesoscale models, because a much larger part of the gravity wave spectrum is resolved and these waves excite large temperature and wind disturbances at higher altitudes (Spiga et al., 2012). Secondly, mesoscale simulations are sometimes performed with lower tops than used in GCMs (50–60 km versus >100 km). This could adversely affect large-scale circulations such as the upper branch of Hadley circulation (Toigo et al., 2002; Spiga and Forget, 2009). With the focus of mesoscale studies usually below ~40 km, this is not necessarily a major problem, although these issues need to be further explored in Martian mesoscale modeling, e.g. to capture the deep consequences of the winter baroclinic zone and the detailed polar warming features identified in MCS measurements (Vasavada et al., 2012). One possible solution is to specify boundary conditions derived from GCM results also at the model top, as is done in MRAMS (e.g. Rafkin et al., 2001) and perhaps other models. As far as the surface boundary is concerned, topography is usually taken into account in a manner consistent with the dynamical core and vertical coordinate system. In σ_z coordinate models, $w = 0$. For models with pressure-based surfaces, the surface moves in z-space with height and a surface pressure tendency must be calculated on a fixed geopotential surface. Surface thermophysical properties (albedo, thermal inertia) intended for mesoscale computations are extracted from high-resolution maps derived from recent spacecraft measurements, mostly on-board the Mars Global Surveyor.

In the process of initialization and definition of boundary conditions, the vertical interpolation of GCM meteorological fields to the terrain-following mesoscale levels must be treated with caution. While deriving the near-surface meteorological fields from GCM inputs, one may address the problem of underlying topographical structures at fine mesoscale horizontal resolution, e.g. a deep crater that is not resolved in the coarse GCM case. A crude extrapolation of the near-surface GCM fields to the mesoscale levels is usually acceptable for terrestrial applications. On Mars, owing to the low density and heat capacity of the Martian atmosphere, the ground temperature is to first order controlled by radiative equilibrium, and thus it is left relatively unaffected by variations of topography (Nayvelt et al., 1997). Further, since the air is heated and cooled through radiation, the air near the surface experiences approximately the same radiative environment regardless of the topographic

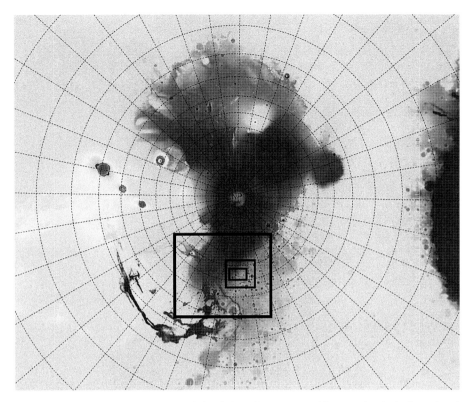

Figure 8.1. A typical grid configuration for a mesoscale model simulation of Mars starts with a superhemispheric mother domain, often centered at one of the poles, in order to minimize the effect of domain boundaries on the thermal tide. Additional grids are spawned within the mother domain to progressively focus on an area of interest, in this case Mawrth Valles. A black and white version of this figure will appear in some formats. For the color version, please refer to the plate section.

elevation. A practical consequence, which renders an extrapolation strategy particularly wrong on Mars, is that the near-surface temperature and wind fields vary much more with the distance from the surface than with the absolute altitude above the aeroid (or equivalently with the pressure level).

8.3.4 Grid Nesting

A feature typically reserved for mesoscale models is the ability to nest computational grids. Within the largest numerical domain of the mesoscale model (often referred to as the mother domain), additional, higher-resolution grids may be embedded (often referred to as nested, child, or spawned grids). Additional nested grids may be embedded within their parent grid (Figure 8.1). Nesting provides a means to better resolve processes in a specified region. Because the memory demand is proportional to the number of grid points, simulations with expansive nested grids or numerous nested grids can cause the simulation to become unwieldy. Also, the integration time step is inversely proportional to spacing. Multiple nested grids with small horizontal grid spacing require increasingly smaller time steps, which can dramatically increase the time it takes to complete a simulation. Parallel processing can help with these issues, but it is not a panacea. Even though models run faster in parallel, memory usage and the rapidly growing size of the output data can still be problematic. Furthermore, as the spacing between model grid points decreases, new, lesser exercised, and often more complicated parameterizations are needed to represent the unresolved physical processes.

Nested grids may be either one-way or two-way (interactive). In the case of one-way nesting, information is supplied to the nested grid only through lateral boundaries, and no information is sent from the nested grid back to the parent. This configuration is acceptable if the impact of higher-resolution circulations within the nested grid is not expected to provide forcing on the larger-scale circulation. In interactive nesting, the solution on the nested grid is averaged to the parent grid and is substituted in place of the parent grid solution over the nested grid domain. Thus, in interactive nested grids, the resolved circulations have a direct impact on the larger-scale parent grid circulations. However, since the boundary conditions of the mother domain are provided by GCM simulations, there is no way for the mesoscale model solution to directly impact the boundary condition information. For example, if the mesoscale model predicts a dust storm, the boundary conditions will not feel the impact of that storm unless it was also predicted in the GCM. In one-way nesting, the circulations on the nested grid can feel the large scale, but the large scale gains no knowledge from the nested grid.

There must be an integer number of nested grid cells within each parent grid cell. A typical nesting ratio between a nested grid and its parent is from 3 to 5. Higher ratio values result in an abrupt change in resolution at the grid boundary and can produce non-physical solutions, spurious wave reflections, and numerical noise due to a large dynamical mismatch at the grid interface.

Values smaller than 3 do not provide a sufficient increase in model resolution to justify the additional cost of adding the nest.

When successive nests are used, care must also be taken to ensure that there is sufficient distance between the child and parent grid boundaries, and that each grid is large enough to allow the expected circulations to develop in less than the advective timescale through the grid. Adequate space between grids is needed to provide sufficient space for the predicted flow coming out of one grid to adjust to the new grid spacing in the nest. The need for a large enough grid is simply to allow enough time for any potentially resolvable circulations to develop. For example, if air enters one side of a grid and there exists a spatially resolvable instability with a timescale of τ, then a grid of length L with atmospheric flow of U could not capture the instability if $U/L < \tau$. It is also best to avoid having a boundary intersect sharp topographical features, as this can also result in spurious results at the boundary, although achieving this can be challenging given Mars' topographical nature.

8.4 KEY MARTIAN MESOSCALE SYSTEMS

Mesoscale systems induce meteorological variability on Mars mostly through thermally driven circulations dominated by surface–atmosphere interactions and waves propagating in the atmosphere. The former kinds of phenomena have received the greatest attention, likely because thermal circulations are the obvious and dominant circulations, much more so than on Earth, that appear in mesoscale simulations, and because they can lift dust or produce clouds, which make the normally invisible circulations visible. The strong thermal circulations are also important for interpreting observed aeolian geomorphology. The direct mechanical interaction of the flow with topographic obstacles produces a range of waves, including lee waves and vertically propagating gravity waves. There is also a range of phenomena that may initially be thermally driven, but then propagate into an environment where additional phenomena, such as bore waves, are triggered.

8.4.1 Near-Surface Mesoscale Phenomena

Regional, diurnal, and seasonal variations of surface temperature are particularly large on Mars (Kieffer et al., 1976; Sutton et al., 1978). In most cases, the low atmospheric density and heat capacity lead to small contributions from sensible heat flux (energy exchange between the atmosphere and surface due to molecular conduction and turbulence) in the Martian surface energy budget (Sutton et al., 1978; Haberle et al., 1993). Consequently, the Martian surface remains close to radiative equilibrium (Nayvelt et al., 1997; Savijärvi and Kauhanen, 2008). From a mesoscale point of view, this surface radiative equilibrium yields intense near-surface circulations, driven either by gradients in topographic elevation or by horizontal variations in soil thermophysical properties.

8.4.1.1 Thermal Circulations Related to Topography

Except for the Viking Landers, Mars Pathfinder, and Mars Phoenix Lander, which are located in relatively flat terrain,

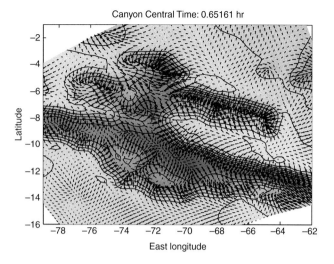

Canyon Central Time: 0.65161 hr

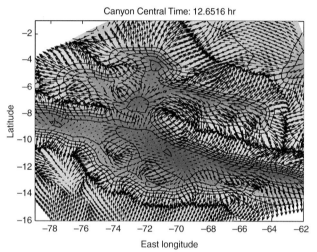

Canyon Central Time: 12.6516 hr

Figure 8.2. The classical mesoscale flow associated with topography is dominated by the thermally driven circulation. In this case, strong katabatic drainage flows develop at night (top) and intense anabatic flows develop during the day (bottom) along the slopes of Valles Marineris. An 8 sol average was used to construct the wind fields. Maximum wind speeds are ~20 m s^{-1}. Shading and contour lines show topography at 1 km intervals. Figure 13 from Tyler et al. (2002).

most of what is known about slope flows has been obtained by mesoscale modeling, by orbital thermal retrievals, and by inference from dust and cloud images. Although direct empirical data have yet to confirm this, strong anabatic and katabatic winds are likely to be ubiquitous in the Martian environment. The ground temperature and near-surface atmospheric temperature are nearly independent of surface elevation on Mars; isotherms tend to follow topography. An immediate consequence of this is the presence of topographically induced thermal circulations driven by local pressure gradients along slopes (Gierasch and Sagan, 1971; Mahrt, 1982; Ye et al., 1990; Parish, 2003). During the day, topographic maxima (hills, crater rims, mountains) are warmer than the surrounding free atmosphere. The thermal contrast is amplified compared to Earth due to the deep boundary layer and very steep near-surface lapse rates; strong upslope circulations develop in response. At night, the opposite will occur (Figure 8.2). Katabatic flows develop, with

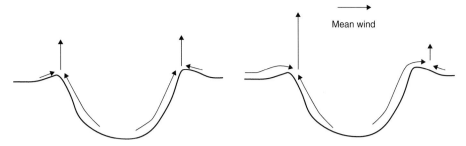

Figure 8.3. The total wind field is roughly the sum of the mean, large-scale wind plus the mesoscale circulations, with the latter mostly a thermally driven circulation associated with topography, as illustrated by this schematic over a crater. Without a mean wind, the circulation is roughly symmetric, with convergence and upwelling over the crater rims (left). When a mean wind is added (right), convergence boundaries and upward velocity can be weakened, such as on the downwind side of the crater, but strengthened on the upwind side. The convergence boundary may also propagate and be displaced downwind.

larger accelerations typically correlated with steeper slopes and strong near-surface temperature inversions (e.g. Spiga, 2011). For a given slope, the wind speeds may be two or three times as large as for Earth (e.g. Blumsack et al., 1973). Upslope, anabatic winds are not necessarily strongest over the steepest slopes. Heating depends on the geometry of the slope and the insolation. Thus, the angle of the slope, as well as the azimuth with respect to the Sun, is important in determining the upslope circulation strength. Radiative cooling is not highly geometry-dependent, so steeper slopes should tend to have stronger katabatic winds, all other things being equal. This usually results in lower velocity for afternoon anabatic winds compared to nighttime katabatic winds (Savijärvi and Siili, 1993), which is also probably related to daytime anabatic flow being deeper than nighttime katabatic flow. The closest terrestrial analog to Mars katabatic winds is in Antarctica under clear sky conditions (Parish and Bromwich, 2007). While the polar regions on Mars are conducive to the formation of strong katabatic winds (Kauhanen et al., 2008; Spiga et al., 2011b), it is not necessarily the location on the planet where those events are the most intense, contrary to Earth.

It is worth noting that the scale and magnitude of the thermal circulations on Mars can be much larger and stronger than those seen on Earth. For example, the daily upslope and downslope circulations associated with the Tharsis Montes likely dominate the meteorology of the region. While such slope-valley circulations are also commonplace on Earth, they do not typically result in regular directional reversals of 20–50 m s⁻¹ in wind speed, as predicted by models. Also, these circulations on Mars can influence regions a thousand or more kilometers away from the source topography. As a result, what is typically considered almost a pure mesoscale circulation on Earth can take on larger-scale characteristics on Mars simply because of the Rossby radius-of-deformation arguments. Further, because the slope flows have the same periodicity as the global thermal tide, the two can be confused. The tide is a global wave response due to direct, large-scale heating of the atmosphere. Large-scale waves triggered by this heating propagate globally and interact dynamically with topography and through wave–wave interactions to produce the net tidal response. In contrast, the slope flows result from differential heating along topographic slopes. The analogy to Earth is that of sea breezes, which result from local changes in heating due to surface properties, and the

atmospheric tide, which, like Mars, results from large-scale heating by the Sun.

All mesoscale models (e.g. Ye et al., 1990; Savijärvi and Siili, 1993; Tyler et al., 2002; Rafkin and Michaels, 2003; Rafkin et al., 2004; Spiga and Forget, 2009) show strong, repeatable slope circulations superimposed on large-scale background winds. This has also been confirmed through simulations with high-resolution GCMs (Spiga and Lewis, 2010). In many cases, the mesoscale circulation dominates over the large-scale one, so that the daily cycle of winds depends mostly on the local topography while the large-scale circulations provide only perturbations. This is particularly true in the tropics outside the reach of transient baroclinic storms, and less so in the high latitudes of the winter hemispheres where large-amplitude traveling waves are present.

The juxtaposition of the large-scale winds and the mesoscale circulations can produce asymmetries in the total circulation in the vicinity of topographic obstacles. In the absence of a mean wind, a uniformly heated hill or crater rim would produce symmetric upslope flow converging at the ridge with rising motion directly above. Large-scale mean winds can cause a propagation and displacement of the convergence boundary downwind of the ridge. In the case of crater rims normal to the large-scale flow, the sum of the mesoscale slope flow and the large-scale flow can favor stronger slope flow and rising motion over one of the rims (Figure 8.3). Anywhere the mesoscale flow opposes the mean motion, there is likely to be stronger convergence and rising motion. Where the mean motion and mesoscale flow are in phase, the winds will be stronger than they would be otherwise.

The only notable exception to the importance of slope flows is during periods of high global dust opacity mixed deeply through the atmosphere. The greenhouse effect of dust under high-opacity conditions tends to make the temperature profile more isothermal (Haberle, 1993). Under these conditions, the thermal contrast between topographic peaks and the surrounding atmosphere is reduced. Pressure gradients are concomitantly reduced and a reduction in winds follows. The reduction in slope winds should act as negative feedback, since lower winds will tend to lift less dust. On the other hand, the large-scale circulation may become more energetic (Haberle et al., 1982), which could counteract the effect. If the high opacity is localized, such as in the early stages of a dust storm, a large temperature gradient between the dust region and the surrounding, relatively clear region can lead to a thermal circulation that accelerates winds and lifts more dust.

To date, slope wind modeling has been focused on the most prominent topographical obstacles on Mars: Valles Marineris and the Tharsis Montes and plateau. Modeling studies of Valles Marineris canyon, where some of the steepest slopes on Mars can be found, predict intense afternoon upslope circulations near the canyon walls. At night, winds reverse and air from the surrounding plains pours into the canyon. The overall structure of the slope winds system around Valles Marineris is basically the same in the independent studies by Tyler et al. (2002), Rafkin and Michaels (2003), Toigo et al. (2003), Richardson et al. (2007), and Spiga and Forget (2009). The amplitudes of the thermally driven Martian slope winds in the Valles Marineris region are in the range 25–35 m s^{-1} in the afternoon and 30–40 m s^{-1} in the night, while their vertical component ranges between about 5 and 10 m s^{-1}. Local maxima of vertical velocity correlate with topographic gradients. Vertical cross-sections of the Valles Marineris canyon circulation along a given latitude also indicate that the near-surface anabatic winds are associated with a compensating downwelling of lesser amplitude in the center of the canyon, a few kilometers above the surface (Rafkin and Michaels, 2003). This tends to squash the convective boundary layer in the center of the canyon, produce a capping inversion, and prevent the radiatively heated air near the surface from rising until it reaches the canyon walls. Trapped close to the surface, the air suffers greater radiative heating than it would under more typical conditions. When the air makes it to the canyon walls, the extra thermal buoyancy is realized as very strong and deep updrafts that rise 10 or more kilometers above the surrounding plateau. The uneven topography of Valles Marineris not only drives powerful slope winds, but also acts as a mechanical obstacle for channeling of the atmospheric flow (Rafkin and Michaels, 2003; Toigo and Richardson, 2003; Spiga and Forget, 2009). The predominant east–west alignment of the canyon may interact with the thermal tide to further accelerate along-canyon winds during the afternoon (Rafkin and Michaels, 2003). Toigo and Richardson (2003) found a similar acceleration, but there was several hours difference in peak winds between the two studies. This may have been due to differences in how the models simulated the phase and amplitude of the global tide. Certainly, the mesoscale circulations can add a significant signal on top of the tidal pressure signal; the details of the mesoscale circulation and the ability of the models to capture the global tide circulation have a direct bearing on the total simulated pressure cycle. As previously mentioned, the location and size of the mother domain may also have an impact on the nature of the tide in the mesoscale models. The true underlying cause of the difference between the phasing of the pressure cycles in the two models remains unresolved, but does highlight the difficulty in validating models over regions where no *in situ* data are available.

Olympus Mons and the Tharsis volcanoes are also preferential sites to study slope winds. Rafkin et al. (2002) showed that modeled anabatic winds over Arsia Mons could reach 40 m s^{-1} and form spiral dust clouds similar to those imaged by the Mars Global Surveyor spacecraft. What is particularly interesting is that the model could reproduce the observed structure with generic climate model output rather than exact initial conditions. This suggests that the combination of correct seasonal insolation and realistic topography is sufficient to quickly overwhelm the inexact initial conditions, and that local effects dominate over the large-scale circulation. Anabatic winds have also been found to play a significant role in forming the summer afternoon clouds over Tharsis volcanoes (Michaels et al., 2006; Spiga and Forget, 2009). Spiga et al. (2011b) studied through mesoscale modeling the katabatic winds on the slopes of Olympus Mons. Figure 8.4 shows temperature and wind profiles in the near-surface atmospheric layer above two locations at the same longitude, one over Olympus flanks and one over surrounding flat plains. Over the flanks of Olympus Mons, the vertical profiles of temperature and wind indicate, to first order, a Prandtl-like slope wind regime, arising from quasi-equilibrium between katabatic acceleration and near-surface friction (e.g. Mahrt, 1982). The katabatic wind layer extends up to ~1 km above the surface over the Olympus slope, with horizontal component reaching 38 m s^{-1} and vertical component reaching ~15 m s^{-1}. As shown by Spiga et al. (2011a), such katabatic winds over Olympus Mons exert a strong thermal influence on the Martian atmosphere and surface, which can overwhelm radiative contributions. Not only does their vertical component result in adiabatic compression and heating of the atmosphere, but their horizontal component enhances the downward sensible heat flux. The latter effect allows the warmer atmosphere obtained through the former effect to heat the surface significantly (the reverse phenomenon, with cooler atmosphere cooling the surface, is also true for anabatic winds). This explains why, according to nighttime measurements, the Martian surface is up to 20 K warmer on slopes than on surrounding plains in the Olympus Mons/Lycus Sulci region, with an apparent correlation between thermal signatures and slope steepness. A corollary is that surface radiative equilibrium does not hold everywhere on Mars, especially over slopes or, as previously discussed, in the bottom of the Valles Marineris canyon. Neglecting the contribution of Martian atmospheric winds in the surface energy budget could have adversely affected thermal inertia retrievals (e.g. Putzig and Mellon, 2007) to the point that artificial (wind-induced) structures correlated with slopes would appear. This underscores how slope winds are a key component of the Martian system, with important geological implications, too.

Although mesoscale studies have focused on mostly the largest topographic features, thermal circulations are forced by topography at all scales. This is evident in the results from studies of MER, Phoenix, and Beagle landing safety studies that contain a variety of smaller topographic features in the modeling domain (e.g. Rafkin and Michaels, 2003; Rafkin et al., 2004; Michaels and Rafkin, 2008). The potential collective importance of these circulations to the global circulation was first pointed out by Rafkin (2003b) and further developed in Rafkin (2012). Although Rafkin (2012) argues that the rising plumes of air above the topographic peaks can penetrate many kilometers above the surrounding convective boundary layer, this new result has yet to be confirmed by observations. The plumes, if they do exist, provide a mechanism by

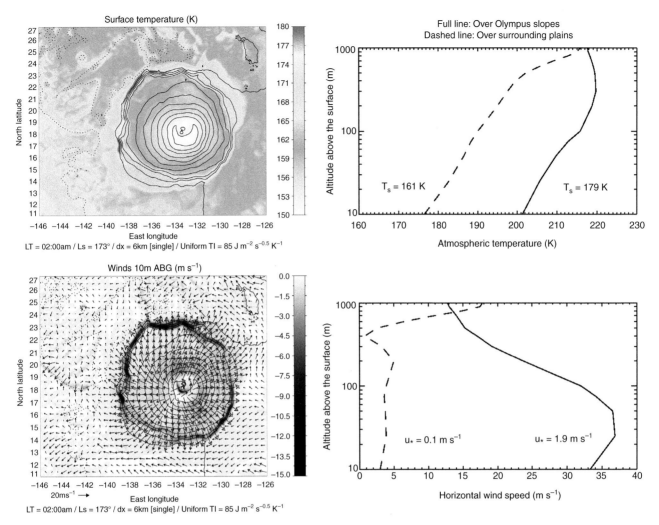

Figure 8.4. (Top left) Surface temperature (K) and (bottom left) winds 10 m above local surface (m s^{-1}), predicted in the Olympus Mons/ Lycus Sulci area with 6 km horizontal grid spacing and assuming uniform soil thermal inertia. Topography is contoured (2 km interval). Vectors indicate wind direction and speed. Vertical velocity (m s^{-1}) is shaded. Vertical profiles of (top right) near-surface temperature (K) and (bottom right) horizontal wind speed (m s^{-1}). The profiles are extracted over the northwestern flank of Olympus Mons and over the plain northward of Olympus Mons. Adapted from figures 4 and 7 of Spiga et al. (2011a). A black and white version of this figure will appear in some formats. For the color version, please refer to the plate section.

which boundary layer material can be transported rapidly to great altitudes. The deepest transport would be associated with the Tharsis Montes, but even generic craters and hills several kilometers in scale might be capable of producing transport to heights in excess of 10 km.

Recent retrievals of vertical dust distribution by the Mars Climate Sounder aboard the Mars Reconnaissance Orbiter reveal extensive and persistent elevated layers of enhanced dust mixing ratio in the tropics and subtropics. These dust layers are difficult to reconcile with transport within the rising branch of the global mean circulation. Rafkin (2012) hypothesized that the deep transport associated with the penetrating plumes associated with topography and local and regional dust storms

might be the primary mechanism by which dust is placed above the convective boundary layer. Condensation of water and cloud scavenging of aerosols is likely to act upon the transported dust, producing even greater structure (especially diurnally and seasonally), but microphysics cannot transport dust to the observed altitudes. Microphysics can only act on dust that is already present. The net effect of upward motion within local deep penetrating circulations and their associated compensating subsidence may also substantially contribute to the overall global mean circulation. Michaels et al. (2006) found that as much as one-third of the upward global mass flux in the mean meridional circulation may take place over the calderas of the four major Tharsis volcanoes.

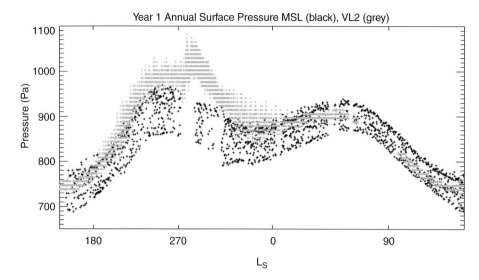

Figure 8.5. A comparison of the annual pressure cycle at MSL and VL2. All VL2 data are plotted (roughly 85 per sol). For MSL, every 10 000th data point is plotted. The jump in pressure shortly after $L_s = 270°$ in the VL2 is caused by a planet-encircling dust storm. Both MSL and VL2 are at nearly the same elevation, so that the data may be compared without an altitude correction. The diurnal variation of pressure is characterized by the spread around a given L_s. Figure is courtesy of M. Mischna.

Near-surface wind measurements to date were acquired on relatively flat terrains, where it is difficult, though not impossible, to separate the circulation caused by large-scale slopes from other contributions, as has been demonstrated by Leovy (1985) with the Viking Landers' data and by Taylor et al. (2008) through the "telltale" experiment on-board Phoenix (Holstein-Rathlou et al., 2010). Savijärvi and Siili (1993) showed through idealized two-dimensional mesoscale simulations that stronger large-scale forcing at the low-latitude Viking 1 site reverses the diurnal turning of winds into backing, while at the higher-latitude Viking 2 site with lower large-scale contribution, the combination of Coriolis and slope accelerations leads to direct veering throughout the day. The fact that the katabatic circulation over moderate Martian slopes is constrained by large-scale circulations, notably thermal tides, is further supported by high-resolution general circulation modeling (Spiga and Lewis, 2010) and three-dimensional mesoscale simulations (Toigo et al., 2002) compared to Viking and Pathfinder observations.

Unfortunately, quantitative measurements of slope winds are not widely available, especially over terrains with uneven topography where those flows are the most prominent. Geological features (dunes or streaks) on Martian craters and mountains are thought to provide indirect evidence of repetitive nighttime katabatic inflow into craters, in good agreement with mesoscale model predictions (Kuzmin et al., 2001; Greeley et al., 2003, 2008; Fenton et al., 2005; Toyota et al., 2011).

It is clear that wind measurements in uneven topographical areas on Mars would be useful to unambiguously detect repetitive, intense, clear-cut slope winds. The Mars Science Laboratory (MSL) carries the Rover Environmental Monitoring System (REMS) with a full complement of fundamental meteorological sensors to measure wind, temperature, and pressure (Gómez-Elvira et al., 2012), and is located in a crater with topography that should produce noticeable slope flows. Unfortunately, the REMS wind sensor was severely damaged during landing and retrieval of winds beyond crude estimates of

speed and direction are not possible. Consequently, if the circulations associated with the complex topography of Gale Crater are to be revealed, it must be done indirectly through other parameters. This also means that direct validation of mesoscale model winds and direct testing of slope wind theory on Mars must be left to future missions.

Pressure is the most robust measurement returned by REMS (Harri et al., 2014). As of writing, MSL has been on the surface of Mars for over one Mars year. Coincidentally, the elevation of the MSL landing site in Gale Crater (~−4500 m) is very close to that of VL2 (~−4495 m), so that direct comparison of pressure data from these two sites can be made without requiring altitude corrections. The annual pressure cycle in Gale Crater is similar to that measured previously by the Viking Landers (Figure 8.5). The seasonal CO_2 cycle dominates the signal with the long-term trends driven by condensation and sublimation of this main atmospheric constituent at high latitudes. An obvious difference between MSL and VL2, however, is that the amplitude of the annual cycle is muted at Gale Crater. The annual VL2 pressure ranged from a mean daily value of ~750 Pa to ~950 Pa. MSL ranges from ~750 Pa to ~900 Pa. This difference is diagnostic of variations in the large-scale latitudinal pressure gradients, possibly driven by the topographic dichotomy (Hourdin et al., 1993; Richardson and Wilson, 2010; Haberle et al., 2014), and asymmetric seasonal insolation associated with orbital eccentricity. Perhaps relatedly, there is less asymmetry in the relative pressure maxima at MSL than at VL2.

At diurnal frequencies, there are additional notable differences between MSL and VL2. The amplitude of the diurnal signal is much larger in Gale Crater; it is the largest variation observed at any location on the planet to date (Haberle et al., 2014). The large amplitude has been attributed to the crater circulation acting constructively with the thermal tide. From a hydrostatic viewpoint, the crater circulation must vent mass out of the crater in the afternoon, and the crater must be a net repository of mass in the night. Presumably, the contribution to the diurnal signal from the thermal tide can be estimated by looking at the pressure signal nearby Gale Crater on a relatively flat topographic region. There are no such pressure observations, but numerical models can provide guidance. Tyler and Barnes (2013) simulated the circulation and pressure cycle of Gale Crater and provided numerical results consistent with a diurnal cycle strongly modified by the crater circulation. Rafkin and Pla-Garcia (2015) suggest the amplified crater pressure signal is not caused by a canonical crater circulation but by an adjustment to hydrostaticity upon which the crater circulation is superimposed.

The results of Tyler and Barnes (2013) also predicted that the afternoon convective boundary layer in the crater should be suppressed due to compensating subsidence associated with upslope circulations. This effect is identical that identified in previous simulations of craters, valleys, and small basins (Rafkin and Michaels, 2003). Although MSL has no direct way of measuring the depth of the boundary layer, imaging of the sky can provide information about the vertical dust distribution in the lower atmosphere. Moores et al. (2014) found optical extinction to be much lower on a line of site to the crater rim than along the line of site to the Sun; the air in the crater has less dust than the atmosphere above. In addition, the paucity of convective vortices and dust devils in Gale Crater are all strongly suggestive of a shallow convective boundary layer.

8.4.1.2 Circulations Related to Thermal Contrasts

Regional-scale thermal contrasts can drive mesoscale circulations. Because of the aforementioned strong radiative forcing, such circulations develop close to the surface where local contrasts of surface temperature develop, related to contrasts in soil thermophysical properties (thermal inertia, albedo). This is the Martian analog to land/sea breeze circulations on Earth, and the underlying physics can be explained through the hypsometric equation that combines hydrostatic equilibrium with the ideal gas law,

$$\frac{\Delta p}{p} = -\frac{\Delta z}{H} \tag{8.5}$$

where $H = RT/g$ is the atmospheric pressure scale height.

Equation (8.5) shows that the thickness of an air layer enclosed within two isobars is greater within a warmer layer than within a cold layer. For instance, consider an area where the albedo is high close to an area where the albedo is low. During daytime, the surface and the atmosphere above the former area are cooler than above the latter area. A few kilometers above the surface, the change in pressure scale height yields a horizontal pressure gradient that causes winds to blow from the warm region to the cold region. The surface pressure then increases in the cold region, which causes a reverse circulation near the surface. This kind of circulation is, therefore, thermally induced.

The most prominent of such circulations on Mars are circulations close to the polar caps. Toigo et al. (2002) were able to show that it is the thermal contrast between the polar cap covered with CO_2 ice with the bare soil at lower latitudes that is the main driver of the strong near-surface winds likely to cause dust lifting at the cap edge. The thermal contrasts between the

icy caps and bare soil extend over a few tens of kilometers. The resulting strong winds are likely the primary mechanism by which near-cap dust storms are triggered, and the dust lifted at the edge of polar caps plays a significant role in the whole Martian dust cycle and the annual dust budget (Cantor et al., 2006). A major advantage to studying polar processes with mesoscale models is that, in contrast to most GCMs to date, the employed map projections do not result in a computational singularity at the pole. Winds on the edge of the caps were predicted by GCMs to be very low during southern spring, while observations indicate that at this season the cap storm activity is at its peak (Toigo et al., 2002). Mesoscale modeling indicates wind stresses compatible with possible dust lifting at the edge of the cap. The influence of topography (through slope winds, see Section 8.4.1.1) is significant but less than that of thermally induced winds. The influence of large-scale condensation flow is found to be weak, but acts constructively to reinforce the thermal cap circulations. A similar conclusion was reached earlier through idealized 2D experiments by Siili et al. (1999).

Another key mesoscale phenomenon defined by thermal contrasts are fronts, sometimes spectacularly associated with hemispheric dust perturbations (Wang et al., 2003). Baroclinic systems are most energetic in the middle and high latitudes of the winter hemisphere. A strong horizontal pressure gradient is associated with the frontal thermal contrast, which may have been detected through mapping surface pressure with the OMEGA spectrometer on-board Mars Express (Spiga et al., 2007). Thus far, the dynamics of frontal activity has not been directly addressed through Martian mesoscale modeling efforts. GCM studies and observations have explored baroclinic activity in general, with little focus on frontal structures and dynamics. Frontal dynamics is an area ripe for future mesoscale modeling efforts.

Tyler and Barnes (2005) identified weak transient eddies (1–1.5% amplitude in pressure) in the northern polar summer that may not be exclusively baroclinic. These eddies are important to address when the release of water vapor from the cap is at a maximum. The summer transient eddies exhibit near-surface wind perturbations of 10–15 m s^{-1}, but the circulations are confined within one scale height above the surface. Tyler and Barnes (2005) proposed that the transient eddies are excited through strong slope winds close to the surface (see previous discussion on topographic circulations). At the end of summer, the influence of baroclinic waves grows and dominates the pressure perturbation signal.

8.4.2 Mesoscale Waves

Gravity waves (hereafter GWs) are mesoscale atmospheric oscillations related to the buoyancy restoring force, which play a key role in the circulation, structure, and variability of planetary atmospheres (Fritts and Alexander, 2003). On Earth, vertically propagating GWs originate in the lower part of the atmosphere by a variety of mechanisms involving topography, fronts, convective cells, jet streams, wind shears, and wave–wave interactions (e.g. Fritts et al., 2006; Spiga et al., 2008). In the 1970s, Mariner 9 and Viking missions revealed that GWs are also ubiquitous in the Martian low-density stable atmosphere (Briggs and Leovy, 1974; Pirraglia, 1976; Pickersgill and

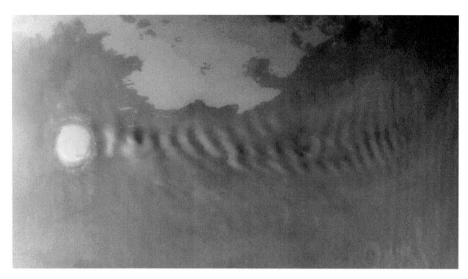

Figure 8.6. An example of a trapped gravity wave train emanating from an ice-filled crater in the northern plains during winter when the westerly winds tend to be strongest. Image credit: MSSS/JPL/NASA.

Hunt, 1979). The crucial role of GWs in transporting energy and momentum and influencing the synoptic circulation and thermal structure has been acknowledged in both planets (cf. Mars global circulation modeling by Forget et al. (1999) and Wilson (1997)). Typical horizontal scales of such phenomena range from thousands of kilometers to a few kilometers.

Since their propagation produces fluctuations in both the temperature and density fields, Martian GWs have been identified from orbit in thermal profiles derived by infrared spectrometry, by radio occultation retrievals (Hinson et al., 1999; Creasey et al., 2006), by density measurements obtained by accelerometers during the aerobraking phases of the Mars Global Surveyor (MGS) and Mars Odyssey (ODY) spacecraft (Keating et al., 1998; Creasey et al., 2006; Fritts et al., 2006), and by MGS Mars Orbiter Laser Altimeter (MOLA) anomalies caused by CO_2 ice wave clouds in the polar night (Pettengill and Ford, 2000; Tobie et al., 2003). All observations indicate that GWs are prominent in the Martian mesosphere at altitudes >60 km, causing temperature and density oscillations over 10%, with vertical wavelengths $\lambda_z \le 10$ km and horizontal wavelengths $\lambda_x \le 200$ km (Fritts et al., 2006; Magalhães et al., 1999). Convective instabilities in the upper part of the Martian atmosphere were detected in Mars Climate Sounder profiles and attributed to GW breakings (Heavens, 2010). The dayglow emission measured on-board Mars Express is suspected to have captured oscillations caused by gravity waves (Spiga et al., 2007; Altieri et al., 2012), as is often observed on Earth (Melo et al., 2006). Gravity waves likely play a role in the formation of some high-altitude CO_2 clouds (Spiga et al., 2012).

Atmospheric "background" conditions, namely wind and stability, cause GWs to be either horizontally trapped or vertically propagating. In the former case, trapped GWs will propagate over great distances in the horizontal dimension and induce temperature perturbations much further from the area of emission. In the latter case, GW fluctuations of temperature, density, and wind increase exponentially with altitude, which makes GWs likely to become unstable at high altitudes, to break (through either convective/static instability, or critical

levels, where wave phase speed becomes comparable with background flow), and exert significant drag on large-scale circulations (Lindzen, 1981; Barnes, 1990), or to be dampened through radiative processes (Eckermann et al., 2011). In both cases, a detectable consequence of GW-induced temperature perturbations is the formation of wave clouds. Trapped gravity wave "trains" have been imaged in water ice cloud structures downstream of topographic obstacles such as craters and volcanoes (Pickersgill and Hunt, 1981; Figure 8.6), especially in mid- to high latitudes at seasons in which westerlies are particularly strong (Wang and Ingersoll, 2002). In polar regions, the cold atmosphere is also conducive to the formation of low-altitude CO_2 ice wave clouds (Tobie et al., 2003), which were initially detected by MOLA as spurious echoes not corresponding to any reasonable topographical structure (Pettengill and Ford, 2000). On the other hand, the upward-propagating GWs that reach the mesosphere without encountering breaking or dissipation induce large temperature perturbations. The mesospheric cold pockets in which high-altitude CO_2 clouds might form were speculated to result from large-scale and mesoscale wave activity (Clancy and Sandor, 1998). Recently, following the observations of mesospheric CO_2 clouds (Chapter 5), idealized mesoscale modeling showed that gravity waves likely play a role in the formation of those clouds (Spiga et al., 2012).

Other sources of gravity waves besides topographical forcing do exist on Mars – which is actually emphasized by results of Creasey et al. (2006). Mars is characterized by powerful convection in the troposphere, related to both boundary layer turbulence and local dust storms. Near-surface super-adiabatic gradients cause intense daytime boundary layer convection, with turbulent updrafts reaching 10 km or more above local surface. The stably stratified free atmosphere above the convective boundary layer is perturbed by those updrafts, which gives rise to internal GW (Rafkin and Michaels, 2003; Spiga and Forget, 2009). Intense vertical circulations could develop inside regional dust storms (sometimes several tens of kilometers wide), which could also trigger gravity waves (Rafkin, 2009). Characteristics and sources of GWs remain, however, to be further assessed on Mars, through both original observational methods and mesoscale modeling (where part of the GW spectra is resolved and not parameterized). By providing new examples of GW phenomena, from an extraterrestrial environment where their existence appears to be both prolific and important, Martian studies could be of wide interest in GW meteorology.

Bore waves are solitary waves closely related to gravity waves (e.g. Crook and Miller, 1985; Rottman and Simpson, 1989; Rottman and Grimshaw, 2003). The top of the strong nocturnal inversions found on Mars effectively divides very stable, dense air from much less dense air above. An even more dense current of air propagating into the inversion can produce a disturbance at the nocturnal inversion interface. The propagation speed and nature of the perturbation depend on gravity wave propagation. Briggs et al. (1977), Hunt et al. (1981) and Pickersgill and Hunt (1981) noted water ice clouds in Viking Orbiter images that seemed to trace out the rising motion and oscillations consistent with a bore wave. Kahn and Gierasch (1982) attributed the cloud features to the closely related hydraulic jump phenomenon. The mesoscale study of Sta. Maria et al. (2006) found the structure and propagation of a modeled disturbance on the slopes of Olympus Mons to be consistent with a bore wave. The modeled waves show seasonality with the strongest disturbances during the summer and weakest during the winter. The summer disturbance also produced strong vertically propagating gravity waves. Bore waves may be ubiquitous on Mars, but will only be visible when sufficient water vapor is present to produce saturation. Further, the best dynamical environment for the generation and propagation of bore waves is found during the dark, early morning hours when visible imagery is useless. Heating from the Sun will quickly erode the nocturnal inversion and destroy bore waves; successful imaging requires fortuitous timing.

Haberle et al. (2014) identified an "evening oscillation" in high-frequency pressure data at Gale Crater. These oscillations had a period of 5–10 min with an amplitude of ~0.5 Pa. The oscillations are consistent with shallow water calculations of gravity wave phase speeds under reasonable assumptions about the nocturnal structure of the atmosphere in the crater.

8.5 CLOUDS AND DUST

8.5.1 Water Ice Clouds

Slope circulations and perturbations of the atmosphere as air flows over topography can produce clouds. Slope flows (Michaels et al., 2006; Spiga and Forget, 2009), mountain (gravity) wave activity, and the thermal tides (Hinson and Wilson, 2004) are all important in generating daytime and nighttime clouds. The downward-propagating thermal wave associated with the tide provides an environment conducive for cloud formation (cooling), while upslope circulations provide a source of water and mountain wave perturbations that further cool the air. Modeling of nighttime water ice cloud cover in areas of complex topography (Wilson et al., 2007) is consistent with observations. Both the slope circulations and the tide (diurnal and semidiurnal) fall in the same frequency domain,

but the two phenomena are quite different. By definition, thermal tides initiate as a global resonance. Nonlinear wave interactions and interactions with topography result in a wide range of wave harmonics. Slope circulations are not global, but regional. The circulations are the result not of a global resonance but of the direct thermal forcing from the topography itself. The overlapping frequency and the large magnitudes of both the thermal tide and slope circulations can make it difficult to disentangle the relative contributions of the two very different forcing mechanisms. During the day, anabatic flows can produce clouds in moist, adiabatically cooled upslope plumes (Rafkin et al., 2002; Michaels et al., 2006). In the case of the Tharsis clouds, the upslope circulations are responsible for providing the source of moisture for both daytime and nighttime clouds. (Downslope nighttime circulations do the opposite: dry air aloft is entrained and carried down the slopes.) Hinson and Wilson (2004) found tidal waves to provide sufficient cooling to produce clouds in the moist air above the Tharsis Montes. Thus, the clouds likely represent a coupling between diurnal mesoscale upslope/downslope water advection and global thermal tides. At higher latitudes, lower-altitude mesoscale water ice clouds are often seen in association with gravity waves generated by an obstacle, within craters and other low-lying areas, and (often intermixed with dust) with baroclinic storms and fronts (e.g. Cantor et al., 2002; Wang and Ingersoll, 2002; Wang and Fisher, 2009). The clouds within high-latitude craters may help produce/maintain localized environmental conditions quite different from what might be expected from a larger-scale climatology.

8.5.2 Carbon Dioxide Ice Clouds

The possibility of CO_2 ice clouds was first outlined by Gierasch and Goody (1968), who determined that CO_2 clouds would only form in the polar night lower atmosphere where air temperatures drop below the relevant frost point (<150 K). Recent observations (e.g. Montmessin et al., 2007; Vincendon et al., 2011) now confirm these clouds can also exist at great altitude (~80 km) in the daytime near the equator. The available evidence strongly suggests that the circulations that produce these clouds and the clouds themselves are within the mesoscale realm.

Compared to water clouds on Earth, CO_2 ice clouds on Mars are novel in that they result from the condensation of the primary constituent of the atmosphere, producing unusual effects such as the local enrichment of non-condensable gases and the direct excitation of sound waves. The primary microphysical difference between CO_2 clouds and water clouds is the mechanism that ultimately limits their nucleation and growth rates. For a minor species, the finite rate of vapor diffusion limits growth, while, for a major species, the finite rate of latent heat removal limits growth. The microphysical theory for the nucleation and growth of these cloud particles (Wood, 1999; Colaprete and Toon, 2003; Määttänen et al., 2005) predicts that homogeneous nucleation is extremely improbable in the contemporary Mars atmosphere. However, heterogeneous nucleation on dust, dust coated in water ice, or water ice particles is possible, but requires a critical supersaturation of >32% for larger ice nuclei. Indirect evidence consistent with CO_2 clouds includes Mars Global Surveyor (MGS) Radio

Science (RS) vertical temperature profiles of the polar night (e.g. Colaprete et al., 2003) that provide compelling evidence for CO_2 supersaturation sufficient for nucleation in the lower atmosphere. Serendipitously, the Mars Orbiter Laser Altimeter (MOLA) aboard MGS, acting as a simple lidar in the polar night, has provided additional evidence suggestive of CO_2 clouds (Pettengill and Ford, 2000; Ivanov and Muhleman, 2001; Colaprete et al., 2003; Neumann et al., 2003). The densest clouds inferred in this way were observed in the south distributed in a wave-2 or wave-3 pattern, further suggesting a possible connection to atmospheric Rossby waves (Ivanov and Muhleman, 2001; Colaprete et al., 2003). MOLA returns from less-dense clouds were observed more frequently in the north polar night. Hayne (2010) used Mars Reconnaissance Orbiter Mars Climate Sounder (MCS) observations to obtain distributions of these suspected CO_2 clouds with altitude, location, and season. The Hayne (2010) MCS data analysis demonstrates that CO_2 clouds appear to be most dense in the lowest 10 km of the polar night atmosphere, although they can extend to altitudes greater than 30 km, and the altitude of the densest clouds decreases equatorward – both are consistent with the prior MOLA observations. Some clouds are correlated with topography, particularly 10–100 km diameter craters (Hayne, 2010), and may form within vertically propagating gravity waves.

The data analysis and atmospheric modeling work of Colaprete et al. (2003, 2008) have demonstrated that significant amounts of CO_2 supersaturation are present in the polar night atmosphere, and that this supersaturation has a magnitude and distribution that permits CO_2 convective clouds driven by latent heat, which are not so dissimilar from large terrestrial cumulus clouds. If so, the CO_2 cloud dynamics are solidly within the mesoscale. Polar night CO_2 clouds may have important roles in the Mars climate system in that they may comprise an important component of the radiative balance of the atmosphere and surface, and will distribute heat and mass in the vertical.

High-altitude (e.g. 80 km) CO_2 clouds form in the near-equatorial region, and are relatively discrete clouds of sufficient opacity to cast shadows on the surface (Montmessin et al., 2007; Vincendon et al., 2011). These clouds generally have a cirrus-like morphology, arranged as east–west streamers without obvious convective traits (Vincendon et al., 2011). Their spatial dimensions and longitudinally clustered distribution suggest that mesoscale processes may be involved in their formation, but the specific processes are unclear. Gravity waves or ageostrophic circulations associated with jets might play a role, as they do for cirrus clouds on Earth.

8.5.3 Dust Storms

Dust storms are perhaps the most widely recognized phenomena on Mars. Large, planet-obscuring dust storms have been observed to occur every few Mars years near perihelion. Regional storms generally occur a few times each year. Smaller, local storms are even more common (Cantor et al., 2001). All of these storms have one thing in common: they each begin at the mesoscale, and the active lifting centers likely remain at the mesoscale throughout the duration of the event. While Mars

may become completely shrouded in dust every few years, dust is not being lifted globally over the entire planet. The 2001 Mars global dust storm was observed in great detail by orbital instrumentation (Smith et al., 2002). The origin of this storm was a series or complex of local dust disturbances in the Hellas Basin area. These disturbances moved slowly east while other transient storms in the southern hemisphere came and went. Within a matter of two weeks, the dust lofted by these storms filled the atmosphere with optically thick dust, which obscured details of what was happening below. What is clear is that the storm, while global in impact, was driven and fed by mesoscale lifting events.

Many large dust storms appear to be initiated when dust lifting along fronts interacts with the thermal tide. When the circulations are phased properly, dust lifted along the front in the northern hemisphere encounters tidally induced winds that rapidly advect dust into the southern hemisphere (Wang et al., 2003). These so-called flushing dust storms are, therefore, a result of the opening of a "tidal gate", which normally remains closed and regionally confines the lifted dust. The timescale of the tidal gate opening may be as short as 10 hours. Thus, although the two interacting systems – a baroclinic disturbance and the tide – may often be considered large-scale features, the result of their interaction may be considered a mesoscale phenomenon.

With the exception of flushing dust storms and the frequent storms along the polar cap, the origin and cause of global storms and smaller-scale regional and local disturbances remains unknown. Some storms may be triggered by slope or thermal circulations, but since these circulations are thought to be relatively repeatable from sol to sol, there must be another factor that triggers the storms at a particular time on a particular sol. Some dust storms have no obvious triggers at all (i.e. they are not near slopes or features that might provide some causality).

An analogy between dust storms and tropical storms on Earth was forwarded by Gierasch and Goody (1973). Tropical thunderstorms are ubiquitous on Earth, but only rarely do they self-organize into mesoscale tropical cyclones (i.e. hurricanes or typhoons). Likewise, small dust storms are ubiquitous on Mars, and only rarely do they organize into more robust and larger disturbances. The analogy goes further. Latent heating within clusters of thunderstorms produces a hydrostatic low pressure, which can accelerate the wind, increasing the flux of moisture into the air, and draw additional moist air into the system. Under the right conditions, this can produce a positive feedback process (wind-induced sensible heat exchange (WISHE)) and the storm can organize and strengthen (Emanuel, 1991). Radiative heating of atmospheric dust on Mars has a diabatic heating effect similar to latent heating in thunderstorms. A hydrostatic low pressure in an initial dust disturbance could accelerate the wind, increase the surface dust flux, and draw dusty air into the system. The simplified analytical solution to this scenario by Gierasch and Goody (1973) demonstrated that such a positive feedback mechanism was at least possible for Mars.

The work of Gierasch and Goody (1973) was largely ignored, if not dismissed, when repeated orbital imagery failed to find obvious signs of any dusty hurricanes. At the same time, few hypotheses have been forwarded to explain the origin and growth of most dust storms. Recent work (Rafkin, 2009, 2012) indicates that the dismissal of a positive radiative dynamic dust feedback may have been premature. The question is not whether dusty hurricanes exist on Mars, but whether a feedback process is operating. A large majority of tropical thunderstorm clusters on Earth fail to ever develop into mature tropical cyclones or develop obvious rotational structures, yet a great number of these same clusters do benefit from WISHE. It is other factors, such as wind shear or cold water, that prevent WISHE from fully developing the systems. Nonetheless, the process is still operating. Tropical depressions and tropical storms may not be hurricanes, but they are organized dynamical mesoscale entities gaining energy from a positive feedback process even if it is not obvious from satellite images.

Even if dusty hurricanes were a reality on Mars, they may not be so easily detected from imagery. Unlike water, dust does not undergo phase changes. Hurricanes are visible mostly because water provides a tracer of upward and downward motion within the dynamical structure. Rising areas are cloudy, sinking areas are clear, and the boundary between the two is sharp. Thus, hurricane eyes and spiral arm bands are obvious in satellite imagery. Geostationary satellites observe the same spot over time and can further provide a time lapse of images that show rotation of the visible features. There is no analogous process for dust that would reveal structure. Dust lofted into the atmosphere would likely conceal whatever structure might be present below the opaque layer. Further, since Mars imagery is not obtained by geostationary satellites, making clear identification of any rotating structures is difficult.

Idealized mesoscale model studies (Rafkin, 2009) demonstrate that, in the absence of other forcing mechanisms, a positive radiative dynamic feedback process should operate on Mars. Like Earth, the process is optimized under conditions where the Coriolis force is large enough to induce rotation and where energy input is large enough to rapidly grow the system. For Mars, there must be sufficient insolation and a readily available reservoir of surface dust that can be lifted to realize diabatic heating from sunlight. This is analogous to the tropical warm waters that feed hurricanes on Earth. Although potential heating is greatest near the equator, the Coriolis force is too weak to produce a balanced rotation (cf. Rossby radius of deformation); latitudes higher than ~15° are required. The higher the latitude, the more favorable the dynamics, but the less favorable the energy input. Optimal growth for idealized dust disturbances was found to occur between approximately 15° and 30° under solstitial solar insolation conditions. As might be expected, more realistic simulations that contain the potential disruptive effects of topography, wind shear, and other large-scale atmospheric circulations have failed to show the development of dusty hurricanes like those envisioned by Gierasch and Goody (1973) and as simulated by Rafkin (2009). What the simulations do show, however, are strong indications that a wind-enhanced interaction of radiation and dust (WEIRD) is occurring in a manner analogous to WISHE on Earth (Rafkin, 2012). Although a dusty hurricane does not develop, the positive feedback process is operating to strengthen the disturbance, invigorate the circulation, and lift dust. WEIRD was found to enhance pressure deficits, increase local winds, increase column opacity, and increase vorticity compared to simulations where the radiative feedback was disallowed. Global dust storms may result on the very rare occasions when all the right amplification factors are in place and when disruptive forcings are small or can be overwhelmed. WEIRD is a hypothesis that can be most easily tested by surface meteorological stations in the vicinity of active dust disturbances. Careful analysis of thermal imaging may also be able to reveal dynamical structure hidden by dust (e.g. Määttänen et al., 2009). A great deal of additional observational, theoretical, and modeling work is still needed to confirm WEIRD or to establish another viable, competing hypothesis.

8.6 AEOLIAN PROCESSES

Wind-driven movement of surface particles is the dominant surface erosion process on contemporary Mars. This process results in the generation and evolution of aeolian geologic features, such as vast areas of circumpolar dunes (e.g. Tsoar et al., 1979), intra-crater dunes (e.g. Fenton et al., 2003, 2005), particulate ripples (e.g. Golombek et al., 2010), yardangs (e.g. Ward, 1979), ventifacts (Bridges et al., 1999), and surface wind streaks (e.g. Thomas and Veverka, 1979; Thomas et al., 1981, 1984; Veverka et al., 1981). Due in large part to the complex topography over much of Mars, many of the relevant flows that control these aeolian surface interactions are influenced by mesoscale or smaller circulations.

Previous comparisons of wind streak orientation with GCM wind fields (e.g. Greeley et al., 1993) show some satisfactory agreement, but also many problematic observations, suggesting that the atmospheric structure and flows that contribute to the formation of wind streaks may be primarily mesoscale in nature or that these features are relics of a past epoch. The importance of mesoscale (or smaller) circulations compared to large-scale circulations in aeolian processes should not be surprising given the strong forcing that topography and surface properties exert on Mars (Figure 8.7). The relatively coarse resolution of GCMs cannot capture the variability of winds that may actually be doing work on the surface (Kuzmin et al., 2001; Greeley et al., 2003). This has immediate consequences for the parameterization of dust lifting in models (and, by extension, evaluation of potential aeolian activity). If GCM winds are below a given lifting threshold, no dust will be lifted. However, mesoscale and smaller circulations can accelerate or overwhelm large-scale winds to produce local or regional areas of lifting. Likewise, if GCM winds exceed a lifting threshold, the implication is that dust is lifted everywhere, whereas local or regional winds may be below the threshold.

Some dust lifting parameterizations do not take into account the subgrid-scale variability of winds within a model grid (e.g. Kahre et al., 2006). Instead, lifting is switched on when the resolved winds exceed a specified threshold. In addition, many models add a second lifting scheme associated with dust devils. Thus, wind lifting due to winds below that which is resolved and which are not dust devils are completely neglected. One method to take into account these hitherto unaccounted for circulations is to assume that the actual winds in a model grid are distributed around

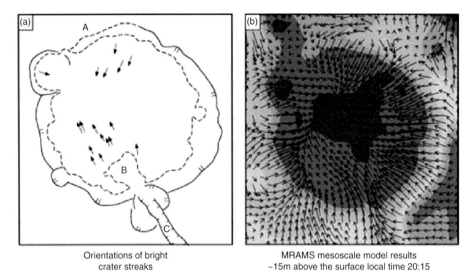

| Orientations of bright crater streaks | MRAMS mesoscale model results ~15m above the surface local time 20:15 |

Figure 8.7. (a) Sketch of Gusev Crater rim (line with tick marks) and floor (dashed line) showing orientations of bright crater streaks (arrows); "A" indicates topographically lowest part of crater rim, and "B" outlines the inferred deltaic deposits from Ma'adim Vallis ("C"). (b) MRAMS simulation (Rafkin and Michaels, 2003) corresponding to hour 20:15, showing wind vectors for the area of Gusev shown in panel (a). This simulation of the nighttime winds shows a correlation with the bright streaks; this is consistent with the model for bright wind streak formation of Veverka et al. (1981) which suggests that the streaks result from the deposition of dust during periods of positive atmospheric static stability, which typically occur during the night. Figure 13 from Greeley et al. (2003).

the model-resolved mean wind. Based on terrestrial observations (e.g. Tuller and Brett, 1984; He et al., 2010), limited Mars observations, and large-eddy simulations (Fenton and Michaels, 2010), a Weibull distribution (Lorenz, 1996; Newman et al., 2002; Fenton and Michaels, 2010) is a reasonable assumption for the shape of the distribution. The width of the distribution is then a measure of the subgrid-scale turbulence, which could include unresolved boundary layer convection (in which dust devils are embedded), as well as any subgrid-scale perturbations of the flow due to unresolved topography and surface properties. In a mesoscale model grid with sufficient resolution (typically on the order of kilometers), the bulk of the variations due to topography and surface properties should be accounted for. Thus, within this mesoscale model grid, variations of wind should be due primarily to turbulence alone. Fenton and Michaels (2010) have shown that the Weibull distributions approximating such boundary layer turbulent wind fluctuations should be a function of both the friction velocity and a shape parameter that is dependent on what process is forcing the turbulence (e.g. free convection versus driven by wind shear). A parameterization based on this notion has been used to simulate dust devil (albedo) tracks that have strong similarities to those observed on Mars (Michaels, 2006).

The temporal aspects of the mesoscale flows that produce aeolian features are also of crucial importance. When the near-surface atmosphere is statically stable (e.g. at night), air is more likely to be deflected horizontally around a modest obstacle. Therefore, aeolian features formed at specific times over the sol may record the dynamical changes in wind direction. Significant mesoscale near-surface wind enhancements may occur only at certain times of year and/or for short periods of time (e.g. minutes or hours; Chojnacki et al., 2011). Integrated over geological timescales, these short periods of strong winds

could do substantial work. Besides the regular and nearly predictable diurnal wind variations, stochastic mesoscale or global-scale processes such as baroclinic systems and dust storms are known to lift dust. Dust streaks are known to be partially or completely obscured following planet-encircling dust storms, but reappear, sometimes in a short period of time. Large swaths of the surface have also been observed to be rapidly cleaned of dust (e.g. Cantor, 2007).

Studies of intra-crater dune fields using mesoscale atmospheric modeling (e.g. Fenton et al., 2003; Hayward et al., 2009; Chojnacki et al., 2011) have resulted in mixed degrees of correlation between wind directions (inferred from aeolian features and modeled) and the estimated aeolian effectiveness (directly proportional to the aerodynamic surface stress) of the model-predicted winds. However, mesoscale atmospheric modeling results for Nili Patera (Michaels, 2011) show unambiguous correlations between observed sand movement (Silvestro et al., 2010), wind direction, aeolian effectiveness, and season.

8.7 MESOSCALE MODEL APPLICATIONS

Aside from the science applications, mesoscale models have become an invaluable tool in support of entry, descent, and landing (EDL) performance assessment for surface-bound spacecraft. Although spacecraft landing systems vary, all are sensitive to some combination of winds, density, or temperature in the lowest few scale heights of the atmosphere. In the case of the Mars Exploration Rovers (MERs), low-level horizontal winds imparted additional kinetic energy that needed to be dissipated by the airbag system (Kass et al., 2003; Rafkin and Michaels, 2003; Toigo and Richardson, 2003). Strong vertical circulations could have caused the rocket descent system either to descend too slowly and run out of fuel, in the case of strong updrafts, or to fail to adequately decelerate, resulting in impact, in the case of strong downdrafts. Above the boundary layer, wind and density had the potential to affect the landing precision by altering the nominal flight trajectory, especially during the parachute descent stage. In particular, strong vertical shear of the horizontal wind had the potential to cant the spacecraft–parachute system (Kass et al., 2003). The Mars Phoenix Lander utilized powered descent in the final stages of landing and was less sensitive than the MERs to horizontal wind, but was still susceptible to impacts from vertical atmospheric motion (Michaels and Rafkin, 2008; Tamppari et al., 2008; Tyler et al., 2008). For the Mars Science Laboratory (MSL), low-level winds were less of a problem, but the interaction of winds and density perturbations with the closed loop flight control system were of concern. Mesoscale models were used to help characterize the environment (Vasavada et al., 2012; Cianciolo et al., 2013). In addition, for the first time, explicit regional and local dust storm events were modeled to assess the impact on EDL performance (e.g. Rafkin, 2012).

Like the science investigations, atmospheric characterization with the mesoscale models is not a true prediction of the atmospheric state for a given time and location. In any planetary atmosphere, chaos will degrade the accuracy of any prediction over time, even if the initial conditions are exact. In the case of Mars, the initial conditions are supplied only by climate models rather than observations, and the simulations are conducted years in advance of EDL. The climate models themselves do not use observed initial conditions, but are spun up from some prescribed initial state until the modeled climate stabilizes. Recent data assimilation efforts (Montabone et al., 2005; Lewis et al., 2007; Hoffman et al., 2010; Rogberg et al., 2010) are permitting GCMs to be more strongly informed by past observations. These techniques may allow for true weather prediction on Mars at some time in the future. Even so, many decisions about EDL must be made too far in advance for such methods to be utilized. There is no possibility for a true prediction, even with data assimilation, months or years in advance of atmospheric entry.

Using an initial condition that matches as closely as possible the observations might be expected to produce the most probable climatological environment. But the probability distribution function of actual circulations may be quite broad. TES and now MCS show that the Mars atmosphere is not exactly repeatable from year to year; it is impossible to specify a singularly correct initial condition from which to begin a forecast. Additionally, TES and MCS temperature retrievals are not in complete agreement for times and places where the observations overlap. The atmospheric circulation is strongly driven by the temperature distribution, and the latitudinal and vertical thermal gradients in particular. Small changes in the thermal distribution, such as might be achieved with reasonable dust profile perturbations, can produce dramatically different mean and wave circulation patterns. A modestly different circulation pattern resulting from nearly the same thermal distribution may be almost as likely as the most probable circulation. Even if the models capture the average mean and wave patterns, it does not mean that the models will predict the phase and magnitude of a particular wave or storm system at a specific place or time.

Whatever uncertainties are associated strictly with the mesoscale models must be added to the inherent uncertainties in the MGCM results. The uncertainty in the MGCM results from errors in physics and numerics as well as errors resulting from the deviation of the model-produced climate from the actual atmospheric state at a given time.

The philosophy of using the models to characterize the landing environment is to provide an envelope of scenarios at least 3σ from the mean or most likely case. To achieve this, mesoscale model data are analyzed statistically to generate a mean and variance of atmospheric structure along the spacecraft flight trajectory. Typically, many consecutive sols of model output are processed covering several hours centered near the local time of atmospheric entry. To the extent that the models capture the sol-to-sol internal variability of the atmospheric circulation, a statistical representation of possible conditions is constructed. The assumption that the models can represent the variability for a fixed dust condition is uncertain.

Like all models, mesoscale models cannot explicitly represent circulations at scales smaller than roughly four times the model grid spacing. Yet, these smaller scales (of order kilometers and less in the horizontal and hundreds of meters to kilometers in the vertical) do impact spacecraft performance. Therefore, it is necessary to add this variability to the model-resolved structures. This can be done by estimating subgrid-scale motions from turbulent parameterizations within the models themselves (e.g. the use of model-predicted subgrid turbulent kinetic energy). It can also be done by using engineering and empirical models to impose a predetermined turbulent spectrum of circulations, or by scaling terrestrial observations (Justus et al., 2002; Kass et al., 2003). Finally, the subgrid-scale fields can be informed from large-eddy simulations. In practice, some combination of all these processes is used.

REFERENCES

Altieri, F., A. Spiga, L. V. Zasova, G. Bellucci, and J.-P. Bibring (2012), Gravity waves mapped by the OMEGA/MEX instrument through O_2 dayglow at 1.27 µm: data analysis and atmospheric modeling, *J. Geophys. Res.*, doi:10.1029/2012JE004065, in press.

Anthes, R. A. (1971), A numerical model of the slowly varying tropical cyclone in isentropic coordinates, *Monthly Weather Review*, 99(8), 617–635.

Antic, S., R. Laprise, B. Denis, and R. de Elía (2006), Testing the downscaling ability of a one-way nested regional climate model in regions of complex topography, *Climate Dynamics*, 26(2), 305–325.

Arakawa, A. (1966), Computational design for long-term numerical integration of the equations of fluid motion: two-dimensional incompressible flow, *J. Computational Phys.*, 1(1).

Barnes, J. R. (1990), Possible Effects of Breaking Gravity Waves on the Circulation of the Middle Atmosphere of Mars, *J. Geophys. Res.*, 95(B2), 1401–1421.

Bennetts, D. A., and B. J. Hoskins (1979), Conditional symmetric instability – a possible explanation for frontal rainbands, *Quarterly Journal of the Royal Meteorological Society*, 105(446), 945–962.

Benson, J. L., P. B. James, B. A. Cantor, and R. Remigio (2006), Interannual variability of water ice clouds over major Martian volcanoes observed by MOC, *Icarus*, 184(2), 365–371.

Blumsack, S. L., P. J. Gierasch, and W. R. Wessel (1973), An Analytical and Numerical Study of the Martian Planetary Boundary Layer Over Slopes, *J. Atmos. Sci.*, 30(1), 66–82.

Bridges, N. T., R. Greeley, A. F. C. Haldemann, et al. (1999), Ventifacts at the Pathfinder landing site, *J. Geophys. Res.*, 104(E4), 8595–8615.

Briggs, G. A., and C. B. Leovy (1974), Mariner Observations of the Mars North Polar Hood, *Bulletin of the American Meteorological Society*, 55(4), 278–296.

Briggs, G., K. Klaasen, T. Thorpe, J. Wellman, and W. Baum (1977), Martian dynamical phenomena during June–November 1976: Viking Orbiter imaging results, *J. Geophys. Res.*, 82(28), 4121–4149.

Cantor, B. (2007), MOC observations of the 2001 Mars planet-encircling dust storm, *Icarus*, 186(1), 60–96.

Cantor, B. A., P. B. James, M. Caplinger, and M. J. Wolff (2001), Martian dust storms: 1999 Mars Orbiter Camera observations, *J. Geophys. Res.*, 106(E10), 23653–23687.

Cantor, B., M. Malin, and K. S. Edgett (2002), Multiyear Mars Orbiter Camera (MOC) observations of repeated Martian weather phenomena during the northern summer season, *J. Geophys. Res.*, 107(E3), doi:10.1029/2001JE001588.

Cantor, B. A., K. M. Kanak, and K. S. Edgett (2006), Mars Orbiter Camera observations of Martian dust devils and their tracks (September 1997 to January 2006) and evaluation of theoretical vortex models, *J. Geophys. Res.*, 111(E12), E12002.

Charney, J. G. (1947), The dynamics of long waves in a baroclinic westerly current, *Journal of Meteorology*, 4(5), 136–162.

Chojnacki, M., D. M. Burr, J. E. Moersch, and T. I. Michaels (2011), Orbital observations of contemporary dune activity in Endeavor Crater, Meridiani Planum, Mars, *J. Geophys. Res.*, 116, E00F19.

Cianciolo, A., B. Cantor, J. Barnes, et al. (2013), Atmosphere Assessment for MARS Science Laboratory Entry, Descent and Landing Operations, Document ID 20140001381, http://ntrs.nas.gov.

Clancy, R. T., and B. J. Sandor (1998), CO_2 ice clouds in the upper atmosphere of Mars, *Geophys. Res. Lett.*, 25(4), 489–492.

Clever, R. M., and F. H. Busse (1992), Three-dimensional convection in a horizontal fluid layer subjected to a constant shear, *Journal of Fluid Mechanics*, 234, 511–527.

Colaitis, A., A. Spiga, F. Hourdin, et al. (2013), A thermal plume model for the Martian convective boundary layer, *J. Geophys. Res.*, 118, 1468–1487, doi:10.1002/jgre.20104.

Colaprete, A., and O. B. Toon (2003), Carbon dioxide clouds in an early dense Martian atmosphere, *J. Geophys. Res.*, 108(E4), 5025.

Colaprete, A., R. M. Haberle, and O. B. Toon (2003), Formation of convective carbon dioxide clouds near the south pole of Mars, *J. Geophys. Res.*, 108(E7), 5081.

Colaprete, A., J. R. Barnes, R. M. Haberle, and F. Montmessin (2008), CO_2 clouds, CAPE and convection on Mars: observations and general circulation modeling, *Planet Space Sci.*, 56(2), 150–180.

Conrath, B. J., J. C. Pearl, M. D. Smith, et al. (2000), Mars Global Surveyor Thermal Emission Spectrometer (TES) observations: atmospheric temperatures during aerobraking and science phasing, *J. Geophys. Res.*, 105(E4), 9509–9519.

Creasey, J. E., J. M. Forbes, and D. P. Hinson (2006), Global and seasonal distribution of gravity wave activity in Mars' lower atmosphere derived from MGS radio occultation data, *Geophys. Res. Lett.*, 33, L01803.

Crook, N. A., and M. J. Miller (1985), A numerical and analytical study of atmospheric undular bores, *Quarterly Journal of the Royal Meteorological Society*, 111(467), 225–242.

Dickinson, R. E., C. P. Lagos, R. E. Newell (1968), Dynamics of the neutral gas in the thermosphere for small Rossby number motions, *J. Geophys. Res.*, 73, 4299–4313, doi:10.1029/JA073i013p04299.

Dimitrijevic, M., and R. Laprise (2005), Validation of the nesting technique in a regional climate model and sensitivity tests to the resolution of the lateral boundary conditions during summer, *Climate Dynamics*, 25(6), 555–580.

Dudhia, J. (1993), A nonhydrostatic version of the Penn State–NCAR mesoscale model: validation tests and simulation of an Atlantic cyclone and cold front, *Monthly Weather Review*, 121(5), 1493–1513.

Eckermann, S. D., J. Ma, and X. Zhu (2011), Scale-dependent infrared radiative damping rates on Mars and their role in the deposition of gravity-wave momentum flux, *Icarus*, 211(1), 429–442.

Emanuel, K. A. (1986), An air–sea interaction theory for tropical cycles. Part 1: Steady-state maintenance, *J. Atmos. Sci.*, 43, 585–604.

Emanuel, K. A. (1991), The theory of hurricanes, *Annual Review of Fluid Mechanics*, 23(1), 179–196.

Emanuel, K. A., M. Fantini, and A. J. Thorpe (1987), Baroclinic instability in an environment of small stability to slantwise moist convection. Part I: Two-dimensional models, *J. Atmos. Sci.*, 44(12), 1559–1573.

Farrell, B. F. (1989), Optimal Excitation of Baroclinic Waves, *J. Atmos. Sci.*, 46(9), 1193–1206.

Fenton, L. K., and T. I. Michaels (2010), Characterizing the sensitivity of daytime turbulent activity on Mars with the MRAMS LES: early results, *Mars*, 5 (Mars Dust Cycle Special Issue), 159–171.

Fenton, L. K., J. L. Bandfield, and A. W. Ward (2003), Aeolian processes in Proctor Crater on Mars: sedimentary history as analyzed from multiple data sets, *J. Geophys. Res.*, 108(E12), 5129.

Fenton, L. K., A. D. Toigo, and M. I. Richardson (2005), Aeolian processes in Proctor Crater on Mars: mesoscale modeling of dune-forming winds, *J. Geophys. Res.*, 110(E6), E06005.

Fisher, J. A., M. I. Richardson, C. E. Newman, et al. (2005), A survey of Martian dust devil activity using Mars Global Surveyor Mars Orbiter Camera images, *J. Geophys. Res.*, 110(E3), E03004.

Forget, F., F. Hourdin, R. Fournier, et al. (1999), Improved general circulation models of the Martian atmosphere from the surface to above 80 km, *J. Geophys. Res*, 104(24), 155–176.

Fritts, D. C., and M. J. Alexander (2003), Gravity wave dynamics and effects in the middle atmosphere, *Rev. Geophys.*, 41(1), 1003.

Fritts, D. C., L. Wang, and R. H. Tolson (2006), Mean and gravity wave structures and variability in the Mars upper atmosphere inferred from Mars Global Surveyor and Mars Odyssey aerobraking densities, *J. Geophys. Res.*, 111(A12), A12304.

Gierasch, P., and R. Goody (1968), A study of the thermal and dynamical structure of the Martian lower atmosphere, *Planet Space Sci.*, 16(5), 615–646.

Gierasch, P. J., and R. M. Goody (1973), A model of a Martian great dust storm, *J. Atmos. Sci.*, 30(2), 169–179.

Gierasch, P., and C. Sagan (1971), A preliminary assessment of Martian wind regimes, *Icarus*, 14(3), 312–318.

Golombek, M., K. Robinson, A. McEwen, et al. (2010), Constraints on ripple migration at Meridiani Planum from Opportunity and HiRISE observations of fresh craters, *J. Geophys. Res.*, 115, E00F08

Gómez-Elvira, J., C. Armiens, L. Castañer, et al. (2012), REMS: the environmental sensor suite for the Mars Science Laboratory Rover, *Space Science Reviews* 170(1–4): 583–640.

Greeley, R., A. Skypeck, and J. B. Pollack (1993), Martian aeolian features and deposits: comparisons with general circulation model results, *J. Geophys. Res.*, 98(E2), 3183–3196.

Greeley, R., R. O. Kuzmin, S. C. R. Rafkin, T. I. Michaels, and R. Haberle (2003), Wind-related features in Gusev Crater, Mars, *J. Geophys. Res.*, 108(E12), 8077.

Greeley, R., P. L. Whelley, L. D. V. Neakrase, et al. (2008), Columbia Hills, Mars: aeolian features seen from the ground and orbit, *J. Geophys. Res.*, 113(E6), E06S06.

Haberle, R. M. (1993), Mars atmospheric dynamics as simulated by the NASA/Ames general circulation model, *J. Geophys. Res.*, 98, 3093–3124.

Haberle, R. M., C. B. Leovy, and J. B. Pollack (1982), Some effects of global dust storms on the atmospheric circulation of Mars, *Icarus*, 50, 322–367.

Haberle, R. M., H. C. Houben, R. Hertenstein, and T. Herdtle (1993), A boundary layer model for Mars: comparison with Viking entry and lander data, *J. Atmos. Sci.*, 50, 1544–1559.

Haberle, R. M., J. Gómez-Elvira, M. de la Torre Juárez, et al. (2014), Preliminary interpretation of the REMS pressure data from the first 100 sols of the MSL mission, *J. of Geophys. Res.*, 119(3), 440–453.

Haltiner, G., and R. T. Williams (1980), *Numerical Prediction and Dynamic Meteorology*, 2nd ed., Wiley.

Harri, A. M., M. Genzer, O. Kemppinen, et al. (2014), Pressure observations by the Curiosity Rover: initial results, *J. of Geophys. Res.*, 119, 82–92.

Harrison, E. J., and R. L. Elsberry (1972), A Method for Incorporating Nested Finite Grids in the Solution of Systems of Geophysical Equations, *J. Atmos. Sci.*, 29(7), 1235–1245.

Hayne, P. O. (2010), Snow clouds on Mars and ice on the Moon: thermal infrared observations and models, Thesis, UCLA, Los Angeles.

Hayward, R. K., T. N. Titus, T. I. Michaels, et al. (2009), Aeolian dunes as ground truth for atmospheric modeling on Mars, *J. Geophys. Res.*, 114(E11), E11012.

He, Y., A. H. Monahan, C. G. Jones, et al.(2010), Probability distributions of land surface wind speeds over North America, *J. Geophys. Res.*, 115(D4), D04103.

Heavens, N. G. (2010), *The impact of mesoscale processes on the atmospheric circulation of Mars*, California Institute of Technology.

Hinson, D. P., and R. J. Wilson (2004), Temperature inversions, thermal tides, and water ice clouds in the atmosphere of Mars, *J. Geophys. Res.*, 109(E01002).

Hinson, D. P., R. A. Simpson, J. D. Twicken, G. L. Tyler, and F. M. Flasar (1999), Initial results from radio occultation measurements with Mars Global Surveyor, *J. Geophys. Res.-Planets*, 104(E11), 26997–27012.

Hoffman, M. J., S. J. Greybush, R. J. Wilson, et al. (2010), An ensemble Kalman filter data assimilation system for the Martian atmosphere: implementation and simulation experiments, *Icarus*, 209(2), 470–481.

Holstein-Rathlou, C., H. P. Gunnlaugsson, J. P. Merrison, et al. (2010), Winds at the Phoenix landing site, *J. Geophys. Res.*, 115(E5) doi:10.1029/2009JE003411.

Hoskins, B. J. (1975), The geostrophic momentum approximation and the semi-geostrophic equations, *J. Atmos. Sci.*, 32, 233–242.

Hoskins, B. J., and A. J. Simmons (1975), A multi-layer spectral model and the semi-implicit method, *Quart. J. R. Meteorol. Soc.*, 101, 637–655.

Hourdin, F., P. Le Van, F. Forget, and O. Talagrand (1993), Meteorological Variability and the Annual Surface Pressure Cycle on Mars, *J. Atmos. Sci.*, 50(21), 3625–3640.

Hunt, G. E., A. O. Pickersgill, P. B. James, and N. Evans (1981), Daily and seasonal Viking observations of Martian bore wave systems, *Nature*, 293(5834), 630–633.

Ivanov, A. B., and D. O. Muhleman (2001), Cloud reflection observations: results from the Mars Orbiter Laser Altimeter, *Icarus*, 154(1), 190–206.

Janjic, Z. I., J. P. Gerrity, and S. Nickovic (2001), An Alternative Approach to Nonhydrostatic Modeling, *Monthly Weather Review*, 129(5), 1164–1178.

Justus, C. G., B. F. James, S. W. Bougher, et al. (2002), Mars-GRAM 2000: a Mars atmospheric model for engineering applications, *Advances in Space Research*, 29(2), 193–202.

Kahn, R., and P. Gierasch (1982), Long Cloud Observations on Mars and Implications for Boundary Layer Characteristics Over Slopes, *J. Geophys. Res.*, 87(A2), 867–880.

Kahre, M. A., J. R. Murphy, and R. M. Haberle (2006), Modelling the Martian dust cycle and surface dust reservoirs with the NASA Ames general circulation model, *Journal of Geophysical Research E: Planets*, 111(6).

Kass, D. M., J. T. Schofield, T. I. Michaels, et al. (2003), Analysis of atmospheric mesoscale models for entry, descent, and landing, *J. Geophys. Res.*, 108(E12), 8090.

Kauhanen, J., T. Siili, S. Järvenoja, and H. Savijärvi (2008), The Mars limited area model and simulations of atmospheric circulations for the Phoenix landing area and season of operation, *J. Geophys. Res.*, 113(E3), E00A14.

Keating, G. M., S. W. Bougher, R. W. Zurek, et al. (1998), The structure of the upper atmosphere of Mars: in situ accelerometer measurements from Mars Global Surveyor, *Science*, 279(5357), 1672–1676.

Kieffer, H. H., S. C. Chase, E. D. Miner, et al. (1976), Infrared thermal mapping of the Martian surface and atmosphere: first results, *Science*, 193(4255), 780–786.

Kuzmin, R. O., R. Greeley, S. C. R. Rafkin, and R. Haberle (2001), Wind-related modification of some small impact craters on Mars, *Icarus*, 153(1), 61–70.

Larsen, S. E., H. E. Jørgensen, L. Landberg, and J. E. Tillman (2002), Aspects of the atmospheric surface layers on Mars and Earth, *Boundary-Layer Meteorology*, 105(3), 451–470.

Leovy, C. (1985), *The General Circulation of Mars – Models and Observations*, Academic Press, Orlando, FL.

Lewis, S. R., P. L. Read, B. J. Conrath, J. C. Pearl, and M. D. Smith (2007), Assimilation of thermal emission spectrometer atmospheric data during the Mars Global Surveyor aerobraking period, *Icarus*, 192(2), 327–347.

Lilly, D. K., and E. L. Petersen (1983), Aircraft measurements of atmospheric kinetic energy spectra, *Tellus A*, 35A(5), 379–382.

Lindzen, R. S. (1981), Turbulence and stress owing to gravity wave and tidal breakdown, *J. Geophys. Res.*, 86(C10), 9707–9714.

Lorenz, R. D. (1996), Martian surface wind speeds described by the Weibull distribution, *J. Spacecraft and Rockets*, 33, 754–756.

Lott, F., and M. Miller (1997), A new sub-grid scale orographic drag parameterization: its formulation and testing, *Q. J. R. Met. Soc.*, 123.

Määttänen, A., H. Vehkamäki, A. Lauri, et al. (2005), Nucleation studies in the Martian atmosphere, *J. Geophys. Res.*, 110(E2), E02002.

Määttänen, A., T. Fouchet, O. Forni, et al. (2009), A study of the properties of a local dust storm with Mars Express OMEGA and PFS data, *Icarus*, 201(2), 504–516.

Magalhães, J. A., J. T. Schofield, and A. Seiff (1999), Results of the Mars Pathfinder atmospheric structure investigation, *J. Geophys. Res.*, 104(E4), 8943–8955.

Mahrer, Y., and R. A. Pielke (1977), A numerical study of the airflow over irregular terrain, *Beitrage zur Physik der Atmosphere*, 50, 98–113.

Mahrt, L. (1982), Momentum Balance of Gravity Flows, *J. Atmos. Sci.*, 39(12), 2701–2711.

Martínez-Alvarado, O., L. Montabone, S. R. Lewis, I. M. Moroz, and P. L. Read (2009), Transient teleconnection event at the onset of a planet-encircling dust storm on Mars, *Ann. Geophys.*, 27(9), 3663–3676.

McWilliams, J. C., and P. R. Gent (1980), Intermediate Models of Planetary Circulations in the Atmosphere and Ocean, *J. Atmos. Sci.*, 37(8), 1657–1678.

Medvedev, A. S., and P. Hartogh (2007), Winter polar warmings and the meridional transport on Mars simulated with a general circulation model, *Icarus*, 186(1), 97–110.

Mellor, G., and T. Yamada (1974), A hierarchy of turbulence closure models for planetary boundary layers, *J. Atmos. Sci.*, 31, 1791–1806.

Melo, S. M. L., O. Chiu, A. Garcia-Munoz, et al. (2006), Using airglow measurements to observe gravity waves in the Martian atmosphere, *Advances in Space Research*, 38(4), 730–738.

MEPAG (2010). *Science Goals, Objectives, Investigations, and Priorities: 2010.* Mars Exploration Program and Assessment Group. Report available at http://mepag.nasa.gov/reports/MEPAG_Goals_Document_2010_v17.pdf.

Michaels, T. I. (2006), Numerical modeling of Mars dust devils: albedo track generation, *Geophys. Res. Lett.*, 33(19), L19S08.

Michaels, T. (2011), Modeling aeolian surface interaction phenomena at Nili and Meroe Paterae, in *Joint AAS Division of Planetary Science and European Planetary Science Conference*, edited, Nantes, France.

Michaels, T. I., and S. C. R. Rafkin (2004), Large-eddy simulation of atmospheric convection on Mars, *Quarterly Journal of the Royal Meteorological Society*, 130(599), 1251–1274.

Michaels, T. I., and S. C. R. Rafkin (2008), Meteorological predictions for candidate 2007 Phoenix Mars Lander sites using the Mars Regional Atmospheric Modeling System (MRAMS), *J. Geophys. Res. Planets*, 113(E3).

Michaels, T. I., A. Colaprete, and S. C. R. Rafkin (2006), Significant vertical water transport by mountain-induced circulations on Mars, *Geophysical Research Letters*, 33(16).

Miller, M. J., P. M. Palmer, and R. Swinbank (1989), Parametrisation and influence of sub-grid scale orography in general circulation and numerical weather prediction models, *Meteorol. Atmos. Phys.*, 40.

Montabone, L., S. R. Lewis, and P. L. Read (2005), Interannual variability of Martian dust storms in assimilation of several years of Mars global surveyor observations, *Advances in Space Research*, 36(11), 2146–2155.

Montabone, L., S. R. Lewis, P. L. Read, and D. P. Hinson (2006), Validation of Martian meteorological data assimilation for MGS/TES using radio occultation measurements, *Icarus*, 185(1), 113–132.

Montmessin, F., B. Gondet, J. P. Bibring, et al. (2007), Hyperspectral imaging of convective CO_2 ice clouds in the equatorial mesosphere of Mars, *J. Geophys. Res.*, 112(E11), E11S90.

Moores, J. E., M. T. Lemmon, H. Kahanpää, et al. (2014), Observational evidence of a suppressed planetary boundary layer in northern Gale Crater, Mars as seen by the Navcam instrument onboard the Mars Science Laboratory Rover, *Icarus*, 249, 129–142.

Moudden, Y., and J. C. McConnell (2005), A new model for multiscale modeling of the Martian atmosphere, GM3, *J. Geophys. Res.*, 110(E4), E04001.

Nayvelt, L., P. J. Gierasch, and K. H. Cook (1997), Modeling and Observations of Martian Stationary Waves, *J. Atmos. Sci.*, 54(8), 986–1013.

Neumann, G. A., D. E. Smith, and M. T. Zuber (2003), Two Mars years of clouds detected by the Mars Orbiter Laser Altimeter, *J. Geophys. Res.*, 108(E4), 5023.

Newman, C. E., S. R. Lewis, P. L. Read, and F. Forget (2002), Modeling the Martian dust cycle 2. Multiannual radiatively active dust transport simulations, *J. Geophys. Res.*, 107(E12), 5124.

Nicholls, M. E., and R. A. Pielke (1994), Thermal compression waves. I: Total-energy transfer, *Quarterly Journal of the Royal Meteorological Society*, 120(516), 305–332.

Ooyama, K. V. (1982), Conceptual evolution of the theory and modeling of the tropical cyclone, *J. Meteor. Soc. Japan*, 6, 369–380.

Panofsky, H. A., and I. van der Hoven (1955), Spectra and cross-spectra of velocity components in the mesometeorological range, *Quarterly Journal of the Royal Meteorological Society*, 81(350), 603–606.

Parish, T. R. (2003), Katabatic Winds, in *Encyclopedia of Atmospheric Sciences*, edited by J. R. Holton, J. Pyle and J. A. Curry, Academic Press.

Parish, T. R., and D. H. Bromwich (2007), Reexamination of the near-surface airflow over the Antarctic continent and implications on atmospheric circulations at high southern latitudes, *Monthly Weather Review*, 135(5), 1961–1973.

Pettengill, G. H., and P. G. Ford (2000), Winter clouds over the North Martian Polar Cap, *Geophys. Res. Lett.*, 27(5), 609–612.

Pickersgill, A. O., and G. E. Hunt (1979), The formation of Martian lee waves generated by a crater, *J. Geophys Res.*, 84(B14).

Pickersgill, A. O., and G. E. Hunt (1981), An Examination of the Formation of Linear Lee Waves Generated by Giant Martian Volcanoes, *J. Atmos. Sci.*, 38(1), 40–51.

Pielke, R. A. (2002), *Mesoscale Meteorological Modeling*, 2nd ed., Academic Press, San Diego.

Pielke, R. A., W. R. Cotton, R. L. Walko, et al. (1992), A comprehensive meteorological modeling system – RAMS, *Meteorology and Atmospheric Physics*, 49(1), 69–91.

Pirraglia, J. A. (1976), Martian atmospheric Lee waves, *Icarus*, 27(4), 517–530.

Putzig, N. E., and M. T. Mellon (2007), Apparent thermal inertia and the surface heterogeneity of Mars, *Icarus*, 191(1), 68–94.

Rafkin, S. C. R. (2003a), The Effect of Convective Adjustment on the Global Circulation of Mars as Simulated by a General Circulation Model, in *6th International Conference on Mars*, edited, Lunar and Planetary Institute, Pasadena, CA.

Rafkin, S. C. R. (Ed.) (2003b), *Reflections on Mars Global Climate Modeling from a Mesoscale Meteorologist*, Granada, Spain.

Rafkin, S. C. R. (2009), A positive radiative-dynamic feedback mechanism for the maintenance and growth of Martian dust storms, *J. Geophys. Res.*, 114(E1), E01009.

Rafkin, S. C. R. (2012), The potential importance of non-local, deep transport on the energetics, momentum, chemistry, and aerosol distributions in the atmospheres of Earth, Mars, and Titan, *Planet. Space Sci.*, 60(1), 147–154.

Rafkin, S. C. R., and T. I. Michaels (2003), Meteorological predictions for 2003 Mars Exploration Rover high-priority landing sites, *J. Geophys. Res.*, 108(E12), 8091, doi:10.1029/2002JE002027.

Rafkin, S. C. R., R. M. Haberle, and T. I. Michaels (2001), The Mars Regional Atmospheric Modeling System: model description and selected simulations, *Icarus*, 151(2), 228–256.

Rafkin, S. C. R., M. R. V. Sta. Maria, and T. I. Michaels (2002), Simulation of the atmospheric thermal circulation of a Martian volcano using a mesoscale numerical model, *Nature*, 419(6908), 697–699.

Rafkin, S. C. R., T. I. Michaels, and R. M. Haberle (2004), Meteorological predictions for the Beagle 2 mission to Mars, *Geophys. Res. Lett.*, 31(1), L01703.

Rafkin, S. C. R, J. Pla-Garcia, M. Kahre, et al. (2016), The Meteorology of Gale Crater as Determined from Rover Environmental Monitoring Station Observations and Numerical Modeling. Part II: Interpretation, *Icarus*, 280, 114–138.

Richardson, M. I., A. D. Toigo, and C. E. Newman (2007), A general purpose, local to global numerical model for planetary atmospheric and climate dynamics, *J. Geophys. Res.*, 112, E09001.

Richardson, M. I., and R. J. Wilson (2010), A topographically forced asymmetry in the Martian circulation and climate, *Nature*, 416, 298–301, doi:10.1038/416298a.

Rogberg, P., P. L. Read, S. R. Lewis, and L. Montabone (2010), Assessing atmospheric predictability on Mars using numerical weather prediction and data assimilation, *Quarterly Journal of the Royal Meteorological Society* 136, 1614–1635

Rossby, C.-G. (1938), On the mutual adjustment of pressure and velocity in certain simple current systems, II, *J. of Marine Res.*, 7.

Rottman, J. W., and R. Grimshaw (2003), *Atmospheric Internal Solitary Waves in Environmental Stratified Flows*, edited by R. Grimshaw, Springer, New York, 61–88.

Rottman, J. W., and J. E. Simpson (1989), The formation of internal bores in the atmosphere: a laboratory model, *Quarterly Journal of the Royal Meteorological Society*, 115(488), 941–963.

Savijärvi, H., and J. Kauhanen (2008), Surface and boundary-layer modelling for the Mars Exploration Rover sites, *Quarterly Journal of the Royal Meteorological Society*, 134(632), 635–641.

Savijärvi, H., and T. Siili (1993), The Martian Slope Winds and the Nocturnal PBL Jet, *J. Atmos. Sci.*, 50(1), 77–88.

Schofield, J. T., J. R. Barnes, D. Crisp, et al. (1997), The Mars Pathfinder Atmospheric Structure Investigation/Meteorology (ASI/MET) Experiment, *Science*, 278(5344), 1752–1758.

Siebesma, A. P., and J. W. M. Cuijpers (1995), Evaluation of Parametric Assumptions for Shallow Cumulus Convection, *J. Atmos. Sci.*, 52(6), 650–666.

Siili, T., R. M. Haberle, J. R. Murphy, and H. Savijärvi (1999), Modelling of the combined late-winter ice cap edge and slope winds in Mars Hellas and Argyre regions, *Planet Space Sci.*, 47(8–9), 951–970.

Siili, T., J. Kauhanen, H. Savijärvi, et al. (2006), Simulations of atmospheric circulations for the Phoenix landing area and season-of-operation with the Mars limited area model (MLAM), paper presented at *Fourth International Conference on Mars Polar Science and Exploration*, Lunar and Planetary Institute, Davos, Switzerland.

Silvestro, S., L. K. Fenton, D. A. Vaz, N. T. Bridges, and G. G. Ori (2010), Ripple migration and dune activity on Mars: evidence for dynamic wind processes, *Geophys. Res. Lett.*, 37(20), L20203.

Skamarock, W. C., and J. B. Klemp (2008), A time-split nonhydrostatic atmospheric model for weather research and forecasting applications, *Journal of Computational Physics*, 227(7), 3465–3485.

Smith, M. D., J. C. Pearl, B. J. Conrath, and P. R. Christensen (2001), Thermal Emission Spectrometer results: Mars atmospheric thermal structure and aerosol distribution, *J. Geophys. Res*, 106, 23929–23945.

Smith, M. D., B. J. Conrath, J. C. Pearl, and P. R. Christensen (2002), Thermal Emission Spectrometer Observations of Martian Planet-Encircling Dust Storm 2001A, *Icarus*, 157(1), 259–263.

Spiga, A. (2011), Elements of comparison between Martian and terrestrial mesoscale meteorological phenomena: katabatic winds and boundary layer convection, *Planet Space Sci.*, 59(10), 915–922.

Spiga, A., and F. Forget (2008), Fast and accurate estimation of solar irradiance on Martian slopes, *Geophys. Res. Lett.*, 35(15), L15201.

Spiga, A., and F. Forget (2009), A new model to simulate the Martian mesoscale and microscale atmospheric circulation: validation and first results., *J. Geophys Res.*, 114(E02009).

Spiga, A., and S. R. Lewis (2010), Martian mesoscale and microscale wind variability of relevance for dust lifting, *International Journal of Mars Science and Exploration*, 5, 146–158.

Spiga, A., F. Forget, B. Dolla, et al. (2007), Remote sensing of surface pressure on Mars with the Mars Express/OMEGA spectrometer: 2. Meteorological maps, *J. Geophys. Res.*, 112(E8), E08S16.

Spiga, A., H. Teitelbaum, and V. Zeitlin (2008), Identification of the sources of inertia–gravity waves in the Andes Cordillera region, *Ann. Geophys.*, 26.

Spiga, A., F. Forget, J.-B. Madeleine, et al. (2011a), The impact of Martian mesoscale winds on surface temperature and on the determination of thermal inertia, *Icarus*, 212(2), 504–519.

Spiga, A., F. Forget, J.-B. Madeleine, et al. (2011b), Elements of comparison between Martian and terrestrial mesoscale meteorological phenomena: katabatic winds and boundary layer convection, *Planet Space Sci.*, 59(10), 915–922.

Spiga, A., F. González-Galindo, M. Á. López-Valverde, and F. Forget (2012), Gravity waves, cold pockets and CO_2 clouds in the Martian mesosphere, *Geophys. Res. Lett.*, 39(2), L02201.

Sta. Maria, M. R. V., S. C. R. Rafkin, and T. I. Michaels (2006), Numerical simulation of atmospheric bore waves on Mars, *Icarus*, 185(2), 383–394.

Sutton, J. L., C. B. Leovy, and J. E. Tillman (1978), Diurnal variations of the Martian surface layer meteorological parameters during the first 45 sols at two Viking Lander sites, *J. Atmos. Sci.*, 35(12), 2346–2355.

Tamppari, L. K., J. Barnes, E. Bonfiglio, et al. (2008), Expected atmospheric environment for the Phoenix landing season and location, *J. Geophys. Res. – Planets*, 113, E00A20, doi:10.1029/2007JE003034.

Taylor, P. A., D. C. Catling, M. Daly, et al. (2008), Temperature, pressure, and wind instrumentation in the Phoenix meteorological package, *J. Geophys. Res.*, 113(E3), E00A10.

Thomas, P., and J. Veverka (1979), Seasonal and secular variation of wind streaks on Mars: an analysis of Mariner 9 and Viking data, *J. Geophys. Res.*, 84(B14), 8131–8146.

Thomas, P., J. Veverka, S. Lee, and A. Bloom (1981), Classification of wind streaks on Mars, *Icarus*, 45(1), 124–153.

Thomas, P., J. Veverka, D. Gineris, and L. Wong (1984), "Dust" streaks on Mars, *Icarus*, 60(1), 161–179.

Tillman, J. E., L. Landberg, and S. E. Larsen (1994), The boundary layer of Mars: fluxes, stability, turbulent spectra, and growth of the mixed layer, *J. Atmos. Sci.*, 51(12), 1709–1727.

Tobie, G., F. Forget, and F. Lott (2003), Numerical simulation of the winter polar wave clouds observed by Mars Global Surveyor Mars Orbiter Laser Altimeter, *Icarus*, 164(33).

Toigo, A. D., and M. I. Richardson (2003), Meteorology of proposed Mars Exploration Rover landing sites, *J. Geophys. Res.*, 108(E12), 8092.

Toigo, A. D., M. I. Richardson, R. J. Wilson, H. Wang, and A. P. Ingersoll (2002), A first look at dust lifting and dust storms near the south pole of Mars with a mesoscale model, *J. Geophys. Res.*, 107(E7), 5050.

Toigo, A. D., M. I. Richardson, S. P. Ewald, and P. J. Gierasch (2003), Numerical simulation of Martian dust devils, *J. Geophys. Res.*, 108(E6), 5047.

Toon, O. B., C. McKay, T. P. Accerman, and K. Santhanam (1989), Rapid calculation of radiative heating rates and photodissociation rates in inhomogeneous multiple scattering atmospheres, *J. Geophys Res.*, 94, 16287–16301.

Toyota, T., K. Kurita, and A. Spiga (2011), Distribution and time-variation of spire streaks at Pavonis Mons on Mars, *Planet Space Sci.*, 59(8), 672–682.

Tsoar, H., R. Greeley, and A. R. Peterfreund (1979), Mars: the north polar sand sea and related wind patterns, *J. Geophys. Res.*, 84(B14), 8167–8180.

Tuller, S. E., and A. C. Brett (1984), The characteristics of wind velocity that favor the fitting of a Weibull distribution in wind speed analysis, *Journal of Climate and Applied Meteorology*, 23(1), 124–134.

Tyler, D., and J. R. Barnes (2005), A mesoscale model study of summertime atmospheric circulations in the north polar region of Mars, *J. Geophys. Res. – Planets*, 110(E6).

Tyler, D. and J. Barnes (2013), Mesoscale modeling of the circulation in the Gale Crater region: an investigation into the complex forcing of convective boundary layer depths, *Mars*, 8, 58–77.

Tyler, D., J. R. Barnes, and R. M. Haberle (2002), Simulation of surface meteorology at the Pathfinder and VL1 sites using a Mars mesoscale model, *J. Geophys. Res. – Planets*, 107(E4).

Tyler, D., Jr., J. R. Barnes, and E. D. Skyllingstad (2008), Mesoscale and large-eddy simulation model studies of the Martian atmosphere in support of Phoenix, *J. Geophys. Res.*, 113(E3), E00A12.

Vasavada, A., A. Chen, J. Barnes, et al. (2012), Assessment of environments for Mars Science Laboratory entry, descent, and surface operations. *Space Science Reviews* 170(1–4): 793–835.

Veverka, J., P. Gierasch, and P. Thomas (1981), Wind streaks on Mars: meteorological control of occurence and mode of formation, *Icarus*, 45(1), 154–166.

Vincendon, M., C. Pilorget, B. Gondet, S. Murchie, and J.-P. Bibring (2011), New near-IR observations of mesospheric CO_2 and H_2O clouds on Mars, *J. Geophys. Res.*, 116, E00J02.

Vinnichenko, N. K. (1970), The kinetic energy spectrum in the free atmosphere – 1 second to 5 years, *Tellus*, 22(2), 158–166.

Wang, H. (2007), Dust storms originating in the northern hemisphere during the third mapping year of Mars Global Surveyor, *Icarus*, 189(2), 325–343.

Wang, H., and A. P. Ingersoll (2002), Martian clouds observed by Mars Global Surveyor Mars Orbiter Camera, *J. Geophys. Res.*, 107(E10), 5078.

Wang, H., M. I. Richardson, R. J. Wilson, et al. (2003), Cyclones, tides, and the origin of a cross-equatorial dust storm on Mars, *Geophys. Res. Lett.*, 30(9), 1488.

Wang, H., and J. A. Fisher (2009), North polar frontal clouds and dust storms on Mars during spring and summer, *Icarus*, 204, 103–113, doi:10.1016/j.icarus,.2009.05.028

Ward, A. W. (1979), Yardangs on Mars: evidence of recent wind erosion, *J. Geophys. Res.*, 84(B14), 8147–8166.

Wilson, R. J. (1997), A general circulation model simulation of the Martian polar warming, *Geophys. Res. Lett.*, 24, 123–126.

Wilson, R. J., and K. P. Hamilton (1996), Comprehensive model simulation of thermal tides in the Martian atmosphere, *J. Atmos. Sci.*, 53, 1290–1326.

Wilson, R. J., G. A. Neumann, and M. D. Smith (2007), Diurnal variation and radiative influence of Martian water ice clouds, *Geophys. Res. Lett.*, 34(2), L02710.

Wing, D. R., and G. L. Austin (2006), Global Mars mesoscale meteorological model, *Icarus*, 185.

Wood, S. (1999), Nucleation and growth of CO_2 ice crystals in the Martian atmosphere, Thesis, UCLA, Los Angeles.

Wyngaard, J. C. (2004), Toward numerical modeling in the "Terra Incognita", *J. Atmos. Sci.*, 61(14), 1816–1826.

Ye, Z. J., M. Segal, and R. A. Pielke (1990), A comparative study of daytime thermally induced upslope flow on Mars and Earth, *J. Atmos. Sci.*, 47(5), 612–628.

The Global Circulation

JEFFREY R. BARNES, ROBERT M. HABERLE, R. JOHN WILSON, STEPHEN R. LEWIS, JAMES R. MURPHY, PETER L. READ

9.1 INTRODUCTION

Mars is a relatively small and dry planet with a very thin atmosphere. These properties are directly reflected in its global atmospheric circulation, which is fundamentally similar to that on Earth in many ways, but very different in other ways. The investigation of the global circulation of the Martian atmosphere has a surprisingly long history, as a result of the basic terrestrial similarities and the remarkably early spacecraft missions to Mars. The tremendous growth in spacecraft observations during the past two decades has made the atmospheric circulation of Mars by far the most fully characterized and understood atmospheric circulation in the Solar System, aside from that of Earth. By comparison with Earth, our knowledge of the global atmospheric circulation of Mars remains limited: there are extremely few direct wind observations, very few good surface pressure observations, and the local time coverage of the global observations is very limited. Observations of the planetary boundary layer (which can be extremely deep by comparison to Earth) on Mars are quite limited at present. It is only with the aid of extensive numerical modeling that our fundamental understanding of the global circulation of the Martian atmosphere has been fleshed out. Data assimilation efforts for Mars are still at a relatively early stage but these have tremendous potential for the future – provided that observations of the global atmosphere continue to be obtained.

The fact that Mars is now a desert planet without oceans and lakes means that the thermal inertia of the surface is small. As a consequence, the magnitude of seasonal changes in the atmospheric circulation is much larger than it is on Earth. The terrestrial analog for seasonal changes in the Mars atmosphere is the middle atmosphere (stratosphere and mesosphere). In the summer hemisphere of Mars the mean meridional (latitudinal) temperature gradient reverses in direct response to the daily insolation, such that temperature increases from low latitudes to the poles – or at least up to very high latitudes where the residual polar ice caps are located. Thus easterly winds tend to prevail at higher altitudes, whereas the winter hemisphere is characterized by very strong westerly winds aloft. With only very small amounts of ozone (see Chapters 4 and 13) and no other solar absorbing gases, the Martian atmosphere lacks a tropopause. Thus the basic circulation regimes tend

to extend to relatively high altitudes: the "lower" atmosphere, the subject of this chapter, is generally defined as extending to altitudes of ~80–100 km. The seasonality of the Martian atmosphere is driven by the obliquity, which at present is very nearly the same as that for Earth, and the orbital eccentricity, which is much larger than Earth's. Perihelion occurs at $L_s \sim 250°$ (close to southern summer solstice), when the solar insolation is ~40% larger than it is at aphelion ($L_s \sim 70°$, close to northern summer solstice)[1]. This means that the atmospheric temperatures go through a strong annual cycle tied to perihelion and aphelion (see Chapter 4). This is the primary reason that the "dust storm season" on Mars occurs during the southern spring and summer seasons, and this then acts to further amplify the annual variation in atmospheric temperatures (see Chapter 10).

The Martian atmosphere is almost entirely composed of CO_2, a greenhouse gas. Given its small total mass, the radiative relaxation time scales in the atmosphere are very short by comparison with Earth. One basic consequence of this is that very large-amplitude thermal atmospheric tides are present on Mars. Dust in the atmosphere increases and alters the thermal forcing of these tides. The planetary boundary layer of Mars is driven in a fundamentally different way than it is on Earth, as a result of the dominance of radiative heating and cooling at low levels above the ground (see Chapter 7). The thinness of the atmosphere causes the daytime mixed layer to be much deeper than on Earth, reaching depths as large as ~10–15 km (the depth of the troposphere on Earth). This effect is magnified over elevated terrain and suppressed over lowlands, causing a regionally varying thermal forcing of the atmosphere. The small size of Mars and the lack of oceans cause the latitudinal extent of the Hadley circulation to be greater than on Earth. It also causes the transient baroclinic eddies to be characterized by planetary zonal scales, as are the largest-amplitude quasi-stationary eddies that are primarily forced by the large topography. These properties represent very significant dynamical differences by comparison to Earth.

Some amount of dust is always present in the Martian atmosphere, and it is a good absorber of solar radiation, and a good absorber and emitter in the infrared. Its presence can generate strong buoyancy during the daytime, which makes it partially analogous to water vapor in Earth's atmosphere. Since the dust is lifted and transported by the winds, and is radiatively active, it adds considerable complexity to the global circulation and the climate. The thermal forcing of the circulation cannot be simulated well without an accurate description of the varying spatial distribution and opacity of the dust. And since this distribution depends directly upon the circulation and its interaction

---

[1] L_s is an angular measure of the planet's orbital position: $L_s = 0°$ corresponds to northern spring equinox; $L_s = 90°$, 180°, and 270° correspond to northern summer solstice, fall equinox, and winter solstice, respectively.

with the surface, the global circulation is intimately coupled to the complexities of the dust cycle (see Chapter 10).

During the last 5–10 years cloud thermal forcing has come to be recognized as an important driver, adding a new layer of complexity to the global circulation and climate system. Water ice clouds, fogs, and hazes can produce strong heating and cooling, almost entirely as a result of their absorption and emission of IR radiation. Even though the atmosphere contains a very small amount of water vapor, the clouds that form can still be optically thick enough to generate very significant heating and cooling rates that can significantly alter the thermal structure and the circulation of the atmosphere. In the aphelion season, modeling results now indicate that cloud thermal forcing may actually be more important than that due to dust in at least some respects.

The dust and water cycles on Mars are strongly coupled. Dust particles act as the ice nuclei for condensation and cloud formation and thus help determine the abundance and distribution of clouds. Conversely, clouds also scavenge dust out of the atmosphere and therefore directly influence the distribution of dust. Carbon dioxide also condenses in the atmosphere, so CO_2 ice particles are a third aerosol species that can impact the thermal forcing of the circulation.

Mars has a unique global mean condensation flow associated with the condensation and sublimation of CO_2 in middle and high latitudes (see Chapter 12). The winds associated with this flow can be of some direct significance near the edges of the rapidly retreating caps in the spring. On a global scale, however, the dynamical effects of the condensation flow are small, as global circulation models (GCMs – also termed global climate models or general circulation models) can capture the basic aspects of the atmospheric circulation in the complete absence of this flow (e.g. Haberle et al., 1993b, 1997). The wintertime polar regions become substantially enriched in non-condensable gases (e.g. Sprague et al., 2007) because of the condensation flow, a phenomenon that has both radiative and dynamical implications that remain largely unexplored.

In this chapter we examine our current understanding of the global atmospheric circulation of Mars. The emphasis is on the knowledge that has been gained since the last comprehensive review of this subject by Zurek et al. (1992). First we discuss how the global circulation has been determined from both observations and models. We then describe the thermal forcing of the circulation, which is very strong compared to Earth. We then systematically discuss the major components of the global circulation: the mean circulation, thermal tides, transient eddies, and stationary eddies. The mean circulation as we define it here includes the overturning circulation, the condensation flow, as well as the western boundary currents – which are not zonally symmetric. We conclude with a brief summary and a look towards future research. There are brief summaries at the end of each section after the following one, to help the reader see the "big picture" along the way.

9.2 DETERMINING THE GLOBAL CIRCULATION

Most of our understanding of the global circulation on Mars has come from the synergy between observations and models – GCMs, in particular. The observational database, although greatly expanded in recent years, is still much too sparse to describe the full spatial and temporal structure of the global circulation. However, the observations we do have are extremely important for constraining the models, because in their absence the basic validity of the models would remain highly uncertain. It is thus only from the combination of observations and modeling that we have been able to determine the basic nature of the global atmospheric circulation on Mars.

9.2.1 Observations

The observations we do have come from Earth-based telescopes, orbital remote sensing, and *in situ* measurements at the surface. While the Earth-based telescopic observations of the late 19th and early 20th centuries revealed the presence of clouds and dust storms moving around the planet (see Chapter 3), it was not until the dawn of the spacecraft era in the mid-1960s, and the subsequent delivery of orbiters and landers, that a picture of the global circulation patterns began to emerge. While virtually every spacecraft mission to Mars has measured some parameter of relevance to the atmosphere, we focus here on those missions whose payloads are the most relevant for global circulation studies. Table 9.1 lists the missions, their operational dates, and the kinds of measurements they made that are most relevant to global atmospheric circulation studies.

To determine the global circulation, wind measurements are the obvious starting point. Unfortunately, the only direct measurements of winds to date come from Earth-based telescopic observations of Doppler-shifted spectral lines (e.g. Clancy et al., 2006; Sonnabend et al., 2006, 2012; Moreno et al., 2009), and landed platforms carrying meteorological payloads (e.g. Hess et al., 1977; Schofield et al., 1997; Gunnlaugsson et al., 2008; Gómez-Elvira et al., 2012). The telescopically derived winds are mostly applicable to the middle atmosphere near 50 km, and since the data are acquired during opposition (when the telescopic resolution is high) they have limited seasonal coverage. Landers sample the near-surface environment at a fixed location. However, given the small number of landers with wind sensors (Viking Lander 1 and 2, Pathfinder, Phoenix, and the Mars Science Laboratory (MSL)), their lack of operational overlap, limited success, and the fact that near-surface winds are strongly controlled by local topography, these datasets have been mainly used to provide constraints on the surface expression of several components of the global circulation; in particular, the thermal tides and the transient eddies. This has been the most applicable when the winds can be combined with simultaneous measurements of surface pressure and temperature.

Interestingly, it is actually the measurement of surface pressures from landed platforms that has provided the richest information on global and large-scale wind systems. Viking, Pathfinder, Phoenix, and MSL each carried pressure sensors. Because the Martian atmosphere is hydrostatic on larger scales (greater than ~1–10 km), surface pressure is a measure of the total mass in the air column, and this is controlled by elevation, temperature, and dynamics. Surface pressure time series contain information on a very wide range of scales of motion, from dust devils, slope flows, and regional circulations, at smaller scales, to transient baroclinic eddies, thermal tides, and the CO_2 cycle, at larger scales (Haberle et al., 2014).

Table 9.1. *Missions to Mars with an atmospheric science component.*

Mission	Dates of operation	Relevant measurements[a]
Mariner 9	Nov 13, 1971 – Oct 27, 1972	T-profiles, Ts, opacity, RS
Viking Orbiter 1	Jun 19, 1976 – Aug 07, 1980	T @ ~25 km, Ts, opacity, RS
Viking Orbiter 2	Aug 07, 1976 – Jul 25, 1978	T @ ~25 km, Ts, opacity, RS
Viking Lander 1	Jul 20, 1976 – Nov 11, 1982	P, T, wind
Viking Lander 2	Sep 03, 1976 – Apr 11, 1980	P, T, wind
MGS	Sep 11, 1997 – Nov 02, 2006	T-profiles, Ts, opacity, RS
Mars Pathfinder	Jul 04, 1997 – Sep 27, 1997	P, T, wind
Mars Odyssey	Oct 24, 2001 – Ongoing	Ts, T @ ~25 km
Mars Express	Dec 25, 2003 – Ongoing	T-profiles, RS
Phoenix	May 25, 2008 – Nov 02, 2008	P, T, wind
MRO	Mar 12, 2006 – Ongoing	P, T, opacity
MSL (Curiosity)	Aug 05, 2012 – Ongoing	P, T, Ts, RH, wind, opacity
Mars Orbiter Mission	Sep 24, 2014 – Ongoing	Ts, opacity

[a] T = temperature, Ts = surface temperature, RS = radio science, P = pressure, RH = relative humidity.

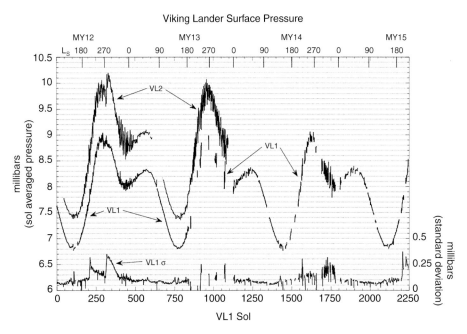

Figure 9.1. Diurnally averaged Viking Lander surface pressures (the lower data curve is for Lander 1 and the higher – and shorter – data curve is for Lander 2), along with the standard deviation of the intra-diurnal surface pressure variations, at the bottom. The time axis on the bottom is VL1 sol number, and on the top it is the areocentric longitude L_s.

Because the Viking Landers and MSL measured the daily variation of surface pressure for more than one Mars year, these multi-year datasets are the most useful for characterizing the diurnal, seasonal, and inter-annual variations of the global wind systems that they can detect, primarily the thermal tides and the transient eddies. Daily averages of the Viking Lander and MSL pressure data, as well as intra-diurnal standard deviations, are shown in Figures 9.1 and 9.2. The pressure data from Pathfinder and Phoenix are also useful, but these missions had comparatively short lifetimes (82 and 152 sols, respectively) and instrument calibration issues (e.g. Haberle et al., 1999).

The large semi-annual variation in daily mean surface pressure shown in Figures 9.1 and 9.2 is a consequence of the seasonal CO_2 cycle examined in detail in Chapter 12. The

asymmetry in this oscillation is due to Mars' orbital eccentricity and the timing of perihelion; southern winter is longer than northern winter, so that the deepest pressure minimum occurs during this season. During fall, winter, and spring at the Viking Lander sites the daily-averaged surface pressures, particularly those at VL2, show sol-to-sol fluctuations due to transient baroclinic eddies (middle-latitude weather systems), which are discussed in detail in Section 9.6. In the MSL data, sol-to-sol fluctuations are much more subdued because the transient eddies do not penetrate to near-equatorial latitudes. The large and seasonally variable standard deviation of the diurnal pressure variation largely reflects the thermal tides that, in sharp contrast to Earth, are a major component of the Martian global circulation. These are discussed in Section 9.5.

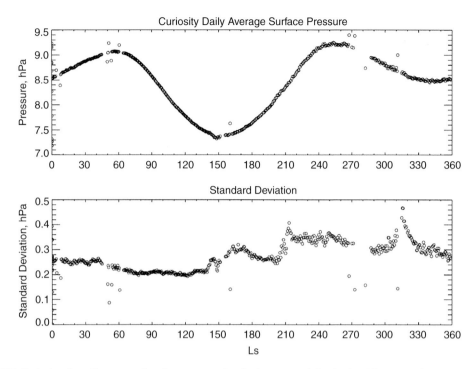

Figure 9.2. The MSL Curiosity diurnally averaged surface pressure data in the top panel, for the first Mars year of observations in Gale Crater. The intra-diurnal standard deviation of the pressure is shown in the lower panel.

While landers provide direct *in situ* measurements of the near-surface environment at fixed locations, orbiters provide a global view of the planet in a relatively short period of time. Naturally, the coverage that a given orbiter can provide depends on its orbital properties. The low, circular, high-inclination, Sun-synchronous orbits of the Mars Global Surveyor (MGS), Mars Odyssey, and Mars Reconnaissance Orbiter (MRO) missions are ideal for daily global mapping, but are limited to two essentially fixed local times. On the other hand, the precessing elliptical orbits of Mariner 9, Viking, and Mars Express allowed the sampling of different local times of day – at the expense of daily global coverage. Taken together, these missions have provided excellent global coverage, but rather limited local time-of-day coverage.

As mentioned above, direct wind measurements from orbiters have yet to be made, though visible imaging systems have been used to track clouds and extract some wind information (e.g. Wang and Ingersoll, 2003). Instead, the main utility of orbiters for global circulation studies comes from their ability to remotely sense the thermal structure of the atmosphere. This has been achieved primarily through radio occultations and infrared sounding. The value of temperature data is that they can be used to constrain global circulation models and construct balanced wind fields (e.g. Smith et al., 2001a,b). Mariner 9, Viking, Mars Global Surveyor (MGS), Mars Express, and the Mars Reconnaissance Orbiter (MRO) each carried payloads capable of sensing temperatures in the lower atmosphere (below ~80 km). Mariner 9, Mars Express, and MGS flew infrared spectrometers that retrieved temperatures from the surface to ~40 km with roughly scale-height resolution (e.g. Hanel et al., 1972; Conrath et al., 2000; Formisano et al., 2004). The Mars Climate Sounder (MCS) on MRO scans the limb to retrieve temperature up to 80 km with half-scale-height resolution (e.g. McCleese et al., 2008). And even though the Viking

Infrared Thermal Mapper (IRTM) and Odyssey's Thermal Emission Imaging System (THEMIS) were not sounding instruments, their measurements of temperatures near the 25 km level provide useful constraints on the circulation (Kieffer et al., 1977; Christensen et al., 2004). Very importantly, most of these retrievals also included dust and cloud opacities to help relate the observed temperature fields to the radiative forcing of the atmosphere.

9.2.2 Global Circulation Models

GCMs have long been an extremely important tool for determining the global circulation on Mars. The output from these models provides a complete description of the three-dimensional structure of the global circulation and how it varies in time. The validity of the simulations depends, of course, on how well they compare with data. This is directly related to how well they represent the physical processes that actually drive the circulation. Over time, the level of sophistication of the representation of physical processes in Mars GCMs has increased considerably. This increase in sophistication has been driven by the increase in observations coming from the spacecraft missions mentioned above (as well as the increasing sophistication of terrestrial GCMs). Compared to the first Mars GCM simulations performed by Leovy and Mintz (1969), when there was extremely little data to compare with, today's GCMs include realistic topography, transport schemes, cloud microphysics, and radiatively active aerosols. They have better horizontal and vertical resolution, higher model tops, more sophisticated grids, and generally simulate the full seasonal cycle for multiple and even many Mars years. This enormous increase in modeling capability is needed to better interpret the vastly larger volume of data that we now possess, but it is essentially the case that the models continue to be "ahead" of the observational data (they simulate far more aspects of the atmosphere than we have observed).

GCMs provide numerical solutions to a set of partial differential fluid dynamical equations based on the principles of conservation of mass, momentum, and heat, and an equation of state. A common assumption, valid for thin atmospheres and large enough horizontal scales, is that the vertical component of the vector momentum equation is approximated extremely well by simple hydrostatic balance. This leads to the "primitive equations" upon which virtually all Mars (and terrestrial) GCMs are based (see e.g. Read and Lewis, 2004). These equations can be discretized and solved on a 3D grid (grid-point models), or expanded in a series of spherical harmonic functions in latitude and longitude (spectral models). For Mars the vertical coordinate is usually chosen to be a normalized pressure coordinate in order to better represent the very large topographical variations.

It is convenient to think of these models in terms of a "dynamical core" that solves the fluid dynamical equations, and a "physics package" that allows the models to simulate the particular processes that are critical for the Mars atmosphere. The dynamical core solver is generic and can be used for any planet that satisfies the assumptions that the governing equations are based on. There are many types of readily available, well-tested, and validated dynamical cores that have been developed for Earth, and are applicable to Mars. While they differ in methodology and conservation properties, they all provide time-dependent solutions to the governing dynamical equations. The physics package, however, is what makes a GCM unique to a specific planet. For Mars this is a primary focus of ongoing development activity for most modeling groups. As more data become available, better representations of physical processes are inevitably needed to improve the model/data comparisons.

9.2.3 Data Assimilation

In the terrestrial weather and climate community, data assimilation techniques (e.g. Lahoz et al., 2010) are commonly used to produce atmospheric "reanalyses", suitable for diagnosing the large-scale atmospheric circulations, which are fully consistent with the atmospheric variables as actually measured. Data assimilation is a combination of observations and numerical models that (a) imposes physical constraints, (b) "absorbs" all of the observational data that are introduced into the model, and (c) enables the estimation of variables that are not actually measured. The model values of variables are "relaxed" or driven towards the observational values, and the result is a hybrid combination of the model and the observational data. Data assimilation offers very significant advantages for the analysis of atmospheric data from Mars, but it presents some large challenges as well (e.g. Banfield et al., 1995, 1996; Lewis and Read, 1995; Zhang et al., 2001; Hoffman et al., 2010; Lee et al., 2011; Greybush et al., 2012). This has now been demonstrated by the assimilation of three Mars years (almost six Earth years) of thermal, dust opacity, and water vapor observations from the Thermal Emission Spectrometer (TES) on MGS (Montabone et al., 2006; Lewis et al., 2007; Lee et al., 2011; Greybush et al., 2012; Steele et al., 2014a). MCS temperature data have also been assimilated (Navarro et al., 2014; Steele et al., 2014b). Atmospheric data assimilation for Mars is different than that for Earth,

primarily because at the present time there are no wind observations and very few surface pressure measurements. The circulation in the models has to be constrained almost entirely by the assimilated temperature observations, along with observations of the dust and clouds. It is not clear at the present time just how well the results of the Mars data assimilations resemble reality – especially at lower levels in the atmosphere (Mooring and Wilson, 2015). The lack of "ground truth" observations with which to assess the assimilations is a major limitation. Additionally, the assimilated observations are far more restricted than for Earth, in sizable part because of their very limited local time coverage. The lack of good local time coverage is considerably more significant for Mars than it would be for Earth because diurnal variations are generally so much larger in the atmosphere of Mars. The thermal forcing due to the radiative effects of dust and water clouds (also CO_2 clouds) is critical for the thermal structure and the circulation, and the current observations do not determine the spatial and temporal distributions of the dust and clouds well enough at present. This essentially leads to a "climate bias" problem in the models being utilized for data assimilation. There is evidence that the strong impact of the aerosols on the temperatures can be used to "update" the aerosol distributions continuously as the assimilation is run (Navarro et al., 2014). The assimilation of the observed radiances, instead of retrieved values, is also something that potentially could improve the assimilation results significantly. Regardless, more extensive and better observational datasets will be critical to the improvement of data assimilation for Mars in the future. Wind and surface pressure observations would be of enormous value for data assimilation for Mars.

9.3 FORCING PROCESSES AND BOUNDARY CONDITIONS

Forcing is used here in the standard way in relation to the atmospheric circulation. The forcing encompasses the "physical" processes that act to drive the circulation, as well as processes that act fundamentally to dissipate it. The former processes include radiative (at both solar and IR wavelengths) heating and cooling, sensible and latent heat transfers from the surface, latent heating in the atmosphere, and chemical heating in the atmosphere. The only latent heating that is important for present-day Mars is associated with CO_2 condensation and sublimation; chemical heating is not important for the large-scale circulation in the lower atmosphere (below ~80–100 km). The latter processes include those taking place in the planetary boundary layer (PBL) that are generally acting to weaken the winds (e.g. surface stresses). Processes in the PBL also can act strongly to force circulations (e.g. slope flows, gravity waves, and stationary eddies) in the presence of topography. Chapter 7 deals with the PBL, and thus the treatment here of this layer is brief and focuses on how the PBL both forces and retards the large-scale circulation. Frictional forces are also present at levels well above the PBL, and these are briefly discussed as well.

At the most fundamental level, forcing could refer to the strictly "external" parameters and properties that determine

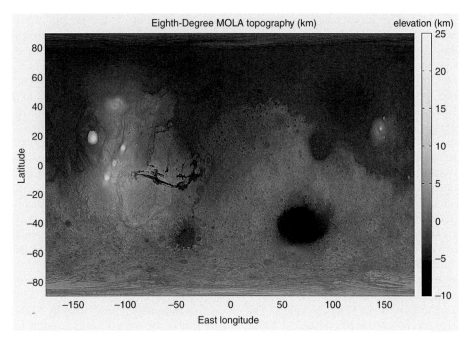

Figure 9.3. Mars topography as determined from the Mars Orbiter Laser Altimeter (MOLA) measurements. The map was produced using 1/8th degree MOLA data.

the global circulation (e.g. the solar flux, the composition of the atmosphere, the size of the planet, etc.). The ultimate goal of modeling atmospheric circulations has to be the ability to simulate the global circulation (with a completely general model) given only the external parameters and properties as an input. There is certainly a very long way to go before such a capability is achieved. But such an achievement would not directly yield much of an understanding of the atmospheric circulation as the dynamical response to the forcing. A framework for understanding the latter can be built upon the simple fact that the circulation generally acts to move the atmosphere away from the thermal state that would exist in the absence of all large-scale circulation. The latter state would be one established by radiative forcing, thermal convection, and latent heating processes – all operating in the vertical direction (within atmospheric columns) in the atmosphere. In any real planetary atmosphere, forced by spatially and temporally varying insolation, this state is incompatible with there being no winds: most basically because of the presence of horizontal pressure gradient forces due to the variations in the vertical temperature profiles in the horizontal directions. The resulting atmospheric circulation and the forcing can be viewed as existing in a quasi-equilibrium state, in which heat and energy transports by the circulation are in balance with the physical forcing processes. The latter strongly depend upon temperature, and in particular the departure of temperature from the state that would exist in the absence of large-scale atmospheric circulation. Where these departures are large, the *net* radiative heating/cooling will be large, thus requiring large dynamical heating/cooling rates for a balance. The actual atmospheric forcing is the result of the complex and highly nonlinear interplay of dynamical and physical processes. For Mars we have realized that this system is more complex than had once been thought, because of the very tight coupling of various dynamical and physical

processes. The presence of three different aerosols (dust, water ice, and CO_2 ice) that can all produce very significant heating and cooling in the atmosphere and on the surface, and that all interact with one another, makes the Mars atmosphere a very complex system. The almost certain existence of dust "reservoirs" on the surface also gives the system a longer timescale memory that it would otherwise not have.

Given the above, and the scarcity of direct observations of the atmospheric forcing, global circulation models have been critical to developing an understanding of the forcing – an understanding that is continuing to evolve in rather significant ways. These models have also been crucial to developing an understanding of the global circulation at large scales. In order to produce good simulations of the circulation, the models need to have realistic amounts and distributions of the three aerosols that can very significantly influence the atmospheric forcing. Different approaches to this have been and are being employed, in a variety of global circulation models. Simulation of the global atmospheric circulation of Mars has advanced greatly over the last 25 years, but it is still at a stage at which all of the key physical processes are generally not fully coupled together in the global models. The spatial resolutions of the global models that contain all of the basic physical processes remains somewhat limited at present, and this almost certainly hinders realistic simulation of some of the key physical processes (e.g. dust lifting and cloud microphysics).

9.3.1 Surface Properties

The exchange of heat and momentum between the surface and atmosphere is a major forcing mechanism for the global circulation. The properties of the surface that control this exchange are its topography, albedo, and thermal inertia. Maps of these properties based on the current observations are shown in Figures 9.3–9.5.

(a)

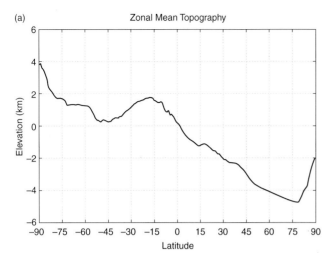

(b)

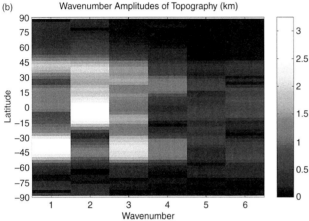

Figure 9.4. (a) Zonally averaged MOLA topography, and (b) the zonal wavenumber amplitudes of the MOLA topography. One-degree topography data were used to produce both panels. A black and white version of this figure will appear in some formats. For the color version, please refer to the plate section.

The topography of the Martian surface is extremely variable, with a maximum elevation difference of almost 30 km between the tops of the Tharsis volcanoes and the bottom of the Hellas Basin. At global scales the most striking feature is the substantial elevation difference between the northern and southern hemispheres (see Figure 9.3). As discussed later, this hemispheric topographic dichotomy, ~4 km on average, has a very significant affect on the global circulation. Zonal variations in elevation are also extremely important. As shown in Figure 9.4, topographic zonal wavenumbers 2 and 3 are prominent in the northern hemisphere subtropical and mid-latitude regions with wavenumber 3 extending to higher latitudes. Wavenumbers 1 and 3 are most prominent in the south, with the former due to the Hellas impact basin and the southern extension of Tharsis. At equatorial latitudes Tharsis and Arabia give rise to a very prominent wavenumber-2 component. Zonal topographic variations provide both mechanical and thermal forcing of the atmosphere. Mechanical forcing is primarily responsible for the wintertime extratropical stationary eddies, while the thermal forcing effects of topography are most important at lower latitudes and in the summer hemisphere.

In addition to its very large topography, the Martian surface is also characterized by continental-scale variations of albedo and thermal inertia (see Figure 9.5). Surface albedos are derived from orbiting bolometers, so that some care must be exercised to separate scattering by airborne dust from that by the surface (e.g. Kieffer et al., 1977; Christensen et al., 2001). Values of thermal inertia have been obtained by fitting thermal models to observed surface temperature variations, and thus the effects of airborne dust must be considered here as well (e.g. Mellon et al., 2000; Christensen et al., 2001; Putzig and Mellon, 2007). The importance of these properties for surface forcing is that the albedo determines how much sunlight is absorbed, and the thermal inertia determines how quickly the surface temperatures respond to energy gain and loss. Surfaces with low (high) albedo have higher (lower) daily mean temperatures, while surfaces with low (high) thermal inertial have larger (smaller) diurnal variations. The albedo and thermal inertia patterns on Mars are somewhat anticorrelated, with the brighter regions generally indicative of dust mantling while the darker regions are more representative of exposed basaltic bedrock of higher thermal conductivity and hence higher thermal inertia. The general pattern of large-scale albedo features is quasi-permanent (e.g. Kieffer et al., 1977; Christensen et al., 2001), but temporal variations at smaller scales due to dust storm activity have been observed (e.g. Pleskot and Miner, 1981; Christensen, 1988; Smith, 2004; Szwast et al., 2006).

9.3.2 Heat Sources and Sinks

9.3.2.1 Radiative Heating

Radiative heating is the dominant forcing process in the Martian atmosphere. This heating originates from the absorption and emission of radiant energy by gases and aerosols. Chapter 6 provides a detailed discussion of radiative transfer in the Martian atmosphere. Here we focus on the atmospheric heating rates that radiative transfer produces, since these are the fundamental driver of the atmospheric circulation.

CO_2 is the primary gas of interest, as water vapor, O_3, CH_4, CO, and other radiatively active gases are present in such small abundances that their heating rates are negligible. For CO_2 the atmosphere is nearly transparent to solar radiation, as it absorbs in narrow bands around 2.7 and 4.0 μm (Savijärvi et al., 2005). Only about 1% of the incoming solar radiation is absorbed by CO_2 in the atmosphere. Consequently, most of the incoming sunlight reaches the surface, where it is absorbed and re-radiated in the thermal infrared. In this region of the spectrum, however, the broad 15 μm vibrational fundamental of CO_2 provides significant opacity. Here, about 10% of the emitted thermal energy is absorbed in the atmosphere, and much of this takes place in the lowest several kilometers. Thus, the aerosol-free Martian atmosphere is largely heated from below by the absorption of infrared radiation coming up from the surface.

Dust and clouds are the most important radiatively active aerosols in the Martian atmosphere. Unlike CO_2, their concentrations, and hence the heating rates they produce, can vary greatly in space and time. Dust absorbs strongly at visible wavelengths

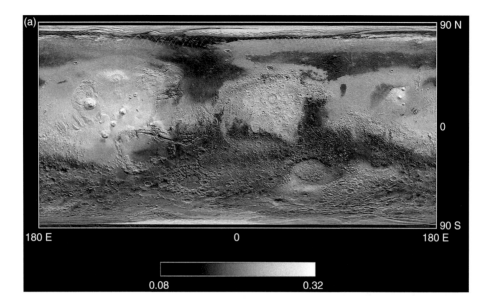

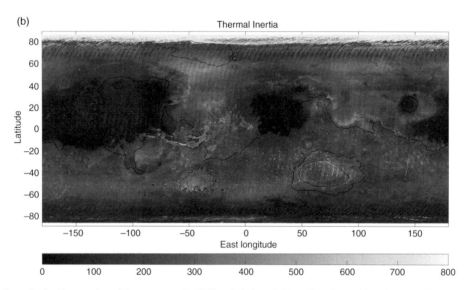

Figure 9.5. (a) Surface albedo (figure adapted from one on the TES website), and (b) surface thermal inertia (taken from Putzig and Mellon (2007)) as determined from TES temperature data. A black and white version of this figure will appear in some formats. For the color version, please refer to the plate section.

such that, when present, even at very small concentrations, it is the dominant absorber. Dust is also active in the infrared due to its broad 9 µm silicate feature (see Chapter 6). For typical (relatively low) dust loadings, the atmosphere absorbs roughly 10% of the solar and infrared radiation incident on the atmosphere (see Chapter 6).

While the great importance of dust heating for the global circulation has long been known (e.g. Haberle et al., 1982), the importance of cloud radiative heating is a relatively recent discovery. The water ice clouds that form the high-altitude, tropical, aphelion cloud belt, for example, absorb upwelling thermal radiation from the ground and warm the atmosphere by as much as 10 K in the time mean (e.g. Wilson et al., 2008b; Madeleine et al., 2012). Without this warming, modeled temperatures are too cold and the mean meridional circulation is weaker than it would be otherwise. In contrast, the polar hood clouds that form

at high latitudes nearer the surface during winter have a cooling effect, which can significantly increase baroclinic wave activity due to the increase in the equator-to-pole temperature gradient (e.g. Hollingsworth et al., 2011; Barnes et al., 2014). Thus, the radiative effects of both clouds and dust are very important drivers of the Martian global atmospheric circulation.

A key aspect of radiative forcing in the Martian atmosphere is how quickly the atmosphere responds to it. A measure of the response time can be obtained by calculating the (e-folding) time it takes for a thermal perturbation to relax to its equilibrium value. Near the surface, this "radiative time constant" or relaxation rate is on the order of a day for a relatively clear Mars atmosphere compared to ~20 days for Earth and ~10^4 days for Venus (Goody and Belton, 1967; Kondratyev and Hunt, 1982; Barnes, 1984; Collins et al., 1995; Hollingsworth and Barnes, 1996; Nayvelt et al., 1997). This very short radiative response

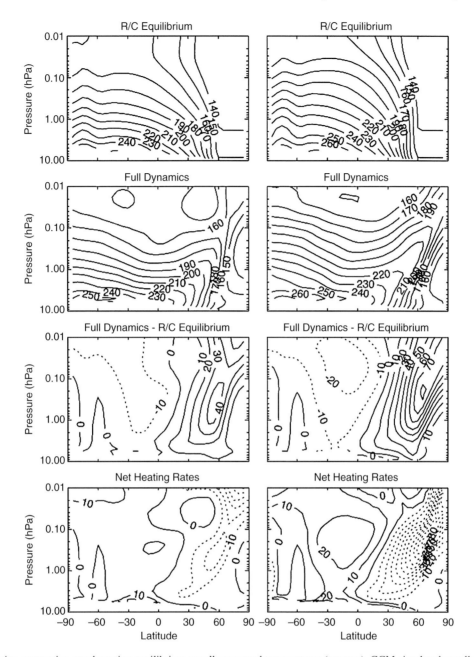

Figure 9.6. Radiative–convective–condensation equilibrium zonally averaged temperatures (top row), GCM-simulated zonally averaged temperatures (second row), the GCM simulated temperatures minus the radiative–convective–condensation temperatures (third row), and the net radiative heating rates (zonally averaged, in K/day) from the GCM simulations (bottom row). The left column is for northern winter solstice and a visible optical depth of 0.3, and the right column is for northern winter solstice and a visible optical depth of 1.0. M.A. Kahre produced the equilibrium thermal fields using a special version of the NASA Ames Mars GCM.

time is the result of the low mass of the Martian atmosphere and the fact that its main constituent is radiatively active. Because dust and clouds are also radiatively active, their presence decreases the response time even further.

In general, the radiative response time on Mars tends to be as short as or shorter than the large-scale dynamical timescales, which is why the thermal structure of much of the atmosphere is relatively close to being in radiative–convective (and condensational) equilibrium. This is particularly true close to the surface where the radiative timescales are very short (<1 day). Importantly, however, there are regions where the dynamical

timescales are comparable to the radiative timescales. This generally occurs at higher altitudes where the radiative timescales can increase and the winds can be much stronger. In these regions the temperatures can depart from radiative equilibrium by very substantial amounts.

These basic concepts are illustrated in Figure 9.6, which shows the temperature fields for radiative–convective–condensational (RCC) equilibrium and for the same conditions with dynamical processes included. The two cases shown are for northern winter solstice, with the first being relatively non-dusty (a visible dust opacity of 0.3) and the second being fairly dusty (an opacity of 1.0). Outside of the winter polar region, the RCC equilibrium temperatures are uniformly warmer in the dusty case than in the non-dusty case. This is mostly due to the absorption of solar radiation by the dust. This warmer RCC equilibrium atmosphere is reflected in the full dynamics simulations, as the dusty atmosphere is almost everywhere warmer than the non-dusty atmosphere. The difference temperature fields directly show the extent to which the circulation is pulling the atmosphere away from the RCC equilibrium state. It can be seen that in both cases the differences are not all that large except in a region in northern middle and high latitudes that largely coincides with the location of the strongest meridional temperature gradients in the full dynamics thermal field. In the non-dusty case the temperatures in this region are more than 40 K warmer than in the RCC equilibrium field, while in the dusty case the temperatures are more than 70 K warmer than in the RCC equilibrium field. The net radiative heating rates must be negative in this region and they are, exceeding 20 K/day in the non-dusty case and 90 K/day in the dusty case. These are indicative of the very strong increase in the strength of the mean meridional circulation with increasing dust loading, as it is being driven by much stronger radiative heating gradients. These cause it to expand substantially in latitude and in depth, as will be discussed in Section 9.4. Comparable dynamically driven temperature changes on Earth (as a fraction of the mean temperature) are only found in the middle atmosphere.

The heating rates that drive the global circulation have not been directly observed, but can be estimated from models – as above. This is typically done by calculating the vertical profiles of solar and infrared fluxes and evaluating their convergences in a layered atmosphere. A variety of methods have been used to calculate fluxes, ranging from simple band models (Gierasch and Goody, 1968; Leovy and Mintz, 1969; Pollack et al., 1981) to full up line-by-line calculations (Crisp, 1990; Pollack et al., 1990). Modern GCMs mostly use a correlated-*k* approach, which is fast, reasonably accurate, and easily includes aerosol scattering (see Mischna et al. (2012) for a comprehensive review of this method). The lines within a gaseous absorption band, normally taken from a database such as HITRAN (Rothman et al., 2013), are re-sorted in probability space to produce a monotonic curve, which can be more easily integrated (by Gaussian quadrature) to obtain an optical depth. For aerosols, opacities depend on single-scattering properties, which can be evaluated with a Mie code given a particle size distribution and the wavelength dependence of the refractive index. The gas and aerosol opacities are then used in a two-stream code to calculate fluxes and heating rates. All of the current Mars GCMs employ radiative codes that perform these calculations, in order to drive the circulation and determine atmospheric temperatures and other variables.

An important aspect of radiative forcing is its diurnal, seasonal, and spatial variability. Diurnal variability is mainly driven by the Mars surface, whose generally low thermal inertia can lead to daily temperature swings as large as ~100 K. These very large diurnal variations force the thermal tides, which are a major component of the global circulation on Mars (see Section 9.5). Under dusty conditions the direct absorption of sunlight in the atmosphere enhances diurnal variations aloft, which can greatly amplify tidal components. Seasonal and latitudinal variations arise from Mars' orbital characteristics. The present Martian obliquity is 25.2°, which makes its seasonal and latitudinal forcing, with one

exception, very similar to Earth's. That exception is due to the much larger eccentricity of the Martian orbit (0.093) compared to Earth (0.017), which leads to a ~40% variation in the solar flux at the top of the atmosphere between perihelion and aphelion as compared to 6% for Earth. Presently, perihelion occurs at $L_s = 250°$ just before winter solstice in the northern hemisphere. Thus, the global circulation is much more vigorous when the planet is near perihelion than at aphelion. Furthermore, because of the lack of extensive reflective clouds and large permanent polar ice caps, surface insolation and hence surface temperatures in the absence of oceans tend to increase toward the pole during summer. The bulk of the summer atmosphere thus has small and reversed meridional temperature gradients, and very little if any transient eddy activity except at very high latitudes near the permanent polar ice caps (as discussed in Section 9.6).

Another unique aspect of radiative forcing on Mars is that, to first order, surface temperatures and their diurnal variations are relatively insensitive to elevation variations. This is essentially due to the fact that surface (ground) temperatures are strongly controlled by the local energy balance between absorbed solar and IR radiation and emitted IR radiation – along with heat conduction below the surface. Locations at the same latitude will thus tend to have very similar surface temperatures if they have similar surface albedo and thermal inertia values. This basic thermal behavior is very non-terrestrial. A consequence of it is that, at given pressure levels in the atmosphere above elevated terrain, it will tend to be warmer over elevated regions and colder over lower regions. Thus elevated regions on Mars tend to create heat sources in the atmosphere while lowland regions tend to produce heat sinks. In the presence of sufficiently deep thermal inversions in the vertical, this relationship will be reversed. This effect creates a strong thermal forcing for large-scale stationary eddies in the tropics and the summer hemisphere.

In the presence of the topographic hemispheric dichotomy, the above effect drives a stronger mean overturning circulation during southern summer than during northern summer (e.g. Richardson and Wilson, 2002). This is very strongly amplified by the orbital eccentricity and the present-day longitude of perihelion such that, when annually averaged, the mean meridional circulation has a southern summer-like structure. The southern summer mean meridional circulation is much stronger than that in northern summer, and in addition to the much stronger insolation near perihelion the greatest reason for this is that the atmospheric dust loading is generally much larger during southern summertime. This provides a much stronger forcing for the mean meridional circulation, as well as the thermal tides. The former is discussed in detail in the next section, while the latter is discussed in Section 9.5.

There is another effect on Mars that can act to drive circulations on a wide range of scales; this is associated with the fact that the planetary boundary layer in the daytime (the mixed layer) tends to be considerably deeper in higher-elevation regions than it is in lower-elevation ones (Hinson et al., 2008b; Spiga et al., 2010; see Chapter 7). The mixed-layer depth in closed low-elevation regions can be further depressed by subsidence forced by rising motions (upslope winds) occurring around the edges of the lowlands (Tyler and Barnes, 2013, 2015). Variations in the mixed-layer depth are also created by albedo and thermal inertia variations. Thus very strong gradients in the mixed-layer depth are present on Mars on a very wide range of horizontal scales. These will tend to be associated with strong horizontal temperature gradients, which can drive circulations on a wide range of scales.

9.3.2.2 Sensible Heat Exchange With the Surface

In addition to radiative exchange with the surface, the atmosphere also exchanges sensible heat (c_pT) with the surface. This exchange is mostly accomplished by turbulence within the PBL. The turbulence can be driven by thermal instabilities (i.e. unstable lapse rates), and/or shear instabilities (strong variations in the wind with height). These are distinguished by the Richardson number (Ri), which is defined as

$$Ri = \frac{g}{\theta} \frac{d\theta/dz}{(du/dz)^2} \qquad (9.1)$$

where g is gravity, θ is potential temperature, u is the wind speed, and z is height. Defined in this way, Ri represents the ratio of the production of turbulent kinetic energy by buoyancy forces to that by shear stresses. Flows are generally considered laminar for $Ri > 0.25$ and turbulent for $Ri < 0.25$. When the atmosphere is thermally stable ($d\theta/dz > 0$) but turbulent ($0 < Ri < 0.25$), wind shear is the dominant factor. As Ri falls below zero, however, the atmosphere becomes unstable and buoyancy forces begin driving the turbulence. For $Ri \ll 0$, as is often the case during the day on Mars, buoyancy is the dominant turbulence production mechanism.

The production of turbulence within the Martian boundary layer is driven by the very strong infrared heating rates in the near-surface atmosphere (e.g. Haberle et al., 1993a; Takahashi et al., 2003; Spiga et al., 2010; also see Chapter 6). During the day, this heating can generate super-adiabatic lapse rates that destabilize the atmosphere and generate convection. Pathfinder observed such lapse rates very near the surface (Schofield et al., 1997), the MER Rovers found them up to several hundred meters (Smith et al., 2006), and orbital remote sensing measurements have detected well-mixed – near-neutral – layers extending at least as high as ~10 km (Hinson et al., 2008b; Spiga et al., 2010). Thus, during the daytime, heat, momentum, and mass (e.g. dust, water vapor, and other trace constituents) are mixed throughout a relatively deep portion of the atmosphere above the surface.

At night, surface temperatures decline, the near-surface atmosphere stabilizes, and an intense surface-based inversion develops – extending upwards to ~1 km (e.g. Smith et al., 2006). In this situation, mixing can only be generated mechanically by shear stresses associated with winds. Nighttime mixed layers are thus very shallow, and this tends to greatly limit the depth through which mixing occurs.

The sensible heat fluxes arising from this strong daily cycle have not been directly measured, but they have been inferred from near-surface temperature data and models. Sutton et al. (1978) used Viking temperature and wind data, a thermal model, and estimates of the surface roughness from imaging data to compute surface sensible heat fluxes from Monin–Obukhov similarity theory. They found maximum surface heat fluxes of ~15–20 W m^{-2} at the Viking Lander 1 site, and somewhat smaller values at the Viking Lander 2 site. These surface heat fluxes would warm a 5 km deep layer by about 7 K per sol. However, in comparison to the upwelling infrared radiation

that provides the major energy input into the lower atmosphere on Mars, these fluxes are relatively small. A daytime surface temperature of 260 K, for example, will radiate ~230–260 W m^{-2} depending on its emissivity. A variety of other studies give similar results for surface sensible heat fluxes. Thus, while surface sensible heat forcing of the Martian atmosphere is not negligible, it is not at all the major energy transfer from the surface. This is in sharp contrast to the Earth, where the sensible and latent heat fluxes provide a primary thermal forcing for the global circulation.

9.3.2.3 Condensation and Latent Heating

Both water vapor and CO_2 condense and sublimate in the Martian atmosphere. Latent heating/cooling occurs and therefore must be considered in relation to thermal forcing. In the case of water vapor in the current atmosphere, the amount condensing and sublimating is quite small and the latent heating is negligible (e.g. Savijärvi, 1995; Montmessin et al., 2004). In the case of CO_2, latent heating can be very significant – both in the atmosphere and on the surface. Condensation of CO_2 in the atmosphere occurs in the winter polar regions and seasonally at very high altitudes in the tropics (e.g. Hinson et al., 1999; Conrath et al., 2000; Colaprete et al., 2003, 2008; Montmessin et al., 2007; Määttänen et al., 2010, 2014; Hu et al., 2012; Hayne et al., 2014). CO_2 begins condensing when temperatures fall below the saturation temperature corresponding to the local CO_2 vapor pressure (the frost-point temperature). Because CO_2 is the major constituent of the atmosphere, latent heating tends to maintain temperatures close to the frost point. The frost-point temperature at 6.1 mbar (near the mean surface pressure) is 147.7 K, and it decreases to 124.1 K at 0.1 mbar (~40 km altitude).

Observations discussed in the above references show that winter polar temperature profiles often follow the CO_2 saturation temperature profile. However, there appear to be some temperature profiles with lapse rates exceeding the CO_2 saturation profile. These are supersaturated profiles that may result from a lack of condensation nuclei (dust); they could potentially lead to CO_2 convection. Colaprete et al. (2003, 2008) discuss this possibility and show that the amount of convective available potential energy in some of the polar profiles is enough to drive relatively deep convection and significantly enhance vertical mixing in the polar night regions. This remains a very interesting and potentially important process, one warranting further study.

CO_2 also condenses directly on the ground. The familiar advance and retreat of the Martian seasonal polar caps are due largely to this process. Winter polar surface temperatures are therefore controlled by latent heat release and are maintained at the frost-point temperature of the local CO_2 vapor pressure. Since the Martian atmosphere is 95% CO_2, to first order this vapor pressure is ~95% of the surface pressure. Given the relatively weak dependence of the frost-point temperature on vapor pressure, the winter polar caps are therefore roughly isothermal, with surface temperatures ~145 K in the north and several degrees cooler in the south. Of course, elevation variations and dynamics will affect the surface pressure, and as winter progresses the non-condensable component of the polar atmosphere is enriched (e.g. Sprague et al., 2007). These factors can result in significant spatial variations of surface temperature within the seasonal ice caps (at least as large as ~5 K).

9.3.3 Momentum Sources and Sinks

9.3.3.1 Interactions With the Surface

The surface is the ultimate source of angular momentum for the global circulation, and its interaction with the atmosphere is therefore important. On Earth, for example, angular momentum is extracted from the surface in the tropical easterly trade belts where it is transported upward and poleward, respectively, by the mean meridional circulation and eddies and then transferred back to the surface in the mid-latitude prevailing westerlies (e.g. Peixoto and Oort, 1992). This momentum exchange between the surface and atmosphere is accomplished through surface wind stresses and mountain torque. An analogous momentum exchange and transfer must occur on Mars, though the condensation flow plays a unique role in the Martian momentum budget (e.g. Haberle et al., 1993b).

Frictional processes within the boundary layer control the surface stress exchange, with the rates determined by the nature of the surface and the characteristics of the boundary layer. The surface roughness, z_0, defined as the height above the surface at which the mean winds vanish, is a key parameter in determining the surface stress τ. For a neutral surface layer (see Chapter 7), the mean wind u and the friction velocity $u^* = (\tau/\rho)^{1/2}$ are related to the surface roughness through

$$u/u^* = \kappa \ln(z/z_0) \qquad (9.2)$$

where ρ is the density, κ is von Kármán's constant (0.35), and z is height. Thus, for a given mean wind speed (at a given height), the friction velocity and hence the surface stress increases with increasing surface roughness. Sutton et al. (1978) estimated maximum friction velocities at the Viking Lander 1 site ranging from 0.4 to 0.6 m s^{-1} assuming a surface roughness of 1 cm. Unfortunately, the spatial variation in Martian surface roughness is not well known, though estimates have been made from rock abundance data (e.g. Hébrard et al., 2012). These estimates show substantial spatial variability in the surface roughness, with values ranging between 0.001 and 2.33 cm. In addition, the surface appears to be smoother than previously assumed, with 84% of the surface having roughness lengths less than the 1 cm value adopted by Sutton et al. (1978) and used in most subsequent modeling work.

As noted in Section 9.3.2.2, the boundary layer can grow very deep during the day (up to 10 km or even deeper) and is typically very shallow at night (<1 km), which means that there is a strong diurnal variation in frictional drag each day. Winds normally increase with height, so that during the daytime, when the boundary layer is deep, momentum is mixed down to the surface and the frictional drag is the strongest. At night the boundary layer is shallow and wind stresses tend to be weaker. Hess et al. (1977) invoked this daily mixing cycle to explain the surface wind behavior at Viking Lander 1 during the early portion of the mission. However, there are certainly exceptions to this general pattern. Nighttime winds on sloping terrains can become very strong and cause maximum stresses, generated mechanically in this instance, to occur at night rather than during the day.

The large topographical relief of the Martian surface can also produce substantial torques on the atmosphere. These torques arise where surface pressure and slopes are correlated. Pollack et al. (1981) found that this so-called "mountain torque" term

was comparable to the surface stress torque in southern mid-latitudes during winter, in early GCM simulations. Thus, both surface stresses and mountain torques play an important role in the angular momentum budget of the Martian atmosphere.

9.3.3.2 Gravity Waves

At spatial scales of a few up to hundreds of kilometers and even larger, topographic variations can force gravity (buoyancy) waves (GWs). These waves can vertically propagate and thus achieve very large amplitudes and break. Breaking (as well as non-breaking) GWs can act to accelerate the large-scale flow and to produce turbulence at altitudes near their breaking level. GWs are also forced by a number of different non-topographic and non-stationary processes (e.g. Théodore et al., 1993; Medvedev et al., 2011). GWs in Mars' atmosphere were first evidenced by images of lee wave clouds (Briggs and Leovy, 1974; Anderson and Leovy, 1978). Observed atmospheric temperatures have been interpreted as providing direct evidence of GWs in the form of vertical temperature oscillations (e.g. Magalhães et al., 1999; Creasey et al., 2006), and in the form of adiabatic and near-adiabatic lapse rates at high altitudes (Heavens et al., 2010). Edmonds et al. (2014) have called into question some of the MCS thermal profiles identified as exhibiting gravity wave effects. Interestingly, the large-magnitude and smaller spatial-scale topographic variances that characterize the southern hemisphere (see Figure 9.3) are not well correlated with the locations of maximum GW activity as determined by the data analyses conducted to date (Creasey et al., 2006).

Gravity waves have been examined in relation to observed cases of winter polar warming (e.g. Barnes, 1990; Collins et al., 1997; Hartogh et al., 2007) and in connection with the vertical extent of the winter westerly jets (e.g. Barnes, 1990; Joshi et al., 1995a; Collins et al., 1997). Surface magnitudes of the forced waves (e.g. McFarlane, 1987; Forget et al., 1999), dissipation of vertically propagating waves (Imamura and Ogawa, 1995; Eckermann et al., 2011), and the atmosphere's vertical thermal and wind structure control the heights to which GWs can propagate and thus induce mean flow accelerations and turbulence. Breaking at lower atmospheric density levels (higher altitudes) results in greater accelerations. Modeled wind accelerations have ranged from about 10 to 1000 m s^{-1} sol^{-1} (e.g. Barnes, 1990; Joshi et al., 1995a; Forget et al., 1999) at altitudes spanning the bottom few scale heights to greater than 100 km (Miyoshi et al., 2011; Medvedev et al., 2011). The spectrum of GWs and their interactions and phase speeds determine the altitudes at which GW effects will be the strongest (e.g. Barnes, 1990; Hartogh et al., 2005; Medvedev et al., 2011). The study of the forcing of the large-scale atmospheric flow by GWs is still in its relatively early stages, in large part because of the lack of good observations of GWs in the atmosphere of Mars. In addition, it is very challenging to determine the actual role of GWs in the global circulation if the thermal forcing due to aerosol heating processes is not sufficiently well determined in the models.

9.3.4 Summary

The atmosphere of Mars is very strongly forced thermally, and this forcing is primarily radiative as a consequence of the low density of the atmosphere and its mostly CO_2 composition. The bulk of the thermal energy transferred from the ground to the atmosphere is carried by IR wavelength photons; the sensible heat transfer is a very secondary (though not unimportant) component of this forcing. The atmosphere is also relatively strongly forced directly, as a result of the absorption of solar radiation by the ever-present atmospheric dust. The basic thermal structure of the atmosphere, and the global circulation, changes dramatically when the atmospheric dust loading becomes substantial. This occurs every year during the dust storm season, which ranges from $L_s \sim 180°$ to $L_s \sim 320°$. The highly eccentric orbit of Mars plays an extremely important role in the thermal forcing of the atmosphere, with the dust storm season being centered on perihelion at $L_s \sim 250°$. The current timing of perihelion enhances the basic hemispheric asymmetry in the seasonal forcing, as a consequence of the topographic dichotomy. The substantially higher elevations in the southern hemisphere lead to a stronger thermal forcing of the mean circulation during the dust storm season (in northern autumn and winter), for a given atmospheric dust loading. Conversely, in the aphelion season ($L_s \sim 0–140°$), the topographic dichotomy acts to weaken the mean circulation even further.

The lack of oceans on Mars causes a very different forcing of the circulation in the summer hemisphere. Atmospheric temperatures actually increase in latitude towards the summer pole, this increase being ended at relatively low levels by the presence of the very cold permanent polar caps at very high latitudes. There is a strongly baroclinic region in the summer hemispheres, but it is located very close to the permanent ice caps. The consequence of this is that most of the summer hemisphere does not experience much if any transient weather activity during a major portion of late spring and summer.

The condensation flow represents a unique atmospheric forcing for Mars in comparison with Earth. This primarily affects the atmospheric thermal structure, constraining the wintertime higher-latitude temperatures to remain at or above the CO_2 frost point. The "direct" impact of the condensation flow on the momentum balance of the atmosphere is only of major significance for the low-level winds near the edges of the seasonal polar caps. Simulations with global models that do not have a CO_2 cycle, but which constrain temperatures to remain at or above the frost point everywhere, demonstrate this, as they produce thermal fields and winds that are very similar to those in GCMs that have full CO_2 cycles.

9.4 THE MEAN CIRCULATION

9.4.1 Fundamental Aspects

The mean circulation describes the large-scale Martian atmospheric thermal state, the winds, and the pressure distribution when averaged over time periods that are long compared to typical diurnal or day-to-day weather variations (meaning more than about 20 days for Mars), but relatively short compared to seasonal changes tied to the position of Mars in its orbit (meaning less than ~100 days). The mean circulation is very often averaged in longitude: zonal averaging. This traditional approach (Peixoto and Oort, 1992) allows one to examine the

basic structure of the temperature field and the winds within the latitude–height plane or a latitude–pressure plane. Zonal averaging has the benefit of allowing the entire atmosphere to be visualized in two-dimensional sections representative of a suitable time interval. Such maps should be interpreted with caution, however, since (a) they may give the impression that the circulation is actually two-dimensional, which it most certainly is not, and (b) the zonal averaging can lead to features that are mostly or even entirely artifacts of the averaging.

In the case of Mars, there are longitudinally varying features of the time-mean circulation that are very closely linked to the zonal-mean circulation. In particular, the Hadley circulation is the dominant structure that is present in the zonally averaged circulation on Mars, and the lower branch of this circulation can exhibit a large amount of zonal asymmetry. This asymmetry takes the basic form of western boundary currents, concentrated near the western edges of lowland regions. Such circulations exist in the terrestrial atmosphere, but they are far more prominent and global in extent on Mars because of the structure of the very large topography. We choose to consider the western boundary currents as being fundamentally a part of the mean circulation rather than to view them as forced stationary eddy (longitudinally varying) circulations. There are strong forced eddy circulations in the atmosphere of Mars, as there are on Earth; these are discussed in Section 9.7.

9.4.2 Basic Theory

If we consider the mean circulation in terms of longitudinal averages of the wind, temperature, and pressure fields, then there are some useful theoretical concepts that provide insight into its structure and variability. These concepts originate from basic principles that mass, angular momentum, and heat are all conserved. It is through these principles that the wind, temperature, and pressure fields are linked. Zurek et al. (1992) reviewed these concepts in some detail, but we briefly summarize them here.

We begin with the concept of geostrophic wind balance. This concept, valid for rapidly rotating planets like Mars, greatly simplifies the momentum equation by assuming that the dominant terms – the pressure gradient and Coriolis terms – are balanced. If $\bar{u}_g$ represents the zonal-mean geostrophic wind (positive for winds blowing from west to east, the overbar denoting a zonal average), it can be shown that

$$\bar{u}_g = -\frac{1}{f}\frac{\partial \Phi}{\partial y} \tag{9.3}$$

where f is the Coriolis parameter, Φ is the zonal-mean geopotential on pressure surfaces, and y is distance in latitude (e.g. Holton and Hakim, 2013). For a hydrostatic atmosphere, the vertical variation of this wind can be shown to be

$$\frac{\partial \bar{u}_g}{\partial \ln p} = -\frac{R}{f}\frac{\partial \bar{T}}{\partial y} \tag{9.4}$$

where R is the gas constant and p is pressure. This is the thermal wind equation and it states that the zonal-mean geostrophic wind increases with height in regions where zonal-mean temperatures decrease with latitude. This is precisely the situation in middle and high latitudes during wintertime, and the thermal wind equation requires the existence of strong westerly jet streams aloft. Furthermore, if the temperature and surface pressure fields (i.e. the surface geostrophic winds) are known, then (9.4) can be vertically integrated to yield the zonal-mean geostrophic wind field. For Mars, the surface pressure field is not known well enough to determine the surface geostrophic winds, so an assumption must be made. Surface winds are generally assumed to be negligible, which is a decent approximation but clearly introduces some uncertainty. A slightly more sophisticated approach that incorporates the centrifugal force in the balance equation leads to the concept of a gradient wind (see e.g. Holton and Hakim, 2013). Given orbital remote sensing measurements of temperature profiles, the thermal and gradient wind concepts provide a simple, yet powerful tool to estimate the mean zonal winds. Their main limitation is that they are not valid near the equator where the Coriolis parameter goes to zero, and for Mars we do not have the needed lower boundary condition because of the lack of accurate surface pressure data.

While zonal-mean zonal winds can be estimated from temperature data, zonal-mean meridional winds cannot because by definition there is no zonal-mean geostrophic meridional wind. (The zonal-mean geostrophic meridional wind would be proportional to the longitudinal gradient of the zonal-mean geopotential or pressure field, but this vanishes by definition of the zonal mean.) Instead, to understand the mean meridional circulation, we need to consider the processes that drive it: thermal forcing due to diabatic heating (e.g. radiation), planetary eddy heat and momentum flux convergences (due to transient eddies, stationary waves, and thermal tides), and mechanical forcing from "frictional" processes (e.g. gravity wave drag). The mean meridional circulation is thus more complicated to understand than the zonal wind field.

The theory of steady, nearly inviscid, Hadley circulations presented by Held and Hou (1980) offers considerable insight into the key factors controlling the mean meridional circulation. This theory assumes that internal friction and eddy forcing are negligible, and that zonal-mean radiative heating and surface friction are the only forcing. (In the case of Mars, the neglect of eddy forcing is almost certainly a much better assumption than it is for Earth.) Furthermore, zonal winds at the surface are assumed to be negligible (but not zero) compared to those in the upper branch of the Hadley cell. Given these assumptions, poleward motion in the upper branch will conserve the absolute angular momentum acquired at the surface in the ascending branch, while zonal winds in the lower branch will have a latitudinal distribution that exerts no net torque on the surface. Because the westerly winds in the upper branch will increase with latitude, thermal wind balance requires an equator-to-pole temperature gradient. However, the increase is limited since, if the circulation extended all the way to the pole, the westerly zonal flow would become infinite and thermal wind balance could no longer be maintained. The meridional extent of the Hadley circulation is thus limited by the ability of diabatic heating mechanisms to produce temperature fields that are in thermal wind balance with the zonal wind fields.

Specifically, the magnitude and latitudinal variation of the heating fields are of crucial importance. The magnitude determines the overall strength of the Hadley circulation since adiabatic cooling and heating in the rising and sinking branches balance these fields; stronger heating and cooling leads to stronger rising and sinking motions and, by mass conservation, stronger poleward and equatorward flows. The latitudinal variation determines the extent of the Hadley circulation since it determines the equator-to-pole temperature gradient, the strength of the westerly winds, and therefore the poleward extent of the Hadley cell. The depth of the Hadley circulation also plays a critical role since deeper circulations produce stronger upper-level zonal winds for a given temperature field. The zonal wind field of a steady, nearly inviscid, zonally symmetric Hadley cell is controlled by angular momentum conservation, while the intensity and extent of its meridional circulation are controlled by the diabatic heating fields.

The nearly inviscid, zonally symmetric theory outlined above is particularly relevant for Mars because radiative heating dominates the forcing. It is, however, important to consider what this theory predicts for off-equatorial heating and hemispherically asymmetric topography, two features that are much more prevalent for Mars than Earth. Schneider (1983) applied this theory to Mars for cases where the latitude of maximum solar heating was not at the equator. He found that the circulation rapidly transitions to a single dominant cross-equatorial Hadley circulation cell whose intensity at solstice is much stronger than at equinox even with the same magnitude of heating. Lindzen and Hou (1988) obtained a similar result for Earth, though in their simulations two cells were always present. For Mars, GCMs consistently predict a single strong cross-equatorial Hadley circulation except close to the equinox seasons. Interestingly, however, the models show seasonal asymmetries in the Hadley circulation that are related to the mean elevation differences between the northern and southern hemispheres (Richardson and Wilson, 2002; Takahashi et al., 2003). Zalucha et al. (2010) extended the nearly inviscid theory to include north–south sloping, zonal-mean topography as exists on Mars. They found that the southern highlands act as elevated heat sources when PBL convection is included in the heating terms and that this is part of the explanation for why the solstice Hadley circulation during southern summer is stronger than that at the opposite solstice season. Thus, the nearly inviscid, zonally symmetric Hadley circulation theory provides a very useful framework for understanding the basic nature of the zonal-mean circulation on Mars.

9.4.3 Observations

Almost all of our observations of the mean circulation in the Mars atmosphere are of temperatures obtained from remote sensing (see Table 9.1). There are very few actual wind observations and we lack enough surface pressure data to construct maps of the pressure or height fields (of constant pressure surfaces). This is the enormous difference between Mars and Earth, where we have very extensive daily observations of the winds and the surface pressures in addition to the temperatures. In some ways the current observational situation for the Mars atmosphere is more similar to that for the oceans on Earth. As discussed in Section 9.2, it is mostly through the extensive use of global circulation models that we have been able to learn about the actual circulation – the winds – in the Mars atmosphere.

Given the basic dynamical scaling that governs large-scale motions on Mars, we can be rather confident (generally) that if the models are able to simulate the observed thermal fields then they will also be simulating the correct wind fields, at least outside of the tropics. In the tropics, the link between the wind and temperature fields is generally not as good as it is outside the tropics.

The mean circulation of the Martian atmosphere varies greatly with the forcing. As discussed in the previous section, the atmospheric forcing varies strongly with season and distance from the Sun. It also varies dramatically with dust loading. At present the "dust storm season" on Mars largely coincides with the period during which Mars is relatively close to perihelion in its orbit, spanning the range defined by $L_s \sim 180$–$340°$. This is certainly not a coincidence, since the circulation intensifies greatly with the much stronger solar forcing during this period. The unusual latitudinal variation of the zonally symmetric component of the topography on Mars (as discussed in Section 9.3) acts to further enhance the asymmetry of the solar forcing and the dust loading, making the mean circulation during southern summer–northern winter stronger than it would otherwise be. It has recently become clear that the radiative effects of the water ice clouds in the atmosphere are much more important for the thermal forcing than previously thought, so these clouds represent a second atmospheric aerosol that is important for the forcing of the mean circulation. CO_2 clouds constitute a third atmospheric aerosol that is also of at least some importance for the forcing of the mean circulation.

9.4.3.1 Thermal Observations

Conrath (1981) was among the first to determine the zonal-mean thermal structure of the Martian atmosphere from spacecraft observations. He analyzed a subset of the 20 000 retrievals of temperature profiles from the Mariner 9 Infrared Interferometer Spectrometer (IRIS) instrument to obtain latitude–height maps of zonal-mean temperatures and zonal winds (from the geostrophic thermal wind equation) for a 40 sol period around $L_s = 330$–$350°$ corresponding to northern hemisphere winter. He also attempted to identify planetary wave features in this dataset, although this was hampered by ambiguities between stationary and propagating features because of the sparse sampling. A portion of this dataset was later analyzed using a more sophisticated retrieval method by Santee and Crisp (1993, 1995) to obtain more or less complete maps of temperature and dust opacity over a 10 sol period ($L_s = 343$–$348°$). These were used by Santee and Crisp (1995) to obtain estimates of gradient zonal winds and eddy fluxes that compare fairly well with more recent detailed measurements and simulations. The Viking Orbiter IRTM obtained further measurements of the global thermal state of the atmosphere, limited to essentially one vertical level centered at ~25 km altitude; the interpretations of these observations are summarized in Zurek et al. (1992). Unfortunately, the subsequent discovery of surface radiation contamination in the atmospheric thermal channel required some revisions of the IRTM observations (Wilson and Richardson, 2000).

A much superior dataset was obtained by the TES experiment on the global mapping MGS Orbiter that operated more or less continuously between 1999 and 2004. Retrievals of temperature profiles as well as column dust and water ice opacities (e.g. Conrath et al., 2000; Smith, 2004) enabled the assembly of a detailed climatology of the seasonally varying, zonal-mean thermal structure and its forcing over the complete annual cycle for the first time (e.g. Smith et al., 2001a,b, 2002; Smith, 2004). More recently, these measurements have been significantly extended in temporal and altitude coverage, and vertical resolution, by the MCS experiment on the global mapping MRO spacecraft (e.g. McCleese et al., 2010). The TES and MCS temperature observations are discussed in Chapter 4. Other sources of temperature measurements include extensive radio occultations from MGS and Mars Express (e.g. Hinson et al., 1999; Pätzold et al., 2004) that provide high-precision and vertical resolution measurements of vertical temperature profiles, though these have very sparse and highly intermittent spatial coverage.

Monthly zonal-mean, average temperature fields from the MCS dataset for Mars Year 30–31 (a non-dust storm year) are shown in Figure 9.7. Near the equinoxes the thermal structure is close to being symmetric about the equator. In the tropics, temperatures decrease with height, whereas polar inversions are readily apparent at the higher latitudes. At the solstices, temperatures generally increase toward the poles in the summer hemispheres – in sharp contrast to the situation on Earth. In the winter hemispheres temperatures can be seen to decrease towards the pole, except at fairly high altitudes. There the temperatures can be seen to reach a maximum at or close to the pole. This polar warming must be produced dynamically, since there is no insolation available to produce it. It should be noted that the northern winter polar warming is stronger than that in southern winter. At lower altitudes the winter polar temperatures are very cold. The dynamically produced, poleward and upward-sloping "tongue" of warm air at high latitudes seen near the solstices in the winter hemisphere is a very basic feature of the atmospheric zonal-mean thermal structure on Mars, as is the poleward tilt with height of the strong latitudinal temperature gradients associated with it.

9.4.3.2 Wind Observations

Early attempts to deduce the Martian mean circulation relied on telescopic observations from Earth, which were of low spatial resolution and were difficult to interpret. Hess (1950) made a very early attempt to map the Martian atmosphere during northern hemisphere summer, based on estimates of surface temperatures and a few cloud-tracked winds observed during oppositions in 1894, 1896 and 1924.

As already noted, the few actual wind measurements that we have come from the landed spacecraft of the Viking, Pathfinder, Phoenix, and MSL missions (see Hess et al., 1977; Schofield et al., 1997; Sullivan et al., 2000; Gunnlaugsson et al., 2008; Gómez-Elvira et al., 2012), cloud tracking studies from orbiters (Kahn, 1983; Wang and Ingersoll, 2003; Määttänen et al., 2010; McConnochie et al., 2010), and Doppler wind measurements using Earth-based telescopes (e.g. Lellouch et al., 1991; Sonnabend et al., 2006, 2012; Cavalié et al., 2008; Moreno et al., 2009). While these datasets provide useful constraints, they have very significant limitations in the context of the zonal-mean circulation.

The landed wind measurements can be difficult to interpret in terms of the mean circulation because of the control of near-surface winds by local topography. The crater environment of the Curiosity Rover is a good example. Accommodation issues (e.g. rover interference on MSL), implementation ambiguities (e.g. wind speeds from the Pathfinder wind sensor), and general instrument limitations (e.g. the Pathfinder wind sock and the Phoenix telltale sensor) further complicate the interpretation. Nevertheless, there have been a few instances where surface winds may be reflecting larger-scale wind systems. Lewis et al. (1999), for example, successfully reproduced the observed diurnal variation of wind directions at the Pathfinder landing site with a GCM, which suggests that the winds observed there were controlled by a circulation system on a scale at least as large as their GCM grid box (~225×225 km at the equator). Also, the Phoenix telltale wind sensor detected a shift toward more westerly winds at the end of the mission that probably indicated a strengthening polar vortex (Holstein-Rathlou et al., 2014). The change of the surface winds to northeasterlies at Viking Lander 2 following the onset of the 1977B global dust storm is consistent with a poleward shift of the sinking branch of the Hadley circulation (Haberle et al., 1982). Aside from these few cases at specific locations and times, however, measured surface winds from landers shed little additional direct light on the nature of the global mean circulation.

A similar situation exists for winds derived from cloud tracking and Doppler techniques. While these winds are less influenced by local topography, the datasets are restricted to specific locations, seasons, and times of day, and as such they are difficult to use for constructing zonal means. Wang and Ingersoll (2003) estimate zonal means from cloud tracking data using MGS imaging observations of the north and south polar regions during selected seasons, but the altitudes are uncertain, a coverage bias remains, and the local time is restricted to near 2 p.m. The Doppler data also have significant spatial and seasonal coverage limitations, being mostly limited to the upper atmosphere at times of opposition.

Proxies for the surface expression of zonal wind systems can be found in surface streaks and dune orientations (e.g. Thomas et al., 1981; Greeley et al., 1993; Geissler, 2005). The so-called Type 1b bright streaks, which develop after major dust storms, are suggestive of a Hadley circulation (Magalhães, 1987), and the vast dune fields surrounding the north residual polar cap suggest that predominantly zonal flows are responsible for their movement (Hayward et al., 2014).

By far, however, the global structure of the zonal-mean circulation is best revealed by temperature data obtained from orbiting spacecraft. MGS and MRO, in particular, have Sun-synchronous, global mapping orbits that cover the atmosphere on a daily basis. (These orbits, however, only yield observations at two local times of day; with a side-scanning IR instrument (MCS), somewhat greater local time coverage has been obtained by MRO.) From these data, zonal-mean zonal winds can be deduced from the zonal-mean thermal structure by assuming gradient wind balance and a near-surface boundary condition. This is typically taken to be a vanishing wind field. An example of such wind fields – derived from MCS data – is shown in McCleese et al. (2010). At the equinoxes, strong westerly jets exist in each hemisphere and the circulation is nearly symmetric about the equator. This symmetry breaks down

rather rapidly away from the equinox seasons, and closer to the solstices strong westerlies characterize the winter hemisphere and easterlies prevail in the summer hemisphere.

The meridional component of the zonal-mean circulation, however, is not a balanced wind field and cannot therefore be approximated from the temperature field. Instead, the mean meridional circulation, which is strongly influenced by frictional processes, eddy mixing, and diabatic heating, is best determined from models, as is illustrated in subsequent sections. However, if the temperature and dust fields are known simultaneously from observations, then radiative heating rates can be calculated, from which the diabatic circulation (see e.g. Holton and Hakim, 2013) can be determined. The diabatic circulation is a good approximation for the mean meridional circulation for Mars because the eddy forcing of the latter is relatively weak, as discussed later in this section. Santee and Crisp (1995) demonstrated this method using Mariner 9 IRIS data and found a vigorous two-cell circulation existed during late northern winter. This is generally consistent with the modeling results discussed in the following sections.

9.4.3.3 Surface Pressure Observations

While the MSL Curiosity Rover now has over two Mars years of pressure data, the most complete observations of the annual pressure cycle are still those obtained by the two Viking Landers in the late 1970s and early 1980s (e.g. Tillman, 1988; Tillman et al., 1993; see Figure 9.1). The difference in pressure between the two landers (both in the northern hemisphere, but separated by over 25° latitude) largely reflects their difference in elevation relative to the mean aeroid. Seasonal variations come from the condensation of atmospheric carbon dioxide on the fall–winter polar cap and sublimation from the spring–summer polar cap, but there is also a component of the pressure gradient arising from the large-scale mean circulation (a latitudinal pressure gradient related to the zonal-mean wind in the winter hemisphere in particular) and a thermal effect as the result of the redistribution of mass between the hemispheres associated with expansion in the summer hemisphere (Hourdin et al., 1995). This latter effect explains the different trajectories of the daily-averaged pressures at MSL and VL2 in spite of their nearly identical elevations (Haberle et al., 2014). To simulate the Viking and MSL pressure curves well thus requires a good simulation of the mean circulation and the atmospheric thermal structure, though the surface pressure in the Mars GCMs is typical tuned to the observations over the course of a Martian year – though only at several model grid points.

9.4.4 Modeling

In recent years there has been a proliferation of GCMs that have been developed to study the Martian atmosphere, driven by the dramatic increase in observational data from the various spacecraft missions. The NASA Ames Mars GCM has the oldest continuous heritage for a current Mars GCM, dating back to the early 1970s. Examples of model data analyses done using previous versions of the Ames GCM can be found in numerous publications (e.g. Haberle et al., 1993b; Barnes et al., 1993, 1996a,b). Examples of seasonal zonal-mean states for Mars' global circulation from other GCMs can also be

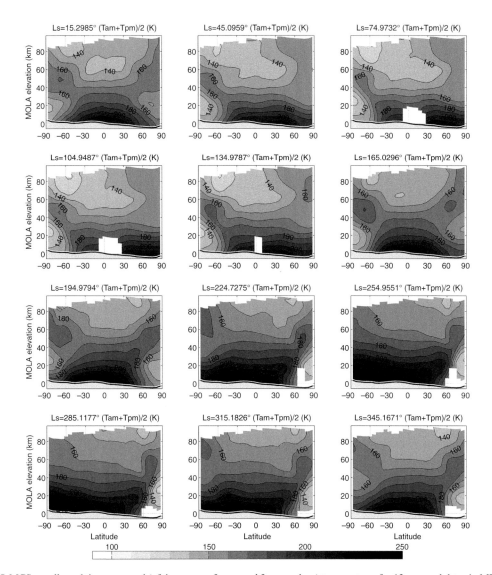

Figure 9.7. MRO MCS zonally and time-averaged (of the average 3 p.m. and 3 a.m. values) temperatures for 12 seasonal dates in MY 30 and 31. The Ls=15.2985° panel is for MY 31, all others are for MY30. White regions have no data. The MCS Science Team provided access to the latest data.

found in the literature (e.g. Wilson and Hamilton, 1996; Forget et al., 1999; Read and Lewis, 2004; Moudden and McConnell, 2005; Medvedev and Hartogh, 2007; Richardson et al., 2007). The Mars Climate Database is discussed in Chapter 4; data produced by the LMD (Laboratoire de Météorologie Dynamique) GCM can be freely sampled using this tool.

In the following section we show results from an assimilation of TES data into a GCM to illustrate the Martian seasonal cycle in more detail. The GCM in this case is the UK spectral version of the joint European LMD Mars GCM (see Forget et al. (1999), though there have been subsequent updates to many of the physical schemes). This particular model uses a reasonably comprehensive set of physical parameterizations appropriate to the neutral lower atmosphere, primarily below about 80–100 km altitude, although the model typically extends several scale heights above this. As such, it provides an adequately realistic representation of the global circulation on Mars from the surface up to levels at ~10^{-1} Pa. TES thermal observations and, crucially, total dust opacity observations are assimilated into the model using the approach described in Lewis et al. (2007) for a period of one Mars year beginning in MY 24 and ending

in early MY 25 (at $L_s = 140°$). This period is chosen as an almost repeatable year with no planet-encircling dust event; a global storm occurred later in MY 25 (after $L_s = 180°$). It should be noted that the mean circulation is most sensitive to the total dust opacity in the model, which is adjusted towards observed values where and when they are available. The results would be broadly similar in a "free-running" GCM with a similar prescribed dust scenario, but the temperatures in the assimilation are closer to the observations because of the assimilation process.

9.4.4.1 Overturning Circulations

(a) **Basic Structure** Figures 9.8–9.10 illustrate the Mars monthly variation of the zonal-mean temperatures, zonal winds, and the mean meridional, mass transport streamfunctions from the assimilation. The temperature plots may be directly compared with thermal observations (e.g. Figure 9.7). In general, the warmest part of the lower atmosphere tends to follow the subsolar latitude, but, unlike the Earth, the warmest region moves to very high latitudes in the summer hemisphere owing

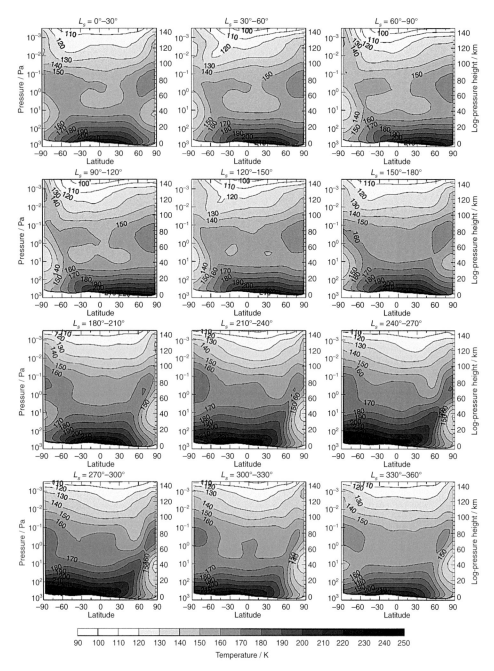

Figure 9.8. Zonally averaged temperatures averaged for 12 periods of 30° of areocentric longitude (adapted from Lewis et al., 1999) spanning the Martian year. The temperatures are from an assimilation using a version of the UK Mars GCM and MGS TES nadir temperature and dust opacity data. It can be noted that these temperature data only extend to ~40 km altitude. The TES data are from MY 24 and MY 25, beginning at $L_s \sim 142°$. The plots were made using a log-pressure (left-hand axis, in Pa) coordinate, with a pseudo-height scale (right-hand axis, in km) provided for approximate guidance only.

to the relatively low surface thermal inertia of the planet and the absence of oceans. The zonal winds are approximately in thermal (gradient) wind balance with these temperature fields, with very strong westerly jets in both hemispheres in middle and high latitudes during wintertime, and easterlies high over the equator and in the summer hemisphere. The strongest zonal-mean winds occur at high latitudes in northern hemisphere winter, when Mars' elliptical orbit takes it closest to the Sun and the thermal gradient from summer to winter hemispheres is the most extreme. At this time there are easterlies

throughout the southern (summer) hemisphere, except close to the surface. The mean meridional mass streamfunction reveals a fairly Earth-like situation at the equinoxes, with rising flow at the equator and descending flow at about 30° latitude away from the equator, quite reminiscent of the classic Hadley cells on Earth. In contrast to Earth's troposphere, however, at the solstices the mean meridional circulation is dominated by a single, equator-crossing Hadley cell with a width of at least 90° of latitude, with the strongest cell occurring close to northern winter solstice and perihelion. It is worth noting that the

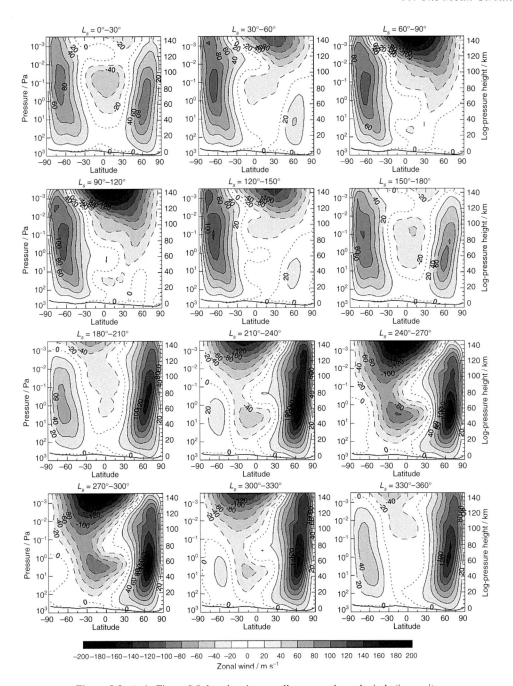

Figure 9.9. As in Figure 9.8, but showing zonally averaged zonal winds (in m s⁻¹).

mean meridional circulation on Mars is not closed as it is on Earth because there is a net source of atmospheric mass at the subliming polar cap and a net sink at the condensing cap. This flow is, however, relatively weak (typically ~1 m s⁻¹ or less; see Read and Lewis, 2004) compared to that in the bulk of the atmosphere.

The Martian Hadley cells are largely confined to "tropospheric" levels below 10 Pa, though there are extensions to "stratospheric" levels at some latitudes, especially around 10–20° on either side of the equator. The near-equinox Hadley cells are not completely symmetric about the equator, as the rising branch is generally centered in the southern hemisphere and extends further poleward in the $L_s = 180$–$210°$

time period than it does at $L_s = 0$–$30°$ (when the rising branch is in the northern hemisphere). This reflects the asymmetry in zonal-mean topography between the northern and southern hemispheres, the stronger thermal forcing at $L_s = 180$–$210°$ as a result of being closer to perihelion, and the greater dust loading at this season. In addition to the Hadley cells, there are Ferrel cells at higher latitudes – but again there are significant differences between the hemispheres. These are most prominent in the northern hemisphere from autumn through spring and are weak in the southern hemisphere. Thermally indirect (with cold air rising and warm air sinking) Ferrel cells are generally associated with the presence of baroclinic eddies that modify the transport of heat and momentum at

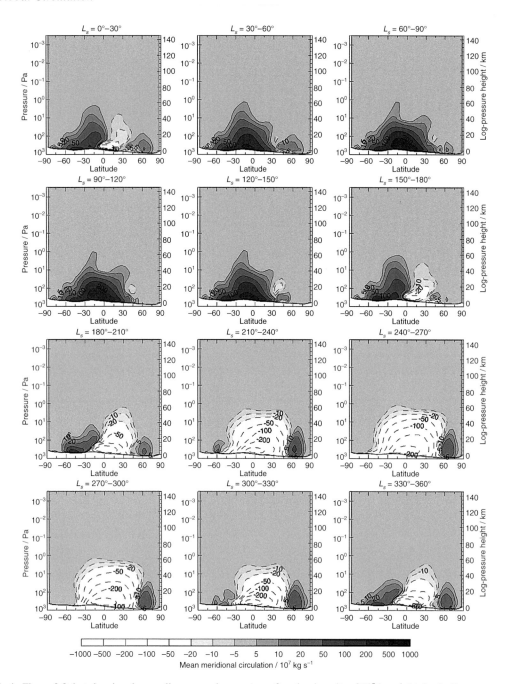

Figure 9.10. As in Figure 9.8, but showing the zonally averaged mass streamfunction in units of 10^7 kg s^{-1}. Light shading represents clockwise circulation and dark shading denotes counterclockwise circulation. Small values of the streamfunction are not contoured or colored, and thus the apparent depth and extent of the circulation cells tend to be smaller than is the case when small values are contoured or colored.

these latitudes (see e.g. Holton and Hakim, 2013). The difference in strength and latitudinal extent of the northern and southern Ferrel cells is indicative of differences in the intensity of baroclinic eddy activity and meridional eddy heat and momentum transports between the north and south. There is generally stronger transient eddy activity in the north, as is discussed in Section 9.6.

The extratropical, winter, westerly jets have peak wind speeds of ~160 m s^{-1} in the northern hemisphere and ~100–120 m s^{-1} in the southern hemisphere. They are located at around 65° latitude in each hemisphere, but maximize at a somewhat

lower level in the north (1–10 Pa) as compared to the south (0.1–1 Pa). Both of the westerly winter jets show a tendency to slope upwards towards the poles, though this is somewhat stronger in the northern hemisphere. In the tropics, westward zonal winds (easterlies) of up to 50 m s^{-1} are found in this assimilation. This is partly because the tropical zonal winds are not generally in thermal (gradient) wind balance, but instead result from a more complex dynamical balance, most likely including the effects of non-axisymmetric circulations such as the thermal tides. In the absence of strong non-axisymmetric disturbances, however, the effect of advection by axisymmetric

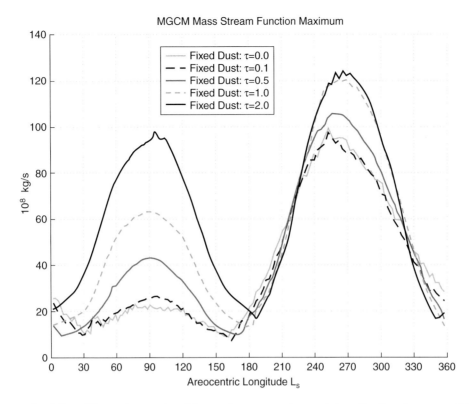

Figure 9.11. The seasonal variation of the maximum value of the zonal-mean, mass transport streamfunction, for a range of dust visible optical depths, as simulated by a GCM. The maximum value typically is located at ~15 km above the surface in the tropics. Adapted from Basu et al. (2006).

transport will generally produce a tendency for tropical east-erly winds (e.g. Held and Hou, 1980), as found at most heights over the equator in Figure 9.9. Some westerly equatorial flow – "superrotation" – is evident in Figure 9.9, indicating that eddy forcing must be playing a role. Hide's theorem (see e.g. Held and Hou, 1980) requires eddy forcing to drive superrotation (in particular, zonal-mean westerly winds at the equator). This forcing appears to be primarily due to the strong thermal tides in the Martian atmosphere.

(b) Seasonal Variability Figures 9.8–9.10 show that the zonal-mean circulation changes very substantially during the year as the seasonal cycle progresses on Mars. Most of the year the zonal-mean circulation is strongly biased towards one hemisphere or the other. In the temperature field, this is reflected in large latitu-dinal excursions of the lower atmospheric temperature maxima almost from one pole to the other, with corresponding changes in the high-altitude polar warmings as the seasons change. At low levels the equator-to-pole temperature contrast almost disappears during summer, actually tending to reverse (such that high lat-itudes are warmer than the equator) in the vicinity of summer solstice. At middle levels (~10 Pa, or ~35–40 km) the cold region over the pole completely disappears near summer solstice as rel-atively direct, continuous insolation warms the atmosphere. This occurs because of the lack of oceans on Mars. At the highest lat-itudes, permanent polar caps are present in both hemispheres – composed of water ice in the north and CO_2 ice in the south.

Zonal winds reflect a (gradient) thermal wind balance at extratropical latitudes, consistent with the thermal fields in Figure 9.7. This leads to easterly (westward) winds in the extra-tropics during summer. Near the equator, the westward flow increases in strength, especially at high altitudes, close to both solstices. Low-level westerly (eastward) flow near the equator, however, tends to be particularly evident in periods close to the equinoxes, when the thermal tides are the strongest close to the equator (e.g. Zurek, 1986; Lewis and Read, 2003).

Perhaps the most remarkable seasonal changes are seen in the variation of the mean meridional circulation, as shown in Figure 9.10. It is apparent that, except close to the times of the equinoxes, the mean meridional circulation is almost entirely dominated by one or the other of the two Hadley cells, depend-ing on the season. The dominant Hadley cell is seen to grow rapidly after each equinox in strength and latitudinal extent, spanning the equator to produce an inter-hemispheric transport circulation that connects the summer and winter hemispheres. This is strongly reminiscent of the behavior found by Lindzen and Hou (1988) for axisymmetric circulations when the subso-lar latitude is moved north or south of the equator.

The mean meridional circulation varies quite strongly in magnitude, as well as in its height and latitude range, with time of year and with the atmospheric dust loading. This is illustrated in Figure 9.11, which shows the maximum absolute value of the mean meridional, mass transport streamfunction as a function of seasonal date for five different fixed (uniform) dust loadings (Basu et al., 2006). This can be compared with peak values at different times of the year with a more realistic varying dust loading in Figure 9.11. The strong mean meridional circulation near to both solstices is evident, as is the much greater variation in the northern hemisphere summer circulation with varying dust loading. The higher dust loadings used for the simulations shown in Figure 9.11 are inappropriate for northern summer on Mars (based on all of the observational data) and consequently the northern summer mean meridional circulation will generally be considerably weaker than that in northern winter, as it is in Figure 9.10.

(c) Transport The mean meridional circulation plays a vital role in the cross-equatorial transport of dust, water vapor, and clouds. Here we consider the role of the seasonally varying mean meridional circulation on (a) the bulk mass transport across the planet, (b) dynamical heat transports (as compared with diabatic processes), and (c) angular momentum transports and the torque balance between the atmosphere and surface.

Despite the emergence of numerous distinct Mars GCMs in the past decade or so, probably the most comprehensive analysis of the above continues to be the early study by Haberle et al. (1993b), using the NASA Ames GCM. This showed that at most seasons – except close to the equinoxes – the meridional mass and heat transports were dominated by the equator-crossing Hadley circulation, with relatively small contributions from stationary and transient eddies. Although this might indicate that eddy transports are relatively weak on Mars, it actually reflects the fact that the mean meridional overturning circulation on Mars is extremely strong compared to that on Earth. Timescales for meridional overturning were estimated by Barnes et al. (1996a) from model simulations of the transport of a passive tracer, and were generally found to be much shorter than for the Earth. These timescales vary significantly with season (Barnes et al., 1996a). Ventilation timescales (defined by the ratio of atmospheric mass to integrated mass flux) ranged from as short as a few days under very dusty northern winter solstice conditions to ~180 days at equinox, with more typical timescales of ~40–60 days under less dusty conditions at other seasons. These can be compared with a typical ventilation timescale of ~180 days for the Earth's stratosphere (e.g. Shia et al., 1989).

A seasonal modulation of the large-scale mass transport on Mars is also evident from measurements of atmospheric argon concentration obtained by the Mars Odyssey Gamma Ray Spectrometer (Sprague et al., 2004, 2007, 2012). The Ar mixing ratio at polar latitudes is found to vary due to the seasonal condensation of CO_2, which modulates the dilution of Ar as the seasonal ice caps advance and retreat. These variations manifest themselves as increased concentrations during the respective fall and winter in each hemisphere; there are very significant differences between the north and south (Sprague et al., 2012). Simulating these variations and their hemispheric asymmetry accurately, even in a statistical sense, appears to pose a significant challenge for the current Mars GCMs (see e.g. Lian et al., 2012) since these tend to predict seasonal Ar variations that are smaller by a factor of 3–4 than observed. Lian et al. (2012) suggested that most current GCMs are somewhat deficient in their representations of tracer transport at high latitudes, something that could have significant implications for the representation of other transport cycles – such as for dust and water – that are extremely important for the climate system of Mars.

A further consequence of the strength of the mean meridional overturning circulation on Mars is the impact on the distribution of atmospheric angular momentum. Barnes and Haberle (1996) examined this in the context of simulations performed with the NASA Ames model. These indicated that, at least under dusty northern winter solstice conditions, the upper-level branch of the mean meridional overturning circulation was sufficiently strong as to allow absolute angular momentum ($M = a\cos\phi(\Omega - \cos\varphi + u)$) to be approximately conserved. There was a near-elimination of the meridional gradients of both M and potential vorticity throughout low and middle latitudes in the north, as well as near-zero potential vorticity. It should be noted that the idealized, nearly inviscid Hadley circulation of Held and Hou (1980) has some of these same properties.

9.4.4.2 Condensation Flow

A unique feature of the Martian atmosphere is a cross-equatorial pole-to-pole flow that is driven by the condensation and sublimation of carbon dioxide ice at higher latitudes (see Chapter 12). Very simple calculations (Read and Lewis, 2004) indicate that this flow can contribute a vertically averaged meridional flow component of ~0.5 m s^{-1}. The Coriolis torque on this meridional flow can produce a zonal wind of ~10 m s^{-1}. Such winds are certainly not large, but they are potentially significant close to the surface. Winds of this magnitude associated with the condensation flow have been found in GCM studies (e.g. Haberle et al., 1993b). As noted above, the existence of a net transfer of mass from the subliming cap in the spring hemisphere to the condensing cap in the autumn hemisphere means that the mean meridional circulation is not "closed" as it is on Earth. In the analysis of GCM data, it is a relatively simple matter to remove the condensation flow so as to obtain the (closed) mean overturning circulation.

On Mars there is a significant atmospheric forcing that results from the removal of the primary gas from the atmosphere to the surface in winter polar latitudes. CO_2 condensation results in the deposition of CO_2 ice at the surface, in the form of either CO_2 "snow" or "frost". The removal of CO_2 gas from any atmospheric column reduces the hydrostatic pressure within the column, with a resultant horizontal pressure gradient between that column and warmer non-condensing columns at lower latitudes. The resulting horizontal pressure gradients drive the pole-to-pole mass flow, which attempts to replace the CO_2 deficit produced by condensation in the autumn/winter seasonal cap region. This mass flow is then reversed when ice sublimes during the subsequent spring season.

The Viking Lander and MSL surface pressure measurements (see Figures 9.1 and 9.2) directly evidence the seasonal variations in atmospheric mass, with minimum pressures occurring during mid- to late northern summer in association with CO_2 ice accumulation in the south seasonal cap; the north has no CO_2 ice at this season. A secondary minimum occurs during late northern winter. Maximum surface pressure occurred during late northern autumn/early northern winter (enhanced in the first Viking year by global dust storm effects), with a secondary maximum during late northern spring. It can be seen that at the near-equatorial MSL site the two pressure maxima are almost

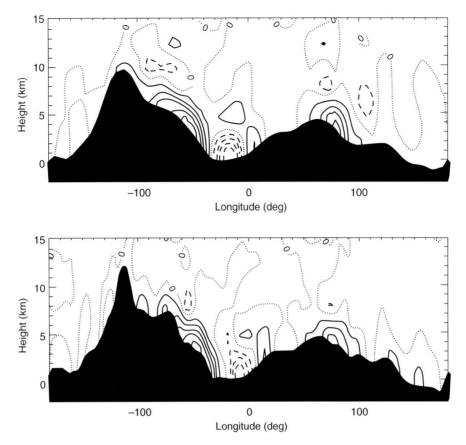

Figure 9.12. Cross-section in longitude and height of the time-averaged meridional wind at the equator for northern summer solstice conditions, in Mars GCM simulations by Joshi et al. (1994). The profile of topography at this latitude is shown in black. The top panel shows a simulation at T21 resolution (~5.6° in longitude) while the bottom panel shows the results for T42 resolution (~2.8°). The contour interval is 5 m s⁻¹, with the zero contour dotted and negative contours dashed.

equal, whereas they are very different at the middle-latitude Viking Lander 2 site.

The initial GCM atmospheric modeling for Mars by Leovy and Mintz (1969) included the thermodynamic aspects of CO_2 condensation/sublimation. Atmospheric mass loss at a rate of 1.1 Pa/day (sol) was calculated for northern winter solstice. With this mass loss confined to high northern latitudes, the resulting meridional pressure gradients force a net poleward flow toward that seasonal cap. Haberle et al. (1993b) showed that the column-integrated mass flow (in GCM simulations) toward the condensing north winter polar cap maximized at ~12 cm s⁻¹, while the maximum flow toward the southern winter cap approached 20 cm s⁻¹. Such velocities are far smaller than the maximum meridional velocity values within the thermally driven Hadley circulation, but the Coriolis torque acting upon the meridional condensation flow can drive zonal winds of ~10–15 m s⁻¹. The influence of these flows upon near-surface polar cap-edge winds can significantly affect their dust lifting ability, most strongly so for the subliming cap (Toigo et al., 2002).

Interestingly, the magnitude and the structure of the condensation flow have not really received very much attention. While GCMs are generally validated through comparisons of the model-generated surface pressure seasonal cycles with available surface pressure measurements, detailed analyses of the condensation mass flow itself have not been pursued. Such

studies, particularly when combined with constraints on the observed spatial and temporal variations of the seasonal caps, could help to provide a more complete understanding of the Martian condensation flow phenomenon.

9.4.4.3 Western Boundary Currents

The time-mean meridional circulation on Mars is not zonally uniform, since it varies strongly with longitude. This is particularly the case for the lower branch of the circulation, which can cross the equator in highly concentrated western boundary currents (WBCs) that lean against the eastward-facing slopes of the major elevated topographic features (Joshi et al., 1994, 1995b). Figure 9.12 illustrates these flows. The time-mean, north–south wind can actually reverse direction at certain longitudes, flow that is almost reminiscent of ocean basin circulations on Earth. The opposite flow at high altitudes (10–60 km) tends to be more uniform and weaker, at least for non-dusty conditions.

Large-scale, north–south-aligned topographic ridges force zonal confinement of meridional flows upon their eastern slope via vorticity conservation and Ekman pumping (Anderson, 1976; Gill, 1980; Joshi et al., 1994, 1995b). Mars possesses such topographic structures in the form of the cross-equatorial Tharsis and Syrtis ridges. On Earth, the "downward-to-the-east" continental slopes provide forcing for oceanic WBCs, while in the atmosphere the east African coast with its prominent mountainous, ridge-like structures also provides a WBC forcing. Mars possesses prominent east–west-sloping topographic ridges spanning the equator in the Tharsis and Syrtis regions (Joshi et al., 1994, 1995b; Wilson and Hamilton, 1996). These provide a forcing capable of intensifying the strong cross-equatorial flow during the solstice seasons.

The existence of strong, equator-crossing, low-level winds in lower latitudes is reflected in dust wind streak markings observed on the Martian surface (see Section 9.4.2.2; Greeley et al., 1992, 1993). There is also some evidence in the evolution of the distribution of dust (Smith, 2004, 2009), and in the longitudes at which some northern dust storms preferentially cross the equator (e.g. Wang, 2007; Hinson et al., 2012). The WBCs are also evidenced in assimilations of regional dust storms such as the Noachis storm of 1997 (Lewis et al., 2007).

While the WBC forcing is in general a time-mean phenomenon, the diurnally varying insolation and resultant thermal tides and upslope/downslope flows can interact with the WBC forcing to enhance the winds at certain local times of day (Joshi et al., 1997). Enhanced dust loading and the resultant decreased PBL depth can augment the mid-afternoon forcing of the WBC, while enhanced vertical mixing in a deeper PBL in the absence of substantial dust opacity can act to reduce the magnitude of the WBC forcing.

9.4.4.4 Eddy Forcing of the Mean Circulation

Departures from the mean circulation in the form of thermal tides, large-scale, and small-scale waves can all play a role in driving the mean circulation through their heat and momentum flux convergences. The mean meridional circulation responds to these convergences so as to maintain thermal wind balance between the zonal-mean zonal wind and temperature fields. The existence of thermally indirect Ferrel cells, for example, illustrates this concept. Eddy heat fluxes on Earth maximize near the jet stream at mid-latitudes and therefore warm (cool) the higher (lower) latitudes and reduce the equator-to-pole temperature gradient. The mean meridional circulation restores thermal wind balance by simultaneously reducing the vertical wind shear and increasing the equator-to-pole temperature gradient. This is accomplished, respectively, by westward (eastward) Coriolis torques generated from equatorward (poleward) motion at upper (lower) levels, and adiabatic heating (cooling) in the descending (ascending) branch.

A convenient way to examine the effects of eddies is through the concept of the transformed Eulerian mean (TEM) circulation (see e.g. Andrews and McIntyre, 1976; Andrews et al., 1987; Holton and Hakim, 2013). In the TEM framework, the thermodynamic effects of the eddy heat flux convergences are subtracted from the Eulerian mean (EM) meridional circulation and what remains is defined as the TEM circulation. The advantage of the TEM is that it accounts for what are often the nearly compensating effects of eddy heating and advection by the mean meridional circulation. Specifically, the vertical component of the TEM circulation is that part of the mean vertical velocity whose thermal effects are not balanced by the horizontal eddy heat flux convergences/divergences. This

component is driven by the diabatic (mostly radiative) heating and can therefore be calculated if the temperature and opacity fields are known (e.g. Santee and Crisp, 1993). (The meridional component of the TEM circulation is forced by eddy heat and momentum fluxes in the mean zonal momentum equation.) The TEM circulation tends to be a reasonably good approximation to the Lagrangian circulation and is therefore a good indicator of the mass transport. In general, if the eddy forcing is weak compared to the mean circulation, the TEM will closely resemble the Eulerian mean circulation; if it is strong, the two circulations can differ very considerably.

For Mars, there is a rather close resemblance of the Eulerian and TEM circulation patterns at some seasons and altitudes. Haberle et al. (1993b) first pointed this out for the northern winter solstice season (see their figures 8 and 44). This would appear to indicate that eddies have relatively little impact on mean atmospheric transport (they still act to produce mixing), at least at altitudes above ~10–15 km. Near the surface, some differences are apparent, indicating that transient eddies and tides do have some effect, although this is generally much smaller relative to the meridional overturning circulation than is the case in the Earth's troposphere and stratosphere (see e.g. Andrews et al., 1987).

The mass streamfunction for the TEM and EM circulations, for northern fall equinox, from a UK data assimilation reanalysis of MGS TES observations, is shown in Figure 9.13. It can be seen that a reversed (and thermally indirect) Ferrel cell circulation is present in the north, at lower altitudes. This is driven by the transient eddies at this season; the transient eddies in the south at this season are too weak to drive any real Ferrel cell there. In the TEM circulation the Ferrel cell in the north is essentially not present, and the Hadley circulation extends further poleward at lower levels. This is at least somewhat reminiscent of Earth's troposphere and middle atmosphere, where generally the transient and stationary eddies (and gravity waves) exert a very strong influence on the Eulerian mean circulation (see e.g. Peixoto and Oort, 1992; Holton and Hakim, 2013; Mitchell et al., 2015).

Nearer the solstice seasons on Mars, the EM and TEM circulations are considerably more similar than in Figure 9.13 (see e.g. Haberle et al., 1993b; Mitchell et al., 2015). Near southern winter solstice, the transient eddies in the south are very weak, such that there may not even be a real Ferrel cell in the EM circulation (see Figure 9.10). The Hadley cell is also stronger than it is near the equinoxes. Near northern winter solstice, the Hadley cell is at its strongest and the transient eddies also are relatively weakened – at lower levels (this is referred to as the solstitial pause in both hemispheres, and will be discussed in Section 9.6). At these seasons the EM and TEM circulations tend to be very similar. The EM circulation reaches extreme strength when the atmosphere becomes very dusty during northern wintertime. The Hadley cell then can become almost global in extent at higher levels in the atmosphere (e.g. Wilson, 1997; Forget et al., 1999). Strong meridional flow at high altitudes towards the north then produces sinking motions and adiabatic warming in the northern (winter) polar region. The EM circulation is extremely strong in this situation, but eddies can still be very important in enabling stronger polar warming to occur, at relatively lower levels. In the modeling studies that

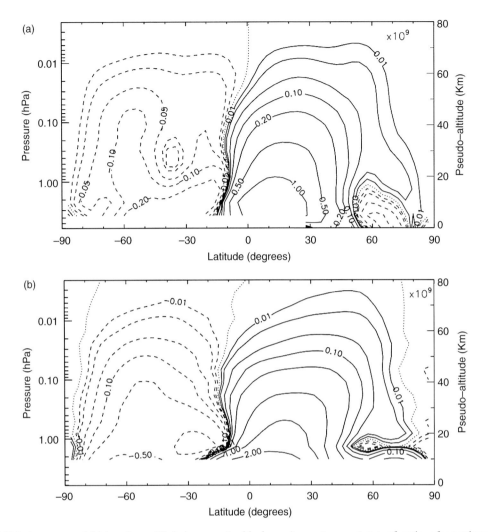

Figure 9.13. (a) Eulerian mean and (b) transformed Eulerian mean (residual mean) mass transport streamfunctions for northern fall equinox. These were produced from an assimilation reanalysis of TES temperature and dust opacity data performed with the UK Mars GCM. The assimilation data span the seasonal interval of $L_s = 165$–$195°$. The solid contours correspond to clockwise circulation and the dashed contours to counterclockwise circulation; the units are 10^9 kg s^{-1}.

have been done of this extreme case (essentially a global dust storm event), it appears that thermal tides play a major role in producing a stronger and lower polar warming (Wilson, 1997; Forget et al., 1999; Kuroda et al., 2009). Planetary waves (transient and/or stationary) and gravity waves may also provide significant forcing of the EM circulation. Tides in the south appear to act to enable dust transport into the southern polar region, something that has been observed to occur in at least some global dust storms (Wilson, 1997).

For the most part, the observed zonal-mean thermal structure does not depart greatly from that produced by the effects of radiative heating/cooling and the mean meridional circulation (see Section 9.3.2.1). However, not all aspects of the thermal structure are consistent with this picture. Following the onset of the 1977B global dust storm around $L_s \sim 270°$, for example, north polar air temperatures from Viking IRTM observations at ~25 km were seen to increase to values well above those expected for radiative equilibrium (e.g. Martin and Kieffer, 1979). Haberle et al. (1982) found that dynamical heating from adiabatic warming in the sinking branch of a dust-heated and greatly expanded

zonally symmetric circulation could explain some but not all the warming, leaving open the possibility of a very substantial role for eddy forcing. Thermal tides are the planetary waves most capable of providing strong eddy forcing under very dusty conditions, and Zurek and Haberle (1988) studied this possibility. They showed that the eddy momentum flux divergences from the classical tides during dusty conditions at this season could provide enough deceleration of the zonal-mean flow to drive a stronger mean meridional circulation. In their simulations the tidal forcing moved the descending branch further poleward, but not far enough to explain the observed polar warming. It now appears that the depth of the domain in these simulations artificially suppressed the expansion of the mean circulation, and prevented the development of an angular-momentum-conserving, pole-to-pole Hadley circulation, which, coupled with strong tidal forcing, could largely explain the observed 1977 polar warming (Wilson, 1997). Barnes and Hollingsworth (1987) had earlier explored the possibility that forced planetary waves could act to drive the polar warming, much as they do for sudden stratospheric warmings on Earth.

Another example of a role for eddy forcing of the mean circulation comes from MCS observations. In particular, McCleese et al. (2008) drew attention to the relatively strong thermal inversions at south polar latitudes above ~40 km altitude appearing during southern winter that appeared to be much stronger than was typical in the GCM simulations at the time. Unlike the polar warming just discussed, this warming is unrelated to dust storms and occurs at a relatively clear time of the year. One possible explanation for this is that upward-propagating planetary waves and thermal tides are strongly dissipated at this time, due to radiative damping and the effects of nonlinear breaking. This can lead to the forcing of enhanced mean downwelling, which then results in stronger adiabatic warming, much as is often observed in Earth's polar mesosphere (where breaking tides and gravity waves are dominant (e.g. Andrews et al., 1987)). Medvedev and Hartogh (2007) examined this feature in GCM simulations with high vertical resolution, and these indicated that eddy dissipation played a key role in modifying and enhancing the mean meridional circulation at high latitudes. This is similar to the so-called "extratropical pump" on Earth (see e.g. Holton et al., 1995). The resulting mean meridional circulation acts to warm the upper polar atmosphere and strengthen the polar thermal inversion at lower levels.

Another possible explanation is that radiatively active clouds, which were largely missing from GCMs at the time, can strengthen the Hadley circulation at this season and therefore increase the adiabatic warming in the descending branch without the need for very strong eddy flux divergences. Subsequent studies have shown that radiatively active clouds during southern winter do drive a stronger Hadley circulation, but the polar inversions that are produced are not as strong as observed (e.g. Steele et al., 2014b). More recently, Kahre et al. (2014) showed that the forcing from radiatively active water clouds also carries dust to higher altitudes, further enhancing the Hadley circulation and resulting in stronger polar inversions. These inversions are still not as strong as the ones observed by MCS during southern winter. It seems very likely that (breaking and nonbreaking) gravity waves play a significant, if not primary, role in producing the polar warming during southern wintertime. The large topographic variations at relatively smaller scales in southern middle and high latitudes (many associated with craters) would certainly seem to favor strong gravity wave activity in the south. By comparison, the northern hemisphere is relatively "smooth" topographically at such scales.

9.4.5 Summary

The Hadley circulations in the atmosphere of Mars are much stronger than those in the terrestrial atmosphere, a fundamental consequence of the very short radiative time constants in the Martian atmosphere. At seasonal dates that are not close to the equinoxes, the lack of oceans means that the latitudes of strongest heating are displaced substantially away from the equator. This causes the meridional extent and strength of the Hadley circulation to increase very considerably, producing a single Hadley cell with a strongly cross-equatorial structure. The cross-equatorial low-level flow is channeled into multiple "western boundary currents" by the very large topography. Only relatively close to the equinoxes does Mars exhibit two roughly comparable Hadley cell circulations, as is always the case in Earth's atmosphere. The strength (and structure) of the mean circulation on Mars varies greatly with the atmospheric dust loading. Under very dusty conditions, the modeling shows that it becomes essentially global in extent at upper levels in the atmosphere. The easterly mean zonal winds at the equator at high altitudes approach the limit for angular-momentum-conserving pole-to-pole flow (assuming vanishing zonal winds at the summer pole).

The mean circulation during the northern wintertime seasons is much stronger than that during the southern wintertime seasons. This is due to three basic factors: the stronger thermal forcing resulting from the current timing of perihelion (at $L_s \sim 250°$, very close to winter solstice); the large hemispheric dichotomy of the zonal-mean topography on Mars; and the fact that the global atmospheric dust loading is generally much greater during the northern wintertime seasons. It must be noted that the last factor exists in sizable part *because* of the first two factors. Present-day Mars has a "dust storm season" and a non-dust storm season (often referred to as the aphelion season), and the mean circulation is generally much weaker during the latter portion of the annual cycle.

The large-scale eddies (stationary, transient, and tidal) are generally weaker in comparison to the mean circulation in the atmosphere of Mars than in the terrestrial atmosphere; this is largely because the mean circulation is so strong in the Martian atmosphere, as the eddy circulations are not weak. Basic measures of the forcing of the mean circulation by eddies reflect this fact, as Figure 9.13 illustrates very clearly. The transformed Eulerian mean circulation does not differ substantially from the Eulerian mean circulation, except at the equinox seasons, when the mean circulation is at its weakest. In this respect, the Martian global atmospheric circulation differs greatly from the terrestrial one.

9.5 THERMAL TIDES

9.5.1 Fundamental Aspects

Thermal tides are the atmospheric response to diurnally varying thermal forcing produced by aerosol heating/cooling (within the atmosphere) and radiative and convective heat transfer from the surface. The thermal tides are planetary-scale gravity waves with periods that are harmonics of the solar day. Hence, slope flows, which vary diurnally but tend to be more local in scale, are not generally considered to be thermal tides. Thermal tides are more prominent throughout the bulk of the Mars atmosphere than they are in any other atmosphere in the Solar System. They are vastly stronger than they are in the bulk of Earth's atmosphere, where they are little more than a scientific curiosity. As a result, the temperature, pressure, and wind fields have a strong and repeatable dependence on local solar time. The tides include westward-propagating, migrating (Sun-synchronous) waves driven in direct response to solar heating, as well as non-migrating, eastward-propagating waves resulting from zonal variations in the thermo-tidal forcing. Zonal modulation of forcing can arise from the longitudinal variations of the surface (topography and surface properties), as well as from radiatively active aerosols (dust and water ice clouds).

Although the thermal tides are present even in clear conditions, when they are forced mostly by surface heating, they are most strongly forced by the dust and clouds that are always present in the Martian atmosphere. Dust heating can be so

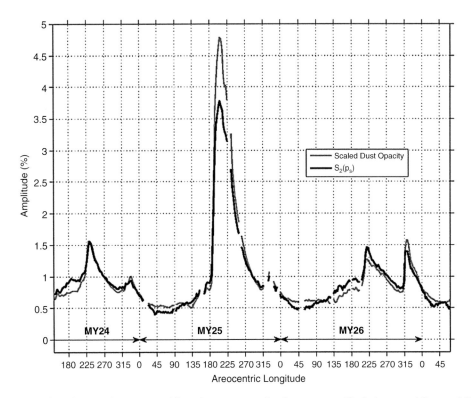

Figure 9.14. The evolution of the Sun-synchronous, semidiurnal component of surface pressure (S_2, darker curve) from an MGCM simulation employing observed TES dust column opacities for MY 24–26. The tidal amplitudes have been normalized by the seasonally varying, diurnal-mean surface pressures. The dust opacity τ (lighter curve) has been scaled to illustrate that the relationship between tide amplitude and column dust opacity is quite well approximated by $S_2 = 1.6\tau + 0.3$. Both the tidal amplitudes and the dust opacities are zonal average values at the equator. Figure adapted from Wilson et al. (2008b).

strong that it provides a very potent positive feedback mechanism for the initiation of dust storms, as first proposed by Leovy et al. (1973). The amplitudes of the tides can provide a direct measure of the thermal forcing of the atmosphere, particularly in the case of the semidiurnal tide. The amplitude of the semidiurnal variation in surface pressure can be directly related to the atmospheric dust opacity, as is shown in Figure 9.14. The diurnally varying winds in the boundary layer associated with tides strengthen with dust loading, and these certainly must play a major role in the development and growth of dust storms. Thermal tides also shape the distribution and diurnal variation of water ice clouds (e.g. Hinson and Wilson, 2004; Benson et al., 2010; Wilson and Guzewich, 2014), whose radiative heating and cooling can in turn amplify and alter the tidal forcing (Hinson and Wilson, 2004; Wilson and Guzewich, 2014; Wilson et al., 2014a). Thus, both dust and clouds play a major role in the excitation of thermal tides on Mars.

9.5.2 Theory

Basic tidal theory was developed almost 50 years ago for Earth's atmosphere, and it can be applied to understand many of the essential aspects of the thermal tides in the Mars atmosphere (e.g. Chapman and Lindzen, 1970). The latitudinal and vertical structure of the tides depends both on the period and structure of the forcing and on the efficiency of the atmospheric response to a given forcing. This response may be usefully considered using classical tidal theory. In classical tidal theory, the primitive equations are linearized about a motionless basic

state with no topography. Assuming the solutions are periodic in time and longitude leads to separate equations for the horizontal and vertical structure. The Laplace tidal equation with the well-known Hough function solutions (see e.g. Chapman and Lindzen, 1970) yields the horizontal structure. The Hough modes most relevant to Mars have been identified in many studies (Zurek, 1976; Bridger and Murphy, 1998; Withers et al., 2003; Guzewich et al., 2012). The vertical structure equation, a second-order differential equation, is solved for each Hough mode – which is associated with an equivalent depth, h_{eq}, which is a separation constant. The atmospheric response is determined by the projection of the thermo-tidal forcing onto these modes. For each s and σ combination (where s = zonal wavenumber and σ = frequency number (1 = diurnal, 2 = semidiurnal, etc.)), there is a complete set of orthogonal Hough modes (identified with a meridional index, n). Thermal tides can propagate vertically when the forcing frequency, $\sigma\Omega$, is greater than f, the Coriolis parameter ($f = 2\Omega \sin \phi$, where Ω is the planetary rotation rate and ϕ is latitude).

For the diurnal tidal components ($\sigma = 1$) in an atmosphere at rest, the transition from vertically propagating behavior to vertically trapped behavior takes place at 30° latitude. The Hough functions for these components cluster into two groups: modes that are relatively confined to tropical latitudes, and modes that tend to be confined to extratropical and polar latitudes. The former are vertically propagating while the latter are vertically trapped. Vertically trapped modes are localized in the immediate region of the thermal forcing, whereas the amplitude of a propagating mode will tend to grow with height above the levels of forcing. This latter behavior is a consequence of the conservation of energy in an atmosphere in which density decreases exponentially with height. Vertically propagating gravity waves have their group and phase velocities directed in opposite directions so that upward group velocity is associated with downward phase velocity. This means that there are tropical tidal maxima/minima (e.g. in temperature) that move downwards in time over the course of a solar day. High-frequency oscillations (motions for which $\sigma\Omega > f$) are characterized by divergent, cross-isobaric flow; while lower-frequency oscillations tend to be quasi-non-divergent, with flow roughly parallel to the isobars so that the motion is quasi-geostrophic. The longer forcing periods provide sufficient time for geostrophic balance to be established and the corresponding vertical motions are relatively weak. The development of vertical motions provides the atmosphere with a means to balance the thermal forcing by adiabatic heating or cooling, an effect that need not be confined to the region of forcing. Hence the dynamic response of the atmosphere can cause temperatures at a given level to deviate substantially from what would be expected on the basis of radiative effects alone.

In general, the longitudinal variation of the atmospheric fields may be decomposed into stationary waves, eastward- and westward-propagating thermal tides, and traveling waves with a broad range of frequencies. The time (t) and longitude (λ) dependence – at a given vertical level and latitude – can be represented as

$$A(\lambda,t) \sim \sum A_{s,\sigma}\cos(s\lambda + \sigma\Omega t - \delta_{s,\sigma}) \qquad (9.5)$$

where s, λ, σ, and t are defined as above, and $\delta_{s,\sigma}$ is the phase of the tidal component. Stationary waves have $\sigma = 0$, and tides, with strictly integer values of σ, propagate both westward ($s = 1, 2, 3, \ldots$) and eastward ($s = -1, -2, -3, \ldots$). The special category of zonally symmetric tides have $s = 0$. Other traveling waves, and in particular the transient eddies (see Section 9.6), are characterized by a spectrum of frequencies ($0 < \sigma < 1$) that are typically relatively small and are thus much less sensitive to local time.

Thermal tides can be separated into migrating and non-migrating components. Migrating tides follow the Sun and are Sun-synchronous, and thus propagate westward with respect to a stationary observer, with an angular phase speed $c = -\sigma\Omega/s$. For the migrating (Sun-synchronous) tides, $c = -\Omega$, which means that $s = \sigma$. Thus, the migrating tides are a special class for which the zonal wavenumber and frequency number are identical. All other tides do not follow the Sun and are hence referred to as non-migrating.

9.5.2.1 Migrating Tides

The diurnal tidal response is dominated by a vertically propagating, Sun-synchronous component (the $\sigma = 1$, $s = 1$, and $n = 1$ Hough mode) in the tropics, which has a vertical wavelength of ~33 km. Tidal modes corresponding to higher n values have more meridional structure and shorter vertical wavelengths. Typically, thermal forcing projects onto a sizable set of these modes, leading to a relatively complex thermal response. The tropical tidal response for vertically distributed heating is dependent on the mix of excited Hough modes, and significant interference leads to a tidal amplitude that does not simply

reflect the aerosol forcing (see e.g. Guzewich et al., 2013). In contrast, the extratropical diurnal tidal response is vertically trapped so that the phase is relatively constant in the vertical in the region of forcing. Moreover, the amplitude of the extratropical tide is related to the strength of the forcing, and is typically dominated by aerosol heating. The phase (18:00–20:00 LT) of the tidal temperature field lags that of the peak radiative forcing by roughly a quarter cycle (6 hours).

For the semidiurnal component of solar forcing, $\sigma\Omega > f$ at all latitudes, and thus the semidiurnal tides are vertically propagating everywhere. The dominant symmetric Hough mode solution for semidiurnal forcing is the $\sigma = 2$, $s = 2$, and $n = 2$ Hough mode. This mode has a very long vertical wavelength (~100–200 km) and a broad meridional scale, so that it responds efficiently to spatially distributed thermal forcing – which is expected to be largely due to dust aerosols for Mars. The Sun-synchronous, semidiurnal surface pressure response, denoted $S_2(p)$, is therefore particularly useful for providing a measure of globally integrated thermal forcing (Zurek, 1981; Zurek and Leovy, 1981; Lewis and Barker, 2005; Wilson et al., 2008b). This is the basis for the close relationship between globally averaged aerosol optical depth and the amplitude of $S_2(p)$ in the Viking Lander 1 surface pressure record (Zurek and Leovy, 1981), as is illustrated in Figure 9.14.

9.5.2.2 Non-Migrating Tides

Westward-propagating, non-migrating, diurnal tides (i.e. $\sigma = 1$, $s = 2, 3, 4, \ldots$) are generally very similar to their migrating counterparts, but have somewhat shorter vertical wavelengths. By contrast, the most prominent components of the eastward-propagating tides are the diurnal Kelvin waves (labeled DK1, DK2, DK3, etc., corresponding to $s = -1, -2, -3, \ldots$), which are meridionally symmetric. DK1 has a vertical structure that corresponds to the equivalent barotropic Lamb wave and it can be resonantly enhanced in the Martian atmosphere (Conrath, 1976; Zurek, 1976, 1988; Hamilton and Garcia, 1986; Wilson and Hamilton, 1996). DK2 and DK3 are vertically propagating modes with expected wavelengths of roughly 80 and 60 km, respectively, and have amplitudes that increase exponentially with height (see Wilson, 2000; Forbes et al., 2002). These properties are those expected for a motionless atmosphere.

Long vertical wavelengths render the diurnal-period Kelvin waves less susceptible to dissipation than the shorter-wavelength westward-propagating modes, allowing them to appear prominently in the upper atmosphere (Forbes and Hagan, 2000; Wilson, 2000, 2002; Forbes et al., 2002; Angelats i Coll, 2005; Withers et al., 2011; Guzewich et al., 2012). Non-migrating, semidiurnal tides also have long vertical wavelengths, and these evidently contribute to the large observed latitude temperature and thermospheric density variations (Bougher et al., 2001; Forbes et al., 2002; Wilson, 2002; Angelats i Coll, 2005; Cahoy et al., 2007; Forbes, 2008; Moudden and Forbes, 2008a,b; 2014; Guzewich et al., 2012; Wolkenberg and Wilson, 2014). The zonal wavenumber-2, semidiurnal Kelvin wave (SK2) may also be resonantly enhanced on Mars. This mode plays a prominent role in surface pressure modulation (Wilson and Hamilton, 1996) and has been identified in atmospheric data that resulted from

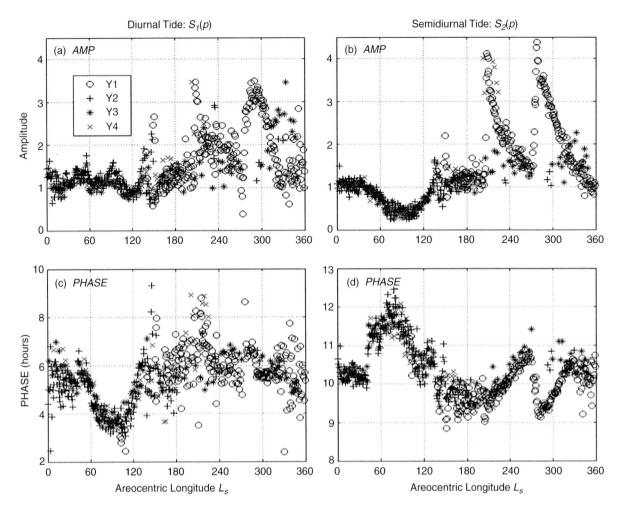

Figure 9.15. The seasonal variation of the (a) diurnal and (b) semidiurnal tidal amplitudes derived from the four-year record of surface pressure at the Viking Lander 1 site (22°N, 312°E). The amplitudes are normalized by the diurnal-mean surface pressures; the values are shown as percentages. The corrected (to true local time) diurnal tidal phase is shown in (c), and the corrected semidiurnal tidal phase is shown in (d). Figure adapted from Wilson et al. (2007). A black and white version of this figure will appear in some formats. For the color version, please refer to the plate section.

near-simultaneous radio science observations by the MGS and MEX spacecraft (Hinson et al., 2008a).

9.5.3 Observations

Direct observations of thermal tides are relatively limited. On the surface, the Viking, Pathfinder, Phoenix, and MSL Landers/Rovers carried meteorological payloads capable of identifying tidal signatures. However, these missions did not overlap and (except for Viking) are single stations, so that spatial structure cannot be determined. Furthermore, the Pathfinder and Phoenix missions were relatively short-lived and their pressure sensors had calibration problems. From orbit, the Mariner 9, Viking, MGS, Mars Express, and MRO spacecraft measured atmospheric temperatures and provided global viewing, but with rather limited local-time-of-day coverage. MGS and MRO, for example, observed at essentially fixed local times due to their Sun-synchronous polar orbits. Nevertheless, these platforms have provided very valuable data that have helped to characterize the nature of thermal tides on Mars.

9.5.3.1 Lander Observations

The Viking Landers measured surface pressures, air temperatures, and winds, and these have been analyzed for their tidal signatures. The local times attributed to the pressure observations in earlier papers (e.g. Wilson and Hamilton, 1996; Bridger and Murphy, 1998) were based on the local solar time appropriate for the first sol after lander touchdown, such that no account was taken of the evolution of the local true solar time as Mars progressed in its highly elliptical orbit. This effect has been shown by Allison (1997) and Allison and McEwen (2000) to yield a variation in true local time of up to 90 minutes over an annual cycle. Consequently, the surface pressure observations from VL1 and VL2 have been reanalyzed (Wilson et al., 2007) and the results for VL1 are shown in Figure 9.15. The amplitudes have been normalized so that the quantity plotted is a percentage of the local diurnally averaged surface pressure. The normalized surface pressure amplitude should remain relatively constant with changes in surface elevation (mean pressure).

The two very large dust storms of 1977 (MY 12) are particularly evident in the semidiurnal tidal record. As noted previously, the amplitude of the Sun-synchronous, equatorial, semidiurnal surface pressure tide is closely correlated with the global mean dust opacity. A large dust storm in MY 15 (1982) is strongly suggested by the VL1 semidiurnal tidal amplitude record. While there are no direct observations of the MY 15 storm, there is some supporting evidence for such an event (Leovy et al., 1985). There is a similar progression of phase and amplitude of the semidiurnal tide at the VL2 site during northern spring and summer.

The Viking measured surface pressures exhibit large-amplitude (up to 4% of the diurnal average pressure) tidal signatures during the two MY 12 dust storms (see Figure 9.15). The main features of the observed semidiurnal surface pressure oscillation at Viking Lander 1 (22.48°N) can be related to direct aerosol radiative heating, at least during dusty periods (Zurek, 1981; Zurek and Leovy, 1981; Leovy et al., 1985). Viking observed surface winds that also indicate large, tidally controlled variations during the two observed dust storms (Murphy et al., 1990). Simulations of the amplitude of the Sun-synchronous, equatorial semidiurnal surface pressure tide (Figure 9.14; Lewis and Barker, 2005; Wilson et al., 2008b) verify the interpretation of a close correlation between this amplitude and the global mean, dust column opacity. This is consistent with the expectation that the semidiurnal migrating tide is dominated by a mode with a broad meridional structure and a long vertical wavelength that efficiently responds to the globally integrated atmospheric dust heating (Wilson and Hamilton, 1996; Bridger and Murphy, 1998). In addition to the migrating tidal signatures present within the Viking Lander pressure measurements, a non-migrating Kelvin wave signal has also been identified (Tillman, 1988; Wilson and Hamilton, 1996).

Apart from dust storm events, the amplitude and phase of both the local diurnal and semidiurnal tide ($S_1(p)$ and $S_2(p)$, respectively) suggest a well-defined seasonal cycle. The VL1 record strongly suggests a high degree of repeatability for four Mars years in the $L_s = 0$–130° season that is consistent with the TES and MCS temperature record (Liu et al., 2003; Smith, 2004, 2008; Wilson et al., 2014b). The semidiurnal tide at VL1 underwent a smooth decrease in amplitude through northern spring followed by an increase after summer solstice. This amplitude behavior is not reflected in the Viking Lander opacity record (Colburn et al., 1988), which indicates a modest decline in opacity from $L_s = 0°$ to roughly 150°. The phase of $S_2(p)$ underwent a clear advance and retardation at both landers as the season progressed through northern summer solstice. Throughout this time the phase was distinctly later than that expected for the Sun-synchronous component of the semidiurnal tide (~09:00 LT).

The daily variability of the surface winds measured by the Viking Landers has been discussed by Hess et al. (1977), Leovy and Zurek (1979), Leovy (1981), Leovy et al. (1985), and Murphy et al. (1990). The interpretation of the wind data is complicated by the contamination of the global tidal signal by regional and local effects induced by topography and other inhomogeneities in the surface (e.g. Blumsack et al., 1973). The diurnal modulation of turbulent mixing may also have a significant effect on the phasing of the diurnal near-surface winds.

Little agreement has been found in terrestrial studies between the observed diurnal cycle of surface winds and that expected from linear tidal theory (e.g. Chapman and Lindzen, 1970). Hess et al. (1977) discussed hodographs of the Viking Lander winds for early northern summer. In this season the VL1 winds exhibited a counterclockwise rotation through the day with a phase such that winds were directed downslope during the morning and upslope during the afternoon. Tidal theory predicts a clockwise rotation of the horizontal surface wind through the day in the northern hemisphere. Hess et al. (1977) explained the observed rotation of the early summer VL1 observations by postulating a diurnal modulation of turbulent mixing in the PBL whereby strong afternoon mixing couples the surface winds to the observed southerly winds aloft. In late summer the observed winds at VL1 began to exhibit clockwise rotation through the day. Since the winds aloft are expected to switch to northerly as the Hadley circulation migrates southward with the subsolar latitude, this change in the diurnal surface wind variation may also be explained by the diurnal modulation of mixing (Murphy et al., 1990). Murphy et al. found that at neither lander was the phase of the diurnal period component of the two horizontal wind components consistent with that expected for a westward-propagating diurnal tide, even during the two MY 12 large dust storms when the tidal forcing was very strong. However, during the late autumn and winter seasons when the aerosol loading was largest, the daily hodographs were dominated by a semidiurnal variation. Leovy and Zurek (1979) and Leovy (1981) showed that the observed VL2 semidiurnal wind components were consistent in phase and amplitude with a Sun-synchronous, semidiurnal pressure signal having the observed strength during these periods. Murphy et al. (1990) demonstrated this to be the case at VL1 as well. GCM simulations of a very dusty atmosphere (Wilson and Hamilton, 1996) are in agreement with these observations.

The Rover Environmental Monitoring Station (REMS) on MSL's Curiosity Rover has measured surface pressures in Gale Crater (4.6°S, 137.4°E). The tidal amplitudes for the first Mars year of operations are shown in Figure 9.16. Normalized diurnal amplitudes are considerably larger than those observed by Viking (VL1). However, the semidiurnal tide has comparable amplitude and seasonal variations at the two sites. While there were no global dust storms during the first Mars year of MSL operations, there were several regional events that show up as a sharp increase in both the diurnal and semidiurnal amplitudes, at $L_s \sim 205°$ and 310°.

9.5.3.2 Orbiter Observations

Mariner 9 obtained the first measurements of thermal tides in the atmosphere of Mars. Longitudinal deviations from mean atmospheric temperatures were found to have amplitudes of ~15 K (Hanel et al., 1972). These deviations were recognized as being related to the westward-propagating, Sun-synchronous, diurnal tide, influenced at that time by the decay of a very strong, global dust storm. Pirraglia and Conrath (1974) estimated the heating necessary to account for observed diurnal temperature differences during this dust storm. This forcing was used to diagnose surface pressure oscillations and diurnally varying winds during the dust storm. Leovy and Zurek (1979) performed calculations that yielded similar results. The

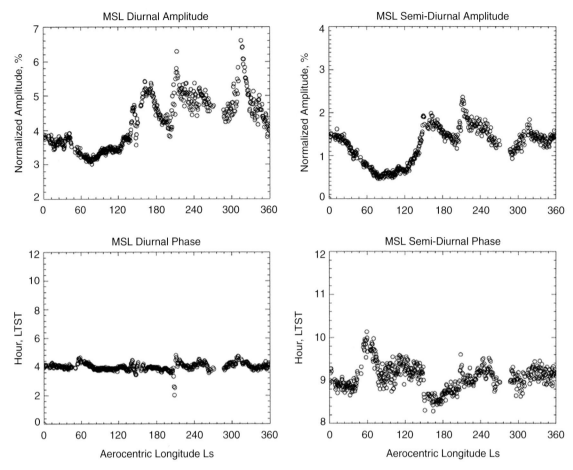

Figure 9.16. MSL tidal amplitudes and phases for the first Mars year of operations, as determined using data from the REMS experiment. The format is the same as that in Figure 9.15. Note that MSL operations began at $L_s \sim 155°$.

observed diurnal temperature range suggested absorption of ~40% of the available insolation (Pirraglia and Conrath, 1974).

The temporal decline in the derived diurnal temperature range at two atmospheric levels (200 and 30 Pa) during the 1971 global storm decay was related to a dust mixing ratio decline, with radiative heating proportional to the forcing provided by the dust's absorption of solar radiation (Conrath, 1975). An expression for the time evolution of the dust produced the same exponential decay at both pressure levels, suggesting a balance between differential particle sedimentation and diffusive mixing. The analytic function describing the vertical dust distribution has been widely used in atmospheric modeling, with the so-called "Conrath parameter" measuring the depth to which dust is essentially uniformly mixed (Conrath, 1975).

The Infrared Thermal Mapper (IRTM) radiometers aboard the two Viking Orbiters monitored the atmosphere over a range of local times (e.g. Martin and Kieffer, 1979), providing observations of the seasonal and diurnal variations of brightness temperature in the 15 μm channel (T_{15}), which is a depth-weighted measure of atmospheric temperatures centered at ~50 Pa (~25 km). Unfortunately, the observed diurnal cycle during relatively clear sky conditions was evidently contaminated by surface radiance due to a "spectral leak" (Wilson and Richardson, 2000), and this made the fitting of tidal amplitudes rather problematic during relatively non-dusty seasons. Revised IRTM temperatures are in

agreement with those observed by TES (Liu et al., 2003) and recent observations of the diurnal cycle of tropical temperature have confirmed the validity of the revision (Sato et al., 2011).

IRTM observations during the first mission year (MY 12) revealed large-amplitude thermal tidal signatures (Figure 9.17) during the course of the 1977a and 1977b dust storms (Martin and Kieffer, 1979; Martin, 1981), under conditions where the influence of the instrument's spectral leak were minimal (Wilson and Richardson, 2000; Liu et al., 2003). Results from a GCM simulation allow a more complete representation of the diurnal cycle of simulated T_{15} (Wilson and Richardson, 1999, 2000), and these simulated temperatures exhibit a substantial semidiurnal variation at tropical latitudes while a large diurnal signal is present at high latitudes. The close agreement between model and observations strongly suggests that the radiative forcing due to the dust field in the simulation is yielding a good agreement with reality. Wilson and Richardson (1999) showed that the simultaneous matching of the observed trends in semidiurnal surface pressure and IRTM temperatures can place useful constraints on the evolving dust distribution, similar to the much simpler fitting described by Conrath (1975).

Atmospheric temperature datasets derived from MGS TES spectra and MRO MCS radiances have both provided characterization of temperature variations correlated with the forcing of thermal tides. In a local solar time reference frame (where $t_{LT} = t + \lambda$), which

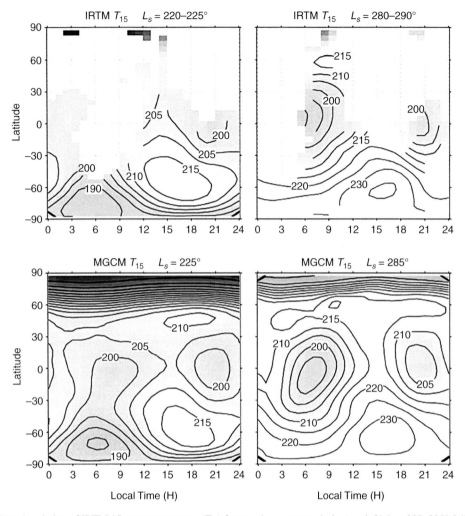

Figure 9.17. The diurnal variation of IRTM 15 μm temperatures (T_{15}) for two dust storm periods: (top left) L_s = 220–225° following the 1977a dust storm and (top right) L_s = 285–290° during the 1977b global dust storm. The bottom panels show the corresponding temperatures from a GFDL (Geophysical Fluid Dynamics Laboratory) GCM simulation. Figure adapted from Wilson and Richardson (2000). The contour interval is 5 K, and the darker regions are colder.

is appropriate for spacecraft observations from Sun-synchronous orbits (MGS, MRO), the longitude–time dependence is given as

$$A(\lambda, t_{LT}) \sim \sum A_{s,\sigma} \cos((s - \sigma)\lambda + \sigma\Omega t_{LT} - \delta_{s,\sigma}) \qquad (9.6)$$

In this reference frame, the migrating (Sun-synchronous) tides ($s = \sigma$) have no longitude dependence. Further, an observed zonal wavenumber m variation can reflect the presence of a stationary wave, $A_{m,0}$, and/or a set of non-migrating tides with $A_{s,\sigma}$, such that $(s-\sigma) = \pm m$ (Forbes and Hagan, 2000; Wilson, 2000; Banfield et al., 2003; Guzewich et al., 2012). For example, an observed wavenumber-2 variation may be due to the presence of diurnal-period, westward- ($A_{3,1}$) and eastward-propagating ($A_{-1,1}$) components, as well as contributions from higher temporal harmonics ($A_{0,2}, A_{4,2}, A_{1,3}, A_{5,3}, \ldots$). These two missions' Sun-synchronous orbits (MGS observed at 2 a.m./2 p.m. while MCS observed at 3 a.m./3 p.m.) preclude direct diagnoses of the amplitudes and phases of tidal thermal structures except during MGS's aerobraking phase, during which a broader range of local time coverage was available (Banfield et al., 2000).

It is natural to define a diurnal average field defined as $T_{avg} = (T_{p.m.} + T_{a.m.})/2$ and a tidal field defined by $T_{diff} = (T_{p.m.} - T_{a.m.})/2$. The 12 hour (half-sol) separation between observations results in the even diurnal harmonics being aliased with the true diurnal mean (Wilson, 2000; Banfield et al., 2003; Lee et al., 2009; Guzewich et al., 2012). In particular, the observed zonally averaged T_{avg} field represents both the true zonal and diurnal-mean temperature field and the contributions from the migrating semidiurnal tide, which can be substantial during dusty periods. For example, simulations and observations suggest an amplitude of the semidiurnal tide in T_{15} of ~10 K during the 1977b dust storm (Figure 9.17) observed by Viking (Wilson and Richardson, 2000), and a similar amplitude is suggested by the UK reanalysis at the peak of the 2001 global dust storm (Wilson et al., 2008a). This tide is phased in the tropics such that temperature maxima occur at 03:00 and 15:00 LT, and so could contribute significantly to the "diurnal mean" (T_{avg}) observed by MCS.

Longitudinally varying fields of T_{avg} and T_{diff} contain aliased signatures of migrating and non-migrating tides (Wilson, 2000;

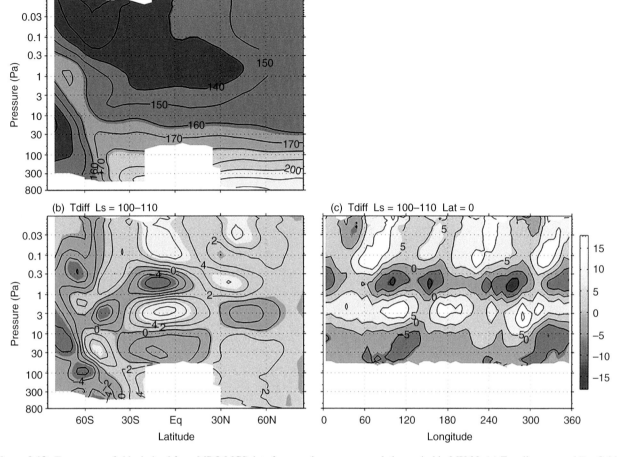

Figure 9.18. Temperature fields derived from MRO MCS data for a northern summer solstice period in MY 30. (a) Zonally averaged T_{avg} field. (b) Zonally averaged T_{diff} field. (c) Longitude structure of the equatorial T_{diff} field showing the presence of non-migrating tidal modes. T_{diff} is the difference between the dayside (3 p.m.) and nightside (3 a.m.) MCS temperatures, and T_{avg} is the average of them. Figure adapted from Wilson et al. (2014b).

Banfield et al., 2000, 2003; Lee et al., 2009; Guzewich et al., 2012), with two or more tidal components being represented by a specific longitudinal structure (zonal wavenumber) in the mean or difference maps. More recently, cross-track instrument viewing by MCS has expanded the local-time-of-day coverage compared to in-track views, and this enhanced coverage enables a more direct determination of the tidal components (Kleinböhl et al., 2013; Wolkenberg and Wilson, 2014).

Figure 9.18(a,b) shows the zonally averaged T_{avg} and T_{diff} fields for $L_s \sim 90°$ from recent MCS data. These results are consistent with those presented by Lee et al. (2009). They showed that the pattern of the maxima and minima is consistent with the diurnal thermal tidal response in a Mars GCM simulation. MCS observations (Wilson, 2012b) show much larger tropical tidal amplitudes due to the much superior vertical resolution and extent of MCS as compared to TES. This is suggested in the analogous figure in Banfield et al. (2003). The extratropical tidal amplitude is quite large in the MCS data (Figure 9.18). Very large tidal amplitudes are also implied by the TES data for the MY 25 dust storm (Smith et. al., 2002; see Chapter 4).

Figure 9.18(c) shows the longitude structure of the equatorial T_{diff} field. The zonal-mean component of this field is the equatorial migrating tide in Figure 9.18(b). There is significant zonal structure, which constitutes the non-migrating thermal tide. Zonal wavenumbers 2–4 are particularly prominent over a range of depths. The zonally modulated tides are of particular interest, as these can be adequately resolved by twice-daily spacecraft coverage (Guzewich et al., 2012). The long vertical wavelength and eastward phase tilt with height suggest the presence of eastward-propagating Kelvin waves.

Wilson (2000) showed that the longitudinal variability of tropical T_{15} synthesized from TES spectra collected during the mapping mission ($L_s = 108–350°$) is dominated by non-migrating thermal tides in low latitudes (30°S–30°N). A similar finding is also described in Banfield et al. (2003). The wavenumber-2 and wavenumber-3 components of tropical temperatures in GCM simulations, yielding ($T_{2p.m.}–T_{2a.m.}$)/2 fields consistent with TES observations, were found to be dominated by eastward-propagating Kelvin modes with long vertical wavelengths (Wilson, 2000).

When viewed at a fixed local time, the diurnal, wavenumber-1, Kelvin wave appears as a wavenumber-2 longitudinal variation. At a local time of 14:00 this Kelvin wave mode will have temperature maxima at longitudes of ~60° and 240°E. Pressure and

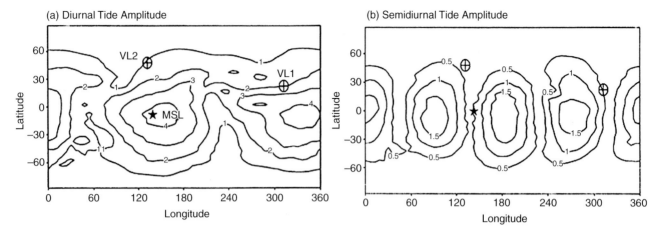

Figure 9.19. The spatial variation of (a) diurnal and (b) semidiurnal surface pressure amplitudes from a GFDL GCM simulation for northern hemisphere summer solstice ($L_s \sim 90°$). Tidal amplitudes are normalized to show the amplitude as a percentage of the local diurnal-mean surface pressure. The locations of the VL1, VL2, and MSL landers are shown. Figure adapted from Wilson and Hamilton (1996).

temperature are in phase for this Kelvin wave mode (DK1); this is referred to as a barotropic tidal structure. An examination of the vertical structure of the TES wavenumber-2 signal (see Wilson, 2000) in the tropics – for 14:00 and 02:00 LT, separately – reveals just such a barotropic vertical structure. Additional contributions from a westward-propagating zonal wavenumber-3 tidal mode (DW3) could appear with the same wavenumber-2 structure, but these are not evident in the data due to the relatively strong vertical smoothing of the TES temperature retrievals. (This is because the DW3 mode is characterized by relatively short vertical wavelengths.) It can be noted that the apparent wavenumber-2 variation in ~130 km level densities derived from MGS accelerometer data (Wilson, 2002) has been attributed to the presence of a Kelvin wave mode (DK1), having a similar phase.

9.5.4 Modeling

Given the limited observational coverage of the thermal tides, the complexities of tidal forcing, and their nonlinear interactions with the mean winds and surface topography, the tides can only be simulated realistically with Mars GCMs. Given the significant advances in the GCMs over the past several decades, a large amount of our present understanding of tides has been derived from GCM studies.

9.5.4.1 Surface Pressure

The interpretation of tidal signals at a given location such as a lander site is complicated by the multiplicity of tidal modes that may be present. So, while the migrating semidiurnal tide is a particularly useful measure of global heating, its isolation in a lander surface pressure record is non-trivial. During dusty periods, this tide is dominant, and this is the reason it can be related directly to the optical depth of the atmosphere. However, during clear periods, interference can be significant, as illustrated in Figure 9.19. The wavenumber-2 modulation of the diurnal tide is expected when eastward- and westward-traveling wave-1 components interfere. Similarly, wavenumber-2 eastward and westward waves produce the wavenumber-4 pattern in the semidiurnal tide.

The pronounced seasonal variation of surface pressure tidal amplitudes (diurnal and semidiurnal) is attributed to

constructive and destructive interference between the Sun-synchronous thermal tides and near-resonant diurnal and semi-diurnal Kelvin waves, as this provides a consistent explanation of the tidal behavior at both Viking Lander sites. The semi-diurnal tide shows a tendency for the normalized amplitudes to equilibrate around 1% during relatively clear periods away from northern summer solstice – when interference with an eastward-propagating semidiurnal Kelvin wave leads to a pronounced decrease in amplitude and an increase in the phase.

Normalized diurnal amplitudes observed by REMS are considerably larger than those observed by Viking, which is a consequence of Curiosity being located at a longitude where eastward and westward modes constructively interfere (see Figure 9.19, and the discussion in Haberle et al. (2014)), and also its location at the bottom of a deep crater where the local slope flow circulations act to further enhance an already large (normalized) diurnal tidal variation (Tyler and Barnes, 2013, 2015; Wilson, 2015).

GCM simulations (Wilson and Hamilton, 1996; Bridger and Murphy, 1998) indicate that the observed $S_2(p)$ phase evolution is not in agreement with that expected for the Sun-synchronous semidiurnal tide at the latitudes of VL1 and VL2. These simulations suggest the presence of resonantly enhanced, eastward-propagating, diurnal and semidiurnal Kelvin waves of zonal wavenumber 1 and 2, respectively. These modes are most strongly forced during the solstice seasons and are most nearly resonant during the relatively cool northern hemisphere late spring and summer (aphelion) seasons. The phase and amplitude behaviors of $S_1(p)$ and $S_2(p)$ are a consequence of interference between the eastward- and westward-propagating components of the diurnal and semidiurnal tides, which results in a zonal modulation of tidal amplitude and phase. The similar phase and amplitude behavior at the two lander sites, which are 180° apart, is consistent with the expectation that such an interference effect should yield a zonal wavenumber-4 pattern for the semidiurnal tide and a wavenumber-2 pattern for the diurnal tide. This is the spatial pattern evident in Figure 9.19, which is for a time of year when the near-resonant Kelvin waves have their maximum amplitudes. Diurnal period eastward and westward zonal wavenumber-1 tides combine to yield the pattern in Figure 9.19(a). Semidiurnal period eastward and westward zonal wavenumber-2 tides combine to yield the pattern in Figure 9.19(b).

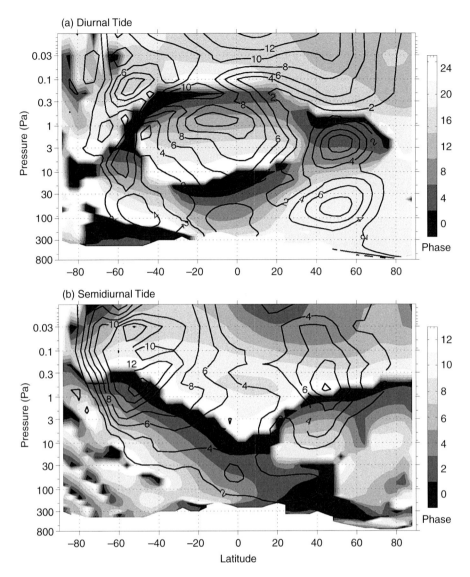

(a) Diurnal Tide

(b) Semidiurnal Tide

Figure 9.20. GCM-simulated temperature amplitudes and phases of the migrating (a) diurnal and (b) semidiurnal tide for early northern summer ($L_s \sim 105°$). Tidal amplitudes are contoured at 2 K intervals and phases are shaded from 00:00 to 24:00 LT for the diurnal tide and from 00:00 to 12:00 LT for the semidiurnal tide. Figure adapted from Kleinböhl et al. (2013). A black and white version of this figure will appear in some formats. For the color version, please refer to the plate section.

9.5.4.2 Temperatures

Figure 9.20 shows the amplitude and phase of the migrating diurnal and semidiurnal components of the thermal tides in a GCM simulation for $L_s = 105°$, a season with low dust loading. A similar figure is shown in Banfield et al. (2003), but with a domain depth appropriate for TES. The simulated diurnal tidal response is dominated by a vertically propagating component in the tropics with a vertical wavelength of ~40 km (Wilson and Richardson, 2000; Takahashi et al., 2006; Kleinböhl et al., 2013). The strong vertical "smoothing" that is present in the TES nadir retrievals means that the observed amplitudes shown in Banfield et al. (2003) are almost certainly reduced significantly compared to the actual atmospheric response. In contrast, the extratropical response is characterized by relatively weak phase variations in height and the TES-observed amplitudes should be more comparable to the actual ones. The tidal components retain a fairly high degree of symmetry about the equator, even for the solstice season forcing. This symmetry is reduced for stronger thermal forcing – dustier conditions – when zonal-mean zonal wind effects become more important (Wilson and Hamilton, 1996; Wilson and Richardson, 2000; Takahashi et al., 2006). The temperature response is strong near the surface, in agreement with simple, one-dimensional model calculations (Gierasch and Goody, 1968; Pollack et al., 1979). In middle latitudes, the phase of $S_1(T)$ shows less variation with height, and the amplitude is largely a direct response to the strength of the aerosol heating. The time of the temperature maximum in these simulations is ~18:00 LT, which is distinctly later than for the tropical temperatures shown in Figure 9.18. The phase of the tropical temperature response varies strongly with height, consistent with a vertically propagating tidal component having a vertical wavelength of ~40 km. At roughly 25 km in height (~0.5 mbar), the local time of the diurnal tidal temperature maximum has advanced to early morning.

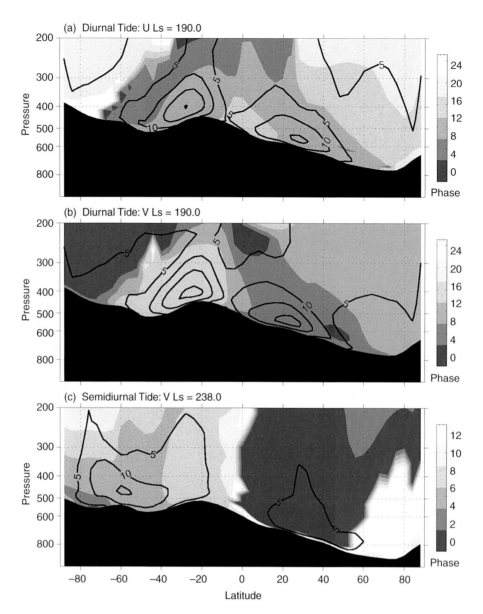

Figure 9.21. Amplitude and phase of GCM-simulated tidal wind fields. The migrating component of the diurnal period zonal and meridional winds for $L_s \sim 190°$ are shown in (a) and (b), respectively. The amplitude contour interval is 5 m s^{-1} and the phase (shading) varies from 00:00 to 24:00 LT. (c) Migrating semidiurnal meridional wind field for $L_s \sim 238°$. The phase varies from 00:00 to 12:00 LT, and the amplitude contour is 5 m s^{-1}. Figure adapted from Wilson (2012a,b). A black and white version of this figure will appear in some formats. For the color version, please refer to the plate section.

9.5.4.3 Winds

Simulations of the diurnal wind cycle at the Viking Lander sites for the summer season have been carried out using rather simple, limited-area models incorporating slope effects (Savijärvi and Siili, 1993; Haberle et al., 1993a). The models allow for an imposed background state but do not account for propagating global-scale tides. These studies have been able to reproduce some of the observed wind characteristics, notably the sense of wind rotation at each site. However, the results show a strong sensitivity to the local topographic slope used in the models.

Hodographs showing the composite diurnal evolution of the simulated, lowest-level winds at the model grid points closest to the VL1 and VL2 sites are presented in Wilson and Hamilton

(1996). The behavior of the simulated winds is in very good agreement with the observations at VL1 (Murphy et al., 1990) and VL2 (Leovy and Zurek, 1979) for dust storm conditions. Wilson and Hamilton (1996) found that simulated low-level diurnal winds are of comparable strength to the time-mean winds. There is an obvious strong influence of the topography on the diurnal winds (as the Sun-synchronous tide by itself would produce a zonally uniform distribution of local-time-of-day winds). There is a tendency for late afternoon upslope (and early morning downslope) winds over much of the surface. A similar pattern holds for all seasons and dust distributions considered.

Figures 9.21–9.23 are derived from a GCM simulation employing the seasonally varying distribution of dust observed by TES during the MY 24/25 year (as in Wilson, 2012a,b). MY

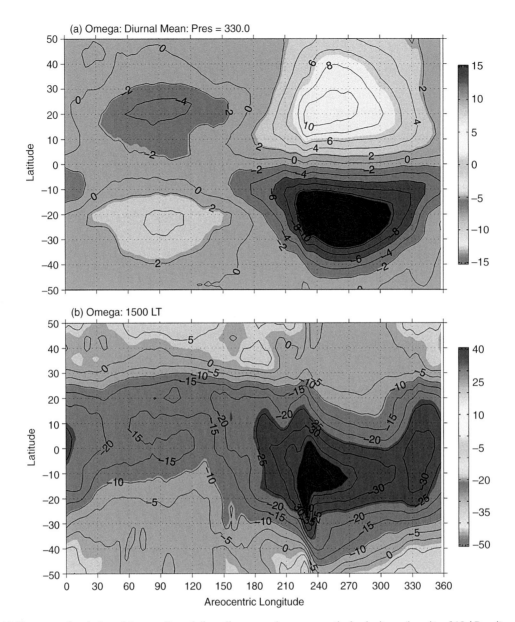

Figure 9.22. (a) The seasonal variation of the zonally and diurnally averaged pressure vertical velocity ω (in units of 10^{-4} Pa s^{-1}) on the 330 Pa pressure level, from a GCM simulation using specified dust column opacity as observed by MGS TES during MY 24/25. (b) As above, but for the pressure vertical velocity at 15:00 LT (no diurnal averaging). Note that the contour intervals and the range of shading differ in the two panels. Figure adapted from Wilson (2012b). A black and white version of this figure will appear in some formats. For the color version, please refer to the plate section.

24 was characterized by significant regional dust lifting that was initiated by flushing storm activity in the north at $L_s \sim 225°$. Subsequent lifting took place in the southern hemisphere, with the main lifting centers migrating southwest into Cimmeria and Promethei by $L_s \sim 236°$. The global opacity peaked at $L_s \sim 238°$ and then continued to decline through the solstice season. A second, somewhat weaker regional dust storm was observed to form at $L_s \sim 330°$.

Figure 9.21(a,b) shows latitude–pressure cross-sections of the amplitude and phase of the migrating component of the diurnal zonal and meridional winds in the lower atmosphere. There is a strong cross-equatorial flow in the lowest 2 km, with a phase such that the northerly (i.e. upslope) flow peaks in the late afternoon. In the absence of topography, the zonal and meridional diurnal tidal winds tend to be symmetric and antisymmetric, respectively, and have distinct, well-separated amplitude maxima at roughly 30°N and 30°S, as suggested in Figure 9.21. This result has been modified by the zonal-mean component of the topography in the model simulation. The tide is phased so that maximum low-level convergence occurs in the afternoon. The diurnally varying wind response typically dominates the diurnal-mean response in model simulations. In the diurnal and zonal mean, the meridional wind simply reflects the lower branch of the Hadley circulation. Tidal amplitudes are a strong function of dust heating, so that surface wind stresses amplify with dust lifting, yielding a significant positive feedback effect for regional dust storm intensification. This is particularly the case for the semidiurnal tide, as shown in

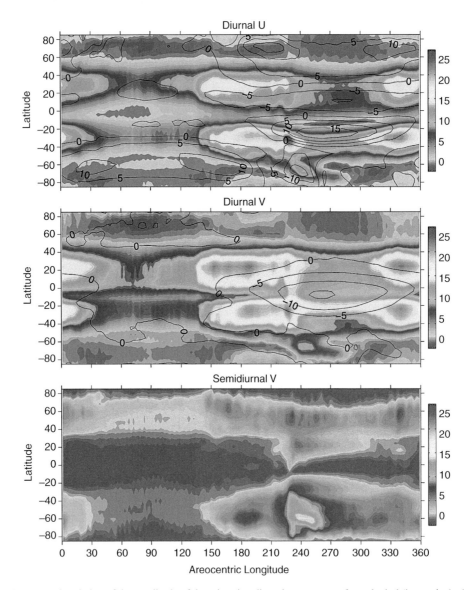

Figure 9.23. (Top) The seasonal variation of the amplitude of the migrating diurnal component of zonal wind (in m s^{-1}, shaded) at ~1 km above ground level, from a GCM simulation with dust opacities appropriate for MY 24/25. The diurnally and zonally averaged zonal wind field is also shown, contoured at intervals of 5 m s^{-1}. (Middle) As in the top panel but for the meridional wind field. (Bottom) As in the top panel but for the semidiurnal component of the meridional wind field. Figure adapted from Wilson (2012a). A black and white version of this figure will appear in some formats. For the color version, please refer to the plate section.

Figure 9.21(c). The semidiurnal wind oscillations are strongest poleward of ~40°N/S.

Of course, there is considerable longitude structure hidden by the zonal averaging, reflecting the additional presence of non-migrating tides that are influenced by zonal variations in the topography. This includes regions in which the phases of the planetary-scale thermal tides interact constructively or destructively with more local slope winds that also have a very strong diurnal variation. The simulated semidiurnal wind fields are much more uniform in longitude than those for the diurnal tide, consistent with the expected domination of semidiurnal variations by the global-scale, Sun-synchronous tide.

Figure 9.22(a) shows the seasonal variation of vertical velocity in the same GCM simulation. The central latitude of the upward branch of the Hadley circulation roughly follows the subsolar latitude and shows a strong hemispheric asymmetry

in intensity. The seasonal evolution of afternoon (15:00 LT) vertical velocity is shown in Figure 9.22(b). It is clear that the afternoon migrating tidal wind response dominates the diurnal-mean (Hadley circulation) response. Again, there is significant longitude structure associated with non-migrating tides. The influence of the dust lifting events in the pre- and post-solstice seasons ($L_s = 230°$ and $330°$) of MY 24 is evident in the tidal response, which is larger than that of the Hadley circulation. Of course, the diurnally integrated impact of the Hadley circulation will be more significant than that of the tides for the long-range vertical and horizontal transport of dust.

The seasonal variations of the migrating tidal amplitudes of the low-level (at ~1 km) wind components are shown in Figure 9.23. There is a clear pattern of tidal amplification that tends to maximize in the equinoctial seasons. This pattern is modulated by the seasonal variations in dust opacity.

The top two panels also show the evolution of the low-level winds associated with the diurnally and zonally averaged (Hadley) circulation at ~1 km. The development of a strong cross-equatorial circulation and intensification of the sub-tropical westerly jet in the south is very clearly evident in the northern winter solstice season. The bottom panel shows the rapid amplification and decay of the semidiurnal tide, whose amplitude is very strongly dependent on atmospheric dust loading.

Modeling studies have generally focused on simulating major dust storms in the solstice season due to the expectation that dust is most efficiently lifted and distributed by the Hadley circulation. However, the observational record indicates that most years are characterized by pre- and post-solstice regional dust lifting. In several years (1977, 1982, and, most notably, 2001) global dust storms have occurred well before the solstice, suggesting that the Hadley circulation does not necessarily play a dominant role in global storm development.

It is notable that the 2001 dust storm intensified at L_s ~ 190° when tidal winds most strongly dominate the Hadley circulation. The low-level, semidiurnal winds peak at mid- to high latitudes in the summer hemisphere with a response that is roughly proportional to column dust opacity. For dusty conditions, the semidiurnal winds are the dominant component of the low-level wind field in middle and high latitudes. Wilson (2012a) has suggested that is quite possible that the migration of dust lifting to high southern latitudes seen in the evolution of many regional dust storms (e.g. Wang and Richardson, 2015) is a consequence of the intensification of the semidiurnal tide.

9.5.5 Summary

Thermal tides are an atmospheric response to diurnally varying thermal forcing (as are slope flow circulations on a very wide range of scales on Mars). While diurnal variability is very evident in a variety of spacecraft observations at the surface and within the atmosphere, it remains very much the case that Mars GCMs are the most comprehensive means of self-consistently relating thermal forcing, temperature response, and the associated circulations. This is true for all of the major components of the global atmospheric circulation of Mars – GCMs are the only way that all (or at least most) of the many facets of these circulations can be filled in and related to each other.

There are some significant issues that remain to be addressed in the future. Better observations of tides would certainly help to improve/constrain the representation of radiative forcing in the Mars GCMs. There is a strong need for a more complete surface network that would enable the isolation of the different tidal component contributions to local tidal signals (see e.g. Haberle and Catling, 1996; Bridger and Murphy, 1998). In particular, sufficient spatial coverage would allow the main components of the migrating and non-migrating tides to be identified and distinguished. A related issue is how representative are local surface pressure variations of large-scale forcing. In particular, Curiosity sits at the bottom of Gale Crater where the local slope flow circulations appear to have a very significant influence on the diurnal pressure cycle. This was first investigated by Tyler and Barnes (2013) with a mesoscale model, and then explored further for highly idealized crater topography in the complete absence of thermal tides (Tyler and Barnes, 2015). Wilson (2015) examined this problem using data from several very high-resolution GCM simulations, and obtained results that were in close agreement with those of Tyler and Barnes. There is a clear enhancement of the diurnal surface pressure range within craters, above that expected in the context of normalization by the higher mean surface pressures. There also appears to be a very significant reduction in the diurnal surface pressure range on (localized) elevated topographic features. Slope flow circulations are absolutely ubiquitous on Mars: in craters (of which there are far more in the southern hemisphere), in basins, in canyons, and in other closed or partially closed topographic depressions on the planet. Strong slope flow circulations exist on "localized" elevated topographic features as well, and these are also very common on Mars.

Incomplete diurnal observations of atmospheric temperatures by orbiting spacecraft in Sun-synchronous orbits (MGS, MRO) significantly limit the information about the strength and distribution of radiative forcing that may be inferred from these observations. For example, results in Kleinböhl et al. (2013) (see Figure 9.20) suggest that semidiurnal tides may achieve large amplitudes even in relatively non-dusty seasons and thus should be accounted for in the interpretation of temperature data. Wilson et al. (2014a) suggest that recent simulations that include radiative heating by water ice clouds imply that these clouds may actually be as, or even more, important than dust is in the forcing of the tropical semidiurnal tide over much of the annual cycle. Currently it is simply not possible to robustly constrain this problem, and thus improved coverage of the diurnal cycle is essential.

Observations of winds will remain rather sparse for the foreseeable future. Ongoing research on the dust cycle, where the parameterization of dust lifting in terms of the surface wind stress is critical (see Chapter 10), will put emphasis on the reliable simulation of near-surface winds. The local effects of topography and thermal stratification are expected to be important. In particular, the transition from "local" slope influence (e.g. at the crater scale) to more planetary scales at which tides – and other circulation components – should dominate will likely be a function of height above the surface. The recent developments in the mesoscale modeling of circulations in Gale Crater (e.g. Tyler and Barnes, 2013, 2015; Wilson, 2015) and other topographic structures (e.g. Valles Marineris) constitute the start of a much more complete investigation of this problem. For very dusty conditions, the near-surface winds will tend to become more strongly dominated by the tides at all scales.

9.6 TRANSIENT EDDIES

9.6.1 Fundamental Aspects

Large-scale, transient eddies – "traveling weather systems" as they are often referred to in Earth's atmosphere – were a well-predicted component of the global circulation of the Mars atmosphere (Hess, 1950; Mintz, 1961; Leovy, 1969; Leovy and Mintz, 1969). Given some of the key basic similarities of Mars and Earth (obliquity, rotation rate, etc.), it was anticipated that such disturbances should be present in the atmosphere of Mars. Observations and modeling going back more than 50 years have established that transient eddies are indeed an extremely important component of the global atmospheric circulation and climate system of Mars. We now know that transient eddies play a very prominent role in the current dust cycle. As is the case for the thermal tides (and the mean circulation), the transient eddies are very strongly affected by the large and unusual topography on Mars. The study of transient eddies in the Mars atmosphere can be viewed as comprising three different periods. During the first period, the Mariner 9 and Viking missions provided strong observational evidence for the existence of transient eddies in the northern fall, winter, and spring seasons. The second period corresponds with an interval following the Viking era, during which there were not any spacecraft missions that obtained useful observational data. During this time, the emphasis shifted strongly to global modeling of the atmosphere, as Mars GCMs became much more capable. The third period of studies of the transient eddies began with the Mars Global Surveyor mission, the first spacecraft mission to acquire a global atmospheric dataset in a regular mapping mode. This period is continuing today with the ongoing MRO mission, although the MRO data have not yet yielded much in the way of new results pertaining directly to the transient eddies.

Extratropical weather systems in the terrestrial atmosphere are generally understood to be fundamentally a consequence of baroclinic instability. This fluid instability generally is present in differentially heated (between low and high latitudes) and rapidly rotating atmospheres; it is sometimes called "slantwise convection". This is a very appropriate term, since the disturbances (eddies) that result from baroclinic instability act to transport heat both horizontally and vertically in the atmosphere, from warmer regions to colder regions. These heat transports are associated with air parcel trajectories that slope very gradually (are nearly flat) in the latitude–height plane. The transient eddies also generally transport momentum, and on Earth these momentum transports are crucial to the maintenance of the extratropical jet streams in the presence of friction. An atmosphere in which there are temperature gradients at constant pressure is a baroclinic atmosphere (an atmosphere without such thermal gradients is termed "barotropic"). The strongest horizontal temperature gradients on large scales in the atmosphere of Mars are those that exist between lower latitudes and the winter polar latitudes. It is these intense thermal gradients that give rise to baroclinic instability. The first Mars GCM experiments (Leovy and Mintz, 1969) were partially motivated by the question of whether or not the atmosphere could transport enough heat into polar latitudes to prevent or reduce the condensation of CO_2 during the wintertime seasons. The answer to this question was "no", though this is primarily a consequence of the very low density of the atmosphere and is not due to the lack of vigor of the transient eddies. On Earth, both the atmosphere and the oceans transport large amounts of heat poleward, and this keeps the high-latitude temperatures considerably warmer than they would otherwise be.

The bulk of this section deals with the Martian transient eddies that can be regarded as direct analogs to middle-latitude weather systems on Earth – highly baroclinic eddies with large amplitudes at and near the ground. There is also discussion of transient eddies in the Mars atmosphere that have a very strongly "upper-level" structure. These eddies are much more prominent in the northern hemisphere than in the southern hemisphere, and our present understanding of their dynamics remains more limited (there being no real counterpart to these eddies on Earth).

9.6.2 Basic Theory

Charney (1947) and Eady (1949) developed the basic linear theory of baroclinic instability in the aftermath of World War II. Weather observations and forecasting efforts had built a foundation for their work, which is one of the singular achievements in atmospheric dynamics. The basic theory addresses the linear instability of a zonal flow that is vertically sheared, which will be the case in the presence of north–south temperature gradients (this is a simple manifestation of the thermal wind relationship of dynamic meteorology – see Section 9.4.2). The theory shows that such a zonal flow will generally be unstable to eddy (zonally asymmetric) disturbances that will propagate in the direction of the mean flow. The unstable disturbances can amplify relatively rapidly for typical north–south temperature gradients, and they act to transport heat both poleward and upward. It is the upward heat transport (warm air rising and cold air sinking) that generates the kinetic energy of the unstable disturbances. Holton and Hakim (2013) and Read and Lewis (2004) present accessible summaries of linear baroclinic instability theory; Read and Lewis specifically discuss baroclinic instability for Mars.

Linear baroclinic instability theory enables predictions to be made for the dominant zonal scales of the weather disturbances, for their propagation speeds, and for their growth rates. It also predicts the basic structures of the disturbances, subject to the limitations of linear dynamics. The earliest studies for Mars made use of the simplest possible linear models to make predictions of these basic properties of transient baroclinic eddies (Mintz, 1961; Leovy, 1969). This work is discussed in more detail below. When the first observations that enabled comparisons with theory and models to be made were obtained by the Viking Meteorology Experiment, an initial focus was placed on assessing how the observed properties of the transient disturbances compared to those predicted by the theory and modeling. The basic answer was that they compared rather well, such that the existence of transient eddies associated with baroclinic instability in the Martian atmosphere was largely confirmed by early analyses of the Viking meteorology data.

In reality, the transient eddies are large-amplitude disturbances that are present in highly complex atmospheric states. Linear theory is certainly not valid, and there are no theories for strongly nonlinear baroclinic instability. One has to turn to numerical modeling to study real atmospheric transient eddies, and the most realistic models for this purpose are GCMs. Given the lack of extensive observations of the transient eddies – as are obtained every day on Earth – GCMs are essential for Mars in order to try to fill in the large gaps in the global observational database (there being essentially no surface pressure and wind observations). Data assimilation is a mathematical and numerical way to do this using GCMs. These models simulate all of the dynamics that is in the basic linear theories, as well as much higher-order dynamics. These dynamics are highly nonlinear. The GCMs also contain representations of the most important physical processes in the atmosphere of Mars (e.g. very strong radiative heating and cooling), along with the global topography and the variable surface thermal properties. It is certainly not an overstatement to say that the GCM studies have contributed enormously to our present understanding of transient eddies in the atmosphere of Mars, as they have for the zonal-mean circulation, the thermal tides, and the quasi-stationary eddies.

9.6.3 Observations

The observations of the transient eddy disturbances fall into two basic categories: observations obtained on the surface by landed instruments, and observations made from orbit by remote sensing instruments. There is almost no overlap between the two types of observations, and this is arguably the greatest deficiency of the observational dataset for Mars. The presentation here of the observations is separated into these two categories. Surface observations of the transient baroclinic eddies are essentially limited to the meteorology measurements obtained by the Viking Landers, while the orbital observations include thermal measurements and imaging. The TES thermal observations have provided a great deal of quantitative information about the transient eddies with regular global coverage. The MGS Radio Science (RS) experiment provided temperature data with very high vertical resolution (~1 km), but these data are restricted to quite narrow latitude bands. The RS experiment also measured surface pressures, and thus allowed the determination of geopotential height on constant-pressure surfaces. This allowed the determination of the meridional geostrophic winds associated with the transient (and quasi-stationary) eddies. It is generally rather difficult to obtain very much quantitative information on the transient eddies from orbital imagery, but this does provide a view of frontal clouds and dust structures associated with the transient eddies. None of the orbital imagery for Mars is synoptic, since different longitude regions are observed at different (universal) times.

9.6.3.1 Surface Observations

Essentially all of the *in situ* observations that helped to determine a number of the key characteristics of the transient eddies were obtained by the Meteorology Experiment on the two Viking Landers. Figure 9.1 shows the Viking Lander pressure data for the entire mission, depicting the sol-average pressure values for both of the landers as well as the values of the pressure standard deviations for each sol for VL1. A remarkable amount of basic information can be obtained from an examination of this figure. The mean pressure at VL1 is lower than that at VL2 largely because of the elevation difference between the two sites, and the large pressure changes on seasonal timescales are due to the condensation and sublimation of CO_2 in the polar regions of Mars during the wintertime seasons (see Section 9.4). The transient eddies manifest themselves in this figure as the much higher frequency "noise" that can be seen

during the wintertime seasons at both of the landers. The strongest weather activity occurred during very late winter at VL2 in the first Viking year – MY 12. In MY 13 it can be seen that the weather activity was again becoming very strong when VL2 ceased operating as late winter was starting. It can be seen that the weather variations at VL1 were substantially weaker than those at VL2, but they were characterized by very similar seasonal variations. A fundamental aspect of the seasonal variations in the transient eddy activity is that in late spring and most of summer it is extremely weak. This was first predicted by the early studies based on linear theory and the first GCM experiments (Mintz, 1961; Leovy, 1969; Leovy and Mintz, 1969). It is a direct consequence of the lack of oceans on Mars, because this allows the north–south temperature gradient in the atmosphere to reverse and weaken during summertime (much as it does in the terrestrial middle atmosphere). On Earth the latitudinal temperature gradient in the troposphere does not reverse in summer because of the very strong thermal buffering provided by the oceans and the extensive permanent ice at high latitudes. Greater cloudiness also contributes to higher latitudes remaining cooler in summer. The latitudinal thermal gradient weakens and is shifted to higher latitudes, but it is still present. On Mars there is still a very strong thermal gradient in summer in high latitudes, but only close to the edges of the permanent polar caps in both hemispheres. Mesoscale modeling studies have shown that transient eddy activity is certainly present in the northern polar region during the summer season (Tyler and Barnes, 2005, 2014).

It can be noted in Figure 9.1 that the weather variations (in surface pressure) were at their very strongest at VL1 in the third Viking year, MY 14, during winter. The magnitudes of the day-to-day pressure variations during this interval were much larger than that at any other time, and they are as large as or even larger than those that characterized the weather activity at VL2 during much of the fall, winter, and spring seasons. It is virtually certain that VL2 would have observed even larger pressure variations at this time had it still been operating. We now know that VL1 was located in a storm zone region on Mars, a region in which the weather activity is considerably stronger than it is in other regions in middle and high latitudes. But the MY 14 winter season was clearly a very unusual one in this regard. Unfortunately, we do not have any analogous extended surface pressure observations post-Viking, except those being obtained now from the Curiosity Rover in the tropics.

An additional important aspect of the transient eddy activity that can be seen in Figure 9.1 is that it can be very strongly altered by dust storm activity. In MY 12 a global dust storm began just after northern winter solstice. It can be seen that, following this seasonal date, the magnitude of the transient eddy activity at both landing sites decreased dramatically. It did not return to the previous levels for quite some time (~50–70 sols). The primary periodicities of the transient eddy disturbances also changed considerably during this interval; this is discussed below. The observations obtained by TES during a global dust storm in MY 25 have enabled us to understand these observations as being in large part due to changes in the vertical and latitudinal structures of the dominant transient eddy disturbances. The changes in the vertical structure are very dramatic and almost certainly are associated with changes in the

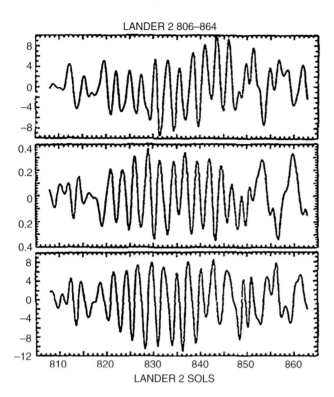

LANDER 2 806–864

LANDER 2 SOLS

Figure 9.24. Viking Lander 2 zonal wind (top panel, in m s⁻¹), pressure (middle, in mbar), and meridional wind (bottom, in m s⁻¹) data that have been low-pass filtered and detrended to remove tidal variations and isolate the variations due to the transient eddies. The data are for a period (L_s ~ 190–226°) during early autumn in MY 13, the second Viking year. Figure adapted from Barnes (1981).

fundamental dynamics of the transient eddies. This is discussed in more detail below.

Figure 9.24 shows a ~55 sol interval of pressure and wind variations observed by Viking Lander 2 in MY 13 (the second Viking year). The data have been binned (25 bins/sol) and then low-pass filtered so as to remove all of the diurnal and subdiurnal variability (see Barnes, 1980, 1981). This interval contains a remarkable sequence of 10 or 11 weather disturbances, with an average period of ~2.7 sols, at this middle-latitude location during early autumn (L_s ~ 190–226°). It is safe to say that a weather sequence like this one would almost certainly never be observed on Earth; we will discuss this aspect (the "regularity") of the data below. At the end of the sequence of ~2.7 sol disturbances, there is a sharp transition to several longer-period disturbances. There is then a 6 sol data gap, followed by a ~125 sol interval of nearly continuous observations (see Barnes, 1981). The first portion of this second interval contains another remarkably regular sequence of disturbances, with an average period of ~2.3 sols. As a result of extensive analyses of the TES and RS atmospheric temperature data, we now know what some of the basic properties of such short-period disturbances are; this is discussed below. Orbital imagery from MGS and MRO has revealed that such short-period disturbances in the northern hemisphere play a very important role in the global dust cycle on Mars.

The phase relationships between the pressures and the winds evidenced by the data in Figure 9.24 are entirely consistent

with those expected for eastward-propagating disturbances centered somewhat to the north of the Viking Lander 2 site (e.g. Barnes, 1980, 1981, 1983). Falling pressure is correlated with winds from the south, and rising pressure with winds from the north. The phase relationships for temperature are as expected for baroclinic weather systems that are acting to transport heat northward: southerly winds are correlated with warm temperatures and northerly winds with cold temperatures. To the extent that the disturbances are quasi-geostrophic (e.g. Holton and Hakim, 2013), the pressure and meridional wind variations can be combined to yield estimates of their zonal wavelengths. In order to do this, the observed winds very close to the surface must be converted into geostrophic winds at the top of the planetary boundary layer (Barnes, 1980, 1981, 1983). Barnes did this by making use of boundary layer parameter values determined by Leovy and Zurek (1979) and Leovy (1981) from analyses of the semidiurnal tidal wind variations that were observed by Viking Lander 2. In application to the sequence of ~2.7 sol disturbances (see Figure 9.24), the non-dimensional zonal wavenumber value that results is ~3.7. This value is essentially consistent with the predictions from the earlier theoretical and modeling studies of baroclinic instability in the Martian atmosphere.

The Viking meteorological data, especially the Lander 2 data, were subjected to analyses to identify the periodicities of the variations associated with transient eddies during the fall, winter, and spring seasons. Ryan et al. (1978), Tillman et al. (1979), Sharman and Ryan (1980), and Niver and Hess (1982) examined the dominant periodicities that were present in the data at Lander 2. The latter two studies looked at only the pressure data, while the first two also looked at wind and temperature data. Ryan et al. made the first estimates of zonal wavenumbers using a very simplified approach. Tillman et al. analyzed observations of a cold frontal passage at Lander 2 in early autumn of MY 12. Barnes (1980, 1981, 1983) analyzed the periodicities present in the pressure, temperature, and winds for five selected intervals in MY 12 and MY 13. Barnes also performed cross-spectral analyses for these intervals, and determined the zonal wavenumbers (and phase speeds) of the dominant transient eddy modes in the data.

The basic picture that emerged from the analyses of the Viking Lander 2 meteorological data was that disturbances with periods of ~2–8 sols were present during the fall, winter, and spring seasons. The wavenumber analyses showed that the longer-period disturbances (~6–8 sols) were of larger zonal scale, corresponding to smaller zonal wavenumbers (~1–2). The shorter-period disturbances (~2–3 sols) were characterized by shorter zonal scales, corresponding to zonal wavenumbers of ~3–5. The longer-period eddies were more prominent in the MY 12 data, while the shorter-period eddies were much stronger in the MY 13 data that were subjected to analysis (Barnes, 1980, 1981, 1983). It was thus apparent that there was significant inter-annual variability in the transient eddy activity. The very large-amplitude pressure variations observed at Lander 1 in MY 14 constitute further evidence of substantial year-to-year changes in the eddy activity.

The dramatic changes that were observed to occur in the transient eddy variations following the onset of the winter solstice global dust storm in MY 12 (see Figure 9.1) are of considerable interest. The magnitudes of the observed variations during the dust storm (an interval more than 50 sols long) were much smaller than those prior to the storm onset, and the variations exhibited a dominant ~6–7 sol periodicity. The zonal wave-number that was determined for this long-period disturbance was ~3 (Barnes, 1980, 1981, 1983), but this value was rather sensitive to the boundary layer parameter values that were used to produce the geostrophic wind estimates. We have since been able to put together a basic picture of the changes in the transient eddies that are associated with large dust storms, from the TES temperature data and modeling studies. This is discussed in more detail below.

Two studies of the Viking Lander 2 pressure data that were done well after the Viking mission produced very interesting and important results. Collins et al. (1996) performed analyses of the pressure data for the entire mission at Lander 2, making use of a singular systems analysis to identify the most prominent eddy modes. Hinson et al. (2012) analyzed the pressure data from both MY 12 and MY 13, and directly compared the results with analyses of both TES and MGS Radio Science (RS) data.

Collins et al. (1996) explicitly characterized a very basic aspect of the eddy activity that can be readily seen in time series of the Viking data – if the data are low-pass filtered to remove the tidal variations, there are transitions between different dominant eddy modes that occur on timescales of ~15–60 sols. In some intervals, two different eddy modes can be present with very comparable amplitudes in surface pressure. The fact that the five different intervals analyzed by Barnes (1980, 1981, 1983) yielded a variety of dominant eddy modes essentially showed that this kind of behavior was occurring. The results obtained by Collins et al. more fully defined the prominence of shorter-period eddies in MY 13 and the much greater prominence of longer-period eddies in MY 12 that Barnes (1980, 1981, 1983) had found.

Hinson et al. (2012) performed analyses of the dominant periodicities and the mode transitions in the Viking Lander 2 pressure data in light of results that he and his collaborators had previously obtained in analyses of MGS RS and TES data. This work will be discussed in the next section. In particular, they found that a short-period wavenumber-3 disturbance almost certainly was dominant during several intervals in autumn and late winter in MY 13. Studies by Hinson and his colleagues had previously shown that short-period wavenumber-3 modes play a key role in much of the dust raising activity associated with transient eddies in the northern hemisphere. This is discussed in the following section.

9.6.3.2 *Orbital Observations*

(a) Eddy Modes – Zonal Wavenumbers and Periods The mapping orbit of MGS allowed the TES temperature data and some of the RS data to be analyzed to yield the zonal wavenumbers and periods of the transient eddy modes. These modes – in the northern hemisphere – had been characterized over somewhat less than two Mars years in the Viking Lander meteorology data, as discussed above. The zonal wavenumbers obtained were estimates, assuming geostrophy and subject to errors due to the estimation of the

meridional geostrophic winds (among other things). With its Sun-synchronous orbit (~12–13 orbits per sol), the TES data are highly asynoptic. Different analysis approaches can be employed to determine the zonal wavenumber and frequency content of such data. Salby (1982a,b) rigorously worked out the wavenumber–frequency resolution of highly asynoptic satellite data, and, in the case of the MGS orbit, zonal wavenumbers extending to 6 can be resolved. The frequency resolution depends upon whether or not both dayside and nightside data are used in the analysis: with both, frequencies extending to roughly one cycle per sol can be resolved. In this case the zonal direction of propagation of the disturbances can be determined. Eastward-traveling waves with a period of 1 sol (diurnal Kelvin wave tidal modes) are resolved, but westward waves with a period of 1 sol (migrating diurnal tides) are not. This is a consequence of the eastward precession of a Sun-synchronous orbit. If data for only one local time per sol are used (as is the case for all of the RS data that can be analyzed for transient eddies), the shortest eddy period that can be resolved is 2 sols, and the direction of zonal propagation cannot be resolved. In this case there are two eddy modes that are always possible, one traveling eastward and one westward. There are an infinite number of other possible modes, but they have much shorter periods (see Hinson and Wilson (2002) for a specific example of this aliasing). Hinson et al. (2012) discuss the aliasing problem for data obtained at one local time, and contrast it with the situation for data obtained at two local times (twice per orbit, at a given latitude). The possibility of aliasing of higher-frequency modes always exists, even for the combined day and night TES data. The fact that very short-period disturbances are not present in the Viking data essentially eliminates this possibility, at least for the near-surface RS data. The fact that the phase relationships between pressure and meridional wind in the Viking Lander 2 data are those expected for large-scale, eastward-traveling waves essentially eliminates the possibility of the dominant eddies in the MGS RS data being westward-propagating. Finally, analyses of the combined dayside and nightside TES data show that most of the variance is indeed eastward-propagating throughout the wintertime seasons.

Different investigators have employed different analysis approaches to examine the MGS TES and RS temperature data. Barnes (2001, 2003a,b, 2006) and Barnes and Tyler (2007) analyzed the TES temperature data using the Fast Fourier Synoptic Mapping (FFSM) technique originally developed by Salby (1982a,b), and later refined for application to terrestrial satellite data by Lait and Stanford (1988). This method transforms the data to rotated coordinates to remove the coupling between longitude and universal time, and then makes use of separate one-dimensional fast Fourier transforms (FFTs) of the day and night data. The FFSM approach has a number of distinct advantages, as discussed by Salby (1982a,b) and Lait and Stanford (1988). The most fundamental one is that the coordinate transform makes it an orthogonal decomposition of the asynoptic data, such that the Nyquist limits are precisely defined. As the name suggests, synoptic maps are easily produced after the wavenumber–frequency spectra are obtained. These maps contain only the fully resolved spectral components. The FFSM technique was developed for application to combined dayside/

nightside data, but it can also readily be applied to "one-sided" data (either day or night only). There are numerous intervals in the TES dataset for which there are not enough data from both the dayside and the nightside orbits in middle and high latitudes, and thus a "one-sided analysis" must be performed.

The various other analyses of the wavenumber and frequency content of the TES and RS data have made use of either a correlation method or a least-squares fitting, to traveling waves, of the data (Hinson and Wilson, 2002; Wilson et al., 2002; Banfield et al., 2004; Hinson, 2006; Wang, 2007; Hinson and Wang, 2010; Wang et al., 2011; Hinson et al., 2012). The greatest advantage of these approaches is that no interpolations of missing data points are required. They are required for the FFSM method since FFTs require uniformly spaced data. The different approaches to analyzing the asynoptic TES and RS data can be viewed as essentially complementary, although each has its own advantages and disadvantages. They have generally yielded very similar results for the dominant wavenumbers and periods in the orbital datasets. For all analyses of one-sided data (and thus all of the RS data analyses), it must simply be assumed that all or most of the variance is indeed associated with eastward-propagating eddies.

The Viking observations of the transient eddy periods were confirmed and greatly extended by the analyses of the TES and RS data. The existence of transient eddy activity in the southern hemisphere during the wintertime seasons was quickly confirmed by the TES data (Hinson and Wilson, 2002); this had been predicted by GCM studies (e.g. Barnes et al., 1993). The dominant periodicities in the TES and RS data are ~2–10 sols at low atmospheric levels; at higher levels (~25 km and above) dominant periodicities as long as ~20–25 sols were found. The longer-period modes (~6–25 sols) are characterized by a dominant zonal wavenumber of 1. The shortest-period modes (~2–3 sols) are generally characterized by a dominant zonal wavenumber of 3, and modes with intermediate periods (~3–5 sols) generally have a dominant wavenumber of 2. The longer-period modes are more prominent in the northern wintertime seasons, while the shortest-period modes are more dominant during the southern wintertime seasons. A very important aspect of the eddy modes revealed by the analyses of the TES and RS data is that typically multiple zonal wavenumbers have modes at the same period/frequency. This is the spectral signature of a storm zone, which is a confinement of the strongest eddy activity to certain longitudinal regions. The storm zones are discussed below.

As had been found to be the case in the Viking meteorology data, the dominant eddy modes varied considerably in time in the TES and RS data. Some of this variation was clearly in direct association with large dust storm activity (large regional and global storms), and some of it was not. Transitions between different dominant transient eddy modes are a very basic aspect of the eddy activity in the Martian atmosphere, one that is linked to dust storm activity, seasonal changes, and the basic nonlinear dynamics of the atmospheric system. Given this behavior, a typical seasonal interval of ~30–60 sols or more will be characterized by at least several different dominant eddy modes. Figure 9.24 evidences this, as observed at the Viking Lander 2 site. One must perform analyses that can identify the dominant periods and wavenumbers in an essentially continuous (in time) manner in order to define seasonal intervals in which

only one eddy mode is dominant. In some intervals, two different modes are present with very comparable amplitudes. For the five selected intervals of Viking data that were analyzed by Barnes (1980, 1981, 1983), no effort was made to break apart the different sub-intervals dominated by different periodicities. Thus two or three major spectral peaks emerged from the analyses of each of the intervals of data. This was also the case in the analyses done by Banfield et al. (2004), as each Mars year was broken up into 12 intervals that were 30° of L_s in length. In most cases only the dominant mode in each interval was characterized. Hinson (2006) and Hinson et al. (2012) isolated intervals that were dominated by different eddy modes, analyzing near-surface data only. The former study used RS data, and thus the identified dominant modes might not be representative of other latitudes in the extratropics. Analyses of the TES data show that different transient eddy modes have their maximum amplitudes at different latitudes (e.g. Barnes, 2003a,b, 2006; Barnes and Tyler, 2007), as well as at different vertical levels.

The biggest surprise in the TES data was almost certainly the existence of very slowly traveling disturbances that exhibited quite large temperature amplitudes at upper levels (~0.5 mbar and above). These disturbances were dominated by zonal wavenumber 1. They were found to be present in the northern hemisphere with quite large amplitudes during seasonal intervals following the onset of large dust storms. Wilson et al. (2002) examined such disturbances in an interval following the start of a large regional dust storm in the northern autumn of MY 24 (at $L_s \sim 230°$). The dominant wave-1 mode had an average period of ~20 sols. Its temperature amplitude at the surface was very small (~2 K) in the TES data, while its upper-level amplitude was very large. As the atmosphere gradually became much less dusty, the wave-1 disturbance weakened and it largely disappeared by $L_s \sim 270°$. Banfield et al. (2004) documented the time evolution of the period and amplitude of this upper-level eddy mode and its replacement by a much shorter-period (~6–7 sols) wave-1 mode around winter solstice. The much shorter-period mode was very similar to the wave-1 modes in other seasonal intervals in northern wintertime, with substantially larger near-surface amplitudes along with very sizable amplitudes at upper levels. During the global dust storm of MY 25, a large-amplitude wave-1 mode with long periods (~15–25 sols) was present in the north (Banfield et al., 2004; Barnes, 2006). This eddy mode did not become prominent until well after the start of the global dust storm at $L_s \sim 185°$, indicating that seasonal changes as well as those due to the dust storm were necessary for its existence. In MY 26 large-amplitude, upper-level, wave-1 modes were prominent following the onset of large regional dust storms in both early autumn and mid-winter (Barnes, 2006). Making use of data assimilation, Lewis et al. (2016) have recently provided a picture of the seasonal variations of the transient eddy activity at both lower and upper levels for all three TES years.

(b) Eddy Structures   Analyses of the TES data allow a determination of the latitudinal and vertical structures of the different transient eddy modes. In an FFSM analysis, the amplitudes and phases of the complex spectral coefficients provide this information given the latitudes and vertical levels of the data analyzed. The same information can also be obtained in the other analysis

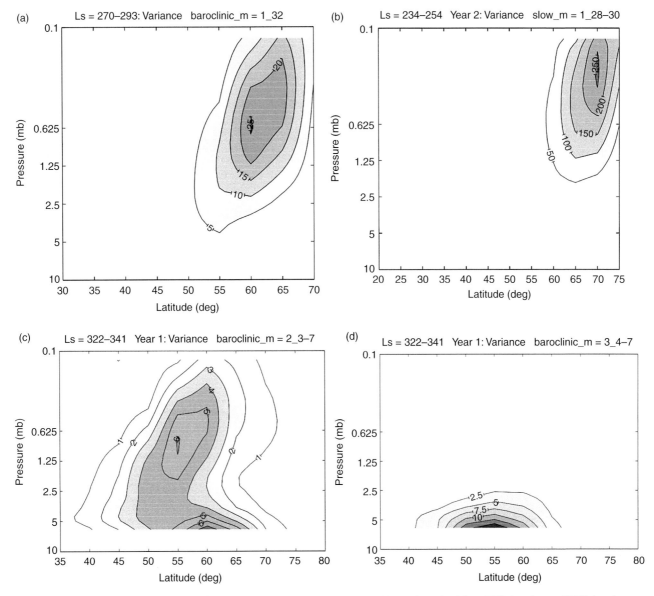

Figure 9.25. (a) Temperature variance (in K²) for a zonal wavenumber-1 transient eddy mode, derived from FFSM analyses of TES data from a seasonal interval in early northern wintertime in MY 24 (L_s ~ 270–293°). The period of this mode is ~7 sols. (b) As (a) but for a period in northern autumn in MY 25 (L_s ~ 234–254°). The period of the mode is ~20–25 sols. (c) As (a) but for a wavenumber-2 mode in a late winter interval (L_s ~ 322–341°) in MY 24. (d) As (a) but for a wavenumber-3 mode. J.R. Barnes performed the FFSM TES data analyses.

approaches. Linear baroclinic instability theory and modeling generally predict that longer-wavelength modes will be deeper in extent than shorter-wavelength modes. The wavenumber spectrum of the energetic transient eddies is greatly "compressed" in the Mars case by comparison with the terrestrial atmosphere, such that zonal wavenumber-1 modes can have extremely deep vertical structures while the wavenumber-3 modes tend to have relatively shallow structures, in *temperature*. Wave-2 modes tend to have an intermediate vertical structure, exhibiting temperature maxima both at the lowest levels and at higher levels. Figure 9.25 shows examples of transient eddy mode (latitude–height) structures for each of the primary zonal wavenumbers (1–3), as determined using FFSM analyses of TES data.

It is very important to note that the transient eddy thermal structures are markedly different from those for other primary atmospheric fields, the horizontal winds, geopotential (or pressure), and surface pressure, in particular. The geopotential and horizontal wind eddy amplitudes will both generally tend to have much deeper vertical structures than the temperature amplitudes do. Modeling studies clearly show this, but it essentially has to be the case on the basis of hydrostatic balance given the thermal structures. The RS data allow a determination of the vertical structure within the very narrow latitude bands of the observations, since the RS retrievals yield values for the surface pressures as well as the temperatures (at very high vertical resolution). Given this, the eddy geopotential field can be determined. With the assumption of geostrophic balance, the eddy meridional wind field can then be determined as well, but not the zonal wind field. An example of a vertical structure determined from the RS data is shown in Figure 9.26. It can be seen that substantial geopotential amplitudes (and thus horizontal winds as well) are present at much greater heights than the sizable temperature amplitudes are, for a wave-3 mode having a rather shallow temperature structure. In the case of the very deep, wave-1, transient eddy modes, large-amplitude geopotential and wind perturbations can extend to very high altitudes in the atmosphere. Modeling studies have indicated that they do; this is discussed briefly in the next section.

The analyses of the TES temperature data also allow synoptic (latitude–longitude) maps of the transient eddy structures to be produced, at various vertical levels. Figure 9.27 shows an example of such a map, produced using an FFSM analysis for an interval in late northern winter in MY 25. A very large-amplitude wavenumber-3 mode is present in this short interval, and the map (at the 6.1 mbar level) shows its horizontal structure at one instant in time early in this interval. Other eddy modes are also present in the interval, but they have much smaller amplitudes than the wave-3 mode; their contributions are incorporated in the map.

(c) Storm Zones GCM modeling studies predicted that storm zones should exist in the Martian northern hemisphere (Hollingsworth et al., 1996, 1997). Almost certainly a consequence primarily of the large low-wavenumber topography in the extratropics, the storm zones are the longitudinal regions in which the transient eddy activity is substantially stronger than in other regions at roughly the same latitudes. Such regions are very prominent in Earth's atmosphere, where they are a consequence of the existence of continents and oceans, as well as topography. The early analyses of the MGS TES data confirmed

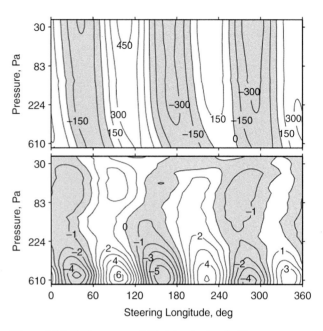

Figure 9.26. Eddy geopotential heights (top) and eddy temperatures (bottom) for a wavenumber-3 transient eddy mode, as derived from MGS RS data in northern wintertime. The cross-sections were derived for a reference frame moving at the wavenumber-3 phase speed (59° of longitude per sol), averaged over the seasonal interval used for the analysis. The "steering longitude" is the east longitude. Contributions from eddy modes at different wavenumbers are present only if their phase speeds are the same as the reference frame phase speed. This means that the eddy modes at different wavenumbers that contribute to a storm zone structure do *not* contribute to these two fields, as they average to zero as a result of having different phase speeds. Figure produced and provided by D.P. Hinson.

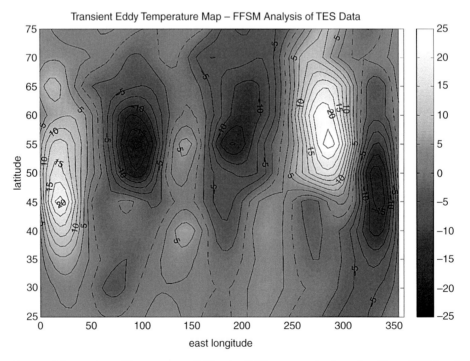

Figure 9.27. A synoptic map of the transient eddy temperature field at the 610 Pa pressure level, produced by an FFSM analysis of TES data for an interval in northern winter of MY 25 ($L_s \sim 326$–$333°$). This map is for a time early in this interval, and it shows a large-amplitude, zonal wavenumber-3, transient eddy mode (other wavenumbers are also present). The contour interval is 2.5 K with the zero contours dashed. The longitudes are east longitudes. J.R. Barnes performed the FFSM analysis of the TES data.

(a)

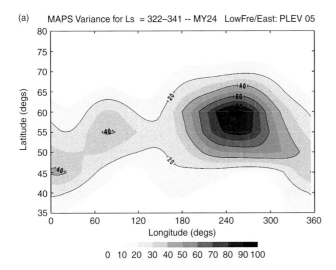

(b)

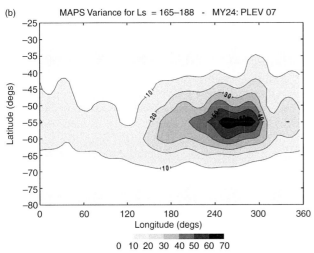

Figure 9.28. Two storm zone plots derived from TES temperature data using FFSM analyses (performed by J.R. Barnes). (a) The storm zone for the 610 Pa level, for an interval in late northern winter in MY 24 ($L_s \sim$ 322–341°). (b) The storm zone for the 370 Pa pressure level, for an interval in late southern winter in MY 24 ($L_s \sim$ 165–188°). In both plots the values shown are the total transient eddy temperature variances (in K²). The longitudes are east longitudes.

that storm zones are present on Mars in both hemispheres. In the northern hemisphere the storm zones – as defined by the transient eddy temperature variance near the surface – generally lie to the west of or in the major lowland regions of Acidalia, Utopia, and Arcadia. In many seasonal intervals during northern autumn and winter, the largest transient eddy variance in the TES data is roughly centered just to the north of Alba Patera and Tempe Terra. In some intervals the strongest eddy activity is located just to the west of and in Utopia. The lowland regions in the north are where the so-called "flushing" dust storms occur during northern autumn and winter. These are discussed below.

In the southern hemisphere there is typically a single major storm zone, which is largely confined to the western hemisphere; the southern eddy activity is typically strongest between approximately 200° and 300°E. Examples of the northern and southern storm zones near the surface are shown in Figure 9.28. In the northern hemisphere the storm zones frequently extend

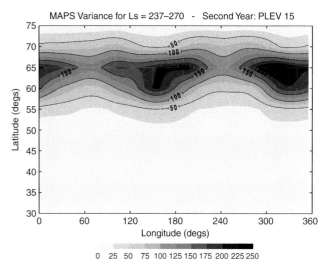

Figure 9.29. Storm zone plot, as in Figure 9.28, for an interval in northern autumn (L_s = 237–270°) in MY 24. The plot is for the 50 Pa pressure level (~25 km altitude).

very strongly to upper levels in the TES data. Figure 9.29 shows an example of an upper-level storm zone structure. The upper-level storm zones tend to lie at the same longitudes as those near the surface, but they are located at higher latitudes – in the vicinity of the greatest latitudinal temperature gradients and thus the maximum zonal wind shears and speeds.

(d) Dust Clouds The MGS MOC and MRO MARCI wide-angle imaging systems have obtained extensive visible imagery that shows dust cloud features that are associated with transient eddies in the northern hemisphere (Wang et al., 2003, 2005; Wang, 2007; Hinson and Wang, 2010; Hinson et al., 2012). In some cases the dust becomes organized into local storms, and these storms actually are able to move into the southern hemisphere in the major lowland regions of the north. Because of its lack of global imaging coverage, Viking never observed this phenomenon. As this dust lifting in the north generally begins in very early autumn, it is a primary contributor to the annual increase in dust opacity in the north at this time. Combined analyses of the MOC imagery and the TES and RS data showed that much of the observed flushing dust storm activity is associated with the presence of large-amplitude, wavenumber-3, transient eddy modes (Wang et al., 2005; Wang, 2007; Hinson and Wang, 2010; Hinson et al., 2012). Such short-period (~2.0–2.7 sols) modes are present during relatively limited seasonal intervals in autumn and winter (see table 1 in Hinson et al., 2012) in MY 24, 25, 26, and 27. Such a mode is shown in Figure 9.27 (as well as in Figure 9.25). Wave-3 modes are favored for dust activity (compared to waves 1 and 2) since they will be associated with stronger meridional winds for given surface pressure perturbations, because of their shorter zonal scales. Hinson and Wang (2010) and Hinson et al. (2012) showed that the stationary eddies can provide a significant boost to the near-surface winds in particular longitudinal regions. Another major circulation component that appears to play a significant role in the flushing dust storms is the diurnal tide, as the ~2 sol periodicity of the eddies appears to allow them to be synchronized with the tidal winds in certain intervals, in the lowlands. The western boundary currents are also strongest near the surface on the western sides of the lowland regions, and are directed southward at these seasons (see Section 9.4).

The flushing dust storms occur in early autumn and also in middle to late winter, during seasonal intervals that at least partially overlap with those in which the wave-3 modes are dominant near the surface. Wang et al. (2013) show the very clear relationship that exists between the two phenomena during MY 24 and MY 26. Between the two seasonal periods there is an extended period during which the near-surface eddy activity is relatively weak; this period is roughly centered on northern winter solstice. This has been termed the "solstitial pause" in lower-level eddy activity (e.g. Lewis et al., 2016). During this extended late autumn and early winter seasonal interval, large-amplitude, upper-level, eddy modes are generally very prominent, as noted above. The TES temperature data show that, when the northern winter atmosphere becomes sufficiently dusty and the seasonal date is not far removed from solstice, the strong transient eddy activity shifts to upper levels and the near-surface activity weakens substantially. The flushing dust storms tend to occur during seasonal intervals in which the near-surface eddy activity is very strong, and that at upper levels is relatively weak. A rather similar minimum in the near-surface transient eddy activity also occurs in southern wintertime, around solstice, but this is not associated with any significant changes in the atmospheric dust loading (e.g. Barnes and Tyler, 2007; Wang et al., 2011; Lewis et al., 2016). In sharp contrast to the northern hemisphere, the very large-amplitude (at upper levels), wave-1, transient eddy modes that are so prominent in the north under dusty conditions do not appear to exist in the south (Barnes and Tyler, 2007; Lewis et al., 2016).

9.6.4 Modeling

Theory and modeling predicted the existence of transient baroclinic eddies in the atmosphere of Mars. The very long gap in successful spacecraft missions to Mars following Viking was marked by the development of considerably more advanced global circulation models, and these were used to study transient eddies along with the other major components of the global circulation. A number of major modeling studies were performed in the decade prior to the MGS mission. Once the MGS TES atmospheric temperature data became available, the focus shifted to the analysis of these data. The models have continued to advance, revealing new aspects of the transient eddy activity. Modeling studies have recently indicated that the radiative effects of water ice clouds are of considerable importance for the transient eddy activity as well as other circulation components. In the absence of new observational data allowing a direct characterization of the transient eddies, modeling has to be used extensively to try to fill in the large gaps that remain in the observational picture. Data assimilation is now allowing the relatively sparse observations to be combined with models in a dynamically self-consistent way, so as to yield a synthesized picture (a model and data hybrid) of the global circulation. In the absence of "ground truth" observations, especially near the ground and everywhere in the case of the winds, the fidelity of the assimilation data products has to remain somewhat uncertain.

9.6.4.1 Early Studies: Linear Instability and GCMs

Some of the earliest modeling work on the transient eddies was referenced in preceding sections. This work consisted of the application of simple linear models to the Mars problem, and the initial Mars GCM studies. Zurek et al. (1992) contains a summary of these early studies, so the discussion here will be brief.

Most of the early modeling studies were linear ones. Mintz (1961) and Leovy (1969) used very simple two-level/two-layer models to get at the most unstable zonal wavenumbers, the phase speeds (periods), and the growth rates of unstable disturbances. They found that the zonal wavenumbers that were most unstable generally were in the range of ~1–4, and they predicted significant seasonal variations in these. In the very simple instability models, this was largely due to variations in the static stability of the wintertime atmosphere. The strong dependence on static stability of the wavenumber of fastest growth in the simple linear models is largely an artifact of the highly simplified vertical eddy structures in these models. This dependence is not present in the vertically continuous, Charney model of baroclinic instability (see e.g. Pedlosky, 1979). Leovy (1969) examined the impact on the linear instability of the very strong radiative damping in the Martian atmosphere, finding that it decreased the unstable growth rates only slightly. Blumsack and Gierasch (1972) made an initial examination of the effects of topography on linearly unstable eddies, showing that it could cause the baroclinic eddies to be strongly stabilized in a simple model depending on the relationship of the topographic slopes to the isentropic slopes.

The use of more realistic linear models began with the study by Gadian (1978). He found maximum instability for zonal wavenumbers of ~3–6, and made estimates of poleward heat fluxes using theoretical parameterizations. Barnes (1983, 1984) employed beta-plane and spherical quasi-geostrophic models to examine linear baroclinic instability in the presence of both radiative and Ekman damping (representing surface friction). Barnes also investigated the effects of zonal-mean topographic slopes on the unstable eddies. The zonal-mean slopes in the northern hemisphere were then known to slope downwards towards the pole in middle latitudes, and such topography was found to be strongly stabilizing. Barnes found that the zonal scales of the disturbances with the largest growth rates did not change very much for a dust storm basic state with greatly increased static stability (as well as stronger radiative damping). But the growth rates for such a state were reduced greatly. The vertical structure of the unstable eddies was found to vary dramatically over the range of strong instability (zonal wavenumbers 2–4). Longer-wavelength disturbances were characterized by much deeper vertical structures, while the short-wavelength ones exhibited much shallower structures. This is a general aspect of the Charney instability problem, but the range of this great variation in depth for Mars is greatly compressed in zonal wavenumber space as noted earlier. This is due largely to the small size of Mars, and it is almost certainly further accentuated by the lack of a tropopause on Mars.

Barnes (1986) performed simulations of nonlinear baroclinic instability with a quasi-geostrophic, beta-plane model that represented a single zonal wavenumber eddy field and a zonal-mean flow. A chief motivation for this study was to examine the influence of very strong radiative damping on the nonlinear equilibration of an unstable disturbance. The simulations showed that the eddy equilibrated fairly rapidly with the mean flow if the radiative damping was Mars-like in strength, but did not for much weaker damping. For strong damping, the equilibrated eddy structure corresponded to that of the most unstable linear mode for the equilibrated zonal-mean state. With the model's strong radiative damping as well as surface friction, this mode was a neutral one, maintaining a constant amplitude and structure. This early study suggested that the strong radiative damping in the Mars atmosphere could play a major role in yielding a more "regular" behavior for the transient eddies than exists on Earth.

Pollack et al. (1981) presented new GCM results, from simulations performed with an improved version of the earlier GCM developed by Leovy and Mintz (1969). The new GCM incorporated topography as well as albedo variations, and had three layers instead of two. This allowed the model to simulate a diurnally varying convective boundary layer. By this time it was well known that the radiative effects of dust were extremely important for the atmosphere, and Pollack et al. (1981) tried to minimize the impact of not including this in the model by focusing on a simulation for southern winter solstice (when the dust loading is relatively low). With topography, large-amplitude quasi-stationary eddies were present in the simulation in the winter extratropics, as well as in the tropics and the summer hemisphere. Transient eddies were prominent in the winter (southern) hemisphere, and their basic properties compared fairly well with those that had been observed by Viking.

9.6.4.2 New GCMs

In the decade preceding the MGS mission, a great amount of effort was invested in advancing global circulation modeling for Mars. Multiple new GCMs were developed, and the pre-existing NASA Ames GCM was greatly improved. Some use was also made of "simplified" GCMs, dynamically the same as the GCMs but containing highly simplified representations of various physical processes.

The Ames GCM was greatly improved by the incorporation of the radiative effects of atmospheric dust, and by much higher vertical resolution and extent (13 model layers, with a model top at ~45–50 km). Barnes et al. (1993) presented results from an extensive study of the transient eddies in a large set of simulations performed with this new Mars GCM. Transient eddies were present in both hemispheres during the autumn, winter, and spring seasons. The periods and wavenumbers were found to be very similar to those determined from the Viking data. The biggest surprise was that the transient eddies in the south were much weaker than those in the north. It was shown that this was in sizable part due to the unique topography in the south, but it also appeared to be at last partly due to the weaker thermal forcing in southern wintertime. Most of the simulations utilized the so-called Mars Consortium topography, which was rather unlike the actual topography (as later determined by the MOLA instrument on MGS) in middle and high latitudes. One or two experiments were performed with an updated topography (the so-called DTM topography), which did not have the very low elevations in southern polar latitudes that the Consortium topography had. Even with this topography the transient eddies were still substantially weaker than those in the north. Barnes et al.

(1993) showed that the GCM transient eddies had relatively deep vertical structures, especially so in geopotential height and kinetic energy. For highly dusty northern winter solstice conditions (a uniform global optical depth of 5), the transient eddy amplitudes near the surface were not much reduced, if at all, in comparison to those obtained for much less dusty conditions. This was in sharp disagreement with the Viking observations during the winter solstice global dust storm of 1977 (MY 12). The westerly jet did not shift further north by more than a very small amount in these GCM simulations. There also was no change in the basic vertical structure of the transient eddies in the very dusty simulations. We now know that these results were largely a consequence of the relatively low height of the model top (~45–50 km).

Also employing the Ames GCM, Hollingsworth et al. (1996, 1997) showed that there was large longitudinal variability in the near-surface transient eddy activity in the northern hemisphere: storm zone structure. An interesting aspect of these results was that the eddy kinetic energy tended to have maxima in the major lowland regions (Acidalia, Utopia, and Arcadia), while the eddy meridional heat flux had its maxima to the west (upstream) of these regions.

Collins et al. (1996) used the fairly new UK and LMD Mars GCMs to examine the transient eddies during the northern wintertime seasons. They performed extended simulations (eight years in length) with the UK model and found that the transient eddies were extremely regular in nature in the absence of a diurnal cycle in the model, but were much less regular with a diurnal cycle in the model. The latter behavior was much more similar to that seen in the Viking meteorology data than the former. The LMD model, which had a much higher top, did not exhibit nearly as regular behavior in the absence of a diurnal cycle. Both models were dominated by zonal wavenumbers 1 and 2, in sharp contrast to the Viking meteorology results as well as the Barnes et al. (1993) GCM results.

9.6.4.3 Modeling and Observations

After the TES atmospheric temperature data began to become available, modeling studies were often targeted at particular aspects of the transient eddies that were revealed by the data analyses. Hinson and Wilson (2002) performed a GCM simulation (using the GFDL (Geophysical Fluid Dynamics Laboratory) model developed by Wilson) for a seasonal interval centered on $L_s \sim 145°$, to examine the southern hemisphere eddies. The simulation was dominated by a wave-3 disturbance having a period of ~2 sols, in good agreement with results obtained from analyses of RS data for this same season, results presented in the same paper. The model eddy activity showed a very pronounced storm zone structure (strongest eddy activity at ~180–330°E), which compared well with that evidenced by the RS data.

Wilson et al. (2002) performed a GCM simulation motivated by TES observations of large-amplitude, wave-1 disturbances at upper levels, following the onset of a large regional dust storm at $L_s \sim 228°$ in MY 24. As discussed above, in the seasonal interval extending almost to winter solstice, a very slowly traveling (with an average period of ~20 sols) wave-1 disturbance was present in the TES data. The temperature amplitude of this mode near the surface was very small (~2 K). After winter solstice, a wave-1 mode with a somewhat different vertical structure and phase speed became dominant. This mode had larger low-level amplitudes and a period of ~7 sols, a very prominent periodicity in the Viking observations. Rather similar wave-1 modes were found to be present in the model data, though the period of the slower mode was ~10 sols and not ~20 sols. The fast wave-1 mode had the basic character expected for an eddy associated with "surface" baroclinic instability, and it appeared to be very similar to the wave-1 modes previously identified by Barnes et al. (1993) in simulations with the NASA Ames GCM. Wilson et al. (2002) showed that the zonal-mean GCM circulation was inertially unstable in lower latitudes at upper levels, and that the meridional wind field of the slow wave-1 mode extended into the southern hemisphere at upper levels. They speculated that the inertial instabilities in the model might be supplying energy to a Rossby wave "free mode" centered in the westerly jet. The possible existence of such eddy modes had been previously suggested by modeling studies done by Hollingsworth and Barnes (1996) and Barnes et al. (1996b). Barnes et al. (1993) had previously shown that inertial instabilities were present in low latitudes in Ames GCM simulations for northern winter solstice.

Considerable modeling effort has been aimed at obtaining a better understanding of the flushing dust storms in northern autumn and winter. Wang et al. (2003) employed the GFDL GCM to attempt to simulate the type of dust features seen in numerous MGS MOC wide-angle images. They simulated dust transport by simply injecting particles into certain regions and modeling their evolution using an offline transport code and the GCM wind fields. The results indicated that the dust field could form curvilinear features resembling those seen in the imagery. At least in some cases the dust could be transported well to the south, even into the southern hemisphere in the Acidalia and Chryse lowlands. Wang et al. (2003) found that only eddy disturbances that entered the lowland regions during the daytime were able to transport dust rapidly southward, and this happened only in early autumn and later winter. They argued that tidal winds and western boundary current winds strongly enhanced the transient eddy winds (southward behind cold fronts), enabling rapid and extended transport of dust to the south. During early autumn and late winter, wave-2 and wave-3 eddies were active, and these had relatively strong near-surface winds. Closer to winter solstice, wave-1 disturbances were prominent in the GCM, and these were characterized by much weaker near-surface winds.

Wilson et al. (2006) used the GFDL GCM to examine the transient eddies in relation to the annual dust cycle and the flushing dust storms. Their model predicted water ice clouds that could be allowed to be radiatively active. It was found that this could very significantly influence the transient eddy activity, through its effects on the zonal-mean thermal structure in middle and high latitudes. This was the first time that the possible impact of such clouds on the transient eddies was noted. This now appears to be extremely important, and is discussed below. Wilson et al. (2006) found that the eddy meridional winds were strongest near the surface during early autumn and late winter in the northern hemisphere, and in late winter and early spring in the south. Strong storm zones existed in the northern lowlands (for the meridional wind variance). The intervals of strongest

low-level meridional wind variance in the north tended to be dominated by zonal wavenumber 3, and these modes had periods of ~2 sols. Wilson et al. also performed an interactive dust lifting simulation, showing an example of a very extended curvilinear dust band associated with a cold front. As in Wang et al. (2003), tidal winds contributed very significantly to the dust lifting and transport in the Chryse Basin – with the storms moving southward during daytime periods.

Hollingsworth and Kahre (2010) carried out a detailed examination of a synoptic event in the northern hemisphere (in late winter) in an interactive dust lifting simulation with the Ames GCM. They found that much of the dust lifting occurred in association with strong nocturnal drainage winds off the Tharsis uplands, and not actually near the cold frontal band. A major factor underlying this may have been that the simulated weather system was a wave-2 mode, and not a wave-3 mode. The latter will tend to produce significantly stronger near-surface winds and thus may also tend to generate more intense frontal zones. The horizontal resolution employed in the modeling was likely also not as high as needed to simulate a very strong frontal band.

Wang et al. (2011) performed a GCM simulation for late southern winter ($L_s \sim 150\text{–}170°$), and showed that a wave-3 mode with a period of ~2 sols was dominant in the model in middle latitudes. This compares well with the analyses of TES data for this season. Barnes and Tyler (2007) showed a near-surface storm zone pattern for surface pressure from a southern wintertime ($L_s \sim 120°$) simulation with a high-resolution mesoscale model, which was very similar to the southern storm zones found in the TES data.

Kavulich et al. (2013) analyzed data from a simulation for northern autumn and winter (using the GFDL GCM) to examine the local dynamics and energetics of the transient eddies. They found that the strongest baroclinic eddy energy conversions (in a time-mean sense) occurred in or just upstream of the Acidalia and Utopia regions, with the eddy kinetic energy maxima located a short distance downstream of these. This is similar to the results obtained by Hollingsworth et al. (1996, 1997). An interesting finding of Kavulich et al. (2013) was that the eddies were extracting kinetic energy from the time-mean flow barotropically (via momentum fluxes) in some regions, and feeding energy back to the mean flow in other regions. The transient eddies in the simulations were globally coherent, but the most intense baroclinic energy conversions were found to be relatively localized in space and time.

Arguably the most interesting and potentially important modeling in the recent era has taken place rather recently. This is global modeling that shows that water ice clouds can be extremely important for the transient eddy activity. Hinson and Wilson (2004) first presented results from a Mars GCM that was able to simulate the radiative effects of water ice clouds. As noted above, Wilson et al. (2006) discussed aspects of the transient eddy activity in a GCM that simulated the water cycle and clouds, and allowed the clouds to be radiatively active. They reported that the latter had a substantial impact on the mean thermal structure in the model, at least in some seasons. The simulated polar hood clouds produced strong IR cooling at lower levels and lower-latitude clouds produced warming at higher levels that acted to enhance the zonal-mean baroclinicity in middle and high latitudes. Wilson (2011) and Hollingsworth et al. (2011) showed that transient eddy activity was very substantially increased in their GCMs during both northern and southern wintertime by the presence of radiatively active clouds. The eddy meridional winds were very strongly increased near the surface in both hemispheres in the Wilson study, in autumn as well as in late winter and early spring.

Barnes et al. (2014) and Rucker (2014) investigated the impact of radiatively active clouds on the dominant transient eddy modes throughout the annual cycle in both hemispheres, by analyzing data from very extended simulations with the Ames GCM. Barnes et al. (2014) noted that the GCM did not produce simulations of the polar hood clouds that agreed well with observational data – the clouds tended to be overly extensive and thick, particularly in the north. The radiatively active clouds had a dramatic impact on the transient eddy activity in both hemispheres. In the north the strength of the wave-3 activity was greatly increased, such that during some seasonal intervals in autumn and late winter the wave-3 modes were dominant. In the absence of radiatively active clouds the wave-3 activity in the GCM simulations was very weak in the north. All of the current GCMs seem to have considerable difficulty in simulating the very strong wave-3 activity in the north that is present in the observational data during certain intervals in early autumn and in middle to late winter. Wang et al. (2013) carried out a GCM study (without radiatively active clouds) that attempted to get better simulations of the wave-3 activity in the north by putting more dust into middle and high latitudes. The added dust was in the form of moving "blobs" having a wave-3 pattern and the same phase speed as the wave-3 eddies in the model. This dust distribution did enhance the wave-3 activity in the model in autumn and later in winter. However, this wave-3 dust as well as a zonally symmetric distribution of added dust were not in good agreement with the TES and MCS dust data (especially the latter data, which show very little dust above the seasonal polar cap). Mulholland et al. (2016) showed that radiatively active clouds played a very important role in producing the solstitial pause in lower-level eddy activity, in both hemispheres. It now appears rather clear that radiatively active water clouds play a very important role in the annual cycle of transient eddy activity in both hemispheres. These clouds may be especially critical to the existence of the strong wave-3 eddies in the northern hemisphere that play such a major role in strong dust storm activity in autumn and in middle to late winter.

9.6.5 Summary

The study of transient eddies in the atmosphere of Mars has been a very fertile area of research for a long time. These eddies were predicted by very early theoretical and modeling work, and their existence was confirmed by analyses of the Viking meteorology data. The MGS orbital mapping mission finally allowed many of the basic properties of the transient eddies to be determined, by analyses of the TES and RS temperature data. The holes in the observational datasets for these eddies are still very large, given the lack of simultaneous surface pressure data over the globe and the lack of wind data. The global models have become more and more sophisticated, and simulations with these have shown that water clouds play a very important role in the annual cycle

of eddy activity, along with variations in atmospheric dust loading. In order to yield good annual and multi-annual simulations of the transient eddies, the global models will have to be able to produce good simulations of the water clouds. The clouds can be very sensitive to the dust in the model, since the dust particles act as ice nuclei. Eventually, fully coupled simulations of the dust and water cycles will be performed, in which the seasonally varying dust loading is not prescribed. Some such simulations have already been carried out. These constitute a considerable challenge to the modeling capabilities for Mars. Reproducing the observed variability in the transient eddy activity could provide some of the strongest "leverage" on the global models because these eddies are rather sensitive to the mean thermal structure of the atmosphere, as well as the distributions of the different aerosols (dust, water ice, and CO_2 ice).

The upper-level transient eddies that are present in the north during very dusty periods in the wintertime seasons are a unique feature of the Martian circulation. On the basis of modeling, these eddies appear to have a strongly barotropic character – drawing kinetic energy directly from the intense zonal-mean zonal flow. These transient eddies appear to represent an alternative "regime" of the middle- and high-latitude circulation in wintertime, since the near-surface eddy amplitudes generally tend to decrease considerably when the upper-level eddies are the most prominent. This is the case in a period roughly centered on winter solstice, and in periods immediately following (usually) the onsets of large regional and global dust storms. The global dust storm of MY 25 is a very interesting case in that the upper-level eddies did not become very prominent until well after its early autumn start. This indicates that the seasonal date is of considerable importance for the upper-level transient eddies, in addition to the atmospheric dust loading.

9.7 STATIONARY EDDIES

9.7.1 Fundamental Aspects

Stationary eddies in the atmosphere of Mars are forced mechanically by topography and thermally by zonal variations in heating. On Mars the latter are produced by the very large topography, as well as by zonal variations in surface albedo and thermal inertia. Zonal variations in dust (at constant pressure) will also provide thermal forcing for stationary eddies to the extent that the dust distribution is quasi-stationary. The transient eddies will also act to force stationary eddies if their heat and momentum fluxes are zonally asymmetric – which they are on Mars, according to the GCM simulations. The topography on Mars is dominated by planetary wavenumbers (1–4), in middle and high latitudes as well as in low latitudes (see Figure 9.4). In addition, Mars has very large zonally symmetric topography as shown in Figure 9.4. This acts to mechanically force stationary eddies in the presence of asymmetric topography (the meridional winds of the stationary eddies force vertical motions as a consequence of the zonally symmetric topographic slopes). The unique hemispheric dichotomy on Mars is associated with zonally symmetric topography that slopes upwards towards the pole in the southern hemisphere, and downwards towards the pole in the bulk of the northern hemisphere.

Planetary-scale stationary eddies on Mars have the basic character of Rossby waves in middle and high latitudes of the winter hemisphere. In this region the zonal-mean zonal winds are westerly from the surface up to very high altitudes (~70–100 km). As discussed in the next section, this means that the wintertime stationary eddies can propagate vertically and attain very large amplitudes at high altitudes. In the tropics and the summer hemisphere of Mars, the zonal flow is mostly easterly. Stationary eddies are vertically trapped by easterly zonal winds, and will thus be confined to lower altitudes. In regions of easterly zonal flow, the stationary eddies do not have the basic character of Rossby waves; in the tropics they will have more of the structure of Kelvin waves, with relatively small meridional winds.

Stationary eddies are a very important component of the global circulation on Mars for a number of reasons. In the winter hemisphere they can transport substantial amounts of heat towards the pole, and they also generally act to transport momentum. Stationary eddies play a major role in the existence of storm zone regions for the transient eddies, in both hemispheres. They act to transport dust and water down the zonal-mean gradient. It appears that they may be important for dust lifting, when the winds associated with them can combine with those of transient eddies (as well as tides, and slope flows) to produce significantly stronger winds near the surface. In the deep middle atmosphere of Mars, the stationary eddies can have very large amplitudes at high altitudes. Along with vertically propagating thermal tides and gravity waves, they thus can play a significant role in coupling the lower atmosphere to the upper atmosphere.

9.7.2 Basic Theory

Stationary eddies are fairly amenable to treatment with steady, linear models. It now appears that such models are able to provide a good basic understanding of stationary eddies in the terrestrial atmosphere (see e.g. Held et al., 2002). The stationary eddies in the atmosphere of Mars should be more linear than their terrestrial counterparts because the zonal-mean zonal flow is much stronger aloft, so it should be the case that the application of basic theory and linear modeling can provide a relatively good basic understanding of the structure and dynamics of the stationary eddies. To date there have not been very many such studies; there have been none at all since the turn of the century.

A very fundamental aspect of the dynamics of stationary eddies is the Charney–Drazin condition, which provides a framework for understanding under what conditions they are able to propagate vertically. As derived from the quasi-geostrophic equations in a beta-plane geometry (see e.g. Holton and Hakim, 2013), this condition implies that the stationary eddies can propagate vertically only if the zonal-mean zonal flow is westerly and weaker than a critical value that depends strongly upon the zonal wavelength of the eddies as well as the vertical and meridional structure of the zonal flow. It is the stationary eddies with the longest zonal scales (lowest wavenumbers) that are most favored for vertical propagation. For Mars, zonal wavenumber-1 eddies should tend to become dominant at sufficiently high levels in the atmosphere in the presence of

strong westerly zonal winds, even when higher wavenumbers have larger amplitudes in the lower atmosphere. The winds in the summer hemisphere on Mars are mostly easterly, except for the subtropical, low-level westerly jet. This is the same as the situation in the terrestrial middle atmosphere. In the presence of easterly winds, vertical wave propagation cannot occur, and the stationary waves cannot have a Rossby wave character.

The Charney–Drazin condition can be generalized to spherical geometry (see e.g. Andrews et al., 1987), in which case stationary eddies are able to propagate both meridionally and vertically depending upon their zonal scales and the structure of the zonal-mean zonal flow. An "index of refraction" for the wave propagation can be defined, such that propagation is possible when this is positive, and the stationary waves will tend to be refracted towards regions of large positive values.

For realistic westerly atmospheric flows, the typical situation on Mars is one in which the lowest-wavenumber stationary eddies can propagate meridionally and vertically in some portion of the atmosphere. At higher levels, even the lowest wavenumbers will become trapped. In the case of Mars, the very strong westerly wintertime jets act to provide "ducts" for the stationary eddies in which wave energy can propagate meridionally and vertically within the jet region. Above some height, the waves will generally become "evanescent" as a result of vertical trapping by the very strong zonal winds. Their amplitudes may continue to increase with height above this height, however. It is the wave energy (per unit mass) that will decrease with height. Stationary eddies that are vertically propagating will produce poleward heat fluxes, and will exhibit a westward phase tilt with height in their geopotential field. That is, at higher levels the troughs and ridges in the wave geopotential field will be shifted to the west compared to the corresponding features near the ground. If the waves are propagating poleward as well as vertically, they will produce equatorward momentum fluxes (there will be poleward momentum fluxes for equatorward wave propagation).

The higher-wavenumber stationary eddies that are vertically trapped will essentially be in phase with the topography in the case of mechanical forcing, such that geopotential/pressure ridges will be present at the location of the topographic ridges and troughs will exist over the topographic lowlands. For stationary eddies that exhibit some increase in amplitude with height, the ridges will be located close to the warmest air and the troughs will be close to the coldest air. For the longest eddies that are vertically propagating, the ridges will be located at least somewhat to the west of the topographic ridges and similarly for the troughs.

In the summer hemisphere, thermal forcing will tend to play a dominant role, except within the subtropical, low-level westerly jet. Since the isotherms tend to follow the topography at lower levels on Mars (the ground temperature being approximately independent of elevation), elevated topographic regions will tend to act as heat sources and lowlands will act as heat sinks if temperature decreases with height. (If temperature increases with height near the surface, the opposite relationship will exist.) Elevated regions will tend to force thermal lows, in which the warm air above the region yields a high-pressure center at higher levels. Such a stationary eddy will be characterized by convergence at low levels and divergence at higher levels, with rising motions at intermediate levels. A lowland region will tend to force the opposite kind of baroclinic circulation, characterized by sinking motions and divergence near the surface, with convergence aloft.

9.7.3 Observations

Early modeling studies predicted that stationary eddies would be present in the atmosphere of Mars, in both the summer and wintertime hemispheres – as discussed in the next section. But there were very few observations made that could detect these circulation systems prior to the MGS mapping mission. The only observational data that could show any evidence of stationary eddies were the atmospheric temperatures obtained by the Mariner 9 IRIS and the Viking IRTM instruments, along with the imaging observations of surface wind streaks obtained by both missions. Conrath (1981) analyzed IRIS temperature data for a late winter period in the north, and found an eddy disturbance with a sizable amplitude at ~0.5–1.0 mbar. The nature of the Mariner 9 orbit made it impossible to determine the zonal wavenumber and the period of this disturbance. A stationary wavenumber-2 eddy was one possibility, along with various transient eddies. We now know that a stationary wavenumber-2 eddy is very prominent during northern wintertime, and it is essentially vertically trapped as discussed below. Conrath inferred a substantial westward phase tilt with height from the data, however, which is not consistent with this component of the stationary eddy field. The phase near the surface is also not what has now been observed for this wave.

Banfield et al. (1996) analyzed the IRTM 15 μm temperature data (before they were corrected for a filter leak) for a period very near $L_s \sim 0°$. They found that a stationary wavenumber-2 disturbance was present in northern middle and high latitudes, as well as a wavenumber-1 disturbance in the south. This is essentially consistent with the TES and MCS data that we now have.

The bright wind streaks found in the Mariner 9 and Viking imaging data offered at least the possibility of providing evidence for stationary eddy circulations; these streaks are thought to form following large dust storms that occur at different seasonal dates in different years. Nayvelt et al. (1997) made use of the Mariner 9 and Viking bright streak data to compare with their linear modeling results, and found that the comparisons were improved when the modeled stationary eddy winds near the surface were included in addition to zonal-mean winds from a NASA Ames GCM simulation for northern winter solstice.

Fully unambiguous detections of stationary waves were made when researchers began to analyze the MGS TES and RS atmospheric temperature data. Hinson et al. (1999) examined RS data obtained prior to the start of the mapping mission for a period in early southern summer. The RS data in southern middle latitudes were sparse, but they clearly indicated that there was a large difference in the temperatures at constant pressure between a region just south of Tharsis and one over Hellas. The difference was entirely as anticipated on the basis of simple considerations for the summer hemisphere: the elevated location south of Tharsis was warm while that over Hellas was cold. Elevated topography was acting as a heat source and a lowland region (the lowest on the planet) was acting as a heat sink. Hinson et al. (2001) performed an analysis of a short segment of pre-mapping RS data acquired at ~65°N at $L_s \sim 75°$, in late

northern spring. These data showed that both wavenumber-1 and wavnumber-2 quasi-stationary waves were present in the time-averaged data, having amplitudes of ~3 K near the surface. The geopotential amplitudes showed no increase with height, except those for wave 2 above ~1.5 mbar. This increase was relatively rapid, which was not as expected for relatively weak westerly winds (considering the Charney–Drazin condition). The RS data had all been obtained at essentially the same local time, and thus the stationary eddy components could be aliased by certain non-migrating tidal components. Modeling that is discussed in the next section showed that the larger geopotential amplitudes at higher levels for zonal wavenumber 2 were almost certainly associated with a diurnal, wavenumber-1, Kelvin wave, tidal mode.

Banfield et al. (2003) carried out a systematic analysis using the TES nadir temperatures for the first full Mars year of mapping data to characterize the stationary eddies in each of 12 "months" defined by 30° intervals of L_s. The results were generally very much in accord with basic theoretical considerations, in that relatively large amplitudes were present during the wintertime seasons in both hemispheres. In northern wintertime zonal wavenumbers 1 and 2 had comparable amplitudes, whereas in southern wintertime wavenumber 1 was generally dominant. In the north both wavenumbers exhibited very little phase change with height, but in the south wavenumber 1 exhibited a significant phase change (a westward tilt) at lower levels. Banfield et al. speculated that this implied poleward heat fluxes, but, in the absence of any information about the geopotential field, this cannot actually be determined. They also noted that there were very sizable changes in the (absolute) phase of wavenumber 1 in the north between autumn and early winter, and that the phase of wave 1 then changed back again later in the winter.

Hinson et al. (2003) analyzed both RS data and TES limb data for a mid-winter period ($L_s \sim 134$–160°) in the south; modeling was also performed, which will be discussed in the next section. This is the most comprehensive study of stationary eddies in a seasonal period that has been performed to date. The seasonal period is essentially the same one that Hinson and Wilson (2002) examined for transient eddy activity. The RS data were at ~68°S latitude; the TES limb data greatly expanded the RS view of the stationary eddies in latitude and in height, extending upwards to ~1 Pa. It was found that large-amplitude, wavenumber-1 and wavenumber-2 eddies were present in the data: a ~7 K maximum amplitude for wave 1, and a ~5 K amplitude for wave 2. Both eddies exhibited maximum amplitudes aloft, and in both cases the region of largest amplitudes shifted strongly towards the pole with increasing height. For both wavenumbers there were upper-level (in the TES limb data, above ~0.1 mbar) secondary maxima in amplitude, located roughly above or somewhat equatorward of the lower-level maxima. In the case of wave 2 the upper-level maximum was larger than that at lower levels. For both waves the region between the two maxima exhibited large phase variations; for wave 2 the amplitudes in this region were quite small (~1 K). The RS data showed that for wave 1 the geopotential and thermal fields were approximately in quadrature with each other, such that there were sizable poleward eddy heat fluxes at lower levels. For wave 2 the phase relationship between geopotential and temperature implied small equatorward heat fluxes at low levels. The stationary wave structures found by Hinson et al. (2003) provide an excellent example of why it is not possible to determine the vertical propagation and the eddy heat fluxes from the eddy temperature structure alone. The same is true for the transient eddies. Geopotential data are required in order to determine this. The lack of geopotential data (due to the lack of surface pressure data) represents a very large hole in the Mars atmospheric database, along with the lack of wind data.

Barnes and Tyler (2007) analyzed TES nadir data for a number of southern wintertime seasonal intervals, and found very similar results for the stationary eddies as in other studies. They noted that the western hemisphere in the south is relatively cold aloft throughout most of the wintertime seasons, while the eastern hemisphere is relatively warm. The warmest region aloft is typically centered on Hellas. The western hemisphere is the location of the southern storm zone at lower levels, as discussed in the previous section on the transient eddies. Colaprete et al. (2005) made use of TES temperature maps for southern winter, produced using the FFSM approach, which showed that the coldest temperatures during much of winter occurred in the western hemisphere at middle and high latitudes; these temperatures were very frequently cold enough to allow CO_2 condensation. In the eastern hemisphere the temperatures were very rarely cold enough for this. The very cold temperatures occurred as a result of the fact that the western hemisphere was cold in a time-mean sense due to the stationary waves, and was also the region in which the strongest transient eddy activity was present. The coldest temperatures occurred when the cold region in a transient eddy disturbance was present in the cold portion of the stationary wave pattern. Colaprete et al. (2005) performed GCM modeling (using a modified version of the Ames GCM that simulated CO_2 clouds and snow) that strongly suggested that this basic hemispheric asymmetry in the atmospheric circulation in the south during wintertime could act to cause the permanent south polar cap to be offset substantially from the pole (as it is).

Wave amplitude plots for stationary waves 1 and 2 are shown in Figures 9.30 and 9.31, for both northern wintertime and southern wintertime seasonal intervals. These were derived by FFSM analyses of the TES nadir data, as described in Section 9.6. The northern examples illustrate that both waves 1 and 2 can have quite large amplitudes at higher levels during winter. At lower levels, as noted, wave 2 is the most prominent – it is very dominant in this example. The southern results are for a seasonal interval that is almost the same as the one in midwinter examined by Hinson et al. (2003), but in MY 25 and not MY 24. It can be seen that the wave-1 and wave-2 amplitude structures are rather similar to those found by Hinson et al., at levels below ~0.1 mbar. Wave 2 has the appearance of a trapped stationary wave that is able to penetrate somewhat up into the westerly zonal jet region. Wave 1 is better able to propagate vertically into the jet region.

Surprisingly, not a great amount of work on stationary eddies has yet been done using the MCS data from MRO. Guzewich et al. (2012) presented several figures showing wave-1 and wave-2 temperature perturbations in the longitude–height plane, for several different bands of latitude. For an interval

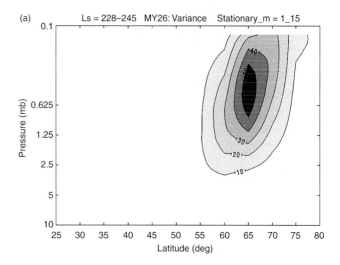

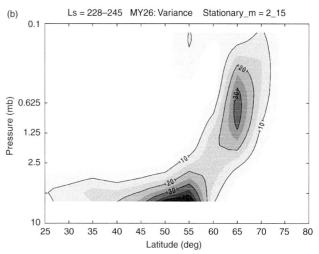

Figure 9.30. Temperature variance plots derived from FFSM analyses of TES data for stationary wavenumber 1: (a) variance for a northern autumn interval ($L_s \sim 228$–$245°$), and (b) variance for a southern winter interval ($L_s \sim 130$–$155°$). The variance values are in K^2. J.R. Barnes performed the FFSM analyses.

centered on southern winter solstice, a large-amplitude zonal wave 1 was present. The wave structure below ~0.1 mbar at ~50°S was largely in accord with that found in the TES data by Banfield et al. (2003) and Barnes and Tyler (2007). Above this level, sizable temperature perturbations were present (extending upwards to ~1 Pa); these were ~180° out of phase with those at lower levels. For a plausible geopotential structure at lower levels – one that would be associated with poleward heat fluxes – this eddy temperature pattern would cause the geopotential perturbations to decay with height above ~0.1 mbar at ~50°S. The wave-1 structure appears to be rather similar to that found in the TES limb data, later in southern winter, by Hinson et al. (2003). Wave 2 in the south, in an autumn period, also exhibited a structure that was similar to the one found by Hinson et al. (2003) in the TES limb data for a midwinter period. Viewed in a longitude–height cross-section at ~50°S, this has the appearance of a "stacked" structure with the temperature perturbations reversing sign between ~1 and ~0.1 mbar. Guzewich et al. (2012) argued that this appeared

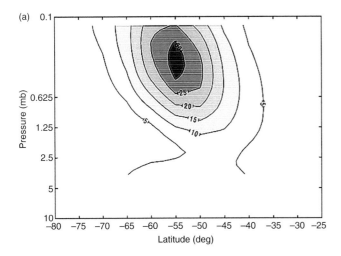

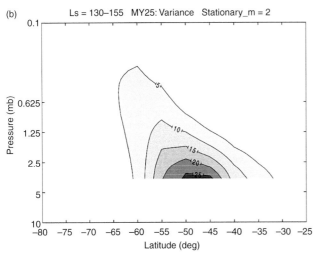

Figure 9.31. As in Figure 9.30, but for stationary wavenumber 2.

to evidence vertical wave propagation, but, as the much more complete study by Hinson et al. (2003) showed, wave 2 is propagating up into the jet core at higher latitudes (as is wave 1). The modeling in Hinson et al. showed rather conclusively that wave 2 was much more strongly vertically trapped than wave 1, as would be expected on the basis of theory. Guzewich et al. also showed stationary wave results for two northern early autumn periods. Wave 1 exhibited sizable amplitudes at ~60°N (as large as ~5 K), extending above 1 Pa; the pattern was similar to that present in the south near solstice, with a "stacked" appearance. This structure is entirely consistent with propagation up and into the jet core at higher latitudes, just as for wave 1 in the south. Wave 2 had the same kind of structure below ~1 mbar (at 50°N), again consistent with some amount of penetration into the jet at higher latitudes aloft. The most interesting aspect of the wave-2 structure in the north is the existence of very sizable temperature perturbations centered at ~0.1 Pa. Wave 2 would not be expected to be able to propagate vertically very strongly during this season (because of the strong zonal winds), so this very high-altitude stationary wave-2 feature could be associated with forcing at relatively high levels in the atmosphere. Breaking waves (planetary or tidal) can act to force stationary eddies; this occurs in the terrestrial middle atmosphere.

It is also possible that this very high-altitude wave-2 feature is due to a non-migrating tidal mode; a zonal wavenumber-4, semidiurnal mode is aliased into the wave-2 stationary wave in the MCS data. The lack of good local-time-of-day coverage can make it difficult to separate certain tidal modes from the stationary waves.

Tellmann et al. (2013) examined Mars Express RS temperature data for a very extended seasonal period in southern wintertime, $L_s \sim 61$–$110°$, by averaging the observations over a very wide latitude region (49–69°S). Stationary temperature perturbations as large as ~10 K were present in the data, with wave 1 dominating at higher levels. The eastern hemisphere was warm at all levels and the western hemisphere was mostly cold, as was found by Barnes and Tyler (2007) and Colaprete et al. (2005). At lower levels there was a geopotential low over Hellas and a high over the southern extension of Tharsis, as would be expected for relatively shorter zonal-scale topographic (mechanical) forcing. The wave-1 geopotential field exhibited a very large amount of vertical phase tilt, ~180°, and a rapid increase in geopotential amplitude at higher levels. Wave 2 displayed a very small, eastward phase tilt with height at lower levels and an equivalent barotropic structure, indicative of vertical trapping by strong westerly winds.

9.7.4 Modeling

Webster (1977) first studied the competing influences of the mechanical forcing produced by the large topography and the thermal forcing associated with it, using a two-layer, linear, primitive equation model. He considered the thermal forcing of topography, recognizing that elevated regions should act as heat sources and low regions as heat sinks as a result of the fact that surface temperatures are independent of elevation on Mars to a very good approximation. This will be the case if temperatures decrease with height. The opposite relationship will exist if temperatures increase with height, such that a simple expression for topographic thermal forcing is proportional to the atmospheric lapse rate and the elevation. Blumsack (1971) had previously made use of such a formulation. Webster (1977) found that mechanical topographic forcing dominated the linear wave response in the winter extratropics, and thermal forcing dominated in low latitudes. The stationary waves in the tropics had much of the basic character of Kelvin wave modes, with zonal winds being prominent and meridional winds weak.

Pollack et al. (1981) performed GCM simulations with topography, using a three-layer version of the earlier Leovy–Mintz Mars GCM; the topography was specified from a combination of Mariner 9 and Earth-based data. A variable surface albedo was also included in the simulations. For southern winter solstice, large-amplitude stationary eddies were present in the southern extratropics. Zonal wavenumbers 1 and 2 were dominant in the stationary eddy patterns. A cyclonic circulation was present at lower levels over Hellas. In the summer subtropics, where low-level westerly winds were present, several smaller-scale, stationary eddies were present. These had a baroclinic structure, with warm, lower-level, low-pressure regions and high-pressure regions aloft. Since these were present over elevated regions, thermal forcing appeared to be of importance for them.

Barnes et al. (1996b) carried out an extensive study of the stationary eddies in a large set of simulations with an improved version of the Ames GCM. The topography in the model was the so-called Consortium dataset; variable albedo and surface thermal inertia datasets were also incorporated in the model. This remains the only comprehensive and detailed study that has been done of the stationary eddies in a Mars GCM. Large-amplitude eddies were present in the wintertime hemisphere of the model, with the geopotential height amplitudes being as large as ~1.5 km at the top of the model (~45 km). The largest amplitudes were present in a southern winter solstice simulation and in a very dusty northern winter solstice case; wavenumber 1 was strongly dominant at high altitudes in both cases. A cyclonic circulation was present at lower levels over Hellas, with another such circulation existing over Argyre. At lower levels in the north a wavenumber-2 pattern was present in wintertime; this was much weaker in the western hemisphere than in the eastern hemisphere. A much stronger and more uniform wave-2 circulation was present at lower levels in a dustier (visible optical depth 1) simulation, and also in a simulation with the so-called "DTM" topography. This topography had larger low-wavenumber amplitudes in the northern extratropics than the Consortium topography did. The stationary eddy temperature perturbations in the winter hemisphere exhibited maxima at ~0.5–0.7 mbar, with values (root mean square (rms) variance) of ~4–12 K at the solstice seasons. The geopotential amplitudes peaked at the model top, as did the stationary eddy kinetic energy. Several aspects of the GCM stationary eddies were very interesting and surprising. One was that, for dusty conditions at northern winter solstice, there were very strong indications of wave propagation across the equator, from the south to the north. The index-of-refraction field was consistent with this, indicating that stationary wavenumber-1 and wavenumber-2 eddies should indeed have been able to propagate across the equator. A second notable aspect was that the amplitude of wavenumber 1 increased very strongly with increasing dust loading in the northern winter solstice simulations. Barnes et al. (1996b) employed rather simplified linear calculations to argue that this could be due to wavenumber 1 being closer to resonance under highly dusty conditions. This was attributed to the winter jet shifting poleward for dusty conditions, and possibly also by changes in the jet strength and structure.

Hollingsworth and Barnes (1996) made use of a linear primitive equation model in spherical geometry to examine the stationary response to the mechanical forcing by the Consortium topography. The model provided quite high resolution in height (2 km) and a very high model top (at 90 km). For a relatively non-dusty northern winter basic state, Hollingsworth and Barnes found that wavenumber 1 was able to propagate vertically to a substantial extent: maximum geopotential amplitudes were located at ~60–70 km. Wavenumber 2 exhibited maximum geopotential amplitudes at ~40 km, while wavenumber 3 was strongly trapped at low levels. Wavenumber 2 dominated the stationary eddy pattern at lower levels (~10–20 km). In contrast, the pattern in the south – for the same basic state – was much more strongly dominated by wavenumber 1, even at lower levels. The maximum geopotential amplitudes at high levels (~50–70 km) were very large for wavenumber 1

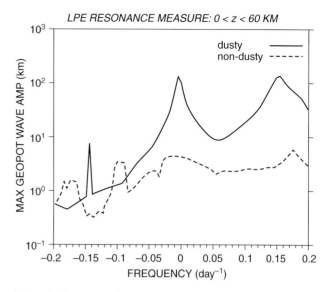

Figure 9.32. Results of zonal wavenumber-1 response calculations with a linear primitive equation model (taken from Hollingsworth and Barnes, 1996), as described in the text. Two different basic state zonal flows were examined, one for relatively non-dusty northern wintertime conditions and the second for very dusty northern wintertime conditions. The vertical axis shows the maximum eddy geopotential height amplitude below 60 km (in km), on a log scale.

(exceeding 2 km). Hollingsworth and Barnes (1996) performed simplified barotropic calculations to show that the stationary eddies were confined in latitude to a much greater extent than they are in the terrestrial atmosphere, and that quite high group velocities enabled the response to isolated topography to be essentially global in the extratropics. This is not the case in the terrestrial atmosphere. Hollingsworth and Barnes also examined the stationary wave response for a dusty northern winter zonal flow. They found that, even in the presence of much stronger dissipation (radiative damping), the wavenumber-1 geopotential amplitudes were very substantially increased (reaching almost 2 km at high altitudes). The wavenumber-2 amplitudes were decreased. The phase relationship of wave 1 with the topography was much different than in the non-dusty case, with the wave being shifted to the west by almost 180° within the "waveguide" region in the jet. On the basis of theory, this indicated that wave 1 could be near resonance (in the presence of strong dissipation) in the linear model. Hollingsworth and Barnes (1996) also performed a "resonant response" calculation using their linear model. They examined a range of low frequencies corresponding to both eastward and westward propagation, forcing the model at the lower boundary. A very substantially enhanced response centered on zero frequency (stationary waves) was obtained for wavenumber 1 for the dusty zonal flow. The stationary response for the non-dusty zonal flow was much weaker. Figure 9.32 shows the linear wave response versus frequency as calculated by Hollingsworth and Barnes (1996) for their non-dusty and dusty basic states. It can be seen that, even with stronger dissipation, the stationary amplitude is much greater in the dusty case than it is in the non-dusty case. The wave-1 amplitudes in the dusty case are extremely large, and could never exist in the atmosphere. But they indicated that resonance in the presence of very strong dissipation was occurring in the linear model.

Nayvelt et al. (1997) also carried out a study using a linear primitive equation model. This model had only nine levels, so it was not well suited to examining vertical wave propagation. Nayvelt et al. focused on the stationary eddy circulations in low latitudes at low levels, and incorporated a simple parameterization for thermal forcing – the same one used by Webster (1977). They made use of a zonal-mean basic state from an Ames GCM simulation for non-dusty northern winter solstice conditions. This state included zonal-mean meridional winds, as well as zonal winds. Nayvelt et al. presented results for isolated mountains, and also their results for the full global topography used in the model (this was derived from the DTM data). The highest model level for which results were shown was 2 mbar, and thus direct comparisons with the Hollingsworth and Barnes (1996) linear results could only be done in a very limited way. A very pronounced wavenumber-2 pattern was present at the 2 mbar level in the north, and this appeared to be linked to the stationary eddy pattern in the tropics and the southern hemisphere. Nayvelt et al. (1997) devoted much of their study to comparing the winds at the lowest model level with the bright streak imaging data, to assess how much the inclusion of the stationary eddy winds improved the comparison. This was discussed in the previous section on observational data.

Two modeling studies of stationary eddies were carried out by Hinson et al. (2001, 2003), for direct comparison with results from MGS RS data analysis studies (as discussed in the previous section). In the first study, the Ames GCM and the GFDL GCM were employed to examine the stationary eddies in high northern latitudes in late spring. Analysis of the Ames GCM data showed that a pattern of temperature perturbations was present (though at somewhat higher latitudes) that was at least somewhat similar to that in the RS data at very low levels. The GFDL GCM simulations showed that a wavenumber-1, diurnal, Kelvin tidal mode could explain the observed wavenumber-2 MGS RS observations at higher levels (above ~0.2 mbar). This mode was aliased in the RS observations because they were obtained at essentially a fixed local time. The observed and simulated eddy patterns at lower levels compared well, except that the simulated wavenumber 2 had a significantly greater vertical extent than the observed wave 2 did (in geopotential). The GCM produced much greater vertical penetration of wavenumber 2 than was present in the observations, probably due to the zonal wind field being different than the actual wind field.

Hinson et al. (2003) represents the best example to date of a focused observational and modeling study of a particular seasonal period: mid-winter in the south (the RS and TES limb observations in this period were discussed in the previous section). The GFDL GCM was used to simulate the stationary eddies in the southern extratropics, for comparison with the data. The GCM simulation was shown to agree remarkably well with the observational data, for both wavenumbers 1 and 2. It extended to significantly greater heights than the data, allowing the full wave structure to be seen. Wave 1 showed considerable vertical propagation in the model, having geopotential amplitudes of ~750 m at the 0.1 Pa level. In contrast, wave 2 was characterized by comparable amplitudes, but at much lower levels (~3–30 Pa). For both waves the maximum geopotential amplitudes were located near the core of the zonal jet. Thus the basic picture was of wave 1 being able to propagate vertically (in the "waveguide" associated with the westerly jet) while wave 2 was able to penetrate vertically as well, but to a substantially lesser extent due to its much shorter zonal scale. The maximum temperature amplitudes for wave 1 were located at the 0.1–0.3 mbar levels in the model and the observations, whereas for wavenumber 2 they were found at the surface. In both cases it is very significant that the contours of the eddy temperature amplitudes in the model were centered poleward (as well as equatorward) of the zonal flow contours, while the geopotential amplitude contours were found to be much more closely aligned with the zonal wind contours. This is a basic aspect of the stationary eddy structure when there is significant vertical penetration and propagation.

9.7.5 Summary

Along with the transient eddies and thermal tides, the forced stationary eddies are a major eddy component of the global circulation on Mars. This is certainly the case for Earth's atmosphere, and, given the very large Mars topography, it would be extremely surprising if it were not also the case for Mars. Compared to the amount of research done to date on tides and transient eddies, the amount of work done on the stationary eddies has not been nearly as substantial. There should be a great deal that can be learned about the stationary eddies from additional linear modeling, as well as from GCM studies. Analyses of the stationary waves in the MCS data have been relatively limited to date; systematic examination of the stationary wave amplitudes in the "middle atmosphere" region in the MCS data is something that very much needs to be done. Stationary waves may be able to propagate upwards to very high levels in the atmosphere, probably high enough to be important for the coupling of the lower and upper atmospheres. They may contribute significantly to the seasonally sustained polar warmings in the middle atmosphere. The details of the dynamical mechanisms by which the stationary eddies act to create storm zone regions for the transient eddies remain rather poorly understood. One can easily list numerous aspects of the dynamics of the stationary eddies and their roles in the dust and water cycles that warrant further study. This directly reflects the relative lack of attention that the stationary eddies have received. The currently available observations of the stationary eddies are much better (more complete) than those of the tides and transient eddies, so there is significantly better data to constrain and motivate modeling work.

9.8 SUMMARY

In this chapter we have examined the progress made in our understanding of the global circulation on Mars, with a focus upon the work done in the post-Viking era. Following Viking, it was established that Hadley cells (overturning circulations) and transient baroclinic eddies, major circulation components in Earth's atmosphere, were very prominent features of the Martian atmospheric circulation. Given the very large topography of Mars, it seemed certain that forced stationary waves had to be very significant components of the global circulation also. Observations obtained by Viking, especially those made by the two landers, very clearly revealed the existence of large-amplitude thermal tides in the atmosphere of Mars. Thermal tides exist in Earth's atmosphere, but at lower levels they are little more than a scientific curiosity. In sharp contrast, the tides in the Martian atmosphere are a primary feature of the local meteorological variations at most locations on the planet. The Viking Lander meteorological data showed that, under very dusty conditions, the tidal amplitudes increased dramatically. The Viking meteorology data also revealed the direct signature of the seasonal condensation and sublimation of CO_2, the primary atmospheric constituent, in the winter polar regions of Mars. Both the thermal tides and the global condensation flow are major components of the global atmospheric circulation on Mars that either do not exist or are completely unimportant in lower atmospheric regions on Earth. On Mars, both owe their existence to the small mass and short radiative response time of the atmosphere, along with the lack of oceans. They directly reflect the fact that Mars is a desert planet with a very thin atmosphere composed primarily of CO_2, a greenhouse gas.

Nothing in this very basic picture of the global circulation has really changed in the years since Viking. However, as a result of the numerous spacecraft missions to Mars during the past two decades, and the increasing sophistication of global circulation models that are now being run with very accurate topographic relief and relatively good values for the atmospheric dust optical parameters, there is a much better – though certainly still incomplete – picture of the nature and variability of the global circulation than we had as recently as the middle of the 1990s. We have discussed this new picture in some detail in this chapter. We now know, for example, that the lower branch of the Hadley cell is channeled into western boundary currents, that sustained wintertime polar warming exists in both hemispheres, that stationary waves are prominent in both hemispheres in winter and summer, that topographically excited, eastward-traveling, tidal modes significantly modulate the spatial distribution of tidal amplitudes throughout the atmosphere, and that transient baroclinic eddies are generally much more vigorous in the northern hemisphere than in the southern hemisphere. Furthermore, we have a good basic picture of how each of these circulation components varies with season and atmospheric dust loading. Despite this, our understanding of how the very thin Martian atmosphere manages to produce global dust storms from time to time remains relatively limited. The biggest puzzle that this unique phenomenon presents is why only a very small fraction of all large regional dust storms grow to become nearly global in extent, something that is rather difficult to understand given that most of the major circulation components tend to increase in strength as the atmospheric dust loading increases. The atmospheric circulation appears to be right on the edge of producing nearly global dust storms every year, during the Mars dust storm season that is centered on perihelion, but in most years this does not actually happen. There are aspects of the atmospheric circulation and the climate system that allow very sizable regional dust storms to occur every year, but that make it much more difficult for global storms to occur. These aspects have not yet been clearly identified and understood.

We also considered the forcing of the global atmospheric circulation, which is accomplished primarily by radiative heating, since latent heat release due to water cloud formation is negligible and sensible heat input at the surface, though significant, is spread out over a relatively deep planetary boundary layer. If not for the presence of aerosols – dust and clouds – the forcing of the global circulation on Mars would be a simple matter of gaseous absorption and emission of IR radiation by CO_2, the primary radiatively active atmospheric (gaseous) constituent. However, dust and clouds are always present on Mars and, because they are radiatively active and readily transported by the winds, they greatly complicate the picture by giving rise to numerous feedbacks that couple the global circulation to the dust and water (and CO_2) cycles in ways that are not fully understood at present. In the context of Mars GCMs, it is now very clear that good simulations of the circulation cannot be produced without a good knowledge of the distribution of clouds and dust in the atmosphere. Current observations do not provide us with that, but even if they did we would certainly want to be able to simulate the fully coupled circulation–aerosol–radiation system that acts to produce the continuously changing atmospheric state on Mars. Achieving the understanding and ability to do this is essential if we are going to be able to understand past (and future) climate change on Mars.

There are certainly some aspects of this coupled system that we do understand, at least in a basic way. We know, for example, that dust heating has a major impact on the strength of the Hadley circulation and the thermal tides, at times expanding the former all the way to the poles because of the nearly inviscid behavior of the atmosphere at high altitudes, and increasing the amplitudes of the latter to a point where the diurnal surface pressure variations can exceed 10% of the mean surface pressure (equivalent to a ~100 mbar surface pressure variation on Earth!). We also know that dust greatly alters the thermal structure of the atmosphere and hence the zonal wind fields, and this in turn strongly affects the transient eddies and forced stationary waves. And during the past decade or so, we have come to realize that radiative heating produced by the relatively (by comparison to Earth) sparse and thin water clouds on Mars is of very considerable importance for multiple components of the global atmospheric circulation. This is the case as a consequence of the small mass of the atmosphere, since very small amounts of water (all that the current cold atmosphere can hold) are able to form clouds that produce very significant IR radiative heating rates.

The radiatively active atmospheric aerosols (dust and clouds) constitute the greatest challenge going forward. We do not yet know the full effects of these aerosols since we do not have the observations needed to fully characterize their sources and sinks, their local time variability, and the lowest portion of the atmosphere where most of the "action" related to the surface–atmosphere exchange of dust, water, and CO_2 takes place. We can certainly hope that future spacecraft missions will close some of these knowledge gaps with orbiters that precess through local time, landed meteorological networks, and boundary layer atmospheric profilers. Missions capable of directly measuring winds throughout the atmosphere would clearly be invaluable to our ongoing efforts to fully characterize and understand the global circulation. This task becomes an extremely challenging one in the almost complete absence of direct wind measurements – and surface pressure data. Given this observational reality, we have come a very long way from the early days of studies of the global atmospheric circulation on Mars. It is fair to say that we have achieved a fairly good first-order understanding of the global circulation through a combination of spacecraft observations (far more than we have for any planet other than Earth) and very extensive numerical modeling of the Martian atmosphere. The next two decades should certainly be very interesting ones for the study of the global atmospheric circulation on Mars. Data assimilation has great promise for the future, but much more extensive and better observations are crucial for increasing our knowledge and understanding of the global circulation of the atmosphere of Mars.

REFERENCES

Allison, M. (1997), Accurate analytic representations of solar time and seasons on Mars with applications to the Pathfinder/Surveyor missions, *Geophys. Res. Lett.*, 24(16), 1967–1970, doi:10.1029/97GL01950.

Allison, M., and M. McEwen (2000), A post-Pathfinder evaluation of areocentric solar coordinates with improved timing recipes for Mars seasonal/diurnal climate studies, *Planet. Space Sci.*, 48(2–3), 215–235, doi:10.1016/S0032-0633(99)00092-6.

Anderson, D.L.T. (1976), The low-level jet as a western boundary current, *Mon. Wea. Rev.*, 104, 907–921, doi:10.1175/1520-0493(1976)104<0907:TLLJAA>2.0.CO;2.

Anderson, E., and C.B. Leovy (1978), Mariner 9 television limb observations of dust and ice hazes on Mars, *J. Atmos. Sci.*, 35, 723–734, doi:10.1175/1520-0469(1978)035<0723:MTLOOD>2.0.CO;2.

Andrews, D.G., and M.E. McIntyre (1976), Planetary waves in horizontal and vertical shear: the generalized Eliassen–Palm relation and the mean zonal acceleration, *J. Atmos. Sci.*, 33(11), doi:10.1175/1520-0469(1976)033<2031:PWIHAV>2.0.CO;2.

Andrews, D.G., F.W. Taylor, and M.E. McIntyre (1987), The influence of atmospheric waves on the general circulation of the middle atmosphere [and discussion], *Philos. Trans. R. Soc. A Math. Phys. Eng. Sci.*, 323(1575), 693–705, doi:10.1098/rsta.1987.0115.

Angelats i Coll, M., F. Forget, M.A. López-Valverde, and F. González-Galindo (2005), The first Mars thermospheric general circulation model: the Martian atmosphere from the ground to 240 km, *Geophys. Res. Lett.*, 32, L04201, doi:10.1029/2004GL021368.

Banfield, D., A.P. Ingersoll, and C.L. Keppenne (1995), A steady-state Kalman filter for assimilating data from a single polar orbiting satellite, *J. Atmos. Sci.*, 52, 737–753, doi:10.1175/1520-0469(1995)052<0737:ASSKFF>2.0.CO;2.

Banfield, D., A.D. Toigo, A.P. Ingersoll, and D.A. Paige (1996), Martian weather correlation length scales, *Icarus*, 119(1), 130–143, doi:10.1006/icar.1996.0006.

Banfield, D., B.J. Conrath, J.C. Pearl, M.D. Smith, and P.R. Christensen (2000), Thermal tides and stationary waves on Mars as revealed by Mars Global Surveyor Thermal Emission Spectrometer, *J. Geophys. Res.*, 105(E4), 9521, doi:10.1029/1999JE001161.

Banfield, D., B.J. Conrath, M.D. Smith, P.R. Christensen, and R.J. Wilson (2003), Forced waves in the Martian atmosphere from MGS TES nadir data, *Icarus*, 161(2), 319–345, doi:10.1016/S0019-1035(02)00044-1.

Banfield, D., B.J. Conrath, P.J. Gierasch, R.J. Wilson, and M.D. Smith (2004), Traveling waves in the Martian atmosphere from MGS TES nadir data, *Icarus*, 170(2), 365–403, doi:10.1016/j.icarus.2004.03.015.

Barnes, J.R. (1980), Time spectral analysis of midlatitude disturbances in the Martian atmosphere, *J. Atmos. Sci.*, 37, 2002–2015, doi:10.1175/1520-0469(1980)037<2002:TSAOMD>2.0.CO;2.

Barnes, J.R. (1981), Midlatitude disturbances in the Martian atmosphere: a second Mars year, *J. Atmos. Sci.*, 38, 225–234, doi:10.1175/1520-0469(1981)038<0225:MDITMA>2.0.CO;2.

Barnes, J.R. (1983), Baroclinic waves in the atmosphere of Mars: observations, linear instability, and finite-amplitude evolution, Ph.D Thesis, University of Washington, Seattle, Washington.

Barnes, J.R. (1984), Linear baroclinic instability in the Martian atmosphere, *J. Atmos. Sci.*, 41, 1536–1550, doi:10.1175/1520-0469(1984)041<1536:LBIITM>2.0.CO;2.

Barnes, J.R. (1986), Finite-amplitude behavior of a single baroclinic wave with multiple vertical modes: effects of thermal damping, *J. Atmos. Sci.*, 43, 58–71, doi:10.1175/1520-0469(1986)043<0058:FABOAS>2.0.CO;2

Barnes, J.R. (1990), Possible effects of breaking gravity waves on the circulation of the middle atmosphere of Mars, *J. Geophys. Res.*, 95(B2), 1401, doi:10.1029/JB095iB02p01401.

Barnes, J.R. (2001), Asynoptic fourier transform analyses of MGS TES data: transient baroclinic eddies, *Bull. Am. Met. Soc.*, 33, 1067.

Barnes, J.R. (2003a), Planetary eddies in the Martian atmosphere: FFSM analysis of TES data, *First Int. Workshop Mars Atmos. Model. Obs.*, Granada, Spain, http://www-mars.lmd.jussieu.fr/granada2003/.

Barnes, J.R. (2003b), Mars weather systems and maps: FFSM analyses of MGS TES temperature data, *Sixth Int. Conf. Mars*, Pasadena, California, www.lpi.usra.edu/meetings/sixthmars2003/abstract-volume.html.

Barnes, J.R. (2006), FFSM studies of transient eddies in the MGS TES temperature data, *Second Int. Workshop Mars Atmos. Model. Obs.*, Granada, Spain, http://www-mars.lmd.jussieu.fr/granada2006/.

Barnes, J.R., and R.M. Haberle (1996), The Martian zonal-mean circulation: angular momentum and potential vorticity structure in GCM simulations, *J. Atmos. Sci.*, 53, 3143–3156, doi:10.1175/1520-0469(1996)053<3143:TMZMCA>2.0.CO;2.

Barnes, J.R., and J.L. Hollingsworth (1987), Dynamical modeling of a planetary wave mechanism for a Martian polar warming, *Icarus*, 71, 313–334.

Barnes, J.R., and D. Tyler (2007), Winter weather on Mars: the unique southern hemisphere, in *Seventh Int. Conf. Mars*, Pasadena, California, www.lpi.usra.edu/meetings/7thmars2007/.

Barnes, J.R., J.B. Pollack, R.M. Haberle, et al. (1993), Mars atmospheric dynamics as simulated by the NASA Ames General Circulation Model: 2. Transient baroclinic eddies, *J. Geophys. Res.*, 98(E2), 3125–3148, doi:10.1029/92JE02935.

Barnes, J.R., T.D. Walsh, and J.R. Murphy (1996a), Transport timescales in the Martian atmosphere: general circulation model simulations, *J. Geophys. Res.*, 101(E7), 16881, doi:10.1029/96JE00500.

Barnes, J.R., R.M. Haberle, J.B. Pollack, H. Lee, and J. Schaeffer (1996b), Mars atmospheric dynamics as simulated by the NASA Ames General Circulation Model: 3. Winter quasi-stationary eddies, *J. Geophys. Res.*, 101, 12753–12776.

Barnes, J.R., M.S. Rucker, and D. Tyler (2014), Transient eddies in the atmosphere of Mars: the crucial importance of water clouds, in *Eighth Int. Conf. Mars*, Pasadena, California, www.hou.usra.edu/meetings/8thmars2014/.

Basu, S., J. Wilson, M. Richardson, and A.P. Ingersoll (2006), Simulation of spontaneous and variable global dust storms with the GFDL Mars GCM, *J. Geophys. Res.*, 111(E9), E09004, doi:10.1029/2005JE002660.

Benson, J.L., D.M. Kass, A. Kleinböhl, et al. (2010), Mars' south polar hood as observed by the Mars Climate Sounder, *J. Geophys. Res.*, 115(E12), E12015, doi:10.1029/2009JE003554.

Blumsack, S.L. (1971), On the effects of topography on planetary atmospheric circulation, *J. Atmos. Sci.*, 28, doi:10.1175/1520-0469(1971)028<1134:OTEOTO>2.0.CO;2.

Blumsack, S.L., and P.J. Gierasch (1972), Mars: the effects of topography on baroclinic instability, *J. Atmos. Sci.*, 29, doi:10.1175/1520-0469(1972)029<1081:MTEOTO>2.0.CO;2.

Blumsack, S.L., P.J. Gierasch, and W.R. Wessel (1973), An analytical and numerical study of the Martian planetary boundary layer over slopes, *J. Atmos. Sci.*, 30, doi:10.1175/1520-0469(1973)030<0066:AAANSO>2.0.CO;2.

Bougher, S.W., S. Engel, D.P. Hinson, and J.M. Forbes (2001), Mars Global Surveyor radio science electron density profiles: neutral atmosphere implications, *Geophys. Res. Lett.*, 28(16), 3091–3094, doi:10.1029/2001GL012884.

Bridger, A.F.C., and J.R. Murphy (1998), Mars' surface pressure tides and their behavior during global dust storms, *J. Geophys. Res.*, 103(E4), 8587, doi:10.1029/98JE00242.

Briggs, G.A., and C.B. Leovy (1974), Mariner observations of the Mars north polar hood, *Bull. Am. Meteorol. Soc.*, 55(4), doi:10.1175/1520-0477(1974)055<0278:MOOTMN>2.0.CO;2.

Cahoy, K.L., D.P. Hinson, and G.L. Tyler (2007), Characterization of a semidiurnal eastward-propagating tide at high northern latitudes with Mars Global Surveyor electron density profiles, *Geophys. Res. Lett.*, 34(15), L15201, doi:10.1029/2007GL030449.

Cavalié, T., F. Billebaud, T. Encrenaz, et al. (2008), Vertical temperature profile and mesospheric winds retrieval on Mars from CO millimeter observations, *Astron. Astrophys.*, 489(2), 795–809, doi:10.1051/0004-6361:200809815.

Chapman, S., and R.S. Lindzen (1970), *Atmospheric Tides – Thermal and Gravitational*, Reidel, Dordrecht.

Charney, J.G. (1947), The dynamics of long waves in a baroclinc westerly current, *J. Meteorol.*, 4(5), 136–162, doi:10.1175/1520-0469(1947)004<0136:TDOLWI>2.0.CO;2.

Christensen, P.R. (1988), Global albedo variations on Mars: implications for active aeolian transport, deposition, and erosion, *J. Geophys. Res.*, 93(B7), 7611, doi:10.1029/JB093iB07p07611.

Christensen, P.R., J.L. Bandfield, V.E. Hamilton, et al. (2001), Mars Global Surveyor Thermal Emission Spectrometer experiment: investigation description and surface science results, *J. Geophys. Res.*, 106(E10), 23823, doi:10.1029/2000JE001370.

Christensen, P.R., B.M. Jakosky, H.H. Kieffer, et al. (2004), The Thermal Emission Imaging System (THEMIS) for the Mars 2001 Odyssey Mission, *Space Sci. Rev.*, 110(1/2), 85–130, doi:10.1023/B:SPAC.0000021008.16305.94.

Clancy, R.T., B.J. Sandor, G.H. Moriarty-Schieven, and M.D. Smith (2006), Mesospheric winds and temperatures from JCMT sub-millimeter CO line observations during the 2003 and 2005 Mars oppositions, *Second Int. Work. Mars Atmos. Model. Obs.*, Granada, Spain, http://www-mars.lmd.jussieu.fr/granada2006/.

Colaprete, A., R.M. Haberle, and O.B. Toon (2003), Formation of convective carbon dioxide clouds near the south pole of Mars, *J. Geophys. Res.*, 108(E7), 5081, doi:10.1029/2003JE002053.

Colaprete, A., J.R. Barnes, R.M. Haberle, et al. (2005), Albedo of the south pole on Mars determined by topographic forcing of atmosphere dynamics, *Nature*, 435(7039), 184–188, doi:10.1038/nature03561.

Colaprete, A., J.R. Barnes, R.M. Haberle, and F. Montmessin (2008), CO_2 clouds, CAPE and convection on Mars: observations and general circulation modeling, *Planet. Space Sci.*, 56(2), 150–180, doi:10.1016/j.pss.2007.08.010.

Colburn, D.S., J.B. Pollack, and R.M. Haberle (1988), *Diurnal Variations in Optical Depth at Mars: Observations and Interpretations*, NASA-TM-100057, A-88067, NAS 1.15:1000057.

Collins, M., S.R. Lewis, and P.L. Read (1995), Regular and irregular baroclinic waves in a Martian general circulation model: a role for diurnal forcing?, *Adv. Sp. Res.*, 16(6), 3–7, doi:10.1016/0273-1177(95)00243-8.

Collins, M., S.R. Lewis, P.L. Read, and F. Hourdin (1996), Baroclinic wave transitions in the Martian atmosphere, *Icarus*, 120(2), 344–357, doi:10.1006/icar.1996.0055.

Collins, M., S.R. Lewis, and P.L. Read (1997), Gravity wave drag in a global circulation model of the Martian atmosphere: parameterisation and validation, *Adv. Sp. Res.*, 19(8), 1245–1254, doi:10.1016/S0273-1177(97)00277-9.

Conrath, B.J. (1975), Thermal structure of the Martian atmosphere during the dissipation of the dust storm of 1971, *Icarus*, 24(1), 36–46, doi:10.1016/0019-1035(75)90156-6.

Conrath, B.J. (1976), Influence of planetary-scale topography on the diurnal thermal tide during the 1971 Martian dust storm, *J. Atmos. Sci.*, 33, 2430–2439, doi:10.1175/1520-0469(1976)033<2430:IOPSTO>2.0.CO;2.

Conrath, B.J. (1981), Planetary-scale wave structure in the Martian atmosphere, *Icarus*, 48(2), 246–255, doi:10.1016/0019-1035(81)90107-X.

Conrath, B.J., J.C. Pearl, M.D. Smith, et al. (2000), Mars Global Surveyor Thermal Emission Spectrometer (TES) observations: atmospheric temperatures during aerobraking and science phasing, *J. Geophys. Res.*, 105(E4), 9509, doi:10.1029/1999JE001095.

Creasey, J.E., J.M. Forbes, and D.P. Hinson (2006), Global and seasonal distribution of gravity wave activity in Mars' lower atmosphere derived from MGS radio occultation data, *Geophys. Res. Lett.*, 33(1), doi:10.1029/2005GL024037.

Crisp, D. (1990), Infrared radiative transfer in the dust-free Martian atmosphere, *J. Geophys. Res.*, 95(B9), 14577, doi:10.1029/JB095iB09p14577.

Eady, E.T. (1949), Long waves and cyclone waves, *Tellus*, 1, 33–52, doi:10.1111/j.2153-3490.1949.tb01265.x.

Eckermann, S.D., J. Ma, and X. Zhu (2011), Scale-dependent infrared radiative damping rates on Mars and their role in the deposition of gravity-wave momentum flux, *Icarus*, 211(1), 429–442, doi:10.1016/j.icarus.2010.10.029.

Edmonds, R.M., J. Murphy, J.T. Schofield, and N.G. Heavens (2014), Considerations on the Presence of Gravity Wave Activity During MCS Limb Staring Observations, *Fifth Int. Work. Mars Atmos. Model. Obs.*, Oxford, UK, http://www-mars.lmd.jussieu.fr/oxford2014/.

Forbes, J.M. (2008), Troposphere–Thermosphere Coupling by Thermal Tides at Earth and Mars, *AGU Spring Meeting Abstracts*, 1, 1.

Forbes, J.M., and M.E. Hagan (2000), Diurnal Kelvin wave in the atmosphere of Mars: towards an understanding of "stationary" density structures observed by the MGS accelerometer, *Geophys. Res. Lett.*, 27(21), 3563–3566, doi:10.1029/2000GL011850.

Forbes, J.M., A.F.C. Bridger, S.W. Bougher, et al. (2002), Nonmigrating tides in the thermosphere of Mars, *J. Geophys. Res.*, 107(E11), 5113, doi:10.1029/2001JE001582.

Forget, F., F. Hourdin, R. Fournier, et al. (1999), Improved general circulation models of the Martian atmosphere from the surface to above 80 km, *J. Geophys. Res.*, 104(E10), 24155, doi:10.1029/1999JE001025.

Formisano, V., D. Grassi, R. Orfei, et al. (2004), The Planetary Fourier Spectrometer (PFS) for Mars Express, In *Mars Express – The Scientific Payload*, Ed. by A. Wilson, Sci. Coord. by A. Chicarro, ESA SP-1240, Noordwijk, Netherlands, ESA Publications Division, 71–94, http://sci.esa.int/mars-express/34885-esa-sp-1240-mars-express-the-scientific-payload/.

Gadian, A.M. (1978), The dynamics of and the heat transfer by baroclinic eddies and large-scale stationary topographically forced long waves in the Martian atmosphere, *Icarus*, 33(3), 454–465, doi:10.1016/0019-1035(78)90184-7.

Geissler, P.E. (2005), Three decades of Martian surface changes, *J. Geophys. Res.*, 110(E2), E02001, doi:10.1029/2004JE002345.

Gierasch, P.J., and R. Goody (1968), A study of the thermal and dynamical structure of the Martian lower atmosphere, *Planet. Space Sci.*, 16(5), 615–646, doi:10.1016/0032-0633(68)90102-5.

Gill, A.E. (1980), Some simple solutions for heat-induced tropical circulation, *Quart. J. Roy. Meteor. Soc.*, 106, 447–462, doi:10.1002/qj.49710644905.

Gómez-Elvira, J., C. Armiens, L. Castaner, et al. (2012), REMS: the environmental sensor suite for the Mars Science Laboratory Rover, *Space Sci. Rev.*, 170(1–4), 583–640, doi:10.1007/s11214-012-9921-1.

Goody, R., and M.J.S. Belton (1967), A discussion of Martian atmospheric dynamics, *Planet. Space Sci.*, 15(2), 247–256, doi:10.1016/0032-0633(67)90193-6.

Greeley, R., N. Lancaster, S. Lee, and P. Thomas (1992), Martian aeolian processes, sediments, and features, In *Mars*, H.H. Kieffer, B.M. Jakosky, C.W. Snyder, and M.S. Mathews, Eds., University of Arizona Press, 730–766.

Greeley, R., A. Skypeck, and J.B. Pollack (1993), Martian aeolian features and deposits: comparisons with general circulation model results, *J. Geophys. Res.*, 98(E2), 3183, doi:10.1029/92JE02580.

Greybush, S.J., R.J. Wilson, R.N. Hoffman, et al. (2012), Ensemble Kalman filter data assimilation of Thermal Emission Spectrometer temperature retrievals into a Mars GCM, *J. Geophys. Res.*, 117(E11), E11008, doi:10.1029/2012JE004097.

Gunnlaugsson, H.P., C. Holstein-Rathlou, J.P. Merrison, et al. (2008), Telltale wind indicator for the Mars Phoenix Lander, *J. Geophys. Res.*, 113, E00A04, doi:10.1029/2007JE003008.

Guzewich, S.D., E.R. Talaat, and D.W. Waugh (2012), Observations of planetary waves and nonmigrating tides by the Mars Climate Sounder, *J. Geophys. Res.*, 117(E3), E03010, doi:10.1029/2011JE003924.

Guzewich, S.D., A.D. Toigo, M.I. Richardson, et al. (2013), The impact of a realistic vertical dust distribution on the simulation of the Martian General Circulation, *J. Geophys. Res. Planets*, 118(5), 980–993, doi:10.1002/jgre.20084.

Haberle, R.M., and D.C. Catling (1996), A Micro-Meteorological mission for global network science on Mars: rationale and measurement requirements, *Planet. Space Sci.*, 44(11), 1361–1383, doi:10.1016/S0032-0633(96)00056-6.

Haberle, R.M., C.B. Leovy, and J.B. Pollack (1982), Some effects of global dust storms on the atmospheric circulation of Mars, *Icarus*, 50(2–3), 322–367, doi:10.1016/0019-1035(82)90129-4.

Haberle, R.M., H.C. Houben, R. Hertenstein, and T. Herdtle (1993a), A boundary-layer model for Mars: comparison with Viking Lander and entry data, *J. Atmos. Sci.*, 50, 1544–1559, doi:10.1175/1520-0469(1993)050<1544:ABLMFM>2.0.CO;2.

Haberle, R.M., J.B. Pollack, J.R. Barnes, et al. (1993b), Mars atmospheric dynamics as simulated by the NASA Ames General Circulation Model: 1. The zonal-mean circulation, *J. Geophys. Res.*, 98(E2), 3093, doi:10.1029/92JE02946.

Haberle, R.M., H.C. Houben, J.R. Barnes, and R.E. Young (1997), A simplified three-dimensional model for Martian climate studies, *J. Geophys. Res.*, 102(E4), 9051, doi:10.1029/97JE00383.

Haberle, R.M., M.M. Joshi, J.R. Murphy, et al. (1999), General circulation model simulations of the Mars Pathfinder atmospheric structure investigation/meteorology data, *J. Geophys. Res.*, 104(E4), 8957, doi:10.1029/1998JE900040.

Haberle, R.M., J. Gomez-Elvira, M. de la Torre Juarez, et al. (2014), Preliminary interpretation of the REMS pressure data from the first 100 sols of the MSL mission, *J. Geophys. Res. Planets*, 119(3), 440–453, doi:10.1002/2013JE004488.

Hamilton, K., and R.R. Garcia (1986), Theory and observations of the short-period normal mode oscillations of the atmosphere, *J. Geophys. Res.*, 91(D11), 11867, doi:10.1029/JD091iD11p11867.

Hanel, R., B.J. Conrath, W. Hovis, et al. (1972), Investigation of the Martian environment by infrared spectroscopy on Mariner 9, *Icarus*, 17(2), 423–442, doi:10.1016/0019-1035(72)90009-7.

Hartogh, P., A.S. Medvedev, T. Kuroda, et al. (2005), Description and climatology of a new general circulation model of the Martian atmosphere, *J. Geophys. Res.*, 110(E11), E11008, doi:10.1029/2005JE002498.

Hartogh, P., A.S. Medvedev, and C. Jarchow (2007), Middle atmosphere polar warmings on Mars: simulations and study on the validation with sub-millimeter observations, *Planet. Space Sci.*, 55(9), 1103–1112, doi:10.1016/j.pss.2006.11.018.

Hayne, P.O., D.A. Paige, N.G. Heavens, and the Mars Climate Sounder Science Team (2014), The role of snowfall in forming the seasonal ice caps of Mars: models and constraints from the Mars Climate Sounder, *Icarus*, 231, 122–130, doi:10.1016/j.icarus.2013.10.020.

Hayward, R.K., L.K. Fenton, and T.N. Titus (2014), Mars Global Digital Dune Database (MGD3): global dune distribution and wind pattern observations, *Icarus*, 230, 38–46, doi:10.1016/j.icarus.2013.04.011.

Heavens, N.G., M.I. Richardson, W.G. Lawson, et al. (2010), Convective instability in the Martian middle atmosphere, *Icarus*, 208(2), 574–589, doi:10.1016/j.icarus.2010.03.023.

Hébrard, E., C. Listowski, P. Coll, et al. (2012), An aerodynamic roughness length map derived from extended Martian rock abundance data, *J. Geophys. Res. Planets*, 117(E4), doi:10.1029/2011JE003942.

Held, I.M., and A.Y. Hou (1980), Nonlinear axially symmetric circulations in a nearly inviscid atmosphere, *J. Atmos. Sci.*, 37(3), 515–533, doi:10.1175/1520-0469(1980)037<0515:NASCIA>2.0.CO;2.

Held, I.M., M. Ting, and H. Wang (2002), Northern winter stationary waves: theory and modeling, *J. Climate*, 15, 2125–2144, doi:10.1175/1520-0442(2002)015<2125:NWSWTA>2.0.CO;2.

Hess, S.L. (1950), Some aspects of the meteorology of Mars., *J. Atmos. Sci.*, 7(1), doi:10.1175/1520-0469(1950)007<0001:SAOTMO>2.0.CO;2.

Hess, S.L., R.M. Henry, C.B. Leovy, J.A. Ryan, and J.E. Tillman (1977), Meteorological results from the surface of Mars: Viking 1 and 2, *J. Geophys. Res.*, 82(28), 4559–4574, doi:10.1029/JS082i028p04559.

Hinson, D.P. (2006), Radio occultation measurements of transient eddies in the northern hemisphere of Mars, *J. Geophy. Res.*, 111, E05002, doi:10.1029/2005JE002612.

Hinson, D.P., and H. Wang (2010), Further observations of regional dust storms and baroclinic eddies in the northern hemisphere of Mars, *Icarus*, 206, 290–305, doi:10.1016/j.icarus.2009.08.019.

Hinson, D.P., and R.J. Wilson (2002), Transient eddies in the southern hemisphere of Mars, *Geophys. Res. Lett.*, 29(7), 1154, doi:10.1029/2001GL014103.

Hinson, D.P., and R.J. Wilson (2004), Temperature inversions, thermal tides, and water ice clouds in the Martian tropics, *J. Geophys. Res.*, 109, E01002, doi:10.1029/2003JE002129.

Hinson, D.P., R.A. Simpson, J.D. Twicken, G.L. Tyler, and F.M. Flasar (1999), Initial results from radio occultation measurements with Mars Global Surveyor, *J. Geophys. Res.*, 104(E11), 26997–27012, doi:10.1029/1999JE001069.

Hinson, D.P., G.L. Tyler, J.L. Hollingsworth, and R.J. Wilson (2001), Radio occultation measurements of forced atmospheric waves on Mars, *J. Geophys. Res. Planets*, 106(E1), 1463–1480, doi:10.1029/2000JE001291.

Hinson, D.P., R.J. Wilson, M.D. Smith, and B.J. Conrath (2003), Stationary planetary waves in the atmosphere of Mars during southern winter, *J. Geophys. Res.*, 108(E1), 5004, doi:10.1029/2002JE001949.

Hinson, D.P., M. Pätzold, R.J. Wilson, et al. (2008a), Radio occultation measurements and MGCM simulations of Kelvin waves on Mars, *Icarus*, 193(1), 125–138, doi:10.1016/j.icarus.2007.09.009.

Hinson, D.P., M. Pätzold, S. Tellmann, B. Häusler, and G.L. Tyler (2008b), The depth of the convective boundary layer on Mars, *Icarus*, 198(1), 57–66, doi:10.1016/j.icarus.2008.07.003.

Hinson, D.P., H. Wang, and M.D. Smith (2012), A multi-year survey of dynamics near the surface in the northern hemisphere of Mars: short-period baroclinic waves and dust storms, *Icarus*, 219(1), 307–320, doi:10.1016/j.icarus.2012.03.001.

Hoffman, M.J., S.J. Greybush, R.J. John Wilson, et al. (2010), An ensemble Kalman filter data assimilation system for the Martian atmosphere: implementation and simulation experiments, *Icarus*, 209(2), 470–481, doi:10.1016/j.icarus.2010.03.034.

Hollingsworth, J.L., and J.R. Barnes (1996), Forced stationary planetary waves in Mars's winter atmosphere., *J. Atmos. Sci.*, 53, doi:10.1175/1520-0469(1996)053<0428:FSPWIM>2.0.CO;2.

Hollingsworth, J.L., and M.A. Kahre (2010), Extratropical cyclones, frontal waves, and Mars dust: modeling and considerations, *Geophys. Res. Lett.*, 37(22), doi:10.1029/2010GL044262.

Hollingsworth, J.L., R.M. Haberle, J.R. Barnes, et al. (1996), Orographic control of storm zones on Mars, *Nature*, 380(6573), 413–416, doi:10.1038/380413a0.

Hollingsworth, J.L., R.M. Haberle, and J. Schaeffer (1997), Seasonal variations of storm zones on Mars, *Adv. Sp. Res.*, 19(8), 1237–1240, doi:10.1016/S0273-1177(97)00275-5.

Hollingsworth, J.L., M.A. Kahre, R.M. Haberle, and F. Montmessin (2011), Radiatively-active aerosols within Mars' atmosphere: implications on the weather and climate as simulated by the NASA ARC Mars GCM, in *Fourth Int. Work. Mars Atmos. Model. Obs.*, Paris, France, http://www-mars.lmd.jussieu.fr/paris2011/.

Holstein-Rathlou, C., H.P. Gunnlaugsson, J.J. Iversen, et al. (2014), Mars wind as seen by the NASA Phoenix Lander telltale, in *Eighth Int. Conf. Mars*, www.hou.usra.edu/meetings/8thmars2014/.

Holton, J.R., and G.J. Hakim (2013), *An Introduction to Dynamic Meteorology*, Fifth Ed., Academic Press.

Holton, J.R., P.H. Haynes, M.E. McIntyre, et al. (1995), Stratosphere-troposphere exchange, *Rev. Geophys.*, 33(4), 403, doi:10.1029/95RG02097.

Hourdin, F., F. Forget, and O. Talagrand (1995), The sensitivity of the Martian surface pressure and atmospheric mass budget to various parameters: a comparison between numerical simulations and Viking observations, *J. Geophys. Res.*, 100(E3), 5501, doi:10.1029/94JE03079.

Hu, R., K. Cahoy, and M.T. Zuber (2012), Mars atmospheric CO_2 condensation above the north and south poles as revealed by radio occultation, climate sounder, and laser ranging observations, *J. Geophys. Res.*, 117, E07002, doi:10.1029/2012JE004087.

Imamura, T., and T. Ogawa (1995), Radiative damping of gravity waves in the terrestrial planetary atmospheres, *Geophys. Res. Lett.*, 22(3), 267–270, doi:10.1029/94GL02998.

Joshi, M.M., S.R. Lewis, P.L. Read, and D.C. Catling (1994), Western boundary currents in the atmosphere of Mars, *Nature*, 367(6463), 548–552, doi:10.1038/367548a0.

Joshi, M.M., B.N. Lawrence, and S.R. Lewis (1995a), Gravity wave drag in three-dimensional atmospheric models of Mars, *J. Geophys. Res.*, 100(E10), 21235, doi:10.1029/95JE02486.

Joshi, M.M., S.R. Lewis, P.L. Read, and D.C. Catling (1995b), Western boundary currents in the Martian atmosphere: numerical simulations and observational evidence, *J. Geophys. Res.*, 100(E3), 5485–5500, doi:10.1029/94JE02716.

Joshi, M.M., R.M. Haberle, J.R. Barnes, J.R. Murphy, and J. Schaeffer (1997), Low-level jets in the NASA Ames Mars general circulation model, *J. Geophys. Res.*, 102(E3), 6511, doi:10.1029/96JE03765.

Kahn, R. (1983), Some observational constraints on the global-scale wind systems of Mars, *J. Geophys. Res.*, 88(A12), 10189, doi:10.1029/JA088iA12p10189.

Kahre, M.A., R.M. Haberle, J.L. Hollingsworth, and R.J. Wilson (2014), Coupling the Mars dust and water cycles: investigating the role of clouds in controlling the vertical distribution of dust during N.H. summer, in *Fifth Int. Work. Mars Atmos. Model. Obs.*, Oxford, UK, http://www-mars.lmd.jussieu.fr/oxford2014/.

Kavulich, M.J., I. Szunyogh, G. Gyarmati, and R.J. Wilson (2013), Local dynamics of baroclinic waves in the Martian atmosphere, *J. Atmos. Sci.*, 70, 3415–3447, doi:10.1175/JAS-D-12-0262.1.

Kieffer, H.H., T.Z. Martin, A.R. Peterfreund, et al. (1977), Thermal and albedo mapping of Mars during the Viking primary mission, *J. Geophys. Res.*, 82(28), 4249–4291, doi:10.1029/JS082i028p04249.

Kleinböhl, A., R.J. Wilson, D. Kass, J.T. Schofield, and D.J. McCleese (2013), The semidiurnal tide in the middle atmosphere of Mars, *Geophys. Res. Lett.*, 40(10), 1952–1959, doi:10.1002/grl.50497.

Kondratyev, K.I., and G.E. Hunt (1982), *Weather and Climate on Planets*, Pergamon Press.

Kuroda, T., A.S. Medvedev, P. Hartogh, and M. Takahashi (2009), On forcing the winter polar warmings in the Martian middle atmosphere during dust storms, *J. Met. Soc. Japan*, 87(5), 913–921, doi:10.2151/jmsj.87.913.

Lahoz, W., B. Khattatov, and R. Menard, Eds. (2010), *Data Assimilation – Making Sense of Observations*, Springer.

Lait, L.R., and J.L. Stanford (1988), Applications of Asynoptic Space–Time Fourier Transform Methods to Scanning Satellite Measurements, *J. Atmos. Sci.*, 45(24), 3784–3799, doi:10.1175/1520-0469(1988)045<3784:AOASFT>2.0.CO;2.

Lee, C., W.G. Lawson, M.I. Richardson, et al. (2009), Thermal tides in the Martian middle atmosphere as seen by the Mars Climate Sounder, *J. Geophys. Res.*, 114, E03005, doi:10.1029/2008JE003285.

Lee, C., W.G. Lawson, M.I. Richardson, et al. (2011), Demonstration of ensemble data assimilation for Mars using DART, MarsWRF, and radiance observations from MGS TES, *J. Geophys. Res.*, 116, E11011, doi:10.1029/2011JE003815.

Lellouch, E., J. Rosenqvist, J.J. Goldstein, S.W. Bougher, and G. Paubert (1991), First absolute wind measurements in the middle atmosphere of Mars, *Astrophys. J.*, 383, 401, doi:10.1086/170797.

Leovy, C.B. (1969), Mars: theoretical aspects of meteorology, *Appl. Opt.*, 8(7), 1279–86, doi:10.1364/AO.8.001279.

Leovy, C.B. (1981), Observations of Martian tides over two annual cycles, *J. Atmos. Sci.*, 38, 30–39, doi:10.1175/1520-0469(1981)038<0030:OOMTOT>2.0.CO;2.

Leovy, C.B., and Y. Mintz (1969), Numerical simulation of the atmospheric circulation and climate of Mars, *J. Atmos. Sci.*, 26(6), doi:10.1175/1520-0469(1969)026<1167:NSOTAC>2.0.CO;2.

Leovy, C.B., and R.W. Zurek (1979), Thermal tides and Martian dust storms: direct evidence for coupling, *J. Geophys. Res.*, 84(B6), 2956, doi:10.1029/JB084iB06p02956.

Leovy, C.B., R.W. Zurek, and J.B. Pollack (1973), Mechanisms for Mars dust storms., *J. Atmos. Sci.*, 30, doi:10.1175/1520-0469(1973)030<0749:MFMDS>2.0.CO;2.

Leovy, C.B., J.E. Tillman, W.R. Guest, and J.R. Barnes (1985), Interannual variability of Martian weather, In *Recent Advances in Planetary Meteorology*, Cambridge Press, 69–84.

Lewis, S.R., and P.R. Barker (2005), Atmospheric tides in a Mars general circulation model with data assimilation, *Adv. Sp. Res.*, 36(11), 2162–2168, doi:10.1016/j.asr.2005.05.122.

Lewis, S.R., and P.L. Read (1995), An operational data assimilation scheme for the Martian atmosphere, *Adv. Sp. Res.*, 16(6), 9–13, doi:10.1016/0273-1177(95)00244-9.

Lewis, S.R., and P.L. Read (2003), Equatorial jets in the dusty Martian atmosphere, *J. Geophys. Res.*, 108, E4, 5034, doi:10.1029/2002JE001933.

Lewis, S.R., M. Collins, P.L. Read, et al. (1999), A climate database for Mars, *J. Geophys. Res.*, 104(E10), 24177, doi:10.1029/1999JE001024.

Lewis, S.R., P.L. Read, B.J. Conrath, J.C. Pearl, and M.D. Smith (2007), Assimilation of thermal emission spectrometer atmospheric data during the Mars Global Surveyor aerobraking period, *Icarus*, 192(2), 327–347, doi:10.1016/j.icarus.2007.08.009.

Lewis, S.R., D.P. Mulholland, P.L. Read, et al. (2016), The solsticial pause on Mars: 1. A planetary wave reanalysis, *Icarus*, 264, 456–464, doi:10.1016/j.icarus.2015.08.039.

Lian, Y., M.I. Richardson, C.E. Newman, et al. (2012), The Ashima/MIT Mars GCM and argon in the Martian atmosphere, *Icarus*, 218(2), 1043–1070, doi:10.1016/j.icarus.2012.02.012.

Lindzen, R.S., and A. Hou (1988), Hadley circulations for zonally averaged heating centered off the equator, *J. Atmos. Sci.*, 45(17), 2416–2427, doi:10.1175/1520-0469(1988)045<2416:HCFZAH>2.0.CO;2.

Liu, J., M.I. Richardson, and R.J. Wilson (2003), An assessment of the global, seasonal, and interannual spacecraft record of Martian climate in the thermal infrared, *J. Geophys. Res.*, 108, E8, 5089, doi:10.1029/2002JE001921.

Määttänen, A., F. Montmessin, B. Gondet, et al. (2010), Mapping the mesospheric CO_2 clouds on Mars: MEx/OMEGA and MEx/HRSC observations and challenges for atmospheric models, *Icarus*, 209(2), 452–469, doi:10.1016/j.icarus.2010.05.017.

Määttänen, A., B. Gondet, F. Montmessin, et al. (2014), Mesospheric CO_2 clouds on Mars: detection, properties and origin, *Eighth Int. Conf. Mars.*, www.hou.usra.edu/meetings/8thmars2014/.

Madeleine, J.-B., F. Forget, E. Millour, T. Navarro, and A. Spiga (2012), The influence of radiatively active water ice clouds on the Martian climate, *Geophys. Res. Lett.*, 39(23), doi:10.1029/2012GL053564.

Magalhães, J.A. (1987), The Martian Hadley circulation: comparison of "viscous" model predictions to observations, *Icarus*, 70(3), 442–468, doi:10.1016/0019-1035(87)90087-X.

Magalhães, J.A., J.T. Schofield, and A. Seiff (1999), Results of the Mars Pathfinder atmospheric structure investigation, *J. Geophys. Res.*, 104(E4), 8943, doi:10.1029/1998JE900041.

Martin, T.Z. (1981), Mean thermal and albedo behavior of the Mars surface and atmosphere over a Martian year, *Icarus*, 45(2), 427–446, doi:10.1016/0019-1035(81)90045-2.

Martin, T.Z., and H.H. Kieffer (1979), Thermal infrared properties of the Martian atmosphere: 2. The 15 μm band measurements, *J. Geophys. Res.*, 84(B6), 2843, doi:10.1029/JB084iB06p02843.

McCleese, D.J., J.T. Schofield, F.W. Taylor, et al. (2008), Intense polar temperature inversion in the middle atmosphere on Mars, *Nat. Geosci.*, 1(11), 745–749, doi:10.1038/ngeo332.

McCleese, D.J., N.G. Heavens, J.T. Schofield, et al. (2010), Structure and dynamics of the Martian lower and middle atmosphere as observed by the Mars Climate Sounder: seasonal variations in zonal mean temperature, dust, and water ice aerosols, *J. Geophys. Res.*, 115, E12016, doi:10.1029/2010JE003677.

McConnochie, T.H., J.F. Bell III, D. Savransky, et al. (2010), THEMIS-VIS observations of clouds in the Martian mesosphere: altitudes, wind speeds, and decameter-scale morphology, *Icarus*, 210(2), 545–565, doi:10.1016/j.icarus.2010.07.021.

McFarlane, N.A. (1987), The effect of orographically excited gravity wave drag on the general circulation of the lower stratosphere and troposphere, *J. Atmos. Sci.*, 44, 1775–1800.

Medvedev, A.S., and P. Hartogh (2007), Winter polar warmings and the meridional transport on Mars simulated with a general circulation model, *Icarus*, 186(1), 97–110, doi:10.1016/j.icarus.2006.08.020.

Medvedev, A.S., E. Yiğit, and P. Hartogh (2011), Estimates of gravity wave drag on Mars: indication of a possible lower thermospheric wind reversal, *Icarus*, 211(1), 909–912, doi:10.1016/j.icarus.2010.10.013.

Mellon, M.T., B.M. Jakosky, H.H. Kieffer, and P.R. Christensen (2000), High-resolution thermal inertia mapping from the Mars Global Surveyor Thermal Emission Spectrometer, *Icarus*, 148(2), 437–455, doi:10.1006/icar.2000.6503.

Mintz, Y. (1961), The general circulation of planetary atmospheres. In *The Atmospheres of Mars and Venus*, 107–146.

Mischna, M.A., C. Lee, and M.I. Richardson (2012), Development of a fast, accurate radiative transfer model for the Martian atmosphere, past and present, *J. Geophys. Res.*, 117(E10), E10009, doi:10.1029/2012JE004110.

Mitchell, D.M., L. Montabone, S. Thomson, and P.L. Read (2015), Polar vortices on Earth and Mars: a comparative study of the climatology and variability from reanalyses, *Quart. J. Roy. Met. Soc.*, 141(687), 550–562, doi:10.1002/qj.2376.

Miyoshi, Y., J.M. Forbes, and Y. Moudden (2011), A new perspective on gravity waves in the Martian atmosphere: sources and features, *J. Geophys. Res.*, 116(E9), E09009, doi:10.1029/2011JE003800.

Montabone L., S.R. Lewis, P.L. Read, and D.P. Hinson (2006), Validation of Martian meteorological data assimilation for MGS/TES using radio occultation measurements, *Icarus*, 185(1), 113–132, doi:10.1016/j.icarus.2006.07.012.

Montmessin, F., F. Forget, P. Rannou, M. Cabane, and R.M. Haberle (2004), Origin and role of water ice clouds in the Martian water cycle as inferred from a general circulation model, *J. Geophys. Res.*, 109(E10), E10004, doi:10.1029/2004JE002284.

Montmessin, F., B. Gondet, J.-P. Bibring, et al. (2007), Hyperspectral imaging of convective CO_2 ice clouds in the equatorial mesosphere of Mars, *J. Geophys. Res.*, 112, E11590, doi:10.1029/2007JE002944.

Mooring, T.A., and R.J. Wilson (2015), Transient eddies in the MACDA Mars reanalysis, *J. Geophys. Res. Planets*, 120, 1671–1696, doi:10.1002/2015JE004824.

Moreno, R., E. Lellouch, F. Forget, et al. (2009), Wind measurements in Mars' middle atmosphere: IRAM Plateau de Bure interferometric CO observations, *Icarus*, 201(2), 549–563, doi:10.1016/j.icarus.2009.01.027.

Moudden, Y., and J.M. Forbes (2014), Insight into the seasonal asymmetry of nonmigrating tides on Mars, *Geophys. Res. Lett.*, 41(7), 2631–2636, doi:10.1002/2014GL059535.

Moudden, Y., and J.C. McConnell (2005), A new model for multiscale modeling of the Martian atmosphere, GM3, *J. Geophys. Res.*, 110(E4), E04001, doi:10.1029/2004JE002354.

Moudden, Y., and J.M. Forbes (2008a), Effects of vertically propagating thermal tides on the mean structure and dynamics of Mars' lower thermosphere, *Geophys. Res. Lett.*, 35, L23805, doi:10.1029/2008GL036086.

Moudden, Y., and J.M. Forbes (2008b), Topographic connections with density waves in Mars' aerobraking regime, *J. Geophys. Res.*, 113, E11009, doi:10.1029/2008JE003107.

Mulholland, D.P., S.R. Lewis, P.L. Read, J.-B. Madaleine, and F. Forget (2016), The solsticial pause on Mars: 2. Modeling and investigation of causes, *Icarus*, 264, 465–477, doi:10.1016/j.icarus.2015.08.038.

Murphy, J.R., C.B. Leovy, and J.E. Tillman (1990), Observations of Martian surface winds at the Viking Lander 1 Site, *J. Geophys. Res.*, 95(B9), 14555, doi:10.1029/JB095iB09p14555.

Navarro, T., F. Forget, E. Millour, and S.J. Greybush (2014), Detection of detached dust layers in the Martian atmosphere from their thermal signature using assimilation, *Geophys. Res. Lett.*, 41, 6620–6626, doi:10.1002/1014GL061377.

Nayvelt, L., P.J. Gierasch, and K.H. Cook (1997), Modeling and observations of Martian stationary waves, *J. Atmos. Sci.*, 54, doi:10.1175/1520-0469(1997)054<0986:MAOOMS>2.0.CO;2.

Niver, D.S., and S.L. Hess (1982), Band-pass filtering of one year of daily mean pressures on Mars, *J. Geophys. Res.*, 87(B12), 10191, doi:10.1029/JB087iB12p10191.

Pätzold, M., F.M. Neubauer, L. Carone, et al. (2004), MaRS: Mars Express Orbiter Radio Science, In *Mars Express – The Scientific Payload*, Ed. by A. Wilson, Sci. Coord. by A. Chicarro, ESA SP-1240, Noordwijk, Netherlands, ESA Publications Division, 71–94, http://sci.esa.int/mars-express/34885-esa-sp-1240-mars-express-the-scientific-payload/.

Pedlosky, J. (1979), *Geophysical Fluid Dynamics*, Springer.

Peixoto, J.P., and A.H. Oort (1992), *Physics of Climate*, American Institute of Physics.

Pirraglia, J.A., and B.J. Conrath (1974), Martian tidal pressure and wind field obtained from the Mariner 9 infrared spectroscopy experiment., *J. Atmos. Sci.*, 31, doi:10.1175/1520-0469(1974)031<0318:MTPAWF>2.0.CO;2.

Pleskot, L.K., and E.D. Miner (1981), Time variability of Martian bolometric albedo, *Icarus*, 45(1), 179–201, doi:10.1016/0019-1035(81)90013-0.

Pollack, J.B., D.S. Colburn, F.M. Flasar, et al. (1979), Properties and effects of dust particles suspended in the Martian atmosphere, *J. Geophys. Res.*, 84(B6), 2929, doi:10.1029/JB084iB06p02929.

Pollack, J.B., C.B. Leovy, P.W. Greiman, and Y. Mintz (1981), A Martian general circulation experiment with large topography, *J. Atmos. Sci.*, 38, 3–29, doi:10.1175/1520-0469(1981)038<0003:AMGCEW>2.0.CO;2.

Pollack, J.B., R.M. Haberle, J. Schaeffer, and H. Lee (1990), Simulations of the general circulation of the Martian atmosphere: 1. Polar processes, *J. Geophys. Res.*, 95(B2), 1447–1473, doi:10.1029/JB095iB02p01447.

Putzig, N., and M.T. Mellon (2007), Apparent thermal inertia and the surface heterogeneity of Mars, *Icarus*, 191(1), 68–94, doi:10.1016/j.icarus.2007.05.013.

Read, P.L., and S.R. Lewis (2004), *The Martian Climate Revisited – Atmosphere and Environment of a Desert Planet*, Springer.

Richardson, M.I., and R.J. Wilson (2002), A topographically forced asymmetry in the Martian circulation and climate., *Nature*, 416(6878), 298–301, doi:10.1038/416298a.

Richardson, M.I., A.D. Toigo, and C.E. Newman (2007), PlanetWRF: a general purpose, local to global numerical model for planetary atmospheric and climate dynamics, *J. Geophys. Res.*, 112(E9), E09001, doi:10.1029/2006JE002825.

Rothman, L.S., I.E. Gordon, Y. Babikov, et al. (2013), The HITRAN2012 molecular spectroscopic database, *J. Quant. Spectroscopy Rad. Transfer*, 130, 4–50, doi:10.1016/j.jqsrt.2013.07.002.

Rucker, M.S. (2014), The effects of clouds on transient baroclinic eddies in a Mars general circulation model, Master's Thesis, Oregon State University, Corvallis, Oregon.

Ryan, J.A., R.M. Henry, S.L. Hess, et al. (1978), Mars meteorology: three seasons at the surface, *Geophys. Res. Lett.*, 5(8), 715–718, doi:10.1029/GL005i008p00715.

Salby, M.L. (1982a), Sampling theory for asynoptic satellite observations. Part I: Space-time spectra, resolution, and aliasing, *J. Atmos. Sci.*, 39(11), 2577–2600, doi:10.1175/1520-0469(1982)039<2577:STFASO>2.0.CO;2.

Salby, M.L. (1982b), Sampling theory for asynoptic satellite observations. Part II: Fast Fourier Synoptic Mapping, *J. Atmos. Sci.*, 39(11), 2601–2614, doi:10.1175/1520-0469(1982)039<2601:STFASO>2.0.CO;2.

Santee, M.L., and D. Crisp (1993), Thermal structure and dust loading of the Martian atmosphere during late southern summer: Mariner 9 revisited, *J. Geophys. Res.*, 98(E2), 3261, doi:10.1029/92JE01896.

Santee, M.L., and D. Crisp (1995), Diagnostic calculations of the circulation in the Martian atmosphere, *J. Geophys. Res.*, 100(E3), 5465, doi:10.1029/94JE03207.

Sato, T.M., H. Fujiwara, Y.O. Takahashi, et al. (2011), Tidal variations in the Martian lower atmosphere inferred from Mars Express Planetary Fourier Spectrometer temperature data, *Geophys. Res. Lett.*, 38(24), doi:10.1029/2011GL050348.

Savijärvi, H. (1995), Mars boundary layer modeling: diurnal moisture cycle and soil properties at the Viking Lander 1 site, *Icarus*, 117(1), 120–127, doi:10.1006/icar.1995.1146.

Savijärvi, H., and T. Siili (1993), The Martian slope winds and the nocturnal PBL jet, *J. Atmos. Sci.*, 50, 77–88, doi:10.1175/1520-0469(1993)050<0077:TMSWAT>2.0.CO;2.

Savijärvi, H., D. Crisp, and A.-M. Harri (2005), Effects of CO_2 and dust on present-day solar radiation and climate on Mars, *Q. J. R. Meteorol. Soc.*, 131(611), 2907–2922, doi:10.1256/qj.04.09.

Schneider, E.K. (1983), Martian great dust storms: interpretive axially symmetric models, *Icarus*, 55(2), 302–331, doi:10.1016/0019-1035(83)90084-2.

Schofield, J.T., J.R. Barnes, D. Crisp, et al. (1997), The Mars Pathfinder atmospheric structure investigation/meteorology (ASI/MET) experiment., *Science*, 278(5344), 1752–1758, doi:10.1126/science.278.5344.1752.

Sharman, R.D., and J.A. Ryan (1980), Mars atmosphere pressure periodicities from Viking observations, *J. Atmos. Sci.*, 37, 1994–2001, doi:10.1175/1520-0469(1980)037<1994:MAPPFV>2.0.CO;2.

Shia, R.-L., Y.L. Yung, M. Allen, R.W. Zurek, and D. Crisp (1989), Sensitivity study of advection and diffusion coefficients in a two-dimensional stratospheric model using excess carbon 14 data, *J. Geophys. Res.*, 94(D15), 18467, doi:10.1029/JD094iD15p18467.

Smith, M.D. (2004), Interannual variability in TES atmospheric observations of Mars during 1999–2003, *Icarus*, 167(1), 148–165, doi:10.1016/j.icarus.2003.09.010.

Smith, M.D. (2008), Spacecraft observations of the Martian atmosphere, *Ann. Rev. Earth Planet. Sci.*, 36(1), 191–219, doi:10.1146/annurev.earth.36.031207.124334.

Smith, M. D. (2009), THEMIS observations of Mars aerosol optical depth from 2002–2008, *Icarus*, 202(2), 444–452, doi:10.1016/j.icarus.2009.03.027.

Smith, M.D., J.C. Pearl, B.J. Conrath, and P.R. Christensen (2001a), One Martian year of atmospheric observations by the thermal emission spectrometer, *Geophys. Res. Lett.*, 28(22), 4263–4266, doi:10.1029/2001GL013608.

Smith, M.D., J.C. Pearl, B.J. Conrath, and P.R. Christensen (2001b), Thermal Emission Spectrometer results: Mars atmospheric thermal structure and aerosol distribution, *J. Geophys. Res.*, 106(E10), 23929, doi:10.1029/2000JE001321.

Smith, M.D., B.J. Conrath, J.C. Pearl, and P.R. Christensen (2002), Thermal Emission Spectrometer observations of Martian planet-encircling dust storm 2001A, *Icarus*, 157(1), 259–263, doi:10.1006/icar.2001.6797.

Smith, M.D., M.J. Wolff, N. Spanovich, et al. (2006), One Martian year of atmospheric observations using MER Mini-TES, *J. Geophys. Res.*, 111(E12), E12S13, doi:10.1029/2006JE002770.

Sonnabend, G., M. Sornig, P.J. Krötz, R.T. Schieder, and K.E. Fast (2006), High spatial resolution mapping of Mars mesospheric zonal winds by infrared heterodyne spectroscopy of CO_2, *Geophys. Res. Lett.*, 33(18), L18201, doi:10.1029/2006GL026900.

Sonnabend, G., M. Sornig, P. Kroetz, and D. Stupar (2012), Mars mesospheric zonal wind around northern spring equinox from infrared heterodyne observations of CO_2, *Icarus*, 217(1), 315–321, doi:10.1016/j.icarus.2011.11.009.

Spiga, A., F. Forget, S.R. Lewis, and D.P. Hinson (2010), Structure and dynamics of the convective boundary layer on Mars as inferred from large-eddy simulations and remote-sensing measurements, *Q. J. R. Meteorol. Soc.*, 136(647), 414–428, doi:10.1002/qj.563.

Sprague, A.L., W.V Boynton, K.E. Kerry, et al. (2004), Mars' south polar Ar enhancement: a tracer for south polar seasonal meridional mixing, *Science*, 306(5700), 1364–7, doi:10.1126/science.1098496.

Sprague, A.L., W.V. Boynton, K.E. Kerry, et al. (2007), Mars' atmospheric argon: tracer for understanding Martian atmospheric circulation and dynamics, *J. Geophys. Res.*, 112(E3), E03S02, doi:10.1029/2005JE002597.

Sprague, A.L., W.V. Boynton, F. Forget, et al. (2012), Interannual similarity and variation in seasonal circulation of Mars' atmospheric Ar as seen by the Gamma Ray Spectrometer on Mars Odyssey, *J. Geophys. Res. Planets*, 117(E4), doi:10.1029/2011JE003873.

Steele, L.J., S.R. Lewis, M.R. Patel, et al. (2014a), The seasonal cycle of water vapor on Mars from assimilation of Thermal Emission Spectrometer data, *Icarus*, 237, 97–115, doi:10.1016/j.icarus.2014.04.017.

Steele, L.J., S.R. Lewis, and M.R. Patel (2014b), The radiative impact of water ice clouds from a reanalysis of Mars Climate Sounder data, *Geophys. Res. Lett.*, 41, 4471–4478, doi:10.1002/2014GL060235.

Sullivan, R., R. Greeley, M. Kraft, et al. (2000), Results of the Imager for Mars Pathfinder windsock experiment, *J. Geophys. Res.*, 105(E10), 24547, doi:10.1029/1999JE001234.

Sutton, J.L., C.B. Levoy, and J.E. Tillman (1978), Diurnal variations of the Martian surface layer meteorological parameters during the first 45 sols at two Viking Lander sites, *J. Atmos. Sci.*, 35, 2346–2355, doi:10.1175/1520-0469(1978)035<2346:DVOTMS>2.0.CO;2.

Szwast, M.A., M.I. Richardson, and A.R. Vasavada (2006), Surface dust redistribution on Mars as observed by the Mars Global Surveyor and Viking Orbiters, *J. Geoph. Res.*, 111(E11), E11008, doi:10.1029/2005JE002485.

Takahashi, Y.O., H. Fujiwara, H. Fukunishi, et al. (2003), Topographically induced north–south asymmetry of the meridional circulation in the Martian atmosphere, *J. Geophys. Res.*, 108(E3), 5018, doi:10.1029/2001JE001638.

Takahashi, Y.O., H. Fujiwara, and H. Fukunishi (2006), Vertical and latitudinal structure of the migrating diurnal tide in the Martian atmosphere: numerical investigations, *J. Geophys. Res.*, 111(E1), E01003, doi:10.1029/2005JE002543.

Tellmann, S., M. Pätzold, B. Häusler, D.P. Hinson, and G.L. Tyler (2013), The structure of Mars lower atmosphere from Mars Express Radio Science (MaRS) occultation measurements, *J. Geophys. Res. Planets*, 118(2), 306–320, doi:10.1002/jgre.20058.

Théodore, B., E. Lellouch, E. Chassefière, and A. Hauchecorne (1993), Solstitial temperature inversions in the Martian middle atmosphere: observational clues and 2-D modeling, *Icarus*, 105(2), 512–528, doi:10.1006/icar.1993.1145.

Thomas, P., J. Veverka, S. Lee, and A. Bloom (1981), Classification of wind streaks on Mars, *Icarus*, 45(1), 124–153, doi:10.1016/0019-1035(81)90010-5.

Tillman, J.E. (1988), Mars global atmospheric oscillations: annually synchronized, transient normal-mode oscillations and the triggering of global dust storms, *J. Geophys. Res.*, 93(D8), 9433, doi:10.1029/JD093iD08p09433.

Tillman, J.E., R.M. Henry, and S.L. Hess (1979), Frontal systems during passage of the Martian north polar hood over the Viking Lander 2 site prior to the first 1977 dust storm, *J. Geophys. Res.*, 84(B6), 2947, doi:10.1029/JB084iB06p02947.

Tillman, J.E., N.C. Johnson, P. Guttorp, and D.B. Percival (1993), The Martian annual atmospheric pressure cycle: years without great dust storms, *J. Geophys. Res.*, 98(E6), 10963, doi:10.1029/93JE01084.

Toigo, A.D., M.I. Richardson, R.J. Wilson, H. Wang, and A.P. Ingersoll (2002), A first look at dust lifting and dust storms near the south pole of Mars with a mesoscale model, *J. Geophys. Res.*, 107(E7), doi:10.1029/2011JE001592.

Tyler, D., and J.R. Barnes (2005), A mesoscale model study of summer-time atmospheric circulations in the north polar region of Mars, *J. Geophys. Res.*, 110(E6), E06007, doi:10.1029/2004JE002356.

Tyler, D., and J.R. Barnes (2013), Mesoscale modeling of the circulation in the Gale Crater region: an investigation into the complex forcing of convective boundary layer depths, *Mars*, 8, 58–77, doi:10.1555/mars.2013.0003.

Tyler, D., and J.R. Barnes (2014), Atmospheric mesoscale modeling of water and clouds during northern summer on Mars, *Icarus*, 237, 388–414, doi:10.1016/j.icarus.2014.04.020.

Tyler, D., and J.R. Barnes (2015), Convergent crater circulations on Mars: influence on the surface pressure cycle and the depth of the convective boundary layer, *Geophys. Res. Lett.*, 42(18), 7343–7350, doi:10.1002/2015GL064957.

Wang, H. (2007), Dust storms originating in the northern hemisphere during the third mapping year of Mars Global Surveyor, *Icarus*, 189(2), 325–343, doi:10.1016/j.icarus.2007.01.014.

Wang, H., and A.P. Ingersoll (2003), Cloud-tracked winds for the first Mars Global Surveyor mapping year, *J. Geophys. Res.*, 108(E9), 5110, doi:10.1029/2003JE002107.

Wang, H., and M.I. Richardson (2015), The origin, evolution, and trajectory of large dust storms on Mars during Mars Years 24–30 (1999–2011), *Icarus*, 251, 112–127, doi:10.1016/j.icarus.2013.10.033.

Wang, H., M.I. Richardson, R.J. Wilson, et al. (2003), Cyclones, tides, and the origin of a cross-equatorial dust storm on Mars, *Geophys. Res. Lett.*, 30(9), 1488, doi:10.1029/2002GL016828.

Wang, H., R.W. Zurek, and M.I. Richardson (2005), Relationship between frontal dust storms and transient eddy activity in the northern hemisphere of Mars as observed by Mars Global Surveyor, *J. Geophys. Res.*, 110(E7), E07005, doi:10.1029/2005JE002423.

Wang, H., A.D. Toigo, and M.I. Richardson (2011), Curvilinear features in the southern hemisphere observed by Mars Global Surveyor Mars Orbiter Camera, *Icarus*, 215(1), 242–252, doi:10.1016/j.icarus.2011.06.029.

Wang, H., M.I. Richardson, A.D. Toigo, and C.E. Newman (2013), Zonal wavenumber three traveling waves in the northern hemisphere of Mars simulated with a general circulation model, *Icarus*, 223(2), 654–676, doi:10.1016/j.icarus.2013.01.004.

Webster, P.J. (1977), The low-latitude circulation of Mars, *Icarus*, 30(4), 626–649, doi:10.1016/0019-1035(77)90086-0.

Wilson, R.J. (1997), A general circulation model simulation of the Martian polar warming, *Geophys. Res. Lett.*, 24(2), 123–126, doi:10.1029/96GL03814.

Wilson, R.J. (2000), Evidence for diurnal period Kelvin waves in the Martian atmosphere from Mars Global Surveyor TES data, *Geophys. Res. Lett.*, 27(23), 3889–3892, doi:10.1029/2000GL012028.

Wilson, R.J. (2002), Evidence for nonmigrating thermal tides in the Mars upper atmosphere from the Mars Global Surveyor Accelerometer Experiment, *Geophys. Res. Lett.*, 29(7), 1120, doi:10.1029/2001GL013975.

Wilson, R.J. (2011), Water ice clouds and thermal structure in the Martian tropics as revealed by Mars Climate Sounder, in *Fourth Int. Workshop Mars Atmos. Model. Obs.*, Paris, France, http://www-mars.lmd.jussieu.fr/paris2011/.

Wilson, R.J. (2012a), The role of thermal tides in the Martian dust cycle, in *Eur. Planet. Sci. Congr.*, Madrid, Spain, http://meetingorganizer.copernicus.org/EPSC2012/EPSC2012-798-1.pdf.

Wilson, R.J. (2012b), Thermal tides as revealed by Mars Climate Sounder, in *Eur. Planet. Sci. Congr.*, Madrid, Spain, http://meetingorganizer.copernicus.org/EPSC2012/EPSC2012-825-1.pdf.

Wilson, R.J. (2015), The impact of planetary-scale thermal forcing and small-scale topography on the diurnal cycle of Martian surface pressure, Abstract P22A-07, presented at *2015 Fall Meeting, AGU*, San Francisco, CA, 14–18 December, http://abstractsearch.agu.org/meetings/2015/FM/P22A-07.html.

Wilson, R.J., and S.D. Guzewich (2014), Influence of water ice clouds on nighttime tropical temperature structure as seen by the Mars Climate Sounder, *Geophys. Res. Lett.*, 41(10), 3375–3381, doi:10.1002/2014GL060086.

Wilson, R.J., and K. Hamilton (1996), Comprehensive model simulation of thermal tides in the Martian atmosphere., *J. Atmos. Sci.*, 53, doi:10.1175/1520-0469(1996)053<1290:CMSOTT>2.0.CO;2.

Wilson, R.J., and M.I. Richardson (1999), Comparison of Mars GCM dust storm simulations with Viking mission observations, *Fifth Int. Conf. Mars*, Pasadena, California, www.lpi.usra.edu/meetings/5thMars99/pdf/sessguid.pdf.

Wilson, R.J., and M.I. Richardson (2000), The Martian atmosphere during the Viking mission. 1: Infrared measurements of atmospheric temperatures revisited, *Icarus*, 145(2), 555–579, doi:10.1006/icar.2000.6378.

Wilson, R. J., D. Banfield, B.J. Conrath, and M.D. Smith (2002), Traveling waves in the northern hemisphere of Mars, *Geophys. Res. Lett.*, 29(14), 1684, doi:10.1029/2002GL014866.

Wilson, R.J., D.P. Hinson, and M.D. Smith (2006), GCM simulations of transient eddies and frontal systems in the Martian atmosphere, *Second Int. Work. Mars Atmos. Model. Obs.*, Granada, Spain, http://www-mars.lmd.jussieu.fr/granada2006/.

Wilson, R.J., S.R. Lewis, and L. Montabone (2007), Thermal tides in an assimilation of three years of Thermal Emission Spectrometer data from Mars Global Surveyor, *Seventh Int. Conf. Mars*, Pasadena, California, www.lpi.usra.edu/meetings/7thmars2007/.

Wilson, R.J., R.M. Haberle, J. Noble, et al. (2008a), Simulation of the 2001 planet-encircling dust storm with the NASA/NOAA Mars general circulation model, *Third Int. Work. Mars Atmos. Model. Obs.*, Williamsburg, Virginia, www.lpi.usra.edu/meetings/modeling2008/.

Wilson, R.J., S.R. Lewis, L. Montabone, and M.D. Smith (2008b), Influence of water ice clouds on Martian tropical atmospheric temperatures, *Geophys. Res. Lett.*, 35(7), doi:10.1029/2007GL032405.

Wilson, R.J., E. Millour, T. Navarro, F. Forget, and M. Kahre (2014a), GCM simulations of aphelion season tropical cloud and temperature structure, *Fifth Int. Work. Mars Atmos. Model. Obs.*, Oxford, UK, http://www-mars.lmd.jussieu.fr/oxford2014/.

Wilson, R.J., S.D. Guzewich, and A. Kleinböhl (2014b), New progress and insights on thermal tides and their forcing from MCS and modeling, *Eighth Int. Conf. Mars*, Pasadena, California, www.hou.usra.edu/meetings/8thmars2014/.

Withers, P., S.W. Bougher, and G.J. Keating (2003), The effects of topographically-controlled thermal tides in the Martian upper atmosphere as seen by the MGS accelerometer, *Icarus*, 164(1), 14–32, doi:10.1016/S0019-1035(03)00135-0.

Withers, P., R. Pratt, J.-L. Bertaux, and F. Montmessin (2011), Observations of thermal tides in the middle atmosphere of Mars by the SPICAM instrument, *J. Geophys. Res.*, 116(E11), E11005, doi:10.1029/2011JE003847.

Wolkenberg, P.M., and R.J. Wilson (2014), Mars Climate Sounder observations of wave structure in the north polar middle atmosphere of Mars during the summer season, *Eighth Int. Conf. Mars*, Pasadena, California, www.hou.usra.edu/ meetings/ 8thmars2014/.

Zalucha, A.M., R.A. Plumb, and R.J. Wilson (2010), An analysis of the effect of topography on the Martian Hadley cells, *J. Atmos. Sci.*, 67(3), 673–693, doi:10.1175/2009JAS3130.1.

Zhang, K.Q., A.P. Ingersoll, D.M. Kass, et al. (2001), Assimilation of Mars Global Surveyor atmospheric temperature data into a general circulation model, *J. Geophys. Res.*, 106(E12), 32863, doi:10.1029/2000JE001330.

Zurek, R.W. (1976), Diurnal tide in the Martian atmosphere, *J. Atmos. Sci.*, 33, 321–337, doi:10.1175/1520-0469(1976)033<0321:DTITMA>2.0.CO;2.

Zurek, R.W. (1981), Inference of dust opacities for the 1977 Martian great dust storms from Viking Lander 1 pressure data, *Icarus*, 45(1), 202–215, doi:10.1016/0019-1035(81)90014-2.

Zurek, R.W. (1986), Atmospheric tidal forcing of the zonal-mean circulation: the Martian dusty atmosphere, *J. Atmos. Sci.*, 43, 652–670, doi:10.1175/1520-0469(1986)043<0652:ATFOTZ>2.0.CO;2.

Zurek, R.W. (1988), Free and forced modes in the Martian atmosphere, *J. Geophys. Res.*, 93(D8), 9452, doi:10.1029/JD093iD08p09452.

Zurek, R.W., and R.M. Haberle (1988), Zonally symmetric response to atmospheric tidal forcing in the dusty Martian atmosphere, *J. Atmos. Sci.*, 45, 2469–2485, doi:10.1175/1520-0469(1988)045<2469:ZSRTAT>2.0.CO;2.

Zurek, R.W., and C.B. Leovy (1981), Thermal tides in the dusty Martian atmosphere: a verification of theory, *Science*, 213(4506), 437–439, doi:10.1126/science.213.4506.437.

Zurek, R.W., J.R. Barnes, R.M. Haberle, et al. (1992), Dynamics of the atmosphere of Mars, In *Mars*, H.H. Kieffer, B.M. Jakosky, C.W. Snyder, and M.S. Mathews, Eds., University of Arizona Press, 835–933.

The Mars Dust Cycle

MELINDA A. KAHRE, JAMES R. MURPHY, CLAIRE E. NEWMAN, R. JOHN WILSON, BRUCE A. CANTOR, MARK T. LEMMON, MICHAEL J. WOLFF

10.1 INTRODUCTION

The presence of suspended dust and its temporal and spatial variability are important components of the Martian climate system. Gierasch and Goody (1972) first recognized that the thermal state of the Martian atmosphere is very sensitive to the quantity and distribution of radiatively active airborne dust and its influence on the low-mass, short-radiative-time-constant atmosphere (Figure 10.1). Airborne dust efficiently absorbs and scatters radiant energy at visible wavelengths, which results in the heating of the dust itself and the atmosphere within which it is suspended. At infrared wavelengths, it efficiently absorbs in the 9 μm silicate band, increases the emissivity of the CO_2 atmosphere at 15 μm, and locally cools or warms the atmosphere depending on the environmental conditions (Toon et al., 1977; Pollack et al., 1979, 1990).

Evidence for the presence of dust within the Martian atmosphere can be traced back to yellow clouds telescopically observed as early as the 18th century (Capen and Martin, 1971; McKim, 1999). The Mariner 9 Orbiter arrived at Mars in November 1971 to find a planet completely enshrouded in airborne dust (Snyder and Moroz, 1992). Since that time, the exchange of dust between the planet's surface and atmosphere and the role of airborne dust on its weather and climate have been studied using observations and numerical models. Observations indicate that dust is present in the Martian atmosphere year-round with seasonally varying abundance. The general year-to-year repeatability of dust loading as a function of season constitutes the dust cycle (Figure 10.1), which is characterized by a low-level dust loading during the "non-dusty season" ($L_s \sim 0–135°$) and a higher dust loading during the "dusty season" ($L_s \sim 135–360°$). Although some aspects of the behavior of dust in the current climate regime are cyclical, there are also behaviors that are more stochastic than repeatable in nature. In order to straightforwardly compare one Mars year to another, we adopt the convention of Clancy et al. (2000) for the "Mars year" numbering system, which defines Mars Year 1 (MY 1) to have begun on April 11, 1955, at planetocentric solar longitude $L_s = 0°$.

Observations and numerical modeling are both critically important for understanding the Martian dust cycle. Observations provide information on where and when dust is present in the Martian atmosphere and on the surface, the nature of the dust itself, and how it affects the thermal state of the atmosphere. Numerical models further our understanding of how and when dust enters the atmosphere, how dust travels through the atmosphere, where and how dust is removed from

the atmosphere, and how dust affects the state of the Martian atmosphere and climate. Because models, particularly global climate models (GCMs), have been used extensively to help interpret observations, numerical modeling results are often presented alongside observations in the following sections.

Dust is lifted from the surface into the atmosphere via processes associated with the exchange of momentum and heat between the atmosphere and the surface at the bottom of the planetary boundary layer (PBL). Two processes – surface wind stress lifting and dust devil (convective vortex) lifting – are thought to be the primary mechanisms of dust lifting on Mars. Surface wind stress lifting generally occurs when the momentum imparted to the surface by near-surface winds exceeds the necessary threshold for sand-sized particles to detach from the surface, travel a short distance, and return to the surface in the process of saltation (Greeley and Iversen, 1985). When these larger particles return to the surface, they impact dust-sized particles, providing enough momentum (in addition to that contributed by the wind alone) to inject the smaller particles into suspension in the atmosphere. While it is also possible to lift dust particles directly into suspension by winds, this process is likely secondary to saltation because very strong winds are required. Dust lifting by convective vortices (i.e. dust devils) occurs when the pressure gradients across the vortex and/or strong surface wind stresses around the vortex act to inject dust from the surface into the atmosphere. These processes are described in further detail in Section 10.5.1.1.

Dust lifting events produce dust clouds/hazes, which can range in spatial scale from local (meters across) to global, and in duration from less than an hour to seasonal timescales. There has been much discussion in the literature regarding the naming and classification convention that is the most descriptive and useful; while the body of literature on this topic can be confusing, it clearly demonstrates that classifying storms can be difficult to do in many cases because there is likely a fairly smooth continuum of dust storm sizes and/or durations. Historically, the largest, longest-lasting dust events have been referred to by different names, including "global dust storms", "planet-encircling dust storms", and "great dust storms". Much of the discussion regarding what term should be used to refer to these storms centers on the fact that they do not result from one global-scale dust lifting event; rather, they are produced from the combination of multiple local and regional dust lifting events. We believe that the term "planet-encircling dust storm" is somewhat unsatisfactory because many regional storms have been observed to encircle the planet at specific latitudes. Thus, we do not choose to adopt that term. We choose instead to

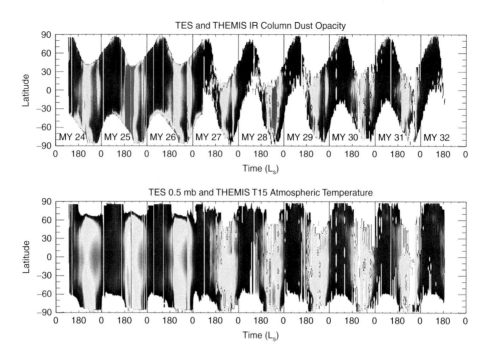

Figure 10.1. Zonally averaged 9 μm column dust opacity (top) and zonally averaged atmospheric temperature for approximately eight of the most recent Martian years, as observed by MGS/TES (MY 24–26) and Mars Odyssey (MY 27–32). The 9 μm column dust opacities range from 0 (black) to >0.5 (red/light gray). Temperatures range from 150 K (black) to 250 K (red/light gray). Data courtesy of M.D. Smith. A black and white version of this figure will appear in some formats. For the color version, please refer to the plate section.

use the term "global dust storm" to refer to the most spatially extended and long-lasting dust storms, because it follows most directly from the progression of scales from small and short-lived to large and long-lived (i.e. local, regional, and global).

Airborne dust affects dynamical processes via its influence on the atmospheric thermal structure. Atmospheric regions that warm or cool due to the distribution of radiatively active dust also experience resulting changes in their mass distribution, which results in modified pressure gradients and hence winds. The altered atmospheric circulation then affects the locations and magnitude of additional dust lifting from the surface, the transport (advection and mixing) of dust within the atmosphere, and the locations where dust is deposited back to the surface (Newman et al., 2002a; Basu et al., 2004, 2006; Kahre et al., 2006, 2008). The response of circulation components (mean overturning circulation, eddies, and condensation flow) to the dust-induced thermal structures can also result in climatically important differences in the fluxes of heat, momentum, dust, water, and non-condensable gases (Haberle et al., 1982; Pollack et al., 1990; Wilson and Hamilton, 1996; Richardson and Wilson, 2002; Montmessin et al., 2004; Lian et al., 2012).

Interactions between the dust, CO_2, and water cycles are important for the current climate of Mars. The dust and CO_2 cycles are coupled through the radiative effects of polar airborne dust, the influence of dust on atmospheric heat transport into the polar regions, the modification of seasonal CO_2 ice cap properties, and the importance of total atmospheric mass, manifested as near-surface gas density, on dust lifting processes. Winter polar atmospheric CO_2 condensation (Hu et al., 2012; Hayne et al., 2012) can be enhanced by the increased emissivity and condensation seed nuclei provided by locally airborne dust, though this may be offset by the increased meridional heat transport resulting from a dusty atmosphere (Pollack et al., 1990). Additionally, dust

falling in the polar region during seasonal cap formation, either directly or as CO_2 snow nuclei, will affect the thermal properties of the CO_2 cap ice (albedo and emissivity) and thus the condensation and sublimation rate of CO_2. The influence of dust on the CO_2 cycle could then feed back on the dust cycle, because the atmosphere–surface exchange of momentum (i.e. surface wind stress) that drives significant dust lifting is directly related to the atmospheric mass (see Section 10.5.1.1).

The dust and water cycles are coupled through cloud condensation processes. Suspended dust particles are thought to provide seed nuclei for heterogeneously nucleated water ice clouds (Michelangeli et al., 1993; Montmessin et al., 2002). Ice-covered dust particles fall at speeds different from those of the dust particles alone, depending on the ratio of dust to ice mass and the density of each material (Rossow, 1978). Thus, cloud formation can change the vertical distribution (and, by extension, the horizontal distribution) of dust and water in the atmosphere. The combined dust and water ice particle exhibits altered radiative properties compared to the dust particle in isolation. Therefore, the radiative effects of water ice clouds affect the thermal and dynamical state of the atmosphere, which in turn alters dust lifting, transport, and sedimentation.

The Martian surface is the primary source and sink for atmospheric dust. Surface evidence of aeolian (wind-driven) processes yields information about the current and past dust cycles. Examples of aeolian signatures that correspond to the current dust climatic state include observed albedo temporal changes and dust depositional streak formation and degradation. Evidence for aeolian processes under past climate regimes is also abundant, including the presence of the polar layered terrain, inactive dunes, and other features that do not appear to be explained by the current wind regime (e.g. Greeley et al., 1993). The spatial distribution of mobile dust particles on the surface today is the result of the integrated effect of aeolian processes over geologic timescales. The timescale required for creating dust-sized particles from the erosion of larger materials is long compared to the orbital obliquity cycle (Golombek et al., 2006), which supports the assumption that, under the current climate system, a finite amount of mobile dust is available. However, there is no *a priori* reason to assume that dust cannot be sequestered into surface reservoirs that are not currently susceptible to aeolian processes. Thus, it is possible that quasi-permanent sinks for dust exist under the current climatic conditions. Such long-term sinks could be activated in the future under different climatic states forced by orbital changes. Alternatively, a continued sequestration of mobile dust could eventually lead to a permanently inactive dust cycle. However, the fact that there is an active dust cycle on Mars today could argue against the existence of permanent sinks over geologic timescales.

Our goal is to present the state of knowledge of the current Martian dust cycle and identify avenues of future investigation that may be most fruitful in expanding this knowledge further. We organize our discussion by focusing on observations that provide characterization of the dust cycle and the physical processes involved, drawing upon insights provided by numerical, laboratory, and field studies. We begin with a brief history of dust cycle observations and climate modeling. We then summarize the state of knowledge regarding atmospheric and surface dust and the seasonal variation of atmospheric dust loading on Mars, including a summary of the morphology and seasonality of dust storms. We next focus on the physical processes involved in dust lifting, removal, and atmospheric transport, and on the state of interactive dust cycle modeling. We conclude with a discussion about the open questions that exist regarding the dust cycle, and possible methods for addressing these questions.

10.2 HISTORY OF OBSERVATIONS AND CLIMATE MODELING

10.2.1 Observations

Observations of the presence of dust in the Martian atmosphere and on its surface date back as far as the early 18th century, when astronomers noted the obscuration of surface features, though the concept of "yellow clouds or hazes" was not introduced until the second half of the 19th century (McKim, 1999). Since then, the monitoring of Martian dust has been carried out with observations made by Earth-based telescopes, instruments in orbit around Mars, and landers and rovers on the Martian surface.

10.2.1.1 Earth-Based Observations

There is a long history of telescopic observations of Mars, beginning with Galileo Galilei, who observed Mars with his primitive telescope in 1609. Although initially observed much earlier, "yellow clouds" that sometimes grew to obscure the entire Martian surface were first suggested to be due to dust near the end of the 19th century by Douglass (1899) and later by Antoniadi (1915). Surface albedo variations were also initially misinterpreted – for example, a "seasonal wave of darkening" each spring was at one time thought to be due to the seasonal growth and decay of vegetation – until McLaughlin (1954) and

Kuiper (1957) suggested that changes in albedo patterns might be explained by the removal and deposition of windblown dust.

Naturally, this lack of understanding of what was being observed caused problems in interpretation, and the early telescopic record of dust storms is difficult to interpret for several other reasons as well:

- high-quality photographs of Mars have only been possible since the early 1900s, and even then the variability of the distance from Earth to Mars has meant great spatial and temporal disparity in the coverage of storms;
- some clouds described as "yellow" (or with no color assigned) would actually have been water or CO_2 ice clouds;
- improvements in visibility were sometimes interpreted as the disappearance of large clouds that had obscured the surface, rather than improvement in seeing for other reasons (such as Mars being closer); and
- surface albedo changes were sometimes thought to indicate clouds, and vice versa.

Despite these difficulties, telescopic observations provide invaluable insight into dust activity before continuous spacecraft observations became available. The British Astronomical Association (BAA) Mars section was formed in 1892 and continues to collect and publish observations of Mars, such as the catalog of McKim (1999), which attempts to interpret all telescopic observations prior to 1993 in light of the aforementioned difficulties. Martin and Zurek (1993) provides another excellent review covering the period 1873–1990. From about the 1950s, visible telescopic observations began to be combined into joint programs, the goal being to coordinate observations by different groups worldwide to ensure adequate longitudinal coverage to keep Mars under near-continuous observation, though with peak activity occurring during oppositions when viewing was optimal. In 1969, the Association of Lunar and Planetary Observers (ALPO) created the International Mars Patrol (IMP), an international network of observers still producing thousands of observations each apparition. A large worldwide observing program, the International Planetary Patrol (IPP), also ran from 1969–1982, partly to provide support for spacecraft missions (Baum, 1973).

10.2.1.2 Orbiters

Mariners 6 and 7 flew by Mars in 1969 (MY 8) during northern autumn on Mars ($L_s = 200°$). Although the TV camera images showed a relatively clear atmosphere, a lack of surface detail suggested the presence of atmospheric dust in the Hellas region (Leovy et al., 1971). Mariners 6 and 7 carried the Infrared Spectrometer (IRS), which covered the infrared spectrum out to approximately 14 μm; however, Mariner 6's mid-IR channel failed, so no data were acquired between 6 and 14 μm. Measured absorption from Mariner 7 in the 9 μm silicate band at midday near the equator indicated the presence of airborne dust (Herr et al., 1998), although quantitative dust opacities have not been derived from these data to date.

Mariner 9, the first spacecraft to obtain measurements from Mars orbit, arrived in November 1971 (MY 9) to find the surface obscured by a fully developed global dust storm. Observations acquired with the Infrared Interferometric Spectrometer (IRIS) exhibited a prominent absorption feature at 9 μm, which was used to monitor the decay of the storm over a several month time interval (Conrath, 1975). The observed decline of atmospheric dust load coincided with an observed cooling of the atmosphere (Conrath, 1975) and the settling of dust to lower altitudes (Hartmann and Price, 1974; Anderson and Leovy, 1978; Fenton et al., 1997). Analysis of IRIS data led to the widely used "Conrath" profile describing the typical vertical dust distribution, which was derived by balancing upward transport due to diffusive mixing against downward gravitational sedimentation (Conrath, 1975). Usually expressed in terms of the "Conrath-ν" parameter, ν, which determines how deeply dust is mixed, we have

$$q(z) = q_0 e^{\nu[1-e^{z/H}]} \qquad (10.1)$$

where q is the dust mixing ratio at altitude z, q_0 is the dust mixing ratio at the surface or a reference altitude, and H is the atmospheric scale height. Although Conrath-ν or modified Conrath-ν (Montmessin et al., 2004) profiles have been used for decades in climate models to simulate the fundamentals of the general circulation, recent observations suggest notable divergence from this profile (see Section 10.3.1.2).

Arriving in 1976 (MY 12), Viking Orbiters 1 and 2 each carried two instruments that provided measurements relevant to dust: the broadband visible camera and the Infrared Thermal Mapper (IRTM). The cameras revealed the temporally varying spatial extent of dust storms, most notably the expansion of the first 1977 global dust storm (the so-called 1977a storm, in MY 12; Thorpe, 1979). The IRTM instruments monitored atmospheric dust by measuring absorption in the 9 μm silicate band, and quantitative column dust opacities were derived from these data (Pleskot and Miner, 1981; Martin, 1986; Christensen, 1988; Liu et al., 2003). The IRTM instruments additionally provided measurements from which properties of the surface such as the albedo and thermal inertia were derived (Kieffer et al., 1977; Christensen, 1982, 1986a,b, 1988; Jakosky, 1986), with the spatio-temporal stability or variability of these properties yielding insight into the location and movement of dust around the planet.

The Mars Global Surveyor (MGS) spacecraft arrived at Mars in 1997 and began science operations in 1999 (MY 24). MGS monitored airborne dust in the thermal infrared through nadir and limb views with the Thermal Emission Spectrometer (TES) and in the visible with the Mars Orbiter Camera (MOC). TES operated for approximately three Mars years, while MOC operated for the entire length of the MGS mission – approximately four Mars years. The instruments on MGS mapped the spatial and temporal patterns of airborne dust and surface properties with unprecedented coverage (Cantor et al., 2001; Christensen et al., 2001; Ruff and Christensen, 2002; Smith, 2004).

Currently, there are three Mars-orbiting spacecraft carrying instruments that monitor dust. Data acquired with the broadband IR Thermal Emission Imaging System (THEMIS) on Mars Odyssey overlapped in time with those obtained with MGS/TES, thus allowing for a continuous record of derived atmospheric dust opacity at 9 μm from the beginning of the MGS mission to the present day (Smith, 2009). The Planetary Fourier Spectrometer (PFS; Grassi et al., 2005; Zasova et al., 2005), the Observatoire pour la Minéralogie, l'Eau, les Glaces et l'Activité (OMEGA, an infrared mapping spectrometer; Määttänen et al., 2009), and the Spectroscopy for the Investigation of

Characteristics of the Atmosphere of Mars Ultraviolet and Infrared Atmospheric Spectrometer (SPICAM; Montmessin et al., 2006; Rannou et al., 2006) instruments on Mars Express (MEX), and the Mars Climate Sounder (MCS; McCleese et al., 2010; Heavens et al., 2011a,b), Compact Reconnaissance Imaging Spectrometer for Mars (CRISM; Smith et al., 2013), and Mars Color Imager (MARCI; Malin et al., 2008; Cantor et al., 2010) instruments on the Mars Reconnaissance Orbiter (MRO) continue to measure the physical properties and distribution of atmospheric dust. The data acquired from these recent and ongoing missions are discussed in more detail in the following sections, as they provide the basis for our evolving understanding of the Martian dust cycle.

10.2.1.3 Landers and Rovers

Observations of dust from the surface of Mars have been acquired from the Viking Landers (VL1, 1976–1980, and VL2, 1976–1982), the Mars Pathfinder Lander (MPF, 1997), the Mars Exploration Rovers (MERs; Spirit, 2004–2010, and Opportunity, 2004–present), the Phoenix Lander (PHX, 2008), and the Mars Science Laboratory Rover (MSL, 2012–present). Observations of the Sun with visible imagers on the landers and rovers have yielded quantitative measurements of the visible line-of-sight dust optical depth at their locations (Colburn et al., 1989; Smith and Lemmon, 1999; Lemmon et al., 2014). Additional quantitative measurements of dust optical depth have been derived from data obtained with the mini-TES instruments on the MERs (Smith et al., 2006). Meteorological data acquired with instruments on VL1, VL2, MPF, PHX, and MSL have been used to characterize the frequency of dust devil passage (Ryan and Lucich, 1983; Murphy and Nelli, 2002; Ringrose et al., 2003; Ellehoj et al., 2010; Kahanpää et al., 2013), and dust devil imaging by Mars Pathfinder and Spirit has been used to characterize dust devil frequency and lifting rates (Metzger et al., 1999, 2000; Ferri et al., 2003; Greeley et al., 2006a,b, 2010).

Both Viking Landers and both MERs operated during planet-encircling dust storms, in 1977 and 2007, respectively, with local visible-band optical depths approaching or exceeding 5. The Viking Landers and MERs operated for more than a Mars year each, at northern tropical and mid-latitudes for Viking, and southern tropical latitudes for MER. Pathfinder and Phoenix operated during northern summer, in northern tropical and polar latitudes, respectively. MSL landed in the southern tropics near the end of southern winter.

The vast increase in the volume and quality of data pertaining to the behavior of dust in the atmosphere and on the surface of Mars since 1997 has yielded a comprehensive description of the current dust cycle. Although observations acquired before MGS and MPF are discussed when appropriate, focus is given here to those obtained between 1997 and the present.

10.2.2 Climate Modeling

Numerical models have been used to investigate aspects of the Martian dust cycle and its effect on climate since the late 1960s (Leovy and Mintz, 1969). Observations made in 1971 by instruments on Mariner 9 of a planet covered in atmospheric dust clouds motivated Gierasch and Goody (1972) to show for the first time that the radiative effects of atmospheric dust could have a significant effect on the thermal state of the atmosphere. Since then, climate models have been designed and utilized to investigate dust cycle phenomena at a range of spatial and temporal scales in one, two, and three dimensions. These models have grown in sophistication, spatial resolution, and simulation duration as more has been discovered about the critical physical processes that govern the dust cycle, and as computers have become significantly more efficient.

There are two general types of Martian dust modeling. The first is based on the need to realistically simulate atmospheric temperatures compared to observations in order to study a variety of aspects of the current Mars climate. These studies typically employ prescribed dust distributions, which allows the modeler to alter the dust distribution in a controlled manner to improve agreement with observed temperatures. The second is to study the processes involved in the dust cycle itself, which requires the explicit inclusion of processes such as dust lifting, transport and removal.

10.2.2.1 Climate Modeling With Prescribed Dust

Three-dimensional global climate models (GCMs) were first utilized in the early 1990s to study the effect of dust on the Martian climate. In the earliest work, the spatial distribution of dust was prescribed and fixed in time and space. In particular, the vertical distribution of dust was typically assumed to be well mixed up to a particular pressure (or altitude). Results from prescribed and fixed dust investigations show great sensitivity of the zonal-mean circulation, transient and stationary eddies, tides, and the patterns and magnitudes of surface wind stresses to the quantity and spatial distribution of atmospheric dust (Pollack et al., 1990, 1993; Barnes et al., 1993; Greeley et al., 1993; Haberle et al., 1993; Hourdin et al., 1993, 1995; Wilson and Hamilton, 1996; Wilson, 1997).

Methodologies for implementing prescribed dust scenarios have become more sophisticated with time. Improving on spatial dust distributions that are constant in time and space, there have been many investigations that employed seasonally and spatially varying prescribed dust loadings. Observed column-integrated dust opacities from instruments aboard landed spacecraft (such as the Viking Landers) and orbiting spacecraft such as MGS/TES and ODY/THEMIS have been used to force GCM simulations (Forget et al., 1999; Lewis et al., 1999; Montmessin et al., 2004; Montabone et al., 2005, 2006; Lewis et al., 2007; Kuroda et al., 2008). Like the earliest prescribed dust GCM studies, these investigations employed prescribed vertical dust profiles (e.g. Conrath or modified Conrath profiles).

Recent work has been carried out to improve the treatment of the vertical distribution in prescribed dust studies with the goal of better simulating the observed temperature structure of the Martian atmosphere. This has been done using two general methods. In the first, models self-consistently predict the vertical distribution of dust while still maintaining column-integrated dust opacities in the model that match observed opacities (Wilson et al., 2008a; Kahre et al., 2010a; Madeleine et al., 2011; Greybush et al., 2012). In the second, observations of the vertical dust distribution by limb sounders such as MRO/MCS or MGS/TES are used to force the models (Guzewich et al., 2013a).

10.2.2.2 Modeling Dust Cycle Processes

The earliest process-focused numerical work on the dust cycle employed one-dimensional theoretical calculations of atmospheric dust based on Mariner 9 observations during the decay phase of the 1971 global dust storm (Hartmann and Price, 1974; Conrath, 1975; Toon et al., 1977; Rossow, 1978; Murphy et al., 1990). Based on the rate at which the atmosphere was observed to clear of dust, estimations were made of dust particle sizes, suspension lifetimes, and the vertical distribution of atmospheric dust.

Two-dimensional axisymmetric models were subsequently developed to investigate how dust was transported by, and how dust affected, the mean overturning circulation. Investigations by Haberle et al. (1982) and Murphy et al. (1993) showed that dust lifted in the southern subtropics during southern summer solstice entered the rising branch of the Hadley cell and heated the atmosphere, which resulted in intensifying and increasing the meridional extent of the Hadley cell. This radiative–dynamic feedback, whereby increasing the atmospheric dust loading increases the strength of the overturning circulation, is a prominent feature in all global climate modeling studies to date and is thought to play an important role in the generation of regional and global-scale dust storms (see Section 10.4.2.3).

The models that incorporate some or all of these processes have also grown in sophistication over time. The first studies to incorporate dust transport in three-dimensional global climate models (GCMs) were Murphy et al. (1995), Wilson and Hamilton (1996), Wilson (1997), and Richardson and Wilson (2002). In these studies, radiatively active dust was injected into the atmosphere from prescribed source regions (negating the need to model the physics of dust lifting explicitly) and the authors investigated dust transport, the effects of atmospheric dust on components of the general circulation, and the effects of dust on the magnitude and patterns of surface wind stresses.

The current state of the art in dust cycle modeling in GCMs is the inclusion of fully interactive dust lifting, transport, and removal of radiatively active dust. The earliest interactive dust cycle studies assumed an unlimited planet-wide supply of surface dust available for lifting (Newman et al., 2002a,b; Basu et al., 2004, 2006; Kahre et al., 2006, 2008). These studies captured some of the observed characteristics of the dust cycle. More recent work has been on the inclusion of finite surface dust reservoirs (Kahre et al., 2005; Wilson and Kahre, 2009; Mulholland et al., 2013; Newman and Richardson, 2015) and on the coupling between the dust and water cycles (Kahre et al., 2011, 2015). Fully interactive GCM dust cycle studies are discussed in detail in Section 10.5.2.

In addition to modeling the dust cycle in global climate models, there has been a growing number of mesoscale and large-eddy simulation (LES) dust cycle modeling studies. Rafkin et al. (2001), Toigo et al. (2003), Michaels and Rafkin (2004), Spiga and Forget (2009), and Gheynani and Taylor (2011) used LES models to simulate the convective vortices that may be capable of lifting dust, although dust lifting is not explicitly included. Michaels (2006) used a mesoscale model to study the formation and behavior of Martian dust devils and their tracks. Rafkin (2009) used a mesoscale model to study the radiative–dynamical feedbacks within local-scale dust disturbances and

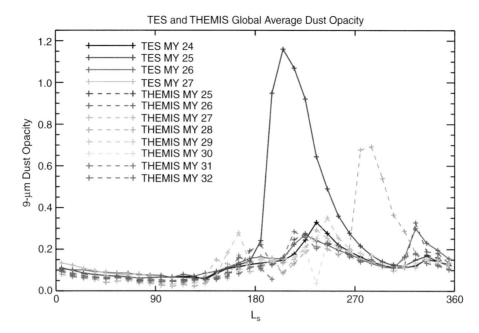

Figure 10.2. Globally averaged 9 μm dust opacity observations from MGS/TES and ODY/THEMIS for multiple Mars years. Data courtesy of M.D. Smith. A black and white version of this figure will appear in some formats. For the color version, please refer to the plate section.

how these feedbacks lead to continued or strengthened dust lifting. Recent investigations have focused on how local and mesoscale circulations may organize dust into elevated layers above the PBL as observed by MRO/MCS and MGS/TES (Rafkin, 2012; Spiga et al., 2013; Section 10.3.1.2).

10.3 THE DISTRIBUTION AND VARIATION OF ATMOSPHERIC AND SURFACE DUST

10.3.1 Atmospheric Dust

As viewed from orbit and from the surface, dust is present in the Martian atmosphere throughout the year (Figures 10.1, 10.2, and 10.3). Although year-to-year variability exists at specific seasons, some aspects of the behavior of dust repeat from one year to the next. The total atmospheric dust loading is generally characterized by low levels during northern hemisphere spring and summer, and increased levels during northern hemisphere autumn and winter (Liu et al., 2003; Smith, 2004, 2008).

10.3.1.1 Overview of Seasonal Patterns

(a) **The Non-Dusty Season ($L_s \sim 0$–$135°$)** During northern spring and the majority of summer, the atmosphere at low to middle latitudes exhibits a 9 μm column dust opacity of ~0.1 (Figure 10.1). Localized higher dust opacities occur along the edges of the seasonal CO_2 caps in both the north and the south, with the highest opacities occurring along the north cap just prior to that cap's disappearance at $L_s \sim 90°$ and along the south cap from the time of its maximum extent through the entirety of its retreat phase. These localized cap-edge high opacities coincide temporally with enhanced visible dust opacities in imaging data from Viking, MGS, MRO, and MEX (Cantor et al., 2001). The morphology and generation mechanisms of these cap-edge dust storms will be discussed in more detail in Section 10.4.2.1.

It is notable that no large regional or global dust storms have been recorded during these seasons (Martin and Zurek, 1993; Zurek and Martin, 1993; Clancy et al., 2000; Cantor et al., 2001; Liu et al., 2003; Smith, 2008; Montabone et al., 2015).

(b) **The Dusty Season ($L_s \sim 135$–$360°$)** Northern autumn and winter are characterized by greater atmospheric dust loadings (Figures 10.1, 10.2, and 10.3). Local, regional, and global dust storms with a variety of timings and sizes are observed (Liu et al., 2003; Wang, 2007; Smith, 2008, 2009; Wang and Richardson, 2015). Although the locations, spatial extents, and seasonality of high-dust-opacity storms can vary significantly from one year to the next, there are repeatable characteristics of storm activity. The following characterization is based on derived dust loadings from MY 24 through MY 31, and it consists of thermal infrared and visible observations from orbit and primarily visible observations from the surface.

GLOBAL DUST STORMS   The most dramatic and thermodynamically significant dust events are global dust storms (Martin and Zurek, 1993; Zurek and Martin, 1993; Smith et al., 2002; Strausberg et al., 2005; Cantor, 2007). Large dust opacities over extended regions accompany mid-level atmospheric temperatures that are more than 20 K warmer than temperatures representative of years in which no significant dust storms are observed (Smith et al., 2002; Smith, 2004, 2008). Since the beginning of the MGS mission (1997), there have been two global dust storms: one in MY 25 (June 2001) and one in MY 28 (June 2007). The MY 25 event was one of the earliest global storms observed to date, beginning at $L_s \sim 185°$ (Smith et al., 2002; Strausberg et al., 2005; Cantor, 2007), whereas the MY 28 event had a more classical solstitial initiation, beginning at $L_s \sim 262°$ (Cantor et al., 2008; Montabone et al., 2011; Wang and Richardson, 2015). Further details and mechanisms proposed for the initiation and growth of these storms are discussed in Section 10.4.2.3.

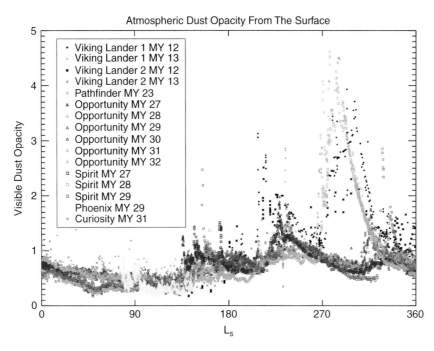

Figure 10.3. Visible dust opacities derived from VL1, VL2, MPF, MERa, MERb, PHX, and MSL. A black and white version of this figure will appear in some formats. For the color version, please refer to the plate section.

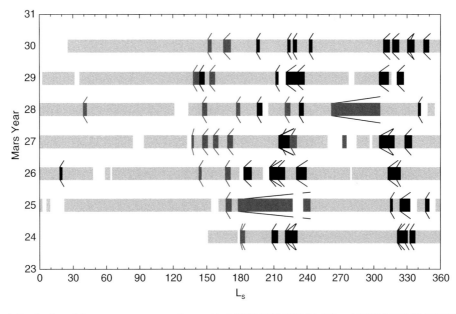

Figure 10.4. Seasonal distribution of dust storm sequences observed in MGS/MOC (MY 24–28) and MRO/MARCI (MY 28–30) Mars Daily Global Maps (MDGM). Light gray bars indicate periods with data but without any dust storm sequences. A vertical line denotes the beginning of a dust storm sequence and two sloped lines indicate the duration of the sequence. Dust storm sequences that originate in the northern hemisphere are black, and those that originate in the southern hemisphere are dark gray. From Wang and Richardson (2015).

OTHER MAJOR FEATURES There are features of the seasonal and spatial atmospheric dust loading that generally occur year after year, particularly in years without a global dust storm (Figure 10.4). We identify four distinct classes of increased dust loading during the dusty season that generate a measurable response in the mid-level atmospheric temperatures (Kass et al., 2016; Wang and Richardson, 2015; Montabone et al., 2015). Although these atmospheric dust features are identifiable in most if not all Mars years since MY 24, there are notable differences in their strength, duration, and

atmospheric thermodynamic response from one year to the next (Smith, 2008; Kass et al., 2016; Wang and Richardson, 2015; Montabone et al., 2015).

Early-Season Activity: $L_s \sim 135-180°$. The first rise in the global atmospheric dust loading results from dust storm activity during the seasonal range from $L_s \sim 135°$ to $L_s \sim 180°$ (Figures 10.1, 10.2, 10.3, and 10.4; Montabone et al., 2015). In most years, these are storms in the southern hemisphere, but in at least one year (MY 29) dust storm activity occurred during this season in the northern hemisphere (Wang and Richardson,

2015). This feature of the annual dust cycle was particularly strong in MY 27 and MY 29. In MY 27, a regional storm produced a dust haze that encircled the planet in the southern hemisphere (Smith, 2009) and raised the visible-wavelength dust opacities observed by the MER Rovers over unity (Lemmon et al., 2014). In MY 29, early-season dust storm activity in both the northern and southern hemispheres resulted in spikes in atmospheric dust loading as seen from both orbit (Montabone et al., 2015) and the surface (Lemmon et al., 2014).

Pre-Solstice Activity: $L_s \sim$ 180–236°. The second (and more prominent) rise in atmospheric dust loading occurs before both perihelion ($L_s = 251°$) and southern summer solstice ($L_s = 270°$). As discussed in more detail in Section 10.4.2.2, this increase in atmospheric dust loading is related to dust storm activity in both the northern and southern hemispheres (Figures 10.1 and 10.4). In the north, dust storms are associated with traveling weather systems (i.e. frontal storms). When conditions are favorable, these storms travel through the low-topographic corridors of Acidalia, Arcadia, and Utopia into the southern hemisphere and in some cases trigger dust storm activity in the southern hemisphere. While many southern hemisphere dust storms appear to be directly related to the arrival of northern hemisphere frontal storms, it is important to note that some southern hemisphere dust storms during this season are isolated events. For example, isolated southern hemisphere dust storms occurred in MY 29 and MY 30 (Wang and Richardson, 2015). Because the rising branch of the Hadley cell is located in the southern hemisphere during this season, dust storm activity in the south quickly leads to increased dust loading throughout the atmosphere and results in mid-level atmospheric warming (Figure 10.1; Kass et al., 2016). While cross-equatorial dust storms have been observed every year from MY 24 to the present (Figure 10.4; Wang et al., 2005; Wang, 2007; Hinson and Wang, 2010; Wang and Richardson, 2015), the degree to which subsequent dust storm activity occurs in the southern hemisphere varies. Prominent cross-equatorial storms and subsequent dust storm activity in the south occurred during MY 24, 26, 27, 29, and 31. The global dust storm in MY 25 either suppressed or obscured this type of dust storm behavior. Significant southern hemisphere dust storm activity triggered by cross-equatorial frontal storms was not observed during this season in MY 28 or 30 (Wang and Richardson, 2015).

Solstitial Activity Near the South Pole: $L_s \sim$ 250–300°. An additional recurring feature in the atmospheric dust loading record occurs near the south pole during the solstitial season and is likely related to the final stages of the retreat phase of the south seasonal CO_2 ice cap (Montabone et al., 2015). While local dust opacities are observed to be quite high (9 μm opacities > 0.5) from orbit (Figure 10.1), this dust storm activity does not produce increased dustiness or a thermodynamic signal outside of the vicinity of the south polar region (Kass et al., 2016; Wang and Richardson, 2015; Montabone et al., 2015). In fact, there is a decrease in dust loading during this season throughout the majority of the atmosphere in years without a global dust storm, so there is not an apparent increase in dustiness as seen by the landers/rovers at low latitudes (Figure 10.3) or in the global mean from orbit (Figure 10.2).

Post-Solstice Activity: $L_s \sim$ 308–336°. The final annual peak in atmospheric dust loading occurs after the southern summer solstitial season. Similar to the pre-solstitial season peak,

this rise in atmospheric dust loading is produced by a series of cross-equatorial dust storms and southern hemisphere dust storms (Figure 10.4). There is a fair amount of variability in the strength of this category of dust storm activity and the atmospheric thermodynamic response amongst the recent recorded Mars years (Figures 10.1, 10.2, and 10.3; Kass et al., 2016; Wang and Richardson, 2015; Montabone et al., 2015). This is a particularly prominent feature in the northern and southern hemispheres in MY 26, 27, and 31. In MY 24, 29, and 30, the increased dust loading was largely confined to the northern hemisphere, with dust extending to subtropical latitudes in the south but not much farther (Figure 10.1).

10.3.1.2 Vertical Distribution

The vertical dust distribution greatly influences the vertical distribution of solar energy deposition and thus atmospheric diabatic heating (Pollack et al., 1979). Because horizontal wind speeds tend to increase with height, the vertical transport of dust influences the rapidity with which dust can be horizontally transported away from its surface source region. Therefore, a full assessment and understanding of the seasonal and spatial variation in the vertical distribution of dust is important for fully characterizing the dust cycle. Observations of line-of-sight dust abundances from the surface looking upwards through different atmospheric air masses provide information regarding the vertical distribution of dust within the bottom 1–3 scale heights (Pollack et al., 1979; Smith et al., 1997; Lemmon et al., 2004), while information on the vertical distribution and extent of dust above the lowest scale height is primarily derived from orbiter observations of the limb (Ajello et al., 1976; Anderson and Leovy, 1978; Jaquin et al., 1986; Montmessin et al., 2006; Cantor, 2007; Clancy et al., 2010; Heavens et al., 2011a,b; Guzewich et al., 2013b).

(a) **Dust in the Lowest Scale Height** Several lines of evidence have suggested that dust is well mixed in the lowest scale height (including the planetary boundary layer) at low and middle latitudes. The definition of "well mixed" in this context is a constant mass mixing ratio given a specified size distribution, composition, and shape of dust particles. Examination of twilight brightness profiles from the Viking Landers (Pollack et al., 1977) and MPF (Smith et al., 1997) imaging cameras and observations of the Sun near the horizon acquired with the MER Pancam instruments (Lemmon et al., 2004, 2014) indicate that the scale height of dust is between 10 and 13 km. Well-mixed dust is inferred from these observations because the derived scale height for dust is consistent with the CO_2 gas scale height. These are not surprising results given the vigor of turbulence in the equatorial and subtropical Martian boundary layer. At Gale Crater, the boundary layer may be depleted of dust due to local-scale circulations (Moores et al., 2014). At higher northern latitudes, observations with the PHX LIDAR instrument are consistent with dust being well mixed up to approximately 4 km (Dickinson et al., 2011; Komguem et al., 2013). At these high latitudes, it is speculated that the scavenging of dust by water ice clouds plays a role in confining dust near the surface. This hypothesis is based on LIDAR observations of cloud fall streaks that suggest that cloud particles fall rapidly from the top of the PBL during the early morning hours (Whiteway et al., 2009).

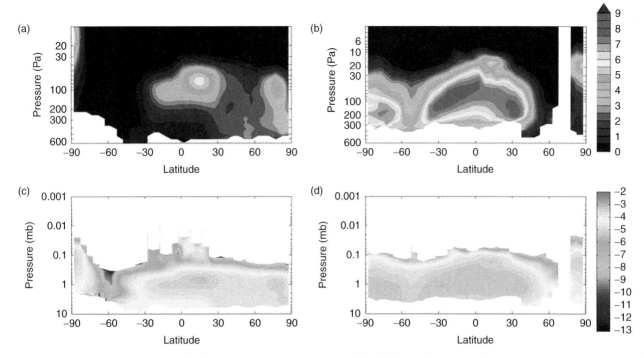

Figure 10.5. (a) Zonal average nightside dust-density-scaled opacity at $L_s = 90°$, MY 30 ($\times 10^4$ m^2 kg^{-1}); (b) zonal average nightside dust-density-scaled opacity at $L_s = 270°$, MY 29 ($\times 10^4$ m^2 kg^{-1}); (c) $\log_{10}$ of zonal average nightside dust-density-scaled opacity at $L_s = 90°$, MY 30 (m^2 kg^{-1}); (d) $\log_{10}$ of zonal average nightside dust-density-scaled opacity at $L_s = 270°$, MY 29 (m^2 kg^{-1}). From Heavens et al. (2011a). A black and white version of this figure will appear in some formats. For the color version, please refer to the plate section.

(b) Dust Above the Lowest Scale Height Information on the vertical extent of atmospheric dust above the lowest scale height has been collected from orbit since Mariner 9. Visible/near-infrared measurements of the limb have been acquired by Mariner 9 (Anderson and Leovy, 1978), Viking (Jaquin et al., 1986), MGS/MOC (Cantor, 2007), MEX/OMEGA (Fouchet et al., 2006), and MRO/CRISM (Smith et al., 2013). Measurements of reflected UV light from the limb have been made from Mariner 9 observations (Ajello et al., 1976) and from MEX/SPICAM observations (Rannou et al., 2006). Thermal infrared limb observations have been gathered from MGS/TES (McConnochie and Smith, 2008; Clancy et al., 2010; Guzewich et al., 2013b) and MRO/MCS (McCleese et al., 2010; Heavens et al., 2011a,b). Limb measurements of dust have also been acquired in a solar occultation geometry by the Phobos spacecraft (Chassefiere et al., 1992) and MEX/SPICAM (Montmessin et al., 2006). Taken together, these data suggest that the vertical extent of dust varies with both season and latitude, with dust extending to the highest altitudes (~60–70 km) at equatorial and subtropical latitudes during the dusty season. Dust is confined nearest to the surface at high latitudes during the non-dusty season and at polar latitudes during northern winter. Prior to the start of data acquisition by MRO/MCS, the available orbiter data were generally consistent with the scenario that dust is generally well mixed up to a particular (but not constant) altitude, with a fall-off in concentration above that altitude.

(c) Elevated Dust Layers MGS/TES and MRO/MCS limb observations have revealed the existence of persistent, large-scale regions that exhibit a local maximum of dust mixing ratio at altitude that is disconnected from the surface (Figure 10.5; McConnochie and Smith, 2008; Heavens et al., 2011a,b, 2014; Guzewich et al., 2013b). Elevated dust layers are observed throughout much of the year at tropical and subtropical latitudes (Heavens et al., 2011a,b, 2014; Guzewich et al., 2013b). While these detached dust layers may have important implications for the Martian climate (as discussed below), we note that they contain only a small fraction of the total column dust.

Detached regions of concentrated dust could have important implications for how dust controls the thermodynamic structure of the atmosphere. Until recently, GCM studies that included fixed dust distributions were based on the vertical prescription developed by Conrath (1975), which does not account for increasing dust concentration with altitude. The thermodynamic implications of prescribed detached dust layers (i.e. non-Conrath-like vertical distributions) are beginning to be explored by ingesting MRO/MCS- or MGS/TES-observed profiles into global climate models. While GCMs have historically captured the bulk character of the thermal structure and general circulation of Mars using Conrath dust profiles, Guzewich et al. (2013a) show that implementing the MGS/TES limb profiles into a GCM produces notable differences in the predicted atmospheric temperatures and in the general circulation.

In addition to the thermodynamic effects, the existence of elevated dust layers has implications for dust transport processes and/or the coupling between the dust and water cycles. While published GCM studies that predict the vertical dust distribution can exhibit elevated dust layers during southern spring and summer (typically associated with the onset phase of a large storm, e.g. when significant dust is carried aloft then equatorward from a southern hemisphere source region in the upwelling branch of the Hadley cell), GCMs generally do not predict elevated layers during northern spring and summer (Richardson and Wilson, 2002; Kahre et al., 2006). This suggests that there are physical processes missing from GCMs attempting to model the dust cycle. One possibility is that cloud scavenging produces detached layers; however, recent water cycle studies do not indicate that elevated layers can be produced by scavenging alone (Kahre et al., 2015; Navarro et al., 2014). The second possibility is that vertical transport processes that are unresolved in GCMs may be of vital importance for producing the observed detached dust layers. Mesoscale modeling of dust transport indicates that local and regional-scale phenomena associated with topographic flows and/or radiative–dynamic feedbacks between dust heating and the circulation can produce detached dust layering (Rafkin et al., 2002; Michaels et al., 2006; Rafkin, 2009; Spiga et al., 2013). Additionally, large-eddy simulation (LES) studies of buoyant thermal plumes that can transport heat, momentum, and aerosols efficiently through the top of the daytime PBL (e.g. Colaïtis et al., 2013) may also lead to an increased understanding of these elevated dust layers. Together, modeling efforts indicate that local-scale and mesoscale circulations are likely critical for producing the observed detached dust layers, although water ice cloud interactions may also play a role. The reader is referred to Chapter 8 for more details on these small-scale phenomena.

10.3.1.3 In Situ Measurements of Surface Dust Accumulation

To date, no direct measurement of the rate of dust accumulation (in terms of either changing thickness or mass flux) upon Mars' surface has been obtained. Indirect dust accumulation rates have been inferred from a number of observations acquired *in situ* by every lander and rover that took measurements from the Martian surface. These techniques include: monitoring the contrast reduction over time due to dust accumulation of the surrounding terrain and of the lander magnetic and calibration targets (Guinness et al., 1979, 1982; Arvidson et al., 1983; Landis and Jenkins, 2000; Johnson et al., 2003; Kinch et al., 2007; Drube et al., 2010); monitoring the degradation of solar panel electrical generation efficiency over time due to accumulated dust (Rover Team, 1997; Arvidson et al., 2004; Stella et al., 2005); and inferring an accumulation rate that is proportional to the local atmospheric column dust opacity (Kinch et al., 2007). The lander/rover deposition estimates from four equatorial/subtropical locations are all in general agreement when the specific seasons and/or atmospheric dust loadings are considered. Representative values for the annual accumulation rates are one to a few dust optical depths, which correspond to several tens of micrometers of accumulated dust when particle size and porosity are accounted for (Johnson et al., 2003; Kinch et al., 2007). Note, however, that deposited dust thickness at VL1 has been reported to be only micrometers in depth (Guinness et al., 1979, 1982). The reported high-latitude deposition rates (Drube et al., 2010) are similar to those reported near the equator (Johnson et al., 2003, Kinch et al., 2007).

10.3.1.4 Properties of Airborne Dust

(a) **Composition** Attempts to discern the composition of Martian atmospheric dust began with some of the first orbiting spacecraft observations of Mars (e.g. Hanel et al., 1972; Toon et al., 1977). As discussed by Hamilton et al. (2005) – see also the review by Smith (2008) and the references therein – the composition of dust comprises mainly three framework silicates: plagioclase, feldspar, and zeolite. Additionally, minor phases of ferric compounds such as goethite and nanophase iron oxide were also identified with a fair degree of confidence. The conservative nature of this mineralogy was driven in no small part by the limited spectral range and quality of the data available to Hamilton et al. (2005), who relied primarily on the structure of the 9 μm band from previous ground-based optical observations. Refractive indices of sufficient quality for the O–H stretch near 3 μm and the absorption feature near 20 μm were then not yet available, though they are now (Wolff et al., 2006, 2009). However, the broad nature of these two features is likely to provide little additional constraint on the dust bulk composition. Mössbauer spectroscopy and X-ray fluorescence measurements of magnetic airfall dust by the MERs shows magnetite, olivine, and ferric oxides presumed to be nanocrystalline phases, suggestive of non-aqueous alteration of basalt (Goetz et al., 2005).

(b) **Particle Sizes** Descriptions of particle sizes of dust aerosols generally focus on the first and second moments of the cross-section-weighted particle size distribution, the *effective radius*, and the *effective variance* as defined by Hansen and Travis (1974):

$$r_{eff} = \frac{\int r \pi r^2 n(r)\, dr}{\int \pi r^2 n(r)\, dr} = \frac{1}{G}\int r \pi r^2 n(r)\, dr \qquad (10.2)$$

$$v_{eff} = \frac{1}{G r_{eff}^2}\int (r - r_{eff})^2 \pi r^2 n(r)\, dr \qquad (10.3)$$

where G is the geometric cross-section of the particles, r is the particle equivalent-sphere radius, and $n(r)dr$ is the number of particles between r and $r+dr$.

A comprehensive tabulation of observationally derived r_{eff} and v_{eff} from work prior to 2003 may be found in Dlugach et al. (2003), with more recent efforts included in Smith (2008). Taken as a whole, these two reviews identify several patterns of note. The general column-averaged size distribution moments in low and moderate dust loading conditions observed by Viking, MPF, MGS/TES, MER/Mini-TES, and MER/Pancam fall typically in the relatively small ranges of $r_{eff} = 1.4$–1.7 μm and $v_{eff} = 0.2$–0.4 (see Figure 6.14). It is notable that the width of the retrieved particle size distribution (v_{eff}) is fairly large, indicating that the dust size distribution is not well represented by a single particle size (i.e. r_{eff}). During the less common high dust loading conditions (e.g. the 2001 global storm), larger sizes have been observed ($r_{eff} = 1.8$–2.5 μm). The size variation results in a seasonal pattern in 9 μm to 880 nm dust opacity ratio, with low values of 0.4 associated with smaller aerosols during low-opacity conditions, and higher values of 0.7 near active dust lifting areas (Lemmon et al., 2014). However, it is important to emphasize that these values represent an average over properties in the vertical column. At present, the retrieval of particle size profiles represents the state of the art in ongoing aerosol studies (i.e. Wolff et al., 2011). Nevertheless, even with the limited nature of previously published efforts, two distinct pictures are revealed. During times of large-scale, highly enhanced dust optical depths, r_{eff} values in the 30–60 km range are very similar to those in the lower portion: $r_{eff} = 1.5–2.0$ μm (Clancy et al., 2010). For more diffuse conditions, r_{eff} decreases from the near-surface (<10 km) values of ~1.4–1.8 μm, to ~1.0 μm near 20 km, to ~0.5 μm in the 30–40 km range, and finally to ≤0.3 μm above 40 km (Montmessin et al., 2006; Rannou et al., 2006; Fedorova et al., 2009; Clancy et al., 2010).

Numerical modeling studies robustly predict that atmospheric transport processes should give rise to the segregation of dust particle sizes within the Martian atmosphere. Two- and three-dimensional models have been used to study how transport within the mean overturning circulation results in the segregation of dust particle sizes. Murphy et al. (1993) used a zonally symmetric circulation model to investigate particle size variability during a simulated large-scale dust storm that originated in the southern middle latitudes at $L_s \sim 270°$. They found that the column-integrated particle size distribution evolved in time after dust was injected into the model, with larger particles falling out of the atmosphere faster than smaller particles. This resulted in a decrease in the effective particle size with time. Kahre et al. (2008) built on this result using a version of the GCM that included interactive dust lifting, transport, and sedimentation. During a simulated global storm at southern hemisphere summer solstice, Kahre et al. (2008) showed that the timescale involved in particles traveling upward in the southern hemisphere and then from the southern hemisphere to the northern hemisphere aloft can be shorter than the sedimentation timescale. This led to a buildup of large particles in an elevated dust layer that stretches across the equator that is driven by the strong Hadley circulation at the southern summer solstice season.

(c) **Radiative Properties** Radiative properties, such as the single-scattering albedo, are difficult to determine observationally without introducing a bias based upon assumptions such as the functional form of the scattering phase function or particle microphysical properties. However, the recent occurrence of multiple operating spacecraft on the surface and in orbit around Mars has provided an opportunity to derive radiative properties from coordinated observing campaigns (e.g. Wolff et al., 2006, 2009). The net result has been a set of refractive indices (see Chapter 6) that allows one to calculate self-consistent radiative and scattering properties of Martian dust particles from first principles. Recently derived solar-band dust single-scattering albedos range from 0.92 to 0.94 (Clancy et al., 2003; Wolff et al., 2006, 2009; Määttänen et al., 2009), which is a distinctly different range than provided by some previous derivations (i.e. Ockert-Bell et al., 1997).

10.3.2 Surface Dust

There is little doubt that the vast majority of dust in the atmosphere originates from and ultimately returns to the surface. Although it is possible that there is a small meteoritic influx of dust at the top of the atmosphere (Flynn, 1992), it is unlikely to be large enough to have a discernible effect on the observed dust cycle. Therefore, observations of the spatial and temporal variations of surface dust reservoirs (i.e. the surface branch of the dust cycle) can provide insight into the location and movement of dust within the Martian surface–atmosphere environment.

10.3.2.1 Location of the Surface Dust Reservoir

The orbital mapping of local surface dust content relies on a basic knowledge of the physical behavior of dust deposits of varying depths. Information regarding the type of materials comprising the top few centimeters of the surface, especially if that surface is dust-covered, is best obtained remotely via the magnitude of the derived surface thermal inertia (TI; $\sqrt{K\rho C_p}$). Regions that exhibit the lowest thermal inertia (i.e. ≤100 J m^{-2} K^{-1} s$^{-1/2}$) are interpreted to be mantled with dust at least a few centimeters thick (Christensen, 1986b). There are three Martian low-thermal-inertia continents – Tharsis, Arabia, and Elysium (Figure 10.6) – which are therefore thought to be dust-covered. Based on observations of rock abundance and thermal inertia, these dust deposits are thought to have depths from a few centimeters to several meters (Christensen, 1986b). While thermal inertia observations are well suited for unambiguously identifying dust-covered regions, they are not well suited for identifying dust-free regions because higher values of thermal inertia have non-unique interpretations and may result from mixtures of fines (i.e. dust), coarser sand-sized particles, bedrock, and ice (Ruff and Christensen, 2002).

Surface albedo also provides useful information about the location of dust on the surface because fine dust particles are highly reflective at visible and near-infrared wavelengths (Sagan et al., 1972, 1973; Veverka et al., 1974; Lecacheux et al., 1991; Bell et al., 1997). Unlike thermal inertia, observations of low surface albedo are well suited to identify regions that are dust-free, because even micrometer-deep layers of dust covering a dark surface will raise the albedo a measurable amount (Wells et al., 1984). Regions such as Syrtis Major and much of Acidalia have albedo values <0.10 and are interpreted to be mostly devoid of surface dust. Although low-albedo surfaces can be used to uniquely identify dust-free surfaces, moderate- and high-albedo surfaces do not uniquely identify dust-covered surfaces. Low-albedo regions have been observed to brighten after dust storms and then darken again (Smith, 2004; Szwast et al., 2006; Cantor, 2007), which has been interpreted as dust deposition on and subsequent removal from normally dust-free surfaces (Sagan et al., 1972, 1973; Veverka et al., 1974; Pleskot and Miner, 1981; Lee, 1987; Lecacheux et al., 1991; Smith, 2004; Szwast et al., 2006). Albedo changes due to specific dust storms are discussed in Section 10.4.2.3.

Additional information regarding the location of dust on the surface of Mars is available from measurements of surface emissivity in a spectral region where the emissivity is sensitive to the size of particles on the surface. Ruff and Christensen (2002) define a dust cover index (DCI) parameter as being the average emissivity value in the spectral range 1350–1400 cm^{-1} (7.1–7.4 μm), and show that it is a reasonable measure of the presence or absence of surface dust. The global pattern of DCI exhibits regions of low and high emissivity (corresponding to high and low dust content, respectively), which are consistent with the low-thermal-inertia and low-albedo regions, respectively (Ruff and Christensen, 2002).

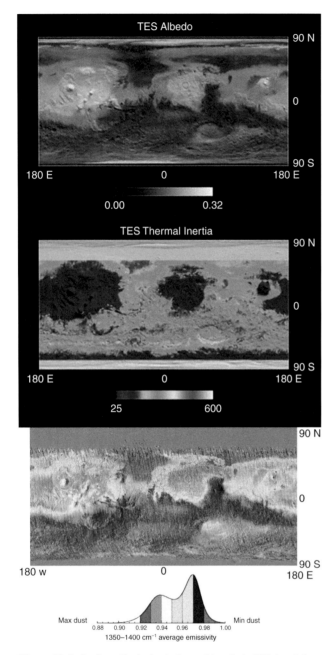

Figure 10.6. Surface albedo (top), thermal inertia (middle) and dust cover index (bottom) as derived from MGS/TES data. From www.mars.asu.edu (top two panels) and Ruff and Christensen (2002) (bottom panel). A black and white version of this figure will appear in some formats. For the color version, please refer to the plate section.

10.4 DUST STORM MORPHOLOGY AND BEHAVIOR

Suspended dust events on scales ranging from meters to global have been observed on Mars. Based on observed spatial and temporal scales and differences in morphology, dust lifting events and their subsequent dust clouds have been divided into two broad categories: dust devils and dust storms. While dust devils are small, short-lived, and have distinct morphologic characteristics, dust storms have a wider range in spatial and temporal scales and morphology. In some cases, linked local and regional storms produce a global (or

nearly global) dust cloud that obscures the large majority of the planet's surface.

10.4.1 Dust Devils

Dust devils are the smallest-scale dust lifting events identified on Mars. Although individual dust devils are local events, their ubiquitous occurrence planet-wide suggests an important role in the Martian dust cycle (Ferri et al., 2003; Fisher et al., 2005; Cantor et al., 2006). Like their terrestrial counterparts, a dust devil can be defined as a thermally driven columnar vortex that is made visible by the lofting of mobile surface dust particles (see Section 10.5.1.1(b)).

10.4.1.1 From Orbit

Active dust devils were first observed from orbit by the Viking Orbiter cameras (Thomas and Gierasch, 1985) and have subsequently been observed in all other orbiter missions: MGS/MOC (Malin and Edgett, 2001; Biener et al., 2002; Cantor et al., 2002, 2006; Fisher et al., 2005; Malin et al., 2010), Mars Odyssey/THEMIS (Cushing et al., 2005; Towner, 2009), MEX/HRSC (Milam et al., 2003; Williams et al., 2004; Greeley et al., 2005; Stanzel et al., 2006, 2008; Reiss et al., 2011), and MRO/MARCI, CTX, and HiRISE (Choi and Dundas, 2011). Orbiter observations show dust devils ranging in size from a few meters across and tens of meters high to over a kilometer across and more than 6 km in height. In some instances, the vortex will also generate either a light- or dark-toned residual surface track extending for many tens to thousands of meters in length.

Dust devils and their light- and dark-toned surface tracks have been observed over a variety of surface materials (Cantor et al., 2006) over latitudes ranging from 80°S to 80°N and from the bottom of the Hellas Basin to the tops of volcanoes (Malin and Edgett, 2001; Cushing et al., 2005; Cantor et al., 2006). Dust devils and tracks have been observed in both hemispheres in all seasons, with peak activity occurring during summer in each hemisphere (Fisher et al., 2005; Cantor et al., 2006). MGS/MOC dust devil monitoring over several Mars years (MY 24–28) indicated that the largest, regularly occurring afternoon active dust devils were observed in northern Amazonis (Malin and Edgett, 2001; Cantor et al., 2002, 2006; Fisher et al., 2005), Syria-Claritas, and eastern Meridiani just west of Schiaparelli Crater, with peak activity on the order of $(1.5–3.9) \times 10^{-3}$ devils/km² (Cantor et al., 2006). Observations of dust devils from orbit to date are limited by the fixed local times of the spacecrafts' orbits. Therefore, information regarding the diurnal variability of dust devil activity must come from surface observations.

10.4.1.2 From the Surface

Instruments on surface-based platforms have also detected dust devils, with distinct patterns of seasonal and diurnal variability.

One or more active dust devils were imaged at the MPF, MER/Spirit, MER/Opportunity, PHX, and MSL sites (Metzger et al., 1999, 2000; Ferri et al., 2003; Greeley et al., 2006a,b, 2010; Ellehoj et al., 2010; Moores et al., 2014). Although cameras on multiple landers have directly imaged dust devils, the longest-lived surface-based camera that imaged enough dust

devils to study seasonal patterns was MER/Spirit (VL1 and VL2 did not image any dust devils, and MER/Opportunity and MSL have imaged only a very limited number). Three Mars years of imaging observations indicate that dust devil activity in Gusev Crater was initiated early in southern hemisphere spring, increased in intensity until late spring, and then generally decreased monotonically through summer (Greeley et al., 2010). An abrupt cessation of dust devil activity was observed during the second year of observations (MY 28) at $L_s \sim 270°$, which coincided with the onset of a global dust storm that raised the local dust optical depth to ~4 (Wolff et al., 2009; Greeley et al., 2010). Dust devil activity at the Spirit site was observed to peak during the early afternoon (Greeley et al., 2010).

Convective vortices (which technically may be referred to as dust devils only if they contain sufficient dust to be visible) create distinctive signatures in lander-observed pressures, temperatures, and winds, as the low-pressure core brings warm surface air toward the center of the vortex (Sinclair, 1969, 1973; Rennó et al., 1998). These signatures have been recorded in the meteorological data of VL1, VL2, MPF, PHX, and, most recently, MSL (Figure 10.7; Ryan and Lucich, 1983; Schofield et al., 1997; Murphy and Nelli, 2002; Ringrose et al., 2003; Ellehoj et al., 2010; Haberle et al., 2014). The intensity, duration, and frequency of observed pressure drops are generally consistent between the various landing sites and with LES models that explicitly simulate Martian convective vortices (Rafkin et al., 2001; Toigo et al., 2003; Michaels and Rafkin, 2004; Gheynani and Taylor, 2011). The typical drop in pressure is approximately 2–3 Pa and lasts on the order of seconds. Convective vortices have been observed to peak during the late morning/early afternoon, between 11:00 and 15:00 hours local time, at VL2, MPF, and PHX (Figure 10.8; Murphy and Nelli, 2002; Ringrose et al., 2003; Ellehoj et al., 2010; Kahanpää et al., 2013; Haberle et al., 2014). Of the convective vortex occurrences recorded during the first 90 sols of operation by the REMS instruments on MSL, only one was associated with a dip in the UV flux recorded at the surface, suggesting that almost none of the recorded vortices in Gale Crater contained dust (Haberle et al., 2014). Haberle et al. (2014) speculate that the vortices in Gale are dust-free because the depth of the PBL is suppressed due to topographically forced dynamics (Tyler and Barnes, 2013).

10.4.2 Dust Storms

Larger-scale dust lifting events and the optically thick dust clouds that they produce are referred to as dust storms. Storms can range in size from a few hundred to many millions of square kilometers, though the surface source regions are likely far less extensive (Cantor, 2007). Dust storm clouds have been observed to exhibit lobate structures indicative of convective mixing (Briggs et al., 1979; James, 1985; Malin et al., 2008). From orbit, active dust lifting centers for significant storm events are difficult to distinguish and are therefore only occasionally observed (Malin et al., 2008), though localized regions of active dust lifting in the shape of tendrils trailing behind larger-scale dust storms have been observed in both hemispheres (Inada et al., 2007; Malin et al., 2008).

Dust storms have historically been classified by size as local (long axis < 2000 km), regional (long axis > 2000 km), and global storms (Martin and Zurek, 1993). Cantor et al. (2001)

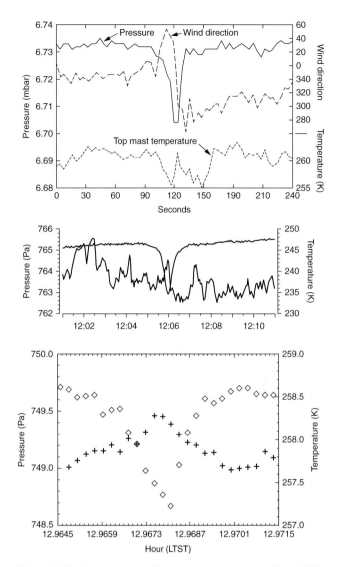

Figure 10.7. Pressure observations of convective vortices from MPF (top), PHX (middle), and MSL (bottom). In the middle panel, the top line represents the measured surface pressure and the bottom line represents measured air temperature. In the bottom panel, the diamond symbols represent the measured surface pressure and the plus symbols represent measured air temperature. From Schofield et al. (1997) (top panel), Ellehoj et al. (2010) (middle panel), and adapted from Haberle et al. (2014) (bottom panel).

used MGS/MOC observations to determine a relationship between the areal extent of a storm and its duration. They noted that some storms that would be classified as regional would at times not be visible from ground-based observations. They therefore refined the definition of local and regional storms to bring in line the historical and ground-based observations. Cantor et al. (2001) refined the definition of local storms as those covering less than 1.6×10^6 km^2 and regional storms as those covering greater than 1.6×10^6 km^2.

10.4.2.1 Local Storms

Local dust storms are the most frequently observed dust storms on Mars. High-resolution global synoptic imaging of Mars, starting with MGS/MOC, has allowed for a continuous, multi-year

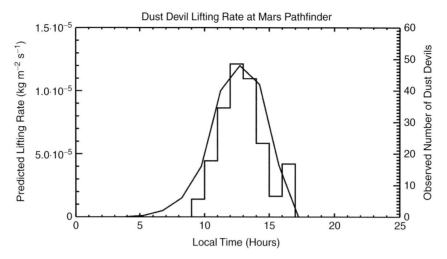

Figure 10.8. Pathfinder diurnal dust devil lifting rate. The solid curve indicates the GCM-predicted dust devil lifting rate (kg m⁻² s⁻¹) as a function of local time at the model grid point nearest the MPF landing site. The histogram indicates the observed diurnal occurrence of dust devils during the Mars Pathfinder mission (Murphy and Nelli, 2002). Adapted from Kahre et al. (2006).

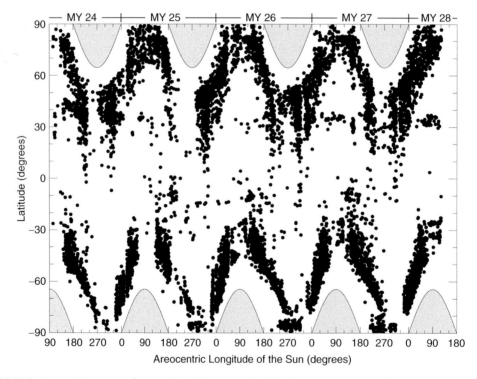

Figure 10.9. MGS/MOC-observed dust storms from multiple Mars years. Each black dot represents one observed dust storm, and the gray shaded region denotes the polar cap/polar hood region.

monitoring effort of the seasons and locations favorable for the occurrence of local dust storms. This growing database of observed dust storm occurrence (Figure 10.9) and morphology yields important insight into possible dust lifting mechanisms and preferred dust source regions on Mars. Like dust devils, the ubiquitous nature of local dust storms makes them important contributors to the dust cycle. Local dust storms are observed in both hemispheres in all seasons, but an annual local dust storm cycle has been observed to exist, with storms developing in specific locations at particular seasons each year (Cantor et al., 2002; Wang et al., 2005; Cantor, 2007): at the edges of the seasonal polar caps and the polar hoods; at the bases of high topographic regions in the northern hemisphere; and in the middle latitudes in both hemispheres (Cantor et al., 2001, 2002, 2010; Cantor, 2007). While the behavior of local storms described below is extensive in temporal and spatial coverage, we note that it is based on observations from a Sun-synchronous orbit and is therefore limited in time-of-day coverage.

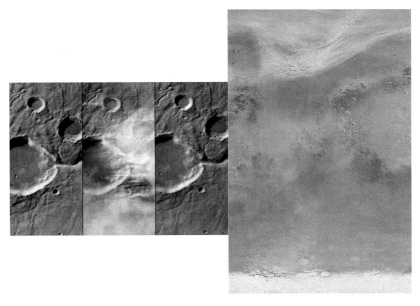

Figure 10.10. Images of a local storm in the Icaria Planum region taken with MRO/CTX (left) and a frontal storm in the Acidalia region taken with MGS/MOC (right). Images courtesy of NASA/JPL/Malin Space Science Systems.

Most local dust storms develop within 10° of latitude of either the receding or advancing seasonal polar cap edge or the corresponding polar hood boundary in both hemispheres. Often forming during the morning hours and expanding rapidly, these storms suggest that the surface winds generated by large-magnitude horizontal surface thermal gradients likely play an important role in lifting dust from the surface (Cantor et al., 2010). In the northern hemisphere, cap-edge local dust storm activity peaks in early local spring ($L_s \sim 10$–$20°$) from ~50° to 70°N along the receding north polar cap (James et al., 1999) and in mid-summer ($L_s \sim 135$–$140°$) from ~70° to 90°N along the residual polar cap edge (Cantor et al., 2010). Local storm activity also peaks along the advancing north polar hood edge between $L_s \sim 170°$ and 190° from 50° to 70°N. In the southern hemisphere, cap-edge local dust storm activity follows the receding south polar cap edge during local spring ($L_s \sim 205$–$250°$) and the advancing polar cap edge during local autumn ($L_s \sim 0$–$40°$; Cantor et al., 2001). Climate models predict high surface wind stresses along the advancing and receding cap edges (with higher values along the receding cap edge) in the locations and seasons at which cap-edge storms are observed, which further supports the idea that these storms are formed via surface wind stress lifting processes (see Section 10.5.1.1).

Correlations between topographic slope and local dust storm occurrence suggest that topographic relief can be important for the initiation of storms. A large number of local storms occur on the northwest flanks of Olympus Mons and Elysium Mons, with strong slope winds hypothesized to produce significant dust lifting (Cantor et al., 2001). The peak period of activity in the region to the northwest of Olympus Mons occurs in north autumn between $L_s \sim 205°$ and 235°, after which the activity declines rapidly, and again in northern winter between $L_s \sim 290°$ and 355°. This suggests that either the circulations that resulted in dust lifting have changed, or that mobile surface dust has been depleted by this season. The latter case would require subsequent surface dust replenishment to enable these

regions to resume as future surface dust sources (Cantor et al., 2001).

Local dust storms also occur in the middle latitudes of Mars, well away from the seasonal cap edge, in both hemispheres. In the northern hemisphere, there is a correlation between low-topography regions and local dust storm activity (Cantor et al., 2001). Northern hemisphere regions of maximum dust storm occurrence include Arcadia, Amazonis, Acidalia, Chryse, and Utopia (Cantor et al., 2001; Wang et al., 2003, 2005). Within these low-lying regions, the location of local dust storm initiation generally migrates southward with advancing season. As northern autumn progresses towards winter ($L_s \sim 180$–$270°$), the latitude of northern hemisphere local storms moves equatorward (Cantor et al., 2001). There is a correlation between the preferred longitudinal corridors of dust storm occurrence at northern middle latitudes with the model-predicted storm tracks (Hollingsworth et al., 1996). The importance of this correlation is discussed further in Section 10.4.2.2. In the southern hemisphere, there is a significant increase in the number of local storms in the Noachis region from $L_s \sim 135°$ to 189°. Local storms in the middle latitudes of both hemispheres exhibit two seasonal peaks in activity: one during local autumn (from $L_s \sim 200°$ to 235° in the north and from $L_s \sim 20°$ to 45° in the south), and one during local winter (from $L_s \sim 310°$ to 340° in the north and from $L_s \sim 135°$ to 179° in the south). There is a dearth of local dust storms during the $L_s \sim 250$–$270°$ seasonal range at northern subtropical and middle latitudes, which is the season of the "mid-winter" suppression of baroclinic eddies. Although GCM simulations show that the seasonal variation in the vigor of the eddy wave activity is sensitive to dust loading, the exact dynamical mechanism behind the mid-winter minimum is not completely understood (Hourdin et al., 1995; Basu et al., 2006; Wilson et al., 2006; Read et al., 2011; Mulholland et al., 2013; Wang et al., 2013).

The morphologies of local dust storms offer clues to storm initiation processes (Figure 10.10). Many local storms exhibit curvilinear features referred to as "streaks", "spirals", and "frontal dust storms" in the literature (Cantor et al., 2001, 2010; Malin et al., 2008; Wang and Fisher, 2009). Such storms occur in both the northern and southern hemispheres and are thought to be associated with frontal weather systems (Wang et al., 2003; Hinson and Wang, 2010). Spiral circulation systems that resemble terrestrial polar vortices are abundant at high northern latitudes during northern spring and summer (Cantor et al., 2002; Wang and Fisher, 2009). Northern autumn and winter frontal dust storms exhibit curvilinear bands that tend to radiate outwards from the pole (Wang et al., 2003). These have been shown to correlate to MGS/TES-observed cold temperature anomalies (Wang et al., 2005), which strengthens the argument that they are associated with baroclinic waves and instabilities (James et al., 1999; Wang et al., 2005; see also Section 10.4.2.2). While many frontal dust storms remain local, some storms initiated by frontal activity grow to regional scales (Cantor et al., 2001; Wang et al., 2005; Cantor, 2007).

10.4.2.2 Regional Storms

While local dust storms are more numerous, regional dust storms increase the atmospheric dust loading over larger areas for longer periods of time and produce significant responses in atmospheric temperatures, particularly once dust is entrained into the Hadley circulation. Regional storms have been observed to occur on the order of 8–35 times per Mars year since MGS began science orbit operations in 1999. More than half of the regional dust storms observed were clearly observed to result from the merging of local storms (Cantor et al., 2001). Although regional storms have been observed throughout most of the Martian year, they occur much more frequently during the dusty season ($L_s \sim 135$–$360°$; Cantor, 2007). Some regional storms exhibit specific characteristics that allow them to be studied as a group, while others are less well understood.

Frontal dust storms are a relatively well-studied class of storm that can develop into regional storms (Figure 10.10). Frontal storms are observed in all seasons on Mars, but those that mostly develop into regional storms occur most frequently during two seasonal windows before and after northern winter solstice ($L_s \sim 200$–$244°$ and $L_s \sim 300$–$355°$), and are likely important contributors to the observed increase in atmospheric dust loading in the northern hemisphere at these seasons (Figure 10.1). It has been demonstrated that frontal dust storms are associated with traveling waves embedded in the strong westerly jet that is present in the northern hemisphere in the autumn, winter, and spring seasons (Cantor et al., 2001; Wang et al., 2003, 2005; Cantor, 2007; Wang, 2007; Hinson and Wang, 2010).

Dust lifted by these transient wave events propagates southward towards and across the equator in north–south-oriented channels defined by low-topography regions: Acidalia into Chryse, Arcadia into Amazonis, and Utopia into Isidis (Wang et al., 2003, 2005; Wang, 2007). Models suggest that these longitudinal corridors are preferred for the return branch of the Hadley cell through the topographically controlled western boundary currents (WBCs) and have been termed the Acidalia, Arcadia, and Utopia storm tracks (Joshi et al., 1995, 1997; Hollingsworth et al., 1996; Wilson and Hamilton, 1996;

Cantor, 2007). Modeling indicates that storm intensification and southward dust transport are most effective when the low-pressure center passes through the longitude of the preferred low-topographic region between 9 a.m. and 7 p.m. local time (Wang et al., 2003). This local time window corresponds to the time of day when the tidal winds are from the north, which allows for the constructive interference between the winds associated with the front itself and those associated with the tide. This interference results in stronger winds potentially capable of lifting more dust into the system and allowing for enhanced southward transport. At other times of day, destructive interference can prohibit dust lifting and weaken or even prohibit southward dust transport. The necessary alignment between the local time of the passage of the low-pressure system with favorable northerly tidal winds has been referred to in the literature (Wang et al., 2003) as a "tidal gate" that allows some frontal storms to transport dust to the equatorial regions and prohibits others from doing so. A major unknown factor regarding these cross-equatorial storms is how much new dust lifting develops and how much dust is transported by the disturbance.

Studies of traveling waves in the MGS/TES observations (Wang et al., 2003, 2005; Hinson and Wang, 2010) and GCM simulations (Basu et al., 2006; Wilson et al., 2006; Wang et al., 2013) have emphasized the role of transient waves that are zonal wave 3 and have a period of 2–2.3 sols. Because southward dust transport can be maximized when the traveling wave is properly synchronized with the tidal winds, a transient wave period of close to 2 sols can result in a succession of waves that continually transport dust southward across the equator. Recent GCM simulations indicate that tidal winds maximize in the subtropics (~ 20–$30°N$ latitude) and during the equinoctial seasons ($L_s \sim 170$–$220°$ and $L_s \sim 330$–$20°$; Wilson, 2012). Thus, it is possible that the cross-equatorial storm period is a convolution of the presence of the tide and traveling wave activity.

As dust storms move southward toward the equator, they can potentially ignite new lifting centers in the northern hemisphere. Once cross-equatorial dust storms travel into the southern hemisphere, observations suggest they can, in some cases, affect the thermal structure of the southern hemisphere, either by developing new lifting centers or by spatially expanding via atmospheric transport processes. During the period from $L_s \sim 225°$ to $230°$ of MY 24, a series of dust storms (cross-equatorial events) that began in the Acidalia and Utopia regions transported dust across the equator (Figure 10.11). Upon reaching subtropical latitudes in the south, further dust lifting is evident in both the MGS/TES opacity and MGS/MOC imagery data. Dust lifted in the southern hemisphere appears to have been entrained into the Hadley circulation, lofting it upward and spreading a haze of dust throughout much of the atmosphere and causing a thermal response up to approximately 40 km (Cantor et al., 2001; Cantor, 2007). Similar regional dust storms occurred in MY 26, both before and after the solstitial season. In these cases, dust traveling across the equator was lofted and spread by the circulation once it reached southern subtropical latitudes (Cantor, 2007). While it appears in the data as though dust lifting must be triggered in the southern hemisphere during these events, the dynamical mechanisms for driving that lifting

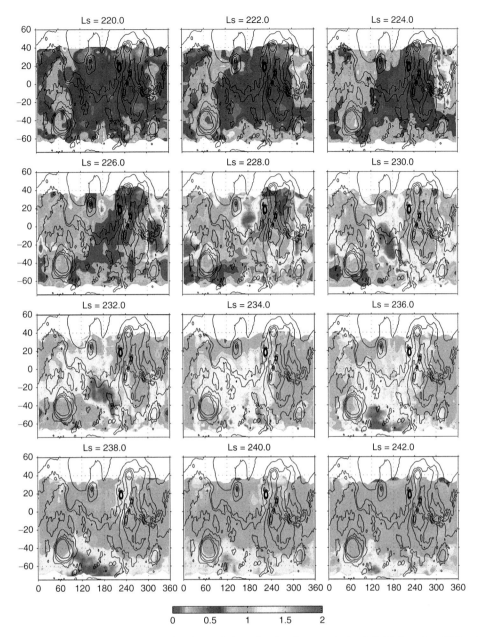

Figure 10.11. MGS/TES-observed 9 μm opacities showing the evolution of regional storms during MY 24 from $L_s = 220°$ to $L_s = 242°$. A black and white version of this figure will appear in some formats. For the color version, please refer to the plate section.

are not well understood. One possibility is that concentrations of dust from the cross-equatorial events lead to enhanced circulation strength via enhanced thermal tides, producing increased surface wind stress-induced dust lifting (Wilson et al., 2008b; Wilson, 2012). However, it is also possible that the timing of some such southern storms is coincidental, rather than being causally related to dust transported from the north.

Another type of dust storm that can develop into a regional storm (or even a global storm) is one that originates along the receding south seasonal polar cap during local spring and summer. Accounting for less than 10% of all regional storms, these storms tend to originate in the Argyre and Hellas regions as the polar cap recedes past these topographic basins. GCMs

predict a maximum in surface wind stress in these regions due to the combined effect of topographic, diurnally varying slope winds, cap-edge circulations driven by latitudinal thermal gradients, and traveling eddies imbedded in the westerly flow (see Section 10.5.2.1).

10.4.2.3 Global Dust Storms

Historically, the largest spatial-scale dust lifting events and the dust clouds that they produce are global dust storms (Martin and Zurek, 1993; Zurek and Martin, 1993; Cantor, 2007). The large dust storms produce substantial visible-band dust optical depths (>3) over large portions of the planet, and have been observed from Earth, from Mars orbit, and from the Martian

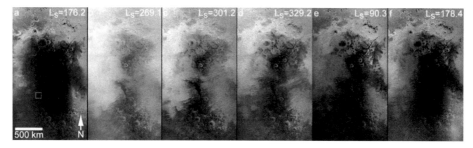

Figure 10.14. Syrtis Major large-scale regional Minnaert albedo ($k = 0.7$) changes observed during a one Mars year period with the red-filter wide-angle MGS/MOC: (a) June 11, 2001; (b) November 10, 2001; (c) January 1, 2002; (d) February 19, 2002; (e) November 4, 2002; and (f) May 3, 2003. From Cantor (2007).

triggered by a single storm that began in Chryse and moved south along the Acidalia storm track at $L_s = 261.9°$ (Cantor et al., 2008; Wang and Richardson, 2015). This storm crossed the equator into the southern hemisphere the next sol, through easternmost Valles Marineris. Although the imaging data are ambiguous due to coverage constraints, it appears as though the storm propagated to the southeast over the next 3 sols into the southern mid-latitudes between 30° and 50°S in the Noachis region (Cantor et al., 2008; Wang and Richardson, 2015). The storm expanded between $L_s = 264.4°$ and 265.1°, with dust extending to the southwest through Argyre and to the seasonal south polar cap edge and to the northeast as far as northern Hellas (Cantor et al., 2008; Wang and Richardson, 2015). The dust cloud continued to expand both westward between 58° and 80°S latitude along the seasonal south polar cap edge and also eastward between 21° and 56°S. The storm became planet-encircling on $L_s = 269.3°$, 13 sols after onset, at southern mid- to high latitudes between 35° and 80°S (Cantor et al., 2008).

By the end of the expansion phase of the storm ($L_s \sim 275.2°$), the atmospheric dust haze encircled the planet from approximately 90°S to 58°N, with peak regional visible-band optical depths of ~4.5 (Smith, 2009; Lemmon et al., 2014). This is consistent with the more limited coverage of the ground-obscuring dust clouds associated with the 2007 storm. Just as with previous planet-encircling storms, the decay of the atmospheric dust haze exhibited an exponential decline. The decay constants at the InSight site were very similar between the 2007 and 2001 events. Like the 2001 storm, the 2007 storm had a number of secondary active lifting centers, observed throughout the course of the dust event, from Meridiani, Syria, Solis Planum, Claritis, Hesperia, Cimeria, Sirenum in the southern hemisphere and Elyisum in the northern hemisphere (Cantor et al., 2008; Smith, 2009). The atmospheric dust loading returned to a seasonally representative level by approximately $L_s \sim 325°$ (Cantor et al., 2008; Smith, 2009).

Vincendon et al. (2009) determined from MEX/OMEGA near-IR observations that the 2007 dust storm did not produce widespread albedo changes but rather only small-scale variations at previous boundaries of bright versus dark terrains. The one exception was the return of the large-scale dark albedo feature, centered at 25°N, 238°W, in Amenthes that was observed by the Viking Orbiters.

10.4.2.4 Mechanisms Controlling Global Dust Storms

The processes that control the genesis, expansion, and ultimately the decay of global dust storms are likely critically dependent on the radiative–dynamic feedbacks (both positive and negative) that occur between atmospheric dust and the atmospheric circulations.

(a) Storm Initiation and Expansion Global dust storms result from the rapid expansion and merging of multiple local and regional storms that generate a dust cloud that is global or near global in extent (Zurek et al., 1992; Cantor, 2007). The local and regional storms involved do not individually differ substantially in morphology from local and regional storms that occur at other times that do not produce a global event. Thus, it is the linkage between them and their combined influence on increasing the atmospheric dust loading that lead to the explosive expansion phase as a global dust storm erupts. Although it has been historically proposed that dust devils could contribute to the generation of global dust storms (Leovy et al., 1973), observations of dust devil occurrence clearly demonstrate an anticorrelation with storm occurrence (Cantor et al., 2006; Malin et al., 2010). Additionally, Montabone et al. (2005) used results from a reanalysis based on MGS/TES temperatures to conclude that a dust devil contribution to storm genesis would have been marginal at best.

(b) Theories for Dust Storm Onset There are several hypotheses for global dust storm initiation and expansion in the literature that are largely based on theoretical arguments (see Zurek et al., 1992). Common to all of these genesis hypotheses are positive radiative–dynamic feedbacks between dust lifting, the heating of the atmosphere due to airborne dust, and the strength of various components of the circulation, which result in the intensification and/or the spatial expansion of dust lifting. The ever-growing volume of observations and the advancement of climate models allow us to address the viability of these theories. Two of the storm genesis hypotheses summarized in Zurek et al. (1992) have been addressed in the more recent literature: the dusty hurricane theory of Gierasch and Goody (1973), and the superposition of planetary-scale circulations theory by Leovy et al. (1973).

DUSTY HURRICANE   In the dusty hurricane theory, it is proposed that the heating of airborne dust would act in a way analogous to the latent heating by condensing water vapor in an Earth hurricane. In this scenario, dust-heated air rises in the center of a vortex, additional air is pulled in at low levels, the vortex spins up and additional dust is lifted (Gierasch and Goody, 1973; Zurek et al., 1992). While this model has

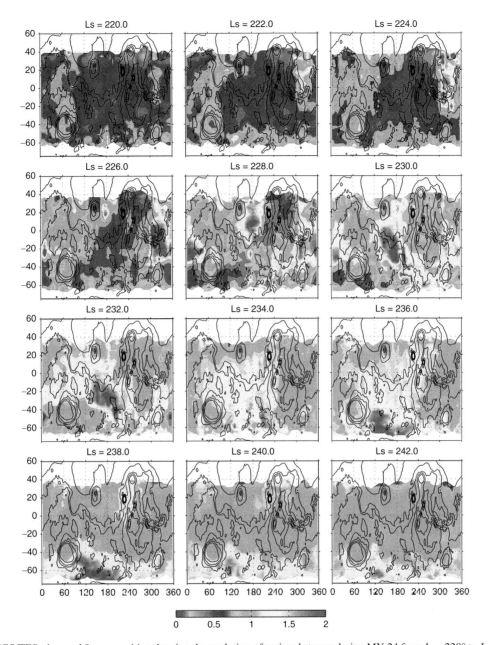

Figure 10.11. MGS/TES-observed 9 μm opacities showing the evolution of regional storms during MY 24 from $L_s = 220°$ to $L_s = 242°$. A black and white version of this figure will appear in some formats. For the color version, please refer to the plate section.

are not well understood. One possibility is that concentrations of dust from the cross-equatorial events lead to enhanced circulation strength via enhanced thermal tides, producing increased surface wind stress-induced dust lifting (Wilson et al., 2008b; Wilson, 2012). However, it is also possible that the timing of some such southern storms is coincidental, rather than being causally related to dust transported from the north.

Another type of dust storm that can develop into a regional storm (or even a global storm) is one that originates along the receding south seasonal polar cap during local spring and summer. Accounting for less than 10% of all regional storms, these storms tend to originate in the Argyre and Hellas regions as the polar cap recedes past these topographic basins. GCMs

predict a maximum in surface wind stress in these regions due to the combined effect of topographic, diurnally varying slope winds, cap-edge circulations driven by latitudinal thermal gradients, and traveling eddies imbedded in the westerly flow (see Section 10.5.2.1).

10.4.2.3 Global Dust Storms

Historically, the largest spatial-scale dust lifting events and the dust clouds that they produce are global dust storms (Martin and Zurek, 1993; Zurek and Martin, 1993; Cantor, 2007). The large dust storms produce substantial visible-band dust optical depths (>3) over large portions of the planet, and have been observed from Earth, from Mars orbit, and from the Martian

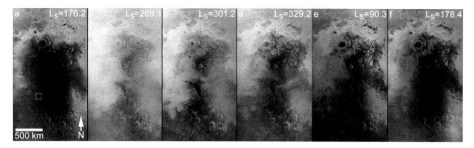

Figure 10.14. Syrtis Major large-scale regional Minnaert albedo ($k = 0.7$) changes observed during a one Mars year period with the red-filter wide-angle MGS/MOC: (a) June 11, 2001; (b) November 10, 2001; (c) January 1, 2002; (d) February 19, 2002; (e) November 4, 2002; and (f) May 3, 2003. From Cantor (2007).

triggered by a single storm that began in Chryse and moved south along the Acidalia storm track at $L_s = 261.9°$ (Cantor et al., 2008; Wang and Richardson, 2015). This storm crossed the equator into the southern hemisphere the next sol, through easternmost Valles Marineris. Although the imaging data are ambiguous due to coverage constraints, it appears as though the storm propagated to the southeast over the next 3 sols into the southern mid-latitudes between 30° and 50°S in the Noachis region (Cantor et al., 2008; Wang and Richardson, 2015). The storm expanded between $L_s = 264.4°$ and 265.1°, with dust extending to the southwest through Argyre and to the seasonal south polar cap edge and to the northeast as far as northern Hellas (Cantor et al., 2008; Wang and Richardson, 2015). The dust cloud continued to expand both westward between 58° and 80°S latitude along the seasonal south polar cap edge and also eastward between 21° and 56°S. The storm became planet-encircling on $L_s = 269.3°$, 13 sols after onset, at southern mid- to high latitudes between 35° and 80°S (Cantor et al., 2008).

By the end of the expansion phase of the storm ($L_s \sim 275.2°$), the atmospheric dust haze encircled the planet from approximately 90°S to 58°N, with peak regional visible-band optical depths of ~4.5 (Smith, 2009; Lemmon et al., 2014). This is consistent with the more limited coverage of the ground-obscuring dust clouds associated with the 2007 storm. Just as with previous planet-encircling storms, the decay of the atmospheric dust haze exhibited an exponential decline. The decay constants at the InSight site were very similar between the 2007 and 2001 events. Like the 2001 storm, the 2007 storm had a number of secondary active lifting centers, observed throughout the course of the dust event, from Meridiani, Syria, Solis Planum, Claritis, Hesperia, Cimeria, Sirenum in the southern hemisphere and Elyisum in the northern hemisphere (Cantor et al., 2008; Smith, 2009). The atmospheric dust loading returned to a seasonally representative level by approximately $L_s \sim 325°$ (Cantor et al., 2008; Smith, 2009).

Vincendon et al. (2009) determined from MEX/OMEGA near-IR observations that the 2007 dust storm did not produce widespread albedo changes but rather only small-scale variations at previous boundaries of bright versus dark terrains. The one exception was the return of the large-scale dark albedo feature, centered at 25°N, 238°W, in Amenthes that was observed by the Viking Orbiters.

10.4.2.4 Mechanisms Controlling Global Dust Storms

The processes that control the genesis, expansion, and ultimately the decay of global dust storms are likely critically dependent on the radiative–dynamic feedbacks (both positive and negative) that occur between atmospheric dust and the atmospheric circulations.

(a) **Storm Initiation and Expansion** Global dust storms result from the rapid expansion and merging of multiple local and regional storms that generate a dust cloud that is global or near global in extent (Zurek et al., 1992; Cantor, 2007). The local and regional storms involved do not individually differ substantially in morphology from local and regional storms that occur at other times that do not produce a global event. Thus, it is the linkage between them and their combined influence on increasing the atmospheric dust loading that lead to the explosive expansion phase as a global dust storm erupts. Although it has been historically proposed that dust devils could contribute to the generation of global dust storms (Leovy et al., 1973), observations of dust devil occurrence clearly demonstrate an anticorrelation with storm occurrence (Cantor et al., 2006; Malin et al., 2010). Additionally, Montabone et al. (2005) used results from a reanalysis based on MGS/TES temperatures to conclude that a dust devil contribution to storm genesis would have been marginal at best.

(b) **Theories for Dust Storm Onset** There are several hypotheses for global dust storm initiation and expansion in the literature that are largely based on theoretical arguments (see Zurek et al., 1992). Common to all of these genesis hypotheses are positive radiative–dynamic feedbacks between dust lifting, the heating of the atmosphere due to airborne dust, and the strength of various components of the circulation, which result in the intensification and/or the spatial expansion of dust lifting. The ever-growing volume of observations and the advancement of climate models allow us to address the viability of these theories. Two of the storm genesis hypotheses summarized in Zurek et al. (1992) have been addressed in the more recent literature: the dusty hurricane theory of Gierasch and Goody (1973), and the superposition of planetary-scale circulations theory by Leovy et al. (1973).

DUSTY HURRICANE In the dusty hurricane theory, it is proposed that the heating of airborne dust would act in a way analogous to the latent heating by condensing water vapor in an Earth hurricane. In this scenario, dust-heated air rises in the center of a vortex, additional air is pulled in at low levels, the vortex spins up and additional dust is lifted (Gierasch and Goody, 1973; Zurek et al., 1992). While this model has

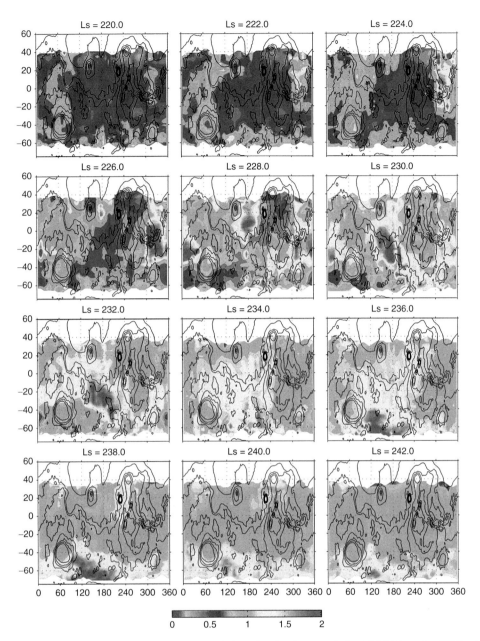

Figure 10.11. MGS/TES-observed 9 μm opacities showing the evolution of regional storms during MY 24 from $L_s = 220°$ to $L_s = 242°$. A black and white version of this figure will appear in some formats. For the color version, please refer to the plate section.

are not well understood. One possibility is that concentrations of dust from the cross-equatorial events lead to enhanced circulation strength via enhanced thermal tides, producing increased surface wind stress-induced dust lifting (Wilson et al., 2008b; Wilson, 2012). However, it is also possible that the timing of some such southern storms is coincidental, rather than being causally related to dust transported from the north.

Another type of dust storm that can develop into a regional storm (or even a global storm) is one that originates along the receding south seasonal polar cap during local spring and summer. Accounting for less than 10% of all regional storms, these storms tend to originate in the Argyre and Hellas regions as the polar cap recedes past these topographic basins. GCMs

predict a maximum in surface wind stress in these regions due to the combined effect of topographic, diurnally varying slope winds, cap-edge circulations driven by latitudinal thermal gradients, and traveling eddies imbedded in the westerly flow (see Section 10.5.2.1).

10.4.2.3 Global Dust Storms

Historically, the largest spatial-scale dust lifting events and the dust clouds that they produce are global dust storms (Martin and Zurek, 1993; Zurek and Martin, 1993; Cantor, 2007). The large dust storms produce substantial visible-band dust optical depths (>3) over large portions of the planet, and have been observed from Earth, from Mars orbit, and from the Martian

Table 10.1. *Observed global dust storms.*

Terrestrial year	Mars year	L_s (deg.)	Type	References
1924	−16	310	Unconfirmed	Antoniadi (1930)
1956	1	249	Confirmed	Miyamoto (1957), Slipher (1962), Martin and Zurek (1993), Zurek and Martin (1993)
1971–1972	9	260	Confirmed	Martin (1974a,b)
1973	10	300	Confirmed	Martin (1974b)
1977	12	204	Confirmed	Briggs et al. (1979), Thorpe (1979), Ryan and Sharman (1981), Zurek and Martin (1993)
1977	12	268	Confirmed	
1979	13	255	Unconfirmed	Leovy (1981), Ryan and Sharman (1981)
1982	15	208	Unconfirmed	Tillman (1988), Zurek and Martin (1993)
2001	25	184.7	Confirmed	Cantor et al. (2002), Smith et al. (2002), Strausberg et al. (2005), Cantor (2007)
2007	28	261.5	Confirmed	Cantor et al. (2008), Smith (2009), Wang and Richardson (2015)

surface. Although the observational record is incomplete due to unfavorable observing geometries for Earth-based observing and non-continuous spacecraft missions to Mars, information about the history of global dust storms has been compiled for the past century (Table 10.1). A total of 10 candidate global storms, seven of which have been confirmed, have been observed in the ~46 Mars years since 1924. Global dust storms have only been observed to begin during Mars' dusty season (L_s ~ 135–360°). Although regional-scale dust clouds occur annually, global storms only occur during some Mars years, with a frequency of approximately one year out of three (Zurek and Martin, 1993; Smith, 2008).

Five confirmed global dust storms have been observed by instruments either in Mars orbit or on the Martian surface – one in 1971 (MY 9; Mariner 9), two in 1977 (MY 12; Viking), one in 2001 (MY 25; MGS), and one in 2007 (MY 28; ODY/MRO/MEX). One additional storm, in 1982 (MY 15), is unconfirmed as having been truly global due to lack of planet-wide observations, but the large semidiurnal tide response it produced in the VL1 pressure observations suggests it was likely global in extent (Tillman, 1988; Zurek and Martin, 1993). Two more confirmed storms – one in 1956 (MY 1) and one in 1973 (MY 10) – are well documented in the ground-based telescopic record (Martin and Zurek, 1993; Zurek and Martin, 1993). There are recurrent patterns that have emerged from the history of these largest-scale dust events (Kahn et al., 1992; Zurek and Martin, 1993; Smith et al., 2002; Cantor, 2007). Five out of eight of these events (1956, 1971, 1973, 1977b, 2007) began within 30° of L_s of the southern hemisphere summer solstice (L_s = 270°), while the remaining three (1977a, 1982, and 2001) began earlier, at L_s = 204°, 208°, and 185°, respectively. Each of these global storms started in the southern hemisphere, and the majority began in either the Noachis–Hellaspontus (to the west of Hellas) region or the Solis Planum–Argyre region.

(a) 2001 Global Dust Storm The 2001 global dust storm is the earliest recorded global dust storm on Mars, with a season of initiation just after the autumnal equinox at L_s ~ 185°. Data were collected on this storm from several ground- and space-based Earth instruments (e.g. Hubble Space Telescope; HST),

as well as multiple instruments on the MGS spacecraft in Mars orbit (i.e. TES and MOC). Together, these data provide unprecedented detail on the development of the storm and its effect on the thermal and dynamical state of the atmosphere.

The 2001 storm began with dust lifting events to the southwest and within the Hellas Basin, which were morphologically similar to those that occur during the recession of the south polar ice cap every spring. Cantor (2007) reported seven precursor storm "pulses" during this season, which have been shown to correspond to traveling eddies passing through the region (Montabone et al., 2008; Wilson et al., 2008a; Noble et al., 2011). This period has been referred to as the precursor stage (Figure 10.12). Based on MGS/TES opacity observations, the storm was initiated in Hesperia (northeast of the Hellas Basin) at L_s = 184.5° (Smith et al., 2002). It is interesting to note that some GCM simulations suggest that it is difficult to transport dust out of the Hellas Basin (Ogohara and Satomura, 2008).

The storm expanded as the centers of lifting spread to the north into the Syrtis Major region in the northern hemisphere, and to the east into the Hesperia and Promethei regions (Strausberg et al., 2005; Cantor, 2007). The dust cloud originating in Hellas spread to the south over the seasonal polar cap, and to the east towards southern Tharsis. Secondary lifting centers became active in Syria, Solis Planum, and Claritas in the southern hemisphere and in Elysium in the northern hemisphere (Strausberg et al., 2005; Cantor, 2007). Although dust was being transported across Hesperia towards the east and had reached the western edge of the Tharsis plateau by L_s = 188°, the imaging and the MGS/TES optical depth data suggest that the activation of a secondary dust lifting center in Syria Planum immediately preceded the rapid storm growth to global encircling scale by L_s ~ 190° (Wilson et al., 2008a). The thermal response to the rapid increase in atmospheric dust is reflected in the MGS/TES-observed equatorial T_{15} temperatures (T_{15} is a brightness temperature roughly representative of the temperature at ~50 Pa; Figure 10.12). As the equatorial 9 μm dust opacity increased from ~0.1 to ~1.75 from L_s ~ 185° to 210°, equatorial T_{15} temperatures rose about 35 K, from ~185 K to ~220 K.

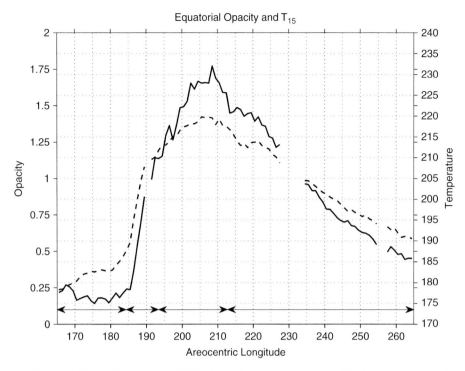

Figure 10.12. Evolution of equatorial 9 μm dust opacity (solid line) and T_{15} temperatures (dashed line) over the course of the 2001 (MY 25) global dust storm. The horizontal arrows denote (from left to right) the approximate L_s ranges of the precursor, expansion, mature, and decay stages of the storm.

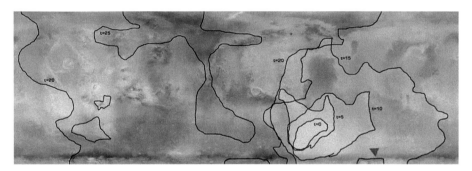

Figure 10.13. MGS/MOC global mosaic overlain by layers of varying opacity indicating the time of dust coverage during the 2001 global dust storm. The $t = 0$ contour corresponds to $L_s = 180°$, with each consecutive contour indicating five Martian days of storm development (i.e. $t = 10$ corresponds to 10 Martian days after the first contour). From Strausberg et al. (2005).

By the end of the expansion phase of the storm (Figure 10.13) – between $L_s \sim 195°$ and $200°$, depending on how the expansion phase is defined (see Strausberg et al., 2005; Cantor, 2007; Wilson et al., 2008a) – the atmospheric dust haze encircled the planet from approximately $60°$S to $60°$N, reached visible optical depths of ~ 5 in some locations, and extended to approximately 60 km above the surface (Smith et al., 2002; Strausberg et al., 2005; Cantor, 2007; Clancy et al., 2010; Guzewich et al., 2014). The decay of the atmospheric dust haze exhibited an exponential decline, with a decay constant that varied from location to location. The decay rate at the VL1 site inferred from MGS/MOC observations was 70–75 sols, which is consistent with decay rates measured at the same location by VL1 following the 1977a storm (Cantor, 2007). The atmospheric dust loading and equatorial T_{15} temperatures returned to a seasonally representative level by approximately $L_s \sim 300°$.

The surface albedo pattern was observed to change significantly in response to the 2001 global dust storm (Smith, 2004; Cantor, 2007). The latitude band between $10°$ and $60°$S exhibited the largest albedo changes, with specifically the low-albedo regions of Syrtis Major, Terra Tyrrhena, Hesperia Planum, and regions to the north and northwest of Argyre measurably brightening by the end of the storm (Figure 10.14). This suggests that the low-albedo regions were free of dust before the 2001 storm but acquired at least a thin veneer of dust due to the global redistribution of dust by the storm. Darkening of these regions (interpreted as dust removal) occurred subsequently, returning the albedo of most of these regions to their pre-storm values by the start of the dusty season the following year (Szwast et al., 2006; Cantor, 2007).

(b) **2007 Global Dust Storm** The 2007 storm rapidly expanded from the Noachis region at $L_s = 264.4°$, but was possibly

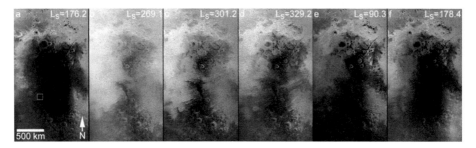

Figure 10.14. Syrtis Major large-scale regional Minnaert albedo ($k = 0.7$) changes observed during a one Mars year period with the red-filter wide-angle MGS/MOC: (a) June 11, 2001; (b) November 10, 2001; (c) January 1, 2002; (d) February 19, 2002; (e) November 4, 2002; and (f) May 3, 2003. From Cantor (2007).

triggered by a single storm that began in Chryse and moved south along the Acidalia storm track at $L_s = 261.9°$ (Cantor et al., 2008; Wang and Richardson, 2015). This storm crossed the equator into the southern hemisphere the next sol, through easternmost Valles Marineris. Although the imaging data are ambiguous due to coverage constraints, it appears as though the storm propagated to the southeast over the next 3 sols into the southern mid-latitudes between 30° and 50°S in the Noachis region (Cantor et al., 2008; Wang and Richardson, 2015). The storm expanded between $L_s = 264.4°$ and 265.1°, with dust extending to the southwest through Argyre and to the seasonal south polar cap edge and to the northeast as far as northern Hellas (Cantor et al., 2008; Wang and Richardson, 2015). The dust cloud continued to expand both westward between 58° and 80°S latitude along the seasonal south polar cap edge and also eastward between 21° and 56°S. The storm became planet-encircling on $L_s = 269.3°$, 13 sols after onset, at southern mid- to high latitudes between 35° and 80°S (Cantor et al., 2008).

By the end of the expansion phase of the storm ($L_s \sim 275.2°$), the atmospheric dust haze encircled the planet from approximately 90°S to 58°N, with peak regional visible-band optical depths of ~4.5 (Smith, 2009; Lemmon et al., 2014). This is consistent with the more limited coverage of the ground-obscuring dust clouds associated with the 2007 storm. Just as with previous planet-encircling storms, the decay of the atmospheric dust haze exhibited an exponential decline. The decay constants at the InSight site were very similar between the 2007 and 2001 events. Like the 2001 storm, the 2007 storm had a number of secondary active lifting centers, observed throughout the course of the dust event, from Meridiani, Syria, Solis Planum, Claritis, Hesperia, Cimeria, Sirenum in the southern hemisphere and Elyisum in the northern hemisphere (Cantor et al., 2008; Smith, 2009). The atmospheric dust loading returned to a seasonally representative level by approximately $L_s \sim 325°$ (Cantor et al., 2008; Smith, 2009).

Vincendon et al. (2009) determined from MEX/OMEGA near-IR observations that the 2007 dust storm did not produce widespread albedo changes but rather only small-scale variations at previous boundaries of bright versus dark terrains. The one exception was the return of the large-scale dark albedo feature, centered at 25°N, 238°W, in Amenthes that was observed by the Viking Orbiters.

10.4.2.4 Mechanisms Controlling Global Dust Storms

The processes that control the genesis, expansion, and ultimately the decay of global dust storms are likely critically dependent on the radiative–dynamic feedbacks (both positive and negative) that occur between atmospheric dust and the atmospheric circulations.

(a) **Storm Initiation and Expansion** Global dust storms result from the rapid expansion and merging of multiple local and regional storms that generate a dust cloud that is global or near global in extent (Zurek et al., 1992; Cantor, 2007). The local and regional storms involved do not individually differ substantially in morphology from local and regional storms that occur at other times that do not produce a global event. Thus, it is the linkage between them and their combined influence on increasing the atmospheric dust loading that lead to the explosive expansion phase as a global dust storm erupts. Although it has been historically proposed that dust devils could contribute to the generation of global dust storms (Leovy et al., 1973), observations of dust devil occurrence clearly demonstrate an anticorrelation with storm occurrence (Cantor et al., 2006; Malin et al., 2010). Additionally, Montabone et al. (2005) used results from a reanalysis based on MGS/TES temperatures to conclude that a dust devil contribution to storm genesis would have been marginal at best.

(b) **Theories for Dust Storm Onset** There are several hypotheses for global dust storm initiation and expansion in the literature that are largely based on theoretical arguments (see Zurek et al., 1992). Common to all of these genesis hypotheses are positive radiative–dynamic feedbacks between dust lifting, the heating of the atmosphere due to airborne dust, and the strength of various components of the circulation, which result in the intensification and/or the spatial expansion of dust lifting. The ever-growing volume of observations and the advancement of climate models allow us to address the viability of these theories. Two of the storm genesis hypotheses summarized in Zurek et al. (1992) have been addressed in the more recent literature: the dusty hurricane theory of Gierasch and Goody (1973), and the superposition of planetary-scale circulations theory by Leovy et al. (1973).

DUSTY HURRICANE   In the dusty hurricane theory, it is proposed that the heating of airborne dust would act in a way analogous to the latent heating by condensing water vapor in an Earth hurricane. In this scenario, dust-heated air rises in the center of a vortex, additional air is pulled in at low levels, the vortex spins up and additional dust is lifted (Gierasch and Goody, 1973; Zurek et al., 1992). While this model has

recently been applied to the growth of local-scale dust disturbances (Rafkin, 2009), it is not likely a viable mechanism for the growth of global-scale storms. One argument against this mechanism is that global storms are not one single dust disturbance, but several local and regional storms combining together to produce a global-scale dust cloud. A second argument against this model for global storms is that there are very few observations of dust storms (of any size larger than a dust devil) that exhibit the predicted spiral structure and those that are observed occur in the northern plains at extratropical and polar latitudes (Gierasch et al., 1979; Hunt and James, 1979; Cantor et al., 2002, 2010).

SUPERPOSITION OF CIRCULATIONS   In the superposition of planetary-scale circulations theory, it is proposed that the interactions between the Hadley cell, tides, and large-scale topographic circulations create optimal conditions for the growth of dust storms. In this scenario, the combination of the season of maximum solar insolation (southern hemisphere summer) and the increased general atmospheric dust loading enhances each of these three circulation components, which can constructively interfere with each other to initiate and grow dust storms (Leovy et al., 1973). Given the recent GCM investigations of the mechanisms controlling the initiation and expansion of global dust storms, it seems very likely that the interaction between different circulation components is indeed key to the explosive growth of global dust storms.

(c) **Numerical Studies of Dust Storm Onset**   Recent GCM studies have focused either on spontaneously predicted storms (using fully interactive dust cycle models) or on specific storms in the observational record (e.g. the 2001 global dust storm).

Before the occurrence of the 2001 global dust storm at L_s ~ 185°, global dust storms were thought to occur within a relatively small seasonal window around southern summer solstice. Thus, the focus was largely on understanding storm development during the solstitial season through the study of radiative–dynamic feedbacks between dust and the general circulation. It has been well established that introducing dust to the atmosphere alters the structure of the Hadley cell throughout the year, but the strength of the response is greatest during the southern summer solstitial period because the solar insolation is at a maximum and the orientation of topographic dichotomy favors south-to-north transport (Richardson and Wilson, 2002; Takahashi et al., 2003; Basu et al., 2006; Zalucha et al., 2010; Wilson, 2012; see Chapter 9). Near southern summer solstice, dust introduced to the atmosphere in the southern hemisphere near the Hadley cell convergence zone absorbs solar energy, heats the atmosphere in the vicinity of the rising branch of the Hadley cell, and results in an increase in the depth and meridional extent of the overturning circulation (Haberle et al., 1982; Murphy et al., 1993; Wilson, 1997, 2012; Newman et al., 2002a; Basu et al., 2004, 2006; Kahre et al., 2006, 2008). In addition to the dust response of the mean overturning component of the general circulation, the eddy (and, in particular, tidal) component of the circulation is also greatly enhanced (Wilson, 2012). The combined effect of the enhanced mean overturning circulation and thermal tides results in an increase in surface wind stresses, which likely leads to the initiation of

additional centers of dust lifting (Murphy et al., 1995; Newman et al., 2002a; Wilson et al., 2008a).

While the response of the Hadley cell circulation likely plays an important role in the expansion of near-solstitial storms, it does not readily explain the explosive expansion of the 2001 global dust storm because of that storm's initiation time (L_s ~ 185°). During this near-equinoctial period, the rising branch of the Hadley cell is in the tropics and the circulation is roughly symmetric about the equator. Thus, the heating in the southern middle latitudes by dust entering the atmosphere in the Hellas region would not directly enhance the Hadley cell. Additionally, daytime vertical velocities associated with the thermal tides dominate over those of Hadley cell circulation at this season, and the nearly linear amplification of the diurnal tide with increasing dust leads to increased surface wind stresses (a strong positive feedback). Thus, it has been proposed that thermal tides may have played a prominent role in the expansion of the 2001 storm (Montabone et al., 2008; Wilson et al., 2008a; Martínez-Alvarado et al., 2009; Wilson, 2012). Montabone et al. (2008) and Martínez-Alvarado et al. (2009), using the UK Reanalysis, found evidence for a phase shift of the diurnal Kelvin wave in response to the initially localized dust lifting in the Hellas region. They suggest that this may communicate a remote change in surface wind stress in the Tharsis region, thereby triggering the dust lifting in the Daedalia–Claritas–Solis Planum region that was crucial for the significant expansion of the storm during the L_s = 188–193° period. Although progress has been made in understanding the 2001 global dust storm, it is clear that more work is necessary.

(d) **Storm Shut-Off and Decay**   Once the majority of dust lifting ceases, the decay of global dust storms sets in. A large and puzzling open question remains regarding why dust lifting stops when the atmosphere is dust-laden and the circulation is likely intense relative to a non-dusty period. There are two general possibilities for the cessation of dust lifting during a global storm: the depletion of surface dust in the dust source regions, or negative radiative–dynamic feedbacks within the atmosphere itself.

The depletion of dust from surface source regions could bring dust lifting to a rapid end. This mechanism could operate in two ways: first, all of the surface dust could be removed from the region; or second, as dust is removed, the remaining dust may be harder to lift for various reasons, raising the effective dust lifting threshold sufficiently to shut down lifting even though surface dust is still present. In either case, for the finite supply of surface dust to be the dominant shutdown mechanism for global dust storms, either the resupply of the predominant source regions for global storms must occur on timescales of at most a few Mars years (since global storms occur in roughly one year in three) or the predominant source regions must be different from one global storm to the next. Although the latter may be true in some cases, there are regions (e.g. Hellas) that have been observed to participate in multiple global dust storms. GCM simulations that include a finite surface dust reservoir indicate that the resupply of source regions on short timescales is challenging because storm source regions are confined spatially, but deposition regions are more spatially extended (see Section 10.5.2.2).

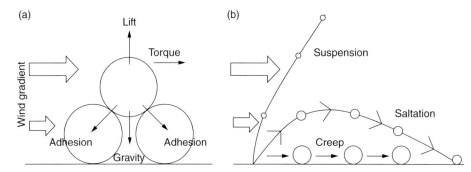

Figure 10.15. (a) The effective forces on particles at the surface, plus the wind field varying with height. (b) The three forms of particle movement: suspension of the lightest particles (kept airborne by turbulence), saltation of particles too heavy to stay aloft back to the surface, and creep of even heavier particles along the surface. Adapted from Merrison et al. (2007).

Pollack et al. (1979) proposed that the increase in static stability of the near-surface atmosphere as the atmospheric dust loading increases past a critical value would suppress near-surface winds enough to bring the surface wind stresses down below the threshold required for lifting. While this mechanism is theoretically possible, simulations of dust storms that include dust lifting based on a surface wind stress threshold parameterization do not predict a negative feedback based on static stability (Haberle et al., 1993; Newman et al., 2002a). In fact, GCM-predicted dust storms that are truly global in extent do not shut off until the Hadley cell circulation decreases in intensity as the season progresses past southern summer solstice (Montabone et al., 2005; Basu et al., 2006; Kahre et al., 2008; see Section 10.5.2.1).

10.5 MODELING THE DUST CYCLE

10.5.1 Theory of Dust Lifting, Removal, and Transport

10.5.1.1 Dust Lifting Theory

The process or processes by which dust can be raised from the surface of Mars has been a topic of great interest for several decades, since Sagan and Pollack (1969) first suggested that windblown dust could account for observed albedo changes on Mars. While dust lifting was already being studied on Earth – largely as a part of sand dune studies (e.g. Bagnold, 1936, 1941, 1956; Owen, 1964) – the greater importance of atmospheric dust loading on Martian weather and climate led to renewed interest in the theory of dust lifting mechanisms from a planetary science perspective (e.g. Sagan and Pollack, 1969; Hess, 1973; Leovy et al., 1973; Sagan and Bagnold, 1975). A large body of work now exists related to dust and sand movement on Earth, Mars, and even Venus and Titan. Thus, only a summary of the major points will be presented here. The primary focus is on the two processes believed to be responsible for the majority of dust lifting from the surface of Mars. The first is dust lifting due to surface wind stress, which can be divided into two broad categories: dust lifting via saltation, and direct suspension of dust particles. The second major process is lifting by convective vortices or "dust devils", for which pressure gradient effects combine with surface wind stress lifting around the vortex core to inject dust through the boundary layer. Additional processes

that may play a role, such as electrical and thermal processes, CO_2 fountaining, and meteoritic impacts, are also briefly discussed.

(a) Lifting by Surface Wind Stress

PHYSICS OF PARTICLE MOVEMENT   The forces on surface particles in a wind field (Figure 10.15a) are the following: gravity (F_g) and adhesion (F_{adh}) resist movement, while aeolian drag can be partitioned into an effective lift force (F_L) and a moment of torque (F_T) in which rolling (or sliding, on smooth surfaces) occurs due to shear stress. Particle detachment occurs when the wind-induced forces balance adhesion plus gravity, i.e. $F_L + F_T = F_g + F_{adh}$. Once detachment occurs, three types of movement are possible, as shown in Figure 10.15(b): suspension, in which (typically fine) particles can remain airborne due to turbulence; saltation, in which gravity dominates and particles fall back to the surface; and creep, in which grains do not leave the surface, but undergo rolling or sliding. By definition, a sand particle is one that falls back to the surface – saltates – rather than going into suspension, i.e. a particle that is too heavy for background vertical winds and turbulence to carry it away from the surface for any significant time, and which instead follows a curving path back to the surface. Dust particles are typically those particles that are carried into suspension and remain aloft for some significant time, though eventually they may return to the surface again at another location.

The shear stress at the surface (hence lift and torque) can be written in terms of a friction or drag velocity, $u^* = (\tau/\rho_{air})^{0.5}$, where τ = stress at the surface, and ρ_{air} is the surface air density. Assuming the near-surface wind speed follows a logarithmic profile with height (e.g. Prandtl, 1935; Bagnold, 1941; Greeley and Iversen, 1985), the wind speed u at a height z above the surface may be written as

$$u = \left(\frac{u^*}{k}\right)\ln\left(\frac{z}{z_0}\right) \qquad (10.4)$$

where k is von Kármán's constant ($k = 0.4$), and z_0 is the aerodynamic surface roughness, the height above the surface at which the wind velocity is essentially zero. Aerodynamic surface roughness is a function of many factors, including topography on a scale of meters and the size of the grains on the surface (Greeley and Iversen, 1987; Greeley et al., 2000).

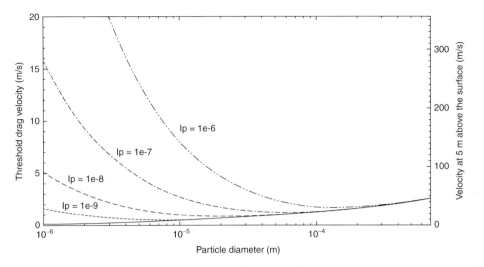

Figure 10.16. The variation of threshold drag velocity as a function of wind magnitudes at 5 m with particle diameter, at a grid point in northern Hellas, using inter-particle cohesions $I_p = 0$ to 1×10^{-6} N m$^{-1/2}$. From Newman et al. (2002b).

The fluid (or static) threshold drag velocity, u_t^*, is defined as the drag velocity at which particles on the surface begin to move continuously, whether by creep, saltation, or suspension. The first significant work on aeolian processes was performed by Bagnold (1936, 1941) to understand the motion of desert sand and dunes. The Bagnold (1941) equation is commonly used to define the threshold drag velocity as

$$u_t^* = A \sqrt{\frac{(\rho_p - \rho_{air})g D_p}{\rho_{air}}} \approx A \sqrt{\frac{\rho_p g D_p}{\rho_{air}}} \qquad (10.5)$$

where ρ_p is the density of the particle material (in practice, much larger than ρ_{air}, enabling the above approximation), g is acceleration due to gravity, and D_p is the mean particle diameter (Lorenz et al., 1995; Roney and White, 2004). Greeley and Iversen (1985) used wind tunnel experiments to produce semi-empirical expressions for the threshold parameter, A, as follows:

$$A = 0.2 \sqrt{\frac{1 + (I_p / \rho_p) g D_p^{2.5}}{1 + 2.5 R_t^*}} \qquad 0.03 < R_t^* < 0.3$$

$$A = 0.129 \sqrt{\frac{1 + (I_p / \rho_p) g D_p^{2.5}}{1.928(R_t^*)^{0.092} - 1}} \qquad 0.3 < R_t^* < 10$$

$$A = 0.12 \sqrt{1 + (I_p / \rho_p) g D_p^{2.5}} \, [1 - 0.0858 e^{-0.0617(R_t^* - 10)}] \qquad R_t^* > 10$$

$$(10.6)$$

Here I_p is the inter-particle cohesion between particles and $R_t^* = u_t^* D_p / \nu$ is the particle friction Reynolds number, where ν is the kinematic viscosity of the air. Since A depends on R_t^*, which in turn depends on u_t^*, this system of equations (10.5) and (10.6) must be solved iteratively. Other formulations for A have also been proposed (e.g. Lu et al., 2005) that provide an even more general description of particle detachment from the surface in a variety of fluids (e.g. air and water).

The wind threshold required for particle movement depends on particle size and inter-particle cohesion. Although smaller particles of a given density are lighter than larger particles (which tends to make smaller particles easier to lift), smaller

particles also experience a stronger inter-particle cohesion effect than larger particles, as cohesion increases with the ratio of surface area to volume (which tends to make smaller particles harder to lift). Due to the competition between these two effects, there is an optimal particle size to be lifted from the surface for a given value of I_p; for example, particles of ~50–150 μm diameter (sand-sized) are most easily lifted for I_p between 10^{-7} and 10^{-6} N m$^{-1/2}$ (Figure 10.16).

The dynamic (or impact) threshold is the drag velocity required to *maintain* saltation once established. The dynamic threshold is typically ~20% lower than the fluid threshold on Earth (Bagnold, 1937; Greeley and Iversen, 1985). However, recent experiments and theoretical studies show a much bigger difference between the two thresholds for Mars conditions (Kok, 2010), up to an order of magnitude for smaller saltating particles (~100 μm diameter), because the lower gravity and vertical drag on Mars allows saltating particles to travel higher and farther, meaning that they are accelerated by the wind for far longer than during a single "hop" on Earth and thus hit the surface with greater force. This produces a hysteresis effect: once saltation is initiated by strong winds, it can be *maintained* by winds an order of magnitude weaker. This potentially huge effect is difficult to incorporate into simple estimates of dust lifting on Mars, and has not yet been fully explored in more complex climate models.

DUST LIFTING BY SALTATING SAND PARTICLES   Sand-sized particles are likely easiest to lift on Mars, making their saltation an important mechanism (sometimes called sandblasting) to raise large amounts of dust (Greeley and Iversen, 1985). In this scenario, sand particles saltate back to the surface, imparting additional momentum that enables the smaller and likely harder-to-raise dust particles to be lifted. A variety of equations have been proposed to describe the vertical dust flux, F_{ws}, based on the horizontal flux of sand during saltation. Particle trajectory, surface wind stress, and wind profile in the saltating layer are considered in the derivation of these expressions (e.g. Bagnold, 1941; Owen, 1964; White, 1979; Westphal et al., 1987; Shao et al., 1993), which tend to depend on either

$(u^*)^3$ or $(u^*)^4$. Many are modified by some function of $u^* - u_t^*$, and all include a constant to be determined by experimental results, or by parameter tuning if applied within a model.

Two forms of vertical dust flux equations have been used in the Mars dust cycle modeling literature. The first is based on White (1979) and Shao et al. (1993) and the second is based on a surface dust lifting prescription developed to represent observations of dust storms in the Sahara (Westphal et al., 1987). In terms of surface wind stress $\tau = \rho u^{*2}$, these two prescriptions for the wind stress dust flux, F_{ws}, for $\tau > \tau^*$ are

$$F_{ws} = \alpha_{ws} \times 2.61 \frac{\tau^{3/2}}{g\sqrt{\rho}} \left(1 - \sqrt{\frac{\tau^*}{\tau}}\right)\left(1 + \sqrt{\frac{\tau^*}{\tau}}\right)^2 \qquad (10.7)$$

and

$$F_{ws} = \alpha_{ws} \times 2.3 \times 10^{-3} \tau^2 \left(\frac{\tau - \tau^*}{\tau^*}\right) \qquad (10.8)$$

where α_{ws} is the tunable lifting efficiency factor, τ is the surface wind stress, τ^* is the threshold surface wind stress, and ρ is the near-surface air density (Newman et al., 2005; Kahre et al., 2006).

DUST LIFTING BY DIRECT SUSPENSION   Recent field and laboratory work has demonstrated the importance of direct suspension of dust particles, particularly in a supply-limited situation (Loosmore and Hunt, 2000; Kjelgaard et al., 2004; Macpherson et al., 2008). The long-term, steady-state dust flux is typically several orders of magnitude smaller than that during sandblasting. However, direct dust lifting events typically begin with stronger, transient lifting of dust, which increases the overall dust contribution. The reason for transient strong injection is not well understood, but in a real setting is unlikely to occur so often that direct lifting can compete with sandblasting in terms of significant dust events (i.e. storms). However, although we expect sandblasting to be a far stronger effect where sand is present, we should also consider what happens in areas without adequate sand, as this may affect the background atmospheric dust abundance on Earth and Mars.

Early calculations of u_t^* for micrometer-sized dust particles and general circulation models of near-surface winds suggested that u^* would be too low on Mars to allow dust-sized particles to be directly lifted by the near-surface wind, at least assuming the same range of inter-particle cohesions given above. However, even if thresholds are higher for dust particles, higher u^* can also be produced by increasing the surface roughness, z_0, which results in higher u^* for a given $u(z)$ via (10.4). Wind tunnel studies (e.g. White et al., 1997; Greeley et al., 2000) incorporating this effect show that even moderately rough surfaces increase u^* enough for dust-sized particles to be lifted directly without saltation required. The effective change in z_0 for a given distribution of roughness elements is typically complicated to assess, but generally there are two effects that combine to enhance lifting: first, dust sitting on individual roughness elements (e.g. small pebbles) sits higher in the boundary layer where the wind shear is stronger than over the smooth surface; and second, turbulence generated around individual elements creates scour zones where dust can be locally entrained.

If dust lifting can occur on Mars in the absence of saltation, the next step is to calculate the dust flux. Initial experiments (e.g. Bagnold, 1954; Chepil and Woodruff, 1963; Shao et al., 1993) using visual measurement showed the dust flux to be negligible following an initial transient burst, but some more recent experiments have shown a non-zero long-term flux proportional to u^{*3} (e.g. Loosmore and Hunt, 2000), though generally the initial burst injects orders of magnitude more dust (Loosmore and Hunt, 2000; Macpherson et al., 2008). Estimating the total flux is likely dependent on the availability, distribution, and placement of dust grains on the surface, as the stronger transient flux may be due to initial removal of grains of a certain size range or positioning.

Estimating the threshold at which direct suspension begins is even more complex. For direct suspension, more so than in sandblasting, the placement of dust particles on the surface appears to be a crucial factor in determining the threshold at which lifting occurs, as it has a huge impact on the highly important cohesive forces between particles. For example, in wind tunnel experiments under Mars conditions (using the MARSWIT facility at NASA Ames), White et al. (1997) found a wide range in threshold for suspension to occur depending on whether the surface was smooth or rough, with some dust particles raised into suspension at similar thresholds to those of much larger saltating particles. In later wind tunnel experiments using Owens (dry) Lake surface soils under Earth conditions, Roney and White (2004) found that dust suspension thresholds varied from 50% to 75% of conventional saltation thresholds.

Due to the complexity of this topic, in particular when the surface soil is inhomogeneous in both its size distribution and placement, further studies of direct suspension are clearly needed before we can fully assess its importance on Mars, in particular whether it could dominate in regions with low sand cover, and whether it has a significant contribution to larger storms and/or the background dustiness.

(b) Lifting by Convective Vortices (Dust Devils)

PHYSICS OF DUST DEVIL FORMATION   Boundary layer convective vortices form when surface heating generates plumes of warm air that rise and interact with cooler air in such a way as to produce initial rotation (e.g. Neakrase and Greeley, 2010a). Once a rotating core forms, the column of warm air rises and stretches vertically, producing an intensification of the rotation due to conservation of angular momentum. Strong upward motion in the vortex core just above the surface produces a pressure drop, leading to additional warm air being pulled in near the surface, further intensifying the vortex until it is self-sustaining, with peak tangential velocities near the surface at the edge of the core. Higher up, however, the vortex is in nearly solid-body rotation with peak tangential velocities at its edge, and with downdrafts and stagnation points dominating inside the core while peak upwelling now coincides with maximum temperatures in the outer section (e.g. Gu et al., 2010). In dusty vortices (dust devils), dust is pulled in at the surface, lofted through the outer section (since little dust is contained in the core aloft, due to downdrafts and centrifugal forces), and then either falls out or is ejected at the vortex top.

Rennó et al. (1998) present a comprehensive thermodynamic theory for the formation of dust devils. In this theory, a convective vortex is viewed as a heat engine, performing mechanical work against frictional dissipation, and driven by heating by warm air pulled into the vortex just above the surface. The mechanical energy made available by the convective heat engine (to drive the vortices), F_{av}, is given by

$$F_{av} \approx \eta F_s \tag{10.9}$$

where η is the thermal efficiency of the heat engine and F_s is the surface sensible heat flux. This heat flux F_s is approximately proportional to the difference between the temperature of the surface and that of the near-surface air, also increasing with drag velocity, while η may be given approximately by $\eta = 1-b$, where

$$b = \frac{p_s^{\chi+1} - p_{top}^{\chi+1}}{p_s - p_{top}(\chi+1)p^{\chi}} \tag{10.10}$$

Here p_s is the ambient surface pressure, p_{top} is the ambient pressure at the top of the convective boundary layer, and χ is the specific gas constant divided by the specific heat capacity at constant pressure. This means that, as p_{top} increases (hence the boundary layer thickness decreases), b increases, and η decreases; in other words, the thicker the convective boundary layer, the greater the efficiency of the dust devil heat engine.

According to the convective heat engine model, the vortex (dust devil) "intensity" as given by the pressure drop is given by

$$\Delta p \approx p_s \left\{ 1 - \exp\left[\left(\frac{\gamma\eta}{\gamma\eta-1} \right) \left(\frac{\eta_H}{\chi} \right) \right] \right\} \tag{10.11}$$

Here γ is the fraction of the total dissipation of mechanical energy that is consumed by friction at the surface (a free parameter sometimes set to 0.5), and η_H is the horizontal thermodynamic efficiency of the dust devil, given by $\eta_H = (T_0 - \bar{T}_s)/\bar{T}_s$, with T_0 the temperature at the center of the vortex at the surface (for which the ground temperature can be taken as the upper bound, as the vortex is a region of strong mixing, which reduces the surface-to-air temperature gradient) and $\bar{T}_s$ the temperature of the near-surface air outside the vortex.

If it is assumed that dust devils are in cyclostrophic balance (i.e. the centrifugal and radial pressure gradient forces balance), the peak tangential velocity is proportional to the square root of the pressure drop:

$$\frac{v^2}{r_m} \simeq \frac{1}{\rho} \frac{\Delta p}{r_m} \tag{10.12}$$

Substituting (10.11) into (10.10), we find that

$$v \approx \sqrt{RT_s \left\{ 1 - \exp\left[\left(\frac{\gamma\eta}{\gamma\eta-1} \right) \left(\frac{\eta_H}{\chi} \right) \right] \right\}} \tag{10.13}$$

Greater thermal efficiency, η, or horizontal thermal efficiency, η_H, will both result in v increasing. This suggests that a taller vortex (as might be expected for a deeper boundary layer, hence larger η) or a wider vortex (as might be expected to result in a larger core-to-boundary temperature gradient, hence larger η_H) should result in a stronger tangential wind.

DUST LIFTING BY CONVECTIVE VORTICES The dust devil intensity (pressure drop) or tangential wind speed are useful characteristics, but do not tell us how much dust can actually be lifted by such vortices. A range of laboratory experiments (Greeley and Iversen, 1985; Neakrase and Greeley, 2010a,b) and terrestrial field experiments (Sinclair, 1973; Balme et al., 2003; Rennó et al., 2004; Ringrose et al., 2007) have been conducted to study dust devils, in part in an attempt to determine the raised dust flux for Earth and Mars conditions.

An interesting question is how much lifting of dust particles is due to the "suction" effect inside the low-pressure dust devil core rather than due to surface wind stress lifting by the strong tangential winds around it. Balme and Hagermann (2006) used laboratory experimental techniques to show that vertical pressure gradient lifting appears far more efficient than drag force lifting. The relative insensitivity to particle size of the suction effect suggests that dust devils could be the primary mechanism responsible for lifting the ubiquitous micrometer-sized dust particles into the Martian atmosphere in low surface wind stress locations and periods, and could potentially provide the dust flux required to maintain the background haze of 0.2–0.3 visible-band optical depths during the non-dusty season (Ferri et al., 2003; Basu et al., 2004; Balme and Greeley, 2006). Neakrase and Greeley (2010a) measured the sediment flux produced by a range of vortices generated by the Arizona State University vortex generator (ASUVG), both for ambient conditions and for Mars conditions (inside NASA's MARSWIT wind tunnel). They found that results were dependent only on the pressure drop or "intensity" of the vortex, rather than on their physical size, suggesting that extrapolating results to far larger events in the field (whether on Earth or Mars) based on the pressure drop alone may be feasible.

Two schemes representing two possible mechanisms for dust lifting by convective vortices have been implemented into GCMs (Newman et al., 2002a,b; Basu et al., 2004; Kahre et al., 2006). The first relates the lifted dust flux to the dust devil activity (10.9) by assuming that the two are proportional, which means that convective vortex dust lifting depends primarily on the sensible heat exchange from the surface to the atmosphere and on the depth of the planetary boundary layer. The second assumes that dust lifting occurs when the tangential frictional wind speed exceeds the threshold frictional wind speed required for saltation. Based on their laboratory work, Greeley and Iversen (1985) presented a semi-empirical formula for tangential wind speed around the vortex core for which a single layer of dust particles ($n = 1$) is lifted:

$$v_{tang}^t = \sqrt{\left(1 + \frac{15}{\rho_d g n D_p} \right) \left(\frac{\rho_d g n D_p}{\rho} \right)} \tag{10.14}$$

Here ρ_d is the particle density, D_p is the particle diameter, n is the number of dust particle layers, and ρ is the air density. Newman et al. (2002b) used this result and (10.13) to predict a "threshold" for convective vortex dust lifting. Equation (10.14) was then solved for n to provide a rough estimate of the number of dust particle layers (hence dust flux) injected when the threshold is exceeded.

(c) Other Processes

ELECTRICAL EFFECTS Dust particles may clump together electrostatically to form a low-density, high-diameter aggregate (e.g. Merrison et al., 2007), thus lowering their surface-area-to-volume ratio. Low-density, high-diameter aggregates are easier to lift than either individual dust grains or sand particles of the same diameter (which would have a higher density and thus weight) because their inter-particle cohesion force decreases (due to their lower surface-area-to-volume ratio) more than their gravitational force increases (due to their higher mass). These aggregates would quickly disintegrate once in the air, releasing individual dust grains and maintaining the large dust abundance observed on Mars (Greeley, 1979; Merrison et al., 2004). White et al. (1997) and Greeley et al. (2000) observed this clumping in wind tunnel experiments. However, electrostatic forces may also increase cohesion between individual particles (Greeley and Leach, 1978).

On Earth, large electric fields are measured in dust devils and dust storms (Stow, 1969; Qu et al., 2004; Rennó et al., 2004; Zhang et al., 2004). These electric fields are most likely caused by charge transfer during collisions between larger (i.e. sand) and smaller (i.e. dust) particles, followed by charge separation of the heavier (positively charged) and lighter (negatively charged) particles via gravitational and aerodynamic forces (Schmidt et al., 1998; Desch and Cuzzi, 2000; Rennó et al., 2003). Strong electric fields may be able to either directly lift surface particles, or at least reduce the amount of force required (i.e. reduce the effective threshold surface wind stress) for them to be raised by the wind. An experimental study by Kok and Rennó (2006) showed that for Earth soils and conditions the effect is greatest for particles with diameters ~50–200 μm, suggesting that charge separation should increase saltation rather than lift dust particles directly. Unlike Earth, however, electrical breakdown of the thin Martian atmosphere occurs at only ~20 kV m^{-1} (Rennó et al., 2003), which may prevent the larger bulk electric fields necessary to produce significant lift, although it is possible that stronger fields may persist within dust devils and add to the other lift effects present there (Farrell et al., 2006).

THERMAL EFFECTS   Light incident on a bed of surface dust is absorbed up to larger depth, but only the top layer can cool via thermal radiation. Hence, temperature increases with depth in a thin dust layer, which is an effect known as the solid-state greenhouse effect (e.g. Niederdörfer, 1933; Kaufmann et al., 2006). Momentum is then transferred between gas molecules and dust particles along the thermal gradient from the warm depths to the cold top layer in a process called "thermophoresis" (Cheremisin et al., 2005). Dust eruptions can occur if this exceeds the combined forces of gravity and cohesion (Wurm and Krauss, 2006; Wurm et al., 2008). Laboratory and microgravity experiments have shown that the incident solar flux on Mars is in the same range as the light flux required for lift to occur, thus could potentially lower the effective threshold for Martian dust lifting (Wurm et al., 2008).

Heating and vaporization of adsorbed CO_2 (Johnson et al., 1975) or H_2O (Huguenin and Clifford, 1979) on dust grains has been suggested to produce additional injection of dust particles. Greeley and Leach (1979) tested H_2O fountaining in the laboratory at low atmospheric pressure and found that dust was sprayed or erupted into the atmosphere through vents or fissures, the production of which also increased surface roughness and increased dust injection via surface wind stress

lifting. High-resolution observations over the last decade have provided increased evidence of CO_2 geysers and associated "dust fountaining" on Mars (e.g. Piqueux et al., 2003; Kieffer, 2007; Thomas et al., 2010; Portyankina et al., 2012), though it remains unclear how much such events contribute to the atmospheric dust load each year.

10.5.1.2 Dust Deposition Theory

The processes that govern the removal of dust from the Martian atmosphere to the surface have been studied in earnest since Mariner 9 provided the first spacecraft observations of the decay phase of a global dust storm (Hartmann and Price, 1974; Conrath, 1975). Since that 1971–1972 event, Mars-orbiting spacecraft have been present to observe the decay of four additional global dust storms (1977a, 1977b, 2001, 2007) as well as thousands of local and regional dust storms (Cantor et al., 2001). *In situ* rates of dust accumulation have also been inferred from measurements obtained by Mars Pathfinder, Mars Phoenix, and the Mars Exploration Rovers (see Section 10.3.2).

Airborne dust is returned to the surface as the result of several processes. The downward mass flux of dust (F_{mass}, kg m^{-2} s^{-1}) crossing the atmosphere–surface boundary can be expressed as

$$F_{mass} = V_d C \qquad (10.15)$$

where V_d is the effective dust deposition velocity and C is the concentration (kg m^{-3}) of atmospheric dust near the surface. The deposition velocity depends on the efficiencies of the removal processes. On Earth, removal processes can broadly be divided into two categories: dry and wet. On Mars, dry deposition likely dominates over wet deposition due to the lack of significant non-gaseous water (Rossow, 1978). For this reason, the following discussion focuses on dry deposition processes. However, modeling studies and the recent observations of large water ice cloud particles near the surface at the Phoenix Lander site argue that wet (frozen) deposition processes may be important at some locations during some seasons (Rodin et al., 1999a,b; Whiteway et al., 2009). The effects of scavenging by water and CO_2 ice clouds are therefore discussed where appropriate.

On Earth, several dry deposition processes play important roles, including gravitational sedimentation, impaction, and Brownian diffusion (Slinn and Slinn, 1980; Ruijgrok et al., 1995). Gravitational sedimentation is the process by which particles settle to the surface under the influence of gravity (see below). Impaction occurs when larger particles that are traveling in an airstream collide with the surface as the stream changes direction, while Brownian diffusion is important for smaller particles that contact the surface as a result of their random motion within a gas. Many of these processes are complexly related to aerosol particle size and density, terrain, and meteorological conditions (Zhang et al., 2001).

A variety of parameterizations have been developed to calculate the deposition velocity for aerosols and pollutants on Earth. A typical approach is to sum the transfer resistances in series. One parameterization that has been utilized for Mars (Murphy et al., 1990) follows from the work of Slinn and Slinn (1980). The deposition velocity (V_d) is calculated from

$$\frac{1}{V_d} = \frac{1}{k_c} + \frac{1}{k_d} + \frac{V_g(R_p)}{k_c k_d} \tag{10.16}$$

where

$$k_c = k_c' + V_g(R_p) \tag{10.17}$$

and

$$k_d = k_d' + V_g(R_p) \tag{10.18}$$

Here V_g is the gravitational sedimentation velocity of particles of radius R_p, and k_c' and k_d' are parameters that account for diffusion and impaction.

Gravitational sedimentation alone typically dominates over the combined effect of all three dry deposition processes for conditions relevant to the Martian atmosphere and dust (Murphy et al., 1990). Specifically, for dust particles between 0.5 and 4 μm in radius, the inclusion of impaction and diffusion makes very little difference to the deposition velocity. For this reason, it is typical to include only gravitational sedimentation in models of dust removal on Mars.

(a) **Gravitational Sedimentation** The gravitational sedimentation velocity (i.e. the "terminal velocity") of a particle is reached when the downward force of gravity and upward drag and buoyancy forces are in balance and no subsequent acceleration is experienced. The gravitational force experienced by a spherical particle is

$$F_{grav} = \frac{4\pi R_p^3 \rho_p g}{3} \tag{10.19}$$

while the drag force opposing gravity is

$$F_{drag} = 6\pi \eta R_p V_g \tag{10.20}$$

In these equations, R_p and ρ_p are the particle's radius and density, respectively, g is gravitational acceleration at the planet's surface, η is the viscosity, and V_g is the particle's gravitational fall velocity. Balancing these equations and solving for velocity (i.e. the terminal velocity) yields the Stokes fall equation,

$$V_g = \left(\frac{2}{9}\right)\frac{\rho_p R_p^2 g}{\eta} \tag{10.21}$$

which is valid within a continuous medium (i.e. where the mean free path of the atmospheric gas, $\lambda = 2\eta/\rho v$, is small compared to the particle radius). The Knudsen number, $Kn = \lambda/R_p$, determines whether the interactions of aerosol particles with gas molecules is best represented by continuum mechanics ($Kn \ll 1$, classical regime) or statistical mechanics ($Kn \gg 1$, kinematic regime). Under the conditions of the classical regime, (10.21) can be used to calculate the terminal velocity of a spherical particle. However, at the temperatures, pressures, and dust particle sizes characteristic of the Martian atmosphere, Kn is often larger than unity. The Cunningham slip-flow correction, $\delta = 1 + \alpha Kn$, is then included to account for a particle's increase in fall velocity (compared to Stokes fall) due to relatively infrequent collisions with gas molecules, modifying (10.21) to become (Rossow, 1978)

$$V_g = \left(\frac{2}{9}\right)\frac{\rho_p R^2 g}{\eta}(1 + \alpha Kn) \tag{10.22}$$

where

$$\alpha = 1.246 + 0.42\exp\left(\frac{-0.87}{Kn}\right) \tag{10.23}$$

The derivation for gravitational velocity assumes spherical dust particles, but the observed scattering properties of airborne Martian dust strongly suggest that suspended Martian dust particles are non-spherical. Therefore, the manner in which modified particle shapes affects fall speed must be considered. The balance of gravitational and drag forces dictate a particle's terminal fall speed, as indicated above. Since a non-spherical shape possesses a greater surface-area-to-volume ratio than an equal-volume spherical particle, drag effects upon the non-spherical particle exceed those upon the spherical particle, resulting in slower terminal velocity values for non-spherical shapes.

Atmospheric dust particles are likely candidates for serving as condensate-cloud seed nuclei for both water and CO_2 ice cloud generation (Michelangi et al., 1993; Colaprete and Toon, 2003; Määttänen et al., 2005). The presence of lower-density ice surrounding a dust particle of a given size will modify the speed at which that dust particle will fall. As ice condenses onto a dust grain, the mass of the combined dust–ice particle will increase, which will tend to increase the particle's sedimentation velocity. However, as the ratio of ice to dust increases, the bulk density of the particle will decrease (relative to a dust particle of the same size), which acts to decrease the particle's sedimentation velocity. To understand how these opposing tendencies affect the overall sedimentation velocity of the dust particle, (10.22) is rearranged using the relationships between dynamic viscosity, Knudsen number, air density and thermal velocity:

$$V_g = \left(\frac{4}{9}\right)\frac{\rho_{cld} R_{cld} g}{Kn \rho_a v_t}(1 + \alpha Kn) \tag{10.24}$$

According to (10.24), $V_g \propto R_{cld}\rho_{cld}$. Since the cloud particle's density scales with R_{cld}^3, the bulk density of the cloud particle quickly approaches the density of ice as ice continues to be added to a dust core. Thus, as the cloud particle first starts to grow, the fall velocity of the dust nuclei is reduced slightly until a minimum is reached, after which the velocity increases as the cloud particle continues to grow. It is notable that cloud particle radii from 1 to ~10 μm (containing 1 μm dust cores) have fall velocities similar to those of the dust particle alone. Large cloud particles (>10 μm) are required to substantially influence the particle fall velocity (i.e. increase it by more than a factor of 2).

10.5.1.3 Dust Transport Theory

Dust grains lifted from the surface travel some distance before returning to the ground. The distance a dust particle travels from its source region depends on the balance between that particle's gravitational fall speed and the efficiency of vertical and horizontal atmospheric transport. It is difficult to directly observe atmospheric transport processes because observed dust storms are the superposition of dust lifting, transport, and removal.

However, observations of particular types of dust events yield insight into how dust moves in the atmosphere.

In principle, the movement of dust through the Martian atmosphere can be fully described by advection due to atmospheric winds. However, modeling the advection of dust with fluid dynamical models is challenging because of the large range of spatial and temporal scales relevant for atmospheric motion (submillimeter and subsecond to global and annual). Meso- and global-scale climate models cannot resolve the smallest spatial and shortest temporal scales, thus small, short-lived processes that contribute to dust transport often need to be parameterized. In GCMs, while horizontal transport is often computed using only resolved winds, vertical transport due to unresolved mechanical and thermal turbulence is often parameterized (in addition to vertical transport by the resolved wind). The situation is further complicated by the fact that model-simulated transport has been shown to be sensitive to the choice of numerical transport scheme and grid structure in GCMs (Lian et al., 2012).

The recent observations of elevated dust layers have led to renewed focus on the processes that govern the vertical transport of dust in the Martian atmosphere. Although the Conrath-ν vertical dust distribution prescription (Section 10.2.1.2) has historically been used extensively in climate models to reasonably simulate the bulk character of the atmospheric thermal and dynamical state of the Martian atmosphere, it is important to note that this prescription is based on dust transport described by diffusion and it cannot physically be employed to describe increasing dust concentration with increasing altitude. Diffusive mixing describes transport from high concentration to low concentration, which is incompatible with a surface source leading to an increasing concentration with altitude. Thus, it seems clear that modifications to the Conrath prescription are needed. While theoretical studies and mesoscale modeling efforts are currently underway to better understand how these elevated layers may develop (e.g. Heavens et al., 2011a; Rafkin, 2012; Spiga et al., 2013), parameterizations for these vertical transport processes have not yet been fully developed for GCMs.

10.5.2 The Simulated Dust Cycle

Global climate modeling efforts to simulate the current Mars dust cycle by including appropriate physical parameterizations for dust lifting, transport, and sedimentation have continued in earnest since the late 1990s. While these investigations have many scientific goals, modeling techniques, and simulated results in common, they also differ notably in the details of their approaches and results. In this section, we describe the results of different "interactive" dust cycle modeling efforts, which means that dust lifting, transport, sedimentation, and the radiative impact of the changing dust distribution are coupled and vary consistently. Simulated results from most (if not all) interactive dust cycle studies show that the general observed behavior of a background, fairly constant dust loading during northern spring and summer and an increased dust loading during southern spring and summer can be well represented by GCMs. However, it is clear that models are not yet able to capture many important aspects of the Mars dust cycle.

Fully interactive dust cycle simulations by various modeling groups (Murphy, 1999; Newman et al., 2002a,b, 2005; Basu et al., 2004, 2006; Kahre et al., 2005, 2006, 2008; Wilson and Kahre, 2009; Mulholland et al., 2013) have all included a combination of convective (dust devil) and surface wind stress dust lifting. The relative contributions of convective and surface wind stress lifting may be found by comparing the observed and predicted seasonal variation in dustiness by either process. Parameterizations of convective lifting imply a much weaker variation with time of year (i.e. a smaller asymmetry between northern and southern summer dust loading) than is observed, suggesting that surface wind stress lifting must be the dominant process overall and account for the peak in dust lifting during southern hemisphere spring–summer, while the relatively smaller rate of convective lifting produces the background dustiness observed to persist at other times.

Surface wind stress lifting schemes typically have two parameters that can be modified ("tuned") to produce more realistic predicted dust cycles: the threshold for lifting, and the lifting efficiency (Equations (10.7) and (10.8)). Typically both are held constant throughout a simulation, though thresholds that vary according to the amount of dust on the surface have also been explored (see discussion of threshold feedbacks below). The threshold for surface wind stress lifting can be calculated within the model simulations based on local conditions (see Newman et al., 2002b; Michaels et al., 2006), but it is often kept constant and used as a tunable parameter. Generally, the surface wind stress threshold and the efficiency factor are varied to yield the most realistic dust cycle possible as defined by a range of criteria chosen by the investigator, examples being the spatial extent of predicted lifting (e.g. Kahre et al., 2006), the seasonal behavior of the predicted dust loading (e.g. Newman et al., 2002a; Basu et al., 2004; Kahre et al., 2006), the range of predicted dust storm size, duration, and location (e.g. Newman et al., 2002a), and/or the inter-annual variability of predicted global dust storms (e.g. Basu et al., 2006). Although different tuning methods have been used, there are several aspects of the predicted spatial and temporal patterns of surface wind stress lifting that are common across most (if not all) interactive dust cycle studies (see Section 10.5.2).

Dust devil lifting schemes typically have a single parameter, the lifting efficiency, which is also held constant throughout a simulation. Investigators have used different methods to tune the dust devil efficiency factor to provide reasonable dust loadings during northern spring and summer. Basu et al. (2004, 2006) tuned the threshold-independent dust devil efficiency factor by comparing predicted atmospheric temperatures to the T_{15} temperatures observed by Viking. This method assumes that the atmospheric dust loading is correct if the atmospheric heating is correct. While this method guarantees a good match to observed temperatures, it assumes that atmospheric dust is the primary radiatively active aerosol responsible for maintaining atmospheric temperatures, which may not always be the case. Recent work has shown that the radiative effects of water ice clouds could also have a non-negligible effect on T_{15} (midlevel, 50 Pa) northern spring and summer (Wilson et al., 2007; Madeleine et al., 2012). Newman et al. (2002a) and Kahre et al. (2005, 2006, 2008) tuned the dust devil efficiency factor by comparing the predicted global and zonal-mean dust opacities to observed dust opacities. Although this method attempts to get at the atmospheric dust content directly, it is sensitive to the dust optical properties used in the model and it does not guarantee the best match to observed atmospheric temperatures. Both methods yield roughly comparable results. In particular, the dust devil (or convective lifting) component in the simulations needs to be fairly weak to maintain an annual cycle simulation with a quiescent northern hemisphere summer.

In the following sections, we focus first on results from studies that assume that there is a planet-wide infinite surface dust reservoir (Section 10.5.2.1). While it is clear, based on observations of surface properties, that dust is not always available to be lifted at all locations on the planet, this group of investigations has advanced our understanding of basic aspects of the dust cycle and has highlighted model weaknesses that need further attention. The implementation of finite surface dust reservoirs is challenging and represents one of the current state-of-the-art research topics in this field. Results from these investigations are presented in Section 10.5.2.2.

10.5.2.1 Infinite Surface Dust Reservoirs

There are several characteristics of the observed dust cycle that modelers attempt to capture in simulations that assume an infinite planet-wide surface dust source. These characteristics include the annual behavior and magnitude of the atmospheric dust loading, the spatial and temporal pattern of dust storm and dust devil onset regions, the variability of dust storm sizes, the variability of atmospheric particle sizes, and the inter-annual variability of global-scale dust storms. As described above, this is done by making choices regarding the dust lifting parameterizations used, the method used for tuning the dust lifting parameterizations, and the properties of the lifted dust particles lifted (e.g. the size distribution). Because no one simulation strategy has yet to capture all of the desired characteristics of the dust cycle, investigators prioritize the above list differently depending on the goals of their particular study.

(a) Patterns of Simulated Dust Lifting

SURFACE WIND STRESS LIFTING   GCMs generally agree on the seasons and locations of maximum and minimum predicted surface wind stresses, although the predicted magnitudes of these stresses vary from one model to the next, likely due to differences in model resolution and physical parameterizations (e.g. planetary boundary layer schemes). Peak surface wind stresses occur at several locations and seasons (Figure 10.17; Wilson et al., 2006): along the edges of the seasonal CO_2 ice caps during the spring cap retreat in both hemispheres; in a zonal band in the southern subtropics during southern summer; in the northern hemisphere mid-latitudes during autumn, winter, and spring; and in a subtropical band in the northern hemisphere during the $L_s \sim 180–210°$ and $L_s \sim 320–340°$ periods.

High surface wind stresses occur along the edges of the seasonal CO_2 ice caps due to the intense and diurnally variable local circulations that develop as the result of the strong thermal contrast between the ice-free and ice-covered surface (Toigo et al., 2002). In the southern hemisphere during the $L_s \sim 150–200°$ period, maximum surface wind stresses are largely localized

to the Hellas and Argyre Basins but they become more zonally uniform later in the season.

Surface wind stresses are enhanced in the southern hemisphere subtropics during southern summer due to strong winds that develop during these seasons as the result of the Coriolis deflection of the lower branch of the Hadley circulation. These southern hemisphere surface wind stresses are strongly modulated by topography and are diurnally variable. Because the strength of the Hadley circulation and the subtropical jet increases with increasing dust loading, there is a strong feedback in surface wind stress at these latitudes.

Strong surface wind stresses, evident in the northern hemisphere during the autumn, winter, and spring ($L_s = 180–350°$) seasons, are associated with the westerly jet that forms in this season. Traveling waves account for much of the strength and variability of surface wind stress in this region. Increasing dust loading leads to a weakening of the surface wind stresses associated with the mid-latitude westerlies. Significantly, the surface wind stresses diminish sharply during the northern hemisphere winter solstice season when the dust loading is increased. This is due to the suppression of baroclinic wave activity. This decline in transient wave activity at solstice was noted at the Viking Lander 2 site following the 1977b global dust storm and is evident in the other GCM simulations and in MGS/TES data (Wang et al., 2005, 2013; Basu et al., 2006; Kavulich et al., 2013; Mullholland et al., 2013)

Surface wind stress dust lifting schemes are based on model-predicted surface wind stresses (see Section 10.5.1.1(a)); thus there is also a fair amount of agreement between models on where surface wind stress dust lifting is predicted to occur. During northern spring and summer, GCM-predicted surface wind stress dust lifting maximizes along the growing south seasonal CO_2 ice cap (Figure 10.17). During northern autumn and winter, predicted surface wind stress dust lifting occurs along the receding south and growing north polar caps, on the slopes of the Tharsis Montes, Alba Patera, and Elysium, in the western boundary currents to the east of Tharsis, Syrtis, and Elysium, and on the flanks of the Hellas and Argyre Basins (Newman et al., 2002b; Basu et al., 2004; Kahre et al., 2005, 2006). Predicted regions of maximum surface wind stress lifting have been compared to observations of dark wind streaks (Thomas et al., 1984; Kahn et al., 1992; Newman et al., 2002b) and catalogs of dust storm onset regions (Martin and Zurek, 1993; Cantor et al., 2001; Newman et al., 2002b, 2005; Haberle et al., 2003; Basu et al., 2004; Kahre et al., 2005, 2006). Models generally compare favorably to these observations, but there are disagreements as well. For example, models tend to predict confined longitudinal zones of high surface wind stress lifting in the south along the receding seasonal CO_2 cap during northern autumn, whereas the observed dust storm events are more zonally symmetric (Newman et al., 2005; Kahre et al., 2006). In this case, changing the surface wind stress threshold and or increasing the model spatial resolution in future studies may improve the agreement.

DUST DEVIL LIFTING   Two parameterizations have been implemented into multiple Mars GCMs to simulate dust devil lifting: the threshold-independent and threshold-dependent schemes outlined in Section 10.5.1. The threshold-independent

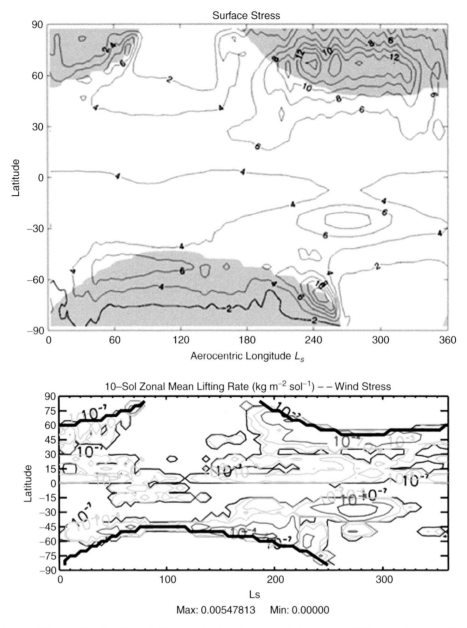

Figure 10.17. Zonal-mean GCM-predicted surface wind stresses (top) and surface wind stress dust lifting rate (bottom). The latitudinal extent of the seasonal CO_2 ice cap is denoted with shading in the top panel and a thick solid black line in the bottom panel. From Wilson et al. (2006) (top) and Kahre et al. (2006) (bottom).

scheme has emerged as the preferred scheme because it has been shown to predict a temporally and spatially smooth dust loading during northern spring and summer (Newman et al., 2002b) and it more cleanly reproduces the seasonal behavior of dust devil activity in both the northern and southern hemispheres when compared to observational surveys of dust devils and dust devil tracks (Fisher et al., 2005; Kahre et al., 2006).

The threshold-independent scheme predicts a spatially ubiquitous source of dust devil lifting during the day (Newman et al., 2002b; Basu et al., 2004; Kahre et al., 2006), which is to be expected as dust lifting via this scheme merely requires a positive surface heat flux and a non-zero planetary boundary layer depth. Kahre et al. (2006) show that the GCM-predicted diurnal peak in dust devil activity at the Mars Pathfinder site

at the time of the Mars Pathfinder mission occurs between 12 p.m. and 1 p.m. local time, which is consistent with the observed peak of dust devil occurrence as derived by measured pressure depressions at MPF (Figure 10.8; Murphy and Nelli, 2002).

The dust devil activity is expected to have a seasonal trend because both the surface-to-atmosphere temperature difference (and thus the heat flux) and the depth of the PBL maximize near the subsolar point. Thus, predicted peak dust devil lifting rates occur during local spring and summer, which is consistent with observations (Newman et al., 2002b; Basu et al., 2004; Fisher et al., 2005; Cantor et al., 2006; Kahre et al., 2006). Preferred southern hemisphere regions are Thaumasia–Solis, Margaritifer Sinus, Mare Tyrrhenum, and Hesperia–Mare Cimmerium,

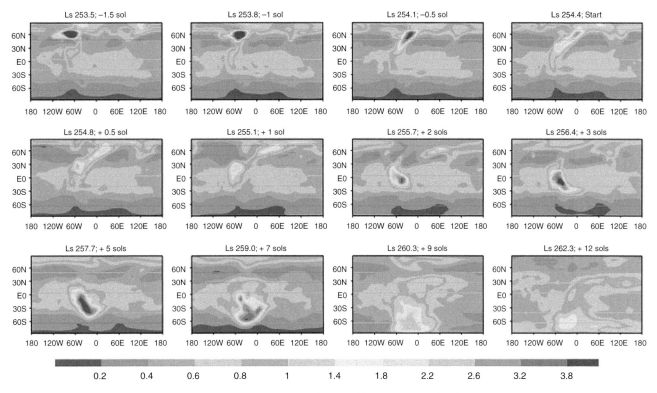

Figure 10.18. GCM-simulated cross-equatorial regional storm that developed in the Chryse region from approximately $L_s = 253°$ to $L_s = 260°$. From Newman et al. (2002a). A black and white version of this figure will appear in some formats. For the color version, please refer to the plate section.

while preferred northern hemisphere regions are Tharsis and to the west of Syrtis Major.

(b) Patterns of Simulated Dust Deposition and Net Dust Accumulation/Deflation The patterns of predicted dust deposition in infinite-reservoir GCM simulations tend to closely mimic the predicted dust lifting patterns due to the fact that a significant fraction of the lifted dust mass falls directly back to the surface before it can be transported away (Basu et al., 2004; Kahre et al., 2006). Thus, preferred regions for dust deposition are often either spatially aligned with surface wind stress lifting regions or offset slightly from these regions. However, since the annual net gain (accumulation) or loss (deflation) in a region depends on the balance between dust lifting and deposition, locations of strongly preferred surface wind stress lifting are losing dust annually, while regions outside of these source regions generally accumulate dust annually (see e.g. figure 18 in Basu et al., 2004). The flanks of Hellas and Argyre, the slopes of the large volcanoes, and the low-topographic areas to the east of Tharsis, Syrtis, and Elysium are net dust deflation regions. As discussed below, these deflation regions are the first to become swept clean of dust and subsequently tend to remain essentially dust-free when finite surface dust reservoirs are implemented into simulations of the dust cycle. Large annual net dust accumulation ($\gtrsim 5$ μm yr^{-1}) is predicted to occur in the center of Hellas, in regions in the Tharsis plateau, and in the north and south polar regions (Basu et al., 2004; Kahre et al., 2006). Smaller net dust accumulation rates of ~ 1–5 μm yr^{-1} occur throughout large regions in the middle and low latitudes. GCM-predicted net dust accumulation rates have

been shown to compare well to estimates of net dust accumulation rates from orbit and from the surface (Basu et al., 2004; Kahre et al., 2006).

(c) Simulated Dust Loading and Dust Storm Behavior The evolution of the model-simulated atmospheric dust loading is the result of dust lifting, transport, and deposition processes. Because airborne dust strongly influences the thermal and dynamical state of the atmosphere, it is necessary to include the feedbacks between lifting, transport, and deposition in order to understand how each process contributes to the dust cycle as a whole. Observations of the atmospheric dust loading show that there is a low-level background dust haze during northern spring and summer and increased atmospheric dustiness during northern autumn and winter (Figure 10.1; see Section 10.3.1.1). Models that include a fully interactive dust cycle are capable of reproducing this general behavior (Newman et al., 2002a,b; Basu et al., 2004; Kahre et al., 2006). There has been a general consensus among these studies that the dust cycle is best represented by the combination of two dust lifting parameterizations: one that provides a ubiquitous, low-level dust source (taken to be dust devils), and one that provides a spatially confined, explosive dust source (taken to be surface wind stress lifting). While the dust devil parameterization provides a smooth, low-level atmospheric dustiness throughout the year, its presence is particularly important during northern spring and summer when threshold-dependent surface wind stress parameterizations (due to their tuning) do not maintain the observed background dust haze (Newman et al., 2002a; Basu et al., 2004; Kahre et al., 2006).

BACKGROUND DUST HAZE DURING NORTHERN SPRING AND SUMMER   During northern spring and summer, GCM-predicted levels of atmospheric dust loading (provided almost entirely by dust devils) are on the order of a couple to a few tenths of a visible-band optical depth at low to middle latitudes, which is consistent with observations (Newman et al., 2002a; Kahre et al., 2005, 2006). Basu et al. (2004) show T_{15} temperatures as an indicator of atmospheric dust loading in lieu of a dust optical depth. Because they require a dust devil lifting parameterization during these seasons to match T_{15} temperatures, it is reasonable to assume that the atmospheric dust loading predicted by their model is consistent with other models. Newman et al. (2002a) demonstrate that there is a negative feedback involved in the dust devil (convective) lifting parameterization: as the atmosphere becomes dustier through dust devil lifting, the surface to near-surface air temperature difference decreases, which reduces the future dust devil lifting rate. This negative feedback limits the amount of dust that the dust devil lifting scheme can provide to the atmosphere, which allows for a relatively constant background dust loading during northern spring and summer.

LOCAL STORMS   Local storms along the advancing and receding seasonal CO_2 cap edges are observed and predicted by models throughout the Martian year (Cantor et al., 2001; Toigo et al., 2002; Basu et al., 2004; Kahre et al., 2006). The simulated storms are the result of increased surface wind stresses due to intense cap-edge circulations (Figure 10.17). During northern spring and summer, these storms remain local and short-lived, but during southern spring and summer, they can grow to become regional in scale (Cantor et al., 2001). While models predict that dust lifting does occur along the cap edges in the appropriate seasons, models tend to underpredict the increase in local dust opacity as a result of these cap-edge storms (Basu et al., 2004; Kahre et al., 2006). Several possible reasons for this exist: first, models may be underpredicting the amount of dust actually lifted in these regions; second, the lifted dust may be returning to the surface too quickly to be reflected in the models as compared to reality, which could be an indication that models do not capture transport processes as well as they should; or third, the numerical smoothing introduced by high-latitude Fourier filtering (required by many GCMs to prevent instabilities there) is averaging out the local high dust abundances that would otherwise be produced (Basu et al., 2004).

REGIONAL STORMS   Model-predicted regional-scale storms exhibit some, but not all, of the characteristics of observed regional dust storms. There are two types of predicted storms that have been examined in detail – frontal storms that lift dust and transport it from the northern hemisphere across the equator into the southern hemisphere, and southern hemisphere storms that tend to originate in the Hellas region. While both of these types of storms begin with local-scale lifting, they can grow into regional dust storms.

Cross-equatorial regional storms that develop in the north during local fall are initialized from localized dust lifting on topographic slopes (e.g. the north slopes of the Tharsis Montes, Alba Patera, or Elysium Mons) near the edge of the growing north seasonal CO_2 cap (Newman et al., 2002a; Basu et al., 2006; Hollingsworth and Kahre, 2010). As low-pressure systems travel eastward past these dust lifting regions, the lifted dust can be concentrated and transported eastward along the leading edge of the frontal system and funneled southward toward the equator through the low-topographic channels of Acidalia, Arcadia, and Utopia (Figure 10.18). As described in Section 10.4.2.2, if the timing of the passage of the low-pressure system aligns properly with the tidal winds, dust lifting can be enhanced along the frontal convergence zone through the constructive interference between the northerly winds associated with the front and the northerly tidal winds, and the dust storm can move across the equator into the southern hemisphere. Aspects of the behavior of these storms have been described in interactive dust cycle studies (Newman et al., 2002a; Basu et al., 2006; Mulholland et al., 2013).

Simulated Hellas regional storms can be split into two categories: early storms that originate on the southwest rim of the Hellas Basin as the seasonal CO_2 ice cap retreats through that latitudinal region from $L_s \sim 175°$ to $L_s \sim 190°$, and late storms that develop on the northern/northeastern rim of the basin as southern summer solstice is approached (Newman et al., 2002a; Basu et al., 2004, 2006; Kahre et al., 2006). The early Hellas storms develop as the interaction between intense cap-edge circulations and slope flows associated with the topographic relief of the basin produce high (threshold-exceeding) surface wind stresses (Siili et al., 1999). Lifted dust tends to be transported to the east by the westerlies that exist in this region at this time of year (Basu et al., 2006). The subsequent evolution of these types of storms differs in different models (or simulations, or simulated years). In some cases, new lifting centers can be ignited as the core of the storm moves away from its original location, and/or the predominant direction of the storm can change from eastward to westward and poleward (Newman et al., 2002a).

Late Hellas storms develop on the northern rim of Hellas between $L_s \sim 240°$ and $L_s \sim 280°$ (Basu et al., 2004; Kahre et al., 2006), due to high surface wind stresses produced by the subtropical westerly jet, combined with slope flows on the edge of the basin and with strong diurnal tidal winds. The Hadley cell circulation deepens and becomes more meridionally extensive as dust enters the atmosphere in the latitudinal region of the upward branch and is rapidly transported upward and away from the source region. The intensified Hadley circulation increases the strength of the westerly jet and combines with the dust-amplified tides to further increase the amount of dust lifted. If lifting remains local to the Hellas region, the storm is prevented from developing from a regional storm into a global dust storm. However, these late-type Hellas regional storms are often the precursors for the simulated global-scale dust storms discussed in the next subsection.

GLOBAL STORMS   Simulated dust storms that are truly global in interactive dust cycle modeling studies typically develop from the late-type Hellas regional storms described above. Specifically, these simulated storms are initiated in the Hellas region, begin between $L_s \sim 240°$ and $280°$, and reach their peak global dust loading after southern summer solstice between $L_s \sim 280°$ and $290°$ (Basu et al., 2004, 2006; Kahre et al., 2006, 2008). This is the period when the Hadley cell circulation is strongest, and the ability of these storms to grow to become global in scale is closely tied to the behavior of the overturning circulation. As the simulated storm grows and the circulation is intensified, secondary lifting regions (in Argyre, for example) are initiated in the same latitudinal belt as the initial storm (i.e. the latitude of the rising branch of the Hadley cell, which is at ~30°S during this period). These additional source regions help the storm become global. The strongly positive radiative–dynamic feedback that facilitates the predicted global-scale behavior of these dust storms is present in all interactive modeling studies to date, and was also firmly established in the early literature by investigators using two- and three-dimensional numerical models (Haberle et al., 1982; Murphy et al., 1993, 1995; Wilson, 1997). Although these predicted global storms reproduce some aspects of observed storms, they do not exhibit some important aspects of the observations. In particular, the predicted storms do not cease as rapidly as observed global dust storms at this (or indeed any) time of year. Dust lifting during simulated storms continues until the strong atmospheric circulation patterns naturally decrease in intensity with the advancing time of year (as Mars approaches northern spring equinox), whereas dust lifting during observed storms appears to cease far more abruptly. This behavior has led to the speculation that surface dust may be depleted, causing a rapid shutdown of dust lifting, which is discussed below.

INTER-ANNUAL VARIABILITY   The large amount of inter-annual variability observed in Martian dust storms is not reproduced by all interactive dust simulations that assume an infinite surface dust reservoir (Newman et al., 2002a; Kahre et al., 2005, 2006). Basu et al. (2004, 2006) showed that a delicate tuning of the surface wind stress lifting threshold and efficiency factor is required to produce a degree of inter-annual variability that is qualitatively consistent with observations. They found that relatively high surface wind stress lifting thresholds and high efficiency factors are necessary to spontaneously develop global-scale storms in some simulated years but not in others. These high threshold–high efficiency factor combinations force the predicted dust cycle to be sensitive to the inherent degree of variability in the model atmosphere by sampling only the high-speed tail of the model-predicted wind distribution. While the predicted inter-annual variability is a desirable outcome of the models, the high thresholds required are very likely unrealistic.

10.5.2.2 Finite Surface Dust Reservoirs

Limiting the availability of surface dust has been proposed as one of the controlling factors of the current dust cycle (Zurek et al., 1992; Pankine and Ingersoll, 2004; Szwast et al., 2006; Cantor, 2007). There are many reasons for including a finite surface dust reservoir into dust cycle models. First, observations of surface properties – albedo in particular – indicate that there is very little to no surface dust in some regions of the planet. Second, these observations also indicate regions where dust appears to be removed entirely from certain source regions during the onset or growth phases of storms, suggesting a potential impact on further growth or the timing of decay. Third, simulations that assume an infinite, planet-wide surface dust reservoir do not reproduce the observed degree of variability of sizes and seasonality of predicted dust storms. For example, truly

global-scale simulated dust storms are all initiated due to lifting on the northern rim of Hellas during a limited L_s range. Fourth, infinite-reservoir simulations do not reproduce the rapid shutdown of lifting in particular locations that is often observed in regional- and global-scale dust storms. Instead, regions of active dust lifting (e.g. Hellas) tend to continue until the circulation weakens due to the changing solar insolation with time of year.

(a) **Simulations Without a Threshold Feedback** The simplest method for implementing finite surface reservoirs is to initialize a fully interactive dust cycle model with a constant amount of surface dust chosen such that dust will be completely depleted from some locations during the course of the simulation. The dust lifting parameterizations can then be tuned (and possibly retuned) as the simulation progresses to evaluate how limiting the surface dust reservoir affects the predicted dust cycle. This simple implementation has proved challenging because models thus far do not predict a sustainable, equilibrated dust cycle.

Kahre et al. (2005) show that, once major dust source regions are depleted, they do not subsequently acquire enough dust to become major source regions in future simulated years if no retuning of the lifting parameter is performed. Instead, the dust cycle eventually "winds down" to a state in which no significant dust storms are predicted and the global dust loading is much lower than the observed dust loading. Mulholland et al. (2013) find similar behavior in simulations that include this simple implementation of surface dust reservoirs: the dust cycle eventually becomes much too quiescent as surface dust is sequestered into regions where little lifting occurs. Newman and Richardson (2015) suggest that retuning of the dust lifting parameters is vital, and point out that there is no reason to expect that the dust lifting parameters that produced the best match to the observed dust cycle for an infinite surface supply should be those most appropriate when only a fraction of the surface has dust available and can act as a source. They thus conclude that a lower threshold and/or higher lifting efficiency must be required to produce realistic dust cycles for an equilibrated surface dust distribution (in which surface dust remains only in locations that gain as much as they receive over several Mars years). However, Newman and Richardson (2015) are unable to find such an equilibrated surface dust distribution via an iterative process in which the dust lifting parameters are incrementally retuned as the surface dust distribution rearranges itself.

(b) **Simulations With a Threshold Feedback** The second method for including a finite surface dust reservoir is to impose a variable threshold that depends on the surface dust amount, which produces a negative feedback on lifting. Pankine and Ingersoll (2002, 2004) explored the issue of inter-annual variability of global dust storms using a highly simplified model that represented the axisymmetric Hadley cell and allowed for heating to depend on dustiness, which was coupled to the surface winds. Pankine and Ingersoll (2002) suggested that a feedback between the value of the threshold friction velocity and the atmospheric dust activity could provide a means of "self-tuning", whereby the threshold lifting value could evolve so that a state of intermittent large storms would characterize the inter-annual variability. They argued that, when a surface experienced net deflation, the lifting threshold would be raised above the original value because the non-erodible surface features would shelter the remaining surface dust. (Note that an argument can be made for the reverse effect, in that this would also lead to increased roughness, which could potentially increase the stress experienced by dust, at least initially.) The presence of a process or circulation that provides for the eventual resupply of dust particles would then result in a system with a lifting threshold that hovers about its critical value, enabling aperiodic dust storm behavior. Pankine and Ingersoll (2004) successfully demonstrated this idea with their low-order model, and an initial test of the concept in a GCM simulation was described in Wilson and Kahre (2009). Mulholland et al. (2013) further developed this idea and showed that the inter-annual variability of global dust storms could be simulated with a reasonable frequency of occurrence. However, in both of these studies, an extra dust resupply term was needed in the form of a continuous decrease in the lifting threshold planet-wide. Without such a factor that operated to resupply major source regions on timescales on the order of several Mars years, GCM simulations behaved similarly to the simulations without a threshold feedback, i.e. the dust cycle "wound down" to an unrealistically quiescent state. Although Mulholland et al. (2013) argue that the resupply factor represents unresolved processes in the model, Newman and Richardson (2015) argue that this factor is unphysical, and that it entirely dominates the more physically justifiable parameterized threshold feedback effect.

10.6 OPEN QUESTIONS

It has been known for decades that dust is critically important for the current (and likely past) climate of Mars, but our understanding of the Martian dust cycle has improved significantly over the past ~15 years. This increase in knowledge has been the result of marked improvements in the observational record and in the sophistication of numerical climate models. However, although we now understand many aspects of the Martian dust cycle, many questions remain regarding many of the physical mechanisms that control it. The following five open questions will likely be the focus of both observations and numerical modeling studies for many years to come.

1. What mechanisms are responsible for maintaining the background dust loading and the elevated dust layers during the non-dusty season?

Background Dust Loading. In the absence of regional and/or global dust storms, dust devils and local storms are candidates for the maintenance of the low-abundance background dust haze during the non-dusty season (L_s ~ 0–135°). During this time period, the majority of local storms are observed near the edges of the retreating north and advancing south seasonal CO_2 ice caps (Figure 10.9). Because these storms occur in regions dominated by sinking atmospheric motion (i.e. not near the subsolar point and the rising branch of the Hadley cell), it is unlikely that the dust lifted in these storms is transported very deeply into the atmosphere and therefore it is unlikely that the dust will be efficiently transported to the lower latitudes. Thus,

many investigators have looked to dust devils as the possible major contributor to the background dust haze. Most studies report that the total dust lifted by dust devils during northern spring and summer is of the same order of magnitude as the estimated global deposition rate (Pollack et al., 1979; Landis and Jenkins, 2000; Ferri et al., 2003; Fisher et al., 2005; Cantor et al., 2006; Greeley et al., 2006b); see Balme et al. (2003) for a lower estimate of the dust devil flux. It therefore seems plausible that dust devils could contribute enough dust to produce the observed background dust haze, though it is currently unknown whether local dust storms could also contribute a non-negligible amount, and if so what the relative fraction may be.

Elevated Dust Layers. The recent observations of elevated dust layers at tropical and subtropical latitudes during northern spring and summer complicate our understanding of the non-dusty season. Although there have been recent mesoscale modeling studies of dust storms that can produce elevated dust layers, as discussed in Section 10.3.1.2 these storms appear to be too infrequent in the observational record to be responsible for the observed dust layers. Other possibilities include transport by buoyant thermal plumes that can transport heat, momentum, and aerosols efficiently through the top of the daytime PBL (and which are currently unrepresented in most models; e.g. see Colaïtis et al., 2013), and topographic effects. Further LES and mesoscale investigations such as these will likely be key to understanding and modeling the vertical distribution of dust during the non-dusty season.

2. How do interactions between the dust, water, and CO_2 cycles affect the dust cycle?

Interactions With the Water Cycle. Interactions between the dust and water cycles through cloud formation could affect the dust cycle in two main ways, as discussed in Section 10.5.1.2: cloud scavenging could influence the vertical distribution of dust and the efficiency of dust deposition on the surface, and cloud radiative effects could alter the thermal and dynamical structure of the atmosphere and surface, thereby potentially affecting the spatial extent, seasonality, and rigor of dust lifting and transport. For example, MRO/MCS-observed profiles of dust and clouds in the tropics during the aphelion season show ice layers residing on top of the dust, which may indicate that clouds are confining the vertical extent of dust. Currently there are many unknowns about how these mechanisms may work, and, by extension, which are crucial in affecting the present (and past) dust cycle. GCMs that include water ice cloud microphysics packages are poised to address several crucial questions, including these: How do water ice clouds control the vertical distribution of atmospheric dust? (For example, do clouds play a role in producing elevated dust layers?) Do the radiative effects of water ice clouds significantly affect where, when, and how much dust is lifted from the surface? And do the interactions between the dust and water cycles play a role in the inter-annual variability of global dust storms?

Interactions With the CO_2 Cycle. While it is clear that dust in the atmosphere and on the surface affects CO_2 condensation and sublimation, it is unclear how feedbacks between the two cycles could affect the dust cycle in particular. For example, it is unknown if year-to-year variability in dust lifting (particularly in the vicinity of the polar caps) can result from small variations that have been observed in the retreat rate of regions of the cap following a regional or global dust storm (e.g. Bonev et al., 2008). While GCMs have historically included the effects of dust to study the CO_2 cycle (e.g. Pollack et al., 1990; Hourdin et al., 1995; Forget et al., 1999; Kahre and Haberle, 2010b), very little focus has been given to understanding how the couplings affect the dust cycle. Addressing these questions with GCMs will likely require the implementation of CO_2 ice cloud microphysics and CO_2 cloud radiative effects in combination with fully interactive dust cycle simulations.

3. What mechanisms control the initiation, growth, and decay of dust storms?

Although it seems clear that feedbacks are likely critically important for the initiation, expansion, and shut-off of dust storms at a range of spatial and temporal scales, exactly how these feedbacks work is not well understood. As discussed in Section 10.4.2.3, positive radiative–dynamic feedbacks between dust lifting, atmospheric heating, and the strengthened atmospheric circulation leading to further dust lifting are likely key to the expansion of dust storms, but it is unclear why some storms grow while others do not. While positive feedbacks can readily be identified in modeling studies for storm growth, it has been difficult to identify negative feedbacks that lead to storm cessation, particularly of regional and global dust storms. It has been hypothesized that the increase in near-surface atmospheric stability accompanying substantial atmospheric dust loads should negatively affect dust lifting, but models have yet to produce this behavior with surface wind stress parameterizations for dust lifting (a negative feedback does exist for dust devil lifting parameterizations, however). Other possibilities for dust storm cessation are the depletion of the surface dust reservoir in major source regions and/or feedbacks between surface dust and the threshold for surface wind stress lifting. The role of the surface dust reservoir is discussed further below.

4. What mechanisms control the inter-annual variability of global dust storms?

The inter-annual variability of global dust storms is one of the most puzzling phenomena of the Mars dust cycle, and – via the influence of dust on the entire circulation – the current Martian climate. The occurrence of one (or, in some cases, more than one) global dust storm in some years but not in others could be due either to processes that operate in the atmosphere or to hysteresis that resides on the surface in the form of the redistributable surface dust reservoir, or both.

Atmosphere. It has been challenging to simulate the inter-annual variability of global dust storms solely due to atmospheric processes because models do not generally display a large degree of inherent inter-annual variability. As discussed in Section 10.5.2.1, increased variability of global dust storms has been produced by utilizing high surface wind stress thresholds that sample the high-speed tail of the model-resolved wind distribution, which is the most likely to exhibit year-to-year variability. However, thresholds required are very likely unrealistically high. Another potential way in which variability could

be introduced into the atmospheric system is via the coupling of the dust, water, and CO_2 cycles, though the effect of these couplings on inter-annual variability has gone largely unexplored to date. Long-duration, fully coupled dust, water, and CO_2 cycle simulations will be needed to understand whether interactions between the cycles produces significant year-to-year variability.

Surface. It is possible that the hysteresis resides on the surface in the form of the redistributable surface dust reservoir. As discussed in Section 10.5.2.2, it has been proposed that the depletion of dust and/or the change in the surface wind stress threshold in critical dust source regions can lead to year-to-year variability of global dust storms. Simulating the dust cycle with finite surface dust reservoirs has proven challenging, likely due to many unknown aspects of the combined surface–atmosphere system. For example, the accurate, self-consist prediction of the spatial and temporal pattern of dust in the atmosphere and on the surface may require both the simulation of an accurate atmospheric circulation (which may not even be possible *a priori* since the dust distribution is so coupled to it) and a perfect knowledge of the current surface dust (and possibly sand) distribution. Additionally, there are uncertainties about the physics of dust lifting itself, including variations in grain type and/or cohesiveness that could affect the surface wind stress threshold both spatially and over time through feedbacks between the threshold and dust lifting and deposition. The many challenges involved in simulating the Martian dust cycle with finite surface dust reservoirs will undoubtedly continue to be addressed in future studies.

5. What is the current global dust budget and how has it evolved over time?

There are several critical and largely unknown questions regarding how the global dust budget is currently evolving, including whether there is an annual net transfer of dust from one hemisphere to the other, whether there are semi-permanent regions of net annual depletion or accumulation, and whether or not the dust cycle is "closed" on seasonal, annual, multi-annual, or orbital change timescales. Observations give us only limited insight into some of these questions. For example, as discussed in Section 10.4.2.3, detailed analysis of changes in the observed surface albedo during years with and without global dust storms indicate that the movement of dust on the surface is complex, with both local and large-scale dust storms moving dust between locations on a range of timescales. Fully interactive dust cycle models must be relied upon to provide the answers to these questions, which highlights the need for continued dust cycle modeling studies and model development. Until models are capable of capturing reality more completely, these significantly important questions will remain open.

REFERENCES

Ajello, J. M., Pang, K. D., Lane, A. L., Hord, C. W., Simmons, K. E., 1976. Mariner 9 ultraviolet spectrometer experiment – bright-limb observations of the lower atmosphere of Mars. *Journal of the Atmospheric Sciences*, 33, 544–552.

Anderson, E., Leovy, C., 1978. Mariner 9 television limb observations of dust and ice hazes on Mars. *Journal of the Atmospheric Sciences*, 35, 723–734.

Antoniadi, E. M., 1915. Report of the section for the observation of Mars, 1909. *Mem. Br. Astron. Assoc.* 20, 25–92.

Antoniadi, E. M., 1930. *La Planète Mars*, 1659–1929, Herman et Cie, Paris. (Translated P. Moore, *The Planet Mars*, Keith Reid, Shaldon, Devon, UK, 1975).

Arvidson, R. E., Guinness, E. A., Moore, H. J., Tillman, J., Wall, S. D. 1983. Three Mars years – Viking Lander 1 imaging observations. *Science*, 222, 463–468.

Arvidson, R. E., Anderson, R. C., Bartlett, P., et al., 2004. Localization and physical property experiments conducted by Opportunity at Meridiani Planum. *Science*, 306 (5702), 1730–1733.

Bagnold, R. A., 1936. The movement of desert sand. *Proceedings of the Royal Society of London. Series A, Mathematical and Physical Sciences*, 157 (892), 594–620.

Bagnold, R. A., 1937. The transport of sand by wind. *The Geographical Journal* 89(5), 409–438.

Bagnold, R. A., 1941. *The Physics of Blown Sand and Desert Dunes.* London: Methuen, 265.

Bagnold, R. A., 1954. Experiments on a gravity-free dispersion of large solid spheres in a Newtonian fluid under shear. *Proceedings of the Royal Society of London. Series A, Mathematical and Physical Sciences*, 225 (1160), 49–63.

Bagnold, R. A., 1956. The flow of cohesionless grains in fluids, *Proc. R. Soc. Lond.*, 249, 235–297.

Balme, M., Greeley, R., 2006. Dust devils on Earth and Mars, *Reviews of Geophysics*, 44 (3).

Balme, M., Hagermann, A., 2006. Particle lifting at the soil-air interface by atmospheric pressure excursions in dust devils. *Geophysical Research Letters*, 33 (19).

Balme, M. R., Whelley, P. L., Greeley, R., 2003. Mars: dust devil track survey in Argyre Planitia and Hellas Basin. *J. of Geophys. Res.*, 108, doi:10.1029/2003JE002096.

Barnes, J. R., Pollack, J. B., Haberle, R. M., et al., 1993. Mars atmospheric dynamics as simulated by the NASA AMES general circulation model. II – Transient baroclinic eddies. *Journal of Geophysical Research*, 98 (E2), 3125–3148.

Basu, S., Richardson, M. I., Wilson, R. J., 2004. Simulation of the Martian dust cycle with the GFDL Mars GCM. *Journal of Geophysical Research*, 109 (E11).

Basu, S., Wilson, J., Richardson, M., Ingersoll, A., 2006. Simulation of spontaneous and variable global dust storms with the GFDL Mars GCM. *Journal of Geophysical Research*, 111 (E9).

Baum, W. A., 1973. The International Planetary Patrol Program: an assessment of the first three years. *Planetary and Space Science*, 21 (9), 1511–1519.

Bell, J. F., Wolff, M. J., James, P. B., et al., 1997. Mars surface mineralogy from Hubble Space Telescope imaging during 1994–1995: observations, calibration, and initial results, *Journal of Geophysical Research*, 102 (E4), 9109–9124.

Biener, K. K., Geissler, P. E., McEwen, A. S., Leovy, C., 2002. Observations of Martian dust devils in MOC wide angle camera images. In *33th Annual Lunar and Planetary Science Conference*, March 11–15, 2002, Houston, TX, abstract no. 2004.

Bonev, B. P., Hansen, G. B., Glenar, D. A., James, P. B., Bjorkman, J. E., 2008. Albedo models for the residual south polar cap on Mars: implications for the stability of the cap under near-perihelion global dust storm conditions, *Planetary and Space Science*, 56 (2), 181–193.

Briggs, G. A., Baum, W. A., Barnes, J., 1979. Viking Orbiter imaging observations of dust in the Martian atmosphere, *Journal of Geophysical Research*, 84 (10), 2795–2820.

Cantor, B. A., 2007. MOC observations of the 2001 Mars planet-encircling dust storm, *Icarus*, 186 (1), 60–96.

Cantor, B. A., James, P. B., Caplinger, M., Wolff, M. J., 2001. Martian dust storms: 1999 Mars Orbiter Camera observations, *Journal of Geophysical Research*, 106 (E10), 23653–23688.

Cantor, B. A., Malin, M., Edgett, K. S., 2002. Multiyear Mars Orbiter Camera (MOC) observations of repeated Martian weather phenomena during the northern summer season, *Journal of Geophysical Research (Planets)*, 107 (E3), doi:10.1029/2001JE001588.

Cantor, B. A., Kanak, K. M., Edgett, K. S., 2006. Mars Orbiter Camera observations of Martian dust devils and their tracks (September 1997 to January 2006) and evaluation of theoretical vortex models, *Journal of Geophysical Research*, 111 (E12), doi:10.1029/2006JE002700

Cantor, B. A., Malin, M. C., Wolff, M. J., et al., 2008. Observations of the Martian atmosphere by MRO-MARCI, an overview of 1 Mars year. In *Third International Workshop on The Mars Atmosphere: Modeling and Observations*, November 10–13, 2008, Williamsburg, Virginia. LPI Contribution No. 1447, 9075.

Cantor, B. A., James, P. B., Calvin, W. M., 2010. MARCI and MOC observations of the atmosphere and surface cap in the north polar region of Mars, *Icarus*, 208 (1), 61–81.

Capen, C. F., Martin, L. J., 1971. The developing stages of the Martian yellow storm of 1971, *Lowell Observatory Bulletin No. 157*, VII, 20, 211–216.

Chassefiere, E., Blamont, J. E., Krasnopolsky, V. A., et al., 1992. Vertical structure and size distributions of Martian aerosols from solar occultation measurements, *Icarus*, 97 (1), 46–69.

Chepil, W. S., Woodruff, N. P., 1963. The physics of wind erosion and its control, *Advances in Agronomy*, 15, 211–302.

Cheremisin, A. A., Vassilyev, Y. V., Horvath, H., 2005. Gravito-photophoresis and aerosol stratification in the atmosphere, *J. Aerosol Sci.*, 36, 1277–1299.

Choi, D. S., Dundas, C. M., 2011. Measurements of Martian dust devils winds with HiRISE. *Geophysical Research Letters*, 38 (24), doi:10.1029/2011GL049806.

Christensen, P. R., 1982. Martian dust mantling and surface composition – interpretation of thermophysical properties, *Journal of Geophysical Research*, 87, 9985–9998.

Christensen, P. R., 1986a. The spatial distribution of rocks on Mars, *Icarus* 68, 217–238.

Christensen, P. R., 1986b. Regional dust deposits on Mars – physical properties, age, and history, *Journal of Geophysical Research*, 91, 3533–3545.

Christensen, P. R., 1988. Global albedo variations on Mars – implications for active aeolian transport, deposition, and erosion, *Journal of Geophysical Research*, 93, 7611–7624.

Christensen, P. R., Bandfield, J. L., Hamilton, V. E., et al., 2001. The Mars Global Surveyor Thermal Emission Spectrometer experiment: investigation description and surface science results, *J. Geophys. Res.*, 106, 23823–23871.

Clancy, R. T., Sandor, B. J., Wolff, M. J., et al., 2000. An intercomparison of ground-based millimeter, MGS TES, and Viking atmospheric temperature measurements: seasonal and interannual variability of temperatures and dust loading in the global Mars atmosphere, *Journal of Geophysical Research*, 105 (E4), 9553–9572.

Clancy, R. T., Wolff, M. J., Christensen, P. R., 2003. Mars aerosol studies with the MGS TES emission phase function observations: optical depths, particle sizes, and ice cloud types versus latitude and solar longitude. *J. of Geophys. Res.* 108, doi:10.1029/2003JE002058.

Clancy, R. T., Wolff, M. J., Whitney, B. A., et al., 2010. Extension of atmospheric dust loading to high altitudes during the 2001 Mars dust storm: MGS TES limb observations. *Icarus*, 207 (1), 98–109.

Colaprete, A., Toon, O. B., 2003. Carbon dioxide clouds in an early dense Martian atmosphere, *Journal of Geophysical Research (Planets)*, 108 (E4), doi:10.1029/2002JE001967.

Colaïtis, A., Spiga, A., Hourdin, F., et al., 2013. A thermal plume model for the Martian convective boundary layer, *Journal of Geophysical Research: Planets*, 118 (7), 1468–1487.

Colburn, D. S., Pollack, J. B., Haberle, R. M., 1989. Diurnal variations in optical depth at Mars, *Icarus* 79, 159–189.

Conrath, B. J., 1975. Thermal structure of the Martian atmosphere during the dissipation of the dust storm of 1971, *Icarus*, 24, 36–46.

Cushing, G. E., Titus, T. N., Christensen, P. R., 2005. THEMIS VIS and IR observations of a high-altitude Martian dust devil, *Geophysical Research Letters*, 32 (23), doi:10.1029/2005GL024478.

Desch, S. J., Cuzzi, J. N., 2000. The generation of lightning in the solar nebula, *Icarus*, 143, 87–105.

Dickinson, C., Komguem, L., Whiteway, J. A., et al., 2011. Lidar atmospheric measurements on Mars and Earth. *Planetary and Space Science*, 59 (10), 942–951.

Dlugach, Z. M., Korablev, O. I., Morozhenko, A. V., et al., 2003. Physical properties of dust in the Martian atmosphere: analysis of contradictions and possible ways of their resolution, *Solar System Research*, 37 (1), 1–19.

Douglass, A. E., 1899. Mars, January 1899. *Popular Astronomy*, 7, 113–117.

Drube, L., Leer, K. Goetz, W., et al., 2010. Magnetic and optical properties of airborne dust and settling rates of dust at the Phoenix landing site. *J. Geophys. Res.* 115, E00E23. doi:10.1029/2009JE003419

Ellehoj, M. D., Gunnlaugsson, H. P., Taylor, P. A., et al., 2010. Convective vortices and dust devils at the Phoenix Mars mission landing site. *J. Geophys. Res.*, 115, E00E16. doi:10.1029/2009JE003413.

Farrell, W. M., Marshall, J. R., Cummer, S. A., Delory, G. T., Desch, M. D., 2006. A model of the ULF magnetic and electric field generated from a dust devil, *Journal of Geophysical Research*, 111 (E11). doi:10.1029/2006JE002689.

Fedorova, A., Korablev, O., Bertaux, J.-L., et al., 2009. Solar infrared occultations by the SPICAM experiment on Mars Express: simultaneous observations of H_2O, CO_2 and aerosol vertical distribution. *Icarus* 200 (1), 96–117.

Fenton, L. K., Pearl, J. C., Martin, T. Z., 1997. Mapping Mariner 9 Dust Opacities, *Icarus*, 130 (1), 115–124.

Ferri, F., Smith, P. H., Lemmon, M., Rennó, N. O., 2003. Dust devils as observed by Mars Pathfinder, *Journal of Geophysical Research*, 108 (E12), doi:10.1029/2000JE001421.

Fisher, J. A., Richardson, M. I., Newman, C. E., et al., 2005. A survey of Martian dust devil activity using Mars Global Surveyor Mars Orbiter Camera images, *Journal of Geophysical Research*, 110 (E3), doi:10.1029/2003JE002165.

Flynn, G. J., 1992. The Contribution of Meteoritic Material to the Dust and Aerosols in the Atmosphere of Mars, *Abstracts of the Lunar and Planetary Science Conference*, 23, 371.

Forget, F., Hourdin, F., Fournier, R., et al., 1999. Improved general circulation models of the Martian atmosphere from the surface to above 80 km, *Journal of Geophysical Research*, 104 (E10), 24155–24176.

Fouchet, T., Bèzard, B., Drossart, P., et al., 2006. OMEGA limb observations of the Martian dust and atmospheric composition. *Second Workshop on Mars Atmosphere Modelling and Observations*, February 27–March 3, Granada, Spain, 223.

Gheynani, B. T., Taylor, P. A., 2011. Large eddy simulation of typical dust devil-like vortices in highly convective Martian boundary layers at the Phoenix Lander site, *Planetary and Space Science*, 59 (1), 43–50.

Gierasch, P. J., Goody, R. M., 1972. The effect of dust on the temperature of the Martian atmosphere, *Journal of Atmospheric Science*, 29, 400–402.

Gierasch, P. J., Goody, R. M., 1973. A model of a Martian great dust storm, *Journal of Atmospheric Science*, 30, 169–179.

Gierasch, P. J., Thomas, P., French, R., Veverka, J., 1979. Spiral clouds on Mars: a new atmospheric phenomenon. *Geophys, Res. Lett.* 6, 405–408.

Goetz, W., Bertelsen, P., Binau, C. S., et al., 2005. *Nature*, 436 (7047), 62–65.

Golombek, M. P., Grant, J. A., Crumpler, L. S., et al., 2006. Erosion rates at the Mars Exploration Rover landing sites and long-term climate change on Mars. *Journal of Geophysical Research*, 111 (E12), doi:10.1029/2006JE002754.

Grassi, D., Ignatiev, N. I., Zasova, L. V., et al., 2005. Methods for the analysis of data from the Planetary Fourier Spectrometer on the Mars Express Mission, *Planetary and Space Science*, 53 (10), 1017–1034.

Greeley, R., 1979. Silt clay aggregates on Mars, *J. Geophys. Res.*, 84, 6248–6254.

Greeley, R., Iversen, J. D., 1985. *Wind as a geological process on Earth, Mars, Venus and Titan*, Cambridge Planetary Science Series, Vol. 4. Cambridge University Press, Cambridge.

Greeley, R., Iversen, J. D., 1987. Measurements of wind friction speeds over lava surfaces and assessment of sediment transport. *Geophysical Research Letters*, 14 (9), 925–928.

Greeley, R., Leach, R., 1978. A preliminary assessment of the effects of electrostatics on aeolian processes, in *Reports of Planetary Geology Program*, 1978–1979, NASA TM-79729, 236–237.

Greeley, R., Leach, R., 1979. "Steam" injection of dust on Mars: laboratory simulation, in *Reports of Planetary Geology Program*, 1978–1979, NASA TM-80339, 304–307.

Greeley, R., Skypeck, A., Pollack, J. B., 1993. Martian aeolian features and deposits – comparisons with general circulation model results, *Journal of Geophysical Research*, 98 (E2) 3183–3196.

Greeley, R., Wilson, G., Coquilla, R., White, B., Haberle, R., 2000. Windblown dust on Mars: laboratory simulations of flux as a function of surface roughness, *Planetary and Space Science*, 48 (12–14), 1349–1355.

Greeley, R., Arvidson, R., Bell, J. F., et al., 2005. Martian variable features: new insight from the Mars Express Orbiter and the Mars Exploration Rover Spirit. *J. Geophys. Res.* 110, E06002, doi:10.1029/ 2005JE002403.

Greeley, R., Arvidson, R. E., Bartlett, P. W., et al., 2006a. Gusev Crater: wind-related features and processes observed by the Mars Exploration Rover, Spirit. *J. Geophys. Res.* 111, E02S09, doi:10.1029/2005JE002491.

Greeley, R., Whelley, P. L., Arvidson, R. E., et al., 2006b. Active dust devils in Gusev Crater, Mars: observations from the Mars Exploration Rover, Spirit. *J. Geophys. Res.* 111, E12S09, doi:10.1029/2006JE002743.

Greeley, R., Waller, D. A., Cabrol, N. A., et al., 2010. Gusev Crater, Mars: observations of three dust devil seasons, *Journal of Geophysical Research*, 115 (E8), doi:10.1029/2010JE003608.

Greybush, S. J., Wilson, R. J., Hoffman, R. N., et al., 2012. Ensemble Kalman filter data assimilation of Thermal Emission Spectrometer temperature retrievals into a Mars GCM, *Journal of Geophysical Research*, 117 (E11) doi:10.1029/2012JE004097

Gu, Z., Wei, W., Zhao, Y., 2010. An overview of surface conditions in numerical simulations of dust devils and the consequent near-surface air flow fields, *Aerosol and Air Quality Research*, 10, 272–281.

Guinness, E. A., Arvidson, R. E., Gehret, D. C., Bolef, L. K., 1979. Color changes at the Viking landing sites over the course of a Mars year, *Journal of Geophysical Research*, 84, 8355–8364.

Guinness, E. A., Leff, C. E., Arvidson, R. E., 1982. Two Mars years of surface changes seen at the Viking Landing sites, *Journal of Geophysical Research*, 87, 10051–10058.

Guzewich, S. D., Toigo, A. D., Richardson, M. I., et al., 2013a. The impact of a realistic vertical dust distribution on the simulation of the Martian general circulation, *Journal of Geophysical Research: Planets*, 118 (5) 980–993.

Guzewich, S. D., Talaat, E. R., Toigo, A. D., Waugh, D. W., McConnochie, T. H., 2013b. High-altitude dust layers on Mars: observations with the Thermal Emission Spectrometer, *J. Geophys. Res. Planets*, 118, 1177–1194, doi:10.1002/jgre.20076.

Guzewich, S. D., Wilson, R. J., McConnochie, T. H., et al., 2014. Thermal tides during the 2001 Martian global-scale dust storm. *Journal of Geophysical Research*, 119(3), doi:10.1002/2013JE004502.

Haberle, R. M., Leovy, C. B., Pollack, J. B., 1982. Some effects of global dust storms on the atmospheric circulation of Mars. *Icarus*, 50, 322–367.

Haberle, R. M., Pollack, J. B., Barnes, J. R., et al., 1993. Mars atmospheric dynamics as simulated by the NASA AMES general circulation model. I – The zonal-mean circulation, *Journal of Geophysical Research*, 98 (E2), 3093–3123.

Haberle, R. M., Murphy, J. R., Schaeffer, J., 2003. Orbital change experiments with a Mars general circulation model, *Icarus*, 161 (1), 66–89.

Haberle, R. M., Gómez-Elvira, J., Torre Juárez, M., et al., 2014. Preliminary interpretation of the REMS pressure data from the first 100 sols of the MSL mission. *Journal of Geophysical Research: Planets*, 119 (3), 440–453.

Hamilton, V. E., McSween, H. Y., Hapke, B., 2005. Mineralogy of Martian atmospheric dust inferred from thermal infrared spectra of aerosols, *Journal of Geophysical Research*, 110, E12, doi:10.1029/2005JE002501.

Hanel, R., Conrath, B., Hovis, W., et al., 1972. Investigation of the Martian environment by infrared spectroscopy on Mariner 9, *Icarus*, 17, 423.

Hansen, J. E., Travis, L. D., 1974. Light scattering in planetary atmospheres, *Space Science Reviews*, 16, 527–610.

Hartmann, W. K., Price, M. J., 1974. Mars: clearing of the 1971 dust storm, *Icarus*, 21, 28.

Hayne, P. O., Paige, D. A., Schofield, J. T., et al., 2012. Carbon dioxide snow clouds on Mars: south polar winter observations by the Mars Climate Sounder, *Journal of Geophysical Research*, 117 (E8).

Heavens, N. G., Richardson, M. I., Kleinböhl, A., et al., 2011a. The vertical distribution of dust in the Martian atmosphere during northern spring and summer: observations by the Mars Climate Sounder and analysis of zonal average vertical dust profiles, *Journal of Geophysical Research*, 116 (E4), doi:10.1029/2010JE003691.

Heavens, N. G., Richardson, M. I., Kleinböhl, A., et al. 2011b. Vertical distribution of dust in the Martian atmosphere during northern spring and summer: high-altitude tropical dust maximum at northern summer solstice, *Journal of Geophysical Research*, 116 (E1).

Heavens, N. G., Johnson, M. S., Abdou, W. A., et al., 2014. Seasonal and diurnal variability of detached dust layers in the tropical Martian atmosphere. *Journal of Geophysical Research*, 119(8), doi:10.1002/2014JE004619.

Herr, K. C., Forney, P. B., Pimentel, G. C., 1998. *Mariner Mars 6/7 Infrared Spectrometers: lab simulation of Mars spectra*, in *29th Annual Lunar and Planetary Science Conference*, March 16–20, 1998, Houston, TX, abstract no. 1518.

Hess, S., 1973. Martian winds and dust clouds, *Planetary and Space Science*, 21 (9), 1549–1557.

Hinson, D. P., Wang, H., 2010. Further observations of regional dust storms and baroclinic eddies in the northern hemisphere of Mars, *Icarus*, 206 (1), 290–305.

Hollingsworth, J. L., Kahre, M. A., 2010. Extratropical cyclones, frontal waves, and Mars dust: modeling and considerations. *Geophys. Res. Letters*, 37, doi:10.1029/2010GL044262.

Hollingsworth, J. L., Haberle, R. M., Barnes, J. R., et al., 1996. Orographic control of storm zones on Mars, *Nature*, 380 (6573), 413–416.

Hourdin, F., Le Van, P., Forget, F., Talagrand, O., 1993. Meteorological Variability and the Annual Surface Pressure Cycle on Mars, *Journal of Atmospheric Sciences*, 50 (21), 3625–3640.

Hourdin, F., Forget, F., Talagrand, O., 1995. The sensitivity of the Martian surface pressure and atmospheric mass budget to various parameters: a comparison between numerical simulations and Viking observations, *Journal of Geophysical Research*, 100 (E3), 5501–5523.

Hu, R., Cahoy, K., Zuber, M. T., 2012. Mars atmospheric CO_2 condensation above the north and south poles as revealed by radio occultation, climate sounder, and laser ranging observations, *Journal of Geophysical Research*, 117 (E7).

Huguenin, R. L., Clifford, S. M., 1979. Mars: origin of the global dust storms. *Bulletin of the American Astronomical Society*, 11, 578.

Hunt, G. E., James, P. B, 1979. Martian extratropical cyclones, *Nature*, v278, 531–532.

Inada, A., Richardson, M. I., McConnochie, T. H., et al., 2007. High-resolution atmospheric observations by the Mars Odyssey Thermal Emission Imaging System, *Icarus*, 192 (2), 378–395.

Jakosky, B. M., 1986. On the thermal properties of Martian fines, *Icarus*, 66, 117–124.

James, P. B., 1985. Martian local dust storms, *Recent advances in planetary meteorology*. Cambridge and New York, Cambridge University Press, 85–99.

James, P. B., Hollingsworth, J. L., Wolff, M. J., Lee, S. W., 1999. North Polar Dust Storms in Early Spring on Mars, *Icarus*, 138 (1), 64–73.

Jaquin, F., Gierasch, P., Kahn, R., 1986. The vertical structure of limb hazes in the Martian atmosphere, *Icarus*, 68, 442–461.

Jianjun, Q., Yan, M., Dong, G., et al., 2004. Wind tunnel simulation experiment and investigation on the electrification of sandstorms, *Sci. China, Ser. D*, 47, 529–539.

Johnson, D. W., Harteck, P., Reeves, R. R., 1975. Dust injection into the Martian atmosphere, *Icarus*, 26, 441–443.

Johnson, J. R., Grundy, W. M., Lemmon, M. T., 2003. Dust deposition at the Mars Pathfinder landing site: observations and modeling of visible/near-infrared spectra, *Icarus*, 163 (2), 330–346, doi:10.1016/S0019-1035(03)00084-8.

Joshi, M. M., Lewis, S. R., Read, P. L., Catling, D. C., 1995. Western boundary currents in the Martian atmosphere: numerical simulations and observational evidence, *Journal of Geophysical Research*, 100 (E3), 5485–5500.

Joshi, M. M., Haberle, R. M., Barnes, J. R., Murphy, J. R., Schaeffer, J., 1997. Low-level jets in the NASA Ames Mars general circulation model, *Journal of Geophysical Research*, 102 (E3), 6511–6524.

Kahanpää, H., de la Torre Juarez, M., Moores, J., et al., 2013. EGU General Assembly 2013, held 7–12 April, 2013 in Vienna, Austria, id. EGU2013-9455.

Kahn, R. A., Martin, T. Z., Zurek, R. W., Lee, S. W., 1992. The Martian dust cycle, in *Mars*, ed. H. Kieffer et al., Univ. Arizona Press, Tucson, 1017–1053.

Kahre, M. A., Haberle, R. M., 2010. Mars CO_2 cycle: effects of airborne dust and polar cap ice emissivity, *Icarus*, 207 (2), 648–653

Kahre, M. A., Murphy, J. R., Haberle, R. M., Montmessin, F., Schaeffer, J., 2005. Simulating the Martian dust cycle with a finite surface dust reservoir, *Geophysical Research Letters*, 32 (20), doi:10.1029/2005GL023495.

Kahre, M. A., Murphy, J. R., Haberle, R. M., 2006. Modeling the Martian dust cycle and surface dust reservoirs with the NASA Ames general circulation model. *J. of Geophys. Res.* 111. doi:10.1029/2005JE002588

Kahre, M. A., Hollingsworth, J. L., Haberle, R. M., Murphy, J. R., 2008. Investigations of the variability of dust particle sizes in the Martian atmosphere using the NASA Ames General Circulation Model. *Icarus*, 195, 576–597.

Kahre, M. A., Wilson, R. J., Hollingsworth, J. L., Haberle, R. M., 2010. Using Assimilation Techniques To Model Mars' Dust Cycle With The NASA Ames And NOAA/GFDL Mars General Circulation Models, American Astronomical Society, DPS meeting #42, #30.19, *Bulletin of the American Astronomical Society*, 42, 1031.

Kahre, M. A., Hollingsworth, J. L., Haberle, R. M., Montmessin, F., 2011. Coupling Mars' dust and water cycles: effects on dust lifting vigor, spatial extent and seasonality, *The Fourth International Workshop on the Mars Atmosphere: Modelling and Observation*, 8–11 February, 2011, Paris, France, 143–146.

Kahre, M. A., Hollingsworth, J. L., Haberle, R. M., Wilson, R. J., 2015. Coupling the Mars dust and water cycles: the importance of radiative-dynamic feedbacks during northern hemisphere summer, *Icarus*, 260, 477–480.

Kass, D. M., Kleinböhl, A., McCleese, D. J., Schofield, J. T. and Smith, M. D., 2016. Interannual similarity in the Martian atmosphere during the dust storm season. *Geophysical Research Letters*, 43(12), 6111–6118.

Kaufmann, E., Kömle, N. I., Kargl, G., 2006. Laboratory simulation experiments on the solid-state greenhouse effect in planetary ices, *Icarus*, 185 (1), 274–286.

Kavulich, M. J., Szunyogh, I., Gyarmati, G., Wilson, R. J., 2013. Local dynamics of baroclinic waves in the Martian atmosphere, *J. Atmos. Sci.*, 70, 3415–3447.

Kieffer, H. H., 2007. Cold jets in the Martian polar caps, *Journal of Geophysical Research*, 112 (E8), doi:10.1029/2006JE002816.

Kieffer, H. H., Martin, T. Z., Peterfreund, A. R., Jakosky, B. M., Miner, E. D., Palluconi, F. D., 1977. Thermal and albedo mapping of Mars during the Viking primary mission, *Journal of Geophysical Research*, 82, 4249–4291.

Kinch, K. M., Sohl-Dickstein, J., Bell, J. F., et al., 2007. Dust deposition on the Mars Exploration Rover Panoramic Camera (Pancam) calibration targets, *Journal of Geophysical Research*, 112 (E6), doi:10.1029/2006JE002807.

Kjelgaard, J. F., Chandler, D. G., Saxton, K. E., 2004. Evidence for direct suspension of loessial soils on the Columbia Plateau, *Earth Surface Processes and Landforms*, 29 (2), 221–236.

Kok, J. F., 2010. An improved parameterization of wind-blown sand flux on Mars that includes the effect of hysteresis, *Geophysical Research Letters*, 37 (12), doi:10.1029/2010GL043646.

Kok, J. F., Rennó, N. O., 2006. Enhancement of the emission of mineral dust aerosols by electric forces, *Geophysical Research Letters*, 33 (19).

Komguem, L., Whiteway, J. A., Dickinson, C., Daly, M., Lemmon, M. T., 2013. Phoenix LIDAR measurements of Mars atmospheric dust, *Icarus*, 223(2), 649–653.

Kuiper, G. P., 1957. Visual Observations of Mars, 1956, *Astrophysical Journal*, 125, 307.

Kuroda, T., Medvedev, A. S., Hartogh, P., Takahashi, M., 2008. Semiannual oscillations in the atmosphere of Mars, *Geophysical Research Letters*, 35 (23), doi:10.1029/2008GL036061.

Landis, G. A., Jenkins, P. P., 2000. Measurement of the settling rate of atmospheric dust on Mars by the MAE instrument on Mars Pathfinder, *Journal of Geophysical Research*, 105 (E1), 1855–1858.

Lecacheux, J., Drossart, P., Buil, C., et al., 1991. CCD images of Mars with the 1 m reflector atop Pic-du-Midi, Proceedings of Colloquium on Phobos-Mars Mission, Paris, France, Oct. 23–27, 1989, A91-29558 11-91. *Planetary and Space Science*, 39, 273–279.

Lee, S. W., 1987. Regional sources and sinks of dust on Mars: Viking observations of Cerberus, SOLIS Planum and Syrtis Major, in *Lunar and Planetary Inst., MECA Symposium on Mars: Evolution of its Climate and Atmosphere*, 71–72.

Lemmon, M. T., Wolff, M. J., Smith, M. D., et al., 2004. Atmospheric Imaging Results from the Mars Exploration Rovers: Spirit and Opportunity, *Science*, 306 (5702), 1753–1756, doi:10.1126/science.1104474.

Lemmon, M. T., Wolff, M. J., Bell, J. F., et al., 2014. Dust aerosol, clouds, and the atmospheric optical depth record over 5 Mars years of the Mars Exploration Rover mission. *Icarus*, doi:10.1016/j.icarus.2014.03.029

Leovy, C. B., 1981. Observations of Martian tides over two annual cycles. *J. Atmos. Sci.*, 38, 30–39.

Leovy, C., Mintz, Y., 1969. Numerical simulation of the atmospheric circulation and climate of Mars, *Journal of Atmospheric Sciences*, 26 (6), 1167–1190.

Leovy, C. B., Smith, B. A., Young, A. T., Leighton, R. B., 1971. Mariner Mars 1969: atmospheric results, *Journal of Geophysical Research*, 76, 297–312.

Leovy, C. B., Zurek, R. W., Pollack, J. B., 1973. Mechanisms for Mars dust storms, *Journal of Atmospheric Science*, 30, 749–762.

Lewis, S. R., Collins, M., Read, P. L., et al., 1999. A climate database for Mars, *Journal of Geophysical Research*, 104 (E10), 24177–24194.

Lewis, S. R., Read, P. L., Conrath, B. J., Pearl, J. C., Smith, M. D., 2007. Assimilation of thermal emission spectrometer atmospheric data during the Mars Global Surveyor aerobraking period, *Icarus*, 192 (2), 327–347.

Lian, Y., Richardson, M. I., Newman, C. E., et al., 2012. The Ashima/MIT Mars GCM and argon in the Martian atmosphere, *ICARUS*, 218, 1043–1070, doi:10.1016/j.icarus.2012.02.012

Liu, J., Richardson, M. I., Wilson, R. J., 2003. An assessment of the global, seasonal, and interannual spacecraft record of Martian climate in the thermal infrared, *Journal of Geophysical Research*, 108 (E8), doi:10.1029/2002JE001921.

Loosmore, G. A., and Hunt, J. R., 2000. Dust resuspension without saltation, *J. Geophys. Res.* 105, 20663–20672.

Lorenz, R. D., Lunine, J. I., Grier, J. A., and Fisher, M. A., 1995. Prediction of aeolian features on planets: application to Titan paleoclimatology, *Journal of Geophysical Research*, 100 (E12), 26377–26386.

Lu, H., Raupach, M. R., and Richards, K. S. (2005), Modeling entrainment of sedimentary particles by wind and water: a generalized approach, *J. Geophys. Res.*, 110, D24114, doi:10.1029/2005JD006418.

Määttänen, A., Vehkamäki, H., Lauri, A., et al., 2005. Nucleation studies in the Martian atmosphere, *Journal of Geophysical Research*, 110 (E2), doi:10.1029/2004JE002308.

Määttänen, A., Fouchet, T., Forni, O., et al., 2009. A study of the properties of a local dust storm with Mars Express OMEGA and PFS data, *Icarus*, 201 (2), 504–516.

Macpherson, T., Nickling, W. G., Gillies, J. A. and Etyemezian, V., 2008. Dust emissions from undisturbed and disturbed supply-limited desert surfaces. *Journal of Geophysical Research: Earth Surface*, 113 (F2) doi:10.1029/2007JF000800.

Madeleine, J.-B., Forget, F., Millour, E., Montabone, L., Wolff, M. J., 2011. Revisiting the radiative impact of dust on Mars using the LMD global climate model, *Journal of Geophysical Research*, 116 (E11), doi:10.1029/2011JE003855.

Madeleine, J.-B., Forget, F., Millour, E., Navarro, T., Spiga, A., 2012. The influence of radiatively active water ice clouds on the Martian climate, *Geophysical Research Letters*, 39 (23), doi:10.1029/2012GL053564.

Malin, M. C., Edgett, K. S., 2001. Mars Global Surveyor Mars Orbiter Camera: interplanetary cruise through primary mission, *Journal of Geophysical Research*, 106 (E10) 23429–23570.

Malin, M. C., Calvin, W. M., Cantor, B. A., et al., 2008. Climate, weather, and north polar observations from the Mars Reconnaissance Orbiter Mars Color Imager, *Icarus*, 194 (2), 501–512.

Malin, M. C., Edgett, K. S., Cantor, B. A., et al., 2010. An overview of the 1985–2006 Mars Orbiter Camera science investigation, *International Journal of Mars Science and Exploration*, 4, 1–60.

Martin, L. J., 1974a. The major Martian dust storms of 1971 and 1973. *Icarus*, 23, 108–115.

Martin, L. J., 1974b. The major Martian yellow storm of 1971. *Icarus*, 22 (2), 175–188.

Martin, L. J., Zurek, R. W., 1993. An analysis of the history of dust activity on Mars, *Journal of Geophysical Research*, 98 (E2) 3221–3246.

Martin, T. Z., 1986. Thermal infrared opacity of the Mars atmosphere, *Icarus*, 66, 2–21.

Martínez-Alvarado, O., Montabone, L., Lewis, S. R., Moroz, I. M., Read, P. L. 2009. Transient teleconnection event at the onset of a planet-encircling dust storm on Mars, *Annales Geophysicae*, 27 (9), 3663–3676.

McCleese, D. J., Heavens, N. G., Schofield, J. T., et al., 2010. Structure and dynamics of the Martian lower and middle atmosphere as observed by the Mars Climate Sounder: seasonal variations in zonal mean temperature, dust, and water ice aerosols, *Journal of Geophysical Research*, 115 (E12), doi:10.1029/2010JE003677.

McConnochie, T. M., Smith, M. D., 2008. Vertically resolved aerosol climatology from the Mars Global Surveyor Thermal Emission Spectrometer (MGS-TES) limb sounding. In *Third International Workshop on the Mars Atmosphere: Modeling and Observations*, Williamsburg, Virginia.

McKim, R. J., 1999. Meeting contribution: recent views of Mars, *Journal of the British Astronomical Association*, 109 (5), 287.

McLaughlin, D. B., 1954. Interpretation of some Martian features, *Publications of the Astronomical Society of the Pacific*, 66 (391), 161.

Merrison, J., Jensen, J., Kinch, K., Mugford, R., Nørnberg, P., 2004. The electrical properties of Mars analogue dust, *Planetary and Space Science*, 52 (4), 279–290.

Merrison, J. P., Gunnlaugsson, H. P., Nørnberg, P., Jensen, A. E., Rasmussen, K. R., 2007. Determination of the wind induced detachment threshold for granular material on Mars using wind tunnel simulations, *Icarus*, 191 (2), 568–580.

Metzger, S. M., Johnson, J. R., Carr, J. R., Parker, T. J., Lemmon, M., 1999. Dust devil vortices seen by the Mars Pathfinder Camera. *Geophys. Res. L.* 26, 2781–2784, doi:10.1029/1999GL008341.

Metzger, S. M., Carr, J. R., Johnson, J. R., Parker, T. J., Lemmon, M. T., 2000. Techniques for identifying dust devils in Mars Pathfinder images, *IEEE Transactions on Geoscience and Remote Sensing*, 38 (2), 870–876, doi:10.1109/36.842015.

Michaels, T. I., 2006. Numerical modeling of Mars dust devils: albedo track generation, *Geophysical Research Letters*, 33 (19), doi:10.1029/2006GL026268

Michaels, T. I., Rafkin, S. C. R., 2004. Large-eddy simulation of atmospheric convection on Mars, *Quarterly Journal of the Royal Meteorological Society*, 130 (599), 1251–1274.

Michaels, T. I., Colaprete, A., Rafkin, S. C. R. (2006), Significant vertical water transport by mountain-induced circulations on Mars, *Geophys. Res. Lett.*, 33, L16201, doi:10.1029/2006GL026562.

Michelangeli, D. V., Toon, O. B., Haberle, R. M., Pollack, J. B., 1993. Numerical simulations of the formation and evolution of water ice clouds in the Martian atmosphere, *Icarus*, 102 (2), 261–285.

Milam, K. A., Stockstill, K. R., Moersch, J. E., et al., 2003. THEMIS characterization of the MER Gusev Crater landing site, *Journal of Geophysical Research*, 108 (E12), doi:10.1029/2002JE002023.

Miyamoto, S., 1957. *The Great Yellow Cloud and the Atmosphere of Mars: Report of Visual Observations During the 1956 Opposition.* Contribution 7.1, Inst. of Astrophys. and Kwasan Obs., Univ. of Kyoto, Japan.

Montabone, L., Lewis, S. R., Read, P. L., 2005. Interannual variability of Martian dust storms in assimilation of several years of Mars global surveyor observations, *Advances in Space Research*, 36 (11) 2146–2155.

Montabone, L., Lewis, S. R., Read, P. L., Hinson, D. P., 2006. Validation of Martian meteorological data assimilation for MGS/TES using radio occultation measurements, *Icarus*, 185 (1), 113–132.

Montabone, L., Martinez-Alvarado, O., Lewis, S. R., Read, P. L., Wilson, R. J., 2008. Teleconnection in the Martian atmosphere during the 2001 planet-encircling dust storm, in *Third International Workshop on The Mars Atmosphere: Modeling and Observations*, held November 10–13, 2008 in Williamsburg, Virginia. LPI Contribution No. 1447, 9077.

Montabone, L., Lemmon, M. T., Smith, M. D., et al., 2011. Reconciling dust opacity datasets and building multi-annual dust scenarios for Mars atmospheric models, in *The Fourth International Workshop on the Mars Atmosphere: Modelling and Observation*, 8–11 February, 2011, Paris, France, 103–105.

Montabone, L., Forget, F., Millour, E., et al., 2015. Eight-year climatology of dust on Mars, *Icarus*, 251, 65–95.

Montmessin, F., Rannou, P., Cabane, M., 2002. New insights into Martian dust distribution and water-ice cloud microphysics, *Journal of Geophysical Research (Planets)*, 107, E6, 4–1, CiteID 5037, doi:10.1029/2001JE001520.

Montmessin, F., Forget, F., Rannou, P., Cabane, M., Haberle, R. M., 2004. Origin and role of water ice clouds in the Martian water cycle as inferred from a general circulation model, *Journal of Geophysical Research*, 109 (E10), doi:10.1029/2004JE002284.

Montmessin, F., Quémerais, E., Bertaux, J. L., et al., 2006. Stellar occultations at UV wavelengths by the SPICAM instrument: retrieval and analysis of Martian haze profiles, *Journal of Geophysical Research*, 111, (E9), doi:10.1029/2005JE002662.

Moores, J., Lemmon, M. T. Kahanpää, H., et al., 2014. Observational evidence of a shallow planetary boundary layer in northern Gale Crater, Mars as seen by the Navcam instrument onboard the Mars Science Laboratory Rover. *Icarus*, 249, 129–142.

Mulholland, D. P., Read, P. L., Lewis, S. R., 2013. Simulating the interannual variability of major dust storms on Mars using variable lifting thresholds, *Icarus*, 223 (1), 344–358.

Murphy, J. R., 1999. The Martian atmospheric dust cycle: insights from numerical model simulations, in *The Fifth International Conference on Mars*, July 19–24, 1999, Pasadena, California, abstract no. 6087.

Murphy, J. R., Nelli, S., 2002. Mars Pathfinder convective vortices: frequency of occurrence, *Geophysical Research Letters*, 29 (23), doi:10.1029/2002GL015214.

Murphy, J. R., Toon, O. B., Haberle, R. M., Pollack, J. B., 1990. Numerical simulations of the decay of Martian global dust storms, *Journal of Geophysical Research*, 95 (30), 14629–14648.

Murphy, J. R., Haberle, R. M., Toon, O. B., Pollack, J. B., 1993. Martian global dust storms – zonally symmetric numerical simulations including size-dependent particle transport, *Journal of Geophysical Research*, 98 (E2), 3197–3220.

Murphy, J. R., Pollack, J. B., Haberle, R. M., et al., 1995. Three-dimensional numerical simulation of Martian global dust storms, *Journal of Geophysical Research*, 100 (E12), 26357–26376.

Navarro, T., Madeleine, J.-B., Forget, F., et al., 2014. Global climate modeling of the Martian water cycle with improved microphysics and radiatively active water ice clouds, *Journal of Geophysical Research: Planets*, doi:10.1002/2013JE004550.

Neakrase, L. D. V., Greeley, R., 2010a. Dust devil sediment flux on Earth and Mars: laboratory simulations, *Icarus*, 206 (1), 306–318.

Neakrase, L. D. V., Greeley, R., 2010b. Dust devils in the laboratory: effect of surface roughness on vortex dynamics, *Journal of Geophysical Research*, 115 (E5), 10.1029/2009JE003465.

Newman, C. E., Richardson, M. I., 2015. The impact of surface dust source exhaustion on the Martian dust cycle, dust storms and interannual variability, as simulated by the MarsWRF General Circulation Model, *Icarus*, 257, doi:10.1016/j.icarus.2015.03.030.

Newman, C. E., Lewis, S. R., Read, P. L., Forget, F., 2002a. Modeling the Martian dust cycle 2. Multiannual radiatively active dust transport simulations, *Journal of Geophysical Research (Planets)*, 107 (E12), doi:10.1029/2002JE001920.

Newman, C. E., Lewis, S. R., Read, P. L., Forget, F., 2002b. Modeling the Martian dust cycle, 1. Representations of dust transport processes, *Journal of Geophysical Research (Planets)*, 107 (E12), doi:10.1029/2002JE001910.

Newman C. E., Lewis, S. R., Read, P. L., 2005. The atmospheric circulation and dust activity in different orbital epochs on Mars. *Icarus*, 174, 135–160.

Niederdörfer, E., 1933. Messungen des Wärmeumsatzes über schneebedecktem Boden, *Meteorol. Z.*, 50, 201–208.

Noble, J., Wilson, R. J., Haberle, R. M., et al. 2011. Comparison of TES FFSM eddies and MOC storms, MY 24–26, in *The Fourth International Workshop on the Mars Atmosphere: Modelling and Observation*, 8–11 February, 2011, Paris, France, 125–128.

Ockert-Bell, M. E., Bell, J. F., Pollack, J. B., McKay, C. P., Forget, F., 1997. Absorption and scattering properties of the Martian dust in the solar wavelengths, *Journal of Geophysical Research*, 102 (E4), 9039–9050.

Ogohara, K., Satomura, T., 2008. Northward movement of Martian dust localized in the region of the Hellas Basin, *Geophysical Research Letters*, 35 (13).

Owen, P. R., 1964. Saltation of uniform grains in air, *Journal of Fluid Mechanics*, 20, 225–242.

Pankine, A. A., Ingersoll, A. P., 2002. Interannual Variability of Martian Global Dust Storms. Simulations with a Low-Order Model of the General Circulation. *Icarus*, 155 (2), 299–323.

Pankine, A. A., Ingersoll, A. P., 2004. Interannual variability of Mars global dust storms: an example of self-organized criticality? *Icarus*, 170 (2), 514–518.

Piqueux, S., Byrne, S., Richardson, M. I., 2003. Sublimation of Mars's southern seasonal CO_2 ice cap and the formation of spiders, *Journal of Geophysical Research*, 108 (E8), doi:10.1029/2002JE002007.

Pleskot, L. K., Miner, E. D., 1981. Time variability of Martian bolometric albedo, *Icarus*, 45, 179–201.

Pollack, J. B., Colburn, D., Kahn, R., et al., 1977. Properties of aerosols in the Martian atmosphere, as inferred from Viking Lander imaging data, *Journal of Geophysical Research*, 82, 4479–4496.

Pollack, J. B., Colburn, D. S., Flasar, F. M., et al., 1979. Properties and effects of dust particles suspended in the Martian atmosphere. *J. of Geophys. Res.* 84, 2929–2945.

Pollack, J. B., Haberle, R. M., Schaeffer, J., Lee, H., 1990. Simulations of the general circulation of the Martian atmosphere. I – Polar processes. *J. of Geophys. Res.* 95, 1447–1473.

Pollack, J. B., Haberle, R. M., Murphy, J. R., Schaeffer, J., Lee, H., 1993. Simulations of the general circulation of the Martian atmosphere. II – Seasonal pressure variations, *Journal of Geophysical Research*, 98 (E2), 3149–3181.

Portyankina, G., Pommerol, A., Aye, K.-M., Hansen, C. J., Thomas, N., 2012. Polygonal cracks in the seasonal semi-translucent CO_2 ice layer in Martian polar areas, *Journal of Geophysical Research*, 117 (E2), doi:10.1029/2011JE003917.

Prandtl, L., 1935. The mechanics of viscous fluids. In W.F. Durand (ed.) *Aerodynamic Theory III*. Berlin: Springer.

Rafkin, S. C. R., 2009. A positive radiative-dynamic feedback mechanism for the maintenance and growth of Martian dust storms, *Journal of Geophysical Research*, 114 (E1), doi:10.1029/2008JE003217.

Rafkin, S. C. R., 2012. The potential importance of non-local, deep transport on the energetics, momentum, chemistry, and aerosol distributions in the atmospheres of Earth, Mars, and Titan. *Planetary and Space Science*, 60 (1), 147–154.

Rafkin, S. C. R., Haberle, R. M., Michaels, T. I., 2001. The Mars Regional Atmospheric Modeling System: model description and selected simulations, *Icarus*, 151 (2), 228–256.

Rafkin, S. C. R., Sta. Maria, M. R. V., Michaels, T. I., 2002. Simulation of the atmospheric thermal circulation of a Martian volcano using a mesoscale numerical model, *Nature*, 419 (6908), 697–699.

Rannou, P., Perrier, S., Bertaux, J.-L., et al., 2006. Dust and cloud detection at the Mars limb with UV scattered sunlight with SPICAM, *Journal of Geophysical Research*, 111 (E9), doi:10.1029/2006JE002693.

Read, P. L., Montabone, L., Mulholland, D. P., et al., 2011. Midwinter suppression of baroclinic storm activity on Mars: observations and models, *The Fourth International Workshop on the Mars Atmosphere: Modelling and Observation*, 8–11 February, 2011, Paris, France, 133–135.

Reiss, D., Zanetti, M., Neukum, G., 2011. Multitemporal observations of identical active dust devils on Mars with the High Resolution Stereo Camera (HRSC) and Mars Orbiter Camera (MOC), *Icarus*, 215 (1), 358–369.

Rennó, N. O., Burkett, M. L., Larkin, M. P., 1998. A simple thermodynamical theory for dust devils, *Journal of Atmospheric Sciences*, 55 (21), 3244–3252.

Rennó, N. O., Wong, A.-S., Atreya, S. K., de Pater, I., Roos-Serote, M., 2003. Electrical discharges and broadband radio emission by Martian dust devils and dust storms, *Geophysical Research Letters*, 30 (22), doi:10.1029/2003GL017879.

Rennó, N. O., Abreu, V. J., Koch, J., et al., 2004. MATADOR 2002: a pilot field experiment on convective plumes and dust devils, *Journal of Geophysical Research*, 109 (E7), doi:10.1029/2003JE002219.

Richardson, M. I., Wilson, R. J., 2002. Investigation of the nature and stability of the Martian seasonal water cycle with a general circulation model, *Journal of Geophysical Research (Planets)*, 107 (E5), doi:10.1029/2001JE001536.

Ringrose, T. J., Towner, M. C., Zarnecki, J. C., 2003. Convective vortices on Mars: a reanalysis of Viking Lander 2 meteorological data, sols 1–60, *Icarus*, 163 (1), 78–87.

Ringrose, T. J., Patel, M. R., Towner, M. C., et al., 2007. The meteorological signatures of dust devils on Mars, *Planetary and Space Science*, 55 (14), 2151–2163.

Rodin, A. V., Clancy, R. T., Wilson, R. J., 1999a. Dynamical properties of Mars water ice clouds and their interactions with atmospheric dust and radiation, *Advances in Space Research*, 23 (9), 1577–1585.

Rodin, A. V., Wilson, R. J., Clancy, R. T., Richardson, M. I., 1999b. The coupled roles of dust and water ice clouds in the Mars aphelion season, in *The Fifth International Conference on Mars*, July 19–24, 1999, Pasadena, California, abstract no. 6235.

Roney, J. A., White, B. R., 2004. Definition and measurement of dust aeolian thresholds, *Journal of Geophysical Research: Earth Surface*, 109 (F1), doi:10.1029/2003JF000061.

Rossow, W. B., 1978. Cloud microphysics – analysis of the clouds of Earth, Venus, Mars, and Jupiter, *Icarus*, 36, 1–50.

Rover Team, 1997. Characterization of the Martian surface deposits by the Mars Pathfinder rover, Sojourner, *Science*, 278 (5344), 1765–1768.

Ruff, S. W., Christensen, P. R., 2002. Bright and dark regions on Mars: particle size and mineralogical characteristics based on Thermal Emission Spectrometer data, *Journal of Geophysical Research (Planets)*, 107 (E12), doi:10.1029/2001JE001580.

Ruijgrok, W., Davidson, C. I., Nicholson, K. W., 1995. Dry deposition of particles, *Tellus*, 47B, 587–601.

Ryan, J. A., Lucich, R. D., 1983. Possible dust devils – vortices on Mars, *Journal of Geophysical Research*, 88, 11005–11011.

Ryan J. A., Sharman, R. D., 1981. Two major dust storms, one year apart: comparison of Viking data, *Journal of Geophysical Research*, 86, 3247–3254.

Sagan, C., Bagnold, R. A., 1975. Fluid transport on Earth and aeolian transport on Mars. *Icarus*, 26. 209–218.

Sagan, C., Pollack, J. B., 1969. Windblown dust on Mars, *Nature*, 223 (5208), 791–794.

Sagan, C., Veverka, J., Fox, P., et al., 1972. Variable features on Mars: preliminary Mariner 9 television results, *Icarus*, 17, 346.

Sagan, C., Veverka, J., Fox, P., et al., 1973. Variable features on Mars, 2, Mariner 9 global results, *Journal of Geophysical Research*, 78 (20), 4163–4196.

Schmidt, D. S., Schmidt, R. A., Dent, J. D., 1998. Electrostatic force on saltating sand, *J. Geophys. Res.*, 103, 8997–9001.

Schofield, J. T., Barnes, J. R., Crisp, D., et al., 1997. The Mars Pathfinder atmospheric structure investigation/meteorology, *Science*, 278 (5344), 1752.

Shao, Y., Raupach, M. R., Findlater, P. A., 1993. Effect of saltation bombardment on the entrainment of dust by wind, *Journal of Geophysical Research*, 98 (D7), 12719–12726.

Siili, T., Haberle, R. M., Murphy, J. R., Savijärvi, H., 1999. Modelling of the combined late-winter ice cap edge and slope winds in Mars Hellas and Argyre regions. *Planetary and Space Science*, 47, 8–9, 951–970.

Sinclair, P. C., 1969. General characteristics of dust devils, *Journal of Applied Meteorology*, 8 (1), 32–45.

Sinclair, P. C., 1973. The lower structure of dust devils, *Journal of Atmospheric Sciences*, 30 (8), 1599–1619.

Slinn, S. A., Slinn, W. G. N., 1980. Predictions for particle deposition on natural waters, *Atmospheric Environment*, 14, 1013–1016.

Slipher, E. C., 1962. *Mars – The Photographic Story*. Northland Press, Flagstaff, AZ.

Smith, M. D., 2004. Interannual variability in TES atmospheric observations of Mars during 1999–2003, *Icarus*, 167 (1), 148–165.

Smith, M. D., 2008. Spacecraft Observations of the Martian atmosphere, *Annual Review of Earth and Planetary Sciences*, 36, 191–219.

Smith, M. D., 2009. THEMIS observations of Mars aerosol optical depth from 2002–2008, *Icarus*, 202 (2), 444–452.

Smith, M. D., Conrath, B. J., Pearl, J. C., Christensen, P. R., 2002. Note: Thermal Emission Spectrometer observations of Martian planet-encircling dust storm 2001A, *Icarus*, 157 (1), 259–263.

Smith, M. D., Wolff, M. J., Spanovich, N., et al., 2006. One Martian year of atmospheric observations using MER Mini-TES, *Journal of Geophysical Research*, 111 (E12), doi:10.1029/2006JE002770.

Smith, M. D., Wolff, M. J., Clancy, R. T., Kleinböhl, A., Murchie, S. L., 2013. Vertical distribution of dust and water ice aerosols from CRISM limb-geometry observations, *Journal of Geophysical Research: Planets*, 118 (2), 321–334.

Smith, P. H., Lemmon, M., 1999. Opacity of the Martian atmosphere measured by the Imager for Mars Pathfinder, *Journal of Geophysical Research*, 104 (E4), 8975–8986, doi:10.1029/1998JE900017.

Smith, P.H., Bell III, J. F., Bridges, N. T., et al., 1997. Results from the Mars Pathfinder Camera. *Science*, 278, 1758–1765, doi:10.1126/science.278.5344.1758.

Snyder, C. W., Moroz, V. I., 1992. Spacecraft exploration of Mars, in *Mars*, ed. H. Kieffer et al., Univ. Arizona Press, Tucson, 71–119.

Spiga, A., Forget, F., 2009. A new model to simulate the Martian mesoscale and microscale atmospheric circulation: validation and first results, *Journal of Geophysical Research*, 114 (E2).

Spiga, A., Faure, J., Madeleine, J.-B., Määttänen, A., Forget, F., 2013. Rocket dust storms and detached dust layers in the Martian atmosphere, *Journal of Geophysical Research: Planets*, 118 (4), 746–767.

Stanzel, C., Pätzold, M., Greeley, R., Hauber, E., Neukum, G., 2006. Dust devils on Mars observed by the High Resolution Stereo Camera, *Geophysical Research Letters*, 33 (11), doi:10.1029/2006GL025816.

Stanzel, C., Pätzold, M., Williams, D. A., et al., 2008. Dust devil speeds, directions of motion and general characteristics observed by the Mars Express High Resolution Stereo Camera, *Icarus*, 197 (1), 39–51.

Stella, P., Ewell, R., Hoskin, J., 2005, Design and performance of the MER (Mars Exploration Rovers) solar arrays, *Proc. IEEE Photovoltaic Spec. Conf.*, 31, 626–630.

Stow, C. D., 1969. Dust and Sand Storm Electrification, *Weather*, 24 (4), 134–144.

Strausberg, M. J., Wang, H., Richardson, M. I., Ewald, S. P., Toigo, A. D., 2005. Observations of the initiation and evolution of the 2001 Mars global dust storm, *Journal of Geophysical Research*, 110 (E2).

Szwast, M. A., Richardson, M. I., Vasavada, A. R., 2006. Surface dust redistribution on Mars as observed by the Mars Global Surveyor and Viking Orbiters, *Journal of Geophysical Research*, 111 (E11), doi:10.1029/2005JE002485.

Takahashi, Y. O., Fujiwara, H., Fukunishi, H., et al., 2003. Topographically induced north–south asymmetry of the meridional circulation in the Martian atmosphere, *Journal of Geophysical Research (Planets)*, 108 (E3), doi:10.1029/2001JE001638.

Thomas, N., Hansen, C. J., Portyankina, G., Russell, P. S., 2010. HiRISE observations of gas sublimation-driven activity in Mars' southern polar regions: II. Surficial deposits and their origins, *Icarus*, 205 (1), 296–310.

Thomas, P. C., Gierasch, P., 1985. Dust devils on Mars, *Science*, 230, 175–177.

Thomas, P. C., Veverka, J., Gineris, D., Wong, L., 1984. Dust streaks on Mars, *Icarus*, 60, 161–179.

Thorpe, T. E., 1979. A history of Mars atmospheric opacity in the southern hemisphere during the Viking extended mission, *Journal of Geophysical Research*, 84, 6663–6683.

Tillman, J. E., 1988. Mars global atmospheric oscillations – annually synchronized, transient normal-mode oscillations and the triggering of global dust storms, *Journal of Geophysical Research*, 93, 9433–9451.

Toigo, A. D., Richardson, M. I., Wilson, R. J., Wang, H., Ingersoll, A. P., 2002. A first look at dust lifting and dust storms near the south pole of Mars with a mesoscale model, *Journal of Geophysical Research (Planets)*, 107 (E7), doi:10.1029/2001JE001592.

Toigo, A. D., Richardson, M. I., Ewald, S. P., Gierasch, P. J., 2003. Numerical simulation of Martian dust devils, *Journal of Geophysical Research Planets*, 108 (E6), doi:10.1029/2002JE002002.

Toon, O. B., Pollack, J. B., Sagan, C., 1977. Physical properties of the particles composing the Martian dust storm of 1971–1972, *Icarus*, 30, 663–696.

Towner, M. C., 2009. Characteristics of large Martian dust devils using Mars Odyssey Thermal Emission Imaging System visual and infrared images, *Journal of Geophysical Research*, 114 (E2), doi:10.1029/2008JE003220.

Tyler, D., Barnes, J. R., 2013. Mesoscale modeling of the circulation in the Gale Crater region: an investigation into the complex forcing of convective boundary layer depths, *Mars*, 58–77. doi:10.1555/mars.2013.0003.

Veverka, J., Sagan, C., Quam, L., Tucker, R., Eross, B., 1974. Variable features on Mars III: comparison of Mariner 1969 and Mariner 1971 photography, *Icarus*, 21, 317.

Vincendon, M., Langevin, Y., Poulet, F., et al., 2009. Yearly and seasonal variations of low albedo surfaces on Mars in the OMEGA/MEx dataset: constraints on aerosols properties and dust deposits, *Icarus*, 200 (2), 395–405.

Wang, H., 2007. Dust storms originating in the northern hemisphere during the third mapping year of Mars Global Surveyor, *Icarus*, 189 (2), 325–343.

Wang, H., Fisher, J. A., 2009. North polar frontal clouds and dust storms on Mars during spring and summer, *Icarus*, 204 (1), 103–113.

Wang, H., Richardson, M. I., 2015. The origin, evolution, and trajectory of large dust storms on Mars during Mars Years 24–30 (1999–2011), *Icarus*, 251, 112–127.

Wang, H., Richardson, M. I., Wilson, R. J., et al., 2003. Cyclones, tides, and the origin of a cross-equatorial dust storm on Mars, *Geophysical Research Letters*, 30 (9), doi:10.1029/2002GL016828.

Wang, H., Zurek, R. W., Richardson, M. I., 2005. Relationship between frontal dust storms and transient eddy activity in the northern hemisphere of Mars as observed by Mars Global Surveyor, *Journal of Geophysical Research*, 110 (E7), doi:10.1029/2005JE002423.

Wang, H., Richardson, M. I., Toigo, A. D., Newman, C. E., 2013. Zonal wavenumber three traveling waves in the northern hemisphere of Mars simulated with a general circulation model, *Icarus*, 223 (2), 654–676.

Wells, E. N., Veverka, J., Thomas, P., 1984. Mars – experimental study of albedo changes caused by dust fallout, *Icarus* 58, 331–338.

Westphal, D. L., Toon, O. B., Carlson, T. N., 1987. A two-dimensional numerical investigation of the dynamics and microphysics of Saharan dust storms, *Journal of Geophysical Research*, 92, 3027–3049.

White, B. R., 1979. Soil transport by winds on Mars, *J. Geophys. Res.*, 84(B9), 4643–4651, doi:10.1029/JB084iB09p04643.

White, B. R., Lacchia, B. M., Greeley, R., Leach, R. N., 1997. Aeolian behavior of dust in a simulated Martian environment, *Journal of Geophysical Research*, 102 (E11), 25629–25640.

Whiteway, J. A., Komguem, L., Dickinson, C., et al., 2009. Mars water-ice clouds and precipitation, *Science*, 325 (5936), 68.

Williams, D. A., Greeley, R., Neukum, G., et al., 2004. Seeing Mars with new eyes: latest results from the High Resolution Stereo Camera on Mars Express, *Geol. Soc. Am. Abstr. Programs*, 36(5), 21.

Wilson, R. J., 1997. Dust transport in the Martian atmosphere as simulated by a general circulation model, *Advances in Space Research*, 19 (8), 1290–1290.

Wilson, R. J., 2012. Martian dust storms, thermal tides, and the Hadley circulation, in *Comparative Climatology of Terrestrial Planets*, held June 25–28, 2012, Boulder, CO. LPI Contribution No. 1675, id.8069.

Wilson, R. J., Hamilton, K., 1996. Comprehensive model simulation of thermal tides in the Martian atmosphere, *Journal of Atmospheric Science*, 53 (9), 1290–1326.

Wilson, R. J., Kahre, M. A., 2009. The role of spatially variable surface dust in GCM simulations of the Martian dust cycle. In *Mars Dust Cycle Workshop*, held September 15–17, 2009, Moffett Field, CA. 108–112. http://spacescience.arc.nasa.gov/mars-climate-modeling-group/documents/mars_dust_cycle_workshop_abstracts.pdf.

Wilson, R. J., Hinson, D., Smith, M. D., 2006. GCM simulations of transient eddies and frontal systems in the Martian atmosphere, *Second Workshop on Mars Atmosphere Modelling and Observations*, February 27–March 3, 2006, Granada, Spain, 154.

Wilson, R. J., Neumann, G. A., Smith, M. D., 2007. Diurnal variation and radiative influence of Martian water ice clouds, *Geophysical Research Letters*, 34 (2), doi:10.1029/2006GL027976.

Wilson, R. J., Haberle, R. M., Noble, J., et al., 2008a. Simulation of the 2001 planet-encircling dust storm with the NASA/NOAA Mars general circulation model. In *Third International Workshop on the Mars Atmosphere: Modeling and Observations*, Williamsburg, Virginia.

Wilson, R. J., Lewis, S. R., Montabone, L., 2008b. Thermal tides in an assimilation of three years of Thermal Emission Spectrometer data from Mars Global Surveyor. In *Third International Workshop on the Mars Atmosphere: Modeling and Observations*, Williamsburg, Virginia.

Wolff, M. J., Smith, M. D., Clancy, R. T., et al., 2006. Constraints on dust aerosols from the Mars Exploration Rovers using MGS overflights and Mini-TES, *Journal of Geophysical Research*, 111 (E12), doi:10.1029/2006JE002786.

Wolff, M. J., Smith, M. D., Clancy, R. T., et al., 2009. Wavelength dependence of dust aerosol single scattering albedo as observed by the Compact Reconnaissance Imaging Spectrometer. *J. of Geophys. Res.* 114, doi:10.1029/2009JE003350.

Wolff, M. J., Clancy, R. T., Smith, M. D., et al., 2011. Deriving vertical profiles of aerosol sizes from TES, American Geophysical Union, Fall Meeting 2011, abstract #P24A-09.

Wurm, G., Krauss, O., 2006. Dust eruptions by photophoresis and solid state greenhouse effects, *Physical Review Letters*, 96 (13), 134301.

Wurm, G., Teiser, J., Reiss, D., 2008. Greenhouse and thermophoretic effects in dust layers: the missing link for lifting of dust on Mars, *Geophysical Research Letters*, 35 (10), doi:10.1029/2008GL033799.

Zalucha, A. M., Plumb, R. A., Wilson, R. J., 2010. An analysis of the effect of topography on the Martian Hadley cells, *Journal of the Atmospheric Sciences*, 67 (3), 673–693.

Zasova, L., Formisano, V., Moroz, V., et al, 2005. Water clouds and dust aerosols observations with PFS MEX at Mars, *Planetary and Space Science*, 53 (10), 1065–1077.

Zhang, H.-F., Wang, T., Qu, J.-J., Yan, M.-H., 2004. An experimental and observational study on the electric effect of sandstorms, *Chin. J. Geophys.*, 47, 53–60.

Zhang, L., Gong, S., Padro, J., Barrie, L. (2001). A size-segregated particle dry deposition scheme for an atmospheric aerosol module. *Atmos. Environ.* 35: 549–560.

Zurek, R. W., Martin, L. J., 1993. Interannual variability of planet-encircling dust storms on Mars, *Journal of Geophysical Research*, 98 (E2), 3247–3259.

Zurek, R. W., Barnes, J. R., Haberle, R. M., et al., 1992. Dynamics of the atmosphere of Mars, in *Mars*, ed. H. Kieffer et al., Univ. Arizona Press, Tucson, 835–933.

11

The Water Cycle

FRANCK MONTMESSIN, MICHAEL D. SMITH, YVES LANGEVIN, MICHAEL T. MELLON, ANNA FEDOROVA

11.1 INTRODUCTION

The study of water on Mars has been a central and ongoing quest of Mars exploration. From the early ground-based measurements of Spinrad et al. (1963) to the recent spaceborne endeavors, attempts to detect water in all possible forms have progressively unveiled the salient features of the seasonal activity of water on Mars. Water, which in the liquid phase is considered to be an essential ingredient for life as we know it, is present only in very low abundances on Mars. Identification of its major reservoirs – the atmosphere, surface ice, and the regolith – yields an estimate of the Martian global water inventory that is far less than Earth's water inventory. Current estimates suggest that, if all these reservoirs were condensed or melted to cover the Martian surface globally, it would produce a layer of liquid water only 20–30 m deep (Smith et al., 1999; see also Chapter 17), far less than the kilometer-deep oceans on Earth. In spite of its scarcity, water has revealed itself as a major actor in the current Mars climate, impacting the latter in many important areas. Mars' climate, in turn, heavily influences the behavior of water. The tenuous and cold CO_2 atmosphere of Mars maintains average thermodynamical conditions significantly below the triple point of water (6 mbar, 273 K), explaining why the latter has only been observed to date in the solid and gaseous phases. Except for some rare transient episodes that may involve liquid water, such as gullies (Malin and Edgett, 2000; Costard et al., 2002), recurring slope lineae (McEwen et al., 2014; Ojha et al., 2015), or deliquescing salts (Rennó et al., 2009; Martín-Torrez et al., 2015), the present-day climate is not favorable for the stability of liquid water at the surface.

There has been an accumulation of evidence that Martian water behaves in a cyclic manner, thereby establishing the paradigm of a water cycle on Mars. It is, however, important to define what is implied by the "cycle" designation. A cycle corresponds to a period of time during which a recurring sequence of phenomena occurs. For Mars, this statement appears valid regardless of the period of time considered. As discussed in Chapter 16 of this book, throughout the Amazonian era, which encompasses the history of Mars for the last couple of billions of years and during which the Mars atmosphere and surface have evolved to be more or less in the same states as they exist today, the Mars water cycle has been submitted to periodic variations imparted by the cyclic changes of the orbital configuration of the planet. Cycling of water on geologic timescales (10^5–10^7 years) nevertheless finds its deepest foundations in the mechanisms that operate today, and which, for the most part, can be observed and subsequently comprehended on a human timescale.

The existence of a water cycle on Mars was deduced from the first multi-annual monitoring of water vapor performed by the Viking Mars Atmospheric Water Detector (MAWD) instrument, which revealed that the same seasonal and spatial pattern qualitatively repeated itself for nearly two consecutive Martian years (Jakosky and Farmer, 1982). After Viking, other missions (Mars Global Surveyor (MGS), Mars Express (MEX), and Mars Reconnaissance Orbiter (MRO)) have confirmed this initial conclusion: seasonal water variations appear to be controlled by exchanges between various reservoirs, achieving on an annual basis a stationary state with some inter-annual differences. The latter statement implies that, even if the budget of Mars water is not necessarily balanced between its reservoirs, i.e. that a permanent net transfer may exist between them, the processes that control the budget evolve in an equilibrated manner. At this point, the following are important questions that have captured the attention of the Mars community for a long time: What and where are the major sources and sinks of water? Is the annual budget of water closed? Are the current locations of the major reservoirs in equilibrium with the present-day climate configuration or do they reflect past water cycle activity? Can we identify dominant processes for the overall Mars water seasonal evolution?

The study of the Mars water cycle is a rapidly evolving area. Every observational piece of evidence, every new theoretical work, has contributed in the last decades to create our contemporary view of the Mars water cycle. The growing field of the Mars water cycle has now gained sufficient maturity to propose key elements to its understanding. Eventually, one will be able to qualitatively and quantitatively characterize all the components of the water cycle: (i) the nature, size, and location of the water reservoirs; (ii) the physical processes driving the exchanges between them and/or affecting the reservoir specifically; and (iii) the magnitude and timing of the mass fluxes. Whereas observations can directly characterize the first component, theoretical tools are definitely required to tackle the other two. Fortunately, the last decades have seen considerable improvements in both the observational and modeling areas, in particular, the reference climatology of atmospheric water that was established by the MGS Thermal Emission Spectrometer (TES) instrument (Smith, 2004), the detection of an ice-rich regolith close to the surface by the Mars Odyssey Gamma Ray Spectrometer (GRS) instrument (Boynton et al., 2002), the detailed characterization of seasonal water ice frosts by the Viking Infrared Thermal Mapper (IRTM), TES (Kieffer et al., 2000; Kieffer and Titus, 2001), MEX Observatoire pour la Minéralogie, l'Eau, les Glaces et l'Activité (OMEGA; Langevin et al., 2007; Appéré et al., 2011) and MRO Compact Reconnaissance Imaging Spectrometers for Mars (CRISM; Smith et al., 2009), as well as the advent and further progress accomplished in parallel by three-dimensional climate models. From these observations and modeling efforts, a consensus view of the present-day Mars water cycle has slowly emerged.

The largest reservoirs of water are found as icy layers either covering the surface or mixed within the regolith. The surface ice reservoir contains approximately 10^6–10^7 times more water than the atmosphere. At the estimated loss rate of water escape to space (10^{-6} kg m^{-2} yr^{-1}; McElroy, 1972; Yung et al., 1988; Kass and Yung, 1999), surface reservoirs will be able to replenish the atmosphere for billions of years. Among the surface reservoirs, the north polar cap plays a pivotal role. It consists of a kilometer-thick water ice dome permanently covering regions poleward of 80°N. The cross-sectional profile of the cap, exposed in many distinct areas, exhibits ubiquitous sequences of alternately bright and dark layers, interpreted as the accumulation over geological timescales of successive water ice deposition and removal events controlled by the Milankovitch cycles of Mars' orbital parameters (Malin et al., 2001; Laskar et al., 2002). The north polar cap is by far more extensive in its exposed part than its southern analog, whose nature and history are more complex to decipher: a thin (meters thick) permanent layer of CO_2 ice prevents the underlying bulk water ice layer from regularly interacting with the atmosphere on a seasonal basis. As a result, the Mars water seasonal cycle is predominantly controlled by the seasonal climate variations at the north pole. The spring–summer season at the north pole, when massive amounts (>10^{12} kg) of water sublime from the cap and are injected into the atmosphere, is followed by a period of extreme dryness subsequent to the onset and further progress of the polar night, where temperature is so cold that water vapor partial pressure falls rapidly towards undetectable amounts. There, water ice frost has been observed to form, with an areal extent tracking seasonally the latitudinal excursion of the polar night vortex boundaries. Once a molecule of water has left its original location of the north polar cap after sublimation, it becomes free to travel across the planet, carried by winds towards the equator, subsequently incorporating into the overturning northern spring–summer Hadley cell that conveys air masses and moisture from the northern to the southern hemisphere. The eccentric Mars orbit imposes significantly (~20 K) warmer conditions at the southern spring–summer pole, making water less prone to reside perennially as frost there. Water vapor is therefore led during the following northern fall–winter season to return to the northern hemisphere through the Hadley cell, to be eventually trapped within the seasonal frost which "creeps" gradually back to the north pole before the next summer sublimation event.

This schematic description, illustrated in Figure 11.1, captures the essence of the seasonal cycle of water. Vast, exposed reservoirs of ice communicate with the atmosphere whose circulation is vigorous enough to transport water from pole to pole and back, thereby closing the water budget on an annual basis. Additional aspects are important to consider as well, such as the possibility for water molecules to be adsorbed and released by the regolith, or the possibility for molecules to condense as

clouds that sediment and/or precipitate to the surface. All these processes contribute to defining the current seasonal evolution of water on Mars.

The goal of this chapter is to provide a comprehensive overview of our knowledge and understanding of the elements of the present-day Mars water cycle depicted in Figure 11.1. The next sections review the history of observations, which have fed the study of Mars water since the first ground-based measurements. The discussion also includes a description, as inferred from observations, of the nature, size, and seasonal evolution of the major reservoirs of water, including the atmosphere (as both vapor and clouds), the perennial and seasonal surface water ice, as well as water stored in the regolith. In the last section, a review of the theoretical work performed to date is given to establish the physical connections between all the reservoirs of water, which constitute the very nature of present-day Mars' water cycle.

11.2 THE SEASONAL RESERVOIRS OF WATER ON MARS

11.2.1 Atmospheric Reservoir

Among all the reservoirs of water that exist on Mars, the atmosphere is the most well observed. A fairly complete dataset has been compiled that provides a temporally regular and spatially extensive survey covering more than nine (non-consecutive) Martian years and allowing comparison of the seasonal behavior of water decades apart. Atmospheric water exists predominantly in the form of vapor, but also in the form of water ice clouds (see Chapter 5) that were first identified by Curran et al. (1973).

11.2.1.1 History of Observations

Establishing the current Mars water cycle paradigm started with the characterization of water in the atmosphere. The first detection was made by Spinrad and colleagues when 11 weak lines of water vapor near 0.82 μm were found in the spectrum of Mars observed by the Mt. Wilson 100-inch refractor on April 12 and 13, 1963 (Spinrad et al., 1963). This important discovery triggered an era of intensive survey: during the 1970s, numerous ground-based observations of Martian water vapor in the same near-infrared band were made (Barker et al., 1970; Barker, 1976), which provided the water vapor column abundance (that is, the total amount of water vapor contained in a column of atmosphere) averaged over the disk of Mars. These first telescopic observations showed a prominent seasonal dependence in atmospheric water vapor, with column abundance varying by a factor of about 2, reaching seasonal maximum value just after each of the two solstices.

The first spacecraft observations of water vapor were made in 1971 and 1972 by the Mariner 9 Infrared Interferometer Spectrometer (IRIS) using the 30 μm rotational band (Conrath et al., 1973). The seasonal coverage of the IRIS observations extended from $L_s = 293°$ to $L_s = 102°$ covering the southern summer and fall seasons. A recent reanalysis of these data showed that, in addition to the previously observed seasonal variation, there was a diurnal variation of water vapor abundance that decreased as the mission progressed (Ignatiev et al.,

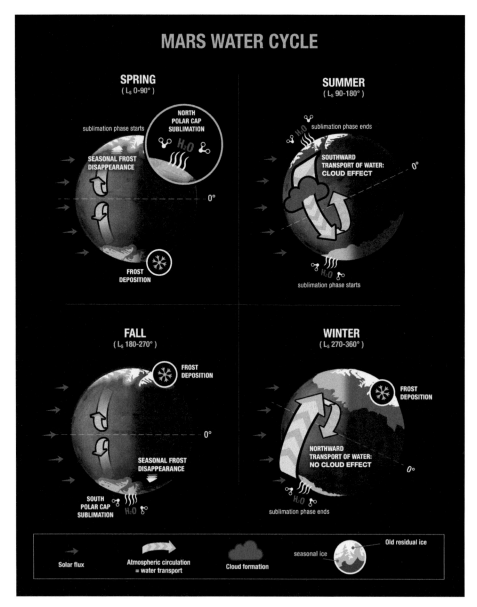

Figure 11.1. Seasonal progression of the present Mars water cycle illustrating the exchange of water between surface and atmospheric reservoirs. A black and white version of this figure will appear in some formats. For the color version, please refer to the plate section.

2002). The Mars 3 mission in 1971–1972 (Moroz and Nadzhip, 1976) and Mars 5 in 1974 (Moroz and Nadzhip, 1975) used the 1.38 μm band to provide measurements of water vapor for several months during the same period and showed good agreement with the Mariner 9 findings.

The first systematic mapping of water vapor with complete spatial and seasonal coverage was obtained by MAWD on-board the Viking 1 and 2 Orbiters using again the 1.38 μm absorption band (Farmer and Laporte, 1972; Farmer et al., 1977). The MAWD observations covered more than one Martian year from June 1976 through April 1979 (Jakosky and Farmer, 1982). For a long time these data formed the classical framework constraining Martian water cycle models (e.g. Davies, 1981; Haberle and Jakosky, 1990; Houben et al., 1997; Richardson and Wilson, 2002a).

After the Viking mission there was a long gap in spacecraft missions to Mars. Only one spacecraft, Phobos 2, reached

Mars during the following decade. Phobos 2 observed the atmosphere using a near-infrared mapping spectrometer for a period of about two months at the beginning of northern spring. Rosenqvist et al. (1992) retrieved detailed maps of water vapor column abundance in the equatorial regions based on the analysis of the 2.56 μm band. These maps revealed spatial variations of up to a factor of 5 that were correlated with topography and surface albedo, while significant diurnal variations were also observed (Titov et al., 1994, 1995).

A large set of ground-based near-infrared observations was collected in 1988 and 1999 (Rizk et al., 1991; Sprague et al., 1996, 2001, 2003, 2006). In addition to the near-infrared data, microwave spectra were observed with the Very Large Array telescope, allowing the retrieval of water vapor columns over a significant latitudinal domain (Clancy et al., 1992, 1996). Millimeter-wavelength ground-based observations of HDO (Encrenaz et al., 1991, 1995; Fouchet et al., 2011), infrared observations from

Mauna Kea (Encrenaz et al., 2005a, 2008a, 2010), Infrared Space Observatory far-infrared spectroscopy (Burgdorf et al., 2000), and data from the Submillimeter Wave Astronomy Satellite (SWAS) reported by Gurwell et al. (2002) further contributed to the growing body of water vapor observations.

An intensive campaign of water vapor measurements from orbital spacecraft began at the end of the 1990s. Smith (2002, 2004, 2008) retrieved the thermal structure, aerosol optical depth, and water vapor column abundance from thermal infrared spectra obtained by TES, and monitored the spatial and seasonal dependence of water vapor for three Martian years between 1997 and 2004. Since 2004, water vapor has been monitored by three different instruments on MEX providing simultaneous observations: OMEGA at 2.56 μm (Encrenaz et al., 2005b, 2008b; Melchiorri et al., 2007; Maltagliati et al., 2009, 2011b), the Planetary Fourier Spectrometer (PFS) at 2.56 μm (Tschimmel et al., 2008) and 30 μm (Fouchet et al., 2007), and the Spectroscopy for Investigation of Characteristics of the Atmosphere of Mars (SPICAM) instrument at 1.38 μm (Fedorova et al., 2006). In addition to global-scale monitoring, solar occultation observations by SPICAM allowed retrieval of the vertical distribution of water vapor (Fedorova et al., 2009; Maltagliati et al., 2011a).

The most recent spacecraft observations were provided by the CRISM instrument using the 2.56 μm band (Smith et al., 2009). CRISM began systematic mapping and monitoring of water vapor in 2007. Observations from CRISM and the three Mars Express instruments have allowed for a direct comparison of the retrievals from the different instruments and have provided a continuation of the water vapor monitoring by the TES instrument, which ceased collecting spectra in September 2004.

From the surface of Mars, atmospheric water vapor was observed for the first time by the Imager for Mars Pathfinder (IMP) camera in 1997 at $L_s = 149$–$154°$ using a filter centered at 0.94 μm (Titov et al., 1999). The dependence of the observed water vapor transmittance with Sun elevation implied that water vapor was not uniformly mixed in the atmosphere but was instead confined to a layer 1–3 km thick near the surface. IMP observations also indicated horizontal inhomogeneity in the layer, but showed no significant morning-to-evening variations of the water vapor abundance. The Mini-TES instrument on-board the Spirit and Opportunity Mars Exploration Rovers was used to retrieve the water vapor column abundance above each of the two rovers for 1.5 Martian years using the 30 μm rotation bands (Smith et al., 2006). The Phoenix Lander in 2008 observed the north polar atmosphere during summer (Tamppari et al., 2010). Water vapor measurements were made using the 0.935 μm filter in a way similar to the Mars Pathfinder experiment (Titov et al., 1999). Besides a measurement of the seasonal cycle of water vapor above the landing site, the Phoenix results have yielded a new understanding of the diurnal variation and vertical distribution of water vapor in the north polar region summer. Water vapor was found to have a more complicated vertical structure than the well-mixed profiles that are often assumed, and the depth of the layer that experiences a diurnal exchange of water with the surface was found to be 0.5–1 km (Tamppari et al., 2010). The most recent results have been obtained by the Curiosity Rover of the Mars Science Laboratory mission whose relative humidity measurements show a very dry environment for Gale Crater but with diurnal exchange with the regolith still possible (Harri et al., 2014; Savijärvi et al., 2016).

11.2.1.2 Comparison Between Instruments

Interpreting results from very different instruments operating under different conditions and using different retrieval algorithms is challenging. For instance, a comparison of the Viking data to earlier ground-based observations (Jakosky and Barker, 1984) showed differences in the southern hemisphere during the summer. The MAWD retrievals suggested a very dry season at the time of the Viking observations, with column abundance about 5–10 pr μm (precipitable micrometers, a unit that will be used hereafter to describe water abundance on Mars; 1 pr μm is equal to 1 g of water per m² equivalent to forming a 1 μm thick layer of condensed water) in the southern hemisphere. On the other hand, Barker et al. (1970) reported a column abundance as high as 40 pr μm for the same location and season using ground-based observations. Two global dust storms occurred during the time of the MAWD observations, and many authors have noted the potential influence of dust aerosol scattering on the MAWD retrievals shown in Figure 11.2 (Davies, 1979; Davies and Wainio, 1981; Jakosky et al., 1988). Actually, Hunten et al. (2000) describe detailed radiative transfer modeling of a combination of CO_2 and water vapor absorption lines to properly account for scattering by aerosols in their water vapor retrievals at 0.82 μm (Sprague et al., 2001, 2003).

Ground-based observations taken after the end of the Viking missions have persistently shown differences with MAWD. In particular, Rizk et al. (1991) observed a more humid southern hemisphere summer than MAWD, a result consistent with Barker et al. (1970). On the other hand, the microwave observations by Clancy et al. (1992) resulted in a water vapor abundance that was roughly one-half that of the MAWD estimate.

Comparison of TES and MAWD water vapor retrievals revealed an overall similarity, but the seasonal distribution obtained by MAWD shows significantly lower water vapor abundance near the south pole during the summer season than later measurements by TES (Smith, 2002). This difference was likely caused by the high dust aerosol loading during the Viking mission leading to the underestimation of water vapor due to the neglect in the MAWD retrieval of the effect of multiple scattering of light by aerosols. Smith (2002) argued that, in contrast to the 1.38 μm water vapor band used by MAWD, the rotation bands in the 28–42 μm spectral region used to retrieve water vapor abundance from TES observations were of sufficiently long wavelength to be significantly less affected by aerosol scattering. The disagreement between MAWD and TES retrievals was further reduced by a reanalysis of the MAWD data taking into account aerosol scattering (Fedorova et al., 2004).

MGS, MEX, and MRO retrievals provide a generally coherent water vapor dataset covering seven Martian years (MY 24–30). However, a more thorough comparison between TES and the PFS, SPICAM, and OMEGA instruments (Korablev et al., 2006; Tschimmel et al., 2008; Maltagliati et al., 2011b) showed systematic differences. Such intercomparisons eventually led to important improvements to the original retrieval algorithms

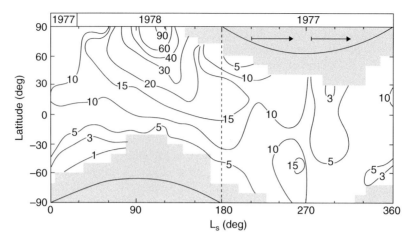

Figure 11.2. The seasonal map of water vapor obtained by MAWD/Viking 1 and 2 (Jakosky and Farmer, 1982).

applied to MAWD (Fedorova et al., 2004, 2010) and TES (Fouchet et al., 2007; Pankine et al., 2010) observations, with the revised water vapor abundances being significantly lower than the original values in both cases.

11.2.1.3 Temporal Variations: Inter-Annual Timescale

Differences between instruments led to the suggestion of inter-annual variability in the Martian water cycle (Jakosky and Barker, 1984; Clancy et al., 1996). While the comparison of the absolute value of water vapor abundance between instruments has proved to be technically challenging, a consistent comparison of the major trends can nonetheless be attempted. Apart from a large and still unexplained variability reported in the southern summer from ground-based measurements (Barker et al., 1970; Jakosky and Barker, 1984; Clancy et al., 1992), all observations of water vapor made from orbit show a consistently higher water vapor abundance near the summer pole and a seasonal evolution of water vapor qualitatively identical from year to year.

A closer examination of water vapor abundances retrieved by TES, shown in Figure 11.3, does in fact reveal inter-annual variations in the details of the spatial distribution. However, these variations are relatively modest when compared to the inter-annual variations observed for atmospheric dust. Smith (2004) notes ~20% more water vapor in the southern hemisphere polar summer maximum in the TES observations for MY 24 as compared to MY 25 and 26, as well as changes in the details of the spatial distribution of water vapor in the north polar region from year to year. Pankine et al. (2010) found that during MY 25 the water vapor column abundance poleward of 80°N latitude was lower than further south along the edge of the cap, while in MY 24 and 26 the water vapor abundance above the residual cap was the same as or higher than that above the edge of the cap. However, despite these differences, Pankine et al. (2010) also noted that the total integrated mass of water vapor in the north polar region was essentially the same for all three years (MY 24–26) observed by TES. Additionally, Pankine et al. (2010) show that details such as the annular structure of water vapor near the north pole in summer were also observed by Viking MAWD and Mars Express PFS.

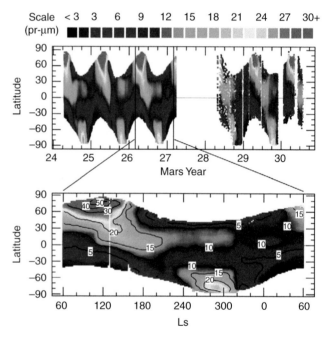

Figure 11.3. An overview of the current annual cycle of water vapor column abundance versus season and latitude over several Mars years. Results from TES (MY 24–27) and CRISM (MY 28–30) are shown. There is a repeatable pattern, with highest water vapor abundance at high latitudes of the summer hemisphere. A black and white version of this figure will appear in some formats. For the color version, please refer to the plate section.

While the existence of significant inter-annual variability cannot be dismissed (although no large inter-annual variation was ever reported by the same instrument), it is likely that the same processes that shape the seasonal and spatial distribution of water vapor operate in a similar way each Martian year and have done so for at least the past 35 years (see further discussion in Section 11.3.2). The two large differences apparent in Figure 11.3, with the decreases in water vapor at MY 25 at $L_s = 180°$ and MY 28 at $L_s = 270°$ are artifacts caused by the difficulties of performing accurate retrievals during the global-scale dust storms at those times (Smith et al., 2002; Smith, 2009).

11.2.1.4 Temporal Variations: Seasonal Timescale

Recent observations now permit a good overall characterization of the distribution of water vapor as a function of season and location, as shown in Figure 11.3. The current climate has a global annually averaged column abundance of about 10 pr μm, with higher abundance at high latitudes in the hemisphere where it is spring or summer. The northern hemisphere summer high-latitude maximum reaches a peak column abundance of roughly 50 pr μm, while the corresponding southern hemisphere summer maximum is weaker and more variable from one Martian year to the next, usually reaching about 25 pr μm.

Despite the modest inter-annual variation in water vapor abundance, the global seasonal trend follows the same general pattern from year to year (see Figure 11.3). The highest water vapor column abundance occurs poleward of 70°N latitude during early summer ($L_s = 110$–$120°$) after sublimation of the bulk of the seasonal polar cap. At this time, water vapor abundances decrease monotonically from north to south with an abundance of roughly 50 pr μm near the pole, 20 pr μm at 15°N latitude, and less than 5 pr μm in the southern hemisphere. After $L_s = 130°$, water vapor abundances rapidly decrease in the north polar region, falling below 10 pr μm by $L_s = 170°$. This decrease in the north is partially balanced by a corresponding increase in water vapor that moves to the south as the season progresses. At 45°N latitude, water vapor reaches a maximum abundance at $L_s = 135°$, while at 30°N latitude maximum water vapor is attained at $L_s = 150°$. By $L_s = 170°$ a well-developed maximum in water vapor develops between the equator and 30°N latitude, which persists throughout northern hemisphere fall and winter until $L_s = 40°$ in the following year when water vapor begins to rapidly increase again throughout the northern hemisphere.

In the southern hemisphere there is a gradual rise in water vapor abundance throughout the southern spring ($L_s = 180$–$270°$) as water vapor is transported southward from the northern hemisphere summertime maximum. In late southern spring ($L_s = 220°$) water vapor increases at high southern latitudes with the sublimation of the southern seasonal polar cap. Maximum southern hemisphere water vapor occurs around $L_s = 290°$. The peak abundance of water vapor at the southern hemisphere summer maximum is somewhat variable, but is generally about one-half that of the maximum value during northern hemisphere summer.

After $L_s = 300°$ water vapor abundance near the south pole decreases, and after $L_s = 330°$ the decrease in water vapor becomes planet-wide. The period between $L_s = 330°$ and $40°$ is the driest time of the year overall, with 5 pr μm or less water vapor column abundance over most of the planet. Water vapor abundance begins to increase significantly once again in the northern hemisphere after $L_s = 40°$, steadily climbing to its peak value once again at $L_s = 120°$, while the abundance of water vapor in the southern hemisphere remains mostly unchanged at a very low levels until southern spring.

As described in more detail in Chapter 5, water ice clouds also have a distinctive seasonal pattern (e.g. Kahn, 1984; Tamppari et al., 2000; Benson et al., 2003; Liu et al., 2003; Smith, 2004, 2009), which is related to the seasonal and spatial variation of atmospheric temperature, and to a lesser extent that of water vapor. The most prominent water ice clouds are those that form a low-latitude belt during the aphelion season. This cloud belt (also known as the aphelion cloud belt) begins to form around $L_s = 0°$ and continues to build in intensity and spatial coverage until around $L_s = 80°$. The clouds have significant optical depth between 10°S and 30°N latitude, with local enhancements over areas of elevated topography, especially the Tharsis volcanoes, Olympus Mons, Elysium, and the Lunae Planum region to the north of the Valles Marineris. The aphelion cloud belt persists until around $L_s = 140°$ at which time it rapidly dissipates, although clouds remain over the volcanoes throughout much of the year. At high latitudes, widespread cloud formation covers the polar regions (the "polar hoods") during the autumn and winter seasons, especially in the northern hemisphere where clouds can reach as far south as 30°N latitude. There is some indication that the polar hood clouds have the highest optical depth near the edges (Liu et al., 2003). The role of clouds in the water cycle is fundamental, as explained in Section 11.3.2.

11.2.1.5 Temporal Variations: Diurnal Timescale

The study of diurnal variability has important implications for water vapor because it allows one to explore the influence of local processes. Water vapor column abundances can vary with local time due to water condensing onto the surface or in the atmosphere, or through exchanges with the regolith. Diurnal changes may potentially occur during the passage of air masses with different origins, but this would require horizontal advection phased with the diurnal cycle, which is not the dominant periodicity of transport.

Several observations indicate that local diurnal variations of water do exist that can reach a factor of 2–3, showing a strong dependence on both location and season (Jakosky et al., 1988; Titov et al., 1995; Hunten et al., 2000; Sprague et al., 2003). Such a variation is equivalent to a change in the absolute abundance of about 10 pr μm, However, the measurements performed to date are still too sparse to reach definite conclusions about the nature of these variations, and some observations yield conflicting trends. For instance, the OMEGA instrument (Melchiorri et al., 2009) observed minor diurnal variations of only a few precipitable micrometers during the day near the south pole, while Maltagliati et al. (2011b) did not find any prominent diurnal signature (at a level of, say, 10 pr μm) from a global comparison between OMEGA (08:00 to 18:00) and TES (14:00) datasets.

The most straightforward evidence for a diurnal activity of water vapor has been established by the landed Phoenix mission, which has operated at 68°N during the northern mid-summer. The near-surface ($z \sim 2$ m) atmospheric humidity measured by the TECP instrument indicated a rapid fall of the partial pressure of water vapor after 18:00, reaching a value of 0.1 Pa at 20:00 that is a factor of 20 smaller than the 2 Pa observed during the rest of the daylight hours (Smith et al., 2009). The SSI camera of Phoenix, supported by co-located sensing by CRISM, has reported diurnal change as large as 15 pr μm on $L_s = 108.3°$. This result cannot be attributed to the sole action of diurnal condensation and/or sublimation of water on the surface and/or in the atmosphere. First, the potential for diurnal fluctuations induced by cloud formation and/or disappearance is too small due to the low water content of the polar clouds (~2 pr μm),

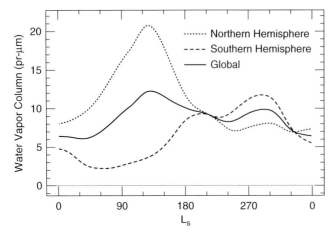

Figure 11.4. The average water vapor abundance averaged over each hemisphere as a function of season for the northern hemisphere (dotted line) and southern hemisphere (dashed line). The global average is also shown as a solid line. An average of 10 pr µm water column is equal to 1.45×10^{12} kg over a hemisphere and 2.90×10^{12} kg over the globe. There is a significant asymmetry between the northern and southern hemispheres.

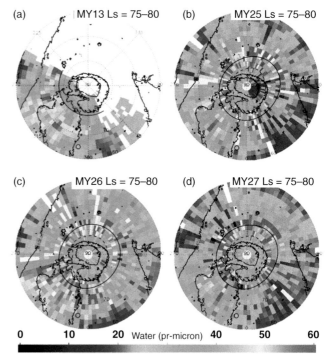

Figure 11.5. From Pankine et al. (2010). Polar maps of water vapor column abundances at $L_s = 75\text{--}80°$: (A) MAWD MY 13; (B) TES MY 25; (C) TES MY 26; (D) TES MY 27. East longitude system with $0°E$ at the bottom. Black contours are MOLA topography at 1 km interval. The boundaries of the H_2O and CO_2 seasonal caps on maps of TES data are shown by black solid and dashed nearly oval curves, respectively. A black and white version of this figure will appear in some formats. For the color version, please refer to the plate section.

as observed by the Phoenix LIDAR (Whiteway et al., 2009; Tamppari et al., 2010). Second, Phoenix did not report any surface frost activity in the late evening, while early-morning CRISM observations showed evidence for a frost layer of no more than 5–10 µm thick. An additional source/sink is therefore needed to account for the 15 pr µm daily changes in water vapor, suggesting a substantial fraction (up to 10 pr µm) controlled by exchanges with the regolith (Tamppari et al., 2010).

11.2.1.6 Spatial Variation: the North Polar Cap Sublimation

The current annual cycle of water vapor as shown in Figures 11.3 and 11.4 is dominated by the rise and fall of water vapor abundance at high northern latitudes during the spring and summer as the north polar seasonal cap sublimates. An analysis of north polar water vapor observed by TES during this important season by Pankine et al. (2009, 2010) provides details about the processes involved. During this period, maximum water vapor abundances form an annulus around the north pole at the edge of the retreating north polar seasonal ice cap. This annulus appears to be caused by the sublimation of surface frost as it is exposed by the retreating seasonal frost (see Figure 11.5). Water vapor can then recondense on the edge of the polar cap, leading to an annulus with increasing water vapor abundance as the cap retreats (Houben et al., 1997; Bass and Paige, 2000; Richardson and Wilson, 2002a; Montmessin et al., 2004; Appéré et al., 2008; Pankine et al., 2010). Observations by TES also show variations in water vapor abundance over small spatial (~100 km) and short temporal (<10 sol) scales, with discrete releases of water vapor from localized sources giving rise to short-lived concentrations of water vapor near the edge of the north polar residual cap (Pankine et al., 2009).

While the highest water vapor concentrations are found in the northern high latitudes during spring and summer, a substantial fraction likely migrates towards lower latitudes as evidenced

by the equatorward expansion of the water vapor concentration contours in the northern hemisphere (see Figure 11.3). Details on transport processes are given in Section 11.3.2.

11.2.1.7 Spatial Variation: Hemispheric Asymmetry

One of the most striking features of the global annual cycle of water vapor abundance shown in Figure 11.3 is the asymmetry between the northern and southern hemispheres. There are several key differences. The summer maximum that appears at polar latitudes in both hemispheres is about twice as strong in terms of maximum water vapor column abundance in the north than it is in the south. A second key difference is the deep minimum in water vapor covering the entire southern hemisphere during the southern fall and early winter. No corresponding minimum exists in the north. A third is the completely different behavior of the decay of the summer maximum between mid-summer and mid-fall. In the northern hemisphere, an area of relatively high water vapor abundance extends increasingly equatorward as the season progresses. Near northern fall equinox ($L_s = 180°$) water vapor appears to cross the equator and continue southward, eventually contributing to the formation of the summer maximum at high southerly latitudes. In stark contrast, the southern hemisphere summer maximum abruptly decays away in middle to late summer, with no indication of enhanced water vapor abundance extending equatorward and little indication of northward transport of water vapor.

Figure 11.4 shows this asymmetry in terms of hemispheric averages of water vapor column abundance as a function of season. There is significantly more water vapor in the northern hemisphere than in the south overall. The amount of northern hemisphere water vapor never reaches the very low levels observed in the southern hemisphere during southern fall and winter ($L_s = 0–120°$) because a moderate amount of water vapor is maintained at low northerly latitudes during the entire year. Although both hemispheres exhibit similarly timed early summer maxima in water vapor, the time dependence of the increase in water vapor leading up to the summer maxima are different for the two hemispheres. In the north, the increase begins rather abruptly at about $L_s = 40°$ and the entire increase occurs between $L_s = 40°$ and $120°$. In the south there is also an increase during the equivalent season ($L_s = 220–300°$) but it is much smaller. The majority of the increase in southern hemisphere water vapor occurs earlier during the winter and early spring and appears to be caused by transport of water vapor from the northern to southern hemisphere. Because of these differences, during the northern summer roughly 85% of all water vapor is in the northern hemisphere, while during the southern summer only 60% of all water vapor is in the southern hemisphere. As explained in Section 11.3.2, this asymmetry is caused by a corresponding seasonal asymmetry in the height at which tropical water ice clouds form. Together with the seasonal variation of the cross-equatorial Hadley circulation, they create a nonlinear pump favoring water transport to the north at the expense of the south.

11.2.1.8 Spatial Variation: Synoptic Scale

The top panel of Figure 11.6 shows the annually averaged water vapor abundance as a function of location on Mars. Immediately apparent is a correlation with topography, with low-altitude regions like Hellas showing relatively large water vapor and high-altitude regions like Tharsis showing relatively small water vapor abundance. Although water vapor is not in general distributed uniformly through the entire atmosphere, for much of the year water vapor is well mixed through enough of the densest part of the atmosphere so that water vapor abundance has some correlation with topography (Smith, 2002).

The effect of topography can be removed by dividing water vapor abundance by a quantity proportional to the surface pressure. The bottom panel of Figure 11.6 shows annually averaged water vapor divided by the quantity ($p_{surf}/610$ Pa), where p_{surf} is the annually averaged surface pressure at a given location in Pa. With this correction, the dependence of water vapor with latitude and longitude that is independent of topography is now apparent. The most obvious feature in this map is the pronounced maximum between 10°S and 40°N latitude, with roughly 50% more water vapor in this latitude band than outside it. Water vapor is maintained year-round at these latitudes.

The bottom panel of Figure 11.6 also shows large longitudinal variations in annually averaged water vapor, with two large regions of high water vapor abundance at low latitudes at 60–150°W longitude (covering the Tharsis region) and 300–10°W longitude (covering Arabia Terra). Although the highlands around Olympus Mons and the Tharsis volcanoes have high water vapor, the volcanoes themselves do not. This dependence of water vapor at low latitudes shows a strong

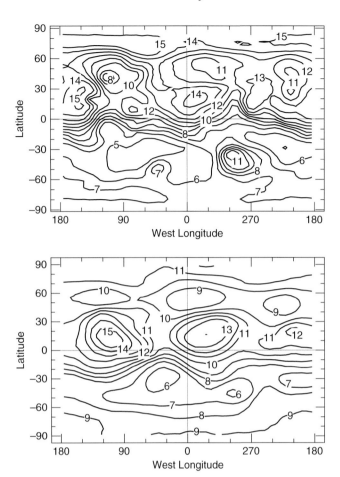

Figure 11.6. Maps of (top) annually averaged water vapor column abundance and (bottom) annually averaged water vapor column abundance divided by ($p_{surf}/610$ Pa) to remove the effect of topography. Values are in pr μm.

positive correlation with albedo (high water vapor in high-albedo regions) and a negative correlation with thermal inertia (Jakosky and Farmer, 1982; Smith, 2002). Water vapor abundance also has a negative correlation with surface pressure that has never been reproduced by models. Albedo, thermal inertia, and surface pressure have direct relations to the average and diurnal variations of surface and subsurface temperatures, the pore space size, and grain size of the regolith, and to the circulation patterns of the atmosphere. These can in turn influence the transport of water vapor and its adsorption and desorption into the regolith. Poleward of 40°N latitude and 60°S latitude there is much less longitude dependence in annually averaged water vapor and less correlation with surface properties. At these latitudes, interactions with the seasonal polar caps become most important.

11.2.1.9 Spatial Variation: the Vertical Distribution and the Issue of Saturation

No extensive climatology exists for the vertical distribution of water, as it does for column abundance. Still, the former is a unique indicator of the relative role of the various sources, sinks, and processes that control Mars' water cycle. Water vapor vertical distribution is the end product of a complex set of phenomena involving convective and/or turbulent mixing, global advection by winds, temperature variations, regolith–atmosphere exchange, and cloud-related processes such as condensation, evaporation, and sedimentation. One major factor constraining water vapor vertical distribution is related to saturation, the altitude of which is controlled by the thermal structure of the atmosphere. During northern spring the average atmospheric temperature in the tropics reaches its annual minimum since the planet is near aphelion. As a result the altitude of water saturation is low and water vapor has been observed to remain below an altitude of 10 km during that season. Subsequently, the saturation altitude in the tropics steadily increases to 40–60 km during southern spring because of the global rise of temperature as Mars approaches perihelion (Jakosky, 1985; Clancy et al., 1996).

It is interesting to note that the water vapor vertical distribution has important implications for dust aerosols. As mentioned above, the major hurdle for the vertical transport of water is the saturation altitude. The saturation ratio S is defined as the ratio of the local water vapor partial pressure to the equilibrium vapor pressure for ice P_{sat}^{ice} at a temperature T. The saturation water vapor pressure is given by (11.1), the reference formulation of Goff and Gratch (1946):

$$\ln P_{sat}^{ice} = 100\left[-9.09718\left(\frac{T_0}{T} - 1\right) - 3.56654\ln\left(\frac{T_0}{T}\right) \right.$$
$$\left. + 0.876793\left(1 - \frac{T_0}{T}\right) + \ln 6.11 \right] \qquad (11.1)$$

with P_{sat}^{ice} given in Pa, and with T_0 being the triple-point temperature (273.16 K). Once the saturation ratio exceeds unity, transformation of water into the solid state, and thereby cloud particle formation, is enabled. However, this condition alone is not sufficient, and a preliminary stage of the phase transition consists of the statistical creation of small icy embryos with a sufficient size to allow further growth by condensation. This stage is known as nucleation. For Martian conditions, homogeneous formation of these critically sized embryos directly from the vapor phase is difficult to achieve because of the enormously high supersaturation (~1000) it requires (Michelangeli et al., 1993; Colaprete et al., 1999; Montmessin et al., 2002; Määttänen et al., 2005). An alternative way for water to nucleate is to make use of the solid substrate provided by the surrounding airborne dust particles, which allows a reduction in the amount of energy needed to establish a mechanically stable interface between the crystal and its environment. This mode of heterogeneous nucleation on dust theoretically requires a saturation ratio of only 1.2–2 and should thus dominate the formation of water ice cloud particles whenever dust is available (Määttänen et al., 2005; Iraci et al., 2010). Nucleation establishes the first major link between the water and the dust cycles. The dust–water connection extends further through the scavenging of dust particles, which follows the fall and sublimation of the dust-containing icy crystals at lower subsaturated altitudes. Hence, the saturation altitude should limit not only the vertical transport of water but also that of dust. Observational clues of this capping effect have been recently provided by the Mars Climate Sounder (Benson et al., 2011).

Davies (1979) examined MAWD data over the Viking Lander 1 site and concluded that water vapor had a uniform distribution with height characterized by 150 ppm of water below 10 km at $L_s = 102°$. The first direct observation of the vertical profile of water vapor was provided by the Auguste solar occultation experiment of Phobos 2 that sounded the near-equatorial troposphere near equinox. These observations demonstrated a sharp decrease in water vapor mixing ratio above 25 km and a nearly constant mixing ratio of about 100 ppm below 20 km. SPICAM results for the end of summer (MY 28, $L_s = 130–160°$) in mid-northern latitudes showed a higher amount of water vapor above the saturation altitude (Fedorova et al., 2009) than expected from a vapor pressure standpoint. Further analysis of the SPICAM observations for the same season during the following Martian year (MY 29) showed additional evidence of water vapor well in excess of the saturation vapor pressure, exceeding it by a factor of up to 10 above 20 km in the northern summer tropics (Maltagliati et al., 2011a, 2013). This may reveal the additional feedback that dust has on water, since the removal of dust by scavenging above the saturation level will necessarily impede the possibility for the remaining water to nucleate and condense there again.

Below the saturation level, water vapor is usually assumed to follow a well-mixed profile, with a uniform mass mixing ratio prevailing down to the surface. Selected vertical profiles of water vapor are shown in Figure 11.7. Using the Phobos 2 Auguste solar occultation data collected near the equator during the $L_s = 1–19°$ period, Rodin et al. (1997) obtained a nearly constant mixing ratio within the 10–25 km altitude range, capped by a sharp decrease above 23–25 km (see Figure 11.7a). In contrast, several observations provide indirect evidence of water being strictly confined to the lowest 3 km of the atmosphere (Titov et al., 1999; Tschimmel et al., 2008). Using PFS observations, Fouchet et al. (2007) found an anticorrelation between water vapor and surface pressure during the northern spring equinox ($L_s = 330–60°$) and northern summer–early fall ($L_s = 90–200°$). They suggested that exchanges with the regolith could cause this anticorrelation and may thus account for up to 3–4 pr μm of the global atmospheric inventory. Smith (2002) found the strongest anticorrelation to occur during northern summer and on an annual basis found a trend for water to be more well mixed in the south than in the north. Several modeling results support confinement of water in the lowest 3 km, with the first hundred meters above the surface controlled by exchanges with the regolith (Zent et al., 1993; Savijärvi, 1995; Pathak et al., 2008).

11.2.2 Surface Reservoir

Surface ice accounts for the major seasonal source and sink of the water cycle, representing the most extensive reservoir accessible to the atmosphere. A new picture of the role of surface ice has emerged since the overview of Thomas et al. (1992), thanks to the discoveries made by recent orbiting missions complemented by new developments in modeling. For a long time, the polar caps of Mars were considered permanent fixtures of the planet. Now the perception has changed by considering the dynamic behavior of water ice at the surface on seasonal timescales.

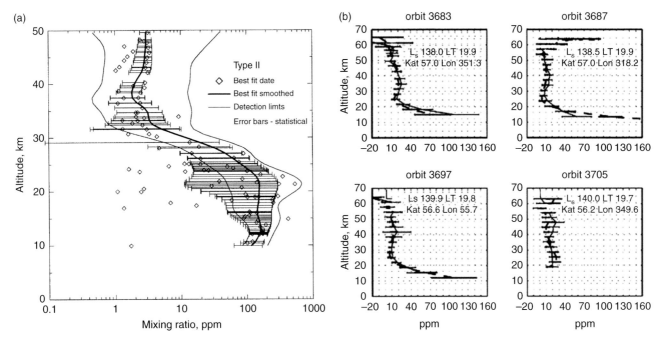

Figure 11.7. Vertical profiles of water vapor mixing ratio. (a) Auguste measurements reported by Rodin et al. (1997). (b) SPICAM profiles analyzed by Fedorova et al. (2009).

By combining the results of the Mars Orbiter Laser Altimeter (MOLA) on-board MGS (Smith et al., 1999) with those of the two radar sounding experiments on MEX (MARSIS; Picardi et al., 2005; Plaut et al., 2007) and MRO (SHARAD; Seu et al., 2007), it has been possible to confirm the upper limit on the contribution of the polar caps to the global water ice inventory by determining a total volume of $(3.2–4.7) \times 10^6$ km^3 for the two polar bulges. The radar results demonstrated that the subsurface ice component (such as observed by the Phoenix Lander and discussed in Section 11.2.3) was not thick enough to contribute significantly to the water inventory except in a few locations (e.g. lobate debris aprons; Plaut et al., 2009). An evaluation from the global coverage achieved by MARSIS demonstrated that the two perennial caps contribute about two-thirds of the global inventory of water ice on Mars (Mouginot et al., 2010). These polar caps formed very early in the history of the planet. However, it is known that water will migrate within a highly dynamic system forced by the chaotic evolution of the obliquity of Mars (Laskar and Robutel, 1993). Perennial water ice deposits play a major role in the water cycle by establishing the driving boundary conditions upon which most of the seasonal activity of water will depend.

11.2.2.1 Perennial Water Ice at High Northern Latitudes

Since the first telescopic observations, the perennial caps were thought to be composed of water ice because of their similarities with Earth's polar caps. For the north cap, this interpretation was confirmed by the thermal infrared observations of IRTM on-board Viking (Kieffer et al., 1976), which demonstrated that the observed temperatures of frost-covered areas were in the 180 K range, incompatible with a CO$_2$ ice substrate. The evolution with time of the albedo of the cap, with a marked decrease close to summer solstice, was considered by Kieffer

(1990) to result either from an increase of the mean grain size or from dust contamination. The dominant role of water ice in surface layers of the north perennial cap was later confirmed by the higher spatial observations of TES (Kieffer and Titus, 2001; Calvin and Titus, 2008). Crater counts led to age estimates of the surface of the perennial cap of a few$\times 10^5$ to 10^6 years (Herkenhoff and Plaut, 2000), confirming that the surface of the north polar cap is a recent geomorphological feature.

Radar experiments on-board MEX (MARSIS) and MRO (SHARAD) demonstrated that the entire north polar layered deposits (NPLD) associated with the polar altimetry protuberance identified by MOLA are mainly composed of water ice. The intricate internal structure, with a large number of closely spaced horizontal scattering surfaces (see Figure 11.8), is attributed to episodes of higher obliquity (Phillips et al., 2008). Recent modeling of the dielectric properties of the NPLD led to the conclusion that the overall dust content of NPLD was lower than 10% (Grima et al., 2009)

Near-infrared imaging spectrometry with OMEGA on-board Mars Express (Langevin et al., 2005) provided constraints on dust contamination and water ice grain size over the northern perennial cap and outlying ice-covered areas, as these parameters strongly impact the shape and strength of the 1.25, 1.5, 2, and 3 μm absorption bands of water ice. Dust contamination was estimated at less than 5 vol.% for ice filling outlying craters such as Korolev and less than 2.5% on the polar cap itself. Maps of the grain size of surface water ice as a function of time were also obtained (see Figure 11.9). The decrease in albedo of the central regions of the cap shortly after solstice observed at visible wavelengths (Bass et al., 2000) was attributed to a marked increase in mean grain size from <100 μm (a value typical of seasonal water ice frost, see below) to ~1 mm (which was associated with perennial ice). These results are fully consistent with modeling of water ice grain evolution as a function of dust contamination

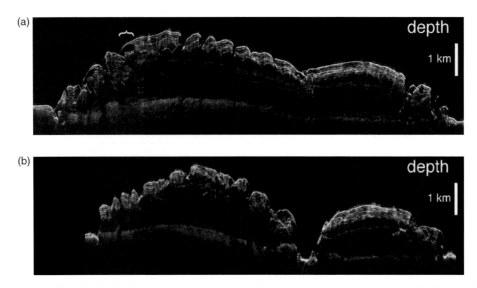

Figure 11.8. Two radargrams from SHARAD/MRO across the north perennial cap from the main lobe on the left to Gemina Lingula on the right. In (a) the radargram crosses just east of Chasma Boreale; in (b) it passes through it. The radargrams indicate that the polar cap is mainly composed of water ice, with a large number of reflecting interfaces. From Putzig et al. (2009).

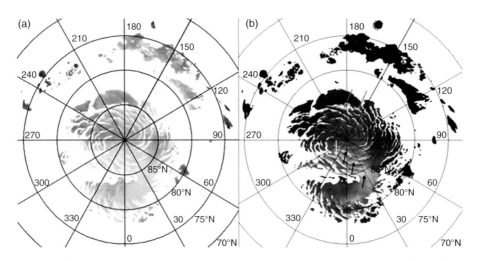

Figure 11.9. Maps of water ice signatures detected by OMEGA/MEX at (a) $L_s = 93$–$98°$ and (b) $L_s = 107$–$110°$ (from Langevin et al., 2005). The gray scale corresponds to different grain sizes, from ~1 mm (dark gray), corresponding to perennial water ice, to less than 100 µm (light gray), corresponding either to the last surviving patches of seasonal frost in central areas of the cap or to deposits of medium-sized water ice grain deposits at the edge of the cap (gray arrows) most likely linked to katabatic winds. The small-grained covered area at 330°E, 87°N (white arrow) corresponds to the "cool bright anomaly" observed by TES (Kieffer and Titus, 2001).

(Barr and Milkovich, 2008), which leads to a maximum size of 1 mm for dust contaminations lower than 2 vol.%. A few areas retain small-sized ice grains during early summer. Some were already identified as "cold bright anomalies" by TES (Kieffer and Titus, 2001). Accumulations of small-sized water ice were observed at the edges of the cap, probably linked to transport processes by katabatic winds. An increase of the albedo of outlying craters such as Korolev (165°E, 73°N) was attributed to a decrease in surface dust contamination due to sublimation.

Later work by Vincendon et al. (2007) studied the evolution of the column density of aerosols and of the surface reflectance spectra of an ice-filled crater obtained by OMEGA. This study demonstrated that albedo variations of ice-covered regions at northern high latitudes were due to a combination of three processes: (i) grain size evolution, in particular the final stages of sublimation of water seasonal frost; (ii) deposition and removal of surface dust; and (iii) changes in the aerosol optical thickness above ice-covered regions.

Aerosol scattering decreases the contrast between bright ice-covered regions and dust-contaminated or ice-free regions. Taking into account this effect impacted only marginally the conclusions of Langevin et al. (2005), as the measured optical thicknesses are low in early summer at high latitudes.

The high spatial resolution (20 m/pixel) provided in the near-infrared by CRISM made it possible to investigate in detail ice deposits in small outlying craters such as Louth (Brown et al.,

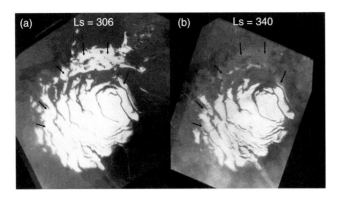

Figure 11.10. South seasonal cap observed by the Mars Orbiter Camera in 2000 (MY 24) at (a) $L_s = 306°$ and (b) $L_s = 340°$ (from James and Cantor, 2001). The black arrows correspond to regions at the edge of the cap or constituting a large outlier which were still very bright at $L_s = 306°$ (albedo ~60%) while presenting a lower albedo (35–45%) at $L_s = 340°$, a period in late southern summer, which corresponds to the minimum extent of the bright areas. There was no global dust storm in the summer of this Martian year.

2008), with coordinated observations by HiRISE providing information at a sub-meter scale on the surface structure of ice deposits. The dust contamination levels (<2% in the center, <4% at the edges of the deposit) and range of grain sizes are consistent with those determined for outlying ice-filled craters, indicating comparable regimes of water ice accumulation.

Observations of the northern perennial cap at visible (Benson and James, 2005), near-infrared, and thermal infrared wavelengths (Kieffer and Titus, 2001; Calvin and Titus, 2008) are now available over many Martian years. The outline and characteristics of the perennial cap appear remarkably stable, with most of the observed variability being attributed to aerosols, clouds, and transient frosting.

11.2.2.2 Perennial Water Ice at High Southern Latitudes

The bright southern perennial cap is much smaller than the northern perennial cap, being restricted to latitudes poleward of 84°S from longitudes 240°E to 30°E at minimum extent ($L_s = 335$–$340°$, Figure 11.10). These bright areas were initially attributed to water ice deposits. However, temperatures <160 K measured by Viking IRTM (Kieffer, 1979) demonstrated that CO_2 ice was the dominant component of surface ice over the south perennial cap. The volume of this CO_2 ice reservoir is estimated to account for only 3% of the total atmospheric inventory and its release into the atmosphere should therefore have a minor influence on Mars' present-day climate (Bibring et al., 2004). Observations at visible wavelengths (James and Cantor, 2001; James et al., 2007, 2010) demonstrated that the final stages of the retreat of CO_2 ice showed significant inter-annual variability. For Martian years with no global dust storm, the extent of CO_2 coverage at $L_s = 305$–$315°$ (Figure 11.10a) is still significantly larger than the minimum extent at $L_s = 340°$ (Figure 11.10b). If a major dust storm occurs during late southern spring, CO_2 coverage at $L_s = 305$–$315°$ has already reached a minimum extent that is very similar to that of years without global dust storms, with little further changes until $L_s = 335$–$340°$.

This major hemispheric composition asymmetry between the northern perennial cap (surface H_2O ice) and the southern perennial cap (CO_2 ice) was attributed to the 7 km altitude difference between the basal areas of the two caps. However, MOLA revealed that the two polar bulges were very similar, with a thickness of several kilometers in the south as well as in the north. Detailed observations of the south perennial cap demonstrated that water ice was present below a thin veneer of CO_2 ice of at most 8 m thickness (Byrne and Ingersoll, 2003; Tokar et al., 2003). Radar investigations by MARSIS (Plaut et al., 2007) confirmed that the south polar layered deposits are dominated by water ice, similarly to the NPLD.

The first success in the search for exposed perennial (summer-surviving) water ice at high southern latitudes was the detection by TES of a small region with temperatures intermediate between that of CO_2 ice and that of dust-covered areas (Titus et al., 2003). The observed temperature in this area was relatively uniform at the 3 km/pixel resolution provided by TES. This supported the proposed interpretation but did not exclude spatial mixing between CO_2 ice-covered and ice-free areas at much smaller scales resulting in such intermediate temperatures. The observations at near-infrared wavelengths by OMEGA shown in Figure 11.11 provided the first spectroscopic identification of water ice deposits at the edge of the CO_2 perennial cap and within a large outlier (see Figure 11.11d). The albedo of these deposits (~45%) indicates a high level of dust contamination, which prevented their identification from observations at visible wavelengths. The H_2O ice area observed by TES was confirmed by OMEGA, but it is among those exhibiting the weakest water ice signatures, and OMEGA observations of another region identified by Titus (2005) as H_2O ice-covered on the basis of its apparent temperature exhibited a spectrum indicating spatial mixing. Spectral signatures of water ice are observed over the perennial cap, indicating significant intimate contamination of CO_2 ice by less than 0.1% of H_2O ice subsequent to a cold trapping mechanism (see Section 11.3.2). Observations of intermediate temperature areas with Mars Odyssey THEMIS (Piqueux et al., 2008) confirmed the OMEGA water ice detections at the edge of the perennial cap and extended the outlier observed by OMEGA. Observations in 2009 (Figure 11.11d) spectrally confirmed the large extent of this water ice outlier. The origin of perennial water ice at high southern latitudes was tentatively attributed by Montmessin et al. (2007) to orbital precession over 25 000 year timescales.

Radiative transfer modeling (Douté et al., 2007) indicates that perennial water ice at high southern latitudes is a component of permafrost. There is no spectrally significant admixture of CO_2 ice. The regions at the edge of the perennial cap are characterized by a dust content of ~25% and a large water ice grain size (~300 μm). These deposits are interpreted as outcrops of the water ice-rich terrains buried under the CO_2 cap. The smaller grain size compared to the north perennial water ice (~1 mm) can be explained by grain evolution with a large dust load (Barr and Milkovich, 2008). The water ice grain size observed in outlying regions is significantly smaller (100–150 μm) than that of water ice at the edge of the cap. Douté et al. (2007) therefore propose that such areas could correspond to surface water ice deposits that survive through southern

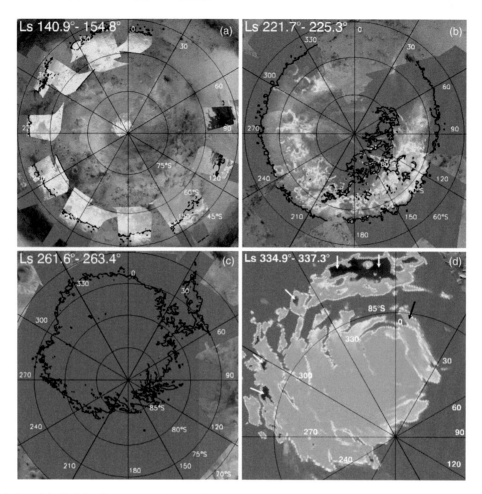

Figure 11.11. Evolution of the H_2O ice signature at 1.5 μm as observed by OMEGA/Mars Express: (a)–(c) during the retreat of the southern seasonal cap, in MY 27, from Langevin et al. (2007); and (d) over the perennial cap, observations in 2009, MY 29, with a more comprehensive coverage than that presented in MY 26 by Bibring et al. (2004). The rainbow color scale ranges from a band strength of 45% (dark blue) to 0% (red). The black line indicates the boundary of the seasonal ice cap as inferred from CO_2 ice spectral signatures. At $L_s = 340°$, the CO_2 ice signatures have reached their minimum extent corresponding to the perennial cap (Figure 11.10b). The perennial CO_2 cap is contaminated by H_2O frost (Bibring et al., 2004; Douté et al., 2007). Regions at the boundary of the cap (white arrows) present an intermediate albedo with no spectral signature of CO_2 frost and strong signatures of H_2O frost. The region identified by TES (black arrow) as H_2O ice-covered on the basis of its intermediate temperature presents weaker spectral signatures of H_2O ice than other regions covered by H_2O frost at the boundary of the CO_2 cap. The full extent of the large outlier (330°E to 10°E, 82°S to 84°S) was first observed by Themis (Piqueux et al., 2008). There is a striking similarity between the map of H_2O ice signatures at $L_s = 335–337°$ and that of bright regions covered by CO_2 frost at $L_s = 306°$ extending beyond the minimum extent of the CO_2 cap (arrows, from Figure 11.10a), which was obtained five Martian years earlier. A black and white version of this figure will appear in some formats. For the color version, please refer to the plate section.

summer. However, an important clue is the almost perfect match noted by Piqueux et al. (2008) between the regions presenting an H_2O ice spectral or thermal signature at the time of minimum extent of CO_2 ice (Figure 11.11d) and those still covered by CO_2 ice at $L_s = 306°$ for Martian years with no global dust storm (Figure 11.10a). In this respect, water ice outliers behave similarly to regions close to the boundary of the CO_2 cap. Transient CO_2 frost at lower latitudes has recently been used as a marker for subsurface water from thermal modeling (Vincendon et al., 2010b). Water ice outliers therefore are most likely related to (but may not be outcrops of) subsurface H_2O ice.

The total area covered by perennial water ice at high southern latitudes is ~20 times smaller than the area covered by perennial water ice in the north. Furthermore, most areas covered by surface water ice in the north are directly exposed to sunlight before the summer solstice while southern perennial water ice remains protected by CO_2 seasonal ice until very late in the summer ($L_s = 325°$), providing the southern exposed perennial ice with a very strong stabilization mechanism. This explains why the sublimation of northern H_2O surface ice completely dominates the seasonal water cycle, as discussed in the previous sections.

11.2.2.3 Seasonal Water Ice Frost in the South

The seasonal evolution of water ice in the southern hemisphere demonstrates the dominant role of the northern polar sources of water in the atmosphere. Given the much higher sublimation temperature of water ice (~180 K) compared to that of CO_2 ice (~145 K) at Martian standard surface pressure, it was expected

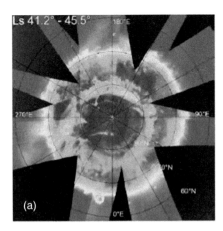 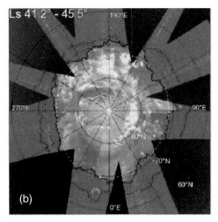 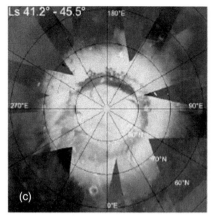

Figure 11.12. Evolution of the northern seasonal deposits in terms of (a) H_2O ice band depth at 1.5 μm, (b) CO_2 ice band depth at 1.43 μm, and (c) reflectance factor at 1.08 μm, for L_s = 41.2–45.5°. From Appéré et al. (2011). The OMEGA albedo maps are displayed over MOC albedo data acquired in summer (Caplinger and Malin, 2001) as a background. CO_2 ice and H_2O ice maps are displayed over MOLA topography (Zuber et al., 1998). Three boundaries are plotted on panel (b): the black outline corresponds to the OMEGA CO_2 ice boundary, the blue outline is the OMEGA H_2O ice boundary, and the white dashed outline is the TES "crocus line" (limit of low temperatures compatible with surface CO_2 ice, averaged over three Martian years of TES observations) computed for the L_s range. A black and white version of this figure will appear in some formats. For the color version, please refer to the plate section.

that water ice frost should lag behind the retreating edge of the seasonal cap (Houben et al., 1997), mostly composed of CO_2 ice with a maximum thicknesses of ~1 m (Smith et al., 2001; Prettyman et al., 2009). Yet a systematic survey of the retreat of the south seasonal cap by TES (Kieffer et al., 2000) did not reveal any extended regions with temperatures that could be interpreted as water ice frost at the edge of the retreating seasonal cap.

The extensive observations of the retreat of the south seasonal cap by OMEGA (Langevin et al., 2007; Schmidt et al., 2009) made it possible to monitor the evolution of spectral properties as a function of L_s. The distribution of these deposits is very consistent with the results of general circulation models (Forget et al., 1999; Montmessin et al., 2004; Colaprete et al., 2005). In southern mid-winter (L_s = 130–136°) water ice frost represents an extended deposit centered on the Hellas Basin, a region that corresponds to extensive cloud activity during that season. Prevailing winds extend this deposit towards the southeast later during southern winter (L_s = 140–155°, Figure 11.11a), sublimation and recondensation of water leading to water ice surface contamination of the outer regions of the seasonal cap except for a "dry spike" west of Hellas (Langevin et al., 2007). In early southern winter, surface H_2O frost deposits have also been observed on pole-facing reliefs as far north as 13°S by OMEGA and CRISM (Vincendon et al., 2010a). Ice cloud activity and contamination of the seasonal cap by surface ice decreases during southern spring. In mid-spring, optically thick water ice frost deposits are restricted to well-defined regions from 100°E to 290°E. These deposits have been observed at the same locations over three Martian years and the reproducibility of the water ice frost cover could be linked to ice grain nucleation and sedimentation in updrafts, as there is a strong correlation with topographic features such as crater rims. CRISM and HiRISE observations made it possible to investigate H_2O frost deposits at a much smaller spatial scale in southern craters (Kereszturi et al., 2011).

The sublimation–recondensation–sedimentation process postulated by Houben et al. (1997) is inhibited after mid-spring, and the seasonal cap shortly before the summer solstice (Figure 11.11c) does not exhibit any spectral signature of H_2O ice. This "dry" situation persists until the retreat begins to uncover the water ice-rich edges of the CO_2 perennial cap after L_s = 310°. These observations confirm the relatively drier climate of the south compared to the north.

11.2.2.4 Seasonal Water Ice Frost in the North

The evolution of the northern seasonal deposits appears simpler than that of the south based on observations at visible wavelengths (James, 1979), with a nearly axisymmetric distribution and a relatively smooth northward motion of the cap edge during retreat (note that observations of the expanding phase are hampered by the polar night at visible and near-IR wavelengths). Observations at visible wavelengths by MOC/MGS (Benson and James, 2005) and MARCI (Cantor et al., 2010) combined with those by OMEGA at 1.05 μm (Appéré et al., 2011; see Figure 11.12) provide a climatology covering many consecutive Martian years. The inter-annual variability of the latitude of the outer boundary at a given L_s is only ~1° rms.

The observations by TES (Kieffer and Titus, 2001) demonstrated that the receding north seasonal cap was surrounded by a bright annulus of water ice-covered regions extending over ~5° in latitude, fully in line with expectations from the sublimation–recondensation conjecture predicted by models (Houben et al., 1997). The edge of this outer part of the seasonal cap dominated by H_2O frost moves northward at a nearly constant rate of ~0.16°/day. At latitudes of 70–75°N, the temperature remains lower than 165 K until L_s = 75°, indicating that CO_2 ice is still present at this stage of the retreat. The advance of the cap starts with water ice frost deposition (L_s = 140° at latitudes of 70–75°N), CO_2 deposition being initiated close to the northern fall equinox. An intriguing feature is the fact that the ring of dark materials surrounding the perennial cap shows up at visible wavelengths from the spring equinox to the end of the retreat, albeit with a much lower albedo contrast than in summer. In early spring the north seasonal cap is still expected to be relatively thick at such latitudes (75–85°N). Kieffer and Titus (2001) interpreted this low albedo as demonstrating that photons travel through a thick translucent slab of CO_2 ice ("cryptic" behavior).

The observations by OMEGA since January 2004 made it possible to monitor the spectral signatures of H_2O and CO_2 ices during the retreat of the northern seasonal frost. The best coverage and time resolution were obtained for MY 27 and 28 (Appéré et al., 2011). These observations confirm through direct spectral identification that the outlying bright ring corresponds to water ice frost. In early spring, the CO_2 ice signature recedes much faster than the edge of the H_2O frost-covered regions, and by $L_s = 45°$ most of the seasonal cap is spectrally dominated by H_2O ice, exhibiting only weak spectral features of CO_2 ice. The disappearance of CO_2 ice signatures in early spring was observed by CRISM over the Phoenix landing site (Cull et al., 2010a,b). In contrast to these authors, Appéré et al. (2011) conclude from radiative transfer modeling that a decrease in CO_2 grain size cannot account for the observed spectral evolution, which requires the formation of an optically thick H_2O ice cover (200 μm or more). This cover is continually renewed by the sublimation–recondensation cycle, moving northward and getting thicker while CO_2 ice continues to sublimate through the overlying layer of water ice.

While the major asymmetry between the exposed perennial reservoirs of ice at the two poles has been acknowledged for a long time, recent observations demonstrate that a strong asymmetry also exists in the composition of the seasonal frosts in mid-spring: the seasonal cap is dominated in the north at $L_s = 45°$ by water ice with a few patches of CO_2 ice directly exposed at the surface. At the corresponding season in the south ($L_s = 225°$, Figure 11.11b), CO_2 ice is directly exposed over most of the seasonal frost, with water ice being observed only over well-defined regions. These asymmetries are an important aspect of surface–atmosphere interactions in the context of the water cycle.

11.2.3 Subsurface Reservoir

11.2.3.1 Forms of Subsurface Water

Subsurface water on Mars today is known to take several forms, as water vapor and ice in the soil pores, and adsorbed thin films on the surfaces of individual soil grains. Additionally, water can be chemically bound within hydrated and hydroxylated minerals and may exist as brines under appropriate vapor pressure and temperature conditions. Stable liquid groundwater, however, if present, will be largely limited to below the permafrost, which currently extends to at least kilometers depth (e.g. Kuzmin, 1983; Clifford, 1993).

Water vapor occurs in the soil pores where the regolith is in diffusive communication with the atmosphere, and wherever ice or adsorbed water occurs in the subsurface, regardless of a connection to the atmosphere. While water vapor represents the smallest subsurface reservoir of water, it is centrally important. Vapor acts as a primary conduit for the transport of water between different regions of the subsurface and in exchange

with the atmosphere (Clifford and Hillel, 1983; Fanale et al., 1986; Mellon and Jakosky. 1993).

Ice exists in the subsurface at depths and in geographic regions where the appropriate stability conditions are met today or were met in the recent past (Leighton and Murray, 1966; Feldman et al., 2008). Such ice, typically referred to as ground ice, has long been predicted to occur in the middle and high Martian latitudes poleward of about 40–50° (e.g. Leighton and Murray, 1966; Farmer and Doms, 1979; Mellon and Jakosky, 1993) and has been confirmed to exist in these same regions (Boynton et al., 2002; Feldman et al., 2002; Mitrofanov et al., 2002). Ground ice, whose spatial distribution is displayed in Figure 11.13, represents the largest identified reservoir of subsurface water. The total amount of water in this reservoir depends on the depth distribution of ice, which is largely unknown. However, assuming a nominal regolith porosity of 40%, the upper meter of high-latitude regolith alone would contain a global equivalent layer of at least 12 cm of water, i.e. more than 10 000 times the atmospheric inventory.

Adsorbed water consists of a thin film of water molecules that are weakly bound to the surface of individual soil grains. When a water vapor molecule collides with a mineral surface it may bounce off that surface or stick, through van der Waals bonding, for a period of time (de Boer, 1968; see also Jakosky, 1985). Conversely, adsorbed molecules will randomly escape the surface, returning to a vapor phase. The balance between sticking and escaping results in an adsorptive equilibrium. While residing in the adsorbed phase, water molecules are able to interact chemically with the minerals and migrate across the mineral surfaces, hopping between adsorption sites. The amount of adsorbed water contained within soil depends strongly on the specific surface area of the soil and only weakly on the soil mineralogy (Anderson and Tice, 1972; Zent and Quinn, 1997). Typically, adsorbed water can compose up to a few percent of the mass of soil, being highest at colder temperatures and higher partial pressures. Adsorbed water acts as a buffer to water vapor transport and affords an important and easily accessible reservoir for atmospheric exchange (e.g. Jakosky, 1983a; Mellon and Jakosky, 1993; Zent et al., 1993; Chevrier et al., 2008). While water ice requires specific pressure–temperature conditions to form, adsorbed water will occur in varying quantities at all temperature and pressure conditions. The amount of this water that is available for exchange with the atmosphere through typical temperature and humidity cycles depends on the surface area and the depth of diffusive penetration, but may be as much as 100 pr μm (atmospheric equivalent water).

In addition to vapor, ice, and adsorbed water, many minerals have been observed that contain chemically bound water, such as phyllosilicates (Milliken et al., 2007; Mustard et al., 2008), hydrated sulfate salts (e.g. Bibring et al., 2005), and perchlorate salts (Hecht et al., 2009; Leshin et al., 2013). Changes in temperature and humidity conditions may release some of this water into the atmosphere over time (Chou and Seal, 2003; Chipera and Vaniman, 2007). However, strongly chemically bound water, such as hydroxyl ions bound within phyllosilicates, is generally inaccessible for atmospheric exchange.

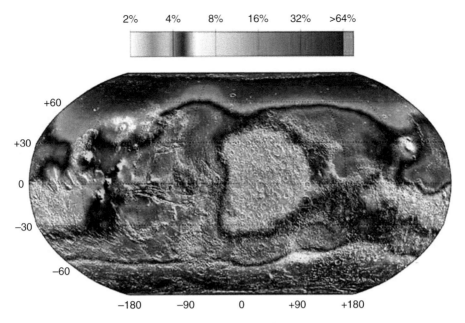

Figure 11.13. A Robinson projection of the water-equivalent hydrogen content of the semi-infinite layer of water-bearing soils derived from the Mars Odyssey GRS spectrometer. From Feldman et al. (2008). Regolith is filled with water ice fills poleward of 60° in both hemispheres. A black and white version of this figure will appear in some formats. For the color version, please refer to the plate section.

11.2.3.2 Subsurface Water: Stability and Distribution

Both ground ice and adsorbed water were predicted to be present on Mars (e.g. Leighton and Murray, 1966; Fanale and Cannon, 1974) before recent spacecraft observations confirmed their existence. Comparisons of ground temperatures and estimates of atmospheric humidity led Leighton and Murray (1966) to suggest ice-rich permafrost would exist at middle and high latitudes. Subsequent studies refined and expanded on this result, examining aspects of local climate conditions, soil properties, geographic variability, and climate change (e.g. Farmer and Doms, 1979; Fanale et al., 1986; Paige, 1992; Mellon and Jakosky, 1993, 1995; Aharonson and Schorghofer, 2006; Chamberlain and Boynton, 2006). Similar conclusions were reached and a consensus has developed that ground ice will persist today at shallow depths of typically ~10 cm below a layer of ice-free soil and poleward of about 40–50° latitude. Most ice-table depths range between 1 and 100 cm, being shallowest in polar regions and deepest near the equator (Mellon et al., 2004). Rarely, extreme depths of a few millimeters to a few meters also occur. Similarly, adsorbed water was predicted to be omnipresent in Martian soil based on laboratory measurements of adsorption on various mineral surfaces at low temperatures and for the expected humidity and temperature conditions found on Mars (Fanale and Cannon, 1971, 1974; Anderson and Tice, 1972; Zent and Quinn, 1995, 1997).

Imaging data from a variety of Mars-orbiting spacecraft have revealed an array of geomorphic evidence of past and current water ice in the subsurface. Large-scale features, such as lobate debris aprons, lineated valley fill, and concentric crater fill, suggest glacial-like deposits of ice (Squyres and Carr, 1986), and many impact craters exhibit lobate ejecta suggesting impact into a ice-rich substrate (Kuzmin et al., 1988), both of which indicate abundant ice-rich deposits at middle and high latitudes. Orbital radar sounding observations have shown that material consistent with thick water ice deposits exists today in some lobate debris aprons (Holt et al., 2008). These deposits are generally deeply buried and may offer limited exchangeability within the present climate; however, longer-term climate changes may periodically have unlocked these deposits. High-resolution images have also shown that the middle- and high-latitude Martian surface is dominated by polygonally patterned ground (Mangold, 2005; Levy et al., 2009). Such patterns are common in terrestrial ice-rich permafrost and form through repeated seasonal contraction cracking of the ice-cemented soil subsurface (Leffingwell, 1915; Lachenbruch, 1962). Their ubiquitous presence on Mars indicates the presence of shallow deposits of ice-rich material (Mellon, 1997; Mellon et al., 2008, 2009).

The gamma ray and neutron spectrometers on-board the Mars Odyssey spacecraft were the first to provide a detection of abundant water in the Martian subsurface (Boynton et al., 2002; Feldman et al., 2002; Mitrofanov et al., 2002). These spectrometers rely on the production of high-energy neutrons in the Martian soil through collisions with galactic cosmic rays and the moderation of neutron energy by hydrogen. Strong hydrogen signals in the middle and high latitudes of Mars are only explained by the presence of shallow deposits of water, which is believed to be ice in this deeply sub-freezing environment. The Phoenix spacecraft landed at a high-latitude location (68°N) where gamma ray and neutron spectrometers indicated high abundances of subsurface water, and where high-resolution images displayed extensive polygonal patterned ground (Arvidson et al., 2008). Direct excavation at the Phoenix site revealed abundant ice-rich ground at a shallow depth consistent with theoretical predictions (Mellon et al., 2009).

Adsorbed water and hydrated minerals have also been observed to occur on Mars. Hydrated sulfate salts were initially reported by Bibring et al. (2005). Milliken et al. (2007) reported hydrated mineral signatures over much of the equatorial and northern latitudes of Mars (from a few to 15 wt.%), with substantially stronger signatures over high northern latitudes. These signatures were attributed to a combination of chemically bound water and adsorbed water. Additionally, diurnal cycles in the dielectric properties of the ice-free soil overburden at the Phoenix landing site (68° N) were consistent with cycles in adsorbed water driven by soil temperature cycles, and regolith–atmosphere exchange of perhaps as much as 10 pr μm (atmospheric equivalent water) in a single diurnal cycle (Zent et al., 2010, 2016). Equatorial signatures of hydrogen in the leakage neutron observations have been attributed to moderate abundances of hydrated magnesium sulfates heterogeneously distributed over the Martian globe (Feldman et al., 2004; Fialips et al., 2005). Perchlorate salt at the Phoenix site at <1 wt.% concentrations may play some role in buffering atmospheric humidity, due to its strong deliquescent properties (Hecht et al., 2009).

11.3 NATURE OF THE PRESENT-DAY WATER CYCLE

11.3.1 Theoretical Studies

11.3.1.1 Previous Work

The use of modeling tools is a long-standing activity in the study of Mars' water cycle. Models provide a theoretical framework in which physical processes can be isolated and further quantified. While the theoretical study of the Mars water cycle was largely inspired by the desire to analyze and understand the MAWD observations of water vapor (Jakosky and Farmer, 1982), the idea that the Mars water cycle could be addressed in terms of physical processes started significantly prior to the MAWD era. Leighton and Murray (1966) made the first theoretical representation of the major seasonal cycles of volatiles, concluding on the likely water composition of the NPC. Leovy (1973) identified the potential importance associated with the cyclic condensation of the CO_2 atmosphere for the fate of water, yet failed to reproduce the observed seasonality of water vapor, leaving the question of carbon dioxide and water interactions unresolved.

As indicated by Richardson and Wilson (2002a), three major questions arose from the MAWD observations: (1) What is the cause of the observed north-to-south asymmetry in water vapor abundances? (2) Does the annual behavior of water correspond to a state of equilibrium, repeating itself year after year? (3) What controls the changes of the bulk abundance of water in the atmosphere? Nearly two decades after Viking, various orbiting experiments have shown that the behavior of water basically reproduces the same seasonal trends year after year and is therefore evolving in a stable manner apart from some evidence of minor inter-annual variability (see Section 11.2.1).

Davies (1981) was the first to investigate the latitudinal and seasonal variability of water vapor observed by MAWD, hypothesizing that exchanges of water were uniquely driven by deposition and sublimation of ice at and from the surface. The model of Davies (1981) used an *ad hoc* representation of

transport where tracer advection was modeled by horizontal diffusion with a prescribed timescale. Although very crude, the modeling work of Davies captured two of the most salient features of the present-day water cycle: (i) the water cycle is essentially controlled by the seasonal variation of temperature in the polar regions; and (ii) the water cycle is in equilibrium with the present-day orbital configuration of Mars, the latter controlling insolation distribution between the polar and the equatorial regions. Later studies using more sophisticated circulation models have essentially confirmed Davies' conclusions.

The first comprehensive assessment of the water cycle performed by Jakosky (1983a,b) followed a similar approach to Davies (1981) but with the inclusion of an active "wettable" regolith in addition to atmospheric transport and surface–atmosphere exchanges (note that a detailed discussion on the regolith is given in Section 11.3.3). Jakosky (1983a,b) acknowledged the central role played by the perennial polar caps in the entire cycle process and also concluded that a near-surface reservoir was necessary for successfully reproducing the seasonality of water vapor, estimating it to account for 10–40% of the atmospheric water variability. His work was nonetheless impaired by a parametric representation of atmospheric transport, a weakness that also affected Davies' work, and which later proved to bias the results significantly. James (1985) specifically investigated the role of the global CO_2 mass fluxes that take place in the polar regions to address the observed desiccated state of the southern hemisphere relative to that of the north. Seasonal condensation and sublimation of CO_2 ice induce a latitudinal pressure gradient that consequently forces meridional advection (20–30% of the total atmospheric mass is involved in this seasonal cycle). According to James (1985), this component of the circulation is made asymmetric between north and south by the eccentricity of Mars' orbit. The larger mass of CO_2 incorporated in the seasonal deposits of the southern hemisphere is released more rapidly than in the north due to Mars orbiting closer to the Sun at that season, thus biasing meridional tracer advection towards a net annual flux of water from south to north, a result that was later confirmed by Richardson and Wilson (2002a).

The first study including a non-parametric representation of atmospheric water transport was published by Haberle and Jakosky (1990), who used a two-dimensional axisymmetric circulation model to analyze the transport mechanism controlling water extraction from its northern pole source region in spring and summer. The conclusion of these authors emphasized two aspects of this critical stage of the water cycle. First, a sea-breeze circulation at the edge of the north residual cap was identified as the sole dynamical mechanism capable of carrying water equatorward. Second, the sea-breeze mechanism was predicted to be too weak in intensity to reproduce the observed summer moistening of the non-polar region. Something was lacking in the model, which Haberle and Jakosky (1990), confirming earlier statements by Jakosky (1983a,b), attributed to the seasonal "breathing" of the mid-latitude regolith through adsorption and desorption of water.

It is now known that all the studies performed to that date were missing a fundamental aspect of the Mars water cycle. The advent of three-dimensional GCMs, with a self-consistent determination of the wind and temperature fields, was the

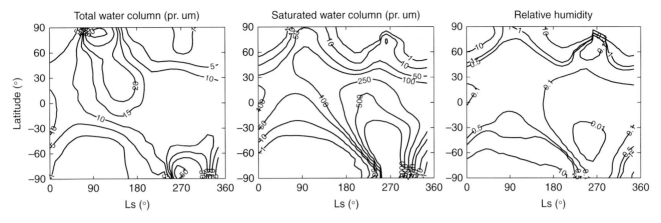

Figure 11.14. GCM simulation results showing (left) the predicted distribution of total water column abundances including the contribution of clouds (scaled to a pressure of 6.1 mbar), (middle) same as left panel except for fully saturated atmospheric columns (only vapor), and (right) the ratio of the left to middle panels, yielding the estimated relative humidity of the atmosphere.

key to a physically valid reproduction of the water vapor seasonal behavior (Houben et al., 1997; Richardson and Wilson, 2002a; Montmessin et al., 2004). With these models, the need to include a regolith as a seasonal water reservoir has become less obvious. Houben et al. (1997) claimed that the regolith had a large role as an exchangeable seasonal reservoir, but their conclusion was contaminated by an error in their model. The overstated importance of the regolith was likely the result of simpler models attempting to compensate for an inadequate transport description. Mars circulation is three-dimensional in essence, and the existence of residual components in the circulation, i.e. traveling and stationary atmospheric waves, accounts for a significant fraction of horizontal transport that parametric or zonally symmetric circulation models have difficulty representing. Additionally, the use of GCMs has allowed an evaluation of the role of water ice clouds. For a long time, clouds were thought to be a minor component of the water cycle. Compared to the Earth or Venus, Mars is certainly less cloudy, but cloud occurrence is phased with key processes of the water cycle that control water exchanges between the polar and nonpolar atmosphere and further exchanges between hemispheres (Richardson et al., 2002; Montmessin et al., 2004), a point that is further emphasized in Section 11.3.2.

11.3.1.2 Modeling Mars' Water Cycle With a General Circulation Model

Because they have brought the most comprehensive representation of Mars' water cycle, the following sections essentially discuss the results of GCM-based water cycle models. Except for a few specific aspects, most GCMs share a common philosophy based on the coupling of an atmospheric dynamics solver with a physics module where processes can be simulated separately for every model grid point (e.g. radiative transfer, turbulence in the boundary layer, CO_2 condensation and/or sublimation, etc.). To simulate Mars' water cycle, several other processes are generally included: condensation and/or sublimation of water ice at/from the surface, atmospheric condensation (to form clouds that can sediment), and advection of atmospheric water species by the GCM-predicted winds. For more detailed descriptions,

the reader is referred to the works of Houben et al. (1997), Richardson and Wilson (2002a), Montmessin et al. (2004), Böttger et al. (2004, 2005), and Navarro et al. (2014). Models are usually initialized with an "infinite" reservoir of water at the north pole and with a representation of the CO_2 residual cap at the south pole. No CO_2 ice is prescribed at the south pole *per se*, but is instead represented by a permanent boundary condition at the surface (Jakosky, 1983a,b; Richardson and Wilson, 2002a; Montmessin et al., 2004; Navarro et al., 2014). The latter is forced to maintain a temperature following that of CO_2 phase change, which is far below the condensation point of water and thus traps the near-surface moisture that flows over this area, a phenomenon that has long been speculated and has now been confirmed by observations (Bibring et al., 2004).

Horizontal resolution is a major limitation of current models. With a typical 5°×5° latitude and longitude resolution, models have difficulty representing major topographic features (e.g. Valles Marineris, the steepest volcanoes) as well as reproducing the detailed structure of the polar and seasonal deposits. In particular, the shape of the north polar cap (NPC) is only approximately represented, yet appears to significantly influence the sublimation processes of water from the pole (Haberle et al., 2011).

Approximately 10–20 years of model simulations are required to obtain an annually repeatable state, which reflects the timescale associated with the building up of a water vapor spatial distribution in balance with transport phenomena. After spin-up, converged fields can be analyzed. The results that are presented in the following sections have been obtained by the same model as the one used by Montmessin et al. (2004) and by numerous other authors (Lefèvre et al., 2004; Levrard et al., 2004; Forget et al., 2006). This model, which has since been updated in Madeleine et al. (2012) and in Navarro et al. (2014) to account for the effects of clouds on radiative transfer, provided the first consistent reproduction of the major water cycle observables, such as the TES water vapor and cloud distributions, producing results qualitatively in line with other GCMs (Richardson and Wilson, 2002a; Böttger et al., 2005). A first subset of results is displayed in Figure 11.14, showing the predicted spatial and temporal distributions of water vapor. The model matches the typical trends and values of the reference

TES dataset (Smith, 2004), with large inputs of water vapor at the summer poles, whereas an otherwise dry atmosphere prevails the rest of the year in the mid- to high-latitude regions.

Contrary to the conclusion of Davies (1979), the Martian atmosphere is far from using its full water-holding capacity, an issue that was emphasized by Richardson and Wilson (2002a). Figure 11.14 shows that, apart from the polar night areas, the relative humidity usually remains below 50%, being 10% on a global and annual average. This result indicates that other processes control the average behavior of atmospheric water, pointing towards the dominant influence of atmospheric transport, as discussed in Section 11.3.2.

11.3.1.3 Surface–Atmosphere Exchanges

An outstanding issue in modeling the water cycle on Mars concerns the theory employed for computing the sublimation flux of water from the surface. These exchanges account for the bulk of the water cycle seasonal activity and therefore deserve special care. However, to date, this problem has been somewhat overlooked, with different authors using different formulations without assessing the potential effect a given choice may have on the results. The most basic representation is to allow surface exchanges with the lowest atmospheric level so as to precipitate any vapor in excess of saturation or to sublime the quantity needed to saturate the first level whenever ice is present at the surface (Davies, 1981; Houben et al., 1997). More sophisticated formulations exist that account for the dynamically unstable configuration of a water-saturated near-surface layer overlaid by denser CO_2 molecules (the free convection model proposed by Ingersoll (1970)) as well as for the turbulent mixing induced by local winds (forced convection; Haberle and Jakosky, 1990). Other sublimation models can be found, such as the formulation of Flasar and Goody (1976), based on a purely turbulent mixing process with water treated as a passive scalar. Such a scheme was used by Montmessin et al. (2004) and Böttger et al. (2004, 2005) and is also used for representing evaporation of water on Earth (Peixoto and Oort, 1992).

Ivanov and Muhleman (2000) stated that free convection should be negligible in the temperature range from 150 to 210 K (though not explained by the authors, it is likely that the vapor pressure of water at those temperatures is too low to affect the molecular weight of the air). These authors have used a different formulation for the sublimation flux, as they assumed that water sublimation should be mainly controlled by forced convection. The major difference with Flasar and Goody (1976) concerns the use of the kinetic theory (where one assumes a Maxwellian distribution for the velocity of the molecules) to determine the upward flux of water. At a temperature of 273 K, kinetics yields a subliming flux that is 1000 times smaller than when computed with other formulations ignoring kinetics. If kinetics affects the regime of forced convection, it should equally affect that of free convection. Yet, the free convection model of Ingersoll, which ignores kinetics, was able to match experimental data (Hecht, 2002). It is likely instead that kinetic processes are only important when sublimation occurs in a pure

vacuum or in a rarefied gas, but not on Mars, where the pressure is sufficient to have a subliming flux of water controlled by Fick's law diffusion.

Many laboratory experiments have been conducted to study the specific Martian conditions for the sublimation of water ice (Hecht, 2002; Sears and Moore, 2005; Bryson et al., 2008; Chittenden et al., 2008) that generally support Ingersoll's approach (Ingersoll, 1970). This agreement would tend to legitimize the use of the free convection theory in conditions of extremely light wind. Hecht (2002) estimates that this regime should dominate for wind speeds lower than 2 m s^{-1}. However, the relevance of laboratory experiments to the real Martian environment is difficult to validate and a rigorous work of intercomparison still needs to be carried out.

Abundant literature exists for the terrestrial case (Brutsaert, 1982), which, because of the complicated nature of evaporation from the surface, requires discretization of the boundary layer in contiguous sublayers ruled by various types of phenomena (pure turbulence, viscous flows, etc.) and where the issue of dynamical closure remains uncertain. In order to determine the flux of sublimation, one needs to consider the evaporative process that is taking place within the very thin layer referred to by Brutsaert (1982) as the interfacial sublayer, and also known as the viscous sublayer. It is defined as the sublayer of the turbulent boundary layer below the dynamic layer where the wind profile is logarithmic. In this sublayer: (i) the flow is not fully turbulent; (ii) the flow is profoundly affected by the nature and the placement of the roughness obstacles; and (iii) the transport of tracers is also dependent on their molecular diffusivity. Despite the diffusive nature of the water transport in this layer, water transport cannot be described by the classical Fick's law. Brutsaert (1982) explains that water diffusion actually takes place into random-lived eddies, whose length and time scales are given by Kolmogorov's theory for microscale turbulence. The solution to this problem has been given by Brutsaert (1982), who obtained various formulations for the evaporation rate depending on the surface roughness.

11.3.2 Key Mechanisms in the Seasonal Behavior of Water

Theoretical studies conducted since Haberle and Jakosky (1990) have provided a renewed understanding of how and why the Mars water cycle behaves the way it is observed. In the current perception of the water cycle, one is able to distinguish between several fundamental processes and components that are the subject of the following discussion. We have chosen to discuss the role of the regolith in a separate section (Section 11.3.3) because it remains the most elusive facet of the water cycle, without a general consensus.

11.3.2.1 A Quest for Equilibrium

The question of how the Martian water cycle evolved into a state of equilibrium has been central to most studies published to date. The idea of equilibrium was however questioned by Jakosky and Farmer (1982), who speculated that the presence of a residual CO_2 ice layer at the surface near the south pole that permanently maintains a "water ice cold trap" surface temperature would necessarily imply an irreversible and permanent

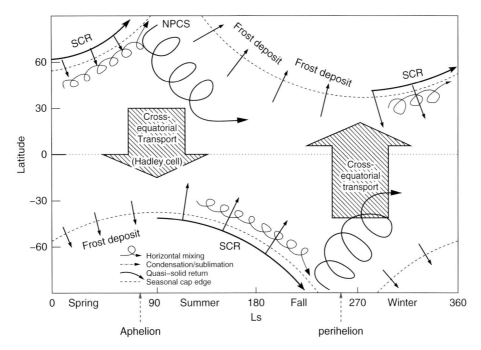

Figure 11.15. Chart describing the main events affecting the Martian water cycle over a year. From Montmessin et al. (2004). NPCS stands for north polar cap sublimation; SCR stands for seasonal cap recession.

deposition of water in that region. Models that have been used to represent this cold trapping effect by the CO_2 residual cap do show a net transfer: approximately 10^{10}–10^{11} kg of water is deposited every year at the surface near the south pole, which represents 1–10% of the seasonal inventory of water vapor (Jakosky, 1983a; Richardson and Wilson, 2002a; Montmessin et al., 2005). Since the condition of equilibrium for the water cycle cannot be met on an annual basis because of this net flux between two of its reservoirs (disregarding the potential net exchanges between the regolith and the atmosphere – see later in this chapter), Richardson and Wilson (2002a) considered that Mars water cycle was evolving in a quasi-equilibrium or in a steady state rather than in a truly equilibrated manner. The lack of major changes in the distribution of water vapor between the Viking and MGS eras, except for the stochastically triggered events of the major dust storms, is nevertheless indicative of physical processes acting quantitatively the same every year. In other words, even if a net flux exists between the reservoirs, their annual derivative is probably close to zero, which is another way to define an equilibrated situation.

Figure 11.15 depicts the major processes involved in the seasonal evolution of the present-day water cycle as inferred from the analysis of GCM experiments. Richardson and Wilson (2002a) provided the basic elements for an understanding of the annual stability of Mars' water cycle. While their model could not quantitatively reproduce the MAWD data (their model was a factor of 2 too wet), their results were in qualitative agreement, thereby establishing the relevance of their description of the mechanisms at work. To first order, one may simplify the problem of equilibration by considering solely the exchanges of water between the poles, which host the bulk of the water inventory and sustain the largest amplitudes of insolation, with the non-polar atmosphere. This simplification is permitted by

assuming that processes are symmetric between the northern and southern hemispheres and that only the communication between the poles and the equator regulates the whole cycle. Although the Richardson and Wilson (2002a) simulations were not run long enough to achieve a true equilibrium, they could still demonstrate via a set of simulations with varying boundary conditions that the present-day water cycle is always seeking to adjust from an initially overly wet or overly dry atmosphere, eventually converging towards a similar state.

The present-day water cycle is dominated by the existence of atmospheric waves, whose effect is to relax latitudinal gradients of heat and tracers (e.g. water vapor) with a seasonally varying strength. During spring and summer, when water vapor is predicted and observed to be at maximum at either pole due to local sublimation, a sluggish meridional mixing is taking place that allows moisture to be advected from the pole and to be exported towards the equator within three specific longitudinal corridors. From the end of summer until next early spring, mid-latitude and polar region climate is driven by the latitudinal and seasonal wandering of the polar vortex, whose latitudinal excursion is at maximum around winter solstice. The winter–spring retraction of the polar vortex is associated with a peculiar phenomenon that Houben et al. (1997) were the first to identify from their modeling work and which is now supported by observations (see Section 11.2.2). A strong horizontal mixing occurs across the vortex boundaries, subsequent to the reinforcement of mid-latitude winter–springtime traveling waves. The polar vortex becomes dynamically porous, allowing poleward intrusions of wet air masses that condense and precipitate their water content as frost. Following this idea, models indicate that the frost that sublimes from the surface of the seasonal cap edge as it gets exposed to sunlight re-forms itself at higher latitudes, gradually creeping back to the pole in a so-called "quasi-solid" return motion (Richardson and Wilson, 2002a). This mechanism explains why the south polar region, where no extensive reservoir of water ice resides, also exhibits a spring–summer maximum in water vapor abundances (Figure 11.15) caused by the seasonal growth of the frost edge, which sublimes entirely once at the pole.

Horizontal mixing is therefore key to the water cycle stability. Early modeling attempts using diffusive approaches were conceptually not so far off the "true" representation of the water cycle, but the introduction of GCMs allowed a self-consistent prediction of the atmospheric circulation and thus a representation of seasonal water advection in a far more reliable way.

11.3.2.2 The Role of Stationary Waves in the Summertime Extraction of Water From the Polar Regions

The water content of the north polar cap (NPC) surpasses the atmospheric reservoir (10 pr μm on average) by a factor $>10^6$. The absence of a comparable exposed reservoir at the south pole, where patches of perennial water ice have nonetheless been discovered (Titus et al., 2003; Bibring et al., 2004; Piqueux et al., 2008) is consistent with the present-day asymmetry of Mars' climate. The 20 K warmer summer season in the south imposes a severe imbalance for the stability of ice between the two poles, favoring on an annual basis storage conditions in the north (Houben et al., 1997; Richardson and Wilson, 2002a; Montmessin et al., 2007). From the early work of Jakosky (1983a,b) until the most recent GCM-led studies, the NPC has always been suspected to play a central role in the Martian water cycle. Exchanges between the NPC and the atmosphere are responsible for the main signature in the seasonal activity of water vapor (Smith, 2002). As described in Section 11.2.1, the annual peak of the globally integrated abundance of water corresponds with the period of maximum exposure for the NPC, the latter releasing into the atmosphere a mass $>2\times10^{12}$ kg of water vapor that is sufficient to double the global atmospheric water content at that season. The existence of the summertime sea-breeze circulation at the north pole identified by Haberle and Jakosky (1990) has been confirmed by 3D models (Richardson, 1999; Tyler and Barnes, 2014) but contributes only a minor fraction to the global water transport. It nonetheless plays two major roles. First, it explains why water vapor abundance is observed to be at maximum at the edge of the cap and not on the cap itself, as a locally convergent circulation develops in the periphery of the NPC. Second, this cap-edge convergence allows sublimed water to be carried within "reach" of the non-symmetric transport mechanisms that convey water to lower latitudes, as explained below.

The conditions within which water is exported from the edge of the NPC to the lower latitudes are now identified. The pioneering work of Haberle and Jakosky (1990) failed to explain the water extraction process on the basis of zonally symmetric dynamical phenomena alone, leading to the biased conclusion that the regolith was accounting for a significant fraction of the seasonal variability of water at mid-latitudes. However, Richardson (1999) specifically investigated the nature of summertime water transport from the north pole. By comparing various transport formulations (diffusion approach, 2D zonally symmetric, and full 3D GCM),

Richardson (1999) was able to establish the 3D nature of the summertime water advection from northern high to mid-latitudes. The surface thermal inertia on Mars (Putzig and Mellon, 2007) exhibits a distinct wave-3 zonal structure in the 45–70°N latitude range that imposes significant surface temperature contrast with longitude. Together with the wave-3 structure of topography associated with Arcadia, Acidalia, and Utopia Planitiae, surface properties force the development of a zonally asymmetric summertime circulation in that latitudinal band. These same regions are known to enhance transient eddy activity in winter, being referred to as the "storm zones" of Mars (Hollingsworth et al., 1996). Water vapor subliming off the NPC is therefore locked into a wave-3 configuration, confining equatorward transport of water within three longitudinal corridors located at 130°W, 20°W, and 270°W (Figure 11.16).

The top panel of Figure 11.16 shows the horizontal wind map (at ~3 km) averaged over the $L_s = 90$–120° northern summer season superimposed with the column abundance of water vapor scaled by a reference pressure of 610 Pa to remove the influence of topography on the water fields. The wind field structure reproduces a distinct wave-3 pattern within the 45°N to 80°N latitudinal range associated with the presence of a strong cyclonic flow sitting westward of the prime meridian at 65°N and dipping into Acidalia Planitia. Two additional stationary eddies centered on the same latitude and located respectively north of Arcadia and the Utopia plains also convey the recently sublimed water to lower latitudes.

Summertime transport of water from the north polar region can be further examined in Figure 11.17, where the meridional flow of water as a function of longitude is plotted for four contiguous latitudinal belts covering the 35°N to 90°N area. The wave-3 advection pattern is evident in this figure and dominates the entire summertime advection process from the north pole to the mid-latitudes, even if the relative contribution of the three modes changes significantly with latitude. As shown by Figure 11.16, poleward of 50°N, the major extraction pathway for water vapor occurs within the southward node of the Acidalia cyclonic flow. Southward of 50°N, the Acidalia flow is blocked by the presence of a strong poleward current skirting along the western flanks of Chryse Planitia, reducing the equatorward advection of water in that area. However, part of the water pushed northward by the boundary current recirculates around Alba Patera (40°N, 110°W) and merges with the southward node of the Arcadia eddy. For this reason, most of the equatorward export in this latitudinal range is accomplished by a nearly pure meridional transport through a narrow longitudinal conduit (140°W to 110°W) funneling water across the western flanks of Tharsis and later spreading it throughout the plateau.

In the 0° to 30°N area, the circulation is dominated by convergent flows between the Arcadia/Tharsis and Acidalia currents, with a strong boundary current carrying water-poor air masses from the south along the eastern flank of the Tharsis bulge, subsequently forcing water to pool in that tropical convergence zone. Water is then incorporated into the northern tropical rising branch of the Hadley circulation, to be further advected across the equator and be transferred to the southern hemisphere (see discussion below).

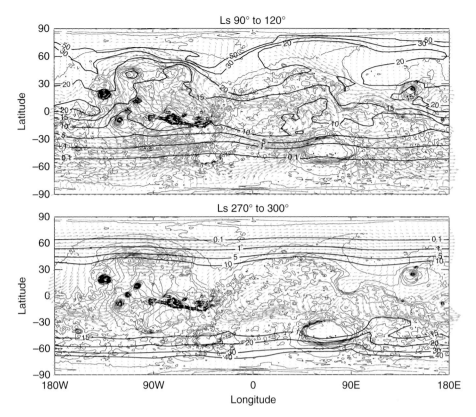

Figure 11.16. Simulation results obtained with the model described in Montmessin et al. (2004) showing maps of water vapor column abundances (scaled to a reference pressure of 6.1 mbar, units in pr μm) averaged over the northern and southern solstice seasons (upper and lower panels). Horizontal wind is indicated by the gray arrows to detail the nature of the near-surface ($z \sim 3$ km) circulation during these periods. A stationary wave-3 pattern ($-130°$, $-20°$, and $90°$ in east longitudes) dominates the northern summer high latitudes, whereas a less-pronounced wave-2 pattern (Argyre and Hellas, 20°W and 120°E) is prevalent in the southern summer polar region. Such structures in the water vapor summertime distribution are corroborated by TES observations (Smith, 2004).

For comparison, Figure 11.16 shows that the southern summer polar region exhibits a smaller degree of zonal structure for water vapor, except for the presence of two "wet filaments" flowing equatorward from the eastern rims of the Argyre (45°W, 45°S) and Hellas (270°W, 45°S) Basins. The relative lack of features in the zonal structure of polar water vapor during the southern summer is indicative of a less effective water extraction process from the south polar region.

11.3.2.3 Exchange of Water Between Hemispheres at the Solstices

The atmospheric circulation of Mars shares many similarities with the terrestrial circulation (see Chapter 9). In particular, the existence of Hadley cells on Mars allows the excess of solar energy deposited in the spring and summer tropics to be redistributed towards the deficient regions of the fall and winter high latitudes. Unlike Earth, however, Mars' circulation is dominated by a single overturning cell at solstices (Haberle et al., 1982, 1993). In addition, circulation cells on Mars can have a theoretically much greater vertical extent than their terrestrial counterpart, since the ozone thermal inversion in the Earth's stratosphere creates a dynamical barrier damping the vertical extension of lower tropospheric flows. The vertical

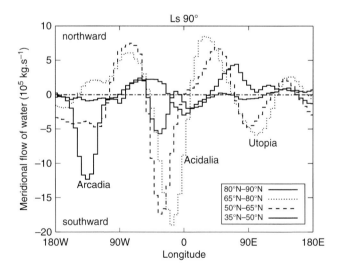

Figure 11.17. Northern summer advection of water from the north polar cap. Meridional flows of water have been obtained from the vertical integration of the simulated water vapor density times the meridional wind and averaged over four latitudinal belts (see legend box). The wave-3 pattern in the southward motion of polar water vapor during northern summer changes with latitude, progressively switching from a dominant contribution of the Acidalia corridor (20°W) at higher latitudes to the Arcadia corridor (120°W) southward of 50°N.

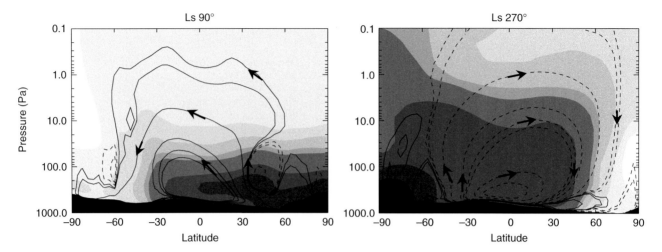

Figure 11.18. Latitude–altitude cross-sections of the time and zonally averaged component of Mars circulation (solid and dashed lines indicate counterclockwise and clockwise orientations, respectively) at both solstices (left and right panels) superimposed with shaded contours of water vapor mass mixing ratios (from 0 ppmv, white, to 500 ppmv, dark gray). The 100, 10, and 1 Pa pressure levels correspond approximately with the 20, 45, and 70 km of altitude levels, respectively.

development of Martian Hadley cells significantly exceeds the tropopause level (near 40 km) and cells are therefore able to mix the atmospheric column over very large depths (see Figure 11.18). Deep mixing is supported by the existence of significant quantities of aerosols and water vapor up to 70 km during southern spring and summer (Jaquin et al., 1986; Fedorova et al., 2009).

Mars has no oceans but instead has a solid surface with a relatively low heat storage capacity, which quickly responds to seasonal changes in insolation and thus maintains close coincidence between the subsolar and the "warmest" areas. Therefore, Martian Hadley cells are subjected to significant seasonal variation. Around solstices ($L_s = 90°$ and $270°$), the Hadley cell is characterized by rising motions in the summer tropics and downwelling in the winter mid-latitudes (see Figure 11.18). Around the equinoxes, the cell splits into an Earth-like configuration with two subcells facing each other about the equator, where a sluggish upwelling motion occurs. The equinoctial circulation is, however, a shorter and much weaker circulation pattern such that most of the cross-equatorial transport is accomplished around the solstices (Forget et al., 1999; Read and Lewis, 2004).

Figure 11.18 shows that the latitude–altitude distribution of water is markedly different between the northern and southern summer solstices and this drives a significant hemispheric asymmetry; this point is discussed in greater detail in Section 11.3.2. According to the model of Montmessin et al. (2004), approximately 10^{12} kg of water are transferred from the north to the south and back every year through the solstitial Hadley cells. Houben et al. (1997) estimated this total to be around $2×10^{12}$ kg, but indicated that this value was heavily dependent upon the action of the regolith within the rising and descending areas of the Hadley cells. Considering the error associated with their regolith representation, we may use the 10^{12} kg figure as a representative estimate of the annual exchange of water between the two hemispheres, i.e. half of the mass of water subliming from the NPC in spring and summer.

11.3.2.4 The Winter–Springtime Return of Water to the Poles

A fundamental aspect of the Mars water cycle concerns its ability to compensate the summertime extraction of water from the poles by an equivalent return flow (minus the permanent loss on the residual CO_2 cap) during the rest of the year. Since the work of Houben et al. (1997), the process by which this occurs is now better understood and has been studied in greater detail thanks to numerous observations (Kieffer et al., 2000; Kieffer and Titus, 2001; Pankine et al., 2010; Langevin et al., 2007; Appéré et al., 2011), as described in Section 11.2.2. Seasonal CO_2 ice caps are places of highly unstable baroclinic disturbances (Leovy, 1973; Barnes, 1981), which participate in the maintenance and regulation of the winter–springtime poleward retreat of water. Based on the observed evolution of the seasonal frost properties in the northern hemisphere, Appéré et al. (2011) indicate that the poleward retreat of the seasonal frosts in the north behave rather symmetrically, characterized by an annulus of water ice extending a few degrees of latitude equatorward of the edge of the CO_2 seasonal frost (see Figure 11.12). From the early phase of the retreat, near $L_s = 320°$, seasonal water ice frost is involved in a sublimation–recondensation mechanism, consisting of poleward redeposition of newly sublimed water. This mechanism operates until the final stages of the recession after $L_s = 70°$. Models appear to be in good qualitative agreement with this description (Houben et al., 1997; Richardson and Wilson, 2002a; Montmessin et al., 2004), even though they do not reproduce important aspects relative to the CO_2 and H_2O ice interactions, in particular the transition between a dominant CO_2–H_2O admixture early in the process towards the formation of stacked layers of either component later on (Appéré et al., 2011).

GCMs indicate that a rather uniform accumulation of water ice (at a rate of a few tenths of a micrometer/day) occurs throughout the surface of the seasonal cap during the retreat phase. This accumulation is forced by the remaining seasonal CO_2 ice located poleward of the water ice annulus, which

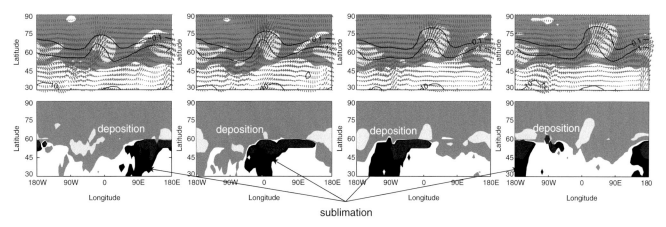

Figure 11.19. Predicted daily evolution of the seasonal frost during northern springtime at $L_s = 10°$. Each plot is separated by an interval of 6 hours from left to right. (Upper panels) Contours of pressure-scaled water vapor column abundances (solid lines, in pr μm), superimposed with contours of total water ice cloud column (filled with gray, 2 and 5 pr μm levels) and near-surface horizontal wind field (arrows). The eastward motion of a transient eddy and its associated cloud can be seen at 60°N. (Lower panels) Surface ice accumulation rate; light gray areas indicate regions where water ice is deposited (two levels at 0.25 and 1 μm/day) whereas dark gray areas indicate the opposite situation, with ice subliming from the surface.

maintains "CO_2 ice cold" surface temperature (~145 K) and therefore acts as a cold trap for the residual humidity of the dry polar atmosphere. However, traveling waves play a complementary and decisive role in the sublimation–recondensation process. As illustrated by Figure 11.19, the seasonal cap edge hosts a variety of propagating disturbances generated by baroclinic instabilities across the boundary of the polar vortex. Enhancement of the transient wave activity during late winter and spring has been also recorded by the Viking Lander meteorological experiments (Tillman et al., 1979).

While the water ice frost sublimation zone essentially tracks the daily westward motion of the Sun, baroclinic waves travel eastward with a dominant three-day period, creating deep poleward intrusion of air masses loaded with the moisture that had just sublimed. Models predict widespread regions of enhanced deposition (>1 μm/day) extending up to 15° north of the sublimation front. These regions are associated with intense cross-frontal mixing between water-rich air masses moving poleward and the cold polar air, giving rise to thick water ice clouds precipitating in the cold equatorward phase of the eddies (see the upper panels of Figure 11.19).

The seasonal cap edge consequently sees a significant and steady increase of its water ice composition during the retreat phase, with an equivalent thickness of water ice growing from 20 μm during the early stage up to >200 μm before the final sublimation event. This is the essence of the mechanism proposed by Houben et al. (1997), which appears to prevail in the seasonal recession of the southern hemisphere frost as well. However, observations have revealed important differences with the north. The retreat of the seasonal frost in the south does not exhibit the presence of a water ice annulus in the periphery of the seasonal CO_2 ice but a distinct longitudinal asymmetry (Kieffer et al., 2000; Langevin et al., 2007) that is reminiscent of the circulation asymmetry forced by the topography of the high southern latitudes (Colaprete et al., 2005). The presence of the deep Hellas Basin depression induces an equatorward expansion of the springtime polar vortex that efficiently pumps in moisture from the mid-latitudes towards

the south pole, leading to concentrated deposition of water ice in an area located southeast of the basin, as revealed by OMEGA and subsequently modeled by GCMs (Langevin et al., 2007).

Although the Houben mechanism provides an effective way for the polar reservoirs to seasonally recycle the water that sublimes during spring and summer, simulations indicate that only half of the water ice that has accumulated during fall and winter will condense again during the retreat, the other half serving to replenish the water vapor content of the atmospheric columns overlying the sublimation zone.

11.3.2.5 Seasonal Changes in the Meridional Advection of Water

Following the method described in Montmessin et al. (2004), which was itself adapted from Peixoto and Oort (1992), it is possible to quantify and isolate the contributions of the various circulation components to the seasonal meridional transport of water. The basic transport decomposition includes the zonally symmetric, time-averaged component (mean circulation) as well as the time and zonal departures from the mean (the components associated with the transient and stationary waves, respectively). A typical expression of water transport is given by

$$\left[\overline{qv}\right] = \left[\overline{q}\right]\left[\overline{v}\right] + \left[\overline{q'v'}\right] + \left[\overline{q^*v^*}\right] \tag{11.2}$$

where v and q are, respectively, the meridional wind velocity and the total water mass mixing ratio (including both vapor and cloud) at a pressure level p. The prime symbol expresses the departure from the time average ($q' = q - \overline{q}$, $v' = v - \overline{v}$) while the star symbol expresses the departure from the zonal average ($q^* = q - [q]$ and $v^* = v - [v]$). Importantly, q and v are, respectively, replaced by Q_{wat} and V in (11.2) to consider the meridional transport of the vertically integrated column of water. To do so, we define $Q_{wat} = \int_0^{p_s} q\,dp\,/\,g$ and $V = \left[\int_0^{p_s} qv\,dp\,/\,g\right]/Q_{wat}$, with g the gravity of Mars and p_s the pressure at the surface.

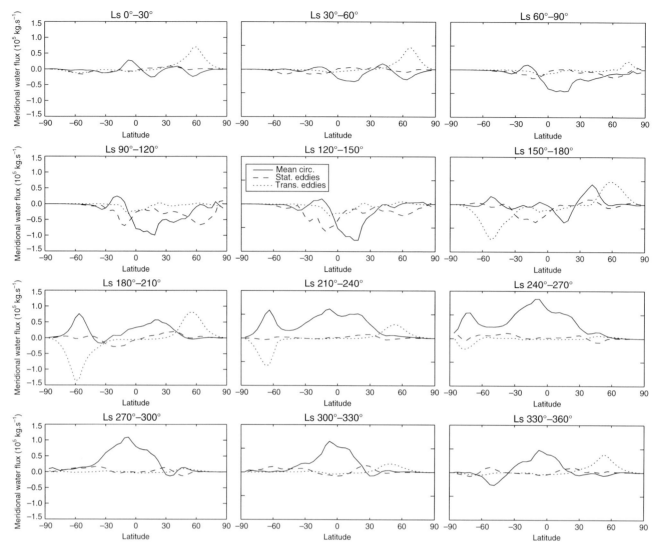

Figure 11.20. Seasonal evolution of the computed meridional flux of water (comprising vapor and clouds, in 10^5 kg s^{-1}) decomposed into the various atmospheric transport components (zonally symmetric circulation, stationary, and transient waves). Southward transport is negative.

The seasonal variation of meridional transport of water, as formulated by the above expressions, is displayed in Figure 11.20. The role of the zonally symmetric circulation is most pronounced at low to mid-latitudes around solstices when the Hadley cells achieve their maximum development. Northern summer solstice is characterized by a net southward transport of water at the equator, with water ice clouds playing a dominant role. The separation of the transport between the vapor and the cloud contributions indicates that no net exchange of water occurs between hemispheres in the form of vapor, and that only cloud advection participates in the southward transfer of water. After $L_s = 120°$, there is a significant increase in the contribution of stationary eddies to the cross-equatorial motion of water. This contribution tends to vanish near the equinox ($L_s = 180°$) when the Hadley circulation reverses itself. At that moment, the southern hemisphere starts to lose water in favor of the northern hemisphere and this configuration will persist until the next equinox ($L_s = 360°$). During that timeframe, the southern summer Hadley cell develops and prevails in the cross-equatorial exchanges (Figure 11.18).

Transport in the extratropical regions is essentially accomplished by the circulation residuals. As detailed in the previous section, a wave-3 stationary weather pattern is responsible for the majority of the equatorial advection of water from the subliming north polar cap during the $L_s = 90°–120°$ period in the 30–80°N latitudinal range. The sea-breeze circulation at the cap shows up as a small contribution to transport in the mean circulation component near the north pole, but is quickly overtaken by the stationary transport pattern.

The Houben transport mechanism is clearly revealed in the plots of Figure 11.19, appearing as the unique poleward transport contribution by transient eddies in the mid-latitudes. This contribution is maximum during the retraction phases of the polar vortex in the winter and spring seasons of both hemispheres ($L_s = 330°$ to 60° in the north and $L_s = 150°$ to 240° in the south). Clouds play a similar role as vapor, since they can be advected in the cold equatorward phase of the disturbances, thus balancing the return of water to the poles by a flux of water in the form of atmospheric ice towards the tropics.

A poleward transient eddy contribution is also found in the mid-latitudes of the northern hemisphere during the $L_s = 150°$ to 210° period. Most of this water will incorporate into the seasonal deposits and will thus reduce the transportable fraction of water in the north. No comparable mechanism is predicted to occur in the south.

Interestingly, the poleward transport of water by transient eddies in the springtime mid-latitudes is balanced by a nearly equivalent equatorward transport controlled by the mean circulation. This reflects the contribution of the meridional flows induced by the disappearance of the seasonal CO_2 frost. As already investigated by James (1985) and Richardson and Wilson (2002a), the current climatic asymmetry of Mars, with a shorter and warmer summer in the south, forces a more intense equatorward advection of water from the receding southern CO_2 cap. Most of the time, the opposite poleward advection by transient eddies wins the transport "battle". However, during the last stages of the southern frost recession ($L_s = 240°$ to 270°), the CO_2 sublimation flow becomes strong enough to subdue the Houben mechanism and to establish a net equatorward advection of water. In contrast, the CO_2 sublimation flow remains persistently too weak during the retreat of the northern seasonal deposits to overcome the poleward transport by transient waves.

11.3.2.6 The Hemispheric Asymmetry and the Role of Clouds

The relatively desiccated state of the southern hemisphere compared to the north has been the motivation for numerous studies in the past. Eventually, the use of GCMs was able to decipher the detailed mechanisms at work in the maintenance of the hemispheric asymmetry of water on Mars. However, the problem of the asymmetry is biased by the fact that the present-day water cycle is influenced by boundary conditions that are history-dependent: the presence of the largest reservoir of water in the north and a small one in the south imposes a severe imbalance for the inter-hemispheric partitioning of water, since the southern hemisphere is essentially dependent on the water "share" allotted by the north. Would this asymmetry be maintained if a comparable reservoir of ice presently existed in the south? Richardson and Wilson (2002a) provide part of the answer. The purpose of their experiment was to study the competition between the two hemispheres for the preservation of their own polar reservoir. They show that such a hypothetical reservoir at the south pole is unstable and would quickly migrate to the north pole.

One obvious source of asymmetry is related to the eccentricity of Mars' orbit, which determines the length and insolation conditions of northern and southern summers, making a >20 K warmer climate during southern summer (phased with perihelion) than during northern summer (phased with aphelion) (see Chapter 4). Together with the enhanced dust activity of the southern summer (see Chapter 10), these two factors conspire to reduce the capacity of the south for retaining atmospheric water. Under the warmer conditions of summer in the south, water is inevitably more inclined to evolve in the vapor phase and thus to be transported elsewhere. This point is illustrated by Figure 11.14(right), which shows that southern summer corresponds with an annual low of relative humidity.

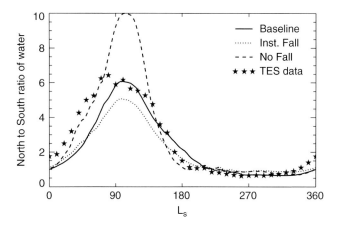

Figure 11.21. Seasonal evolution of the north-to-south ratio of water vapor as given by different versions of the model. Adapted from Montmessin et al. (2004). Baseline, predicted cloud particle size; Inst. Fall, cloud particle fall forced instantaneous; and No Fall, non-sedimenting cloud particles. Models are compared with TES observations. The contrast of humidity between the north and the south is clearly the strongest around aphelion ($L_s = 70°$).

Two mechanisms play a critical role in this context. The first has been proposed by Clancy et al. (1996) and has since then been referred to as the "Clancy effect". It is based on the cold aphelion climate imposing a saturation altitude of water near or below the returning branch of the northern summer Hadley cell. The validity of this mechanism was at first rejected by Houben et al. (1997), but has been later supported by Richardson and Wilson (2002a) and Montmessin et al. (2004). Above the saturation level (the hygropause), water vapor subsists in very low quantities and contributes negligible amounts to the southward upper branch of the Hadley cell, making the cross-equatorial flow of vapor to the south less efficient during northern spring and summer. This effect explains why the GCMs predict the absence of a net meridional flux of vapor around $L_s = 90°$, as discussed in the previous section. Water vapor in excess of saturation in principle condenses into clouds, which is observed to occur in the equatorial region during aphelion and is known as the aphelion cloud belt, a major seasonal manifestation of Mars' climate (Clancy et al., 1996; Smith, 2004; see discussion in Section 11.2.1 and Chapter 5). Recent observations suggest, however, that water vapor may exist in excess of saturation (Maltagliati et al., 2011a), which may allow some residual transport in the vapor phase, a point not addressed by climate models so far. In any case, the formation of clouds near the hygropause controls the bulk amount of the water migration from the northern hemisphere to the south. Clouds can be advected by winds, although less efficiently than vapor because of their gravitational fall. Figure 11.21 shows a set of GCM experiments extracted from Montmessin et al. (2004), indicating that the north-to-south ratio of water around aphelion ($L_s = 70°$) is entirely dependent on the fall speed of the cloud particles lofted in the aphelion cloud belt. The baseline GCM, which predicts particle radius around 3–5 μm similar to observations (Clancy et al., 2003), reproduces quantitatively the TES measurements, whereas the two end-member cases show that the north-to-south ratio can be modulated by a factor with a specific surface area of about 9.8 m^2 g^{-1} and extrapolated from laboratory temperatures (>250 K) to Mars soil temperatures (as low as 150 K). Zent and Quinn (1997) made new measurements of palagonite samples and showed that the Fanale and Cannon isotherm overestimates the adsorption of water per unit surface area by about a factor of 30 at Mars environmental conditions. While adsorption depends only secondarily on grain mineralogy, it depends very strongly on the specific surface area, which will likely vary considerably across the surface of Mars. Generally speaking, coarse-grained or poorly chemically weathered basalts (such as might be found in sand dunes) will exhibit low surface areas (perhaps <10 m^2 g^{-1}), while weathering products such as phyllosilicate smectite clays will have much higher surface areas (approaching 1000 m^2 g^{-1}) due to the layered-sheet mineral structure (Zent and Quinn, 1997).

11.3.3.3 Regolith–Atmosphere Exchange: Theoretical and Observational Limitations

At present, the role of diurnal and seasonal exchange of water between the regolith and atmosphere is not well understood. Existing studies contrast in their conclusions regarding the magnitude of the exchange of adsorbed water and the importance of the regolith. Additionally, many of the studies discussed above have utilized measurements of Fanale and Cannon (1971, 1974) when modeling the adsorptive capacity of the Martian regolith and its temperature dependence, which have been subsequently shown to be inaccurately extrapolated to Martian conditions (Zent and Quinn, 1997). Numerous new measurements of adsorption have been conducted for a variety of Mars-analog materials (e.g. Zent and Quinn, 1995, 1997; Zent et al., 2001; Bryson et al., 2008; Chevrier et al., 2008; Jänchen et al., 2009; Beck et al., 2011). Care should be taken in incorporating appropriate adsorption isotherms in water cycle models. Additionally, Martian soils and their adsorptive capacity are likely to vary substantially with geographic region and depth. For example, regions of higher surface albedo and lower thermal inertia represent vast deposits of fine-grained soils (commonly referred to as "dust") perhaps smaller than 40 μm in diameter (Palluconi and Kieffer, 1981; Mellon et al., 2000). Finer-grained soils will exhibit higher surface area. Likewise, areas with higher concentrations of smectite clays will exhibit high surface area and greater adsorbed water abundance.

The role of shallow ground ice as a reservoir for atmospheric exchange has only been minimally considered. Mellon and Jakosky (1995) found that about 40 g cm^{-2} exchanges with the atmosphere (and presumably polar caps) over an obliquity-oscillation timescale of 10^5 years. If this amount were expelled at a constant rate over half an obliquity cycle, it roughly translates to 8 μm cm^{-2} yr^{-1}. Key diffusive properties of porosity and tortuosity have been revised since these calculations (e.g. Sizemore and Mellon, 2008), so this estimate should be taken as a qualitative result. Nonetheless, Böttger et al. (2005) indicated a similar scale of exchange between ground ice and the atmosphere on seasonal timescales. At present, it is not clear how much water exchanges from the ground ice in the current epoch, as either a steady loss or gain due to inter-annual variability or orbitally driven climate change.

The role of ground ice in the diurnal and seasonal water cycle depends strongly on the burial depth of the ice table and diffusion through the ice-free overburden. Analysis of Mars Odyssey data suggests ice-table depths that are typically twice as deep as theoretical models predict (e.g. Mellon et al., 2004). However, this difference is attributed to the presence of rocks and slopes on the Martian surface and their nonlinear effects on leakage neutrons (Sizemore and Mellon, 2006), such that the ice-table depth in the soil areas between rocks is as predicted. In general, there is strong evidence that the ice-table depth locally and globally exists in diffusive equilibrium with the current atmosphere, or perhaps a slightly more humid climate than is presently measured (Mellon et al., 2004, 2009; Mellon and Feldman, 2006). In addition, the ice-table depth will vary regionally with latitude (e.g. Leighton and Murray, 1966) and with soil thermal properties (Paige, 1992). Depths can vary from millimeters to meters, but are most often a few to a few tens of centimeters (Mellon et al., 2004).

Some minerals observed on Mars contain chemically bound water that can be released (or incorporated) as environmental conditions change. Such water has not been considered in modeling the atmospheric water cycle. Sulfate salts, such as kieserite (MgSO$_4$·H$_2$O), hexahydrite (MgSO$_4$·6H$_2$O), and epsomite (MgSO$_4$·7H$_2$O), have been detected in mono- and polyhydrated forms (Bibring et al., 2005). These minerals are believed to be stable on Mars (Chou and Seal, 2003; Feldman et al., 2004; Fialips et al., 2005; Chipera and Vaniman, 2007). Changes in ground temperature and atmospheric humidity may cause transitions between stable states, though it is possible that reaction kinetics may be too slow to play a significant role on seasonal or shorter timescales (Chipera and Vaniman, 2007). In addition, Hecht et al. (2009) suggested that perchlorate salts (a strongly deliquescent mineral) may play some role in buffering atmospheric humidity, an effect that has not been previously considered. Detailed modeling of the role of hydrated minerals is needed to understand their potential impact on the Martian water cycle.

11.4 SUMMARY AND PERSPECTIVES

11.4.1 An Integrated Picture of the Present-Day Water Cycle

This chapter provides a comprehensive overview of our understanding of Mars' water cycle, including a description of its various atmospheric, surface, and subsurface reservoirs. Some of the major questions raised in the introduction have been answered, whereas others remain uncertain. The main seasonal sources and sinks of water have been identified: seasonal and perennial ices have been mapped and characterized, and atmospheric water has been tracked seasonally in its vapor and condensed phases. On average, the seasonal cycle of water on Mars is characterized by a factor of 2 variation of the overall water vapor abundance, forced by the seasonal sublimation and condensation of about 3×10^{12} kg of surface ice. A third of this mass (10^{12} kg) is exchanged seasonally between the two hemispheres during the solstitial seasons, implying that a major fraction (two-thirds) of water annually recycles in the northern hemisphere. Models show that >85% of atmospheric water

A poleward transient eddy contribution is also found in the mid-latitudes of the northern hemisphere during the $L_s = 150°$ to 210° period. Most of this water will incorporate into the seasonal deposits and will thus reduce the transportable fraction of water in the north. No comparable mechanism is predicted to occur in the south.

Interestingly, the poleward transport of water by transient eddies in the springtime mid-latitudes is balanced by a nearly equivalent equatorward transport controlled by the mean circulation. This reflects the contribution of the meridional flows induced by the disappearance of the seasonal CO_2 frost. As already investigated by James (1985) and Richardson and Wilson (2002a), the current climatic asymmetry of Mars, with a shorter and warmer summer in the south, forces a more intense equatorward advection of water from the receding southern CO_2 cap. Most of the time, the opposite poleward advection by transient eddies wins the transport "battle". However, during the last stages of the southern frost recession ($L_s = 240°$ to 270°), the CO_2 sublimation flow becomes strong enough to subdue the Houben mechanism and to establish a net equatorward advection of water. In contrast, the CO_2 sublimation flow remains persistently too weak during the retreat of the northern seasonal deposits to overcome the poleward transport by transient waves.

11.3.2.6 The Hemispheric Asymmetry and the Role of Clouds

The relatively desiccated state of the southern hemisphere compared to the north has been the motivation for numerous studies in the past. Eventually, the use of GCMs was able to decipher the detailed mechanisms at work in the maintenance of the hemispheric asymmetry of water on Mars. However, the problem of the asymmetry is biased by the fact that the present-day water cycle is influenced by boundary conditions that are history-dependent: the presence of the largest reservoir of water in the north and a small one in the south imposes a severe imbalance for the inter-hemispheric partitioning of water, since the southern hemisphere is essentially dependent on the water "share" allotted by the north. Would this asymmetry be maintained if a comparable reservoir of ice presently existed in the south? Richardson and Wilson (2002a) provide part of the answer. The purpose of their experiment was to study the competition between the two hemispheres for the preservation of their own polar reservoir. They show that such a hypothetical reservoir at the south pole is unstable and would quickly migrate to the north pole.

One obvious source of asymmetry is related to the eccentricity of Mars' orbit, which determines the length and insolation conditions of northern and southern summers, making a >20 K warmer climate during southern summer (phased with perihelion) than during northern summer (phased with aphelion) (see Chapter 4). Together with the enhanced dust activity of the southern summer (see Chapter 10), these two factors conspire to reduce the capacity of the south for retaining atmospheric water. Under the warmer conditions of summer in the south, water is inevitably more inclined to evolve in the vapor phase and thus to be transported elsewhere. This point is illustrated by Figure 11.14(right), which shows that southern summer corresponds with an annual low of relative humidity.

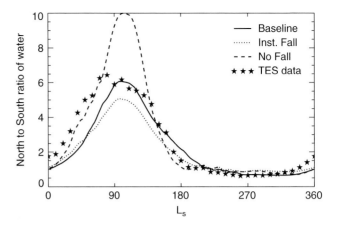

Figure 11.21. Seasonal evolution of the north-to-south ratio of water vapor as given by different versions of the model. Adapted from Montmessin et al. (2004). Baseline, predicted cloud particle size; Inst. Fall, cloud particle fall forced instantaneous; and No Fall, non-sedimenting cloud particles. Models are compared with TES observations. The contrast of humidity between the north and the south is clearly the strongest around aphelion ($L_s = 70°$).

Two mechanisms play a critical role in this context. The first has been proposed by Clancy et al. (1996) and has since then been referred to as the "Clancy effect". It is based on the cold aphelion climate imposing a saturation altitude of water near or below the returning branch of the northern summer Hadley cell. The validity of this mechanism was at first rejected by Houben et al. (1997), but has been later supported by Richardson and Wilson (2002a) and Montmessin et al. (2004). Above the saturation level (the hygropause), water vapor subsists in very low quantities and contributes negligible amounts to the southward upper branch of the Hadley cell, making the cross-equatorial flow of vapor to the south less efficient during northern spring and summer. This effect explains why the GCMs predict the absence of a net meridional flux of vapor around $L_s = 90°$, as discussed in the previous section. Water vapor in excess of saturation in principle condenses into clouds, which is observed to occur in the equatorial region during aphelion and is known as the aphelion cloud belt, a major seasonal manifestation of Mars' climate (Clancy et al., 1996; Smith, 2004; see discussion in Section 11.2.1 and Chapter 5). Recent observations suggest, however, that water vapor may exist in excess of saturation (Maltagliati et al., 2011a), which may allow some residual transport in the vapor phase, a point not addressed by climate models so far. In any case, the formation of clouds near the hygropause controls the bulk amount of the water migration from the northern hemisphere to the south. Clouds can be advected by winds, although less efficiently than vapor because of their gravitational fall. Figure 11.21 shows a set of GCM experiments extracted from Montmessin et al. (2004), indicating that the north-to-south ratio of water around aphelion ($L_s = 70°$) is entirely dependent on the fall speed of the cloud particles lofted in the aphelion cloud belt. The baseline GCM, which predicts particle radius around 3–5 μm similar to observations (Clancy et al., 2003), reproduces quantitatively the TES measurements, whereas the two end-member cases show that the north-to-south ratio can be modulated by a factor of up to 3 depending on the cloud sedimentation properties. While the model suggests that advection in the upper branch of the cell is essentially made of cloud mass instead of vapor, the comparison between the various simulations suggests that cloud sedimentation sequesters water in the northern summer tropics. A direct consequence of this sequestration is the presence of an anvil-shaped vertical distribution of water vapor in the northern summer tropics (Figure 11.18) at around 10–20 km altitude, showing substantial water vapor enrichment below the condensation level. This anvil shape is caused by cloud particles lofted above the hygropause that sediment and release their water content after sublimation. The asymmetry of the Clancy effect lies in the absence of a comparable sequestration effect in the southern summer tropics. At that season, the saturation level is located at a much higher altitude (>40 km) as a result of the warmer perihelion climate. This point is well illustrated by the comparison of the solstitial water vapor vertical distributions displayed in Figure 11.18. Adding to the climate asymmetry, Richardson and Wilson (2002b) have demonstrated that the north-to-south topography dichotomy of Mars forces a more intense southern summer Hadley circulation compared to that of the northern summer, regardless of the orbit eccentricity and of the dust loading arguments. The reason for this asymmetry is related to the near-independence of Mars' surface temperatures on elevation (see Chapter 9 for details). Thus, this bias further adds to the current climate conditions already favoring the accumulation of water in the northern hemisphere.

Two other mechanisms, already mentioned in the previous section, further participate in forcing an asymmetry. Figure 11.20 shows that a large influx of water to the north pole occurs soon after the sublimation season of the NPC, following the onset of the polar vortex and associated eddies. This late summer transport, which has no counterpart in the south, supplies an early return mechanism of water to the north pole and therefore helps the northern hemisphere to retain its water instead of losing it to the southern hemisphere. This early return in the north combines with the above Hadley circulation arguments to reinforce the hemispheric dichotomy, as does the CO_2 sublimation flows from the seasonal caps (although this aspect was considered minor by Richardson and Wilson (2002a)).

The only two factors acting in favor of the southern hemisphere are the more sluggish summertime south polar circulation, which impedes the ability of the south pole to export its newly sublimed water, and the shortness of its summer. However, they are no match for the environment and present-day climate arguments, which are otherwise largely favorable to the north. The climatic contrast between northern and southern summers is certainly the major source of present-day asymmetry, since phase changes respond nonlinearly to temperature changes by virtue of the Clausius–Clapeyron law (see (11.1)). Montmessin et al. (2007) have shown that, in the recent past (~25 000 years ago), when Mars' orbital variations led to an opposite climatic configuration with a shorter and warmer summer in the north, the present-day asymmetry was likely reversed.

The above discussion emphasizes the key role played by clouds in the seasonal transport of water. The two major stages of the water cycle where cloud advection is important are the aphelion cross-equatorial transport and the non-summer return process in the mid- to high latitudes of both hemispheres. The net overall effect of clouds is to provide additional mobility to water and therefore to increase the moistening capability of the water cycle. Their average microphysical properties, with a radius of a few micrometers, allow them to be maintained aloft on the order of days in the tropics and on the order of weeks at high latitudes during the non-summer period. This timescale is critical, as it permits cloud advection over significant distances (>1000 km) and thus allows water to reach regions that are otherwise unattainable in the vapor phase. Increasing the mobility of water increases its ability to be exported farther away from the source polar region, thereby reducing the efficiency of the return mechanism to the poles (Montmessin et al., 2004). The net overall consequence of clouds is to induce a global moistening of the Martian atmosphere by up to a factor of 2, even though the average cloud fraction remains below 15% of the atmospheric inventory.

11.3.3 The Role of the Regolith

11.3.3.1 Previous Work

Early studies of the water cycle on Mars already began to consider subsurface reservoirs of water and exchange with the atmosphere (e.g. Leovy, 1973; Flasar and Goody, 1976). With the Viking spacecraft measurements of the seasonal cycles in the global distribution of atmospheric water, interest in the role of the regolith increased in an attempt to explain the observations (e.g. Jakosky and Farmer, 1982). In particular, the roles of adsorbed water (Fanale and Cannon, 1971, 1974), and ground ice (Leighton and Murray, 1966) have been the focus of attention. Flasar and Goody (1976) modeled the diurnal cycle of water, including diffusion into the ground and exchange with subsurface reservoirs of ice and adsorbed water. They concluded that adsorbed water in the topmost centimeter of the regolith can play a significant role in drying the atmosphere in the evenings, but that it is unlikely that significant ice will form.

Based on the Viking MAWD results, Jakosky and Farmer (1982) noted that the springtime increase in water vapor in the northern mid-latitudes preceded exposure of polar ice and suggested that regolith sources of water were the most likely cause. Jakosky (1983a,b) examined the adsorptive capacity of the regolith on depth scales available for diffusive exchange and suggested that 10–40% of the seasonal water cycle could result from exchange with the regolith. Zent et al. (1993) also examined diurnal cycles of water in the planetary boundary layer. Their results suggest that diurnal exchange is small (~1 pr µm), but that the regolith is important in establishing the column abundance of water vapor. Jakosky et al. (1997) re-examined Viking MAWD data and found a factor of 2 depletion in water vapor at night. Modeling the diurnal cycle, they showed that including the effects of adsorbed water produced a similar depletion and concluded that the regolith plays a major role in the diurnal cycle. Houben et al. (1997) modeled the seasonal transport of water and concluded that regolith adsorption in the topmost 10 cm plays a significant role. Richardson and Wilson (2002a) also examined the seasonal water cycle using a general circulation model and, pointing out an error in Houben et al.'s calculations, concluded that the MAWD observations could also be explained without regolith adsorption.

In a way similar to Richardson and Wilson (2002a), Böttger et al. (2004, 2005) studied the water cycle with a GCM coupled to a regolith model. They concluded that at high latitudes the seasonal adsorption and desorption of water in the shallow regolith is important, but that the diurnal cycle is not reproduced by regolith adsorption. This effect becomes especially important at times of planet-encircling dust storms, as the reduction of diurnal temperature fluctuations at the surface results in an increase in the adsorptive capacity of the regolith, providing more consistent results for the water column abundances in the southern summer hemisphere (Böttger et al., 2004). They also found that shallow high-latitude ground ice exchanges with the atmosphere and suggested that ground ice in the southern hemisphere would be currently experiencing a gradual loss to be deposited in the northern hemisphere permafrost, mainly due to the current pole-to-pole asymmetry in the atmospheric water content (e.g. Jakosky and Farmer, 1982). While the latter may be true, it should be noted that integrations of ground ice accumulations and loss over orbital cycles indicate the reverse situation, that ground ice is currently being lost from the northern permafrost and accumulating in the southern permafrost on interannual timescales due to precession of the season of perihelion (Mellon and Jakosky, 1995; Zent, 2008). In addition, Mellon et al. (2004), in examining ground ice stability relative to Mars Odyssey results, reported the timescales for the loss of ground ice to the atmosphere as a function of burial depth and temperature. Their results suggest that ground ice would need to be shallower than about 10 cm to begin exchanging significant amounts of water (>1 pr μm) with the atmosphere on seasonal timescales and shallower than 1 cm to play a role in diurnal exchange.

The direct observations of near-surface humidity and regolith water content by the Phoenix spacecraft at a northern high latitude indicate that a substantial amount of water (perhaps as much as 10 pr μm) may be exchanged between the upper centimeters of the regolith and the atmosphere on diurnal timescales (Zent et al., 2010). Uncertainty remains in the magnitude of this exchange due to uncertainty in understanding the response of soil dielectric properties to specific changes in adsorbed water content.

In general, at the current time, disagreement still remains about the role of the regolith (in terms of both adsorbed water and ice) in the water cycle for both diurnal and seasonal timescales. Although most investigators suggest that the regolith plays a substantial role in the seasonal water cycle, doubt remains as to whether the regolith is necessary to explain the observations. The role of the regolith in the diurnal cycle is even less clear, though observations would suggest it is important, particularly at higher latitudes. Additionally, while the role of adsorbed water continues to be considered in detail by modelers, the potential role of ground ice has only minimally been addressed.

11.3.3.2 Water Movement and Regolith–Atmosphere Exchange: What Is the Relevant Physics?

Water in the soil subsurface can be quite dynamic in response to diurnal, seasonal, and longer changes in the Martian climate.

Changes in subsurface temperature and atmospheric absolute humidity are the primary drivers of water exchange between the atmosphere and regolith. The rate of this exchange relative to the rates of seasonal or climatic change is critically important to understanding the role of regolith reservoirs of water. These rates are also highly dependent on regolith properties and climate conditions. Surface and subsurface ground temperatures drive much of the process and vary diurnally, seasonally, and with depth (see, for examples, Kieffer et al., 1977; Mellon et al., 2000). The molecular diffusion of water vapor in the soil pores is the dominant transport process. However, surface diffusion, capillary flow, and advection may also play some, perhaps minor, role in special circumstances.

At the sub-freezing temperatures typically found on Mars, the exchange of water between the atmosphere and regolith subsurface occurs predominantly through molecular diffusion of water vapor. Following Fick's first law, a diffusive flux, N, of water vapor molecules is driven by a divergence in the density in water vapor molecules, n, between the atmosphere and regolith or between different regions within the regolith, expressed in the vertical dimension as

$$N = -D \frac{\varepsilon}{\tau} \frac{\partial n}{\partial z} \qquad (11.3)$$

(see Mellon and Jakosky, 1993, and references therein). The effective diffusion coefficient, D, includes both Knudsen flow (gas molecules primarily collide with soil grains) and normal flow (collisions between gas molecules dominate). The diffusion coefficient D is a function of the regolith's pore radius, the ambient pressure, temperature, and the gas physical properties. The porosity ε and the tortuosity τ are regolith properties. Tortuosity represents the winding path that a diffusing gas molecule takes through a soil and is best determined experimentally. Values of 3–5 have historically been used in Mars studies (e.g. Smoluchowski, 1968; Clifford and Hillel, 1983; Mellon and Jakosky, 1993); however, recent measurements by Sizemore and Mellon (2008) suggest that a value closer to 1.5 is appropriate for Martian soils, resulting in more rapid transport of water vapor.

Evaluation of (11.3) at the surface gives the exchanging flux of water vapor between the surface and atmosphere. The vapor density, n, in the subsurface is controlled by equilibrium with non-vapor phases at regolith temperatures. For example, when, and at depths where, ice is present, the vapor density is given by equilibrium with that ice. When ice is absent, equilibrium with surface adsorption controls the vapor density, which in itself depends on the vapor density. Proper phase partitioning is important to understanding the amount of water in a vapor phase available for diffusive transport (e.g. Mellon and Jakosky, 1993). The presence of hydrated minerals, soluble salts, and brines may also play a role in phase partitioning.

In the absence of ice, the exchange of water between the regolith and atmosphere (the rate of diffusive migration of water into and out of the regolith) depends primarily on the adsorption isotherm (the relationship between adsorbed water content and the temperature and vapor density). The climate studies described above commonly use the adsorption isotherm of water on basalt powder as measured by Fanale and Cannon (1971, 1974). These measurements were for samples with a specific surface area of about 9.8 m² g⁻¹ and extrapolated from laboratory temperatures (>250 K) to Mars soil temperatures (as low as 150 K). Zent and Quinn (1997) made new measurements of palagonite samples and showed that the Fanale and Cannon isotherm overestimates the adsorption of water per unit surface area by about a factor of 30 at Mars environmental conditions. While adsorption depends only secondarily on grain mineralogy, it depends very strongly on the specific surface area, which will likely vary considerably across the surface of Mars. Generally speaking, coarse-grained or poorly chemically weathered basalts (such as might be found in sand dunes) will exhibit low surface areas (perhaps <10 m² g⁻¹), while weathering products such as phyllosilicate smectite clays will have much higher surface areas (approaching 1000 m² g⁻¹) due to the layered-sheet mineral structure (Zent and Quinn, 1997).

11.3.3.3 Regolith–Atmosphere Exchange: Theoretical and Observational Limitations

At present, the role of diurnal and seasonal exchange of water between the regolith and atmosphere is not well understood. Existing studies contrast in their conclusions regarding the magnitude of the exchange of adsorbed water and the importance of the regolith. Additionally, many of the studies discussed above have utilized measurements of Fanale and Cannon (1971, 1974) when modeling the adsorptive capacity of the Martian regolith and its temperature dependence, which have been subsequently shown to be inaccurately extrapolated to Martian conditions (Zent and Quinn, 1997). Numerous new measurements of adsorption have been conducted for a variety of Mars-analog materials (e.g. Zent and Quinn, 1995, 1997; Zent et al., 2001; Bryson et al., 2008; Chevrier et al., 2008; Jänchen et al., 2009; Beck et al., 2011). Care should be taken in incorporating appropriate adsorption isotherms in water cycle models. Additionally, Martian soils and their adsorptive capacity are likely to vary substantially with geographic region and depth. For example, regions of higher surface albedo and lower thermal inertia represent vast deposits of fine-grained soils (commonly referred to as "dust") perhaps smaller than 40 μm in diameter (Palluconi and Kieffer, 1981; Mellon et al., 2000). Finer-grained soils will exhibit higher surface area. Likewise, areas with higher concentrations of smectite clays will exhibit high surface area and greater adsorbed water abundance.

The role of shallow ground ice as a reservoir for atmospheric exchange has only been minimally considered. Mellon and Jakosky (1995) found that about 40 g cm⁻² exchanges with the atmosphere (and presumably polar caps) over an obliquity-oscillation timescale of 10⁵ years. If this amount were expelled at a constant rate over half an obliquity cycle, it roughly translates to 8 μm cm⁻² yr⁻¹. Key diffusive properties of porosity and tortuosity have been revised since these calculations (e.g. Sizemore and Mellon, 2008), so this estimate should be taken as a qualitative result. Nonetheless, Böttger et al. (2005) indicated a similar scale of exchange between ground ice and the atmosphere on seasonal timescales. At present, it is not clear how much water exchanges from the ground ice in the current epoch, as either a steady loss or gain due to inter-annual variability or orbitally driven climate change.

The role of ground ice in the diurnal and seasonal water cycle depends strongly on the burial depth of the ice table and diffusion through the ice-free overburden. Analysis of Mars Odyssey data suggests ice-table depths that are typically twice as deep as theoretical models predict (e.g. Mellon et al., 2004). However, this difference is attributed to the presence of rocks and slopes on the Martian surface and their nonlinear effects on leakage neutrons (Sizemore and Mellon, 2006), such that the ice-table depth in the soil areas between rocks is as predicted. In general, there is strong evidence that the ice-table depth locally and globally exists in diffusive equilibrium with the current atmosphere, or perhaps a slightly more humid climate than is presently measured (Mellon et al., 2004, 2009; Mellon and Feldman, 2006). In addition, the ice-table depth will vary regionally with latitude (e.g. Leighton and Murray, 1966) and with soil thermal properties (Paige, 1992). Depths can vary from millimeters to meters, but are most often a few to a few tens of centimeters (Mellon et al., 2004).

Some minerals observed on Mars contain chemically bound water that can be released (or incorporated) as environmental conditions change. Such water has not been considered in modeling the atmospheric water cycle. Sulfate salts, such as kieserite ($MgSO_4 \cdot H_2O$), hexahydrite ($MgSO_4 \cdot 6H_2O$), and epsomite ($MgSO_4 \cdot 7H_2O$), have been detected in mono- and polyhydrated forms (Bibring et al., 2005). These minerals are believed to be stable on Mars (Chou and Seal, 2003; Feldman et al., 2004; Fialips et al., 2005; Chipera and Vaniman, 2007). Changes in ground temperature and atmospheric humidity may cause transitions between stable states, though it is possible that reaction kinetics may be too slow to play a significant role on seasonal or shorter timescales (Chipera and Vaniman, 2007). In addition, Hecht et al. (2009) suggested that perchlorate salts (a strongly deliquescent mineral) may play some role in buffering atmospheric humidity, an effect that has not been previously considered. Detailed modeling of the role of hydrated minerals is needed to understand their potential impact on the Martian water cycle.

11.4 SUMMARY AND PERSPECTIVES

11.4.1 An Integrated Picture of the Present-Day Water Cycle

This chapter provides a comprehensive overview of our understanding of Mars' water cycle, including a description of its various atmospheric, surface, and subsurface reservoirs. Some of the major questions raised in the introduction have been answered, whereas others remain uncertain. The main seasonal sources and sinks of water have been identified: seasonal and perennial ices have been mapped and characterized, and atmospheric water has been tracked seasonally in its vapor and condensed phases. On average, the seasonal cycle of water on Mars is characterized by a factor of 2 variation of the overall water vapor abundance, forced by the seasonal sublimation and condensation of about 3×10^{12} kg of surface ice. A third of this mass (10^{12} kg) is exchanged seasonally between the two hemispheres during the solstitial seasons, implying that a major fraction (two-thirds) of water annually recycles in the northern hemisphere. Models show that >85% of atmospheric water

exists in the form of vapor, the remaining fraction being water in the form of clouds.

Closure of the water cycle on an annual basis is driven by rapid atmospheric transport processes and by interactions between the surface and the atmosphere through seasonal ice sublimation and re-formation. This implies that the present-day water cycle reflects equilibrium with the current climatic conditions, which favor the storage of water in the northern hemisphere.

The most important processes in the water cycle are the exchanges between the northern polar cap, by far the largest exposed reservoir of water on the planet, and the atmosphere, which controls the spring–summer extraction and the winter–spring return of water to the poles. In this picture, atmospheric transport can be separated into two main components: the zonally symmetric Hadley circulation as the prime mechanism for water exchange between the two hemispheres, and planetary waves (the residual, non-zonal component of circulation), which control the poleward and equatorward fluxes at mid- and high latitudes of both hemispheres.

Important to these exchanges are clouds. Water ice clouds are the unique atmospheric conveyors of water in the regions and at the seasons where low temperatures impose very low water vapor abundance. This occurs in the northern spring and summer tropics when clouds are predicted to unlock a net transfer of 10^{12} kg water to the southern hemisphere through the upper branch of the solstitial Hadley cell. Similarly, clouds delay the winter–spring return of water to the polar regions since they can be advected by the equatorward motions of the traveling disturbances that develop at the edges of the polar vortices, ending up with a significant buildup (a factor of 2) in the annually averaged concentration of water in the tropics.

There remains considerable uncertainty as to the exact role played by the exchanges between the regolith and the atmosphere. The fact that many water cycle models have been able to reproduce the salient features of the seasonal cycle of water vapor without the action of a regolith leaves the question of the influence of the seasonal storage and release of water in to and out from the subsurface still open. As explained in the previous section, regolith representation is hampered by the lack of knowledge of some major parameters, in particular the regolith adsorptive capacity, whose spatial variability is hard to constrain without relevant *in situ* measurements. While the role of the regolith appears fundamental for the evolution of water on geologic timescales, its influence on the diurnal and seasonal fate of water remains to be quantified.

11.4.2 The Seasonal Water Cycle: a Mars Climate Regulator?

The Mars water cycle is likely responsible for several fundamental climatic feedbacks. In the form of vapor, water has little effect on the radiative budget of the planet and is usually neglected. However, when condensed as frost or as clouds, water can modulate the local radiative properties of the surface and of the atmosphere. On the surface, seasonal and perennial ices change the thermophysical properties of the exposed layers of the soil, changing in turn the seasonal cycles of deposition and sublimation of ice.

One of the most studied impacts of the water cycle on Mars' climate is the direct radiative forcing of water ice clouds (Wilson et al., 2007, 2008; Haberle et al., 2011; Madeleine et al., 2012; Navarro et al., 2014). Wherever and whenever clouds form, they can significantly affect the thermal structure through thermal infrared emission and increased reflectivity in the visible part of the solar spectrum. Wilson et al. (2008) showed the tight coupling between water ice cloud radiative cooling and thermal tide vertical propagation in the tropics, deepening temperature inversions in the troposphere by up to 20 K and increasing surface temperature at night (Wilson et al., 2007). Recent GCM simulations confirm the need to include cloud radiative feedbacks to match the observed seasonality of tropical and polar atmospheric temperature profiles (Haberle et al., 2011; Madeleine et al., 2012; Navarro et al., 2014).

However, an even more important climatic effect of clouds may arise from their interactions with suspended dust. As previously described (see Section 11.2.1 and Chapter 5 for further details), water ice cloud formation occurs via nucleation on mineral airborne dust, the primary radiative agent of the Martian atmosphere. As the icy crystals grow, they fall more rapidly and carry with them their dust cores. This induces a dust "cleansing" or "scavenging" effect and directly perturbs the local radiative budget and subsequently the thermal structure of the atmosphere. The nonlinear nature of these interactions, due to the exponential relationship of condensation with temperature, is the likely source of metastable climatic states (Rodin et al., 1999).

As such, the water cycle occupies a nodal position in Mars climate, connecting all major climatic variables in an exponential way. It is therefore tempting to associate this nonlinear perturbation with the existence of climatic inter-annual variability that is evidenced by the non-periodic occurrence of planet-encircling dust storms (see Chapter 10). Climatic interactions with dust and with CO_2 (such as proposed by Jakosky and Haberle (1990)) might by revealed with future improvements in water cycle models.

11.4.3 The Future of Observations and Models

Monitoring of water in almost all its forms during the last decade has facilitated considerable advances in our understanding of Mars' water cycle. Currently orbiting spacecraft (MEX and MRO) continue their observations of water vapor and clouds, providing insights into areas where substantial knowledge gaps still exist. For instance, the first results collected by these missions on the vertical distribution of water vapor will be completed to permit partial reconstruction of the 3D variability of water, a poorly known aspect of the water cycle. Future missions, such as the ExoMars Trace Gas Orbiter, will be particularly well suited to carry on this task. However, unless an ambitious network of stations lands on Mars with the appropriate payload, the true nature of regolith–atmosphere exchange and the role of subsurface reservoirs will remain difficult to determine.

Climate modelers now acknowledge the need to include water-cycle-related climatic feedbacks (e.g. cloud effect on radiative transfer, dust nucleation and scavenging) that were neglected in the original models. This can only improve our understanding of the controls on the water cycle. The community is now making the conceptual transition from "Do I have the right climate model to study the water cycle?" to "I cannot correctly model the climate without the water cycle." Obviously, such ambitious undertakings come with increased complexity. In this regard, several areas require further study to solidify their theoretical foundations, such as the coupling of dust and water in the atmosphere and on the surface, which subsequently affects the radiative properties and hence the climate. However, the growing body of observational constraints is far from being fully analyzed, and sufficient material is now at the disposal of the community to further the knowledge of the present-day Mars water cycle.

REFERENCES

Aharonson, O., and N. Schorghofer, Subsurface ice on Mars with rough topography, *J. Geophys. Res.*, 111, E11007, 1–10. 2006.

Anderson, D. M., and A. R. Tice, Predicting unfrozen water contents in frozen soils from surface area measurements, *Highway Res. Rec.*, 393, 12–18, 1972.

Appéré, T., B. Schmitt, A. Pommerol, et al., Spatial and temporal distributions of the water ice annulus during recession of the northern seasonal condensates on Mars, *3rd International Workshop on the Mars Atmosphere: Modeling and Observations*, Williamsburg, Virginia, 2008.

Appéré, T., B. Schmitt, Y. Langevin, et al., Winter and spring evolution of northern seasonal deposits on Mars from OMEGA on Mars Express, *J. Geophys. Res*, 116, E05001, 2011.

Arvidson, R. E., D. Adams, G. Bonfiglio, et al., Mars Exploration Program 2007 Phoenix landing site selection and characteristics, *J. Geophys. Res.*, 113, E00A03, 2008.

Barker, E. S., Martian atmospheric water observations: 1972–74 apparition, *Icarus*, 28, 247–268, 1976.

Barker, E. S., R. A. Schorn, A. Worszczyk, R. G. Tull, and S. J. Little, Mars: detection of atmospheric water vapor during the southern hemisphere spring and summer season, *Science*, 170, 1308–1310, 1970.

Barnes, J. R., Midlatitude disturbances in the Martian atmosphere: a second Mars year, *J. Atmos. Sci.*, 38, 225–234, 1981.

Barr, A. C., and S. M. Milkovich, Ice grain size and the rheology of the Martian polar deposits, *Icarus*, 194, 513–518, 2008.

Bass, D. S., and D. A. Paige, Variability of Mars' North polar water ice cap. II. Analysis of Viking IRTM and MAWD data, *Icarus*, 144, 397–409, 2000.

Bass, D. S., K. E. Herkenhoff, and D. A. Paige, Variability of Mars' North Polar Water Ice Cap. I. Analysis of Mariner 9 and Viking Orbiter Imaging Data, *Icarus*, 144, 382–396, 2000.

Beck, P., A. Pommerol, B. Schmitt, and O. Brissaud, Kinetics of water adsorption on minerals and the breathing of the Martian regolith, *J. Geophys. Res.*, 115, E10011, 2011.

Benson, J. L., and P. B. James, Yearly comparisons of the Martian polar caps: 1999–2003 Mars Orbiter Camera observations. *Icarus*, 174 (2), 513–523, 2005.

Benson, J. L., B. P. Bonev, P. B. James, et al., The seasonal behavior of water ice clouds in the Tharsis and Valles Marineris regions of Mars: Mars Orbiter Camera observations. *Icarus*, 165, 34–52, 2003.

Benson, J. L., D. M. Kass, and A. Kleinböhl, Mars' north polar hood as observed by the Mars Climate Sounder, *J. Geophys. Res.*, 116, E03008, doi:10.1029/2010JE003693, 2011.

Bibring, J.-P., Y. Langevin, F. Poulet, et al., Perennial water ice identified in the south polar cap of Mars, *Nature*, 428, 627–630, 2004.

Bibring, J.-P., Y. Langevin, A. Gendrin, et al., Mars surface diversity as revealed by the OMEGA/Mars Express observations, *Science*, 307, 1576–1581, 2005.

Böttger, H. M., S. R. Lewis, R. L. Read, and F. Forget, The effect of a global dust storm on simulations of the Martian water cycle, *Geophys. Res. Let.*, 31(22), L22702, 2004.

Böttger, H. M., S. R. Lewis, R. L. Read, and F. Forget, The effects of the Martian regolith on GCM water cycle simulations, *Icarus*, 177, 174–189, 2005.

Boynton, W. V., W. C. Feldman, S. W. Squyres, et al., Distribution of hydrogen in the near surface of Mars: evidence for subsurface ice deposits. *Science*, 297, 81–85, 2002.

Brown A. J., S. Byrne, L. L. Tornabene, and T. Roush, Louth Crater: evolution of a layered water ice mound, *Icarus*, 196 (2), 433–445, 2008.

Brutsaert, W., *Evaporation Into the Atmosphere*, Kluwer Academic, Norwell, MA, 1982.

Bryson K. L., V. Chevrier, D. W. G. Sears, and R. Ulrich, Stability of ice on Mars and the water vapor diurnal cycle: experimental study of the sublimation of ice through a fine-grained basaltic regolith, *Icarus*, 196, 446–458, 2008.

Burgdorf, M. J., T. Encrenaz, E. Lellouch, et al., ISO observations of Mars: an estimate of the water vapor vertical distribution and the surface emissivity, *Icarus*, 145, 79–90, 2000.

Byrne, S., and A. P. Ingersoll, A Sublimation Model for Martian South Polar Ice Features, *Science* 299 (February): 1051–53, 2003.

Calvin W. M., and T. N. Titus, Summer season variability of the north residual cap of Mars as observed by the Mars Global Surveyor Thermal Emission Spectrometer (MGS-TES), *Planet. Space Sci.*, 56, 212–226, 2008.

Cantor B. A., P. B. James, and W. M. Calvin, MARCI and MOC observations of the atmosphere and surface cap in the north polar region of Mars, *Icarus*, 208, 61–81, 2010.

Caplinger, M. A., and M. C. Malin, Mars Orbiter Camera geodesy campaign, *J. Geophys. Res.*, 106 (23), 595–23, 606, doi:10.1029/2000JE001341, 2001.

Chamberlain, M. A., and W. V. Boynton, Response of Martian ground ice to orbit-induced climate change, *J. Geophys. Res.*, 112, E06009, 2006.

Chevrier, V., D. R. Ostrowski, and D. W. G. Sears, Experimental study of the sublimation of ice through an unconsolidated clay layer: implications for the stability of ice on Mars and the possible diurnal variations in atmospheric water, *Icarus*, 196, 459–476, 2008.

Chipera, S. J., and D. T. Vaniman, Experimental stability of magnesium sulfate hydrates that may be present on Mars, *Geochimica et Cosmochimica Acta*, 71, 241–250, 2007.

Chittenden, J. D., V. Chevrier, L. A. Roe, et al., Experimental Study of the Effect of Wind on the Stability of Water Ice on Mars, *Icarus*, 196, 477, doi:10.1016/j.icarus.2008.01.016, 2008.

Chou, I. M. and R. R. Seal, Determination of epsomite–hexahydrite equilibria by the humidity-buffer technique at 0.1 MPa with implications for phase equilibria in the system $MgSO_4$–H_2O, *Astrobiology*, 3, 619–630, 2003.

Clancy, R. T., A. W. Grossman, and D. O. Muhleman, Mapping Mars water vapor with the Very Large Array, *Icarus*, 100, 48–59, 1992.

Clancy, R. T., A. W. Grossman, M. J. Wolff, et al., Water vapor saturation at low altitudes around Mars aphelion: a key to Mars climate?, *Icarus*, 122, 36–62, 1996.

Clancy, R. T., M. J. Wolff, and P. R. Christensen, Mars aerosol studies with the MGS TES emission phase function observations: optical depths, particle sizes, and ice cloud types versus latitude and solar longitude, *J. Geophys. Res.*, 108(E9), 5098, doi:10.1029/2003JE002058, 2003.

Clifford, S. M., A model for the hydrological and climatic behavior of water on Mars, *J. Geophys. Res.*, 98, 10973–11016, 1993.

Clifford, S. M., and D. Hillel, The stability of ground ice in the equatorial region of Mars, *J. Geophys. Res.*, 88, 2456–2474, 1983.

Colaprete, A., O. B. Toon, and J. A. Magalhaes, Cloud formation under Mars Pathfinder conditions, *J. Geophys. Res.*, 104, 9043–9053, 1999.

Colaprete A., J. R. Barnes, R. M. Haberle, et al, Albedo of the south pole on Mars determined by topographic forcing of atmosphere dynamics. *Nature*, 435, 184–188, 2005.

Conrath, B., R. Curran, R. Hanel, et al., Atmospheric and surface properties of Mars obtained by infrared spectroscopy on Mariner 9, *J. Geophys. Res.*, 78(20), 4267–4278, 1973.

Costard, F., F. Forget, N. Mangold, and J. P. Peulvast, Formation of recent Martian debris flows by melting of near-surface ground ice at high obliquity, *Science*, 295, 110–113, 2002

Cull S., R. E. Arvidson, M. Mellon, et al. Seasonal H_2O and CO_2 ice cycles at the Mars Phoenix landing site: 1. Prelanding CRISM and HiRISE observations, *Journal of Geophysical Research*, 115, E00D17, 2010a.

Cull S., R. E. Arvidson, R. V. Morris, et al., Seasonal ice cycle at the Mars Phoenix landing site: 2. Postlanding CRISM and ground observations, *J. Geophys. Res.*, 115, E00E19, 2010b.

Curran, R. J., B. J. Conrath, R. A. Hanel, V. G. Kunde, and J. C. Pearl, Mars: Mariner 9 spectroscopic evidence for H_2O ice clouds, *Science*, 182, 381–383, 1973.

Davies, D. W., The vertical distribution of Mars water vapor, *J. Geophys. Res.*, 84, 2875–2879, 1979.

Davies, D. W., and L. A. Wainio, Measurements of water vapor in Mars' Antarctic, *Icarus*, 45, 216–230, 1981.

Davies, D. W., The Mars water cycle, *Icarus*, 45, 398–414, 1981.

de Boer, J. H., *The Dynamic Character of Adsorption*, Oxford University Press, London, 1968.

Douté S., B. Schmitt, Y. Langevin, et al., South pole of Mars: nature and composition of the icy terrains from Mars Express OMEGA observations, *Planet. Space Sci.*, 55, 113–133, 2007.

Encrenaz, T., E. Lellouch, J. Rosenqvist, et al. The atmospheric composition of Mars: ISM and ground-based observational data. *Ann. Geophysicae*, 9, 797–803, 1991.

Encrenaz, T., E. Lellouch, J. Cernicharo, G. Paubert, and S. Gulkis, A tentative detection of the 183-GHz water vapor line in the Martian atmosphere: constraints upon the H_2O abundance and vertical distribution, *Icarus*, 113, 110–118, 1995.

Encrenaz, T., B. Bezard, T. Owen, et al., Infrared imaging spectroscopy of Mars: H_2O mapping and determination of CO_2 isotopic ratios, *Icarus*, 179(1), 43–54, 2005a.

Encrenaz, T., R. Melchiorri, T. Fouchet, et al., A mapping of Martian water sublimation during early northern summer using OMEGA/Mars Express, *Astron. Astrophys.*, 441, 9–12, 2005b.

Encrenaz, T., T. K. Greathouse, M. J. Richter, et al., Simultaneous mapping of H_2O and H_2O_2 on Mars from infrared high-resolution imaging spectroscopy. *Icarus*, 195(2), 547–556, 2008a.

Encrenaz, T., T. Fouchet, R. Melchiorri, et al., Study of the Martian water vapor over Hellas using OMEGA and PFS aboard Mars Express, *Astronomy and Astrophysics*, 484(2), 547–553, 2008b.

Encrenaz, T., T. K. Greathouse, B. Bézard, et al., Water vapor map of Mars near summer solstice using ground-based infrared spectroscopy, *Astronomy and Astrophysics*, 520, A33, 2010.

Fanale, F. P., and W. A. Cannon, Adsorption on the Martian regolith, *Nature*, 230, 502–504, 1971.

Fanale, F. P., and W. A. Cannon, Exchange of adsorbed H_2O and CO_2 between the regolith and atmosphere of Mars caused by changes in surface insolation, *J. Geophys. Res.*, 79, 3397–3402, 1974.

Fanale, F. P., J. R. Salvail, A. P. Zent, and S. E. Postawko, Global distribution and migration of subsurface ice on Mars, *Icarus*, 67, 1–18, 1986.

Farmer, C. B., and P. E. Doms, Global seasonal variations of water vapor on Mars and the implications for permafrost, *J. Geophys. Res.*, 84, 2881–2888, 1979.

Farmer, C. B., and D. D. LaPorte, The detection and mapping of water vapor in the Martian atmosphere. *Icarus*, 16, 34–46, 1972.

Farmer, C. B., D. W. Davies, A. L. Holland, D. D. LaPorte, and P. E. Doms, Mars: water vapor observations from the Viking Orbiters. *J. Geophys. Res.*, 82 (28), 4225–4248, 1977.

Fedorova, A. A., A. V. Rodin, and I. V. Baklanova, MAWD observations revisited: seasonal behavior of water vapor in the Martian atmosphere, *Icarus*, 171(1), 54–67, 2004.

Fedorova, A., O. Korablev, J.-L. Bertaux, et al., Mars water vapor abundance from SPICAM IR spectrometer: seasonal and geographic distributions, *J. Geophys. Res.*, 111, E09S08, doi:10.1029/2006JE002695, 2006.

Fedorova, A. A., O. I. Korablev, J.-L. Bertaux, et al., Solar infrared occultation observations by SPICAM experiment on Mars-Express: simultaneous measurements of the vertical distributions of H_2O, CO_2 and aerosol, *Icarus*, 200(1), 96–117, 2009.

Fedorova, A. A., S. Trokhimovsky, O. Korablev, and F. Montmessin, Viking observation of water vapor on Mars: revision from up-to-date spectroscopy and atmospheric models, *Icarus*, 208, 156–164, 2010.

Feldman, W. C., W. V. Boynton, R. L. Tokar, et al., Global distribution of neutrons from Mars: results from Mars Odyssey. *Science*, 297, 75–78, 2002.

Feldman W. C., M. T. Mellon, S. Maurice, et al., Hydrated states of $MgSO4$ at equatorial latitudes on Mars, *Geophys. Res. Lett.*, 31(16) Art. L16702, 2004.

Feldman, W. C., M. T. Mellon, O. Gasnault, S. Maurice, and T. H. Prettyman, Volatiles on Mars: scientific results from the Mars Odyssey Neutron Spectrometer, in *The Martian Surface: Composition, Mineralogy, and Physical Properties*, J. F. Bell, ed., Cambridge University Press, London, 125–148, 2008.

Fialips, C. I., J. W. Carey, D. T. Vaniman, et al., Hydration state of zeolites, clays, and hydrated salts under present-day Martian surface conditions: can hydrous minerals account for Mars Odyssey observations of near-equatorial water-equivalent hydrogen?, *Icarus*, 178, 74–83, 2005.

Flasar, F. M., and R. M. Goody, Diurnal behavior of water on Mars, *Planet. Space Sci.*, 24, 161–181, 1976.

Forget F., F. Hourdin, R. Fournier, et al., Improved general circulation models of the Martian atmosphere from the surface to above 80 km. *J. Geophys. Res.*, 104, 24155–24176, 1999.

Forget, F., R. M. Haberle, F. Montmessin, B. Levrard, and J. W. Head, Formation of Glaciers on Mars by Atmospheric Precipitation at High Obliquity, *Science*, 311 (January): 368–71. doi:10.1126/science.1120335, 2006.

Fouchet, T., E. Lellouch, N.I. Ignatiev, et al., Martian water vapor: Mars Express PFS/LW observations, *Icarus*, 190, 32–49, 2007.

Fouchet, T., R. Moreno, E. Lellouch, et al., Interferometric millimeter observations of water vapor on Mars and comparison with Mars Express measurements, *Planetary and Space Science*, 59, 683–690, 2011.

Goff, J. A., and S. Gratch, Low-pressure properties of water from –160 to 212 F, *Transactions of the American Society of Heating and Ventilating Engineers*, 25–164, New York, 1946.

Grima C., W. Kofman, J. Mouginot, et al., North polar deposits of Mars: extreme purity of the water ice, *Geophys. Res. Lett.*, 36, L03203, 2009.

Gurwell, M. A., E. A. Bergin, G. J. Melnick, et al., Submillimeter wave astronomy satellite observations of the Martian atmosphere: temperature and vertical distribution of water vapor, *Astrophys. J.*, 539, L143–L146, 2002.

Haberle, R. M., and B. M. Jakosky, Sublimation and transport of water from the North residual polar cap on Mars. *J. Geophys. Res.*, 95(B2), 1423–1437, 1990.

Haberle, R. M., C. B. Leovy, and J. B. Pollack, Some effects of global dust storms on the atmospheric circulation of Mars, *Icarus*, 50, 322–367, 1982.

Haberle, R. M., J. B. Pollack, J. R. Barnes, et al., Mars atmospheric dynamics as simulated by the NASA Ames General Circulation Model: 1. The zonal-mean circulation, *J. Geophys. Res.*, 98(E2), 3093, doi:10.1029/92JE02946, 1993.

Haberle, R. M., F. Montmessin, M. A. Kahre, et al., Radiative effects of water ice clouds on the Martian seasonal water cycle. *4th International Workshop on the Mars Atmosphere: Modeling and Observations*, Extended Abstracts, 223–226, Paris, France, 8–11 February 2011.

Harri, A.-M., M. Genzer, O. Kemppinen, et al., Mars Science Laboratory relative humidity observations: initial results., *J. Geophys. Res. Planets*, 119(9), 2132–2147, doi:10.1002/2013JE004514, 2014.

Hecht, M. H., Metastability of liquid water on Mars, *Icarus*, 156, 373–386, 2002.

Hecht, M. H., S. P. Kounaves, R. C. Quinn, et al., Detection of perchlorate and the soluble chemistry of Martian soil: findings from the Phoenix Mars Lander, *Science*, 325(64), 2009.

Herkenhoff, K. E., and J. J. Plaut, Surface ages and resurfacing rates of the polar layered deposits on Mars. *Icarus*, 144, 243–53, 2000.

Hollingsworth, J. L., R. M. Haberle, J. R. Barnes, et al., Orographic control of storm zones on Mars, *Nature*, 380(6573), 413–416, doi:10.1038/380413a0, 1996.

Holt, J. W., A. Safaeinili, J. J. Plaut, et al., Radar sounding evidence for buried glaciers in the southern mid-latitudes of Mars, *Science*, 322, 1235–1238, 2008.

Houben, H., R. M. Haberle, R. E. Young, and A. P. Zent, Modeling the Martian seasonal water cycle, *J. Geophys. Res.*, 102 (E4), 9063–9083, 1997.

Hunten, D. M., A. L. Sprague, and L. R. Doose, Correction for dust opacity of Martian atmospheric water vapor abundances, *Icarus*, 147, 42–48, 2000.

Ignatiev, N. I., L. V. Zasova, V. Formisano, D. Grassi, and A. Maturilli, Water vapor abundance in Martian atmosphere from revised Mariner 9 IRIS data, *Adv. Space Rev.*, 29(2), 157–162, 2002.

Ingersoll, A. P., Mars: occurrence of liquid water, *Science*, 168, 972–973, doi:10.1126/science.168.3934.972, 1970.

Iraci, L. T., B. D. Phebus, B. M. Stone, and A. Colaprete, Water ice cloud formation on Mars is more difficult than presumed: laboratory studies of ice nucleation on surrogate materials, *Icarus*, 210(2), doi:10.1016/j.icarus.2010.07.020, 2010.

Ivanov, A. B., and D. O. Muhleman, The role of sublimation for the formation of the northern ice cap: results from the Mars Orbiter Laser Altimeter, *Icarus*, 144, 436–448, 2000.

Jakosky, B. M., The role of seasonal reservoirs in the Mars water cycle. I. Seasonal exchange of water with the regolith, *Icarus*, 55, 1–18, 1983a.

Jakosky, B. M., The role of seasonal reservoirs in the Mars water cycle. II. Coupled models of the regolith, the polar caps, and atmospheric transport, *Icarus*, 55, 19–39, 1983b.

Jakosky, B. M., The seasonal cycle of water on Mars, *Space Science Reviews*, 41, 131–200, 1985.

Jakosky, B. M., and E. S. Barker, Comparison of ground-based and Viking Orbiter measurements of Martian water vapor: variability of the seasonal cycle, *Icarus*, 57, 322–334, 1984.

Jakosky, B. M., and C. B. Farmer, The seasonal and global behavior of water vapor in the Mars atmosphere: complete global results of the Viking atmospheric water detector experiment, *J. Geophys. Res.*, 87, 2999–3019, 1982.

Jakosky, B. M., and R. M. Haberle, Year-to-year instability of the Mars south polar cap, *J. Geophys. Res.* 95, 1359. doi:10.1029/JB095iB02p01359, 1990.

Jakosky, B. M., R. W. Zurek, and M. R. LaPointe, The observed day-to-day variability of Mars atmospheric water vapor, *Icarus*, 73, 80–90, 1988.

Jakosky, B. M., A. P. Zent, and R. W. Zurek, The Mars water cycle: determining the role of exchange with the regolith, *Icarus*, 130, 87–95, 1997.

James P. B., Recession of Martian north polar cap – 1977–1978 Viking observations, *J. Geophys. Res.*, 84, 8332–8334, 1979.

James, P. B., The Martian hydrologic cycle – effects of CO_2 mass flux on global water distribution, *Icarus*, 64, 249–264, 1985.

James P. B., and B. A. Cantor, Mars Orbiter Camera observations of the Martian south polar cap in 1999–2000, *J. Geophys. Res.*, 106, 23635–23652, 2001.

James P. B., P. C. Thomas, M. J. Wolff, and B. P. Bonev, MOC observations of four Mars years variations in the south polar residual cap of Mars, *Icarus*, 192, 318–326, 2007.

James P. B., P. C. Thomas, and M. C. Malin, Variability of the south polar cap of Mars in Mars Years 28 and 29, *Icarus*, 208, 82–85, 2010.

Jänchen, J., R. V. Morris, D. L. Bish, M. Janssen, and U. Hellwig, The H_2O and CO_2 adsorption properties of phyllosilicate-poor palagonitic dust and smectites under Martian environmental conditions, *Icarus*, 200, 463–467, 2009.

Jaquin, F., P. Gierasch, and R. Kahn, The vertical structure of limb hazes in the Martian atmosphere, *Icarus*, 68, 442–461, 1986.

Kahn, R., The Spatial and Seasonal Distribution of Martian Clouds and Some Meteorological Implications, *J. Geophys. Res.*, 89, 6671–88, 1984.

Kass, D. M., and Y. L. Yung, Water on Mars: isotopic constraints on exchange between the atmosphere and surface, *Geophys. Res. Let.*, 26(24) 3653–3656, 1999.

Kereszturi, A., M. Vincendon, and F. Schmidt, Water ice in the dark dune spots of Richardson Crater on Mars, *Planet. Space Sci.*, 59, 26–42, 2011.

Kieffer, H. H., Mars south polar spring and summer temperatures: a residual CO_2 frost, *J. Geophys. Res.*, 84, 8263–8288, 1979.

Kieffer, H. H., H_2O grain size and the amount of dust in Mars' residual north polar cap, *J. Geophys.Res.*, 95(B2), 1481–1493, 1990.

Kieffer, H. H., and Titus, T. N., TES mapping of Mars' north seasonal cap, *Icarus*, 154, 162–180, 2001.

Kieffer H. H., S. C. Chase, T. Z. Martin, E. D. Miner, and F. D. Palluconi, Martian North Pole summer temperature: dirty water ice, *Science*, 194, 1341–1344, 1976.

Kieffer, H. H., T. Z. Martin, A. R. Peterfreund, et al., Thermal and albedo mapping of Mars during the Viking primary mission, *J. Geophys. Res.*, 82, 4249–4291, 1977.

Kieffer, H. H., T. N. Titus, K. F. Mullins, and P. R. Christensen, Mars south polar spring and summer behavior observed by TES: seasonal cap evolution controlled by frost grain size, *J. Geophys. Res.*, 105(E4), 9653–9700, 2000.

Korablev, O., N. Ignatiev, A. Fedorova, et al., Water in Mars atmosphere: comparison of recent data sets. *2nd International Workshop on the Mars Atmosphere: Modeling and Observations*, Granada, Spain, 2006.

Kuzmin, R. O., *The Cryolithosphere of Mars*, Izdatel'stvo Nauka, Moscow, 1983.

Kuzmin, R. O., N. N. Bobina, E. V. Zabalueva, and V. P. Shashkina, Structural inhomogeneities of the Martian cryolithosphere, *Solar System Res.*, 22, 195–212, 1988.

Lachenbruch, A. H., Mechanics of thermal contraction cracks and ice-wedge polygons in permafrost, *Geol. Soc. Am. Spec. Paper*, 70, 69, 1962.

Langevin, Y., F. Poulet, J.-P. Bibring, et al., Summer evolution of the north polar cap of Mars as observed by OMEGA/Mars Express, *Science*, 307, 1581–1584, 2005.

Langevin Y., J.-P. Bibring, F. Montmessin, et al., Observations of the south seasonal cap of Mars during recession in 2004–2006 by the OMEGA visible/near-infrared imaging spectrometer on board Mars Express. *J. Geophys. Res.*, 112, CiteID E08S12, 2007.

Laskar, J., and P. Robutel, The chaotic obliquity of the planets, *Nature*, 361, 608–612, 1993.

Laskar, J., B. Levrard, and J. F. Mustard, Orbital forcing of the Martian polar layered deposits, *Nature*, 419, 375–377, 2002.

Lefèvre, F., S. Lebonnois, F. Montmessin, and F. Forget, Three-dimensional modeling of ozone on Mars, *J. Geophys. Res.*, 109, E07004, doi:10.1029/2004JE002268, 2004.

Leffingwell, E. K., Ground-ice wedges: the dominant form of ground-ice on the north coast of Alaska, *J. Geol.*, 23, 635–654, 1915.

Leighton, R. B., and B. C. Murray, Behavior of carbon dioxide and other volatiles on Mars, *Science*, 153, 136–144, 1966.

Leovy, C. B., Exchange of water vapor between the atmosphere and surface of Mars, *Icarus*, 18, 120–125, 1973.

Leshin, L. A., P. R. Mahaffy, C. R. Webster, et al., Volatile, isotope, and organic analysis of Martian fines with the Mars Curiosity Rover, *Science*, 341(6153), 1238937, doi:10.1126/science.1238937, 2013.

Levrard B., F. Forget, F. Montmessin, and J. Laskar, Recent ice-rich deposits formed at high latitudes on Mars by sublimation of unstable equatorial ice during low obliquity, *Nature*, 431, 1072–1075, 2004.

Levy, J., J. Head, and D. Marchant, Thermal contraction crack polygons on Mars: classification, distribution, and climate implications from HiRISE observations, *J. Geophys. Res.*, 114, E01007, 2009.

Liu, J., M. I. Richardson, and R. J. Wilson, An assessment of the global, seasonal, and interannual spacecraft record of Martian climate in the thermal infrared, *J. Geophys. Res.*, 108, doi:10.1029/2002JE001921, 2003.

Määttänen, A., H. Vehkamaki, A. Lauri, et al., Nucleation studies in the Martian atmosphere, *J. Geophys. Res.*, 110, E02002, doi:10.1029/2004JE002308, 2005.

Madeleine, J.-B., F. Forget, E. Millour, T. Navarro, and A. Spiga, The Influence of Radiatively Active Water Ice Clouds on the Martian Climate, *Geophys. Res. Let.*, 39, doi:10.1029/2012GL053564, 2012.

Malin, M. C., and K. S. Edgett, Evidence for recent groundwater seepage and surface runoff on Mars, *Science*, 288, 2330–2335, 2000.

Malin, M. C., M. A. Caplinger, and S. D. Davis, Observational evidence for an active surface reservoir of solid carbon dioxide on Mars, *Science*, 294, 2330–2335, 2001.

Maltagliati, L., D. V. Titov, T. Encrenaz, et al., Observations of atmospheric water vapor above the Tharsis volcanoes on Mars with the OMEGA/MEx imaging spectrometer, *Icarus*, 194(1), 53–64, 2009.

Maltagliati, L., F. Montmessin, A. Fedorova, et al., Evidence of water vapor in excess of saturation in the atmosphere of Mars, *Science*, 333, 1868–1872, 2011a.

Maltagliati, L., D.V. Titov, T. Encrenaz, et al., Annual survey of water vapor behavior from the OMEGA mapping spectrometer onboard Mars Express, *Icarus*, 213(2), 480–495, 2011b.

Maltagliati, L., F. Montmessin, O. Korablev, et al., Annual Survey of Water Vapor Vertical Distribution and Water-Aerosol Coupling in the Martian Atmosphere Observed by SPICAM/MEx Solar Occultations, *Icarus*, 223, 942–62. doi:10.1016/j.icarus.2012.12.012, 2013.

Mangold, N., High latitude patterned ground on Mars: classification, distribution and climate control, *Icarus*, 174, 336–359, 2005.

Martín-Torres, F. J., M.-P. Zorzano, P. Valentín-Serrano, et al., Transient liquid water and water activity at Gale Crater on Mars, *Nat. Geosci.*, 8(5), 357–361, doi:10.1038/ngeo2412, 2015.

McElroy, M. B., Mars: an evolving atmosphere, *Science*, 175, 443–445, doi:10.1126/science.175.4020.443, 1972.

McEwen, A. S., C. M. Dundas, S. S. Mattson, et al., Recurring slope lineae in equatorial regions of Mars, *Nat. Geosci.*, 7(1), 53–58, doi:10.1038/ngeo2014, 2014.

Melchiorri, R., T. Encrenaz, T. Fouchet, et al., Water vapor mapping on Mars using OMEGA/Mars Express. Water vapor mapping on Mars using OMEGA/Mars Express, *Planetary and Space Science*, 55(3), 333–342, 2007.

Melchiorri R., T. Encrenaz, P. Drossart,et al., OMEGA/Mars Express: south pole region, water vapor daily variability, *Icarus*, 201, 102–112, 2009.

Mellon, M. T., Small-scale polygonal features on Mars: seasonal thermal contraction cracks in permafrost, *J. Geophys. Res.*, 102, 25617–25628, 1997.

Mellon, M. T., and W. C. Feldman, The global distribution of Martian subsurface ice and regional ice stability, 37th *Lunar and Planet. Sci. Conf.*, Houston, 2006.

Mellon, M. T., and B. M. Jakosky, Geographic variations in the thermal and diffusive stability of ground ice on Mars, *J. Geophys. Res.*, 98, 3345–3364, 1993.

Mellon, M. T., and B. M. Jakosky, The distribution and behavior of Martian ground ice during past and present epochs, *J. Geophys. Res.*, 100(E11), 781–11, 799, 1995.

Mellon M. T., B. M. Jakosky, H. H. Kieffer, and P. R. Christensen, High-resolution thermal inertia mapping from the Mars Global Surveyor Thermal Emission Spectrometer, *Icarus*, 148, 437–455, 2000.

Mellon, M. T, W. C. Feldman, and T. H. Prettyman, The presence and stability of ground ice in the southern hemisphere of Mars, *Icarus*, 169, 324–340, 2004.

Mellon, M. T., W. V. Boynton, W. C. Feldman, et al., Ice-table depth and ice characteristics in Martian permafrost at the proposed Phoenix landing site., *J. Geophys. Res.*, 113, E00A25, 2008.

Mellon, M. T., R. E. Arvidson, H. G. Sizemore, et al., Ground ice at the Phoenix landing site: stability state and origin, *J. Geophys. Res.*, 114, E00E07, 2009.

Michelangeli, D. V., O. B. Toon, R. B. Haberle, and J. B. Pollack, Numerical simulations of the formation and evolution of water ice clouds in the Martian atmosphere, *Icarus*, 100, 261–285, 1993.

Milliken, R. E., J. F. Mustard, F. Poulet, et al., Hydration state of the Martian surface as seen by Mars Express OMEGA: 2. H_2O content of the surface, *J. Geophys. Res.*, 112, E08S07, 2007.

Mitrofanov, I., D. Anfimov, A. Kozyrev, et al., Maps of Subsurface Hydrogen from the High Energy Neutron Detector, Mars Odyssey, *Science*, 297, 78–81, 2002.

Montmessin, F., P. Rannou, and M. Cabane, New insights into Martian dust distribution and water-ice cloud microphysics, *J. Geophys. Res.*, 107(E6), 5037, doi:10.1029/2001JE001520, 2002.

Montmessin, F., F. Forget, P. Rannou, M. Cabane, R.M. Haberle, Origin and role of water ice clouds in the Martian water cycle as inferred from a general circulation model, *J. Geophys. Res.*, 109, doi:10.1029/2004JE002284, 2004.

Montmessin, F., T. Fouchet, and F. Forget, Modeling the annual cycle of HDO in the Martian atmosphere, *J. Geophys. Res.*, 110, E03006, doi:10.1029/2004JE002357, 2005.

Montmessin F., R. M. Haberle, F. Forget, et al., On the origin of perennial water ice at the South Pole of Mars: a precession-controlled mechanism?, *J. Geophys. Res*, 112, E08S17, 2007.

Moroz, V. I., and A. E. Nadzhip, Preliminary measurement results of the water vapor content in the planetary atmosphere from measurements onboard the Mars 5 spacecraft, *Cosmic Research*, 13(N1), 28–30, translation, 1975.

Moroz, V. I., and A. E. Nadzhip, Water vapor in the atmosphere of Mars based on measurements on board Mars 3, *Cosmic Research*, 13, N5, 658–670, translation, 1976.

Mouginot, J., A. Pommerol, W. Kofman, et al., The 3–5 MHz global reflectivity map of Mars by MARSIS/Mars Express: implications for the current inventory of subsurface H_2O, *Icarus*, 210, 612–625, 2010.

Mustard, J. F., S. L. Murchie, S. M. Pelkey, et al., Hydrated silicate minerals on Mars observed by the Mars Reconnaissance Orbiter CRISM instrument, *Nature*, 454, 305–309, 2008.

Navarro, T., J.-B. Madeleine, F. Forget, et al., Global climate modeling of the Martian water cycle with improved microphysics and radiatively active water ice clouds, *J. Geophys. Res. Planets*, 119(7), 1479–1495, doi:10.1002/2013JE004550, 2014.

Ojha, L., M. B. Wilhelm, S. L. Murchie, et al., Spectral evidence for hydrated salts in recurring slope lineae on Mars, *Nature Geosci.*, 8, 829–832, doi:10.1038/NGEO2546, 2015.

Paige, D. A., The thermal stability of near-surface ground ice on Mars, *Nature*, 356, 43–45, 1992.

Palluconi, F. D., and H. H. Kieffer, Thermal inertia mapping of Mars from 60°S to 60°N, *Icarus*, 45, 415–426, 1981.

Pankine, A. A., L. K. Tamppari, M. D. Smith, Water vapor variability in the north polar region of Mars from Viking MAWD and MGS TES datasets, *Icarus*, 204, 87–102, 2009.

Pankine, A. A., L. K. Tamppari, and M. D. Smith, MGS TES observations of the water vapor above the seasonal and perennial ice caps during northern spring and summer, *Icarus*, 210, 58–71, 2010.

Pathak, J., D. V. Michelangeli, L. Komguem, J. Whiteway, and L. K. Tamppari, Simulating Martian boundary layer water ice clouds and the lidar measurements for the Phoenix mission, *J. Geophys. Res.*, 113, CiteID E00A05, doi:10.1029/2007JE002967, 2008.

Peixoto, J. P., and A. H. Oort, *Physics of Climate*, American Institute of Physics, New York, 1992.

Phillips R. J., M. T. Zuber, S. E. Smrekar, et al., Mars north polar deposits: stratigraphy, age and geodynamical response, *Science*, 320, 1182–1185, 2008.

Picardi G., J. J. Plaut, D. Biccari, et al., Radar soundings of the subsurface of Mars, *Science*, 310, 1925–1928, 2005.

Piqueux, S., C. S. Edwards, and P. R. Christensen, Distribution of the ices exposed near the south pole of Mars using Thermal Emission Imaging System (THEMIS) temperature measurements, *J. Geophys. Res.*, 113, E08014, doi:10.1029/2007JE003055, 2008.

Plaut, J. J., G. Picardi, A. Safaeinili, et al., Subsurface radar sounding of the south polar layered deposits of Mars, *Science*, 316, 92–96, 2007.

Plaut J. J., A. Safaeinili, J. W. Holt, et al., Radar evidence for ice in lobate debris aprons in the mid-northern latitudes of Mars, *Geophys. Res. Lett.*, 36, L02203, 2009.

Prettyman, T. H., W. C. Feldman, and T. N. Titus, Characterization of Mars's seasonal caps using neutron spectroscopy, *J. Geophys. Res.*, 114, E08005, 2009.

Putzig, N., and M. Mellon, Apparent Thermal Inertia and the Surface Heterogeneity of Mars, *Icarus*, 191, 68–94. doi:10.1016/j.icarus.2007.05.013, 2007.

Putzig, N. E., R. J. Phillips, B. A. Campbell, et al., Subsurface structure of Planum Boreum from Mars Reconnaissance Orbiter Shallow Radar soundings, *Icarus*, 204, 443–57, doi:10.1016/j.icarus.2009.07.034, 2009.

Read, P. L. and S. R. Lewis, *The Martian Climate Revisited: Atmosphere and Environment of a Desert Planet*, Springer-Praxis Books, 2004.

Rennó, N. O., B. J. Bos, D. Catling, et al., Possible physical and thermodynamic evidence for liquid water at the Phoenix landing site, *J. Geophys. Res.*, 114, E00E03, doi:10.1029/2009JE003362, 2009.

Richardson, M. I., A general circulation model study of the Mars water cycle, *Ph.D. thesis*, Univ. of Calif., Los Angeles, 1999.

Richardson, M. I., and R. J. Wilson, Investigation of the nature and stability of the Martian seasonal water cycle with a general circulation model, *J. Geophys. Res.*, 197, E5, 5031, 2002a.

Richardson, M. I., and R. J. Wilson, A topographically forced asymmetry in the Martian circulation and climate, *Nature*, 416, 298–301, 2002b.

Richardson, M. I., R. J. Wilson, and A. V. Rodin, Water ice clouds in the Martian atmosphere: general circulation model experiments with a simple cloud scheme, *J. Geophys. Res.*, 107(E9), 5064, doi:10.1029/2001JE001804, 2002.

Rizk, B., W. K. Wells, D. M. Hunten, et al., Meridional Martian water abundance profiles during the 1988–1989 season, *Icarus*, 90, 205–213, 1991.

Rodin, A. V., O. I. Korablev, V. I. Moroz, Vertical distribution of water in the near-equatorial troposphere of Mars: water vapor and clouds, *Icarus*, 125(1), 212–229, 1997.

Rodin, A.V., R. T. Clancy, R. J. Wilson, and M. Richardson, Dynamical properties of Mars water ice clouds and their interactions with atmospheric dust and radiation, *Adv. Space Res.*, 23, 1577–1585, 1999.

Rosenqvist, J., P. Drossart, M. Combes, et al., Minor constituents in the Martian atmosphere from the ISM/Phobos Experiment, *Icarus*, 98, 254–270, 1992.

Savijärvi, H., Mars boundary layer modeling: diurnal cycle and soil properties at the Viking Lander 1 Site, *Icarus*, 117, 120–27, 1995.

Savijärvi, H., A.-M. Harri, and O. Kemppinen, The diurnal water cycle at Curiosity: role of exchange with the regolith, *Icarus*, 265, 63–69, doi:10.1016/j.icarus.2015.10.008, 2016.

Schmidt F., S. Douté, B. Schmitt, et al., Albedo control of seasonal south polar cap recession on Mars, *Icarus*, 200, 374–394, 2009.

Sears, D. W. G., and S. R. Moore, On laboratory simulation and the evaporation rate of water on Mars, *Geophys. Res. Lett.*, 32, L16202, doi:10.1029/2005GL023443, 2005.

Seu, R., R. J. Phillips, D. Biccari, et al., SHARAD sounding radar on the Mars Reconnaissance Orbiter, *J. Geophys. Res.*, 112, E05S05, 2007.

Sizemore, H. G. and M. T. Mellon, Effects of soil heterogeneity on Martian ground-ice stability and orbital estimates of ice table depth, *Icarus*, 185, 358–369, 2006.

Sizemore, H. G., and M. T. Mellon, Laboratory characterization of the structural properties controlling dynamical gas transport in Mars-analog soils, *Icarus*, 197, 606–620, 2008.

Smith D. E., M. T. Zuber, S. C. Solomon, et al., The global topography of Mars and implications for surface evolution, *Science*, 284, 1495–1503, 1999.

Smith D. E., M. T. Zuber, and G. A. Neumann, Seasonal variations of snow depth on Mars, *Science*, 294, 2141–2146, 2001.

Smith, M. D., The annual cycle of water vapor as observed by the Thermal Emission Spectrometer, *J. Geophys. Res.*, 107, doi:10.1029/2001JE001522, 2002.

Smith, M. D., Interannual variability in TES atmospheric observations of Mars during 1999–2003, *Icarus*, 167, 148–165, 2004.

Smith, M. D., Mars water vapor climatology from MGS/TES, *abstract from the Mars Water Cycle Workshop*, 21–23 April 2008, Paris, France, 2008.

Smith, M. D., THEMIS observations of Mars aerosol optical depth from 2002–2008, *Icarus*, 202, 444–452, 2009.

Smith, M. D., B. J. Conrath, J. C. Pearl, and P. R. Christensen, Thermal Emission Spectrometer observations of Martian planet-encircling dust storm 2001A, *Icarus*, 157, 259–263, 2002.

Smith, M. D., M. J. Wolff, N. Spanovich, et al., One Martian year of atmospheric observations using MER Mini-TES, *J. Geophys. Res.*, 111, E12S13, doi:10.1029/2006JE002770, 2006.

Smith, M. D., M. J. Wolff, R. T. Clancy, and S. L. Murchie, Compact Reconnaissance Imaging Spectrometer observations of water vapor and carbon monoxide, *J. Geophys. Res.*, 114, doi:10.1029/2008JE003288, 2009.

Smoluchowski, R., Mars: retention of ice, *Science*, 159, 1348–1350, 1968.

Spinrad, H., G. Münch, and L. D. Kaplan, The detection of water vapor on Mars, *Astrophys. J.*, 137, 1319–1321, 1963.

Sprague, A. L., D. M. Hunten, R. E. Hill, B. Rizk, and W. K. Wells, Martian water vapor, 1988–1995, *J. Geophys. Res.*, 101(E10), 23229–23241, 1996.

Sprague, A. L., D. M. Hunten, R. E. Hill, L. R. Doose, and B. Rizk, Water vapor abundances over Mars north high latitude regions: 1996–1999, *Icarus*, 154, 183–189, 2001.

Sprague, A. L., D. M. Hunten, L. R. Doose, and R. E. Hill, Mars atmospheric water vapor abundance: 1996–1997. *Icarus*, 163(1), 88–101, 2003.

Sprague, A. L., D. M. Hunten, L. R. Doose, et al., Mars atmospheric water vapor abundance: 1991–1999, Emphasis 1998–1999, *Icarus*, 184(2), 372–400, 2006.

Squyres, S. W., and M. H. Carr, Geomorphic evidence for the distribution of ground ice on Mars, *Science*, 231, 249–252, 1986.

Tamppari, L. K., R. W. Zurek, and D. A. Paige, Viking-era water-ice clouds. *J. Geophys. Res.*, 105, 4087–4107, 2000.

Tamppari, L. K., D. Bass, B. Cantor, et al., Phoenix and MRO coordinated atmospheric measurements, *J. Geophys. Res.*, 115(E12), E00E17, doi:10.1029/2009JE003415, 2010.

Thomas, P. C., S. W. Squyres, K. E. Herkenhoff, A. D. Howard, and B. C. Murray, Polar deposits of Mars, in *Mars*, edited by H. H. Kieffer et al., 767–795, Univ. of Arizona Press, Tucson, 1992.

Tillman, J. E., R. M. Henry, and S. L. Hess, Frontal systems during passage of the Martian north polar hood over the Viking Lander 2 site prior to the first 1977 dust storm, *J. Geophys. Res.*, 84, 2947–2955, 1979.

Titov, D. V., M. I. Moroz, A. V. Grigoriev, et al., Observations of water vapour anomaly above Tharsis volcanoes on Mars in the ISM (Phobos-2) experiment, *Planet. Space Sci.*, 42, 1001–1010, 1994.

Titov, D. V., J. Rosenqvist, V. I. Moroz, A. V. Grigoriev, G. Arnold, Evidences of the regolith-atmosphere water exchange on Mars from the ISM (Phobos-2) infrared spectrometer observations, *Adv. Space Res.*, 16(6), 23–33, 1995.

Titov, D. V., W. J. Markiewicz, N. Thomas, et al., Measurements of the atmospheric water vapor on Mars by the Imager for Mars Pathfinder, *J. Geophys. Res.*, 104, E4, 9019–9026, 1999.

Titus, T. N., Thermal infrared and visual observations of a water ice lag in the Mars southern summer, *Geophys. Res. Lett.*, 32, L24204, 2005.

Titus, T. N., H. H. Kieffer, and P. R. Christensen, Exposed water ice discovered near the South Pole of Mars, *Science*, 299, 1048–1051, 2003.

Tokar, R. L., R. C. Elphic, W. C. Feldman, et al., Mars odyssey neutron sensing of the south residual polar cap, *Geophys. Res. Lett.*, 30(13), 10–1, 2003.

Tschimmel, M., N. I. Ignatiev, D. V. Titov, et al., Investigation of water vapor on Mars with PFS/SW on Mars Express, *Icarus*, 195, 557–575, 2008.

Tyler, D., and J. R. Barnes, Atmospheric mesoscale modeling of water and clouds during northern summer on Mars, *Icarus*, 237, 388–414, doi:10.1016/j.icarus.2014.04.020, 2014.

Vincendon, M., Y. Langevin, F. Poulet, J.-P. Bibring, B. Gondet, Recovery of surface reflectance spectra and evaluation of the optical depth of aerosols in the near-IR using a Monte Carlo approach: application to the OMEGA observations of high-latitude regions of Mars, *J. Geophys. Res.*, 112, E08S13, 2007.

Vincendon M., F. Forget, and J. Mustard, Water ice at low to midlatitudes on Mars, *J. Geophys. Res.*, 115, E10001, 2010a.

Vincendon, M., J. Mustard, F. Forget, et al., Near-tropical subsurface ice on Mars, *Geophys Res. Lett.*, 37, L01202, 2010b.

Whiteway, J. A., L. Komguem, C. Dickinson, et al., Mars water-ice clouds and precipitation, *Science*, 325, 68, 2009.

Wilson, R. J., G. A. Neumann, and M. D. Smith, Diurnal variation and radiative influence of Martian water ice clouds, *Geophys. Res. Lett.*, 34, L02710, doi:10.1029/2006GL027976, 2007.

Wilson, R. J., S. R. Lewis, L. Montabone, and M. D. Smith, Influence of water ice clouds on Martian tropical atmospheric temperatures, *Geophys. Res. Lett.*, 35, L07202, doi:10.1029/2007GL032405, 2008.

Yung, Y. L., J. Wen, J. P. Pinto, K. K. Pierce, and M. Allen, HDO in the Martian atmosphere – implications for the abundance of crustal water, *Icarus*, 76, 146–59, 1988.

Zent, A. P., 2008. A historical search for habitable ice at the Phoenix landing site, *Icarus*, 196, 285–408, 2008.

Zent, A. P., and R. C. Quinn, Simultaneous adsorption of CO_2 and H_2O under Mars-like conditions and applications to the evolution of the Martian climate, *J. Geophys. Res.*, 100, 5341–5349, 1995.

Zent, A. P. and R. C. Quinn, Measurement of H_2O adsorption under Mars-like conditions: effects of adsorbent heterogeneity, *J. Geophys. Res.*, 102, 9085–9095, 1997.

Zent, A. P., R. M. Haberle, H. C. Houben, and B. M. Jakosky, A coupled subsurface-boundary layer model of water on Mars, *J. Geophys. Res.*, 98, 3319–3337, 1993.

Zent, A. P., D. J. Howard, and R. C. Quinn, H_2O adsorption on smectites, Application to the diurnal variations of H_2O in the Martian atmosphere, *J. Geophys. Res*, 106, 14667–14674, 2001.

Zent, A. P., M. Hecht, D. Cobos, et al., Initial results from the Thermal and Electrical Conductivity Probe (TECP) on Phoenix, *J. Geophys. Res.*, 115, E00E14, 2010.

Zent, A. P., M. H. Hecht, T. L. Hudson, S. E. Wood, and V. F. Chevrier, A revised calibration function and results for the Phoenix mission TECP relative humidity sensor, *J. Geophys. Res. Planets*, 121(4), 626–651, doi:10.1002/2015JE004933, 2016.

Zuber, M. T., D. E. Smith, S. C. Solomon, et al., Observations of the north polar region of Mars from the Mars Orbiter Laser Altimeter, *Science*, 282, 2053–2060, doi:10.1126/science.282.5396.2053, 1998.

12

The CO$_2$ Cycle

TIMOTHY N. TITUS, SHANE BYRNE, ANTHONY COLAPRETE, FRANÇOIS FORGET,
TIMOTHY I. MICHAELS, THOMAS H. PRETTYMAN

12.1 INTRODUCTION

The CO$_2$ cycle on Mars involves several interactions between the atmosphere and the surface. CO$_2$ gas composes ~95% of the Martian atmosphere (Owen et al., 1977; Mahaffy et al., 2013), and approximately 25–30% of the atmosphere is cycled through the seasonal caps annually (Tillman et al., 1993; Forget and Pollack, 1996; Kelly et al., 2006; Prettyman et al., 2009). The exact mass distribution between the CO$_2$ gas and solid (ice) phases varies not only seasonally, but also over much longer timescales. Changes in obliquity, season of perihelion, and other orbital parameters affect the latitudinal and seasonal distributions of insolation, which in turn affect aspects of the CO$_2$ cycle, including the amount and extent of seasonal ice (e.g. Mischna et al., 2003). The seasonal condensation and sublimation of CO$_2$ at high latitudes controls atmospheric circulation on a global scale and is thus a critical aspect of the Martian climate (Hourdin et al., 1993, Forget et al., 1998).

While the CO$_2$ cycle is a dominant driving force of the Martian climate, the climate is also affected by the H$_2$O cycle (Chapter 11) and the dust cycle (Chapter 10). These two cycles interact and modify the CO$_2$ cycle on both the regional and the local scales. Subsurface H$_2$O ice, which is widespread in the Mars polar regions (e.g. Bandfield and Feldman, 2008), can behave much like a thermal capacitor by storing up heat in the summer and reducing net CO$_2$ ice accumulation in the fall and winter (e.g. Haberle et al., 2008). Dust can change both the emissivity and the albedo of CO$_2$ ice and snow, thus changing the rate at which the ice sublimes (e.g. Bonev et al., 2002, 2008). While this chapter focuses on the CO$_2$ cycle, the discussion of dust and H$_2$O is unavoidable due to these interactions. Furthermore, surface topography affects the CO$_2$ cycle by creating atmospheric gravity waves (Chapter 8) that can alter the type and amount of CO$_2$ ice deposition. This occurs on both the local level, e.g. orographic lifting resulting in CO$_2$ snow downwind from a crater (Forget et al., 1998; Colaprete and Toon, 2002; Tobie et al., 2003), and the regional scale, e.g. the southern Tharsis plateau along with the Hellas and Argyre Basins create significantly differing climates in the eastern and western hemispheres of the south polar region (Colaprete et al., 2005).

A thorough understanding of all these processes is needed if we are to understand the Mars of today and, perhaps more importantly, the Mars of the past. A variety of modeling techniques (see Chapter 9) have been used to reconstruct Mars' past climate (Chapters 16 and 17). The results of these models, while useful, must be interpreted with caution, as much of the physics has been either parameterized or adjusted to match present-day observations. There is no guarantee that these parameterizations are invariant for the entirety of Mars history.

Throughout this chapter, great care has been taken to distinguish between H$_2$O and CO$_2$ ices and between the various descriptors of the Martian polar caps. For example, those portions of the polar caps that retain intact volatiles (H$_2$O in the north; CO$_2$ and H$_2$O in the south) throughout the summer are referred to as residual caps. The parts of the polar caps that undergo complete sublimation are referred to as seasonal caps. The composition of the seasonal caps is based on temperature, i.e. the seasonal cap is referred to as CO$_2$ seasonal cap as long as the surface temperature remains near the CO$_2$ frost point, even in the absence of CO$_2$ ice spectral features. This distinction is important because a thermally thin, but optically thick in the visible and near-infrared, cold trapped layer of H$_2$O ice can hide the spectral signature of CO$_2$ ice, while remaining at or near CO$_2$ frost-point temperatures.

This chapter is laid out in six sections. Section 12.1 has provided an introduction to the Mars CO$_2$ cycle. Section 12.2 discusses the Mars polar energy balance and how the CO$_2$ cycle is modified by the other atmospheric cycles. A brief historical overview is included to provide the reader with the importance of the use of energy balance models to better understand and quantify the Martian CO$_2$ cycle (see also Chapter 3). Section 12.3 provides a discussion of the instrumental capabilities and observables that allow for the monitoring of the Mars CO$_2$ cycle. Section 12.4 is a discussion of the active processes, e.g. the growth and retreat of the polar seasonal caps. Section 12.5 discusses the permanent CO$_2$ deposits and long-term reservoirs. Finally, Section 12.6 is a summary of the current knowledge and a discussion of unresolved issues and some enigmas.

12.2 POLAR ENERGY BALANCE

The polar caps were first observed in 1784 by William Herschel and were assumed to be composed of H$_2$O snow and ice, just as on Earth. For example, in 1845 Ormsby MacKnight Mitchel observed an albedo feature believed to be a temporary springtime polar lake (Blunck, 1977, 1982). While most scientists accepted this theory of H$_2$O composition without question, a few suggested, as early as the 19th century, that the polar caps could not be composed of H$_2$O (e.g. Stoney, 1898). Their argument was based on Mars being smaller than the Earth, and therefore that it would not have the gravitational field to prevent water from escaping into space. Leighton and Murray (1966) used a simple energy balance model to show that water ice could not be the major constituent of the polar caps, and suggested that CO_2 ice was the most likely candidate. This proposition was later confirmed by Mariner observations (Neugebauer et al., 1971). It has been nearly a half century since Leighton and Murray published their iconic paper, and the analysis tool used to understand the Mars CO_2 cycle – the exchange of CO_2 between the atmosphere, the surface, and the subsurface – continues to be some variation of the energy balance model (which includes conservation of mass, momentum, etc.), which includes Mars general circulation models or Mars global climate models (MGCMs) – see Chapter 9.

The amount, rate, and distribution of CO_2 condensation and sublimation are determined by the balance of several mass (e.g. exchange between CO_2 reservoirs) and energy sources and sinks, including insolation (exposure to sunlight), net radiative loss to space, the latent heat of deposition (predominantly from CO_2 ice), summertime heat storage in the regolith, and atmospheric storage and transport of energy and mass.

Since Leighton and Murray (1966), many models have been developed to simulate the formation of the Martian polar caps by examining their energy balance. This includes simple "surface energy balance models" (e.g. Briggs, 1974; Davies et al., 1977; James and North, 1982; Paige and Wood, 1992; Wood and Paige, 1992), and the more complete global climate models designed to simulate all aspects of the climate, including the advection of heat by the atmosphere and the contribution from CO_2 ice clouds and precipitation (Pollack et al., 1990, 1993; Hourdin et al., 1993, 1995; Forget et al., 1998; Haberle et al., 2008; Guo et al., 2009). Such models were able to reproduce many aspects of the observed CO_2 cycle and in particular the annual variations in atmospheric pressure observed by the Viking Landers (e.g. Figure 4.17; Hess et al., 1977, 1979). However, until 2008, most studies concluded that it was necessary to use artificially low emissivity values (near 0.5) for the modeled caps in order to decrease the net infrared cooling and thus reduce the deposition rates by a factor of about 2. Otherwise, the predicted polar caps would become much too massive and the modeled seasonal pressure cycle would exhibit too large an amplitude. The use of this relatively low CO_2 ice emissivity was justified by the fact that solid CO_2 is extremely transparent in the thermal infrared between 20 and 30 μm (e.g. Hansen, 2013), and that small-grain deposits originating from atmospheric condensation could exhibit an emissivity well below unity (Forget et al., 1995). However, thermal infrared observations performed in the polar night (Forget and Pollack, 1996; Kieffer et al., 2000; Kieffer and Titus, 2001) confirmed that in most places the seasonal polar caps are made of large grains or are contaminated by dust or water ice and exhibit an emissivity higher than 0.8. Clearly, some physical process that limits the condensation rate was either not taken into account or not properly treated by the models. Part of the solution for this enigma came from the Mars Odyssey and Phoenix missions, which revealed the presence of a high-thermal-inertia ice-rich layer below a few centimeters of soil in the Martian high latitudes (e.g. Boynton et al., 2002; Feldman et al., 2002, 2004; Mitrofanov et al., 2002; Tokar et al., 2002, 2003; Mellon et al., 2004; Prettyman et al., 2004). Global climate models had assumed that these regions contain a sand-like

lower-thermal-inertia material, as deduced from observations of the diurnal temperature cycle, which is sensitive only to the first centimeters. In reality, at seasonal scales, the ice layer is able to store a significant amount of heat during summer and release it in fall and winter, thus significantly reducing the CO_2 condensation rate (e.g. Haberle et al., 2008).

The simplest form of the polar energy balance was used by Paige and Ingersoll (1985):

$$F_{rad} + F_{horiz} + F_{cond} = S_{atm} + S_{CO2} \qquad (12.1)$$

where F_{rad} is the net radiation at the top of the atmosphere, F_{horiz} is the net horizontal heat flux, F_{cond} is the net vertical conduction of heat from the surface, S_{atm} is the rate of change of total potential energy storage of the atmospheric column, and S_{CO2} is the latent heat of fusion for CO_2. The net radiation at the top of the atmosphere is simply the insolation incident upon the atmosphere minus the radiative losses to space:

$$F_{rad} = (1 - A_{top})\frac{S_0}{U^2}\mu_0 - \varepsilon\sigma T^4 \qquad (12.2)$$

where A_{top} is the effective albedo at the top of the atmosphere, S_0 is the solar insolation at 1 AU, U is the distance of Mars from the Sun in AU, and μ_0 is the cosine of the solar incidence angle. The amount of latent heat from the CO_2 ice phase to the gas phase is simply

$$S_{CO2} = -L_{CO2}\frac{dM_{CO2}}{dt} \qquad (12.3)$$

$$\frac{dM_{CO2}}{dt} = \frac{1}{L_{CO2}}\left\{ \varepsilon\sigma T^4 - (1 - A_{top})\frac{S_0}{U^2}\langle\mu_0\rangle \right\} \qquad (12.4)$$

The term $\langle\mu_0\rangle$ is the diurnal mean of the cosine of the solar incidence angle, where negative values have been set to zero. This final expression assumes no lateral transfer of heat in the atmosphere and no vertical conduction of heat into and out of the regolith.

This equation (graphically represented in Figure 12.1) is at the heart of the CO_2 cycle exchange between the atmospheric gas phase and the surface ice phase. Changes in ice emissivity, ε, modify the amount of thermal emission and therefore affect the rate at which CO_2 ice accumulates. The temperature, T, is actually affected by changes in the atmospheric pressure, as the equilibrium frost temperature of CO_2 is pressure-dependent (see Table 12.1). The top-of-the-atmosphere Bond albedo, A_{top}, is affected by how dusty the ice is and how cloudy the atmosphere is. CO_2 ice sublimation rates are largely determined by the effective Bond albedo.

12.2.1 Climate Change on Mars

With the possible exception of buried CO_2 ice deposits in the south polar layered deposits (see Section 12.5.2), the largest known reservoir of CO_2 is the Mars atmosphere. (See Table 12.2 for comparisons of reservoirs and their respective timescales.) The atmospheric mass (and therefore air pressure) changes seasonally by ~25% (when compared to the maximum atmospheric pressure) as the seasonal polar caps grow and retreat (e.g. Tillman et al., 1993; Forget and Pollack, 1996;

Table 12.1 *Physical properties of CO_2*

Property	Phase	Value or equation	Units	Reference
Density	Solid	1606	kg m^{-3}	Kieffer et al. (2000)
Heat capacity	Solid	$459.6+1.3585T$	kJ kg^{-1} K^{-1}	H.H. Kieffer (personal communication)
Thermal conductivity	Solid	$k = 93.4/T$	W s^{-1} m^{-1} K^{-1}	Kravchenko and Krupskii (1986)
	Gas	$k = -12.0817+1.39898T$	W s^{-1} m^{-1} K^{-1}	Vesovic et al. (1990)
Latent heat of sublimation	Transition	$6.52308\times10^5 - 371.28T$	J kg^{-1}	Mullins et al. (1963)
$\ln(p) = a - b/T$, with $a = 27.9546$ and $b = 3182.48$, p in Pa, and T in K				James et al. (1992)

Table 12.2 *Table of CO_2 reservoirs, the estimated timescale for recycling (in Earth years), and the relative size of the reservoir compared to the current size of the atmosphere.*

Timescale (Earth years) for atmospheric interaction	Reservoir	Amount (relative to current atmosphere) (%)	References
1.9 (= 1 Mars year)	Seasonal caps	~25	Leighton and Murray (1966), Tillman et al. (1993), Kelly (2006)
100–1000	South polar residual cap: CO_2 overlying H$_2$O ice	~3	Byrne and Ingersoll (2003), Bibring et al. (2004), Thomas et al. (2009)
51,000 (argument of perihelion)			Titus et al. (2003), Montmessin et al. (2007)
120,000 (obliquity)	Buried south polar CO_2 ice	~80	Phillips et al. (2011)
	Adsorbed within regolith	100–1000	Fanale and Salvail (1994), Manning et al. (2006)
Millions	Clathrates in polar layered deposits	~0	Mellon (1996), Nye et al. (2000)
Several billion	Carbonates	?	

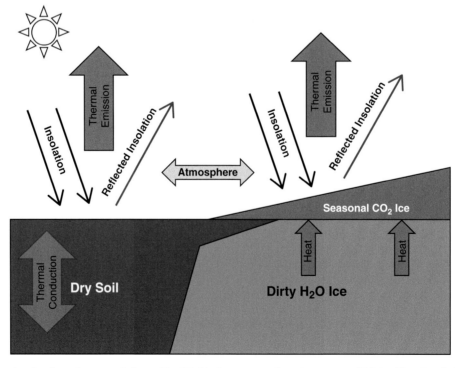

Figure 12.1. Cartoon showing the polar energy balance. The thin black arrows are the solar energy, which is either absorbed or reflected (thin gray arrows) back into space. The two thick open arrows pointing up are the radiative heat loss to space. The thick open arrows in the regolith are the heat conduction and the release of the stored solar heat back into surface ice. The sideways arrow is the heat transported by the atmosphere.

Kelly et al., 2006; Prettyman et al., 2009). The atmosphere also responds to changes in the release or absorption of CO_2 from other sources or sinks. The mass of the south polar residual cap (SPRC) has been reported to change on the annual to centennial scale (e.g. Thomas et al., 2013). However, these changes may correspond to a change in atmospheric pressure of only a few pascals and are therefore difficult to detect (e.g. Byrne and Ingersoll, 2003; Haberle et al., 2009; Haberle and Kahre, 2010). Based on GCM results, the magnitude of seasonal atmospheric pressure variations is most pronounced with changes in obliquity, which occur over much longer timescales (e.g. Toon et al., 1980; François et al., 1990; Kieffer and Zent, 1992; Mellon and Jakosky, 1995; Haberle et al., 2003; Newman et al., 2005).

Evidence that liquid water once flowed across the surface of Mars, cutting out river channels (e.g. Baker, 1979; Carr, 1979) and lake deposits (e.g. Carr, 1983; Head and Pratt, 2001; Cabrol et al., 2002), suggests that Mars had a significantly more massive atmosphere in the distant past (Noachian). The question of where the excess CO_2 went has yet to be determined. Some of the CO_2 may be buried within the south polar layered deposits (SPLD) (Phillips et al., 2011) or stored as carbonates (e.g. Bandfield et al., 2003; Ehlmann et al., 2008, 2011; Brown et al., 2010b; Michalski and Niles, 2010). At least some of the CO_2 must have been lost to space (e.g. Leblanc and Johnson, 2001, 2002; Kass and Yung, 1995) and therefore is no longer available as a potential reservoir. The purpose of the Mars Atmosphere and Volatile Evolution (MAVEN) mission is to study the upper atmosphere and determine the potential loss rates over the last 4.5 Ga (Jakosky, 2008).

12.2.2 Astronomical Effects

The predominant driver of the Mars CO_2 cycle is the amount and distribution of solar radiation. Therefore, any variations in the orbital or axial rotation parameters that modify solar insolation may have a dramatic impact on the CO_2 cycle. As detailed in Chapter 16, compared to the Earth, Mars experiences large variations of its orbital parameters (eccentricity, season of perihelion) and especially of its obliquity, which directly impacts the insolation at high latitudes. As detailed in Laskar et al. (2004) and as shown in Figures 16.1 and 16.2, Mars' obliquity can vary widely and somewhat erratically. In most periods, obliquity first varies with a period of about $\sim 10^5$ years with a peak-to-peak amplitude of 10–20°. However, the mean value of this oscillation can evolve on a 10^6–10^7 year timescale, so that obliquity is thought to have ranged between 0° and more than 60° in the past (the current obliquity is 25.2°).

Several studies have been performed to estimate the changes in surface temperature and CO_2 ice cap processes induced by the obliquity variations (Toon et al., 1980; François et al., 1990; Kieffer and Zent, 1992; Mellon and Jakosky, 1995; Haberle et al., 2003; Newman et al., 2005). These studies have shown that, at obliquities higher than today, the region of polar night expands toward the equator and that the latitudinal extension of the seasonal CO_2 cap increases. For a 60° obliquity experiment, the winter caps extend all the way to the equator (Haberle et al., 2003) and the seasonal CO_2 pressure variations are strongly enhanced. However, at obliquities a few degrees higher than today, the increase in mean polar insolation will likely induce the sublimation of any perennial CO_2 ice cap. At obliquities

lower than today, the extension of the seasonal CO_2 ice caps should be proportionally smaller. It is also expected that the fraction of the atmosphere constantly held as CO_2 ice in the perennial polar cap will increase (Ward et al., 1974), while decreasing the mean atmospheric pressure. For example, long-term GCM simulations performed by Newman et al. (2005) predict a global annual-average pressure of 350 Pa and 40 Pa for an obliquity of 15° and 5°, respectively (compared to $\sim$610 Pa today). However, the processes controlling the evolution of the permanent CO_2 ice caps at low obliquity remain poorly understood, notably because the variations of the CO_2 ice cap albedo in past conditions is difficult to predict. Furthermore, Kreslavsky and Head (2005, 2011) have suggested that at low obliquity the permanent CO_2 ice deposits should have formed on steep pole-facing slopes at moderately high latitudes rather than at the poles. They even found possible geological evidence of this in the form of CO_2 ice glaciers in three localized sites near 70°N latitude (Kreslavsky and Head, 2011).

12.2.3 Other Processes That Modify the CO_2 Cycle

While solar insolation is the driver of the CO_2 cycle, several properties of the atmosphere, surface, and subsurface can modify, store, or delay the absorption of solar radiation. For surface CO_2 ice, these properties include albedo, emissivity, and latent heat. For the atmosphere, these properties include opacity (due to both absorption and scattering), lateral transport of heat, convection, and latent heat. For surfaces that are ice-free for a fraction of a year, the properties of thermal inertia, albedo, and emissivity exert the greatest influence in modifying CO_2 ice accumulation.

12.2.3.1 Albedo and Emissivity

The albedo and emissivity of ice are determined by composition and grain size. The albedo of the CO_2 ice determines the fraction of solar insolation that is absorbed, and therefore sets the sublimation rate. The effective emissivity of the ice controls how much heat is radiatively lost and primarily regulates the condensation rate (James et al., 1992). The albedo and emissivity of CO_2 ice are dependent upon grain size. Fine-grained CO_2 ice (e.g. snow) has a high albedo at solar wavelengths and low emissivity at 25 μm (e.g. Hansen, 1997; Kieffer et al., 2000; Appéré et al., 2011; see Figure 12.2). Although the major constituent of seasonal ice is CO_2, the ice is contaminated by small amounts of water and dust. In particular, water frosts and dust can form thin layers on top of the CO_2 ice, changing both the albedo and the emissivity.

As a general rule, the seasonal caps are brighter in late spring and summer than they are in early spring. This is true for both the northern and the southern caps, with the notable exception of the so-called "cryptic region" (Titus et al., 1998; Kieffer et al., 2000). The "cryptic region" is a region of the southern seasonal polar cap that consistently remains dark throughout spring sublimation (discussed in more detail in Section 12.4.4). Possible causes for this spring brightening include: (1) a change in the ice properties with increased solar insolation (e.g. less dust contamination, additional cracks); (2) geometrical photometric effects; and (3) cold trapping of water vapor into frost (Paige, 1985; Houben et al., 1997). The north polar region is generally

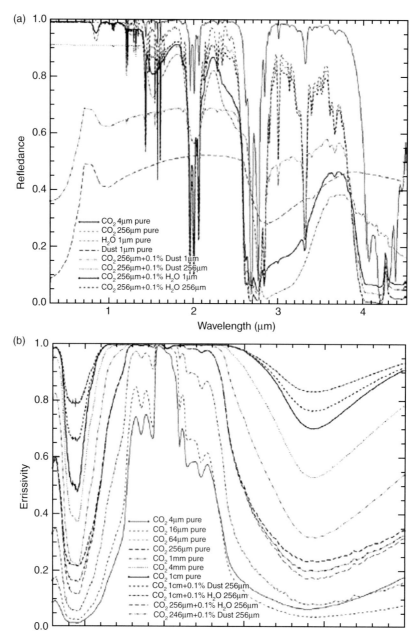

Figure 12.2. Model spectra of CO$_2$ ice: (A) figure 25.1 from Titus et al. (2008); (b) figure 25.2 from Titus et al. (2008).

"wetter" than the south polar region, and migration of H$_2$O ice is observed along the edge of the retreating northern seasonal cap (Kieffer and Titus, 2001; Schmitt et al., 2005a,b; Titus, 2005a; Wagstaff et al., 2008). While the Mars Express Observatoire pour la Minéralogie, l'Eau, les Glaces et l'Activité (OMEGA) near-IR imaging spectrometer observed H$_2$O as a component of the southern retreating seasonal cap (e.g. Langevin et al., 2005), those observations do not reveal large-scale H$_2$O ice lags on the edge of the cap. Possible H$_2$O ice lags are seen at the end of the southern cap's retreat, but these could also be an artifact of subpixel mixing of remnant CO$_2$ ice and dust (Titus, 2005b; Douté et al., 2007). The second possible process, photometric effects, has not been fully explored. For most studies, CO$_2$ ice has simply been assumed to be Lambertian (Paige, 1985; Paige and Ingersoll, 1985; James et al., 2005).

12.2.3.2 Atmospheric Dust Opacity

During the condensation phase (autumn and winter), atmospheric dust can modify the condensation mode by increasing the atmospheric thermal infrared radiative cooling (Pollack et al., 1990) and also by providing condensation nuclei for the precipitation of H$_2$O and CO$_2$ snow (e.g. Hayne et al., 2014). This would suggest that southern spring and summer near-global dust storms could potentially enhance atmospheric condensation and snowfall in the northern hemisphere. There seems to be a correlation between global dust storm events and an increase in low-emissivity deposition of CO$_2$ snow (e.g. Forget and Pollack, 1996; Cornwall and Titus, 2009, 2010).

During the sublimation phase (spring), atmospheric dust can alter sublimation rates. Bonev et al. (2008) demonstrated that an increase in atmospheric aerosols during the sublimation phase of the southern cap can modify the sublimation rates by redistributing the solar radiation from the visible wavelengths to the thermal infrared wavelengths. This redistribution occurs through the absorption of sunlight by the atmospheric dust, which increases the dust temperature, and then the absorbed energy is re-emitted in the thermal infrared. Because less solar (visible) radiation reaches the surface, the sublimation rate for the darker CO_2 decreases. The brighter CO_2 ice, which normally only absorbs a small portion of the solar insolation, absorbs more of the increased thermal downwelling radiation from the atmospheric aerosols, thus increasing its sublimation rate.

12.2.3.3 Atmospheric Clouds

The presence of both H_2O and CO_2 ice clouds can reduce the net accumulation of surface CO_2, but in very different ways: CO_2 ice clouds have low emissivities and/or result in the deposition of low-emissivity snowfall on the surface, thus reducing the net radiative losses to space. Water ice clouds increase the emissivity of the middle atmosphere, which is often warmer than the surface. They thus tend to warm the surface through downward infrared radiation. More details on clouds can be found in Chapter 5.

12.2.3.4 Ice-Free Surface Albedo, Emissivity, and Thermal Inertia

The albedo, emissivity, and thermal inertia of the volatile-free surfaces in the polar region also modify the CO_2 cycle. High-albedo materials will absorb less sunlight than darker materials, and thus will remain colder throughout the summer, thereby inducing volatiles to condense at an earlier date. Surfaces with higher emissivity will radiate heat more efficiently and therefore will remain cooler throughout the summer. Low-thermal-inertia surfaces, which heat up and cool down quickly, will maintain a cooler diurnal average temperature and will also exhibit early volatile condensation. Finally, the vertical distribution of thermal inertia can have a dramatic impact on the net accumulation of CO_2 ice (Haberle et al., 2008). High-thermal-inertia material (e.g. H_2O ice) located near the surface but covered by lower-thermal-inertia material can act as a thermal capacitor, storing heat during the summer and releasing that heat in the autumn. This release of stored heat causes basal sublimation of the CO_2 ice and thus reduces the net accumulation of ice. With the discovery of near-surface H_2O ice by the GRS suite of instruments (Boynton et al., 2002), the spatially dependent heat storage of the regolith has become a component of the CO_2 cycle energy balance that must be considered in thermal models and GCMs.

12.2.3.5 Atmospheric Transport

The final contribution to the energy balance discussed here is atmospheric heat capacity and the lateral transport of atmospheric heat. The heat capacity of the atmosphere is small, and thus its contribution can generally be neglected. However, the lateral transport of heat is not well constrained and could be quite large, depending on the location and season (Pollack et al., 1990; Hourdin et al., 1995; Haberle et al., 2004). A more detailed discussion of heat transport by the atmosphere is covered in Chapter 6.

12.2.4 Mesoscale Models and the CO₂ Cycle

The application of mesoscale atmospheric models to discern aspects of the CO_2 cycle on Mars has been relatively limited thus far. Two-dimensional modeling studies of southern hemisphere locations by Siili et al. (1997, 1999) showed that, although the speed of the sublimation flow itself is not great, it can combine with mesoscale thermal contrasts at the CO_2 ice edge and/or topographic slopes to produce mesoscale circulations of tens of meters per second near the surface. Such circulations would yield aerodynamic surface stresses of the approximate magnitude needed to lift dust from the surface, potentially resulting in the observed cap-edge dust storms. Extending that work into three dimensions, Toigo et al. (2002) additionally found that sublimation flow exerts its greatest effect upon near-surface cap-edge circulations when the receding seasonal CO_2 ice cap is intermediate in size (e.g. mid- to late southern spring). Tyler and Barnes (2005) examined the effects of topography on katabatic winds and the mixing of H_2O vapor via eddy currents during the northern summer. They found that the northern atmosphere transitions to its winter state as early as $L_s = 150°$. In order to ascertain the state of the atmosphere that would be encountered by the NASA Phoenix Lander in the high northern latitudes of Mars during late spring, two mesoscale atmospheric models (Michaels and Rafkin, 2008; Tyler et al., 2008) were run with various Thermal Emission Spectrometer (TES) and Mars Orbiter Camera (MOC) based parameterizations specific to the location and state of the receding seasonal cap (and its water-ice-dominated transition regions). While Tyler et al. (2008) used surface properties for the cap that were homogeneous at the subgrid scale (less than approximately a few kilometers), Michaels and Rafkin (2008) employed an approach where TES-based maps of the cap's albedo and thermal inertia were transformed into varying subgrid-scale components of CO_2 ice, water ice, and dust (or bare ground). The asymmetries of the north polar residual cap topography and the variegated edge of the receding seasonal cap (and its associated thermal contrast with bare ground) were shown by Tyler et al. (2008) to be primary sources of complex modulations for the relatively weak high-latitude baroclinic storms extant in the late northern spring. Michaels and Rafkin (2008) found similar behavior. Spiga et al. (2011) found that katabatic winds flowing over CO_2 ice on mesoscale topographic slopes would enhance surface–atmosphere turbulent heat transfer at those locations, with a tendency to increase the sublimation rate in the spring and limit the deposition of CO_2 ice in the autumn–winter. Mesoscale atmospheric modeling (Spiga et al., 2012) has also indicated that mesoscale gravity waves may have a key role in forming CO_2 ice clouds that have been observed within the mesosphere of Mars (e.g. González-Galindo et al., 2011).

12.3 CO₂ DETECTION, MONITORING, AND OBSERVATIONS

As discussed in the previous section, several factors affect the CO_2 cycle. These various factors are observed and monitored through a variety of remote sensing techniques and (for atmospheric pressure) *in situ* techniques. In the visible, CO_2 ice is usually assumed to be bright, but this is not always the case. First, CO_2 ice can have a range of albedo depending on grain size or porosity and amount of trace contaminants (e.g. dust or H_2O ice). Second, it is difficult to distinguish between H_2O and CO_2 volatiles using only visible wavelengths. The best diagnostic tools to identify CO_2 ice make use of several spectral absorption features in the near-infrared (e.g. 1.4 μm, 2.2 μm) and the thermal infrared using both brightness temperature and spectral transparency features (25 μm). Several spacecraft missions over the last five decades have provided key observations, as outlined in Table 12.3.

12.3.1 *In Situ* Observations

Surface air pressure measurements have now been conducted by landed spacecraft at five separate sites on Mars: Viking Landers 1 and 2 (see Figure 4.17), Pathfinder, Phoenix, and the currently operating Mars Science Laboratory (MSL). Haberle and Kahre (2010) tabulate the operating dates of the first four and discuss important details and caveats of their measurement accuracies. A Tavis pressure transducer on each of the Viking Landers operated for more than 3 Mars years at the VL1 site and ~1.5 Mars years at the VL2 location (e.g. Chamberlain et al., 1976; Tillman, 1988). The Pathfinder ASI/MET package included a more modern Tavis magnetic reluctance diaphragm sensor that measured pressure during the mission (~0.13 Mars year in duration) with improved measurement cadence and sensitivity versus Viking (Schofield et al., 1997; Seiff et al., 1997). The Phoenix Lander included a Väisälä pressure sensor that operated for ~0.25 Mars year (Taylor et al., 2010). The MSL is currently taking air pressure measurements on Mars using a Väisälä-based custom sensor (Gomez-Elvira et al., 2012).

12.3.2 Remote Sensing Observations

12.3.2.1 *Solar and Thermal Radiometers*

The use of radiometers has been widely used to understand the surface of Mars. Radiometers have higher signal-to-noise ratio than spectrometers and have provided the best constraints on polar energy balance. The Mariner 1969 Infrared Radiometer (IRR) experiment (Herr and Pimentel, 1969; Herr et al., 1972) was a two-band thermal radiometer which first determined that the surface temperature of the southern seasonal cap was ~150 K, consistent with a CO_2 ice composition (Neugebauer et al., 1969). The Viking Orbiters' Infrared Thermal Mapping (IRTM) experiments (Kieffer et al., 1972) were a follow-up to the IRR. The Viking IRTM solar bolometers had a spatial resolution that ranged from about 18 to 114 km, and were used to monitor the polar caps at the regional scale, confirming that both seasonal caps were primarily composed of CO_2 ice and that the southern residual cap was also composed of CO_2 ice. The Mars Global Surveyor (MGS) Thermal Emission Spectrometer (TES) (Christensen et al., 1992) was composed of three subsystems, two of which were radiometers (one solar and one thermal). TES had three sub-instruments that consisted of a 3×2 solar bolometer array and a 3×2 thermal bolometer array. The spatial resolution of TES was ~3–5 km and allowed for continuous monitoring of the seasonal and residual caps for nearly a

decade. The latest of the radiometers sent to Mars is the Mars Reconnaissance Obiter (MRO) Mars Climate Sounder (MCS) (McCleese et al., 2007). MCS has a spatial resolution of ~1 km nadir and ~5 km vertical. While radiometers are not technically imagers, the observations can often be used to form low-resolution "mosaics", as demonstrated by Kieffer et al. (2000).

12.3.2.2 *Visible Imaging and Spectroscopy*

With the notable exception of MAVEN, all orbital spacecraft have been sent to Mars with a visible camera. Even with the arrival of a suite of non-visible imagers and spectrometers, visible imaging remains an important technique for monitoring and understanding polar processes, and therefore the CO_2 cycle.

Visible imagers have been the most prevalent observational tools for monitoring the polar caps. These range from the lowest-resolution MGS Mars Orbiter Camera (MOC) Wide Angle global imaging (~7.5 km/pixel at the center of the image) (Malin et al., 1998) and MRO Mars Color Imager (MARCI) (Malin et al., 2001a) (1–10 km/pixel), used to monitor the polar caps at the regional to global scale, to the highest-resolution MRO High-Resolution Imaging Science Experiment (HiRISE) images (~25 cm/pixel), that allow detailed monitoring of smaller-scale features (McEwen et al., 2007), such as the fans in the cryptic region (see Section 12.4).

The first images of the surface of Mars were sent back by Mariner 4 (M4) in 1965, covering less than 1% of the surface. Mariners 6 (M6) and 7 (M7) (Leighton et al., 1969) provided an uncanny glimpse of the polar caps. Both Viking Orbiters had cameras that were used to monitor the polar caps (James et al., 1992). Comparisons between M9 and Viking images showed inter-annual variation of the southern perennial cap (James et al., 1992). MGS MOC provided both high-spatial-resolution images (Narrow Angle) and global monitoring (Wide Angle) of polar surface processes. MOC Narrow Angle images were the first images of sufficient spatial resolution to identify several polar surface processes, such as the "Swiss cheese" terrain (Thomas, et al., 2000) and spots and fans in the polar ice (Malin et al., 1998). MOC Wide Angle was used to monitor the recession of the seasonal caps (James et al., 2000; Benson and James, 2005), while the Mars Odyssey Thermal Emission Imaging System (THEMIS) (Christensen et al., 2004) has a subsystem capable of visible imaging (five colors) at ~18 m/pixel resolution. THEMIS VIS has provided the highest spatial resolution with global coverage. Mars Express (MEX) High Resolution Stereo Camera (HRSC) (Jaumann et al., 2007) has also imaged the polar regions of Mars. HRSC is specifically defined with multiple cameras for stereo imaging. MRO has three cameras, HiRISE, Context Camera (CTX) (Malin et al., 2007), and the Mars Reconnaissance Orbiter Mars Color Imager (MARCI) (Bell et al., 2009). All of these cameras have provided insights into the surface portion of the CO_2 cycle.

12.3.2.3 *UV Imaging and Spectroscopy*

The main contribution of UV spectroscopy to the understanding of the Martian CO_2 cycle has been the detection and characterization of mesospheric ice clouds associated with atmospheric temperatures below the CO_2 condensation point, thus inferred to

Table 12.3 *Key observations made by the following spacecraft missions: Mariner 7 (M7), Mariner 9 (M9), Viking Orbiters (VO), Viking Landers (VL), Mars Pathfinder (MPF), Mars Global Surveyor (MGS), Mars Odyssey (MOdy), Mars Express (MEX), Mars Reconnaissance Orbiter (MRO), and Mars Science Laboratory (MSL).*

	Mission	Key observable(s) relevant to CO_2 cycle
Laser altimeter		
MOLA	MGS	Polar night CO_2 ice clouds; thickness of CO_2 ice
Gamma-ray/neutron detectors		
GRS	MOdy	Seasonal variation of atmospheric argon and the thickness of CO_2 ice; ice table characteristics (water-equivalent hydrogen abundance and ice table depth)
NS	MOdy	Seasonal variation of the mixing ratio of atmospheric non-condensable gases (effectively N_2 and Ar) and the thickness of CO_2 ice; ice table characteristics (water-equivalent hydrogen abundance and ice table depth)
HEND	MOdy	Seasonal variation of the thickness of CO_2 ice; ice table characteristics (water-equivalent hydrogen abundance and ice table depth)
Radar sounder		
SHARAD	MRO	Subsurface CO_2 deposits
Radio occultation		
RS	MGS	Atmospheric CO_2 supersaturation (with respect to ice) vertical profiles
MaRS	MEX	Atmospheric CO_2 supersaturation (with respect to ice) vertical profiles
Infrared and near-infrared spectrometers/imagers		
IRR, IRIS, IRTM	M7, M9, VO	Surface CO_2 ice identification and its properties
TES	MGS	Growth/recession/change of CO_2 ice caps; CO_2 ice properties
THEMIS	MOdy	Growth/recession/change of CO_2 ice caps
OMEGA	MEX	Growth/recession/change of CO_2 ice caps; disambiguation of impure CO_2 ice; CO_2 ice clouds
CRISM	MRO	Surface CO_2 ice properties; CO_2 ice clouds; phenomena related to CO_2 jets
MCS	MRO	Growth/recession/change of CO_2 ice caps; CO_2 ice cap properties, such as grain size and chemical composition; polar night CO_2 ice clouds
Visible imagers		
Orbiter cameras	M7, M9, VO	Growth/recession/change of CO_2 ice caps
MOC-WA	MGS	Growth/recession/change of CO_2 ice caps
MOC-NA	MGS	Growth/recession/change of CO_2 ice caps; phenomena related to CO_2 jets
HRSC	MEX	Growth/recession/change of CO_2 ice caps
MARCI	MRO	Growth/recession/change of CO_2 ice caps
HiRISE	MRO	Growth/recession/change of CO_2 ice caps; phenomena related to fans and CO_2 jets
Landed (*in situ*) instrumentation		
air pressure	VL, MPF, MSL	Magnitude/timing of global CO_2 sublimation/deposition

be CO_2 ice clouds. As an example, the Mars Express SPICAM-UV (Bertaux et al., 2000), which is an instrument similar to MAVEN's IUVS, observed CO_2 ice clouds near an altitude of 100 km (Montmessin et al., 2006). More details about CO_2 ice clouds are covered in Section 12.4.2 and in Chapter 5.

12.3.2.4 Near-Infrared and Short-Wave Infrared Imaging and Spectroscopy

There have been three instruments capable of observing Mars in this spectral range: Mars Express OMEGA and Planetary Fourier Spectrometer (PFS), and the MRO CRISM. Of these three instruments, OMEGA provides a regional to global view, while CRISM provides the highest spatial resolution. There are several spectral features in the short-wave infrared that are diagnostic of both CO_2 ice and CO_2 gas, e.g. 1.435 μm and

2.28 μm. The 1.4 μm feature is present for both the gas and ice phases of CO_2. Forget et al. (2007) and Toigo et al. (2013) used OMEGA and CRISM 2 μm band observations to monitor the CO_2 atmospheric pressure and found similar results to the Viking Orbiter measurements. While near-infrared and short-wave infrared spectroscopy have provided constraints on the CO_2 component of the atmosphere, their main contribution has been in monitoring and characterizing the seasonal and residual polar caps. A detailed discussion about this is presented in Section 12.4.3. Langevin et al. (2007) and Cull et al. (2010a,b) provide good descriptions of ice retrieval methods.

12.3.2.5 Thermal Imaging and Spectroscopy

Thermal instruments have proven quite useful for monitoring the polar seasonal caps and tracking volatile composition.

Thermal infrared observations can track both the advance and the retreat of the seasonal cap, as opposed to visible cameras, which depend on reflected light. The presence of CO_2 ice in vapor pressure equilibrium can be thermally inferred, even when the near-infrared (NIR) spectral signatures are obscured by the presence of an optically thick layer of H_2O ice.

M6 and M7 Infrared Spectroscopy (IRS) experiments provided high-quality infrared spectra between 1.9 and 14.4 μm. The Mariner 1971 (M9) Infrared Spectroscopy (IRS) experiment (Hanel et al., 1970, 1972) provided follow-on observations. These observations remained the only thermal spectroscopy of the Martian surface for nearly three decades. MGS TES expanded the thermal studies of both Mariner spectrometers and Viking IRTM radiometers. Because the TES instrument included an interferometer, detailed studies of CO_2 grain size could be conducted (e.g. Kieffer et al., 2000; Kieffer and Titus, 2001; Titus et al., 2001; Hansen, 2013). In addition to MGS TES, another spectrometer, MEX Planetary Fourier Spectrometer (PFS), provided additional data of the polar caps (Hansen et al., 2005; Giuranna et al., 2008).

The M01 THEMIS instrument included not only visible imaging, but also thermal imaging. THEMIS provides eight unique narrow-band filters and is capable of imaging at 100 m resolution. This instrument package provides a unique capability to simultaneously image both temperature and albedo variations within the same scene. This capability was used to monitor and constrain the polar processes involved in the southern springtime cap (Kieffer et al., 2006).

12.3.2.6 Orbital Observations With Nuclear Spectroscopy

The 2001 Mars Odyssey Gamma Ray Spectrometer instrument suite (Boynton et al., 2004) provided a wealth of new information about the Martian surface and atmosphere. Nuclear spectroscopy data with global coverage were acquired from an approximately 400 km altitude (2 h period), circular polar mapping orbit. Data were accumulated for six Mars years (from April 2002 to present time of writing), with occasional losses due to solar energetic particle events, spacecraft safings, and other anomalies. The instrument suite consists of three subsystems: a gamma ray spectrometer (GRS), a neutron spectrometer (NS), and a high-energy neutron detector (HEND). As of 2014, acquisition of neutron spectroscopy data (NS and HEND) was ongoing; however, the GRS subsystem was taken out of service in 2009. Gamma ray spectroscopy data are available through September 2009, providing over three Mars years of useful data.

The nuclear emissions measured by these instruments were produced by the decay of radioelements within the regolith of Mars and by the interaction of galactic cosmic rays (GCR) with the Martian surface and atmosphere (e.g. Prettyman, 2014). Gamma rays and neutrons produced by GCR interactions sampled the entire atmospheric column and surface materials to depths of a few decimeters. A steady rain of GCRs falls on Mars uniformly from all directions. Consequently, GRS instruments were able to obtain full global coverage, independent of solar illumination.

The spatial resolution for all three GRS subsystems was on the order of 10° of arc length on the surface (about 600 km).

At this coarse spatial scale, the GRS instruments characterized the distribution of CO_2 ice in the seasonal caps and monitored changes in the composition of the atmosphere near the poles. The field of view of the instruments was such that the entire surface of Mars was covered every day; however, the statistics of counting generally required binning of data on longer timescales in order to detect and quantify atmospheric variations. For some measurements, timescales on the order of 10° of L_s could be sampled, enabling the investigation of seasonal processes.

12.3.2.7 Radar Observations

Radar has not been useful in the monitoring the seasonal CO_2 cycle because the seasonal CO_2 ice layers on the surface are too thin (~1 m at the thickest) to be visible to radar. However, radar has proven to be a useful tool in the search for buried CO_2 reservoirs. The unexpected discovery of buried CO_2 ice deposits under the south polar layered deposits (Phillips et al., 2011) has remained somewhat controversial and is discussed in depth in Section 12.5.2.

12.3.2.8 Radio and Gravity Science

Radio occultation measurements through the Martian atmosphere provided constraints on the atmospheric vertical temperature profile. These observations confirmed theoretical expectations that the radiative cooling of the atmosphere during the polar night would be buffered by condensing atmospheric carbon dioxide, thus maintaining atmospheric temperatures at the carbon dioxide saturation temperature (Hinson et al., 1999; Hu et al., 2012). However, these observations also revealed portions of the polar night temperatures falling below the saturation temperature (Colaprete et al., 2008). The magnitude of supersaturation observed was consistent with what is required to nucleate carbon dioxide cloud particles on dust grain (Glandorf et al., 2002). Radio-occultation-derived temperatures, along with TES-derived temperatures, were also used to identify a standing wave that dominated the southern pole's nighttime circulation (Hinson et al., 2003). This standing wave, controlled by the large southern hemisphere topography (namely Hellas Basin), is thought to be responsible for the offset of the southern pole perennial cap (Colaprete et al., 2005).

12.4 ACTIVE PROCESSES AND SEASONAL RESERVOIRS

The seasonal CO_2 cycle is a prominent feature of global atmospheric circulation on Mars. The atmosphere of Mars is primarily CO_2 (about 95 vol.% on average), with the balance consisting of Ar and N_2 (about 1.9 vol.% each,), trace gases, water vapor, and dust (Owen et al., 1977; Mahaffy et al., 2013; Chapter 4). The present obliquity of Mars (about 25.2°) results in seasonal variations in illumination similar to that of Earth (e.g. Head et al., 2003). During autumn and winter in both hemispheres, the temperature of the polar atmosphere drops to the condensation temperature of CO_2 (about 145 K at 4 mbar), resulting in the deposition of CO_2 on the surface at high latitudes (e.g. Leighton and Murray, 1966; Titus et al., 2001). The CO_2 ice sublimes as the seasonal cap is exposed to sunlight during spring. The edges of the seasonal CO_2 caps can reach latitudes down to 50° (James et al., 1992). In the spring sunlight, the caps appear as prominent, bright features that have long been observed from Earth (e.g. Herschel, 1784).

Seasonal frost has been implicated in a variety of processes that currently affect the Martian surface. Mass movements and gully formation have been shown to be common on northern polar dunes during times when they are covered with seasonal frost (Hansen et al., 2011). Additionally, activity in the mid-latitude gullies located on both dunes and crater walls has been found to be restricted to times when seasonal CO_2 frost is present (Dundas et al., 2010; Diniega et al., 2010, 2013). Although it is not well understood exactly how the seasonal frost drives these activities, they could possibly be explained by either explosive eruptions of gas that sublimated from the bottom side of transparent ice slabs (as occurs in the south polar jets), or avalanching of an unstable accumulation of granular frost (Hugenholtz, 2008). CO_2 frost avalanching has also been observed on steep northern polar scarps from $L_s = 8°$ to 48° over three consecutive Martian years (Russell et al., 2008) and may play a role in accelerating the retreat of these scarps.

The active seasonal CO_2 reservoirs can be divided between the surface ice and the atmosphere. The surface CO_2 ice can further be divided between the northern hemisphere, the southern hemisphere, and the high-altitude equatorial volcanic regions. Under the current epoch, these high-altitude equatorial deposits, which are part of a diurnal cycle, have negligible impact on the seasonal CO_2 cycle and are usually neglected. The southern hemisphere can further be subdivided, as there are distinct local climate differences that affect both condensation and subsequent sublimation of volatiles. This section will first discuss the atmospheric CO_2 reservoir, then that of the surface volatiles, and then will cover the three distinct polar climate zones and the higher elevations of the Tharsis volcanic region.

12.4.1 Characterization of the Atmosphere

Continuous, global monitoring of atmospheric processes and surface–atmosphere interactions by a fleet of orbiting spacecraft instruments supplement data acquired for decades from Earth-based observations, robotic landers, and orbiters. Our understanding of the Martian climate is guided by general circulation models (GCMs), with model predictions validated against *in situ* and remote sensing data (Pollack et al., 1990, 1993; Haberle et al., 1993; Hourdin et al., 1995; Murphy et al., 1995; Forget et al., 1999). For example, using the NASA Ames Research Center (ARC) GCM, the albedo and emissivity of the seasonal caps were "tuned" in order to fit Viking Lander pressure data, providing estimates of global surface pressure and the amount of CO_2 cycled through the seasonal caps (e.g. Haberle et al., 2008). GCM-based estimates indicate that 25% of the atmosphere is cycled through the seasonal caps, resulting in a meridional transport of mass and energy that strongly influences global and regional weather patterns.

However, Haberle and Kahre (2010) used GCMs to compare atmospheric pressure measurements from the two Viking Landers, Pathfinder, and Phoenix. Their analysis suggests that the atmosphere may have increased by approximately 10 Pa

between the Viking and Phoenix eras. However, 10 Pa is less than the uncertainty in the Phoenix data. The Viking data have an uncertainty of ~4 Pa and an elevation uncertainty equivalent to ~3 Pa. Interpolations in locations between the gridded models and actual location of the landers and rovers can potentially add another 4 Pa of uncertainly. The combined uncertainties exceed the estimated increase of 10 Pa, so additional observations will be needed to confirm this conclusion. If the mean atmospheric mass is increasing, the most likely source of the extra CO_2 is the south polar residual cap. The possibility of secular climate change has implications for the stability of the perennial CO_2 ice deposits (see Section 12.5.1).

CO_2 ice clouds are also observed at high altitudes in the Martian tropics at both equinoxes between +30° and –30° latitude, and at pressures less than 0.005 mbar (about 60–70 km). These high-altitude tropical clouds contain much smaller particle sizes than their polar counterparts. Mesospheric clouds appear in TES limb data (Clancy et al., 2000, 2004), SPICAM occultation data (Montmessin et al., 2006), Mars Orbiter Camera (MOC) limb images (Clancy et al., 2007), and MCS limb observations (Sefton-Nash et al., 2013). TES spectral fits estimate that hazes above 60 km were composed of either water ice clouds, with an effective mean particle radius of ≤1.0 μm, or CO_2 ice clouds, with a mean particle radius of ≤1.5 μm (Clancy et al., 2007). Montmessin et al. (2006) analyzed SPICAM ultraviolet stellar occultation data for the hazes and determined the mean radius of particles to be 0.3 μm. SPICAM observed some very high-level hazes at altitudes greater than 90 km with a mean particle radius of 0.1 μm; the temperatures of these hazes were simultaneously retrieved from SPICAM, and the hazes appear to have occurred where atmospheric temperatures were below the CO_2 saturation temperature. In several instances of these high-altitude haze observations, derived temperatures were consistent with CO_2 supersaturations between 10% and 100%. This association suggests that hazes at very high altitudes are composed of CO_2 ice. Direct spectral identification of these high-altitude hazes as CO_2 ice is reported in Montmessin et al. (2007). Perhaps one of the most interesting aspects of a supersaturated CO_2 atmosphere is the observation of spontaneous nucleation, which correlate to cold pockets caused by gravity waves (Määttänen et al., 2005, 2007; Listowski et al., 2013, 2014). The cloud particle size estimates and altitudes matter because they provide insights into the formation, maintenance, and decay mechanisms of these clouds. It also indicates which optical property paradigm (changes significantly with particle size) should be used to estimate their radiative transfer effects (e.g. opacity).

Perhaps the most striking atmospheric discovery made by Mars Odyssey's nuclear instruments was the magnitude and character of seasonal changes in the composition of Mars' polar atmosphere (Sprague et al., 2004). The deposition of CO_2 ice at high latitudes results in the flow of atmosphere towards the poles. In the southern hemisphere, a strong, coherent polar winter vortex forms, limiting meridional mixing of gases from the polar region to lower latitudes. The removal of CO_2 from the atmosphere results in the concentration of non-condensable gases (e.g. N_2, Ar, and CO) at high latitudes. During fall and early winter, a factor of 6 enhancement in the column abundance of non-condensable gases (relative to the average measured by

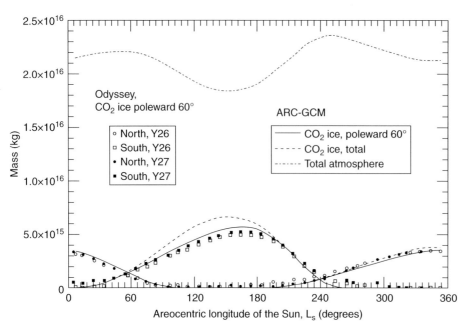

Figure 12.3. The inventory of seasonal CO$_2$ ice as measured by Odyssey's neutron spectrometer (NS; Prettyman et al., 2009) is compared to GCM calculations, for which seasonal cap properties were adjusted to fit Viking Lander pressure measurements (Haberle et al., 2008). Epithermal neutron measurements for MY 26 and 27 (April 2002–January 2006) were analyzed to determine the mass of CO$_2$ ice poleward of 60° in both hemispheres. The NS measurements are consistent with the GCM calculations, which suggests that Mars' climate has not changed significantly since the Viking era (MY 11–12) (Prettyman et al., 2009). The uncertainty in the NS measurements is greater than 10 Pa. The average mass of Mars' atmosphere is about 2.2×10^{16} kg, which varies by about 25% due to seasonal condensation and sublimation of CO$_2$ frost at high latitudes.

Viking Lander 2) was independently measured by Odyssey's gamma ray and neutron spectrometers (Figure 12.3) (Sprague et al., 2004, 2007; Prettyman et al., 2009). This enhancement of inert gases relative to the CO$_2$ gas is consistent with thermal infrared measurements from MCS (Hayne et al., 2012). The gamma-ray spectrometer observed variations in the prominent gamma rays produced by neutron capture with atmospheric Ar; whereas the neutron spectrometer measured variations in thermal neutron absorption by N$_2$ and Ar. Eddy mixing results in the diffusion of N$_2$ and Ar away from the poles. The fact that the peak enhancement in non-condensable gases occurs well before the reversal of the condensation flow (maximum CO$_2$ ice deposition, Figure 12.4) indicates changes in eddy mixing processes and/or the strength of the winter vortex. Despite the presence of a strong winter vortex in the northern hemisphere, enrichment of non-condensable gases is relatively weak and chaotic (Feldman et al., 2003; Sprague et al., 2012), which points to fundamental differences in the magnitude of various dynamical processes acting in the northern and southern hemispheres (Colaprete et al., 2008). Initial GCM simulations of non-condensable gas enrichment were unable to fully reproduce the magnitude of the observed enhancement of non-condensable gases (Nelli et al., 2007; Colaprete et al., 2008); however, using a state-of-the-art dynamical core and tracer transport scheme (Lian et al., 2012) and/or incorporation of improved models of CO$_2$ cloud microphysics (Colaprete et al., 2008) results in better fits to non-condensable gas column abundances in both hemispheres.

12.4.2 Condensation and the Polar Night

During the polar night, the entire polar surface is covered by CO$_2$ ice, the thickness and density of which is a function of location and season. However, because the majority of the polar caps are shrouded in darkness during the late fall and winter, these regions (during these seasons) are the least observed parts of Mars. In addition to the lack of reflected sunlight (which makes visible and near-infrared observations useless), a polar

hood of atmospheric H$_2$O ice (essentially a cloud bank) further obscures the view of the condensing polar cap. MGS MOLA (D. Smith et al., 2001b) was an exception for observing the surface ice as it had its own light source at 1 μm. Measuring surface albedo during the polar night was not a primary function of MOLA; so much of the surface reflectivity data collected by MOLA was either saturated due to the presence of high-albedo fine-grained CO$_2$ ice (Ivanov, 2002) or obscured by clouds. (Neumann et al., 2003).

Even thermal imaging and spectroscopy have had difficulty in observing many of the polar night processes due to extremely low signal-to-noise ratio, and at the CO$_2$ frost temperature (T ~ 145 K) thermal emission decreases dramatically shortward of 11 μm. The Viking Orbiter IRTM instruments had channels at 20 μm and 11 μm that provided the first ability to monitor polar night processes. The MGS TES spectrometer had sufficient signal-to-noise ratio in the spectral range 11–40 μm to monitor polar processes for nearly four Mars years. MEX PFS has been able to track the cap advance and recession of the southern cap (Giuranna et al., 2008), demonstrating that the cap advance was approximately double the rate of the seasonal cap retreat. Mars Odyssey THEMIS and MRO MCS continue to actively monitor both surface and atmospheric process. Mars Odyssey THEMIS has limited ability to observe polar night processes because the longest-wavelength filter that can observe the surface through the CO$_2$ atmosphere is 12.6 μm, which can identify CO$_2$ surface ice, but has insufficient signal-to-noise ratio to differentiate grain size variations caused by the various deposition processes. At the time of writing, MCS has collected three complete Mars years of observation of the seasonal ice caps.

The Mars Odyssey Gamma Ray Spectrometer (GRS) suite of instruments is also able to peer through the veil of darkness. GRS measurements of the inert gas mixing ratio is covered in Section 12.4.1, while the discussion of CO$_2$ ice deposition and sublimation is discussed in Section 12.4.3.

During the polar night, the polar atmosphere is dynamically separated from the rest of the atmosphere by the polar vortex.

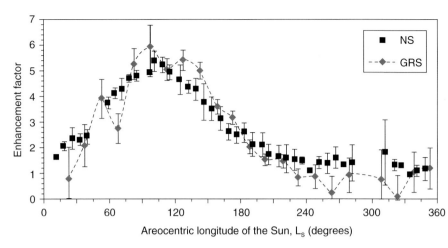

Figure 12.4. Mars' southern atmosphere is enriched in non-condensable gases (N_2 and Ar) as CO_2 ice is deposited on the surface during fall and winter. The non-condensable gas "enhancement factor" separately measured by gamma ray and neutron spectroscopy (labeled, respectively, GRS and NS) poleward of 75°S is shown (Sprague et al., 2004, 2007; Prettyman et al., 2009). The plot is reproduced from Prettyman et al. (2009). The enhancement factor is defined as the ratio of the column abundance of non-condensable gases to the average abundance measured by Viking Lander 2.

While the polar vortex is a barrier to atmospheric exchange at lower altitudes, above ~60 km transport into the polar region is less restricted (see Haberle et al., 1993; Forget et al., 1999), thus allowing the injection of dust during periods of global dust storms. This is mainly due to the downwelling branch of the winter hemisphere Hadley circulation branch (Conrath et al., 1973; M. Smith et al., 2001; McCleese et al., 2008).

The condensation of CO_2 ice occurs either through direct surface condensation (Ditteon and Kieffer, 1979) or via snowfall from clouds (Pollack et al., 1990; Colaprete and Toon, 2002; Eluszkiewicz et al., 2008) often induced by orographic lifting (Forget et al., 1998; Cornwall and Titus, 2009; Hayne et al., 2012) or gravity waves (Tobie et al., 2003). Kuroda et al. (2013) used a GCM with an updated CO_2 condensation and sublimation scheme, which replicates the zonally averaged results from MRO MCS. Their simulation suggests that half of the northern seasonal cap could be formed from CO_2 snow, while the remaining half is formed from condensation directly onto the surface. Observations from MCS have estimated that ~3–20% of the south polar seasonal cap is formed from snowfall (Hayne et al., 2014). Hayne et al. (2012), also using MCS observations, concluded that the larger regions of low emissivity could only be explained by the presence of long-lived regional CO_2 ice clouds, combined with surface CO_2 snow. The smaller, transient regions of low emissivity are still most likely due to surface CO_2 snow formed by orographic lifting as suggested by Cornwall and Titus (2009, 2010).

It is generally believed that within the polar night a balance exists between diabatic processes such as radiative cooling and latent heating from condensing CO_2. This assumption, that CO_2 is always in vapor pressure equilibrium with its solid phase, manifests itself in Mars general circulation models (MGCMs) in such a way as to never allow the atmospheric temperature to dip below the saturation temperature of CO_2. However, this assumption becomes invalid if there is no solid phase present. A variety of observations from Mars Global Surveyor (MGS) radio science (RS), the MGS Thermal Emission Spectrometer (TES), the Mars Reconnaissance Orbiter (MRO) Mars Climate Sounder (MCS) and Mars Express (MEX) SPICAM have demonstrated this assumption to be largely inaccurate for significant portions of the polar night atmosphere and in the mesosphere (e.g. Forget et al., 2009). Figure 12.5 shows an example of supersaturation (where temperatures deviate from the CO_2

saturation temperature) as derived from TES observations. Furthermore, significant regions of the Martian mesosphere have also been observed to have temperatures well below the condensation temperatures of CO_2. This supersaturation is widespread and not uncommon and the observed temperature profiles suggest unstable regions containing significant amounts of potential convective energy.

The process of nucleation describes the initial formation of a crystal spontaneously without a substrate (homogeneous nucleation) or on a dust grain or similar aerosol (heterogeneous nucleation) (see Wood, 1999; Ortega et al., 2011). The homogeneous nucleation of most vapors requires very high levels of saturation, as the energy barrier involved with forming clusters of molecules is high. If a pre-existing surface is involved, as is the case with heterogeneous nucleation, the surface energy per molecule is greatly reduced and the vapor saturation level needed to nucleate decreases by as much as two orders of magnitude from that of homogeneous nucleation. In either case, however, some amount of supersaturation is typically needed before nucleation rates become significant. Recent studies have shown a similar requirement for CO_2 cloud nucleation (Glandorf et al., 2002). Understanding the nucleation process is critical to determining the characteristics of CO_2 clouds, including when and where they form, and their constituent particle sizes and number density. The latent heat associated with the potential level of condensation at supersaturation temperatures can result in local warming and possible atmospheric instability. Convection resulting from the release of latent heat is commonplace within the Earth's atmosphere; however, this potential for convection within the Martian polar night is a relatively new concept, being developed only since the observations of CO_2 supersaturation temperatures within the Martian atmosphere. The convective available potential energy (CAPE) is a measure of the potential air parcel buoyancy associated with warming that results from the latent heat release from condensation that occurs within the parcel. On Earth, water vapor is a minor constituent, and thus CAPE is largely limited by the availability of water vapor. On Mars, however, the condensable gas is the primary gas and thus not limited by diffusion or the availability of vapor. Figure 12.6 shows CAPE values for the north and south poles derived from TES observations. Significant quantities of CAPE is seen in the polar night early in either hemisphere's winters, and especially

Temperature (K)

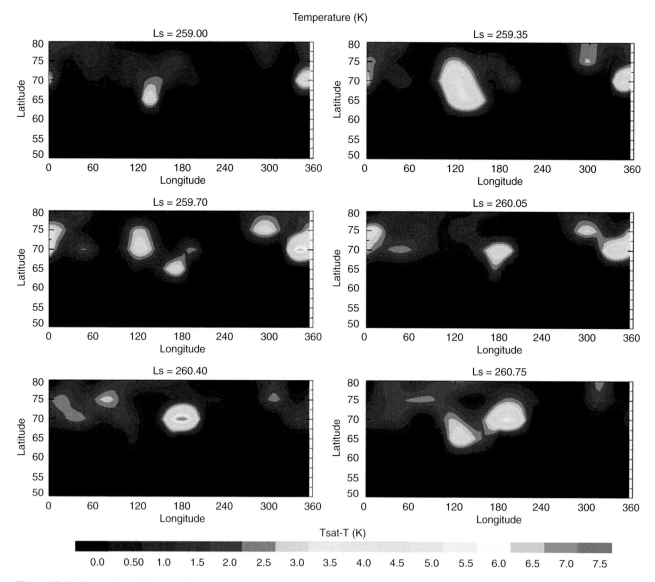

Figure 12.5. The number of degrees of temperature (derived from TES observations at approximately 1.37 mbar) below carbon dioxide saturation, $T_{sat} - T$ (K), for six representative periods during the northern winter. A black and white version of this figure will appear in some formats. For the color version, please refer to the plate section.

nearest the polar night boundary, were baroclinic activity may have an important role in creating pockets of supersaturation and CAPE. This CAPE, if realized through nucleation of significant numbers of small ice nuclei, can result in convection and increased vertical mixing. This vertical mixing has been shown to have an important effect on polar temperatures and cloud/dust distributions (Colaprete et al., 2008).

12.4.3 Surface Volatiles, e.g. Seasonal Caps

There are four distinct regions on Mars where CO_2 ice resides on the surface: the north and south polar regions, subtropical poleward-facing slopes (Vincendon et al., 2010), and the highest altitudes of Tharsis (Cushing and Titus, 2008). The highest-altitude regions of the Tharsis region (e.g. the caldera of Arsia Mons) are extreme environments, even by Mars standards. The diurnal surface temperature can change by a factor of 2, and,

for at least some fraction of a Mars year, the volcanic calderas have CO_2 ice deposited nightly (Cushing and Titus, 2008). The poleward-facing slopes of the subtropical regions that remain in shadow for long periods of time can have surface temperatures at the CO_2 ice frost point. Vincendon et al. (2010) observed the presence of small amounts of CO_2 ice in these regions. While the amount of CO_2 ice deposited in the subtropical shadow regions is small under the current epoch, the presence of CO_2 ice at low latitudes does suggest that, under differing climatic conditions, the polar caps could potentially extend to the low latitudes.

The two polar regions form three distinct climate zones: (1) the northern region (N), characterized by the presence of H$_2$O ice; (2) the portion of the southern polar cap that includes the Mountains of Mitchel (~50°E) extending west to ~230°E (SW), which is characterized by bright albedo and enhanced CO_2 snowfall; and (3) the southern polar cap that extends east of the Mountains of Mitchel, through the cryptic region to ~230°E

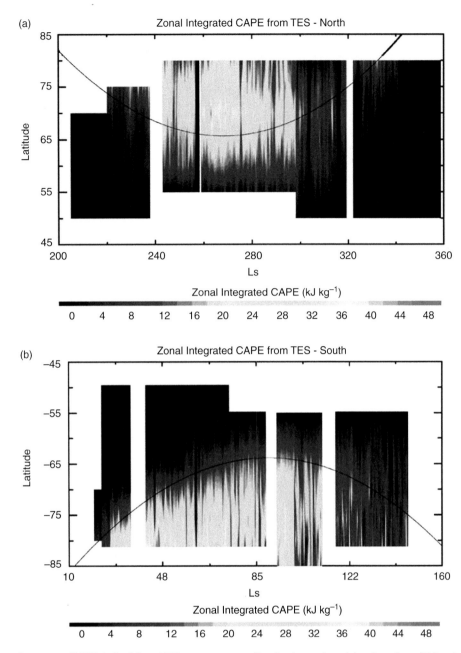

Figure 12.6. The zonally average CAPE derived from TES temperature profiles for the northern (a) and southern (b) hemispheres. The black curve shows the approximate location of the polar night boundary. A black and white version of this figure will appear in some formats. For the color version, please refer to the plate section.

(SE), characterized by dark albedo, dust, and slab ice. This division between the southwest and the southeast polar regions is the same as used by Schmidt et al. (2010). It has been suggested that these two climates zones be referred to as Prometheus and Argentea, a reference to their geographical locations.

12.4.3.1 Mass and Density of the Seasonal Polar Caps

The mass of the polar caps is an important parameter for understanding the Martian climate. The distribution of seasonal CO_2 ice provides information about the interaction between the atmosphere, topography, and surface properties. The density of the CO_2 ice constrains several important polar processes, including deposition mechanisms and subsequent densification.

Calculations and remote sensing data that have been analyzed to determine seasonal cap characteristics are reviewed here, as is the use of neutron spectroscopy to determine CO_2 cap thickness. Several approaches have been used to estimate the amount of CO_2 that condenses onto the seasonal cap – see James et al. (1992) for earlier estimates. The most robust of these have been GCMs that were tuned to match the atmospheric pressure curves from the Viking Landers (Chamberlain et al., 1976; Tillman et al., 1993). While these models often use unrealistic albedo and emissivity values for the seasonal caps, they do match the pressure curves, and therefore must match the net global accumulation; however, they do not necessarily match local distributions. Other methods use energy balance considerations. For example, Kieffer et al. (2000) computed the peak annual southern mass at several locations from thermal and albedo observations throughout the sublimation season, yielding 800–1200 kg m^{-2} (polar cap sublimation and accumulation budgets are often expressed in mass per unit area; e.g. for an assumed density of 1000 kg m^{-3}, a budget of 1200 kg m^{-2} would correspond to a thickness of 1.2 m). Kieffer and Titus (2001) used energy balance during both condensation and sublimation seasons in the north to estimate the zonal distribution of cap mass, not only finding peak budgets of 1100 kg m^{-2} near the pole, but also finding the condensation budget to be several hundred kg m^{-2} larger than the sublimation budget near 70°N, indicating that substantial zonal heat transport is likely to occur during the spring. This analysis, however, neglected the effects of heat storage in the regolith. The polar CO_2 condensation creates an observable increase in the planet's rotation rate by conservation of angular momentum (Folkner et al., 1997; Yoder and Standish, 1997). A combination of gravity and differential elevation measurements were used to estimate an average seasonal cap density of 910±230 kg m^{-3} (D. Smith et al., 2001a)

Detailed analysis of perturbations to Mars gravity field as a function of season was conducted by Smith et al. (2009). They estimated that the maximum mass of the northern seasonal cap was 3.5×10^{15} kg and that of the southern cap was 6.5×10^{15} kg.

The mass of CO_2 within the seasonal caps was subsequently measured by Mars Odyssey's GRS instrument suite. The thickness of the cap was determined directly from observations of the attenuation of gamma rays produced in the regolith by the overlying CO_2 ice (Kelly et al., 2006), and by measurements of neutron counting rates, which systematically increase as CO_2 ice is deposited onto the surface (Litvak et al., 2005, 2006; Prettyman et al., 2009). Measurements of the amount of CO_2 in the residual caps confirm GCM-based analyses of Viking Lander pressure data shown in Figure 12.3 and by Kelly et al. (2006). Furthermore, the inventory and spatial distribution of seasonal CO_2 ice, as determined by the Odyssey GRS suite, was found to be consistent with seasonal, zonal gravity measurements (Konopliv et al., 2011). Inter-annual differences in the CO_2 cap inventory were found to be small, within analytical uncertainties (Kelly et al., 2006; Prettyman et al., 2009). A comparison of epithermal neutron counting rates binned poleward of 80° in both hemispheres for MY 26–29 further indicates that accumulation of CO_2 ice at high latitudes is generally consistent from year to year in the current obliquity regime (Maurice et al., 2011). Based on measurements from Mars Odyssey instruments and modeling, aspects of the climate that would affect the seasonal CO_2 cycle have not significantly changed since the Viking era; however, improvements in data reduction and monitoring over many Mars years would be necessary to detect subtle changes such as a gradual loss of southern perennial CO_2 ice (Haberle and Kahre, 2010; Malin et al., 2001b). An episodic disappearance of large portions of the perennial cap, as suggested by Jakosky and Haberle (1990), could be detected by the Mars Odyssey's instruments (Tokar et al., 2003; Prettyman et al., 2004).

Depths and thicknesses of layers determined by nuclear spectroscopy are reported as column abundances or areal densities (g cm^{-2}). Consequently, if the geometric thickness of a layer can be estimated from geophysical constraints or measured (via laser altimetry or photoclinometry), then the bulk density of the layer can be determined. For example, the density of the seasonal CO_2 caps was determined by combining column abundances measured by the GRS instruments and geometric thicknesses determined by the Mars Orbital Laser Altimeter (e.g. Aharonson et al., 2004). Results indicate that the seasonal ice density is much lower than the theoretical density of CO_2 ice and that the density changes with time, which may point to variations in the emplacement mechanisms (snow versus direct condensation) and alteration of the surface over time (Hecht, 2008).

Measurements of density are less reliable than measurements of mass, owing to uncertainties in linear thickness measurements and comparisons between regionally averaged frost values (from nuclear spectroscopy) and local measurements provided by MOLA. Density estimates for the northern seasonal cap vary from 500 to 1100 kg m^{-3} (e.g. D. Smith et al., 2001a; Aharonson et al., 2004) depending on how the estimate was determined and over which latitude region. The southern seasonal cap appears to have a density of 1000 kg m^{-3}. The density of the seasonal cap is likely to be inconsistent, with variations that correspond to latitude, season, and mechanisms of the local depositional history (e.g. direct condensation versus snowfall). The density of the CO_2 ice can also increase over time due to compaction and sintering. Impurities, such as dust or H_2O, can also affect density. Finally, the presence of noble gases (or any trapped gas that does not completely condense) can create bubbles inside the CO_2 ice, thus increasing its porosity and lowering the ice's effective density (Pilorget et al., 2011).

As discussed in the beginning of this section, the south polar region has two climate zones. A logical question is that, if these two climate zones have differing CO_2 ice deposition mechanisms, does the CO_2 ice density reflect these differences? Using MOLA data, Jian and Ip (2009) found no difference in seasonal ice depths between the SE and SW climate zones. Even if the ice column depths are symmetric between these two climate zones, differences in density could still be present, which suggests there are processes that can increase the density of porous ice. More investigation is needed.

12.4.3.2 Seasonal Cap Edges

The most easily observable parameter that describes the polar seasonal caps is their size, which has been measured since the days of Herschel (1784). However, telescopic observations of the dynamic nature of the seasonal caps have been restricted to the recession phase of the caps. During the period of seasonal cap growth (or cap advance), Mars is tilted away from the Sun (and, consequently, the Earth), thus placing the cap edge at unfavorable viewing angles. In addition, the cap edge is close to the terminator and is often obscured by clouds or haze. Fischbacher et al. (1969) compiled recession curves for the polar caps from 1905 to 1965, using photographic material archived at the Lowell Observatory in Flagstaff, Arizona. A number of other observers used telescopic data to characterize the cap recessions from the early 1960s through to the present. Spacecraft observations began in 1969 with Mariner 7 and continue to the present day. Compilations of both telescopic and spacecraft observations of the polar caps can be found in James et al. (1992), Hansen (1999), and Benson and James (2005).

The advance and retreat of the polar cap from year to year may contain many clues to help elucidate a number of physical

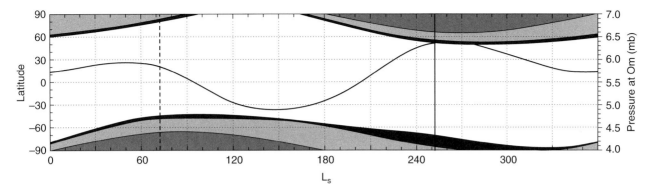

Figure 12.7. Polar cap-edge behavior as a function of season. The dark gray areas represent the polar night. The light gray regions represent the minimum extent of the seasonal cap, and the black regions represent the maximum extent. The thickness of the black region also represents that amount of asymmetry in the seasonal cap. The north polar seasonal cap remains mostly symmetric while the southern cap becomes asymmetric during maximum extent and during the cap recession after $L_s = 180°$. The dashed and solid vertical lines are aphelion and perihelion, respectively. The wavy gray line is the atmospheric pressure curve. This figure is based on cap estimates from Titus and Cushing (2014) and is a modified version of figure 1 from Piqueux et al. (2015).

processes that are not well understood. For example, summer-time heat storage in the regolith could delay the onset of the seasonal CO_2 cap. The onset of the seasonal cap could also be directly affected by the thermal inertia of the near-surface desiccated regolith and place constraints on the depth of the ice table.

Parameterizations of the seasonal cap edges provide useful constraints on atmospheric GCMs and mesoscale models (as shown in Figure 12.7). Resolving the longitudinal differences of the cap edges as the seasonal caps advance and retreat constrains the times when zonal means are appropriate and when longitudinal asymmetries make zonal means invalid. Figure 12.7 illustrates one such parameterization, where both the advance and the retreat of the seasonal caps have been modeled as a series of sines and cosines. These types of parameterizations can also be used when modeling other observational data that have large spatial resolutions, such as GRS and NS data. By knowing where the cap edge should be located, instruments with coarse spatial resolution can correct for subpixel mixing caused by their large instrument point-spread functions that convolve both frosted and frost-free areas.

The H_2O cycle can modify the CO_2 cycle in two major ways: H_2O ice contamination, which alters the CO_2 ice albedo and emissivity (e.g. Warren et al., 1990); and perennial near-surface H_2O ice tables, which act as thermal capacitors, storing heat throughout the summer and releasing it during the fall and winter. H_2O ice contamination either can be intimately mixed with the CO_2 ice, or can condense as a layer on top of the CO_2 ice. The H_2O ice layer can be either optically thin or optically thick. Very small amounts of water ice, either as a part of the CO_2 ice or as an optically thin layer, can have a strong spectral signature. Optically thick H_2O ice will effectively obscure CO_2 ice spectral features (Appéré et al., 2011), but the temperatures will remain near those of CO_2 ice. Furthermore, the presence of H_2O frost can brighten the surface, thus slowing sublimation in the spring.

The dust cycle also impacts the CO_2 cycle by changing both the albedo and emissivity of the CO_2 surface ice. The most direct effect is that mixing of dust within surface CO_2 ice will

decrease the albedo while increasing the effective emissivity. Atmospheric dust, if present during the polar night, can provide condensation nuclei and thus increase the amount of snowfall (e.g. Cornwall and Titus, 2009). Atmospheric dust, if present during retreat of the seasonal cap (such as often occurs during global dust storms in the southern spring), redistributes the wavelength regime of visible and thermal radiation on the surface (Bonev et al., 2002). Atmospheric dust will absorb insolation at the visible wavelengths and re-emit the energy in the thermal. This results in dark regions subliming slower (as they mostly absorb in the visible) and bright regions subliming faster (as they mostly absorb in the thermal).

12.4.3.3 Northern Seasonal Cap Edge

The north polar seasonal cap (NPSC) has been less rigorously studied than the south, both because Mars is typically farther from Earth during Martian northern winter and also due to the consistent formation of the northern "polar hood", which typically obscures the cap between $L_s = 150°$ and $L_s = 30°$ (Benson et al., 2011). Consequently, the advance of the cap has not been well characterized until the 21st century. Kieffer and Titus (2001) used zonal means to observe variations in surface temperature and visible bolometric albedo according to season using MGS TES. The TES thermal observations show a nearly perfect symmetric advance; that is, condensation at consistent latitude across all longitudes, with the most northern edge of the seasonal cap occurring between longitudes 245°E and 265°E and the most southern edge of the seasonal cap occurring between 280°E and 30°E (Titus, 2005a). The advance of the northern cap typically leads the advance of the edge of polar night by 10° of latitude (Figure 12.7). Brown et al. (2012) used CRISM observations to monitor the cap edge near the end of winter ($L_s = 304–0°$). At this time, the northern seasonal cap extended to 50°N latitude, and was nearly symmetric at $L_s = 304°$.

The retreat of the northern seasonal cap is also nearly symmetric, having been well characterized at visible wavelengths by both telescopic and spacecraft observations. The northern spring retreat is also nearly symmetric in both visual and thermal observations, and follows the same small asymmetries as seen in the advance (e.g. Kieffer and Titus, 2001). The latitude difference between the maximum and minimum extent of the thermal cap is generally between 3° and 5°, regardless of whether the cap is advancing or retreating.

Benson and James (2005) document changes in northern spring recession rates between 2000 and 2002. These changes occurred in the same longitude ranges as where Titus (2005a) discusses similar asymmetries, with the slower retreat occurring between longitudes 300°E and 30°E and the faster retreat between 230°E and 270°E. In addition to four Martian years of seasonal observations by TES and MOC, the northern seasonal cap was observed in detail by OMEGA in 2004 and 2006 (MY 27–28). During winter of this year, the CO$_2$ ice was initially transparent (Appéré et al., 2011), which suggests slab CO$_2$ ice. A transparent (or at least translucent) CO$_2$ slab ice appears to be the normal state for much of the early springtime cap (e.g. Kieffer and Titus, 2001).

Brown et al. (2012) used CRISM observations to define four distinct phases of the cap recession: the pre-sublimation phase ($L_s = 304–0°$), the early spring phase ($L_s = 0–25°$), the asymmetric retraction phase ($L_s = 25–62°$), and the stable phase ($L_s = 62–92°$). The pre-sublimation phase presents the maximum extent of the winter cap. While the cap remains relatively stable during this period, a small amount of sublimation does occur at the lower latitudes. During the early spring phase, the cap is retreating and remains mostly symmetric. During the asymmetric retraction phase, NIR spectroscopic observations of the cap-edge retreat observed the cap to become asymmetric (Brown et al., 2012). While cap-edge tracking using NIR spectroscopy shows an asymmetric retreat, thermal observations do not (Kieffer and Titus, 2001). This suggests that the CO$_2$ ice cap remains symmetrical, while an annulus of H$_2$O ice that surrounds the cap may be asymmetric (see the discussion in the next section). The stable phase is the late spring period when all apparent CO$_2$ ice spectral signatures have disappeared. While the spectral signature of CO$_2$ ice may have disappeared, the H$_2$O ice temperatures appear to remain buffered to near summer solstice (Kieffer and Titus, 2001; Wagstaff et al., 2008).

In addition to the temporal phases discussed by Brown et al. (2012), the northern seasonal cap has distinct zonal regions. Giuranna et al. (2007b) used MEX-PFS observations at $L_s = 40°$N to discuss five concentric regions of the seasonal cap: Region I was the H$_2$O ice (20 µm grain size) that extended from 65°N to 76°N; Region II was dust-free CO$_2$ ice (5 mm grain size) that extended from 76°N to 79°N; Region III was dusty CO$_2$ ice (3.7 mm grain size) that extended from 79°N to 81°N; Region IV was nearly dust-free CO$_2$ ice (3 mm grain size) that extended from 81°N to 84°N; and Region V was dust-free CO$_2$ ice (5 mm) that extended from 84°N to the pole.

12.4.3.4 H$_2$O Ice Annulus

The bright ring at intermediate temperatures (180 K) observed by TES (Kieffer and Titus, 2001) and THEMIS (Wagstaff et al., 2008) in early spring was verified by OMEGA as resulting from H$_2$O frost (Bibring et al., 2005; Appéré et al., 2011),

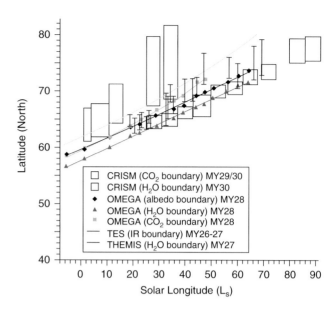

Figure 12.8. Comparison of CRISM MY 30 H$_2$O and CO$_2$ cap-edge results to cap-edge retreat rates of OMEGA (Appéré et al., 2011), TES (Titus, 2005a), and THEMIS (Wagstaff et al., 2008). Figure is figure 5d from Brown et al. (2012).

with a sublimation front that lags by up to 4–6° in latitude south of the CO$_2$ ice sublimation front. Contamination of H$_2$O ice into the CO$_2$ ice is ubiquitous in the northern seasonal cap at all stages of its evolution. H$_2$O ice dominates the spectral signatures over most of the seasonal cap after mid-spring ($L_s \sim 45°$). The importance of the H$_2$O ice annulus that surrounds the retreating CO$_2$ cap and the cold trapping of optically thick layers of H$_2$O ice on top of the CO$_2$ ice are important to consider when estimating the CO$_2$ surface ice component of the CO$_2$ cycle. The H$_2$O annulus is not disguisable in the visible from the adjacent CO$_2$ ice cap, and so the visible cap appears larger than the actual seasonal CO$_2$ ice cap. In addition, the region of the CO$_2$ ice cap adjacent to the H$_2$O annulus is often covered by an optically thick layer of H$_2$O ice that obscures the presence of the underlying CO$_2$ ice. These two processes must be considered when estimating the seasonal cap edge and the mass of the seasonal CO$_2$ ice. Brown et al. (2012) demonstrated this "watery" cap edge when they compared visible, NIR spectroscopy, and thermal cap-edge detections as a function of season (Figure 12.8).

In addition to the presence of an H$_2$O ice annulus that tracks the seasonal cap retreat, occasional brightening of the entire seasonal cap occurs (Kieffer and Titus, 2001). This temporary brightening of the cap (referred to as "flashing") typically lasts for a day, and then returns to normal albedo. This type of flashing is most likely due to a sudden cold trapping of water vapor across the entire cap, resulting in a thin and temporary layer of bright water frost. This relatively slow type of flashing should not be confused with the "flashing" observed telescopically by many amateur astronomers (e.g. Wilson, 1937; Wells and Hale, 1971; Haas, 2003), which occurs very quickly and is probably the reflection of sunlight off of ice crystals in high-altitude Martian clouds, or possibly specular reflections off isolated pockets of surface ice.

12.4.3.5 South Polar Seasonal Cap Edge

During the southern fall, the advance of the southern seasonal cap is nearly symmetric, similar to the northern seasonal cap. Both thermal (Giuranna et al., 2007a; Titus et al., 2008; Titus and Cushing, 2014) and NIR (Brown et al., 2010a) observations suggests that CO_2 ice condensation phase begins around $L_s = 340°$. Only a 2–4° difference in the latitude of the cap edge has been observed during this phase (Titus and Cushing, 2014), with the most northern extension generally occurring near longitude 300°E. A continuous cover of surface frost extends to between 50°S and 55°S and reaches as far north as 40°S in the relatively low-elevation Hellas and Argyre Basins, which are covered preferentially by the frost as expected from energy balance considerations. Localized seasonal CO_2 ice frost deposits can form at even lower latitudes on pole-facing slopes. Using OMEGA and CRISM spectroscopic observations, Vincendon et al. (2010) detected the presence of winter CO_2 frost deposits on such slopes up to 34°S, and showed that high-thermal-inertia buried water ice in the subsurface is required to explain the absence of CO_2 frost at lower latitude up to 25°S. During this period, the polar hood forms around $L_s = 10°$ and begins to track the cap edge until about $L_s = 70°$ (Benson et al., 2010). Benson et al. (2010) discuss the south polar hood, which tracks the polar cap edge. The polar hood reappears in the interior of the cap around $L_s = 100°$.

Brown et al. (2010a) used CRISM observations to define four distinct phases of the cap recession: the pre-sublimation phase ($L_s = 160–200°$), the cryptic phase ($L_s = 200–240°$), the asymmetric retraction phase ($L_s = 240–290°$), and the stable cap phase ($L_s = 290–340°$). During the pre-sublimation phase, the cap edge is retreating, but CO_2 ice condensation may still occur at high latitudes. It is at the end of this phase that the polar hood disappears for the remainder of the Mars year (Benson et al., 2010).

The cryptic phase begins around $L_s = 200°$ with the rapid appearance of the cryptic region (Kieffer et al., 2000; Section 12.4.4) between $L_s = 195°$ and 210° (Brown et al., 2010a). CRISM observations of the cryptic region during this time also indicate the presence of pure water ice. During this phase, the southern seasonal cap remains symmetric in shape. At approximately $L_s \sim 220°$, the cryptic region (Kieffer et al., 2000; Section 12.4.4) has reached maximum albedo contrast with the rest of the seasonal cap.

The retreat of the southern cap has been known to be asymmetric since early telescopic observations, but was often reported to begin earlier than is now known from thermal and NIR spacecraft observations. It is during this asymmetric retraction phase that the cryptic region quickly sublimes away and the Mountains of Mitchel become detached. Observations by MGS, Mars Odyssey, Mars Express, and MRO have shown that the asymmetric sequence repeats consistently from year to year. The MOC observations of the global recession in 1999, 2001, and 2003 suggest that the overall seasonal cap is insensitive to global-scale dust storms. However, Bonev et al. (2002, 2008) showed evidence for local variations due to the redistribution of absorbed solar energy. During large dust storms, when the direct absorption of solar energy is reduced and thermal downwelling radiation is increased due to increased atmospheric opacity,

bright regions of the cap have higher-than-normal sublimation rates, while darker regions have lower-than-normal sublimation rates. Inevitably, this means that the cryptic region lasts longer and the bright features like the "Mountains of Mitchel" disappear earlier. The stable cap phase ($L_s = 290–340°$) is the period of minimal cap size. Any remaining CO_2 ice is considered part of the south polar residual cap (SPRC) and is discussed in Section 12.5.2.

12.4.4 The Cryptic Region, Fans, Cold CO_2 Jets, and Dunes

12.4.4.1 The Southern Seasonal Cap Cryptic Region

The cryptic (meaning "obscure" or "camouflaged") region is a dark and cold region within the southern seasonal cap that has been identified either as "dark and dirty" dry ice (Titus et al., 1998; Kieffer et al., 2000) or as translucent dry ice (Kieffer et al., 2006; Kieffer, 2007). While these authors were the first to recognize the cryptic region as dark CO_2 ice, they were not the first to have observed the region. Observations of this dark seasonal feature date back to 1845, when first seen by the American astronomer O. M. Mitchel (Blunck, 1977, 1982). This southern springtime dark region was also observed by several astronomers through the late 19th and early 20th centuries. In 1892, Barnard, Hussey, Schaeberle, and Young observed the cryptic region from $L_s \sim 210°$ to 229°. The cryptic region was again observed in 1894 by Barnard, Pickering, and Lowell ($L_s \sim 206–237°$). Fournier observed and drew an illustration ($L_s \sim 235°$) of the cryptic region in 1909. In 1924, Antoniadi (1930) observed this region, which he called Depressio Magna (~80°S, 270°W), or Big Depression. Antoniadi even speculated that this dark region inside the bright seasonal cap might be a polar lake. In addition to the observation of Depressio Magna, Antoniadi also observed a smaller dark patch, which he referred to as Depressio Parva or Little Depression (Blunck, 1977, 1982). These two dark patches correlate with the location of the cryptic region as observed by both the Viking Orbiter Infrared Thermal Mapper (IRTM) and MGS TES (Titus et al., 1998, 2008; Kieffer et al., 2000). Depressio Magna, or what is presently referred to as the cryptic region, can still be seen in the best ground-based images taken of Mars during favorable oppositions (Titus et al., 2008).

The two Viking Orbiter IRTM instruments were the first thermal IR instruments to observe the southern retreating cap (Kieffer, 1979). While observed by the IRTM, the cryptic region was not fully recognized as a region of dark CO_2 ice and was relegated to a footnote in a paper by Kieffer (1979). With the arrival of MGS, three new instruments were observing the cryptic phenomenon: a high-resolution optical camera (MOC); a thermal IR spectrometer (TES, a thermal spectrometer and bolometer, and a solar bolometer with ~3×6 km resolution); and MOLA, which, when operated as a reflectometer, could map the cap albedo at a wavelength near 1 μm.

The TES thermal and solar bolometer observations revealed a region of the springtime southern cap that was cold and dark (Titus et al., 1998; Kieffer et al., 2000). TES thermal spectra of this region showed none of the 25 μm transparency band signature that is typically observed in fine-grained CO_2 frost and snow, suggesting that the cryptic region was composed of either coarse-grained or slab CO$_2$, or dirty dry ice (Kieffer et al., 2000). While thermal IR spectral observations could not determine the form of CO$_2$ ice that composed the cryptic region, other observations provided indirect evidence that supported the concept of a coarse-grained translucent slab. For example, MOC images, with resolution as high as ~1.5 m/pixel, revealed that the cryptic region had a cornucopia of bizarre albedo features that have been referred to as a "zoo" (Kieffer et al., 2000, 2006; Kieffer, 2007). Sub-kilometer dark areas were observed within the seasonal frost. Also discovered were fields of dark, round spots with parallel-oriented tails ("fans"), fields of spots with individual or collective medium-toned halos, and fields of dark (later to become relatively light) radial and ragged branching araneiform patterns ("black spiders"), usually centered on the narrow ends of fans (Piqueux et al., 2003). Some of the spots must form in the dark, as they are observed in pre-polar dawn images (Aharonson, 2004; Kieffer, 2007).

Initial interpretations, based solely on imaging, of these dark surfaces within the polar frost were that they were defrosted surfaces and should be warm (Malin et al., 1998). However, the TES instrument observed low temperatures in the cryptic region for weeks after the formation of the spots and fans, indicating that solid CO$_2$ was still present at the surface (Kieffer et al., 2006). A small diurnal temperature variation of ~5 K indicated the possible presence of either a thin layer of surface dust or a layer of near-surface atmospheric dust (Kieffer et al., 2000; Titus and Kieffer, 2001).

The THEMIS simultaneous visual and thermal imaging of these features with 100 m resolution (or better) confirmed the hypothesis that the spots and surrounding terrain remained at solid CO$_2$ temperatures for much of the spring. Once the spots began to warm, however, they grew and coalesced quickly, forming large defrosted regions (Kieffer et al., 2006). All the albedo features disappeared when the ground warmed, except for the araneiforms, which are observed as topographical depressions when the cap is gone (Hansen et al., 2010).

After the arrival of Mars Express and MRO, NIR and short-wave mid-IR hyperspectral imaging observations became available. The OMEGA spectrometer has provided additional information about the cryptic region through the analysis of H$_2$O and CO$_2$ ice absorption features (Langevin et al., 2006, 2007). Spectral modeling has proven to be very successful in the interpretation of OMEGA observations (e.g. Langevin et al., 2006, 2007; Titus et al., 2007). As an example, the evolution of the observed spectrum from the cryptic region during the southern spring ($L_s = 197°$) could be interpreted first as an intimate mixture of dust (0.7 wt.%) and H$_2$O ice (0.06 wt.%) within the CO$_2$ ice (Langevin et al., 2006). Later ($L_s = 223°$), there was evidence for extensive contamination of the surface layers by 7 wt.% of dust but no longer any spectrally detectable H$_2$O. Finally, two weeks before the area became completely free of ices ($L_s = 242°$), the spectra were consistent with areal mixing of 25% ice-free areas and 75% ice-covered areas within each pixel (Langevin et al., 2006). Such sub-kilometer-scale ice-free and ice-covered areas are a consistent feature of regions close to the subliming cap edge at all stages of the retreat of the southern seasonal cap.

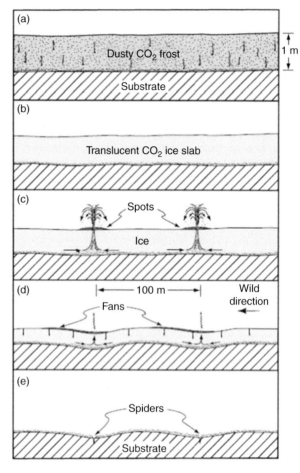

Figure 12.9. A model for the formation of polar spots, fans, and araneiforms. CO$_2$ ice formed during winter (a) self-cleans and anneals to form a translucent slab (b). Jets form as the pressure builds at the base of the slab, carrying sand and dust aloft to form dark spots (c). Dark material is transported downwind, forming fans (d). Because of the near-equality in sublimation rates between spot and spot-free regions, the entire upper surface lowers evenly (d). The high-velocity gas flow at the base of the ice erodes the araneiforms whose center lies beneath a spot (d). Eventually the ice is completely removed and the dark granular material is back on the surface (e). This is figure 4 from Kieffer et al. (2006).

12.4.4.2 The Cold Jet Hypothesis

A series of unanticipated observations by the MGS TES and MOC, with confirmation by THEMIS (Kieffer et al., 2006) and OMEGA (Langevin et al., 2006), has led to a complex model for surface processes in the seasonal polar cap (Figure 12.9). A possible explanation pivots on the physical and optical properties of solid CO$_2$, which has a very small absorption coefficient in the solar wavelength region and becomes opaque to thermal radiation within a few millimeters thickness (Hansen, 1997, 2005; Kieffer, 2007). H$_2$O and dust are also involved in this proposed explanation that is summarized by a brief description of the annual sequence based on the discussion by Kieffer (2000), and modeled in detail by Kieffer (2007), Pilorget et al. (2011), and Portyankina et al. (2010). The venting model is the best candidate to explain the spots, fans, and araneiforms observed at different times and places in the seasonal cap.

Deep in the Martian polar night there is some CO_2 snowfall, but most of the surface is covered by solid CO_2 that is deposited in the form of a uniform, continuous, non-scattering slab with embedded dust and H_2O ice grains. While the exact ratio of snowfall to direct condensation is being debated, analysis by Piqueux et al. (2003) has shown that much of the southern cap has a dark albedo during early spring, suggesting either dirty ice or slab ice. Pre-polar-dawn seasonal heat flow (heat stored in the regolith from the summer) sublimes some CO_2 ice from the bottom of the slab, and results in some "vents" forming and releasing this gas (Aharonson, 2004; Pilorget et al., 2011).

Following seasonal sunrise, in some areas the seasonal CO_2 ice brightens due to fracturing or surficial frosting (either CO_2 or H_2O cold trapping), but in other areas (specifically the cryptic region) the dark ice slab persists. Within the dark dirty CO_2 ice regions, the solar energy is largely absorbed by embedded dust grains (which either burrow downward or escape upward), cleaning the CO_2 slab. The largest dust grains would burrow down to the surface underlying the ice, while the smallest dust grains would blow away with the sublimation winds (Kieffer, 2007). Small holes created near the surface quickly anneal, forming a translucent slab (Kieffer, 2007). Sunlight then penetrates to the bottom of the slab, warming the soil and subliming ice from the bottom. The pressure from the CO_2 ice sublimation levitates the ice slab and may force gas into the subsurface (Kieffer, 2007). Vents develop in the slab, which allows the high-pressure gas to escape. As this sub-slab gas converges toward the vents, the gas scours the soil surface along ragged channels (araneiforms). Laboratory experiments using glass beads in a Hele-Shaw cell demonstrate that Martian araneiforms can be formed by these jets/vents (de Villiers et al., 2012). If the gas velocity is sufficiently high, the moving gas will capture and transport dust within the flow. Dust entrained in the jetting gas falls out and downwind to form "fans".

As this process continues throughout the early spring, the dust that forms the fans starts to form an optically thick layer on the surface of the ice. OMEGA spectroscopic observations suggest that by mid-spring ($L_s \sim 220°$) the surface of the CO_2 ice appears to be strongly contaminated by dust, with few photons making it to the underlying surface, suggesting that most venting activity has ceased. Dust may also be deposited on the surface shortly after equinox by storm systems originating from the Hellas region (Colaprete et al., 2005; Langevin et al., 2006).

As the slab thins and the sub-slab pressure drops, either gas enriched in water vapor is released from the subsurface to recondense as water frost inside the sub-ice channels (Kieffer, 2007) or some of the CO_2 recondenses from adiabatic cooling as the pressure is released (Titus et al., 2007). In either case, the araneiforms appear to brighten. As soon as the CO_2 slab is gone, the remaining water frost warms and evaporates. Only the topographic ghosts of the araneiforms persist through the summer. Thus, seasonal reworking of the soil below the ice slab may help to explain the uniformity and youthful appearance of the terrain upon which the vents form (Piqueux and Christensen, 2008a).

12.4.4.3 Cryptic Region Location

Why is the cryptic region located where it is on the seasonal cap? Initial attempts to find correlations of the cryptic region's location with local geologic features failed (Kieffer et al., 2000). A majority of the cryptic region lies over the PLDs, but it also extends over other geologic units. Intra-seasonal and inter-annual variations in the shape of the cryptic region indicate that atmospheric dynamics might play a role. Indeed, this hypothesis has gained additional support from mesoscale modeling that has demonstrated that regional topography (primarily the Hellas Basin) sets up an $N = 1$ zonal standing wave during the southern winter (Colaprete et al., 2005). This standing wave appears to cause two distinct microclimates to form over the southern polar region: the western hemisphere (longitudes 210–30°E, south and southwest of Argyre) is colder, cloudier, and has more snowfall, while the eastern hemisphere (longitudes 30–210°E, south and southeast of Hellas) is slightly warmer and wetter (higher water vapor abundance). The cryptic region lies in the "warmer and wetter" microclimate, where there is less snowfall and enhanced radiative loss to space due to increased surface emissivity. Oddly enough, this would suggest that the cryptic region would have additional CO_2 ice accumulation, but analysis from Jian and Ip (2009) suggests that this is not the case. The greater availability of H_2O ice in the eastern hemisphere (the Prometheus climate zone), while significantly smaller than the northern hemisphere, may be key to the formation of jets and fans. Experiments conducted by Laufer et al. (2013) found that pure CO_2 ice quietly sublimes, while CO_2 ice slabs contaminated by small amounts of H_2O did exhibit venting action.

The detailed analysis of the CO_2 cycle, at the regional and local scales, suggests that the surface is modified, e.g. formation of araneiforms or reworking of the surface dust through fan formation. While these small-scale processes may not directly impact the bulk Mars atmosphere, their presence under modern climate conditions may hold clues as to past processes that formed the polar layered deposits (PLD). If this is the case, then the processes that form the cryptic region today may have formed the PLD in the past.

12.4.4.4 Jets and Fans in the North

This same process, the solar heating of the regolith through a translucent slab of CO_2 ice, has been used to explain the observations of fans on springtime dunes in the north and the south (Hansen et al., 2013; Pommerol et al., 2013; Portyankina et al., 2013). The vents are typically found in three locations where the slab ice appears to be weakest: (1) the dune crest, the boundary between the dunes and inter-dune substrate, and through polygonal cracks in the ice (Hansen et al., 2013). Just as in the southern cryptic region, the formation of fans (and thus jet activity) is largely restricted to early spring (Pommerol et al., 2013).

12.5 LONG-TERM CO₂ RESERVOIRS AND THEIR ROLES ON PRESENT-DAY MARS

Much recent attention has focused on the atmospheric and seasonal frost CO_2 reservoirs, and the geologic changes they have driven in the recent past. Atmospheric CO_2 can mobilize surface dust through direct lifting and/or dust devils and change the albedo pattern of the planet as a whole, which can drive climate change of a sort (Geissler, 2005; Fenton et al., 2007). Consideration of the timing of these albedo measurements relative to global dust storms suggests that it is these discrete events that dominate regional brightening and darkening on Mars (Szwast et al., 2006). Additionally, Szwast et al. (2006) show that locations experiencing a seasonal CO$_2$ mantle have their albedo values reset on an annual basis. Inter-annual HiRISE observations have ended decades of uncertainty by showing that dunes and ripples of sand-sized material are also actively migrating under current conditions (Silvestro et al., 2010; Bridges et al., 2012).

Although the geologic effects of the atmosphere and seasonal frost may accumulate over time, the reservoirs themselves are reset each Martian year. However, CO$_2$ is also stored in other reservoirs that interact with the atmosphere–frost system on considerably longer timescales than annually. CO$_2$ has been speculated to exist as adsorbed onto regolith grains or as clathrates. The potential magnitude of these reservoirs is difficult to assess, as neither has actually been demonstrated to exist. CO$_2$ may also be locked up as carbonates, and therefore no longer part of the modern CO$_2$ cycle. Carbonates have been detected within surface rocks (Ehlmann et al., 2008, 2011; Brown et al., 2010; Sutter et al., 2012); however, these detections are not widespread. The total amount of previously gaseous CO$_2$ that may be currently bound up in rocks is difficult to estimate and it has been suggested that most of the carbonates on Mars are deeply buried and visible only in areas where deep rocks have been uplifted, such as impact crater central peaks (Michalski and Niles, 2010).

12.5.1 Residual Caps and Polar Layered Deposits

Apart from the seasonal frost, two main reservoirs of CO$_2$ ice are known on Mars. The current southern residual cap has been known to be composed of CO$_2$ for some time (Kieffer, 1979). It is volumetrically minor compared with the current atmosphere–frost system (Byrne and Ingersoll, 2003; Tokar et al., 2003; Prettyman et al., 2004) and has areas at its margins and in its interior where underlying water ice shows through (Titus et al., 2003; Bibring et al., 2004; Hansen et al., 2005). Beneath this water ice, a buried reservoir of CO$_2$ ice has recently been discovered by the Shallow Radar (SHARAD) sounder (Phillips et al., 2011) that rests upon the south polar layered deposits with a volume estimated to be comparable to the current atmosphere–frost system. These two reservoirs are described in more detail in the following sections.

The north and south polar layered deposits were originally thought to not contain significant quantities of CO$_2$ ice. Thermal arguments (Mellon, 1996) preclude deeply buried CO$_2$ ice or clathrate (however, this was based on planetary heat flow estimates now thought to be too high (Phillips et al., 2008)). The measured strength of CO$_2$ ice (Durham et al., 1999) is low enough to preclude the preservation of a topographic structure with multi-kilometer relief over timescales similar to the age of the SPLD (Nye et al., 2000). The CO$_2$ reservoir discovered by Phillips et al. (2011) sits on top of the SPLD (rather than being at its base) and so is thermodynamically stable. It is also only

a few hundred meters thick and may be much younger than the SPLD, and so circumvents the mechanical constraints of Nye et al. The north polar layered deposits are thought to be much younger (Koutnik et al., 2002; Banks et al., 2010); however, the lack of any evidence of flow in radar-imaged internal layers argues for a strong (and so CO$_2$-poor) composition (Karlsson et al., 2011). A complete review of the polar layered deposits from a geological perspective can be found in Byrne (2009).

12.5.2 The Residual South Polar Cap

It has been known from centuries of telescopic observations that the seasonal cap in the southern hemisphere retreats to leave a much smaller bright deposit, displaced from the pole by about 200 km, that persists throughout the summer. Early spacecraft thermal observations showed that this cap was, like the seasonal caps, composed of CO$_2$ ice (Kieffer, 1979). Several unanswered questions concerning this residual CO$_2$ cap immediately presented themselves. Why is it centered 200 km from the south pole? Why would CO$_2$ accumulate at the higher-elevation south pole rather than the north, where higher atmospheric pressures favor the formation of CO$_2$ ice? Indeed, the residual cap not only sits at the higher pole, but also sits at the highest topographic point in the south polar region. Finally, the size of this reservoir was of pressing interest in that the Leighton and Murray (1966) model suggested atmospheric pressures were buffered by a large surface reservoir.

Modeling work has shown that the stability of this residual CO$_2$ cap critically depends on its high albedo (Jakosky and Haberle, 1990) and that two stable states (frosted or defrosted) exist. Blackburn et al. (2010) predicted the disappearance of the residual cap within a few Martian years when using a very low albedo in the spring and early summer. However, such a rapid change is unlikely given the length of time we have been observing this feature. If the residual cap were to become defrosted, then heat stored in the subsurface would ensure less CO$_2$ accumulation the following winter and another full defrosting the following year. This may have happened in some locations at the periphery of the cap. Titus et al. (2003) and Bibring et al. (2004) showed exposed water ice at the edges of the cap, whereas further away the terrain is mantled by a thermally insulating and optically thick material. Small windows to the underlying terrain within the perennial cap also show this underlying water ice (Byrne and Ingersoll, 2003; Bibring et al., 2004), so a reasonable interpretation is that recent retreat of the perennial CO$_2$ cover has exposed this peripheral water ice (Piqueux et al., 2008). The timing of such a retreat is unknown, although Piqueux and Christensen (2008b) note only minimal changes in the polar cap's extent since the Mariner 9 mission (in 1972). Telescopic observations in 1969 of enhanced water vapor above the residual cap in late southern summer have led many to speculate that much of the cap was removed that year (Barker et al., 1970; Jakosky and Barker, 1984). Mariner 9 observations showed a relatively variegated late summer cap one Mars year later (possibly the result of a global dust storm that had just cleared); however, by the time the Viking missions arrived (two Mars years after Mariner 9), the cap appeared uniform and brighter (James et al., 2001). Blackburn et al. (2010) estimated that the current SPRC is subliming at a rate of 0.4 m of thickness per year; if this rate continues, the "permanent" CO_2 ice will disappear in 6–7 Mars years. This estimate is not consistent with other observations that show little change (or even expansion) in the areal extent of the cap (e.g. Winfree and Titus, 2007). However, a cyclic disappearance and re-emergence of the SPRC, as suggested by Byrne et al. (2008), would be consistent with most observations.

Large-grained CO_2 ice, such as that seen in the cryptic terrain (Kieffer and Titus, 2001), has a lower albedo than that of a snow-pack composed of smaller grains. The high albedo of the perennial CO_2 ice throughout the year implies the presence of fresh frost up until winter condensation begins anew, and so the cap is undergoing net accumulation. Although highly repetitive by terrestrial standards, Martian meteorology does not repeat exactly each year, and so net accumulation may not be the case every year. Observations by Mariner 9 in the wake of a global dust storm showed the cap albedo to be much lower and variegated than seen by subsequent missions (James et al., 2007, and references therein). Modeling of the radiative effects of atmospheric dust (Bonev et al., 2008) confirms that dusty years can result in a net loss of material. Paradoxically, the seasonal and residual caps were brighter in the year following the 2001 global dust storm (Byrne et al., 2008), which suggests that atmospheric dust may also aid subsequent condensation, perhaps by providing high-altitude condensation nuclei during polar night. The off-pole location of the cap may also owe itself to meteorological effects.

High-resolution imagery of the residual cap showed a CO_2 ice slab that was punctuated by flat-floored quasi-circular pits with diameters from a few meters up to a kilometer (Thomas et al., 2000). Repeat imagery of these pits showed that their walls darken when defrosted during the late summer (Malin and Edgett, 2001) and retreat by several meters per Martian year (Malin et al., 2001b). These expanding pits have been subdivided into populations that penetrate thick (~8 m) and thin (~2 m) slabs, and these populations have faster- and slower-than-average expansion rates, respectively (Thomas et al., 2005). In most locations, only (rapidly shrinking) remnants of the original 8 m thick CO_2 slab remain; however, near 0°E a broad expanse of this thick ice is preserved and contains a population of pits with a narrow size distribution (Byrne and Ingersoll, 2003). Their measured expansion rates and sizes suggest they began forming at this location close to 1940. However, most of the residual cap presents a complicated picture where overlapping slabs of CO_2 ice of different thicknesses contain pits of different sizes (Thomas et al., 2005). Piqueux and Christensen (2008b) have attempted to use historical telescopic and spacecraft data to reconstruct a history of the residual cap that involves at least two major episodes of erosion.

The growth of these pits presents somewhat of a conundrum because their lateral expansion should rapidly (when compared with orbital element variations) consume the remaining CO_2 ice. Although this expansion prompted initial suggestions of climate change, it seems more likely that this ice loss is at least partly offset by net condensation onto the bright surfaces. Byrne et al. (2008) modeled an accumulating CO_2 landscape and found surface roughness to steadily increase until pits could spontaneously form. Once formed, these pits always expand due to the high insolation absorbed by their steep walls. By controlling surface roughness through allowing CO_2 snowfall to drift into surface hollows, accumulation of a new CO_2 slab can proceed.

Accumulation of these slabs and their subsequent destruction by expanding pits combine to yield a cyclic polar cap (over century timescales). An alternative model, suggested by Line and Ingersoll (2010), allows for penetration of sunlight into a semi-transparent accumulating CO_2 ice slab. Energy absorbed within the slab causes internal sublimation and increases porosity until collapse features form on the surface. Once formed, these pits grow as before. Interestingly, both approaches yield repeating polar caps over 100 year timescales; both would interpret different parts of the current residual cap to be at different phases in their life cycle; and neither requires any climate change to explain the current observations.

12.5.3 A Buried Polar CO₂ Reservoir

Although the residual cap is a small reservoir and accounts for probably no more than a few percent of the current atmosphere (Byrne and Ingersoll, 2003; Thomas et al., 2009), there exists evidence for a larger south polar CO_2 reservoir comparable in mass to the current atmosphere (Phillips et al., 2011). Radar sounding by the SHARAD instrument has revealed several reflection-free zones beneath the south polar layered deposits. One of these zones occurs directly beneath the current residual cap and an intervening water ice layer. SHARAD also detected layered material below this reflection-free zone where the individual strata exhibited a large amount of relief in time-delay space. In order for these layers to appear flat in physical-depth space (as the better resolved north polar layers do (Phillips et al., 2008)), it is necessary for the overlying reflection-free material to have a dielectric constant that implies pore-free CO_2 ice. This reflection-free zone closely corresponds to a geologic unit (Aa3) mapped on the basis of surface morphology (Kolb et al., 2006; Tanaka et al., 2007), which contains troughs and rimless pits interpreted as collapse features. Extrapolating the thickness of the reflection-free zone over the extent of Aa3 yields a mass of CO_2 equivalent to ~80% of the current atmosphere. The effect of this reservoir on the general atmospheric circulation would only be modest. Seasonal caps would become slightly larger and more long-lived, a result that would drop mean global temperatures and which is partly offset by a weak increase in the greenhouse effect (Phillips et al., 2011). However, the effect of this additional atmospheric CO_2 on the thermodynamics of water could be significant, by raising mean pressures to ~10 mbar, so that melting becomes possible (given an energy source) over much more of the planet's surface.

The emplacement of this unit has not yet been thoroughly investigated. It is expected that low obliquities may cause atmospheric collapse (Newman et al., 2005) at the poles; however, such ice could be sublimated back into the atmosphere in subsequent periods of high obliquity (assuming the CO_2 ice deposit is not capped by an importable impermeable layer of water ice). The last obliquity low that was not preceded by an obliquity high occurred about ~350 kyr ago (Laskar et al., 2004), although the obliquity only dipped to ~20°. Lower obliquities have been reached over the past million years (most notably ~800 kyr ago when it fell to 15°) (Laskar et al., 2004), but were always followed by obliquity highs. However, in contrast to most of the past few million years, obliquity variations have not been very strong in the past ~200 kyr, so argument of perihelion variations may still have exerted some control over the deposition of this material.

There are few independent constraints on the age of this deposit. It has been suggested on the basis of GCM modeling that the water ice layer that separates this deposit from the residual cap was laid down ~20 kyr ago (Montmessin et al., 2007), which would make material in the reflection-free zone at least that old. The surface of the rest of the south polar layered deposits is thought to be tens of millions of years in age based on crater counts (Koutnik et al., 2002; Plaut, 2005), which provides a loose upper limit.

12.6 CURRENT STATE OF KNOWLEDGE AND OUTSTANDING QUESTIONS

For nearly half of a century, a suite of spacecraft and instruments have greatly improved our understanding of the Martian CO_2 cycle. However, with each increment of new understanding, there come a flood of additional questions. There are as many unresolved issues in Martian polar science as ever, and new data seem to only deepen certain enigmas. In addition to the questions presented here, other unresolved Mars polar issues are discussed in Titus and Michaels (2009) and Clifford et al. (2013).

Current Understanding of the CO₂ Cycle

Annual Cycle and Inter-Annual Variation

- Some 25–30% of the atmosphere cycles through the seasonal caps with little global inter-annual variation. Local inter-annual variations do occur, but these variations tend to cancel each other out, e.g. the increase in sublimation of bright CO_2 ice during dust storms is canceled by the decrease in sublimation rate of the dark CO_2 regions (Bonev et al., 2002).
- The CO_2 ice that forms the seasonal polar caps is deposited by both direct surface deposition and accumulation of CO_2 snow. Once the CO_2 ice is on the surface, the effective grain size (or more appropriately porosity for slab ice) can evolve through a variety of processes.
- The Mars seasonal caps have ranges of albedo and emissivity that vary both spatially and temporally. These variations are due to changes in effective grain size and concentrations of contaminants (e.g. H_2O ice, dust).
- The northern seasonal cap, as viewed in the visible, is larger than the portion of the seasonal cap that is in thermal contact with CO_2 ice. This is due to the formation of an H_2O ice annulus that surrounds the CO_2 ice cap and varies in zonal width from 60 km to 420 km (Kieffer and Titus, 2001; Wagstaff et al., 2008; Appéré et al., 2011).

Outstanding Questions on the CO₂ Cycle

Mysteries of CO₂ Ice Properties Under Mars Conditions

- Past experimental work has focused on the optical constants of CO_2 ice under Mars conditions (e.g. Hansen, 1997). While these optical constants are now adequately known, other

areas require new laboratory studies. These include further study of CO_2 ice behavior, e.g. how grain size changes under polar night conditions and under solar illumination, or more studies concerning the physics of CO_2 jets from solid greenhouse effects.

Mysteries of the Present CO₂ Cycle in a Typical Martian Year

- We are still very much in the dark about what happens in the polar night; this includes much of the details of the condensation process.
- What are the processes that determine condensation mode?
- What role do CO_2 ice clouds play in the CO_2 cycle and where do the clouds form?
- What are the processes that cause density changes in surface CO_2 ice?
- What is the role of the polar hood? While the polar hood is composed of water ice clouds, do these clouds modify the energy balance of the CO_2 cycle by reducing CO_2 condensation and reducing radiative losses to space?

Mysteries of the Inter-Annual Variability in the CO₂ Cycle

- How do global dust storms affect the CO_2 cycle?
- What climatic information (if any) is stored in the residual cap?
- What is the nature of the CO_2 ice deposits ("Swiss cheese", "slab ice", "cryptic regions", etc.) and what are the detailed processes responsible for their evolution?
- Is the SPRC eroding, and, if so, where is the CO_2 going and on what timescale?
- How much CO_2 ice covers the southern residual cap, and how often is the underlying H_2O ice cap exposed?

Mysteries of the Effects of Shifting Orbital Elements on the CO₂ Cycle

- Understanding seasonal and longer-scale obliquity variations and their connections to the fine laminations of the polar layered deposits.
- Further validation, characterization, and modeling of the SPLD buried CO_2 ice deposit is needed. Several questions exist, e.g. how old is the buried CO_2 ice unit? Is this unit stable under modern climatic conditions?

Need for Future Missions

While our understanding of the Martian CO_2 cycle has increased during the golden age of Mars exploration, a continued effort is needed to monitor the CO_2 cycle with both instruments that maintain a continuous history with legacy datasets and those with new capabilities. Current instrumentation used to monitor the CO_2 cycle are part of an aging fleet of spacecraft (M01, MEX, MRO). MAVEN, the newest spacecraft to arrive at Mars, is not designed for polar monitoring due to its instrumentation and orbit.

The polar night is of particular interest because half of the CO_2 cycle is occurring in the dark. Legacy instruments such as TES or MCS will monitor some activity, but cannot measure albedo. Flying an imaging lidar system where the lasers are tunable to the spectral features of CO_2 and H_2O ices would provide the next step in understanding (Brown et al., 2014). Additional suggestions for future instrumentation are covered by Titus and Michaels (2009) and Titus et al. (2010a,b).

ACKNOWLEDGMENTS

We would like to thank Adrian Brown, Paul Hayne, and Michael Smith for their insightful and extremely helpful reviews. We would also like to thank Glen Cushing and Laz Kestay for their insightful review comments as part of the USGS internal review process. Discussions with Hugh Kieffer were quite helpful. We would also like to recognize the late Gary Hansen, whose life work has contributed so much to the understanding of CO_2 ice on Mars.

REFERENCES

Aharonson, O. (2004) Sublimation at the base of a seasonal CO_2 slab on Mars, 35th Lunar and Planetary Science Conference, March 15–19, 2004, League City, Texas, abstract no. 1918.

Aharonson, O., Zuber, M. T., Smith, D. E., et al. (2004) Depth, distribution, and density of CO_2 deposition on Mars, *J. Geophys. Res.*, 109, E05004, doi:10.1029/2003JE002223.

Antoniadi, E. M. (1930) *The Planet Mars*, Trans. Patrick Moore, Devon, UK: Keith Reid Ltd., 1975.

Appéré, T., Schmitt, B., Langevin, Y., et al. (2011) Winter and spring evolution of northern seasonal deposits on Mars from OMEGA on Mars Express, *J. Geophys. Res.*, 116, E5, E05001.

Baker, V. R. (1979) Erosional processes in channelized water flows on Mars, *Journal of Geophysical Research*, 84, 7985–7993.

Bandfield, J. L. and W. C. Feldman (2008) Martian high latitude permafrost depth and surface cover thermal inertia distributions, *J. Geophys. Res.*, 113, E08001.

Bandfield, J. L., Glotch, T. D., Christensen, P. R. (2003) Spectroscopic identification of carbonate minerals in the Martian dust, *Science*, 301, 5636, 1084–1087.

Banks, M. E., Byrne, S., Galla, K., et al. (2010) Crater population and resurfacing of the Martian north polar layered deposits, *J. Geophys. Res.*, 115, E8, E08006.

Barker, E. S., Schorn, R. A., Woszczyk, A., et al. (1970) Mars: detection of atmospheric water vapor during the southern hemisphere spring and summer season, *Science*, 170, 3964, 1308–1310.

Bell, J. F., Wolff, M. J., Malin, M. C., et al. (2009) Mars Reconnaissance Orbiter Mars Color Imager (MARCI): instrument description, calibration, and performance, *J. Geophys. Res.*, 114, E8, CiteID E08S92.

Benson, J. L. and James, P. B. (2005) Yearly comparisons of the Martian polar caps: 1999–2003 Mars Orbiter Camera observations, *Icarus*, 174, 513–523, doi:10.1016/j.icarus.2004.08.025.

Benson, J., Kass, D. M., Kleinböhl, A., et al. (2010) Mars' south polar hood as observed by the Mars Climate Sounder, *J. Geophys. Res.*, 115, E12015, doi:10.1029/2009JE003554.

Benson, J. L., Kass, D. M., Kleinböhl, A. (2011) Mars' north polar hood as observed by the Mars Climate Sounder, *J. Geophys. Res.*, 116, E3, E03008.

Bertaux, J.-L., Fonteyn, D., Korablev, O., et al. (2000) The study of the Martian atmosphere from top to bottom with SPICAM light on Mars Express, *Planetary and Space Science*, 48, 12–14, 1303–1320.

Bibring, J.-P., Langevin, Y., Poulet, F., et al. (2004) Perennial water ice identified in the South polar cap of Mars, *Nature*, 428, 627–30.

Bibring, J.-P., Langevin, Y., Gendrin, A., et al. (2005) Mars surface diversity as revealed by the OMEGA/Mars Express observations, *Science*, 307, 1576–1581, doi:10.1126/science.1108806.

Blackburn, D. G., Bryson, K. L., Chevrier, V. F., et al. (2010) Sublimation kinetics of CO_2 ice on Mars, *Planetary and Space Science*, 58(5), 780–791. doi:10.1016/j.pss.2009.12.004

Blunck, J., (1977) *Mars and Its Satellites: A Detailed Commentary on the Nomenclature*, Exposition Press, Hicksville, N.Y.

Blunck, J., (1982) *Mars and Its Satellites: A Detailed Commentary on the Nomenclature*, 2nd edn., Exposition Press, Hicksville, NY.

Bonev, B. P., James, P. B., Bjorkman, J. E., and Wolff, M. J. (2002) Regression of the Mountains of Mitchel polar ice after the onset of a global dust storm on Mars, *Geophys. Res. Lett.*, 29, 21, 2017, doi:10.1029/2002GL015458.

Bonev, B. P., Hansen, G. B., Glenar, D. A., et al. (2008) Albedo models for the residual south polar cap on Mars: implications for the stability of the cap under near-perihelion global dust storm conditions, *Planetary and Space Science*, 56, 2, 181–193.

Boynton, W. V., Feldman, W. C., Squyres, S. W., et al. (2002) Distribution of hydrogen in the near surface of Mars: evidence for subsurface ice deposits, *Science*, 297, 5578, 81–85.

Boynton W., Feldman W., Mitrofanov I., et al. (2004) The Mars Odyssey gamma-ray spectrometer instrument suite, *Space Science Reviews*, 110, 1, 37–83.

Bridges, N. T., Ayoub, F., Avouac, J.-P., et al. (2012) Earth-like sand fluxes on Mars, *Nature*, 485, 7398, 339–342.

Briggs, G. A. (1974) The nature of the residual Martian polar caps, *Icarus*, 23, 167–191.

Brown, A. J., Calvin, W. M., McGuire, P. C., et al. (2010a) Compact Reconnaissance Imaging Spectrometer for Mars (CRISM) south polar mapping: first Mars year of observations, *J. Geophys. Res.*, 115, E00D13, doi:10.1029/2009JE003333.

Brown, A. J., Hook, S. J., Baldridge, A. M., et al. (2010b) Hydrothermal formation of clay–carbonate alteration assemblages in the Nili Fossae region of Mars, *Earth and Planetary Science Letters*, 297, 174–182.

Brown, A. J., Calvin, W. M., and Murchie, S. L. (2012) Compact Reconnaissance Imaging Spectrometer for Mars (CRISM) north polar springtime recession mapping: first 3 Mars years of observations, *J. Geophys. Res.*, 117, E00J20.

Brown, A. J., Michaels, T. I., Byrne, S., et al. (2014) The case for a modern multiwavelength, polarization-sensitive LIDAR in orbit around Mars, *Journal of Quantitative Spectroscopy and Radiative Transfer*, 153, 131–143.

Byrne, S. (2009) The polar deposits of Mars, *Annual Review of Earth and Planetary Sciences*, 37, 1, 535–560.

Byrne, S. and Ingersoll, I. (2003) A Sublimation Model for Martian South Polar Ice Features, *Science*, 299, Issue 5609, 1051–1053.

Byrne, S., Zuber, M. T., and Neumann, G. A. (2008) Interannual and seasonal behavior of Martian residual ice-cap albedo, *Planetary and Space Science*, 56, 2, 194–211.

Cabrol, N. A., Grin, E. A., Fike, D. (2002) Gusev Crater: a landing site for MER A, in *33rd Annual Lunar and Planetary Science Conference*, March 11–15, Houston, Texas, abstract no. 1142.

Carr, M. H. (1979) Formation of Martian flood features by release of water from confined aquifers, *Journal of Geophysical Research*, 84, 2995–3007.

Carr, M. (1983) Stability of streams and lakes on Mars, *Icarus*, 56, 476–495.

Chamberlain, T. E., Cole, H. L., Dutton, R. G., et al. (1976) Atmospheric measurements on Mars: the Viking meteorology experiment, *American Meteorological Society Bulletin*, 57, 1094–1104. doi:10.1175/1520-0477(1976)057.

Christensen, P. R., Anderson, D. L., Chase, S. C., et al. (1992) Thermal emission spectrometer experiment – Mars Observer mission, *J. Geophys. Res.*, 97, E5, 7719–7734.

Christensen, P. R., Jakosky, B. M., Kieffer, H. H., et al. (2004) The Thermal Emission Imaging System (THEMIS) for the Mars 2001 Odyssey Mission, *Space Science Reviews*, 110, 1, 85–130.

Clancy, R. T., Sandor, B. J., Wolff, M. J., et al. (2000) An intercomparison of ground-based millimeter, MGS TES, and Viking atmospheric temperature measurements: seasonal and interannual variability of temperatures and dust loading in the global Mars atmosphere, *J. Geophys. Res.*, 105, E4, 9553–9572.

Clancy, R. T., Sandor, B. J., Moriarty-Schieven, G. H. (2004) A measurement of the 362 GHz absorption line of Mars atmospheric H$_2$O$_2$, *Icarus*, 168, 1, 116–121, doi:10.1016/j.icarus.2003.12.003.

Clancy, R. T., Wolff, M. J., Whitney, B. A., et al. (2007) Mars equatorial mesospheric clouds: global occurrence and physical properties from Mars Global Surveyor Thermal Emission Spectrometer and Mars Orbiter Camera limb observations, *J. Geophys. Res*, 112, E4, CiteID E04004, doi:10.1029/2006JE002805.

Clifford, S. M., Yoshikawa, K., Byrne, S., et al. (2013) Introduction to the fifth Mars Polar Science special issue: key questions, needed observations, and recommended investigations, *Icarus*, 225, 2, 864–868.

Colaprete, A. and Toon, O. B. (2002) Carbon dioxide snow storms during the polar night on Mars, *J. Geophys. Res. (Planets)*, 107, E7, 5-1–5-16, doi:10.1029/2001JE001758.

Colaprete, A., Barnes, J. R., Haberle, R. M., et al. (2005) Albedo of the south pole on Mars determined by topographic forcing of atmosphere dynamics, *Nature* 435, 184–8.

Colaprete, A., Barnes, J. R., Haberle, R. M., and Montmessin, F. (2008) CO$_2$ clouds, CAPE and convection on Mars: observations and general circulation modeling. *Planetary and Space Science* 56, 2, 150–180.

Conrath, B., Curran, R., Hanel, R., et al. (1973) Atmospheric and surface properties of Mars obtained by infrared spectroscopy on Mariner 9, *Journal of Geophysical Research*, 78, 20, 4267–4278.

Cornwall, C. and Titus, T. N. (2009) Spatial and temporal distributions of Martian north polar cold spots before, during, and after the global dust storm of 2001, *J. Geophys. Res.*, 114, E2, CiteID E02003.

Cornwall, C.; Titus, T. N. (2010) A comparison of Martian north and south polar cold spots and the long-term effects of the 2001 global dust storm, *J. Geophys. Res.*, 115, E6, CiteID E06011.

Cull, S., Arvidson, R. E., Mellon, M., et al. (2010a) Seasonal H$_2$O and CO$_2$ ice cycles at the Mars Phoenix landing site: 1. Prelanding CRISM and HiRISE observations, *J. Geophys. Res*, 115, 1, CiteID E00D16.

Cull, S., Arvidson, R. E., Morris, R. V., et al. (2010b) Seasonal ice cycle at the Mars Phoenix landing site: 2. Postlanding CRISM and ground observations, *J. Geophys. Res.*, 115, 1, E00E19.

Cushing, G. and Titus, T. (2008) MGS-TES thermal inertia study of the Arsia Mons caldera, *J. Geophys. Res.*, 113, E6, E06006.

Davies, D. W.; Farmer, C. B.; Laporte, D. D. (1977) Behavior of volatiles in Mars' polar areas – a model incorporating new experimental data, *J. Geophys. Res.*, 82, 3815–3822.

de Villiers, S., A. Nermoen, B. Jamtveit, et al. (2012) Formation of Martian araneiforms by gas-driven erosion of granular material, *Geophys. Res. Lett.*, 39, L13204, doi:10.1029/2012GL052226.

Diniega, S., Byrne, S., Bridges, N. T., et al. (2010) Seasonality of present-day Martian dune-gully activity, *Geology*, 38, 11, 1047–1050.

Diniega, S., Hansen, C. J., McElwaine, J. N., et al. (2013) A new dry hypothesis for the formation of Martian linear gullies, *Icarus*, 225, 1, 526–537.

Ditteon, R. and Kieffer, H. H. (1979) Optical properties of solid CO$_2$: application to Mars, *J. Geophys. Res.*, 84, 8294–8300, doi:10.1029/JB084iB14p08294.

Douté, S., Schmitt, B., Langevin, Y., Bibring, J.-P., et al. (2007) South pole of Mars: nature and composition of the icy terrains from Mars Express OMEGA observations, *Planet. Space Sci.*, 55, 113–133.

Dundas, C. M., McEwen, A. S., Diniega, S., et al. (2010) New and recent gully activity on Mars as seen by HiRISE, *Geophys. Res. Lett.*, 37, 7, CiteID L07202.

Durham, W. B., Kirby, S. H., and Stern, L. A. (1999) Steady-state flow of solid CO2: preliminary results, *Geophys. Res. Lett.*, 26, 3493–3496.

Ehlmann, B. L., Mustard, J. F., Murchie, S. L., et al. (2008) Orbital identification of carbonate-bearing rocks on Mars, *Science*, 322, 1828–1832.

Ehlmann, B. L., Mustard, J. F., Murchie, S. L., et al. (2011) Subsurface water clay mineral formation during the early history of Mars, *Nature*, 479, 53–60.

Eluszkiewicz, J., Moncet, J.-L., Shephard, M. W., et al. (2008) Atmospheric and surface retrievals in the Mars polar regions from the Thermal Emission Spectrometer measurements, *J. Geophys. Res.*, 113, E10, E10010.

Fanale, F. P., and Salvail, J. R. (1994) Quasi-periodic atmosphere–regolith–cap CO2 redistribution in the Martian past, *Icarus*, 111, 2, 305–316.

Feldman, W. C., Boynton, W. V., Tokar, R. L., et al. (2002) Global distribution of neutrons from Mars: results from Mars Odyssey, *Science* 297, 5578, 75–78, doi:10.1126/science.1073541.

Feldman, W., Prettyman, T., Boynton, W., et al. (2003) CO$_2$ frost cap thickness on Mars during northern winter and spring, *J. Geophys. Res.*, 108, E9, 5103.

Feldman, W. C., Prettyman, T. H., Maurice, S., et al. (2004) Global distribution of near-surface hydrogen on Mars, *J. Geophys. Res.*, 109, E09006, doi:10.1029/2003JE002160.

Fenton, L. K., Geissler, P. E., and Haberle, R. M. (2007) Global warming and climate forcing by recent albedo changes on Mars, *Nature*, 446, 7136, 646–649.

Fischbacher, G. E., Martin, L. J., and Baum, W. A. (1969) Martian Polar Cap Boundaries. Final Report A, Contract 951547, Jet Propulsion Laboratory, Pasadena, CA. Planetary Research Center, Lowell Observatory, Flagstaff, AZ.

Folkner, W. M., Yoder, C. F., Yuan, D. N., Standish, E. M., Preston, R. A. (1997) Interior structure and seasonal mass redistribution of Mars from radio tracking of Mars Pathfinder, *Science*, 278, 5344, 1749.

Forget, F. and Pollack, J. B. (1996) Thermal infrared observations of the condensing Martian polar caps: CO$_2$ ice temperatures and radiative budget, *J. Geophys. Res.*, 101, E7, 16865–16880.

Forget, F., Hansen, G. B., Pollack, J. B. (1995) Low brightness temperatures of Martian polar caps: CO$_2$ clouds or low surface emissivity? *J. Geophys. Res.*, 100, E10, 21219–21234.

Forget, F., Hourdin, F., and Talagrand, O. (1998) CO$_2$ Snowfall on Mars: simulation with a general circulation model, *Icarus*, 131, 2, 302–316.

Forget F., Hourdin F., Fournier R., et al. (1999) Improved general circulation models of the Martian atmosphere from the surface to above 80 km, *J. Geophys. Res., Planets*, 104(E10), 24155–24175.

Forget, F., Spiga, A., Dolla, B., et al. (2007) Remote sensing of surface pressure on Mars with the Mars Express/OMEGA spectrometer: 1. Retrieval method, *J. Geophys. Res.*, 112, E8, E08S15.

Forget, F., Montmessin, F., Bertaux, J.-L., et al. (2009) The density and temperatures of the upper Martian atmosphere measured by stellar occultations with Mars Express SPICAM, *J. Geophys. Res.*, 114, E01004 doi:10.1029/2008JE003086.

François, L. M., Walker, J. C. G., Kuhn, W. R. (1990) A numerical simulation of climate changes during the obliquity cycle on Mars, *J. Geophys. Res.*, 95, 14761–14778.

Geissler, P. E. (2005) Three decades of Martian surface changes, *J. Geophys. Res.*, 110, E2, E02001.

Giuranna, M., Formisano, V., Grassi, D. and Maturilli, A. (2007a) Tracking the edge of the south seasonal polar cap of Mars, *Planetary and Space Science*, 55, 10, 1319–1327.

Giuranna, M., Hansen, G., Formisano, V., et al. (2007b) Spatial variability, composition and thickness of the seasonal north polar cap of Mars in mid-spring, *Planetary and Space Science*, 55, 1328–1345.

Giuranna, M., Grassi, D.; Formisano, V., et al. (2008) PFS/MEX observations of the condensing CO_2 south polar cap of Mars, *Icarus*, 197, 386–402.

Glandorf, D., Colaprete, A., Toon, O. B., and Tolbert, M. (2002) CO_2 snow on Mars and early Earth, *Icarus*, 160, 66–72.

Gomez-Elvira, J., Armiens, C., Castañer, L., et al. (2012) REMS: the environmental sensor suite for the Mars Science Laboratory Rover, *Space Sci. Rev.*, 170, 583–640, doi:10.1007/s11214-012-9921-1.

González-Galindo, F., Määttänen, A., Forget, F., and Spiga, A. (2011) The Martian mesosphere as revealed by CO_2 cloud observations and General Circulation Modeling, *Icarus*, 216, 10–22.

Guo, X., Lawson, W. G., Richardson, M. I., Toigo, A. (2009) Fitting the Viking Lander surface pressure cycle with a Mars General Circulation Model, *J. Geophys. Res*, 114, Issue E7, CiteID E07006.

Haas, W. H. (2003) Flashes on Mars observed in 1937 and some random remarks, *J. Assoc. Lunar Planet. Observers, The Strolling Astronomer*, 45, 43–45.

Haberle R. M. and Kahre M. A. (2010) Detecting secular climate change on Mars, *Mars*, 5, 68–75, doi:10.1555/mars.2010.0003.

Haberle R. M., Pollack J. B., Barnes J. R., et al. (1993) Mars atmospheric dynamics as simulated by the NASA Ames General Circulation Model: 1. The zonal-mean circulation. *J. Geophys. Res.*, *Planets*, 98, E2, 3093–3123.

Haberle, R. M., Murphy, J. R. and Schaeffer, J. (2003) Orbital change experiments with a Mars general circulation model, *Icarus*, 161, 1, 66–89.

Haberle, R. M., Mattingly, B., and Titus, T. N. (2004) Reconciling different observations of the CO2 ice mass loading of the Martian north polar cap, *Geophys. Res. Lett.*, 31, L05702.

Haberle R. M., Forget F., Colaprete A., et al. (2008) The effect of ground ice on the Martian seasonal CO_2 cycle, *Planetary and Space Science*, 56, 2, 251–255.

Haberle, R. M., Kahre, M. A., Malin, M., and Thomas, P. C. (2009) The disappearing south residual cap on Mars: where is the CO2 going? in *Third International Workshop on Mars Polar Energy Balance and the CO2 Cycle*, July 21–24, Seattle, Washington. LPI Contribution No. 1494, 19–20.

Hanel, R. A., Conrath, B. J., Hovis, W. A., et al. (1970) Infrared Spectroscopy Experiment for Mariner Mars 1971, *Icarus*, 12, 1, 48–62.

Hanel, R., Conrath, B., Hovis, W., et al. (1972) Investigation of the Martian environment by infrared spectroscopy on Mariner 9, *Icarus*, 17, 423–442.

Hansen, C. J., Thomas, N., Portyankina, G. et al. (2010) HiRISE observations of gas sublimation-driven activity in Mars' southern polar regions: I. Erosion of the surface, *Icarus*, 205, 1, 283–295.

Hansen, C. J., Bourke, M., Bridges, N. T., et al. (2011) Seasonal erosion and restoration of Mars' northern polar dunes, *Science*, 331, 6017, 575.

Hansen, C. J., Byrne, S., Portyankina, G., et al. (2013) Observations of the northern seasonal polar cap on Mars: I. Spring sublimation activity and processes, *Icarus*, 225, 2, 881–897.

Hansen, G. B. (1997) The infrared absorption spectrum of carbon dioxide ice from 1.8 to 333 micrometers, *J. Geophys. Res.*, 102, 21569–21587.

Hansen, G. B. (1999) Control of the radiative behavior of the Martian polar caps by surface CO_2 ice: evidence from Mars Global Surveyor measurements, *J. Geophys. Res.*, 104, 16471–16486.

Hansen, G. B. (2005) Ultraviolet to near-infrared absorption spectrum of carbon dioxide ice from 0.174 to 1.8 mm, *J. Geophys. Res.*, 110, E11003, doi:10.1029/2005JE002531.

Hansen, G. B. (2013) An examination of Mars' north seasonal polar cap using MGS: composition and infrared radiation balance, *Icarus*, 225, 869–880.

Hansen, G. B., Giuranna, W., Formisano, V., et al. (2005) PFS-MEX observation of ices in the residual south polar cap of Mars, *Planet. Space Sci.*, 53, 1089–1095.

Hayne, P. O., Paige, D. A., Schofield, J. T., et al. (2012) Carbon dioxide snow clouds on Mars: south polar winter observations by the Mars Climate Sounder, *J. Geophys. Res.*, 117, E08014, doi:10.1029/2011JE004040.

Hayne, P. O., Paige, D. A., Heavens, N. G. (2014) The role of snowfall in forming the seasonal ice caps of Mars: models and constraints from the Mars Climate Sounder, *Icarus*, 231, 122–130.

Head, J. W., Pratt, S. (2001) Extensive Hesperian-aged south polar ice sheet on Mars: evidence for massive melting and retreat, and lateral flow and ponding of meltwater, *Journal of Geophysical Research*, 106, E6, 12275–12300.

Head J. W., Mustard J. F., Kreslavsky M. A., et al. (2003) Recent ice ages on Mars, *Nature*, 426, 6968, 797–802.

Hecht, M. H. (2008) The texture of condensed CO_2 on the Martian polar caps *Planet. Space Sci.*, 56, 246–250.

Herr, K. C. and Pimentel, G. C. (1969) Infrared absorptions near three microns recorded over the polar cap of Mars, *Science*, 166, 3904, 496–499.

Herr, K. C., Forney, P. B., and Pimentel, G. C. (1972) Mariner Mars 1969 infrared spectrometer, *Applied Optics*, 11, 3, 493–501.

Herschel, W. (1784) On the remarkable appearances at the polar regions of the planet Mars, the inclination of its axis, the position of its poles, and its spheroidical figure; With a few hints relating to its real diameter and atmosphere, *Philos. Trans. R. Soc. London*, 74, 233–273.

Hess, S. L., Henry, R. M., Leovy, C. B., et al. (1977) Meteorological results from the surface of Mars – Viking 1 and 2, *J. Geophys. Res.*, 82, 4559–4574.

Hess, S. L., Henry, R. M., and Tillman, J. E. (1979) The seasonal variation of atmospheric pressure on Mars as affected by the south polar cap, *J. Geophys. Res.*, 84, 2923–2927.

Hinson, D. P., Flasar, F. M., Simpson, R. A., et al. (1999) Initial results from radio occultation measurements with Mars Global Surveyor, *J. Geophys. Res.*, 104, 26997–27012.

Hinson, D. P., Wilson, R. J., Smith, M. D. and Conrath, B. J. (2003) Stationary planetary waves in the atmosphere of Mars during southern winter, *J. Geophys. Res.*, 108, doi:10.1029/2002JE001949.

Houben, H., Haberle, R. M., Young, R. E., and Zent, A. P. (1997) Modeling the Martian seasonal water cycle, *J. Geophys. Res.*, 102, 9069–9084.

Hourdin, F., Van, P. L., Forget, F., and Talagrand, O. (1993) Meteorological variability and the annual surface pressure cycle on Mars, *Journal of Atmospheric Sciences*, 50, 21, 3625–3640. doi:10.1175/1520-0469(1993)050.

Hourdin F., Forget F., and Talagrand O. (1995) The sensitivity of the Martian surface pressure and atmospheric mass budget to various parameters: a comparison between numerical simulations and Viking observations, *Journal of Geophysical Research*, 100, E3, 5501–5523.

Hu, R., Cahoy, K., Zuber, M. T. (2012) Mars atmospheric CO$_2$ condensation above the north and south poles as revealed by radio occultation, climate sounder, and laser ranging observations, *J. Geophys. Res.*, 117, E7, CiteID E07002.

Hugenholtz, C. H. (2008) Frosted granular flow: a new hypothesis for mass wasting in Martian gullies, *Icarus*, 197, 65–72.

Ivanov, A. B. (2002) Some aspects of the Martian climate in the Mars Orbiter Laser Altimeter (MOLA) investigation. Part I. Evolution of the polar residual ice caps. Part II. Polar night clouds. Part III. Interpretation of the MOLA reflectivity measurement in terms of the surface albedo and atmospheric opacity. PhD Thesis, California Institute of Technology, Source DAI-B 62/11.

Jakosky, B. (2008) The Mars Atmosphere and Volatile Evolution (MAVEN) Mars Scout Mission, in *Third International Workshop on The Mars Atmosphere: Modeling and Observations*, held November 10–13, 2008 in Williamsburg, Virginia. LPI Contribution No. 1447, 9036.

Jakosky, B. M., and Barker, E. S. (1984) Comparison of ground-based and Viking Orbiter measurements of Martian water vapor–variability of the seasonal cycle, *Icarus*, 57, 322–334.

Jakosky, B. M. and Haberle, R. M. (1990) Year-to-year instability of the Mars south polar cap, *J. Geophys. Res: Solid Earth*, 95, B2, 1359–1365.

James, P. B. and North, G. R. (1982) The seasonal CO$_2$ cycle on Mars – an application of an energy balance climate model, *J. Geophys. Res.*, 87, 10271–10283.

James, P. B., Kieffer, H. H., and Paige, D. A. (1992) The seasonal cycle of carbon dioxide on Mars. In *Mars* (ed. H. H. Kieffer et al.) University of Arizona Press, 934–968.

James, P. B., Cantor, B. A., Malin, M. C., et al. (2000) The 1997 spring regression of the Martian south polar cap: Mars Orbiter Camera observations, *Icarus*, 144, 2, 410–418.

James, P. B., Cantor, B. A., and Davis, S. (2001) Mars Orbiter Camera observations of the Martian south polar cap in 1999–2000, *Journal of Geophysical Research*, 106, E10, 23635–23652.

James, P. B., Bonev, B. P., and Wolff, M. J. (2005) Visible albedo of Mars' south polar cap: 2003 HST observations, *Icarus*, 174, 2, 596–599.

James, P. B., Thomas, P. C., Wolff, M. J., Bonev, B. P. (2007) MOC observations of four Mars year variations in the south polar residual cap of Mars, *Icarus*, 192, 2, 318–326.

Jaumann, R., Neukum, G., Behnke, T., et al. (2007) The high-resolution stereo camera (HRSC) experiment on Mars Express: instrument aspects and experiment conduct from interplanetary cruise through the nominal mission, *Planet. Space Sci.*, 55, 7–8.

Jian, J.-J., and Ip, W.-H. (2009) Seasonal patterns of condensation and sublimation cycles in the cryptic and non-cryptic regions of the south pole, *Advances in Space Research*, 43, 1, 138–142.

Karlsson, N. B., Holt, J. W., and Hindmarsh, R. C. A. (2011) Testing for flow in the north polar layered deposits of Mars using radar stratigraphy and a simple 3D ice-flow model, *Geophys. Res. Lett.*, 38, 24204.

Kass, D. M., and Yung, Y. L. (1995) Loss of atmosphere from Mars due to solar wind-induced sputtering, *Science*, 268, 5211, 697–699.

Kelly, E. J. (2006) Seasonal polar carbon dioxide frost on Mars: spatiotemporal quantification of carbon dioxide utilizing 2001 Mars Odyssey gamma ray spectrometer data, PhD dissertation, University of Arizona, AAT 3206184. DAI-B 67/01.

Kelly N., Boynton, W., Kerry, K., et al. (2006) Seasonal polar carbon dioxide frost on Mars: CO$_2$ mass and columnar thickness distribution, *J. Geophys. Res.*, 111(E3), E03S07.

Kieffer, H. H. (1979) Mars south polar spring and summer temperatures: a residual CO$_2$ frost, *J. Geophys. Res.*, 84, 8263–88.

Kieffer, H. H. (2000) Annual punctuated CO$_2$ slab-ice and jets on Mars, *International Conference on Mars Polar Science and Exploration*, 93.

Kieffer, H. H. (2007) Cold jets in the Martian polar caps, *J. Geophys. Res.*, 112, E08005, doi:10.1029/2006JE002816.

Kieffer, H. H., and Zent, A. P. (1992) Quasi-periodic climate change on Mars, in *Mars* (A93-27852 09-91), 1180–1218.

Kieffer, H. H. and Titus, T. (2001) TES mapping of Mars' north seasonal cap, *Icarus*, 154, 162–180.

Kieffer, H. H., Neugebauer, G., Munch, G., Chase, S. C., Jr., and Miner, E. (1972) Infrared Thermal Mapping Experiment: the Viking Mars Orbiter, *Icarus*, 16, 1, 47–56.

Kieffer, H. H., Titus, T. N., Mullins, K. F., and Christensen, P. R. (2000) Mars south polar spring and summer behavior observed by TES: seasonal cap evolution controlled by frost grain size, *J. Geophys. Res.*, 105, 9653–9699.

Kieffer, H. H., Christensen, P. R., and Titus, T. N. (2006) CO$_2$ jets formed by sublimation beneath translucent slab ice in Mars' seasonal south polar ice cap, *Nature*, 442, 793–796.

Kolb, E. J. and Tanaka, K. L. (2006) Accumulation and erosion of south polar layered deposits in the Promethei Lingula region, Planum Australe, Mars, *International Journal of Mars Science and Exploration*, 2, 1–9.

Kolb, E. J., Tanaka, K. L., Greeley, R., et al. (2006) The residual ice cap of Planum Australe, Mars: new insights from the HRSC experiment, in *37th Annual Lunar and Planetary Science Conference*, March 13–17, League City, Texas, Abstract No. 2408.

Konopliv, A. S., Asmar, S. W., Folkner, W. M., et al. (2011) Mars high resolution gravity fields from MRO, Mars seasonal gravity, and other dynamical parameters, *Icarus*, 211, 401–428.

Koutnik, M., Byrne, S., and Murray, B. (2002) South polar layered deposits of Mars: the cratering record, *J. Geophys. Res. (Planets)*, 107, 5100.

Kravchenko, Y. G., and Krupskii, I. N. (1986) Thermal conductivity of solid N$_2$O and CO$_2$, *Sov. J. Low Temp. Phys.*, 12, 1, 46–48.

Kreslavsky, M. A., and Head, J. W. (2005) Mars at very low obliquity: atmospheric collapse and the fate of volatiles, *Geophysical Research Letters*, 32, 12, L12202.

Kreslavsky, M. A., and Head, J. W. (2011) Carbon dioxide glaciers on Mars: products of recent low obliquity epochs (?), *Icarus*, 216, 1, 111–115.

Kuroda, T., Medvedev, A. S., Kasaba, Y., and Hartogh, P. (2013) Carbon dioxide ice clouds, snowfalls, and baroclinic waves in the northern winter polar atmosphere of Mars, *Geophys. Res. Lett.*, 40, 8, 1484–1488, doi:10.1002/grl.50326.

Langevin, Y., Poulet, F., Bibring, J.-P., et al. (2005) Summer evolution of the north polar cap of Mars as observed by OMEGA/Mars Express, *Science*, 307, 1581–1583.

Langevin, Y., Douté, S., Vincendon, M., et al. (2006) No signature of clear CO$_2$ ice from the "cryptic" regions in Mars' south seasonal cap, *Nature*, 442, 790–792.

Langevin, Y., Bibring, J.-P., Montmessin, F., et al. (2007) Observations of the south seasonal cap of Mars during recession in 2004–2006 by the OMEGA visible/near-infrared imaging spectrometer on board Mars Express, *J. Geophys. Res.*, 112, E08S12, doi:10.1029/2006JE002841.

Laskar, J., Correia, A. C. M., Gastineau, M., Joutel, F., Levrard, B., Robutel, P. (2004) Long term evolution and chaotic diffusion of the insolation quantities of Mars, *Icarus*, 170, 2, 343–364.

Laufer, D., Bar-Nun, A., Pat-El, I., and Jacovi, R. (2013) Experimental studies of ice grain ejection by massive gas flow from ice and implications to Comets, Triton and Mars, *Icarus*, 222, 2013, 73–80.

Leblanc, F., and Johnson, R. E. (2001) Sputtering of the Martian atmosphere by solar wind pick-up ions, *Planetary and Space Science*, 49, 6, 645–656.

Leblanc, F., and Johnson, R. E. (2002) Role of molecular species in pickup ion sputtering of the Martian atmosphere, *Journal of Geophysical Research (Planets)*, 107, E2, 5–1, doi:10.1029/2000JE001473.

Leighton, R. B., and Murray, B. C. (1966) Behavior of carbon dioxide and other volatiles on Mars, *Science*, 153, 3732, 136–144.

Leighton, R. B., Horowitz, N. H., Murray, B. C., et al. (1969) Mariner 6 and 7 television pictures: preliminary analysis, *Science*, 166, 3901, 49–67.

Lian, Y., Richardson, M. I., Newman, C. E., et al. (2012) The Ashima/MIT Mars GCM and argon in the Martian atmosphere, *Icarus*, 218, 2, 1043–1070.

Line, M. R., and Ingersoll, A. P. (2010) Can the solid state greenhouse effect produce ~100 year cycles in the Mars south polar residual CO_2 ice cap? American Geophysical Union, Fall Meeting 2010, abstract #P53F-07.

Listowski, C., Määttänen, A., Riipinen, I., Montmessin, F., and Lefèvre, F. (2013) Near-pure vapor condensation in the Martian atmosphere: CO_2 ice crystal growth, *J. Geophys. Res. (Planets)*, 118, 2153–2171, doi:10.1002/jgre.20149.

Listowski, C., Määttänen, A., Montmessin, F., Spiga, A., and Lefèvre, F., (2014) Modeling the microphysics of CO_2 ice clouds within wave-induced cold pockets in the Martian mesosphere, *Icarus*, 237, 239–261.

Litvak M., Mitrofanov I., Kozyrev A., et al. (2005) Modeling of Martian seasonal caps from HEND/ODYSSEY data, *Advances in Space Research*, 36, 11, 2156–2161.

Litvak M., Mitrofanov I., Kozyrev A., et al. (2006) Comparison between polar regions of Mars from HEND/Odyssey data, *Icarus*, 180, 1, 23–37.

Määttänen, A., Vehkamäki, H., Lauri, A., et al. (2005) Nucleation studies in the Martian atmosphere, *J. Geophys. Res. (Planets)*, 110, E02002, doi:10.1029/2004JE002308.

Määttänen, A., Vehkamäki, H., Lauri, A., Napari, I., Kulmala, M., 2007. Two-component heterogeneous nucleation kinetics and an application to Mars, *Journal of Chemical Physics*, 127, 134710. doi:10.1063/1.2770737.

Mahaffy, P. R., Webster, C. R., Atreya, S. K., et al. (2013) Abundance and Isotopic Composition of Gases in the Martian Atmosphere from the Curiosity Rover, *Science*, 341, 6143, 263–266.

Malin, M. C., and Edgett, K. S. (2001) Mars Global Surveyor Mars Orbiter Camera: interplanetary cruise through primary mission, *Journal of Geophysical Research*, 106, E10, 23429–23570.

Malin, M. C., Carr, M. H., Danielson, G. E., et al. (1998) Early views of the Martian surface from the Mars Orbiter Camera of Mars Global Surveyor, *Science*, 279, 5357, 1681.

Malin, M. C., Calvin, W., Clancy, R. T., et al. (2001a) The Mars Color Imager (MARCI) on the Mars Climate Orbiter, *J. Geophys. Res.*, 106, E8, 17651–17672.

Malin, M. C., Caplinger, M. A., and Davis, S. D. (2001b) Observational evidence for an active surface reservoir of solid carbon dioxide on Mars, *Science*, 294, 5549, 2146–2148.

Malin, M. C., Bell, J. F., Cantor, B. A., et al. (2007) Context Camera Investigation on board the Mars Reconnaissance Orbiter, *J. Geophys. Res.*, 112, E5, CiteID E05S04.

Manning, C. V., McKay, C. P., and Zahnle, K. J. (2006) Thick and thin models of the evolution of carbon dioxide on Mars, *Icarus*, 180, 1, 38–59.

Maurice S., Feldman, W., Diez, B., et al. (2011) Mars Odyssey neutron data: 1. Data processing and models of water-equivalent-hydrogen distribution, *J. Geophys. Res.*, 116, E11, E11008.

McCleese, D. J., Schofield, J. T., Taylor, F. W., et al. (2007) Mars Climate Sounder: an investigation of thermal and water vapor structure, dust and condensate distributions in the atmosphere, and energy balance of the polar regions, *J. Geophys. Res.*, 112, E5, E05S06.

McCleese, D. J., Schofield, J. T., Taylor, F. W., et al. (2008) Intense polar temperature inversion in the middle atmosphere on Mars, *Nature Geoscience*, 1, 11, 745–749.

McEwen, A. S., Eliason, E. M., Bergstrom, J. W., et al. (2007) Mars Reconnaissance Orbiter's High Resolution Imaging Science Experiment (HiRISE), *J. Geophys. Res.*, 112, E5, E05S02.

Mellon, M. T. (1996) Limits on the CO_2 content of the Martian polar deposits, *Icarus*, 124, 1, 268–279.

Mellon, M. T. and Jakosky, B. M. (1995) The distribution and behavior of Martian ground ice during past and present epochs, *J. Geophys. Res.*, 100, E6, 11781–11799.

Mellon, M. T., Feldman, W. C., and Prettyman, T. H. (2004) The presence and stability of ground ice in the southern hemisphere of Mars, *Icarus*, 169, 2, 324–340.

Michaels, T. I. and Rafkin, S. C. R. (2008) Meteorological predictions for candidate 2007 Phoenix Mars Lander sites using the Mars Regional Atmospheric Modeling System (MRAMS), *J. Geophys. Res.*, 113, E00A07, doi:10.1029/2007JE003013.

Michalski, J.R. and Niles, P.B. (2010) Deep crustal carbonate rocks exposed by meteoritic impact on Mars, *Nature Geoscience*, 3, 751–755.

Mischna, M. A., Richardson, M. I., Wilson, R. J., and McCleese, D. J. (2003) On the orbital forcing of Martian water and CO_2 cycles: a general circulation model study with simplified volatile schemes, *J. Geophys. Res.*, 108, E6, CiteID 5062, doi:10.1029/2003JE002051.

Mitrofanov, I., Anfimov, D., Kozyrev, A., et al. (2002) Maps of subsurface hydrogen from the High Energy Neutron Detector, Mars Odyssey, *Science*, 297, 5578, 78–81.

Montmessin, F., Bertaux, J. L., Quemerais, E., et al. (2006) Subvisible CO_2 ice clouds detected in the mesosphere of Mars, *Icarus*, 183, 403–410.

Montmessin, F., Haberle, R. M., Forget, F., et al. (2007) On the origin of perennial water ice at the south pole of Mars: a precession-controlled mechanism? *J. Geophys. Res. (Planets)*, 112, 8.

Mullins, J. C., Kirk, B. S., and Ziegler, W. T. (1963) *Calculations of the vapor pressure and heats of vaporization and sublimation of liquids and solids, especially below one atmosphere. V. Carbon monoxide and carbon dioxide.* Tech Rep. No. 2, Project A-663, Engineering Experiment Station, Georgia Inst. of Tech., August.

Murphy, J. R., Pollack, J. B., Haberle, R. M., et al. (1995) Three-dimensional numerical simulation of Martian global dust storms. *J. Geophys. Res.*, 100, E12, 26357–26376.

Nelli, S. M., Murphy, J. R., Sprague, A. L., et al. (2007) Dissecting the polar dichotomy of the non-condensable gas enhancement on Mars using the NASA Ames Mars General Circulation Model. *J. Geophys. Res.: Planets*, 112(E8).

Neugebauer, G., Münch, G., Chase, S. C., Jr., et al. (1969) Mariner 1969: preliminary results of the Infrared Radiometer Experiment, *Science*, 166, 3901, 98–99.

Neugebauer, G., Münch, G., Kieffer, H., Chase, S. C., Jr., and Miner, E. (1971) Mariner 1969 Infrared Radiometer results: temperatures and thermal properties of the Martian surface, *Astronomical Journal*, 76, 719.

Neumann, G. A., Smith, D. E. and Zuber, M. T. (2003) Two Mars years of clouds detected by Mars Orbiter Laser Altimeter, *J. Geophys. Res.*, 108, doi:10.1029/2002JE001849.

Newman, C. E., Lewis, S. R. and Read, P. L. (2005) The atmospheric circulation and dust activity in different orbital epochs on Mars, *Icarus*, 174, 135–160.

Nye, J., Durham, W. B., Schenk, P. M., and Moore, J. M. (2000) The instability of a south polar cap on Mars composed of carbon dioxide, *Icarus*, 144, 2, 449–455.

Ortega, I. K., Määttänen, A., Kurtén, T., Vehkamäki, H. (2011) Carbon dioxide–water clusters in the atmosphere of Mars, *Computational and Theoretical Chemistry*, 965, 353–358.

Owen T., Biemann K., Rushneck D., et al. (1977) The composition of the atmosphere at the surface of Mars. *J. Geophys. Res.*, 82, 28, 4635–4639.

Paige, D. A. (1985) The annual heat balance of the Martian polar caps from Viking observations, Ph.D. dissertation, California Institute of Technology.

Paige, D. A., and Ingersoll, A. P. (1985) Annual heat-balance of Martian polar caps: Viking observations, *Science*, 228, 1160–1168.

Paige, D. A., and Wood, S. E. (1992) Modeling the Martian seasonal CO₂ cycle 2. Interannual variability, *Icarus*, 99, 1, 15–27.

Phillips, R. J., Zuber, M. T., Smrekar, S. E., et al. (2008) Mars north polar deposits: stratigraphy, age, and geodynamical response, *Science*, 320, 1182.

Phillips, R. J., Davis, B. J., Tanaka, K. L., et al. (2011) Massive CO₂ ice deposits sequestered in the south polar layered deposits of Mars, *Science*, 332, 6031, 838.

Pilorget, C., Forget, F., Millour, E., et al. (2011) Dark spots and cold jets in the polar regions of Mars: new clues from a thermal model of surface CO₂ ice, *Icarus*, 213, 131–149.

Piqueux, S., and Christensen, P. R. (2008a) North and south subice gas flow and venting of the seasonal caps of Mars: a major geomorphological agent, *J. Geophys. Res.*, 113, E6, E06005.

Piqueux, S., and Christensen, P. R. (2008b) Deposition of CO₂ and erosion of the Martian south perennial cap between 1972 and 2004: implications for current climate change, *J. Geophys. Res.*, 113, E02006, doi:10.1029/2007JE002969.

Piqueux, S., Byrne, S., and Richardson, M. I. (2003) Sublimation of Mars's southern seasonal CO₂ ice cap and the formation of spiders, *J. Geophys. Res.*, 108, 5084, doi:10.1029/2002JE002007.

Piqueux, S., Edwards, C. S., and Christensen, P. R. (2008) Distribution of the ices exposed near the south pole of Mars using Thermal Emission Imaging System (THEMIS) temperature measurements *J. Geophys. Res.*, 113, E8, E08014.

Piqueux, S., Byrne, S., Kieffer, H. H., Titus, T. N., and Hansen, C. J. (2015) Enumeration of Mars years and seasons since the beginning of telescopic exploration, *Icarus*, 251, 332–338.

Plaut, J. J. (2005) An inventory of impact craters on the Martian south polar layered deposits, *36th Annual Lunar and Planetary Science Conference*, 2319.

Pollack, J. B., Haberle R. M., Schaeffer J., and Lee, H. (1990) Simulations of the general circulation of the Martian atmosphere: 1. Polar processes, *J. Geophys. Res: Solid Earth*, 95, B2, 1447–1473.

Pollack, J. B., Haberle R. M., Murphy J. R., Schaeffer J., and Lee, H. (1993) Simulations of the general circulation of the Martian atmosphere: 2. Seasonal pressure variations, *J. Geophys. Res.: Planets*, 98, E2, 3149–3181.

Pommerol, A., Appéré, T., Portyankina, G., et al. (2013) Observations of the northern seasonal polar cap on Mars III: CRISM/HiRISE observations of spring sublimation, *Icarus*, 225, 2, 911–922.

Portyankina, G., Markiewicz, W. J., Thomas, N., Hansen, C. J., and Milazzo, M. (2010) HiRISE observations of gas sublimation-driven activity in Mars' southern polar regions: III. Models of processes involving translucent ice, *Icarus*, 205, 1, 311–320.

Portyankina, G., Pommerol, A., Aye, K.-M., et al. (2013) Observations of the northern seasonal polar cap on Mars II: HiRISE photometric analysis of evolution of northern polar dunes in spring, *Icarus*, 225, 2, 898–910.

Prettyman, T. H. (2014) Remote sensing of chemical elements using nuclear spectroscopy, in *Encyclopedia of the Solar System*, Elsevier, 1161–1183.

Prettyman, T. H., Feldman, W. C., Mellon, M. T., et al. (2004) Composition and structure of the Martian surface at high southern latitudes from neutron spectroscopy, *J. Geophys. Res.* 109, E5, E05001.

Prettyman, T. H., Feldman, W. C., and Titus, T. N. (2009) Characterization of Mars' seasonal caps using neutron spectroscopy, *J. Geophys, Res.*, 114, E8, E08005.

Russell, P., Thomas, N., Byrne, S., et al. (2008) Seasonally active frost-dust avalanches on a north polar scarp of Mars captured by HiRISE, *Geophys. Res. Lett.*, 35, 23, CiteID L23204.

Schmitt, B., Douté, S., Langevin, Y., et al. (2005a) Northern seasonal condensates on Mars by OMEGA/Mars Express, in *Annu. Lunar Planet. Sci. Conf. XXXVI*, League City, Texas, Abstract No. 2326, March 14–18.

Schmitt, B., Douté, S., Langevin, Y., et al. (2005b) Spring sublimation of the seasonal condensates on Mars from OMEGA/Mars Express, in *Fall AGU Meeting*, Abstract No. P23C-02.

Schmidt, F., Schmitt, B., Douté, S., et al. (2010) Sublimation of the Martian CO₂ Seasonal South Polar Cap, *Planetary and Space Science*, 58, 10, 1129–1138.

Schofield, J. T., Barnes, J. R., Crisp, D., et al. (1997) The Mars Pathfinder atmospheric structure investigation/meteorology, *Science*, 278, 1752. doi:10.1126/science.278.5344.1752

Sefton-Nash, E., Teanby, N. A., Montabone, L., et al. (2013) Climatology and first-order composition estimates of mesospheric clouds from Mars Climate Sounder limb spectra, *Icarus*, 223, 710–721.

Seiff, A., Tillman, J. E., Murphy, J. R., et al. (1997) The atmosphere structure and meteorology instrument on the Mars Pathfinder Lander, *J. Geophys. Res.*, 102, E2, 4045–4056.

Siili, T., Haberle, R. M., and Murphy, J. R. (1997) Sensitivity of Martian Southern polar cap edge winds and surface stresses to dust optical thickness and to the large-scale sublimation flow, *Advances in Space Research*, 19, 1241–1244, doi:10.1016/S0273-1177(97)00276-7.

Siili, T., Haberle, R. M., Murphy, J. R., and Savijärvi, H. (1999) Modelling of the combined late-winter ice cap edge and slope winds in Mars Hellas and Argyre regions, *Planetary and Space Science*, 47, 951–970, doi:10.1016/S0032-0633(99)00016-1.

Silvestro, S., Fenton, L. K., Vaz, D. A., Bridges, N. T., and Ori, G. G. (2010) Ripple migration and dune activity on Mars: evidence for dynamic wind processes, *Geophys. Res. Lett.*, 37, 20, CiteID L20203.

Smith, D. E., Zuber, M. T., and Neumann, G. A. (2001a) Seasonal variations of snow depth on Mars, *Science*, 294, 2141–2146.

Smith, D. E., Zuber, M. T., Frey, H. V., et al. (2001b) Mars Orbiter Laser Altimeter: experiment summary after the first year of global mapping of Mars, *J. Geophys. Res.*, 106, 23689–23722.

Smith, D. E., Zuber, M. T., Torrence, M. H., et al. (2009) Time variations of Mars' gravitational field and seasonal changes in the masses of the polar ice caps, *J. Geophys. Res.*, 114, E05002, doi:10.1029/2008JE003267.

Smith, M. D., Pearl, J. C., Conrath, B. J., and Christensen, P. R. (2001) Thermal Emission Spectrometer results: Mars atmospheric thermal structure and aerosol distribution, *Journal of Geophysical Research*, 106, E10, 23929–23945.

Spiga, A., Forget, F., Madeleine, J.-B., et al. (2011) The impact of Martian mesoscale winds on surface temperature and on the determination of thermal inertia, *Icarus*, 212, 504–519.

Spiga, A., González-Galindo, F., López-Valverde, M.-Á., and Forget, F. (2012) Gravity waves, cold pockets and CO₂ clouds

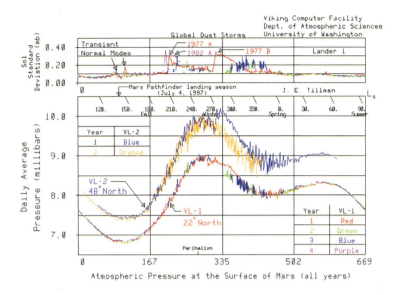

Figure 2.1 The Viking Lander pressure curves measured on the surface at 23°N and 48°N by the Viking Lander 1 and 2 meteorological sensors, respectively. The daily (sol) average and the sol standard deviations are shown. The bottom axis is given in sols, dated from the arrival of Viking Lander 1; the upper axis of the lower panel gives the time of year in L_s, the areocentric longitude of the Sun. The effects of topography, latitude, weather, and even a nearly global dust event are shown (see text). Figure provided courtesy of James Tillman, a veteran of the Viking mission and Viking Lander Meteorology Team. A black and white version of this figure will appear in some formats.

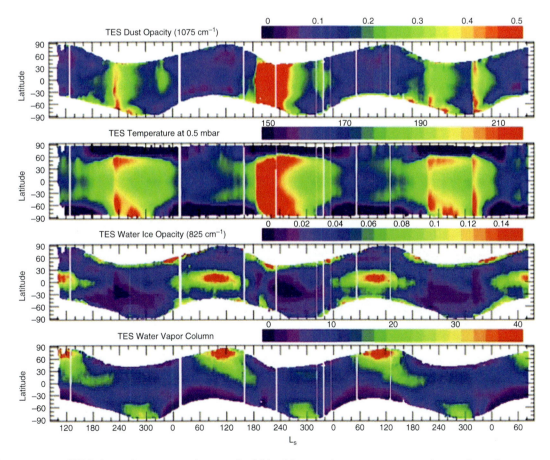

Figure 3.4 Three years of TES observations were used to map the fields of dust opacity, temperature, water ice opacity, and water vapor as a function of season and latitude. The concentration of dust during the season around perihelion is evident. While the inter-annual general repeatability is evident, variability is evidenced by the global dust event that started at equinox in the second Mars year and by the large dust event that occurred at $L_s \sim 310°$ in year 3 but not in the first two years. The figure nicely illustrates the correlation between water vapor near the north pole and the aphelion cloud belt. A black and white version of this figure will appear in some formats.

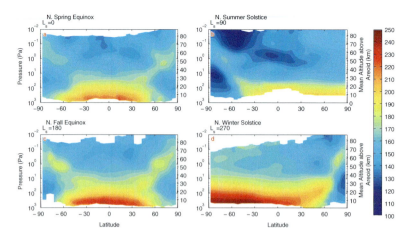

Figure 4.6 Typical zonally averaged daytime (~3:00 p.m. local time) temperatures as a function of latitude and pressure (or height above the surface) as observed by Mars Reconnaissance Orbiter MCS at four seasons throughout the Martian year. Figure after McCleese et al. (2010) and used with permission. A black and white version of this figure will appear in some formats.

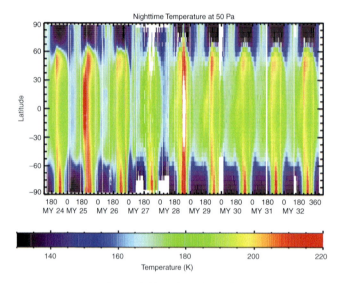

Figure 4.15 Nighttime zonally averaged temperatures at 50 Pa (~25 km above the surface) shown as a function of season (L_s) and latitude covering nine Martian years. Shown is a combination of retrieved temperatures from the Mars Global Surveyor TES (MY 24–27), Mars Odyssey THEMIS (MY 27–present), and Mars Reconnaissance Orbiter MCS (MY 28–present) instruments. A black and white version of this figure will appear in some formats.

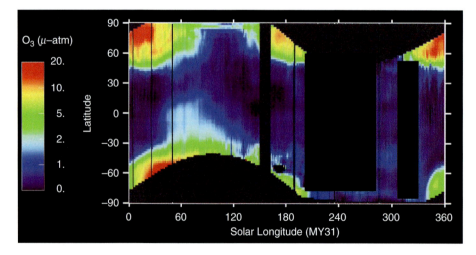

Figure 4.20 The seasonal and latitudinal variation of ozone column abundance in micrometer atmospheres (μm atm) retrieved from MARCI observations during MY 31. Periods between $L_s = 210°$ and $330°$ when high-dust-opacity contaminated ozone retrievals have been excluded. Values are zonal averages. A black and white version of this figure will appear in some formats.

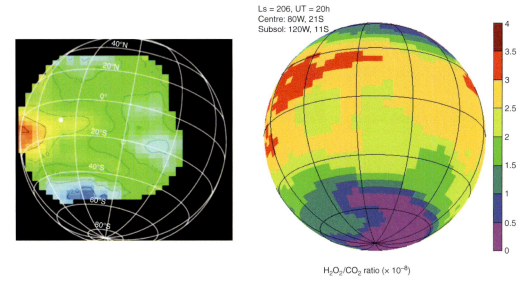

Ls = 206, UT = 20h
Centre: 80W, 21S
Subsol: 120W, 11S

H_2O_2/CO_2 ratio ($\times 10^{-8}$)

Figure 4.21 (Left) Map of the H_2O_2/CO_2 mixing ratio, derived from the line depth ratio of weak H_2O_2 and CO_2 transitions around 8.0 mm, obtained from TEXES data recorded in June 2003 (MY 26, $L_s = 206°$). The H_2O_2 mixing ratio is indicated in multiples of 10^{-8}; contours are separated by 0.5×10^{-8}. The subsolar point is indicated by a white dot. (Right) Map of the H_2O_2 mixing ratio, in the same units, as modeled by the GCM developed at the Laboratoire de Météorologie Dynamique under the conditions of the observations. Figure is from Encrenaz et al. (2004), and is used with permission. A black and white version of this figure will appear in some formats.

Figure 5.1 Hubble Space Telescope imaging of Mars was obtained during its 1997 opposition (James et al., Space Telescope Science Institute, Press release). The distinctive aphelion cloud belt (ACB), composed of water ice clouds, encompasses the northern subtropics of Mars around the cold aphelion (northern summer) portion of the Mars orbit. A black and white version of this figure will appear in some formats.

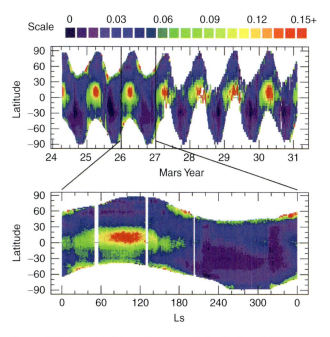

Figure 5.12 The global (zonally averaged) distribution of 12 µm cloud absorption optical depth is presented versus latitude and L_s for multiple Mars years, corresponding to the 1999–2011 period. The period MY 26 is expanded for better viewing of a typical annual behavior. The period MY 24–26 incorporates MGS TES measurements (Smith, 2004), whereas the period MY 27–31 incorporates MO THEMIS measurements (Smith, 2009). The ACB and PH cloud structures apparent in this figure exhibit modest inter-annual variations. Figure provided by Michael Smith, Goddard Space Flight Center. A black and white version of this figure will appear in some formats.

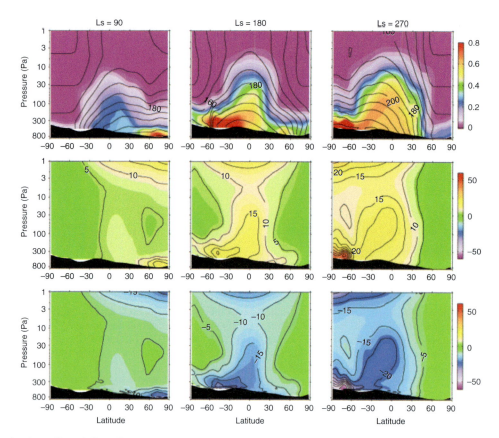

Figure 6.38 Simulated zonally and diurnally averaged radiative/convective equilibrium temperatures and heating rates using specified aerosol distributions in three seasons: $L_s = 90°$ (left), $L_s = 180°$ (center), and $L_s = 270°$ (right). (Top row) Dust is shown as a pressure-normalized visible optical depth (0.67 μm, Pa⁻¹) and the temperature is contoured at 10 K intervals. (Middle row) Solar heating rates (deg sol⁻¹) corresponding to the dust aerosol fields shown in the top row. (Bottom row) Long-wave heating rates. The equatorial column-integrated dust optical depth is ~0.3, 0.6, and 0.7 for the $L_s = 90°$, 180°, and 270° simulations, respectively. A black and white version of this figure will appear in some formats.

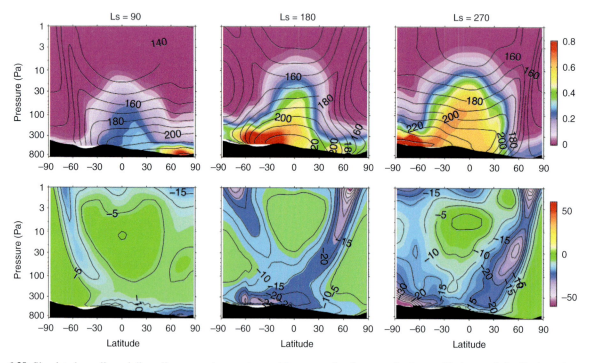

Figure 6.39 Simulated zonally and diurnally average temperature and long-wave heating rates for the specified aerosol distributions shown in Figure 6.38. (Top row) Dust is shown as a pressure-normalized visible optical depth (0.67 μm, Pa⁻¹) and the temperature is contoured at 10 K intervals. (Bottom row) Long-wave heating rates (deg sol⁻¹) corresponding to the dust aerosol fields shown in the top row. A black and white version of this figure will appear in some formats.

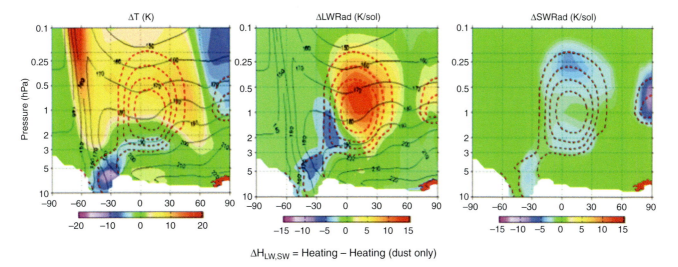

$$\Delta H_{LW,SW} = Heating - Heating (dust only)$$

Figure 6.40 (Left) Zonally and diurnally averaged aphelion season ($L_s = 110°$) temperature from a GCM simulation with radiatively active water ice clouds. The temperature field is displayed with contours at 10 K intervals. The (color) shading illustrates the temperature difference against a baseline simulation without water ice clouds. The mass mixing ratio of the simulated cloud (dashed red contours) is shown at intervals of 10 ppm. (Center) The change in LW heating attributable to water ice clouds ($\Delta LW = LW - LW_{dust\ only}$). (Right) As for center panel, but for SW heating. A black and white version of this figure will appear in some formats.

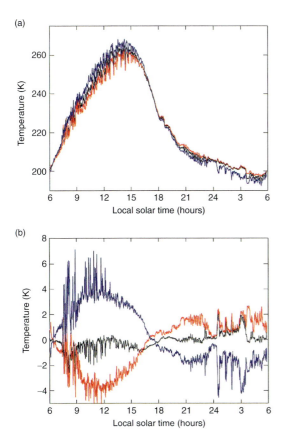

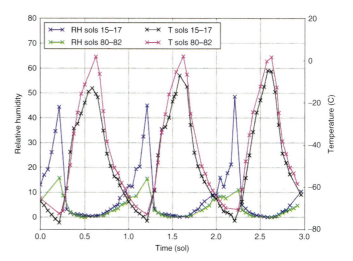

Figure 7.10 MSL REMS-H 1.5 m relative humidity and temperature observations during sols 15–17 ($L_s \sim 158°$) and 80–82 ($L_s \sim 196°$). The first period is cooler and moister; the second is warmer and drier. Figure adapted from Harri et al. (2014a) with permission. A black and white version of this figure will appear in some formats.

Figure 7.8 (a) The diurnal variation of atmospheric temperature at the Pathfinder landing site measured by the top (red), middle (black), and bottom (blue) mast thermocouples from 06:00 LT on sol 25 to 06:00 LT on mission sol 26. The T sampling interval was 4 s throughout this period, but the plots use 2 min running means for clarity (averaging reduces the amplitude and frequency of the fluctuations present in the raw data). (b) The data plotted as T deviations from the mean of all three thermocouples. Adapted from Schofield et al. (1997) with permission. A black and white version of this figure will appear in some formats.

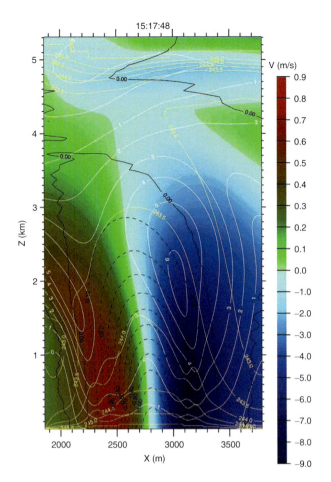

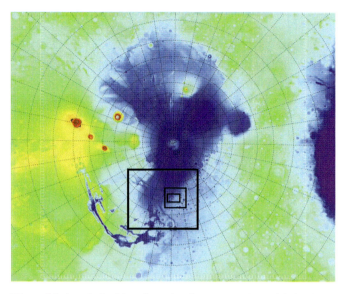

Figure 8.1 A typical grid configuration for a mesoscale model simulation of Mars starts with a superhemispheric mother domain, often centered at one of the poles, in order to minimize the effect of domain boundaries on the thermal tide. Additional grids are spawned within the mother domain to progressively focus on an area of interest, in this case Mawrth Valles. A black and white version of this figure will appear in some formats.

Figure 7.16 High-resolution simulation ($\Delta x = 10$ m) of the "no wind" simulation dust devil of Toigo et al. (2003). Here is plotted a vertical slice through the center of the dust devil. Background color shows the tangential wind speed. Black contours show the pressure perturbation (in Pa), reaching a maximum difference near the surface of about 1 Pa less than the background. Yellow contours show potential temperature (in K), and the warm core of the dust devil. White contours show upward wind velocity (in m s^{-1}). Upward wind velocity peaks at the walls of the dust devil, and the decrease in upward velocity can be seen in the center of the dust devil core. Figure adapted from Toigo et al. (2003) with permission. A black and white version of this figure will appear in some formats.

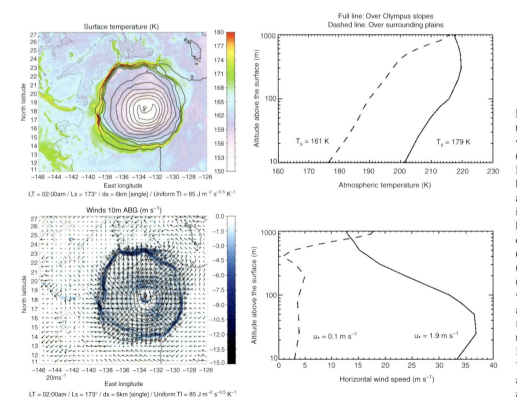

Figure 8.4 (Top left) Surface temperature (K) and (bottom left) winds 10 m above local surface (m s^{-1}), predicted in the Olympus Mons/Lycus Sulci area with 6 km horizontal grid spacing and assuming uniform soil thermal inertia. Topography is contoured (2 km interval). Vectors indicate wind direction and speed. Vertical velocity (m s^{-1}) is shaded. Vertical profiles of (top right) near-surface temperature (K) and (bottom right) horizontal wind speed (m s^{-1}). The profiles are extracted over the northwestern flank of Olympus Mons and over the plain northward of Olympus Mons. Adapted from figures 4 and 7 of Spiga et al. (2011a). A black and white version of this figure will appear in some formats.

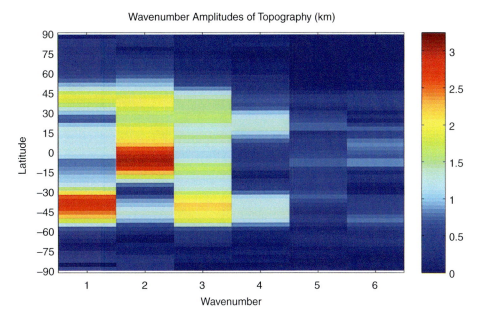

Figure 9.4 (b) the zonal wavenumber amplitudes of the MOLA topography. One-degree topography data were used to produce both panels. A black and white version of this figure will appear in some formats.

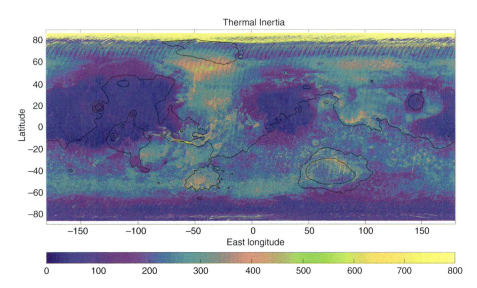

Figure 9.5 (b) surface thermal inertia (taken from Putzig and Mellon (2007)) as determined from TES temperature data. A black and white version of this figure will appear in some formats.

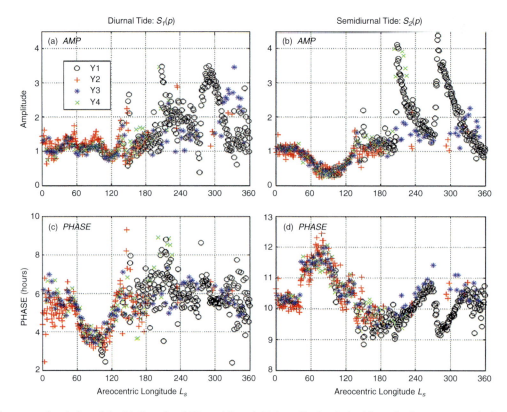

Figure 9.15 The seasonal variation of the (a) diurnal and (b) semidiurnal tidal amplitudes derived from the four-year record of surface pressure at the Viking Lander 1 site (22°N, 312°E). The amplitudes are normalized by the diurnal-mean surface pressures; the values are shown as percentages. The corrected (to true local time) diurnal tidal phase is shown in (c), and the corrected semidiurnal tidal phase is shown in (d). Figure adapted from Wilson et al. (2007). A black and white version of this figure will appear in some formats.

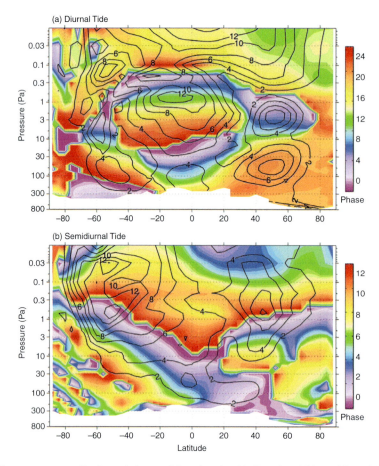

Figure 9.20 GCM-simulated temperature amplitudes and phases of the migrating (a) diurnal and (b) semidiurnal tide for early northern summer ($L_s \sim 105°$). Tidal amplitudes are contoured at 2 K intervals and phases are shaded from 00:00 to 24:00 LT for the diurnal tide and from 00:00 to 12:00 LT for the semidiurnal tide. Figure adapted from Kleinböhl et al. (2013). A black and white version of this figure will appear in some formats.

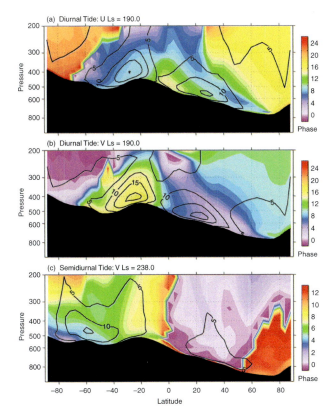

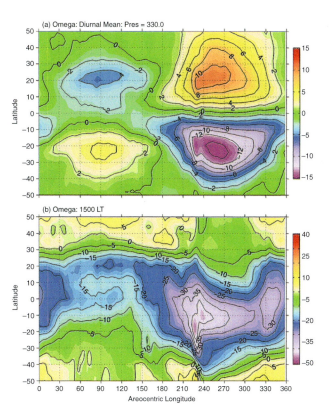

Figure 9.21 Amplitude and phase of GCM-simulated tidal wind fields. The migrating component of the diurnal period zonal and meridional winds for $L_s \sim 190°$ are shown in (a) and (b), respectively. The amplitude contour interval is 5 m s^{-1} and the phase (shading) varies from 00:00 to 24:00 LT. (c) Migrating semidiurnal meridional wind field for $L_s \sim 238°$. The phase varies from 00:00 to 12:00 LT, and the amplitude contour is 5 m s^{-1}. Figure adapted from Wilson (2012a,b). A black and white version of this figure will appear in some formats.

Figure 9.22 (a) The seasonal variation of the zonally and diurnally averaged pressure vertical velocity ω (in units of 10^{-4} Pa s^{-1}) on the 330 Pa pressure level, from a GCM simulation using specified dust column opacity as observed by MGS TES during MY 24/25. (b) As above, but for the pressure vertical velocity at 15:00 LT (no diurnal averaging). Note that the contour intervals and the range of shading differ in the two panels. Figure adapted from Wilson (2012b). A black and white version of this figure will appear in some formats.

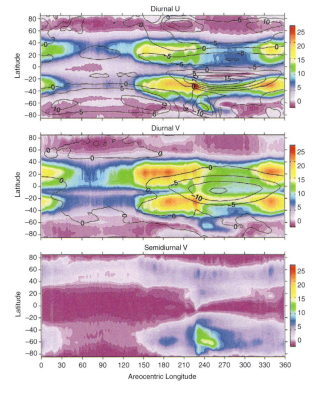

Figure 9.23 (Top) The seasonal variation of the amplitude of the migrating diurnal component of zonal wind (in m s^{-1}, shaded) at ~1 km above ground level, from a GCM simulation with dust opacities appropriate for MY 24/25. The diurnally and zonally averaged zonal wind field is also shown, contoured at intervals of 5 m s^{-1}. (Middle) As in the top panel but for the meridional wind field. (Bottom) As in the top panel but for the semidiurnal component of the meridional wind field. Figure adapted from Wilson (2012a). A black and white version of this figure will appear in some formats.

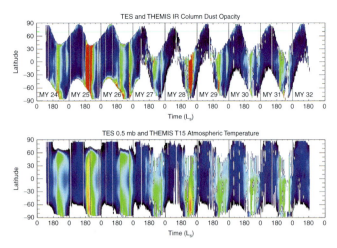

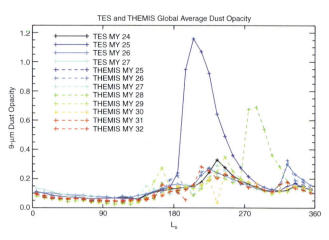

Figure 10.1 Zonally averaged 9 µm column dust opacity (top) and zonally averaged atmospheric temperature for approximately eight of the most recent Martian years, as observed by MGS/TES (MY 24–26) and Mars Odyssey (MY 27–32). The 9 µm column dust opacities range from 0 (black) to >0.5 (red/light gray). Temperatures range from 150 K (black) to 250 K (red/light gray). Data courtesy of M.D. Smith. A black and white version of this figure will appear in some formats.

Figure 10.2 Globally averaged 9 µm dust opacity observations from MGS/TES and ODY/THEMIS for multiple Mars years. Data courtesy of M.D. Smith. A black and white version of this figure will appear in some formats.

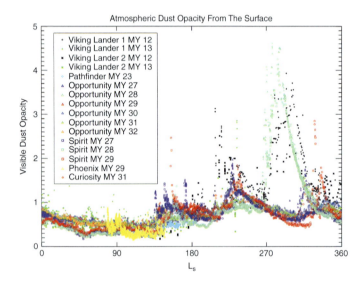

Figure 10.3 Visible dust opacities derived from VL1, VL2, MPF, MERa, MERb, PHX, and MSL. A black and white version of this figure will appear in some formats.

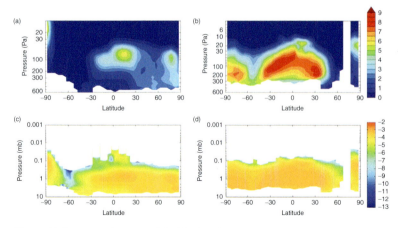

Figure 10.5 (a) Zonal average nightside dust-density-scaled opacity at $L_s = 90°$, MY 30 ($\times10^4$ m^2 kg^{-1}); (b) zonal average nightside dust-density-scaled opacity at $L_s = 270°$, MY 29 ($\times10^4$ m^2 kg^{-1}); (c) $\log_{10}$ of zonal average nightside dust-density-scaled opacity at $L_s = 90°$, MY 30 (m^2 kg^{-1}); (d) $\log_{10}$ of zonal average nightside dust-density-scaled opacity at $L_s = 270°$, MY 29 (m^2 kg^{-1}). From Heavens et al. (2011a). A black and white version of this figure will appear in some formats.

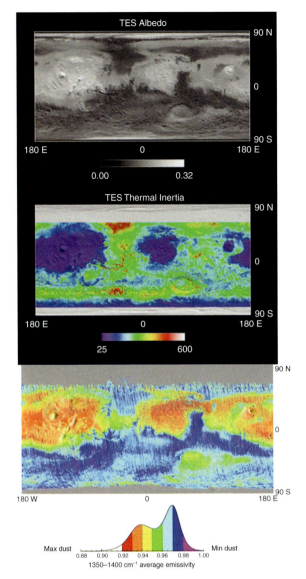

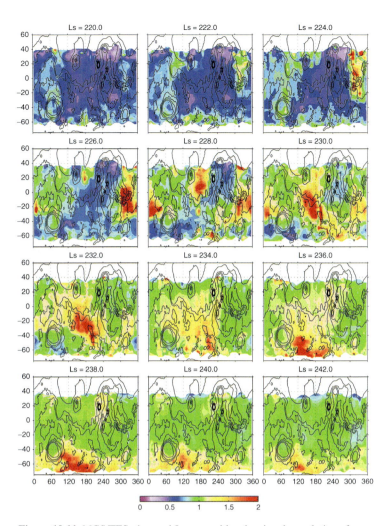

Figure 10.11 MGS/TES-observed 9 μm opacities showing the evolution of regional storms during MY 24 from $L_s = 220°$ to $L_s = 242°$. A black and white version of this figure will appear in some formats.

Figure 10.6 Surface albedo (top), thermal inertia (middle) and dust cover index (bottom) as derived from MGS/TES data. From www.mars.asu.edu (top two panels) and Ruff and Christensen (2002) (bottom panel). A black and white version of this figure will appear in some formats.

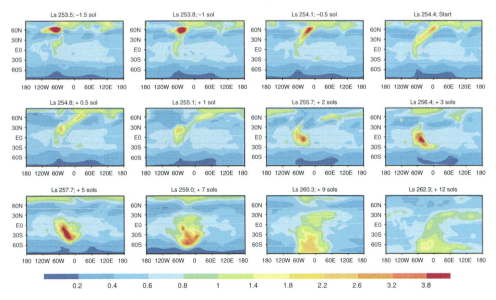

Figure 10.18 GCM-simulated cross-equatorial regional storm that developed in the Chryse region from approximately $L_s = 253°$ to $L_s = 260°$. From Newman et al. (2002a). A black and white version of this figure will appear in some formats.

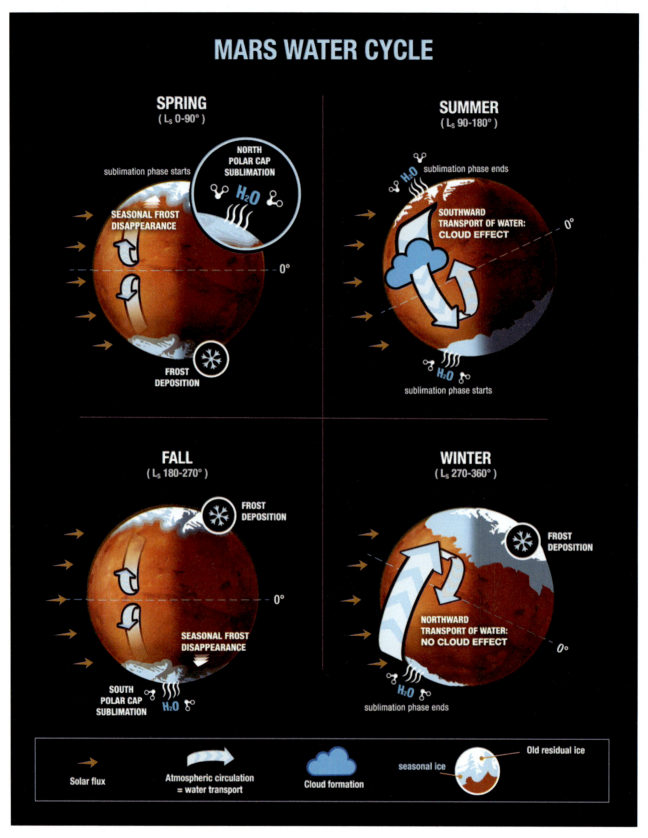

Figure 11.1 Seasonal progression of the present Mars water cycle illustrating the exchange of water between surface and atmospheric reservoirs. A black and white version of this figure will appear in some formats.

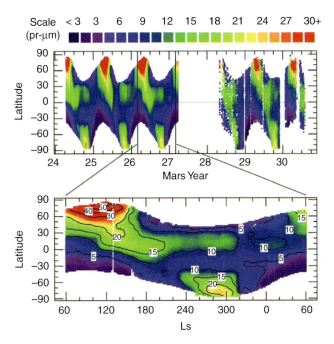

Figure 11.3 An overview of the current annual cycle of water vapor column abundance versus season and latitude over several Mars years. Results from TES (MY 24–27) and CRISM (MY 28–30) are shown. There is a repeatable pattern, with highest water vapor abundance at high latitudes of the summer hemisphere. A black and white version of this figure will appear in some formats.

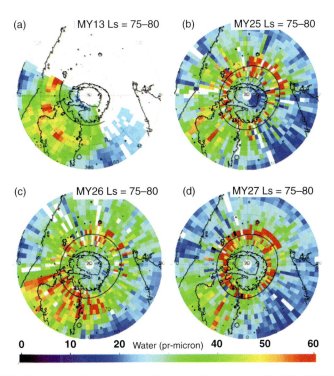

Figure 11.5 From Pankine et al. (2010). Polar maps of water vapor column abundances at $L_s = 75$–$80°$: (A) MAWD MY 13; (B) TES MY 25; (C) TES MY 26; (D) TES MY 27. East longitude system with $0°E$ at the bottom. Black contours are MOLA topography at 1 km interval. The boundaries of the H_2O and CO_2 seasonal caps on maps of TES data are shown by black solid and dashed nearly oval curves, respectively. A black and white version of this figure will appear in some formats.

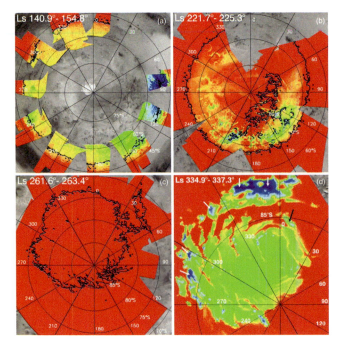

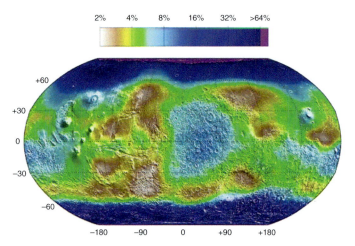

Figure 11.13 A Robinson projection of the water-equivalent hydrogen content of the semi-infinite layer of water-bearing soils derived from the Mars Odyssey GRS spectrometer. From Feldman et al. (2008). Regolith is filled with water ice fills poleward of 60° in both hemispheres. A black and white version of this figure will appear in some formats.

Figure 11.11 Evolution of the H_2O ice signature at 1.5 μm as observed by OMEGA/Mars Express: (a)–(c) during the retreat of the southern seasonal cap, in MY 27, from Langevin et al. (2007); and (d) over the perennial cap, observations in 2009, MY 29, with a more comprehensive coverage than that presented in MY 26 by Bibring et al. (2004). The rainbow color scale ranges from a band strength of 45% (dark blue) to 0% (red). The black line indicates the boundary of the seasonal ice cap as inferred from CO_2 ice spectral signatures. At $L_s = 340°$, the CO_2 ice signatures have reached their minimum extent corresponding to the perennial cap (Figure 11.4). The perennial CO_2 cap is contaminated by H_2O frost (Bibring et al., 2004; Douté et al., 2007). Regions at the boundary of the cap (white arrows) present an intermediate albedo with no spectral signature of CO_2 frost and strong signatures of H_2O frost. The region identified by TES (black arrow) as H_2O ice-covered on the basis of its intermediate temperature presents weaker spectral signatures of H_2O ice than other regions covered by H_2O frost at the boundary of the CO_2 cap. The full extent of the large outlier (330°E to 10°E, 82°S to 84°S) was first observed by Themis (Piqueux et al., 2008). There is a striking similarity between the map of H_2O ice signatures at $L_s = 335–337°$ and that of bright regions covered by CO_2 frost at $L_s = 306°$ extending beyond the minimum extent of the CO_2 cap (arrows, from Figure 11.3a), which was obtained five Martian years earlier. A black and white version of this figure will appear in some formats.

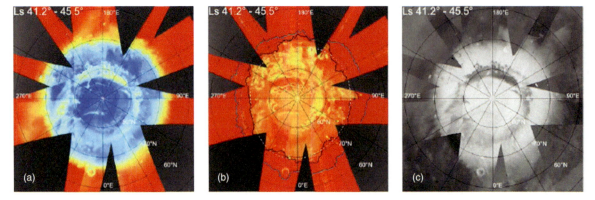

Figure 11.12 Evolution of the northern seasonal deposits in terms of (a) H_2O ice band depth at 1.5 μm, (b) CO_2 ice band depth at 1.43 μm, and (c) reflectance factor at 1.08 μm, for $L_s = 41.2–45.5°$. From Appéré et al. (2011). The OMEGA albedo maps are displayed over MOC albedo data acquired in summer (Caplinger and Malin, 2001) as a background. CO_2 ice and H_2O ice maps are displayed over MOLA topography (Zuber et al., 1998). Three boundaries are plotted on panel (b): the black outline corresponds to the OMEGA CO_2 ice boundary, the blue outline is the OMEGA H_2O ice boundary, and the white dashed outline is the TES "crocus line" (limit of low temperatures compatible with surface CO_2 ice, averaged over three Martian years of TES observations) computed for the L_s range. A black and white version of this figure will appear in some formats.

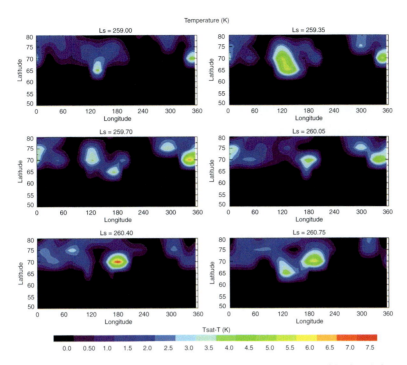

Figure 12.5 The number of degrees of temperature (derived from TES observations at approximately 1.37 mbar) below carbon dioxide saturation, $T_{sat}-T$ (K), for six representative periods during the northern winter. A black and white version of this figure will appear in some formats.

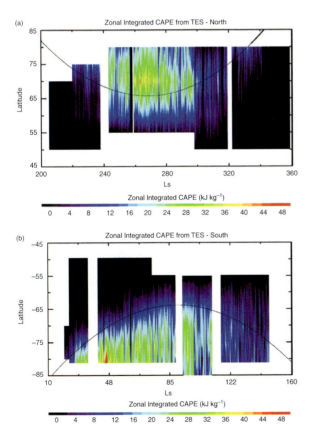

Figure 12.6 The zonally average CAPE derived from TES temperature profiles for the northern (a) and southern (b) hemispheres. The black curve shows the approximate location of the polar night boundary. A black and white version of this figure will appear in some formats.

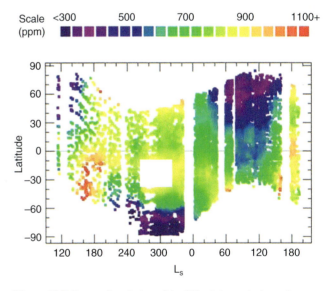

Figure 13.5 Seasonal evolution of the CO mixing ratio (ppmv) as observed between September 2006 and January 2009 by the CRISM instrument on the MRO spacecraft. The artifact caused by extreme dust loading between $L_s = 270°$ and $305°$ (white square) is masked out. From Smith et al. (2009). A black and white version of this figure will appear in some formats.

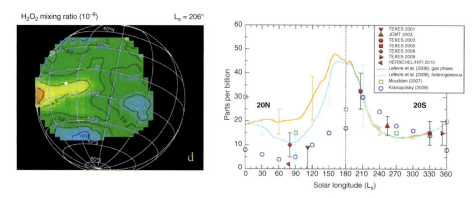

Figure 13.7 Observations of H_2O_2 on Mars. (Left) First mapping of the H_2O_2 mixing ratio (units of 10^{-8}) obtained in June 2003 at $L_s = 206°$ (Encrenaz et al., 2004). Spatial variations are thought to be related to variations in H_2O amount. (Right) Seasonal cycle of H_2O_2 on Mars. Observations in the infrared (TEXES) and models refer to 20°N for $L_s = 0$–180° and to 20°S for $L_s = 180$–360°, in order to match as best as possible the observing conditions induced by the axial tilt of the planet. Submillimeter observations using the James Clerk Maxwell Telescope (JCMT) by Clancy et al. (2004) refer to the entire disk. The points determined from Herschel at $L_s = 77°$ (Herschel/HIFI 2010) and TEXES 2001 at $L_s = 112°$ are upper limits. GCM simulations updated from Lefèvre et al. (2008) ignore (yellow curve) or include (blue curve) heterogeneous chemistry. Error bars for the GCM represent variability (2σ). Model values by Krasnopolsky (2009) and Moudden (2007) are indicated by blue and green circles, respectively. Updated from Encrenaz et al. (2012). A black and white version of this figure will appear in some formats.

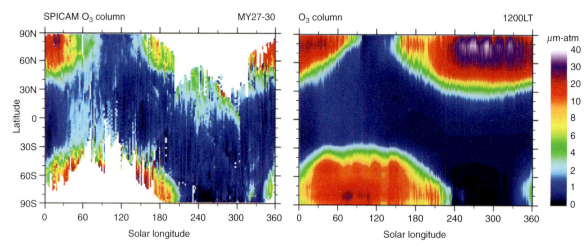

Figure 13.10 Seasonal evolution of the daytime ozone column (μm atm). (Left) Ultraviolet measurements from the SPICAM instrument on Mars Express averaged over MY 27–30 (updated from Perrier et al., 2006). (Right) Three-dimensional simulation by the LMD global climate model with photochemistry described by Lefèvre et al. (2004), with updated kinetics and improved water cycle. A black and white version of this figure will appear in some formats.

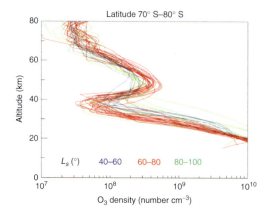

Figure 13.11 Nighttime O_3 vertical profile (molecule cm^{-3}) observed from 2004 to 2011 by the SPICAM instrument on Mars Express between 70°S and 80°S and covering $L_s = 40$–100°. In the southern hemisphere polar night, a secondary maximum peaking at 50 km is detected above the near-surface O_3 layer that is typical of polar regions in winter. Dust opacity prevents the observation below ~20 km. From Montmessin and Lefèvre (2013). A black and white version of this figure will appear in some formats.

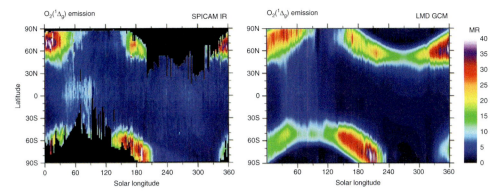

Figure 13.14 Seasonal evolution of the vertically integrated $O_2(^1\Delta_g)$ dayglow at 1.27 μm, in units of megarayleigh (1 MR = 10^{12} photon cm^{-2} s^{-1} $(4\pi$ $sr)^{-1}$ = 4.5×10^{15} cm^{-2} = 1.7 μm atm for the dayglow at 1.27 μm). (Left) Satellite observations averaged over 2004–2012 with the infrared channel of SPICAM on Mars Express (updated from Fedorova et al., 2006a). (Right) Three-dimensional simulation by the LMD global climate model with photochemistry described by Lefèvre et al. (2004), with updated kinetics and improved water cycle. Calculations were made with an $O_2(^1\Delta_g)$ quenching rate by CO_2 equal to 0.25×10^{-20} cm^3 s^{-1}. A black and white version of this figure will appear in some formats.

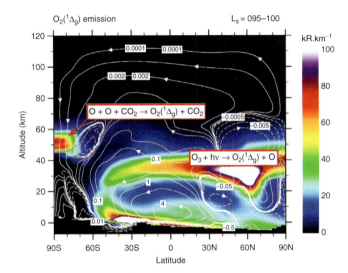

Figure 13.16 A schematic of the vertical and latitudinal distribution of the $O_2(^1\Delta_g)$ emission (zonally averaged, in kR km^{-1}) as simulated by the LMD general circulation model for $L_s = 95$–$100°$. Two primary pathways exist for the production of $O_2(^1\Delta_g)$. (1) Over solar-illuminated latitudes, the photolysis of ozone leads at this season to $O_2(^1\Delta_g)$ emission in a layer located above the hygropause between 25 and 40 km. A strong $O_2(^1\Delta_g)$ emission also outweighs the quenching by CO_2 near the surface in the southern hemisphere. This results from the dry surface conditions (see Figure 13.1) and subsequent large ozone densities at this season. (2) In the polar night, the three-body association of atomic oxygen transported downwards over the winter pole leads to $O_2(^1\Delta_g)$ emission between 45 and 60 km. The meridional streamfunction (10^9 kg s^{-1}) is superimposed in white contour lines. They visualize the downwelling branch of the winter hemisphere Hadley circulation, bringing large amounts of O atoms produced in the upper atmosphere towards lower altitudes into the polar night. From Clancy et al. (2012). A black and white version of this figure will appear in some formats.

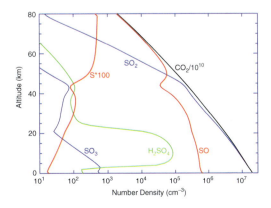

Figure 13.18 Vertical distribution of sulfur species (molecule cm^{-3}) for mean conditions below 80 km, assuming $SO_2 = 0.1$ ppbv at the surface. These results are obtained by inclusion of the SO_x chemistry in the 1D model of Krasnopolsky (2010). This chemistry involves updated data on photolysis of SO_2 and H_2SO_4, reaction between SO_3 and H_2O, and heterogeneous reaction between SO_2 and H_2O_2 on ice particles. Uptake of H_2SO_4 on water ice aerosols also occurs between 20 and 45 km and at night below 3 km. This explains the broad gas-phase H_2SO_4 peak between 3 and 20 km where uptake does not occur. A black and white version of this figure will appear in some formats.

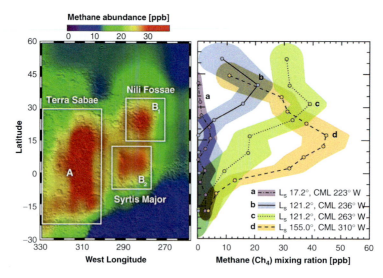

Figure 13.19 Methane mixing ratio (ppbv) observed from Earth by Mumma et al. (2009). (Left) Regions where methane appeared enhanced during summer 2003 (L_s = 121–155°). (Right) Latitudinal and temporal variability of methane. Profile a was obtained in February 2006 at L_s = 17.2°. Profiles b, c, and d were obtained in January–March 2003 between L_s = 121° and 155°. A black and white version of this figure will appear in some formats.

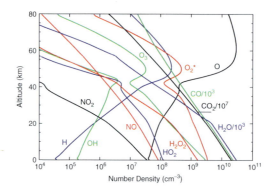

Figure 13.21 Density profiles (molecule cm^{-3}) calculated by the global mean photochemical model of Krasnopolsky (2010), using the chemical reaction rates listed in Table 13.2. O_2^* is $O_2(^1\Delta_g)$. The surface pressure is set to 6.1 hPa and the H_2O vertical column is 9.5 pr µm. The H_2O density profile is restricted by the saturation value. A downward flow of O and CO from the dissociation of O_2 and CO_2 above 80 km is also imposed at the upper boundary. The calculated abundances of the major photochemical products are 1600 ppmv for O_2, 120 ppmv for CO, 20 ppmv for H_2, 0.9 µm atm for O_3, 1.5 MR or 2.5 µm atm for $O_2(^1\Delta_g)$, and 7.6 ppbv for H_2O_2. These values are in reasonable agreement with the observations, with the exception of CO, which is underestimated by a factor of 8. A black and white version of this figure will appear in some formats.

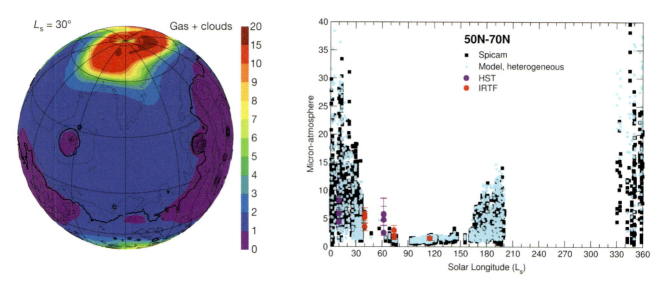

Figure 13.24 Ozone column (µm atm) calculated by the LMD general circulation model. (Left) Synoptic distribution in northern spring (L_s = 30°) with a photochemical scheme taking into account heterogeneous reactions of HO_x on water ice clouds. The observer is facing the 180° meridian at local noon, and the black contours correspond to Mars topography. (Right) Comparison of the same GCM simulation with the ozone columns measured in the 50–70°N latitude band by SPICAM (Perrier et al., 2006), HST (Clancy et al., 1999), and IRTF (Fast et al., 2006). From Lefèvre et al. (2008). A black and white version of this figure will appear in some formats.

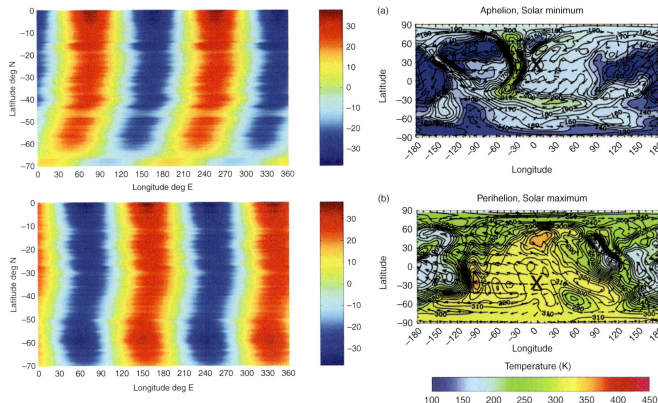

Figure 14.6 Accelerometer mass density wave-2 components (from wave-1, -2, -3 fit) at 130 km taken from MGS ($L_s = 28–93°$) and MRO ($L_s = 36–100°$) sampling: (top) dayside (LT = 13:00 to 17:00) MGS densities in the southern winter hemisphere, and (bottom) nightside (LT = 02:00–03:00) MRO densities in the southern winter hemisphere. Amplitudes are given as percent differences (see color bars) from the mean. Notice the 180° phase shift of these day versus night wave-2 amplitudes. From Keating et al. (2008). A black and white version of this figure will appear in some formats.

Figure 14.14 MTGCM simulated exobase temperatures and superimposed horizontal neutral winds for: (a) $L_s = 90°$, $F_{10.7} = 70$ and (b) $L_s = 270°$, $F_{10.7} = 200$ conditions, corresponding to solar cycle plus seasonal extremes. The subsolar points are indicated by the crosses (local noon meridian and solar declination latitude). Dayside (low SZA) temperatures range from ~170 to 310 K; nightside temperatures drop to as low as ~120–130 K. Maximum horizontal winds reach: (a) ~390 m s^{-1} to (b) ~550 m s^{-1}. The average exobase altitudes are: (a) ~165 km and (b) 195 km. Taken from Valeille et al. (2009). A black and white version of this figure will appear in some formats.

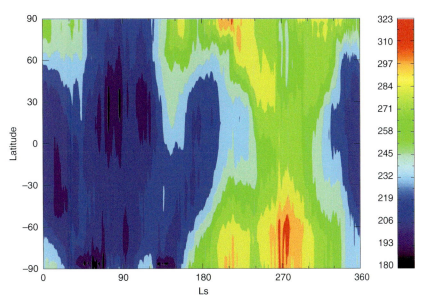

Figure 14.15 Zonal-mean temperatures (K) at the exobase predicted by the LMD-MGCM as a function of latitude and season. Figure taken from González-Galindo et al. (2009a). A black and white version of this figure will appear in some formats.

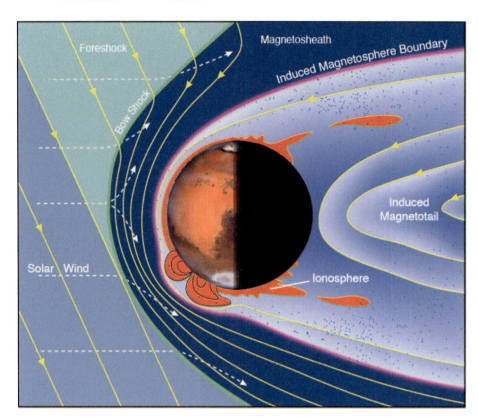

Figure 15.1 Schematic of the Martian plasma interaction region. Various plasma boundaries and regions are described in Section 15.2.1. Image courtesy of S. Bartlett. A black and white version of this figure will appear in some formats.

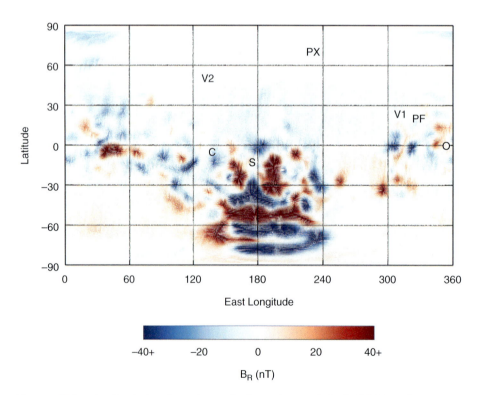

Figure 15.8 Geographic map of Mars showing the location and strength of crustal magnetic fields. Whiskers of average magnetic field strength at 400 km above the Martian nightside are colored according to the strength of the radial (vertical) field component. The length of each whisker is proportional to the log of the horizontal field component. The locations of landed spacecraft are indicated by initials: V1 and V2 for Viking 1 and 2, PF for Pathfinder, S and O for Spirit and Opportunity, PX for Phoenix, and C for Curiosity. A black and white version of this figure will appear in some formats.

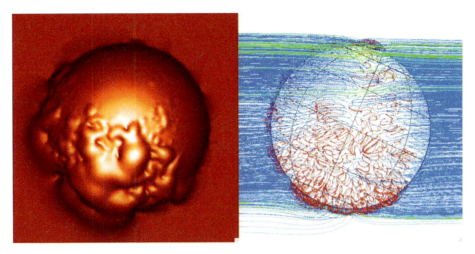

Figure 15.9 The results of calculations showing crustal field influences on the local plasma environment. (Left) The location at which incident solar wind pressure is balanced by a combination of ionospheric thermal pressure and crustal field magnetic pressure is shown as an isosurface. (Right) The configuration of magnetic field lines near Mars shows varied topology of closed (red), open (blue), and unconnected (green) magnetic field. Both images have 180°E as the central longitude. From Brain (2006). A black and white version of this figure will appear in some formats.

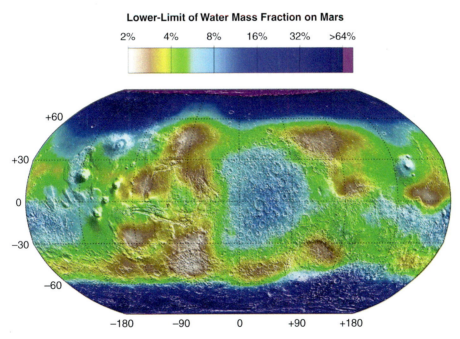

Figure 16.7 A map of apparent, average water content in the subsurface (%) estimated from neutron spectroscopy (Feldman et al., 2004). Fast neutrons are generated by cosmic rays, and the higher the concentration of hydrogen nuclei, the faster the neutrons thermalize, leaving fewer neutrons to escape at epithermal energies. Blue corresponds to a high concentration of water ice in the subsurface. The actual water ice concentration in the ice-rich layer below the dry sediment layer at high latitude should be significantly higher than these apparent values (Diez et al., 2008). Image credit: NASA/JPL/Los Alamos National Laboratory. A black and white version of this figure will appear in some formats.

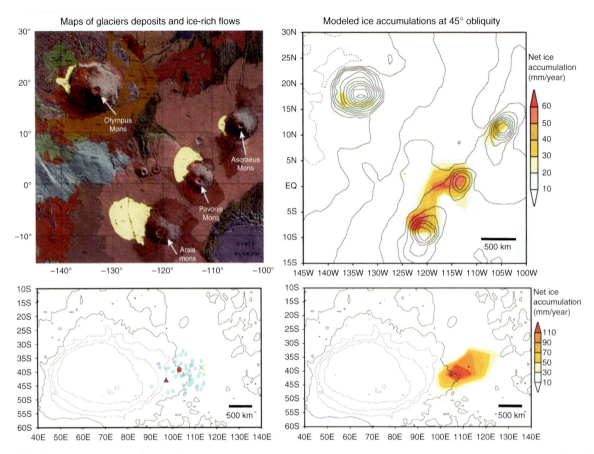

Figure 16.9 (Top left) Geologic map of the Tharsis region showing the location of fan-shaped deposits of Amazonian age (white-yellow) located on the northwest slopes of the Tharsis Montes and Olympus Mons. (Top right) Net surface water ice accumulation (in millimeters per Martian year) in the Tharsis region simulated by the LMD (Laboratoire de Météorologie Dynamique) GCM model with 45° obliquity and assuming that surface water ice is present on the northern polar cap. Superimposed MOLA topography contours are 2000 m apart. (Bottom left) Topographic map of the Hellas Basin with contours every 2 km showing the location of various glacier-like landforms such as lobate debris aprons (blue circles and purple triangle) and debris-covered glaciers (red dot, green cross). (Bottom right) Net surface water ice accumulation predicted in the same area by an LMD GCM simulation performed with 45° obliquity and assuming that surface water ice is initially present only in the southern polar cap. Figures from Forget et al. (2006). A black and white version of this figure will appear in some formats.

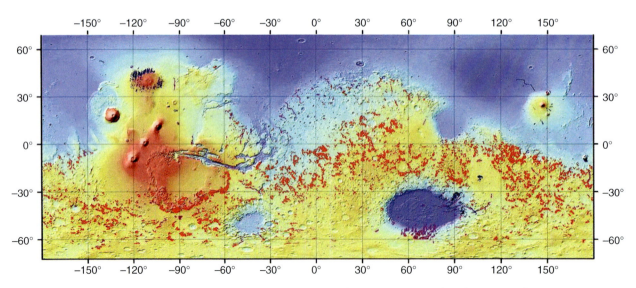

Figure 17.2 The global distribution of valleys. Valleys superimposed on Noachian terrain are shown in red. Younger valleys are in other colors. The Noachian valleys are mainly in terrain over 0 km in elevation. The absence of valleys in the low-lying Noachian of northwest Arabia (blue) has been attributed to the former presence of an ocean at the time that the valleys formed. The scarcity of valleys in the high-standing Noachian between Hellas and Argyre remains unexplained. From Hynek et al. (2010). A black and white version of this figure will appear in some formats.

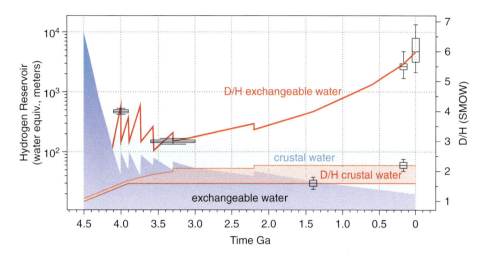

Figure 17.8 Time evolution of D/H (red line, right axis) and water reservoirs (blue shadings, left axis). Open box symbols are D/H SNC data with error bars representing ranges. Layered horizontal rectangles are D/H measurements from ALH 84001 (~3.9–4.1 Ga) and Gale clays (~3.0–3.5 Ga). Pale red hatched region represents D/H in the crustal reservoir. Jaggedness in some of the curves schematically illustrates how the system responds to catastrophic release of new water to the surface, such as outflow channels that release long-buried water from the deep crust. The general trend of D/H ratios growing more quickly before ~3 Ga than after is based on the assumption that H escape was faster then. A black and white version of this figure will appear in some formats.

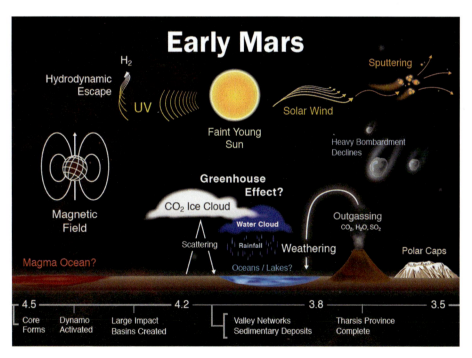

Figure 17.11 Cartoon summary of the processes acting on early Mars. Graphic courtesy of Christina Olivas. A black and white version of this figure will appear in some formats.

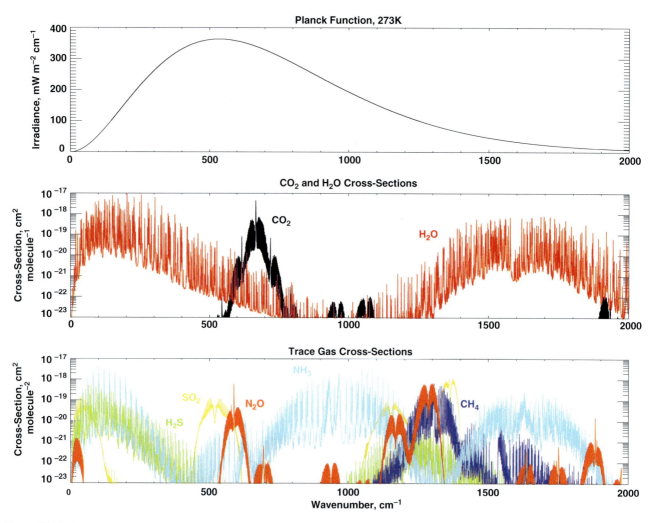

Figure 17.13 (Top panel) The Planck function for a 273 K surface. (Middle and bottom panels) The infrared absorption cross-sections for the gases listed in Table 17.4. A black and white version of this figure will appear in some formats.

in the Martian mesosphere, *Geophys. Res. Lett.*, 39, L02201, doi:10.1029/2011GL050343.

Sprague, A. L., Boynton, W. V., Kerry, K. E., et al. (2004) Mars' south polar Ar enhancement: a tracer for south polar seasonal meridional mixing. *Science*, 306, 5700, 1364–1367.

Sprague, A. L., Boynton, W. V., Kerry, K. E., et al. (2007) Mars' atmospheric argon: tracer for understanding Martian atmospheric circulation and dynamics, *J. Geophys. Res.*, 112, E3, E03S02.

Sprague, A. L., Boynton, W. V., Forget, F., et al. (2012) Interannual similarity and variation in seasonal circulation of Mars' atmospheric Ar as seen by the Gamma Ray Spectrometer on Mars Odyssey, *J. Geophys. Res.*, 117, E4, CiteID E04005.

Stoney, G. J. (1898) Of atmospheres upon planets and satellites, *Astrophys. J.*, 7, 25.

Sutter, B., Boynton, W. V., Ming, D. W., et al. (2012) The detection of carbonate in the Martian soil at the Phoenix Landing site: a laboratory investigation and comparison with the Thermal and Evolved Gas Analyzer (TEGA) data, *Icarus*, 218, 290–296.

Szwast, M. A., Richardson, M. I., and Vasavada, A. R. (2006) Surface dust redistribution on Mars as observed by the Mars Global Surveyor and Viking orbiters, *Journal of Geophysical Research*, 111, E11, E11008.

Tanaka, K. L., Kolb, E. J., and Fortezzo, C. (2007) Recent advances in the stratigraphy of the polar regions of Mars, *Seventh International Conference on Mars*, 1353, 3276.

Taylor, P. A., Kahanpää, H., Weng, W., et al. (2010) On pressure measurement and seasonal pressure variations during the Phoenix mission, *J. Geophys. Res.*, 115, E00E15. doi:10.1029/2009JE003422.

Thomas, P. C., Malin, M. C., Edgett, K. S., et al. (2000) North–south geological difference between the residual polar caps on Mars, *Nature*, 404, 161–4.

Thomas, P. C., Malin, M. C., James, P. B., et al. (2005) South polar residual cap of Mars: features, stratigraphy, and changes, *Icarus*, 174, 535–59, 2005.

Thomas, P. C., James, P. B., Calvin, W. M., et al. (2009) Residual south polar cap of Mars: stratigraphy, history, and implications of recent changes, *Icarus*, 203, 2, 352–375.

Thomas, P. C., Calvin, W. M., Gierasch, P., et al. (2013) Time scales of erosion and deposition recorded in the residual south polar cap of Mars, *Icarus*, 225, 2, 923–932.

Tillman, J. E. (1988) Mars global atmospheric oscillations: annually synchronized, transient normal mode oscillations and the triggering of global dust storms, *J. Geophys. Res.*, 93, 9433–9451.

Tillman, J. E., Johnson, N. C., Guttorp, P., and Percival, D. B. (1993) The Martian annual atmospheric pressure cycle – years without great dust storms, *J. Geophys. Res.*, 98, 10963–10971. doi:10.1029/93JE01084.

Titus, T. N. (2005a) Mars polar cap edges tracked over 3 full Mars years, in *Annu. Lunar Planet. Sci. Conf. XXXVI*, Houston, TX: Lunar and Planetary Institute, Abstract #1993, March 14–18.

Titus, T. N. (2005b) Thermal infrared and visual observations of a water ice lag in the Mars southern summer, *Geophys. Res. Lett.*, 32, L24204, doi:10.1029/2005GL024211.

Titus, T. N., and Kieffer, H. H. (2001) IR spectral properties of dust and ice at the mass south polar cap, American Astronomical Society, DPS Meeting #33, Abstract #19.15, *Bull. Am. Astron.*, 33, 1071.

Titus, T. N. and Michaels, T. I. (2009) Determining priorities for future Mars polar research, *Eos, Transactions American Geophysical Union*, 90, 40, 351–351.

Titus, T. N., and Cushing, G. E. (2014) Monitoring the Mars polar caps during Mars years 24–28, in *45th Lunar and Planetary Science Conference*, 17–21 March, The Woodlands, Texas. LPI Contribution No. 1777, 2177.

Titus, T. N., Kieffer, H. H., and Mullins, K. F. (1998) TES observations of the south pole, *American Astronomical Society, DPS meeting #30*, #20.05, *Bull. Amer. Astron. Soc.*, 30, 1049.

Titus, T. N., Kieffer, H. H., Mullins, K. F., and Christensen, P. R. (2001) TES premapping data: slab ice and snow flurries in the Martian north polar night, *J. Geophys. Res.*, 106, E10, 23181–23196.

Titus, T. N., Kieffer, H. H., and Christensen, P. R. (2003) Exposed water ice discovered near the south pole of Mars, *Science*, 299, 1048–1051.

Titus, T. N., Kieffer, H. H., Langevin, Y., et al. (2007) Bright fans in Mars cryptic region caused by adiabatic cooling of CO_2 gas jets, American Geophysical Union, Fall Meeting 2007, abstract #P24A-05.

Titus, T. N., Calvin, W. M., Kieffer, H. H., et al. (2008) Martian polar processes, in *The Martian Surface – Composition, Mineralogy, and Physical Properties*. Ed. Jim Bell, III. Cambridge University Press, Cambridge, 578.

Titus, T. N., Prettyman, T. H., Brown, A., et al. (2010a) Mars Ice Condensation and Density Orbiter, paper presented at LPSC XXXXI, LPI, Houston, TX.

Titus, T. N., Prettyman, T. H., Michaels, T., et al. (2010b) Mars Polar Science for the Next Decade Decadal Survey, White Paper, http://www8.nationalacademies.org/ssbsurvey/publicview.aspx, 6.

Tobie, G., Forget, F., and Lott, F. (2003) Numerical simulation of the winter polar wave clouds observed by Mars Global Surveyor Mars Orbiter Laser Altimeter, *Icarus*, 164, 1, 33–49.

Toigo, A. D., Richardson, M. I., Wilson, R. J., et al. (2002) A first look at dust lifting and dust storms near the south pole of Mars with a mesoscale model, *J. Geophys. Res.*, 107, E7, doi:10.1029/2001JE001592.

Toigo, A. D., Smith, M. D., Seelos, F. P., and Murchie, S. L. (2013) High spatial and temporal resolution sampling of Martian gas abundances from CRISM spectra, *J. Geophys. Res. Planets*, 118, 89–104, doi:10.1029/2012JE004147.

Tokar, R. L., Feldman, W. C., Prettyman, T. H., et al. (2002) Ice concentration and distribution near the south pole of Mars: synthesis of Odyssey and global surveyor analyses, *Geophys. Res. Lett.*, 29, 1904, doi:10.1029/2002GL015691.

Tokar, R., Elphic, R., Feldman W., et al. (2003) Mars Odyssey neutron sensing of the south residual polar cap, *Geophys. Res. Lett.*, 30, 13, 1677.

Toon, O. B., Pollack, J. B., Ward, W., Burns, J. A., and Bilski, K. (1980) The astronomical theory of climatic change on Mars, *Icarus*, 44, 552–607. doi:10.1016/0019-1035(80)90130-X.

Tyler, D., and Barnes, J. R. (2005) A mesoscale model study of summertime atmospheric circulations in the north polar region of Mars, *J. Geophys. Res. Planets*, 110, E06007.

Tyler, D. Jr., Barnes, J. R., and Skyllingstad, E. D. (2008) Mesoscale and large-eddy simulation model studies of the Martian atmosphere in support of Phoenix, *J. Geophys. Res.*, 113, E00A12, doi:10.1029/2007JE003012.

Vesovic, V., Wakeham, W. A., Olchowy, G. A., et al. (1990) The transport properties of carbon dioxide, *Journal of Physical and Chemical Reference Data*, 19, 3, 763–808.

Vincendon, M., Mustard, J., Forget, F., et al. (2010) Near-tropical subsurface ice on Mars., *Geophys. Res. Lett.*, 37, 1, L01202.

Wagstaff, K. L., Titus, T. N., Ivanov, A. B., Castaño, R., and Bandfield, J. L. (2008) Observations of the north polar water ice annulus on Mars using THEMIS and TES, *Planet. Space Sci.*, 56, 256–265., doi:10.1016/j.pss.2007.08.008.

Ward, W. R., Murray, B. C., and Malin, M. C. (1974) Climatic variations on Mars: 2. Evolution of carbon dioxide atmosphere and polar caps, *Journal of Geophysical Research*, 79, 24, 3387.

Warren, S. G., Wiscombe, W. J., and Firestone, J. F. (1990) Spectral albedo and emissivity of CO$_2$ in Martian polar caps – model results, *J. Geophys. Res.*, 95, 14717–14741.

Wells, E. H., and Hale, D. P. (1971) Flashes on Mars observed in 1937 and some random remarks, *Nature*, 232, 324–325.

Wilson, L. J. (1937) Apparent flashes seen on Mars, *Pop. Astron.*, 45, 430.

Winfree, K. N., and Titus, T. N. (2007) Trends in the south polar cap of Mars, in *Seventh International Conference on Mars*, July 9–13, Pasadena, CA, LPI Contribution No. 1353, 3373.

Wood, S. E. (1999) Nucleation and growth of CO$_2$ ice crystals in the Martian atmosphere, Thesis, UCLA, Los Angeles.

Wood, S. E., and Paige, D. A. (1992) Modeling the Martian seasonal CO$_2$ cycle 1: Fitting the Viking Lander pressure curves, *Icarus*, 99, 1, 1–14.

Yoder, C. F., and Standish, E. M. (1997) Martian precession and rotation from Viking Lander range data, *J. Geophys. Res.*, 102, E2, 4065–4080.

13

Atmospheric Photochemistry

FRANCK LEFÈVRE, VLADIMIR KRASNOPOLSKY

13.1 INTRODUCTION

The bulk atmosphere of Mars is almost entirely composed of CO_2, with small amounts of N_2 and Ar, and trace amounts of H_2O (Table 13.1). As a result, the only processes that can initiate Martian photochemistry are the photolysis of CO_2 and H_2O by ultraviolet solar light. Therefore, Martian photochemistry can be summarized by the interactions between oxygenated products of CO_2 and hydrogen products of H_2O. This picture is much simpler than that of other planetary atmospheres of our Solar System. For instance, contrary to Earth or Titan, nitrogen chemistry plays a minor role on Mars. And unlike Earth and Venus, chlorine and sulfur chemistries have not been identified. Also, Mars lacks the richness of the organic chemistry observed in Earth's troposphere, the giant planets, and Titan. Yet, despite this apparent simplicity, it must be recognized that Martian photochemistry is still not completely understood. As we will show in this chapter, significant progress has been made in our understanding of the short-lived species that determine the equilibrium state of the Martian atmosphere, such as ozone (O_3) or hydrogen peroxide (H_2O_2). However, for longer-lived species the situation is less satisfactory. In particular, the abundance of CO – the product of CO_2 photolysis – has been a challenge and remains unsolved for chemical models. The fact that such a basic problem persists after decades of research may be indicative that an important process is missing or largely inaccurate in our understanding of the Mars photochemical system.

It was revealed in the 1960s (e.g. Kliore et al., 1965; Belton and Hunten, 1966; Kaplan et al., 1969) that Mars could maintain an almost pure atmosphere of CO_2 (95.5%), with only trace amounts of its photodissociation products CO (volume mixing ratio of ~9×10^{-4}), O (~10^{-3} at 100 km), and O_2 (1.4×10^{-3}) (Table 13.1). At the time of that discovery, the remarkable stability of Mars CO_2 was in striking contradiction with the known kinetics of a pure CO_2 atmosphere exposed to sunlight. Indeed, solar ultraviolet radiation penetrates all the way to the surface in the thin Martian atmosphere and photodissociates CO_2 into carbon monoxide and atomic oxygen in its ground state, $O(^3P)$, or its excited state, $O(^1D)$:

$$CO_2 + h\nu \rightarrow CO + O(^3P) \qquad \lambda < 205\,\text{nm} \qquad (13.1)$$

$$CO_2 + h\nu \rightarrow CO + O(^1D) \qquad \lambda < 167\,\text{nm} \qquad (13.2)$$

For clarity, $O(^3P)$ will be denoted as O in the remainder of this chapter. Once CO_2 is photodissociated, it is difficult to restore it since the reverse reaction

$$CO + O + M \rightarrow CO_2 + M \qquad (13.3)$$

is spin-forbidden and therefore very slow (M represents a third body, usually CO_2 on Mars). Thus, the photolysis of CO_2 should in principle lead to a large buildup of CO and O. In addition, copious amounts of O_2 should be formed by the collision of oxygen atoms, which is about a hundred times more efficient than reaction (13.3):

$$O + O + M \rightarrow O_2 + M \qquad (13.4)$$

The net result of CO_2 photolysis is therefore

$$2CO_2 + h\nu \rightarrow 2CO + O_2 \qquad (13.5)$$

This suggests a production of CO and O_2 at a ratio of 2 : 1, only compensated by the slow reaction (13.3) and by the photolysis of O_2, which prevents O_2 building up indefinitely:

$$O_2 + h\nu \rightarrow O + O \qquad \lambda < 240\,\text{nm} \qquad (13.6)$$

A theoretical model of the photochemistry of a pure CO_2 atmosphere can be constructed from the set of reactions (13.1)–(13.6). The results of such a model are however in disagreement with the observations. The obtained mixing ratios of CO and O_2 are two orders of magnitude larger than the measured quantities (0.08 and 0.04, respectively, as calculated for instance by Nair et al. (1994)), and their 2 : 1 ratio is also opposite to the ~1 : 2 observed CO : O_2 ratio. These large differences were puzzling until the classic papers of McElroy and Donahue (1972) and Parkinson and Hunten (1972) showed that the composition of the Mars atmosphere could be precisely controlled by the "odd-hydrogen" species (H, OH, HO_2, known as HO_x) produced by the photolysis of water vapor. Although HO_x are only present in trace amounts of a few parts per billion (ppbv), we will show in Section 13.2 that these short-lived species can catalytically convert CO into CO_2 at a rate that is orders of magnitude faster than the direct reaction (13.3). In fact, this process is so efficient that photochemical models now have the opposite problem of underestimating CO relative to the observations. In Section 13.3 we will present the chemistry of oxygen and some of the large advances accomplished during the last decade in the observational knowledge of these compounds. The role of nitrogen species, which is poorly constrained by observations, will be described in Section 13.4. We will examine the possibility of sulfur and chlorine chemistry in Section 13.5. In Section 13.6, we will discuss the discovery of low levels of methane claimed by several teams since 2004, and also detected at the surface of Mars by the Curiosity Rover.

Table 13.1. *Observed chemical composition of the Martian atmosphere.*[a]

Species	Mixing ratio	Comments and references
CO_2	0.955–0.960	Kuiper (1949), global and annual mean pressure 6.1 mbar, Kliore et al. (1973), Mahaffy et al. (2013)
N_2	0.019–0.027	Owen et al. (1977), Mahaffy et al. (2013)
Ar	0.016–0.019	Owen et al. (1977), Mahaffy et al. (2013)
Ne	1–2.5 ppmv	Owen et al. (1977), Bogard et al. (2001)
Kr	0.3–0.36 ppmv	Owen et al. (1977), Bogard et al. (2001)
Xe	50–80 ppbv	Owen et al. (1977), Bogard et al. (2001)
H	$(3–30)\times10^4$ cm^{-3}	at 250 km for $T_\infty \approx 300$ and 200 K, Anderson and Hord (1971), Anderson (1974), Chaufray et al. (2008)
O	0.005–0.02	at 125 km, Strickland et al. (1972), Stewart et al. (1992), Chaufray et al. (2009)
O_2	$(1.2–1.4)\times10^{-3}$	Barker (1972), Carleton and Traub (1972), Trauger and Lunine (1983), Hartogh et al. (2010a), Mahaffy et al. (2013)
CO	300–1600 ppmv	Kaplan et al. (1969), Krasnopolsky (2007), Billebaud et al. (2009), Smith et al. (2009), Hartogh et al. (2010b), Sindoni et al. (2011)
H_2O	0–70 pr μm	Spinrad et al. (1963), Smith (2004), Fedorova et al. (2006b), Fouchet et al. (2007), Melchiorri et al. (2007), Tschimmel et al. (2008), Smith et al. (2009), Sindoni et al. (2011)
O_3	0–60 μm atm	Barth and Hord (1971), Barth et al. (1973), Clancy et al. (1999), Fast et al. (2006), Lebonnois et al. (2006), Perrier et al. (2006)
$O_2(^1\Delta_g)$	0.6–35 μm atm [b]	Noxon et al. (1976), Fedorova et al. (2006a), Krasnopolsky (2013)
He	10 ppmv	Krasnopolsky et al. (1994), Krasnopolsky and Gladstone (2005)
H_2	17 ppmv	Krasnopolsky and Feldman (2001)
H_2O_2	0–40 ppbv	Clancy et al. (2004), Encrenaz et al. (2004, 2012)
CH_4	0–40 ppbv	Formisano et al. (2004), Krasnopolsky et al. (2004), Mumma et al. (2009), Geminale et al. (2011), Krasnopolsky (2012)
	0.7–7 ppbv	Webster et al. (2015)
C_2H_6	<0.2 ppbv	Krasnopolsky (2012), Villanueva et al. (2013)
H_2CO	<3 ppbv	Krasnopolsky et al. (1997), Villanueva et al. (2013)
CH_3OH	<6.9 ppbv	Villanueva et al. (2013)
H_2S	<100 ppbv	Maguire (1977)
HCl	<0.2 ppbv	Hartogh et al. (2010a)
SO_2	<0.3 ppbv	Encrenaz et al. (2011), Krasnopolsky (2012)
NO	<1.7 ppbv	Krasnopolsky (2006b)
HCN	<2.1 ppbv	Villanueva et al. (2013)

[a] First detections along with the latest data are given; ppmv, parts per million; ppbv, parts per billion; pr μm, precipitable micrometers.

[b] $O_2(^1\Delta_g)$ dayglow at 1.27 μm is measured in megarayleigh (MR); 1 MR = 1.67 μm atm of $O_2(^1\Delta_g)$ for this airglow.

Finally, we will conclude (Section 13.7) with a summary of the current strengths and weaknesses of photochemical models and the outstanding issues in our quantitative understanding of Martian photochemistry.

We will only address in this chapter the photochemistry of the low and middle atmosphere up to about 100 km. The active chemistry taking place in the thermosphere is specifically addressed in Chapter 14.

13.2 HYDROGEN CHEMISTRY

In the lower atmosphere of Mars, the primary sources of odd hydrogen (HO_x) are the photolysis of water vapor,

$$H_2O + h\nu \rightarrow OH + H \qquad \lambda < 200 \text{ nm} \qquad (13.7)$$

and its oxidation by atomic oxygen in its excited state $O(^1D)$,

$$H_2O + O(^1D) \rightarrow OH + OH \qquad (13.8)$$

The latter process plays an important role in the lowest scale height near the surface. Above that level and up to the condensation level of water vapor (known as the hygropause), the photolysis of H_2O in (13.7) largely dominates the production of HO_x. This photolysis occurs in the same spectral domain as CO_2 and is therefore highly dependent on the overhead column of CO_2 (and, near the surface, to dust loading). For example, at a solar zenith angle of 60°, a mean Sun distance of 1.52 AU, and a visible dust opacity of 0.6, the H_2O photolysis frequency increases from 4×10^{-11} s^{-1} near the Martian surface (at 6 hPa) to 2×10^{-8} s^{-1} at 40 km. Even at this latter level, the timescale of the H_2O photolysis remains longer than one terrestrial year. Yet, this slow photolysis is sufficient to produce the few parts

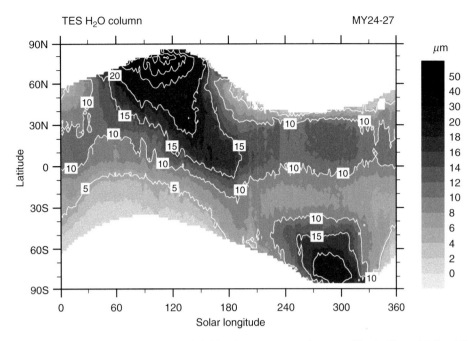

Figure 13.1. Seasonal evolution of the water vapor column (precipitable micrometers, pr μm) measured by the Thermal Infrared Spectrometer (TES) on-board Mars Global Surveyor (Smith, 2004). Data are averaged over years 1999–2004. Contour intervals are 5, 10, 15, 20, 30, 40, 50, and 60 pr μm.

per billion (ppbv) of HO_x that contribute, as we will show in the following, to the chemical stability of the Mars atmosphere.

The seasonal evolution of water vapor on Mars is now well documented by a large dataset of satellite observations. At high latitudes, sublimation and condensation processes near the pole lead to the alternation of a strong spring–summer seasonal maximum (30–50 pr μm) and a very dry (<1 pr μm) vortex in winter (Figure 13.1). Atmospheric transport propagates a fraction of this signal towards mid- to low latitudes where H_2O seasonal variations are less, around a mean value of ~10 pr μm (see Chapter 11 for more details). This general pattern is only representative of the vertically integrated H_2O column. Because of the strong altitude dependence of the H_2O photolysis in (13.7), it is very important to consider how H_2O is distributed along the vertical to understand the seasonal variations of the production rate of HO_x. The vertical distribution of H_2O is known to vary by a large extent with latitude, season, local time, and the local meteorological situation. For instance, ground-based (Clancy et al., 1996) and satellite (Maltagliati et al., 2011, 2013) measurements have shown that orbital changes in temperature at low and middle latitudes lead to a dramatic rise of the hygropause from 10 km around aphelion ($L_s = 71°$) up to 50 km around perihelion ($L_s = 251°$). This behavior of H_2O is well reproduced by three-dimensional (3D) models (see Figure 13.2; Montmessin et al., 2004). Even above the hygropause, models indicate that the warm, dusty atmosphere of perihelion can carry about a thousand times more H_2O than during the dust-free, cold aphelion period. The difference in the altitude of water vapor saturation – or sometimes supersaturation (Maltagliati et al., 2011) – and the amount of H_2O allowed to cross that cold trap have a dramatic impact on the amount of HO_x produced in the middle and upper atmosphere. This is illustrated in Figure 13.2, where global climate model simulations predict a two-orders-of-magnitude increase of odd-hydrogen species HO_x from aphelion to perihelion.

In fact, the dryness of the middle and upper atmosphere at aphelion is such that the role of H_2O as a source of HO_x may be severely reduced. In these conditions, an important source of HO_x then becomes the oxidation of H_2 by $O(^1D)$:

$$H_2 + O(^1D) \rightarrow OH + H \tag{13.9}$$

This process is less efficient than H_2O photolysis. However, in the dry conditions of aphelion, reaction (13.9) is the dominant HO_x production term at all altitudes above ~30 km. $O(^1D)$ is produced by photolysis of ozone in the lower and middle atmosphere, which is weakly dependent on the solar zenith angle. Reaction (13.9) also plays also a major role in the production of HO_x in the dehydrated regions at the sunlit edge of the polar night in winter. In the upper atmosphere, $O(^1D)$ is one product of CO_2 photolysis by ultraviolet radiation (reaction (13.2)), which becomes more and more available as altitude increases and the CO_2 opacity decreases. Thus, similar to H_2O photolysis in (13.7), the efficiency of reaction (13.9) at producing hydrogen radicals increases with altitude. It results from both processes that the total HO_x mixing ratio is a maximum in the upper atmosphere, with amounts varying from ~100 to ~10000 ppbv at 100 km, depending on the amount of H_2O at this level (Figure 13.3). At high altitude, HO_x is entirely in the form of H. The OH formed in (13.7)–(13.9) is indeed rapidly converted in (13.11) to H by the abundant O atoms that result from CO_2 photolysis and the quenching of $O(^1D)$:

$$CO_2 + h\nu \rightarrow CO + O(^1D) \tag{13.2}$$

$$O(^1D) + CO_2 \rightarrow O + CO_2 \tag{13.10}$$

$$OH + O \rightarrow H + O_2 \tag{13.11}$$

In the lower atmosphere, less O is available for reaction (13.11) and the main loss process of OH is in this case the reaction with CO, which forms CO_2:

$$CO + OH \rightarrow CO_2 + H \tag{13.12}$$

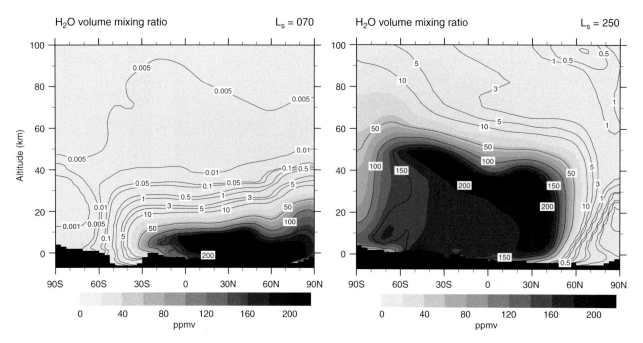

Figure 13.2. Zonally averaged distribution of water vapor calculated by the LMD (Laboratoire de Météorologie Dynamique) global climate model at aphelion $L_s = 70°$ (left) and perihelion $L_s = 250°$ (right). The simulation was performed with an updated version of the model described by Montmessin et al. (2004).

The hydrogen atom H obtained in (13.12) reacts in turn with O_2:

$$H + O_2 + M \rightarrow HO_2 + M \qquad (13.13)$$

The rate of this three-body reaction strongly increases with pressure. This explains why H has a very short lifetime (~1 s) in the lower atmosphere and is rapidly removed by reaction (13.13) to make the hydroperoxy radical HO_2 the dominant form of HO_x near the surface (Figure 13.3). Between both regimes (H-dominated at high altitudes, HO_2-dominated at low altitudes), OH is usually the main HO_x species in the middle atmosphere.

The production of HO_2 in (13.13) is at photochemical equilibrium with the fast loss reaction with the O atoms produced by CO_2 photolysis,

$$HO_2 + O \rightarrow OH + O_2 \qquad (13.14)$$

which recycles the OH initially lost in reaction (13.12).

This sequence of reactions (13.12)–(13.14) can therefore be represented by the following cycle, first proposed by McElroy and Donahue (1972):

Cycle I

$$CO + OH \rightarrow CO_2 + H \qquad (13.12)$$

$$H + O_2 + M \rightarrow HO_2 + M \qquad (13.13)$$

$$HO_2 + O \rightarrow OH + O_2 \qquad (13.14)$$

Net: $CO + O \rightarrow CO_2$

The net result of cycle I is the re-formation of CO_2 from its photolysis product CO. According to model simulations (Nair et al., 1994; Yung and DeMore, 1999; Krasnopolsky, 2010), it

contributes to about 85% of the recycling of CO into CO_2 in the Mars atmosphere. It is important to note that odd hydrogen is not lost in these cycles and therefore acts as a catalyst. From Table 13.2 it can be seen that the column-integrated loss rate of CO determined by (13.12) is about 100 times larger than the production rate of HO_x determined by H_2O photolysis. Thus, every HO_x radical produced by H_2O is used about 100 times in cycle I (and its variants presented in the following) to re-form CO_2 before it is returned to a less reactive form of hydrogen. The reaction (13.12) is the major process of formation of the CO–O bond that balances the breaking of this bond by the photolysis of CO_2. However, the production and loss rates of CO are not in local equilibrium. In the middle and upper atmosphere, the lack of OH makes reaction (13.12) inefficient and CO is produced without any significant loss (Figure 13.4). In contrast, between the surface and the hygropause, the CO loss rate by reaction (13.12) exceeds its production. CO has therefore a slow negative trend in the lower atmosphere, where its lifetime is about five terrestrial years. This long lifetime does not authorize chemical variations of CO with local time or even at the seasonal timescale. Yet, observations show a pronounced seasonal cycle of the CO mixing ratio at high latitudes (Figure 13.5), with an enhancement of more than a factor of 2 in winter and a depletion of the same order of magnitude in summer (Krasnopolsky, 2003a, 2007; Encrenaz et al., 2006; Smith et al., 2009). This phenomenon is not of chemical origin. It is a response to the condensation–sublimation cycle of CO_2 to and from the seasonal ice caps, which leads to enrichment–depletion of all non-condensable gases in polar regions, as also seen in observations of argon (Sprague et al., 2004).

The global photochemical equilibrium of the Martian atmosphere therefore results from a subtle balance between the CO production (CO_2 loss), occurring throughout the atmosphere

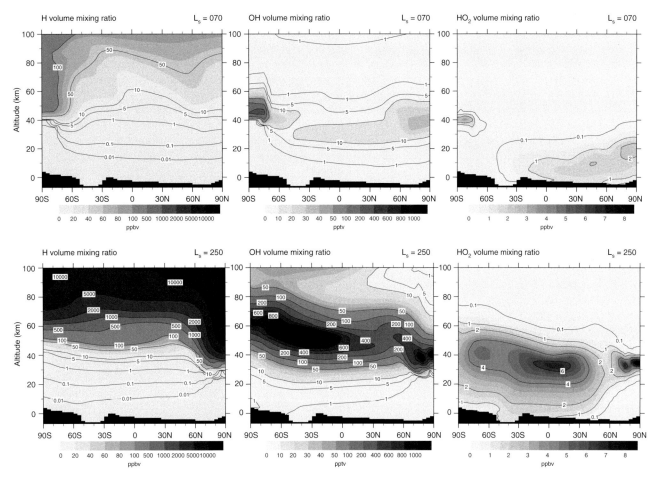

Figure 13.3. Zonally averaged distribution of odd-hydrogen species calculated by the LMD global climate model for daytime conditions at aphelion $L_s = 70°$ (top row) and perihelion $L_s = 250°$ (bottom row). (Left) Atomic hydrogen H (ppbv). Strong production of H by reactions (13.7)–(13.9) and weak loss by reactions (13.18) and (13.19) lead to a maximum at high altitudes. Note the two-orders-of-magnitude increase from aphelion to perihelion. At both periods, downward transport also brings H-enriched air towards lower altitudes above the winter pole. (Center) Hydroxyl radical OH (parts per trillion, pptv). OH peaks at the top at the hygropause at mid- to low latitudes. Maxima at high latitudes are due to large quantities of H and subsequent reaction (13.28). (Right) Hydroperoxy radical HO_2 (ppbv). From the model of Lefèvre et al. (2004) after two Martian years of simulation. Long-lived species O_2, CO, and H_2 are chemically integrated but were initialized to their observed mean mixing ratio (1400 ppmv, 800 ppmv, and 15 ppmv, respectively) at the beginning of the model run.

during the day, and the CO loss (CO_2 production) that occurs between the surface and the hygropause. Vertical mixing maintains a roughly constant CO mixing ratio in the lower atmosphere up to ~50 km. Measurements of Martian CO carried out from space or from the Earth converge towards a mean value of 900–1100 ppmv at middle latitudes (Krasnopolsky, 2007; Billebaud et al., 2009; Hartogh et al., 2010a; Sindoni et al., 2011). The most complete CO dataset, obtained with the CRISM instrument on-board Mars Reconnaissance Orbiter (Smith et al., 2009), indicates a lower globally averaged value of ~700 ppmv, with strong seasonal variations at high latitudes (Figure 13.5).

The amount of OH available for reaction (13.12) obviously plays a critical role in the equilibrium value of CO. As OH is a product of H_2O photolysis, it is essential to represent properly the strong variations of H_2O over the Martian year in models aimed at a quantitative understanding of Martian photochemistry. Clancy and Nair (1996) illustrated this problem by using a one-dimensional (1D) model forced by a stepwise annual cycle

of the water vapor profile. They showed that the variations of the condensation level of H_2O between perihelion and aphelion is the source of up to a two-orders-of-magnitude variation in HO_x density at 20–40 km, implying that the rate of the CO_2 recycling also varies considerably over one annual cycle.

Another indirect product of H_2O photolysis is the hydrogen peroxide H_2O_2 formed by the self-reaction of HO_2 radicals:

$$HO_2 + HO_2 \rightarrow H_2O_2 + O_2 \qquad (13.15)$$

Hydrogen peroxide is recognized to be a powerful oxidizer of organic material, as confirmed by laboratory experiments (Quinn and Zent, 1999; Gough et al., 2011). Mineral processing of H_2O_2 in the Martian surface could also lead to highly reactive hydrogen radicals or superoxides (Atreya et al., 2011) and might explain the apparent absence of organics as revealed by the Viking experiments (Biemann et al., 1976). It is evident from reaction (13.15) that the production of H_2O_2 is quadratically dependent on the concentration of HO_2 (and hence of H_2O). Models therefore predict H_2O_2 to be essentially confined to the

Table 13.2. *Main reactions in the lower and middle atmosphere of Mars, their rate coefficients, and column rates in the globally averaged model of Krasnopolsky (2010). The photolysis frequencies and column rates are dayside mean values. Velocity of H_2 at the upper boundary is that of the diffusion-limited flux. As suggested by Zahnle et al. (2008), a surface sink of O_3 and H_2O_2 is assumed to balance the escape of hydrogen to space and maintain the atmosphere's redox budget. It is possible to check the balance for each species in the model by comparing its column production and loss rates in the table. For example, H_2 is formed by $H+HO_2$ with a rate of 7.91×10^8 cm^{-2} s^{-1} and lost in the reactions with $O(^1D)$ and OH, with rates of 1.04×10^8 and 4.69×10^8 cm^{-2} s^{-1}, respectively. The difference between the production and loss, 2.18×10^8 cm^{-2} s^{-1}, escapes to space (V_{H2}). The escape of H_2 is balanced by the surface loss of oxygen in H_2O_2 and O_3, so that $3.43\times10^7+3\times6.13\times10^7 = 2.18\times10^8$ cm^{-2} s^{-1}. Balance of all other species may be checked in the same way.*[a]

Reaction	Rate coefficient	Column rate
$CO_2 + h\nu \rightarrow CO + O$	—	1.49×10^{12}
$CO_2 + h\nu \rightarrow CO + O(^1D)$	—	2.36×10^{10}
$O_2 + h\nu \rightarrow O + O$	—	1.70×10^{11}
$O_2 + h\nu \rightarrow O + O(^1D)$	—	1.11×10^{10}
$H_2O + h\nu \rightarrow H + OH$	—	1.76×10^{10}
$HO_2 + h\nu \rightarrow OH + O$	2.6×10^{-4}	4.23×10^{10}
$H_2O_2 + h\nu \rightarrow OH + OH$	4.2×10^{-5}	5.68×10^{10}
$O_3 + h\nu \rightarrow O_2(^1\Delta) + O(^1D)$	3.4×10^{-3}	6.38×10^{12}
$O(^1D) + CO_2 \rightarrow O + CO_2$	$7.4 \times 10^{-11}e^{120/T}$	6.42×10^{12}
$O(^1D) + H_2O \rightarrow OH + OH$	2.2×10^{-10}	1.50×10^{9}
$O(^1D) + H_2 \rightarrow OH + H$	1.1×10^{-10}	1.04×10^{8}
$O_2(^1\Delta) + CO_2 \rightarrow O_2 + CO_2$	10^{-20}	4.94×10^{12}
$O_2(^1\Delta) \rightarrow O_2 + h\nu$	2.24×10^{-4}	1.46×10^{12}
$O + CO + CO_2 \rightarrow CO_2 + CO_2$	$2.2 \times 10^{-33}e^{-1780/T}$	5.15×10^{7}
$O + O + CO_2 \rightarrow O_2 + CO_2$	$1.2 \times 10^{-32}(300/T)^2$	2.55×10^{10}
$O + O_2 + CO_2 \rightarrow O_3 + CO_2$	$1.4 \times 10^{-33}(300/T)^{2.4}$	6.45×10^{12}
$H + O_2 + CO_2 \rightarrow HO_2 + CO_2$	$1.7 \times 10^{-31}(300/T)^{1.6}$	1.93×10^{12}
$O + HO_2 \rightarrow OH + O_2$	$3 \times 10^{-11}e^{200/T}$	1.64×10^{12}
$O + OH \rightarrow O_2 + H$	$2.2 \times 10^{-11}e^{120/T}$	3.64×10^{11}
$CO + OH \rightarrow CO_2 + H$	1.5×10^{-13}	1.64×10^{12}
$H + O_3 \rightarrow OH + O_2$	$1.4 \times 10^{-10}e^{-470/T}$	6.17×10^{10}
$H + HO_2 \rightarrow OH + OH$	7.3×10^{-11}	2.36×10^{10}
$H + HO_2 \rightarrow H_2 + O_2$	$1.3 \times 10^{-11}(T/300)^{0.5}e^{-230/T}$	7.91×10^{8}
$H + HO_2 \rightarrow H_2O + O$	1.6×10^{-12}	5.18×10^{8}
$OH + HO_2 \rightarrow H_2O + O_2$	$4.8 \times 10^{-11}e^{250/T}$	1.71×10^{10}
$HO_2 + HO_2 \rightarrow H_2O_2 + O_2$	$2.3 \times 10^{-13}e^{600/T}$	5.75×10^{10}
$OH + H_2O_2 \rightarrow HO_2 + H_2O$	$2.9 \times 10^{-12}e^{-160/T}$	7.08×10^{8}
$OH + H_2 \rightarrow H_2O + H$	$3.3 \times 10^{-13}(T/300)^{2.7}e^{-1150/T}$	4.69×10^{8}
$O + O_3 \rightarrow O_2 + O_2$	$8 \times 10^{-12}e^{-2060/T}$	4.83×10^{7}
$OH + O_3 \rightarrow HO_2 + O_2$	$1.5 \times 10^{-12}e^{-880/T}$	1.30×10^{7}
$HO_2 + O_3 \rightarrow OH + O_2 + O_2$	$10^{-14}e^{-490/T}$	1.77×10^{8}
$H_2O_2 + O \rightarrow OH + HO_2$	$1.4 \times 10^{-12}e^{-2000/T}$	6.76×10^{6}
$NO_2 + h\nu \rightarrow NO + O$	0.0037	7.26×10^{10}
$NO_2 + O \rightarrow NO + O_2$	$5.6 \times 10^{-12}e^{180/T}$	1.78×10^{10}
$NO + HO_2 \rightarrow NO_2 + OH$	$3.5 \times 10^{-12}e^{250/T}$	9.04×10^{10}
diffusion-limited flux of H_2	$V_{H2} = 1.14 \times 10^{13}/[M]_{79km}$	2.18×10^{8}
loss of H_2O_2 at the surface	$V_{H2O2} = 0.02$	3.43×10^{7}
loss of O_3 at the surface	$V_{O3} = 0.02$	6.13×10^{7}
photolysis of O_2 above 80 km[a]	$V_{O2} = 0.32$	1.20×10^{10}

[a] Photolysis frequencies are in s^{-1}. Second- and third-order reaction rate coefficients are in cm^3 s^{-1} and cm^6 s^{-1}, respectively. Velocities V are in cm s^{-1}. Photolysis frequencies for HO_2, H_2O_2, O_3, and NO_2 refer to the lower atmosphere and are calculated for $\lambda > 200$ nm at 1.517 AU. Column reaction rates are in cm^{-2} s^{-1}. They are multiplied by $(1+z/R)^2$ at each altitude z to account for Mars sphericity (R being the Mars radius) and integrated from 1 to 79 km.

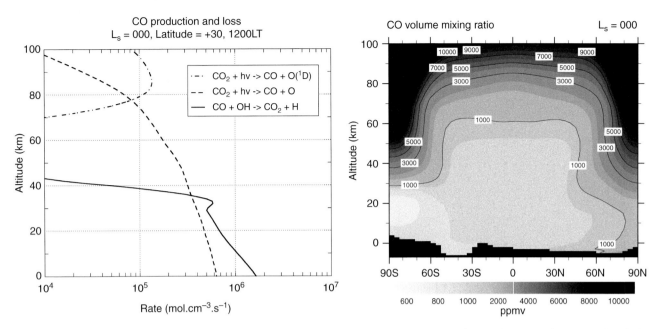

Figure 13.4. (Left) CO production and loss rates calculated at local noon and $L_s = 0°$, for a latitude of $30°$. (Right) Zonally averaged distribution of carbon monoxide CO (ppmv) calculated by the LMD (Laboratoire de Météorologie Dynamique) global climate model for daytime conditions at northern spring equinox ($L_s = 0°$). At high altitudes, strong CO_2 photolysis and weak loss by reaction with OH lead to a large increase in CO. At this season, downward transport brings CO-enriched air towards lower altitudes above both poles. From the model of Lefèvre et al. (2004) after a uniform initialization of CO = 800 ppmv as measured by CRISM at low to mid-latitudes and $L_s = 0°$ (Smith et al., 2009), followed by two Martian years of simulation.

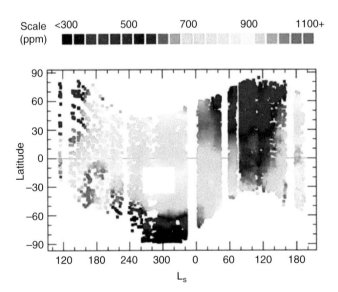

Figure 13.5. Seasonal evolution of the CO mixing ratio (ppmv) as observed between September 2006 and January 2009 by the CRISM instrument on the MRO spacecraft. The artifact caused by extreme dust loading between $L_s = 270°$ and $305°$ (white square) is masked out. From Smith et al. (2009). A black and white version of this figure will appear in some formats. For the color version, please refer to the plate section.

first 20–30 km above the surface where HO_2 is most abundant (Figure 13.6). During the day, the formation of H_2O_2 is close to photochemical equilibrium with its loss by photolysis,

$$H_2O_2 + h\nu \rightarrow OH + OH \qquad \lambda < 350 \text{ nm} \qquad (13.16)$$

which occurs with a typical timescale of ~6 hours. This is significantly slower than the loss processes of H, OH, and HO_2 that occur in a few seconds to minutes in the lower atmosphere. H_2O_2 can therefore accumulate from HO_2, and act as a temporary reservoir of HO_x. H_2O_2 also contributes to the re-formation of CO_2 and to the stability of the Mars atmosphere through the following variant of cycle I, obtained when the HO_2 produced in (13.13) reacts with itself to form H_2O_2 rather than with O:

Cycle II

$$2(CO + OH \rightarrow CO_2 + H) \qquad (13.12)$$

$$2(H + O_2 + M \rightarrow HO_2 + M) \qquad (13.13)$$

$$HO_2 + HO_2 \rightarrow H_2O_2 + O_2 \qquad (13.15)$$

$$H_2O_2 + h\nu \rightarrow OH + OH \qquad (13.16)$$

Net: $CO + O_2 \rightarrow CO_2$

Photochemical models estimate that this cycle initially proposed by Parkinson and Hunten (1972) contributes to about 8% of the total restoration of CO_2 from CO (Nair et al., 1994; Yung and DeMore, 1999; Krasnopolsky, 2010).

The fact that H_2O_2 is a measurable species that provides quasi-direct quantitative information on the amount of HO_x makes it a desirable target for observers. After years of searching, H_2O_2 was finally detected in 2003 by two independent measurements performed from Earth (Clancy et al., 2004; Encrenaz et al., 2004). These data as well as those obtained in the following years (Encrenaz et al., 2008, 2012; Hartogh

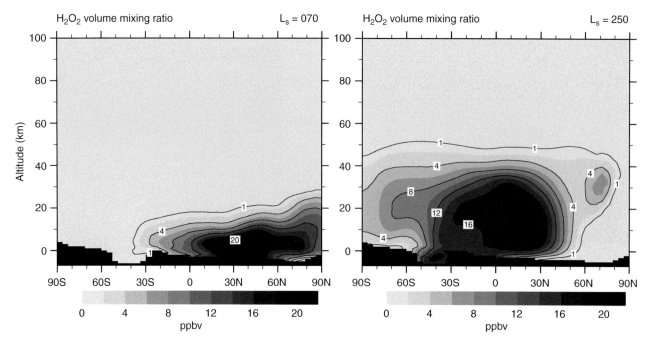

Figure 13.6. Zonally averaged distribution of hydrogen peroxide H_2O_2 (ppbv) calculated by the LMD global climate model for daytime conditions at aphelion $L_s = 70°$ (left) and perihelion $L_s = 250°$ (right). The distribution of H_2O_2 is tightly correlated to that of its source HO_2. From the model of Lefèvre et al. (2004).

et al., 2010a) indicate a substantial seasonal evolution of the H_2O_2 mixing ratio from less than 10 ppbv around aphelion to more than 30 ppbv at $L_s = 210°$ (Figure 13.7). The pronounced H_2O_2 minimum observed at aphelion occurs, surprisingly, when its primal source H_2O is largest. This phenomenon contradicts the expected H_2O–HO_x correlation and cannot be explained by models only considering the known gas-phase chemistry. However, it can be somewhat improved by models including heterogeneous chemical reactions on the water ice clouds that form in this season (Krasnopolsky, 2006a, 2009; Lefèvre et al., 2008). The H_2O_2 dataset thus may provide a clue that heterogeneous processes (discussed in Section 13.7) play a role in lowering the amount of HO_x and the oxidation rate of CO at certain times of the year, even though the very low upper limits for H_2O_2 determined by Encrenaz et al. (2002) and Hartogh et al. (2010a) are still challenging to reproduce and will require further investigation.

The cycles that oxidize CO into CO_2 terminate when the catalyzer HO_x is irreversibly returned to a long-lived form of hydrogen. The most efficient reaction of this type is

$$OH + HO_2 \rightarrow H_2O + O_2 \qquad (13.17)$$

which re-forms the stable species H_2O and limits the amount of HO_x that can exist in the atmosphere. Reaction (13.17) is therefore of fundamental importance for a quantitative understanding of Martian atmospheric chemistry. Unfortunately, this has been a difficult reaction to study. The recommended reaction rate is still subject to a large uncertainty at Martian temperatures (Sander et al., 2011). A minor contribution to the removal of HO_x also exists via the reaction between H and HO_2, forming either molecular hydrogen or water vapor:

$$H + HO_2 \rightarrow H_2 + O_2 \qquad (13.18)$$

$$H + HO_2 \rightarrow H_2O + O \qquad (13.19)$$

Most of the production of H_2 in (13.18) is balanced by its loss reactions with OH (80% of the loss) and $O(^1D)$ (20%):

$$H_2 + OH \rightarrow H_2O + H \qquad (13.20)$$

$$H_2 + O(^1D) \rightarrow OH + H \qquad (13.9)$$

The chemistry of H_2 is very slow in the lower atmosphere, where its lifetime is of the order of 300 terrestrial years. H_2 can therefore be transported passively to high altitudes where the molecule is essentially destroyed by ionospheric reactions. These processes produce H atoms, which may then escape to space. H_2 was detected in the upper atmosphere of Mars by Krasnopolsky and Feldman (2001) using the Far Ultraviolet Spectroscopic Explorer. Extrapolation of the observed abundances to the middle and lower atmosphere resulted in a mixing ratio of 17 ppmv (Table 13.1). Finally, odd hydrogen may also be removed by condensation of H_2O_2 and precipitation of solid H_2O_2 onto the surface. This process is calculated to occur at $T < 180$ K for an H_2O_2 mixing ratio of ~10 ppbv and a pressure of 6 hPa (Kong and McElroy, 1977b; Lindner, 1988).

13.3 OXYGEN CHEMISTRY

On Earth, the reactive forms of oxygen O, $O(^1D)$, and O_3 result from the photolysis of O_2, which is itself produced in large amounts (21% of the total composition) by photosynthesis. In the absence of such process on Mars, O_2 has to be formed photochemically and is present in much smaller quantities. The first Earth-based measurements of Martian O_2 used the lines of the O_2

H$_2$O$_2$ mixing ratio (10^{-8})   L$_s$ = 206°

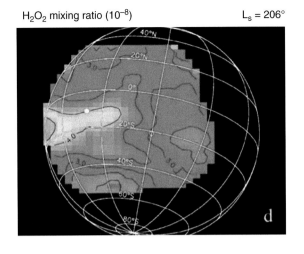

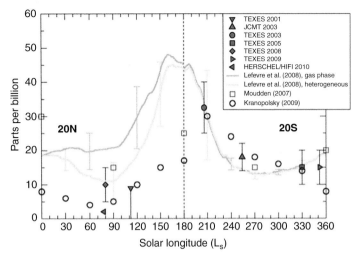

Figure 13.7. Observations of H$_2$O$_2$ on Mars. (Left) First mapping of the H$_2$O$_2$ mixing ratio (units of 10^{-8}) obtained in June 2003 at L_s = 206° (Encrenaz et al., 2004). Spatial variations are thought to be related to variations in H$_2$O amount. (Right) Seasonal cycle of H$_2$O$_2$ on Mars. Observations in the infrared (TEXES) and models refer to 20°N for L_s = 0–180° and to 20°S for L_s = 180–360°, in order to match as best as possible the observing conditions induced by the axial tilt of the planet. Submillimeter observations using the James Clerk Maxwell Telescope (JCMT) by Clancy et al. (2004) refer to the entire disk. The points determined from Herschel at L_s = 77° (Herschel/HIFI 2010) and TEXES 2001 at L_s = 112° are upper limits. GCM simulations updated from Lefèvre et al. (2008) ignore (yellow curve) or include (blue curve) heterogeneous chemistry. Error bars for the GCM represent variability (2σ). Model values by Krasnopolsky (2009) and Moudden (2007) are indicated by blue and green circles, respectively. Updated from Encrenaz et al. (2012). A black and white version of this figure will appear in some formats. For the color version, please refer to the plate section.

band at 762 nm and indicated a mixing ratio of 1.2×10^{-3} (Barker, 1972; Carleton and Traub, 1972; Trauger and Lunine, 1983). Observations performed in the submillimeter range from the Herschel/HIFI instrument have derived a mean O$_2$ mixing ratio of 1.4×10^{-3} (Hartogh et al., 2010a). The only *in situ* measurement of O$_2$ was performed by the Sample Analysis at Mars (SAM) suite of instruments on-board the Curiosity Rover (Mahaffy et al., 2013). The O$_2$ mixing ratio reported from SAM is 1.45×10^{-3}, in very good agreement with the previous Herschel observations.

As mentioned in Section 13.1, the most evident pathway to form O$_2$ in a CO$_2$ atmosphere is the recombination of O atoms via

$$O + O + M \rightarrow O_2 + M \qquad (13.4)$$

On Mars, this reaction plays an important role in the production of O$_2$ above 50 km but represents only ~10% of the formation rate of the O–O bond integrated over the whole atmosphere. Due to the presence of OH in the middle and lower atmosphere, a much more efficient process to form the O–O bond is the reaction

$$OH + O \rightarrow O_2 + H \qquad (13.11)$$

which is responsible for ~90% of the O$_2$ production on Mars. O$_2$ can therefore be considered as an indirect product of the CO$_2$ and H$_2$O photolysis, since these processes provide the reactants O and OH of reaction (13.11). This gives another illustration of the importance of the HO$_x$ radicals in regulating Martian atmospheric chemistry. Reactions (13.14), (13.15), (13.17), and (13.18) producing O$_2$ from the reactant HO$_2$ are not considered as production mechanisms of O$_2$ since one O$_2$ molecule is used in (13.13) to form HO$_2$.

The main (75%) loss process of O$_2$ occurs through reaction (13.13) with H, but this reaction does not break the O–O bond and O$_2$ is immediately returned by reaction (13.14) between HO$_2$ and O. The primary process breaking the O–O bond is the O$_2$ photolysis:

$$O_2 + h\nu \rightarrow O(^1D) + O \qquad \lambda < 175 \text{ nm} \qquad (13.21)$$

$$O_2 + h\nu \rightarrow O + O \qquad 175 \text{ nm} < \lambda < 240 \text{ nm} \qquad (13.22)$$

Although the O$_2$ photolysis represents a minor channel (<5%) in terms of instantaneous loss of O$_2$, this mechanism accounts for ~50% of the total destruction of the O–O bond in the Mars atmosphere. Other important processes to consider are the photolysis of HO$_2$ and H$_2$O$_2$, as well as the reaction

$$H + HO_2 \rightarrow OH + OH \qquad (13.23)$$

These three processes all produce OH and account for ~30% of the breaking of the O–O bond. The remaining ~20% is theoretically accomplished by the reaction between NO and HO$_2$, if one assumes a mean amount of 0.6 ppbv of nitrogen oxides in the low atmosphere (see Section 13.4). From these various loss processes, the mean lifetime of O$_2$ in the Mars atmosphere is estimated to be about 60 years (Krasnopolsky, 2010). O$_2$ is therefore expected to be uniformly mixed by atmospheric transport, and should be subject only to slow seasonal variations induced by the condensation and sublimation cycle of CO$_2$ in polar regions.

Due to the low abundance of O$_2$, the O$_2$ photolysis in (13.21) and (13.22) is not like on Earth the main contributor to the production of what is generally referred as the "odd-oxygen" family, O(^{1}D)+O+O$_3$, denoted O$_x$. On Mars this role is played at

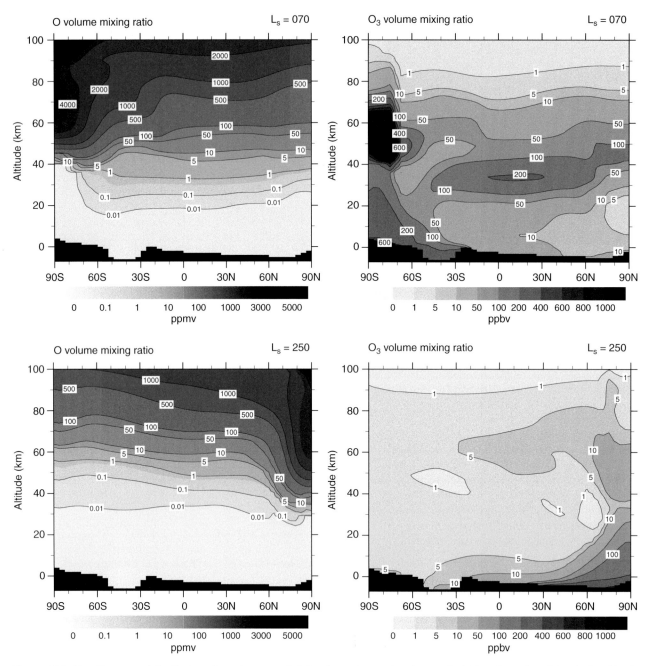

Figure 13.8. Zonally averaged distribution of atomic oxygen O (left column, in ppmv) and ozone O_3 (right column, in ppbv) calculated by the LMD global climate model for daytime conditions, near aphelion $L_s = 70°$ (top row) and near perihelion $L_s = 250°$ (bottom row). From the model of Lefèvre et al. (2004) after two Martian years of simulation. Long-lived species O_2, CO, and H_2 are chemically integrated but were initialized to their observed mean value (1.4×10^{-3}, 800 ppmv, and 15 ppmv, respectively) at the beginning of the model run.

~80% by the CO_2 photolysis, which produces rapidly increasing amounts of O atoms with altitude. At 100 km, the O mixing ratio may reach values of several thousands of ppmv that approach those of CO (Figure 13.8), the other direct product of CO_2 photolysis.

The loss process of O atoms depends on the altitude. In the thermosphere and down to ~70 km, the only significant sink of O is reaction (13.4). This reaction is slow and authorizes a chemical lifetime of several months for the O_x family that is dominated by O in the upper atmosphere (Figure 13.9). O atoms may therefore be transported quasi-passively by the

Hadley circulation from their region of production into the polar night, where they are brought towards lower altitudes by the downwelling motion above the winter pole (Figure 13.8). In the middle and lower atmosphere, the main sink of O atoms is the well-known three-body reaction with O_2 leading to the formation of ozone:

$$O + O_2 + M \rightarrow O_3 + M \qquad (13.24)$$

This reaction is a loss of O but is null in terms of O_x since O_3 is produced. In the sunlit portion of the atmosphere, the inverse process returning O from O_3 within the O_x family is the O_3

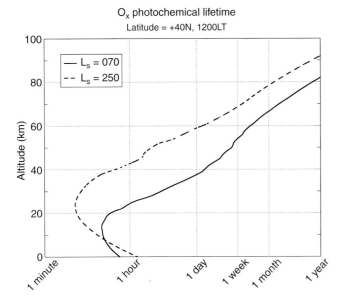

Figure 13.9. Mean photochemical lifetime of the O_x family $(O+O(^1D)+O_3)$ calculated by the model of Lefèvre et al. (2004) at 40°N and local noon for $L_s = 70°$ (solid curve) and $L_s = 250°$ (dashed curve). Rise of the hygropause and the increase in upper atmosphere H_2O from aphelion to perihelion cause a dramatic reduction of the O_x lifetime above 20 km.

photolysis, which occurs at a timescale shorter than 5 minutes at all altitudes:

$$O_3 + h\nu \rightarrow O + O_2 \qquad \text{~90\% for } \lambda > 320 \text{ nm} \qquad (13.25)$$

$$O_3 + h\nu \rightarrow O(^1D) + O_2(^1\Delta_g) \qquad \text{~90\% for } \lambda < 305 \text{ nm} \qquad (13.26)$$

The $O(^1D)$ produced in (13.26) is then rapidly quenched by CO_2 in reaction (13.10). Thus, the reactions of exchange between O, $O(^1D)$, and O_3 are fast and each of those species is at photochemical equilibrium during the day within the O_x family. For ozone, the daytime steady-state expression for production and loss leads to the following ratio of abundances:

$$\frac{O_3}{O} = \frac{k_{24}[O_2][CO_2]}{J_{O3}} \qquad (13.27)$$

Here k_{24} is the reaction rate (cm^6 s^{-1}) of reaction (13.24) and J_{O3} is the photolysis frequency (s^{-1}) of ozone. This ratio is of the order of ~1 at 25 km and of ~50–100 near the surface. O_3 is therefore the dominant O_x species in the lower atmosphere. It is also the only O_x species that is directly measurable by remote sensing, thanks to its strong absorption bands both in the ultraviolet (255 nm) and in the infrared (9.6 μm). These properties allow relatively easy access to the abundance of O_3 in the Martian atmosphere, provided that the observation is performed outside the Earth's ozone-rich atmosphere, or that the terrestrial absorption can be eliminated by Doppler shift. Early in the space age, Martian O_3 was searched for and discovered by the Mariner ultraviolet spectrometers in 1969 and 1971–1972 (Barth and Hord, 1971; Barth et al., 1973). The first positive measurements were of O_3 columns of 10–50 μm atm and were only found in the polar regions. This must be compared to the

Earth's globally averaged O_3 column of ~3 mm atm. With such low quantities, ozone on Mars does not offer an efficient shield comparable to the Earth's ozone layer against solar ultraviolet radiation and all wavelengths larger than 200 nm can reach the Martian surface. The ultraviolet spectrometer of Mariner also revealed that ozone is strongly variable on Mars. While positive detections were obtained at high latitudes in winter–spring, ozone was below the detection limit of the instrument (3 μm atm) equatorward of 45° latitude during any season. Lane et al. (1973) noted a rise in ozone measured by Mariner 9 when water vapor froze out of the atmosphere. This fact was reproduced by the modeling studies that followed in the late 1970s (e.g. Liu and Donahue, 1976; Kong and McElroy, 1977b; Krasnopolsky and Parshev, 1979; Shimazaki and Shimizu, 1979), which highlighted the crucial role played by H_2O and its photolysis products HO_x in the chemistry and variations of ozone. This role is easy to understand since model calculations show that the strongest sink of O_x in the lower and middle atmosphere is

$$HO_2 + O \rightarrow OH + O_2 \qquad (13.14)$$

This reaction suppresses irreversibly the O atoms otherwise used in (13.24) to form O_3. It is also one of key steps of the cycle I described in Section 13.2. The efficiency of this catalytic cycle is such that the chemical lifetime of O_x determined by reaction (13.14) is, during the day, shorter than 1 hour at low to middle latitudes below 25 km (Figure 13.9). Ozone, the main form of O_x at those altitudes, is therefore quite sensitive to the amount of HO_2 and is expected to be anticorrelated to H_2O, the source of HO_x. This tight coupling makes O_3 a useful tracer of the odd-hydrogen chemistry that stabilizes the CO_2 atmosphere of Mars, and O_3 measurements offer a powerful constraint for photochemical models.

There has been much progress in recent years in the knowledge of the climatology of O_3 on Mars, which is the short-lived species best described by the observations. Latitudinal cross-sections of the O_3 column at several Mars seasons have been obtained from the Earth using infrared heterodyne spectroscopy (Espenak et al., 1991; Fast et al., 2006, 2009) and from the Hubble Space Telescope (HST) in the ultraviolet range (Clancy et al., 1999). Since 2004, the SPICAM ultraviolet spectrometer on-board Mars Express has also been providing a near-complete coverage of the O_3 column distribution (Perrier et al., 2006), as well as the first observations of the O_3 nighttime distribution (Lebonnois et al., 2006). All these measurements have clearly identified the anticorrelation between O_3 and H_2O that is predicted by the photochemical theory. A significant example is provided by the O_3 column climatology obtained by SPICAM in three Martian years of observation (Figure 13.10), which is most of the time inversely related to the H_2O column abundance as measured by TES (Figure 13.1). Compared to Earth, the very strong seasonal and spatial variability of ozone (a factor of ~100 as opposed to ~3 on our planet) is a striking feature of Martian photochemistry. As for Mariner, the maximum abundances of ozone measured by SPICAM are near 40 μm atm and are found at high latitudes in winter, when the absence of water vapor and lack of sunlight prevent almost completely the production of HO_x. In these conditions, the chemical lifetime of ozone may increase from its typical value of approximately an hour up to several days, and ozone can accumulate inside the polar vortex.

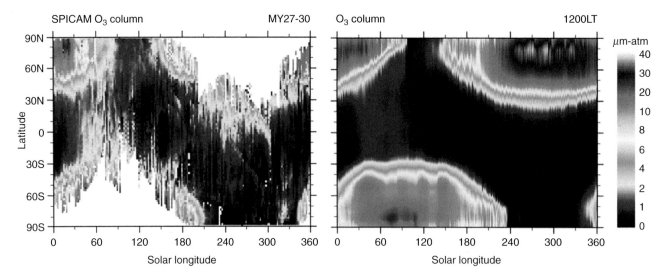

Figure 13.10. Seasonal evolution of the daytime ozone column (μm atm). (Left) Ultraviolet measurements from the SPICAM instrument on Mars Express averaged over MY 27–30 (updated from Perrier et al., 2006). (Right) Three-dimensional simulation by the LMD global climate model with photochemistry described by Lefèvre et al. (2004), with updated kinetics and improved water cycle. A black and white version of this figure will appear in some formats. For the color version, please refer to the plate section.

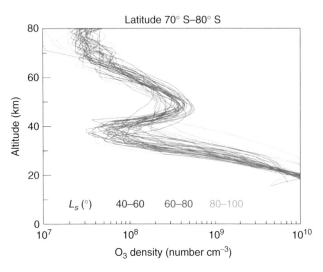

Figure 13.11. Nighttime O_3 vertical profile (molecule cm^{-3}) observed from 2004 to 2011 by the SPICAM instrument on Mars Express between 70°S and 80°S and covering $L_s = 40–100°$. In the southern hemisphere polar night, a secondary maximum peaking at 50 km is detected above the near-surface O_3 layer that is typical of polar regions in winter. Dust opacity prevents the observation below ~20 km. From Montmessin and Lefèvre (2013). A black and white version of this figure will appear in some formats. For the color version, please refer to the plate section.

This is especially true at polar low altitudes (Figure 13.8), where model simulations predict that ozone is not reached by the ozone-destroying HO_x produced by the descent of H atoms from the thermosphere (Figure 13.3). This model calculation is also supported by the observations of the vertical distribution of O_3 performed by SPICAM in stellar occultation mode, showing that most of the polar ozone (in terms of number density) is confined below 20 km in winter (Lebonnois et al., 2006). Recently, a further analysis of SPICAM polar night data also revealed

the presence of a secondary O_3 layer between 30 and 60 km that is only present in the southern hemisphere (Montmessin and Lefèvre, 2013; Figure 13.11). This layer is well reproduced by GCM simulations (Figure 13.8) and shows O_3 mixing ratios larger than anywhere else (0.5–1.5 ppmv) on the planet at this time of the year. Its formation is related to the deep vertical downwelling of O-rich air masses by the Hadley circulation, visible in Figure 13.8, which promotes the formation of ozone by reaction (13.24). The fact that no elevated O_3 polar layer is observed in the northern hemisphere can be explained by the much larger H amounts that are present near the north pole at that time of year (see Figure 13.3). As the reaction

$$H + O_3 \rightarrow OH + O_2 \qquad (13.28)$$

is the main ozone loss process above 20 km in the polar night, the destruction of ozone is about 100 times stronger above the northern winter pole than above its southern counterpart, which prevents a significant buildup of polar ozone in the middle atmosphere. The OH radicals produced in (13.28) are vibrationally excited and are a source of near-infrared emission (Meinel band) in the upper terrestrial mesosphere (Bates and Nicolet, 1950). Recently this OH emission was detected for the first time on Mars from CRISM limb observations of polar night (Clancy et al., 2013a). The emission was observed in a distinct layer between 40 and 55 km. GCM simulations show that the OH polar nightglow is stimulated by the descent of H atoms from the thermosphere, which provides a further indication of the deep Hadley solstitial circulation of the Mars atmosphere.

In terms of integrated ozone vertical column, in the polar vortices only the large concentrations of ozone located near the surface are significant. In these conditions the photochemistry is slow and atmospheric transport plays a major role. Three-dimensional model simulations have shown (Lefèvre et al., 2004) that the confinement of large O_3 column amounts inside the polar vortex and the dynamical disturbances in the shape of the vortex caused by wave activity can explain the considerable

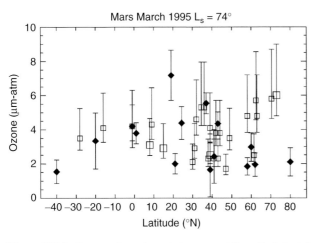
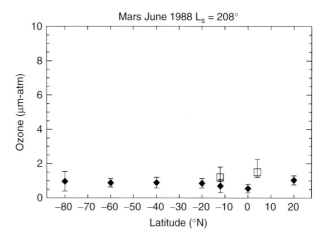

Figure 13.12. Ozone column (µm atm) measured from Earth by infrared heterodyne spectroscopy (filled symbols; Fast et al., 2006) and from the Hubble Space Telescope Faint Object Spectrograph in the ultraviolet (open symbols; Clancy et al., 1999): (left) March 1995, $L_s = 74°$; (right) June 1988, $L_s = 208°$. From Fast et al. (2006).

day-to-day variability in ozone measured at high latitudes by Mariner (Traub et al., 1979), SPICAM (Perrier et al., 2006), and the MARCI imager on Mars Reconnaissance Orbiter (Clancy et al., 2016).

The ozone annual cycle at low latitudes is of much more limited amplitude than at high latitudes. Ultraviolet measurements performed with the HST (Clancy et al., 1999), later confirmed by infrared observations from the Earth (Fast et al., 2006), gave evidence of a broad maximum of 2–6 µm atm centered on Mars aphelion, which contrasts with O_3 columns of ~1 µm atm shortly after equinox (Figure 13.12). At low latitudes SPICAM and MARCI (Clancy et al., 2016) also show an ozone maximum around aphelion (although weaker than in the HST and Earth-based data) as well as substantial quantities between $L_s = 240°$ and 300° likely associated with the southward transport of O_3-rich air masses from the northern polar vortex (Figure 13.10). Overall the O_3 at low to mid-latitudes is not clearly anticorrelated to the amount of H_2O as measured by TES (Figure 13.1). Indeed, the variation in ozone is mainly caused by changes in the vertical distribution of H_2O rather than in the H_2O integrated column. As already mentioned in Section 13.2, the wide vertical excursion of the H_2O saturation altitude over the Martian year cause dramatic changes in the modeled HO_x content of the middle atmosphere (Figure 13.3). Around aphelion, the HO_x production is much reduced above the low hygropause and so is the O_x loss, as illustrated in Figure 13.9, where the photochemical lifetime of O_x at $L_s = 70°$ reaches several hours to one day above 25 km. This leads to the buildup of a prominent seasonal O_3 layer above 25 km, first hypothesized by Clancy and Nair (1996) and reproduced by 3D simulations (Lefèvre et al., 2004). Figure 13.8 shows however that the quantitative agreement with the O_3 observations at low latitudes is still imperfect. It is likely that a better match requires a very accurate description of the H_2O condensation and sedimentation processes around the hygropause level, which are difficult to represent in the models. With only a few uncertain O_3 profiles obtained at the same season ($L_s = 10–15°$) from the Mars 5 (Krasnopolsky and Parshev, 1979) and Phobos 2 (Blamont and Chassefière, 1993) spacecraft, the modeled evolution of O_3 at

low to middle latitudes could not be proved until the extensive O_3 profiling campaign by SPICAM provided a clear validation of the proposed mechanism (Lebonnois et al., 2006). Using the technique of stellar occultation, the instrument confirmed the presence of the ozone layer predicted by models between 25 and 50 km but essentially limited to the season $L_s = 30–110°$ bracketing the Mars aphelion (Figure 13.13). Outside of this period, no ozone is found in the altitude range ($z > 20$ km) probed by the SPICAM instrument, meaning that the ozone detected in nadir viewing is confined near the surface.

The photochemistry of oxygen on Mars can also be investigated by the observation of the dayglow emitted at 1.27 µm by O_2 in its excited state $O_2(^1\Delta_g)$. This dayglow was discovered on Mars by Noxon et al. (1976) and observed in three latitude bands by Traub et al. (1979). The dayglow is excited by photolysis of O_3 in (13.26) with a yield of $O_2(^1\Delta_g)$ close to one. The deactivation of the excited $O_2(^1\Delta_g)$ molecule towards the ground electronic state $O_2(X^3\Sigma_g)$ occurs by emission of a photon at 1.27 µm with a relaxation time of ~1.2 h:

$$O_2(^1\Delta_g) \rightarrow O_2(X^3\Sigma_g) + h\nu \tag{13.29}$$

At low altitudes (i.e. large densities), deactivation of $O_2(^1\Delta_g)$ also occurs by collision with CO_2:

$$O_2(^1\Delta_g) + CO_2 \rightarrow O_2(X^3\Sigma_g) + CO_2 \tag{13.30}$$

The rate constant of this quenching by CO_2 is currently only constrained by an upper limit equal to 2×10^{-20} cm³ s⁻¹ (Sander et al., 2011). Because $O_2(^1\Delta_g)$ is essentially produced by the O_3 photolysis, the dayglow intensity in (13.29) reflects the abundance of O_3 at the altitudes where the quenching in (13.30) plays a minor role. This altitude range is usually 20–40 km and is just that of the variations of the H_2O condensation level discussed in Section 13.2, which makes the $O_2(^1\Delta_g)$ dayglow a sensitive tracer of the interactions between the oxygen and hydrogen species. Using a long-slit high-resolution spectrograph, Krasnopolsky (1997) initiated a long-term ground-based campaign of mapping observations of the $O_2(^1\Delta_g)$ dayglow (Krasnopolsky and Bjoraker, 2000; Novak et al., 2002; Krasnopolsky, 2003b, 2007, 2013). At low latitudes, the

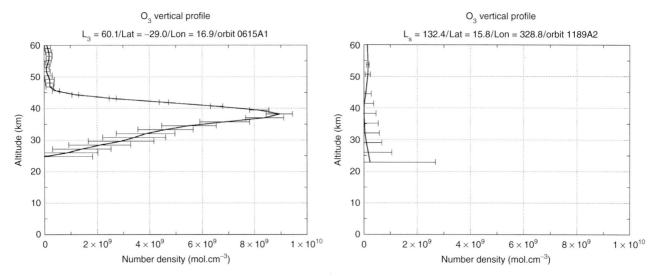

Figure 13.13. Examples of O_3 nighttime vertical profiles (molecule cm^{-3}) measured at low latitudes. The data were obtained by stellar occultation with the SPICAM instrument on-board Mars Express. (Left) Near aphelion at $L_s = 60°$ (29°S, 17°E). (Right) In northern summer at $L_s = 132°$ (16°N, 328°E). In both cases dust opacity prevents the observation below ~25 km. From Lebonnois et al. (2006).

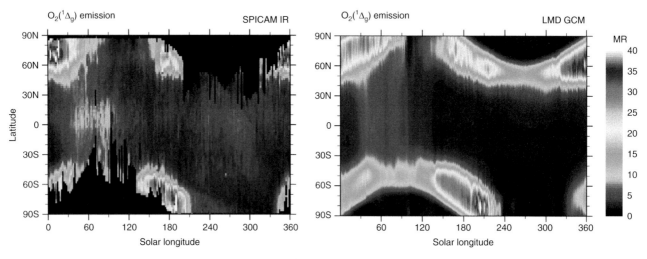

Figure 13.14. Seasonal evolution of the vertically integrated $O_2(^1\Delta_g)$ dayglow at 1.27 μm, in units of megarayleigh (1 MR = 10^{12} photon cm^{-2} s^{-1} (4π sr)$^{-1}$ = 4.5×10^{15} cm^{-2} = 1.7 μm atm for the dayglow at 1.27 μm). (Left) Satellite observations averaged over 2004–2012 with the infrared channel of SPICAM on Mars Express (updated from Fedorova et al., 2006a). (Right) Three-dimensional simulation by the LMD global climate model with photochemistry described by Lefèvre et al. (2004), with updated kinetics and improved water cycle. Calculations were made with an $O_2(^1\Delta_g)$ quenching rate by CO_2 equal to 0.25×10^{-20} cm^3 s^{-1}. A black and white version of this figure will appear in some formats. For the color version, please refer to the plate section.

measurements show a decreasing $O_2(^1\Delta_g)$ dayglow from ~6 MR (megarayleigh) near aphelion to ~1 MR near perihelion, whereas the dayglow in the subpolar regions is highly variable, with intensities up to 20 MR. Since the arrival of Mars Express at Mars in 2004, the infrared channel of SPICAM has provided the first continuous monitoring of the $O_2(^1\Delta_g)$ dayglow (Fedorova et al., 2006a). The 2004–2012 average of the SPICAM observations in Figure 13.14 shows an obvious correlation with the O_3 observations by the same instrument in Figure 13.10. The brightest airglow is observed in the sunlit portions of the winter polar vortices, where the O_3 is usually located near the surface but in quantities so large that the radiative deactivation of $O_2(^1\Delta_g)$ can outweigh the effect

of quenching by CO_2. At low latitudes, the maximum $O_2(^1\Delta_g)$ emission around aphelion is clearly visible and is caused by the photolysis of the seasonal maximum of O_3 observed above the hygropause. The comparison of the SPICAM observations with the Earth-based measurements of Krasnopolsky (2003b) shows a very good agreement if one takes into account the difference in the local time of the observations. Other observations of the $O_2(^1\Delta_g)$ dayglow from Mars Express have been performed by Altieri et al. (2009) with the OMEGA instrument. Because of its mapping capabilities, OMEGA has provided the first snapshots of the $O_2(^1\Delta_g)$ latitude–longitude distribution. The observations show an emission of up to 30 MR at high southern latitudes, with a clear maximum around

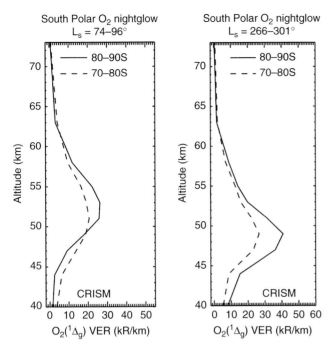

Figure 13.15. The $O_2(^1\Delta_g)$ nightglow observed over both polar regions by CRISM on Mars Reconnaissance Orbiter. (Left) Averaged $O_2(^1\Delta_g)$ volume emission profiles (kR km^{-1}) over $L_s = 74–96°$ for latitude bins of 70–80°S (dashed lines) and 80–90°S (solid lines). (Right) Averaged $O_2(^1\Delta_g)$ volume emission profiles (kR km^{-1}) over $L_s = 266–301°$ for latitude bins of 70–80°N (dashed lines) and 80–90°N (solid lines). From Clancy et al. (2013b).

local noon. This very bright $O_2(^1\Delta_g)$ dayglow can only result from the photolysis of large amounts of O_3 and is therefore consistent with direct observations of O_3 and chemical models, which all indicate the presence of strong O_3 maxima in the dry winter polar vortices.

Another pathway to form $O_2(^1\Delta_g)$ is the termolecular association of O atoms by reaction (13.4). Indeed, this reaction produces O_2 in several excited states, which leads ultimately to the production of $O_2(^1\Delta_g)$ with a quantum yield estimated to 0.7–0.75 on Mars and Venus (Crisp et al., 1996; Krasnopolsky, 2011). The emission at 1.27 μm that follows has been observed for a long time on the nightside of Venus (e.g. Connes et al., 1979; Gérard et al., 2009). On Mars, the $O_2(^1\Delta_g)$ emission at 1.27 μm caused by the association of O atoms can also be searched for at night, when the $O_2(^1\Delta_g)$ emission triggered by the ozone photolysis is shut off. The first detections of the $O_2(^1\Delta_g)$ nightglow on Mars were reported from the observations of the OMEGA (Bertaux et al., 2012), SPICAM (Fedorova et al., 2012), and CRISM (Clancy et al., 2012, 2013b) spectrometers on Mars Express and MRO. All instruments show that the $O_2(^1\Delta_g)$ nightglow at 1.27 μm is a quasi-permanent feature of latitudes beyond 70° in winter that occurs in a distinct layer located at 40–60 km altitude in the polar night region (Figure 13.15). Once integrated vertically, the nighttime $O_2(^1\Delta_g)$ emission caused by the reaction O+O+M is about one order of magnitude weaker (0.1–0.4 MR) than the typical values of the daytime $O_2(^1\Delta_g)$ emission caused by O_3 photolysis. Three-dimensional models show that the Martian $O_2(^1\Delta_g)$ nightglow is a direct consequence

of the polar convergent flow associated with the strong wintertime Hadley circulation (Figure 13.16). Above the winter pole, the O atoms produced at high altitudes and solar-illuminated latitudes are transported by the downwelling branch of the Hadley cell until the density is large enough to trigger the three-body reaction (13.4) and the subsequent $O_2(^1\Delta_g)$ nightglow at 1.27 μm. This process continues until complete exhaustion of O atoms at lower altitudes, which explains why the emission is confined to a relatively narrow altitude layer. Near equinox, the $O_2(^1\Delta_g)$ nightglow can even be observed simultaneously over both polar regions, when the Hadley circulation splits into two branches and a downward transport of O atoms occurs over high latitudes of both hemispheres (Clancy et al., 2012, 2013b). Three-dimensional simulations of the LMD GCM show that the mechanisms leading to the $O_2(^1\Delta_g)$ nightglow can be well reproduced near solstice and equinox, but the model tends to overestimate the emission by 25% and the modeled seasonal variations between these two periods are still poorly simulated (Clancy et al., 2012). These problems reflect an inaccurate description of the polar night upper atmosphere circulation, which is notoriously difficult to reproduce precisely in 3D models, and for which observations of tracers like nighttime $O_2(^1\Delta_g)$ provide powerful constraints.

13.4 NITROGEN CHEMISTRY

The presence of molecular nitrogen in the atmosphere of Mars was discovered in 1976 by the Viking experiments (Nier et al., 1976; Owen et al., 1977), which determined a mixing ratio of 0.027 for N_2. This figure remained unchallenged until the SAM suite on Curiosity recently revealed a mixing ratio of 0.0189 for N_2 (Mahaffy et al., 2013). This 30% smaller proportion of N_2 is believed to be more accurate than the previous Viking determination and will have to be taken into account in future model simulations. However, the strong N–N bond makes N_2 a relatively inert species. Breaking of this bond and activation of nitrogen chemistry occurs at high altitude ($z > 120$ km) in the Mars atmosphere by photon and electron impact dissociation, as well as by a few ionospheric reactions not detailed here. Dissociation of N_2 in the extreme ultraviolet leads to the formation of atomic nitrogen in the ground state and in the metastable state N(^{2}D):

$$N_2 + h\nu \rightarrow N + N(^2D) \qquad 80 < \lambda < 100 \text{ nm} \qquad (13.31)$$

The N(^{2}D) atoms are quenched by O and CO or react with CO_2 to form NO:

$$N(^2D) + CO_2 \rightarrow NO + CO \qquad (13.32)$$

Energetic electrons formed by photoionization of CO_2, N_2, O, and CO contribute to N production at high altitude via

$$N_2 + e^- \rightarrow N + N(^2D) + e^- \qquad (13.33)$$

Thus, the odd-nitrogen species N and NO are the main products of nitrogen chemistry in the upper atmosphere. At night, the reaction between N and O may excite the ultraviolet nightglow of the NO γ and δ bands, as observed by SPICAM (Bertaux et al., 2005):

O$_2$($^1\Delta_g$) emission   L$_s$ = 095–100

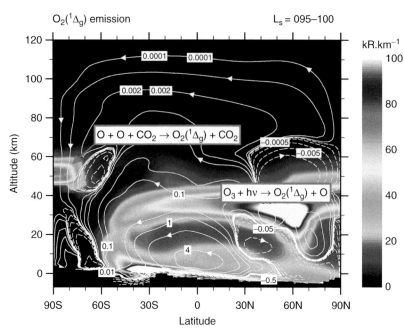

Figure 13.16. A schematic of the vertical and latitudinal distribution of the O$_2$($^1\Delta_g$) emission (zonally averaged, in kR km^{-1}) as simulated by the LMD general circulation model for L_s = 95–100°. Two primary pathways exist for the production of O$_2$($^1\Delta_g$). (1) Over solar-illuminated latitudes, the photolysis of ozone leads at this season to O$_2$($^1\Delta_g$) emission in a layer located above the hygropause between 25 and 40 km. A strong O$_2$($^1\Delta_g$) emission also outweighs the quenching by CO$_2$ near the surface in the southern hemisphere. This results from the dry surface conditions (see Figure 13.1) and subsequent large ozone densities at this season. (2) In the polar night, the three-body association of atomic oxygen transported downwards over the winter pole leads to O$_2$($^1\Delta_g$) emission between 45 and 60 km. The meridional streamfunction (10^9 kg s^{-1}) is superimposed in white contour lines. They visualize the downwelling branch of the winter hemisphere Hadley circulation, bringing large amounts of O atoms produced in the upper atmosphere towards lower altitudes into the polar night. From Clancy et al. (2012). A black and white version of this figure will appear in some formats. For the color version, please refer to the plate section.

$$N + O \rightarrow NO + h\nu \qquad (13.34)$$

During the day, NO is removed by predissociation in the δ(0–0) and (1–0) bands at 190 and 182 nm:

$$NO + h\nu \rightarrow N + O \qquad (13.35)$$

The sink of odd nitrogen produced in reactions (13.31) and (13.33) is the recombination by

$$N + NO \rightarrow N_2 + O \qquad (13.36)$$

If the production of N in the upper atmosphere is larger than that of NO, then the downward flux of odd nitrogen mainly consists of N atoms. In the lower atmosphere, a rapid loss process of N is

$$N + HO_2 \rightarrow NO + OH \qquad (13.37)$$

This reaction is followed by (13.36), which altogether limit N and NO to small amounts by returning them to the stable species N$_2$. If the downward flux of NO is greater than that of N, then the excess of NO initiates a more complicated chemistry (Yung et al., 1977; Krasnopolsky, 1993, 1995; Nair et al., 1994) and NO can persist in larger amounts in the lower atmosphere where its photolysis in (13.35) is slow. In these conditions an important reaction becomes

$$NO + HO_2 \rightarrow NO_2 + OH \qquad (13.38)$$

This reaction efficiently converts HO$_2$ into one hydroxyl radical OH available to oxidize CO by reaction (13.12). In addition, it causes the breaking of the O–O bond and the subsequent release of O through the rapid photolysis of NO$_2$ at visible wavelengths:

$$NO_2 + h\nu \rightarrow NO + O \qquad \lambda < 420 \text{ nm} \qquad (13.39)$$

The above reactions (13.38) and (13.39) introduce the following cycle, which illustrates the possible coupling between the nitrogen, hydrogen, and oxygen chemistries in the Mars lower atmosphere:

Cycle III

$$NO + HO_2 \rightarrow NO_2 + OH \qquad (13.38)$$

$$NO_2 + h\nu \rightarrow NO + O \qquad (13.39)$$

$$CO + OH \rightarrow CO_2 + H \qquad (13.12)$$

$$H + O_2 + M \rightarrow HO_2 + M \qquad (13.13)$$

Net: $\qquad CO + O_2 \rightarrow CO_2 + O$

As for cycles I and II, the net effect of this cycle is the reformation of CO$_2$ from CO but here both HO$_x$ and nitrogen oxides NO$_x$ (NO and NO$_2$) are used as catalysts. Model studies estimate the contribution of cycle III to the total production of CO$_2$ to be 5–10% (Nair et al., 1994; Yung and DeMore, 1999; Krasnopolsky, 2010). Another consequence of cycle III is the formation of O atoms, which facilitates the production of ozone. This is a well-known effect of this cycle in the polluted troposphere on Earth. The termination of cycle III occurs when NO$_2$ returns to a more stable form of nitrogen such as nitric acid HNO$_3$ or peroxynitric acid HO$_2$NO$_2$:

$$NO_2 + OH + M \rightarrow HNO_3 + M \qquad (13.40)$$

$$NO_2 + HO_2 + M \rightarrow HO_2NO_2 + M \qquad (13.41)$$

HNO$_3$ and HO$_2$NO$_2$ are however photolyzed more rapidly on Mars than they are on Earth, and consequently play a lesser role as reservoirs of NO$_x$. For a total NO$_x$ amount of 0.6 ppbv in the first scale height above the surface, one-dimensional models predict a mixing ratio of the order of 1 pptv for HNO$_3$ and of 10 pptv for HO$_2$NO$_2$ (e.g. Yung and DeMore, 1999).

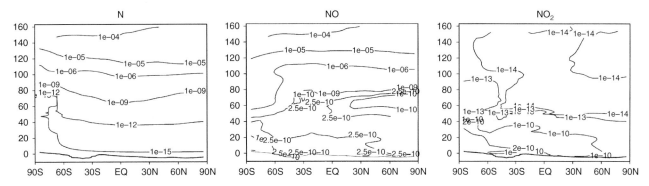

Figure 13.17. Latitude–altitude distribution of the mixing ratio (zonal mean) of the main nitrogen species N, NO, and NO_2 at $L_s = 90°$, calculated by the general circulation model of Moudden and McConnell (2007).

As mentioned above, the overall importance of the nitrogen chemistry in the Mars lower atmosphere depends on a fine balance between the production of N and NO in the thermosphere, whose kinetics is not known within the required accuracy (Nair et al., 1994; Krasnopolsky, 1995). NO was detected down to 120 km by the Viking mass spectrometers (Nier and McElroy, 1977), which suggests a downward flow of NO. However, the rather bright NO nightglow observed by SPICAM (Bertaux et al., 2005; Cox et al., 2008) in the flow descending to the middle atmosphere (60–80 km) suggests that N is more abundant than NO in the thermosphere. At night, the airglow produced in reaction (13.34) is immediately followed by the recombination of odd nitrogen in (13.36), which prevents the supply of NO to the lower atmosphere. Up to now the only attempt to detect NO in the lower atmosphere was made using its strongest absorption lines at 5.3 μm and resulted in an upper limit of 1.7 ppbv (Krasnopolsky, 2006b). The nitrogen chemistry on Mars is therefore poorly constrained by observational data. In any case, a precise estimation of its importance requires better kinetic data, models extending up to the region of production of odd nitrogen ($z > 150$ km), and a good representation of transport from the thermosphere down to the lower atmosphere. This latter point can be better accomplished with a general circulation model with full nitrogen chemistry. Up to now the only simulation of this kind was performed by Moudden and McConnell (2007), who obtained ~0.4 ppbv of NO_x in the lowest 20 km (Figure 13.17). This number is close to the ~0.6 ppbv previously obtained with one-dimensional models (Nair et al., 1994; Krasnopolsky, 1995).

13.5 SULFUR AND CHLORINE CHEMISTRY

On Earth, sulfur dioxide SO_2 is the most abundant species outgassed by volcanoes after CO_2 and H_2O. Although there is no evidence of active volcanism on Mars today, localized outgassing sources cannot be excluded, and the presence of SO_2 in the atmosphere might be indicative of such activity. However, despite many attempts, SO_2 remains up to now undetectable on Mars. The two most recent searches for SO_2, both performed from the Earth in the thermal infrared, are those of Encrenaz

et al. (2011) and Krasnopolsky (2012), who both determined an upper limit of 0.3 ppbv. This value is well below the value of ~10 ppbv from which SO_2 can have a significant impact on the CO_2 chemical stability of Mars (Krasnopolsky, 1993). The influence of SO_2 on the Martian photochemistry at the global scale is therefore negligible. Nevertheless, it is interesting to calculate the theoretical loss rate of SO_2, as it may be converted to a chemical lifetime and an upper limit on the production of SO_2 at the surface. Krasnopolsky (2005) developed a 1D model of sulfur chemistry for that purpose. After release of SO_2 in the atmosphere, photochemical equilibrium is established between SO_2 and its photolysis products SO and S:

$$SO_2 + h\nu \rightarrow SO + O \tag{13.42}$$

$$SO + h\nu \rightarrow S + O \tag{13.43}$$

Photolysis in (13.42) and (13.43) does not constitute a net sink of SO_2 since S is rapidly re-formed to SO,

$$S + O_2 \rightarrow SO + O \tag{13.44}$$

and SO_2 is recycled from SO by

$$SO + O_2 \rightarrow SO_2 + O \tag{13.45}$$

$$SO + OH \rightarrow SO_2 + H \tag{13.46}$$

$$SO + HO_2 \rightarrow SO_2 + OH \tag{13.47}$$

Reactions (13.45)–(13.47) maintain SO_2 as the dominant sulfur species up to ~45 km (Figure 13.18). Above that level, their efficiency decreases and SO becomes the primary sulfur species.

The irreversible loss of SO_2 occurs mainly through the reaction with OH, which initiates the following oxidation chain:

$$SO_2 + OH + M \rightarrow HSO_3 + M \tag{13.48}$$

$$HSO_3 + O_2 \rightarrow SO_3 + HO_2 \tag{13.49}$$

$$SO_3 + H_2O + H_2O \rightarrow H_2SO_4 + H_2O \tag{13.50}$$

The end product of the chain, sulfuric acid H_2SO_4, can then either condense (Wong et al., 2003) or be trapped at the surface of water ice aerosols (Krasnopolsky, 2005). Thus the precipitation of a solid form of H_2SO_4 to the surface is the most likely removal process of sulfur in the Mars atmosphere. The

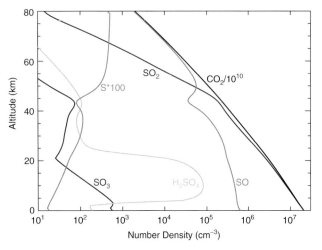

Figure 13.18. Vertical distribution of sulfur species (molecule cm^{-3}) for mean conditions below 80 km, assuming $SO_2 = 0.1$ ppbv at the surface. These results are obtained by inclusion of the SO_x chemistry in the 1D model of Krasnopolsky (2010). This chemistry involves updated data on photolysis of SO_2 and H_2SO_4, reaction between SO_3 and H_2O, and heterogeneous reaction between SO_2 and H_2O_2 on ice particles. Uptake of H_2SO_4 on water ice aerosols also occurs between 20 and 45 km and at night below 3 km. This explains the broad gas-phase H_2SO_4 peak between 3 and 20 km where uptake does not occur. A black and white version of this figure will appear in some formats. For the color version, please refer to the plate section.

SO_2 lifetime is determined by reaction (13.48). It was estimated by Krasnopolsky (2005) to be of the order of two terrestrial years, which results in an upper limit to the production of SO_2 of 17 000 t yr^{-1} to sustain a globally averaged mixing ratio of 1 ppbv. The new upper limit of 0.3 ppbv determined by Encrenaz et al. (2011) and Krasnopolsky (2012) implies that the maximum production of SO_2 must be reduced accordingly to about 5000 t yr^{-1}, which is smaller than the volcanic source of SO_2 on Earth by a factor of ~1500 (Haywood and Boucher, 2000). This fact, along with the absence of any "hot spots" or endogenic heat sources in the THEMIS images (Christensen et al., 2003), suggests that outgassing at the surface of Mars must be very weak, if it exists at all. It is also noteworthy that SO_2 is a relatively long-lived species on Mars. This means that it can be transported by the winds far from its source and be diluted by atmospheric mixing. Thus, a possible detection of SO_2 in the future would be indicative of outgassing somewhere at the surface of the planet but not necessarily in the proximity of the region of detection.

In addition to sulfur, volcanic eruptions on Earth are capable of injecting large amounts of chlorine into the atmosphere, usually in the form of hydrochloric acid HCl. If present on Mars, HCl might therefore be indicative of some outgassing activity. The discovery of small amounts of perchlorate (ClO_4) in the Martian soil by the Phoenix Lander (Hecht et al., 2009) and the Curiosity Rover (Glavin et al., 2013) also raises the possibility of chlorine delivered to the atmosphere by the surface or the airborne dust. Whatever the source is, HCl should be the dominant chlorine species on Mars since any Cl atom released in the lower atmosphere is rapidly converted to the stable species HCl by reaction with HO_2:

$$Cl + HO_2 \rightarrow HCl + O_2 \qquad (13.51)$$

However, as for SO_2, HCl has up to now never been detected in the Mars atmosphere. Recently the upper limit of 2 ppbv determined from the Earth by Krasnopolsky et al. (1997) was lowered to 0.2 ppbv from Herschel/HIFI submillimetric observations (Hartogh et al., 2010a). With this amount of HCl, the predicted mixing ratio of Cl is less than 1 pptv in the lower atmosphere. The production of odd chlorine on Mars is therefore at least three orders of magnitude smaller than that of odd hydrogen. On the basis of the current data in hand, this makes chlorine a negligible contributor as a potential catalyst of Martian atmospheric chemistry.

13.6 THE METHANE CONTROVERSY

Between 2004 and 2012, four independent teams reported detections of low levels (10–60 ppbv) of methane on Mars. If true, these would constitute the first observations of an organic compound on that planet and would be the main novelty of recent years in the inventory of minor species in its atmosphere. However, as we will show in the following, the claims for the presence of methane have been highly controversial over the last decade. Recently, the most robust search for methane was performed on the surface of Mars by the Tunable Laser Spectrometer (TLS) on Curiosity. The latest TLS measurements indicate a background CH_4 level of 0.7 ppbv and have identified a pulse of 7 ppbv only observed over a period of two months (Webster et al., 2015).

The first report of methane detection on Mars was by Krasnopolsky et al. (2004), who observed 10±3 ppbv of methane in January 1999 at $L_s = 88°$ using the Fourier Transform Spectrometer (FTS) at the Canada–France Hawaii Telescope. The field of view covered a significant part of the Martian disk. The same year, Formisano et al. (2004) detected varying methane amounts between 0 and 30 ppbv over a few orbits of the PFS spectrometer on Mars Express. Geminale et al. (2011) expanded on this using the same instrument and method over a six-year baseline. Their results revealed substantial seasonal variations of methane and local enhancements of up to 70 ppbv at high northern latitudes in summer. The global mean value derived from PFS is about 15 ppbv. Using the CSHELL spectrometer at the Infrared Telescope Facility (IRTF, Hawaii), Mumma et al. (2009) observed in January–March 2003 a strong local maximum (~50 ppbv) of methane at low latitudes over the Syrtis Major region (Figure 13.19). No significant levels of methane were found in their observations performed three years later in January–February 2006. Krasnopolsky (2012) reprocessed with a refined analysis his CSHELL observations of February 2006 at $L_s = 10°$ and found about 10 ppbv of methane over the Valles Marineris region and ~3 ppbv outside this region. His observations in December 2009 at $L_s = 20°$ and March 2010 at $L_s = 70°$ showed no detection, with an upper limit of 8 ppbv. Villanueva et al. (2013) also used the CSHELL spectrometer in January 2006 but only derived an upper limit of 7.8 ppbv, in contrast to the 10 ppbv detected by Krasnopolsky (2012) one month later. Villanueva et al. (2013) did not detect methane in their later observations of November 2009 and

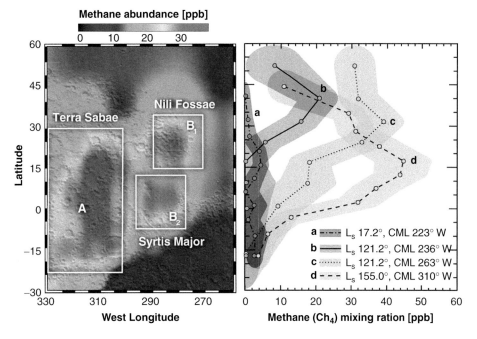

Figure 13.19. Methane mixing ratio (ppbv) observed from Earth by Mumma et al. (2009). (Left) Regions where methane appeared enhanced during summer 2003 (L_s = 121–155°). (Right) Latitudinal and temporal variability of methane. Profile a was obtained in February 2006 at L_s = 17.2°. Profiles b, c, and d were obtained in January–March 2003 between L_s = 121° and 155°. A black and white version of this figure will appear in some formats. For the color version, please refer to the plate section.

April 2010. All the above studies measured CH_4 in its absorption band at 3.3 μm. Using the band at 8 μm, Fonti and Marzo (2010) performed a statistical analysis of the Thermal Emission Spectrometer (TES) spectra and extracted a weak signal attributed to methane absorption and showing seasonally variable methane at the ~10–30 ppbv level.

It is fair to say that the claimed detections of such low levels of methane were all obtained at the limits of the instrumental capabilities. PFS and TES spacecraft instruments do not have the required sensitivity and spectral resolution for an unambiguous identification of CH_4 (Zahnle et al., 2011). Detection of methane is made by summing thousands of spectra, which does not suppress – and can even increase – instrumental effects and systematic errors. Despite their much greater spectral resolution, ground-based observations are not easier because the Martian methane must be viewed through the Earth's atmosphere, which contains ~10^4 more methane molecules above the observer than what is retrieved on Mars. The measurements must therefore exploit the Doppler shift of the Martian lines when Mars is approaching or receding from Earth. However, even in these conditions, the claimed detections of methane are close to the noise level. In any case, the very different geographical distributions and seasonal variations of methane obtained from space (Fonti and Marzo, 2010; Geminale et al., 2011) and from the Earth (Mumma et al., 2009) are puzzling and cast doubt on their veracity.

Despite these reservations, the observational claims for methane on Mars between 1999 and 2006 naturally raise the question of sources and sinks. Regarding the source, methane is not produced photochemically. It is therefore tantalizing to relate its existence to a possible past or extant life on the planet, since more than 90% of methane on Earth has a biological origin. Krasnopolsky (2006c) calculated that methane from impacts of comets, meteorites, and interplanetary dust is insignificant and argued that the lack of volcanism, hot spots, and SO_2 (which is more abundant than CH_4 in terrestrial outgassing) favors the hypothesis of a biogenic origin for Martian methane. On the other hand, Keppler et al. (2012) argued that carbonaceous micrometeorites might be in sufficient quantity on the surface of Mars to become a significant abiogenic source of methane when exposed to ultraviolet radiation. Alternatively, Atreya et al. (2007) proposed that methane could be produced abiogenically by hydrothermal processes such as serpentinization, i.e. the reaction between iron-bearing silicates and water, producing H_2, which in turn reacts with CO_2 to form CH_4. On Mars this would require however the existence of shallow aquifers and an active hydrogeochemistry, which have yet to be discovered.

Regarding the sink, the primary photochemical loss of methane occurs through its oxidation by OH or O(^{1}D), and its photolysis by the Lyman α line above ~80 km altitude. These processes lead to a global chemical lifetime estimated to be 300–340 terrestrial years (Summers et al., 2002; Krasnopolsky et al., 2004; Lefèvre and Forget, 2009), to which corresponds a tiny source of 260 t yr^{-1} to maintain a steady-state value of 10 ppbv. This may be compared with the terrestrial value of 582×10^6 t yr^{-1} (Ciais et al., 2013). With a lifetime of ~300 years, methane on Mars is expected to be homogeneously mixed by atmospheric transport. Lefèvre and Forget (2009) showed that reproducing the methane observations by Mumma et al. (2009) with a GCM requires a methane lifetime shorter than ~200 days, and hence an unknown sink that is at least 600 times faster than the loss derived from the current kinetic data used by the atmospheric chemistry community. The fact that such a strong oxidizing process would have been overlooked would be a surprise, since conventional models do a rather good job at reproducing short-lived chemical species on Mars (e.g. O_3 and H_2O_2), as shown in previous sections.

Several studies explored the chemical or physical processes that could explain a fast methane loss but none of them is supported by current laboratory data. For instance, methane could be destroyed by electrochemical processes triggered by the strong electric fields generated during dust storms (DeLory et al., 2006; Farrell et al., 2006). A large-scale electrochemical production of oxidants in the atmosphere is disputed by the calculations of Krasnopolsky (2006a) and seems difficult to reconcile with current observations of H_2O_2, O_3, and CO (Lefèvre and Forget, 2009). However, locally, the excess of H_2O_2 produced in dust storms could lead to its precipitation out of the atmosphere onto the Martian surface (Atreya et al., 2006, 2007). Methane would then be scavenged by large amounts of H_2O_2 or other superoxides embedded in the regolith. This possibility is for the moment not supported by laboratory work, which shows no apparent oxidation of methane with H_2O_2 on analogs of Martian soil or perchlorates (Gough et al., 2011). In addition, kinetic data on reactions of CH_4 with metal oxide and superoxide ions are extremely slow at Martian temperatures (Krasnopolsky, 2006c). Laboratory work also shows that wind-driven agitation of quartz crystals could produce active sites that sequester methane (Knak Jensen et al., 2014), but it is difficult to extrapolate the efficiency of these highly idealized experiments to the real Martian atmosphere. Another speculative possibility is that the decomposition of methane condensed as clathrates would be a source of episodic release of CH_4 (e.g. Chastain and Chevrier, 2007; Chassefière, 2009). However, this does not solve the problem of the methane sink (Zahnle et al., 2011). According to laboratory work (Trainer et al., 2010), trapping of CH_4 on polar ice analogs (including clathrates) appears anyway to be negligible in the modern Martian conditions.

Because of the potential implications of the presence of methane and in the light of the controversial observational dataset described above, the first *in situ* measurements promised by TLS on Curiosity and its far superior detection capabilities have been eagerly awaited. The first atmospheric samples collected by TLS in the Gale Crater (4°S, 137°E) spanned an eight-month period in spring–summer. By combining all of the individual measurements, it was concluded that methane was not detected, having an upper limit of only 1.3 ppbv (Webster et al., 2013). In a subsequent analysis, Webster et al. (2015) presented the entire TLS dataset reprocessed over a period of almost one Martian year (605 sols). The results indicate detection of methane at two levels of abundance (Figure 13.20). A background CH_4 level of 0.7±0.2 ppbv, based on high-precision methane-enriched experiments, is observed during the first eight months and last four months of the dataset. Between these two periods, four sequential measurements of TLS indicate a pulse of 7±2 ppbv of methane over two months ($L_s = 56$–$82°$). If true, these results raise a number of questions that currently have no answer. First, the only way to reconcile the last Earth-based reports of methane (2003–2006) and the low background level initially measured by Curiosity would be a near-complete disappearance of methane in less than 10 years. There is currently no confirmed

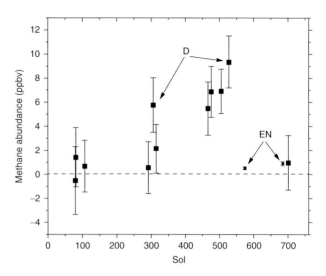

Figure 13.20. Methane mixing ratio (ppbv) measured by TLS on the Curiosity Rover versus Martian sol. Sol 1 was on August 6, 2012. All measurements were made at night, except the two marked "D" ingested during the day. The values with smaller error bars labeled EN were retrieved from the "methane enrichment" runs. From Webster et al. (2015).

mechanism that could explain a loss of methane that is in such stark contradiction with its theoretical 300-year lifetime. Then, it is difficult to conceive a release process that is sufficiently restricted in time to explain the methane pulse detected by TLS. For instance, both serpentinization (occurring a few kilometers below the surface) and the action of ultraviolet light on organics (widespread on the surface) are slow and continuous processes that have in principle no preferred particular time. Even in the case of episodic methane release, transport and mixing by the winds would also rapidly smooth out any plume of methane emitted by a localized source, unless that emission occurred very close to Curiosity. Because it challenges so much our current understanding of the atmospheric chemistry and physics on Mars, the discovery of variable amounts of methane by TLS needs to be confirmed by further detections. Future searches for methane by Curiosity and the Trace Gas Orbiter are therefore a high priority to unravel a mystery that may impact Mars science in a fundamental way.

13.7 RECONCILING OBSERVATIONS AND MODELS

13.7.1 One-Dimensional Annually Averaged Models

As for the Earth's atmosphere, photochemical models are essential tools for understanding the composition and chemical processes at work in the Martian atmosphere. One-dimensional globally averaged models were especially important when observational data were scarce and global climate models (GCMs) did not exist. Their first major success was the demonstration of the catalytic effect of the hydrogen chemistry by McElroy and Donahue (1972) and Parkinson and Hunten (1972), who could provide a qualitative explanation for the surprisingly low amounts of observed CO and O_2 on Mars. After four decades of progress and many new observational constraints, this initial discovery has never been called into question. However, our quantitative understanding of the long-lived chemical species on Mars remains poor. In particular, the observed equilibrium value of CO (900–1100 ppmv) has been historically difficult to reproduce, and this issue has still not been solved in current long-term simulations of the Martian photochemistry.

It was assumed in the early 1970s that Martian atmospheric temperatures decreased with height with a steep gradient of -5 K km^{-1} down to 150 K above 15 km. In these conditions, water vapor was restricted to the first 5 km above the surface and photochemical models tended to produce too little HO$_x$, which led to large CO amounts relative to the observations. To counteract this effect, models adopted unrealistically large H_2O_2 amounts (Parkinson and Hunten, 1972) or very strong eddy mixing coefficients (McElroy and Donahue, 1972; Liu and Donahue, 1976; Kong and McElroy, 1977a) to speed up CO oxidation and to re-form CO_2 fast enough in the simulations. Later, it was recognized that the room-temperature value of the CO_2 absorption cross-section had to be reduced under Martian conditions to account for its temperature dependence (Lewis and Carver, 1983; Anbar et al., 1993a). This has the effect of increasing the amount of ultraviolet radiation available to photodissociate water vapor, creating the opposite problem of an overproduction of HO$_x$ and too little CO in the models. With temperature-dependent CO_2 cross-sections and the standard kinetics, the typical CO mixing ratio calculated by current one-dimensional models is 120 ppmv (Nair et al., 1994; Krasnopolsky, 2010), which is about eight times smaller than the observed value of 900–1100 ppmv (Section 13.2). Figure 13.21 provides an example of the photochemical composition calculated by such models (Krasnopolsky, 2010). The abundance of short-lived species (O_3, $O_2(^1\Delta_g)$, H_2O_2) is in good agreement with the observations, but this success still relies on a CO amount that is largely underestimated (120 ppmv). To solve the general issue of CO in globally averaged models, Nair et al. (1994) modified the rates of the key reactions CO+OH and HO$_2$+OH within their experimental uncertainties and reached ~500 ppmv of CO in their model. However, subsequent laboratory studies and stratospheric measurements on Earth have not supported the suggested modifications. Krasnopolsky (1995) in turn proposed to decrease the efficiency of the HO$_x$ chemistry by reducing the H_2O photolysis by a factor of 2 to account for uncertainties in the measured H_2O absorption cross-sections, and obtained a CO abundance of ~500 ppmv. Water vapor is indeed essentially photolyzed in the narrow interval 190–200 nm where its absorption cross-section was not at that time well constrained by measurements, especially at low temperatures. The later laboratory data by Chung et al. (2001) indicated a moderate decrease (-20%) with temperature of the H_2O cross-section between 295 K and 250 K and 187–189 nm. Unfortunately, measurements at temperatures more representative of the altitudes at which H_2O is photolyzed ($T < 200$ K) are still missing.

The above models only consider gas-phase chemistry. On Earth, heterogeneous reactions between gaseous species and solid particles (such as ice or mineral dust) also play an important role in controlling the composition of the atmosphere and are a standard feature in most photochemical models. In view

of the ubiquity or water ice clouds and dust in the atmosphere of Mars, there are good reasons to examine the possibility of heterogeneous processes also active on that planet. The reactions that have received particular interest on Mars are the interactions of gaseous OH, HO$_2$, and H_2O_2 with water ice and dust particles, since they may provide a sink for HO$_x$ and contribute to throttle down the recycling rate of CO_2 from CO. In order to have an effect on the HO$_x$ budget, the uptake of these species on the solid phase must be irreversible (i.e. reactive) or, if reversible, must be sufficiently strong to lead to a significant removal from the gas phase. The irreversible losses of OH and HO$_2$ on water ice have been identified in the laboratory, with reaction probabilities of ~0.03 and ~0.025, respectively (Cooper and Abbatt, 1996). The uptake of H_2O_2 on water ice is reversible and the partitioning coefficient between the gas and solid phases is still uncertain (Crowley et al., 2010). By analogy with what occurs on mineral dust surfaces on Earth, the uptake of HO$_x$ or other species on Martian dust may also be important. However, in the absence of real samples, the uptake coefficients derived on terrestrial dust particles are still highly speculative when applied to Mars. Kong and McElroy (1977a) made the first attempt to consider quantitatively the heterogeneous destruction of HO$_x$, but only took into account the adsorption on the regolith at the Martian surface. With more comprehensive 1D models including gas–surface interactions on water ice clouds and airborne dust, Anbar et al. (1993b) demonstrated the potential importance of heterogeneous chemistry in lowering the abundance of HO$_x$, while Krasnopolsky (1993) noted a doubling of CO in his model that led to improved agreement with the observations. Both authors emphasized, however, the lack of observational constraints on critical parameters such as the HO$_x$ reaction probabilities on dust and ice, which were largely unknown at that time, or the altitude distribution of the cloud and dust particles. In a more recent study, Krasnopolsky (2006a) investigated the impact of heterogeneous chemistry with an updated kinetics and ice vertical distribution. He showed that it was possible to fit the observed CO only on the condition that the probabilities for heterogeneous loss of HO$_x$ on ice are 4–30 times larger than those currently recommended for atmospheric modeling on Earth (Sander et al., 2011). In addition, the seasonal–latitudinal variations of O_3, $O_2(^1\Delta_g)$, and H_2O_2 calculated with the same extreme heterogeneous parameters strongly disagree with the observations, which is problematic. This result suggests that heterogeneous chemistry is not a simple and unique solution to the low CO amounts predicted by Martian photochemical models. An alternative way to decrease oxidizing species (and hence to increase CO) may be to consider their dry deposition at the surface, as done for O_3 and H_2O_2 in the 1D model of Zahnle et al. (2008). With this assumption, their model could reach a CO value of 470 ppmv. This is larger than the typical value produced by 1D gas-phase models but is still underestimated by a factor of 2 relative to the observations. Even with the inclusion of dry deposition, it is only at the cost of extreme conditions of dryness throughout the atmosphere and the Martian year that Zahnle et al. (2008) could reach a CO value of 750 ppmv in reasonable agreement with the observed CO.

All the above calculations are based on annually and globally averaged models. These models are very useful to estimate

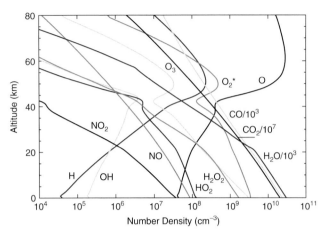

Figure 13.21. Density profiles (molecule cm^{-3}) calculated by the global mean photochemical model of Krasnopolsky (2010), using the chemical reaction rates listed in Table 13.2. O$_2$* is O$_2$($^1\Delta_g$). The surface pressure is set to 6.1 hPa and the H$_2$O vertical column is 9.5 pr μm. The H$_2$O density profile is restricted by the saturation value. A downward flow of O and CO from the dissociation of O$_2$ and CO$_2$ above 80 km is also imposed at the upper boundary. The calculated abundances of the major photochemical products are 1600 ppmv for O$_2$, 120 ppmv for CO, 20 ppmv for H$_2$, 0.9 μm atm for O$_3$, 1.5 MR or 2.5 μm atm for O$_2$($^1\Delta_g$), and 7.6 ppbv for H$_2$O$_2$. These values are in reasonable agreement with the observations, with the exception of CO, which is underestimated by a factor of 8. A black and white version of this figure will appear in some formats. For the color version, please refer to the plate section.

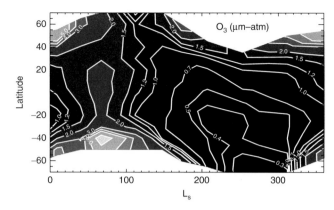

Figure 13.22. Ozone column (μm atm) calculated by the 1D model of Krasnopolsky (2009) at different latitudes and seasons. The model includes an irreversible uptake coefficient of 3×10^{-4} for H$_2$O$_2$ on ice. This map can be compared to the SPICAM and ground-based observations of ozone displayed in Figures 13.10 and 13.12, respectively.

the first-order distribution of chemical species as a function of height and are the only tools available to examine the fate of the long-lived species, which require simulations of several decades (for CO and O$_2$) to thousands of years (for H$_2$) before reaching equilibrium. A basic problem of globally averaged models, however, is their inability to represent the dramatic latitudinal and vertical variations of water vapor over the Martian year, which are now well documented by satellite observations (Figure 13.1). Since H$_2$O is the main source of HO$_x$ species, the oxidation rate of CO and re-formation rate of CO$_2$ obviously experience very large variations in time and space over the Martian year that cannot be easily represented in an annually globally averaged sense. In addition, averaged models cannot account properly for the large observed variations in temperature, water ice, and dust, for which the response in terms of gas-phase kinetics and heterogeneous chemistry is known to be highly nonlinear.

13.7.2 One-Dimensional Time-Dependent Models

A first step in developing more complex models that avoid the problems mentioned above is to constrain the existing 1D models with varying profiles of temperature and water vapor. By nature, this approach can only represent local conditions and still neglects the changes due to horizontal transport. Time-dependent 1D models are, however, very convenient to study the fast chemistry by simulating the distribution of short-lived species that respond quickly to the local abundance of H$_2$O. Owing to its short lifetime (see Section 13.3), ozone is

obviously a good candidate: this was recognized by Shimazaki (1981), who used a uniform but seasonally varying profile of water vapor constrained by Viking data to reproduce with some success the Mariner 9 observations of high-latitude ozone. Clancy and Nair (1996) calculated the seasonal changes in photochemistry using seasonally varying water vapor profiles constrained from direct observations or temperature profiling (Clancy at al., 1996). They could predict a large decrease in the low-latitude O$_3$ amount between aphelion and perihelion, a trend that was confirmed several years later by O$_2$($^1\Delta_g$) Earth-based observations (Krasnopolsky, 2003b) and by the SPICAM nighttime O$_3$ measurements (Lebonnois et al., 2006; Figure 13.13). However, the CO mixing ratio obtained in their simulation was still largely underestimated and only 16% larger than that calculated without seasonal variations of temperature and H$_2$O (Nair et al., 1994). More recently, Krasnopolsky (2006a, 2009) modeled the seasonal and altitudinal variations of the photochemistry with a 1D model using constant values of the long-lived species CO, O$_2$, and H$_2$ as constrained by observations, whereas the seasonal evolution of H$_2$O, water ice clouds, and dust amount was adjusted to the TES observations (Smith, 2004). This study took advantage of the much improved observational dataset of short-lived species in recent years to investigate the latitudinal–seasonal evolution of O$_3$, H$_2$O$_2$, and O$_2$($^1\Delta_g$). Figure 13.22 displays the results of these 1D simulations carried out for ~100 seasonal–latitudinal points. It shows that good agreement between modeled and observed ozone can be obtained in the tropics and mid-latitudes provided that the long-lived species and H$_2$O are maintained in the model to their observed values (some results for H$_2$O$_2$ are shown in Figure 13.7). The model cannot accurately represent trace gases when the effects of transport become important, as is the case in the polar night or at high latitudes in winter, where O$_3$ is underestimated (Figure 13.22).

One-dimensional time-dependent models have also been used to study the variations of Martian chemical species with local time (Krasitski, 1978; Shimazaki, 1981; García Muñoz et al., 2005; Krasnopolsky, 2006a). García Muñoz et al. (2005)

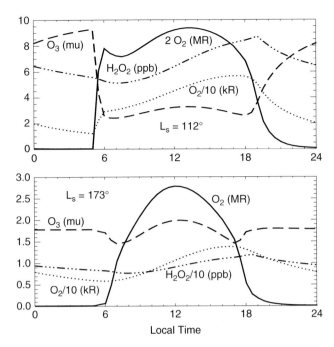

Figure 13.23. Modeled diurnal variations of the O_3 column (μm atm), H_2O_2 (ppbv), the $O_2(^1\Delta_g)$ dayglow at 1.27 μm (MR, solid line), and the $O_2(^1\Delta_g)$ airglow at 1.27 μm from the reaction $O + O + M$ (kR, dotted line): top, results at $L_s = 112°$; bottom, results at $L_s = 173°$. From Krasnopolsky (2006a).

examined the diurnal cycle of the emissions of $O_2(^1\Delta_g)$ and of the vibrationally excited levels of OH. During the day, the peak of the $O_2(^1\Delta_g)$ dayglow produced by the O_3 photolysis (reaction (13.26)) occurs shortly after noon in their model, in rather good agreement with the Earth-based observations (Krasnopolsky, 2003b) and the $O_2(^1\Delta_g)$ images obtained from OMEGA (Altieri et al., 2009). At this time of the day, their modeled intensities are 0.55 MR or 0.95 MR depending on the value adopted for the quenching rate of $O_2(^1\Delta_g)$ in (13.26), taken as $k_{CO2} = 2 \times 10^{-20}$ cm^3 s^{-1} or $k_{CO2} = 1 \times 10^{-20}$ cm^3 s^{-1}, respectively. These intensities calculated at 20°N are rather low compared those observed from Mars Express (Figure 13.14) and by Krasnopolsky (2013). A similar study was performed by Krasnopolsky (2006a) for $L_s = 112°$ and 173° at the subsolar latitudes of 22°N and 3°N, respectively, who used the profiles of temperature, water vapor, and dust measured in the same conditions by TES (Smith, 2004). With these constraints, the calculated values of the $O_2(^1\Delta_g)$ dayglow at 1.27 μm (Figure 13.23) are significantly larger than in García Muñoz et al. (2005) and in rather good agreement with the observations. At $L_s = 112°$, the model predicted the nighttime ozone layer observed in this season by SPICAM (Figure 13.13), which results from the absence of photolysis and the rapid conversion of O atoms to O_3 by reaction (13.24) at dusk. These results were obtained with the same assumptions regarding the heterogeneous loss of H_2O_2 as mentioned above for Krasnopolsky (2009). García Muñoz et al. (2005) and Krasnopolsky (2006a) also examined the emission of $O_2(^1\Delta_g)$ produced at night. As mentioned earlier, the main source of O_2 excitation in these conditions is the termolecular association of oxygen atoms in reaction (13.4).

Both studies reported a nighttime emission of 10–30 kR at low latitude, in agreement with the upper limit of 40 kR derived by Krasnopolsky (2013) and weaker by an order of magnitude than the polar $O_2(^1\Delta_g)$ nightglow detected by OMEGA (Bertaux et al., 2012), SPICAM (Fedorova et al., 2012), and CRISM (Clancy et al., 2012).

13.7.3 Multi-Dimensional Models

With the sizable increase in the observational dataset of atmospheric constituents obtained in the last decade, multi-dimensional models have become essential tools for interpreting the interactions between chemistry and transport in all the environmental conditions (e.g. L_s, season, local time, dust, clouds) for which measurements are now available. The first 3D model that included a simple chemistry (CO_2 photolysis and O recombination) was the Mars thermospheric GCM developed by Bougher et al. (1990) but its lower boundary was at ~100 km. Moreau et al. (1991) then studied the meridional distribution of chemical species down to the surface with a two-dimensional (zonally averaged) model using a simplified water vapor field and dynamics. It is only recently that the increase in computer power has allowed the development of full GCMs of Mars with photochemistry, similar to those commonly used for the study of the Earth's atmospheric chemistry. In addition to an adequate representation of atmospheric transport, Martian GCMs are able to provide a realistic description of the 3D field of water vapor and its variations at all scales, which is a crucial advantage for constraining properly the fast chemistry of the lower atmosphere. A drawback is that the long computation times restrict 3D chemical simulations to a few Martian years. This duration is sufficient to investigate the fast chemistry of O_3, H_2O_2, or $O_2(^1\Delta_g)$, but is not long enough to examine the evolution of long-lived species (CO, O_2, and H_2), which are usually initialized to their observed value. Lefèvre et al. (2004) coupled a photochemical module to the LMD general circulation model and characterized the 3D variations of ozone in the atmosphere of Mars. Their model could reproduce successfully the Mariner observations of O_3 but showed a significant underestimation of the O_3 maximum observed at low latitudes in the aphelion season (Section 13.3). This issue noted with a pure gas-phase chemical scheme was later confirmed by a stringent comparison between the same LMD GCM and the complete SPICAM dataset of ozone (Lefèvre et al., 2008). In their study, Lefèvre et al. (2008) demonstrated that GCM simulations that include the recommended HO_x uptake on water ice clouds could lead to much improved quantitative agreement with ozone measurements without violating current laboratory constraints (Figure 13.24). At the same time, the outstanding disagreement between the modeled H_2O_2 and the low upper limit observed by Encrenaz et al. (2002) around aphelion was reduced, allowing a better quantitative agreement with the H_2O_2 dataset throughout the Martian year (Figure 13.7). Despite these improvements, the latest version of the LMD GCM in Figure 13.10 still tends to underestimate the low-latitude O_3 column when compared to the SPICAM and MARCI observations around aphelion. This problem reflects more the difficulties of GCMs in reproducing precisely the effect of the aphelion cloud belt on the water vapor vertical profile, rather than being a gap in our understanding

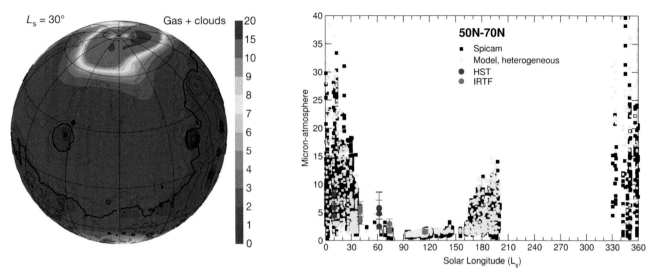

Figure 13.24. Ozone column (μm atm) calculated by the LMD general circulation model. (Left) Synoptic distribution in northern spring ($L_s = 30°$) with a photochemical scheme taking into account heterogeneous reactions of HO_x on water ice clouds. The observer is facing the 180° meridian at local noon, and the black contours correspond to Mars topography. (Right) Comparison of the same GCM simulation with the ozone columns measured in the 50–70°N latitude band by SPICAM (Perrier et al., 2006), HST (Clancy et al., 1999), and IRTF (Fast et al., 2006). From Lefèvre et al. (2008). A black and white version of this figure will appear in some formats. For the color version, please refer to the plate section.

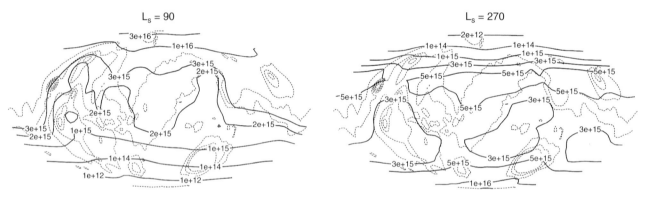

Figure 13.25. Column-averaged mixing ratio of H_2O_2 calculated by the GM3 general circulation model (Moudden, 2007): (left) $L_s = 90°$; (right) $L_s = 270°$. The seasonal evolution of H_2O_2 calculated by this model is compared to observations in the right panel of Figure 13.7.

of the photochemistry. The same pattern is observed in the comparison between the LMD GCM and the $O_2(^1\Delta_g)$ dayglow observations (Figure 13.14). This is expected since the source of the $O_2(^1\Delta_g)$ dayglow is the O_3 photolysis, but this also underlines the good consistency of the SPICAM data obtained in two different wavelength ranges. As in Krasnopolsky (2009), the uncertain quenching rate of $O_2(^1\Delta_g)$ in (13.30) had to be reduced in the LMD GCM to agree with SPICAM at high latitudes. In Figure 13.14 a slow value of 0.25×10^{-20} cm^3 s^{-1} was adopted, well under the currently recommended upper limit of 2×10^{-20} cm^3 s^{-1} (Sander et al., 2011).

A 3D study was also presented by Moudden and McConnell (2007) who added a chemistry module to the 3D Mars Global Multiscale Model (GM3). A notable feature of their model is the comprehensive description of the nitrogen chemistry illustrated in Figure 13.17. The seasonal evolution of ozone given by GM3 agrees qualitatively with the observations considered in their study, and is consistent with the results of the LMD GCM.

The quantitative analysis reveals, however, an overestimation of ozone in the perihelion season, which can probably be attributed to the constant dust optical depth of 0.5 used in GM3, leading to too cold and dry conditions for that period. Moudden (2007) also used GM3 to characterize the 3D evolution of H_2O_2 and showed the extreme variations of this compound at high latitudes related to those of H_2O. At low to mid-latitudes, the H_2O_2 columns calculated by Moudden (2007) (Figure 13.25) are in reasonable agreement with the current observations (Figure 13.7).

In recent years 3D models have been increasingly used to analyze the measurements of the reactive species O_3, $O_2(^1\Delta_g)$, or H_2O_2. In such simulations the short-lived chemical species are discussed against a background of long-lived species CO, H_2, or O_2 that is usually set to its observed value. Under these conditions, current 3D models indicate that the fast chemistry on Mars can be rather well reproduced quantitatively with the standard set of gas-phase and heterogeneous reactions used for terrestrial studies. However, this first step towards a global quantitative understanding of Martian photochemistry should be strengthened by a better knowledge of some critical parameters. The precise amount of water vapor present in the middle and upper atmosphere is still insufficiently known due to the lack of observational data, but also because of the inevitable simplification of microphysical processes leading to water ice condensation and sedimentation in models. New laboratory data are also needed to reduce the uncertainties on some important reactions governing the chemistry of HO_x radicals, on the possible heterogeneous reactions on Martian solid surfaces, or on the temperature dependence of the CO_2 absorption cross-section below 200 K and that of H_2O below 250 K.

Because of the needed computation time, a current limitation of 3D models is that they have not been integrated over sufficiently long periods to investigate the long-term equilibrium value of the slowly evolving species CO, O_2, or H_2. Therefore, the significant underestimation of Martian CO predicted by 1D globally averaged models is still an outstanding issue, and the classical problem of Mars' photochemical stability cannot therefore be considered fully understood. With the increase in computer power, it is obvious that 3D models, which are the best-suited tools to reproduce the interactions between chemistry and transport on long timescales, will soon address this challenge.

REFERENCES

Altieri, F., Zasova, L., D'Aversa, E., et al. (2009) O_2 1.27 μm emission maps as derived from OMEGA/MEx data, *Icarus*, 204, 499–511.

Anbar, A. D., Allen, M., and Nair, H. A. (1993a) Photodissociation in the atmosphere of Mars: impact of high-resolution temperature-dependent CO_2 cross-section measurements, *J. Geophys. Res.*, 98, 10925–10931.

Anbar, A. D., Leu, M. T., Nair, H. A., et al. (1993b) Adsorption of HO_x on aerosol surfaces: implications for the atmosphere of Mars, *J. Geophys. Res.*, 98, 10933–10940.

Anderson, D. E. (1974) Mariner 6, 7, and 9 ultraviolet spectrometer experiment: analysis of hydrogen Lyman-alpha data, *J. Geophys. Res.*, 79, 1513–1518.

Anderson, D. E., and Hord, C. W. (1971) Mariner 6 and 7 ultraviolet spectrometer experiment: analysis of hydrogen Lyman alpha data, *J. Geophys. Res.*, 76, 6666–6671.

Atreya, S. K., Wong, A.-S., Rennó, N. O., et al. (2006) Oxidant enhancement in Martian dust devils and storms: implications for life and habitability, *Astrobiology*, 6, 439–450.

Atreya, S. K., Mahaffy, P. R., and Wong, A. S. (2007) Methane and related trace species on Mars: origin, loss, implications for life and habitability, *Planet. Space Sci.*, 55, 358–69.

Atreya, S. K., Witasse, O., Chevrier, V. F., et al. (2011) Methane on Mars: current observations, interpretation, and future plans, *Planet. Space Sci.*, 59. 133–136.

Barker, E. S. (1972) Detection of molecular oxygen in the Martian atmosphere, *Nature*, 238, 447–448.

Barth, C. A., and Hord, C. W. (1971) Mariner 6 and 7 ultraviolet spectrometer experiment: topography and polar cap, *Science*, 173, 197–201.

Barth, C. A., Hord, C. W., Stewart, A. I., et al. (1973) Mariner 9 ultraviolet spectrometer experiment: seasonal variation of ozone on Mars, *Science*, 179, 795–796.

Bates, D. R., and Nicolet, M. (1950) The photochemistry of atmospheric water vapor, *J. Geophys. Res.*, 55, 301–327.

Belton, M. J. S., and Hunten, D. M. (1966) The abundance and temperature of CO_2 in the Martian atmosphere, *Astrophys. J.*, 145, 454–467.

Bertaux, J. L., Leblanc, F., Perrier, S., et al. (2005) Nightglow in the upper atmosphere of Mars and implications for atmospheric transport, *Science*, 307, 567–569.

Bertaux, J. L., Gondet, B., Lefèvre, F., et al. (2012) First detection of O_2 1.27 μm nightglow emission at Mars with OMEGA/MEX and comparison with general circulation model predictions, *J. Geophys. Res.*, 117, doi:10.1029/2011JE003890.

Biemann, K., Owen, T., Rushneck, D. R., et al. (1976) Search for organic and volatile inorganic components in two surface samples from the Chryse Planitia region of Mars, *J. Geophys. Res.*, 82, 4641–4658.

Billebaud, F., Brillet, J., Lellouch, E., et al. (2009) Observations of CO in the atmosphere of Mars with PFS onboard Mars Express, *Planet. Space Sci.*, 57, 1446–1457.

Blamont, J. E., and Chassefière, E. (1993) First detection of ozone in the middle atmosphere of Mars from solar occultation measurements, *Icarus*, 104, 324–336.

Bogard, D. D., Clayton, R. N., Marti, K., et al. (2001) Martian volatiles: isotopic composition, origin, and evolution, *Space Sci. Rev.*, 96, 425–58.

Bougher, S. W., Roble, R. G., Ridley, E. C., et al. (1990) The Mars thermosphere 2. General circulation with coupled dynamics and composition, *J. Geophys. Res.*, 95, 14811–14827.

Carleton, N. P., and Traub, W. A. (1972) Detection of molecular oxygen on Mars, *Science*, 177, 988–992.

Chassefière, E. (2009) Metastable methane clathrate particles as a source of methane to the Martian atmosphere, *Icarus*, 204, 137–144.

Chastain, B. K., and Chevrier, V. (2007) Methane clathrate hydrates as a potential source for Martian atmospheric methane, *Planet. Space Sci.*, 55, 1246–1256.

Chaufray, J. Y., Bertaux, J. L., Leblanc, F., et al. (2008) Observation of the hydrogen corona with SPICAM on Mars Express, *Icarus*, 195, 598–613.

Chaufray, J. Y., Leblanc, F., Quémerais, E., et al. (2009) Martian oxygen density at the exobase deduced from O I 130.4-nm observations by Spectroscopy for the Investigation of the Characteristics of the Atmosphere of Mars on Mars Express, *J. Geophys. Res.*, 114, E02006.

Christensen, P. R., Bandfield, J. L., Bell, J. F., et al. (2003) Morphology and composition of the surface of Mars: Mars Odyssey THEMIS results, *Science*, 300, 2056–2061.

Chung, C. Y., Chew, E. P., Cheng, B. M., et al. (2001) Temperature dependence of absorption cross-section of H_2O, HDO, and D_2O in the spectral region 140–193 nm, *Nucl. Instr. Meth. Phys. Res. A*, 467, 1572–1576.

Ciais, P., Sabine, C., Bala, G., et al. (2013) Carbon and other biogeochemical cycles. In *Climate Change 2013: The Physical Science Basis. Contribution of Working Group I to the Fifth Assessment Report of the Intergovernmental Panel on Climate Change*, eds T. F. Stocker, D. Qin, G.-K. Plattner, et al. Cambridge University Press, Cambridge, UK.

Clancy, R. T., and Nair, H. (1996) Annual (aphelion–perihelion) cycles in the photochemical behavior of the global Mars atmosphere, *J. Geophys. Res.*, 101, 12785–12790.

Clancy, R. T., Grossman, A. W., Wolff, M. J., et al. (1996) Water vapor saturation at low altitudes around Mars aphelion: a key to Mars climate?, *Icarus*, 122, 36–62.

Clancy, R. T., Wolff, M. J., and James, P. B. (1999) Minimal aerosol loading and global increases in atmospheric ozone during the 1996–1997 Martian northern spring season, *Icarus*, 138, 49–63.

Clancy, R. T., Sandor, B. J., and Moriarty-Schieven, G. H. (2004) A measurement of the 362 GHz absorption line of Mars atmospheric H_2O_2, *Icarus*, 168, 116–121.

Clancy, R. T., Sandor, B. J., Wolff, M. J., et al. (2012) Extensive MRO CRISM observations of 1.27 μm O_2 airglow in Mars polar night and their comparison to MRO MCS temperature profiles and LMD GCM simulations, *J. Geophys. Res.*, 117, doi:10.1029/2011JE004018.

Clancy, R. T., Sandor, B. J., García-Muñoz, A., et al. (2013a) First detection of Mars atmospheric hydroxyl: CRISM near-IR measurement versus LMD GCM simulation of OH Meinel band emission in the Mars polar winter stratosphere, *Icarus*, 226, 272–281.

Clancy, R. T., Sandor, B. J., Wolff, M. J., et al. (2013b) Correction to "Extensive MRO CRISM observations of 1.27 μm O_2 airglow in Mars polar night and their comparison to MRO MCS temperature profiles and LMD GCM simulations", *J. Geophys. Res.*, 118, doi:10.1002/jgre.20073.

Clancy, R. T., Wolff, M. J., Lefèvre, F., et al. (2016) Daily global mapping of Mars ozone column abundances with MARCI UV band imaging, *Icarus*, 266, 112–133.

Connes, P., Noxon, J. F., Traub, W. A., and Carleton, N. P. (1979) $O_2(^1\Delta)$ emission in the day and night airglow of Venus, *Astrophys. J.*, 233, 29–32, doi:10.1086/183070.

Cooper, P. L., and Abbatt, J. P. D. (1996) Heterogeneous interactions of OH and HO_2 radicals with surfaces characteristic of atmospheric particulate matter, *J. Phys. Chem.*, 100, 2249–2254.

Cox, C., Saglam, A., Gérard, J. C., et al. (2008) Distribution of the ultraviolet nitric oxide Martian night airglow: observations from Mars Express anc comparisons with a one-dimensional model, *J. Geophys. Res.*, 113, doi:10.1029/2007JE003037.

Crisp, D., Meadows, V. S., Bézard, B., et al. (1996) Ground-based near-infrared observations of the Venus nightside: 1.27 μm $O_2(^1\Delta_g)$ airglow from the upper atmosphere, *J. Geophys. Res.*, 101, 4577–4593.

Crowley, J. N., Ammann, M., Cox, R. A., et al. (2010) Evaluated kinetic and photochemical data for atmospheric chemistry: Volume V – Heterogeneous reactions on solid substrates, *Atmos. Chem. Phys.*, 10, 9059–9223.

Delory, G. T., Farrell, W. M., Atreya, S. K., et al. (2006) Oxidant enhancement in Martian dust devils and storms: storm electric fields and electron dissociative attachment, *Astrobiology*, 6, 451–462.

Encrenaz, T., Greathouse, T. K., Bézard, B., et al. (2002) A stringent upper limit of the H_2O_2 abundance in the Martian atmosphere, *Astron. Astrophys.*, 396, 1037–1044.

Encrenaz, T., Bézard, B., Greathouse, T. K., et al. (2004) Hydrogen peroxide on Mars: evidence for spatial and seasonal variations, *Icarus*, 170, 424–429.

Encrenaz, T., Fouchet, T., Melchiorri, R., et al. (2006) Seasonal variations of the Martian CO over Hellas as observed by OMEGA/Mars Express, *Astron. Astrophys.*, 459, 265–270.

Encrenaz T., Greathouse, T. K., Richter, M. J., et al. (2008) Simultaneous mapping of H_2O and H_2O_2 on Mars from high-resolution imaging spectroscopy, *Icarus*, 195, 547–555.

Encrenaz, T., Greathouse, T. K., Richter, M. J., et al. (2011) A stringent upper limit to SO_2 in the Martian atmosphere, *Astron. Astrophys.*, 530, A37.

Encrenaz, T., Greathouse, T. K., Lefèvre, F., and Atreya, S. K. (2012) Hydrogen peroxide on Mars: observations, interpretation, and future plans, *Planet. Space. Sci.*, 68, 3–17.

Espenak, F., Mumma, M. J., Kostiuk, T., et al. (1991) Ground-based infrared measurements of the global distribution of ozone in the atmosphere of Mars, *Icarus*, 92, 252–262.

Farrell, W. M., Marshall, J. R., Cummer, S. A., Delory, G. T., and Desch, M. D. (2006) A model of the ULF magnetic and electric field generated from a dust devil, *J. Geophys. Res.*, 111, E11, E11004, doi:10.1029/2006JE002689.

Fast, K., Kostiuk, T., Espenak, F., et al. (2006) Ozone abundance on Mars from infrared heterodyne spectra. I. Acquisition, retrieval, and anticorrelation with water vapor, *Icarus*, 181, 419–431.

Fast, K., Kostiuk, T., Lefèvre, F., et al. (2009) Comparison of HIPWAC and Mars Express SPICAM observations of ozone on Mars 2006–2008 and variation from 1993 IRHS observations, *Icarus*, 203, 20–27.

Fedorova, A., Korablev, O., Perrier, S., et al. (2006a) Observations of O_2 1.27 μm dayglow by SPICAM IR: seasonal distribution for the first Martian year of Mars Express, *J. Geophys. Res.*, 111, doi:10.1029/2006JE002694.

Fedorova, A., Korablev, O., Bertaux, J. L., et al. (2006b) Mars water vapor abundance from SPICAM IR spectrometer: seasonal and geographic distributions, *J. Geophys. Res.*, 111, doi:10.1029/2006JE002695.

Fedorova A., Lefèvre, F., Guslyakova, S., et al. (2012) The O_2 nightglow in the Martian atmosphere by SPICAM onboard of Mars-Express, *Icarus*, 219, 596–608.

Fonti, S., and Marzo, G. A. (2010) Mapping the methane on Mars, *Astron. Astrophys*, 512, A51.

Formisano, V., Atreya, S. K., Encrenaz, T., et al. (2004) Detection of methane in the atmosphere of Mars, *Science*, 306, 1758–1761.

Fouchet, T., Lellouch, E., Ignatiev, N. I., et al. (2007) Martian water vapor: Mars Express PFS/LW observations, *Icarus*, 190, 32–49.

García Muñoz, A., McConnell, J. C., McDade, I. C., et al. (2005) Airglow on Mars: some model expectations for the OH Meinel bands and the O_2 IR atmospheric band, *Icarus*, 176, 75–95.

Geminale, A., Formisano, V., and Sindoni, G. (2011) Mapping methane in Martian atmosphere with PFS-MEX data, *Planet. Space Sci.*, 59, 137–148.

Gérard, J.-C., Cox, C., Soret, L., et al. (2009) Concurrent observations of the ultraviolet nitric oxide and infrared O_2 nightglow emissions with Venus Express, *J. Geophys. Res.*, 114, doi:10.1029/2009JE003371.

Glavin, D. P., Freissinet, C., Miller, K. E., et al. (2013) Evidence for perchlorates and the origin of chlorinated hydrocarbons detected by SAM at the Rocknest aeolian deposit in Gale Crater, *J. Geophys. Res.*, 118, 1955–1973.

Gough, R. V., Turley, J. J., Ferrell, G. R., et al. (2011) Can rapid loss and high variability of Martian methane be explained by surface H_2O_2?, *Planet. Space Sci.*, 59, 238–246.

Hartogh, P., Jarchow, C., Lellouch, E., et al. (2010a) Herschel/HIFI observations of HCl, H_2O_2, and O_2 in the Martian atmosphere – initial results, *Astron. Astrophys.*, 521, doi:10.1051/0004-6361/201015160.

Hartogh, P., Blecka, M. I., Jarchow, C., et al. (2010b) First results on Martian carbon monoxide from Herschel/HIFI observations, *Astron. Astrophys.*, 521, doi:10.1051/201015159.

Haywood, J., and Boucher, O. (2000) Estimates of the direct and indirect radiative forcing due to tropospheric aerosols: a review, *Rev. Geophys.*, 38, 513–543.

Hecht, M. H., Kounaves, S. P., Quinn, R. C., et al. (2009) Detection of Perchlorate and the Soluble Chemistry of Martian Soil at the Phoenix Lander Site, *Science*, 325, 64–67.

Kaplan, L. D., Connes, J., and Connes, P. (1969) Carbon monoxide in the Martian atmosphere, *Astrophys. J.*, 157, 187–192.

Keppler, F., Vigano, I., McLeod, A., et al. (2012) Ultraviolet-radiation-induced methane emissions from meteorites and the Martian atmosphere, *Nature*, 486, 93–96.

Kliore, A. J., Cain, D. L., Levy, G. S., et al. (1965) Occultation experiment: result of the first direct measurement of Mars' atmosphere and ionosphere, *Science*, 149, 1243–1245.

Kliore, A. J., Fjeldbo, G., Seidel, B. L., et al. (1973) S band radio occultation measurements of the atmosphere and topography of Mars with Mariner 9: extended mission coverage of polar and intermediate latitudes, *J. Geophys. Res.*, 78, 4331–4351.

Knak Jensen, S. J., Skibsted, J., Jakobsen, H. J., et al. (2014) A sink for methane on Mars? The answer is blowing in the wind, *Icarus*, 236, 24–27.

Kong, T. Y., and McElroy, M. B. (1977a) Photochemistry of the Martian atmosphere, *Icarus*, 32, 168–189.

Kong, T. Y., and McElroy, M. B. (1977b) The global distribution of O_3 on Mars, *Planet. Space Sci.*, 25, 839–857.

Krasitski, O. P. (1978) A model for the diurnal variations of the composition of the Martian atmosphere, *Cosmic Res.*, 16, 434–442.

Krasnopolsky, V. A. (1993) Photochemistry of the Martian atmosphere (mean conditions), *Icarus*, 101, 313–332.

Krasnopolsky, V. A. (1995) Uniqueness of a solution of a steady state photochemical problem: applications to Mars, *J. Geophys. Res.*, 100, 3263–3276.

Krasnopolsky, V. A. (1997) Photochemical mapping of Mars, *J. Geophys. Res.*, 102, 13313–13320.

Krasnopolsky, V. A. (2003a) Spectroscopic mapping of Mars CO mixing ratio: detection of north–south asymmetry, *J. Geophys. Res.*, 108, doi:10.1029/2002JE001926.

Krasnopolsky, V. A. (2003b) Spectroscopy of Mars O_2 1.27 µm dayglow at four seasonal points, *Icarus*, 165, 315–325.

Krasnopolsky, V. A. (2005) A sensitive search for SO_2 in the Martian atmosphere: implications for seepage and origin of methane, *Icarus*, 178, 487–492.

Krasnopolsky, V. A. (2006a) Photochemistry of the Martian atmosphere: seasonal, latitudinal, and diurnal variations, *Icarus*, 185, 153–170.

Krasnopolsky, V. A. (2006b) A sensitive search for nitric oxide in the lower atmospheres of Venus and Mars: detection on Venus and upper limit for Mars, *Icarus*, 182, 80–91.

Krasnopolsky, V. A. (2006c) Some problems related to the origin of methane on Mars, *Icarus*, 180, 359–367.

Krasnopolsky, V. A. (2007) Long-term spectroscopic observations of Mars using IRTF/CSHELL: mapping of O_2 dayglow, CO, and search for CH_4, *Icarus*, 190, 93–102.

Krasnopolsky, V. A. (2009) Seasonal variations of photochemical tracers at low and middle latitudes on Mars: observations and models, *Icarus*, 201, 564–569.

Krasnopolsky, V. A. (2010) Solar activity variations of thermospheric temperatures on Mars and a problem of CO in the lower atmosphere, *Icarus*, 207, 638–647.

Krasnopolsky, V. A. (2011) Excitation of the oxygen nightglow on the terrestrial planets, *Planet. Space Sci.*, 59, 754–766.

Krasnopolsky, V. A. (2012) Search for methane and upper limits to ethane and SO_2 on Mars, *Icarus*, 217, 144–152.

Krasnopolsky, V. A. (2013) Night and day airglow of oxygen at 1.27 µm on Mars, *Planet. Space Sci.*, 85, 243–249, 2013.

Krasnopolsky, V. A., and Bjoraker, G. L. (2000) Mapping of Mars $O_2(^1\Delta)$ dayglow, *J. Geophys. Res.*, 105, 20179–20188, 2000.

Krasnopolsky, V. A., and Feldman, P. D. (2001) Detection of molecular hydrogen in the atmosphere of Mars, *Science*, 294, 1914–1917.

Krasnopolsky, V. A., and Gladstone, G. R. (2005) Helium on Mars and Venus: EUVE observations and modeling, *Icarus*, 176, 395–407.

Krasnopolsky, V. A., and Parshev, V. A. (1979) Ozone photochemistry of the Martian lower atmosphere, *Planet. Space Sci.*, 27, 113–120.

Krasnopolsky, V. A., Bowyer, S., Chakrabarti, S., et al. (1994) First measurement of helium on Mars: implications for the problem of radiogenic gases on the terrestrial planets, *Icarus*, 109, 337–351.

Krasnopolsky, V. A., Bjoraker, G. L., Mumma, M. J., et al. (1997) High-resolution spectroscopy of Mars at 3.7 and 8 µm: a sensitive search for H_2O_2, H_2CO, HCl, and CH_4, and detection of HDO, *J. Geophys. Res.*, 102, 6525–6534.

Krasnopolsky, V. A., Maillard, J. P., and Owen, T. C. (2004) Detection of methane in the Martian atmosphere: evidence for life?, *Icarus*, 172, 537–547.

Kuiper, G. P. (1949) Survey of planetary atmospheres, in *The Atmospheres of the Earth and Planets* (Kuiper, G. P. Ed), Chicago Press, Chicago.

Lane, A. L., Barth, C. A., Hord, C. W., et al. (1973) Mariner 9 ultraviolet spectrometer experiment: observations of ozone on Mars, *Icarus*, 18, 102–108.

Lebonnois, S., Quémerais, E., Montmessin, F., et al. (2006) Vertical distribution of ozone on Mars as measured by SPICAM/Mars Express using stellar occultations, *J. Geophys. Res.*, 111, E09S05, doi:10.1029/2005JE002643.

Lefèvre, F., and Forget, F. (2009) Observed variations of methane on Mars unexplained by known atmospheric chemistry and physics, *Nature*, 460, 720–723.

Lefèvre, F., Lebonnois, S., Montmessin, F., et al. (2004) Three-dimensional modeling of ozone on Mars, *J. Geophys. Res.*, 109, E07004, doi:10.1029/2004JE002268.

Lefèvre, F., Bertaux, J. L., Clancy, R. T. (2008) et al., Heterogeneous chemistry in the atmosphere of Mars, *Nature*, 454, 971–975.

Lewis, J. S., and Carver, J. H. (1983) Temperature dependence of the carbon dioxide photoabsorption cross section between 1200 and 170 Angstroms, *J. Quant. Spectrosc. Radiat. Transfer*, 30, 297–309.

Lindner, B. L. (1988) Ozone on Mars: the effects of clouds and airborne dust, *Planet. Space Sci.*, 36, 125–144.

Liu, S. C., and Donahue, T. M. (1976) The regulation of hydrogen and oxygen escape from Mars, *Icarus*, 28, 231–246.

Maguire, W. C. (1977) Martian isotopic ratios and upper limits for possible minor constituents as derived from Mariner 9 infrared spectrometer data, *Icarus*, 32, 85–97.

Mahaffy, P. R., Webster, C. R., Atreya, S. K., et al. (2013) Abundance and isotopic composition of gases in the Martian atmosphere from the Curiosity Rover, *Science*, 341, 263–266.

Maltagliati, L., Montmessin, F., Fedorova, A., et al. (2011) Evidence of water vapor in excess of saturation in the atmosphere of Mars, *Science*, 333, 1868–1870.

Maltagliati, L., Montmessin, F., Korablev, O., et al. (2013) Annual survey of water vapor vertical distribution and water-aerosol coupling in the Martian atmosphere observed by SPICAM/Mex solar occultations, *Icarus*, 223, 942–962.

McElroy, M. B., and Donahue, T. M. (1972) Stability of the Martian atmosphere, *Science*, 177, 986–988.

Melchiorri, R., Encrenaz, T., Fouchet, T., et al. (2007) Water vapor mapping on Mars using OMEGA/Mars Express, *Planet. Space Sci.*, 55, 333–342.

Montmessin, F., and Lefèvre, F. (2013) Transport-driven formation of a polar ozone layer on Mars, *Nature Geo.*, 6, 930–933.

Montmessin, F., Forget, F., Rannou, P., et al. (2004) Origin and role of water ice clouds in the Martian water cycle as inferred from a general circulation model, *J. Geophys. Res.*, 109, E10004, doi:10.1029/2004JE002284.

Moreau, D., Esposito, L. W., and Brasseur, G. (1991) The chemical composition of the dust-free Martian atmosphere: preliminary results of a two-dimensional model, *J. Geophys. Res.*, 96, 7933–7945.

Moudden, Y. (2007) Simulated seasonal variations of hydrogen peroxide in the atmosphere of Mars, *Planet. Space Sci.*, 55, 2137–2143.

Moudden, Y., and McConnell, J. C. (2007) Three-dimensional on-line modeling in a Mars general circulation model, *Icarus*, 188, 18–34.

Mumma, M. J., Villanueva, G. L., Novak, R. E., et al. (2009) Strong release of methane on Mars in northern summer 2003, *Science*, 323, 1041–1045.

Nair, H., Allen, M., Anbar, A. D., et al. (1994) A photochemical model of the Martian atmosphere, *Icarus*, 111, 124–150.

Nier, A. O., and McElroy, M. B. (1977) Composition and structure of Mars' upper atmosphere: results from the neutral mass spectrometers on Viking 1 and 2, *J. Geophys. Res.*, 82, 4341–4348.

Nier, A. O., Hanson, W. B., Seiff, A., et al. (1976) Composition of the Martian atmosphere: preliminary results from Viking 1, *Science*, 193, 786–788.

Novak, R., Mumma, M. J., DiSanti, M. D., et al. (2002) Mapping of ozone and water in the atmosphere of Mars near the 1997 aphelion, *Icarus*, 158, 14–23.

Noxon, J. F., Traub, W. A., Carleton, N. P., et al. (1976) Detection of O_2 dayglow emission from Mars and the Martian ozone abundance, *Astrophys. J.*, 207, 1025–1030.

Owen, T., Biemann, K., Rushnek, D. R., et al. (1977) The composition of the atmosphere at the surface of Mars, *J. Geophys. Res.*, 82, 4635–4639.

Parkinson, T. D., and Hunten, D. M. (1972) Spectroscopy and aeronomy of O_2 on Mars, *J. Atmos. Sci.*, 29, 1380–1390.

Perrier, S., Bertaux, J. L., Lefèvre, F., et al. (2006) Global distribution of total ozone on Mars from SPICAM/MEX UV measurements, *J. Geophys. Res.*, 111, E09S06, doi:10.1029/2006JE002681.

Quinn, R. C., and Zent A. P. (1999) Peroxide-modified titanium dioxide: a chemical analog of putative Martian soil oxidants, *Origins Life Evol. Biosphere*, 29, 59–72.

Sander, S. P., Abbatt, J. P. D., Barker, J. R., et al. (2011) *Chemical Kinetics and Photochemical Data for Use in Atmospheric Studies, Evaluation Number 17*, Jet Propulsion Laboratory, Pasadena.

Shimazaki, T. (1981) A model of temporal variations in ozone density in the Martian atmosphere, *Planet. Space Sci.*, 29, 21–33.

Shimazaki, T., and Shimizu, M. (1979) The seasonal variation of ozone density in the Martian atmosphere, *J. Geophys. Res.*, 84, 1269–1276.

Sindoni, G., Formisano, V., and Geminale, A. (2011) Observations of water vapour and carbon monoxide in the Martian atmosphere with the SWC of PFS/MEX, *Planet. Space Sci.*, 59, 149–162.

Smith, M. D. (2004) Interannual variability in TES atmospheric observations of Mars during 1999–2003, *Icarus*, 167, 148–165.

Smith, M. D., Wolff, M. J., Clancy, R. T., et al. (2009) Compact Reconnaissance imaging spectrometer observations of water vapor and carbon monoxide, *J. Geophys. Res.*, 114, doi:10.1029/2008JE003288.

Spinrad, H., Münch, G., and Kaplan, L. D. (1963) The detection of water vapor on Mars, *Astrophys. J.*, 137, 1319–1321.

Sprague, A. L., Boynton, W. V., Kerry, K. E., et al. (2004) Mars' south polar Ar enhancement: a tracer for south polar seasonal meridional mixing, *Science*, 306, 1364–1367.

Strickland, D.J., Thomas, G. E., and Sparks, P. R. (1972) Mariner 6 and 7 ultraviolet spectrometer experiment: analysis of the O I 1304 and 1356 Å emissions, *J. Geophys. Res.*, 77, 4052–4058.

Stewart, A. I. F., Alexander, M. J., Meier, R. R., et al. (1992) Atomic oxygen in the Martian thermosphere, *J. Geophys. Res.*, 97, 21–102.

Summers, M. E., Lieb, B. J., Chapman, E., et al. (2002) Atmospheric biomarkers of subsurface life on Mars, *Geophys. Res. Lett.*, 29, doi:10.1029/2002GL015377.

Trainer, M. G., Tolbert, M. A., McKay, C. P. and Toon, O. B. (2010) Limits on the trapping of atmospheric CH_4 in Martian polar ice analogs, *Icarus*, 208, 192–197.

Traub, W. A., Carleton, N. P., Connes, P., et al. (1979) The latitude variation of O_2 dayglow and O_3 abundance on Mars, *Astrophys. J.*, 229, 846–850.

Trauger, J. T., and Lunine, J. I. (1983) Spectroscopy of molecular oxygen in the atmosphere of Venus and Mars, *Icarus*, 55, 272–281.

Tschimmel, M., Ignatiev, N. I., Titov, D. V., et al. (2008) Investigation of water vapor on Mars with PFS/SW of Mars Express, *Icarus*, 195, 557–575.

Villanueva, G. L., Mumma, M. J., Novak, R. E., et al. (2013) A sensitive search for organics (CH_4, CH_3OH, H_2CO, C_2H_6, C_2H_2, C_2H_4), hydroperoxyl, nitrogen compounds (N_2O, NH_3, HCN) and chlorine species on Mars using ground-based high-resolution infrared spectroscopy, *Icarus*, 223, 11–27.

Webster, C. R., Mahaffy, P. R., Atreya, S. K., et al. (2013) Low upper limit to methane abundance on Mars, *Science*, 342, 355–357.

Webster, C. R., Mahaffy, P. R., Atreya, S. K., et al. (2015) Mars methane detection and variability at Gale Crater, *Science*, 347, 415–427.

Wong, A. S., Atreya, S. K., and Encrenaz, T. (2003) Chemical markers of possible hot spots on Mars, *J. Geophys. Res.*, 108, E4, 5026, doi:10.1029/2002JE002003.

Yung, Y. L., and DeMore, W. B. (1999) *Photochemistry of Planetary Atmospheres*, Oxford University Press, Oxford/New York.

Yung, Y. L., Strobel, D. F., Kong, T. Y., et al. (1977) Photochemistry of nitrogen in the Martian atmosphere, *Icarus*, 30, 26–41.

Zahnle, K., Haberle, R. M., Catling, D. C., et al. (2008) Photochemical instability of the ancient Martian atmosphere, *J. Geophys. Res.*, 113, E11004, doi:10.1029/2008JE003160.

Zahnle, K., Freedman, R. S., and Catling, D. C. (2011) Is there methane on Mars?, *Icarus*, 212, 493–503.

Upper Neutral Atmosphere and Ionosphere

STEPHEN W. BOUGHER, DAVID A. BRAIN, JANE L. FOX, FRANCISCO GONZALEZ-GALINDO,
CYRIL SIMON-WEDLUND, PAUL G. WITHERS

14.1 INTRODUCTION AND SCOPE

The Mars thermosphere (~100–200 km) is an intermediate atmospheric region strongly impacted by coupling from below with the lower atmosphere (via gravity waves, planetary waves and tides, and dust storms) and coupling from above with the exosphere and ultimately the Sun (via solar radiation and solar wind particles) (see Bougher, 1995; Bougher et al., 2002, 2015a). The thermospheric layer extends from the top of the middle atmosphere (mesopause) to the beginning of space (exobase). The thermosphere is also characterized by a transition region called the homopause (~115–130 km), below which atomic and molecular constituents are well mixed (see Stewart, 1987; Bougher et al., 2000). Above, in the heterosphere, individual species separate according to their unique scale heights via molecular diffusion. Mars' thermosphere is dominated by CO_2 and its dissociation products O and CO, as well as N_2. Finally, the thermosphere is the topmost completely bound layer of the atmosphere. Above, in the exosphere, collisions no longer dominate and the lightest species may form an extended hot corona and escape to space (e.g. Valeille et al., 2009, 2010).

The Mars thermosphere–ionosphere system can change dramatically over time since it is controlled by two highly variable components of the Sun's energy output, solar radiation (~0.1–200 nm) and the solar wind (e.g. Bougher et al., 2002, 2009a). The amount of soft X-ray (0.1–5 nm) and extreme ultraviolet (EUV, 5–110 nm) solar radiation most responsible for heating the Mars thermosphere (and forming its ionosphere) varies significantly over time. These temporal variations result from the changing heliocentric distance (~1.38–1.67 AU), the planet's obliquity (determining the local season), and the changing solar radiation itself. Both solar rotation (~27-day periodic changes in the solar output) and solar cycle (~11-year periodic overall changes in solar output) variations of the solar X-ray and EUV fluxes are significant (up to factors of ~3–100), producing dramatic variations in global thermospheric temperatures, composition, and winds (e.g. Bougher et al., 2000, 2002, 2009a; Forbes et al., 2008), as well as ionospheric densities (e.g. Fox, 2004a). For example, as a result of solar cycle and orbital variations in energy inputs, the Mars upper atmosphere "breathes" (i.e. expands from warming and contracts from cooling) as the Martian seasons and the solar cycle advance. In fact, the solar EUV fluxes received at Mars vary by ±22% throughout the Martian year, solely the result of the changing heliocentric distance.

The Mars dayside thermosphere is characterized by temperatures that increase dramatically with altitude above a minimum at the mesopause (~100–120 km) (Forget et al., 2009;

McDunn et al., 2010). This temperature increase is largely due to solar EUV–UV heating, while approaching an asymptotic maximal value at the exobase (~160–220 km). This deposited heat is conducted downward toward the mesopause (via molecular thermal conduction), where it is partially radiated away to space by significant CO_2 15 μm emission. Finally, global-scale winds transport atomic and molecular constituents around the planet, and redistribute the heat produced on the dayside, causing an intricate feedback of global composition, temperature, and wind fields (Bougher et al., 2000, 2006, 2009b).

Embedded within the thermosphere is the ionosphere, a weakly ionized plasma. The primary charged particles that make up the Mars ionosphere (e.g. O_2^+, CO_2^+, O^+, CO^+, and NO^+) are formed by: (1) solar EUV fluxes and subsequent photoelectrons that ionize local neutral thermospheric species (e.g. CO_2, N_2, CO, O, etc.); (2) precipitating particles (e.g. suprathermal electrons) that ionize these same neutral species (especially on the nightside); and (3) subsequent ion–neutral photochemical reactions (e.g. Fox, 2004a,b, 2009). Since Mars has a negligible intrinsic magnetic field, the variable solar wind (including its particles and interplanetary magnetic field) interacts with the Mars near-space environment (including the thermosphere, ionosphere, and exosphere), resulting in ionization, neutral heating, pickup ion escape, ion outflow, hot species escape, and ion sputtering (see Chassefière and Leblanc, 2004). A detailed review of these escape processes is provided in Chapter 15.

The Mars thermosphere–ionosphere system also constitutes an important atmospheric reservoir that regulates present-day escape processes from the planet (see Bougher et al., 2015a). Characterization of this system, and its spatial and temporal (e.g. solar cycle, seasonal, diurnal) variability, is crucial to determining escape rates. For example, the Mars thermosphere regulates escape processes by: (1) absorbing and redistributing solar EUV and UV energy, (2) filtering and absorbing upward-propagating waves of various scales, (3) mediating interactions between the solar wind and the ionosphere, and (4) controlling the transport of species from the lower atmosphere through eddy diffusion and large-scale dynamics, thereby providing a means of regulating atmospheric escape.

Computational models are fundamental tools for the study of planetary atmospheres, and in particular for the upper atmosphere of Mars, where observations are scarce. While some portion of our knowledge of this region comes from observational data, usually from orbit, models allow for a deeper insight into the physical processes that produce observed atmospheric structures. Models also complement observational data, helping to overcome their usually limited spatial and temporal coverage, and are used to predict atmospheric conditions and guide future observations. In this chapter we will focus on the models currently used to study the Martian upper atmosphere and their present status, achievements, and their capabilities and limitations. Thus, we will necessarily neglect many models developed in the past that produced important advances in our knowledge of this atmospheric region, but that are no longer used today (see references in Bougher et al., 2008).

In this chapter, we will review the structure and dynamics of the Martian neutral upper atmosphere and ionosphere, and their variability, using spacecraft datasets prior to the Mars Atmosphere and Volatile Evolution (MAVEN) mission. In addition, recent airglow and auroral emissions, including their spatial and temporal variations, will be discussed. A survey of current neutral upper atmosphere and ionosphere numerical models will be given in order to illustrate the power of these tools to interpret, assimilate, and better utilize the available datasets. Finally, a summary of outstanding problems will be provided to anticipate the scientific return from MAVEN.

14.2 NEUTRAL UPPER ATMOSPHERE STRUCTURE AND VARIABILITY

14.2.1 Thermal Structure and Energy Balance

What processes determine Mars thermospheric temperatures and their variability, and how do these variations and the underlying processes contrast with those for the Venus thermosphere?

Since Mars and Venus are both CO_2-dominated planets, the energy balance processes for these thermospheres are generally thought to be similar. These processes include: (1) solar EUV–UV heating (and its variation with solar cycle, solar rotation, heliocentric distance, and local solar declination), (2) molecular thermal heat conduction, (3) CO_2 15 μm cooling, (4) horizontal advection, and (5) adiabatic heating and cooling associated with global dynamics.

However, the relative importance of these processes may be different on each planet. According to Bougher et al. (1999b, 2009b), the primary Mars thermospheric energy balance on the dayside occurs between EUV heating and thermal heat conduction, with CO_2 cooling playing a tertiary role. In addition, Mars adiabatic cooling, due to rising motions on the dayside in the global thermospheric circulation, plays a progressively more important role as the solar cycle advances from minimum to maximum. For Venus, modeling suggests that the primary dayside thermospheric balance occurs between EUV heating and CO_2 cooling, with molecular thermal conduction and global winds playing a minor role. Bougher et al. (1999b) suggest that these differences are largely the result of fundamental planetary parameters for Venus and Mars (e.g. heliocentric distance and the resulting impact of unique atomic O abundances on the magnitude of CO_2 cooling). Last, for both planets, solar wind precipitating particles (suprathermal) may provide episodic heating of the thermosphere. We will briefly examine each of these Mars heating and cooling processes in turn.

Solar EUV photons are absorbed primarily by CO_2 in the Martian lower thermosphere, and by atomic O at higher altitudes approaching 200 km. Maximum EUV absorption occurs where

the CO_2 optical depth at these wavelengths equals unity. This is ~120–130 km near the subsolar point (see Paxton and Anderson, 1992). This level also corresponds to the maximum per volume EUV heating rate and peak photoionization of CO_2. Solar UV fluxes longward of ~110.0 nm are absorbed at altitudes below ~100 km, and are thus important for photodissociation and associated chemistry, but are of secondary importance (compared to the near-IR) for local solar heating. In terms of commonly used units of K/day, Mars dayside peak EUV heating rates occur at ~150–170 km (see Bougher et al., 1999b).

The detailed calculation of neutral gas EUV heating rates from first principles is a major undertaking, best conducted using comprehensive one-dimensional (1D) models that address energy partitioning (e.g. Fox and Dalgarno, 1979). In particular, large (computationally time-consuming) upper atmosphere general circulation models (GCMs) typically make use of detailed EUV and UV heating efficiencies calculated separately by these 1D models (see Bougher et al., 1999b, 2008). Fox and Dalgarno (1979) describe Mars EUV and UV heating efficiency calculations, for which thermospheric EUV efficiencies of 20–25% and UV efficiencies of 22% are obtained. Initial heating results from dissociative recombination of O_2^+ (at and above the ionosphere peak) and from photodissociation and quenching of metastable species in the lower thermosphere. Significant heat is lost due to (1) the excitation of CO_2 vibrational levels by hot atoms, molecules, and ions, and (2) the subsequent radiative energy loss via CO_2 15 μm emission before quenching can occur (see Bougher et al., 1999b). Altitude-independent EUV and UV heating efficiency values of ~19–22% are commonly used in thermospheric GCMs (see Bougher et al., 2004, 2006, 2009b; Huestis et al., 2008; González-Galindo et al., 2009a,b, 2010).

The relative importance of CO_2 15 μm emission in providing cooling in the Martian thermosphere is dependent upon several factors. First of all, inelastic collisions of atomic oxygen atoms and CO_2 molecules are especially effective in exciting $CO_2(v = 2)$ vibrational states, after which the excitation may be quenched either by another collision (with some molecule or atom) or by emission of 15 μm radiation (see Bougher et al., 1994, 1999b; López-Valverde et al., 1998; Huestis et al., 2008). The latter yields cooling at thermospheric heights, where the timescale for 15 μm emission is shorter than that for collisional quenching. The overall importance of this 15 μm cooling process depends on: (a) the atomic O abundance, (b) the CO_2 abundance, and (c) the collision excitation rate coefficient. To be consistent with the generally accepted way of describing this process in the literature, we refer to the rate coefficient for collisional de-excitation of $CO_2(v = 2)$ and call it the CO_2–O quenching rate coefficient (see Huestis et al., 2008). It is assumed that the CO_2–O quenching rate is directly linked to the collision excitation rate coefficient by detailed balance (e.g. Schunk and Nagy, 2009). This O–CO_2 quenching rate has been the subject of considerable debate for many years; i.e. values measured in the laboratory and retrieved by fitting spacecraft atmospheric observations vary by a factor of 3–4 (see review of Huestis et al. (2008) and references therein). However, this rate coefficient has recently been measured in the laboratory at room temperatures (300 K) and found to be ~1.5×10^{-12} cm^3 s^{-1}, in contrast to values of ~3.0×10^{-12} cm^3 s^{-1} which have commonly been used in recent GCM model simulations that seek to compare the heat budgets and thermospheric temperatures of Venus, Earth, and Mars (see Bougher et al., 1999b, 2000, 2008; González-Galindo et al., 2009a,b). For Mars, this discrepancy (between laboratory and GCM required CO_2–O quenching rates) is attributed to uncertainties in the upper atmosphere atomic oxygen abundance, which has never been directly measured, but has been merely inferred from UV airglow and ion composition measurements (e.g. Stewart et al., 1972, 1992; Strickland et al., 1973; Hanson et al., 1977; Chaufray et al., 2009). Thus, quantification of the role of CO_2 15 μm cooling in the Mars thermosphere requires: (a) proper characterization of atomic oxygen abundances (i.e. GCM dayside O abundances may need to be higher), (b) application of hydrodynamic models that assume appropriate EUV–UV heating efficiencies, and (c) the utilization of a modern O–CO_2 quenching rate (~1.5×10^{-12} cm^3 s^{-1} near 300 K). Meanwhile, uncertainties in calculated CO_2 cooling rates up to a factor of 2 are likely.

Molecular thermal conduction redistributes heat in an effort to reduce the temperature gradient. On the Mars dayside, the temperature gradient is generally positive upward, so EUV thermal energy deposited above ~130 km is conducted downward toward the mesopause (~90–110 km) where CO_2 15 μm emission radiates this heat to space. The role of global winds on the thermal structure is typically approximated in 1D numerical models by parameterization of eddy heat conduction. However, multi-dimensional hydrodynamic models (see Section 14.5.2) are better suited to this task, but still require a parameterization for turbulent heat conduction not resolved by the model grid (e.g. Bougher et al., 2008).

The role of thermospheric global winds in controlling the temperature structure is predicted to change dramatically with the solar cycle and Martian seasons (see Bougher et al., 1999b, 2000, 2009b, 2015a). Generally, solar EUV-driven winds diverge and are subject to upwelling (providing cooling) in the mid-afternoon region near the subsolar point; conversely, winds converge and descend (providing heating) on the nightside and at high latitudes. It is likely that Martian thermospheric winds play an increasingly important role in regulating temperatures as the solar cycle (solar minimum to maximum) and seasons (aphelion to perihelion) advance (Bougher et al., 1999b, 2009b). For instance, without this "dynamical thermostat", Martian exobase temperatures on the dayside would be considerably warmer during perihelion/solar maximum conditions than those currently estimated from limited data.

However, the magnitude of Martian thermospheric winds also depends on other dynamical forcing mechanisms. These include: (a) momentum deposition by upward-propagating gravity waves; (b) upward-propagating tides and planetary waves (and their nonlinear interactions); and (c) the effects of dust storms (e.g. Smith, 2004) and subsequent acceleration of the winds throughout the entire atmosphere (e.g. Forbes et al., 2002; Bougher et al., 2011a; Medvedev et al., 2011, 2012). Dust storm impacts on thermospheric winds were observed during the regional Noachis dust storm in 1997 (Baird et al., 2007).

The Martian upper atmosphere thermal structure is poorly constrained by a limited number of measurements at selected locations, seasons, and periods scattered throughout the solar cycle (e.g. Müller-Wodarg et al., 2008; Bougher et al., 2015a). The upper atmosphere vertical thermal structure has been characterized by: (a) Viking Landers 1 and 2 entry accelerometer-derived temperatures (Seiff and Kirk, 1977); (b) Viking Landers 1 and

2 Upper Atmosphere Mass Spectrometer (UAMS)-derived temperatures (Nier and McElroy, 1977); (c) Mars Global Surveyor (MGS), Mars Odyssey, and Mars Reconnaissance Orbiter (MRO) aerobraking accelerometer-derived temperatures from mass density scale heights (e.g. Keating et al., 1998, 2003, 2006, 2008); and (d) Mars Express SPICAM temperatures obtained by stellar occultation (Forget et al., 2009; McDunn et al., 2010). Figure 14.1 illustrates MRO nightside and MGS dayside temperature profiles derived from accelerometer data near aphelion/solar minimum to moderate conditions (from Keating et al., 2008). Dayside exospheric (isothermal) temperatures are reached above ~160 km, approaching 190–200 K; corresponding nightside values approach ~135–150 K above ~150 km. These low-latitude temperatures reveal rather small diurnal contrasts for "extreme" near-solar-minimum/aphelion conditions, the lower limit of what is likely over the solar cycle and Martian seasons. Predicted heat balances corresponding to these measured temperature profiles will be addressed in Section 14.5.2.1.

Winter polar warming, a feature of the lower thermospheric temperature structure, was first observed by accelerometers, but not predicted beforehand (see Keating et al., 2003, 2008; Bougher et al., 2006). Figure 14.2 illustrates latitudinal variations in temperatures at 120 km near aphelion (northern summer, southern winter) and perihelion (southern summer, northern winter) conditions as observed by MGS and Odyssey, respectively. As MGS approaches the southern winter pole, moving from the dayside to the nightside, lower thermospheric temperatures decrease from ~130–140 K (low–middle dayside latitudes) to ~90–100 K (near the winter pole), before warming again slightly onto the nightside (up to ~110 K). By contrast, Odyssey temperatures increased sharply as the spacecraft crossed the northern winter pole (from 120 to 160 K), before declining to ~100 K at low–middle latitudes on the nightside.

This seasonal variation in the magnitude of winter polar warming is attributed to the seasonal changes in the global thermospheric circulation during solstices, which is predicted to be stronger near perihelion, thereby providing enhanced dynamical heating resulting from convergent and descending winds in northern winter polar latitudes (Bougher et al., 2006). The magnitude of this winter polar warming is also likely regulated by seasonal changes in momentum deposition associated with gravity waves, planetary waves, and tides (e.g. Bell et al., 2007; González-Galindo et al., 2009b; Bougher et al., 2011a; Medvedev et al., 2011).

The combined solar cycle and seasonal variations in Martian dayside upper thermosphere (i.e. exosphere) temperatures have been the subject of considerable analysis and study since the first Mariner 6, 7, and 9 ultraviolet spectrometer (UVS) measurements (1969–1972), and up to recent Mars Express SPICAM UVS measurements (2004–present) (e.g. Stewart, 1987; Bougher et al., 1999b, 2000, 2009b, 2010, 2015a; Keating et al., 2003, 2008; Leblanc et al., 2006a; Withers, 2006; Forbes et al., 2008; González-Galindo et al., 2009a; Huestis et al., 2010; Krasnopolsky, 2010; Stiepen et al., 2015). The significant Mars eccentricity demands that both the solar cycle and seasonal variations in near-exobase temperatures (T_{exo}) be considered together (e.g. Bougher et al., 2000).

Near solar minimum, dayside low solar zenith angle (SZA) T_{exo} values have been extracted from the Viking 1 entry science

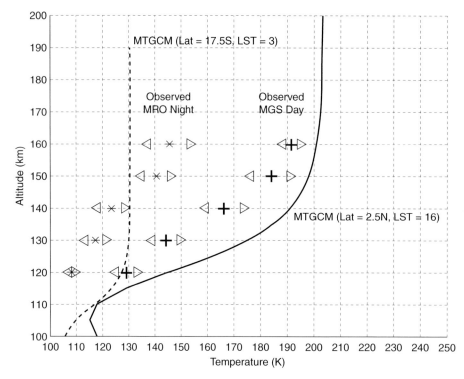

Figure 14.1. MRO nightside and MGS dayside temperature profiles derived from accelerometers. Symbols (+ and *) denote averaged dayside- and nightside-derived temperatures; triangles denote the one-sigma error bars corresponding to these mean values. Observed exospheric temperatures are ~190–200 K (dayside) and 140–150 K (nightside) for these sampling periods. Coupled MGCM–MTGCM (see Section 14.5.2.1) simulated profiles for similar conditions are provided. From Keating et al. (2008).

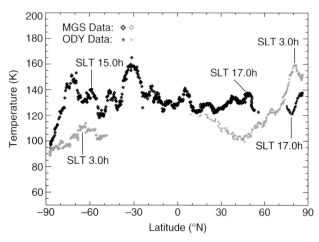

Figure 14.2. MGS ($L_s = 35$–$95°$) and Odyssey ($L_s = 260$–$310°$) accelerometer density-derived temperatures traversing the dayside to nightside and across the winter pole at 120 km. From Bougher et al. (2006). Northern winter polar warming is significant at this altitude during perihelion conditions (~40–60 K) and negligible in the southern winter polar region during local aphelion conditions.

datasets (~186 K) (Nier and McElroy, 1977; Seiff and Kirk, 1977), MGS accelerometer density scale heights (~190–200 K) (see Figure 14.1; also Withers, 2006; Keating et al., 2008), and from Mars Express SPICAM UVS airglow (CO_2^+ doublet) scale heights (~201±10 K) (Leblanc et al., 2006a; Bougher et al., 2010). These limited spacecraft measurements are presently used to characterize T_{exo} for near-solar-minimum, aphelion conditions at Mars. Conversely, solar moderate to

maximum, near-perihelion T_{exo} values are currently estimated from Mariner 6 and 7 CO Cameron emissions (~315 K) (e.g. Stewart, 1972), and Mariner 9 CO Cameron emissions (~325 K) (Stewart et al., 1972). Most recently, MGS drag measurements near ~390 km (Forbes et al., 2008; Bougher et al., 2009b) revealed T_{exo} approaching ~300 K for these extreme solar and orbital conditions.

Figure 14.3 illustrates T_{exo} values extracted from MGS drag measurements over 1999–2005, nearly spanning the solar cycle for all Mars seasons (Forbes et al., 2008). Superimposed are coupled MGCM–MTGCM model (see Section 14.5.2.1) T_{exo} simulated values for corresponding conditions, showing a similar trend. Some doubt was recently raised about whether these MGS drag estimates of T_{exo}, especially near solar minimum conditions, are consistent with the underlying densities retrieved (Krasnopolsky, 2010). Nevertheless, a composite estimate of the "extreme" solar cycle plus seasonal variation of dayside exospheric temperatures of about ~100–140 K is estimated for Mars at the present time. By contrast, solar cycle variation of Venus dayside temperatures is observed to be about 70–80 K (Kasprzak et al., 1997). The weaker variation for Venus is attributed to the dominant role of CO_2 15 µm cooling in balancing EUV heating, thereby acting as a "thermostat" to regulate thermospheric temperatures (e.g. Bougher et al., 1999b).

Recent investigations were conducted (e.g. Bougher et al., 2010; Huestis et al., 2010; Krasnopolsky, 2010; Stiepen et al., 2015) to identify the best airglow emission for reliable extraction of Mars exospheric temperatures (e.g. CO (a–X) Cameron bands at 180–260 nm, CO_2^+ (B–X) ultraviolet doublet near 289 nm, or O I at 297.2 nm). In this regard, a reanalysis of

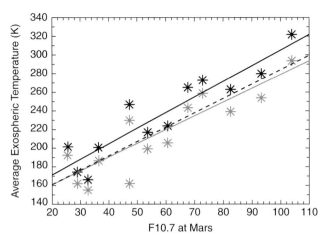

Figure 14.3. Simulated dayside (low SZA) exospheric temperatures over the solar cycle are compared with MGS drag temperatures. The dashed curve corresponds to the MGS drag-derived temperatures over the solar cycle (Forbes et al., 2008). Black (gray) asterisks correspond to MGCM–MTGCM (see Section 14.5.2) simulations using a 22% (19%) EUV–UV heating efficiency. The black (gray) curve corresponds to the linear least-squares fit to the 22% (19%) model data points (black and gray asterisks). The $F_{10.7}$ index at Mars (*x*-axis) is a shorthand to indicate that Earth-measured solar EUV fluxes (minimum, moderate, maximum) are scaled by the Mars heliocentric distance as a function of season for Mars model application. From Bougher et al. (2009b).

Mariner and Mars Express ultraviolet airglow datasets is being carried out to improve the estimates of T_{exo} values from solar minimum to maximum conditions in advance of new datasets anticipated from the MAVEN mission (see Section 14.4).

14.2.1.1 One-Dimensional Models of the Thermal and Radiative Balance

The determination of the thermal state and the remote sensing of the atmosphere require precise radiative models. One of these models, specially designed for the upper atmosphere of Mars, has been developed by the Instituto de Astrofísica de Andalucía (IAA/CSIC, Granada, Spain) as part of an ongoing project to build a 1D non-stationary radiative–convective–composition model (Rodrigo et al., 1990; López-Valverde et al., 2000, 2006). Its core radiative transfer model, specially suited to study non-local thermodynamic equilibrium (NLTE) situations (López-Valverde and López-Puertas, 1994a,b), is based on long experience in modeling these processes in the upper terrestrial atmosphere (López-Puertas and Taylor, 2001) and produces two types of results: populations of vibrational states of CO_2 and CO, and heating/cooling rates in the IR (1–20 μm) by their ro-vibrational bands. The model has been applied both to the radiative balance and the thermal structure of the Martian upper atmosphere (López-Puertas and López-Valverde, 1995; López-Valverde et al., 2000) and to the remote sensing of mesospheric CO_2 and CO emissions at Mars (López-Valverde et al., 2005; Formisano et al., 2006) and at Venus (Roldán et al., 2000; López-Valverde et al., 2007; Gilli et al., 2009).

In addition to the IR radiative balance, the IAA model includes a scheme for the calculation of UV heating and a photochemical module including ionospheric reactions (González-Galindo

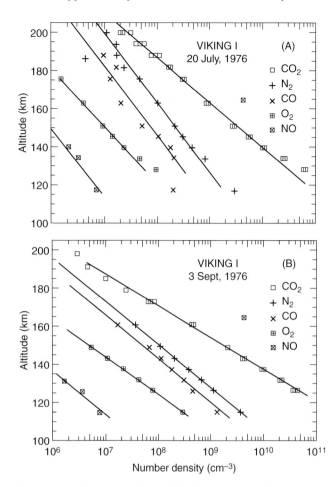

Figure 14.4. Number densities of CO_2, N_2, CO, O_2, and NO determined from measurements made by the Upper Atmospheric Mass Spectrometers (UAMS) on the Viking 1 and 2 Landers. Ar was also measured, but is not shown here. From Nier and McElroy (1977).

et al., 2005, 2008). The IAA model has also been used as a benchmark for the development of parameterizations of important physical processes. In particular, López-Valverde and López-Puertas (2001) developed a parameterization for the CO_2 NLTE thermal cooling rate, and López-Valverde et al. (1998) proposed a scheme for the NLTE solar heating rate by CO_2; both schemes are being used by most GCMs studying the Martian mesosphere and thermosphere. González-Galindo et al. (2005) presented fast schemes to calculate the UV heating and the photochemistry of the major species in the upper atmosphere which are used in the Laboratoire de Météorologie Dynamique (LMD) Mars general circulation model (MGCM) in conjunction with a more sophisticated 1D photochemical model for the lower atmosphere (see Section 14.5.2.2).

14.2.2 Density Structure and Composition

Prior to the MAVEN mission, the composition of the Mars thermosphere, specifically neutral species densities, had only been directly measured with the UAMS instruments on-board the descending Viking 1 and 2 Landers (e.g. Nier and McElroy, 1977). Figure 14.4 illustrates Viking 1 and 2 densities for mid-afternoon (SZA ~ 44°) low to mid-latitude locations,

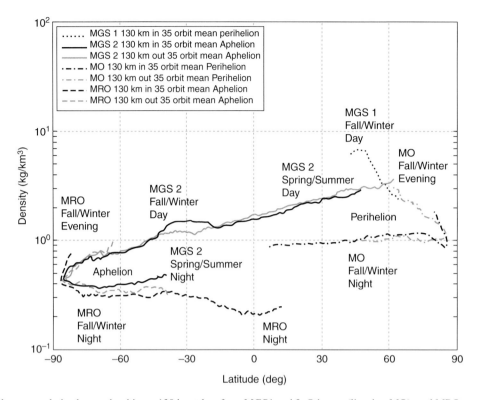

Figure 14.5. Accelerometer-derived mass densities at 130 km taken from MGS1 and 2, Odyssey (listed as MO), and MRO sampling. Inbound and outbound (orbit portion) aeropass measurements are plotted to illustrate the relative stability of thermospheric climatology over several weeks between repeated spacecraft sampling periods. Curves are illustrated to show diurnal, seasonal, and inter-annual trends in density variations. Longitude variations are removed by taking 35-orbit running means. From Keating et al. (2006).

Table 14.1. *MGS1 (phase 1), MGS2 (phase 2), Odyssey, and MRO aerobraking sampling parameters. Temporal (seasonal, solar cycle) and spatial (latitude, solar local time) coverage at 130 km is summarized for inbound legs of spacecraft aeropasses.*

Aerobraking campaigns (interval)	Vertical structures (number)	Seasonal coverage L_s (degrees)	10.7 cm flux at Earth, $F_{10.7}$ (s.f.u.)[a]	Latitude (degrees)	Local solar time (hours)
MGS1 (1997–1998)	200	180–300	70–90	~35N–65N	18–11
MGS2 (1998–1999)	600	34–92	100–150	50N–80S to 80S–30S	17–15 to 1–2
Odyssey (2001–2002)	300	260–310	175–200	70N–80N to 80N–15N	18–17 to 2–3
MRO (2006)	450	35–99	80–100	80S–85S to 80S–12N	20–18 to 2–3

[a] Solar flux units.

corresponding to solar minimum, near aphelion conditions. The major species is CO_2, followed by N_2 and CO. Atomic O could not be measured by the UAMS; instead, ionosphere model calculations have estimated that the O density exceeds the CO_2 density at ~200 km (e.g. Hanson et al., 1977). This uncertain abundance of O is a major unknown for the Mars upper atmosphere; in particular, systematic measurements of atomic O (both spatially and temporally) are needed to constrain the heat budget and chemistry of the dayside thermosphere (see Sections 14.2.1 and 14.5.2).

By contrast, mass densities of the Mars thermosphere have been repeatedly measured by accelerometers on-board MGS, Odyssey, and MRO spacecraft (e.g. Keating et al., 1998, 2003, 2006, 2008; Bougher et al., 1999a, 2006; Tolson et al., 1999, 2005, 2007, 2008; Withers et al., 2003; Withers, 2006; Zurek et al., 2015). The sampling periods (seasons, solar cycle) and locations (latitude, local time) of these accelerometer measurements are summarized in Table 14.1. Each of these nearly polar orbiting spacecraft sampled different longitudes as the planet rotated. Different spacecraft accelerometers permitted density recovery over various altitude ranges: ~115–160 km (MGS1), ~100–160 km (MGS2), ~95–140 km (Odyssey), and ~95–170 km (MRO). It is clear that the accelerometer sampling of mass densities is quite sparse, providing only a rudimentary database for characterizing the upper atmosphere structure and its variations over the solar cycle and Mars seasons. Still, this accelerometer database provides the best record of Mars' upper atmosphere prior to MAVEN.

Figure 14.5 illustrates accelerometer-derived mass densities at 130 km taken from MGS1 and 2 (phases 1 and 2), Odyssey (listed as MO), and MRO sampling (Keating et al., 2006). This reference altitude was chosen since MGS2, Odyssey, and MRO aeropasses all contain regular sampling at 130 km, thus facilitating a comparison of the thermospheric structure during these aerobraking campaigns. This altitude is generally ~20–30 km above periapsis.

MGS2 densities show the transition from spring–summer (northern hemisphere) to fall–winter (southern hemisphere) during the daytime; densities are shown to decrease by nearly a factor of 10 (from 60°N to 85°S latitude) in accord with underlying temperatures, which become cooler toward the local winter pole. The MGS2 data also show the variation in the southern hemisphere between dayside and nightside densities, with a substantial drop in densities on the nightside related to even cooler temperatures. Four Martian years later, the MRO spacecraft overlaps in seasonal coverage for similar latitudes and mostly nightside local times (see Table 14.1). These MGS2 and MRO densities (near aphelion conditions) are quite similar, suggesting that year-to-year thermospheric density variations near aphelion are small.

Odyssey density data are also shown in the northern hemisphere near perihelion. These measurements are made in the dayside winter polar region and on the nightside approaching the equator. These winter polar densities (near perihelion) are nearly a factor of 2.5 larger than corresponding MGS2 and MRO winter polar densities (near aphelion). The inflation of the entire atmosphere (by ~15 km) with the changing seasons (see Section 14.3.3) is responsible. In addition, the Odyssey measurements are obtained near solar maximum of the solar cycle (see Table 14.1). Comparing measurements near the equator, substantial density variation is shown between solar maximum/perihelion (Odyssey) and solar minimum/aphelion (MRO) conditions on the nightside.

The differences in inbound and outbound densities are very small even though the simultaneous latitudes of the measurements differ substantially between inbound and outbound aeropasses. This indicates that the climatology over these time intervals, of the order of weeks, is stable for the three accelerometer campaigns.

Mass densities derived from accelerometer measurements also reveal large-amplitude Sun-synchronous longitudinal (planetocentric) variations at altitudes of ~100–160 km (e.g. Keating et al., 1998; Forbes and Hagan, 2000; Forbes et al., 2002; Withers et al., 2003). These density structures are produced by vertically propagating non-migrating solar thermal tides that are excited near Mars' surface and propagate into the thermosphere; these longitude features only appear as stationary waves from an orbit that precesses slowly in local time (e.g. Forbes and Hagan, 2000; Forbes et al., 2002; Wilson, 2002; Angelats i Coll et al., 2004; Moudden and Forbes, 2008).

Excitation of these waves is commonly attributed to topographic modulation of near-surface solar heating (e.g. Forbes et al., 2002). However, secondary contributors to the excitation of these non-migrating tides (e.g. surface thermal inertia and albedo) are also being investigated (Moudden and Forbes, 2008, 2010). Studies have reported the dominance of wave-2 and wave-3 tidal components in the Sun-synchronous aerobraking maps (e.g. Forbes and Hagan, 2000; Forbes et al., 2002; Withers et al., 2003; Angelats i Coll et al., 2004). An example is provided in Figure 14.6, which shows wave-2 amplitudes at 130 km taken from MGS ($L_s = 28$–93°) and MRO ($L_s = 36$–100°) sampling on the dayside and nightside, respectively (Keating et al., 2008). The 180° phase shift (at low to mid-latitudes) is consistent with the strong influence of a diurnal Kelvin wave. Moudden and Forbes (2010) presented a new interpretation of these MGS and MRO longitudinal variations in terms of planetary wave–tidal interactions. They found that a significant amount of the density variability in the aerobraking region can be attributed to the effect of tidal modulation by planetary waves in the 5–20 day period range.

Lastly, about four Martian years of MGS electron reflectometry (ER) data were employed to derive neutral mass densities at ~185 km on the nightside near 2 a.m. local time (Lillis et al., 2010). Density measurements were extracted specifically from April 1999 to November 2006. Seasonally repeating features include: (a) overall expansion and contraction of the nighttime thermosphere with heliocentric distance, and (b) much lower densities at the aphelion winter pole compared to the perihelion winter pole. Inter-annual differences are also observed, and may be related to changing dust opacities or solar extreme ultraviolet (EUV) fluxes.

14.3 IONOSPHERIC STRUCTURE AND VARIABILITY

How does the structure of the Martian ionosphere compare with that of the Earth? What processes control these ionospheric layers? Why is O_2^+ the dominant ion at Mars and not CO_2^+, as might be expected for a CO_2 atmosphere?

14.3.1 Spacecraft Measurements of the Ionosphere

Electron density profiles from the ionosphere of Mars have been obtained by many radio occultation (RO) experiments since the U.S. Mariner 4 flyby in 1964 (e.g. Fjeldbo and Eshleman, 1968; Lindal et al., 1979; Barth et al., 1992; Mendillo et al., 2003). From 1998 to 2005, the Mars Global Surveyor (MGS) Radio Science Subsystem (RSS) returned thousands of high-latitude electron density profiles in the SZA range 71–89° (e.g. Hinson et al., 1999). Also, the radio science experiment MaRS on the European Mars Express (MEX) Orbiter returned ~500 electron density profiles during five Earth occultation seasons spanning the period from April 2004 to September 2008 (e.g. Pätzold et al., 2005). The altitude resolution for RO measurements is about 1–2 km and the sensitivity is generally in the range $(3$–$8) \times 10^3$ cm^{-3}. The altitude range of an RO profile depends on the sensitivity of the instrument, with the lowest- and highest-altitude data acquired around 80–100 km and 200–350 km, respectively.

MEX also carries the Mars Advanced Radar for Subsurface and Ionospheric Sounding (MARSIS) instrument, which

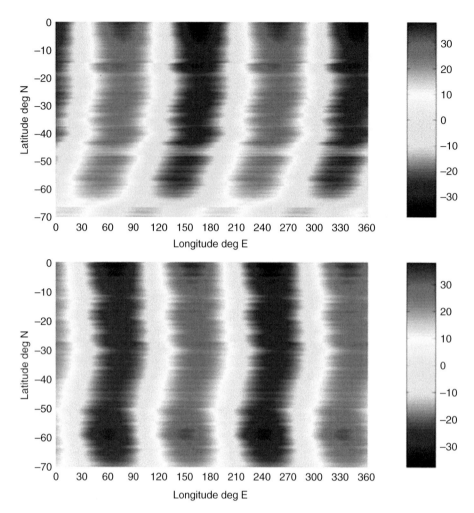

Figure 14.6. Accelerometer mass density wave-2 components (from wave-1, -2, -3 fit) at 130 km taken from MGS (L_s = 28–93°) and MRO (L_s = 36–100°) sampling: (top) dayside (LT = 13:00 to 17:00) MGS densities in the southern winter hemisphere, and (bottom) nightside (LT = 02:00–03:00) MRO densities in the southern winter hemisphere. Amplitudes are given as percent differences (see color bars) from the mean. Notice the 180° phase shift of these day versus night wave-2 amplitudes. From Keating et al. (2008). A black and white version of this figure will appear in some formats. For the color version, please refer to the plate section.

measures the topside ionospheric electron density profile and the peak density by vertical sounding, with a nominal sensitivity of about 1.24×10^2 cm^{-3} (Gurnett et al., 2008). In practice, however, electron densities below $\sim 5 \times 10^3$ cm^{-3} are not detected, and the altitude resolution is ~ 7 km, which is comparable to the neutral scale height (e.g. Gurnett et al., 2005, 2008; Morgan et al., 2008; Picardi, 2008). This experiment can access regions of the ionosphere that RO experiments cannot for geometrical reasons: the deep nightside ionosphere (SZA greater than 125°) and the region near the subsolar point (SZA less than 45°). In addition, MARSIS also obtains *in situ* electron densities along the orbit with a much better sensitivity (~ 1.0 cm^{-3}) (e.g. Gurnett et al., 2010). These local electron densities are derived from the excitation of electron plasma oscillations, and are limited to altitudes above ~ 275 km.

Prior to the MAVEN mission, the only existing *in situ* data on the composition of the Martian ionosphere consisted of the two altitude profiles of the densities of O_2^+, CO_2^+, and O^+ from about 110 to 300 km as measured by the Retarding Potential Analyzers (RPA) on the Viking 1 and 2 Landers near 44° SZA

at low solar activity in July and August 1976 (e.g. Hanson et al., 1977). A recent review of Mars ionosphere structure and modeling studies is found in Haider et al. (2011).

14.3.2 Introduction to Ionospheric Regions: Terrestrial Example

On Earth, the major ionospheric regions are designated by the letters D, E, and F; the F region is divided into F_1 and F_2 regions. Figure 14.7, from Bauer and Lammer (2004), shows the layers in the terrestrial ionosphere and their sources. Each of these regions is formed by a different process. This nomenclature has been adopted, but without universal acceptance, for other planets as well (e.g. Fox, 2004a). It is often challenging to determine the process responsible for an observed ionospheric feature on another planet reliably enough to apply this nomenclature in the absence of modeling.

The F region, where the ultimate source of ions is solar EUV photons in the 10–100 nm wavelength range, is topmost. The column density of the absorbing species in the F region is of

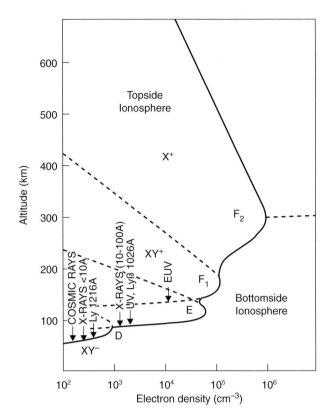

Figure 14.7. Layers in the terrestrial ionosphere (from Bauer and Lammer, 2004).

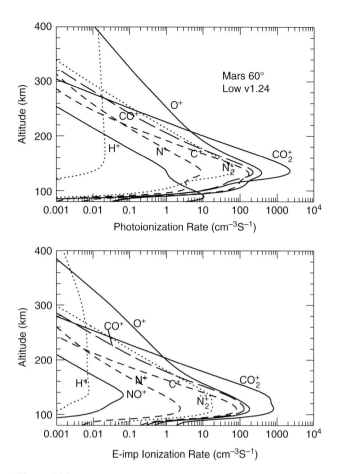

Figure 14.8. Major-ion production rate profiles (SZA ~ 60°) by photoionization and electron impact ionization, from Martian model simulations appropriate to low-solar-activity conditions.

the order of $(5–10)×10^{16}$ cm^{-2}. The F$_1$ subregion is in photochemical equilibrium and is dominated by molecular ions. By contrast, in the overlying F$_2$ subregion, transport processes take over from chemical control at the F$_2$ peak, which is dominated by atomic ions. Below the F region lies the E region, where the ultimate source of ions is absorption of solar soft X-ray photons in the 1–10 nm range. The column densities of the absorbing species in the E region are of the order of $1×10^{18}$ cm^{-2}. The lower E region may also contain ions produced by meteoroid ablation. In the low-lying D region, the ultimate source of ions (both positive and negative) is absorption of solar hard X-ray photons with wavelengths less than 1 nm. Besides direct photoionization, photoelectron impact ionization also produces ions throughout the ionosphere. Its production rate maximizes just below the height of peak photoionization rate, or deeper still for the more energetic electron population.

There are two important potential transitions with altitude in the ionosphere. The first transition illustrates the dominance by molecular ions in the D, E, and F$_1$ regions to atomic ions in the F$_2$ region. The dominant molecular ions in the D, E, and F$_1$ regions are not necessarily the ionized form of the dominant neutral. Instead, the first ions produced are chemically transformed to "terminal" molecular ions, which are, in general, those with low ionization potentials (IP). In the ionospheres of the terrestrial planets (i.e. Venus, Earth, and Mars), the terminal ions generally are O$_2^+$ (IP = 12.07 eV) and NO$^+$ (IP = 9.26 eV) (e.g. Fox, 2006). Second is the transition from photochemical equilibrium at low altitudes, where direct and chemical production mechanisms are balanced locally by

chemical loss, to higher regions where plasma transport plays a significant role in the distribution of ions. At low altitudes, the loss of plasma is controlled by the dissociative recombination of molecular ions, whereas at high altitudes it is controlled by downward transport of atomic ions by diffusion and subsequent conversion into short-lived molecular ions. This second transition occurs roughly where the time constant for destruction by chemical reactions, $\tau_c = 1/L$, is equal to the time constant for loss by diffusion, τ_D. The specific loss rate L is given by $\sum_i k_i n_i$, where k_i is the rate coefficient for the reaction of the atomic ion with neutral species i, and n_i is its number density. The time constant for diffusion, τ_D, is usually approximated as H^2/D_a, where H is the scale height of the background atmosphere and D_a is the ambipolar diffusion coefficient of the atomic ion (see Bauer and Lammer, 2004; Schunk and Nagy, 2009).

14.3.3 Ionospheric Regions on Mars

In Figure 14.8, we present altitude profiles of the production rates of the major ions by photoionization and electron impact ionization for a low-solar-activity 60° SZA model of the Martian ionosphere (see Section 14.3.4 for ionospheric model details). The major ion produced by photoionization is CO$_2^+$ and the rate of peak ionization occurs at an altitude of 136 km,

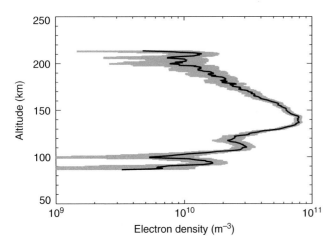

Figure 14.9. MGS radio occultation profile 5045K56A.EDS exhibits three clear layers: the M2 layer at 140 km, the M1 layer at 110 km, and the meteoric layer at 90 km. The profile was measured at latitude 79.9°N, longitude 316.0°E, 9.9 h LST, $L_s = 160.1°$, and SZA = 73.2°. The nominal profile is the solid line and the one-sigma uncertainties in the electron densities are marked by the shaded region. From Withers et al. (2008).

while O^+ production dominates above about 200 km. These ions are transformed by many ion–neutral reactions, such as

$$CO_2^+ + O \rightarrow O_2^+ + CO \qquad (14.1)$$

$$O^+ + CO_2 \rightarrow O_2^+ + CO \qquad (14.2)$$

$$O_2^+ + N \rightarrow NO^+ + O \qquad (14.3)$$

The terminal molecular ions are destroyed mainly by dissociative recombination (DR), for example:

$$O_2^+ + e^- \rightarrow O + O \qquad (14.4)$$

The shape of the dayside ionosphere of Mars is dominated by layers in the E and F_1 regions, where the F_1 layer (sometimes called the M2 layer) is the strongest (primary) ionospheric peak and the E layer (sometimes called the M1 layer) is weaker (secondary) and more variable in shape and magnitude. An episodic meteoric ion layer is sometimes observed near 90 km. Figure 14.9 shows evidence of these three distinct Martian ionospheric layers in a single MGS RO profile (Withers et al., 2008). The D and F_2 regions have not been observed to form local maxima in electron density at Mars. The O^+ density profile measured by the Viking RPAs does, however, show an F_2 peak (see Section 14.3.4) precisely where the transition from chemical- to transport-controlled conditions occurs.

14.3.3.1 The F_1 Region

On Mars, the major electron density peak is found in the F_1 region (M2 region). Unlike the analogous terrestrial ionospheric peak, the F_1 layer and peak are clearly visible in all dayside RO profiles of Mars. The peak appears in the altitude range ~120–180 km, and rises with SZA, somewhat like that of a Chapman layer (Zhang et al., 1990b).

In fact, several investigators have invoked Chapman theory and associated variants to study variations in the dayside

peak electron densities and their altitudes with SZA. The magnitude of the peak electron density for a given solar activity at a SZA of χ is equal to $n_0(\cos \chi)^b$, where n_0 is the subsolar maximum electron density and b is 0.5 for an ideal Chapman layer. Some investigators have treated b as a free parameter, with values close to 0.5 indicating a "Chapman-like" dependence. For example, Zhang et al. (1990a) used the available RO profiles from the Mariner 9 and Viking Orbiters to predict a subsolar maximum electron density in the range $(1.5–2.3) \times 10^5$ cm^{-3} at an altitude of ~120 km, with $b = 0.57$. The subsolar electron peak density was measured by MARSIS to decrease from a value of ~1.8×10^5 to 1.4×10^5 cm^{-3} as the solar activity decreased from moderately low to low during the period August 2005 to July 2007 (Morgan et al., 2008). They also observed the average altitude of the subsolar peak as about 128 km, which is slightly higher than previous predictions from extrapolations of RO data (Withers, 2009).

The altitude of the dayside peak for a given SZA appears at approximately unit optical depth for EUV photons. This implies a constant column density, which is approximately at a constant pressure level, and therefore is coupled to spatial and temporal variations of temperature and pressure in the lower atmosphere. The height of the F_1 ion peak varies ~15 km with season (Stewart, 1987; Zhang et al., 1990a) as a result of the change in heliocentric distance from aphelion ($L_s \sim 70$–90°) to perihelion ($L_s \sim 250$–270°). The observed low SZA F_1 peak heights typically range from ~120 to 135 km. In addition, Bougher et al. (2001, 2004) investigated the altitude behavior of the F_1 ion peak derived from the MGS RSS occultations as a function of longitude. The dataset was characterized by large zenith angles and moderate solar conditions at aphelion. They showed that the peak heights exhibit wave-2 and wave-3 oscillations, with 1σ variations of ~10 km about a mean value of 134–135 km. This variability, which is observed to be consistent at aphelion from year to year, is related to planetary and tidal wave activity in the lower atmosphere (Forbes et al., 2002; Moudden and Forbes, 2010).

14.3.3.2 The F_2 Region

At Earth, the F_2 peak appears near 300 km, and exhibits a maximum electron density of the order of 10^6 cm^{-3}, which is also an absolute maximum for the entire ionosphere (e.g. Kelley, 2009; Schunk and Nagy, 2009). On Mars, by contrast to Earth, a well-defined F_2 peak is not seen in the electron density profiles. The ion density profiles derived from the Viking RPA show that O_2^+ is the major ion over the altitude range 110–280 km, with smaller densities of the other major ion produced, CO_2^+. The maximum O^+ density measured by Viking is in the range 600–750 cm^{-3} near 225 km at low solar activity. The major chemical loss process for O^+ at its peak is reaction (14.2) above. The O^+ density at its peak is smaller than the density of O_2^+ over the entire range measured by the Viking 1 and 2 Landers, and in 1D models, extending to 300 km (e.g. Krasnopolsky, 2002; Fox, 2004a). Despite a peak occurring in the O^+ density profile, no visible maximum has been seen at the corresponding heights in any measured electron density profile, suggesting that O^+ is a minor ion compared with O_2^+. Multi-dimensional magnetohydrodynamic (MHD) models and hybrid codes predict that the O^+ densities do exceed those of O_2^+ at altitudes above ~200 km (see Chapter 15); the reason for this should be further investigated (e.g. Ma et al., 2004; Brecht and Ledvina, 2006; Fox, 2009, and references therein).

The topside scale height of O_2^+ in the Martian ionosphere as measured by the RPA on Viking 1 was observed to be smaller than expected for diffusive equilibrium conditions (Hanson et al., 1977). Simulated ion density profiles could, however, reproduce the observations if a loss process for ions was imposed on the topside of the model. Both Chen et al. (1978) and Fox et al. (1996; see also Fox, 2009) found that imposing an upward velocity on the O_2^+ profile of about 1 km s^{-1} reproduced the Viking ion density profiles. Shinagawa and Cravens (1989) suggested that the loss process was the divergence of the horizontal ion fluxes, by analogy to Venus. Thus these "eroded" profiles may be characteristic of an upward and anti-sunward flow of ions, which may then converge and flow downward on the nightside or escape from the planet. Fox (2009) modeled the low- and high-solar-activity ionospheres to compare to Viking 1 and Mariner 6 and 7 profiles, and computed the ion fluxes implied by the measurements and the maximum fluxes that could be imposed at the top of the models for the major ions. The predicted hemispheric averaged upward fluxes of $O^+ + O_2^+$ ions were shown to vary with solar activity over the range (5–13)$\times 10^7$ cm^{-2} s^{-1}, and were much less than the predicted maximum upward fluxes. Three-dimensional MHD models simulate the upward and anti-sunward ion flow and are shown to accurately capture these "eroded" Viking ion density profiles (e.g. Ma et al., 2004). See further discussion in Chapter 15.

14.3.3.3 The E Region

Photoionization by soft X-rays in the E region (M1 region) produces very high-energy photoelectrons (roughly 110 eV to 1.2 keV) and Auger electrons. The energetic electrons may in turn produce further ionization, with one ion–electron pair being produced for every ~34 eV by which the electron energy exceeds the thermal energy (e.g. Fox et al., 2008). In fact, the main source of ionization in the E region is electron impact ionization; the ions CO_2^+, O^+, and CO^+ are mostly produced by ionization and dissociative ionization of CO_2, as shown in Figure 14.8. As discussed above, these ions are transformed chemically into terminal molecular ions, mostly O_2^+ and NO^+, which are then destroyed by DR (14.4). The E region is in photochemical equilibrium.

The E region is visible in almost all dayside RO profiles as a shoulder or a separate peak below the F_1 peak. The analogous nature of the terrestrial E and the Martian M1 peaks, and of the terrestrial F_1 and the Martian M2 peaks, has been firmly established by the simultaneous observations of the variability and pressure levels of occurrence of the electron density profile of the M1 and M2 peaks on Mars and the E and F_1 peaks, respectively, on Earth (Rishbeth and Mendillo, 2004). In addition, Mendillo et al. (2006) showed that the M1 layer responds to solar flares in the same way as the E region in the terrestrial ionosphere (see Section 14.3.3).

14.3.3.4 Meteoric Layer

In the lower E region on Earth, there is a more or less constant layer of meteoric ions, such as Mg^+, Fe^+, Na^+, Si^+, and their oxides, such as MgO^+ and MgO_2^+. Metal atoms are usually formed from ablation of meteoric particles near 1 μbar. On the terrestrial planets, the ablated atoms may then be ionized by photoionization or by charge transfer from NO^+ or O_2^+ (see the reviews by Grebowsky et al. (2002) and by Plane (2003)).

The chemistry of meteoric ions on Mars has been studied by Pesnell and Grebowsky (2000), by Molina-Cuberos et al. (2003), and most recently by Whalley and Plane (2010). The latter authors have measured the most important reactions of Mg species in the Martian atmosphere, and have argued that nearly all the metallic ions on Mars should be Mg^+, because the reactions of O with MgO_2^+ and MgO^+ re-form the atomic metal ions faster than the molecular ions are neutralized by DR.

A distinct Martian ionospheric layer, probably composed of meteoric ions, can be seen near 80 km in RO profiles from Mariner 7 and 9, MGS, and MEX (Fjeldbo et al., 1970; Kliore et al., 1972; Pätzold et al., 2005; Withers et al., 2008). Withers et al. (2008) found a layer width of about 10 km, mean altitude of 91.7 km, and mean density of 1.33$\times 10^4$ cm^{-3} in MGS profiles.

14.3.4 Episodic Dayside Ionosphere Variability

Withers (2009) has reviewed the response of the dayside Martian ionosphere to various external and internal factors. Here we focus upon episodic external (e.g. solar flares) and internal (e.g. dust storm, crustal-field-induced) forcing that results in perturbations to the more regular solar cycle and seasonal variations of the ionospheric structure.

Since the dayside F_1 peak occurs at approximately a constant pressure level, its altitude is also impacted by episodic variations of temperature and pressure in the lower atmosphere. It is observed that electron density peak altitudes rise in the presence of regional or global dust storms in the lower atmosphere, as was the case for the Mariner 9 primary mission and the early part of the Viking mission (Kliore et al., 1972; Zhang et al., 1990a). For the Mariner 9 primary mission, F_1 peak heights were reported in the range from ~134 to ~154 km for an SZA of 50–60° (Kliore et al., 1972), which is ~20–30 km higher than average.

Solar flares are marked by a sudden increase in the photon fluxes at soft and hard X-ray wavelengths, resulting in dramatic changes in the E layer. For example, MGS radio occultation electron density profiles were acquired within minutes of a solar flare on April 15, 2001 (X14.4 class flare) and April 26, 2001 (M7.8 class flare) (Mendillo et al., 2006). The electron densities above the F_1 layer did not change by more than the measurement uncertainties, yet the electron densities at lower altitudes were significantly increased (doubled at 100 km).

The crustal magnetic field strength at ionospheric altitudes can exceed 1000 nT (Brain et al., 2003; Arkani-Hamed, 2004). The geographic distribution of Mars' crustal field is briefly discussed in Chapter 15. The most striking effects of the magnetic field on the ionospheric densities are seen over strong and vertical magnetic fields. Nielsen et al. (2007) and Duru et al. (2008) used MARSIS data to show that peak and topside densities are twice as large as usual over such regions. Ness et al. (2000) also found suggestions of enhanced electron densities in Mariner 9 radio occultation data above strong and vertical crustal fields. Alternatively, Krymskii et al. (2003) found that peak electron densities are higher than expected within mini-magnetospheres. Mini-magnetospheres (e.g. Mitchell et al., 2001) are defined as regions where both ends of magnetic field lines that thread the ionosphere intersect with the surface, rather than being open to space.

For both the vertical field (cusp-like) regions and minimagnetospheres, the cause of the increased electron densities has been attributed to elevated electron temperatures (Krymskii et al., 2003), which reduce loss rates due to dissociative recombination (Schunk and Nagy, 2009). Increases in electron temperature by a factor of 3.2–7.2 were inferred by Nielsen et al. (2007). High electron temperatures in vertical field regions have been attributed to a two-stream plasma instability involving inflowing solar wind plasma and those in minimagnetospheres to the trapping of hot photoelectrons. Finally, Withers et al. (2005) showed that a small number of MGS radio occultation profiles contain unusually large changes in electron density over short vertical distances. These anomalous profiles are preferentially located over regions of strong vertical crustal magnetic fields.

14.3.5 Nightside Ionosphere

The nightside ionosphere of Mars was discovered by the Soviet spacecraft Mars-4, which showed a peak density of 7×10^3 cm^{-3} at 107 km in a pre-dawn RO electron density profile at 127° SZA on February 10, 1974 (Kolosov et al., 1975). The radio occultation experiments on the Viking 1 and 2 Orbiters detected nightside ionospheric peaks in 40% of measurements for which the SZA was greater than 95°, and the average peak density was 5×10^3 cm^{-3} (Zhang et al., 1990b). Zhang et al. showed three examples of nightside electron density profiles, for 104°, 117°, and 124° SZA, which exhibited peaks at 180, 150, and 150 km, respectively.

Model simulations reveal that the Martian lower ionosphere is sunlit at least to 95° SZA, and that the upper ionosphere is sunlit at least to 100° SZA (e.g. Fox and Yeager, 2006). For the nightside orbits in which a well-defined density peak is not observed, the electron densities may either be below the limit of detection or may represent "disappearing ionospheres" by analogy to Venus (Cravens et al., 1982).

The sources of the nightside ionosphere are assumed to be similar to those at Venus: transport of ions from the dayside or precipitation of energetic particles. Bertaux et al. (2005b) discovered a region of enhanced "auroral" UV emissions that appear to be localized to a region of large, nearly vertical crustal magnetic fields (see Section 14.4 and Chapter 15). Charged particles, probably electrons, are assumed to precipitate into the thermosphere, producing both emissions and ionization (Bertaux et al., 2005b). Fillingim et al. (2007) modeled electron precipitation-induced patches of ionization on the nightside for two electron spectra: a typical tail spectrum as measured by the MGS Magnetometer and Electron Reflectometer (MAG/ER), and an accelerated spectrum. Assuming photochemical equilibrium, which may not be appropriate (e.g. Fox et al., 1993), Fillingim et al. (2007) derived density profiles for these two cases, which exhibited peaks of 1.7×10^4 and 5.7×10^4 cm^{-3} at altitudes of 166 and 156 km, respectively. Fillingim et al. (2007) proposed that ionization would be localized and patchy, due to precipitation of solar wind electrons in regions of open (nearly radial) magnetic field lines, and the absence of ionization in regions of closed (or horizontal) magnetic fields. Safaeinili et al. (2007) came to

a similar conclusion, when they analyzed MEX MARSIS data to determine the total electron content (TEC) of the nightside ionosphere, which they defined as regions where the SZAs are greater than 100°. They found that the TEC was enhanced over small regions where the crustal magnetic field had a large radial component, and postulated that in these regions the magnetic field lines were connected to the solar wind.

Lillis et al. (2009, 2011) also used an electron transport code that includes crustal magnetic fields to model ion production rates by precipitating solar wind electrons with various pitch angle distributions. They quoted a primary peak ionization rate of ~10 cm^{-3} s^{-1} at 160 km, and a lower peak ionization rate of 1 cm^{-3} s^{-1} at ~130 km. A photochemical equilibrium model showed that the maximum electron density at the primary peak was of the order of 1×10^4 cm^{-3}, with a lower shoulder appearing near 130 km with a maximum electron density of about 2×10^3 cm^{-3}. By contrast, invoking measurements of the Analyzer of Space Plasmas and Energetic Atoms (ASPERA-3) instrument on MEX, Frahm et al. (2006) and Dubinin et al. (2008) proposed that the precipitating electrons were photoelectrons with energies in the 30–100 eV range that were transported from the dayside to the nightside, possibly along magnetic field lines (e.g. Liemohn et al., 2007).

One of the most comprehensive studies to date of the Martian nightside ionosphere is that by Němec et al. (2010). They used the MARSIS instrument on MEX to measure the topside electron density profile and peak electron density on the nightside of Mars. They found nightside ionospheres in both northern and southern hemispheres, with a detection limit of ~5×10^3 cm^{-3}. For an SZA less than 125°, in regions of small or no magnetic fields, the occurrence rate of detectable plasma densities decreased with increasing SZA, from which they inferred that ion transport from the dayside may play a role in production of the ionosphere. This suggestion is supported by the measurements of the trans-terminator flow by the ASPERA-3 experiment on MEX by many investigators, including Fränz et al. (2010, and references therein). MHD models of the Martian interaction with the solar wind also predict large anti-sunward fluxes of ions, with about half escaping from the planet (e.g. Ma and Nagy, 2007; Fox, 2009). Alternatively, electron precipitation regions of enhanced ionization in the deep nightside were also identified in MARSIS datasets (e.g. Němec et al., 2011).

14.3.6 Mars Ion and Electron Temperatures

Near the Mars F_1 peak (also called the M2 peak) and below, where there are sufficient collisions, the neutral, ion, and electron temperatures (T_n, T_i, and T_e, respectively) are in thermal equilibrium. Slightly above the peak, T_e departs from T_n and rapidly increases with altitude to values that are typically thousands of kelvins. At higher altitudes (170–180 km) T_i departs from T_n and approaches the values of T_e. Prior to the MAVEN mission, measurements of T_e and T_i profiles were limited to those returned from the RPA on the Viking Landers during solar minimum conditions (Hanson and Mantas, 1988). Several investigators have attempted to simulate the observed values of T_e and T_i for Mars, with mixed success (e.g. Chen et al., 1978; Johnson, 1978; Rohrbaugh et al., 1979; Choi et al., 1998). However, improved knowledge of T_e values is especially

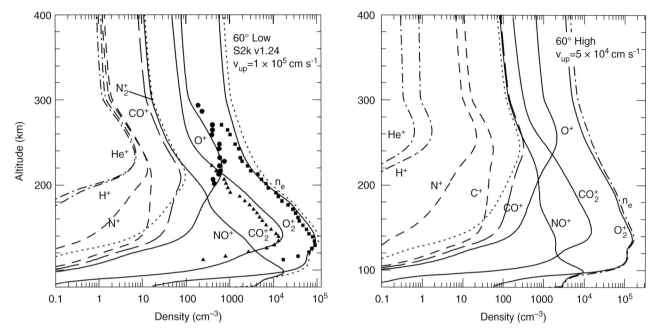

Figure 14.10. Major-ion density profiles for both (left) low- and (right) high-solar-activity models (at 60° SZA) of the Martian ionosphere. Viking 1-measured ion densities (O_2^+, O^+, CO_2^+) are superimposed for low-solar-activity conditions (Hanson et al., 1977).

important since, in the photochemical equilibrium region, the rate coefficients of dissociative recombination reactions (which depend on T_e) control the loss of molecular ions, and therefore the molecular ion densities.

14.3.7 One-Dimensional Models of the Neutral–Ion Composition

These models are devoted to the study of the number densities of the different atmospheric neutral–ion species and their variations due to chemistry and vertical transport by solving their continuity equations, typically making use of a fixed temperature and/or pressure profile. They are computationally fast, which makes them very well suited for the study of very minor species and the long-term evolution and stability of the atmosphere. On the other hand, they do not consider horizontal motions, which are known to affect the distribution of the atmospheric species and prevents the use of these models for studying localized phenomena in which transport plays a significant role, such as the NO and O_2 nightglow.

One of these photochemical models has been developed by V.A. Krasnopolsky (e.g. Krasnopolsky, 1986). This model has been extensively used in the last 20 years for the study of the composition of the atmosphere of Mars and also those of Venus, Titan, and other bodies. In its Martian version, it is a versatile model used both for studies of the lower and middle atmosphere (Krasnopolsky, 2006), and for the study of the upper atmosphere and ionosphere (Krasnopolsky, 2002, 2010). For the latter studies, the most recent version of this model is used to compute the concentration profiles of 11 neutral species and 18 ions between 80 and 400 km, taking into account 86 reactions. Boundary conditions for the abundances of different species are fixed. Formation of ions by photoionization, by photodissociative ionization, and by photoelectrons is included, as

well as vertical transport by molecular and eddy diffusion and ambipolar diffusion for the ions. This model has been used to simulate the composition of the upper atmosphere and ionosphere under different solar conditions, paying special attention to the abundance of deuterated species and the processes that can produce a loss of water (Krasnopolsky, 2002). In addition, this model can be used to address exospheric temperatures and their solar cycle variations (e.g. Krasnopolsky, 2010).

Another 1D photochemical model has been developed by J.L. Fox and co-workers and has been used to study the neutral upper atmosphere and the ionosphere of Mars. Steady-state concentrations of nine minor neutral species, three additional neutral species, and 14 ions are calculated utilizing about 220 reactions at altitudes between 80 and 400 km. The background atmosphere in this model is composed of nine neutral species. Fixed profiles of neutral, ion, and electron temperatures are inputs to the model and taken from measurements or outputs from previous GCM simulations. Boundary conditions for all the species are set at the top and the bottom of the model. The model includes formation of ions by photodissociation, photoionization, photodissociative ionization, electron impact dissociation, electron impact ionization, and electron impact excitation. Vertical transport by molecular and eddy diffusion, as well as ambipolar diffusion for the ions, are also taken into account.

This model has been used for multiple studies of the Martian upper atmosphere and ionosphere, including the study of the effects of enhanced fluxes of solar X-rays in the ionosphere (Fox, 2004a), the importance of dissociative recombination in the atmospheres of Mars and Venus (Fox, 2005), the morphology of the Martian ionosphere and its comparison with MGS data (Fox and Yeager, 2006), and the escape of oxygen atoms (Fox and Hać, 2009, 2010). Figure 14.10 illustrates the predicted major-ion density profiles for both high and low solar activity (60° SZA). For low-solar-activity conditions, the Viking

Table 14.2. *Mechanisms leading to airglow. X represents any thermospheric species, while A and B are atoms or fragments forming the more complex molecule AB. The energetic electrons in processes 3, 4, and 5 may be photoelectrons, precipitating auroral electrons, or secondary electrons.*

Process	Mechanism	Description	Abbreviation
1	$X + h\nu \rightarrow X^{+*} + e$	Photoionization and excitation	PI
2	$AB^* + h\nu \rightarrow A^* + B$	Photodissociative excitation	PD
3	$X + e^* \rightarrow X^* + e$	Electron impact excitation	EE
4	$X + e^* \rightarrow X^{+*} + 2e$	Electron impact ionization and excitation	EI
5	$AB + e^* \rightarrow A^* + B + e$	Electron impact dissociative excitation	ED
6	$AB^+ + e \rightarrow A^* + B$	Dissociative recombination	DR
7	$A + h\nu \rightarrow A^* \rightarrow A + h\nu$	Resonant scattering by atoms	RS
8	$AB(v') + h\nu'' \rightarrow AB^*(v'') + h\nu''$	Fluorescent scattering by molecules	FS
9	$A^* \rightarrow A^{*''} + h\nu$	Radiative cascade from upper states	RC
10	$A + B \rightarrow AB^{**} \rightarrow AB^* + h\nu$	Radiative recombination	CL
11	$O + O + CO_2 \rightarrow O_2^* + CO_2$	Three-body recombination	TBR

1 measured ion densities (O_2^+, CO_2^+, O^+) are superimposed. The model F_1 peak is seen to range from about 1.1×10^5 cm^{-3} at 136 km to 1.7×10^5 cm^{-3} at 137 km, from low to high solar activity, and consists mostly of O_2^+, with smaller densities of CO_2^+. This compares reasonably well to the electron density peaks retrieved from MARSIS for 60° SZA (Morgan et al., 2008). The magnitude of the predicted low-solar-activity peak is in good agreement with the fits to the data, but the model peak altitude is somewhat higher than the measured RO peaks at 60°, which vary from 120 to 132 km (e.g. Zhang et al., 1990a). A good agreement with Viking 1 data is obtained when an upward flux of ions at the top of the model (400 km) of 10^5 cm s^{-1} is imposed (see Figure 14.10, left), which contributes to constrain the rate of escape of ions from the Martian atmosphere. The E peak appears near 116–117 km, with values of $\sim 6.6 \times 10^4$ cm^{-3} at low solar activity and 1.2×10^5 cm^{-3} at high solar activity.

A Boston University group has also developed a Mars ionosphere model, which has been used to study how much ionospheric variability is caused by solar variability (Martinis et al., 2003), how ionospheric conditions might affect radio navigation systems (Mendillo et al., 2004), simultaneous RO profiles from different hemispheres (Mendillo et al., 2011), and the effects of solar flares (Lollo et al., 2012).

14.4 AIRGLOW – AURORA AND VARIABILITY

What processes control the production of airglow and auroral emissions at Mars? How are these emission features used to diagnose the photochemistry, energetics, and dynamics of the Mars upper atmosphere?

Upper atmospheres of planets emit photons in the UV, visible, and IR regions of the electromagnetic spectrum. These emissions may be classified as airglow, which includes dayglow or nightglow, or aurorae. Dayglow is luminosity that arises ultimately from the more or less direct interaction of solar photons and photoelectrons with atmospheric gases during the daytime. Nightglow encompasses emissions that arise from reactions of species that

either originate on the nightside or are transported from the dayside to the nightside. Aurorae are emissions that arise from the interaction of energetic particles other than photoelectrons with atmospheric gases. The source of these emissions is usually the transition of the excited electronic state of a gas to a lower state, which may or may not be the ground state. Hence the airglow is not only a direct tracer of the atmospheric neutral composition, but is also linked to energy transport, dynamics, and photochemistry of planetary upper atmospheres (Slanger and Wolven, 2002). For Mars, both airglow and aurora involve interactions of particles and photons with CO_2, CO, O, O_2, NO, N_2, and CO_2^+.

On the dayside of Mars, dayglow arises predominantly through processes 1–9 shown in Table 14.2, which are discussed in detail in Fox and Dalgarno (1979), Fox (1992), Paxton and Anderson (1992), and Slanger and Wolven (2002).

On the Martian nightside, the main source of nightglow is chemiluminescence that arises from recombination of N and O atoms that are mainly produced by photodissociation, electron impact dissociation, dissociative ionization, or chemical reactions on the dayside. These atoms are transported to the nightside by the atmospheric circulation (Bertaux et al., 2005a). The recombination can be radiative, in reactions such as $N + O \rightarrow NO^* \rightarrow NO + h\nu$, which produces emissions in the NO δ and γ bands, or three-body reactions such as $O + O + M \rightarrow O_2^* + M$, where M is mainly CO_2. The O_2^* can denote any of a number of energetically excited states of oxygen, such as the $O_2(a^1\Delta_g)$ state, which emits a 1.27 μm photon in the (0,0) transition to the ground state $O_2(X^3\Sigma_g^-)$. This emission has been seen from the ground on the dayside (e.g. Slanger and Wolven, 2002), and has recently been detected on the nightside by the visible and IR spectrometer OMEGA on-board Mars Express (Bertaux et al., 2011). The visible b–X atmospheric bands, and the visible Herzberg band systems, also arise from three-body recombination of O atoms.

The Martian auroral emissions are created by the impact of particles, including electrons and protons (see Galand and Chakrabarti, 2002) precipitating along crustal magnetic field lines into the thermosphere, where they produce energetic secondary electrons. Processes 3–5 and 9 of Table 14.2 describe these emissions.

Table 14.3. *Typical limb brightness (in rayleighs) and excitation sources of the main UV (Mariner and Mars Express) and IR (Mars Express) emissions on the dayside and nightside and for auroral emissions. The neutral or ion species involved in the excitation mechanisms are in parentheses; the symbols for sources are defined in Table 14.2. Energetic electrons can either be photoelectrons or precipitating auroral electrons.*

Wavelength (nm)	Species and associated transition	Sources for excited state[a]	Brightness (R)		Missions
			Day	Night[b]	
121.6	H I Lyman α	RS (H)	10^4	100‡	Mariner, MEX
130.2–130.4–130.6	$O(^3S–^3P)$ triplet	PD (CO_2), PD (CO), ED (CO_2), ED (CO), RS (O)	500	—	Mariner, MEX
135.6–135.8	$O(^5S_2–^3P_{1,2})$ doublet	ED (CO_2), EE (O)	200	—	Mariner, MEX
135–170	$CO(A^1\Pi–X^1\Sigma^+)$	PD (CO_2), ED (CO_2), DR (CO_2^+), EE (CO), FS (CO)	10^2–10^3	—	Mariner, MEX
150–650	$N_2(A^3\Sigma_u^+–X^1\Sigma_g^+)$ Vegard–Kaplan	EE (N_2), RC (N_2*)	100–450 (0,5) and (0,6) bands	—	MEX
180–250	$CO(a^3\Pi–X^1\Sigma^+)$ Cameron	PD (CO_2), ED (CO_2), DR (CO_2^+), EE (CO)	10^4–10^5	10^3†	Mariner, MEX
190–260	$NO(A^2\Sigma^+–X^2\Pi)$ and $(C^2\Pi–X^2\Pi)$	CL (N,O), RC (NO*)	—	100‡	MEX
210–250	$CO^+(B^2\Sigma^+–X^2\Sigma^+)$	PI (CO_2), EI (CO_2), PI (CO), EI (CO)	<200	—	MEX upper limit
288–289	$CO_2^+(^2\Sigma_u^+–^2\Pi_g^+)$	PI (CO_2), EI (CO_2)	10^3–10^4	150†	Mariner, MEX
310–340	$CO_2^+(^2\Pi_u–^2\Pi_g^+)$ Fox–Duffendack–Barker	PI (CO_2), EI (CO_2)	10^4	—	Mariner
297.2	$O(^1S–^3P)$ trans-auroral line	PD (CO_2), ED (CO_2), EE (O), DR (O_2^+)	10^3–10^4	0–20†	Mariner, MEX
1270	$O_2(a^1\Delta_g–X^3\Sigma_g^-)$	TBR (O,O,CO_2) night, PD (O_3) day	(1–30)×10^6 nadir	14×10^6‡	MEX

[a] See Table 14.2 for abbreviations.

[b] Aurorae (†) or nightglow (‡).

14.4.1 Airglow Observations

A number of space missions have measured the Martian airglow, and all the emissions have been in the UV and IR.

14.4.1.1 The Mariner Missions (1969–1972)

UV spectral measurements of the Martian atmosphere were first recorded by the NASA Mariner 6, 7, and 9 missions between 1969 and 1972 in the 110–350 nm wavelength range (Barth et al., 1971, 1972). The Mariner 6 and 7 flybys corresponded to a period of high solar activity, but Mariner 9 orbited the planet at moderate solar activity. All these measurements showed similar features. A Mariner 9 spectrum, which corresponds to the sum of 120 individual spectra, is presented in Figure 14.11. The most prominent features were identified by Barth et al. (1972) and are listed in Table 14.3.

Through comparison of the Mariner spectra with laboratory measurements, McConnell and McElroy (1970) and Barth et al. (1971) determined that airglow features are mainly produced between 100 and 200 km altitude by photon and photoelectron excitation mechanisms. These mechanisms, listed in Tables 14.2 and 14.3, form the basis of our current understanding. In addition, Stewart et al. (1972) found a correlation between the daily 10.7 cm solar flux at Mars ($F_{10.7}$) and the Cameron band intensities recorded by Mariner 9. They derived exospheric temperatures of 315 and 325 K from the scale heights of the emission for Mariner 6 and 7, and Mariner 9, respectively. From the intensity of the O I (130.4 nm) line, Strickland et al. (1973) suggested that the abundance of O was less than 1% at 135 km altitude.

14.4.1.2 The Mars Express Mission (2004–Present): SPICAM

The dual UV–IR spectrometer SPICAM on-board Mars Express, inserted into a near-polar orbit around Mars in 2003, consists of a charge-coupled device (CCD) imaging spectrometer with a spectral range of 110–320 nm in the UV and 1.0–1.7

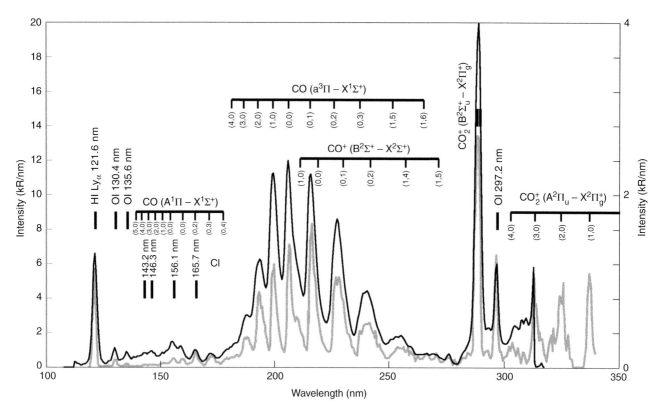

Figure 14.11. Comparison of 120 averaged Martian dayglow spectra as recorded by SPICAM (solar minimum conditions) and by Mariner 9 (solar moderate conditions). The left vertical scale (gray) corresponds to Mariner 9 intensities; the right vertical scale (black) corresponds to SPICAM intensities. The observed features are similar. SPICAM viewed the sunlit limb during orbit 1321 (January 27, 2005) between 100 and 150 km altitude. The Mariner signal is four times brighter on average as a result of higher solar activity. The Mariner spectrum, adapted from Barth et al. (1972), is shown on this figure over the entire wavelength range by combining the spectra from the two separate wavelength channels.

μm in the IR (Bertaux et al., 2006). Observations have been performed in nadir or limb viewing configurations. Up to the present, SPICAM has recorded airglow and auroral emissions for nearly 9000 orbits.

Figure 14.11 shows a comparison of the dayglow spectra measured by the SPICAM UV and the Mariner 9 spectrometer. Similar features are seen for both datasets. The most prominent emissions are the CO(a–X) Cameron bands (180–260 nm), H Lyman α, and the CO_2^+(B–X) ultraviolet doublet (near 289 nm). Since the SPICAM measurements were carried out during solar minimum, in which $F_{10.7}$ ranged from 75 to 135, the intensities were smaller by a factor of ~4 than those of Mariner 9. Leblanc et al. (2006a) showed that SPICAM could detect features such as O I (130.4 nm), O I (135.6 nm), C I (156.1 nm), and C I (165.7 nm). They also reported the first detection of the N_2 (0,5), (0,6), and (0,7) Vegard–Kaplan bands in the range 220–290 nm.

From individual limb profiles measured by SPICAM, it is possible to calculate the altitude profiles of the emissions by integrating directly over wavelengths or by fitting a synthetic spectrum to the data (Simon et al., 2009). Such profiles, shown in Figure 14.12, can be studied as a function of solar longitude L_s, which represents seasonal variations. In their statistical study, Simon et al. (2009) showed the existence of two seasonal maxima in brightness located at L_s ~ 140° and L_s ~ 290°, for the three main emissions: the CO Cameron bands (180–260 nm), the CO_2^+(B–X) ultraviolet doublet (near 289 nm) and the O

I (297.2 nm) line. These maxima correspond to larger neutral densities, which were simultaneously measured by SPICAM using stellar occultation techniques (Forget et al., 2009), and are probably related to seasonal dust storms at Mars. The total brightness and peak altitudes of the Cameron bands and the CO_2^+ ultraviolet doublet are 118 kR and 121 km, and 21 kR and 119 km, respectively, as shown in the recent study of SPICAM limb observations in the 90–180° L_s range (Cox et al., 2010). The variations of these intensities were shown to depend on solar zenith angle and solar activity, as predicted by models (e.g. Simon et al., 2009; Cox et al., 2010). Extraction of averaged exospheric temperatures from Mariner and SPICAM airglow scale heights is discussed in Section 14.2.1.

Recent studies using SPICAM have also focused on the possible existence of a second low-altitude peak in the O I (297.2 nm) emission (Huestis et al., 2010; Gronoff et al., 2012b). Following theoretical predictions (e.g. Fox and Dalgarno, 1979), an upper peak is observed near 130 km altitude, and is due to CO_2 dissociation by EUV photons with $\lambda < 100$ nm, while the lower peak at 90 km altitude results from the dissociation of CO_2 by Ly α photons. More observations with a systematic subtraction of the scattered solar spectrum at low altitudes are needed to confirm the existence of the doubled-peaked O I (297.2 nm) emission.

The Martian nightglow was first detected by Bertaux et al. (2005a) using SPICAM and is mainly the result of the chemiluminescence of the NO($A^2\Sigma^+$–$X^2\Pi$) and NO($C^2\Pi$–$X^2\Pi$)

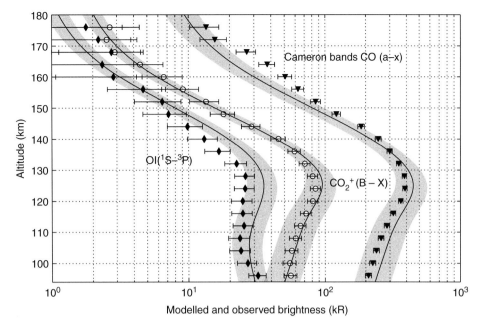

Figure 14.12. Altitude profiles of the Cameron CO, CO_2^+, and O I (297.2 nm) emissions for $L_s = 147°$, $F_{10.7} = 120$ (symbols) as measured by SPICAM and comparison to kinetic transport model calculations with propagated uncertainties (solid lines). This composite plot is based upon corrected individual profiles presented in Gronoff et al. (2012b), with a correction factor of $1/4\pi$.

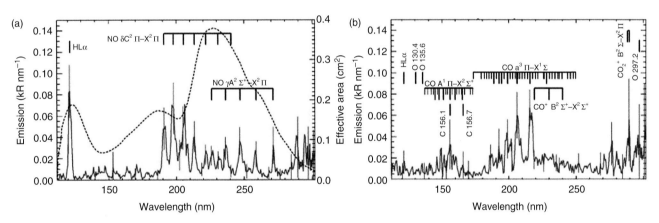

Figure 14.13. (a) Martian nightglow and (b) average auroral spectra recorded in limb viewing geometry by SPICAM at the highest spectral resolution. These spectra were obtained on August 11, 2004. The NO δ and γ and Lyman α nightglow in (a) were subtracted in (b) to produce the auroral spectrum, in which the CO(a–X) and CO_2^+(B–X) emissions clearly appear. The dashed line in (a) represents the effective area of the detector in cm^2 (right scale), i.e. the sensitivity of SPICAM as a function of wavelength. Error bars of 1σ are indicated for a few spectral lines as thin vertical segments. From Bertaux et al. (2005b).

transitions forming the γ and δ bands of the nitric oxide molecule in the 190–270 nm wavelength range. As previously discussed, the NO emissions arise when $N(^4S)$ and $O(^3P)$, produced on the dayside, are transported by the global atmospheric circulation to the nightside, where they recombine to produce NO in the $C^2\Pi(v = 0)$ state, which radiates to the ground state forming the δ bands; the C state may also radiate to the $A^2\Sigma^+(v = 0)$ state, which in turn radiates to the ground state in the γ bands (Cox et al., 2008). Figure 14.13(a) shows the nightglow spectrum observed by SPICAM.

A list of the main emissions and their peak limb brightness is given in Table 14.3. The reported intensities indicate only the observed brightness. The intensities have been shown to vary

with a number of parameters, including but not limited to season, solar activity, solar zenith angle, geographic location, and atmospheric composition.

14.4.2 Models of Airglow Emissions and Comparison to Observations

Understanding the UV emissions at Mars requires use of data from observations and concomitant modeling. Modeling the airglow of Mars has been a major effort for decades (e.g. Barth et al., 1971) and requires an understanding of the main sources of each emission and the associated photochemistry. Detailed modeling has led to successful predictions of the mechanisms and intensities of observed emissions (see Slanger and Wolven, 2002). Table 14.3 summarizes the known main excitation production mechanisms for several observed bright emissions.

Competition between radiative relaxation and quenching by atmospheric gases is of great importance in the understanding of these emissions. For instance, in addition to the fact that the relative importance of the two possible mechanisms for the formation of the green line is not yet secure (see for instance Gronoff et al., 2008), the ratio $r = A_{557.7}/A_{297.2}$ is still the subject of debate (e.g. Slanger et al., 2006, 2008) and its value depends upon whether it is measured in the laboratory ($r \sim 20$), observed in the atmosphere ($r \sim 9.8$), or computed theoretically ($r \sim 16$).

Measurement uncertainties of cross-sections vary from a few percent up to at least 75% depending on the mechanism. Such large uncertainties certainly call for new laboratory measurements. A few cross-section and chemistry databases are available online (www.nist.gov/physlab/data/ionization/; http://webbook.nist.gov; and http://amop.space.swri.edu/ from the SouthWest Research Institute). However, no database has specifically been designed for aeronomy studies. Hence, a comprehensive database for planetary aeronomy calculations, called AtMoCiAD (for Atomic and Molecular Cross-sections for Ionization and Airglow Database), is being developed and utilized, with a focus on these uncertainties (e.g. Simon Wedlund et al., 2011; Gronoff et al., 2012a).

A long-standing effort in modeling airglow emissions has been made using different techniques to calculate the energy deposition due to photons, photoelectrons, and suprathermal electrons. These efforts include the continuous slowing-down approximation (e.g. Fox and Stewart, 1991) or discrete local energy deposition techniques (Fox and Dalgarno, 1979; Fox, 2004a,b), multi-stream discrete ordinate methods (e.g. Mantas and Hanson, 1979; Haider et al., 2006; Simon et al., 2009) and Monte Carlo (MC) approaches (Shematovich et al., 2008). Both multi-stream and MC methods calculate solutions of the Boltzmann kinetic equation.

14.4.2.1 Photochemical Equilibrium Model (Fox and Dalgarno, 1979)

One of the first comprehensive photochemical emission models (Fox and Dalgarno, 1979) used a discrete local approximation for electron energy deposition (e.g. Dalgarno and Lejeune, 1971; Cravens et al., 1975), which incorporated a procedure for determining the equilibrium velocity distribution of the secondary electrons. Electron fluxes were then derived by dividing the electron production by the loss terms at a given energy. The background neutral model was based on the UAMS measurements of the Viking 1 Lander (e.g. Nier and McElroy, 1977). The authors modeled production mechanisms and altitude profiles of the emissions of several species, including CO, O, CO_2^+, C, and N_2. They reported the integrated overhead intensities of the emissions corresponding to Viking conditions. This model still serves as reference for subsequent calculations.

14.4.2.2 Kinetic/Fluid Trans-Mars Model (Simon et al., 2009)

Based on the terrestrial ionosphere model TRANSCAR (Lilensten and Blelly, 2002), the Trans-Mars model is a 1D multi-stream electron kinetic/fluid transport code that describes the dayside and auroral ionosphere of Mars from 80 to 500 km altitude. The kinetic part of the model provides a solution to the Boltzmann equation for suprathermal electrons in eight scattering angles and yields electron, ion, and excited state production rates. The fluid part calculates the moments of the distribution function, i.e. ion and electron densities and temperatures, and ion velocities, so that dissociative recombination can be consistently taken into account. Altitude profiles of the main airglow emissions are computed following the full photoemission description of Gronoff et al. (2008) and Simon et al. (2009). The inputs to the model are elastic and inelastic cross-sections (EUV photon and electron impact), chemical rate coefficients, the EUV solar flux parameterized by $F_{10.7}$, and thermospheric altitude profiles of neutral species and temperatures given by global circulation models such as the MTGCM (see Section 14.5.2.1).

This model has been updated (Witasse et al., 2002; Simon et al., 2009; Gronoff et al., 2012a) to take into account the latest inelastic cross-sections and to account for different viewing geometries. A new version named Aeroplanets is currently being developed to unify the models on several planetary objects (e.g. Venus, Earth, Mars, Jupiter, and Titan).

14.4.2.3 Monte Carlo Transport Model (Shematovich et al., 2008)

The direct simulation Monte Carlo (DSMC) technique is the basis of the model of Shematovich et al. (2008), which finds a stochastic solution to the Boltzmann equation between 75 and 250 km altitude. This model was first developed for Earth (Shematovich et al., 1994) and has been subsequently extended to Mars (Shematovich et al., 2008; Cox et al., 2010). Inputs to this model are elastic and inelastic cross-sections (EUV photon and electron impact), chemical reactions, solar EUV fluxes, and the neutral background model (densities and temperatures) from the MTGCM.

Both of the aforementioned recent models are consistent with each other, and the investigators conclude that satisfactory agreement with the SPICAM observations can be obtained after taking into account the uncertainties about the neutral atmosphere and cross-sections, as shown in Figure 14.12. Simon et al. (2009) divided the Viking CO_2 densities by a factor of 3 (consistent with the higher atmospheric dust load for the Viking versus the Mars Express observing periods) in order to reduce the energy deposition peak altitude. In addition, the electron impact dissociative excitation cross-sections (to produce the Cameron bands) were divided by a factor of 2 to obtain a satisfactory fit. Similarly, Cox et al. (2010) reported an overestimate of the Cameron band intensities by a factor of 1.74 and of the CO_2^+ intensities by a factor of 1.41. Since the DSMC model is able to reproduce the measured intensity variations with SZA and $F_{10.7}$, they ascribed this discrepancy mostly to uncertainties in the excitation cross-sections.

Several visible oxygen emissions, such as the 557.7 nm green line O(^{1}S–^{1}D), the 630.0 nm red line O(^{1}D–^{3}P), and the 844.6 nm line O(^{3}P*–^{3}S°), have never been observed. The Trans-Mars model for Viking conditions (solar minimum) predicts overhead dayglow intensities of 15×10^3 R, 230 R, and 50 R for these three emissions, respectively.

14.4.2.4 *Implications for Neutral Density Retrievals: O and CO₂*

Though *in situ* observations of CO_2, N_2, O_2, CO, and NO were made with the Viking Landers in 1976 (Nier and McElroy, 1977), the neutral upper atmosphere and its variability are still ill-constrained (see Section 14.2). The study and inversion of airglow emissions is an efficient tool to retrieve neutral densities, and has been extensively used at Earth using both theoretical models (e.g. Strickland and Donahue, 1970) and data analysis (e.g. DeMajistre et al., 2004). Since the Mariner missions, models have been used to determine, for example, the altitude profile of O (Strickland et al., 1972, 1973; Stewart et al., 1992).

Retrieval techniques use several prominent dayglow UV emissions such as O I (130.4 nm) and O I (297.2 nm). At Mars, O I (130.4 nm) emission is mainly formed by photodissociation of CO_2 and solar resonant scattering (Fox and Dalgarno, 1979; Stewart et al., 1992). Chaufray et al. (2009) used a Monte Carlo radiative transfer model to calculate an atomic oxygen density of about 10^7 cm^{-3} at the exobase for low solar zenith angles. The [O]/[CO₂] mixing ratio at 135 km was found to be between 0.6% and 1.2% using SPICAM data, versus abundances of 0.5–1% derived by Strickland et al. (1973) for Mariner 6, 7, and 9 and 0.6–1.0% by Stewart et al. (1992). Fox and Dalgarno (1979) derived an O mixing ratio of ~2% from the CO_2^+ and O_2^+ ion density profiles measured by the Retarding Potential Analyzer on Viking 1 (Hanson et al., 1977).

Limb profiles of the O I (297.2 nm) line measured by SPICAM were successfully inverted by Gronoff et al. (2012b) to retrieve the density of CO_2 and its variability. This study also showed that the upper atmosphere of Mars undergoes seasonal density variations of up to an order of magnitude, in agreement with the studies of Forget et al. (2009) and McDunn et al. (2010).

14.4.3 Auroral Observations

Highly localized UV aurora-like emissions have been observed at Mars (Bertaux et al., 2005b), associated with strong, radially oriented crustal magnetic fields. Figure 14.13(b) shows a typical auroral spectrum in limb geometry from which the background average nightglow has been subtracted. Auroral-intensified emissions are similar to dayglow features, e.g. Cameron bands and CO_2^+ UV doublet with limb intensities of 1.5–2 kR and 150 R, respectively. Typical limb brightnesses for these two auroral emissions are listed in Table 14.3.

Ten auroral events were identified between July 2004 and March 2006, of which eight were in nadir viewing geometry. All events occurred near the Martian crustal fields, in locations often associated with open magnetic field lines. Auroral brightnesses are highly variable and are on average weaker than dayglow by a factor of ~15. At the time of the nadir auroral emission events, the Mars Express MARSIS instrument recorded high total electron content (TEC ~10^{11} cm^{-2}) while the ASPERA-3/ELS electron sensor measured an increase of the total incident particle energy flux by an order of magnitude (Leblanc et al., 2008).

Models of the first observations of auroral emission suggested a lower-energy (<100 eV) electron source (Leblanc

et al., 2006b) compared to the high-altitude auroral-like charged particle distributions measured by Mars Global Surveyor or ASPERA-3 (Brain et al., 2006; Lundin et al., 2006). It was proposed that transport of photoelectrons from the Martian dayside could provide the source. While this mechanism may operate at Mars, the connection between the auroral emission events and the more energetic electrons became more apparent when the emission observed in nadir viewing was correlated with *in situ* observations by the ASPERA-3/ELS of peaked electron energy distributions as well as localized increases in ionospheric TEC near the emission. Further, the ratios between the Cameron band and CO_2^+(B–X) emissions were recalculated, allowing a higher-energy electron source in the model. These results suggest a link in cusp-like magnetic field structures between the *in situ* particle observations made at high altitude and the response of the ionosphere below. The mechanisms that may drive the Martian aurora are discussed further in Chapter 15.

14.5 NEUTRAL UPPER ATMOSPHERE AND IONOSPHERE MODELING

What are the advantages of using multi-dimensional models to address the coupled energetics, chemistry, and dynamics of the Martian thermosphere–ionosphere system? What is the status of this GCM modeling, including the capabilities and limitations of key models being used today?

A variety of models of the Martian upper atmosphere have been created in the last few decades, from the first 1D models designed to analyze radio occultation electron density profiles (McElroy, 1967; Fox and Dalgarno, 1979) and the first Martian GCM (Leovy and Mintz, 1969), to the present efforts to create complex multi-dimensional models coupling the thermosphere and ionosphere with the exosphere and the magnetosphere. These models use different approaches to include all or some of the physical processes important in this atmospheric region. They are commonly used to determine the thermal structure and/or the atmospheric composition.

Two key factors motivate the development of computational models of ever increasing complexity: (a) the refinement and the increasing amount of observational data, which require more sophisticated and precise models to address more specific and strongly focused phenomena; and (b) the application of new numerical formulations for physical processes that provide improved accuracy on both the local and global levels. Each of these objectives is facilitated by the increase of computational power.

14.5.1 Modern One-Dimensional Codes

One-dimensional models only consider atmospheric variations in the vertical dimension. While they have the obvious drawback of neglecting the horizontal interchanges of matter and the general circulation, they are computationally faster and, more important, they allow for a detailed treatment of physical processes without making approximations and/or parameterizations typical of multi-dimensional models. This makes these 1D models a perfect test bed to examine "state-of-the-art" formulations of specific processes and phenomena, to analyze the sensitivity of the atmospheric state to different parameters and processes, and to develop and test fast schemes and parameterizations for their use in multi-dimensional models. As a consequence, most of these 1D models are strongly specialized, focusing usually on very specific processes, and they usually make assumptions regarding important atmospheric parameters that they cannot simulate. Modern 1D models of the thermal and radiative balance, the neutral–ion composition, and the atmospheric emissions have been presented in Sections 14.2.1.1, 14.3.7, and 14.4.2, respectively.

14.5.2 Modern Multi-Dimensional Codes

General circulation models (also called global climate models) decompose the atmosphere into a 3D grid (usually a longitude–latitude–altitude grid), and can be schematically divided into two parts: a dynamical core that solves the equations of fluid dynamics, in most cases the primitive equations of meteorology (for more details, see Bougher et al., 2008; Schunk and Nagy, 2009); and a set of physical processes, usually included in the form of parameterizations, that force the dynamics and determine the thermal state and the composition of the atmosphere. Most of these models are adapted from pre-existing terrestrial GCMs, for which unique radiative transfer and photochemical codes appropriate for Martian conditions have to be added.

GCMs allow for self-consistent simulations of the upper atmosphere, as they include the couplings and interactions between temperature fields, chemistry, radiation, and dynamics, in some cases including also the coupling with the lower atmosphere. Their two main limitations are caused by the high demand for computer power, which forces (a) parameterization of some of the physical processes considered, and (b) the use of a relatively low horizontal resolution (typically a few hundreds of kilometers at the equator), preventing them from resolving processes with typical scales smaller than the grid size. However, the continuous increase of computational power and the adoption of massively parallel architectures is contributing to minimize these problems.

Two approaches have been used thus far to capture the physics of the entire Martian atmosphere (ground to exobase) in GCM frameworks: (1) coupling of separate lower and upper atmosphere codes, and (2) adoption of single framework "whole atmosphere" model codes (~0–250 km). Each approach has advantages and limitations that we will discuss (below) in connection with specific GCM tools being used today.

14.5.2.1 Coupled MGCM–MTGCM (Hydrodynamic): 0–250 km

The Mars Thermospheric General Circulation Model (MTGCM) is a finite-difference primitive equation model that self-consistently solves for time-dependent neutral temperatures, neutral–ion densities, and three-component neutral winds over the Mars globe (e.g. Bougher et al., 1999a,b, 2000, 2002, 2004, 2006; Bell et al., 2007). The modern MTGCM code contains prognostic equations for the major neutral species (CO_2, CO, N_2, and O), selected minor neutral species (Ar, NO, $N(^4S)$,

and O_2), and several photochemically produced ions (O_2^+, CO_2^+, O^+, CO^+, and NO^+). All fields are calculated on 33 pressure levels above 1.32 μbar, corresponding to altitudes from roughly 70 to 300 km (at solar maximum conditions), with a 5° resolution in latitude and longitude. The vertical coordinate is log pressure, with a vertical spacing of 0.5 scale heights. Key adjustable parameters that can be varied for MTGCM cases include the $F_{10.7}$ index (a proxy measured at the ground for solar EUV/UV fluxes), the heliocentric distance, and solar declination corresponding to Mars seasons. A fast NLTE 15 μm cooling scheme is implemented in the MTGCM, dynamically dependent upon simulated atomic O abundances, along with corresponding near-IR heating rates (e.g. Bougher et al., 2006). These inputs are based upon detailed 1D NLTE model calculations for the Mars atmosphere (López-Valverde et al., 1998).

A comprehensive dayside photochemical ionosphere is formulated for the MTGCM, including the five major ions. Key ion–neutral reactions and rates are taken from Fox and Sung (2001); empirical electron and ion temperatures are adapted from the Viking mission. Nightside ions are not yet simulated, but will require either day-to-night ion drifts from modern MHD models (e.g. Ma et al., 2004; Dong et al., 2015), or energetic electron precipitation sources of nightside ionization (e.g. Fillingim et al., 2007).

The coupling of separate lower and upper atmosphere codes is implemented to link the MTGCM to the lower atmosphere. The MTGCM is driven from below by the NASA Ames Mars MGCM code (Haberle et al., 1999) at the 1.32 μbar level (near 60–80 km). This detailed coupling allows both the migrating and non-migrating tides to cross the MTGCM lower boundary and the effects of the expansion and contraction of the Mars lower atmosphere (due to dust heating) to extend to the thermosphere. Key prognostic variables are passed upward every time step from the MGCM to the MTGCM at the 1.32 μbar level at every MTGCM grid point: temperatures, zonal and meridional winds, and geopotential heights. These two climate models are each run with a 2 minute time step, with the MGCM exchanging fields with the MTGCM at this frequency. No downward coupling is presently activated between the MTGCM and the MGCM. However, the impacts of lower atmosphere dynamics upon the upper atmosphere are dominant, and most important for reproducing observed thermospheric densities and temperatures (see Bell et al., 2007).

This coupled model configuration has been validated using an assortment of spacecraft observations, including thermosphere, ionosphere, and mesosphere datasets from MGS, Odyssey, MRO, and Mars Express (Bougher et al., 1999a,b, 2000, 2004, 2006, 2009b; Lillis et al., 2010; McDunn et al., 2010). For example, recent studies show that dayside (low SZA) exospheric temperature variations predicted by the MTGCM over the solar cycle and Mars seasons are in reasonably good agreement with available aerobraking, orbital drag, and UV airglow measurements (e.g. Keating et al., 1998, 2003, 2008; Leblanc et al., 2006a; Mazarico et al., 2007; Forbes et al., 2008; Bougher et al., 2009b, 2010; Krasnopolsky, 2010). Exobase neutral horizontal winds and temperatures for solar maximum/perihelion and solar minimum/aphelion conditions are illustrated in Figure 14.14. In general, the global wind patterns simulated by the MTGCM

(a)

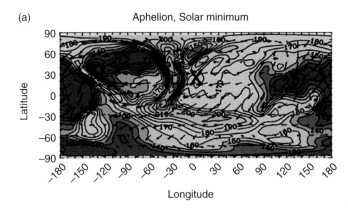

(b)

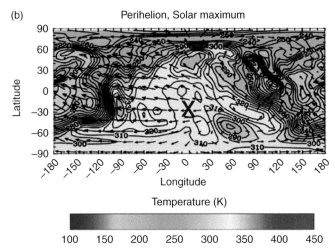

Figure 14.14. MTGCM simulated exobase temperatures and superimposed horizontal neutral winds for: (a) $L_s = 90°$, $F_{10.7} = 70$ and (b) $L_s = 270°$, $F_{10.7} = 200$ conditions, corresponding to solar cycle plus seasonal extremes. The subsolar points are indicated by the crosses (local noon meridian and solar declination latitude). Dayside (low SZA) temperatures range from ~170 to 310 K; nightside temperatures drop to as low as ~120–130 K. Maximum horizontal winds reach: (a) ~390 m s^{-1} to (b) ~550 m s^{-1}. The average exobase altitudes are: (a) ~165 km and (b) 195 km. Taken from Valeille et al. (2009). A black and white version of this figure will appear in some formats. For the color version, please refer to the plate section.

reveal strong summer-to-winter inter-hemispheric circulations that are consistent with observed NO nightglow emissions (Bertaux et al., 2005a) and changing polar warming features (Bougher et al., 2006). Simulated day-to-night global winds yield nightside thermospheric densities in accord with those derived from MGS electron reflectometry (Lillis et al., 2010). Middle atmosphere (~80–130 km) mass density and temperature profiles predicted by the MGCM–MTGCM are in good agreement with corresponding Mars Express SPICAM stellar occultation profiles over most Mars seasons (McDunn et al., 2010). Finally, observed thermosphere–ionosphere tidal features are reproduced in the simulated MTGCM dayside mass density and electron density fields as a function of longitude (Bougher et al., 2001, 2004).

GCMs can also provide information about underlying processes and parameters not available from observations. One example is the MGCM–MTGCM study of the thermal budgets

predicted to be responsible for the observed dayside MGS aerobraking temperatures illustrated in Figure 14.1. These dayside heat balances (not shown here, but presented in Bougher et al. (2009b)) indicate that EUV heating is largely balanced at this location by molecular thermal conduction (above ~130 km). Both dynamical cooling (due to upwelling and divergent winds) and CO_2 15 μm cooling also contribute to a lesser degree. More research is needed to confirm these heating and cooling terms and their relative importance. In particular, gravity wave momentum deposition and its impacts on thermospheric winds should be systematically investigated (e.g. Bougher et al., 2009b, 2011a; Medvedev et al., 2011, 2012).

The major strengths of the coupled MGCM–MTGCM codes are that: (1) the detailed neutral energetics and large-scale dynamics of both the Martian lower and upper atmospheres are being captured in the coupled model framework; (2) the model simulations conducted thus far reproduce most of the observed features of the thermosphere; and (3) the incorporated neutral–ion chemistry (above ~80 km) enables a realistic dayside ionosphere (below 200 km) to be simulated according to observations. The primary limitation of the coupled MGCM–MTGCM framework is that linking two separate codes across an interface is not seamless, i.e. an exact match of thermal and dynamical processes across the interface is not possible.

This primary MGCM–MTGCM limitation is being overcome with the development and validation of a new single framework "whole atmosphere" model code (~0–250 km). The new Mars Global Ionosphere–Thermosphere Model (M-GITM) combines the terrestrial GITM framework (e.g. Ridley et al., 2006; Deng et al., 2008) with Mars fundamental physical parameters, ion–neutral chemistry, and key radiative processes in order to capture the basic observed features of the thermal, compositional, and dynamical structure of the Mars atmosphere from the ground to the exosphere (Bougher et al., 2008, 2011b,c, 2015b; Pawlowski and Bougher, 2010, 2012; Pawlowski et al., 2011). The 3D global GITM framework uses an altitude-based vertical coordinate and allows for the relaxation of the assumption of hydrostatic equilibrium. The formulations and subroutines required for incorporation into the new M-GITM code have largely been taken from existing Mars MGCM and MTGCM codes. Previous coupled MGCM–MTGCM simulations are in fact being used to assist with M-GITM validation studies.

M-GITM presently captures solar cycle and seasonal trends in the upper atmosphere that are consistent with pre-MAVEN observations, yielding significant periodic changes in the temperature structure, the species density distributions, and the large-scale global wind system. Archival simulations spanning the full range of applications of the current M-GITM code, including 12 model runs for various solar cycle and seasonal conditions, are described in Bougher et al. (2015b). These simulations will serve as a benchmark against which to compare episodic variations (e.g. due to solar flares and dust storms) and non-orographic gravity wave impacts (e.g. on both the global circulation and temperatures) in future M-GITM studies.

14.5.2.2 LMD-MGCM (Hydrodynamic): 0–240 km

The Laboratoire de Météorologie Dynamique (LMD) MGCM solves the primitive equations of hydrodynamics in a sphere

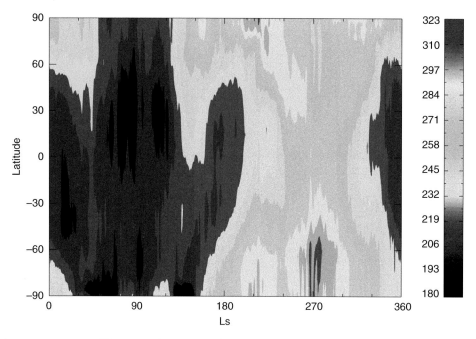

Figure 14.15. Zonal-mean temperatures (K) at the exobase predicted by the LMD-MGCM as a function of latitude and season. Figure taken from González-Galindo et al. (2009a). A black and white version of this figure will appear in some formats. For the color version, please refer to the plate section.

using a finite-difference scheme based on the terrestrial LMDZ model (Sadourny and Laval, 1984). It originally extended up to about 80 km of altitude (Forget et al., 1999), including all the relevant processes in the lower atmosphere, such as a realistic CO_2 condensation scheme, a water cycle (Montmessin et al., 2004), a photochemical model of the lower atmosphere (Lefèvre et al., 2004), and a number of subgrid-scale processes. Now it has become a ground-to-exosphere model, extending from the surface of the planet to an altitude of about 240 km. This allows the model to self-consistently study the coupling between the lower and the upper atmosphere, essential for the determination of the structure of the upper atmosphere. While it is usually run with a 5.625°×3.75° longitude–latitude grid, it is possible to stretch the grid to zoom over particular areas. In the vertical, it uses a hybrid coordinate system, with sigma coordinates in the lower atmosphere and pressure coordinates in the upper layers (González-Galindo et al., 2009a). Fifty vertical levels are used, with an uneven sampling to allow for a higher vertical resolution close to the surface, giving a vertical resolution of about 7 km (close to half a scale height) in the thermosphere.

To simulate the upper atmosphere, the LMD-MGCM includes a parameterization of the IR radiative transfer under NLTE conditions, based on López-Valverde and López-Puertas (2001) and López-Valverde et al. (1998) and recently improved to account for the effects of oxygen variability (González-Galindo et al., 2013), a fast scheme to calculate the UV heating including the day-to-day variability of the solar flux (González-Galindo et al., 2005, 2013), a photochemical model especially developed for the study of the rarified Martian upper atmosphere (González-Galindo et al., 2005), which considers the chemistry of the C, O, H, and N families and ionospheric reactions (González-Galindo et al., 2013), schemes for the molecular diffusion and viscosity (Angelats i Coll et al., 2005; González-Galindo et al.,

2009a, 2013), and a multi-fluid dynamical core for the transport of plasma (Chaufray et al., 2014).

Comparisons with observed features – such as the longitudinal variations of thermospheric densities obtained during MGS aerobraking (Angelats i Coll et al., 2004), the thermospheric polar warming observed during Mars Odyssey aerobraking (González-Galindo et al., 2009b), the NO nightglow observed by SPICAM (González-Galindo et al., 2008), the thermospheric densities derived from MGS electron reflectometry (Lillis et al., 2010), the SPICAM temperature and density profiles (Forget et al., 2009), and the MGS RSS electron densities (González-Galindo et al., 2013) – have validated the model and identified areas of needed improvement. Figure 14.15 shows the seasonal variability of predicted zonal-mean exospheric temperatures, which is in agreement with measurements (González-Galindo et al., 2009a), although the use of a fixed atomic oxygen concentration in the 15 μm cooling scheme produced an overestimation of temperatures in the lower thermosphere (Forget et al., 2009), a problem now overcome with the implementation of an improved cooling scheme (González-Galindo et al., 2013). The importance of a correct representation of the photochemistry to obtain realistic thermospheric temperatures (González-Galindo et al., 2005) and the effects of the orographic gravity waves on the thermospheric temperatures and winds (Angelats i Coll et al., 2005) have also been tested using this model. An intercomparison with the coupled MGCM–MTGCM shows that both models predict similar thermospheric temperatures when similar inputs (e.g. solar EUV–UV fluxes, cross-sections, heating efficiencies, near-IR heating rates) are used (González-Galindo et al., 2010). The effects of solar variability (González-Galindo et al., 2013) and the importance of plasma transport (Chaufray et al., 2014) over the ionosphere have also been evaluated using this model. The ground-to-exosphere LMD-MGCM has been used to build the Mars Climate Database (Lewis et al., 1999), a compilation of outputs of the model for several dust and solar condition scenarios, which is freely available to the community and now includes the upper atmosphere.

The major strengths of the LMD-MGCM model are: (1) it is a ground-to-exosphere model, allowing for a self-consistent description of the whole atmospheric range; (2) it can be integrated for long periods of time, allowing for multi-annual simulations including the day-to-day variability of the solar flux and the dust load; and (3) it includes sophisticated radiative transfer and chemical schemes that have been validated against multiple datasets. However, it does not include the effects of non-orographic gravity waves, shown to play a significant role in the heat balance of the upper atmosphere (Medvedev et al., 2011, 2015).

14.5.2.3 UCL MarsTIM (Hydrodynamic): 60–250 km

The University College London (UCL) Martian Thermosphere Ionosphere Model (MarsTIM) simulates the Martian upper atmosphere above a pressure level of 0.883 Pa (approximately 60 km of altitude). The mathematical formulation and the dynamical core are based on a previous GCM for the thermosphere of Titan (Müller-Wodarg et al., 2000). The primitive equations are solved using a finite-difference scheme on a longitude–latitude–pressure grid with variable resolution.

The modeled thermosphere is described by seven neutral species (CO_2, N_2, O, Ar, NO, O_2, and CO). Major components (CO_2, N_2, and O) are subject to mutual diffusion following the treatment of Fuller-Rowell and Rees (1980), while minor species are considered to be in diffusive equilibrium above the homopause. No neutral chemistry is considered, although a set of neutral–ion reactions is included to simulate the Martian ionosphere. The energy balance includes the IR heating and cooling rates in NLTE, calculated following López-Valverde and López-Puertas (2001), the UV heating, for which the scheme developed by Müller-Wodarg et al. (2000) is used, and thermal conduction.

Although this model does not include the lower atmosphere (e.g. early MarsTIM version of Winchester and Rees (1995)), some tests to couple it with the Mars Climate Database (e.g. Lewis et al., 1999) have been made, which showed the effects of the lower atmosphere over the upper atmosphere and allowed for a reasonable agreement with existent datasets (Moffat-Griffin et al., 2007). This model has also been used, together with the kinetic electron transport Trans-Mars model (described in Section 14.4.2.2), to calculate the production of secondary ions in the ionosphere (Nicholson et al., 2009).

The strengths of this MarsTIM model include: (1) a detailed energy balance treatment for the upper atmosphere; (2) a modern neutral diffusion scheme for thermospheric species; and (3) a novel application making use of the Trans-Mars model for electron transport. The primary limitation is the lack of self-consistent coupling with the lower atmosphere, which is important to simulate the variable structure of the thermosphere–ionosphere system.

14.5.2.4 Canadian GM3 (Non-Hydrostatic, Hydrodynamic): 0–160 km

The Global Mars Multiscale Model (GM3) of York University, Toronto (Moudden and McConnell, 2005) is the adaptation to

Mars of the Global Environmental Multiscale (GEM) model, developed at the Meteorological Service of Canada for weather forecasting purposes. The dynamical core, inherited from the GEM model, is a grid-point model that uses a semi-implicit time integration and a semi-Lagrangian advection calculation for the resolution of the non-hydrostatic equations of fluid dynamics. The model has a variable horizontal resolution capability, allowing even for mesoscale calculations without a considerable reduction of the integration time step. The model extends from the surface up to a level with pressure around 10^{-8} mbar (around 170 km), with 101 levels in the vertical, described using a hybrid coordinate system. MOLA data are used to include the Martian topography.

In the lower atmosphere, the radiative effects of dust and CO_2 are included following Forget et al. (1999) and Hourdin (1992); and in the upper atmosphere, NLTE effects are taken into account using the parameterization developed by López-Valverde and López-Puertas (1994a,b, 2001). A chemical model of the neutral atmosphere, including 23 species and 65 reactions between them, a scheme to simulate the water cycle, and a molecular diffusion scheme have also been included in the model (Moudden and McConnell, 2007). For the UV heating, a simplified scheme using a single wavelength and flux is presently used.

This model has been shown to produce mesospheric temperatures in agreement with Pathfinder entry profiles, and has been used to study the seasonal and dust-induced variability of temperatures and winds below about 170 km (Moudden and McConnell, 2005). The distribution of chemical species and their productions and losses have also been studied with this model, finding that the predicted concentrations of CO and O in the lower thermosphere are in good agreement with previous models and observations (Moudden and McConnell, 2007). Another application of the GM3 model has been the study of the dynamics of the Martian upper atmosphere, and in particular the wave activity at these altitudes and its connection with lower altitudes (Moudden and Forbes, 2008, 2010; Forbes and Moudden, 2009).

The strengths of this GM3 model include: (1) its unique dynamical core; (2) the variable horizontal resolution capability, allowing even for mesoscale simulations; (3) a detailed photochemical model capturing both lower and upper atmosphere neutral reactions; and (4) a modern radiation scheme. However, a more sophisticated UV heating scheme would be necessary for detailed thermospheric studies (Moudden and McConnell, 2005), which limits the validity of this model to altitudes below ~160 km. In addition, GM3 requires the addition of ion–neutral chemistry to address the ionosphere.

14.5.2.5 Max Planck Institute GCM (Hydrodynamic, 0–160 km)

The GCM developed at the Max Planck Institute is a redesigned version of the terrestrial Cologne Model of the Middle Atmosphere. Initially (Hartogh et al., 2005), its dynamical core was a grid-point solver for the primitive equations of hydrodynamics under the hydrostatic approximation. In later versions (Medvedev et al., 2011), however, a spectral dynamical core has substituted the grid-point solver. The first version of the model extended up to 130 km. Later extensions have raised its top to about 160 km (Medvedev and Yigit, 2012), using 67 vertical hybrid levels.

In the lower atmosphere, the radiative effects of dust are taken into account using 19 representative wavelengths. For CO_2, an LTE radiation scheme based on the *k*-distribution method is used in the lower atmosphere. In the upper atmosphere, an optimized version of the exact NLTE code of Kutepov et al. (1998) and Gusev and Kutepov (2003), employing the method of accelerated lambda inversion, is implemented. Both methods produce similar results at the transition between them (60–70 km). A scheme including 37 spectral intervals between 5 and 105 nm is employed to simulate the UV heating. The most novel aspect of this model is the inclusion of a spectral nonlinear parameterization of the effects of gravity waves with non-zero phase speed, based on a terrestrial model (Yigit et al., 2008). This scheme accounts for the vertical propagation, refraction, saturation, breaking, and dissipation of the waves, and computes their momentum deposition and heating/cooling rates.

The Max Planck Institute model has been used to study the gravity-wave-induced heating and cooling in the upper atmosphere (Medvedev and Yigit, 2012), finding a significant effect. The thermal effects of gravity waves have been recently compared with those produced by a modification of the atomic oxygen profile used in the NLTE cooling scheme (Medvedev et al., 2015), finding that the gravity-wave-induced heating/cooling can be dominant in the polar regions. This model has also been used to simulate the effects of global dust storms in the Martian upper atmosphere (Medvedev et al., 2013).

The strengths of the Max Planck Institute model include: (1) the inclusion of a parameterization of the thermal and dynamical effects of non-orographic gravity waves; (2) a spectral dynamical core; and (3) an independent NLTE radiative transfer code. However, the model does not include a photochemical scheme, preventing it from simulating the composition of the upper atmosphere.

14.5.3 Coupling of Thermosphere–Ionosphere, Exosphere, and Magnetosphere Models to Address the Integrated System

Both 1D models and GCMs are useful for the study of the collisional atmosphere, up to the exobase (placed between 160 and 220 km, following Valeille et al. (2010)). Although they have been used sometimes for the study of the exosphere (e.g. Krasnopolsky, 2010), their results in those layers should be considered with caution, as the formalism used to represent the vertical transport is strictly valid only when collisions dominate, that is, below the exobase. Boqueho and Blelly (2005) introduced a unified treatment of the atmosphere below and above the exobase that extends the validity of their model above the exobase. Another important phenomena that can limit the validity of ionospheric models above about 200 km is the interaction with solar wind and the effects of magnetic fields. These processes are negligible below about 200 km, in the photochemically dominated region, but not above. Present efforts to couple thermosphere–ionosphere models with models of the exosphere and the magnetosphere, which would avoid these limitations, are discussed in Chapter 15.

14.6 OUTSTANDING PROBLEMS

Many of the outstanding problems of the Mars thermosphere–ionosphere (presented in this chapter) will be addressed with new datasets and modeling activities as a result of the MAVEN mission (2013–2016). The overview in this chapter on our pre-MAVEN understanding of the Mars thermosphere–ionosphere system is therefore warranted.

The composition of the Mars thermosphere, specifically neutral species densities, has only been directly measured with the UAMS instruments on-board the descending Viking Landers. This paucity of data is aggravated by the fact that atomic O could not be measured *in situ* by the UAMS. Instead, ionosphere model calculations have been used to estimate O abundances, and models have been employed to determine abundances from Mariner and Mars Express UV airglow measurements. Despite these efforts, the abundance of O is still a major unknown for the Mars upper atmosphere (Bougher et al., 2015a). Systematic measurements of neutral upper atmosphere densities (both spatially and temporally) are sorely needed for both the dayside and nightside thermosphere, and will be obtained by MAVEN instruments (e.g. Jakosky et al., 2015; Mahaffy et al., 2015). In particular, atomic O measurements will be obtained and used to constrain the heat budget and chemistry of the dayside thermosphere. These measurements can also be used to extract the solar cycle plus seasonal variations of dayside thermospheric and exospheric structure, including temperatures, a key element in volatile escape calculations. In addition, thermospheric neutral wind measurements are presently lacking. However, when such winds are combined with simultaneous neutral density and temperature measurements, they enable 3D models to investigate the intricate feedbacks of energetic, chemical, and dynamical processes in the upper atmosphere.

Many basic ionospheric properties and processes are poorly constrained by observations. Data on plasma composition and energetics are essentially limited to the two Viking Lander descent datasets, and reliable data on plasma motion are entirely absent. Systematic measurements of ion density profiles above ~200 km are needed, and will be obtained by MAVEN instruments (e.g. Jakosky et al., 2015; Mahaffy et al., 2015). Additional observational constraints spanning ranges of latitude, SZA, and solar cycle will also be provided. The localized and inhomogeneous crustal magnetic fields of Mars, which are unique in the Solar System, are expected to cause the ionospheric electrodynamics of Mars to differ substantially from those of Earth. They are likely to have significant influence on ionospheric structure, dynamics, and energetics.

Airglow studies all show the need for reliable and up-to-date databases of chemical physics parameters (i.e. cross-sections) with constrained uncertainties in order to develop a complete quantitative understanding of airglow processes. Finally, interactive coupling of thermosphere–ionosphere models with models of the exosphere and magnetosphere, along with systematic measurements of the thermosphere–ionosphere–exosphere and solar wind interaction region, are needed to evaluate the relative importance of various volatile loss mechanisms at Mars (e.g. Dong et al., 2015).

REFERENCES

Angelats i Coll, M., Forget, F., López-Valverde, M. A., et al. (2004) Upper atmosphere of Mars up to 120 km: Mars Global Surveyor data analysis with the LMD general circulation model, *J. Geophys. Res.*, 109, E01011, doi:10.1029/2003JE002163.

Angelats i Coll, M., Forget, F., López-Valverde, M. A., and González-Galindo, F. (2005) The first Mars thermospheric general circulation model: the Martian atmosphere from the ground to 240 km, *Geophys. Res. Lett.*, 32, L04201, doi:10.1029/2004GL021368.

Arkani-Hamed, J. (2004) A coherent model of the crustal magnetic field of Mars, *J. Geophys. Res.*, 109, E09005, doi:10.1029/2004JE002265.

Baird, D. T., Tolson, R., Bougher, S. W., and Steers, B. (2007) Zonal wind calculation from MGS Accelerometer and rate data, *AIAA J. Spacecraft and Rockets*, 44 (6), 1180–1187.

Barth, C. A., Hord, C. W., Pearce, J. B., et al. (1971) Mariner 6 and 7 ultraviolet spectrometer experiment: upper atmosphere data, *J. Geophys. Res.*, 76, 2213–2227.

Barth, C. A., Stewart, A. I., Hord, C. W., and Lane, A. L. (1972) Mariner 9 ultraviolet spectrometer experiment: Mars airglow spectroscopy and variations in Lyman alpha, *Icarus*, 17, 457–468.

Barth, C. A., Stewart, A. I. F., Bougher, S. W., et al. (1992) Aeronomy of the current Martian atmosphere, In *Mars* (Kieffer, H. H., Jakosky, B. M., Snyder, C. W., and Matthews, M. S. Eds), University of Arizona Press, Tucson, 1054–1089.

Bauer, S. J., and Lammer, H. (2004) *Planetary Aeronomy: Atmosphere Environments in Planetary Systems*, Springer, Berlin.

Bell, J. M., Bougher, S. W., and Murphy, J. R. (2007) Vertical dust mixing and the interannual variations in the Mars thermosphere, *J. Geophys. Res.*, 112, E12002, doi:10.1029/2006JE002856.

Bertaux, J.-L., Leblanc, F., Perrier, S., et al. (2005a) Nightglow in the upper atmosphere of Mars and implications for atmospheric transport, *Science*, 307, 566–569, doi:10.1126/science.1106957.

Bertaux, J.-L., Leblanc, F., Witasse, O., et al. (2005b) Discovery of an aurora on Mars, *Nature*, 435, 790–794.

Bertaux, J.-L., Korablev, O., Perrier, S., et al. (2006) SPICAM on Mars Express: observing modes and overview of UV spectrometer data and scientific results, *J. Geophys. Res.*, 111, E10S90, doi:10.1029/2006JE02690.

Bertaux, J.-L., Gondet, B., Bibring, J.-P., et al. (2011) First Detection of O_2 Recombination Nightglow Emission at 1.27 μm in the Atmosphere of Mars with Omega/MEX and Comparison with Model in *The Fourth International Workshop on the Mars Atmosphere: Modeling and observations*, Paris, France.

Boqueho, V., and Blelly, P.-L. (2005) Contributions of a multimoment multispecies approach in modeling planetary atmospheres: example of Mars, *J. Geophys. Res.*, 110, A01313, doi:10.1029/2004JA010414.

Bougher, S. W. (1995) Comparative thermospheres: Venus and Mars, *Adv. Space Res.*, 15 (4), 21–25.

Bougher, S. W., Hunten, D. M., and Roble, R. G. (1994) CO_2 cooling in terrestrial planet thermospheres, *J. Geophys. Res.*, 99, 14609–14622.

Bougher, S. W., Keating, G., Zurek, R., et al. (1999a) Mars global surveyor aerobraking: atmospheric trends and model interpretation, *Adv. Space Res.*, 23, 1887–1897, doi:10.1016/S0273-1177(99)00272-0.

Bougher, S. W., Engel, S., Roble, R. G., and Foster, B. (1999b) Comparative terrestrial planet thermospheres 2. Solar cycle variation of global structure and winds at equinox, *J. Geophys. Res.*, 104, 16591–16611, doi:10.1029/1998JE001019.

Bougher, S. W., Engel, S., Roble, R. G., and Foster, B. (2000) Comparative terrestrial planet thermospheres 3. Solar cycle variation of global structure and winds at solstices, *J. Geophys. Res.*, 105, 17669–17692, doi:10.1029/1999JE001232.

Bougher, S. W., Engel, S., Hinson, D. P., and Forbes, J. M. (2001) Mars Global Surveyor Radio Science electron density profiles: neutral atmosphere implications, *Geophys. Res. Lett.*, 28, 3091–3094, doi:10.1029/2001GL012884.

Bougher, S. W., Roble, R. G., and Fuller-Rowell, T. J. (2002) Simulations of the upper atmospheres of the terrestrial planets, In *Atmospheres in the Solar System, Comparative Aeronomy* (Mendillo, M., Nagy, A. F., and Waite, J. H., Jr. Eds), AGU Monograph #130, American Geophysical Union, Washington, D.C., 261–288.

Bougher, S. W., Engel, S., Hinson, D. P., and Murphy, J. R. (2004) MGS Radio Science electron density profiles: interannual variability and implications for the Martian neutral atmosphere, *J. Geophys. Res.*, 109, E03010, doi:10.1029/2003JE002154.

Bougher, S. W., Bell, J. M., Murphy, J. R., et al. (2006) Polar warming in the Mars thermosphere: seasonal variations owing to changing insolation and dust distributions, *Geophys. Res. Lett.*, 33, L02203, doi:10.1029/2005GL024059.

Bougher, S. W., Blelly, P.-L., Combi, M., et al. (2008) Neutral upper atmosphere and ionosphere modeling, *Space Sci. Rev.*, 139, 107–141, doi:10.1007/s11214-008-9401-9.

Bougher, S. W., Valeille, A., Combi, M. R., and Tenishev, V. (2009a) Solar cycle and seasonal variability of the Martian thermosphere-ionosphere and associated impacts upon atmospheric escape, SAE Technical Paper #2009-01-2386, SAE International.

Bougher, S. W., McDunn, T. M., Zoldak, K. A., and Forbes, J. M. (2009b) Solar cycle variability of Mars dayside exospheric temperatures: model evaluation of underlying thermal balances, *Geophys. Res. Lett.*, 36, L05201, doi:10.1029/2008GL036376.

Bougher, S. W., Simon, C., Gronoff, G., et al. (2010) Exospheric temperatures at Mars derived from SPICAM dayglow measurements, in *2010 Fall AGU Meeting*, San Francisco, California.

Bougher, S. W., McDunn, T., Murphy, J., et al. (2011a) Coupling of Mars lower and upper atmosphere revisited: impacts of gravity wave momentum deposition on upper atmosphere structure, in *The Fourth International Workshop on the Mars Atmosphere: Modeling and Observations*, Paris, France.

Bougher, S. W., Ridley, A., Pawlowski, D., et al. (2011b) Development and validation of the ground-to-exosphere Mars GITM code: solar cycle and seasonal variations of the upper atmosphere, in *The Fourth International Workshop on the Mars Atmosphere: Modeling and Observations*, Paris, France.

Bougher, S. W., Pawlowski, D. J., and Murphy, J. R. (2011c) Toward an understanding of the time-dependent responses of the Martian upper atmosphere to dust storm events, in *2011 Fall AGU Meeting*, San Francisco, California.

Bougher, S. W., Cravens, T. E., Grebowsky, J., and Luhmann, J. L. (2015a) The aeronomy of Mars: characterization by MAVEN of the upper atmosphere reservoir that regulates volatile escape, *Space Science Rev.*, 195, 423–456, doi:10.1007/s11214-014-0053-7.

Bougher, S. W., Pawlowski, D., Bell, J. M., et al. (2015b) Mars Global Ionosphere–Thermosphere Model: solar cycle, seasonal and diurnal variations of the Mars upper atmosphere, *J. Geophys. Res.*, 120, 311–342, doi:10.1002/2014JE004715.

Brain, D. A., Bagenal, F., Acuña, M. H., and Connerney, J. E. P. (2003) Martian magnetic morphology: contributions from the solar wind and crust, *J. Geophys. Res.*, 108, 1424, doi:10.1029/2002JA009482.

Brain, D. A., Halekas, J. S., Peticolas, L. M., et al. (2006) On the origin of aurorae on Mars, *Geophys. Res. Lett.*, 33, L01201, doi:10.1029/2005GL024782.

Brecht, S. H., and Ledvina, S. A. (2006) The solar wind interaction with the Martian ionosphere/atmosphere, *Space Sci. Rev.*, 126, 15–38.

Chassefière, E., and Leblanc, F. (2004) Mars atmospheric escape and evolution: interaction with the solar wind, *Planet. Space Sci.*, 52, 1039–1058.

Chaufray, J. Y., Leblanc, F., Quémerais, E., and Bertaux, J. L. (2009) Martian oxygen density at the exobase deduced from O I 130.4-nm observations by Spectroscopy for the Investigation of the Characteristics of the Atmosphere of Mars on Mars Express, *J. Geophys. Res.*, 114, E02006.

Chaufray, J.-Y., González-Galindo, F., Forget, F., et al. (2014) Three-dimensional Martian ionosphere model: II. Effect of transport processes due to pressure gradients, *J. Geophys. Res.*, 119, 1614–1636, doi:10.1002/2013JE004551.

Chen, R. H., Cravens, T. E., and Nagy, A. F. (1978) The Martian ionosphere in light of the Viking observations, *J. Geophys. Res.*, 83, 3871–3876.

Choi, Y. W., Kim, J., Min, K. W., et al. (1998) Effect of the magnetic field on the energetics of Mars ionosphere, *Geophys. Res. Lett.*, 25, 2753–2756.

Cox, C., Saglam, A., Gérard, J.-C., et al. (2008) Distribution of the ultraviolet nitric oxide Martian night airglow: observations from Mars Express and comparisons with a one-dimensional model, *J. Geophys. Res.*, 113, E08012, doi:10.1029/2007JE003037.

Cox, C., Gérard, J.-C., Hubert, B., et al. (2010) Mars ultraviolet dayglow variability: SPICAM observations and comparison with airglow model, *J. Geophys. Res.*, 115, E04010, doi:10.1029/2009JE003504.

Cravens, T. E., Victor, G. A., and Dalgarno, A. (1975) The absorption of energetic electrons by molecular hydrogen gas, *Planet. Space Sci.*, 23, 1059–1070.

Cravens, T. E., Brace, L. H., Taylor, H. A., et al. (1982) Disappearing ionospheres on the nightside of Venus, *Icarus*, 51, 271–282.

Dalgarno, A., and Lejeune, G. (1971) The absorption of electrons in atomic oxygen, *Planet. Space Sci.*, 19, 1653–1667.

DeMajistre, R., Paxton, L. J., Morrison, D., et al. (2004) Retrievals of nighttime electron density from Thermosphere Ionosphere Mesosphere Energetics and Dynamics (TIMED) mission Global Ultraviolet Imager (GUVI) measurements, *J. Geophys. Res.*, 109, A05305, doi:10.1029/2003JA010296.

Deng, Y., Richmond, A. D., Ridley, A. J., and Liu, H.-L. (2008) Assessment of the non-hydrostatic effect on the upper atmosphere using a general circulation model (GCM), *Geophys. Res., Lett.*, 35, L01104, doi:10.1029/2007GL032182.

Dong, C. F., Bougher, S. W., Ma, Y., et al. (2015) Solar wind interaction with the Martian upper atmosphere: crustal field orientation, solar cycle, and seasonal variations, *J. Geophys. Res.*, 120, doi:10.1002/2015JA020990.

Dubinin, E., Modolo, R., Fraenz, M., et al. (2008) Plasma environment of Mars as observed by simultaneous MEX-ASPERA-3 and MEX-MARSIS observations, *J. Geophys. Res.*, 113, A10217, doi:10.1029/2008JA013355.

Duru, F., Gurnett, D. A., Morgan, D. D., et al. (2008) Electron densities in the upper ionosphere of Mars from the excitation of electron plasma oscillations, *J. Geophys. Res.*, 113, A07302, doi:10.1029/2008JA013073.

Fillingim, M. O., Peticolas, L. M., Lillis, R. J., et al. (2007) Model calculations of electron precipitation induced ionization patches on the nightside of Mars, *Geophys. Res. Lett.*, 34, L12101, 0.1029/2007GL029986.

Fjeldbo, G., and Eshleman, V. R. (1968) The atmosphere of Mars analyzed by integral inversion of the Mariner IV occultation data, *Planet. Space Sci.*, 16, 1035–1059.

Fjeldbo, G., Kliore, A., and Seidel, B. (1970) The Mariner 1969 occultation measurements of the upper atmosphere of Mars, *Radio Sci.*, 5, 381–386.

Forbes, J. M., and Hagan, M. E. (2000) Diurnal Kelvin wave in the atmosphere of Mars: towards an understanding of "stationary" density structures observed by the MGS Accelerometer, *Geophys. Res. Lett.*, 27, 21, doi:10.1029/2000GL011850.

Forbes, J. M., and Moudden, Y. (2009) Solar terminator wave in a Mars general circulation model, *Geophys. Res. Lett.*, 36, L17201, doi:10.1029/2009GL039528.

Forbes, J. M., Bridger, A. F. C., Bougher, S. W., et al. (2002) Nonmigrating tides in the thermosphere of Mars, *J. Geophys. Res.*, 107, 5113, doi:10.1029/2001JE001582.

Forbes, J. M., Lemoine, F. G., Bruinsma, S. L., et al. (2008) Solar flux variability of Mars' exosphere densities and temperatures, *Geophys. Res. Lett.*, 35, L01201, doi:10.1029/2007GL031904.

Forget, F., Hourdin, F., Fournier, R., et al. (1999) Improved general circulation models of the Martian atmosphere from the surface to above 80 km, *J. Geophys. Res.*, 104, 24155–24175.

Forget, F., Montmessin, F., Bertaux, J.-L., et al. (2009) Density and temperatures of the upper Martian atmosphere measured by stellar occultations with Mars Express SPICAM, *J. Geophys. Res.*, 114, E01004, doi:10.1029/2008JE003086.

Formisano, V., Maturilli, A., Giuranna, M., et al. (2006) Observations of non-LTE emission at 4–5 microns with the Planetary Fourier Spectrometer aboard the Mars Express mission, *Icarus*, 182, 51–67, doi:10.1016/j.icarus.2005.12.022.

Fox, J. L. (1992) Airglow and aurora in the atmospheres of Venus and Mars, in *Venus and Mars: Atmospheres, Ionospheres, and Solar Wind Interactions*, Geophys. Monogr. Ser., 66, American Geophysical Union, Washington, D.C., 191–222.

Fox, J. L. (2004a) Response of the Martian thermosphere/ionosphere to enhanced fluxes of solar soft X rays, *J. Geophys. Res.*, 109, A11310, doi:10.1029/2004JA010380.

Fox, J. L. (2004b) CO_2^+ dissociative recombination: a source of thermal and nonthermal C on Mars, *J. Geophys. Res.*, 109, A08306, doi:10.1029/2004JA010514.

Fox, J. L. (2005) Effects of dissociative recombination on the composition of planetary atmospheres, *Journal of Physics: Conference Series*, 4, 32–37.

Fox, J. L. (2006) Aeronomy, In *Atomic, Molecular and Optical Physics Handbook* (Drake, G. W. F. Ed), 2nd edn., American Institute of Physics Press, Woodbury, New York, 1259–1292.

Fox, J. L. (2009) Morphology of the dayside ionosphere of Mars: implication for ion outflows, *J. Geophys. Res.*, 114, E12005, doi:10.1029/2009JE003432.

Fox, J. L., and Dalgarno, A. (1979) Ionization, luminosity, and heating of the upper atmosphere of Mars, *J. Geophys. Res.*, 84, 7315–7333.

Fox, J. L., and Ha, A. B. (2009) Photochemical escape of oxygen from Mars: a comparison of the exobase approximation to a Monte Carlo method, *Icarus*, 204, 527–544, doi:10.1016/j.icarus.2009.07.005.

Fox, J. L., and Ha, A. B. (2010) Isotope fractionation in the photochemical escape of O from Mars, *Icarus*, 208, 176–191, doi:10.1016/j.icarus.2010.01.019.

Fox, J. L., and Stewart, A. I. F. (1991) The Venus ultraviolet aurora: a soft electron source, *J. Geophys. Res.*, 96, 9821–9828.

Fox, J. L., and Sung, K. Y. (2001) Solar activity variations of the Venus thermosphere-ionosphere, *J. Geophys. Res.*, 106, 21305–21336, doi:10.1029/2001JA000069.

Fox, J. L., and Yeager, K. E. (2006) Morphology of the near-terminator Martian ionosphere: a comparison of models and data, *J. Geophys. Res.*, 111, A10309, doi:10.1029/2006JA011697.

Fox, J. L., Brannon, J. F., and Porter, H. S. (1993) Upper limits to the nightside ionosphere of Mars, *Geophys. Res. Lett.*, 20, 1339–1342.

Fox, J. L., Zhou, P., and Bougher, S. W. (1996) The thermosphere/ionosphere of Mars at high and low solar activities, *Adv. Space Res.*, 17 (11), 203–218.

Fox, J. L., Galand, M. I., and Johnson, R. E. (2008) Energy deposition in planetary atmospheres by charged particles and solar photons, *Space Sci. Rev.*, 139, 3–62.

Frahm, R. A., Sharber, J. R., Winningham, J. D., et al. (2006) Locations of atmospheric photoelectron energy peaks within the Mars Environment, *Space Sci. Rev.*, 126, 389–402.

Fränz, M., Dubinin, E., Nielsen, E., et al. (2010) Transterminator ion flow in the Martian ionosphere, *Planet. Space Sci.*, 58, 1442–1454.

Fuller-Rowell, T. J., and Rees, D. (1980) A three dimensional, time-dependent, global model of the thermosphere, *J. Atmos. Sci.*, 37, 2545.

Galand, M., and Chakrabarti, S. (2002) Auroral processes in the solar system, In *Atmospheres in the Solar System, Comparative Aeronomy* (Mendillo, M., Nagy, A. F. and Waite, J. H., Jr. Eds), AGU Monograph 130, American Geophysical Union, Washington, D.C., 55–76.

Gilli, G., López-Valverde, M. A., Drossart, P., et al. (2009) Limb observations of CO_2 and CO non-LTE emissions in the Venus atmosphere by VIRTIS/Venus Express, *J. Geophys. Res.*, 114, E00B29, doi:10.1029/2008JE003112.

González-Galindo, F., López-Valverde, M. A., Angelats i Coll, M., and Forget, F. (2005) Extension of a Martian general circulation model to thermospheric altitudes: UV heating and photochemical models, *J. Geophys. Res.*, 110, E09008, doi:10.1029/2004JE002312.

González-Galindo, F., Gilli, G., López-Valverde, M. A., et al. (2008) Nitrogen and Ionospheric Chemistry in the Thermospheric LMD-MGCM in *Third International Workshop on The Mars Atmosphere: Modeling and Observations*, LPI Contribution No. 1447, 9007.

González-Galindo, F., Forget, F., López-Valverde, M. A., Angelats i Coll, M., and Millour, E. (2009a) A ground-to-exosphere Martian general circulation model: 1. Seasonal, diurnal, and solar cycle variation of thermospheric temperatures, *J. Geophys. Res.*, 114, E04001, doi:10.1029/2008JE003246.

González-Galindo, F., Forget, F., López-Valverde, M. A., and Angelats i Coll, M. (2009b) A ground-to-exosphere Martian general circulation model: 2. Atmosphere during solstice conditions – thermospheric polar warming, *J. Geophys. Res.*, 114, E08004, doi:10.1029/2008JE003277.

González-Galindo, F., Bougher, S. W., López-Valverde, M. A., Forget, F., and Murphy, J. (2010) Thermal and wind structure of the Martian thermosphere as given by two general circulation models, *Planet. Space Sci.*, 58, 1832–1849, doi:10.1016/j.pss.2010.08.013.

González-Galindo, F., Chaufray, J.-Y., López-Valverde, M.A., et al. (2013) Three-dimensional Martian ionosphere model: I. The photochemical ionosphere below 180 km, *J. Geophys. Res.*, 118, 2105–2123, doi:10.1002/jgre.20150

Grebowsky, J. M., Moses, J. I., and Pesnell, W. D. (2002) Meteoric material – an important component of planetary atmospheres, In *Atmospheres in the Solar System: Comparative Aeronomy*, Geophysical Monograph 130, AGU, Washington, DC, doi:10.1029/130GM15.

Gronoff, G., Lilensten, J., Simon, C., et al. (2008) Modelling the Venusian airglow, *Astron. Astrophys.*, 482, 1015–1029.

Gronoff, G., Simon Wedlund, C., Mertens, C. J., and Lillis, R. J. (2012a) Computing uncertainties in ionosphere-airglow models. I. Electron flux and species production uncertainties for Mars, *J. Gephys. Res.*, 117, A04306, doi:10.1029/2011JA016930.

Gronoff, G., Simon Wedlund, C., Mertens, C. J., et al. (2012b) Computing uncertainties in ionosphere-airglow models. II. The Martian airglow, *J. Gephys. Res.*, 117, A05309, doi:10.1029/2011JA017308.

Gurnett, D. A., Kirchner, D. L., Huff, R. L., et al. (2005) Radar soundings of the ionosphere of Mars, *Science*, 310, 1929–1933.

Gurnett, D. A., Huff, R. L., Morgan, D. D., et al. (2008) An overview of radar soundings of the Martian ionosphere from the Mars Express spacecraft, *Adv. Space Res.*, 41, 1335–1346.

Gurnett, D. A., Morgan, D. D., Duru, F., et al. (2010) Large density fluctuations in the Martian ionosphere as observed by the Mars Express radar sounder, *Icarus*, 206, 83–94.

Gusev, O.A., and Kutepov, A.A. (2003) Non-LTE gas in planetary atmospheres, in *Stellar Atmosphere Modeling*, ASP Conference series, 288, 318–330.

Haberle, R. M., Joshi, M. M., Murphy, J. R., et al. (1999) General circulation model simulations of the Mars Pathfinder atmospheric structure investigation/meteorology data, *J. Geophys. Res.*, 104, 8957–8974, doi:10.1029/1998JE900040.

Haider, S. A., Seth, S. P., Choksi, V. R., and Oyama, K. I. (2006) Model of photoelectron impact ionization within the high latitude ionosphere at Mars: comparison of calculated and measured electron density, *Icarus*, 185, 102–112.

Haider, S. A., Mahajan, K. K., and Kallio, E. (2011) Mars ionosphere: a review of experimental results and modeling studies, *Rev. of Geophys.*, 49, RG4001, doi:10.1029/2011RG000357.

Hanson, W. B., and Mantas, G. P. (1988) Viking electron temperature measurements – evidence for a magnetic field in the Martian ionosphere, *J. Geophys. Res.*, 93, 7538–7544.

Hanson, W. B., Sanatani, S., and Zuccaro, D. R. (1977) The Martian ionosphere as observed by the Viking Retarding Potential Analyzers, *J. Geophys. Res.*, 82, 4351–4363.

Hartogh, P., Medvedev, A.S., Kuroda, T., et al. (2005) Description and climatology of a new general circulation model of the Martian atmosphere, *J. Geophys. Res.*, 110, E11008, doi:10.1029/2005JE002498.

Hinson, D. P., Simpson, R. A., Twicken, J. D., et al. (1999) Initial results from radio occultation measurements with Mars Global Surveyor, *J. Geophys. Res.*, 104, 26997–27012.

Hourdin, F. (1992) A new representation of the absorption by the CO_2 15-microns band for a Martian general circulation model, *J. Geophys. Res.*, 97, 18319–18335.

Huestis, D. L., Bougher, S. W., Fox, J. L., et al. (2008) Cross sections and reaction rates for comparative aeronomy, *Space Sci. Rev.*, 139, 63–106, doi:10.1007/s11214-008-9383-7.

Huestis, D. L., Slanger, T. G., Sharpee, B. D., and Fox, J. L. (2010) Chemical origins of the Mars ultraviolet dayglow, *Faraday Discuss.*, 147, 307–322.

Jakosky, B. M., Lin, R. P., Grebowsky, J. M., et al. (2015) The Mars Atmosphere and Volatile Evolution (MAVEN) mission to Mars, *Space Sci. Rev.*, 195, 3–48, doi:10.1007/s11214-015-0139-x.

Johnson, R. E. (1978) Comment on the ion and electron temperatures in the Martian upper atmosphere, *Geophys. Res. Lett.*, 5, 989–992.

Kasprzak, W. T., Keating, G. M., Hsu, N. C., et al. (1997) Solar activity behavior of the thermosphere, in *Venus II: Geology, Geophysics, Atmosphere, and Solar Wind Environment*, eds S. W. Bougher, D. M. Hunten, and R. J. Philips, Tucson, AZ, University of Arizona Press, 225.

Keating, G. M., Bougher, S. W., Zurek, R. W., et al. (1998) The structure of the upper atmosphere of Mars: in situ accelerometer measurements from Mars Global Surveyor, *Science*, 279, 1672–1676.

Keating, G. M., Theriot, M., Tolson, R., et al. (2003) Brief Review on the Results Obtained with the MGS and Mars Odyssey 2001 Accelerometer Experiments in *Mars Atmosphere: Modeling and Observations Workshop*, Granada, Spain.

Keating, G. M., Bougher, S. W., Theriot, M. E., et al. (2006) Atmospheric Structure from Mars Reconnaissance Orbiter Accelerometer Measurements in *Proceedings of European Planetary Science Congress*, Berlin, Germany.

Keating, G. M., Bougher, S. W., Theriot, M. E., and Tolson, R. H. (2008) Properties of the Mars Upper Atmosphere Derived from Accelerometer Measurements in *Proceedings of 37th COSPAR Scientific Assembly 2008 and 50th Anniversary*, Montreal, Canada.

Kelley, M. C. (2009) *The Earth's Ionosphere: Plasma Physics and Electrodynamics*, 2nd edn. Academic Press, New York.

Kliore, A. J., Cain, D. L., Fjeldbo, G., et al. (1972) The atmosphere of Mars from Mariner 9 radio occultation measurements, *Icarus*, 17, 484–516.

Kolosov, M. A., Iakovlev, G. D., Iakovleva, O. I., et al. (1975) Results of investigations of the atmosphere of Mars by the method of radio transillumination by means of the automatic interplanetary stations "Mars-2", "Mars-4", and "Mars-6", *Cosmic Research*, 13 (1), 46–50 (transl. *Kosmicheskie Issledovaniia*, 13, 54–59).

Krasnopolsky, V. A. (1986) *Photochemistry of the Atmospheres of Mars and Venus*, Springer, New York.

Krasnopolsky, V. A. (2002) Mars' upper atmosphere and ionosphere at low, medium, and high solar activities: implications for evolution of water, *J. Geophys. Res.*, 107, 5128–5139.

Krasnopolsky, V. A. (2006) Photochemistry of the Martian atmosphere: seasonal, latitudinal, and diurnal variations, *Icarus*, 185, 153–170.

Krasnopolsky, V. A. (2010) Solar activity variations of thermospheric temperatures on Mars and a problem of CO in the lower atmosphere, *Icarus*, 207, 638–647.

Krymskii, A. M., Breus, T. K., Ness, N. F., et al. (2003) Effect of crustal magnetic fields on the near terminator ionosphere at Mars: comparison of in situ magnetic field measurements with the data of radio science experiments on board Mars Global Surveyor, *J. Geophys. Res. – Space Physics*, 108, 1431–1444.

Kutepov, A. A., Gusev, O. A., and Ogibalov, V. P. (1998) Solution of the non-LTE problem for molecular gas in planetary atmospheres: superiority of accelerated lambda iteration, *J. Quant. Spectrosc. Radiat. Transf.*, 60, 199–220.

Leblanc, F., Chaufray, J. Y., Lilensten, J., Witasse, O., and Bertaux, J.-L. (2006a) Martian dayglow as seen by the SPICAM UV spectrograph on Mars Express, *J. Geophys. Res.*, 111 (9), E09S11, doi:10.1029/2005JE002664.

Leblanc, F., Witasse, O., Winningham, J., et al. (2006b) Origins of the Martian aurora observed by Spectroscopy for Investigation of Characteristics of the Atmosphere of Mars (SPICAM) on board Mars Express, *J. Geophys. Res.*, 111, AO9313, doi:10.1029/2006JA011763.

Leblanc, F., Witasse, O., Lilensten, J., et al. (2008) Observations of aurorae by SPICAM ultraviolet spectrograph on board Mars Express: simultaneous ASPERA-3 and MARSIS measurements, *J. Geophys. Res.*, 113, 8311, doi:10.1029/2008JA013033.

Lefèvre, F., Lebonnois, S., Montmessin, F., and Forget, F. (2004) Three-dimensional modeling of ozone on Mars, *J. Geophys. Res.*, 109, E07004, doi:10.1029/2004JE002268.

Leovy, C. B., and Mintz, Y. (1969) Numerical simulation of the weather and climate of Mars, *J. Atmos. Phys.*, 26, 1169–1190.

Lewis, S. R., Collins, M., Read, P. L., et al. (1999) A climate database for Mars, *J. Geophys. Res.*, 104, 24177–24194.

Liemohn, M. W., Ma, Y., Nagy, A. F., et al. (2007) Numerical modeling of the magnetic topology near Mars auroral observations, *Geophys. Res. Lett.*, 34, 24202, doi:10.1029/2007GL031806.

Lilensten, J., and Blelly, P. L. (2002) The TEC and F2 parameters as tracers of the ionosphere and thermosphere, *J. Atmos. Sol. Terr. Phys.*, 64, 775–793.

Lillis, R. J., Fillingim, M. O., Peticolas, L. M., et al. (2009) Nightside ionosphere of Mars: modeling the effects of crustal magnetic fields and electron pitch angle distributions on electron impact ionization, *J. Geophys. Res.*, 114, E11009, doi:10.1029/2009JE003379.

Lillis, R. J., Bougher, S. W., González-Galindo, F., et al. (2010) Four Martian years of nightside upper thermospheric mass densities derived from electron reflectometry: method extension and comparison with GCM simulations, *J. Geophys. Res.*, 115, E07014, doi:10.1029/2009JE003529.

Lillis, R. J., Fillingim, M., and Brain, D. A. (2011) Three-dimensional structure of the Martian nightside ionosphere: predicted rates of impact ionization from Mars Global Surveyor Magnetometer and Electron Reflectometer measurements of precipitating electrons, *J. Geophys. Res.*, 116, A12317, doi:10.1029/2011JA016982.

Lindal, G. F., Hotz, H. B., Sweetnam, D. N., et al. (1979) Viking radio occultation measurements of the atmosphere and topography of Mars – data acquired during 1 Martian year of tracking, *J. Geophys. Res.*, 84, 8443–8456.

Lollo, A., Withers, P., Fallows, K., et al. (2012) Numerical simulations of the ionosphere of Mars during a solar flare, *J. Geophys. Res.*, 117, A05314, doi:10.1029/2011JA017399.

López-Puertas, M., and López-Valverde, M. A. (1995) Radiative energy balance of CO_2 non-LTE infrared emissions in the Martian atmosphere, *Icarus*, 114, 113–129.

López-Puertas, M., and Taylor, F. W. (2001) *Non-LTE Radiative Transfer in the Atmosphere*, World Scientific, Singapore.

López-Valverde, M. A., and López-Puertas, M. (1994a) A non-local thermodynamic equilibrium radiative transfer model for infrared emissions in the atmosphere of Mars. 1: Theoretical basis and nighttime populations of vibrational levels, *J. Geophys. Res.*, 99, 13093–13115.

López-Valverde, M. A., and López-Puertas, M. (1994b) A non-local thermodynamic equilibrium radiative transfer model for infrared emissions in the atmosphere of Mars. 2: Daytime populations of vibrational levels, *J. Geophys. Res.*, 99, 13117–13132.

López-Valverde, M. A., and López-Puertas, M. (2001) A fast computation of radiative heating rates under non-LTE in a CO_2 atmosphere, In *IRS 2000: Current Problems in Atmospheric Radiation* (Smith, W., and Timofeyev, V. Eds), Deepak Publishing, Hampton, Virginia.

López-Valverde, M. A., Edwards, D. P., López-Puertas, M., and Roldán, C. (1998) Non-local thermodynamic equilibrium in general circulation models of the Martian atmosphere 1. Effects of the local thermodynamic equilibrium approximation on thermal cooling and solar heating, *J. Geophys. Res.*, 103, 16799–16812.

López-Valverde, M. A., Haberle, R. M., and López-Puertas, M. (2000) Non-LTE radiative mesospheric study for Mars Pathfinder entry, *Icarus*, 146, 360–365.

López-Valverde, M. A., López-Puertas, M., López-Moreno, J. J., et al. (2005) Analysis of CO_2 non-LTE emissions at 4.3μm in the Martian atmosphere as observed by PFS/Mars Express and SWS/ISO, *Planet. Space Sci.*, 53, 1079–1087.

López-Valverde, M. A., González-Galindo, F., and Forget, F. (2006) 1-D and 3-D modeling of the upper atmosphere of Mars, in *2nd Workshop on Mars Atmosphere Modeling and Observations*, Granada, Spain.

López-Valverde, M. A., Drossart, P., Carlson, R., et al. (2007) Non-LTE infrared observations at Venus: from NIMS/Galileo to VIRTIS/Venus Express, *Planet. Space Sci.*, 55, 1757–1771.

Lundin, R., Winningham, D., Barabash, S., et al. (2006) Plasma acceleration above Martian magnetic anomalies, *Science*, 311, 980–983.

Ma, Y., and Nagy, A. F. (2007) Ion escape fluxes from Mars, *Geophys. Res. Lett.*, 34, L08201, 10.1029/2006GL029208.

Ma, Y., Nagy, A. F., Sokolov, I. V., and Hansen, K. C. (2004) Three-dimensional, multispecies, high spatial resolution MHD studies of the solar wind interaction with Mars, *J. Geophys. Res.*, 109, A07211, doi:10.1029/2003JA010367.

Mahaffy, P. R., Benna, M., King, T., et al. (2015) The Neutral Gas and Ion Mass Spectrometer on the Mars Atmosphere and Volatile Evolution mission, *Space Sci. Rev.*, 195, 49–73, doi:10.1007/s11214-014-0091-1.

Mantas, G. P., and Hanson, W. B. (1979) Photoelectron fluxes in the Martian ionosphere, *J. Geophys. Res.*, 84, 369–385.

Martinis, C. R, Wilson, J. K., and Mendillo, M. J. (2003) Modeling day-to-day ionospheric variability on Mars, *J. Geophys. Res.*, 108, 1383, doi:10.1029/2003JA009973.

Mazarico, E., Zuber, M. T., Lemoine, F. G., and Smith, D. E. (2007) Martian exospheric density using Mars Odyssey radio tracking data, *J. Geophys. Res.*, 112, E05014, doi:10.1029/2006JE002734.

McConnell, J. C., and McElroy, M. B. (1970) Excitation processes for Martian dayglow, *J. Geophys. Res.*, 75, 7290–7293.

McDunn, T. L., Bougher, S. W., Murphy, J., et al. (2010) Simulating the density and thermal structure of the middle atmosphere (80–130 km) of Mars using the MGCM–MTGCM: a comparison with MEX/SPICAM observations, *Icarus*, 206, 5–17.

McElroy, M. B. (1967) The upper atmosphere of Mars, *Astrophys. J.*, 150, 1125–1138.

Medvedev, A. S. and Yigit, E. (2012) Thermal effects of internal gravity waves in the Martian upper atmosphere, *Geophys. Res. Lett.*, 39, L05201, doi:10.1029/2012GL050852.

Medvedev, A. S., Yigit, E., Hartogh, P., and Becker, E. (2011) Influence of gravity waves on the Martian atmosphere: general circulation modeling, *J. Geophys. Res.*, 116 (E10004), 14–32.

Medvedev, A. S., Yigit, E., Kuroda, T., et al. (2013) General circulation modeling of the Martian upper atmosphere during global dust storms, *J. Geophys. Res.*, 118, 2234–2246, doi:10.1002/2013JE004429.

Medvedev, A. S., González-Galindo, F., Yigit, E., et al. (2015) Cooling of the Martian thermosphere by CO_2 radiation and gravity waves: an intercomparison study with two general circulation models, *J. Geophys. Res.*, 120, doi:10.1002/2015JE004802.

Mendillo, M., Smith, S., Wroten, J., et al. (2003) Simultaneous ionospheric variability on Earth and Mars, *J. Geophys. Res.*, 108, 1432, doi:10.1029/2003JA009961.

Mendillo, M. J., Pi, X., Smith, S., et al. (2004) Ionospheric effects upon a satellite navigation system at Mars, *Radio Science*, 39, RS2028, doi:10.1029/2003RS002933.

Mendillo, M., Withers, P., Hinson, D., et al. (2006) Effects of solar flares on the ionosphere of Mars, *Science*, 311, 1135–1138.

Mendillo, M. J., Lollo, A., Withers, P., et al. (2011) Modeling Mars' ionosphere with constraints from same-day observations by Mars Global Surveyor and Mars Express, *J. Geophys. Res.*, 116, A11303, doi:10.1029/2011JA016865.

Mitchell, D. L., Lin, R. P., Mazelle, C., et al. (2001) Probing Mars' crustal magnetic field and ionosphere with the MGS Electron Reflectometer, *J. Geophys. Res.*, 106, 23419–23428.

Moffat-Griffin, T., Aylward, A. D., and Nicholson, W. (2007) Thermal structure and dynamics of the Martian upper atmosphere at solar minimum from global circulation model simulations, *Ann. Geophys.*, 25, 2147–2158.

Molina-Cuberos, G. J., Witasse, O., Lebreton, J.-P., et al. (2003) Meteoric ions in the atmosphere of Mars, *Planet. Space Sci.*, 51, 239–249.

Montmessin, F., Forget, F., Rannou, P., et al. (2004) Origin and role of water ice clouds in the Martian water cycle as inferred from a general circulation model, *J. Geophys. Res.*, 109, E10004, doi:10.1029/2004JE002284.

Morgan, D. D., Gurnett, D. A., Kirchner, D. L., et al. (2008) Variation of the Martian ionospheric electron density from Mars Express radar soundings, *J. Geophys. Res.*, 113, A09303, doi:10.1029/2008JA013313.

Moudden, Y., and Forbes, J. M. (2008) Effects of vertically propagating thermal tides on the mean structure and dynamics of Mars' lower thermosphere, *Geophys. Res. Lett.*, 35, L23805, doi:10.1029/2008GL036086.

Moudden, Y., and Forbes, J. M. (2010) A new interpretation of Mars aerobraking variability: planetary wave–tide interactions, *J. Geophys. Res.*, 115, E09005, doi:10.1029/2009JE003542.

Moudden, Y., and McConnell, J. C. (2005) A new model for multiscale modeling of the Martian atmosphere, GM3, *J. Geophys. Res.*, 110, E04001, doi:10.1029/2004JE002354.

Moudden, Y., and McConnell, J. C. (2007) Three-dimensional on-line chemical modeling in a Mars general circulation model, *Icarus*, 188, 18–34, doi:10.1016/j.icarus.2006.11.005.

Müller-Wodarg, I. C. F., Yelle, R. V., Mendillo, M., et al. (2000) The thermosphere of Titan simulated by a global three-dimensional time-dependent model, *J. Geophys. Res.*, 105, 20833–20856, doi:10.1029/2000JA000053.

Müller-Wodarg, I. C. F., Strobel, D. F., Moses, J. I., et al. (2008) Neutral atmospheres, *Space Sci. Rev.*, 139, doi:10.1007/s11214-008-9404-6.

Němec, F., Morgan, D. D., Gurnett, D. A., and Duru, F. (2010) Nightside ionosphere of Mars: radar soundings by the Mars Express spacecraft, *J. Geophys. Res.*, 115, E12009, 10.1029/2010JE003663.

Němec, F., Morgan, D. D., Gurnett, D. A., and Brain, D. A. (2011) Areas of enhanced ionization in the deep nightside ionosphere of Mars, *J. Geophys. Res.*, 116, E06006, 10.1029/2011JE003804.

Ness, N. F., Acuña, M. H., Connerney, J. E. P., et al. (2000) Effects of magnetic anomalies discovered at Mars on the structure of the Martian ionosphere and solar wind interaction as follows from radio occultation experiments, *J. Geophys. Res.*, 105, 15991–16004.

Nicholson, W. P., Gronoff, G., Lilensten, J., et al. (2009) A fast computation of the secondary ion production in the ionosphere of Mars, *MNRAS*, 400, 369–382.

Nielsen, E., Fraenz, M., Zou, H., et al. (2007) Local plasma processes and enhanced electron densities in the lower ionosphere in magnetic cusp regions on Mars, *Planet. Space Sci.*, 55, 2164–2172.

Nier, A. O., and McElroy, M. B. (1977) Composition and structure of Mars' upper atmosphere: results from the Neutral Mass Spectrometers on Viking 1 and 2, *J. Geophys. Res.*, 82, 4341–4349.

Pätzold, M., Tellmann, S., Häusler, B., et al. (2005) A sporadic third layer in the ionosphere of Mars, *Science*, 310, 837–839.

Pawlowski, D. J., and Bougher, S. W. (2010) Ground to exobase modeling of the Martian atmosphere using M-GITM, in *2010 Fall AGU Meeting*, Abstract #P52A-08, San Francisco, California.

Pawlowski, D. J., and Bougher, S. W. (2012) Comparative aeronomy: the effects of solar flares at Earth and Mars, in *Comparative Climatology of Terrestrial Planets Conference*, Boulder, Colorado.

Pawlowski, D. J., Bougher, S. W., and Chamberlain, P. (2011) Modeling the response of the Martian upper atmosphere to solar flares, in *2011 Fall AGU Meeting*, San Francisco, California.

Paxton, L. J., and Anderson, D. E. (1992) Far ultraviolet remote sensing of Venus and Mars. In *Venus and Mars: Atmospheres, Ionospheres and Solar Wind Interactions* (Luhmann, J. G., Tatrallyay, M., and Pepin, R. O. Eds), Geophysical Monograph #66, American Geophysical Union, Washington, D.C., 113–189.

Pesnell, W. D., and Grebowsky, J. (2000) Meteoric magnesium ions in the Martian atmosphere, *J. Geophys. Res.*, 105, 1695–1708.

Picardi, G. (2008) An overview of radar soundings of the Martian ionosphere from the Mars Express spacecraft, *Adv. Space Res.*, 41, 1335–1346.

Plane, J. M. C. (2003) Atmospheric chemistry of meteoric metals, *Chem. Rev.*, 103, 4963–4984.

Ridley, A. J., Deng, Y., and Tóth, G. (2006) The global ionosphere thermosphere model, *J. Atmos. Solar-Terr. Phys.*, 68, 839–864.

Rishbeth, H., and Mendillo, M. (2004) Ionospheric layers of Mars and Earth, *Planet. Space Sci.*, 52, 849–852.

Rodrigo, R., García-Alvarez, E., López-González, M. J., and López-Moreno, J. J. (1990) A nonsteady one-dimensional theoretical model of Mars' neutral atmospheric composition between 30 and 200 km, *J. Geophys. Res.*, 95, 14795–14810.

Rohrbaugh, R. P., Nisbet, J. S., Bleuler, E., and Herman, J. R. (1979) The effect of energetically produced O_2^+ on the ion temperatures of the Martian thermosphere, *J. Geophys. Res.*, 84, 3327–3338.

Roldán, C., López-Valverde, M. A., López-Puertas, M., and Edwards, D. P. (2000) Non-LTE infrared emissions of CO_2 in the atmosphere of Venus, *Icarus*, 147, 11–25.

Sadourny, R., and Laval, K. (1984) January and July performance of the LMD general circulation model, In *New Perspectives in Climate Modeling* (Berger, A., and Nicolis, C. Eds), Elsevier, Amsterdam, 173–197.

Safaeinili, A., Kofman, W., Mouginot, J., et al. (2007) Estimation of the total electron content of the Martian ionosphere using radar sounder surface echoes, *Geophys. Res. Lett.*, 34, L23204, doi:10.1029/2007GL032154.

Schunk, R., and Nagy, A. (2009) *Ionospheres: Physics, Plasma Physics, and Chemistry*, 2nd edition. Cambridge University Press, New York.

Seiff, A., and Kirk, D. B. (1977) Structure of the atmosphere of Mars in summer at mid-latitudes, *J. Geophysical Res.*, 82, 4364–4378.

Shematovich, V. I., Bisikalo, D. V., and Gérard, J. C. (1994) A kinetic model of the formation of the hot oxygen geocorona. 1: Quiet geomagnetic conditions, *J. Geophys. Res.*, 99, 23217–23228.

Shematovich, V. I., Bisikalo, D. V., Gérard, J.-C., et al. (2008) Monte Carlo model of electron transport for the calculation of Mars dayglow emissions, *J. Geophys. Res.*, 113, 2011, doi:10.1029/2007JE002938.

Shinagawa, H., and Cravens, T. E. (1989) A one-dimensional multispecies magnetohydrodynamic model of the dayside ionosphere of Mars, *J. Geophys. Res.*, 94, 6506–6516.

Simon, C., Witasse, O., Leblanc, F., Gronoff, G., and Bertaux, J.-L. (2009) Dayglow on Mars: kinetic modelling with SPICAM UV limb data, *Planet. Space Sci.*, 57, 1008–1021.

Simon Wedlund, C., Gronoff, G., Lilensten, J., Ménager, H., and Bartélemy, M. (2011) Comprehensive calculation of the energy per ion pair or W values for five major planetary upper atmospheres, *Ann. Geophys.*, 29, 187–195.

Slanger, T. G., and Wolven, B. C. (2002) Airglow processes in planetary atmospheres, In *Atmospheres in the Solar System: Comparative Aeronomy* (Mendillo, M., Nagy, A., and Waite, J. H. Eds), Geophys. Monog. Ser. 130, American Geophysical Union, Washington, D.C., 77–93.

Slanger, T. G., Cosby, P. C., Sharpee, B. D., et al. (2006) O(^{1}S–^{1}D,^{3}P) branching ratio as measured in the terrestrial nightglow, *J. Geophys. Res.*, 111, 12318, doi:10.1029/2006JA011972.

Slanger, T. G., Cravens, T. E., Crovisier, J., et al. (2008) Photoemission phenomena in the solar system, *Space Sci. Rev.*, 139, 267–310.

Smith, M. D. (2004) Interannual variability in TES atmospheric observations of Mars during 1999–2003, *Icarus*, 167, 148–165.

Stewart, A. I. (1972) Mariner 6 and 7 ultraviolet spectrometer experiment: implications of CO_2^+, CO and O airglow, *J. Geophysical Res.*, 77, 54–68, doi:10.1029/JA077i001p00054.

Stewart, A. I. F. (1987) Revised time dependent model of the Martian atmosphere for use in orbit lifetime and sustenance studies, *LASP-JPL Internal Report, NQ-802429*, Jet Propulsion Lab, Pasadena, California.

Stewart, A. I., Barth, C. A., Hord, C. W., and Lane, A. L. (1972) Mariner 9 ultraviolet spectrometer experiment: structure of Mars's upper atmosphere, *Icarus*, 17, 469–474.

Stewart, A. I., Alexander, M. J., Meier, R. R., et al. (1992) Atomic oxygen in the Martian thermosphere, *J. Geophys. Res.*, 97, 91–102.

Stiepen, A., Gerard, J.-C., Bougher, S. W., et al. (2015) Mars thermospheric scale height: CO Cameron and CO_2^+ dayglow observations from Mars Express, *Icarus*, 245, 295–305.

Strickland, D. J., and Donahue, T. M. (1970) Excitation and radiative transport of OI 1304 Å resonance radiation – I: The dayglow, *Planet. Space Sci.*, 18, 661–689.

Strickland, D. J., Thomas, G. E., and Sparks, P. R. (1972) Mariner 6 and 7 ultraviolet spectrometer experiment: analysis of the O I 1304- and 1356-Å emissions, *J. Geophys. Res.*, 77, 4052–4068.

Strickland, D. J., Stewart, A. I., Barth, C. A., et al. (1973) Mariner 9 ultraviolet spectrometer experiment: Mars atomic oxygen 1304-Å emission, *J. Geophys. Res.*, 78, 4547–4559.

Tolson, R. H., Keating, G. M., Cancro, G. J., et al. (1999) Application of accelerometer data to Mars Global Surveyor aerobraking operations, *J. Spacecraft and Rockets*, 36 (3), 323–329.

Tolson, R. H., Dwyer, A. M., Hanna, J. L., et al. (2005) Application of accelerometer data to Mars aerobraking and atmospheric modeling, *J. Spacecraft and Rockets*, 42 (3), 435–443.

Tolson, R. H., Keating, G. M., Zurek, R. W., et al. (2007) Application of accelerometer data to atmospheric modeling during Mars aerobraking operations, *J. Spacecraft and Rockets*, 44 (6), 1172–1179.

Tolson, R. H., Bemis, E., Hough, S., et al. (2008) Atmospheric modeling using accelerometer data during Mars Reconnaissance Orbiter aerobraking operations, *J. Spacecraft and Rockets*, 45 (3), 511–518.

Valeille, A., Combi, M. R., Bougher, S. W., et al. (2009) Three-dimensional study of Mars upper thermosphere/ionosphere and hot oxygen corona: 2. Solar cycle, seasonal variations and evolution over history, *J. Geophys. Res.*, 114, E11006, doi:10.1029/2009JE003389.

Valeille, A., Combi, M. R., Tenishev, V., et al. (2010) A study of suprathermal oxygen atoms in Mars upper thermosphere and exosphere over the range of limiting conditions, *Icarus*, 206, 18–27.

Whalley, C. L., and Plane, J. M. C. (2010) Meteoric ion layers in the Martian atmosphere, *Faraday Discussions*, 147, 349–368.

Wilson, R. J. (2002) Evidence for non-migrating thermal tides in the Mars upper atmosphere from the Mars Global Surveyor Accelerometer Experiment, *Geophys. Res. Lett.*, 29 (7), doi:10.1029/2001GL013975.

Winchester, C., and Rees, D. (1995) Numerical models of the Martian coupled thermosphere and ionosphere, *Adv. Space Res.*, 15, 51.

Witasse, O., Dutuit, O., Lilensten, J., et al. (2002) Prediction of a CO_2^{2+} layer in the atmosphere of Mars, *Geophys. Res. Lett.*, 29, 1263, doi:10.1029/2002GL014781.

Withers, P. G. (2006) Mars Gobal Surveyor and Mars Odyssey accelerometer observations of the Martian upper atmosphere during aerobraking, *Geophys. Res. Lett.*, 33, L02201, doi:10.1029/2005GL024447.

Withers, P. G. (2009) A review of observed variability in the dayside ionosphere of Mars, *Adv. Space Res.*, 44, 277–307.

Withers, P. G., Bougher, S. W., and Keating, G. M. (2003) The effects of topographically controlled thermal tides in the Martian upper atmosphere as seen by the MGS Accelerometer, *Icarus*, 164, 14–32.

Withers, P. G., Mendillo, M., Risbeth, H., et al. (2005) Ionospheric characteristics above Martian crustal magnetic anomalies, *Geophys. Res. Lett.*, 32, L16204, doi:10.1029/2005GL023483.

Withers, P., Mendillo, M., Hinson, D. P., and Cahoy, K. (2008) Physical characteristics and occurrence rates of meteoric plasma layers detected in the Martian ionosphere by the Mars Global Surveyor Radio Science Experiment, *J. Geophys. Res.*, 113, A12314, 110.1029/2008JA013636.

Yigit, E., Aylward, A. D., and Medvedev, A. S. (2008) Parameterization of the effects of vertically propagating gravity waves for thermosphere general circulation models: sensitivity study, *J. Geophys. Res.*, 113, D19106, doi:10.1029/2008JD010135.

Zhang, M. H. G., Luhmann, J. G., Kliore, A. J., and Kim, J. (1990a) A post-Pioneer Venus reassessment of the Martian dayside ionosphere as observed by radio occultation methods, *J. Geophys. Res.*, 95, 14829–14839.

Zhang, M. H. G., Luhmann, J. G., and Kliore, A. J. (1990b) An observational study of the nightside ionospheres of Mars and Venus with radio occultation methods, *J. Geophys. Res.*, 95, 17095–17102.

Zurek, R. W., Tolson, R. H., Baird, D., Johnson, M. Z., and Bougher, S. W. (2015) Application of MAVEN Accelerometer and attitude control data to Mars atmospheric characterization, *Space Sci. Rev.*, 195, 303–317, doi:10.1007/s11214-014-0095-x.

15

Solar Wind Interaction and Atmospheric Escape

DAVID A. BRAIN, STANISLAV BARABASH, STEPHEN W. BOUGHER, FIRDEVS DURU,
BRUCE M. JAKOSKY, RONAN MODOLO

15.1 INTRODUCTION

15.1.1 Motivation

Mars and its atmosphere are embedded in a constantly flowing stream of charged particles from the Sun. This solar wind forms the upper boundary for the Martian atmosphere, but is not spatially distinct. Rather, it intermingles with the uppermost atmospheric layers, leading to complex and sometimes unexpected feedbacks in the Martian plasma interaction over a region of space that is many times the volume of the planet itself. The solar wind is fast-moving and dynamic, typically traveling a Mars diameter in 15–20 s and changing on timescales as short as minutes. It is also tenuous, with just a few particles per cubic centimeter, and a mass encountering Mars over the past four billion years that is on the order of just 1% of the total present-day atmospheric mass. But because the solar wind is flowing and carries the interplanetary magnetic field (IMF), it is able to influence the atmosphere in important ways.

Because it lacks a significant global dynamo magnetic field to shield the atmosphere today, the solar wind induces magnetic fields in the conducting Martian ionosphere, forming a Venus-like induced magnetosphere in many locations (Figure 15.1). Mars is smaller than Venus, however, and as a consequence its neutral exosphere extends, comet-like, to high altitudes relative to the radius of the planet, where it can interact directly with the incident solar wind. Mars also possesses strong crustal magnetic fields that form localized regions that act in many respects like small-scale versions of the terrestrial dayside magnetosphere, shielding the atmosphere from the solar wind in some locations and creating small-scale regions of access between the solar wind and atmosphere in others. With elements of the obstacles at Venus, comets, and Earth, the Martian obstacle to the solar wind responds to changes in the solar wind and IMF, the variable upper atmosphere (which responds chiefly to solar ultraviolet photon flux), and the orientation of crustal fields (which rotate with the planet) with respect to incident flowing plasma.

The Martian environment provides a compelling laboratory for the study of space plasma physics. A number of plasma boundaries and regions form near Mars, some with limited physical extent relative to fundamental plasma length scales, meaning that kinetic (non-fluid) effects can play an important role in deflecting the solar wind around the planet. A variety of plasma processes occur in an induced magnetospheric environment fundamentally different from the global magnetospheres (like Earth) where such processes are most often measured *in situ*. Charged particles from the atmosphere strongly interact with the incident solar wind and thereby influence the global structure of the interaction region, providing an example of the consequences of the mixing of two plasma populations. Finally, the Martian plasma interaction region spans both collisionless (far from the planet) and collisional (near and below the exobase) regimes, so that different limiting physical assumptions are required to adequately model the environment in different locations.

An important aspect of unmagnetized planetary interactions is the role they may play in atmospheric evolution. Mineralogical evidence suggests that liquid water was abundant on early Mars, and geological evidence tells us that it was present at the surface for sufficiently long periods of time to reshape the terrain we observe today. The present Martian atmosphere lacks sufficient greenhouse warming to support liquid water at the surface. Thus, we infer that the atmosphere has undergone fundamental change over the past several billion years (detailed evidence is presented in Chapter 17). Support for this inference can be drawn from measured atmospheric isotope ratios, which show that light isotopes have been preferentially removed over time. Light isotopes are more readily accessible at the top of the atmosphere, and easier to strip away. Thus the isotope ratios indicate that atmospheric escape to space has played an important role in the evolution of the atmosphere over time. Solar ultraviolet photons and the solar wind both provide energy sources for the upper atmosphere, heating it, ionizing it, and driving chemistry and dynamics. Upper atmospheric particles (neutrals or ions) that receive sufficient energy to overcome the gravitational pull of the planet (~5 km s^{-1} or ~2 eV for an oxygen atom) escape to space via a variety of physical mechanisms – some involving plasma processes and some not. Atmospheric escape to space is observed to operate today, and may have been the dominant atmospheric loss process at Mars over the past ~3.5 billion years. The community is presently working to understand whether it contributed significantly to the evolution of the Martian climate over time. In this sense, Mars also serves as a nearby laboratory for interactions of distant stars with planetary atmospheres unprotected by a global magnetic field.

15.1.2 Organization

This chapter describes the present understanding of the solar wind interaction and atmospheric escape to space at Mars, and identifies the major unanswered questions facing the community over the next several years. The reader is advised to consult previous reviews of the Martian plasma interaction by Luhmann et al. (1992a), Nagy et al. (2004), and Brain (2006), and reviews of evolution and escape of terrestrial atmospheres

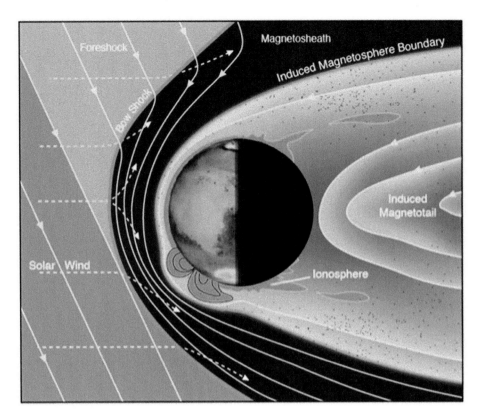

Figure 15.1. Schematic of the Martian plasma interaction region. Various plasma boundaries and regions are described in Section 15.2.1. Image courtesy of S. Bartlett. A black and white version of this figure will appear in some formats. For the color version, please refer to the plate section.

by Hunten (1982), Chassefière and Leblanc (2004), and Lammer et al. (2008). This summary is written at a time when a wealth of observations of the Martian upper atmosphere and plasma environment has been collected over more than a solar cycle by Mars Global Surveyor (MGS) and Mars Express (MEX), and new results are just being made available from the Mars Atmosphere and Volatile Evolution (MAVEN) mission. More results are expected over the next several years. Models are capable of simulating near-Mars space in ever-increasing detail. However, the models are not well constrained by data. The community still awaits complete observations of the entire system – from drivers to upper atmospheric reservoirs to escape – that will help to evaluate its importance in the evolution of the Martian climate over the history of the Solar System. This includes measurements made from the densest regions of the ionosphere all the way out to the unperturbed solar wind, over a long enough period of time to capture the full range of external and Mars variations that occur at present.

In Section 15.2 we describe data and models that have been used to understand the global Martian plasma interaction region and atmospheric escape. Section 15.3 summarizes the current understanding of the Martian plasma interaction, starting with a steady-state description and then discussing time-dependent effects and the influence of crustal magnetic fields. In Section 15.4 we provide an overview of atmospheric escape from Mars, including context and reservoirs, a description of neutral and ion loss processes, and estimates of loss rates. In Section 15.5 we summarize the main points of this chapter, and review the outstanding questions and tasks facing future investigators and missions. This

chapter attempts to avoid specialized space plasma physics jargon, and/or to provide enough context that the meaning of certain terms is obvious for readers unfamiliar with plasmas.

15.2 AVAILABLE DATA AND MODELS

15.2.1 Relevant Observations

Table 15.1 summarizes the relevant spacecraft measurements made at Mars over nearly five decades, through 2011. The Mariner and Mars missions of the 1960s and 1970s crossed into the Martian plasma interaction region, establishing Mars as a weakly magnetized obstacle to the solar wind, though the exact nature of this obstacle was debated for many years afterward. Building upon these early measurements, three spacecraft missions have chiefly been responsible for shaping the current understanding of the Martian solar wind interaction: Phobos 2, Mars Global Surveyor (MGS), and Mars Express (MEX). A fourth mission, MAVEN, is presently in the process of advancing this understanding.

The Phobos 2 mission in the late 1980s approached much closer to the planet (~850 km altitude during five initial elliptical orbits) than previous orbiting missions, with a full complement of particle and field instruments to measure magnetic fields, plasma waves, and ions and electrons at a variety of energies. Measurements were made from both five elliptical and ~110 circular orbits (~6000 km altitude) before contact with the spacecraft was lost (Sagdeev and Zakharov, 1989). The

Table 15.1. *Spacecraft missions to Mars that have provided measurements relevant for describing the plasma environment and atmospheric escape.*

Mission	Dates	Selected key results
Mariner 4, 6, 7, and 9	1965–1972	Detected bow shock and magnetosheath First ionospheric density profiles
Mars 2, 3, and 5	1971–1974	Measured bow shock, sheath, and inner magnetosphere
Viking 1 and 2	1976	*In situ* upper atmospheric profiles (density, composition, temperature) Measurement of atmospheric nitrogen isotopes indicating atmospheric escape
Phobos 2	1989	First sampling of central wake Measured escaping ions
Mars Global Surveyor	1997–2006	Discovered crustal fields Sampled ionosphere down to/below main peak
Mars Odyssey and Mars Reconnaissance Orbiter	2001–present	Upper atmosphere density profiles
Mars Express	2004–present	Discovered localized aurora Detected ionopause Confirmed pressure balance near Mars
Rosetta	2007	Distant bow shock crossing
Mars Science Lander	2012	Measurement of atmospheric argon isotopes indicating atmospheric escape
MAVEN	2014–present	Simultaneous measurements of drivers, reservoirs, and escape

mission made the first detailed measurements of the plasma environment behind Mars, and provided the first estimates of the escape rate of ions from the atmosphere.

MGS was equipped with a magnetometer and suprathermal electron instrument (MAG/ER). It spent nearly two years in a precessing elliptical orbit, and seven more in a circular mapping orbit fixed in local time. These orbital configurations allowed MGS to sample the entire plasma interaction region, reach altitudes low enough that strong crustal fields were discovered, and build statistics about a small slice of the plasma interaction region over more than half a solar cycle (see review by Brain, 2006).

MEX began returning observations from Mars nearly two years before MGS was lost, and continues to make observations through 2013 as this chapter was written. MEX measures suprathermal ions, suprathermal electrons, and energetic neutral atoms (ENAs) with the Analyzer of Space Plasmas and Energetic Atoms (ASPERA) instrument. It probes the ionosphere and samples magnetic field magnitude and thermal electron density *in situ* using the Mars Advanced Radar for Subsurface and Ionospheric Sounding (MARSIS). The SPICAM ultraviolet (and infrared) spectrometer instrument measures ultraviolet signatures from the atmosphere that are relevant for understanding the structure and composition of the thermosphere, ionosphere, and exosphere. MEX has a precessing elliptical orbit, allowing it to visit all parts of the plasma interaction region. Mars Express has detected aurora near Martian crustal fields (Bertaux et al., 2005), provided new estimates of ion escape rates (Barabash et al., 2007), identified the Martian ionopause (Duru et al., 2009), provided upper limits on ENAs produced at Mars (Futaana et al., 2006), and begun to piece together the detailed physics that structures the interaction region and powers atmospheric escape (e.g. Dubinin

et al., 2011). These results have been partially constrained by the lack of an on-board vector magnetometer to organize relevant plasma measurements, although the two-year overlap period with the MGS mission helped to provide some magnetic context before it was lost (Fedorov et al., 2006).

Most recently, the Rosetta spacecraft flew by Mars en route to a comet, recording the most distant bow shock crossings behind Mars of any previous spacecraft, and making measurements of plasma boundaries simultaneous with MEX (Edberg et al., 2009a).

In addition to *in situ* sampling of the particles and fields in the plasma environment, a few missions (Mariner 9, MGS, MEX) have used radio occultation techniques to extract information about upper atmospheric and ionospheric densities at intermediate solar zenith angles at the base of this interaction region – see recent reviews by Withers (2009) and Haider et al. (2011). Measurements of the neutral thermosphere and exosphere are also highly relevant, since these regions serve as the reservoirs for escaping particles and planetary ions that participate in the plasma interaction. These measurements are discussed in Chapter 14. More detailed information about the missions and instrumentation to make relevant plasma measurements at Mars can be found in reviews by Luhmann et al. (1992a), Barabash and Lundin (2006), and Brain (2006) and references therein.

Most recently, the MAVEN spacecraft arrived at Mars in September 2014 and began measuring the interaction of the Martian upper atmosphere with its space environment. MAVEN carries nine science instruments designed to measure the drivers of atmospheric escape (solar EUV flux, solar wind, IMF), the reservoirs for escape (neutrals and charged particles in the thermosphere, ionosphere, and exosphere), and escaping atmospheric particles and processes (neutral and charged) (Jakosky et al., 2015a; Lillis et al., 2015). Preliminary results from MAVEN have just been published at the time of delivery of this chapter for publication. Several notable results are incorporated where possible in this chapter. Additional key early results are summarized in Jakosky et al. (2015b).

Conclusions about the importance of atmospheric escape to space over time come primarily from measurements of atmospheric isotopes. These measurements, as well as other measurements indicating the conditions on early Mars, are described in detail in Chapter 17.

Insight into the Martian plasma interaction and escape processes has not come exclusively from spacecraft missions to Mars. There are many similarities between the plasma interactions at Mars and Venus, for example, and to first order the Martian interaction is often considered to be Venus-like. Perhaps the most relevant non-Mars missions, therefore, have been Pioneer Venus Orbiter and Venus Express, which have carried more complete particle and field instrument suites than any Mars mission other than perhaps the short-lived, high-pericenter Phobos mission. For a recent review of the Martian plasma interaction in context with Venus (and other bodies), consult Bertucci et al. (2011).

15.2.2 Relevant Modeling Approaches

Models are important tools for placing spacecraft measurements in context (both spatially and in the parameter space of conditions that drive the system), and for probing cause and effect in the physics underlying the plasma interaction and atmospheric escape. Individual model *results* are discussed throughout this chapter, where appropriate. Here we describe the different modeling *approaches* that are employed when studying plasma and atmospheric escape near Mars. Models fall into two general categories: global "system in a box" models for the entire interaction region from the unperturbed solar wind down to (and sometimes including) the Martian ionosphere and thermosphere; and non-global models developed to describe a subset of the regions or processes within the system.

Global models seek to describe the entire plasma interaction region near Mars, and have been used for decades. The earliest models (e.g. Dryer and Heckman, 1967; Spreiter et al., 1970) used gas dynamic assumptions, where the solar wind is treated as a fluid flowing past a blunt obstacle. The magnetic field carried by the solar wind does not modify the flow, and as fluid plasma parcels are diverted around the obstacle the field drapes around the planet. These models were used to compare to early spacecraft observations of the Martian bow shock in order to infer the nature and size of the obstacle, and to assess the draping structure of magnetic fields around the obstacle (see discussion and references in Luhmann et al. (1992a)).

Today, such models employ either magnetohydrodynamic (MHD) or hybrid assumptions. MHD models implement equations for the motion of electrically conducting *fluid* subject to electromagnetic forces, while hybrid models strive to include ion kinetic effects by tracking *individual* ion motion (still treating the electrons as a neutralizing fluid). All currently active models for Mars include electrons, solar wind protons, and two or more planetary ion species (such as O^+, O_2^+, or CO_2^+). MHD models can track the various species as a single fluid or

as multiple fluids, and can assume force-free magnetic fields ($E - v \times B = 0$) or include additional terms in Ohm's law (such as the $j \times B$, or Hall, term). Hybrid modelers can make similar choices about which terms to include from Ohm's law, and all hybrid models for Mars at present include the Hall term. There is some question about which model type is most appropriate for the Martian plasma interaction. On the one hand, the small scale sizes in the Martian interaction region relative to the scales over which ions gyrate around magnetic field lines suggest that the kinetic effects captured by hybrid models should be important. On the other hand, fluid models appear to adequately reproduce many characteristics of the interaction region. They are particularly well suited for the collisional regions of the upper atmosphere and ionosphere, and often have higher spatial resolution that can better capture the details of the interaction near the planet. It may also be that plasma waves in the interaction region perform a function analogous to collisions in a non-magnetic fluid, making the motion of plasma past the planet more fluid-like than it might be otherwise.

The hybrid and MHD plasma models currently employed for Mars have many differences in implementation. But they differ most in their treatment of the inner boundary to the simulations. Some models (e.g. Harnett and Winglee, 2006; Kallio et al., 2010) assume the structure and state of the ionosphere at the model inner boundary, while others simulate the ionosphere's formation from background neutral profiles provided from thermospheric models (e.g. Modolo et al., 2006; Ma and Nagy, 2007; Terada et al., 2009; Najib et al., 2011; Brecht and Ledvina, 2012; Dong et al., 2014). Comparison of the results obtained from the different models run for identical input conditions suggest that the differences in model physics, implementation, and boundary conditions can have significant effects on simulation results (Brain et al., 2010b). Ledvina et al. (2008) and Kallio et al. (2011) provide a thorough discussion of the physics and implementation choices available to global plasma modelers.

Several models that cannot be classified as global MHD or hybrid have proved useful for probing the motion of plasma near Mars. Test particle models, for example, trace the motion of millions or billions of individual particles subject to Lorentz motion in the magnetic and electric fields output from the global models in order to probe the angular distributions of ions in certain regions, or the likely source regions of escaping ions or sputtered neutrals (e.g. Luhmann and Kozyra, 1991; Fang et al., 2010a,b; Curry et al., 2013). Electron transport models follow the motion of electrons along individual flux tubes near Mars in order to identify possible source regions of auroral-like electrons identified in crustal field cusps or the consequences of these electrons when they impact the collisional atmosphere (e.g. Liemohn et al., 2003; Lillis et al., 2011). And ion tracing codes have been employed to describe ion heating, transport, and escape in a two-dimensional (2D) slice of the ionosphere (Andersson et al., 2010).

Most of the model approaches described above have been used to simulate the escape of planetary *ions* from Mars. But, as we discuss in Section 15.4, ion escape may be only a small fraction of the total atmospheric escape. Other modeling approaches have been employed to expressly simulate neutral escape processes. These include both one- (1D) and three-dimensional

(3D) models and calculations for photochemical escape and sputtering, most recently using Monte Carlo techniques (Fox and Haċ, 2009; Valeille et al., 2009a,b; Yagi et al., 2012).

15.3 PLASMA ENVIRONMENT

15.3.1 Steady-State Interaction

Mars is a member of the class of Solar System objects that lack an intrinsic global magnetic field, but, through interaction with their surroundings, generate a magnetic field. This magnetic field is capable of diverting a large fraction of incident plasma before it can collide with the object's atmosphere or surface. Such objects (including Venus, Titan, comets, and Europa) form what is termed an *induced* magnetosphere.

The main obstacle to incident plasma in an induced magnetosphere is fundamentally the same as the obstacle in an intrinsic magnetosphere such as Earth's: magnetic fields divert the plasma around the planet. Crustal magnetic fields contribute to the deflection in some regions at Mars. In other regions, the IMF carried by the solar wind induces currents (via Faraday's law) in the conducting ionosphere that generate magnetic field (via Ampere's law). The shielding magnetic fields deflect solar wind particles around the obstacle (via the Lorentz force), forming a plasma wake behind the planet. This is achieved in a collisionless environment. Note that Faraday's law requires that the IMF constantly vary for the above to work. Otherwise ionospheric currents would not be induced and the IMF may penetrate directly into the upper atmosphere, carrying the solar wind with it (see Luhmann, 1995). In reality the IMF is quite variable (e.g. Russell, 2001), but its magnetic field is still occasionally able to diffuse into the ionosphere. Additionally, ionospheric currents may also be induced by the electric field carried by the solar wind (Podgornyi et al., 1982), in a direction perpendicular to those created via Faraday's law. Analyses of the magnetic field orientation in the ionosphere of Venus suggest that both mechanisms contribute (Luhmann, 1992; Dubinin et al., 2013).

A variety of distinct plasma boundaries and regions result from the interaction described above (see review by Nagy et al., 2004). They are shown in the cartoon in Figure 15.1, summarized in Table 15.2, and described in further detail here, beginning with the unperturbed solar wind and proceeding in a direction away from the Sun, from day to night. Subsequent sections discuss time variability and crustal magnetic field influence on the plasma interaction.

15.3.1.1 Upstream Region

Spacecraft measurements show that the solar wind at ~1.5 AU has density ~1–3 cm^{-3} and travels with a median speed of 300–400 km s^{-1} (Fränz et al., 2007). Typical solar wind proton and electron temperatures are 10–20 eV and ~2–10 eV, respectively. The upstream IMF has a typical field strength of ~3 nT, and is oriented ~56° to the Mars–Sun line, on average, according to the expected Parker spiral direction (Brain et al., 2003). Based on the parameters provided in Table 15.2, the typical solar wind flow at ~1.5 AU is super-Alfvénic (Alfvén speed ~50 km s^{-1}) and super-magnetosonic (Mach number ~6), with a dynamic pressure of 0.5–0.8 nPa (about 45% and 22% of the typical dynamic pressure at Earth and Venus, respectively). Yet this flow has slowed considerably by the time it approaches close to and is diverted around Mars in the underlying plasma interaction region.

The Martian exosphere extends upstream of the shock, and exospheric neutrals can be ionized (primarily through charge exchange and photoionization) and accelerated by the solar wind. These exospheric "pickup ions" have a number of observable effects on their local environment. First, their interaction with the passing solar wind generates electromagnetic plasma waves (at the local gyrofrequency) which can be detected by spacecraft magnetometers (Russell et al., 1990). The precise generation mechanism is still debated (Delva et al., 2011), as is the role of energization and transport of exospheric neutrals to large planetary distances before being ionized (Wei and Russell, 2006). Second, pickup ions can be efficiently reflected from the bow shock, allowing them to gain additional energy from the solar wind before passing through the shock (Dubinin et al., 2006a). Third, pickup ions may act to "mass-load" the solar wind upstream from the shock, slowing it by ~5% (Kotova et al., 1997). The term "mass-loading" refers to a situation where slow-moving mass is added to a flowing plasma, decelerating it via conservation of momentum. Finally, exospheric pickup ions can be directly measured *in situ* by spacecraft ion instruments (Rahmati et al., 2015).

Similar upstream solar wind deceleration, though less extreme, has been observed at Earth, which lacks an extended neutral exosphere. There it is thought that solar wind ions reflected from the shock cause the mass-loading (Bame et al., 1980). This occurs in a region magnetically connected to the bow shock – the foreshock. The Martian foreshock is populated by plasma waves generated at the shock, similar to the foreshock regions at most planets (Brain et al., 2002). Several ion populations have been identified in the foreshock, including newly created planetary pickup ions, ions traveling along magnetic field lines, and ions reflected from the shock (Yamauchi et al., 2012); the latter population is influenced in part by the curvature of the bow shock (Richer et al., 2012). The foreshock also contains outwardly propagating regions of hot plasma propagating upstream similar to those observed at Earth (Øieroset et al., 2001). These "hot flow anomalies" result from the interaction of discontinuities in the IMF with the shock, and can cause significant plasma boundary motion in the terrestrial magnetosphere (Eastwood et al., 2008a).

15.3.1.2 Bow Shock

Figure 15.1 shows that a collisionless bow shock is the outermost plasma boundary in the Martian interaction region. A bow shock forms when there is an underlying obstacle that diverts the incident solar wind flow, and as such may be used to infer the size of the obstacle. It marks the transition from supersonic to subsonic solar wind flow. The bow shock lies closest to Mars near the subsolar point[1] at ~1.6 R_M from the planet's

---

[1] More precisely, the shock should lie closest to the planet at the sub-flow point. Even when the flow from the Sun is strictly radial, the aberration of the solar wind in the reference frame of Mars rotates the entire plasma interaction region by a few degrees. Here we use "subsolar" and "sub-flow" interchangeably.

Table 15.2. *Characteristics of the main regions and plasma boundaries near Mars.*

Feature	Location[a]	Characteristics
Upstream solar wind	Exterior to bow shock	~400 km s^{-1} protons and electrons with ~3 nT magnetic field. Dominated by dynamic pressure
Foreshock	Upstream region magnetically connected to the shock	Filled with waves, incident solar wind, and particles reflected/ launched from the shock
Bow shock	~1.6 R_M from Mars center near subsolar point ~2.6 R_M near terminator	Boundary below which incident plasma becomes subsonic. Located close to Mars, relative to a magnetized planet
Magnetosheath	Between shock and IMB, extending into the wake	Filled with mostly shocked solar wind plasma. Dominated by thermal plasma pressure, and filled with ULF waves. Magnetic field strength increases throughout as plasma velocity decreases due to mass-loading
Induced magnetosphere boundary (IMB)	~1.3 R_M from Mars center near subsolar point ~2.1 R_M near terminator	Boundary separating region dominated by solar wind plasma from region dominated by planetary plasma. Marked by changes in ion composition, electron energy spectra, and magnetic field strength, oscillations, and draping. Likely the obstacle to the solar wind that sets the location of the bow shock. Extends to nightside where it is referred to as wake boundary
Induced magnetosphere	Region between IMB and ionopause	Filled with planetary plasma, and dominated by magnetic pressure. Less turbulent than the magnetosheath. Field lines "drape" around the ionosphere
Ionopause	~450 km dayside altitude, increasing near the terminator and decreasing or disappearing on nightside	Location below which thermal plasma pressure becomes dominant. Marked by rapid increase in thermal electron densities with decreasing altitude
Ionosphere	Region below ionopause	Region of cold thermal planetary plasma created by photoionization of upper atmospheric neutrals. Very tenuous on the planetary nightside
Induced magnetotail	Region below wake boundary (nightside extent of IMB)	Nightside extent of induced magnetosphere, filled with planetary plasma. Consists of two lobes of oppositely directed weak magnetic field
Tail current sheet	Interface between magnetotail tail lobes	Plasma sheet filled with energetic planetary ions

[a] R_M = Mars radius.

center, and flares away from the planet so that it lies at ~2.6 R_M at the terminator (Trotignon et al., 2006; Edberg et al., 2008, and references therein). The shock distance is small relative to the size of the planet, similar to the shock at Venus. Mars is smaller than Venus, so that the shock is also physically small compared to relevant plasma length scales. For instance, the Martian magnetosheath has thickness on the order of a solar wind proton gyroradius (the size of a charged particle's helical motion about a magnetic field line), so that solar wind ions may have insufficient room to thermalize before encountering the obstacle (Moses et al., 1988). Observations suggest that the shock is asymmetric in the plane containing the IMF as well as the orthogonal plane containing the solar wind electric field (the "north–south" and "dawn–dusk" directions in MSE[2] coordinates) (Zhang et al., 1991b; Vignes et al., 2002). Proposed causes of the asymmetries include mass-loading effects, kinetic effects of incident solar wind ions, and differences in the angle the IMF makes with the shock surface in different regions (i.e. quasi-parallel versus quasi-perpendicular effects, where parallel and perpendicular refer to the angle between the local magnetic field and the normal to the shock surface). Global plasma models (especially those that include kinetic effects) might be used to address this issue. Hybrid simulations of the Venus and Martian plasma interaction, however, do not reproduce the observed asymmetries (Brecht, 1990; Modolo and Chanteur, 2008). The shock has structure similar to that observed at other planets, including foot, ramp, and overshoot regions, which have characteristics that scale with fundamental plasma parameters (Tatrallyay et al., 1997).

---

[2] MSE refers to the "Mars–solar–electric field" coordinate system, where the center of Mars is the origin, the x-axis points toward the Sun, and the z-axis points in the direction of E_{sw}. Similarly, "Mars–solar–orbital" or "MSO" coordinates have the x-axis point toward the Sun and the z-axis point northward out of Mars' orbital plane.

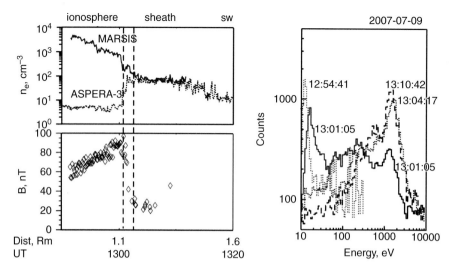

Figure 15.2. Mars Express measurements of the change in plasma parameters as the spacecraft crosses the IMB. (Top left) Measured suprathermal (ASPERA-3) electron densities increase outside the IMB, while thermal densities (MARSIS) decrease. (Bottom left) Magnetic field strength decreases abruptly outside the IMB for this orbit. (Right) Ion spectra change from predominantly planetary ions below the IMB (before 13:01:05) to predominantly solar wind after. From Dubinin et al. (2008a).

15.3.1.3 Magnetosheath and Induced Magnetosphere Boundary

The magnetosheath is a region dominated by shocked solar wind plasma. It is characterized by a strong thermal plasma pressure component; solar wind ions are slowed as they pass through the shock, and the magnetosheath plasma is hotter and denser than the upstream plasma. The flow continues to slow as it travels deeper into the magnetosheath, conserving momentum as new planetary ions are added from the extended exosphere via mass-loading. As the plasma slows, the magnetic field that it carries increases, "piling up" around the planetary obstacle and extending around to the nightside of the planet to form the boundary to the induced magnetotail (Bertucci et al., 2003). The small spatial scale of the sheath relative to typical solar wind gyroradii indicates that kinetic plasma processes may be important there. Indeed, the sheath is filled with signatures of low-frequency plasma waves that can be interpreted using kinetic plasma wave theory (Espley, 2004). Some waves are likely generated locally in the sheath due to interactions between the different ion populations; others may be generated at the bow shock and propagate inward along magnetic field lines.

The boundary at the bottom of the magnetosheath, where the solar wind is at last completely deflected by the planetary obstacle to its flow (see Figure 15.1), has been a topic of some controversy. Different spacecraft missions and instruments have observed it in different ways (based on measurements of ions, electrons, and magnetic field) and it has proved difficult to determine whether there is a single boundary in the plasma, or several that are physically distinct from each other. But most investigators would agree that this boundary region (which we refer to here as the induced magnetosphere boundary, or IMB) marks a narrow transition from a region dominated by solar wind protons to one dominated by planetary plasma. The boundary also marks: a transition from a region dominated by thermal pressure to one dominated by magnetic pressure; a sharp decrease in solar wind proton fluxes; a change in ion composition; an increase in magnetic field strength; a decrease in magnetic field fluctuations; a change in plasma wave characteristics; an increase in the "draping" of the magnetic field; and a change in the suprathermal electron energy spectrum (Bertucci et al., 2003) (Figure 15.2). These signatures are closely spaced (suggesting they may all be associated with the same boundary). Further, it is not yet fully understood which signatures help to form or maintain the IMB and which signatures result from its existence – though it seems clear that the presence of significant planetary ion densities are required for the IMB to form.

Insight into the nature of this boundary should come from three sources. First, simulations have reached the point at which they are able to spatially resolve this boundary and probe its physics. Already, we have learned from simulations that, in the gas dynamic limit, the IMB acts as the planetary obstacle to the solar wind flow (Crider et al., 2004). Simulations coupled with theory have also shown how the IMB is naturally maintained due to the Lorentz forces acting on external solar wind ions (deflecting them around the IMB) and internal planetary ions (keeping them inside) in the presence of electric fields in the Martian system (Simon et al., 2007), while those planetary ions that are outside are removed. This mechanism predicts that the boundary should be more distinct on the side of Mars opposite the direction of the electric field in the solar wind. Second, a similar boundary is seen at other induced magnetospheres, such as those of Venus, comets, and Titan. Insight gained from studies of these objects should be applicable to Mars (see review by Bertucci et al., 2011). Finally, new measurements from spacecraft containing a complete suite of particle and field instruments are planned over the next several years, which should allow, for the first time, simultaneous measurement of all the identified signatures of the IMB.

15.3.1.4 Induced Magnetosphere and Ionopause

Below the IMB lies the induced magnetosphere, a region of quiescent but strong magnetic field, dominated by planetary

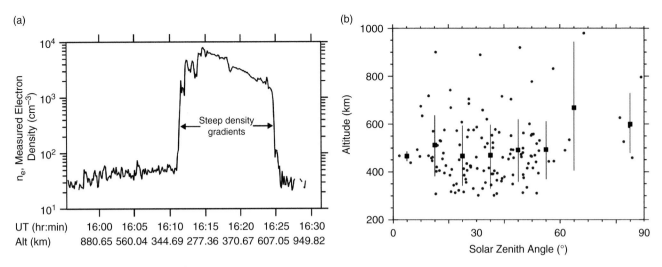

Figure 15.3. Signatures of the Martian ionopause. (a) Large and sudden increases in thermal electron density measured by MARSIS at the top of the ionosphere. (b) Average ionopause height as a function of solar zenith angle. From Duru et al. (2009).

ions. Though the density of solar wind ions markedly decreases in this region, the total plasma density must increase because the momentum in the plasma is conserved by planetary ions that are colder than the shocked solar wind. Observations and simulations for Venus and Mars confirm this expectation (Lundin et al., 1991; Bauske et al., 1998). Some of the planetary ions in the induced magnetosphere are carried away from the planet and escape – it is the region from which the majority of ion escape is believed to occur. Other planetary ions are carried toward the planet itself, and precipitate near the exobase (~200 km altitude). The magnetic field is typically higher in the induced magnetosphere than in surrounding regions (excepting crustal field regions), transported inward through the IMB. Since the obstacle for solar wind protons appears to be the IMB, strong currents are likely to run throughout the induced magnetosphere. However, there have been no spacecraft with instruments capable of directly measuring currents in this region. Though plasma oscillations with small amplitude have been observed there (e.g. Bertucci et al., 2003), they have a variety of observed characteristics and have proved difficult to identify (Espley, 2004).

At the base of the induced magnetosphere lies the ionosphere – a region of relatively cold plasma that, on the dayside, is created chiefly by photoionization of thermospheric species. Ionospheric structure and composition are described in detail in Chapter 14. Here we describe the upper boundary to the ionosphere. Two definitions for this boundary have been employed based on observations. The first relies on the fact that the photoionization of atmospheric CO_2 and atomic oxygen produces electrons with specific energy (Mitchell et al., 2000). The presence of these photoelectrons is used as a tracer of the ionosphere (Mitchell et al., 2000, 2001; Dubinin et al., 2006b; Frahm et al., 2007), and the location between regions where photoelectrons are evident and where they are not is termed the "photoelectron boundary" (Nagy et al., 2004). In some observations this boundary is located very close to the IMB, and in others there is a clear separation between the two boundaries (Mitchell et al., 2001). Observational identification of the

photoelectron boundary relies on the absence of other electron populations with comparable flux in the relevant energy range, such as shocked solar wind electrons. Since electrons travel quickly along magnetic flux tubes, the photoelectron boundary can therefore be thought of as a topological boundary separating flux tubes containing significant quantities of solar wind electrons from flux tubes containing mostly ionospheric photoelectrons.

At Venus, the upper boundary to the ionosphere is called the ionopause, defined as the location where ionospheric electron density decreases rapidly or (with similar results) as the location where ionospheric thermal pressure is balanced by the magnetic pressure in the induced magnetosphere – see discussion in Phillips et al. (1988). At Mars, farther from the Sun and with less robust atmospheric photoionization compared to Venus, the ionospheric thermal pressure is typically insufficient to balance the magnetic pressure from above, and so the field penetrates into the ionosphere, dominating the pressure there (Dubinin et al., 2008b). Steep gradients in local electron density similar to those observed at Venus are sometimes evident in MEX observations, indicating a transient ionopause (Duru et al., 2009) (Figure 15.3). In most cases, the photoelectron boundary and ionopause coincide. In others, the ionopause is situated at higher altitudes. The average ionopause altitude is almost constant on the dayside at around 450 km, and may increase near the terminator.

15.3.1.5 Dayside Pressure Balance

Before proceeding to a discussion of nightside plasma regions, it is appropriate to mention that a convenient way to distinguish the different dayside plasma regions near Mars is to consider that the total pressure at any given location in the Martian system should be conserved. Upstream from the planet, the solar wind is dominated by dynamic pressure ($\sim \rho v^2$). The solar wind is slowed and heated at the bow shock, so that thermal plasma pressure ($\sim nkT$) becomes dominant in the magnetosheath. Colder planetary plasma is added to the system in significant

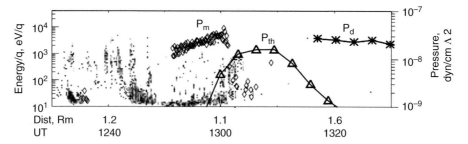

Figure 15.4. Pressure balance throughout the dayside Martian plasma interaction region, demonstrated by MEX measurements. Magnetic (P_m), thermal (P_{th}), and dynamic (P_d) pressure terms are shown as a function of time for a single spacecraft trajectory passing near the Martian subsolar point. The spacecraft passed from deep inside the interaction region at ~12:50 UT to upstream from the bow shock at 13:30. The background shading shows heavy-ion fluxes as a function of time and energy (magnitude scale not shown). From Dubinin et al. (2008b).

quantities below the IMB, and the plasma flow is still slow. Therefore, the dominant pressure term in the induced magnetosphere is magnetic pressure (~$B^2/2\mu_0$ from induced magnetic fields (or IMF that has diffused below the IMB). Magnetic pressure continues to dominate below the ionopause, except for periods when the ionosphere is unmagnetized (at these times ionospheric thermal pressure dominates). The exception at Mars is when and where its crustal fields contribute to the magnetic pressure near the boundary (further discussed in Section 15.3.3). Though the concept of pressure balance was used for many years to explain the differences between regions in induced magnetospheres using theory (Zhang et al., 1991a) and simulation (Liu et al., 1999), it was only recently demonstrated using observations that pressure is in reality at least approximately conserved near Mars (Dubinin et al., 2008b) (Figure 15.4).

15.3.1.6 Nightside Ionosphere

(The reader should also see Chapter 14.) The Martian nightside ionosphere is quite tenuous, similar to Venus during solar minimum conditions, with typical peak densities on the order of 10^4 cm^{-3} and temperatures near the exobase of ~0.1 eV (Fowler et al., 2015), depending upon the location relative to crustal fields. The total electron content of the nightside ionosphere has been mapped using MEX MARSIS measurements (Safaeinili et al., 2007), and is on the order of a few×10^{10} cm^{-2}. The primary source for ionospheric plasma deep on the nightside is believed to be charged particle impact; transport from the dayside can supply some ionospheric plasma in nightside regions near the terminator (Němec et al., 2010).

15.3.1.7 Magnetotail and Current Sheet

Moving upward from the ionosphere on the nightside, an induced magnetotail forms in the Martian wake region from IMF field lines draped around (and through) the dayside ionosphere and stretched anti-sunward behind the planet. The magnetic field in the tail is weak (~10 nT), though it is still stronger than the unperturbed IMF (Yeroshenko et al., 1990). The draped field forms two lobes with opposite magnetic polarity, separated by a current sheet. The current sheet and tail lobes are surrounded by the wake boundary, a nightside extension of the IMB.

The orientation of the tail current sheet is determined primarily by the orientation of the upstream IMF; for an IMF

parallel to the Martian equatorial plane, the current sheet should be oriented in the noon–midnight meridian. Well-defined current sheets consistent with an induced magnetotail configuration are sometimes evident as low as ~400 km on the nightside, though their non-uniform geographic distribution suggests that crustal fields may play some role in their presence at low altitudes (Halekas et al., 2006). Simulations and perhaps some observations suggest that crustal fields certainly have a strong influence on the morphology of the Martian tail region, especially at low altitudes (Ma et al., 2002; Luhmann et al., 2015).

The tail contains spatially structured plasma populations (Kallio et al., 1995; Fedorov et al., 2006; Harada et al., 2015a). Throughout the region inside the wake boundary, planetary ions dominate spacecraft measurements and simulations (Figure 15.5). The current sheet (also called plasma sheet) is filled with planetary ions, accelerated to keV energies or more. These ions are more energetic and have higher fluxes in the "northern" (in MSE coordinates) portion of the plasma sheet, due to efficient acceleration from ion pickup processes (Barabash et al., 2007). Outside of the plasma sheet, planetary ions tend to have lower energy in the central tail, and higher energies toward the periphery (Kallio et al., 1995). Planetary ions are gradually accelerated to speeds in excess of 100 km s^{-1} over the first few Martian radii downstream, and each species reaches approximately the same energy (so that they travel at different speeds) (Nilsson et al., 2010). Global hybrid and test particle simulations also show a spatial variation of planetary ions in the wake region, with a large, energetic, pickup ion "plume" apparent in the "northern" hemisphere, and a lower energy channel of escaping ions in the "southern" portion of the plasma sheet (Brecht and Ledvina, 2006; Fang et al., 2008). The existence of the ion plume as an escape channel was recently observationally verified by Dong et al. (2015b). The spatial variation suggests that different acceleration mechanisms are important in different locations.

15.3.2 Time-Dependent Effects

Though the basic structure of the plasma interaction region can be described using a steady state, it is highly time-dependent in reality. The external conditions constantly and chaotically vary, and the inner boundary of the interaction region rotates with the

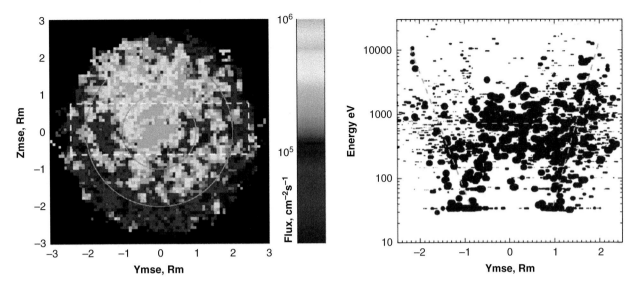

Figure 15.5. Structure of planetary ion fluxes in the Martian magnetotail. (Left) Flux of planetary heavy ions measured by MEX, projected into a plane parallel to the Martian terminator and organized by the solar wind convection electric field (which points in the $+z$ direction here). (Right) Energies of ions measured between the dashed white lines at left, as a function of the location across the magnetotail. From Fedorov et al. (2008).

planet. These changes are reflected throughout the entire system. Many of the most interesting phenomena that occur near Mars result from the constant adjustment of the environment to these ever-changing conditions.

15.3.2.1 Response to Drivers

A number of drivers have been identified that determine the configuration of the plasma environment at any given time. Each of the main drivers is discussed below.

Interplanetary Magnetic Field. The IMF can be characterized by its orientation, strength, and degree of variability. The IMF clock angle (angle in the MSO y–z plane, approximately perpendicular to the solar wind flow direction) determines the direction of the solar wind electric field, since $E_{sw} = -v_{sw} \times B_{IMF}$, and v_{sw} is typically directed approximately in the MSO $-x$ direction. The clock angle therefore influences the trajectory of pickup ions, and is one of the main drivers of asymmetry in the IMB and bow shock. The IMF cone angle (angle with respect to the Mars–Sun line) determines the location of the foreshock region, and which regions of the shock are quasi-parallel versus quasi-perpendicular (which can influence the structure and waves in the magnetosheath). Cone angle has been demonstrated at Venus to control the nature of plasma waves in the magnetosheath (e.g. Du et al., 2010), and may control the pathways for ion escape as well (Masunaga et al., 2011). Hybrid simulations by Modolo et al. (2012) using time-dependent input conditions demonstrated that a rotation in the IMF causes the entire interaction region to rotate in response (as expected), but with a "lag" of as much as 2 minutes in some regions (Figure 15.6). The typical Parker spiral configuration of the magnetic field causes asymmetric draping of the IMF around the obstacle, creating "dawn–dusk" (in MSE coordinates) asymmetries throughout the system. The strength of the IMF influences the plasma environment more indirectly by influencing the speed at which a disturbance can travel along magnetic field lines (the Alfvén speed), and in turn the Alfvénic

and magnetosonic Mach numbers that help to determine the shape of the bow shock. Finally, the variability in the IMF affects the degree to which shielding currents can be generated in the ionosphere (and consequently the degree to which magnetic field can diffuse inward). Sudden changes in IMF may also briefly enhance bulk atmospheric escape rates in induced magnetospheres (Ong et al., 1991), discussed in Section 15.3.

Solar Wind. The solar wind density and velocity determine the incident dynamic pressure at Mars. Since there is pressure balance throughout the dayside interaction region, the density and velocity play a large role in determining the size of the different plasma boundaries, the plasma temperature in the magnetosheath, and the strength of the magnetic field in the induced magnetosphere. At Venus, the solar wind pressure determines whether the ionosphere is magnetized (Russell and Vaisberg, 1983); a similar result is expected for Mars, but has not yet been demonstrated. The solar wind density and temperature also determine the sound speed, which in turn influences the magnetosonic Mach number (and therefore the bow shock).

Solar Photon Flux. Solar extreme ultraviolet (EUV) photons are the primary source of the ionosphere, which is the inner boundary of the plasma interaction region. Ionospheric densities and total electron content have been clearly correlated with EUV flux (see discussion in Chapter 14). The EUV flux therefore determines the ionospheric thermal pressure, which is ultimately responsible for balancing the incident pressure of the solar wind. Mars' orbital eccentricity is large enough that a constant solar output would still result in considerable variation in EUV fluxes at Mars; differences in the ionosphere have been observed and predicted during different Martian seasons (see discussion in Chapter 14). EUV flux has also been demonstrated to have effects that reach beyond the ionosphere. The altitudes of the IMB and bow shock have been successfully correlated with EUV flux (Edberg et al., 2009b), though attempts to correlate the ionopause–photoelectron boundary location with EUV flux have been unsuccessful (Mitchell et al., 2001).

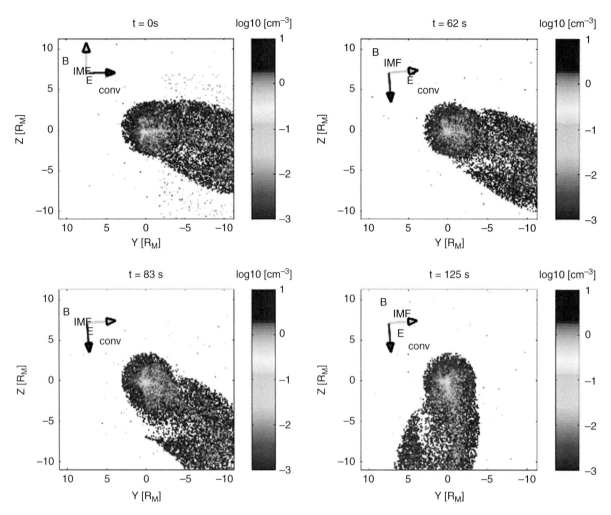

Figure 15.6. The configuration of O$^+$ ions in the Martian magnetotail from a simulation by Modolo et al. (2012). In the time-dependent simulation, a 90° rotation in the IMF was introduced by 62 s (top right), but the tail did not fully reconfigure until ~125 s (bottom right).

Mars Orientation. The orientation of Mars would not be a major factor in controlling variability in its plasma environment if it lacked crustal magnetic fields. However, the crustal fields rotate with the planet once every ~24.7 hours, and are strong enough that they influence the altitude of all the major plasma boundaries at Mars. Their effects are numerous, and are discussed in more detail in Section 15.3.3.

It is evident from the above descriptions that the drivers influence every aspect of the Martian system, including the location of plasma boundaries, the configuration of magnetic fields, the inner ionospheric boundary, and the trajectory and flux of escaping planetary ions. Simulations are especially well suited to studies of the effects and relative importance of the individual drivers (Brecht and Ledvina, 2006; Harnett and Winglee, 2006; Modolo et al., 2006; Ma and Nagy, 2007; Kallio et al., 2008; Fang et al., 2010a; Najib et al., 2011; Dong et al., 2014). Spacecraft observations have also yielded a number of observed effects, and quantified them (recent references include Lundin et al. (2008) and Nilsson et al. (2011)). Studies of plasma boundary locations, in particular, are plentiful; recent studies have focused on quantifying the effect that different drivers have on their location in isolation (Edberg et al., 2009b).

15.3.2.2 Shear-Related Processes

The response of the Martian system to the drivers described above is time-dependent only because the drivers themselves vary as a function of time. If the drivers did not vary, or if they only varied episodically, then the Martian system would reach equilibrium, as simulations with stable input conditions frequently demonstrate. However, there is a robust suite of dynamical processes that occur at Mars that do not require variations in the boundary conditions to operate. Many of these processes are caused by shear between different plasma regions.

One of the most frequently discussed shear-related process for unmagnetized planets is the Kelvin–Helmholtz (KH) instability (Wolff et al., 1980), which results when there is velocity shear between two fluids. KH instability has been reported at Earth's magnetopause (Hasegawa et al., 2004), and should also grow at the Martian ionopause and at crustal magnetic fields (Penz et al., 2005). Ultimately, the growth of these instabilities would lead to the mixing of plasmas across the shear boundary and, in the case of Mars or Venus, to detached ionospheric clouds. Though ionospheric plasma has been detected above the ionopause at both Venus and Mars (Brace et al., 1982; Cloutier

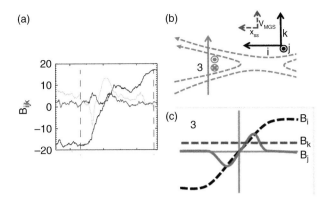

Figure 15.7. Evidence for magnetic reconnection at Mars. (a) MGS vector magnetic field observations, expressed in a minimum variance coordinate system, show a current sheet with embedded bipolar magnetic field signatures. This can be explained if the spacecraft passed through a current sheet along the trajectory shown in (b), sampling Hall magnetic fields in the reconnection exhaust region and yielding measurements similar to those diagrammed in (c). From Halekas et al. (2009).

et al., 1999), it is difficult to verify that the clouds are detached using only one spacecraft; it is possible that the observations could be explained by a boundary in motion or with an irregular shape, or by the external field rotations mentioned earlier as proposed in Ong et al. (1991). However, both theory and simulations show that shear instabilities should operate at Mars and Venus, efficiently removing atmospheric plasma (Terada et al., 2002; Penz et al., 2005). Density fluctuations observed in the Martian ionosphere are consistent with a KH instability formation mechanism (Gurnett et al., 2010), though the nearby field orientations sometimes used as a hallmark of KH instability have not been simultaneously measured.

A shear-related process of considerable interest in space plasmas is magnetic reconnection. In a simplified view, reconnection can occur when the magnetic field on either side of a plasma boundary or layer is oppositely aligned. In this case, the "frozen-in" condition of magnetic field to plasma can be violated over small spatial scales (in a "diffusion region"), allowing magnetic field lines to reconfigure and the isolated plasma populations to mix. In reality the magnetic field on either side of the plasma boundary does not have to be antiparallel, so that reconnection can occur in a variety of magnetic morphologies. Reconnection is thought to be a source of efficient charged particle acceleration, converting magnetic potential energy to particle kinetic energy, and is thought to play an important role in plasma dynamics throughout the Solar System and Universe. The first *in situ* measurement of a reconnection diffusion region outside Earth's magnetosphere was reported at Mars based on magnetic field observations in the nightside current sheet (Eastwood et al., 2008b). Since this first report, many similar events have been identified (Halekas et al., 2009), indicating that reconnection is a common process at Mars (Figure 15.7). The extent to which crustal magnetic fields play a role in initiating reconnection is currently unknown; certainly reconnection is made more likely in some regions by the presence of an additional source of magnetic field that can interact with the IMF (Krymskii et al., 2002). However, reconnection may also

occur in regions that lack crustal field influence, such as the tail current sheet (reported in Harada et al., 2015b), or across the ionopause after a rotation in the IMF. Reconnection may play a role in particle acceleration and atmospheric escape near Mars.

A final notable shear-related process is the formation of magnetic flux ropes near Mars. Flux ropes (coiled magnetic field structures) were first observed in the ionosphere of Venus (Russell and Elphic, 1979). They are believed to form through shear interaction across the Venus ionopause that acts to twist magnetic flux tubes, possibly involving a KH instability. Flux ropes are ubiquitous structures in space plasmas, and often form near magnetic boundaries. They are capable of transporting plasma across boundaries, making them a topic of some interest in the study of atmospheric escape at unmagnetized planets. At Mars, three distinct types of magnetic flux rope have been reported. First, small-scale Venus-like ionospheric flux ropes are present in the Martian ionosphere, though less frequently than at Venus (Cloutier et al., 1999; Vignes et al., 2004). Second, small "magnetic islands" were reported by Eastwood et al. (2008b) near reconnection sites in the nightside current sheet. Such islands are simulated and observed in reconnection "exhaust" regions. Finally, large "plasmoid" structures have been reported downstream from strong crustal magnetic fields (Brain et al., 2010a), and may be detached crustal magnetic flux tubes analogous to plasmoids from the magnetosphere seen in Earth's magnetotail. All of these structures have the same essential twisted field morphology, making it problematic to distinguish them from each other in observations (Briggs et al., 2011). However, all are also hallmarks of a dynamic magnetic environment at Mars.

15.3.2.3 Extreme Events

Intense changes in the external drivers of the Martian plasma environment can occur for hours or days. These "solar storms" or "space weather events" can consist of a combination of increases in solar photon fluxes from solar flares, the sudden arrival of an interplanetary shock and the compressed solar wind plasma and field that follows it, the rotation and increase in the strength of the IMF due to passing coronal mass ejection (CME) drivers, and injections of charged solar energetic particles (SEPs) that are generated at solar flares and at the CME shocks. All four effects have observed consequences for Earth's magnetosphere and atmosphere (Baker et al., 2008). And all four have potentially greater effect at an unmagnetized planet with a thinner atmosphere, such as Mars. Solar storms are relatively infrequent today, with a few large storms per solar cycle and perhaps tens of more moderate storms. However, observations of stars similar to but younger than the Sun suggest that solar storms were more frequent and more intense earlier in Martian history (see discussion in Khodachenko et al., 2007). Therefore, study of their effects on Mars is of interest both for their present impact on the Martian atmosphere, and for the accumulated effect they may have had over time.

Solar storms have been detected *in situ* at Mars by both spacecraft instruments and engineering systems. But separating the effects on the Martian environment of storm-related photons, charged particles, and magnetic field can be difficult, and unambiguous interpretation of the effects of solar storms can be problematic in some cases. Measurements of the effects of flare-related photons are perhaps the easiest to interpret. For example, radio occultation results confirm that large X-ray flares cause a clear enhancement in ionospheric density below the main peak of as much as 200% (Mendillo et al., 2006). See further details in Chapter 14.

The effects of solar energetic particles on the plasma environment and upper atmosphere are harder to discern, because they may arrive together with the energetic photons from flares and/or with the solar wind compressions and magnetic fields of a CME. But storm-related SEPs have been identified in observations from five spacecraft at Mars: Phobos 2, MGS, Mars Odyssey, Mars Express, and MAVEN (Crider et al., 2005; McKenna-Lawlor et al., 2005; Futaana et al., 2008; Jakosky et al., 2015c). Both MGS and MEX relied on an indirect method of energetic particle detection: increased background count rates from their charged particle instruments, resulting from SEPs penetrating the aluminum housing of the instrument and causing counts. Using these observations, more than 80 SEP events have been identified over all portions of the solar cycle, including relatively well-known events such as the "Bastille Day event" in 2000, and the "Halloween storms" in 2003. Observations suggest that impacting energetic particles may increase ionization rates in the upper atmosphere. Radar pulses reflected from the Martian surface "disappear" during SEP events (even in darkness, where solar photons should have no influence); the disappearance is interpreted as resulting from a layer of SEP-induced impact ionization between the spacecraft and the ground (Morgan et al., 2006). And an excess of suprathermal electrons are observed streaming away from the atmosphere during large storm events; careful analysis shows that the most likely explanation is electrons produced during the impact of SEPs (Lillis et al., 2012). Additionally, deposition of incident storm-related suprathermal electrons in the thermosphere have been observed to cause widespread diffuse aurora (Schneider et al., 2015). One puzzling issue is that individual profiles of ionospheric electron density do not contain any evidence that SEPs observably increase ionization in the atmosphere (Uluşen et al., 2012). In fact, the profiles suggest that a depletion in the topside ionosphere may occur during these events. These conflicting results have yet to be satisfactorily reconciled.

Large CMEs also pass Mars, carrying enhanced magnetic field with them. The period of elevated magnetic field can last for days after the CME shock has passed, and the entire plasma interaction region is affected by the elevated external pressure (Crider et al., 2005). During this period the ionosphere is compressed and becomes more strongly magnetized (influencing charged particle transport), and the solar wind electric field increases (allowing stronger acceleration of more planetary ions) (Jakosky et al., 2015c). Two results may be attributable to the increased magnetic field, though it is not entirely certain. First, increased plasma wave activity was observed in the ionosphere during a large storm event, possibly near the oxygen gyrofrequency (Espley, 2005). Such activity may result from increased mobilization of planetary ions. Second, the measured ion escape rate during even moderate solar storms may increase by an order of magnitude or more (Futaana et al., 2008; Jakosky et al., 2015c). Model calculations show that the energy input into the upper atmosphere by SEP events is smaller than that from typical EUV flux except for only the largest SEP events (Leblanc et al., 2002). Mars Express ion observations during less extreme corotating interaction region (CIR) passages show that the pressure pulses associated with CIRs increase escape fluxes by factors of anywhere between ~2.5 and >10 (Dubinin et al., 2009a; Edberg et al., 2010).

Global plasma modeling groups have undertaken simulations to determine the influence of solar storm conditions on the plasma environment, focusing primarily on atmospheric ion escape rates in the present epoch and in the past (Harnett and Winglee, 2006; Ma and Nagy, 2007; Terada et al., 2009; Curry et al., 2015; Dong et al., 2015b). Together, they find that escape rates should increase during solar storms. The magnitude of the effect may range from factors of a few to factors of 100 or more depending upon the severity and duration of the disturbed conditions.

Finally, the effects of solar storms should not be limited to the plasma environment and ionosphere of Mars. SEPs range in energy from tens of keV up to hundreds of MeV or more. The most energetic SEPs are capable of reaching the Martian surface, and influencing the surface–atmosphere interactions (via neutron production in the crust), atmospheric chemistry, and the radiation environment for short periods (see discussion in Schwadron et al., 2010; Hassler et al., 2012).

15.3.3 Crustal Field Influence

Three Solar System bodies are known to possess localized magnetized regions in their crust: Earth, the Moon, and Mars[3]. Of these, the crustal magnetic fields at Mars (Figure 15.8) cause the largest signatures at spacecraft altitude, and have the most pronounced effect on their plasma environment. The Martian crustal fields are found over roughly two-thirds of the planet, but are strongest in the vicinity of the oldest portions of the Martian crust – in the southern highlands, particularly in the Terra Cimmeria/Sirenum region (130–230°E longitude). Moderately strong crustal field sources also appear to be preferentially distributed within ~20° of the crustal dichotomy boundary. The magnetization is believed to have been acquired in the presence of an ancient global dynamo magnetic field (Acuña et al., 1998) that shut off more than four billion years ago (Lillis et al., 2008). Today, crustal fields are as strong as 1000 nT near the main peak of the Martian ionosphere, compared to typical induced ionospheric magnetic fields of ~30 nT (see discussion in Brain, 2006). Though crustal fields clearly have local effects, it is being increasingly recognized that they may also influence particles and fields at great distances from their location. In this sense, crustal fields may have a global influence at Mars.

15.3.3.1 Magnetic Pressure and Topology

The presence of crustal fields at Mars modifies the canonical Venus or comet-like plasma interaction in two basic ways. First,

---

[3] The MESSENGER mission to Mercury has revealed localized magnetic signatures from a region near Mercury's north pole, which may also result from crustal magnetization (Purucker et al., 2012).

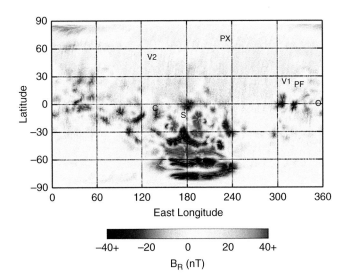

Figure 15.8. Geographic map of Mars showing the location and strength of crustal magnetic fields. Whiskers of average magnetic field strength at 400 km above the Martian nightside are colored according to the strength of the radial (vertical) field component. The length of each whisker is proportional to the log of the horizontal field component. The locations of landed spacecraft are indicated by initials: V1 and V2 for Viking 1 and 2, PF for Pathfinder, S and O for Spirit and Opportunity, PX for Phoenix, and C for Curiosity. A black and white version of this figure will appear in some formats. For the color version, please refer to the plate section.

crustal fields add magnetic pressure to the Martian system at ionospheric altitudes (and above), which helps to counter the pressure of the solar wind. Crustal fields replace or augment the fields formed by induced ionospheric shielding currents in some locations, forming "mini-magnetospheres" that resemble a global magnetosphere in many respects. However, boundary layers and regions of current are observed and simulated to overlie mini-magnetospheres (Luhmann et al., 2002; Harnett and Winglee, 2003), formed from the interaction of the shocked solar wind with the crustal fields. The pressure from the crustal fields alters the shape of the obstacle to the solar wind, perturbing various plasma boundaries upward locally (Figure 15.9, left). Crustal fields influence the location of all main plasma boundaries at Mars, though their influence on the bow shock is relatively weak (Mitchell et al., 2001; Crider et al., 2002; Edberg et al., 2009b; Duru et al., 2010). There is even some evidence that, when crustal fields are situated near the subsolar point, the IMB is pushed outward globally on both the dayside (Brain et al., 2005) and the nightside (as the wake boundary) (Verigin et al., 2001).

Second, crustal fields provide an additional source of magnetic field in the system to interact with the IMF, thereby enabling three different basic topologies for any given flux tube: closed, open, and unconnected (Figure 15.9, right). Closed field lines connect at both ends to the planet, and should shield the upper atmosphere from magnetized charged particles, preventing impact ionization in some regions. Open field lines connect to the planet at only one end, and form cusp-like regions analogous to terrestrial magnetospheric cusp regions. Such regions can provide a conduit for energy and particle exchange between the solar wind and the ionosphere (Krymskii

et al., 2002). Unconnected field lines connect at either end to the IMF, and are the draped field lines usually envisioned in a Venus-like interaction.

All three topologies have been encountered and mapped geographically at 400 km altitudes, using pitch angle distributions of suprathermal electrons (Brain et al., 2007). There are some locations on the nightside above strong crustal fields that the observations indicate are always closed, though we only have robust measurements from a single altitude and local time. Other locations near crustal fields are much more interesting because they are sometimes observed to be open, and other times observed to be closed. Therefore, the magnetic field topology near Mars changes in certain situations. The topology changes have so far been correlated with solar wind pressure and with IMF direction. Changes with pressure may "push" closed field regions below the altitude of the spacecraft so that they are not measured. Changes in IMF direction may facilitate magnetic reconnection, so that some cusp regions alternate between open and closed.

15.3.3.2 Influence on Particle Motion

Crustal magnetic fields should change the direction of charged particles via the Lorentz force. The influence of crustal fields on particle motion is much clearer in observations of electrons than it is for ions. This is not surprising, because electrons have smaller gyroradii than ions, so are more tightly bound to magnetic field lines. Crustal fields have been demonstrated to shield magnetosheath electrons from the ionosphere in closed field regions, while permitting them to access the upper atmosphere and ionosphere directly in cusps (Brain et al., 2005). Crustal fields also modify the angular distribution of electrons by virtue of the magnetic mirror force (or the first adiabatic invariant), which decreases (and eventually reverses) the velocity component of an electron along a magnetic field line as it moves into a region of higher field strength. In this manner, one-sided and two-sided "loss cones" are formed in pitch angle distributions on open and closed field lines, respectively (Brain et al., 2007).

The influence of crustal fields on ion motion is less certain. From the perspective of the canonical picture of the Martian plasma obstacle, if the crustal fields take the place of induced magnetic fields in deflecting the solar wind around the planet, then they should alter the trajectory of solar wind ions. However, observational studies of ion measurements have yielded no clear influence by crustal fields on solar wind ion motion (Dieval et al., 2012). The observational results for *planetary* ion motion are somewhat more promising. Early observations by Mars Express showed evidence for energized planetary ions at low altitudes on the Martian dayside, presumably accelerated away from the planet in cusp regions (Lundin et al., 2004). However, follow-up statistical studies are divided on whether there is a clear influence on planetary ions (Nilsson et al., 2006, 2011; Dubinin et al., 2008d; Lundin et al., 2011).

15.3.3.3 Aurora

A special case of the influence of crustal fields on charged particle motion is discrete aurora – a topic reviewed by Brain and Halekas (2012) and not to be confused with the diffuse aurora

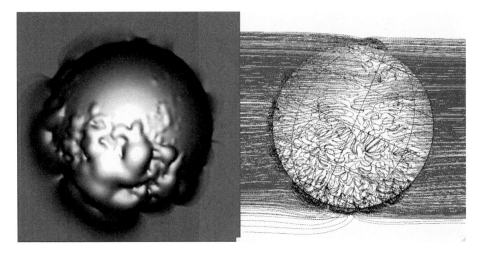

Figure 15.9. The results of calculations showing crustal field influences on the local plasma environment. (Left) The location at which incident solar wind pressure is balanced by a combination of ionospheric thermal pressure and crustal field magnetic pressure is shown as an isosurface. (Right) The configuration of magnetic field lines near Mars shows varied topology of closed (red), open (blue), and unconnected (green) magnetic field. Both images have 180°E as the central longitude. From Brain (2006). A black and white version of this figure will appear in some formats. For the color version, please refer to the plate section.

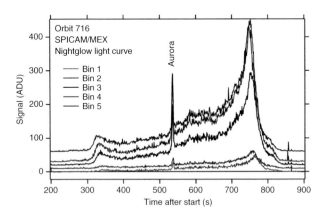

Figure 15.10. First detection of aurora on Mars, by Mars Express. The signal intensity from the SPICAM instrument is shown as a function of time during a nightside limb observation (approaching the terminator at ~750 s). The intensity spike at ~530 s corresponds to auroral emission in a crustal magnetic field cusp. From Bertaux et al. (2005).

mentioned in an earlier section. Ultraviolet auroral emission at Mars was reported soon after Mars Express arrived at Mars with the appropriate instrumentation (Figure 15.10; Bertaux et al., 2005). The emission in nightside crustal cusp regions is discussed in more detail in Chapter 14.

Aurorae on Earth and other planets result when charged particles in the magnetosphere are accelerated downward along magnetic field lines, colliding with and exciting particles in the atmosphere. But the processes that initiate aurorae are still debated today. Electrons with peaked energy distributions that resemble "inverted V" terrestrial auroral electron distributions have been identified in the nightside plasma environment by both MGS and MEX, predominantly near magnetic cusps (Brain, 2006; Lundin et al., 2006). The peaks in energy range from ~200 eV up to 4 keV in a few rare cases. Energized

planetary ion distributions are observed by MEX over the same altitude range (~350–8000 km) and at the same times that auroral-like electron distributions are observed (Lundin et al., 2006). The total energy of the ion and electron peaks is conserved for a given event, with ions carrying more energy at low altitudes, and electrons carrying more energy at high altitudes. These observations are consistent with the existence of an electric field, aligned along cusp magnetic field lines, accelerating electrons downward and ions upward. Such "field-aligned current" regions are commonly believed to be responsible for the particle acceleration in terrestrial auroral regions. The manner in which they are established, however, is still a topic of debate. An additional piece of evidence supporting the idea that potential structures are established on the Martian nightside is the detection of magnetic perturbations perpendicular to the ambient field direction in cusp regions (Brain et al., 2006). Field-aligned currents should cause such perturbations.

At least four possible processes have been proposed to explain Martian aurorae. It seems possible that each of them could operate at Mars, but none of them seem individually capable of explaining all of the observations. Acceleration of electrons in quasi-static electric fields, mentioned above, is one of the leading explanations for the observations. But it is not fully understood how such regions form near Mars, nor is it well understood how the currents close in the Martian ionosphere. Dubinin et al. (2008c) noted that the conductivity along crustal field flux tubes in the Martian ionosphere should be higher than at Earth due to smaller magnetic field strengths, so that only small electric fields and/or narrow auroral flux tube regions should be supported, perhaps only intermittently. However, auroral structures lasting weeks have been observed in some cases (Dubinin et al., 2009b). A second candidate process is transport of ionospheric photoelectrons from the dayside of the planet, along magnetic field lines and into cusp regions (Leblanc et al., 2006). This process was initially proposed to explain the relative intensity of auroral emission at different wavelengths. However, not all examples of auroral emission are consistent with this explanation, and the cross-sections for such calculations are not well known. A third candidate is acceleration of electrons in current sheets at Mars. Many of the most energetic peaked electron distributions recorded by MGS occurred in current sheets (Halekas et al., 2008), and the electron angular distributions observed by MEX are more consistent with acceleration in current sheets than in field-aligned current regions (Dubinin et al., 2009b). However, it has not been established how or whether current sheets on the Martian nightside connect to cusp regions, and whether the observed current sheets are connected to the tail plasma sheet. Another process that may accelerate electrons to auroral energies is magnetic reconnection, which has been reported on the Martian nightside. There may be reconnection that occurs in a quasi-steady situation where strong magnetic shear is present, but also in the course of the changes in connections between Martian and external fields as the solar wind varies and as the planet rotates. However, there are no observations that directly link reconnection with aurorae, and reconnection cannot be responsible for the energization of upward-moving ions at the same location as downward-moving electrons.

More concrete answers about the causes of aurorae at Mars may come from global modeling with improved spatial resolution, or (perhaps more likely) dedicated local models. Future measurements that combine *in situ* charged particles, measurements of the local magnetic and electric fields, and observations of the ionosphere at the base of cusp flux tubes would also prove invaluable. Some progress has been made in this area, with Mars Express observations of auroral emission below the spacecraft, coupled with an increase in ionospheric total electron content and *in situ* measurements of auroral-like electron distributions (Duru et al., 2006; Nielsen et al., 2007; Dubinin et al., 2009b; Němec et al., 2011). Such observations will allow us to correlate the processes accelerating particles above the atmosphere, with their effects in the ionosphere.

15.3.3.4 Influence on Ionosphere

At an umagnetized planet lacking crustal fields, such as Venus, the magnetic field in the dayside ionosphere either is very small (e.g. during solar maximum when ionospheric pressure exceeds solar wind pressure) or is largely horizontal (e.g. during solar minimum). Horizontal magnetic fields inhibit vertical transport of plasma, although gradient drifts and collisional drag may still allow some transport. Crustal magnetic fields create regions of strong vertical magnetic field in some locations (cusp regions), allowing more efficient vertical transport of plasma, and particle exchange between the ionosphere and the plasma environment. In other locations strong horizontal crustal magnetic fields shield the ionosphere, as discussed above, reducing particle impact ionization and heating by magnetosheath electrons and related particle escape. This situation creates heterogeneities in the structure of the ionosphere. Such non-uniformity has been observed on both the dayside and nightside as vertical "walls" of ionization near the boundaries of cusp regions reported based on Mars Express MARSIS radar observations (Duru et al., 2006; Nielsen et al., 2006; Němec et al., 2011), and as abrupt changes in topside electron density profiles obtained from radio occultation measurements (Withers, 2005). Additional evidence from the nightside comes from maps of total electron content, which show that regions of strong horizontal crustal magnetic field have lower electron content (Safaeinili et al., 2007), implying that impact ionization sources for the nightside ionosphere are inhibited in closed field regions.

This non-uniform impact ionization on the nightside creates a patchy ionosphere (Figure 15.11), with some regions having

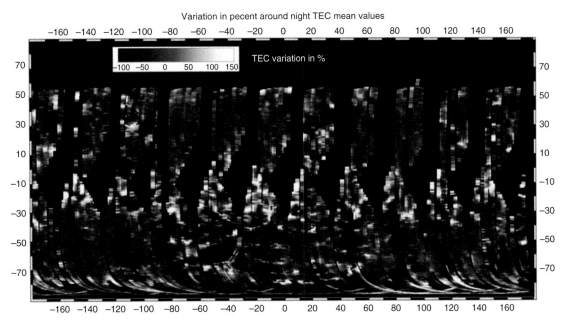

Figure 15.11. Geographic map of variability in total electron content (TEC) in the nightside ionosphere of Mars. Large variability is seen in equatorial regions, but also near the edges of strong crustal fields in the southern hemisphere, corresponding to regions of mostly vertical magnetic field. From Safaeinili et al. (2007).

densities three orders of magnitude higher than nearby regions (Lillis et al., 2011). This non-uniformity should create large pressure gradients that cause plasma transport across magnetic field lines (e.g. horizontal currents). Upper atmospheric winds in the vicinity of crustal fields can also drive horizontal currents, which contribute to local heating (via collisions) of the upper atmosphere (Fillingim et al., 2010). Additional plasma transport enabled by crustal fields can occur when a closed flux tube has one foot point in sunlight, and the other in shadow. In this case, photoelectrons generated in the sunlit portion of the flux tube may be transported along the flux tube and deposited in the nightside ionosphere (Uluşen and Linscott, 2008), or may escape into the wake if the field is open (Liemohn et al., 2006).

Crustal fields should also influence ionospheric conductivity, because both the Pedersen (parallel to the electric field) and the Hall (perpendicular to both the electric and magnetic fields) conductivities are inversely proportional to the local magnetic field strength. The vertical conductivity structure on the dayside of the planet was computed in different regions by Opgenoorth et al. (2010). They found that conductivities near strong crustal fields were reduced by one to two orders of magnitude relative to unmagnetized regions. Further, while unmagnetized regions have two peaks in Pedersen conductivity (with a peak in Hall conductivity in between), magnetized regions have only one peak in Pedersen conductivity overlaying a peak in Hall conductivity. This situation should give rise to conductivity gradients at the edges of crustal field regions, affecting the locations where currents flow.

15.4 ATMOSPHERIC ESCAPE

15.4.1 Source and Loss Processes for Atmospheres

Atmospheric "escape to space" is a loosely defined term referring to the results of a set of physical processes that provide particles in the thermosphere, ionosphere, and exosphere with sufficient energy to escape the planet. Figure 15.12 summarizes these processes. All particles escaping via these mechanisms must pass through the plasma environment. They are energized in upper atmospheric regions, which both have a strong influence on the plasma environment and are affected by that environment. It is natural to study escape to space as an isolated research topic, but the state of the atmosphere at any given moment is the product of an entire suite of source and loss processes that have acted over time. Therefore, it is important also to understand the atmosphere as a system, and to determine how the different processes act together and influence each other. The motivation for doing so is a better understanding of the evolution of the Martian climate over its history – a topic that is described in more detail in Chapter 17.

A number of source and loss processes have been identified for terrestrial planet atmospheres, and are well described in a variety of works (e.g. Hunten, 1992). Three main source processes contribute particles to the atmosphere: delivery by impacts, outgassing from the interior, and release from the surface and subsurface (including chemical reactions and sublimation from polar caps). The first two processes should have been most important early in Martian history, with ever-decreasing significance as both the impact flux and the internal heat of the planet declined over time. Release from the surface and subsurface, such as seasonal exchange of CO_2 between the polar caps and atmosphere and variations in atmospheric abundance associated with changes in Mars' obliquity (see Chapter 16), is thought to be climatically important over shorter timescales.

Four main loss processes have influenced the atmosphere: hydrodynamic escape, impact erosion, incorporation into the surface and subsurface, and escape to space. Hydrodynamic escape occurs when a light species escapes in sufficient abundance that is equivalent to a net upward wind, and drags heavier species along with it through collisions. This process is usually enabled by high solar EUV flux or another form of heating. It should have been significant for Mars and other terrestrial planets, but only during the first few hundred million years after formation (Zahnle and Kasting, 1986). Ejection to space of atmospheric particles by solid-body impacts was an important loss mechanism after hydrodynamic escape ceased, and before the end of the late heavy bombardment (Melosh and Vickery, 1989; Brain and Jakosky, 1998). Impactors carry volatiles, however, so can also add particles to an atmosphere (Cameron, 1983). The polar caps, surface, and subsurface are capable of removing CO_2 and other species from the atmosphere over a variety of timescales and through a variety of processes (see Chapters 16 and 17). While exchange with the surface and subsurface is largely reversible, escape to space permanently removes particles from the atmosphere. It is likely to have been the dominant permanent loss process over the past ~3.8 billion years, and is the focus of the remainder of this chapter.

There is strong evidence that all of the source and loss processes mentioned above have operated at Mars (Jakosky and Jones, 1994; Jakosky and Phillips, 2001). The main question is their relative importance as a function of time and integrated over Martian history. Uncertainty about the "initial state" of the Martian atmosphere further complicates studies of climate evolution. Liquid water was once stable on the surface of Mars for sufficiently long periods of time to alter its landscape and geochemistry (e.g. Carr and Clow, 1981; Craddock and Howard, 2002; Hoke and Hynek, 2009), and a straightforward inference is that Mars once possessed a thicker atmosphere that provided sufficient greenhouse warming to keep surface water liquid. This conclusion has been increasingly re-examined in recent years, based on a number of factors discussed in detail in Chapter 17. They include difficulties with modeling sufficient greenhouse warming from an early thick atmosphere, geochemical analyses of surface minerals and meteorites, and the development of alternative explanations for the formation of certain geomorphologic features that have been interpreted to indicate the evidence of a warmer ancient atmosphere. Given the renewed debate in the Mars science community, it is important that the contributions from each of the source and loss processes be evaluated and quantified, including the contributions from escape to space.

Evidence for the prominent role of atmospheric "escape to space" in climate evolution comes primarily from measurements of isotope ratios in the Martian atmosphere, which show that the atmosphere is enriched in heavy isotopes of several species (see the discussion in Jakosky and Phillips (2001) and

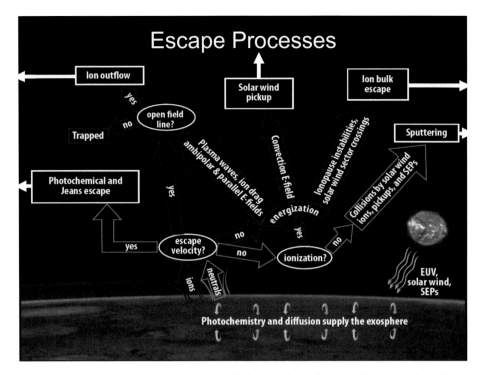

Figure 15.12. Schematic of contemporary escape processes (rectangular boxes) relevant for atmospheric neutrals and ions, presented as a "decision tree". Solar inputs shown at the lower right contribute to the energization of atmospheric particles.

Chapter 17). Of the many processes described above, escape to space alone is capable of fractionating the atmosphere in this way. Such escape occurs above the homopause at Mars, in regions where each atmospheric species diffusively separates. As a result, the upper atmosphere contains a greater abundance of lighter isotopes, which are then preferentially energized and escape, leaving the atmosphere enriched in the heavier isotopes.

15.4.2 Reservoirs for Escape

Escape to space removes both neutral and ionized species from one of three atmospheric "reservoirs", described below. All three regions are also discussed in Chapter 14.

Thermosphere. This region of neutral particles extends from the mesopause (~100 km) up to the exobase (~200 km), with temperature that increases as a function of altitude as a result of EUV input from the Sun. The homopause at ~120 km separates the lower, well-mixed thermosphere from a large region where each species is in a separate diffusive equilibrium. The most abundant thermospheric species are CO_2, CO, O, and N_2. The thermosphere is a collisional region. Any neutral particle that reaches the top of the thermosphere with escape energy should be removed from the atmosphere if it has an outward trajectory. Particles reaching an altitude where collisions are rare and having energy less than the escape energy will populate an extended exosphere or corona.

Exosphere. The exosphere is a region of neutral particles above the thermosphere, extending from the exobase (~200 km) upward to altitudes as high as 30 000 km (Feldman et al., 2011). The dominant species in the exosphere at low altitudes (≲300 km) is oxygen; above ~300 km the dominant species is hydrogen. The exosphere is a non-collisional region (by

definition), and so any upward-directed particle within it that has velocity greater than the escape velocity for Mars is likely to be removed from the atmosphere. The lower boundary to the exosphere is referred to as the exobase – the location where the mean free path for collisions is equal to the atmospheric scale height. Particles with insufficient energy to escape will return to the exobase on ballistic trajectories unless they are first ionized. In reality, the exobase is not a sharp boundary; collisions still occur above the exobase, and particles below the exobase can still escape. Recently it was pointed out, for example, that the assumption that particles scatter isotropically in the region below the classical exobase is likely to be incorrect. Instead, colliding particles should strongly forward-scatter, with the result that much of the escape flux from the neutral atmosphere may come from well *below* the classically defined exobase near 200 km (Fox and Hać, 2009).

Ionosphere. This region of ionized atmospheric particles extends from the exosphere down into the lower atmosphere. The ionosphere at Mars has its main density peak in the thermosphere, near ~115–130 km at the subsolar point; this peak is produced primarily by photoionization of thermospheric neutrals by solar EUV. Other sources of ionization include particle impact (e.g. the auroral and entering magnetosheath electrons discussed earlier) and charge exchange with precipitating ions. The dominant species in the ionosphere are O_2^+ at low altitudes (≲300 km) and O^+ at higher altitudes. Unlike neutrals, ionospheric particles can be accelerated by magnetic and electric fields. Some ions reaching the exobase from below may be accelerated away from the planet by the passing solar wind, even if they initially lack escape energy. Others may be accelerated and reimpact the collisional atmosphere at the exobase. More details about the ionosphere can be found in Chapter 14 and Section 15.2.1.

The reservoirs for escape are not isolated from each other, and the boundaries between them are not sharp. Particles are readily exchanged between the reservoirs via ionization and recombination, or transition through the exobase region. Therefore, when evaluating atmospheric escape at Mars from any one reservoir, caution must be taken to identify the fraction of escaping particles that are added to the other reservoirs rather than being removed from the planet. Evaluation of the net escape to space is made challenging by the fact that different sets of physical processes are necessary to adequately describe each region, and because the observations required to evaluate escape from each reservoir are different. At present, no spacecraft has made all the necessary observations in concert. And no simulation has been developed that captures escape from all three reservoirs self-consistently. Integrated modeling studies are underway, linking the thermosphere–ionosphere–exosphere regions and the solar wind environment in an effort to couple the reservoirs for escape and thereby calculate self-consistent neutral and ion escape rates (see Chapter 14 for more detail).

15.4.3 Neutral Escape Processes

With rare exceptions, uncharged or neutral particles that escape the Martian atmosphere are energized in the thermosphere. This is because any particle that reaches the exobase region with sufficient velocity on an outward trajectory will escape the planet unless it first collides with another particle (which is rare, by definition, in the exosphere) or is ionized (in which case, it can either escape or precipitate/return to the atmosphere, depending on its subsequent behavior). The required *escape velocity* is determined from considering a balance between the gravitational potential energy of a given particle and the kinetic energy required to overcome it. Escape velocity depends upon altitude and the mass of the planet, but does not depend upon the mass of the escaping particle. In practice, a particle's energy is more likely to be directly measured than its velocity. Since escape velocity does not depend on mass, the required *escape energy* from a given location is mass-dependent.

Three main escape processes have been identified for neutrals: Jeans escape, photochemical escape, and sputtering. Jeans (or thermal) escape results when a portion of the thermal distribution for an atmospheric species exceeds the energy necessary for escape. Neutral temperatures near the exobase are sufficiently low (~200–350 K) that only species with small mass (H, D, and He) can escape via this mechanism in significant quantity. A ~250 K temperature corresponds to ~0.02 eV for H, which has escape energy of ~0.1 eV. Since less massive species are removed more efficiently, Jeans escape also contributes to fractionation and can increase the D/H ratio of the atmosphere. The D/H ratio observed in the Martian atmosphere today is about five times the terrestrial value (Owen et al., 1988; Krasnopolsky et al., 1997), indicating that substantial escape of hydrogen to space has occurred over Martian history. Estimates of thermal escape of hydrogen have been made by several authors based on spectroscopic ground-based observations as well as early Mariner measurements (e.g. Anderson and Hord, 1971; McElroy, 1972; Yung et al., 1988; Krasnopolsky et al., 1997).

Photochemical escape is a process where exothermic chemical reactions provide atmospheric species with sufficient velocity to escape. These reactions typically involve dissociative recombination of an ionized molecule with a nearby electron, resulting in two fast neutral atoms. For example, O_2^+ recombines with an electron, and the energy released splits the O_2 molecule to form two fast oxygen atoms. Similar reactions occur for CO_2^+, N_2^+, and CO^+, and photochemical loss can be significant for O, N, and C. Photochemical escape fluxes depend upon ionospheric molecular densities near the exobase, as well as electron density and temperature.

Photochemical loss, like Jeans escape, has not been observed directly at Mars, but can be inferred from comparison of upper atmospheric measurements to the results of model calculations. A large number of models for photochemical escape of different species under different conditions have also been created, using observed thermospheric and ionospheric profiles as input (McElroy et al., 1977; Lammer and Bauer, 1991; Fox, 1993; Krasnopolsky, 1993; Nagy et al., 2001; Fox and Hać, 2009). Such models suggest that photochemical escape is the dominant loss process for neutrals heavier than hydrogen today (Lammer et al., 2009).

Atmospheric sputtering is a process by which energetic incident particles (including ionospheric particles accelerated by the solar wind) collide with thermospheric and ionospheric particles near the exobase. Momentum and energy are exchanged between particles during the Coulomb collisions, and some of the "target" atmospheric particles can be ejected to space. Each incident particle may have substantial energy, and therefore be responsible for the ejection of many atmospheric particles. Watson et al. (1980) first considered whether sputtering by impacting solar wind protons would be important at unmagnetized planets, but Luhmann and Kozyra (1991) and then Johnson and Luhmann (1998), proposed that the O^+ pickup ions, the expected dominant species on the dayside, would be the more effective sputterers. There are no unambiguous observations that sputtering is actively occurring at Mars or contributes significantly to the present-day atmospheric escape rate. To this point most results have been confined to theoretical calculations (Luhmann and Kozyra, 1991; Leblanc and Johnson, 2002), which agree that typical sputtered escape fluxes are small relative to other processes. However, estimates of loss rates due to sputtering for more extreme conditions (Luhmann et al., 1992b) suggest it may have greater impact under certain circumstances. Recent observations have begun to quantify the incident flux of particles that could cause sputtering (Leblanc et al., 2015).

The measurements of isotopic ratios of atmospheric gases provide compelling evidence that loss to space has occurred. Above the homopause, diffusive mixing is slower than gravitational separation, so each gas will take on its own independent scale height based on its mass. As a result, heavier gases do not extend to as high an altitude as lighter ones, and are relatively depleted at the altitudes from which escape occurs. The ratios of stable isotopes are particularly valuable, as they tell us about specific gases. The enrichment of $^{15}N/^{14}N$ indicates that roughly 90% of the atmospheric nitrogen has escaped to space over time (Jakosky et al., 1994). Although originally attributed to photochemical processes, it is not possible to determine the loss mechanism from the isotope ratio alone, and sputtering is a viable process as well. The enrichment of $^{38}Ar/^{36}Ar$ is especially important, though, as argon does not participate in chemical or

Table 15.3. *Contemporary global escape rates (in units of 10^{24} s^{-1}) of relevant species for three processes that remove neutral particles from the Martian atmosphere, updated from Lammer et al. (2008). Rates are given as a range based on the available literature: [1] range computed from Anderson and Hord (1971), Chaufray et al. (2008), and Feldman et al. (2011); [2] Krasnopolsky and Feldman (2001); [3] refer to Table 15.4 for reference list; [4] range computed from Fox and Bakalian (2001) and Nagy et al. (2001); [5] Fox and Dalgarno (1983); and [6] Leblanc and Johnson (2002).*

	H	H_2	O	CO	CO_2	C	N
Jeans escape	129–300 [1]	3.3 [2]					
Photochemical escape			7–210 [3]			0.02–6.4 [4]	0.19–0.72 [5]
Sputtering			0.34 [6]	0.037 [6]	0.05 [6]	0.09 [6]	

photochemical processes; it is a particularly good indicator of sputtering loss. The 20–30% enrichment in this ratio provides compelling evidence for the sputtering process having operated, and suggests that as much as 90% of the atmospheric argon has been lost to space since the atmosphere was last in its initial state (Jakosky et al., 1994; Hutchins et al., 1997). Ratios such as D/H, $^{38}Ar/^{36}Ar$, $^{13}C/^{12}C$, $^{15}N/^{14}N$, and $^{18}O/^{16}O$ have been used to infer that Mars lost 50–90% of its atmosphere by the escape processes discussed in this section (Jakosky and Phillips, 2001). Measurements of the various isotope ratios, and their interpretation, are discussed in Chapter 17.

Table 15.3 summarizes the present-day atmospheric escape rates for neutral particles over a range of solar conditions. From the table we see that the global loss rate of neutrals by all processes is a few$\times 10^{26}$ particles, dominated by Jeans escape of hydrogen and photochemical escape of atomic oxygen. The escape rates for neutrals are not measured directly, however. Instead, measurements of the atmospheric reservoirs (e.g. their densities and scale heights) are used as the basis of calculations of escape rates. These simulations range from 1D globally averaged models of the ionosphere and thermosphere to detailed 3D coupled simulations of the thermosphere and exosphere (Fox, 1993; Chaufray et al., 2007; Valeille et al., 2009a,b; Yagi et al., 2012). Improved knowledge of these reservoirs may change the predicted escape rates significantly.

The large range of calculated values for photochemical escape of atomic oxygen is partly due to difference in approach, but also due to solar and seasonal variability. Table 15.4 demonstrates this point by providing some of the main estimates of neutral oxygen escape over the past decade, divided into low and high EUV flux time periods. Hot O escape rates vary by a factor of 2–3 over the solar cycle. Furthermore, seasonal variation (produced due to the fact that the Martian orbit has a significant eccentricity) of a factor of 1.5–2 is computed by the latest Monte Carlo simulations for which thermosphere–ionosphere 3D model inputs are used (Valeille et al., 2009a,b; Yagi et al., 2012).

15.4.4 Ion Escape Processes

Any atmospheric particle that is ionized above the exobase can be accelerated, usually anti-sunward, by ambient electric fields, including those in the solar wind. Although some of these are "returned" to the atmosphere (precipitate at the exobase),

a significant fraction is expected to avoid that fate. Thus, the ionosphere, exosphere, and upper thermosphere all serve as reservoirs for escaping ions[4]. Escaping ions were first identified at Mars by the Phobos spacecraft (Lundin et al., 1989; Verigin et al., 1991), and have been subsequently measured by Mars Express (Barabash et al., 2007) and MAVEN (e.g. Brain et al., 2015). Further, several sophisticated global models of the interaction of the solar wind with the Martian atmosphere are now capable of simulating ion motion and predicting escape rates under various input conditions using several approaches (see Brain et al., 2010b).

A number of processes have been identified by which atmospheric ions can escape Mars. Different authors classify these processes in different ways – some according to the physical mechanism responsible for the acceleration, and some according to the region of origin or pathway that ions take as they escape. That there are many different escape mechanisms that have been discussed indicates the inherent difficulty in distinguishing the physical processes responsible for ion escape using observations.

For the purposes of our discussion here, we classify the ion loss processes into three categories. Escape rates for each ion loss process (and total ion escape rates) are provided in Table 15.5.

Ion pickup refers to the situation where a neutral particle is ionized (via photons, electron impact, or charge exchange) and accelerated away from the planet by a motional electric field ($E = -v \times B$). Since the atmosphere can be considered to be collisional below the exobase, it is reasonable to assume that ion pickup occurs primarily for ionized exospheric neutrals (though some ionized thermospheric neutrals near the exobase region may escape via pickup as well). Test particle calculations suggest that the loss of atmospheric ions via the pickup process can be significant (Luhmann, 1990; Fang et al., 2010a). Both test particle and global hybrid simulations show that the most energetic pickup ions should escape the planet in a "plume" originating above the hemisphere for which the solar wind convection electric field points away

---

[4] Technically, particles in the thermosphere or exosphere that are ionized and entrained in the passing plasma flow briefly become ionospheric particles before escaping.

Table 15.4. *Solar cycle variation in estimated photochemical escape rates of atomic oxygen. Rates are given in units of 10^{25} s^{-1}.*

Reference	Equinox (low)	Equinox (high)	Ratio
Fox and Haċ (2009)	14.4	21.0	1.46
Valeille et al. (2009a,b)	6.0	19.0	3.17
Chaufray et al. (2007)	1.0	4.0	4.0
Cipriani et al. (2007)	2.1	5.0	2.38
Krest'yanikova and Shematovitch (2005)	0.7	11.0	16.9
Hodges (2002)	4.4	18.0	4.10
Kim et al. (1998), corrected by Nagy et al. (2001)	3.4	8.5	2.5
Average	4.57	12.36	2.70

Table 15.5. *Contemporary global escape rates (in units of 10^{24} s^{-1}) of relevant species for three processes that remove ionized particles from the Martian atmosphere, updated from Lammer et al. (2008). Rates are given as a range based on the available literature – citations are for end-member estimates, only: [1] Fang et al. (2008, 2010a); [2] Lundin et al. (2009); [3] Nilsson et al. (2011); [4] Brain et al. (2010b); [5] Najib et al. (2011); [6] Brecht and Ledvina (2010); [7] Ma and Nagy (2007); and [8] Harnett and Winglee (2006). References [2]–[4] are based on observations, while the others are based on simulation results.*

	Ion pickup	Ion outflow	Bulk escape	All processes
H$^+$		2.0 [2]		
H$_2$$^+$		0.12 [2]		
O$^+$	0.54–3.7 [1]	1.0–2.1 [2, 3]		0.16–52 [5, 6]
O$_2$^{1}		0.67–1.4 [2, 3]		0.19–26 [7, 8]
CO$_2$$^+$		0.2–0.35 [2, 3]		0.11–0.21 [5, 7]
All heavy ions			0.79–1.6 [4]	

from the planet – a picture that has recently been confirmed with observations (Dong et al., 2015a). Estimates of loss via pickup are shown in Table 15.5. Charge exchange is sometimes treated as a loss process separate from ion pickup, because the collision between particles that results in ionization also provides energy to the planetary particle that can be sufficient for escape. However, even if the collision provided no energy, charge exchange occurring above the exobase should result in acceleration of a newly born ion in an ambient motional electric field. In some cases, charge exchange can result in the loss of two planetary particles; if an accelerated planetary ion encounters an exospheric neutral, then both may escape the system – the target particle as an ion and the impactor as an energetic neutral atom. Charge exchange has recently been observed as a mechanism allowing solar wind particles to access low altitudes; solar wind protons charge exchange with the exosphere at high altitude and travel toward Mars, where they charge exchange again and become ionized (Halekas et al., 2015).

Ion outflow refers to the acceleration of low-energy particles out of the ionosphere via plasma heating and outward-directed electric fields. In this case the ion acceleration can occur below the exobase where collisions maintain a more fluid-like

behavior. A number of processes are referred to collectively as outflow in the terrestrial literature, including wave heating, polar wind, and auroral outflow (see André and Yau, 1997; Moore and Horwitz, 2007; Yau et al., 2007). These processes should all have analogs in the Martian ionosphere and crustal magnetic fields. Ions escaping via outflow are often thought of as "cold" in contrast to pickup ions, although pickup ions can also be cold in regions where the bulk plasma velocity (v) and magnetic field (B) are nearly aligned (resulting in a small $v \times B$ pickup acceleration). Outflow processes (as distinct from ion pickup) were first discussed for Mars by Lundin et al. (1990), who estimated a combined loss rate from outflow and pickup of ~2×10^{25} O$^+$ ions per second. Mars Express has observed significant fluxes of cold ions escaping from Mars (Lundin et al., 2009; Nilsson et al., 2012), with average temperatures lower than 10 eV close to the planet that gradually increase downstream from Mars. Various outflow mechanisms have been evaluated using a combination of observational and theoretical considerations. Some calculated estimates suggest that the loss via wave heating and auroral outflow can be significant (Ergun et al., 2006), while others provide more conservative estimates (Dubinin et al., 2009b). Simulations suggest that moderate amounts of wave heating are sufficient to account for the observed ionospheric density profile for O^+, and that related outflow processes can account for a substantial portion of ion loss (Andersson et al., 2010).

Bulk escape refers to any process for which coherent portions of the ionosphere are detached via magnetic and/or velocity shear processes and accelerated away from the planet. For example, shear-related surface instabilities (such as the KH instability) could form at the ionopause of an induced magnetosphere, steepening into waves that eventually detach, releasing ionospheric particles to the passing plasma flow (Terada et al., 2002; Penz et al., 2004). While KH instability has been reported at Earth's magnetopause (Hasegawa et al., 2004), observations at Mars have proved difficult to interpret unambiguously (Duru et al., 2009; Gurnett et al., 2010; Halekas et al., 2011). Plasmoid-style flux ropes may also remove ionospheric plasma from crustal magnetic field regions (Brain et al., 2010a). Loss rates via bulk plasma escape are poorly constrained at present by observations or by models. The presence of viscous (i.e. bulk) processes and detached plasma clouds have been reported for both Venus and Mars (Brace et al., 1982; Grard et al., 1992; Cloutier et al., 1999), but a reliable estimate of the amount of atmosphere removed as clouds at Mars has not been obtained. Brain et al. (2010a) estimate that as much as 10% of the total ion loss could occur as plasmoids from reconnection processes. Global plasma simulations for Mars have not yet isolated the contribution from bulk processes, although Terada et al. (2002) concluded that KH instability could contribute significantly to ion loss from the Venus atmosphere.

An appealing alternative classification scheme for ion escape processes is rooted in MHD physics. In a sense, all ions are accelerated away from Mars by electric fields, and it is convenient to think of ion escape processes in terms of the type of electric field responsible for their removal (see discussion in Dubinin et al., 2011). A simplified version of Ohm's law describes the most important electric field terms that influence ion motion:

$$\mathbf{E} = -\mathbf{v} \times \mathbf{B} + \frac{1}{ne}\mathbf{J} \times \mathbf{B} + \frac{1}{ne}\nabla \mathbf{P}_e \qquad (15.1)$$

Here E is the total electric field, v is the plasma bulk velocity, $\mathbf{B}$ is the magnetic field, $\mathbf{J}$ is the current density, $\mathbf{P}_e$ is the electron pressure tensor, n is the plasma number density, and e is the electron charge. The three terms on the right-hand side of the equation are the motional electric field, the Hall electric field, and the electron pressure gradient. The Ohmic and electron inertial terms have been neglected. There is varying overlap of these three terms with the ion escape processes described earlier. For example, the $v \times B$ term is the convection electric field usually used to describe ion pickup. Ion outflow can be caused both by the Hall electric field and by ambipolar electric fields resulting from electron pressure gradients. Both Mars Express observations and global simulations have been used to evaluate the importance of these three terms in different locations near Mars (Dubinin et al., 2011; Lundin, 2011). These results indicate that $v \times B$ acceleration is most important at high altitudes and that $J \times B$ acceleration is particularly important near the magnetic poles and in the plasma sheet where the magnetic fields are most tightly draped. Acceleration via pressure

gradients is likely to be important near crustal magnetic field cusps and around the terminator where photoproduction sharply decreases.

Finally, it is worth noting that evaluating the processes responsible for atmospheric ion escape using spacecraft measurements is inherently difficult. Escaping ions are measured *in situ*, along with the ambient magnetic and electric fields (if possible). However, local measurements cannot always or easily indicate which of the above-described mechanisms provided the escape energy to an observed distribution of escaping ions, or whether the ions in the observed distribution will ultimately escape the Martian system (or reimpact the atmosphere). For example, all ions behave as pickup ions at sufficiently large distances from Mars, regardless of whether they were accelerated in auroral regions of the ionosphere or the high-altitude exosphere. The large number of empty cells in Table 15.5 underscores this point. Further, several processes may combine to provide an ion with sufficient energy to escape the planet. Reliable models are therefore crucial to evaluating the relative importance of each of the loss processes.

15.4.5 Total Atmospheric Loss

There is no doubt that the many physical processes described above remove particles from the Martian atmosphere. A major unanswered question is whether these processes have a significant impact on the planet's volatile inventory or climate. To address this question, it is first necessary to evaluate the total atmospheric loss rate at present, assess how it varies with the relevant drivers, and estimate the loss over Martian history.

Estimates of the atmospheric loss rates today are obtained from a combination of models and measurements. Escaping neutral particles have not been detected directly; instead, loss rates are inferred from measured properties of the Martian upper atmosphere and exosphere, and have been modeled using a variety of techniques. Ion loss rates are inferred from statistics of *in situ* measurements, and predicted using global models. Referring to Tables 15.3 and 15.5, a number of different species are removed from the atmosphere, with the highest (reported) loss rates associated with species deriving from CO_2, H_2O, and N_2. At present, it appears that the escape of neutral particles may exceed the escape of ionized species by an order of magnitude. Net ion loss rates (independent of the process responsible for escape) are best measured downstream from the planet, since all escaping ions end up entrained in the passing solar wind. Mars Express has measured planetary ions at various distances downstream from the planet, and infers loss rates of $\sim 2 \times 10^{24}$ ions/s, dominated by O^+ (Figure 15.13) (Nilsson et al., 2011). Table 15.5 shows a range of loss rates for each heavy-ion species. Though rates have been derived in the literature based on both spacecraft data analysis and global plasma modeling, the models account for both the maximum and minimum published rates. However, these estimates are limited by the orbit sampling geometry of the spacecraft, and may not fully include all of the escaping subpopulations (e.g. the pickup ion "plume" apparent in the work of Luhmann and Schwingenschuh (1990), Brecht and Ledvina (2006), Fang et al. (2008), and others).

Loss rates should vary in response to drivers from the Sun (photons) and solar wind (incident particles and fields). Monte

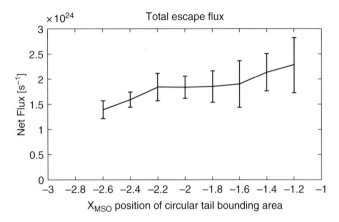

Figure 15.13. Average heavy-ion escape flux as a function of distance downstream from Mars as measured by ASPERA-3. Adapted from Nilsson et al. (2011).

Carlo models of the loss of hot atomic oxygen suggest that loss rates vary by a factor of 2–3 over the solar cycle (Kim et al., 1998; Valeille et al., 2009a). Variations in response to changes in the solar wind or IMF should mostly influence sputtering loss rates (since sputtering results from ions precipitating at the exobase), which are thought to be considerably smaller than loss rates via photochemical or thermal processes under present normal conditions. Ion loss rates have been demonstrated to vary with EUV flux and solar wind pressure based on Mars Express measurements (Lundin et al., 2008; Nilsson et al., 2010, 2011). Loss rates reported by Mars Express during solar minimum are lower than loss rates measured by Phobos 2 during solar maximum by a factor of ~2–10. Global plasma models have explored changes in ion loss rates with various parameters, and generally confirm the trends evident in the observations: as solar EUV or solar wind pressure increases, so do the net planetary heavy-ion loss rates (Ma et al., 2004; Modolo et al., 2005; Brecht and Ledvina, 2006; Harnett and Winglee, 2006; Ma and Nagy, 2007; Najib et al., 2011). However, the model results vary in terms of both the relative and absolute amounts of ion loss as a function of driving conditions, and even whether the loss of rates of individual species increase or decrease as conditions change. Since the models are rarely run for identical input conditions, the results are difficult to compare. At present, the community is engaged in an activity to compare the results of many different models run for identical solar minimum and solar maximum conditions. Initial efforts for solar nominal conditions are reported in Brain et al. (2010b) and show variations of an order of magnitude or more in heavy-ion escape rates (Figure 15.14).

Since Martian crustal magnetic fields influence the shape of the planetary solar wind obstacle, the orientation of Mars with respect to the incident plasma flow should also influence loss rates – at least for escaping ions. The magnitude of this effect is not yet clear, however. Global multi-fluid plasma simulations showed little variation (~15% or smaller) in global escape rates for different orientations of crustal fields with respect to the flow (Harnett and Winglee, 2006), while MHD simulations demonstrated variations of a factor of 4 for different orientations (Ma et al., 2004). A test particle model showed variations of ~10% (Fang et al., 2010b). Models similarly disagree about the change in loss rates when crustal fields are added to a simulation relative

to an "unmagnetized" case: relative changes vary over several orders of magnitude, from <10% (Harnett and Winglee, 2006), to ~30% (Ma et al., 2002), to factors of ~4 (Fang et al., 2010a) and ~30 (Brecht and Ledvina, 2012). Differences between the various estimates could result from the use of different input conditions, different physical assumptions in the models, or different spatial resolutions employed in the models. The latter may be particularly important because crustal fields should not only prevent the solar wind from accessing the Martian ionosphere in certain regions, but also facilitate atmospheric escape along open flux tubes that connect to the magnetosheath and solar wind. Models lacking sufficient resolution to capture these effects may poorly estimate the influence of crustal fields on ion loss rates. Observations suggest that the strong crustal field regions in the southern hemisphere of Mars have lower escape rates than the weakly or unmagnetized northern hemisphere (Lundin et al., 2011; Nilsson et al., 2011; Dubinin et al., 2012).

The presence of crustal fields at Mars, and their likely formation in the presence of a global dynamo magnetic field, suggests that the Martian atmosphere was less exposed to solar wind-related stripping processes early in its history. While the timing of the shut-off of the Martian dynamo is debated, it likely persisted until ~4.1 Ga or beyond (Lillis et al., 2008). During this period, much of the atmosphere was likely protected from the solar wind in the same way that Earth's atmosphere is shielded in regions far from the polar cusps. Even after the cessation of the dynamo, crustal fields may have covered most or all of the Martian surface initially, and been gradually erased as the magnetized crust in different regions was heated above the Curie temperature of the rocks by impact or volcanic events. Thus, assuming that the presence of a crustal field inhibits loss, ion escape rates (and, indirectly, photochemical and sputtering rates) would have been reduced by a global "blanket" of crustal fields that was erased over time.

The results described above establish that escape to space occurs today, and varies with relevant drivers from the Sun and solar wind. But is this escape important to consider when studying the evolution of the Martian atmosphere? If we take the measured escape rates and assume they have remained unchanged over the last ~4 billion years, there is a range of possible answers to this question. Based on oxygen escape rates of ~10^{25} s^{-1} inferred from Phobos observations, Lundin et al. (1990) concluded that all of the oxygen atoms in the Martian atmosphere (including those bound in molecules like CO_2 and H_2O) could be removed in 100 million years[5]. However, 10^{25} s^{-1} has turned out to be near the high end of the range of published escape rates for most species, and the Lundin et al. (1990) estimate only considered particles presently in the ~7 mbar atmosphere today (and not the substantial quantity of oxygen atoms in reservoirs in contact with the atmosphere, such as the polar caps). Luhmann et al. (1992b) and then Leblanc and Johnson (2002) estimated that 150–200 mbar of atmospheric O and CO_2 could be removed via sputtering processes over Martian history, considering changing solar EUV fluxes. This quantity is substantially greater than the current atmospheric abundance,

---

[5] The escape rate of hydrogen is calculated to be significantly larger, so that hydrogen and oxygen are lost at a ratio of somewhere between 4 : 1 and 20 : 1. Thus, for the Martian atmosphere to retain its oxidation state over time, there must be a sink of oxygen (likely at the surface). See discussion in Fox and Hać (2009).

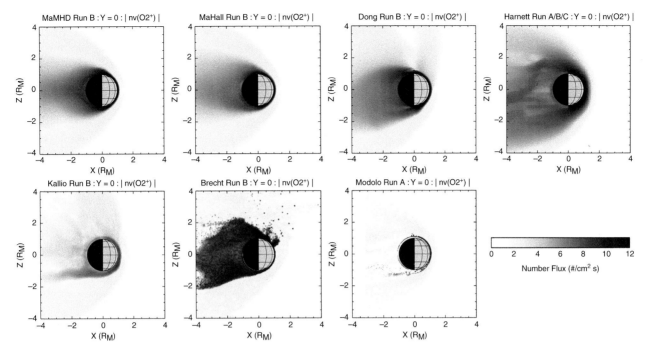

Figure 15.14. Modeled O_2^+ flux as a function of location near Mars, as simulated by seven different global plasma simulations for identical input conditions. After Brain et al. (2010b).

but likely to be smaller than the amount needed to account for an early atmosphere with substantial greenhouse warming. However, the two estimates provided above are for a subset of the species and processes believed to strip atmospheric particles to space, and many other estimates have been made over the past few decades that span a very large range in total loss.

Further, the escape rates measured and modeled for conditions at Mars today are not likely to be indicative of the loss rates over Martian history. Earlier in Solar System history the many drivers of atmospheric escape to space are believed to have been stronger (Luhmann et al., 1992b; Ayres, 1997; Ribas et al., 2005; Wood et al., 2005), so that loss rates were higher billions of years ago. For this reason, quantitative determination of how today's loss rates vary as a function of solar wind drivers is critical to being able to extrapolate loss rates back in time. With knowledge of this variation in hand, models can be employed to constrain the total atmospheric loss at earlier epochs. Several efforts to assess the total loss over history have been made to date. One of the earliest efforts, by Luhmann et al. (1992b), calculated that ~140 mbar of CO_2 could have been removed via sputtering. More recently, Barabash et al. (2007) used the Modolo et al. (2005) and Ma et al. (2004) global plasma models, constrained by Mars Express ion loss measurements, to infer that only 1–43 mbar of CO_2 have been removed from the atmosphere via ion escape processes. However, this estimate only considered the present-day variations between solar minimum and solar maximum conditions, rather than an estimate of conditions at early Mars. Manning et al. (2011) tackled this issue by parameterizing the results of several simulations by Ma et al. (including results obtained for extreme solar wind and solar EUV flux) to infer that 6–110 mbar of CO_2 have been removed from the atmosphere in the past 3.85 Ga. The two estimates above correspond to the loss of a global ocean of

water having depth of 0.04–6.0 m. These estimates can be compared to estimates of ~12 m in the past 3.5 Ga (by Lammer et al. (2003) using a combination of models and previously published work for the different loss processes), ~8 m in the first 150 million years of Martian history (by Terada et al. (2009) using a global MHD model), and ~30 m since the end of hydrodynamic escape (by Krasnopolsky and Feldman (2001) using measurements of atmospheric molecular hydrogen in combination with measurements of D/H fractionation in atmospheric reservoirs). The above estimates are only a few examples of the many that have been published in recent decades – with estimates of the total atmospheric loss and early water budget that vary widely.

There are many uncertainties associated with the various methods of assessing atmospheric escape to space, both at present and (especially) over Martian history. Isotope measurements require accurate knowledge of the initial state of the Martian atmosphere, and when it was last achieved. For example, large impacts may partially "reset" an atmosphere's isotope ratios by providing unfractionated gas to the atmosphere. Extrapolation of present-day loss rates back in time requires accurate knowledge of the time history of the EUV and solar wind drivers of loss, which are currently obtained from observations (sometimes in conjunction with models) of Sun-type stars. The extrapolations further require accurate knowledge of the state of the atmosphere as a function of time; many of the current estimates non-self-consistently assume the present-day atmosphere has existed throughout Martian history. The timing of the shut-off of the Martian dynamo magnetic field will influence which loss processes are active and efficient at different times in Martian history. And there are many uncertainties in measurements of the present-day loss, both in terms of net loss rates of neutrals and ions, and in terms of the relative importance and response to drivers of the several processes that contribute to the net loss. Given the many evident uncertainties, there is a clear need for a comprehensive and contemporaneous set of measurements of all the processes acting today, together with any evidence that can be obtained regarding the early EUV, stellar wind, and activity histories of Sun-like stars.

15.5 SUMMARY AND OUTSTANDING QUESTIONS

15.5.1 Summary

The Martian thermosphere, exosphere, and ionosphere form the tenuous upper boundary for the Martian atmosphere – the region through which all energy and material from the Sun that encounters Mars must either pass or be deflected. The Martian upper atmosphere, ionosphere, and crustal magnetic fields interact directly with the incident solar wind plasma, forming a highly dynamic induced magnetosphere that influences the structure and variability of the upper atmosphere, and results in a rich array of plasma processes. Further, the energy input to the atmosphere from the Sun and solar wind energizes atmospheric particles, giving some of them sufficient velocity to escape the planet.

The Martian solar wind interaction has been studied *in situ* by spacecraft since 1965, though three recent missions (Phobos 2, Mars Global Surveyor, and Mars Express) have contributed the most comprehensive measurements that shape our current understanding of the physical processes operating in the interaction region, as well as atmospheric escape rates. The observations have been complemented by a variety of models that span a range of physical assumptions ranging from gas dynamic, to MHD, to hybrid and test particle models. Increases in computational capability in recent years has enabled model advances that include increased spatial and temporal resolution, incorporation of crustal fields into simulations, and time-dependent solar wind drivers.

Together, the observations and models have revealed a Venus-like induced magnetosphere, where the flowing solar wind drives currents in the Martian ionosphere that generate magnetic fields that deflect incident solar wind particles. The Martian solar wind interaction differs from that at Venus because of the presence of strong crustal magnetic fields, because the low Martian gravity permits an extended exosphere that interacts directly with the unperturbed solar wind, and because the smaller size of Mars and weaker interplanetary field there make the ion gyroradii larger relative to the scale size of the entire system. A number of plasma regimes form near Mars, including a magnetosheath dominated by shocked solar wind, an induced magnetosphere where planetary ions modify the plasma flow, an ionosphere, and an induced magnetotail that includes a wake and a plasma sheet that transport planetary ions downstream from the planet. On the dayside of the planet, solar wind dynamic pressure is converted into thermal plasma pressure in the magnetosheath, then magnetic pressure in the induced magnetosphere region. This pressure is balanced by ionospheric thermal pressure during some times, and the solar wind magnetic field diffuses into the ionosphere at others.

The morphology of the solar wind interaction region is time-variable. There are asymmetries with respect to the solar wind convection electric field, and the IMF cone angle. As the IMF varies, so does the entire interaction region. Solar wind dynamic pressure changes expose more or less of the atmosphere to the erosion processes described here. Further, the ionosphere varies as solar EUV fluxes vary, and consequently the Martian solar wind obstacle. In addition to variation in the basic structure of the interaction region with different drivers, a number of shear-related dynamical processes occur near Mars, including magnetic reconnection and its associated ionospheric plasmoids (observed) and surface instabilities (predicted and modeled), resulting in detached plasma clouds (observed). Extreme events such as solar storms also modify the interaction region, resulting in enhanced ionization, enhanced wave activity and ionospheric escape, and compression of the inner magnetosphere.

Crustal magnetic fields are sufficiently strong that they locally modify the solar wind obstacle, raising the altitude of the induced magnetosphere. They shield the atmosphere locally from incident ionizing particles, but also provide Earth-like cusp regions where currents and particles can flow between the ionosphere and solar wind. Crustal magnetic field lines are not static, and open and close, and expand and contract in response to changes in the IMF orientation and strength, as well as solar wind pressure. Ultraviolet auroral emission has been observed in cusp regions, along with incident electrons and outflowing ions consistent with field-aligned current regions.

The close proximity of the solar wind to the Martian ionosphere, as well as the considerable extent of the exosphere relative to the size of the planet, leads to a number of atmospheric escape processes for both neutral and ionized particles. This stands in contrast to planets with intrinsic magnetospheres, such as Earth, where atmospheric escape is generally confined to magnetic cusp regions. At Mars, particles may escape from most regions of the upper atmosphere, provided they: (1) have escape energy, (2) are at or above a region where collisions become rare, referred to as the exobase region, and (3) have a trajectory that carries them away from the planet. Particles may escape from the upper thermosphere, the exosphere, and the ionosphere.

Today, neutral particles may have sufficient energy to escape on their own by virtue of their thermal distribution (Jeans escape), may receive the energy necessary to escape from sunlight and subsequent chemical reactions (photochemical escape), or may receive the energy to escape from collisions with particles entering the atmosphere from above (e.g. by sputtering). Thermal escape is most relevant for hydrogen today, while the other processes remove mostly heavier atmospheric species such as oxygen, carbon, and nitrogen. Escape rates of neutral particles have been indirectly estimated using a combination of measurements of the atmospheric reservoirs for escape and models.

Escaping ionized particles are more difficult to classify because they may be accelerated to escape energy by a combination of processes. In general, we refer to (i) ion pickup, where a newly ionized particle is accelerated by a motional ($v \times B$) electric field, (ii) ion outflow (sometimes referred to as cold ionospheric outflow), where ionospheric plasma pressure gradients, ambipolar electric fields, and wave heating remove particles directly from the ionosphere via a number of mechanisms, and (iii) bulk plasma escape, where a portion of the ionosphere is detached by a shear-related process and accelerated downstream in the ambient plasma flow. Ion escape has been measured directly at Mars, though estimates of total loss rates from these *in situ* measurements vary by an order of magnitude or more. Different global simulations have a similar (or perhaps larger variation) in total ion loss rates.

The total atmospheric escape from both ions and neutrals varies with the same drivers observed to modify the solar wind interaction region. This variation can be extrapolated to estimate the total atmospheric loss over Martian history using scaling arguments, or by tuning simulations to match the present-day variation. Current efforts to estimate the total atmospheric loss yield predictions that vary from a few mbar up to several hundred mbar or more. The global water layer that has been removed is anywhere from a few centimeters up to a few tens of meters. Better constraints are clearly needed on the amount of atmosphere that has been removed to space.

Although atmospheric escape driven by processes related to the Sun and solar wind provide only one contribution to Martian atmospheric evolution, isotope ratios suggest that such processes have played an important role in the evolution of the climate. But these processes must be considered in context with other atmospheric source and loss processes (for example, volcanism is a source, and sequestration in or below the surface is a loss) if we are to successfully trace the history of the Martian atmosphere through time.

15.5.2 Outstanding Questions

The community of scientists interested in atmosphere–solar wind interactions and atmospheric escape to space has made great strides over the past several decades, progressing from fundamental discoveries about the nature of the Martian obstacle to the solar wind, to sophisticated measurements and models of the interconnected system of atmospheric reservoirs and drivers. A number of outstanding questions and needs have been raised directly or indirectly in the preceding sections (and decades). The most urgent (in our opinion) questions are provided here in an unprioritized list for the interested reader.

- What physical assumptions are required to successfully simulate the Martian plasma interaction? Do the dominant processes vary from region to region?
- What are the mechanisms responsible for the formation of the induced magnetosphere? How is it maintained and how does it adapt to changing solar wind conditions?
- On what timescales and how often does the solar wind magnetic field diffuse into the Martian ionosphere, and what are the consequences for the atmosphere?
- How and where do ionospheric currents flow close at Mars, and do they matter?
- Do magnetic and velocity shear-related surface instabilities occur at the Martian ionopause? How can they be distinguished in observations? What are their consequences?
- What mechanisms are responsible for the Martian aurora, and how frequently (and strongly) does it occur? Is it visible from the Martian surface? Does it affect the atmosphere other than locally?
- What are the escape rates of neutrals and ions from Mars, and how do they vary with the external drivers? Does the escape of neutral particles always exceed the escape of ions

from Mars under present-day conditions? What is the relative importance of ion pickup, cold ion outflow, and bulk plasma escape, and do they ever exceed neutral escape?
- What effects do crustal magnetic fields have on atmospheric escape rates?
- Is atmospheric escape at Mars diffusion-limited, or supply-limited?
- To what extent do conditions in the lower atmosphere (including dust storms) influence escape?
- How has atmospheric escape varied over Martian history? What is the net amount of atmosphere that has been removed by processes related to the Sun and solar wind? Is solar activity and its interplanetary consequences important to consider when studying atmospheric escape over long periods of time?

Three parallel research paths are likely to produce answers to these questions in the coming years. First, continued measurements (by Mars Express and MAVEN, for example) should answer or significantly constrain many of the questions posed above. Second, ever-advancing model capabilities, especially in the area of time dependence, allow increasingly detailed explorations of the global interaction of the solar wind with Mars. Finally, synthesis of results for the Martian interaction with results for analogous bodies (Venus, comets, Pluto, Titan) hold the potential to teach us how the atmospheres of unmagnetized bodies everywhere respond to their space environment.

ACKNOWLEDGMENTS

We acknowledge useful feedback from J. Luhmann and another referee for this chapter, as well as discussions with J. Deighan, W. Petersen, and R. Lillis.

REFERENCES

Acuña, M., J. E. P. Connerney, P. A. Wasilewski, et al. (1998), Magnetic field and plasma observations at Mars: initial results of the Mars global surveyor mission, *Science*, 279(5357), 1676–1680.

Anderson, D. E. J., and C. W. Hord (1971), Mariner 6 and 7 Ultraviolet Spectrometer experiment: analysis of hydrogen Lyman-alpha data, *J. Geophys. Res.*, 76(2), 6666, doi:10.1029/JA076i028p06666.

Andersson, L., R. E. Ergun, and A. I. F. Stewart (2010), The Combined Atmospheric Photochemistry and Ion Tracing code: reproducing the Viking Lander results and initial outflow results, *Icarus*, 206(1), 120–129, doi:10.1016/j.icarus.2009.07.009.

André, M., and A. Yau (1997), Theories and observations of ion energization and outflow in the high latitude magnetosphere, *Space Sci. Rev.*, 80(1), 27–48, doi:10.1023/A:1004921619885.

Ayres, T. R. (1997), Evolution of the solar ionizing flux, *J. Geophys. Res.*, 102(E), 1641–1652, doi:10.1029/96JE03306.

Baker, D. N., S. G. Kanekal, J. P. McCollough, et al. (2008), Adverse space weather at the solar cycle minimum, *American Geophysical Union Meeting*, Abstract No. SH31C-05.

Bame, S. J., J. R. Asbridge, W. C. Feldman, et al. (1980), Deceleration of the solar wind upstream from the earth's bow shock and the origin of diffuse upstream ions, *J. Geophys. Res.*, 85, 2981–2990, doi:10.1029/JA085iA06p02981.

Barabash, S., and R. Lundin (2006), ASPERA-3 on Mars Express, *Icarus*, 182(2), 301–307, doi:10.1016/j.icarus.2006.02.015.

Barabash, S., A. Fedorov, R. Lundin, and J.-A. Sauvaud (2007), Martian atmospheric erosion rates, *Science*, 315(5811), 501–503, doi:10.1126/science.1134358.

Bauske, R., A. F. Nagy, T. I. Gombosi, et al. (1998), A three-dimensional MHD study of solar wind mass loading processes at Venus: effects of photoionization, electron impact ionization, and charge exchange, *J. Geophys. Res.*, 103(A), 23625–23638, doi:10.1029/98JA01791.

Bertaux, J.-L., F. Leblanc, O. Witasse, et al. (2005), Discovery of an aurora on Mars, *Nature*, 435(7), 790–794, doi:10.1038/nature03603.

Bertucci, C., C. Mazelle, D. H. Crider, et al. (2003), Magnetic field draping enhancement at the Martian magnetic pileup boundary from Mars Global Surveyor observations, *Geophys. Res. Lett.*, 30(2), 1099, doi:10.1029/2002GL015713.

Bertucci, C., F. Duru, N. Edberg, et al. (2011), The induced magnetospheres of Mars, Venus, and Titan, *Space Sci. Rev.*, 162(1), 113–171, doi:10.1007/s11214-011-9845-1.

Brace, L. H., R. F. Theis, and W. R. Hoegy (1982), Plasma clouds above the ionopause of Venus and their implications, *Planetary and Space Science*, 30, 29–37, doi:10.1016/0032-0633(82)90069-1.

Brain, D. A. (2006), Mars Global Surveyor measurements of the Martian solar wind interaction, *Space Sci. Rev.*, 126(1), 77–112, doi:10.1007/s11214-006-9122-x.

Brain, D., and J. S. Halekas (2012), Aurora in Martian mini magnetospheres, *Auroral Phenomenology and Magnetospheric Processes: Earth and Other Planets, Geophysical Monograph Series*, 197, 123–132, doi:10.1029/2011GM001201.

Brain, D. A., and B. M. Jakosky (1998), Atmospheric loss since the onset of the Martian geologic record: combined role of impact erosion and sputtering, *J. Geophys. Res.*, 103(E), 22689–22694, doi:10.1029/98JE02074.

Brain, D. A., F. Bagenal, M. H. Acuña, et al. (2002), Observations of low-frequency electromagnetic plasma waves upstream from the Martian shock, *J. Geophys. Res. – Space*, 107(A), 1076, doi:10.1029/2000JA000416.

Brain, D. A., F. Bagenal, M. H. Acuña, and J. E. P. Connerney (2003), Martian magnetic morphology: contributions from the solar wind and crust, *J. Geophys. Res.*, 108(A), 1424, doi:10.1029/2002JA009482.

Brain, D. A., J. S. Halekas, R. Lillis, et al. (2005), Variability of the altitude of the Martian sheath, *Geophys. Res. Lett.*, 32(1), 18203, doi:10.1029/2005GL023126.

Brain, D. A., J. S. Halekas, L. M. Peticolas, et al. (2006), On the origin of aurorae on Mars, *Geophys. Res. Lett.*, 33(1), 01201, doi:10.1029/2005GL024782.

Brain, D. A., R. J. Lillis, D. L. Mitchell, J. S. Halekas, and R. P. Lin (2007), Electron pitch angle distributions as indicators of magnetic field topology near Mars, *J. Geophys. Res.*, 112(A), 09201, doi:10.1029/2007JA012435.

Brain, D. A., A. H. Baker, J. Briggs, et al. (2010a), Episodic detachment of Martian crustal magnetic fields leading to bulk atmospheric plasma escape, *Geophys. Res. Lett.*, 37(1), 14108, doi:10.1029/2010GL043916.

Brain, D., S. Barabash, A. Boesswetter, et al. (2010b), A comparison of global models for the solar wind interaction with Mars, *Icarus*, 206(1), 139–151, doi:10.1016/j.icarus.2009.06.030.

Brain, D. A., J. P. McFadden, J. S. Halekas, et al. (2015), The spatial distribution of planetary ion fluxes near Mars observed by MAVEN, *Geophys. Res. Lett.*, 42, 9142–9148, doi:10.1002/2015GL065293.

Brecht, S. H. (1990), Magnetic asymmetries of unmagnetized planets, *Geophys. Res. Lett.*, 17(9), 1243–1246, doi:10.1029/GL017i009p01243.

Brecht, S. H., and S. A. Ledvina (2006), The solar wind interaction with the Martian ionosphere/atmosphere, *Space Sci. Rev.*, 126(1), 15–38, doi:10.1007/s11214-006-9084-z.

Brecht, S. H., and S. A. Ledvina (2010), The loss of water from Mars: numerical results and challenges, *Icarus*, 206(1), 164–173, doi:10.1016/j.icarus.2009.04.028.

Brecht, S. H., and S. A. Ledvina (2012), Control of ion loss from Mars during solar minimum, *Earth, Planets and Space*, 64(2), 165–178, doi:10.5047/eps.2011.05.037.

Briggs, J., D. A. Brain, M. L. Cartwright, J. P. Eastwood, and J. S. Halekas (2011), A statistical study of flux ropes in the Martian magnetosphere, *Planetary and Space Science*, 59(1), 1498–1505, doi:10.1016/j.pss.2011.06.010.

Cameron, A. G. W. (1983), Origin of the atmospheres of the terrestrial planets, *Icarus*, 56, 195–201, doi:10.1016/0019-1035(83)90032-5.

Carr, M. H., and G. D. Clow (1981), Martian channels and valleys – their characteristics, distribution, and age, *Icarus*, 48, 91–117, doi:10.1016/0019-1035(81)90156-1.

Chassefière, E., and F. Leblanc (2004), Mars atmospheric escape and evolution; interaction with the solar wind, *Planetary and Space Science*, 52(1), 1039–1058, doi:10.1016/j.pss.2004.07.002.

Chaufray, J. Y., R. Modolo, F. Leblanc, et al. (2007), Mars solar wind interaction: formation of the Martian corona and atmospheric loss to space, *J. Geophys. Res.*, 112(E9), doi:10.1029/2007JE002915.

Chaufray, J. Y., J. L. Bertaux, F. Leblanc, and E. Quemerais (2008), Observation of the hydrogen corona with SPICAM on Mars Express, *Icarus*, 195(2), 598–613, doi:10.1016/j.icarus.2008.01.009.

Cipriani, F., F. Leblanc, and J.-J. Berthelier (2007), Martian corona: nonthermal sources of hot heavy species, *J. Geophys. Res.*, 112(E), 07001, doi:10.1029/2006JE002818.

Cloutier, P. A., C. C. Law, D. H. Crider, et al. (1999), Venus-like interaction of the solar wind with Mars, *Geophys. Res. Lett.*, 26(1), 2685–2688, doi:10.1029/1999GL900591.

Craddock, R. A., and A. D. Howard (2002), The case for rainfall on a warm, wet early Mars, *J. Geophys. Res. – Planets*, 107(E), 5111, doi:10.1029/2001JE001505.

Crider, D. H., M. H. Acuña, J. E. Connerney, et al. (2002), Observations of the latitude dependence of the location of the Martian magnetic pileup boundary, *Geophys. Res. Lett.*, 29(8), 11–1, doi:10.1029/2001GL013860.

Crider, D. H., D. A. Brain, M. H. Acuña, et al. (2004), Mars Global Surveyor Observations of Solar Wind Magnetic Field Draping Around Mars, *Space Sci. Rev.*, 111(1), 203–221, doi:10.1023/B:SPAC.0000032714.66124.4e.

Crider, D. H., J. Espley, D. A. Brain, et al. (2005), Mars Global Surveyor observations of the Halloween 2003 solar superstorm's encounter with Mars, *J. Geophys. Res.*, 110(A), doi:10.1029/2004JA010881.

Curry, S. M., M. Liemohn, X. Fang, D. Brain, and Y. Ma (2013), Simulated kinetic effects of the corona and solar cycle on high altitude ion transport at Mars, *J. Geophys. Res. – Space*, 118(6), 3700–3711, doi:10.1002/jgra.50358.

Curry, S. M., J. G. Luhmann, Y. J. Ma, et al. (2015), Response of Mars O^+ pickup ions to the 8 March 2015 ICME: inferences from MAVEN data-based models, *Geophys. Res. Lett.*, 42, 9095–9102, doi:10.1002/2015GL065304.

Delva, M., C. Mazelle, and C. Bertucci (2011), Upstream Ion Cyclotron Waves at Venus and Mars, *Space Sci. Rev.*, 162(1), 5–24, doi:10.1007/s11214-011-9828-2.

Dieval, C., E. Kallio, S. Barabash, et al. (2012), A case study of proton precipitation at Mars: Mars Express observations and hybrid simulations, *J. Geophys. Res.*, 117(A), 06222, doi:10.1029/2012JA017537.

Dong, C., S. W. Bougher, Y. Ma, et al. (2014), Solar wind interaction with Mars upper atmosphere: results from the one-way coupling between the multifluid MHD model and the MTGCM model, *Geophys. Res. Lett.*, 41(8), 2708–2715, doi:10.1002/2014GL059515.

Dong, C., Y. Ma, S. W. Bougher, et al. (2015a), Multifluid MHD study of the solar wind interaction with Mars' upper atmosphere during the 2015 March 8th ICME event, *Geophys. Res. Lett.*, 42, 9103–9112, doi:10.1002/2015GL065944.

Dong, Y., X. Fang, D. A. Brain, et al. (2015b), Strong plume fluxes at Mars observed by MAVEN: an important planetary ion escape channel, *Geophys. Res. Lett.*, 42, 8942–8950, doi:10.1002/2015GL065346.

Dryer, M., and G. R. Heckman (1967), On the hypersonic analogue as applied to planetary interaction with the solar plasma, *Planet. Space Sci.*, 15, 515–546.

Du, J., T. L. Zhang, W. Baumjohann, et al. (2010), Statistical study of low-frequency magnetic field fluctuations near Venus under the different interplanetary magnetic field orientations, *J Geophys Res-Space*, 115(A), 12251, doi:10.1029/2010JA015549.

Dubinin, E., M. Fraenz, J. Woch, et al. (2006a), Hydrogen exosphere at Mars: pickup protons and their acceleration at the bow shock, *Geophys. Res. Lett.*, 33(2), 22103, doi:10.1029/2006GL027799.

Dubinin, E., M. Fränz, J. Woch, et al. (2006b), Plasma Morphology at Mars. Aspera-3 Observations, *Space Sci Rev*, 126(1), 209–238, doi:10.1007/s11214-006-9039-4.

Dubinin, E., R. Modolo, M. Fraenz, et al. (2008a), Plasma environment of Mars as observed by simultaneous MEX-ASPERA-3 and MEX-MARSIS observations, *J. Geophys. Res.*, 113(A), 10217, doi:10.1029/2008JA013355.

Dubinin, E., R. Modolo, M. Fraenz, et al. (2008b), Structure and dynamics of the solar wind/ionosphere interface on Mars: MEX-ASPERA-3 and MEX-MARSIS observations, *Geophys. Res. Lett.*, 35(1), 11103, doi:10.1029/2008GL033730.

Dubinin, E., G. Chanteur, M. Fraenz, and J. Woch (2008c), Field-aligned currents and parallel electric field potential drops at Mars. Scaling from the Earth' aurora, *Planetary and Space Science*, 56(6), 868–872, doi:10.1016/j.pss.2007.01.019.

Dubinin, E., G. Chanteur, M. Fraenz, et al. (2008d), Asymmetry of plasma fluxes at Mars. ASPERA-3 observations and hybrid simulations, *Planetary and Space Science*, 56(6), 832–835, doi:10.1016/j.pss.2007.12.006.

Dubinin, E., M. Fraenz, J. Woch, et al. (2009a), Ionospheric storms on Mars: impact of the corotating interaction region, *Geophys. Res. Lett.*, 36(1), 01105, doi:10.1029/2008GL036559.

Dubinin, E., M. Fraenz, J. Woch, S. Barabash, and R. Lundin (2009b), Long-lived auroral structures and atmospheric losses through auroral flux tubes on Mars, *Geophys. Res. Lett.*, 36(8), 08108, doi:10.1029/2009GL038209.

Dubinin, E., M. Fraenz, A. Fedorov, et al. (2011), Ion Energization and Escape on Mars and Venus, *Space Sci. Rev.*, 162(1), 173–211, doi:10.1007/s11214-011-9831-7.

Dubinin, E., M. Fraenz, J. Woch, et al. (2012), Upper ionosphere of Mars is not axially symmetrical, *Earth*, 64(2), 113–120, doi:10.5047/eps.2011.05.022.

Dubinin, E., M. Fraenz, J. Woch, et al. (2013), Toroidal and poloidal magnetic fields at Venus. Venus Express observations, *Planetary and Space Science*, 87, 19–29, doi:10.1016/j.pss.2012.12.003.

Duru, F., D. A. Gurnett, T. F. Averkamp, et al. (2006), Magnetically controlled structures in the ionosphere of Mars, *J. Geophys. Res.*, 111(A12), doi:10.1029/2006JA011975.

Duru, F., D. A. Gurnett, R. A. Frahm, et al. (2009), Steep, transient density gradients in the Martian ionosphere similar to the ionopause at Venus, *J. Geophys. Res.*, 114(A), 12310, doi:10.1029/2009JA014711.

Duru, F., D. D. Morgan, and D. A. Gurnett (2010), Overlapping ionospheric and surface echoes observed by the Mars Express radar sounder near the Martian terminator, *Geophys. Res. Lett.*, 37(2), 23102, doi:10.1029/2010GL045859.

Eastwood, J. P., D. G. Sibeck, V. Angelopoulos et al. (2008a), THEMIS observations of a hot flow anomaly: solar wind, magnetosheath, and ground-based measurements, *Geophys. Res. Lett.*, 35(17), doi:10.1029/2008GL033475.

Eastwood, J. P., D. A. Brain, J. S. Halekas, et al. (2008b), Evidence for collisionless magnetic reconnection at Mars, *Geophys. Res. Lett.*, 35(2), doi:10.1029/2007GL032289.

Edberg, N. J. T., M. Lester, S. W. H. Cowley, and A. I. Eriksson (2008), Statistical analysis of the location of the Martian magnetic pileup boundary and bow shock and the influence of crustal magnetic fields, *J. Geophys. Res.*, 113(A8), doi:10.1029/2008JA013096.

Edberg, N. J. T., U. Auster, S. Barabash, et al. (2009a), Rosetta and Mars Express observations of the influence of high solar wind pressure on the Martian plasma environment, *Ann Geophys-Germany*, 27(12), 4533–4545.

Edberg, N. J. T., D. A. Brain, M. Lester, et al. (2009b), Plasma boundary variability at Mars as observed by Mars Global Surveyor and Mars Express, *Ann. Geophys.*, 27(9), 3537–3550, doi:10.5194/angeo-27-3537-2009.

Edberg, N. J. T., H. Nilsson, A. O. Williams, et al. (2010), Pumping out the atmosphere of Mars through solar wind pressure pulses, *Geophys. Res. Lett.*, 37(3), 03107, doi:10.1029/2009GL041814.

Ergun, R. E., L. Andersson, W. K. Peterson, et al. (2006), Role of plasma waves in Mars' atmospheric loss, *Geophys. Res. Lett.*, 33(1), 14103, doi:10.1029/2006GL025785.

Espley, J. R. (2004), Observations of low-frequency magnetic oscillations in the Martian magnetosheath, magnetic pileup region, and tail, *J. Geophys. Res.*, 109(A7), doi:10.1029/2003JA010193.

Espley, J. R. (2005), Low-frequency plasma oscillations at Mars during the October 2003 solar storm, *J. Geophys. Res.*, 110(A9), doi:10.1029/2004JA010935.

Fang, X., M. W. Liemohn, A. F. Nagy, et al. (2008), Pickup oxygen ion velocity space and spatial distribution around Mars, *J. Geophys. Res.*, 113(A), 02210, doi:10.1029/2007JA012736.

Fang, X., M. W. Liemohn, A. F. Nagy, J. G. Luhmann, and Y. Ma (2010a), Escape probability of Martian atmospheric ions: controlling effects of the electromagnetic fields, *J. Geophys. Res.*, 115(A), 04308, doi:10.1029/2009JA014929.

Fang, X., M. W. Liemohn, A. F. Nagy, J. G. Luhmann, and Y. Ma (2010b), On the effect of the Martian crustal magnetic field on atmospheric erosion, *Icarus*, 206(1), 130–138, doi:10.1016/j.icarus.2009.01.012.

Fedorov, A., E. Budnik, J.-A. Sauvaud, et al. (2006), Structure of the Martian wake, *Icarus*, 182(2), 329–336, doi:10.1016/j.icarus.2005.09.021.

Fedorov, A., C. Ferrier, J.-A. Sauvaud, et al. (2008), Comparative analysis of Venus and Mars magnetotails, *Planetary and Space Science*, 56(6), 812–817, doi:10.1016/j.pss.2007.12.012.

Feldman, P. D, A, J. Steffl, J. W. Parker, et al. (2011), Rosetta-Alice observations of exospheric hydrogen and oxygen on Mars, *Icarus*, 214(2), 394–399, doi:10.1016/j.icarus.2011.06.013.

Fillingim, M. O., L. M. Peticolas, R. J. Lillis, et al. (2010), Localized ionization patches in the nighttime ionosphere of Mars and their electrodynamic consequences, *Icarus*, 206(1), 112–119, doi:10.1016/j.icarus.2009.03.005.

Fowler, C. M., L. Andersson, R. E. Ergun, et al. (2015), The first in situ electron temperature and density measurements of the Martian nightside ionosphere, *Geophys. Res. Lett.*, 42, 8854–8861, doi:10.1002/2015GL065267.

Fox, J. (1993), On the escape of oxygen and hydrogen from Mars, *Geophys. Res. Lett.*, 20(17), 1747–1750.

Fox, J. L., and F. M. Bakalian (2001), Photochemical escape of atomic carbon from Mars, *J. Geophys. Res.*, 106(A), 28785–28796, doi:10.1029/2001JA000108.

Fox, J. L., and A. Dalgarno (1983), Nitrogen escape from Mars, *Journal of Geophysical Research*, 88, 9027–9032, doi:10.1029/JA088iA11p09027.

Fox, J. L., and A. B. Hać (2009), Photochemical escape of oxygen from Mars: a comparison of the exobase approximation to a Monte Carlo method, *Icarus*, 204(2), 527–544, doi:10.1016/j.icarus.2009.07.005.

Frahm, R. A., J. R. Sharber, J. D. Winningham, et al. (2007), Locations of atmospheric photoelectron energy peaks within the Mars environment, *Space Sci. Rev.*, 126(1–4), 389–402, doi:10.1007/s11214-006-9119-5.

Fränz, M., E. Dubinin, E. Roussos, et al. (2007), Plasma Moments in the Environment of Mars, *Space Sci Rev*, 126(1–4), 165–207, doi:10.1007/s11214-006-9115-9.

Futaana, Y., S. Barabash, A. Grigoriev, et al. (2006), First ENA observations at Mars: ENA emissions from the Martian upper atmosphere, *Icarus*, 182(2), 424–430, doi:10.1016/j.icarus.2005.09.019.

Futaana, Y., S. Barabash, M. Yamauchi, et al. (2008), Mars Express and Venus Express multi-point observations of geoeffective solar flare events in December 2006, *Planetary and Space Science*, 56(6), 873–880, doi:10.1016/j.pss.2007.10.014.

Grard, R., A. Skalsky, C. Nairn, J. G. Trotignon, and K. Schwingenschuh (1992), Waves and cold plasmas near Mars, *Advances in Space Research*, 12, 243–249.

Gurnett, D. A., D. D. Morgan, F. Duru, et al. (2010), Large density fluctuations in the Martian ionosphere as observed by the Mars Express radar sounder, *Icarus*, 206(1), 83–94, doi:10.1016/j.icarus.2009.02.019.

Haider, S. A., K. K. Mahajan, and E. Kallio (2011), Mars ionosphere: a review of experimental results and modeling studies, *Rev. Geophys.*, 49(4), 4001, doi:10.1029/2011RG000357.

Halekas, J. S., D. A. Brain, R. J. Lillis, et al. (2006), Current sheets at low altitudes in the Martian magnetotail, *Geophys. Res. Lett.*, 33(1), 13101, doi:10.1029/2006GL026229.

Halekas, J. S., D. A. Brain, R. P. Lin, J. G. Luhmann, and D. L. Mitchell (2008), Distribution and variability of accelerated electrons at Mars, *Advances in Space Research*, 41(9), 1347–1352, doi:10.1016/j.asr.2007.01.034.

Halekas, J. S., J. P. Eastwood, D. A. Brain, et al. (2009), In situ observations of reconnection Hall magnetic fields at Mars: evidence for ion diffusion region encounters, *J. Geophys. Res.*, 114(A), 11204, doi:10.1029/2009JA014544.

Halekas, J. S., D. A. Brain, and J. P. Eastwood (2011), Large-amplitude compressive "sawtooth" magnetic field oscillations in the Martian magnetosphere, *J. Geophys. Res.*, 116(A), 07222, doi:10.1029/2011JA016590.

Halekas, J. S., et al. (2015), MAVEN observations of solar wind hydrogen deposition in the atmosphere of Mars, *Geophys. Res. Lett.*, 42, 8901–8909, doi:10.1002/2015GL064693.

Harada, Y., J. S. Halekas, J. P. McFadden, et al. (2015a), Marsward and tailward ions in the near-Mars magnetotail: MAVEN observations, *Geophys. Res. Lett.*, 42, 8925–8932, doi:10.1002/2015GL065005.

Harada, Y., J. S. Halekas, J. P. McFadden, et al. (2015b), Magnetic reconnection in the near-Mars magnetotail: MAVEN observations, *Geophys. Res. Lett.*, 42, 8838–8845, doi:10.1002/2015GL065004.

Harnett, E. M., and R. M. Winglee (2003), 2.5-D fluid simulations of the solar wind interacting with multiple dipoles on the surface of the Moon, *J. Geophys. Res. – Space*, 108(A), 1088, doi:10.1029/2002JA009617.

Harnett, E. M., and R. M. Winglee (2006), Three-dimensional multi-fluid simulations of ionospheric loss at Mars from nominal solar wind conditions to magnetic cloud events, *J. Geophys. Res.*, 111(A), 09213, doi:10.1029/2006JA011724.

Hasegawa, H., M. Fujimoto, T. D. Phan, et al. (2004), Transport of solar wind into Earth's magnetosphere through rolled-up Kelvin–Helmholtz vortices, *Nature*, 430(7), 755–758, doi:10.1038/nature02799.

Hassler, D. M. et al. (2012), The Radiation Assessment Detector (RAD) investigation, *Space Sci. Rev.*, 170(1), 503–558, doi:10.1007/s11214-012-9913-1.

Hodges, R. R. (2002), The rate of loss of water from mars, *Geophys. Res. Lett.*, 29(3), 1038, doi:10.1029/2001GL013853.

Hoke, M. R. T., and B. M. Hynek (2009), Roaming zones of precipitation on ancient Mars as recorded in valley networks, *J. Geophys. Res.*, 114(E), 08002, doi:10.1029/2008JE003247.

Hunten, D. M. (1982), Thermal and nonthermal escape mechanisms for terrestrial bodies, *Planetary and Space Science*, 30, 773–783, doi:10.1016/0032-0633(82)90110-6.

Hunten, D. M. (1992), Evolution of the atmosphere of Venus and Mars, in *Venus and Mars: Atmospheres, Ionosphers, and Solar Wind Interactions*, Proceedings of the Chapman Conference, Balatonfured, Hungary, June 4–8, 1990 (A92-50426-21-91).

Hutchins, K. S., B. M. Jakosky, and J. G. Luhmann (1997), Impact of a paleomagnetic field on sputtering loss of Martian atmospheric argon and neon, *J. Geophys. Res.*, 102(E), 9183–9190, doi:10.1029/96JE03838.

Jakosky, B. M., and J. H. Jones (1994), Evolution of water on Mars, *Nature*, 370(6), 328–329, doi:10.1038/370328a0.

Jakosky, B. M., and R. J. Phillips (2001), Mars' volatile and climate history, *Nature*, 412(6), 237–244.

Jakosky, B. M., R. O. Pepin, R. E. Johnson, and J. L. Fox (1994), Mars atmospheric loss and isotopic fractionation by solar-wind-induced sputtering and photochemical escape, *Icarus*, 111, 271–288, doi:10.1006/icar.1994.1145.

Jakosky, B. M., J. M. Grebowsky, J. M. Luhmann, et al. (2015a), The Mars Atmosphere and Volatile Evolution (MAVEN) mission, *Space Sci Rev*, 21, doi:10.1007/s11214-015-0139-x.

Jakosky, B. M., J. M. Grebowsy, J. G. Luhmann, and D. A. Brain (2015b), Initial results from the MAVEN mission to Mars, *Geophys. Res. Lett.*, 42, 8791–8802, doi:10.1002/2015GL065271.

Jakosky, B. M. et al. (2015c), MAVEN observations of the response of Mars to an interplanetary coronal mass ejection, *Science*, 350(6), 0210, doi:10.1126/science.aad0210.

Johnson, R. E., and J. G. Luhmann (1998), Sputter contribution to the atmospheric corona on Mars, *J. Geophys. Res.*, 103, 3649, doi:10.1029/97JE03266.

Kallio, E., H. Koskinen, S. Barabash, C. M. C. Nairn, and K. Schwingenschuh (1995), Oxygen outflow in the Martian magnetotail, *Geophys. Res. Lett.*, 22(1), 2449–2452, doi:10.1029/95GL02474.

Kallio, E., R. A. Frahm, Y. Futaana, A. Fedorov, and P. Janhunen (2008), Morphology of the magnetic field near Mars and the role of the magnetic crustal anomalies: dayside region, *Planetary and Space Science*, 56(6), 852–855, doi:10.1016/j.pss.2007.12.002.

Kallio, E., K. Liu, R. Jarvinen, V. Pohjola, and P. Janhunen (2010), Oxygen ion escape at Mars in a hybrid model: high energy and low energy ions, *Icarus*, 206(1), 152–163, doi:10.1016/j.icarus.2009.05.015.

Kallio, E., J.-Y. Chaufray, R. Modolo, D. Snowden, and R. Winglee (2011), Modeling of Venus, Mars, and Titan, *Space Sci. Rev.*, 162(1), 267–307, doi:10.1007/s11214-011-9814-8.

Khodachenko, M. L. et al. (2007), Coronal mass ejection (CME) activity of low mass M stars as an important factor for the habitability of terrestrial exoplanets. I. CME impact on expected magnetospheres of Earth-like exoplanets in close-in habitable zones, *Astrobiology*, 7(1), 167–184, doi:10.1089/ast.2006.0127.

Kim, J., A. F. Nagy, J. L. Fox, and T. E. Cravens (1998), Solar cycle variability of hot oxygen atoms at Mars, *J. Geophys. Res.*, 103(A), 29339–29342, doi:10.1029/98JA02727.

Kotova, G. A., M. I. Verigin, N. M. Shutte, et al. (1997), Planetary heavy ions in the magnetotail of Mars – results of the TAUS and MAGMA experiments aboard PHOBOS, *Advances in Space Research*, 20, 173, doi:10.1016/S0273-1177(97)00529-2.

Krasnopolsky, V. A. (1993), Photochemistry of the Martian atmosphere (mean conditions), *Icarus*, 101, 313–332, doi:10.1006/icar.1993.1027.

Krasnopolsky, V. A., and P. D. Feldman (2001), Detection of molecular hydrogen in the atmosphere of Mars, *Science*, 294(5), 1914–1917, doi:10.1126/science.1065569.

Krasnopolsky, V. A., G. L. Bjoraker, M. J. Mumma, and D. E. Jennings (1997), High-resolution spectroscopy of Mars at 3.7 and 8 μm: a sensitive search of H_2O_2, H_2CO, HCl, and CH_4, and detection of HDO, *J. Geophys. Res.*, 102(E), 6525–6534, doi:10.1029/96JE03766.

Krest'yanikova, M. A., and V. I. Shematovich (2005), Stochastic models of hot planetary and satellite coronas: a photochemical source of hot oxygen in the upper atmosphere of Mars, *Sol. Syst. Res.*, 39(1), 22–32, doi:10.1007/s11208-005-0002-9.

Krymskii, A. M., T. K. Breus, N. F. Ness, et al. (2002), Structure of the magnetic field fluxes connected with crustal magnetization and topside ionosphere at Mars, *J. Geophys. Res. – Space*, 107(A), 1245, doi:10.1029/2001JA000239.

Lammer, H., and S. J. Bauer (1991), Nonthermal atmospheric escape from Mars and Titan, *Journal of Geophysical Research*, 96, 1819–1825, doi:10.1029/90JA01676.

Lammer, H., F. Selsis, I. Ribas, et al. (2003), Atmospheric loss of exoplanets resulting from stellar X-ray and extreme-ultraviolet heating, *Astrophysical Journal*, 598(2), L121–L124, doi:10.1086/380815.

Lammer, H., J. F. Kasting, E. Chassefière, et al. (2008), Atmospheric escape and evolution of terrestrial planets and satellites, *Space Sci. Rev.*, 139(1), 399–436, doi:10.1007/s11214-008-9413-5.

Lammer, H., J. F. Kasting, E. Chassefière, et al. (2009), Atmospheric escape and evolution of terrestrial planets and satellites, *Comparative Aeronomy*, 2, 399, doi:10.1007/978-0-387-87825-6_11.

Leblanc, F., and R. E. Johnson (2002), Role of molecular species in pickup ion sputtering of the Martian atmosphere, *J. Geophys. Res. – Planet*, 107(E), 5010, doi:10.1029/2000JE001473.

Leblanc, F., J. G. Luhmann, R. E. Johnson, and E. Chassefiere (2002), Some expected impacts of a solar energetic particle event at Mars, *J. Geophys. Res. – Space*, 107(A), 1058, doi:10.1029/2001JA900178.

Leblanc, F., O. Witasse, J. Winningham, et al. (2006), Origins of the Martian aurora observed by Spectroscopy for Investigation of Characteristics of the Atmosphere of Mars (SPICAM) on board Mars Express, *J. Geophys. Res.*, 111(A), 09313, doi:10.1029/2006JA011763.

Leblanc, F., R. Modolo, S. Curry, et al. (2015), Mars heavy ion precipitating flux as measured by Mars Atmosphere and Volatile Evolution, *Geophys. Res. Lett.*, 42, 9135–9141, doi:10.1002/2015GL066170.

Ledvina, S. A., Y. J. Ma, and E. Kallio (2008), Modeling and simulating flowing plasmas and related phenomena, *Space Sci. Rev.*, 139(1), 143–189, doi:10.1007/s11214-008-9384-6.

Liemohn, M. W., D. L. Mitchell, A. F. Nagy, et al. (2003), Comparisons of electron fluxes measured in the crustal fields at Mars by the MGS magnetometer/electron reflectometer instrument with a B field-dependent transport code, *J. Geophys. Res.*, 108(E), 5134, doi:10.1029/2003JE002158.

Liemohn, M. W. et al. (2006), Numerical interpretation of high-altitude photoelectron observations, *Icarus*, 182(2), 383–395, doi:10.1016/j.icarus.2005.10.036.

Lillis, R. J., H. V. Frey, and M. Manga (2008), Rapid decrease in Martian crustal magnetization in the Noachian era: implications for the dynamo and climate of early Mars, *Geophys. Res. Lett.*, 35(1), 14203, doi:10.1029/2008GL034338.

Lillis, R. J., M. O. Fillingim, and D. A. Brain (2011), Three-dimensional structure of the Martian nightside ionosphere: predicted rates of impact ionization from Mars Global Surveyor magnetometer and electron reflectometer measurements of precipitating electrons, *J. Geophys. Res.*, 116(A), 12317, doi:10.1029/2011JA016982.

Lillis, R. J., D. A. Brain, G. T. Delory, et al. (2012), Evidence for superthermal secondary electrons produced by SEP ionization in the Martian atmosphere, *J. Geophys. Res.*, 117(E), 03004, doi:10.1029/2011JE003932.

Lillis, R. J., D. A. Brain, S. W. Bougher, et al. (2015), Characterizing atmospheric escape from Mars today and through time, with MAVEN, *Space Sci. Rev.*, 195, 357–422.

Liu, Y., A. F. Nagy, C. P. T. Groth, et al. (1999), 3D multi-fluid MHD studies of the solar wind interaction with Mars, *Geophys. Res. Lett.*, 26(1), 2689–2692, doi:10.1029/1999GL900584.

Luhmann, J. G. (1990), A model of the ion wake of Mars, *Geophysical Research Letters*, 17, 869–872, doi:10.1029/GL017i006p00869.

Luhmann, J. G. (1992), Pervasive large-scale magnetic fields in the Venus nightside ionosphere and their implications, *Journal of Geophysical Research*, 97, 6103–6121, doi:10.1029/92JE00514.

Luhmann, J. G. (1995), Plasma interactions with unmagnetized bodies, *Introduction to Space Physics*.

Luhmann, J., and J. U. Kozyra (1991), Dayside pickup oxygen ion precipitation at Venus and Mars – spatial distributions, energy deposition and consequences, *J. Geophys. Res.*, 96, 5457–5467.

Luhmann, J. G., and K. Schwingenschuh (1990), A model of the energetic ion environment of Mars, *Journal of Geophysical Research*, 95, 939–945, doi:10.1029/JA095iA02p00939.

Luhmann, J. G., C. T. Russell, L. H. Brace, and O. L. Vaisberg (1992a), The intrinsic magnetic field and solar-wind interaction of Mars, In *Mars*, 1090–1134.

Luhmann, J. G., R. E. Johnson, and M. H. G. Zhang (1992b), Evolutionary impact of sputtering of the Martian atmosphere by O^+ pickup ions, *Geophysical Research Letters*, 19, 2151–2154, doi:10.1029/92GL02485.

Luhmann, J. G., M. H. Acuña, M. Purucker, C. T. Russell, and J. G. Lyon (2002), The Martian magnetosheath: how Venus-like? *Planetary and Space Science*, 50(5), 489–502, doi:10.1016/S0032-0633(02)00028-4.

Luhmann, J. G., C. Dong, Y. Ma, et al. (2015), Implications of MAVEN Mars near-wake measurements and models, *Geophys. Res. Lett.*, 42, 9087–9094, doi:10.1002/2015GL066122.

Lundin, R. (2011), Ion acceleration and outflow from Mars and Venus: an overview, *Space Sci. Rev.*, 162(1), 309–334, doi:10.1007/s11214-011-9811-y.

Lundin, R., H. Borg, B. Hultqvist, A. Zakharov, and R. Pellinen (1989), First measurements of the ionospheric plasma escape from Mars, *Nature*, 341, 609–612, doi:10.1038/341609a0.

Lundin, R., A. Zakharov, R. Pellinen, et al. (1990), ASPERA/Phobos measurements of the ion outflow from the Martian ionosphere, *Geophysical Research Letters*, 17, 873–876, doi:10.1029/GL017i006p00873.

Lundin, R., O. Norberg, E. M. Dubinin, N. Pisarenko, and H. Koskinen (1991), On the momentum transfer of the solar wind to the

Martian topside ionosphere, *Geophysical Research Letters*, 18, 1059–1062, doi:10.1029/90GL02604.

Lundin, R., S. Barabash, H. Andersson, et al. (2004), Solar wind-induced atmospheric erosion at Mars: first results from ASPERA-3 on Mars Express, *Science*, 305(5), 1933–1936, doi:10.1126/science.1101860.

Lundin, R., D. Winningham, S. Barabash, et al. (2006), Plasma acceleration above Martian magnetic anomalies, *Science*, 311(5), 980–983, doi:10.1126/science.1122071.

Lundin, R., S. Barabash, A. Fedorov, et al. (2008), Solar forcing and planetary ion escape from Mars, *Geophys. Res. Lett.*, 35(9), 09203, doi:10.1029/2007GL032884.

Lundin, R., S. Barabash, M. Holmström, et al. (2009), Atmospheric origin of cold ion escape from Mars, *Geophys. Res. Lett.*, 36(1), 17202, doi:10.1029/2009GL039341.

Lundin, R., S. Barabash, M. Yamauchi, H. Nilsson, and D. Brain (2011), On the relation between plasma escape and the Martian crustal magnetic field, *Geophys. Res. Lett.*, 38(2), 02102, doi:10.1029/2010GL046019.

Ma, Y.-J., and A. F. Nagy (2007), Ion escape fluxes from Mars, *Geophys. Res. Lett.*, 34(8), 08201, doi:10.1029/2006GL029208.

Ma, Y., A. F. Nagy, K. C. Hansen, et al. (2002), Three-dimensional multispecies MHD studies of the solar wind interaction with Mars in the presence of crustal fields, *J. Geophys. Res. – Space*, 107(A), 1282, doi:10.1029/2002JA009293.

Ma, Y., A. F. Nagy, I. V. Sokolov, and K. C. Hansen (2004), Three-dimensional, multispecies, high spatial resolution MHD studies of the solar wind interaction with Mars, *J. Geophys. Res.*, 109(A), 07211, doi:10.1029/2003JA010367.

Manning, C. V., Y. Ma, D. A. Brain, C. P. McKay, and K. J. Zahnle (2011), Parametric analysis of modeled ion escape from Mars, *Icarus*, 212(1), 131–137, doi:10.1016/j.icarus.2010.11.028.

McElroy, M. B. (1972), Mars: an evolving atmosphere, *Science*, 175, 443–445.

McElroy, M. B., T. Y. Kong, and Y. L. Yung (1977), Photochemistry and evolution of Mars' atmosphere – a Viking perspective, *J. Geophys. Res.*, 82, 4379–4388.

McKenna-Lawlor, S. M. P., M. Dryer, C. D. Fry, et al. (2005), Predictions of energetic particle radiation in the close Martian environment, *J. Geophys. Res.*, 110(A), 03102, doi:10.1029/2004JA010587.

Melosh, H. J., and A. M. Vickery (1989), Impact erosion of the primordial atmosphere of Mars, *Nature*, 338, 487–489, doi:10.1038/338487a0.

Mendillo, M., P. Withers, D. Hinson, H. Rishbeth, and B. Reinisch (2006), Effects of Solar Flares on the Ionosphere of Mars, *Science*, 311(5), 1135–1138, doi:10.1126/science.1122099.

Mitchell, D. L., R. P. Lin, H. Rème, et al. (2000), Oxygen Auger electrons observed in Mars' ionosphere, *Geophys. Res. Lett.*, 27(1), 1871–1874, doi:10.1029/1999GL010754.

Mitchell, D. L., R. P. Lin, C. Mazelle, et al. (2001), Probing Mars' crustal magnetic field and ionosphere with the MGS Electron Reflectometer, *J. Geophys. Res.*, 106(E), 23419–23428, doi:10.1029/2000JE001435.

Modolo, R., and G. M. Chanteur (2008), A global hybrid model for Titan's interaction with the Kronian plasma: application to the Cassini Ta flyby, *J. Geophys. Res.*, 113(A), 01317, doi:10.1029/2007JA012453.

Modolo, R., G. M. Chanteur, E. Dubinin, and A. P. Matthews (2005), Influence of the solar EUV flux on the Martian plasma environment, *Ann. Geophys. – Germany*, 23(2), 433–444, doi:10.5194/angeo-23-433-2005.

Modolo, R., G. M. Chanteur, E. Dubinin, and A. P. Matthews (2006), Simulated solar wind plasma interaction with the Martian exosphere: influence of the solar EUV flux on the bow shock and the magnetic pile-up boundary, *Ann. Geophys. – Germany*, 24(12), 3403–3410.

Modolo, R., G. M. Chanteur, and E. Dubinin (2012), Dynamic Martian magnetosphere: transient twist induced by a rotation of the IMF, *Geophys. Res. Lett.*, 39(1), 01106, doi:10.1029/2011GL049895.

Moore, T. E., and J. L. Horwitz (2007), Stellar ablation of planetary atmospheres, *Rev. Geophys.*, 45(3), 3002, doi:10.1029/2005RG000194.

Morgan, D. D., D. A. Gurnett, D. L. Kirchner, et al. (2006), Solar control of radar wave absorption by the Martian ionosphere, *Geophys. Res. Lett.*, 33(13), doi:10.1029/2006GL026637.

Moses, S. L., F. V. Coroniti, and F. L. Scarf (1988), Expectations for the microphysics of the Mars-solar wind interaction, *Geophysical Research Letters*, 15, 429–432, doi:10.1029/GL015i005p00429.

Nagy, A. F., M. W. Liemohn, J. L. Fox, and J. Kim (2001), Hot carbon densities in the exosphere of Mars, *J. Geophys. Res.*, 106(A), 21565–21568, doi:10.1029/2001JA000007.

Nagy, A. F., D. Winterhalter, K. Sauer, et al. (2004), The plasma environment of Mars, *Space Sci. Rev.*, 111(1), 33–114, doi:10.1023/B:SPAC.0000032718.47512.92.

Najib, D., A. F. Nagy, G. Tóth, and Y. Ma (2011), Three-dimensional, multifluid, high spatial resolution MHD model studies of the solar wind interaction with Mars, *J. Geophys. Res. – Space*, 116(A), A05204, doi:10.1029/2010JA016272.

Němec, F., D. D. Morgan, D. A. Gurnett, and F. Duru (2010), Nightside ionosphere of Mars: radar soundings by the Mars Express spacecraft, *J. Geophys. Res.*, 115(E), 12009, doi:10.1029/2010JE003663.

Němec, F., D. D. Morgan, D. A. Gurnett, and D. A. Brain (2011), Areas of enhanced ionization in the deep nightside ionosphere of Mars, *J. Geophys. Res.*, 116(E), 06006, doi:10.1029/2011JE003804.

Nielsen, E., H. Zou, D. A. Gurnett, et al. (2006), Observations of vertical reflections from the topside Martian ionosphere, *Space Sci. Rev.*, 126(1–4), 373–388, doi:10.1007/s11214-006-9113-y.

Nielsen, E., D. Morgan, D. Kirchner, J. Plaut, and G. Picardi (2007), Absorption and reflection of radio waves in the Martian ionosphere, *Planetary and Space Science*, 55(7–8), 864–870, doi:10.1016/j.pss.2006.10.005.

Nilsson, H., E. Carlsson, H. Gunell, et al. (2006), Investigation of the influence of magnetic anomalies on ion distributions at Mars, *Space Sci. Rev.*, 126(1–4), 355–372, doi:10.1007/s11214-006-9030-0.

Nilsson, H., E. Carlsson, D. A. Brain, et al. (2010), Ion escape from Mars as a function of solar wind conditions: a statistical study, *Icarus*, 206(1), 40–49, doi:10.1016/j.icarus.2009.03.006.

Nilsson, H., N. Edberg, G. Stenberg, and S. Barabash (2011), Heavy ion escape from Mars, influence from solar wind conditions and crustal magnetic fields, *Icarus*, doi:10.1016/j.icarus.2011.08.003.

Nilsson, H., G. Stenberg, Y. Futaana, et al. (2012), Ion distributions in the vicinity of Mars: signatures of heating and acceleration processes, *Earth*, 64(2), 135–148, doi:10.5047/eps.2011.04.011.

Øieroset, M., D. L. Mitchell, T. D. Phan, R. P. Lin, and M. H. Acuña (2001), Hot diamagnetic cavities upstream of the Martian bow shock, *Geophys. Res. Lett.*, 28(5), 887–890, doi:10.1029/2000GL012289.

Ong, M., J. G. Luhmann, C. T. Russell, R. J. Strangeway, and L. H. Brace (1991), Venus ionospheric "clouds" – relationship to the magnetosheath field geometry, *Journal of Geophysical Research*, 96, 11133, doi:10.1029/91JA01100.

Opgenoorth, H. J., R. S. Dhillon, L. Rosenqvist, et al. (2010), Day-side ionospheric conductivities at Mars, *Planetary and Space Science*, 58(10), 1139–1151, doi:10.1016/j.pss.2010.04.004.

Owen, T., J. P. Maillard, C. de Bergh, and B. L. Lutz (1988), Deuterium on Mars – the abundance of HDO and the value of D/H, *Science*, 240, 1767–1770.

Penz, T., N. V. Erkaev, H. K. Biernat, et al. (2004), Ion loss on Mars caused by the Kelvin–Helmholtz instability, *Planetary and Space Science*, 52(13), 1157–1167, doi:10.1016/j.pss.2004.06.001.

Penz, T., I. L. Arshukova, N. Terada, et al. (2005), A comparison of magnetohydrodynamic instabilities at the Martian ionopause, *Advances in Space Research*, 36(1), 2049–2056, doi:10.1016/j.asr.2004.11.039.

Phillips, J. L., J. G. Luhmann, W. C. Knudsen, and L. H. Brace (1988), Asymmetries in the location of the Venus ionopause, *Journal of Geophysical Research*, 93, 3927–3941, doi:10.1029/JA093iA05p03927.

Podgornyi, I. M., E. M. Dubinin, P. L. Israelevich, and C. P. Sonett (1982), Comparison of measurements of electromagnetic induction in the magnetosphere of Venus with laboratory simulations, *Moon and the Planets*, 27, 397–406, doi:10.1007/BF00929994.

Purucker, M. E., C. L. Johnson, R. M. Winslow, et al. (2012), Evidence for a crustal magnetic signature on Mercury form MESSENGER magnetometer observations, in *43rd Lunar and Planetary Science Conference*, March 19–23, The Woodlands, TX, LPI Contribution No. 1659.

Rahmati, A., D. E. Larson, T. E. Cravens, et al. (2015), MAVEN insights into oxygen pickup ions at Mars, *Geophys. Res. Lett.*, 42, 8870–8876, doi:10.1002/2015GL065262.

Ribas, I., E. F. Guinan, M. Güdel, and M. Audard (2005), Evolution of the solar activity over time and effects on planetary atmospheres. I. High-energy irradiances (1–1700 Å), *Astrophysical Journal*, 622(1), 680–694, doi:10.1086/427977.

Richer, E., G. M. Chanteur, R. Modolo, and E. Dubinin (2012), Reflection of solar wind protons on the Martian bow shock: investigations by means of 3-dimensional simulations, *Geophys. Res. Lett.*, 39(1), 17101, doi:10.1029/2012GL052858.

Russell, C. T. (2001), Solar wind and interplanetary magnetic field: a tutorial, *Space Weather*, 125, 73–89, doi:10.1029/GM125p0073.

Russell, C. T., and R. C. Elphic (1979), Observation of magnetic flux ropes in the Venus ionosphere, *Nature*, 279, 616–618, doi:10.1038/279616a0.

Russell, C. T., and O. Vaisberg (1983), The interaction of the solar wind with Venus, in *Venus*, Tucson, AZ, University of Arizona Press, 873–940 (A83-37401 17-91).

Russell, C. T., J. G. Luhmann, K. Schwingenschuh, W. Riedler, and Y. Yeroshenko (1990), Upstream waves at Mars – PHOBOS observations, *Geophysical Research Letters*, 17, 897–900, doi:10.1029/GL017i006p00897.

Safaeinili, A., W. Kofman, J. Mouginot, et al. (2007), Estimation of the total electron content of the Martian ionosphere using radar sounder surface echoes, *Geophys. Res. Lett.*, 34(2), 23204, doi:10.1029/2007GL032154.

Sagdeev, R. Z., and A. V. Zakharov (1989), Brief history of the Phobos mission, *Nature*, 341(6243), 581–585, doi:10.1038/341581a0.

Schneider, N. M., J. I. Deighan, S. K. Jain, et al. (2015), Discovery of diffuse aurora on Mars, *Science*, 350(6261), doi:10.1126/science.aad0313.

Schwadron, N. A., L. Townsend, K. Kozarev, et al. (2010), Earth–Moon–Mars Radiation Environment Module framework, *Space Weather*, 8(1), doi:10.1029/2009SW000523.

Simon, S., A. Boesswetter, T. Bagdonat, and U. Motschmann (2007), Physics of the Ion Composition Boundary: a comparative 3-D hybrid simulation study of Mars and Titan, *Ann. Geophys. – Germany*, 25(1), 99–115.

Spreiter, J. R., A. L. Summers, and A. W. Rizzi (1970), Solar wind flow past nonmagnetic planets – Venus and Mars, *Planet. Space Sci.*, 18, 1281–1299.

Tatrallyay, M., G. Gévai, I. Apáthy, et al. (1997), Magnetic field overshoots in the Martian bow shock, *J. Geophys. Res.*, 102(A), 2157–2164, doi:10.1029/96JA00073.

Terada, N., S. Machida, and H. Shinagawa (2002), Global hybrid simulation of the Kelvin–Helmholtz instability at the Venus ionopause, *J. Geophys. Res. – Space*, 107(A), 1471, doi:10.1029/2001JA009224.

Terada, N., Y. N. Kulikov, H. Lammer, et al. (2009), Atmosphere and water loss from early Mars under extreme solar wind and extreme ultraviolet conditions, *Astrobiology*, 9(1), 55–70, doi:10.1089/ast.2008.0250.

Trotignon, J. G., C. Mazelle, C. Bertucci, and M. H. Acuña (2006), Martian shock and magnetic pile-up boundary positions and shapes determined from the Phobos 2 and Mars Global Surveyor data sets, *Planetary and Space Science*, 54(4), 357–369, doi:10.1016/j.pss.2006.01.003.

Uluşen, D., and I. Linscott (2008), Low-energy electron current in the Martian tail due to reconnection of draped interplanetary magnetic field and crustal magnetic fields, *J. Geophys. Res.*, 113(E), 06001, doi:10.1029/2007JE002916.

Uluşen, D., D. A. Brain, J. G. Luhmann, and D. L. Mitchell (2012), Investigation of Mars' ionospheric response to solar energetic particle events, *J. Geophys. Res.*, 117(A), 12306, doi:10.1029/2012JA017671.

Valeille, A., M. R. Combi, S. W. Bougher, V. Tenishev, and A. F. Nagy (2009a), Three-dimensional study of Mars upper thermosphere/ionosphere and hot oxygen corona: 2. Solar cycle, seasonal variations, and evolution over history, *J. Geophys. Res. – Planets*, 114, E11006, doi:10.1029/2009JE003389.

Valeille, A., V. Tenishev, S. W. Bougher, M. R. Combi, and A. F. Nagy (2009b), Three-dimensional study of Mars upper thermosphere/ionosphere and hot oxygen corona: 1. General description and results at equinox for solar low conditions, *J. Geophys. Res. – Planets*, 114, E11005, doi:10.1029/2009JE003388.

Verigin, M. I., N. M. Shuttle, A. A. Galeev, et al. (1991), Ions of planetary origin in the Martian magnetosphere (Phobos 2/TAUS experiment), *Planet. Space Sci.*, 39, 131–137, doi:10.1016/0032-0633(91)90135-W.

Verigin, M. I., G. A. Kotova, A. P. Remizov, et al. (2001), Evidence of the influence of equatorial Martian crustal magnetization on the position of the planetary magnetotail boundary by phobos 2 data, *Advances in Space Research*, 28(6), 885–889, doi:10.1016/S0273-1177(01)00510-5.

Vignes, D., M. H. Acuña, J. E. P. Connerney, et al. (2002), Factors controlling the location of the Bow Shock at Mars, *Geophys. Res. Lett.*, 29(9), 42–1, doi:10.1029/2001GL014513.

Vignes, D., M. H. Acuña, J. E. P. Connerney, et al. (2004), Magnetic flux ropes in the Martian atmosphere: global characteristics, *Space Sci. Rev.*, 111(1), 223–231, doi:10.1023/B:SPAC.0000032716.21619.f2.

Watson, C. C., P. K. Haff, and T. A. Tombrello (1980), Solar wind sputtering effects in the atmospheres of Mars and Venus, In *Lunar and Planetary Science Conference*, 11, 2479–2502.

Wei, H. Y., and C. T. Russell (2006), Proton cyclotron waves at Mars: exosphere structure and evidence for a fast neutral disk, *Geophys. Res. Lett.*, 33(23), doi:10.1029/2006GL026244.

Withers, P. (2005), Ionospheric characteristics above Martian crustal magnetic anomalies, *Geophys. Res. Lett.*, 32(16), doi:10.1029/2005GL023483.

Withers, P. (2009), A review of observed variability in the dayside ionosphere of Mars, *Advances in Space Research*, 44(3), 277–307, doi:10.1016/j.asr.2009.04.027.

Wolff, R. S., B. E. Goldstein, and C. M. Yeates (1980), The onset and development of Kelvin–Helmholtz instability at the Venus ionopause, *J. Geophys. Res.*, 85, 7697–7707, doi:10.1029/JA085iA13p07697.

Wood, B. E., H. R. Müller, G. P. Zank, J. L. Linsky, and S. Redfield (2005), New mass-loss measurements from astrospheric

Lyα absorption, *Astrophysical Journal*, 628(2), L143–L146, doi:10.1086/432716.

Yagi, M., F. Leblanc, J. Y. Chaufray, et al. (2012), Mars exospheric thermal and non-thermal components: seasonal and local variations, *Icarus*, 221(2), 682–693, doi:10.1016/j.icarus.2012.07.022.

Yamauchi, M., Y. Futaana, A. Fedorov, et al. (2012), Ion acceleration by multiple reflections at Martian bow shock, *Earth*, 64(2), 61–71, doi:10.5047/eps.2011.07.007.

Yau, A. W., T. Abe, and W. K. Peterson (2007), The polar wind: recent observations, *Journal of Atmospheric and Solar-Terrestrial Physics*, 69(1), 1936–1983, doi:10.1016/j.jastp.2007.08.010.

Yeroshenko, Y., W. Riedler, K. Schwingenschuh, J. G. Luhmann, and M. Ong (1990), The magnetotail of Mars – PHOBOS observations, *Geophysical Research Letters*, 17, 885–888, doi:10.1029/GL017i006p00885.

Yung, Y. L., J. S. Wen, J. P. Pinto, et al. (1988), HDO in the Martian atmosphere – implications for the abundance of crustal water, *Icarus*, 76(1), 146–159, doi:10.1016/0019-1035(88)90147-9.

Zahnle, K. J., and J. F. Kasting (1986), Mass fractionation during transonic escape and implications for loss of water from Mars and Venus, *Icarus*, 68, 462–480, doi:10.1016/0019-1035(86)90051-5.

Zhang, T. L., J. G. Luhmann, and C. T. Russell (1991a), The magnetic barrier at Venus, *Journal of Geophysical Research*, 96, 11145, doi:10.1029/91JA00088.

Zhang, T. L., K. Schwingenschuh, H. Lichtenegger, W. Riedler, and C. T. Russell (1991b), Interplanetary magnetic field control of the Mars bow shock – evidence for Venus like interaction, *Journal of Geophysical Research*, 96, 11265, doi:10.1029/91JA01099.

Recent Climate Variations

FRANÇOIS FORGET, SHANE BYRNE, JAMES W. HEAD, MICHAEL A. MISCHNA, NORBERT SCHÖRGHOFER

16.1 INTRODUCTION

As detailed in previous chapters, the present-day Mars climate is a complex system in which the atmospheric dynamics are coupled with the dust cycle (mineral dust is lifted by the wind and controls the radiative properties of the atmosphere), the CO_2 cycle (the carbon dioxide atmosphere condenses seasonally at high latitudes to form polar caps, inducing surface pressure variations over all the planet), and the water cycle (water vapor is transported by the atmosphere and forms clouds, hazes, and frost which affect the radiative balance). This climate is highly variable in space and time. For instance, global dust storms can shroud the planet and strongly warm the atmosphere some years, but not every year.

The purpose of this chapter is to explore how this climate system may vary over longer timescales, ranging between about 100 and 10^8 years. There are two reasons why we can believe that the Martian climate has strongly varied in the past and will strongly vary in the future on such timescales. First, the climate on Mars depends on the orbital and rotation parameters of the planet, and in particular its obliquity (the inclination of Mars' axis of rotation to its orbit plane). In the Earth's case, such oscillations are small ($\pm 1.3°$ for obliquity), but they are thought to have played a key role in the glacial and interglacial climate cycles (Imbrie and Imbrie, 1979). In the Martian case, calculations have shown that Mars' obliquity can vary widely and somewhat erratically, between 0° and more than 60° (Laskar et al., 2004). Such large variations must induce considerable climate variations. Second, while water ice is currently unstable at the surface of Mars outside the polar regions, Mars is partly covered by landforms resulting from the local accumulation of ice, such as debris-covered glaciers and ice mantles. These landforms are thought to have formed in the geologically recent past, when the climate system was not very different than today. The kilometers-thick polar ice caps themselves also appear to have formed relatively recently. They exhibit thousands of layers that are most likely related to periodic climate changes.

In this chapter, we first review our current knowledge of the variations of Mars obliquity and orbital parameters, and their expected impact on insolation (incoming solar radiation), surface temperatures, the CO_2 cycle, and atmospheric pressure. This is followed by a discussion of the induced change on atmospheric circulation and the dust cycle. We then focus on water (in all phases), with an overview of the geological evidence for climate variations and past ice ages on Mars followed by a description of their interpretation by climate modeling studies. We also discuss the possibility of liquid water in the past millions of years on Mars. Finally, we mention the possibility of very recent climate change in the past hundred years.

16.2 VARIATION OF MARS' ORBITAL AND SPIN PARAMETERS

Like all objects in the Solar System, Mars' orbital and spin motions slowly evolve as a consequence of gravitational interaction with the other bodies in the system. In the case of Mars, these dynamical variations are especially vigorous because of resonances. This results in substantial changes in the surface distribution of insolation and heating over many timescales. While tightly coupled, the dominant orbital parameters, i.e. obliquity, eccentricity, and the solar longitude of perihelion, each independently modify the timing, distribution, and magnitude of the solar radiation incident on the Martian surface. These three parameters, known in the terrestrial literature as Milankovitch parameters, are discussed in turn.

16.2.1 Obliquity

Presently, the obliquity of Mars is 25.2°, quite similar to Earth's present value of 23.5°. Thus, the seasons on Mars undergo much the same annual cycle as those on Earth. Unlike Earth, however, Mars experiences dramatic oscillations in obliquity with time (Figures 16.1 and 16.2). These oscillations are attributable to secular precession of both the spin axis and the orbital planes of Mars convolved with the periodic oscillation of the inclination of the Martian orbital plane. The magnitude of these shifts was first identified by Ward (1973, 1974) through identification of these secular spin–orbit resonances that amplify Martian polar tilt. Two distinct cycles are observable in the obliquity history of Mars. The shorter, ~10^5 year, cycle is a consequence of the differential spin axis and orbital plane precession rates. This more rapid oscillation is modulated on a ~10^6 year cycle by slow oscillations in the inclination of the Martian orbital plane. On Earth, resonance between orbital and spin axis precession is quite weak, and obliquity varies with time by no more than 1–2°. The greater strength of the resonance between these terms on Mars produces oscillations with amplitudes that reach 10° or more.

Laskar and Robutel (1993) and Touma and Wisdom (1993) performed a more precise numerical integration of the equations of motion of the planets and used this to identify the inherent chaotic nature of Martian obliquity, preventing meaningful extrapolation

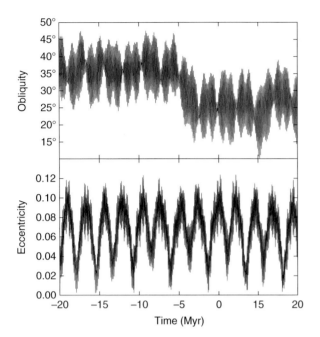

Figure 16.1. Martian obliquity and eccentricity history for the past 20 Myr, and projected forward 10 Myr. Data from Laskar et al. (2004).

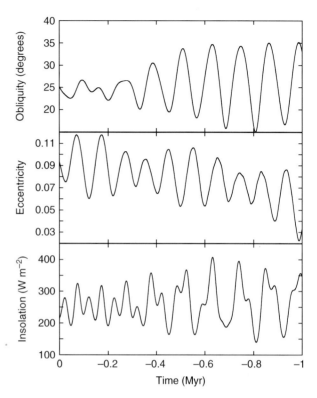

Figure 16.2. Martian obliquity, eccentricity, and north polar insolation at the summer equinox ($L_s = 90°$) in the past one million years. From Laskar et al. (2002). The north pole insolation is a key parameter controlling the stability of water ice of the northern polar cap.

of Martian spin–orbit history beyond ~5–10 Myr. Obliquity could be mapped back ~10 Myr. A robust behavior seen in their results is the secular rise in obliquity as one goes back beyond about 4 Ma. Prior to this point in time, the longer-term mean Martian obliquity was closer to 35°, with excursions of 10° in both directions. Obliquity, therefore, has been as high as 45° within the past 10 Myr. A careful analysis of Figure 16.2 reveals that, at present, Mars is found within a "node" of the obliquity cycle, where obliquity oscillations have been minimal for the past 300 kyr.

Most recently, Laskar et al. (2004), using an improved numerical integration scheme and improved input parameters, extended the precise knowledge of Martian obliquity back to 20 Ma. A statistical study of possible obliquity scenarios over longer timescales (both 250 Ma and 4 Ga) was also performed, and while these analyses do not provide exact values for obliquity history this far in the past, production of a probability density function of likely obliquities is nonetheless an informative way of considering plausible past climate scenarios. Laskar et al. (2004) found that, within the past 250 Myr, the obliquity of Mars may have varied between 0° and 66°, with a mean obliquity of about 35°. Over the full 4–5 Ga age of the Solar System, there are even more extreme departures, with possible excursions from 0° to 82° about a mean value of 38°. This high mean value implies that the present climate of Mars is different than the long-term norm.

16.2.2 Eccentricity

Eccentricity is a measure of the ellipticity of a planet's orbit. Changes to a planet's eccentricity will modify the planet–Sun distance, and thus modify the magnitude and relative duration of the seasonal variations in temperature. Higher eccentricities result in aphelion being more distant from the Sun, with a correspondingly reduced insolation and longer and cooler seasons. The present Mars–Sun distance at aphelion is ~21% larger than

at perihelion and Mars receives about one-third less sunlight during this period. Eccentricity is the only spin–orbit parameter that modifies the Sun–Mars distance, and is therefore the only parameter that adjusts the total planetary insolation. Martian eccentricity is comparatively high, at 0.093 (exceeded only by the eccentricity of Mercury), although within the past 20 Ma it has ranged from ~0.0 to 0.13 (Laskar et al., 2004), with a short-term period of 95 000 years modulated within a longer 2 Myr envelope (Figure 16.1). Long-term calculations by Laskar et al. (2004) obtain a mean eccentricity value over the past 4–5 Ga of 0.069.

16.2.3 Season of Perihelion

Tightly coupled to the influence of eccentricity is the orbital position (or "solar longitude", L_s) of perihelion, which determines the time of year, or season, of closest approach to the Sun. The timing of perihelion is linked to the precession of the spin axis, and is a secular cycle taking ~51 000 years (the difference between the true motion of the perihelion and the true motion of the equinox). When perihelion occurs at solstice, the summer hemisphere will experience maximum insolation, but for a shorter time. Presently, perihelion of Mars occurs at $L_s = 251°$, near northern winter–southern summer solstice ($L_s = 270°$). This results in mild seasons in the north and strong seasons in the south, with short northern winters and long southern winters. As the season of perihelion precesses, the hemispheric influences will reverse, with warmer summers in the north and colder summers in the south.

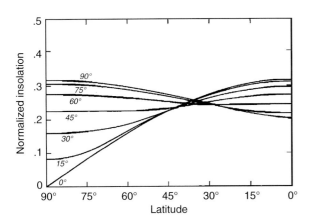

Figure 16.3. Average annual insolation as a function of latitude for various values of obliquity. Insolation normalized to the solar constant at Mars mean distance from the Sun. From Ward (1974).

16.2.4 Impact on Insolation

The *total average insolation* upon the planet is regulated by the distance of Mars relative to the Sun, and thus varies very little, only with eccentricity. Its distribution, however, is controlled by the combined effects of obliquity and timing of perihelion.

The *annual mean insolation* at a given latitude mostly varies with obliquity, as illustrated on Figure 16.3. The effect of obliquity is strongest in the polar regions but remains limited below 50° latitude. At the pole, the annual mean insolation varies with the sine of obliquity, and for obliquities larger than about 54°, the polar regions receive even more insolation than the tropics on average.

The *seasonal variations of insolation* are directly controlled by the variations of Mars' orbit and obliquity (Figure 16.4). In both space and time, the contrast between summer and winter strongly increases with obliquity. The polar night at winter solstice extends to a latitude of about 90° − obliquity. At summer solstice, for an obliquity larger than about 18°, the summer poles do receive more insolation on daily average than the rest of the planet because the Sun is perpetually above the horizon. Convolution of the orbital and eccentricity histories over the past million years (Laskar et al., 2002) shows a nearly threefold change in the amount of insolation at the north pole, for example, with it being a minimum when both obliquity and eccentricity are relatively low (Figure 16.2) and a maximum when both values are high. This affects the stability of ice at the poles, as discussed below in Section 16.6.2.

16.3 VARIATION OF TEMPERATURE AND PRESSURE

16.3.1 Surface Temperatures and Seasonal CO_2 Ice Caps

With an atmosphere thinner than 1000 Pa, the Martian surface is almost in radiative equilibrium (except when CO_2 ice is present). Surface temperatures primarily depend on incident radiation, albedo, and thermal inertia. Assuming that the atmosphere is relatively clear, surface temperatures can be estimated with some accuracy with energy balance models and global climate models (Toon et al., 1980; François et al. 1990; Kieffer and

Zent, 1992; Mellon and Jakosky, 1995; Nakamura and Tajika, 2002; Haberle et al., 2003; Newman et al., 2005). Figure 16.5 shows the variations of zonal-mean surface temperatures as a function of season and latitude assuming present-day Mars conditions, but with obliquities ranging between 0° and 60° (NASA Ames Mars global climate model (GCM); Haberle et al., 2003).

At low obliquity (<20°), the seasonal variations are small and the poles are cold all year long since the Sun is always low in the sky. As discussed below, it is likely that thick permanent CO_2 ice caps will form.

At high obliquity (>30°), seasons are more obvious. At high latitude, daily mean temperature can exceed 0°C in summer. At 60° obliquity, the equator-to-pole differences reaches 100 K at summer solstice. The latitudinal extension of the CO_2 cap strongly increases with obliquity. For the 60° obliquity experiment, the winter caps extend all the way to the equator. Thus, virtually the entire surface of the planet has a CO_2 ice covering at some time during the year. Such a CO_2 ice cover tends to cool the surface on average, notably because its albedo is higher than that of bare ground if present as frost or snow[1]. As a result, in the simulations of Haberle et al. (2003), even at 60° obliquity the poles remain cooler on average than the equator in spite of the fact that they receive more insolation at the top of the atmosphere. Furthermore, the increasing extent of the bright seasonal CO_2 ice caps results in a decrease of the global mean annual temperatures with increasing obliquity (Haberle et al., 2003). This is probably true only if one assumes that obliquity only impacts the insolation and that the CO_2 cycle and the Martian atmosphere remain the same. In reality, it is likely that the amount of airborne dust will change, that the surface pressure will vary with obliquity, and that water ice clouds will play a major role. The possible impact of these phenomena on surface temperatures is detailed below.

16.3.2 Pressure Variations

The mass of CO_2 in the Martian atmosphere (and thus the surface pressure) probably varies significantly in response to the insolation and temperature changes induced by the variations of obliquity. Estimating the extent of these variations is a key question of Martian climatology.

16.3.2.1 Atmospheric Collapse at Low Obliquity (<20°)

It is expected that the fraction of the atmosphere constantly held as CO_2 ice, and which is presently near the south pole (see Chapter 12), will increase (Ward et al., 1974). The mean annual atmospheric pressure is controlled through vapor-pressure equilibrium by the mean annual temperature of the existing CO_2 ice reservoirs. Using the Clausius–Clapeyron relation (see Chapter 12), one can show that, if the reduced low-obliquity insolation decreases this temperature from 142 K (current temperature at the south pole; Paige and Ingersoll, 1985) to 140, 130, and 120 K, the mean CO_2 pressure will decrease to 73%, 15%, and 1.6% of its present value, respectively. For instance, long-term

---

[1] CO_2 ice can also form a transparent slab layer (see Chapter 12). The conditions that favor this remain poorly understood.

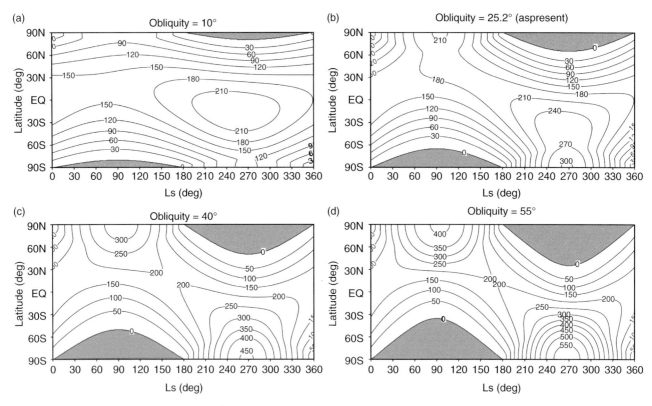

Figure 16.4. Daily-averaged insolation (W m^{-2}) as a function of season for various obliquities assuming the present eccentricity (0.0934) and longitude of perihelion (251°).

GCM simulations performed by Newman et al. (2005) predict a final global mean pressure of 350 Pa and 40 Pa for an obliquity of 15° and 5°, respectively (compared to ~610 Pa today). Similarly, Figure 16.6 shows the very large variations of the atmospheric pressure during a possible epoch of very low obliquities in the past, as calculated by Kreslavsky and Head (2005).

Such an evolution is often described as an atmospheric collapse (Kieffer and Zent 1992; Haberle et al., 1994; Kreslavsky and Head, 2005). When almost all atmospheric CO_2 is condensed, a residual Ar–N$_2$ atmosphere (with possibly some CO) remains, with a mean pressure of about 30 Pa. The atmosphere still has the properties of a real atmosphere rather than an exosphere: the molecular free path is still much smaller than the characteristic atmospheric height. Such an atmosphere still shields the surface from the majority of solar wind protons and high-speed interplanetary dust particles. This means that the processes of regolith formation and maturation typical for atmosphere-free bodies do not occur on Mars during these periods (Kreslavsky and Head, 2005). Another consequence of this pressure drop is that it causes a decrease in the thermal conductivity of porous regolith materials. Wood and Griffiths (2007) suggested that this could lead to a significant increase of subsurface temperatures, up to 20 or 30 K, as the planetary heat flow becomes trapped below a more insulating upper layer. They speculated that, in some conditions, this could possibly melt the subsurface ice present at high latitudes.

If the CO_2 atmosphere collapses onto the present residual CO_2 ice cap, in theory its thickness should increase by several tens of meters. In reality, several physical processes may play a role and alter this scenario. First, insolation is not the only parameter controlling the CO_2 ice cap evolution. For instance, the stability of the present-day south residual cap depends on its very high summer albedo, itself controlled by poorly known physical processes, which may be affected by the change of insolation and cap thickness. They may create a negative feedback, which may prevent atmospheric collapse at moderate obliquity. For instance, a lower summer insolation may result in more dust and fewer cracks within the CO_2 ice (James et al., 1992), therefore a lower albedo, and a polar cap less stable than today. Second, one must take into account the ability of CO_2 ice glaciers to flow and spread (CO_2 ice is known to be much softer than H$_2$O ice at the same temperature) (Durham et al., 1999).

Third, using a simple season-resolved energy balance model, Kreslavsky and Head (2005) have suggested that at low obliquity the permanent CO_2 ice deposits should have formed on steep pole-facing slopes at moderately high latitudes rather than at the poles. They also found that, when the obliquity returns to values similar to today, such slope CO_2 ice glaciers may have survived until almost the present-day values. Can we find geological evidence left by such glaciers? Kreslavsky and Head (2011) found three localized sets of small arcuate ridges associated with slopes in the northern polar area of Mars (~70°N latitude). They showed that they were morphologically similar to sets of drop moraines left by episodes of advance and retreat of cold-based glaciers, but that comparison with other glacial features on Mars suggested that these features differ in important aspects from those associated with water ice flow. Instead, they interpreted these features to be due to perennial accumulation and flow of solid carbon dioxide during recent periods of very low obliquity.

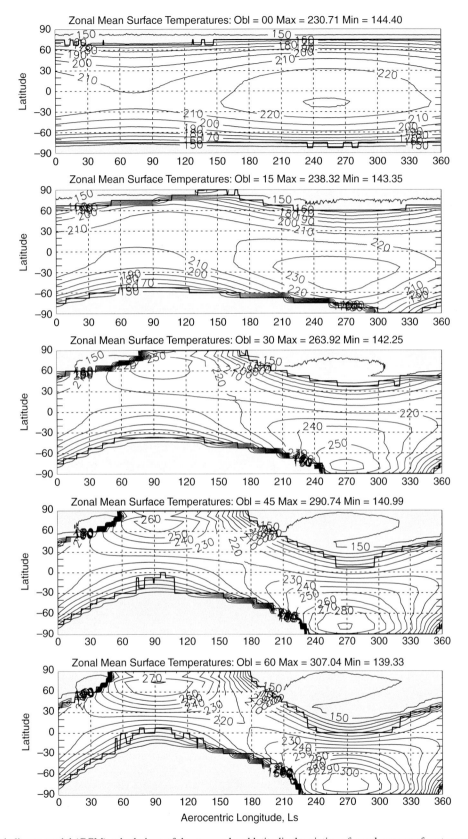

Figure 16.5. Global climate model (GCM) calculations of the seasonal and latitudinal variation of zonal-mean surface temperatures assuming present-day Mars conditions, but with various obliquities. Contour intervals are 5 K. Thick solid lines mark the boundary of the CO_2 seasonal polar caps. The present eccentricity and longitude of perihelion are assumed. The atmospheric visible dust opacity is set to 0.3 at the 610 Pa level and the amount of CO_2 in the atmosphere–seasonal cap system is assumed to be as at present (724 Pa). Figure from Haberle et al. (2003), showing the second year of simulations after changing the obliquity from present conditions.

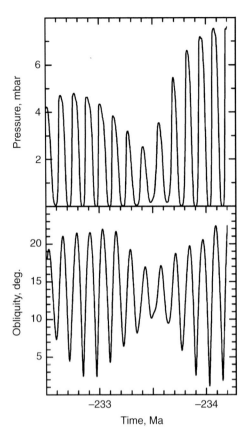

Figure 16.6. An example of a ~2 Ma long history of obliquity oscillations during a low-obliquity epoch, as simulated by Kreslavsky and Head (2005), with a CO_2 ice albedo of 0.65. The obliquity data are taken from one of the calculation series from Laskar et al. (2004). Adapted from Kreslavsky and Head (2005).

16.3.2.2 *Behavior at High Obliquity (>30°)*

(a) Impact of the CO_2 Seasonal Cycle Even assuming that the total amount of CO_2 available in the atmosphere–seasonal cap system remains the same as today, obliquity can affect the surface pressure. As mentioned above, seasonal caps will be larger at high obliquities. This means an increase in the amplitude of the seasonal CO_2 cycle, and a reduction in the mean annual surface pressure (Toon et al., 1980; Haberle et al., 2003; Mischna et al., 2003). While the fall and spring pressure maxima are relatively unaffected by obliquity, the winter minima are significantly lowered (e.g. figure 8 in Haberle et al., 2003). With an obliquity of 60°, the global mean surface pressure in southern mid-winter is estimated to reach 300 Pa compared to 500 Pa today.

(b) Sublimation of Perennial CO_2 Ice Caps In reality, however, it is very likely that the total amount of CO_2 available in the atmosphere–seasonal cap system will significantly increase with obliquity. At obliquity a few degrees higher than today, the increase in mean polar insolation will likely induce the sublimation of the current perennial CO_2 ice cap that we can observe overlying a thin water ice layer (Bibring et al., 2004) near the south pole. This cap is relatively thin and small, though (<5 m thick, ~90 000 km²) (Thomas et al., 2009), and its sublimation

would release only a few tens of pascals. However, observations by the MRO's Shallow Radar (SHARAD) have revealed a buried deposit of CO_2 ice within the south polar layered deposits of Mars with a volume of 9500–12 500 km³, about 30 times that previously estimated for the south pole residual cap (Phillips et al. 2011). If released into the atmosphere at times of high obliquity, this CO_2 reservoir would approximately double the atmospheric pressure. This is probably a lower limit, since it is not excluded that other CO_2 deposits may be sequestered in the polar layered deposits.

(c) Desorption of CO_2 From the Regolith It has also been suggested that surface pressure may be increased by CO_2 released from the regolith. In the pores of the Martian regolith, as for any gas in contact with a solid interface, CO_2 molecules can be physically bound to the mineral surfaces through the process known as adsorption. This reservoir of CO_2 has been described as an "ocean" of adsorbed CO_2 (Fanale and Salvail, 1994), in equilibrium with the atmosphere. The amount of adsorbed CO_2 on a mineral surface weakly depends on pressure (at least above 200 Pa), but strongly increases with decreasing temperature. Therefore, in theory, most of the adsorbed CO_2 is currently stored in the high-latitude subsurface where the annual mean temperatures are lower. As the obliquity of Mars increases, these temperatures increase. After several thousands of years or more, the corresponding thermal waves can reach depths of several tens or several hundreds of meters, inducing desorption of CO_2 and its diffusion into the atmosphere. Many models of the evolution of the atmosphere–regolith–cap system with obliquity have been published (Fanale and Cannon, 1974, 1978; Toon et al., 1980; Fanale et al., 1982, 1986; Zent et al., 1987; François et al., 1990; Fanale and Salvail, 1994; Zent and Quinn, 1995; Armstrong et al., 2004). The earlier studies suggested that the desorbed CO_2 could double the present-day atmosphere (for obliquity of 60°), with pressure increasing linearly with obliquity for values larger than present (see review in Kieffer and Zent, 1992). However, estimating how much CO_2 may be released at high obliquity remains difficult because many parameters remain poorly known: (1) the adsorption properties of the regolith materials, (2) the depth and pore volume of the regolith, and (3) the diffusion time of CO_2 in the Martian regolith. For instance, the most recent modeling studies of the atmosphere–regolith–cap system (Fanale and Salvail, 1994, Armstrong et al., 2004) have found that the pressure may actually not increase with obliquity because of the fact that, while desorption occurs at high latitude at certain depths, it is balanced by increased adsorption at other latitudes and depths. It is also difficult to take into account the presence of water in the system. The subsurface high-latitude water ice discovered by Mars Odyssey and observed by Phoenix (see Section 16.5.1 below) may fill the available pore space and preclude surface transfer from any adsorbed CO_2 in lower layers. Moreover, in the remaining open pores, water molecules compete with CO_2 molecules for a finite number of adsorption sites and could inhibit the adsorption of CO_2 (Carter and Husain, 1974). Zent and Quinn (1995) combined laboratory measurements and numerical modeling and found that, currently, H_2O does not substantially displace adsorbed CO_2, implying that the adsorbed CO_2 inventory is about the same either with or without water. They speculated that at higher obliquity the increase in water vapor (see below) could displace large amounts of CO_2 and induce a pressure increase even larger than mentioned above. However, modeling of this effect showed that it was small, because high-latitude groundwater ice buffers the partial pressure of H_2O in the pores (Zent and Quinn, 1995).

(d) Resulting Pressure at High Obliquity and Implication
Undoubtedly, calculating the pressure increase at high obliquity remains difficult. A conservative estimate is to assume a doubling of the atmosphere–seasonal cap inventory resulting from the sublimation of the perennial CO_2 ice cap, above an unknown threshold obliquity, probably ranging between the present-day obliquity and 35° (Phillips et al., 2011). This atmospheric pressure will be modulated by the seasonal CO_2 cycle. Its amplitude should be less than 500 Pa. This can be deduced from NASA Ames Mars GCM simulations reported by Phillips et al. (2011), performed at 35° obliquity assuming the present inventory (710 Pa) plus 500 Pa. It was found that most of the additional 500 Pa of CO_2 ended up in the atmosphere and that the annual mean seasonal cap masses increased by only about 80 Pa. Surface temperatures decreased only slightly (~0.7 K) because the CO_2 ice was on the ground for a longer period. Phillips et al. (2011) concluded that the two main implications of the pressure increase would be: (1) to exceed the triple-point pressure at the surface in more locales, potentially allowing liquid water to persist without boiling (see Section 16.7); and (2) to enable higher wind stresses, leading to increased frequency and intensity of dust storms.

16.4 ATMOSPHERIC CIRCULATION AND THE DUST CYCLE

16.4.1 Changes in the General Circulation

The changes in the atmospheric dynamics induced by the modification of insolation resulting from the variation of obliquities or longitude of perihelion can be explored using general circulation models (Fenton and Richardson, 2001; Haberle et al., 2003; Newman et al., 2005; Forget et al., 2006).

It has been found that varying the areocentric longitude of perihelion has a definite impact on the circulation. It is stronger when perihelion aligns with solstice rather than equinox, and especially with northern solstice (Newman et al., 2005), due to a bias from the Martian topography (the Mars' global north–south elevation difference favors a dominant southern summer Hadley circulation (Richardson and Wilson, 2002)).

The influence of obliquity is much stronger, since it directly affects the latitudinal contrast in insolation and temperatures. This thermal contrast is the primary cause of atmospheric movements, and in particular of the meridional overturning Hadley cell formed between the spring–summer and the fall–winter hemispheres (see Chapter 9). All GCMs show a strong intensification of the Hadley circulation with increasing obliquity. This results in a strong increase of the surface wind speeds in the Hadley cell return flow (trade winds, western boundary current, and the monsoon-like summer tropics westerlies; see Chapter 9). The Hadley circulation dominates over baroclinic, thermal contrast (e.g. cap edge), and topographic flows.

16.4.2 Dust Lifting and Dust Storms

As detailed in Chapter 10, dust lifting depends on wind velocity, atmospheric density (pressure), surface roughness, and particle size. In particular, the amount of lifted dust is thought to be proportional to surface pressure and – beyond a threshold – to the cube of the wind velocity (White, 1979). As orbital and eccentricity variations can significantly affect pressure and surface winds, they can modify the dust cycle.

At obliquities lower than today, with a lower surface pressure and much weaker general circulation winds, the dust storm activity is expected to be reduced, and the Martian atmosphere less dusty.

At high obliquities, GCM simulations show that surface winds and thus the dust lifting potential increases sharply with obliquity. It is greatest at times of high obliquity when perihelion coincides with northern summer solstice (Haberle et al., 2003; Newman et al., 2005). The expected increase in pressure should also contribute to increasing the lifting and transport potential.

Haberle et al. (2003) found a strong correlation between the deflation potential (the thickness of dust that would be removed over a given period of time) and surface thermal inertia: regions of high deflation potential correspond to regions of high thermal inertia (high rock abundance), and regions of low deflation potential correspond to regions of low thermal inertia (high dust/sand abundance). While the regions of preferred lifting (high deflation potential) expand somewhat with increasing obliquity and dust loading, the central parts of Tharsis, Arabia, and Elysium showed no tendency for significant lifting at any obliquity or longitude of perihelion. They concluded that these regions might therefore be very old and represent net long-term sinks for atmospheric dust (see also Christensen, 1986).

When active dust transport is included in GCMs, a wind stress lifting parameter selected to produce realistic results for the current obliquity produces huge amounts of lifting for obliquities of 35°, due to the strong positive feedbacks between atmospheric dust loading, the atmospheric circulation, and wind stress lifting (Newman et al., 2005). Can we conclude that at higher obliquities Martian meteorology would be marked by abundant and intense dust storms? This is not certain because the amount of dust lifted by the winds depends not only on the wind velocities, but also on the availability of dust. As on Mars today (see Chapter 10), the "recharging" of dust sources may limit the frequencies of large dust storms and the mean dust loading. At high obliquities, water ice may also play a major role due to the increased intensity of the water cycle (see Section 16.6), which can create a long-term sink for dust. Surface ice may limit the availability of dust on a significant part of the planet and dust may be much more scavenged from the atmosphere by ice atmospheric condensation and precipitation.

16.5 GEOLOGICAL EVIDENCE OF CLIMATE VARIATIONS AND PAST ICE AGES ON MARS

The Martian surface displays many geological features that reveal significant changes in the Martian water cycle and climate in the "recent" past (i.e. during the Amazonian era, possibly two billion years old, but in many cases much more recent). This is because most of these landforms are related to glacial processes (glaciers, water ice mantling, water ice-rich layered deposits, etc.). Their formation across Mars is difficult to reconcile with the present-day climate, since currently water ice does not last exposed on the surface for more than a few months outside the polar regions (see Chapter 11). Such non-polar ice deposits imply major climate changes.

Studying these landforms can address several scientific questions.

(a) They can reveal the actual climate and geological processes at work. Recent Mars history is dominated by a global hyperarid, very cold, low atmospheric pressure environment and glacial conditions at non-polar latitudes. Interpretations of this environment on the basis of our terrestrial experience is not straightforward, although we are assisted by an understanding of glacial and periglacial conditions in areas that are polar analogs to Mars (such as the Antarctic dry valleys; Marchant and Head, 2007; Head and Marchant, 2014), and an understanding of the behavior of polar water ice under different insolation conditions, using GCMs, as detailed below.

(b) As mentioned above, while a robust solution for the spin-axis/orbital parameter history of Mars has been developed for the most recent ~20 Ma (Laskar et al., 2004), it cannot be mapped further back into the past due to the chaotic nature of the solutions. Thus, the documentation and dating of the Martian polar and non-polar deposits may provide clues to climate and orbital history (e.g. Head et al., 2009).

(c) Estimates of the total water abundance on Mars and the nature and magnitude of sources and sinks throughout its history have been a matter of controversy for decades; total water abundance estimates span several orders of magnitude (see summary in Carr, 1996). Because the nature of Mars' earliest history and the processes operating during that time are so poorly known, one strategy is to start with the present environment, when climate conditions are more well known, and work backward in time, using the geologic record as a measure of the presence, location, and state of water (Head and Marchant, 2009). This inventory will eventually permit us to assess the migration paths and behavior of water during long-term climate change and to document changes in the nature and volumetric significance of the water cycle in the past geologic history of Mars (Carr and Head, 2015).

In the following sections we identify some of the major non-polar water ice-related deposits and assess their significance compared to the major surface reservoir on Mars, the current polar caps.

16.5.1 The Latitude-Dependent Mantle and Recent Ice Ages

Numerous lines of evidence show the presence of a geologically very young (a few million years at most), several meters-thick, water ice-rich mantling deposit at mid- to high latitudes

on Mars. This mantling deposit consists of multiple surface layers whose characteristics vary primarily with latitude. It is observed down to ~30° latitude in both hemispheres. These layers are interpreted to have been ice-rich when formed, and their deposition and modification driven by spin-axis/orbital parameter-induced climate change (Head et al., 2003). The lines of evidence are as follows.

(a) Remote neutron and gamma-ray spectroscopy has mapped the elemental composition of the uppermost meter of the Martian regolith (Boynton et al., 2002; Mitrofanov et al., 2002; Feldman et al., 2002, 2004, 2007, 2008; Prettyman et al., 2004; Litvak et al., 2006). Poleward of about 60° latitude, the abundance of hydrogen nuclei is so high that H_2O must be present in amounts well in excess of that permitted by filling pore spaces alone. Figure 16.7 shows a map of epithermal (i.e. partially moderated) neutron counts, where low values are indicative of H_2O.

(b) At 68°N, the Phoenix Lander exposed buried water ice a few centimeters below the soil surface beneath the descent engine (Smith et al., 2009) and in trenches (Arvidson et al., 2009; Mellon et al., 2009; Cull et al., 2010). Some of the trenches revealed ice-cemented soil, but others exhibited light-toned, almost pure water ice.

(c) At lower latitudes, fresh, post-mantle impact craters have excavated essentially pure buried water ice (Byrne et al., 2009); observations of ejecta and sublimation rates point to very high ice abundances, in excess of pore ice alone.

Additionally, the observed surface morphology provides suggestive evidence for the volatile-rich nature of this mantling deposit.

(d) Poleward of 55–60°, in correlation with the groundwater ice detected by the Mars Odyssey Neutron Spectrometer, the terrain is characterized by bumpy polygon-like features interpreted to be contraction-crack polygons, thought to mark the presence of shallow ice-rich deposits undergoing thermal cycling (Mangold et al., 2004; Mangold, 2005; Levy et al., 2009). Solar insolation-related polygon asymmetry supports the broader presence of this buried water ice (Levy et al., 2008).

(e) Between 30° and 60° latitude, a distinct pitted mantle texture is interpreted to be the dissected remnant of a former ice-rich dust deposit (Mustard et al., 2001); associated viscous-flow features and gullies are seen in microenvironments in this zone (Milliken et al., 2003). The degradation and dissection of the deposit point to recent climate change (Mustard et al., 2001; Head et al., 2003), perhaps reflecting return of mid-latitude water ice to polar regions during the recent phase of lower obliquity (Section 16.6.2).

The mechanisms responsible for the origin of the water ice-rich layer currently observed poleward of 60° latitude (and which appear to have extended down to 30° latitude in the recent past) are being debated. The physical processes at work are detailed in Section 16.6.3. Regardless, these data compellingly point to climate-driven water ice and dust mobilization and emplacement during very recent periods of higher obliquity.

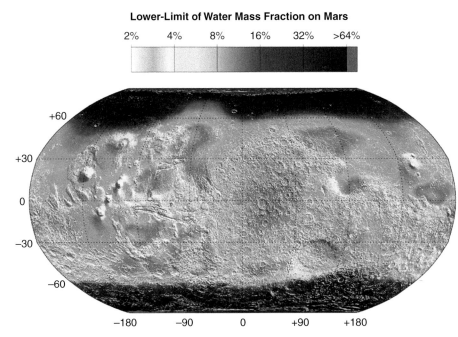

Figure 16.7. A map of apparent, average water content in the subsurface (%) estimated from neutron spectroscopy (Feldman et al., 2004). Fast neutrons are generated by cosmic rays, and the higher the concentration of hydrogen nuclei, the faster the neutrons thermalize, leaving fewer neutrons to escape at epithermal energies. Blue corresponds to a high concentration of water ice in the subsurface. The actual water ice concentration in the ice-rich layer below the dry sediment layer at high latitude should be significantly higher than these apparent values (Diez et al., 2008). Image credit: NASA/JPL/Los Alamos National Laboratory. A black and white version of this figure will appear in some formats. For the color version, please refer to the plate section.

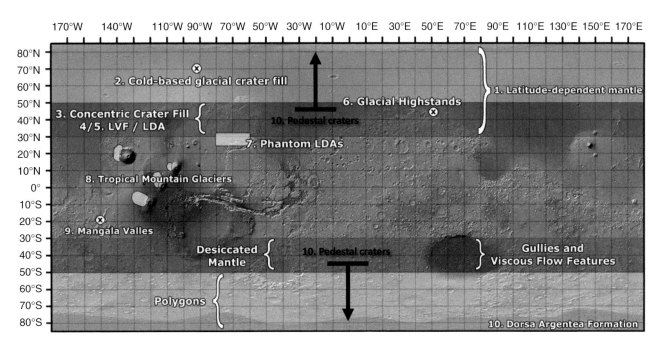

Figure 16.8. Location of various types of non-polar water ice-related deposits. See text and Head and Marchant (2009). "LVF" means "lineated valley fill"; "LDA" means "lobate debris aprons".

16.5.2 Evidence for Older, More Extensive, Lower-Latitude Ice-Related Deposits

While the recent latitude-dependent mantle is interpreted to have formed during the last few million years (when mean obliquity was near the current value), several types of older, non-polar water ice-related deposits have been discovered in specific locations on Mars (see map in Figure 16.8).

(a) *Mid-latitude lobate debris aprons.* Lobate debris aprons are rock-covered lobes which surround the base of many mid-latitude massifs, extending out to about 20 km (Squyres,

1978, 1979; Squyres et al., 1992; Pierce and Crown, 2003; Chuang and Crown, 2005). Their shape and morphology (which include, in particular, surface lineations) point to a debris-covered glacier mode of origin for many lobate debris aprons (Head and Marchant, 2006; Kress and Head, 2008; Head et al., 2010), recently confirmed by SHARAD data, showing nearly pure water ice underlying a meters-thick surface lag (Holt et al., 2008; Plaut et al., 2009).

(b) *Mid-latitude lineated valley fill and plateau glaciation.* Lineated valley fill is a furrowed and ridged texture that characterizes many valleys in the mid-latitudes of Mars.

Several hypotheses have been suggested for their origin, but recent data strongly support some earlier interpretations (Lucchitta, 1984) that debris-covered glacial flow formed the lineated valley fill (Head et al., 2010) and together with lobate debris aprons formed regional valley glacial land systems (Head et al., 2006a,b, 2010; Fastook et al., 2011, 2014).

(c) *Mid- to high-latitude concentric crater fill.* Many medium- to large-sized craters in the mid- to high latitudes contain fill that displays a series of concentric ridges parallel to the rim crest, suggestive of flow. Recent studies of their morphology and structure (Dickson et al., 2010; Levy et al., 2010) show that such concentric fill are also debris-covered glaciers within the craters (Kreslavsky and Head, 2006; Levy et al., 2010; Fastook and Head, 2014).

(d) *Mid-latitude ice highstands.* The discovery of the different types of debris-covered glaciers described above has raised the question of the presence of larger and more extensive water ice sheets in the past. Several studies have revealed evidence of highstands of ice (e.g. perched lobes, trimlines, moraines) (Dickson et al., 2008, 2010); these features suggest that almost a kilometer of water ice has been lost from lineated valley fill in some areas, evidence for much more extensive glaciation than that represented by the remaining deposits themselves.

(e) *Pedestal and excess-ejecta craters.* Pedestal craters are craters perched on a decameters-thick pedestal interpreted to be a remnant ice–dust-rich deposit armored from erosion by the cratering event. Recent analysis of pedestal craters has shown their widespread distribution at latitudes poleward of about 30° in both hemispheres (Kadish et al., 2009, 2010; Kadish and Head, 2011a,b) and reveals strong evidence for ice below pedestal protective veneers (Kadish et al., 2008a). Excess ejecta craters (larger craters whose ejecta is thought to have buried water ice-rich deposits) show characteristics (Black and Stewart, 2008) that further support regional mid- to high-latitude ice cover during the Amazonian and thicknesses of several tens to over a hundred meters (Kadish et al., 2010). Some adjacent areas show evidence for related ice-rich deposits of comparable thickness (Pedersen and Head, 2010). Together, these data support the presence of very extensive, decameters thick, water ice deposits at mid- to high latitudes in the late Amazonian (Kadish et al., 2010), prior to the most recent ice age deposits.

(f) *Low mid-latitude phantom lobate debris aprons.* Evidence for the former presence of water ice-rich deposits surrounding massifs at latitudes even lower than the lobate debris aprons suggests former ice lobes (Hauber et al., 2008) at latitudes <30°.

(g) *Tropical mountain glaciers.* Huge fan-shaped deposits on the NW flanks of Olympus Mons and the Tharsis Montes (Figure 16.9) were previously interpreted to have formed from a wide range of processes (landslides, volcanism, tectonism, glaciation, etc.) (Zimbelman and Edgett, 1992). Recent understanding of the process of cold-based glaciation on Mars (e.g. Marchant and Head, 2007) has led to the interpretation that these features are of cold-based glacial origin and that they represent the former presence of huge tropical mountain glaciers that formed during the Amazonian during periods of high obliquity (Head and Marchant, 2003; Head et al., 2005; Shean et al., 2005, 2007; Milkovich et al., 2006; Kadish et al., 2008b; Scanlon et al., 2014, 2015).

In summary, analysis of Amazonian-aged, non-polar, ice-related deposits suggests that throughout the Amazonian the climate system was able to cause the migration of ice from the poles to form equatorial and mid-latitude glaciers, and mid- to high-latitude regional blankets of snow and ice. Furthermore, abundant evidence suggests that significant quantities of water ice remain sequestered in association with these ancient ice-rich deposits. These deposits have probably been sequestered by development of sublimation lags, sealing them off from significant participation in the current annual water cycle. This sequestration represents a process of removal of water from the active system and a corresponding decrease in the total volume involved in the active water cycle with time (Levy et al., 2014).

16.5.3 The Polar Layered Deposits: Archive of Recent Climate Changes?

Each of the two polar regions of Mars is home to an extensive ice sheet that is several hundred kilometers across and up to 3–4 km thick. Known as the polar layered deposits (PLD; or NPLD/SPLD for the northern/southern poles), these deposits have long been thought to be an atmospherically deposited mixture of dust and water ice (e.g. Cutts, 1973). Radar transparency indicates that the volume fraction of dust is small, ~2% for the NPLD (Phillips et al., 2008; Grima et al., 2009) and ~10% for the SPLD (Picardi et al., 2005; Plaut et al., 2007). Modeling of a gravity anomaly associated with the SPLD indicates a density of 1200–1300 kg m^{-3}, suggesting a volumetric dust content of ~15% (Zuber et al., 2007; Wieczorek, 2008). In bulk, both PLD are dominated by water ice although with a higher dust content in the SPLD.

The SPLD sits on top of the heavily cratered southern highlands; while the NPLD (along with a lower sand-rich basal unit) sits on top of the Vastitas Borealis Formation, which is typically smooth and flat at the multi-kilometer scale. Radar data show that the basal topography is minimally affected by the presence of the PLD and resembles these surrounding units (Picardi et al., 2005; Plaut et al., 2007; Phillips et al., 2008). There is no evidence of lithospheric deflection in response to PLD loading, a result that was surprising in the northern case, where a thinner lithosphere had been expected (Phillips et al., 2008). Measurement of the basal and surface topographies allow for an accurate characterization of PLD volume of 0.8/1.6 million km^3 for the NPLD/SPLD (Plaut et al., 2007; Putzig et al., 2009). An older "basal unit" underlying the NPLD has been identified by its geomorphology. It is about 0.45 million km^3 (Selvans et al., 2010) in volume and, in contrast to the overlying NPLD, contains alternating sand-rich and ice-rich layers (Byrne and Murray, 2002; Edgett et al., 2003; Fishbaugh and Head, 2005). Erosion of this basal unit currently supplies sand-size particles to the circumpolar erg.

The Dorsa Argentea Formation extends in part beneath the SPLD (Kolb and Tanaka, 2006) and is hypothesized to be the

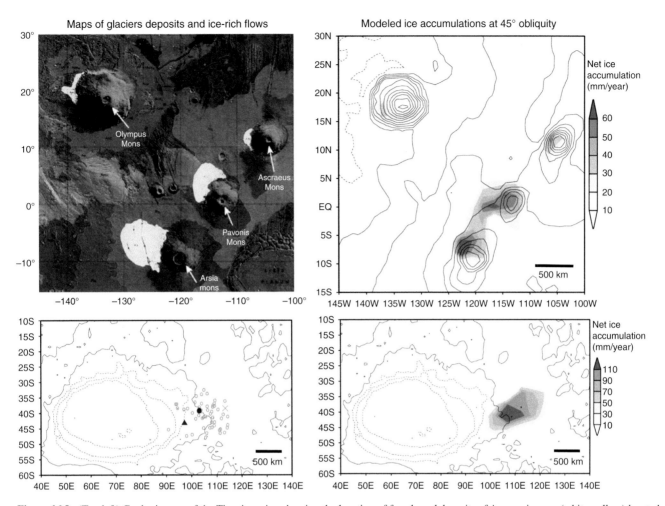

Figure 16.9. (Top left) Geologic map of the Tharsis region showing the location of fan-shaped deposits of Amazonian age (white-yellow) located on the northwest slopes of the Tharsis Montes and Olympus Mons. (Top right) Net surface water ice accumulation (in millimeters per Martian year) in the Tharsis region simulated by the LMD (Laboratoire de Météorologie Dynamique) GCM model with 45° obliquity and assuming that surface water ice is present on the northern polar cap. Superimposed MOLA topography contours are 2000 m apart. (Bottom left) Topographic map of the Hellas Basin with contours every 2 km showing the location of various glacier-like landforms such as lobate debris aprons (blue circles and purple triangle) and debris-covered glaciers (red dot, green cross). (Bottom right) Net surface water ice accumulation predicted in the same area by an LMD GCM simulation performed with 45° obliquity and assuming that surface water ice is initially present only in the southern polar cap. Figures from Forget et al. (2006). A black and white version of this figure will appear in some formats. For the color version, please refer to the plate section.

devolatized remnants of a Hesperian PLD-like ice sheet (Head and Pratt, 2001), possibly melted through local volcanic activity (Ghatan and Head, 2002) or melting by elevated geotherms (Fastook et al., 2012) beneath a thick deposit.

Both the north polar basal unit and the Dorsa Argentea Formation can be clearly distinguished from the overlying PLD in the radar data (Plaut et al., 2007; Phillips et al., 2008; Putzig et al., 2009).

The PLD take their name from the exposures of internal layering visible in numerous interior troughs and bounding scarps. Spiraling interior troughs typically expose a few hundred meters of stratigraphy on their equatorward-facing walls, which have slopes of only 5–10° (Ivanov and Muhleman, 2000). Below these troughs in the NPLD, stratigraphic discontinuities visible in radar data indicate that the troughs have been present for an appreciable fraction of NPLD accumulation and have migrated poleward during this period (Smith and Holt, 2010). Bounding scarps, which also expose layers, can have slopes up to 67° (at MOLA scales). Ongoing mass wasting on NPLD bounding scarps (Russell et al., 2008) has been observed, as well as indications of past landslides on SPLD scarps (Murray et al., 2001). Two large canyons, Chasma Australe in the SPLD and Chasma Boreale in the NPLD, are up to hundreds of kilometers long, several kilometers deep and penetrate to the PLD basement materials (Fishbaugh and Head, 2002; Tanaka et al., 2008). The SPLD contain other, similar but smaller, canyons (Kolb and Tanaka, 2006).

Layers as thin as 0.1 m have been observed in trough exposures; however, these slopes are mantled by a dust-colored layer, interpreted as a sublimation lag, that may obscure even thinner layers (Herkenhoff et al., 2007). Exposed layers may be relatively thick "marker beds" or thinner layers that occur in sets (Fishbaugh et al., 2010) and layer sequences can be correlated in images from trough to trough (Fishbaugh and Hvidberg,

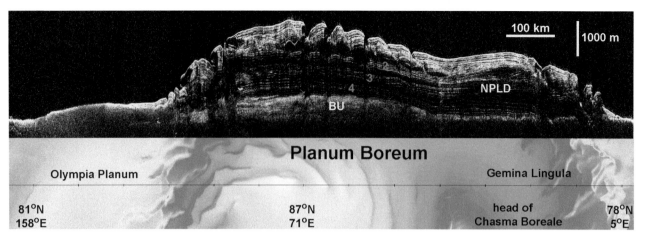

Figure 16.10. (Top) A radargram from Mars Reconnaissance Orbiter radar SHARAD showing the "basal unit" (BU) and a cross-section of the north polar layered deposits (NPLD) revealing numerous layers which are signatures of past climate processes (Phillips et al., 2008). (Bottom) Corresponding ground track shown on a digital elevation model (DEM) derived from Mars Orbiter Laser Altimeter (MOLA) data. Elevation range is approximately −4.5 km (green) to −2 km (white). Image credit: NASA/JPL-Caltech/University of Rome/SwRI.

2006). Unconformities are visible in stratigraphic exposures (Tanaka, 2005), especially at the edges of the NPLD, indicating that episodes of erosion punctuated accumulation. Away from these exposures, radar data (Figure 16.10) show that NPLD layers are mostly continuous over the entire deposit (Phillips et al., 2008; Putzig et al., 2009; Holt et al., 2010). Four distinct packets of radar reflections are visible in the NPLD, separated by sections that appear massive (to radar waves). All of the interior troughs expose layers from the uppermost packet while the bounding scarps may also expose lower strata. In the SPLD radar data, few distinct layers are visible (Seu et al., 2007) and the returned power is dominated instead by diffuse scattering. Geologic mapping has shown the SPLD to comprise several units separated by unconformities (Kolb and Tanaka, 2006).

Despite their icy composition and multi-kilometer thickness, there is no unambiguous evidence of features related to ice flow. Winebrenner et al. (2008) showed that the Gemina Lingula region of the NPLD has a shape expected of a flowing ice sheet with a restricted accumulation zone; however, it has also been shown that sublimation can produce the overall dome-like shape of the NPLD (Ivanov and Muhleman, 2000). A lack of significant ice flow may be due to the low ice temperatures and, to a lesser extent, to the low Martian gravity. Using different plausible thermal histories, Hvidberg (2003) and Pathare and Paige (2005) derived NPLD-trough closure times, driven by ice flow, that differ by over an order of magnitude. Thus the role of ice flow cannot be accurately predicted without knowing the details of the thermal and mass-balance history.

The surface of the NPLD is covered with the moderately high-albedo water ice that forms the northern residual ice cap (Kieffer et al., 1976). Although dust-free, this residual ice cap is large-grained (Kieffer, 1990; Langevin et al., 2005), most likely the result of thermal metamorphism. At high resolution, the late summer surface has bright frosted mounds, often arranged in quasi-linear strings, with intervening defrosted hollows (Thomas et al., 2000). This homogeneous texture repeats periodically over most of the deposit with a characteristic wavelength of 10–20 m (Milkovich et al., 2011).

The surface of the SPLD is mostly blanketed by material whose albedo, thermal inertia, and color are indistinguishable from those of dust (Vasavada et al., 2000). This material is often interpreted as a sublimation lag. A small section of the SPLD is covered with CO_2 ice up to hundreds of meters thick. As detailed in Section 16.3.2, the quantity of CO_2 stored in this reservoir is close to that in the current atmosphere (Phillips et al., 2011). A smaller CO_2 ice reservoir overlies this one, forming the southern residual ice cap (a discontinuous slab of CO_2 ice 2–8 m thick) that amounts to no more than a few percent of the current atmosphere (Byrne and Ingersoll, 2003; Thomas et al., 2009). These two CO_2 ice reservoirs are separated by intervening water ice (Titus et al., 2003; Bibring et al., 2004) that is thought to have been deposited ~20 kyr ago when the season of perihelion was opposite to that of today (see Section 16.6.1).

16.6 MODELING WATER CYCLE VARIATIONS TO UNDERSTAND PAST ICE AGES ON MARS

Can we understand the formation of the ice-related landforms described above using models of the Martian climate and water cycle?

On present-day Mars, the atmospheric water cycle is primarily controlled by the interaction between the atmosphere and the water ice exposed at the surface of the northern polar layered deposits (see Chapter 11). In summer, the ice is warmed by the Sun and sublimes. For a few months, the northern polar region becomes a source of water vapor that is transported away by the atmosphere. During the rest of the year, most of this water ultimately comes back to the northern polar cap through various transport mechanisms. The cycle is closed and near equilibrium (Farmer and Doms, 1979; Richardson and Wilson, 2002; Montmessin et al., 2004). To first order, the general characteristics of the water cycle, as revealed by the measurements of column water vapor and clouds, are well reproduced by GCMs on the basis of universal physical equations. On this basis, much has been learned by running climate models designed to simulate the details of the present-day climate, but with changes in the orbital and obliquity parameters reflective of those expected in the past.

16.6.1 Variation With Perihelion

Presently, Mars perihelion occurs at $L_s = 251°$, near southern summer solstice ($L_s = 270°$). As a result, the northern pole receives 30% less insolation than the southern pole at summer solstice, when water ice is exposed. All things being equal, this favors the stabilization of water ice in the north rather than in the south: using the Laboratoire de Météorologie Dynamique (LMD) GCM, Montmessin et al. (2007) explored the impact of the reversal of perihelion season, which occurred as recently as 21 000 and 75 000 years ago (see Section 16.2.3). They found that in these conditions water ice at the north pole was no longer stable and accumulated instead near the south pole with rates as high as 1 mm yr^{-1}. This could have led to the formation of a meters-thick circumpolar water ice mantle, and explain the origin of the small water ice sheet currently observed near the south pole (Bibring et al., 2004). As perihelion slowly shifted back to the current value, southern summer insolation intensified and the water ice layer became unstable. The layer recessed poleward until the residual CO_2 ice cover eventually formed on top of it and protected water ice from further sublimation. The southern accumulation of water ice at reversed perihelion should be limited to a few meters at each cycle since, on average, the topographic asymmetry favors the accumulation of ice in the northern polar regions. On the one hand, models suggest that Mars' global north–south elevation difference favors a dominant southern summer Hadley circulation (responsible for that inter-hemispheric transport of water) independently of perihelion timing (Richardson and Wilson, 2002). On the other hand, the topography may prevent the formation of a northern "dusty season" and the related atmospheric warming at reversed perihelion (Montmessin et al., 2007). Thus water may be more cold trapped on average in the northern hemisphere than in the southern hemisphere.

16.6.2 Variation With Obliquity

16.6.2.1 Reducing the Obliquity

Assuming that the main ice reservoirs interacting with the atmosphere are the polar caps (ignoring exchange with the regolith), at obliquities lower than today it can be expected that less water vapor will be present in the atmosphere because of reduced polar summer insolation. Simulations performed with the LMD GCM with the current surface pressure, as in Forget et al. (2006), but assuming an obliquity of 15° instead of today's 25.2°, yield column abundances reaching at most 15 pr μm (precipitable micrometers) at summer at high latitude and about 10 pr μm in the tropics, compared to 75 pr μm and 15 pr μm currently. The abundances are further reduced by a factor of 2 in a simulation with obliquity set to 10°. Moreover, as discussed in Section 16.3.2, it is likely that at such low obliquities the surface pressure will be much reduced and the atmospheric

water content even lower. Less atmospheric water vapor affects the stability of subsurface water ice (see below) and possibly the composition of the atmosphere, since less-abundant radicals will be produced by the photolysis of water vapor (see Chapter 13).

16.6.2.2 Increasing the Obliquity

As predicted by Toon et al. (1980) and Jakosky and Carr (1985) on the basis of energy balance calculations, GCM simulations performed assuming the same climate system and initial water ice reservoirs (northern polar cap) as today, but assuming higher obliquities, have all shown that the amount of water vapor involved in the seasonal water cycle readily increases with obliquity because of the increase in polar summer temperatures (Richardson and Wilson, 2002; Mischna et al., 2003; Levrard et al., 2004; Forget et al., 2006). Typically, at 45° obliquity, the column water abundance reaches 2000–3000 pr μm above the northern residual polar cap around summer solstice (compared to 75 pr μm on present-day Mars) and about 100–300 pr μm in the summer tropics (compared to ~15 pr μm currently) (Richardson and Wilson, 2002; Mischna et al., 2003; Levrard et al., 2004; Forget et al., 2006).

For moderate obliquities, in spite of the increased absolute humidity, water ice remains unstable outside the polar regions and all the water transported by the atmosphere is effectively recycled back to the polar caps, as it is today. However, above an uncertain obliquity threshold between about 35° and 45° (depending on the chosen orbit, the assumed dust loading, as well as the climate model itself), models show that the water ice will start to accumulate outside the polar regions until the polar reservoir is depleted[2]. Mischna et al. (2003) modeled that the ice tends to accumulate preferentially in regions of high thermal inertia or high topography, whereas Forget et al. (2006) found that water ice accumulation (reaching 30–70 mm yr^{-1}) occurs in four localized area on the flanks of the Tharsis Montes, Olympus Mons, and Elysium Mons (Figure 16.9). After a few thousand years, such accumulations would form glaciers several hundred meters thick. The difference between the two models can be attributed to the cloud microphysics: Mischna et al. (2003) assumed constant cloud particle size set to 2 μm and thus prevented most precipitation, whereas Forget et al. (2006) included a simplified cloud microphysics which allowed the ice particles to grow with condensation and thus to precipitate where intense condensation was predicted.

The location of the glaciers predicted by Forget et al. (2006) can be compared to the location of the glacier-related deposits mapped in the Tharsis region (Head and Marchant, 2003) (Figure 16.9). The agreement is remarkable, with maximum deposition predicted on the western flanks of Arsia and Pavonis

---

[2] This can occur well before complete sublimation of the reservoir because of the possible formation of a protective sublimation dust lag that prevents the exhaustion of the polar glacier. If the polar ice is not available to sublime, or is slowly subliming through the protective dust lag, the abundance of water vapor at very high obliquity could have been relatively low (Mischna and Richardson, 2005).

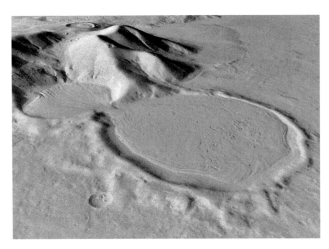

Figure 16.11. Debris-covered glaciers to the east of the Hellas Basin (257°W, 39.2°S). Three-dimensional stereo view of two craters filled with sediments, apparently very rich in water ice (Head et al., 2005). The ice has flowed from the smaller crater (9 km across) into the lower, larger one (16 km). Image obtained by Mars Express HRSC. Credits: ESA/DLR/FU Berlin (G. Neukum).

where the largest deposits are observed and lower deposition on the flanks of Ascraeus and Olympus. During that season, large amounts of water ice tend to condense out on the western side of the volcanoes because of strong westerly winds blowing upslope. In such an upward flow, the summertime water-rich air is strongly adiabatically cooled by 10–20 K. Water condenses and forms ice particles of 20–50 μm in diameter that sediment onto the surface (compared to 6–8 μm in present-day Tharsis clouds). This model permitted the formulation of flow models for the tropical mountain glaciers (Fastook et al., 2008).

Forget et al. (2006) also performed high-obliquity simulations assuming that water ice was available in the south polar region (between 90°S and 80°S) rather than in the north, a likely condition with reversed perihelion, as mentioned above. Under such conditions, water ice accumulation still occurs in the southern part of Tharsis, but the highest rates are now predicted to be in the eastern Hellas region, corresponding closely to where a unique concentration of ice-related landforms is observed (Figure 16.11). Interestingly, the process leading to ice precipitation differs from the one occurring on the volcanoes. In eastern Hellas, almost all the ice is accumulated during a 90 day period around southern summer solstice. At that time, the southern ice cap sublimes and releases large amounts of water vapor to the polar atmosphere. This water vapor is not easily transported toward the equator because the south polar region is isolated by a mid-latitude westward summer vortex, except near eastern Hellas. There, the deep Hellas Basin forces a stationary planetary wave that results in a strong northward flow that transports large amounts of water out of the polar region. The moist and warm polar air meets colder air coming from northern Hellas, and the subsequent cooling results in strong condensation and precipitation.

In the model, the formations of glaciers in eastern Hellas as well as on the flank of the tropical volcanoes directly results from the control of Mars meteorology by topographic effects. Similarly, global climate models simulating present-day Mars

can easily predict the location of observed water ice clouds. This is thus a robust result, which is probably not model-dependent.

16.6.2.3 Returning from High Obliquity to Lower Obliquity

Using the LMD model similar to that of Forget et al. (2006) (but at lower resolution), Levrard et al. (2004) showed that, when Mars returns to lower obliquity conditions, the low- and mid-latitude glaciers formed at high obliquity become unstable, water ice partially sublimes and tends to accumulate in both hemispheres above 60° latitude. Once water is no longer available from the low- and mid-latitude glaciers, water then tends to return to the poles (where it is now), but probably leaves some water ice under a residual lag deposit of dust. They suggest that such a process could explain the presence of the ice-rich mantling observed by geomorphology and detected by the GRS instrument aboard Mars Odyssey (see Section 16.5.1). Madeleine et al. (2009) extended these calculations, taking into account the possibility that the atmosphere may be relatively dusty at high obliquity (see Section 16.4.2). They showed that, during periods of moderate obliquity (25–35°) and high dust opacity (1.5–2.5), if water ice deposited on the flanks of the Tharsis volcanoes at higher obliquity is still available for sublimation, broad-scale glaciation in the northern mid-latitudes occurs, especially in the Deuteronilus–Protonilus Mensae region (0–80°E, 30–50°N), where large concentrations of lobate debris aprons and lineated valley fills are observed. They proposed that high atmospheric dust contents increase its water vapor holding capacity, thereby moving the saturation region to the northern mid-latitudes. Precipitation events are then controlled by topographic forcing of stationary planetary waves and transient weather systems, producing surface water ice distribution and amounts that are consistent with the geological record. Moreover, not only is the modeled accumulation maximum in the regions where glacier-like landforms have been observed, but it is also found that, everywhere poleward of ~50° latitude, some water ice could have accumulated. This could explain the origin of the ice-rich mantling detected by the GRS instrument, but at higher obliquity than in Levrard et al. (2004), which is in better agreement with the expected obliquities at the estimated time of the formation of the mantling.

16.6.2.4 Taking Into Account the Radiative Effect of Clouds

While water ice clouds play only a secondary role in the present-day climate, they should have been much thicker at high obliquity or with non-polar sources. Preliminary GCM simulations performed by Haberle et al. (2011) and Madeleine et al. (2011) suggested that their effect may completely change previous GCM results. In particular, if one assumes that the cloud particle radii remain near 10 μm, clouds could induce a greenhouse effect able to significantly warm the planet at high obliquity. Madeleine et al. (2014) analyzed in detail the impact of clouds at 35° obliquity with the LMD GCM, including an improved microphysics (Figure 16.12). They found that the primary effect of clouds is to warm the atmosphere by absorbing both solar radiation and infrared radiation emitted by the

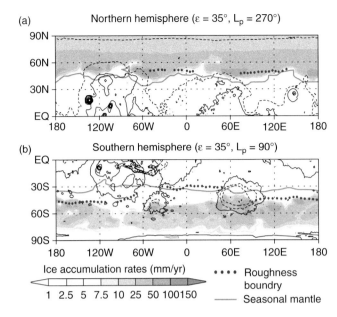

(a)

Northern hemisphere ($\varepsilon = 35°$, $L_p = 270°$)

(b)

Southern hemisphere ($\varepsilon = 35°$, $L_p = 90°$)

Ice accumulation rates (mm/yr)

1 2.5 5 7.5 10 25 50 100 150

• • • • Roughness boundry

——— Seasonal mantle

Figure 16.12. Net surface water ice accumulation (in millimeters per Martian year) simulated at 35° obliquity and 0.1 eccentricity by the LMD GCM, including the radiative effect of water ice clouds. Surface water ice is assumed to be initially present on the northern polar cap, as today. (a) With perihelion at $L_s = 270°$; (b) with perihelion at $L_s = 90°$. The boundary of the latitude-dependent mantle (LDM) is indicated by the thick dotted line, based on the roughness map of Kreslavsky and Head (2000). The thick solid line shows the 1 cm contour of the annual maximum ice thickness. Figure from Madeleine et al. (2014).

surface. This induces a positive feedback since the atmosphere can then hold much more water vapor before saturation, and form thicker clouds. An increase in the annual mean atmospheric temperature of ~50 K was simulated in the tropics at an altitude of ~20 km. In such conditions, modeled precipitation of ice/snow on the surface at mid-latitudes could reach ~10 cm at the end of winter, several orders of magnitude more than on present-day Mars. If the summer atmosphere is dusty enough (Madeleine et al., 2014) or if the snow albedo is high enough (e.g. 0.7; Forget et al., 2014), a fraction of this surface snow could survive the summer and accumulate an ice sheet year after year without any exposed glaciers in the tropics. This could explain why the latitude-dependent mantle appears to have formed relatively recently in the past two millions years (Head et al., 2003), a period during which the obliquity often reached 35°.

16.6.2.5 Remaining Modeling Issues

Global climate model studies have demonstrated that the present Mars climate system forced by orbit and obliquity oscillations is able to mobilize large amounts of water ice around the planet. Overall, the agreement between observed ice sheets, glacier landform locations, and model predictions points to an atmospheric origin for the ice and permits a better understanding of the details of the formation of Martian glaciers. However, one must be careful when interpreting these simulations quantitatively, because the models used in the studies cited above may

have neglected physical processes that could have played a key role in the past. These include the following.

- *Radiative effect of water vapor.* This is neglected in present-day Mars models. As in the case of clouds, it may influence our results under past conditions, given the higher water vapor holding capacity of the atmosphere.
- *Dust lifting and coupling* with the cloud microphysics, as well as scavenging of dust by water ice particles.
- *Coalescence of ice crystals* induced by high precipitation events.
- *Physics of the ice deposits.* Latent heat exchange induced by sublimation or melting of the deposits, heating within the ice layer by absorption of solar radiation (Clow, 1987), and the protective effect of a dust lag (Jakosky et al., 1993; Mischna and Richardson, 2005).
- *Influence of exchange with the subsurface.* Processes like adsorption/desorption and condensation/sublimation in the porous soil may affect the water cycle. Although buried ice is separated from the atmosphere by a layer of soil, vapor freely exchanges with the atmosphere, and it is part of the water cycle in the long term. In terms of area, subsurface ice is by far the largest ice reservoir on Mars (Figure 16.7).

16.6.3 Origin of the Subsurface Water Ice

In this section we discuss the possible condensation of water ice in the subsurface rather than its accumulation on the surface addressed in the previous section. The accumulation of such subsurface ice from mid-latitudes to the poles (see Section 16.5.1) had long been predicted, beginning with Leighton and Murray (1966). Even in non-polar locations where water ice is currently unstable at the surface (because of the high daytime and summertime temperatures, which cause its sublimation), below a few centimeters or tens of centimeters, subsurface temperature remains close to the seasonal or annual mean, and water ice can be stable (Farmer and Doms, 1979). When the vapor density above ice, averaged throughout the year, is equal to the atmospheric average vapor density, then the net annual exchange of water with the atmosphere will be zero. Figure 16.13 illustrates the concept of exchange of water vapor between the subsurface water ice and the atmosphere. Temperature amplitudes decrease rapidly with depth, as does the time-averaged saturation vapor pressure. For an annual mean near-surface vapor density of 2×10^{19} molecules m^{-3}, typical of present-day Mars, the mean annual surface temperature needs to be less than about 193 K for water ice to persist indefinitely. The lower the mean surface temperature, the shallower the equilibrium depth (Schörghofer, 2007b, 2008).

Water ice can grow in-place in the void spaces of the soil by diffusion and re-sublimation, which leads to ice-cemented soil or "pore ice", a process uncommon on Earth. Pore ice has been theoretically predicted by Mellon and Jakosky (1993) and reproduced in the laboratory by Hudson et al. (2009). Surface temperature oscillations, either diurnal or seasonal, can pump water vapor into the ground. The effect is due to the decaying temperature amplitude with depth and the nonlinear dependence of saturation vapor pressure on temperature, which leads to a net downward vapor flux (Figure 16.13e). Although forming

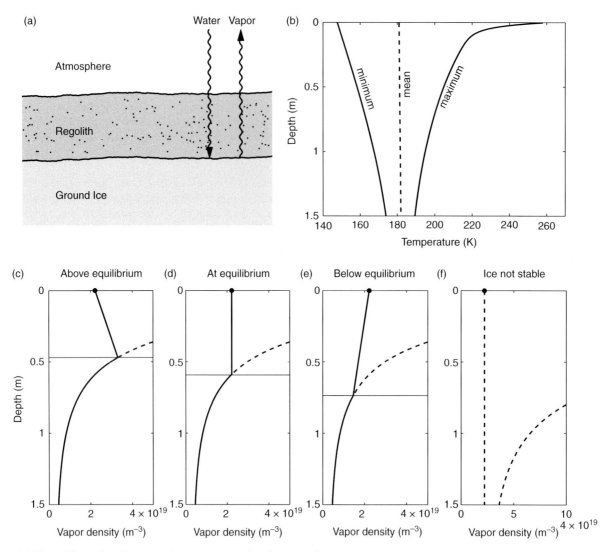

Figure 16.13. (a) Illustration of vapor exchange between subsurface water ice and the atmosphere. (b) Vertical profile of maximum, mean, and minimum subsurface temperatures over one Mars year. In this example, the diurnal skin depth is 3.2 cm, and the seasonal skin depth is 0.84 m. (c)–(f) Long-term average vapor density profiles are represented by thick solid lines. The assumed mean annual vapor density on the surface is 2.2×10^{19} molecules m^{-3}. (c) Water is lost to the atmosphere. The mean vapor density increases linearly from the surface to the ice table. (d) At equilibrium, the long-term average vapor density of water ice and the atmosphere balance each other. (e) Water ice forms from atmospherically derived vapor. (f) If the temperature is too high, ice is unstable.

in the absence of a liquid phase, the ice can contain atmospheric inclusions (bubbles) (Hudson et al., 2009).

Mellon and Jakosky (1993) produced the first theoretical maps of subsurface water ice stability. The subsurface ice found in 2002 by the Mars Odyssey GRS closely matches the theoretical predictions for flat ground (Figure 16.7). This is a rare example of climate model predictions preceding an observational discovery. Studies by Mellon et al. (2004), Schörghofer and Aharonson (2005) and Diez et al. (2008) have all confirmed a relation between the distribution of water ice in the uppermost meter and vapor equilibrium prediction, for the southern and northern hemispheres. Ice-rich permafrost has also been detected on pole-facing slopes in the shallow subsurface (<1 m depth) at latitudes as low as 25° in the southern hemisphere (Vincendon et al., 2010). The ice was revealed by its high thermal inertia, which affects the observed surface distribution of

seasonal CO_2 frost. In that case also, the stability of subsurface water ice on pole-facing slopes had been well predicted by Aharonson and Schörghofer (2006).

What was not anticipated, however, is the fact that the icy layer in both hemispheres is much more ice-rich than pore ice alone would allow, as explained in Section 16.5.1. More precisely, the H_2O content of the ice-cemented soil expected to form through diffusion is limited by the available pore space, typically about 40 vol.% for natural soils. This is much less than observed by the neutron spectrometer, which provided a constraint for the H_2O fraction as 60±10 mass% or 70–85 vol.% in the southern hemisphere (Prettyman et al., 2004). The Phoenix Mars Lander explored this water ground ice *in situ* at 68°N latitude through a dozen trench complexes and landing thruster pits (Smith et al., 2009). As expected, shallow ground ice was found to be abundant under a layer of relatively loose ice-free soil with

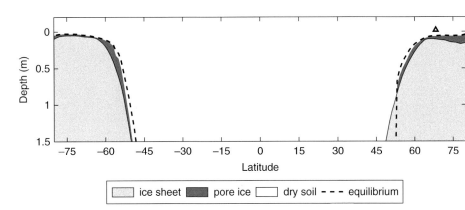

Figure 16.14. Subsurface water ice distribution according to zonally averaged subsurface ice model calculations. In this scenario an ice sheet consisting of 97% water ice and 3% dust was emplaced 4.5 Ma ago and evolved by subsurface–atmosphere exchange. The triangle indicates the latitude of the Phoenix Lander site. The dashed lines indicate the equilibrium depth for the present-day atmosphere. Adapted from Schörghofer and Forget (2012).

a mean depth of 4.6 cm (Mellon et al., 2009). Ice-cemented soil comprised about 90% of the icy material exposed by trenching, but light-toned, almost pure water ice was also observed.

What process may have led to the formation of such an ice-rich layer? This remains debated. It has been suggested that cryoturbation, thin film migration, and ice lensing may all be able to produce water ice contents in excess of porosity through soil ice segregation. However, on Earth, the formation of ice lenses requires the thawing of the permafrost, and the feasibility of this mechanism under Martian conditions remains uncertain. Alternatively, as mentioned above, the ice sheet could be the remnant of past ice ages induced by variations in obliquity (Head et al., 2003; Mischna et al., 2003; Levrard et al., 2004). As mentioned above, the ice would have slowly accumulated on the surface along with a few percent of dust. As observed on Earth, during its recession, the dusty water ice would have buried itself beneath the material it left behind, forming a "sublimation lag" or "refractory mantle". A problem with this scenario is that it seems inconsistent with the observations of rocks and boulders in the high northern latitudes, including near the Phoenix landing site. However, the emplacement age of the boulders at the landing site is not sufficiently constrained to rule out recent emplacement of a massive water ice sheet. Moreover, one could speculate that the very slow growth of the ice sheet may be able to progressively lift the boulders through ice–rock interaction. For instance, boulders meters in diameter are consistently found to be concentrated at or near the cracks that define polygonal networks, indicating a mobilization process (see e.g. Orloff et al., 2011) that may keep the boulder on top of the ice.

16.6.4 History of the Subsurface Ice Mantle

With orbital variations, surface temperatures and the amount of water in the Martian atmosphere change significantly. Thus the extent and depth of the subsurface water ice is expected to vary over time (Mellon and Jakosky, 1995). The timescale of the response has been scrutinized theoretically (Mellon and Jakosky, 1993; Hudson et al., 2007; Schörghofer, 2007b) and in laboratory experiments (Chevrier et al., 2007; Hudson et al., 2007; Hudson and Aharonson, 2008; Sizemore and Mellon, 2008), and is generally faster than orbital cycles.

Geomorphological evidence for recent and episodic desiccation is ample. As mentioned above, imaging and altimetry have revealed the presence of sedimentary mantling deposits that are layered, meters thick, and latitude-dependent (Kreslavsky and Head, 2002; Head et al., 2003). They occur in both hemispheres from mid-latitudes to the poles. The layer is interpreted to be water ice-rich and to have undergone degradation recently. Kostama et al. (2006) identified latitudinal trends in polygon texture and impact crater evidence for a latitudinal age progression of mantle properties. The inferred mean crater retention age of the mantle is ~0.1 Ma. The spatial distribution of craters suggests an age difference between the highest latitudes (younger) and some lower-latitude regions (older).

A quantitative model of subsurface water ice evolution was developed by Schörghofer (2007a). It time-integrates the depletion and accumulation of subsurface ice as a result of subsurface–atmosphere vapor exchange with orbital history. The subsurface water ice distribution obtained from one such model calculation is shown in Figure 16.14 (Schörghofer and Forget, 2012). This example assumes a massive global ice sheet formed by precipitation 4.5 Ma ago. Ice retreats during dry low-obliquity periods and pore ice forms during humid periods of high obliquity, thus creating a three-layered vertical structure. At the Phoenix landing site, there is, in this scenario, a thin layer of ice-cemented soil underlain by massive water ice, consistent with the high volumetric ice content determined by neutron spectroscopy (Feldman et al., 2008). Equatorward of the equilibrium boundary, the ice sheet retreats continuously, and may not have reached equilibrium with the present-day atmosphere yet. Equilibrium or near-equilibrium between the atmospheric water vapor and the subsurface water ice explains the ice in the uppermost meter of the subsurface, but deeper-lying ice requires more time to adjust.

In fact, small recent impacts have excavated ice at latitudes as low as 43.3°N, slightly further equatorward than expected for present-day equilibrium (Figure 16.15; Byrne et al., 2009). The new impact craters excavated material that has a brightness and color suggestive of water ice, and near-infrared spectra confirmed this composition. Repeated imaging showed fading over several months, as expected for sublimating ice. Thermal models suggest clean ice rather than ice in soil pores (Byrne et al., 2009). Is this the remnant of an ice mantle deposit now out of equilibrium? This is possible, but other explanations can be envisioned. For instance, the models defining the stability extract the near-surface humidity from measured column abundances assuming that the water vapor is vertically well mixed. In reality water vapor could be confined near the surface. This would favor the stability of subsurface ice at lower latitude.

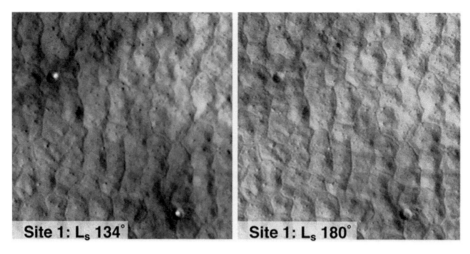

Figure 16.15. Recent impacts into pure water ice (HiRISE false color data; Byrne et al., 2009); latitude is 46.3°N. Panels are 75 m across. Icy material exposed by impacts at $L_s = 134°$ seems to have completely sublimed over 140–200 days.

Variations in subsurface content are important for the global H_2O budget. Simple geometry reveals that a 1° movement of a water ice layer margin in latitude (around 60° latitude) corresponds to 0.6×10^6 km^2, comparable to the area of the north polar cap. Contractions of the subsurface water ice presumably lead to the deposition of a layer on the polar cap, but coupled polar cap–subsurface ice models are not yet available.

Orbital variations may not be the only cause for redistribution of subsurface water ice. Albedo changes, known to occur on short timescales, can lead to notable surface temperature changes (Schörghofer and Aharonson, 2005).

16.6.5 Formation of the Polar Layered Deposits

The premise that climatically modulated deposition of ice and dust accumulated to form the polar layered deposits (PLD) has existed since their initial discovery (Murray et al., 1972; Cutts, 1973) and remains basically sound. Additional complexity has been included over the past four decades driven in large part by new datasets of high-resolution imaging, topography, and surface-penetrating radar. In parallel with these observational advances, solutions for the orbital history of Mars have improved to the point where the past 10–20 Myr (Figure 16.1) can be confidently quantified (a period that may exceed the age of the NPLD, although not the SPLD).

When did the polar layered deposits form? Constraints on the age of the lowermost strata of the PLD exist only in that they superpose geologic units that are billions of years in age (Kolb and Tanaka, 2006; Tanaka et al., 2008). Constraints on the uppermost surfaces are more informative. Crater counts on the SPLD indicate a surface age of tens of millions years (Herkenhoff and Plaut, 2000; Koutnik et al., 2002). The size–frequency distribution of craters on the northern deposits is more complex and craters are much more sparse. Banks et al. (2010) discovered ~100 craters (all less than 400 m across) in the inter-trough regions and found that they were best described as a population where crater lifetime is roughly proportional to crater diameter. Thus, there is no one "age" that can characterize the surface, but all the craters currently visible could have

been formed in the past 10–20 kyr (a period 1000 times shorter than recorded in the SPLD crater population).

As discussed above, climate models suggest that obliquities larger than about 40° produce intense summer conditions resulting in net ablation of polar water ice each Martian year (Mischna et al., 2003; Forget et al., 2006; Levrard et al., 2007), leading to PLD removal in a few tens of thousands of years. Such values of obliquity occurred as recently as about 5 Myr ago (Figure 16.1). Although some interpret this as meaning that the age of the NPLD must be less than ~5 Myr, the surface age of the SPLD (tens of millions of years) immediately indicates that the situation is more complex. Sublimation of dusty ice is expected to produce a lag deposit that can thermally insulate underlying material (Hofstadter and Murray, 1990; Mischna and Richardson, 2005). The generation and effectiveness of these dust lags is likely a key controlling factor in PLD formation, but they remain relatively uninvestigated. Levrard et al. (2007) attempted to include their effects in a model relating orbital element variation to polar climate and mass balance. In their model, lags were assumed to have no effect until 1 m thick, at which point they reduced the ablation rate by a factor of 10; however, these lags were unable to preserve the NPLD prior to 4 Myr ago in their model. Radar and gravity data suggest that the SPLD are several times dustier than the NPLD. It is possible that the SPLD are dusty enough to produce protective lags whereas the NPLD are not, a difference that may reconcile the surface ages and climate models.

Low obliquities drive polar ice accumulation (at the expense of mid-latitude ground ice) and high obliquities drive polar ice loss; as discussed in Section 16.6.1, the accumulation should be favored on one of the poles at the expense of the other depending on the eccentricity and the season of perihelion, with the north pole favored on average because of the topographic asymmetry.

Accumulation rates have also been found to be spatially non-uniform. Holt et al. (2010) showed that the large NPLD re-entrant, Chasma Boreale, was created as the NPLD accumulated both to its north and south and that another (similarly large) canyon was completely filled in after experiencing much higher accumulation rates than surrounding terrain.

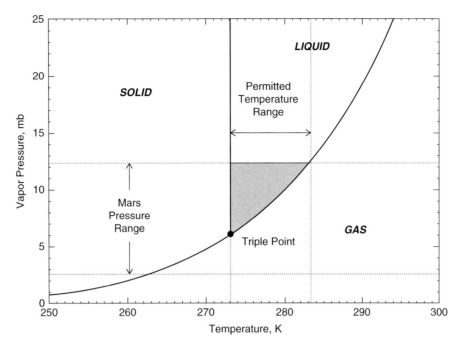

Figure 16.16. Water phase diagram illustrating range of Martian surface pressures and permitted temperature range over which liquid water can be transiently stable. From Haberle et al. (2001).

16.7 LIQUID WATER IN THE PAST MILLIONS OF YEARS

At present, liquid water is not stable at the Martian surface (Ingersoll, 1970); however, periodic conditions may have permitted episodes of surface liquid water in recent Martian history – within the past tens of millions of years. Throughout this period, the mass of the Martian atmosphere has remained largely the same, with only modest deviations from its present state. Thus, in terms of liquid water at the surface, we distinguish "recent" Martian history to be that generally of the late Amazonian when the planet was very much the dry, barren desert we are familiar with today. The presence of liquid water during ancient Martian history (more than three billions years ago) is discussed separately in Chapter 17.

16.7.1 Surface Liquid Water on Mars

16.7.1.1 Pure Water

The global mean water vapor content on Mars is ~10 pr μm, equivalent to an atmospheric vapor pressure of about 0.1 Pa. The surface pressure ranges between 200 and 1000 Pa (2–10 mbar), comprising predominantly CO_2, meaning that water has an average mixing ratio of only a few hundredths of one percent. The global mean surface pressure has a value near the triple-point pressure of water[3] (610 Pa), the point above which liquid water can exist without boiling (Figure 16.16). However, this triple-point pressure is defined in terms of the water vapor

pressure only, and so liquid water is actually far from being stable on Mars. Water can exist in its liquid state only transiently under the very limited conditions described below. First, the local surface (or overburden) pressure must exceed the triple-point pressure of water. Second, surface temperatures must fall within a restricted range dictated by the local surface pressure (Figure 16.17). Third, a source of liquid water must be present and available to melt. Fourth, enough energy must be available to overcome evaporative cooling (Ingersoll, 1970).

Given these conditions, transient liquid water may be possible in the present day due to the wide range of surface temperatures experienced at the surface, but its existence requires that the surface temperature fall between the melting point of water ice and the boiling point of liquid water dictated by the local surface pressure. At the present day, more than half of the surface exceeds 0°C sometime during the Mars year, albeit only during brief periods of the day. For pressures above the triple-point pressure, water may be stable against boiling, but will nevertheless very quickly be lost to the dry Martian atmosphere, first by freezing, followed by sublimation.

In any case, on present-day Mars, ice is extremely unlikely to be available when the temperature gets above 0°C; it has already sublimed well before. Unlike most terrestrial environments, the frost-point temperature (~200 K) is usually significantly below the melting point (273 K). Water ice on Mars is generally only found at temperatures below the frost point, because it is lost to the atmosphere as vapor at temperatures above it. In the time period it takes to warm from the frost-point temperature toward the melting point, significant loss occurs.

Understandably, models do not predict formation of liquid water in the present-day climate, unless an additional mechanism or assumption is invoked, such as aquifers, a solid-state greenhouse effect, a "dirty snowpack", or salts (e.g. Clow, 1987; Heldmann and Mellon, 2004; Heldmann et al., 2007).

---

[3] This is probably a coincidence. However, it has been suggested that the surface pressure stays close to the triple point of water because at higher pressure aqueous chemistry depletes atmospheric CO_2 by forming carbon-containing sedimentary rocks as on Earth (Kahn, 1985).

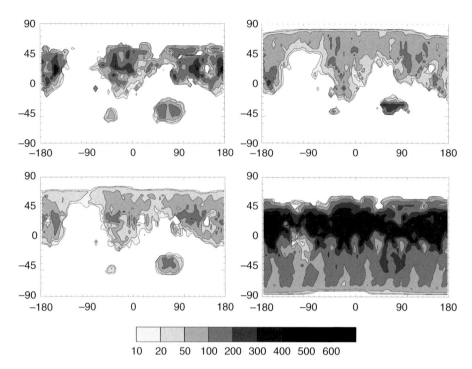

Figure 16.17. Surface map of those locations where both pressure and temperature conditions are satisfied for transient liquid water. (Top left) Present-day Mars. (Top right) Mars with a reversed season of perihelion ($L_s = 71°$). (Bottom left) Mars with obliquity of 45°. (Bottom right) Mars with mean surface pressure 2× present-day. Contours represent the number of days in the Martian year where conditions are met for some portion of the day. Created from the numerical simulations detailed in Richardson and Mischna (2005).

16.7.1.2 *Liquid Brines*

In lieu of pure water, a brine solution can reduce the triple-point pressure, and correspondingly reduce the temperature and overburden pressure required for melting. The presence of salts in the Martian soil has been long established (Clark and Van Hart, 1981, and references therein), and the formation of saline brines is a plausible means to permit liquid water at temperatures below the freezing point of pure water. The discovery of high levels of perchlorate (~1 wt.%) in the Phoenix soil (Hecht et al., 2009; Kounaves et al., 2010) and by Curiosity (Glavin et al., 2013) has rekindled consideration of the importance of such solutions to withstand freezing at reduced surface temperatures. Perchlorate salts are of particular interest because of their very low eutectic temperature (as low as 205 K), which approaches the present-day mean global temperature of Mars (see e.g. Chevrier et al., 2009). More commonly observed chloride and sulfate salts have more modest effects on the melting temperature, reducing it by 5–20 K. So while it is possible that a reduction in freezing-point temperature may have facilitated the formation of liquid on the Martian surface to a degree, it was likely not the sole process. Modeling results by Haberle et al. (2001) explore the potential role of salts on expanding regions of possible surface water. For an assumed NaCl brine (eutectic point 251 K), nearly the entire planet experiences conditions favorable for melting at some point during the year. At 251 K, the equilibrium vapor pressure of the solution is ~120 Pa – well below the lowest values obtained on the present-day Martian surface. Hence, for such a brine

solution, boiling would be avoided globally. The evaporation rate is also reduced. However, it must be noted that such brine solutions at their eutectic point (which contains 23 wt.% NaCl) would be highly viscous and quite unlike the liquid solutions one typically envisions (more akin to honey rather than fluid water). Such high-concentration brines are proposed to be at the origin of the "recurring slope lineae" observed on Martian tropical slopes (Ojha et al., 2015).

16.7.2 Why Surface Liquid Water May Have Been More Likely in the Recent Past

16.7.2.1 *Impact of Perihelion*

Currently, the present combination of Martian topography and orbital orientation provides for very limited locations where both the pressure and temperature conditions are satisfied for liquid water to form if water ice is available. Locations that satisfy the pressure requirement are predominantly found in the northern lowlands, which have mean surface pressures of ~700–1000 Pa. However, peak temperatures during northern summer are suppressed due to solstice occurring during the aphelion season when Mars is furthest from the Sun. By reversing the timing of perihelion, peak summer temperatures occur in the north, satisfying both the pressure and temperature restrictions in much of the northern hemisphere. Thus, as recently as 25 000 years ago, the northern hemisphere may have experienced a widespread distribution of potential liquid water events (Richardson and Mischna, 2005).

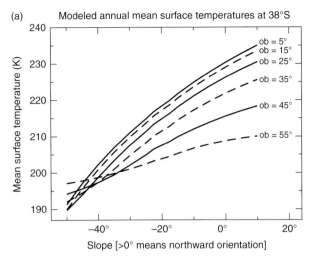

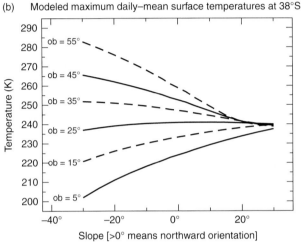

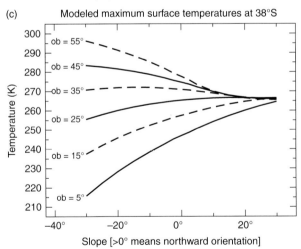

Figure 16.18. (a) Annual mean, (b) maximum daily mean, and (c) absolute maximum surface temperature (K) simulated at 38°S (the latitude of major icy deposits in eastern Hellas) using a one-dimensional version of the LMD Mars global climate model (Forget et al., 1999, 2006) coupled with the slope insolation model of Spiga and Forget (2008). A circular orbit is assumed. The surface albedo and thermal inertia are set to 0.2 (0.6 when CO_2 ice is present) and 1000 J m^{-2} K^{-1} s$^{-1/2}$ (SI units), respectively. The atmospheric surface pressure is 610 Pa and the column dust optical depth is 0.2. Figure adapted from Hartmann et al. (2014). Note that, at high obliquity, poleward-facing slopes are coldest on average, but warmest in summer.

16.7.2.2 *Impact of Obliquity*

Excursion to high obliquities may favor the formation of liquid water for the following reasons:

- As explained in Section 16.3.2, the release of the south polar CO_2 ice reservoir and possible CO_2 desorption may double the surface pressure if the obliquity is increased by a few degrees. In such conditions, on most of the planet the pressure would be above the triple point of water, and liquid water theoretically possible.
- As discussed in Section 16.6.2, the water cycle at higher obliquities (or when returning from high obliquities with tropical and mid-latitude glaciers out of equilibrium) is much more intense, allowing the accumulation of water ice in areas where summer afternoon temperature can reach the melting point.
- The changes in insolation geometry can enhance the heating of slopes in mid- and high latitudes. In particular, as shown by Costard et al. (2002), at high obliquity poleward-facing slopes at mid- and high latitudes can be both the coldest places on an annual average sense (a condition that favors the accumulation of water ice) and the warmest places during summer, in both the seasonal and instantaneous sense (Figure 16.18). Diurnal mean temperatures above 0°C are predicted on poleward-facing slopes at mid- and high latitudes. The corresponding thermal waves could have melted any ice present down to a few tens of centimeters.

16.7.3 Evidence for Liquid Water in the Recent Martian Past: Gullies and Fluvial Valleys

16.7.3.1 *Gullies*

Martian gullies are found within crater walls at both low and mid-latitudes especially on poleward-facing slopes (Malin and Edgett, 2000). They are estimated to have formed in the past million years. Since their discovery, they have attracted considerable attention because they resemble terrestrial debris flows and fluvial channels formed by the action of liquid. In many cases, their sinuosities, the connections of their channels, and the occurrence of levees suggest that they were formed by flows with a significant proportion of liquid, or at least that the flow was fluidized by a gas. Early studies speculated a source below the surface, either subsurface aquifers (Malin and Edgett, 2000) or melting of near-surface groundwater ice from geothermal heating (Hartmann, 2001). However, there is no clear association between the location of debris flows and the general distribution of putative recent geothermal activity, and in many cases the involvement of aquifers seems unlikely, for instance at the top of an isolated peak where gullies have been observed to originate (Dickson et al., 2007). An alternative, outlined by Costard et al. (2002), suggests melting of groundwater ice by exposure to temperatures above 273 K during periods of high obliquity. As mentioned above, they found that above-freezing temperatures can occur at high obliquities in the near-surface of pole-facing slopes on Mars, and that such temperatures are only predicted at latitudes and for slope orientations corresponding to where the gullies have been observed on Mars. This latter theory, further expressed in Morgan et al. (2010) and Schon and Head (2011), argues that the presence of alcove-less gullies (without an obvious source for the water) is consistent with gully formation through degradation and melting of ice-rich mantling deposits (see Section 16.5.1) most likely deposited in response to excursion to high obliquity (Section 16.6.2). By examining the surface heat balance at typical gully locations, Williams et al. (2009) extended the scenario and argued for a seasonal surface snow (rather than near-surface ground ice) source of gully meltwater.

It is possible that some occurrences of Martian gullies are not related to liquid water. Frequent monitoring of the Martian surface has revealed that some gully channel formation is ongoing on present-day Mars (Reiss et al., 2010; Dundas et al., 2010, 2012, 2015; Diniega et al., 2013), notably on sand dunes. The timing of the observed gully activity appears to occur during the period when seasonal CO_2 frost is present and defrosting, pointing to a role of the CO_2 condensation–sublimation cycle, which could destabilize and gas-fluidize the debris flow through a process with no terrestrial analog (Pilorget and Forget, 2016). It remains debated whether all gullies are formed through such dry ice processes, or whether the most Earth-like gullies are the consequence of processes involving liquid water as on Earth.

16.7.3.2 Larger Flows?

The very limited and local nature of the gully systems suggests that the liquid water only briefly formed and traveled for short distances before refreezing or evaporating. Have larger events involving more liquid water occurred on Mars in its recent history? Two geological features can be mentioned here. First, Dickson et al. (2009) described fluvial valley systems of more than tens of kilometers in size within Lyot Crater that appeared to be relatively recent (middle or possibly late Amazonian). They suggested that their existence could be explained by the fact that the interior of Lyot Crater is an optimal microenvironment for the melting of water ice and water flows, since its very low elevation leads to the highest surface pressure of Mars (outside the Hellas Basin), and temperature conditions at its location in the northern mid-latitudes are sufficient for melting during periods of high obliquity. Second, among the outflow channels apparently carved by catastrophic flood events (induced by geophysical events rather than climate processes), some are recent enough to have played a role in the recent Mars environment. In particular, Athabasca Valley seems to have formed in association with Cerberus Fossae (Burr et al., 2002; Plescia, 2003), possibly as recently as tens of millions of years ago. The released water is thought to have flowed down a 10–30 km wide valley for tens to hundreds of kilometers. Its evaporation may have led to significant precipitation and the formation of glaciers in other locations on Mars.

16.8 CLIMATE VARIATIONS IN THE PAST HUNDREDS OF YEARS

One of the most compelling aspects of studying climate on Mars is that the climate is always changing. The large orbital element and spin variations (Figures 16.1 and 16.2) can be seen to persist into the future as well as the past. Coincidentally,

we are in a period of Mars' history where the obliquity is not rapidly changing and does not have particularly extreme values. This may make the current climate more prone to effects that are usually secondary, such as precession of the timing of perihelion.

16.8.1 The Northern Residual Water Ice Polar Cap

Nowhere is this uncertainty better manifested than in the current exchange of water between the northern residual cap (NRC) and the rest of the planet. As explained above, over the past 10 kyr, the theoretical expectation is that the northern polar cap should have accumulated water at the expense of other reservoirs such as the mid-latitude ground ice or the south polar regions (Chamberlain and Boynton, 2007; Levrard et al., 2007; Montmessin et al., 2007; Schörghofer, 2007a,b). The impact cratering record shows that NRC impact craters have been infilled with water ice and hidden over the past 20 kyr (Banks et al., 2010), which suggests recent NRC accumulation in general. However, visual and hyperspectral data show the NRC is composed of clean large-grained water ice (Kieffer, 1990; Langevin et al., 2005). Ice grain size increases with time, so the observation of large ice grain sizes implies that old water ice is being exposed during the summer, which would mean that the NRC is currently undergoing a net loss of material. In addition to this, the observation of new ice-exposing impacts at latitudes lower than anticipated shows that retreat of mid-latitude ice has recently been slower than expected (Byrne et al., 2009; Schörghofer and Forget, 2012).

In short, we may have had a recent polar accumulation period of the NRC (lasting ~10 kyr) coincident with a recent downturn in north polar insolation (Figure 16.2) that has just ended (allowing older large-grained ice to be exposed by ablation). Accumulation may have stopped decades, centuries, or a millennium ago; however, it is not clear why polar accumulation should have recently stopped at all, as polar insolation continues to decrease.

If we had happened to be exploring Mars at almost any other time in its history, we would be able to see clear evidence of either vigorous accumulation or ablation of the polar ice deposits. Instead, we are left with a rather ambiguous situation where usually small effects can become dominant. Indeed, even the extent of the NRC varies inter-annually (Malin and Edgett, 2001; Hale et al., 2005), although these variations are limited to a few percent of its area and seem to reverse on timescales of years (Byrne et al., 2008b).

16.8.2 The Southern Residual CO_2 Ice Polar Cap

Although current orbitally driven climate change is slow by historical standards, things are not static. The south polar residual cap (SRC), discussed earlier as an available reservoir of CO_2 ice, has quasi-circular pits (Thomas et al., 2000) that have been observed to expand by several meters from year to year (Malin et al., 2001). This initially prompted speculation of climate change as this ablated CO_2 was entering the atmosphere, increasing the mean pressure (indeed, given the rate that the pits are expanding, they should completely erode the SRC in a few centuries). However, the extremely bright flat surfaces of the SRC indicate that they are actively accumulating mass all year.

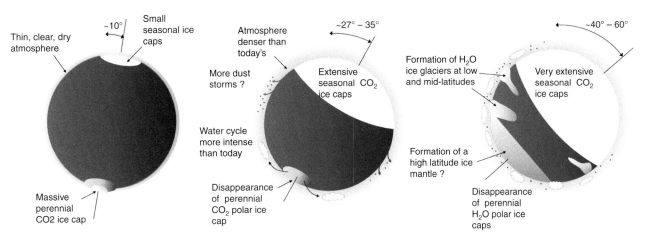

Figure 16.19. The effects of obliquity on the Martian climate: (a) low obliquity (0° to 20°); (b) high obliquity (~27° to 35°); and (c) very high obliquity (40° to 60°).

Thus, it is not clear whether the SRC as a whole is currently gaining or losing mass. Pressure data from MSL may ultimately reveal the long-term mass balance of the SRC (for details, see Haberle and Kahre, 2010). Models of landscape evolution suggest that these expanding pits may erode the CO_2 ice slabs that they are embedded in even as new slabs of CO_2 ice are accumulating in the pit centers and surrounding terrain (Byrne et al., 2008a). The current state of expanding pits and accumulating flat surfaces in the SRC may have persisted much longer than the observed changes in the SRC imply.

16.8.3 Dust and Albedo Changes

Other forms of climate change may also exist in the form of shifting dust patterns. Dust forms a high-albedo, low-thermal-inertia covering that obscures a substantial portion of the planet, affecting surface temperatures and the circulation patterns they generate. Comparing Viking and MGS imagery, Geissler (2005) reported surface albedo changes over a substantial fraction of the planet due to dust movement. These changes were thought to occur over several decades. Fenton et al. (2005) investigated the climatic effects of these changes and concluded that surface air was currently warming. However, these results have been invalidated by Szwast et al. (2006), who showed that the reported inter-annual albedo variations were more likely seasonal, pre-/post-dust storm variations, and that there was no evidence of changes over longer timescales.

16.9 CONCLUSION

Recent geological features, such as glacier-related landforms, desiccating icy mantles, and layered polar deposits, prove that the Martian climate must have changed in the recent geological past. Their presence and their characteristics cannot be explained with the present-day climate.

The primary cause of these climate changes must have been the variations of Mars' orbital and spin parameters, which are more pronounced on Mars than on the Earth. Numerical modeling of the climate system that we know today (atmosphere, surface, subsurface), but performed taking into account the variations of these parameters, have shown that the induced changes in the water cycle were sufficient to move water ice around the planet. To first order, these models have been successful in predicting the location and characteristics of the geologically recent ice-related landforms. More generally, such models and other theoretical considerations can be used to envision the planet's responses to the orbital and spin parameters. For instance, the effects of obliquity variations are summarized in Figure 16.19. At obliquities lower than today's, the atmosphere should be relatively clear, dry and thin, with a large part of the current atmospheric CO_2 frozen in the polar regions. Conversely, at high obliquities, a thicker, somewhat dustier and more humid atmosphere is expected. This extends the stability of subsurface ice to latitudes lower than today, and, at high enough obliquities, allows the formation of non-polar ice glaciers.

The details of the past climate remains theoretical. On the one hand, it is possible that one or several physical processes that we have neglected in our calculations played a role in the climate that occurred in the past. As explained in Section 16.6.2, further modeling work is necessary to fully represent all the changes in the coupled systems. A striking example is the role of thick clouds at high obliquities. In some of the most recent models, the clouds are found to produce a strong greenhouse effect, warming surface temperatures by many tens of degrees. In other models, the effect is minimal. The discrepancies may be related to how the models represent cloud microphysical processes. Thus, just as for Earth, clouds will be a challenge for future GCM research. On the other hand, several major questions related to the available observations remain puzzling. For instance, the origin of the high-latitude subsurface water ice (which seems relatively pure in many locations) is highly debated, as explained in Section 16.6.3. Also, it remains difficult to relate the apparent ages of the various non-polar ice deposits as well as the observed strata in the polar caps to the actual calculation of Mars orbital and spin variations. There is no doubt that many answers to these questions, as well as many new enigmas will come in the near future thanks to new observations and models.

REFERENCES

Aharonson, O., and Schörghofer, N. (2006) Subsurface ice on Mars with rough topography, *J. Geophys. Res.*, 111(E10), 11007.

Armstrong, J. C., Leovy, C. B., and Quinn, T. (2004) A 1 Gyr climate model for Mars: new orbital statistics and the importance of seasonally resolved polar processes, *Icarus*, 171, 255–271.

Arvidson, R. E., Bonitz, R. G., Robinson, M. L., et al. (2009) Results from the Mars Phoenix Lander Robotic Arm experiment, *Journal of Geophysical Research (Planets)*, 114, 5685.

Banks, M. E., Byrne, S., Galla, K., et al. (2010) Crater population and resurfacing of the Martian north polar layered deposits, *Journal of Geophysical Research (Planets)*, 115, 8006.

Bibring, J., Langevin, Y., Poulet, F., et al. (2004) Perennial water ice identified in the south polar cap of Mars, *Nature*, 428, 627–630.

Black, B. A., and Stewart, S. T. (2008) Excess ejecta craters record episodic ice-rich layers at middle latitudes on Mars, *Journal of Geophysical Research (Planets)*, 113, 2015.

Boynton, W. V., Feldman, W. C., Squyres, et al. (2002) Distribution of hydrogen in the near surface of Mars: evidence for subsurface ice deposits, *Science*, 297, 81–85.

Burr, D. M., Grier, J. A., McEwen, A. S., and Keszthelyi, L. P. (2002) Repeated aqueous flooding from the Cerberus Fossae: evidence for very recently extant, deep groundwater on Mars, *Icarus*, 159, 53–73.

Byrne, S. and Ingersoll, A. P. (2003) A sublimation model for Martian south polar ice features, *Science*, 299, 1051–1053.

Byrne, S., and Murray, B. C. (2002) North polar stratigraphy and the paleo-erg of Mars, *Journal of Geophysical Research (Planets)*, 107, 5044.

Byrne, S., Russell, P. S., Fishbaugh, K. E., et al. (2008a) Explaining the persistence of the southern residual cap of Mars: HiRISE data and landscape evolution models, *LPSC abstracts*, 39, 2252.

Byrne, S., Zuber, M. T., and Neumann, G. A. (2008b) Interannual and seasonal behavior of Martian residual ice-cap albedo, *Planet. Space Sci.*, 56, 194–211.

Byrne, S., Dundas, C. M., Kennedy, M. R., et al. (2009) Distribution of mid-latitude ground ice on Mars from new impact craters, *Science*, 325, 1674–1676.

Carr, M. H. (1996) *Water on Mars*, New York, Oxford University Press.

Carr, M. H., and Head, J. W. (2015) Martian surface/near-surface water inventory: sources, sinks, and changes with time, *Geophys. Res. Lett.*, 42, doi:10.1002/2014GL062464.

Carter, J. W., and Husain, H. (1974) The simultaneous adsorption of carbon dioxide and water vapour by fixed beds of molecular sieves, *Chem. Eng. Sci.*, 29, 267–273.

Chamberlain, M. A., and Boynton, W. V. (2007), Response of Martian ground ice to orbit-induced climate change, *J. Geophys. Res.*, 112 (E6), E06009.

Chevrier, V., Sears, D. W. G., Chittenden, J. D., et al. (2007) Sublimation rate of ice under simulated Mars conditions and the effect of layers of mock regolith JSC Mars-1, *Geophys. Res. Lett.*, 34, L02203.

Chevrier, V., Hanley, J., and Altheide, T. S. (2009) Stability of perchlorate hydrates and their liquid solutions at the Phoenix landing site, Mars, *Geophysical Research Letters*, 36, L10202.

Christensen, P. R. (1986) Regional dust deposits on Mars: physical properties, age, and history, *J. Geophys. Res.*, 91, 3533–3545.

Chuang, F. C., and Crown, D. A. (2005) Surface characteristics and degradational history of debris aprons in the Tempe Terra/Mareotis fossae region of Mars, *Icarus*, 179, 24–42.

Clark, B. C., and Van Hart, D. C. (1981) The salts of Mars, *Icarus*, 45, 370–378.

Clow, G. D. (1987) Generation of liquid water on Mars through the melting of a dusty snowpack, *Icarus*, 72, 95–127.

Costard, F., Forget, F., Mangold, N., and Peulvast, J. P. (2002) Formation of recent Martian debris flows by melting of near-surface ground ice at high obliquity, *Science*, 295, 110–113.

Cull, S., Arvidson, R. E., Mellon, M. T., et al. (2010) Compositions of subsurface ices at the Mars Phoenix landing site, *Geophys. Res. Lett.*, 37, L24203.

Cutts, J. A. (1973) Nature and origin of layered deposits of the Martian polar region, *J. Geophys. Res.*, 78, 4231–4249.

Dickson, J. L., Head, J. W., and Kreslavsky, M. A. (2007) Martian gullies in the southern mid-latitudes of Mars: evidence for climate-controlled formation of young fluvial features based upon local and global topography, *Icarus*, 188, 315, doi:10/1016/j.icarus.2006.11.020.

Dickson, J. L., Head, J. W., and Marchant, D. R. (2008) Late Amazonian glaciation at the dichotomy boundary on Mars: evidence for glacial thickness maxima and multiple glacial phases, *Geology*, 36, 411–414.

Dickson, J. L., Fassett, C. I., and Head, J. W. (2009) Amazonian-aged fluvial valley systems in a climatic microenvironment on Mars: melting of ice deposits on the interior of Lyot Crater, *Geophys. Res. Lett.*, 36, 8201.

Dickson, J. L., Head, J. W., and Marchant, D. R (2010) Kilometer-thick ice accumulation and glaciation in the northern mid-latitudes of Mars: evidence for crater-filling events in the Late Amazonian at the Phlegra Montes, *Earth and Planetary Science Letters*, 294, 332–342.

Diez, B., Feldman, W. C., Maurice, S. et al. (2008) H layering in the top meter of Mars, *Icarus*, 196, 409–421.

Diniega, S., Hansen, C. J., McElwaine, J. N., et al. (2013) A new dry hypothesis for the formation of Martian linear gullies, *Icarus*, 225, 526–537.

Dundas, C. M., McEwen, A. S., Diniega, S., Byrne, S., and Martinez-Alonso, S. (2010) New and recent gully activity on Mars as seen by HiRISE, *Geophys. Res. Letters*, 37, L07202.

Dundas, C. M., Diniega, S., Hansen, C. J., Byrne, S., and McEwen, A. S. (2012) Seasonal activity and morphological changes in Martian gullies, *Icarus*, 220, 124–143

Dundas, C. M., Diniega, S., and McEwen, A. S. (2015) Long-term monitoring of Martian gully formation and evolution with MRO/HiRISE, *Icarus*, 251, 244–263.

Durham, W. B., Kirby, S. H., and Stern, L. A. (1999) Steady-state flow of solid CO_2: preliminary results, *Geophysical Research Letter*, 26, 3493–3496.

Edgett, K. S., Williams, R. M. E., Malin, M. C., Cantor, B. A., and Thomas, P. C. (2003) Mars landscape evolution: influence of stratigraphy on geomorphology in the north polar region, *Geomorphology*, 52, 289–297.

Fanale, F. P., and Cannon, W. A. (1974) Exchange of adsorbed H_2O and CO_2 between the regolith and atmosphere of Mars caused by changes in surface insolation, *J. Geophys. Res.*, 79, 3397–3402.

Fanale, F. P., and Cannon, W. A. (1978) Mars – the role of the regolith in determining atmospheric pressure and the atmosphere's response to insolation changes, *J. Geophys. Res.*, 83, 2321–2325.

Fanale, F. P., and Salvail, J. R. (1994) Quasi-periodic atmosphere-regolith-cap CO_2 redistribution in the Martian past, *Icarus*, 111, 305–316.

Fanale, F. P., Salvail, J., Banerdt, W. B., and Saunders, R. S. (1982) Mars: the regolith–atmosphere–cap system and climate change, *Icarus*, 50, 381–407.

Fanale, F. P., Salvail, J. R., Zent, A. P., and Postawko, S. E. (1986) Global distribution and migration of subsurface ice on Mars, *Icarus*, 67, 1–18.

Farmer, C. B., and Doms, P. E. (1979) Global seasonal variation of water vapor on Mars and the implications of permafrost, *J. Geophys. Res.*, 84, 2881–2888.

Fastook J. L., and Head, J. W. (2014) Amazonian mid- to high-latitude glaciation on Mars: supply-limited ice sources, ice accumulation patterns, and concentric crater fill glacial flow and ice sequestration, *Planet. Space Sci.*, 91, 60–76.

Fastook, J. L., Head, J. W., Marchant, D. R., and Forget, F. (2008) Tropical mountain glaciers on Mars: altitude-dependence of ice accumulation, accumulation conditions, formation times, glacier dynamics, and implications for planetary spin-axis/orbital history, *Icarus*, 198, 305, doi:10.1016/j.icarus.2008.08.008.

Fastook J. L., Head, J. W, Forget, F., Madeleine, J.-B., and Marchant, D. R. (2011) Evidence for Amazonian northern mid-latitude regional glacial landsystems on Mars: glacial flow models using GCM-driven climate results and comparisons to geological observations, *Icarus*, 216, 23–39, doi:10.1016/j.icarus.2011.07.018.

Fastook, J. L., Head, J. W., Marchant, D. R., Forget, F., and Madeleine, J.-B. (2012) Early Mars climate near the Noachian-Hesperian boundary: independent evidence for cold conditions from basal melting of the south polar ice sheet (Dorsa Argentea Formation) and implications for valley network formation, *Icarus*, 219, 25–40.

Fastook, J. L., Head, J. W., and Marchant, D. R. (2014) Formation of lobate debris aprons on Mars: assessment of regional ice sheet collapse and debris-cover armoring, *Icarus*, 228, 54–63, doi:10.1016/j.icarus.2013.09.025.

Feldman, W. C., Boynton, W. V., Tokar, R. L., et al. (2002) Global distribution of neutrons from Mars: results from Mars Odyssey, *Science*, 297, 75–78.

Feldman, W. C., Prettyman, T. H., Maurice, S., et al. (2004) Global distribution of near-surface hydrogen on Mars, *Journal of Geophysical Research (Planets)*, 109, 9006.

Feldman, W. C., Mellon, M. T., Gasnault, O., et al. (2007) Vertical distribution of hydrogen at high northern latitudes on Mars: the Mars Odyssey Neutron Spectrometer, *Geophys. Res. Lett.*, 34, L05201.

Feldman, W. C., Bandfield, J. L., Diez, B., et al. (2008) North to south asymmetries in the water-equivalent hydrogen distribution at high latitudes on Mars, *J. Geophys. Res.*, 113, E08006.

Fenton, L. K., and Richardson, M. I. (2001) Martian surface winds: insensitivity to orbital changes and implications for aeolian processes, *Journal of Geophysical Research (Planets)*, 106, 32885.

Fenton, L. K., Toigo, A. D., and Richardson, M. I. (2005) Aeolian processes in Proctor Crater on Mars: mesoscale modeling of dune-forming winds, *Journal of Geophysical Research (Planets)*, 110(E9), 6005.

Fishbaugh, K. E., and Head, J. W. (2002) Chasma Boreale, Mars: topographic characterization from Mars Orbiter Laser Altimeter data and implications for mechanisms of formation, *Journal of Geophysical Research (Planets)*, 107(E3), 5013.

Fishbaugh, K. E., and Head, J. W. (2005) Origin and characteristics of the Mars north polar basal unit and implications for polar geologic history, *Icarus*, 174, 444–474.

Fishbaugh, K. E., and Hvidberg, C. S. (2006) Martian north polar layered deposits stratigraphy: implications for accumulation rates and flow, *Journal of Geophysical Research (Planets)*, 111, 6012.

Fishbaugh, K. E., Hvidberg, C. S., Byrne, S., et al. (2010) First high-resolution stratigraphic column of the Martian north polar layered deposits, *Geophys. Res. Lett.*, 37, 7201.

Forget, F., Hourdin, F., Fournier, et al. (1999) Improved general circulation models of the Martian atmosphere from the surface to above 80 km, *J. Geophys. Res.*, 104, 24155–24176.

Forget, F., Haberle, R. M., Montmessin, F., Levrard, B., and Head, J. W. (2006) Formation of glaciers on Mars by atmospheric precipitation at high obliquity, *Science*, 311, 368–371.

Forget, F., Madeleine, J.-B., Head, J. W., et al. (2014) What does obliquity do to the climate? In *Eighth International Conference on Mars*, LPI Contribution 1791, 1318.

François, L. M., Walker, J. C. G., and Kuhn, W. R. (1990) A numerical simulation of climate changes during the obliquity cycle on Mars, *J. Geophys. Res.*, 95, 14761–14778.

Geissler, P. E. (2005) Three decades of Martian surface changes, *Journal of Geophysical Research (Planets)*, 110(E9), 2001.

Ghatan, G. J., and Head, J. W. (2002) Candidate subglacial volcanoes in the south polar region of Mars: morphology, morphometry, and eruption conditions, *Journal of Geophysical Research (Planets)*, 107, 5048.

Glavin, D. P., Freissinet, C., Miller, K. E., et al. (2013) Evidence for perchlorates and the origin of chlorinated hydrocarbons detected by SAM at the Rocknest aeolian deposit in Gale Crater, *J. Geophys. Res.*, 118, 1955–1973.

Grima, C., Kofman, W., Mouginot, J., et al. (2009) North polar deposits of Mars: extreme purity of the water ice, *Geophys. Res. Lett.*, 36, 3203.

Haberle, R. M., and Kahre, M. A. (2010) Detecting secular climate change on Mars, *The Mars Journal*, 4, 68–75, doi:10.1555/mars.2010.0003

Haberle, R. M., Tyler, D., McKay, C. P., and Davis, W. L. (1994) A model for the evolution of CO_2 on Mars, *Icarus*, 109, 102–120.

Haberle, R. M., McKay, C. P., Schaeffer, J., Cabrol, N., et al. (2001) On the possibility of liquid water on present-day Mars, *J. Geophys. Res.*, 106, 23317–23326.

Haberle, R. M., Murphy, J. R., and Schaeffer, J. (2003) Orbital change experiments with a Mars general circulation model, *Icarus*, 161, 66–89.

Haberle, R. M., Montmessin, F., Kahre, M. A., et al. (2011) Radiative effects of water ice clouds on the Martian seasonal cycle. in *The Fourth International Workshop on the Mars Atmosphere: Modelling and Observations*, Paris, France.

Hale, A. S., Bass, D. S., and Tamppari, L. K. (2005) Monitoring the perennial Martian northern polar cap with MGS MOC, *Icarus*, 174, 502–512.

Hartmann, W. K. (2001) Martian seeps and their relation to youthful geothermal activity, *Space Science Review*, 96, 405–410.

Hartmann W. K., Ansan, V., Berman, D. C., Mangold, N., and Forget, F. (2014) Comprehensive analysis of glaciated Martian crater Greg, *Icarus*, 228, 96–120.

Hauber, E., van Gasselt, S., Chapman, M. G., and Neukum, G. (2008) Geomorphic evidence for former lobate debris aprons at low latitudes on Mars: indicators of the Martian paleoclimate, *Journal of Geophysical Research (Planets)*, 113, 2007.

Head, J. W., and Marchant, D. R. (2003) Cold-based mountain glaciers on Mars: western Arsia Mons, *Geology*, 31(7), 641.

Head, J. W. III, and Marchant, D. R. (2006) Modification of the walls of a Noachian crater in northern Arabia Terra (24E, 39N) during northern mid-latitude Amazonian glacial epochs on Mars: nature and evolution of lobate debris aprons and their relationships to lineated valley fill and glacial systems. In *37th Annual Lunar and Planetary Science Conference*, Lunar and Planetary Inst. Technical Report, 37, 1126.

Head, J. W., and Marchant, D. R. (2009). Inventory of ice-related deposits on Mars: evidence for burial and long-term sequestration of ice in non-polar regions and implications for the water budget and climate evolution. In *Lunar and Planetary Institute Science Conference Abstracts*, 40, 1356.

Head, J. W., and Marchant, D. R. (2014) The climate history of early Mars: insights from the Antarctic McMurdo Dry Valleys hydrologic system, *Antarctic Science*, 26, 774–800, doi:10.1017/S0954102014000686.

Head, J. W., and Pratt, S. (2001) Extensive Hesperian-aged south polar ice sheet on Mars: evidence for massive melting and retreat, and lateral flow and ponding of meltwater, *J. Geophys. Res.*, 106, 12275–12300.

Head, J. W., Mustard, J. F., Kreslavsky, M. A., Milliken, R. E., and Marchant, D. R. (2003) Recent ice ages on Mars, *Nature*, 426, 797–802.

Head, J. W., Neukum, G., Jaumann, R., et al. (2005) Tropical to mid-latitude snow and ice accumulation and glaciation on Mars, *Nature*, 434, 346–351.

Head, J. W., Marchant, D. R., Agnew, M. C., Fassett, C. I., and Kreslavsky, M. A. (2006a) Extensive valley glacier deposits in the northern mid-latitudes of Mars: evidence for late Amazonian obliquity-driven climate change, *Earth and Planetary Science Letters*, 241, 663–671.

Head, J. W., Nahm, A. L., Marchant, D. R., and Neukum, G. (2006b) Modification of the dichotomy boundary on Mars by Amazonian mid-latitude regional glaciation, *Geophys. Res. Lett.*, 33, 8.

Head, J. W., Marchant, D. R., Forget, F., et al. (2009) Deciphering the late Amazonian history of Mars: assessing obliquity predictions with geological observations and atmospheric general circulation models, in *40th Lunar and Planet. Sci. Conf.*, March 23–27, The Woodlands, TX, Abstract No. 1349.

Head, J. W., Marchant, D. R., Dickson, J. L., Kress, A. M., and Baker, D. M. (2010) Northern mid-latitude glaciation in the Late Amazonian period of Mars: criteria for the recognition of debris-covered glacier and valley glacier land system deposits, *Earth and Planetary Science Letters*, 294, 306–320.

Hecht, M. H., Kounaves, S. P., Quinn, R. C., et al. (2009) Detection of perchlorate and the soluble chemistry of Martian soil at the Phoenix Lander site, *Science*, 325, 64–67.

Heldmann, J. L., and Mellon, M. T. (2004) Observations of Martian gullies and constraints on potential formation mechanisms, *Icarus*, 168, 285–304.

Heldmann, J. L., Carlsson, E., Johansson, H., Mellon, M. T., and Toon, O. B. (2007) Observations of Martian gullies and constraints on potential formation mechanisms. II. The northern hemisphere, *Icarus*, 188, 324–344.

Herkenhoff, K. E., and Plaut, J. J. (2000) Surface ages and resurfacing rates of the polar layered deposits on Mars, *Icarus*, 144, 243–253.

Herkenhoff, K. E., Byrne, S., Russell, P. S., Fishbaugh, K. E., and McEwen, A. S. (2007) Meter-scale morphology of the north polar region of Mars, *Science*, 317, 1711.

Hofstadter, M. D., and Murray, B. C. (1990) Ice sublimation and rheology – implications for the Martian polar layered deposits, *Icarus*, 84, 352–361.

Holt, J. W., Safaeinili, A., Plaut, J., et al. (2008) Radar sounding evidence for buried glaciers in the southern mid-latitudes of Mars, *Science*, 322, 1235.

Holt, J. W., Fishbaugh, K. E., Byrne, S., et al. (2010) The construction of Chasma Boreale on Mars, *Nature*, 465, 446–449.

Hudson, T. L., and Aharonson, O. (2008) Diffusion barriers at Mars surface conditions: salt crusts, particle size mixtures, and dust, *J. Geophys. Res.*, 113, E09008.

Hudson, T. L., Aharonson, O., Schörghofer, N., et al. (2007) Water vapor diffusion in Mars subsurface environments, *J. Geophys. Res.*, 112(E5), E05016.

Hudson, T. L., Aharonson, O., and Schörghofer, N. (2009) Laboratory experiments and models of diffusive emplacement of ground ice on Mars, *J. Geophys. Res.*, 114, E01002.

Hvidberg, C. S. (2003) Mass balance processes on the north polar cap on Mars. In *EGS–AGU–EUG Joint Assembly*, 2996.

Imbrie, J., and Imbrie, K. P. (1979) *Ice Ages: Solving the Mystery*. Harvard University Press.

Ingersoll, A. P. (1970) Mars: occurrence of liquid water, *Science*, 168, 972–973.

Ivanov, A. B., and Muhleman, D. O. (2000) The role of sublimation for the formation of the northern ice cap: results from the Mars Orbiter Laser Altimeter, *Icarus*, 144, 436–448.

Jakosky, B. M., and Carr, M. H. (1985) Possible precipitation of ice at low latitudes of Mars during periods of high obliquity, *Nature*, 315, 559–561.

Jakosky, B. M., Henderson, B. G., and Mellon, T. M. (1993) The Mars water cycle at other epochs: recent history of the polar caps and layered terrain, *Icarus*, 102, 286–297.

James, P. B., Kieffer, H. H., and Paige, D. A. (1992) The seasonal cycle of carbon dioxide on Mars, in *Mars*, eds H. H. Kieffer, B. M. Jakosky, C. B. Snyder, and M. S. Matthews, University of Arizona Press, Tucson, AZ, 934–968.

Kadish, S. J., and Head, J. W. (2011a) Impacts into non-polar ice-rich paleodeposits on Mars: excess ejecta craters, perched craters and pedestal craters as clues to Amazonian climate history, *Icarus*, 215, 34–46.

Kadish, S. J., and Head, J. W. (2011b) Preservation of layered paleodeposits in high-latitude pedestal craters on Mars, *Icarus*, 213, 443–450.

Kadish, S. J., Head, J. W., Barlow, N. G., and Marchant, D. R. (2008a) Martian pedestal craters: marginal sublimation pits implicate a climate-related formation mechanism, *Geophys. Res. Lett.*, 35, 16104.

Kadish, S. J., Head, J. W., Parsons, R. L., and Marchant, D. R. (2008b) The Ascraeus Mons fan-shaped deposit: volcano ice interactions and the climatic implications of cold-based tropical mountain glaciation, *Icarus*, 197, 84–109.

Kadish, S. J., Barlow, N. G., and Head, J. W. (2009) Latitude dependence of Martian pedestal craters: evidence for a sublimation-driven formation mechanism, *Journal of Geophysical Research (Planets)*, 114, 10001.

Kadish, S. J., Head, J. W., and Barlow, N. G. (2010) Pedestal crater heights on Mars: a proxy for the thicknesses of past, ice-rich, Amazonian deposits, *Icarus*, 210, 92–101.

Kahn, R. (1985) The evolution of CO_2 on Mars, *Icarus*, 62, 175–190.

Kieffer, H. H. (1990) H_2O grain size and the amount of dust in Mars' residual north polar cap, *J. Geophys. Res.*, 95, 1481–1493.

Kieffer, H. H., and Zent, A. P. (1992) Quasi-periodic climate change on Mars, in *Mars*, eds H. H. Kieffer, B. M. Jakosky, C. B. Snyder, and M. S. Matthews, University of Arizona Press, Tucson, AZ, 1180–1218.

Kieffer, H. H., Chase, S. C., Miner, E. D., et al. (1976) Infrared thermal mapping of the Martian surface and atmosphere: first results, *Science*, 193, 780–786.

Kolb, E. J., and Tanaka, K. L. (2006) Accumulation and erosion of south polar layered deposits in the Promethei Lingula region, Planum Australe, Mars, *International Journal of Mars Science and Exploration*, 2, 1–9.

Kostama, V.-P., Kreslavsky, M. A., and Head, J. W. (2006) Recent high-latitude icy mantle in the northern plains of Mars: characteristics and ages of emplacement, *Geophys. Res. Lett.*, 33, L11201.

Kounaves, S. P., Hecht, M. H., Kapit, J. et al. (2010) Wet chemistry experiments on the 2007 Phoenix Mars Scout Lander mission: data analysis and results, *Journal of Geophysical Research (Planets)*, 115, E00E10.

Koutnik, M., Byrne, S., and Murray, B. (2002) South polar layered deposits of Mars: the cratering record, *Journal of Geophysical Research (Planets)*, 107, 5100.

Kreslavsky, M. A., and Head, J. W. (2000) Kilometer-scale roughness of Mars: results from MOLA data analysis, *Journal of Geophysical Research (Planets)*, 105, 26695.

Kreslavsky, M. A., and Head, J. W. (2002) Mars: nature and evolution of young latitude-dependent water-ice-rich mantle, *Geophys. Res. Lett.*, 29(15), 1719.

Kreslavsky, M. A., and Head, J. W. (2005) Mars at very low obliquity: atmospheric collapse and the fate of volatiles. *Geophys. Res. Lett.*, 32, 12202.

Kreslavsky, M. A., and Head, J. W. (2006) Modification of impact craters in the northern plains of Mars: implications for Amazonian climate history, *Meteoritics and Planetary Science*, 41, 1633–1646.

Kreslavsky, M. A., and Head, J. W. (2011) Carbon dioxide glaciers on Mars: products of recent low obliquity epochs?, *Icarus*, 216, 111–115.

Kress, A. M., and Head, J. W. (2008) Ring-mold craters in lineated valley fill and lobate debris aprons on Mars: evidence for subsurface glacial ice, *Geophys. Res. Lett.*, 35, 23206.

Langevin, Y., Poulet, F., Bibring, J.-P., et al. (2005) Summer evolution of the north polar cap of Mars as observed by OMEGA/Mars Express, *Science*, 307, 1581–1584.

Laskar, J., and Robutel, P. (1993) The chaotic obliquity of the planets, *Nature*, 361, 608–612.

Laskar, J., Levrard, B., and Mustard, J. F. (2002) Orbital forcing of the Martian polar layered deposits, *Nature*, 419, 375–377.

Laskar, J., Correia, A. C. M., and Gastineau, M. (2004) Long term evolution and chaotic diffusion of the insolation quantities of Mars, *Icarus*, 170, 343–364.

Leighton, R. R., and Murray, B. C. (1966) Behavior of carbon dioxide and other volatiles on Mars, *Science*, 153, 136–144.

Levrard, B., Forget, F., Montmessin, F., and Laskar, J. (2004) Recent ice-rich deposits formed at high latitudes on Mars by sublimation of unstable equatorial ice during low obliquity, *Nature*, 431, 1072–1075.

Levrard, B., Forget, F., Montmessin, F., and Laskar, J. (2007) Recent formation and evolution of northern Martian polar layered deposits as inferred from a Global Climate Model, *Journal of Geophysical Research (Planets)*, 112, 6012.

Levy, J. S., Head, J. W., Marchant, D. R., and Kowalewski, D. E. (2008) Identification of sublimation-type thermal contraction crack polygons at the proposed NASA Phoenix landing site: implications for substrate properties and climate-driven morphological evolution, *Geophys. Res. Lett.*, 35, 4202.

Levy, J., Head, J., and Marchant, D. (2009) Thermal contraction crack polygons on Mars: classification, distribution, and climate implications from HiRISE observations, *Journal of Geophysical Research (Planets)*, 114, 1007.

Levy, J., Head, J. W., and Marchant, D. R. (2010) Concentric crater fill in the northern mid-latitudes of Mars: formation processes and relationships to similar landforms of glacial origin, *Icarus*, 209, 390–404.

Levy, J. S., Fassett, C. I., Head, J. W., Schwarts, C., and Wateters, J. L. (2014) Sequestered glacial ice contribution to the global Martian water budget: geometric constraints on the volume of remnant, midlatitude debris-covered glaciers, *J. Geophys. Res.*, 119, 1–9, doi:10.1002/2014JE004685

Litvak, M. L., Mitrofanov, I. G., Kozyrev, A. S., et al. (2006) Comparison between polar regions of Mars from HEND/Odyssey data, *Icarus*, 180, 23–37.

Lucchitta, B. K. (1984) Ice and debris in the fretted terrain, Mars, *Journal of Geophysical Research Supplement*, 89, 409.

Madeleine, J.-B., Forget, F., Head, J. W., et al. (2009), Amazonian northern mid-latitude glaciation on Mars: a proposed climate scenario, *Icarus*, 203, 390–405.

Madeleine, J.-B., Forget, F., and Millour, E. (2011) Modeling radiatively active water ice clouds: impact on the thermal structure and water cycle. In *The Fourth International Workshop on Mars Atmosphere: Modelling and observation*, Paris, France.

Madeleine, J.-B., Head, J. W., Forget, F., et al. (2014) Recent ice ages on Mars: the role of radiatively active clouds and cloud microphysics, *Geophys. Res. Lett.*, 41, 4873–4879

Malin, M. C., and Edgett, K. S. (2000) Sedimentary Rocks of Early Mars, *Science*, 290, 1927–1937.

Malin, M. C., and Edgett, K. S. (2001) Mars Global Surveyor Mars Orbiter Camera: interplanetary cruise through primary mission, *J. Geophys. Res.*, 106, 23429–23570.

Malin, M. C., Caplinger, M. A., and Davis, S. D. (2001) Observational evidence for an active surface reservoir of solid carbon dioxide on Mars, *Science*, 294, 2146–2148.

Mangold, N. (2005) High latitude patterned grounds on Mars: classification, distribution and climatic control, *Icarus*, 174, 336–359.

Mangold, N., Maurice, S., Feldman, W. C., Costard, F., and Forget, F. (2004) Spatial relationships between patterned ground and ground ice detected by the Neutron Spectrometer on Mars, *Journal of Geophysical Research (Planets)*, 109, 8001.

Marchant, D. R., and Head, J. W. (2007) Antarctic dry valleys: microclimate zonation, variable geomorphic processes, and implications for assessing climate change on Mars, *Icarus*, 192, 187–222.

Mellon, M. T., and Jakosky, B. M. (1993) Geographic variations in the thermal and diffusive stability of ground ice on Mars, *J. Geophys. Res.*, 98(E2), 3345–3364.

Mellon, M. T., and Jakosky, B. M. (1995) The distribution and behavior of Martian ground ice during past and present epochs, *J. Geophys. Res.*, 100(E6), 11781–11799.

Mellon, M. T., Feldman, W. C., and Prettyman, T. H. (2004) The presence and stability of ground ice in the southern hemisphere of Mars, *Icarus*, 169, 324–340.

Mellon, M. T., Arvidson, R. E., Sizemore, H. G., et al. (2009) Ground ice at the Phoenix landing site: stability state and origin, *J. Geophys. Res.*, 114, E00E07.

Milkovich, S. M., Head, J. W., and Marchant, D. R. (2006) Debris-covered piedmont glaciers along the northwest flank of the Olympus Mons scarp: evidence for low-latitude ice accumulation during the late Amazonian of Mars, *Icarus*, 181, 388–407.

Milkovich, S. M., Byrne, S., and Russell, P. S. (2011) Variations in surface texture of the north polar residual cap of Mars, in *Fifth Mars Polar Science Conf.*, LPI Contributions, No. 1623, 6029.

Milliken, R. E., Mustard, J. F., and Goldsby, D. L. (2003) Viscous flow features on the surface of Mars: observations from high-resolution Mars Orbiter Camera (MOC) images, *Journal of Geophysical Research (Planets)*, 108(E6), 5057.

Mischna, M. A., and Richardson, M. I. (2005) A reanalysis of water abundances in the Martian atmosphere at high obliquity, *Geophys. Res. Lett.*, 23, L03201.

Mischna, M. A., Richardson, M. I., Wilson, R. J., and McCleese, D. J. (2003) On the orbital forcing of Martian water and CO_2 cycles: a general circulation model study with simplified volatile schemes, *Journal of Geophysical Research (Planets)*, 108(E6), 16–1.

Mitrofanov, I., Anfimov, D., Kozyrev, A., et al. (2002) Maps of subsurface hydrogen from the High Energy Neutron Detector, Mars Odyssey, *Science*, 297, 78–81.

Montmessin, F., Haberle, R. M., Forget, F., et al. (2007) On the origin of perennial water ice at the south pole of Mars: a precession-controlled mechanism? *Journal of Geophysical Research (Planets)*, 112(E11), 8.

Montmessin, F., Forget, F., Rannou, P., Cabane, M., and Haberle, R. M. (2004) Origin and role of water ice clouds in the Martian water cycle as inferred from a general circulation model, *J. Geophys. Res.*, 109(E10), E10004, doi:10.1029/2004JE002284.

Morgan, G. A., Head, J. W., Forget, F., Madeleine, J.-B., and Spiga, A. (2010) Gully formation on Mars: two recent phases of formation suggested by links between morphology, slope orientation and insolation history, *Icarus*, 208, 658–666.

Murray, B. C., Soderblom, L. A., Cutts, J. A., et al. (1972) Geological framework of the south polar region of Mars, *Icarus*, 17, 328.

Murray, B., Koutnik, M., Byrne, S., et al. (2001) Preliminary geological assessment of the northern edge of Ultimi lobe, Mars south polar layered deposits, *Icarus*, 154, 80–97.

Mustard, J. F., Cooper, C. D., and Rifkin, M. K. (2001) Evidence for recent climate change on Mars from the identification of youthful near-surface ground ice, *Nature*, 412, 411–414.

Nakamura, T., and Tajika, E. (2002) Stability of the Martian climate system under the seasonal change condition of solar radiation, *J. Geophys. Res.*, 107, 5094.

Newman, C. E., Lewis, S. R., and Read, P. L. (2005) The atmospheric circulation and dust activity in different orbital epochs on Mars, *Icarus*, 174, 135–160.

Ojha, L., Wilhelm, M. B., Murchie, S. L., et al. (2015) Spectral evidence for hydrated salts in recurring slope lineae on Mars, *Nature Geoscience*, 8, 11, 829–832.

Orloff, T., Kreslavsky, M., Asphaug, E., and Korteniemi, J. (2011) Boulder movement at high northern latitudes of Mars, *Journal of Geophysical Research (Planets)*, 116, 11006.

Paige, D. A., and Ingersoll, A. P. (1985) Annual heat balance of Martian polar caps: Viking observations, *Science*, 228, 1160–1168.

Pathare, A. V., and Paige, D. A. (2005) The effects of Martian orbital variations upon the sublimation and relaxation of north polar troughs and scarps, *Icarus*, 174, 419–443.

Pedersen, G. B. M., and Head, J. W. (2010) Evidence of widespread degraded Amazonian-aged ice-rich deposits in the transition between Elysium Rise and Utopia Planitia, Mars: guidelines for the recognition of degraded ice-rich materials, *Planet. Space Sci.*, 58, 1953–1970.

Phillips, R. J., Zuber, M. T., Smrekar, S. E., et al. (2008) Mars north polar deposits: stratigraphy, age, and geodynamical response, *Science*, 320, 1182.

Phillips, R. J., Davis, B. J., Tanaka, K. L., et al. (2011) Massive CO_2 ice deposits sequestered in the south polar layered deposits of Mars, *Science*, 332, 838.

Picardi, G., Plaut, J. J., Biccari, D., et al. (2005) Radar soundings of the subsurface of Mars. *Science*, 310, 1925–1928.

Pierce, T. L., and Crown, D. A. (2003) Morphologic and topographic analyses of debris aprons in the eastern Hellas region, Mars, *Icarus*, 163, 46–65.

Pilorget, C., and Forget, F. (2016) Formation of gullies on Mars by debris flows triggered by CO_2 sublimation, *Nature Geoscience*, 9, 65–69.

Plaut, J. J., Picardi, G., Safaeinili, A., et al. (2007) Subsurface radar sounding of the south polar layered deposits of Mars, *Science*, 316, 92.

Plaut, J. J., Safaeinili, A., Holt, J. W., et al. (2009) Radar evidence for ice in lobate debris aprons in the mid-northern latitudes of Mars, *Geophys. Res. Lett.*, 36, 2203.

Plescia, J. B. (2003) Cerberus Fossae, Elysium, Mars: a source for lava and water, *Icarus*, 164, 79–95.

Prettyman, T. H., Feldman, W. C., Mellon, M. T., et al. (2004) Composition and structure of the Martian surface at high southern latitudes from neutron spectroscopy, *J. Geophys. Res.*, 109, E05001.

Putzig, N. E., Phillips, R. J., Campbell, B. A., et al. (2009) Subsurface structure of Planum Boreum from Mars Reconnaissance Orbiter Shallow Radar soundings, *Icarus*, 204, 443–457.

Reiss, D., Erkeling, G., Bauch, K. E., and Hiesinger, H. (2010) Evidence for present day gully activity on the Russell Crater dune field, Mars, *Geophys. Res. Letters*, 37, L06203.

Richardson, M. I., and Wilson, R. J. (2002), A topographically forced asymmetry in the Martian circulation and climate, *Nature*, 416(6878), 298–301, doi:10.1038/416298a.

Richardson, M. I., and Mischna, M. A. (2005) Long-term evolution of transient liquid water on Mars, *Jour. Geophys. Res. (Planets)*, 110, E03003.

Russell, P., Thomas, N., Byrne, S., et al. (2008) Seasonally active frost-dust avalanches on a north polar scarp of Mars captured by HiRISE, *Geophys. Res. Lett.*, 35, 23204.

Scanlon, K. E., Head, J. W., Wilson, L., and Marchant, D. R. (2014) Volcano-ice interactions in the Arsia Mons tropical mountain glacier deposits, *Icarus*, 237, 315–339.

Scanlon, K. E., Head, J. W., and Marchant, D. R. (2015) Volcanism-induced, local wet-based glacial conditions recorded in the late Amazonian Arsia Mons tropical mountain glacier deposits, *Icarus*, 250, 18–31.

Schon, S. C., and Head, J. W. (2011) Keys to gully formation processes on Mars: relation to climate cycles and sources of meltwater, *Icarus*, 213, 428–432.

Schörghofer, N. (2007a) Dynamics of ice ages on Mars, *Nature*, 449(7159), 192–194.

Schörghofer, N. (2007b) Theory of ground ice stability in sublimation environments, *Phys. Rev. E.*, 75, 041201.

Schörghofer, N. (2008) Temperature response of Mars to Milankovitch cycles, *Geophys. Res. Lett.*, 35, L18201.

Schörghofer, N., and Aharonson, O. (2005) Stability and exchange of subsurface ice on Mars, *J. Geophys. Res.*, 110(E5), E05003.

Schörghofer, N., and Forget, F. (2012) History and anatomy of subsurface ice on Mars, *Icarus*, 220, 1112–1120.

Selvans, M. M., Plaut, J. J., Aharonson, O., and Safaeinili, A. (2010) Internal structure of Planum Boreum, from Mars advanced radar for subsurface and ionospheric sounding data, *Journal of Geophysical Research (Planets)*, 115, 9003.

Seu, R., Phillips, R. J., Alberti, G., et al. (2007) Accumulation and erosion of Mars south polar layered deposits, *Science*, 317, 1715.

Shean, D. E., Head, J. W., and Marchant, D. R. (2005) Origin and evolution of a cold-based tropical mountain glacier on Mars: the Pavonis Mons fan-shaped deposit, *Journal of Geophysical Research (Planets)*, 110(E9), 5001.

Shean, D. E., Head, J. W., Fastook, J. L., and Marchant, D. R. (2007) Recent glaciation at high elevations on Arsia Mons, Mars: implications for the formation and evolution of large tropical mountain glaciers, *Journal of Geophysical Research (Planets)*, 112, 3004.

Sizemore, H. G., and Mellon, M. T. (2008) Laboratory characterization of the structural properties controlling dynamical gas transport in Mars-analog soils, *Icarus*, 197, 606–620.

Smith, I. B., and Holt, J. W. (2010) Onset and migration of spiral troughs on Mars revealed by orbital radar, *Nature*, 465, 450–453.

Smith, P. H., Tamppari, L. K., Arvidson, R. E., et al. (2009) H_2O at the Phoenix Landing Site, *Science*, 325(5936) 58–61.

Spiga, A., and Forget, F. (2008) Fast and accurate estimation of solar irradiance on Martian slopes, *Geophys. Res. Lett.*, 35, L15201.

Squyres, S. W. (1978) Martian fretted terrain – flow of erosional debris, *Icarus*, 34, 600–613.

Squyres, S. W. (1979) The distribution of lobate debris aprons and similar flows on Mars, *J. Geophys. Res.*, 84, 8087–8096.

Squyres, S. W., Clifford, S. M., Kuzmin, R. O., Zimbelman, J. R., and Costard, F. M. (1992) Ice in the Martian regolith, in *Mars* (Kieffer, H. H., Jakosky, B. M., Snyder, C. W., and Matthews, M. S. Eds), University of Arizona Press, Tucson.

Szwast, M. A., Richardson, M. I., and Vasavada, A. R. (2006) Surface dust redistribution on Mars as observed by the Mars Global Surveyor and Viking orbiters, *J. Geophys. Res.*, 111(E11), E11008, doi:10.1029/2005JE002485.

Tanaka, K. L. (2005) Geology and insolation-driven climatic history of Amazonian north polar materials on Mars, *Nature*, 437, 991–994.

Tanaka, K. L., Rodriguez, J. A. P., Skinner, J. A., et al. (2008) North polar region of Mars: advances in stratigraphy, structure, and erosional modification, *Icarus*, 196, 318–358.

Thomas, P. C., Malin, M. C., Edgett, K. S., et al. (2000) North–south geological differences between the residual polar caps on Mars, *Nature*, 404, 161–164.

Thomas, P. C., James, P. B., Calvin, W. M., Haberle, R., and Malin, M. C. (2009) Residual south polar cap of Mars: stratigraphy, history, and implications of recent changes, *Icarus*, 203, 352–375.

Titus, T. N., Kieffer, H. H., and Christensen, P. R. (2003) Exposed water ice discovered near the south pole of Mars, *Science*, 299, 1048–1051.

Toon, O. B., Pollack, J. B., Ward, W., Burns, J. A., and Bilski, K. (1980) The astronomical theory of climatic change on Mars, *Icarus*, 44, 552–607.

Touma, J., and Wisdom, J. (1993) The chaotic obliquity of Mars, *Science*, 259, 1294–1297.

Vasavada, A. R., Williams, J.-P., Paige, D. A., et al. (2000) Surface properties of Mars' polar layered deposits and polar landing sites, *J. Geophys. Res.*, 105, 6961–6970.

Vincendon, M., Mustard, J., Forget, F., et al. (2010) Near-tropical subsurface ice on Mars, *Geophys. Res. Lett.*, 37, L01202.

Ward, W. R. (1973) Large-scale variations in the obliquity of Mars, *Science*, 181, 260–262.

Ward, W. R. (1974) Climate variations on Mars. 1. Astronomical theory of insolation, *J. Geophys. Res.*, 79, 3375–3386.

Ward, W. R., Murray, B. C., and Malin, M. C. (1974) Climatic variations on Mars. 2. Evolution of carbon dioxide atmosphere and polar caps, *J. Geophys. Res.*, 79, 3387–3395.

White, B. R. (1979) Soil transport by winds on Mars, *J. Geophys. Res.*, 84, 4643–4651.

Wieczorek, M. A. (2008) Constraints on the composition of the Martian south polar cap from gravity and topography, *Icarus*, 196, 506–517.

Williams, K. E., Toon, O. B., Heldmann, J. L., and Mellon, M. T. (2009) Ancient melting of mid-latitude snowpacks on Mars as a water source for gullies, *Icarus*, 200, 418–425.

Winebrenner, D. P., Koutnik, M. R., Waddington, E. D., et al. (2008) Evidence for ice flow prior to trough formation in the Martian north polar layered deposits, *Icarus*, 195, 90–105.

Wood, S. E., and Griffiths, S. D. (2007) Mars subsurface warming at low obliquity. In *Seventh International Conference on Mars*, LPI Contributions, 1353, 3387.

Zent, A. P., and Quinn, R. C. (1995) Simultaneous adsorption of CO_2 and H_2O under Mars-like conditions and application to the evolution of the Martian climate, *J. Geophys. Res.*, 100, 5341–5349.

Zent, A. P., Fanale, F. P., and Postawko, S. E. (1987) Carbon dioxide – adsorption on palagonite and partitioning in the Martian regolith, *Icarus*, 71, 241–249.

Zimbelman, J. R., and Edgett, K. S. (1992) The Tharsis Montes, Mars – comparison of volcanic and modified landforms, *Lunar and Planetary Science Conference Proceedings*, 22, 31–44.

Zuber, M. T., Phillips, R. J., Andrews-Hanna, J. C., et al. (2007) Density of Mars south polar layered deposits, *Science*, 317, 1718.

The Early Mars Climate System

ROBERT M. HABERLE, DAVID C. CATLING, MICHAEL H. CARR, KEVIN J. ZAHNLE

17.1 INTRODUCTION

Today Mars is a cold, dry, desert planet. The atmosphere is thin and liquid water is not stable at the surface. But there is evidence that very early in its history it was warmer and wetter. Since Mariner 9 first detected fluvial features on its ancient terrains, researchers have been trying to understand what climatic conditions could have permitted liquid water to flow on the surface. Though the evidence is compelling, the problem is not yet solved.

The main issue is coping with the faint young Sun. During the period when warmer conditions prevailed ~3.7–4.1 Ga[1], the Sun's luminosity was ~25% less than it is today. How can we explain the presence of liquid water on the surface of Mars under such conditions? A similar problem exists for Earth, which would have frozen over under a faint Sun even though the evidence suggests otherwise.

Attempts to solve the "faint young Sun paradox" rely on greenhouse warming from an atmosphere with a different mass and composition than we see today. This is true for both Mars and Earth. However, it is not a straightforward solution. Any greenhouse theory must (a) produce the warming and rainfall needed, (b) have a plausible source for the gases required, (c) be sustainable, and (d) explain how the atmosphere evolved to its present state. These are challenging requirements, and judging from the literature they have yet to be met.

In this chapter we review the large and growing body of work on the early Mars climate system. We take a holistic approach that involves many disciplines, since our goal is to present an integrated view that touches on each of the requirements listed in the preceding paragraph. We begin with a brief discussion of geological and mineralogical eras in Section 17.2.1, and the major events in early Mars' history (Section 17.2.2). We then step through the observational evidence of the early climate system, which comes from the geology (Section 17.3.1), mineralogy (Section 17.3.2), and isotopic (Section 17.3.3) data. Each of the datasets presents a consistent picture of a warmer and wetter past with a thicker atmosphere. How much warmer and wetter and how much thicker are matters of debate, but conditions then were certainly different than they are today.

We then discuss the origin and evolution of the early atmosphere (Section 17.4), from accretion and core formation to the end of the late heavy bombardment, including estimates of the volatile inventory, outgassing history, and potential loss

mechanisms. This sets the stage for a comprehensive look at the climate system of early Mars (Section 17.5) and the attempts to solve the faint young Sun problem. We take some time to review the basic physics involved and then discuss the different ideas highlighting their strengths and weaknesses. We conclude in Section 17.6 with a summary and a discussion of potentially promising avenues of future research.

17.2 OVERVIEW OF EARLY MARS

17.2.1 Geological and Mineralogical Eras

The geological history of Mars has been grouped into three epochs: Noachian, Hesperian, and Amazonian. These eras are mainly classified on the basis of crater populations linked to a model that converts them to absolute ages (e.g. Hartman and Neukum, 2001). The oldest recognizable terrains, those from the Noachian era, preserve geologic features that date from the period of heavy bombardment, which ended ~3.7 Ga. How far back in time the geologic record extends is unclear because of uncertainties in the early cratering history, but the oldest Noachian surfaces visible in imaging data are probably ~4.1 Ga in age (Werner, 2008). Some workers therefore identify a pre-Noachian period that extends from the time of formation of the crust to ~4.1 Ga (Frey, 2003). Vague circular structures, remnants of large impact basins, probably date from this earliest era. Noachian terrains have crater densities and fluvial features similar to those in the Noachis Terra region of the southern highlands. Hesperian terrains, named after those in Hesperia Planum, are of intermediate age and date from the end of the Noachian to ~2.9–3.3 Ga. Volcanic plains dominate Hesperian terrains, but this does not necessarily imply an increase in volcanic activity, since much of the evidence for earlier volcanism would have been destroyed by the high impact rates (Greeley and Schneid, 1991). Amazonian terrains are the youngest and date from the end of the Hesperian to present. These terrains were shaped during a period of declining geologic and volcanic activity.

Bibring et al. (2006) suggested a parallel timeline classification based on the detection of phyllosilicates and sulfates by Mars Express. Phyllosilicates require liquid water to form, and sulfates are found in sedimentary environments. Phyllosilicates require alkaline conditions and are present mostly on the oldest terrains. Sulfates, on the other hand, require acidic environments and appear to have formed after the phyllosilicates. However, mapping on smaller spatial scales shows that phyllosilicates

---

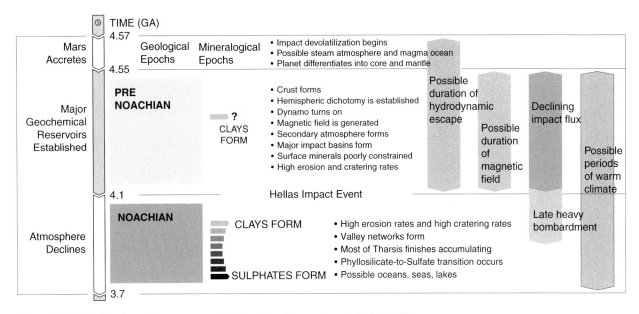

Figure 17.1. Timeline for major events on early Mars. Graphic courtesy of Christina Olivas.

and sulfates are found together in Meridiani Planum (Wiseman et al., 2008; Wray et al., 2009a), Gusev Crater (Wang et al., 2006), Gale Crater (Milliken et al., 2010), Cross Crater (Swayze et al., 2008), Mawrth Vallis (Farrand et al., 2009), and are interbedded in Columbus Crater in Terra Sirenum (Wray et al., 2009b). This may indicate that aqueous environments during the Noachian varied locally with different pH, water/rock ratios, salinity, and temperature. Today, the surface is dominated by anhydrous iron oxides, which can form in the absence of liquid water. Thus, Bibring et al. (2006) suggested three epochs: the Phyllosian, when neutral waters weathered basalts, the Theilkian, when a transition occurred to more acidic environments, and the Siderikian, marked by a final transition to a dry environment. These epochs roughly correspond to the Noachian, Hesperian, and Amazonian epochs.

17.2.2 Timeline of Major Events

Figure 17.1, which is based on material drawn from Fassett and Head (2010), Carr and Head (2010), and Werner and Tanaka (2011), is a summary of the major events that occurred on early Mars. Note that the timing is somewhat uncertain, and the relative chronologies, though plausible, are also subject to revision. The intent here is to provide a context for the discussions in later sections where details and references are provided.

Mars Accretes (4.57–4.55 Ga). Mars accreted to its present size within the first 20 Ma of Solar System history, perhaps even sooner. During this period, the planet differentiated into a core, mantle, and crust. During accretion the devolatilization of impactors probably created an atmosphere mostly composed of water (steam). Accretion may have been fast enough to produce a magma ocean. Hydrodynamic escape powered by the high extreme ultraviolet (EUV) fluxes of the young Sun probably resulted in massive loss of volatiles.

Pre-Noachian (4.55–4.1 Ga). As the planet cooled, convection within the core generated a dynamo and a magnetic field that held off the solar wind, thereby suppressing the loss of

atmospheric constituents by solar wind stripping. The global hemispheric dichotomy was established. Weathering of surface materials may have begun during this period, but the alteration products are poorly constrained. The Tharsis volcanic province began forming and many major impact basins formed. This period ends with the Hellas impact event, which we take to occur at 4.1 Ga, but which may have been as late as 3.9 Ga, depending on crater modeling assumptions. By this time the dynamo ceased, the magnetic field collapsed, and the atmosphere was exposed to solar wind erosion. If there was a period of hydrodynamic escape, it likely ended late in this period.

Noachian (4.1–3.7 Ga). This is the period where the ancient geologic record is best preserved. And it is for this period that warm and wet conditions are thought to have prevailed. During this period, the valley networks formed, global-scale layered terrains developed, erosion and cratering rates were high, and oceans, lakes, and/or seas may have been present. The atmosphere at that time may have been thick enough to provide enough greenhouse warming to drive an active hydrological cycle with rainfall and runoff. Impacts may have created episodic warm environments. By the end of this period, surface waters became more acidic, sulfates began forming, the atmosphere thinned, and erosion rates declined. Note that the time evolution of surface minerals is a general trend and that local exceptions can be found (e.g. phyllosilicates above and below sulfates). Also note that valley network formation may have continued into the early Hesperian (Fassett and Head, 2010, 2011).

17.3 EVIDENCE FOR A DIFFERENT ATMOSPHERE AND CLIMATE SYSTEM

17.3.1 Geological Evidence

The morphology of the Martian surface strongly supports the supposition that surface conditions during the Noachian were, at least episodically, both warmer and wetter than they were

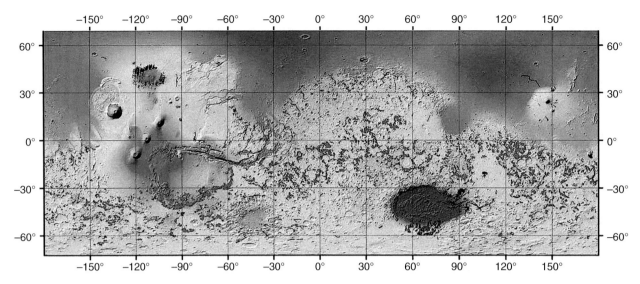

Figure 17.2. The global distribution of valleys. Valleys superimposed on Noachian terrain are shown in red. Younger valleys are in other colors. The Noachian valleys are mainly in terrain over 0 km in elevation. The absence of valleys in the low-lying Noachian of northwest Arabia (blue) has been attributed to the former presence of an ocean at the time that the valleys formed. The scarcity of valleys in the high-standing Noachian between Hellas and Argyre remains unexplained. From Hynek et al. (2010). A black and white version of this figure will appear in some formats. For the color version, please refer to the plate section.

subsequently. Valley networks and associated features, such as deltas and lakes, common in terrains that survive from this early era, almost certainly resulted from precipitation and the subsequent flow of water across the surface. Erosion rates were much higher during this early era than they were later, and the mineralogical evidence discussed in the next section supports moist surface conditions. Despite these conclusions, considerable uncertainty persists as to how frequent, intense, and sustained the fluvial episodes were, what enabled the fluvial episodes, and whether there was ever a stable, global hydrologic system in which precipitation, runoff, infiltration, and groundwater flow were in quasi-equilibrium with evaporation and sublimation from large bodies of water and ice.

The valley networks have long been viewed as strong evidence of precipitation on early Mars (e.g. Carr, 1981; Baker, 1982) and, as we acquire better imagery and altimetry, the evidence becomes more compelling. The global distribution of valley networks is shown in Figure 17.2. In planimetric form the valleys resemble terrestrial river valleys. They are almost exclusively incised into Noachian or Lower Hesperian terrains, although there are rare younger exceptions (e.g. Mangold et al., 2004). They branch upstream and are typically 1–4 km wide and 50–200 m deep. They tend to have V-shaped cross-sections in their upper reaches and rectangular cross-sections downstream. While a few are over 2000 km long, the vast majority are less than 200 km long and drain into local lows, rather than regional lows such as the northern plains and the Hellas Basin. Drainage densities (stream length per unit area) vary considerably with location and the resolution of the observation from extremely low values up to 1 km^{-1}, which is well within the terrestrial range (Craddock and Howard, 2002; Hynek and Phillips, 2003; Hynek et al., 2010; Ansan and Mangold, 2013). The apparent low drainage densities observed in early imagery, amphitheater heads of tributaries, common in prominent valleys such as

Nirgal Vallis, and rectangular cross-sections suggested to many early workers that groundwater sapping had played a major role in the formation of many of the valleys (Pieri, 1980; Carr and Clow, 1981; Baker et al., 1990; Gulick, 1998), although all acknowledged that precipitation and/or hydrothermal circulation were needed to recharge the groundwater system to enable sustained or episodic flow. However, better imaging and altimetry now show that dense, area-filling networks indicative of surface runoff, such as shown in Figure 17.3, are common throughout the Noachian terrains (Hynek et al., 2010). Precipitation followed by surface runoff, coupled with infiltration and groundwater seepage, must have occurred at least episodically in the Noachian (Craddock and Howard, 2002; Irwin and Howard, 2002; Hynek and Phillips, 2003; Stepinski and O'Hara, 2003; Howard et al., 2005; Carr, 2006). Dimensions of channels within some valleys, meander lengths, and boulder sizes suggest peak discharges of 10^2–10^4 m^3 s^{-1} (Moore et al., 2003; Jerolmack et al., 2004; Howard et al., 2005; Irwin et al., 2005), comparable to discharges of typical terrestrial rivers and substantially smaller than the discharges of the large flood features that formed mostly later in Mars' history.

A distinction must be made between the general, pervasive degradation of the ancient highlands and the formation of most of the valley networks. During the Noachian there was widespread erosion of crater rims and other high ground, and partial infilling of lows such as craters (Figure 17.4). Several impact craters as large as several hundred kilometers across, with original rim heights of hundreds of meters and depths of a few kilometers, that are superimposed on Hellas ejecta have, for example, been almost completely filled and had their rims almost completely eroded away so that they are barely visible. However, the observed valley networks appear to have contributed little to that degradation. They were incised into the degraded landscape, but were not its cause. Poorly dissected

Figure 17.3. Warrego Vallis (42°S, 267°E). The dense area-filling drainage pattern is compelling evidence for precipitation. The scene is 120 km across. (THEMIS Mosaic.)

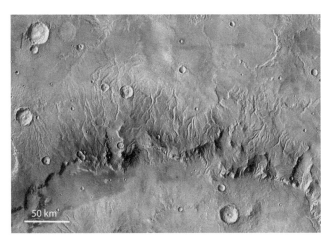

Figure 17.4. The rim of Huygens Crater (10°S, 55°E). The scene is typical of dissected Noachian terrain. Valleys on higher ground drain into local lows, which are mostly undissected. Barely discernible crater rims (upper right and upper left) indicate that extensive erosion occurred prior to the formation of the crisply defined valleys. (THEMIS/MOLA.)

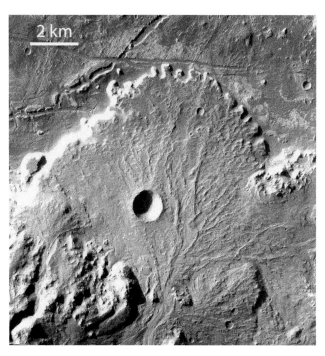

Figure 17.5. Delta in Holden Crater (26°S, 256°E). A stream, having cut through the south rim of the crater, deposited its sediment load to form what was likely a delta on Holden's floor. Former distributary channels, now evident in inverted relief, can be traced across the surface of the delta. (THEMIS.)

Noachian areas such as northwest Arabia and the region between Hellas and Argyre are just as degraded as the dissected areas. The pristine-appearing valleys incised into the highland terrains appear to result from a late Noachian to early Hesperian episode of intense incision (Howard et al., 2005; Irwin et al., 2005). Degraded, difficult-to-discern valleys throughout the highlands (e.g. Baker and Partridge, 1986) may be traces of fluvial activity prior to the late episode.

The Noachian terrains are clearly much more eroded than younger terrains. While Hesperian craters as small as a few kilometers across generally preserve all their primary impact features, even delicate textures on their ejecta, Noachian impact craters several hundreds of kilometers across mostly have highly eroded rims and partly filled interiors. The contrast implies a precipitous drop in erosion rates at the end of the Noachian. The number of fresh-appearing craters with well-preserved ejecta patterns on Noachian terrains is comparable to the number on the oldest Hesperian terrains, which also suggests that high erosion rates persisted until the end of the Noachian and then rapidly declined (Craddock and Maxwell, 1993). Golombek and Bridges (2000) and Golombek et al. (2006), in summarizing all the data available on Martian erosion rates, concluded that, although Noachian erosion rates were 2–5 orders of magnitude higher than they were subsequently, they still appear to have been lower than average terrestrial rates, being comparable to those of old flat cratons on Earth, or about 5 μm yr^{-1}.

The valleys form an immature drainage system with low basin concavities, undissected interfluves between drainage basins, and poor correlation of basin circularity with elevation within the basins (Stepinski and O'Hara, 2003). Stream profiles are poorly graded and closely follow the regional slopes (Howard et al., 2005). Either erosion by the incised valleys was not sufficiently sustained to allow regional integration of the valleys, or disruption by impacts and other events prevented it. The terminal episode during which most of the valleys formed appears to have ended abruptly because associated fans and deltas are undissected (Figure 17.5). During the late incision phase and during the period of degradation that preceded it, eroded debris was mostly deposited in local lows where layered deposits are common. Large drainage basins comparable to the Amazon and Mississippi that would have resulted in transport of erosional debris to regional lows did not develop. Only modest amounts

0.5 km

Figure 17.6. Layers in Terby Crater (28°S, 162°E). Many upland craters have flat floors and are partly to almost completely filled with what are likely sediments. Where the interior deposits are exposed, they are commonly seen to be finely layered, as seen here. One possibility is that the layers accumulated in intra-crater lakes and that sedimentation rates were modulated by the obliquity cycle. (MOC.)

of erosional debris ($<3 \times 10^4$ km³) have, for example, accumulated in Hellas despite a drainage area of 1.5×10^7 km², several kilometers of relief, and pervasive erosion of its rim.

Lakes were likely common throughout Noachian terrain while the valleys were forming (Cabrol and Grin, 1999, 2010), as expected of a poorly graded landscape undergoing fluvial erosion. Most valleys terminate in closed local depressions such as craters or inter-crater lows, where at least transient closed lakes likely formed. Seemingly fine-grained, horizontally layered, easily erodible sediments underlie most of these low areas (Figure 17.6). Chlorine-rich deposits found in local lows within the Noachian uplands may be the result of evaporation of lakes (Osterloo et al., 2008). Temporary, late Noachian to early Hesperian lakes are also suggested by the MER findings in Meridiani. Some of the sulfate-rich deposits found there may have been deposited in transient inter-dune lakes, and subsequently altered as a result of oscillations in the local groundwater table (Grotzinger et al., 2005). Intermittent lakes in Gale Crater possibly lasting for thousands to millions of years (Grotzinger

et al., 2015) further support the notion that lakes were common in the early Hesperian. Almost 300 lows have been identified within the highlands that have both inlet and outlet valleys indicating the former presence of an open basin lake, i.e. one that overflowed (Fassett and Head, 2008). Balancing estimated stream discharges against measured basin area and estimated evaporation rates, Matsubara et al. (2011) inferred that some of the larger open lakes would have taken thousands of years to overflow. However, most basins are closed. From the fact that most local basins did not overflow, coupled with estimates of peak discharges from channel dimensions and computer simulations of fluvial erosion of cratered landscapes, Howard (2007), Barnhart et al. (2009), and Matsubara et al. (2011) concluded that the valley network system did not result from a few deluge-type events, such as might be caused by large impacts, but rather from modest fluvial events over extended periods of time. The model times required to simulate the observed landscapes depend on the frequency and magnitude of the fluvial events, but times could be as long as millions of years (Howard, 2007). Similarly, hydrologic modeling suggests that the Meridiani and other layered lacustrine deposits in Arabia formed by groundwater upwelling over millions of years. The fine-scale regular layering of many sediments in putative lake deposits suggests that inflow into the lakes was controlled by regularly changing conditions such as might be induced by obliquity changes.

Many deltas have been observed where valleys intersect local lows. A particularly striking example in Eberswalde Crater has many of the characteristics of terrestrial deltas and dimensions of its component parts that suggest discharge of several hundred cubic meters per second, comparable to terrestrial rivers (Malin and Edgett, 2003; Moore et al., 2003). The time taken to build the delta is not narrowly constrained but is estimated to be in the range of thousands to millions of years (Moore et al., 2003). Over 50 deltas situated along the plains' upland boundary, close to the −1.8 km contour, have been attributed to sedimentation along the edge of a northern ocean (Di Achille and Hynek, 2010). In most cases, however, neither fans nor deltas are observed where a valley enters a low, possibly because the sediment load is generally too fine-grained (Howard, 2007).

Whether there were ever oceans on Mars is one of the planet's most controversial issues (Parker et al., 1989, 1993; Baker et al., 1991; Head et al., 1999; Baker, 2001; Clifford and Parker, 2001; Carr and Head, 2003). Discussion has focused mainly on the possibility of post-Noachian oceans mainly because they could have resulted from the large post-Noachian floods, and because any evidence for oceans would be better preserved for the post-Noachian era than for the Noachian. However, the Noachian is the time for which we have the best evidence for warm conditions under which oceans might be present. Moreover, large bodies of water were likely needed to provide the precipitation that eroded the valley networks. Clifford and Parker (2001) argue from estimates of the global inventory of water and the thermal conditions implied by the valley networks that possibly one-third of the planet was covered by oceans during parts of the Noachian, and have tentatively identified shorelines that could be Noachian in age. The presence of deltas along a possible shoreline has already been mentioned. In addition, Moore and Wilhelms (2001) suggested that two well-defined breaks in slope within the Hellas Basin could be shorelines. Despite these suggestions, the prospect for finding compelling geomorphic evidence of former Noachian oceans is poor, since such evidence, if it ever existed, would be vulnerable to erasure by burial and erosion. Nevertheless, if Noachian Mars had a large inventory of water and if warm conditions prevailed as indicated by the fluvial land lacustrine features, then bodies of water would have accumulated in lows such as the northern basin and Hellas.

An "icy highlands" model has recently been proposed as an alternative to the "warm wet ocean" model (Wordsworth et al., 2013; see Section 17.5.4). In this scenario, the near-surface inventory of water accumulates as snow and ice in high-altitude terrains. Episodic melting as a result of volcanism, impacts (Segura et al., 2008), and/or spin axis/orbital perturbations (Head and Marchant, 2014) provides meltwater to cut the valleys. The model is attractive in that it does not require sustained global warming but only local, episodic events.

17.3.1.1 Summary

The geologic evidence for precipitation in the Noachian and flow of water across the surface is compelling. The dimensions of the valleys, lakes, and associated features such as deltas suggest rivers with discharges comparable to terrestrial rivers. The erosion rates, although poorly constrained, are estimated to have been similar to arid and semi-arid regions on Earth where water is the erosive agent. But not knowing the flow frequency and the time over which conditions were favorable to fluvial activity, the amount of water that flowed across the surface is poorly constrained. Nevertheless, simulations of the evolution of the Noachian landscape suggest that times of millions of years and precipitation rates equivalent to arid to semi-arid regions on Earth were required to achieve the observed degraded landscape (Howard, 2007). If true, then evaporation from large bodies of water must have occurred to provide precipitation and drive a global hydrologic cycle.

17.3.2 Mineralogical Evidence

The surface of Mars is predominantly basalt. When basaltic minerals react with water, the resulting solution can produce phyllosilicates, silica, sulfates, carbonates, iron oxides, and other alteration products (e.g. Burns, 1987; Catling, 1999; Christensen et al., 2001). Consequently, the water that flowed on the surface of early Mars, as demonstrated in Section 17.2.1, should have produced such minerals. Indeed, they have been detected, as summarized in Table 17.1. The question, however, is: Were they formed in a sustained warmer and wetter climate with an active hydrological cycle, or were they formed by other means, such as impact events or subsurface hydrothermal alteration, which are independent of the climate system?

17.3.2.1 Carbonates

Carbonates have been of particular interest since they are expected to form in warm wet conditions produced by thick CO_2 atmospheres (Kahn, 1985; Pollack et al., 1987). Such atmospheres were thought to produce the greenhouse effect needed to explain the fluvial features (see Section 17.5.2). Carbonates form when CO_2-rich water reacts with basaltic

materials to release various cations (principally Fe^{2+}, Mg^{2+}, Ca^{2+}, K^+, and Na^+) and bicarbonate anions (HCO_3^-). When the resulting solution concentrates, bicarbonate combines with the cations to produce minerals such as siderite, calcite, or magnesite. Used up in this way, for example, 1 bar of CO_2 would produce a planet-wide cover of ~20 m of pure calcite[2].

Carbonates are present on Mars, but they do not outcrop globally (Niles et al., 2013). Occurrences include small volumes (<1 vol.%) in Martian meteorites (Bridges et al., 2001), layered carbonates overlying olivine-rich units in Nili Fossae (Ehlmann et al., 2008a; Ehlmann and Edwards, 2014), layered carbonates excavated by an impact in Syrtis Major (Michalski and Niles, 2010), magnesite outcrops in Gusev Crater (Morris et al., 2010), a 2–5 wt.% magnesite component of the global Martian dust (Bandfield et al., 2003), and 3–5 wt.% carbonate in soil at the Phoenix landing site of mixed Ca/Mg/Fe content (Boynton et al., 2009; Sutter et al., 2012). The area of carbonate outcrops detected in the near-infrared is ~10^5 km^2, assuming that they extend under associated olivine units. Given an average thickness of ~100 m and assuming 100% calcite, the CO_2 they contain would be only ~3 mbar.

The absence of global outcropping has been attributed to the presence of sulfuric or sulfurous acids that would have suppressed their formation (Fairén et al., 2004; Bullock and Moore, 2007; Halevy et al., 2007). On early Mars, a slow rain of microscopic sulfurous and sulfuric acid droplets could have been generated from atmospheric oxidation of volcanic SO_2 hydrated by water vapor (Settle, 1979; Smith et al., 2014). While carbonate suppression by acidic waters is valid for steady-state conditions, in reality weathering fluids in contact with basalts become more alkaline in time. So carbonate formation is unlikely to have been suppressed everywhere at all times.

Do the known carbonates tell us anything about past climates? Notably, the Nili Fossae and Gusev Crater carbonates are magnesite-dominated. In Nili Fossae, magnesite is associated with the probable alteration of olivine-rich ejecta around Isidis (Mustard et al., 2008). On Earth, magnesite commonly forms through hydrothermal alteration of Mg-rich igneous rocks. Consequently, the association of magnesite with olivine probably tells us more about past geothermal activity than past climate.

Unlike the carbonates in Nili Fossae, however, those excavated from the deep crust have spectral signatures of siderite or possibly calcite (Michalski and Niles, 2010). This may be a glimpse of very ancient, deeply buried sedimentary carbonates from a warmer climate or it could simply be evidence of ancient subsurface hydrothermal alteration. In either case, ferrous iron in siderite implies a reducing environment, unlike the modern oxidizing surface environment, which can favor Mg-rich carbonates (Hausrath and Olsen, 2013).

The carbonates in global dust and Phoenix soils may not have any connection to the Noachian climate. While the magnesite in global dust could have a contribution from the erosion of deposits such as those in Nili Fossae or in Gusev (Ehlmann et al., 2008a),

---

[2] Depth = $(P_s M_{CaCO3})/(g M_{CO2} \rho_{CaCO3})$, where surface pressure $P_s = 10^5$ Pa, $M_{CaCO3} = 0.1$ kg mol^{-1}, $M_{CO2} = 0.044$ kg mol^{-1}, g is gravity, and density $\rho_{CaCO3} = 2710$ kg m^{-3}.

Table 17.1 *Alteration minerals reported from orbital or* in situ *measurements on Mars. (Adapted and updated from Ehlmann (2010) and Murchie et al. (2009a).)*

Class	Minerals	Chemistry	Locations on Mars
Non-silicates			
Carbonates	Calcium carbonate	$CaCO_3$	Soil at Phoenix site; large impact crater east of Syrtis Major
	Magnesite	$MgCO_3$	Nili Fossae; Gusev Crater; global dust
	Siderite	$FeCO_3$	Large impact crater east of Syrtis Major
Sulfates	Gypsum	$CaSO_4 \cdot 2H_2O$	Juventae Chasma[a]; Iani Chaos; north circumpolar dunes
	Bassanite	$CaSO_4 \cdot 0.5H_2O$	Mawrth Vallis
	Mono- and poly-hydrated sulfates	$(Fe,Mg)SO_4 \cdot nH_2O$, e.g. kieserite, $MgSO_4 \cdot H_2O$	Valles Marineris region: Ius, Hebes, Capri, Candor
			Melas, Juventae Chasma, Iani Chaos, Aram Chaos, Meridiani, Gale Crater
	Jarosite	$KFe_3^{3+}(SO_4)_2(OH)_6$	Mawrth Vallis, Meridiani
		$(H_3O)Fe_3^{3+}(SO_4)_2(OH)_6$	Melas chasma
	Alunite	$KAl_3(SO_4)_2(OH)_6$	Terra Sirenum craters
Chlorides	Metal chlorides	type not identified	Terra Sirenum plains
Perchlorates	No mineral names exist	Mg_2ClO_4, $(Na,K)ClO_4$	Soil at Phoenix site; north polar cap
Iron oxides	Hematite	Fe_2O_3	Meridiani; Aram Chaos; Candor, Melas, Juventae, Tithonium, and Eos Chasmata
	Goethite	$FeOOH$	Gusev Crater rocks
Sheet silicates			
Phyllosilicates	Fe/Mg smectites	$(Ca,Na)_{0.3-0.5}(Fe,Mg,Al)_{2-3}$ $(Si,Al)_4O_{10}(OH)_2$, e.g. Fe-rich nontronite $Na_{0.3}Fe_2^{3+}(Si,Al)_4O_{10}(OH)_2$ and Mg-rich saponite $Ca_{1/6}$ $Mg_3(Al,Si)_4O_{10}(OH)_2$	Mawrth Vallis; Nili Fossae; Holden, Eberswalde, Jezero, Terby, and Gale Craters; walls of Valles Marineris; Terra Sirenum; Meridiani; southern highlands
	Kaolinite group	$Al_2Si_2O_5(OH)_4$ (kaolinite), e.g. halloysite $Al_2Si_2O_5(OH)_4 \cdot (0-2)H_2O$	Mawrth Vallis; Nili Fossae
	Montmorillonite	$(Na,Ca)_{0.33}(Al,Mg)_2(Si_4O_{10})$ $(OH)_2$	Mawrth Vallis
	Al,K phyllosilicates	$KAl_2AlSi_3O_{10}(OH)_2$	Nili Fossae
	Chlorite	$(Mg,Fe^{2+})_5Al(Si_3Al)O_{10}(OH)_8$	Large northern plains or southern highland craters
	Serpentine	$(Mg,Fe)_3Si_2O_5(OH)_4$	Claritas Rise; Nili Fossae; highlands impact craters; regional olivine-rich unit near Isidis
	Prehnite	$Ca_2Al(AlSi_3O_{10})(OH)_2$	Large northern plains craters; southern highlands
Framework silicates			
Silica	Opaline silica	$SiO_2 \cdot 2H_2O$	Gusev Crater; Valles Marineris Hesperian plains; Ius/Melas Chasmata; Noctis Labyrinthus; Mawrth Vallis
Analcime (analcite)	Analcime/analcite	$NaAlSi_2O_6 \cdot H_2O$	Highlands bordering Isidis or Hellas; Terra Cimmeria

[a] Gypsum is reported by Gendrin et al. (2005) in Juventae Chasma but disputed by others (Bishop et al., 2009; Kuzmin et al., 2009).

it could also have been produced relatively recently. Laboratory experiments show that a few weight percent of carbonate could form on the surface of basaltic fines in "wet" conditions of only 10^2–10^3 monolayers of water over about 1 Ga, although the exact timescale depends on the specific surface area ($m^2 g^{-1}$) of the fines (Stephens, 1995a,b). Such conditions could have been produced during the Amazonian at times of high obliquity (see Chapter 16). Furthermore, the soil at the Phoenix site is likely derived from the ejecta material of the 600 Ma old Heimdall Crater, which has also been exposed to periodic wet conditions (Boynton et al., 2009).

17.3.2.2 Sulfates

Sulfates are abundant on Mars. Near-infrared spectroscopy shows that hydrated sulfates exist in light-toned layered deposits (LLDs) (Gendrin et al., 2005; Bibring et al., 2006; Murchie et al., 2009a), while *in situ* analyses demonstrate that sulfate is ubiquitous in the soil (Yen et al., 2005; Kounaves et al., 2010). Astonishingly, sulfates outcrop as mountain-sized LLDs inside low-latitude chasms and canyons, often in association with chaotic terrain (Catling et al., 2006; Glotch and Rogers, 2007; Bishop et al., 2009; Lichtenberg et al., 2010; Roach et al., 2010a,b). On the ground, the Opportunity Rover provided a close-up view of LLDs in Meridiani Planum (Squyres et al., 2006), where millimeter-scale spherules of hematite (Weitz et al., 2006) are embedded in sulfate-rich sandstones.

The Meridiani LLD sandstones date to the Noachian–Hesperian boundary ~3.7 Ga (Hynek and Phillips, 2008) and are composed of a mixture of sulfates, hydrated silicates, iron oxides, and basaltic debris (Clark et al., 2005). Regionally, they are several hundred meters thick and cover >500 000 km^2. The meter-scale cross-bedding seen in the Burns formation and Victoria Crater (Squyres et al., 2009) indicate that these sedimentary stacks formerly consisted of migrating dunes (Grotzinger et al., 2005).

Squyres et al. (2006) proposed the following scenario for the formation of the Meridiani LLDs. Playa deposits of sulfates were wind-eroded and redeposited as sand dunes (McLennan and Grotzinger, 2008). Groundwater subsequently penetrated the sandstone. Hematite concretions formed when the dissolved iron in the groundwater precipitated on timescales as short as ~10^3 years, constrained by the typical size and volume fraction of concretions (Sefton-Nash and Catling, 2008). Hydrothermal conditions may also have existed (Golden et al., 2008). The centimeter-scale cross-laminations in the upper meter of the Burns formation, the so-called "festoons", suggest gentle subaqueous flow (Grotzinger et al., 2005). Thus, groundwater upwelled to the surface and ponded between dunes. Inter-dune ponding requires a rise in the water table but not direct rainfall (McLennan and Grotzinger, 2008). However, if such hydrology is responsible, the late Noachian or Lower Hesperian climate might require significant rainfall to charge groundwater aquifers.

The similarity of Valles Marineris LLDs to the sulfate-bearing LLDs in Meridiani Planum may indicate a common formation process. In western Candor Chasma, for example, deposits are dominated by dust and basaltic sand, accompanied by hydrated sulfates and crystalline ferric minerals (Bibring et al., 2007; Murchie et al., 2009b). Often, polyhydrated sulfates sit stratigraphically above layers of kieserite and ferric oxides (Roach et al., 2010a). The similarity to Meridiani LLDs suggests a regional process whereby upwelling groundwater modified aeolian sands, dust, or volcanic ash. In this case, the sulfates could derive from acid weathering of basalt followed by formation of evaporites, which for Meridiani, at least, must have been eroded and redeposited by the wind.

Alternatively, sulfates could form by dry deposition as they do in the Atacama desert on Earth (Michalski et al., 2004). Oxygen and sulfur isotopes in Atacama sulfates are mass-independently fractionated, which indicates an atmospheric origin (Farquhar et al., 2000). Sulfur in Martian meteoritic sulfates is similarly fractionated (Farquhar et al., 2007) and likely originated from photochemical conversion of atmospheric SO_2 to submicrometer particles and subsequent dry deposition. In some cases, the atmospheric sulfate apparently was subsequently reduced to sulfide in the meteorite source region (Greenwood et al., 2000). Whether similar isotopic signatures exist in Martian sulfates is currently unknown, but can potentially be addressed by MSL.

The presence of abundant sulfates has implications for Mars' volcanic history. The high abundance of sulfur in the Martian meteorites relative to water suggests that sulfur gases were important in the pre-Noachian and Noachian when Mars was most volcanically active, assuming that the meteorite compositions can be used to infer likely volcanic gas compositions, as discussed by Wänke and Dreibus (1994). If sulfur emissions exceeded water vapor, Martian volcanoes would have effectively spewed out a gas mixture equivalent to concentrated sulfuric acid. Geochemical models suggest that Martian volcanoes should have released gases with sulfur contents 10–100 times that of gases emitted on Earth (Gaillard and Scaillet, 2009). The total amount of oxidized sulfur emitted from Tharsis would have been equivalent to an approximately 20–60 m thick layer of globally distributed sulfates, which is consistent with the estimated global inventory (Righter et al., 2009). We discuss the potential role of sulfur-bearing volcanic gases on the early Mars climate system in Section 17.5.2.

17.3.2.3 Phyllosilicates

Phyllosilicates[3] are products of basaltic weathering and require high water activity in near-neutral or alkaline conditions to form (Ehlmann et al., 2013). They were originally thought to be present in airborne dust from Mariner 9 observations (Toon et al., 1977) and later more definitively identified in Noachian surface materials by Mars Express (Bibring et al., 2005, 2006; Poulet et al., 2005; Ehlmann et al., 2011). However, only ~3% of Noachian surfaces have hydrous minerals such as phyllosilicates or hydrous carbonates (Carter et al., 2012). Phyllosilicates constitute a large group of minerals such as kaolinites or

---

[3] The term "clay" alone refers to particulate materials with grain size less than 1/256 mm or 4 μm. The term "clay mineral" is generally taken to mean phyllosilicates, including aluminosilicate minerals with a crystal structure that derives from phyllosilicates.

smectites that are defined by their sheet arrangements[4]. Phyllosilicates have also been identified in Martian meteorites (e.g. Gooding, 1992) and confirmed by *in situ* measurements of mudstone material by MSL.

There are three stratigraphic classes of phyllosilicates: (i) deep phyllosilicates, (ii) layered phyllosilicates, and (iii) phyllosilicates in sedimentary basins (Murchie et al., 2009a). The deep phyllosilicates are excavated by impact craters with diameters of tens of kilometers in the Noachian southern highlands (Ehlmann et al., 2009; Wray et al., 2009b) or northern lowlands (Michalski and Niles, 2010). They are also found in layers outcropping from underneath Hesperian units (Mustard et al., 2009). The minerals in these phyllosilicates include chlorite, prehnite, and analcime, which are typically formed under hydrothermal conditions. Prehnite, for example, forms under temperatures of 200–350°C and lithostatic pressures below ~2 kbar. However, the presence of prehnite in impact ejecta and its thermal stability suggests that prehnite existed at depth before excavation (Fairén et al., 2010).

Layered phyllosilicates are laterally extensive in Mawrth Vallis and Nili Fossae and consist of Al phyllosilicates overlying kaolinite and Fe/Mg smectites in light-toned outcrops (Ehlmann et al., 2009). In Mawrth Vallis, the Al phyllosilicate layer can also include opaline silica, montmorillonite, and kaolinite (Bishop et al., 2008). Because smectite units drape underlying topography, possible precursors could be layers of volcanic ash that were altered by water. Fe–Mg smectites are also exposed by small ~100 m scale craters in the southern highlands (Wray et al., 2009b). In some places where ultramafic rocks have been altered – including Nili Fossae, some southern highland craters, and western Arabia – serpentine is detected (Ehlmann et al., 2010). This mineral forms in hydrothermal warm waters up to ~400°C in chemically reducing conditions.

Finally, the sedimentary phyllosilicates preserved in the basins of crater interiors have a diverse mineralogy. Fe–Mg phyllosilicates are observed in dozens of inter-crater basins in Sirenum and in northern Noachis (Wray et al., 2009b). In some places, such as Cross Crater in Terra Sirenum, Al phyllosilicates are also found along with sulfate sediments containing alunite and silica, indicating acidic alteration (Swayze et al., 2008).

Whether phyllosilicates have implications for the climate depends on their formation environment. The coexistence of Mg-rich smectites and carbonates suggests an equilibrium $p\mathrm{CO_2}$ of 1–10 mbar, which is comparable to the present surface pressure (Chevrier et al., 2007). Consequently, if the phyllosilicates formed in equilibrium with the atmosphere under warmer, wetter conditions, Chevrier et al. (2007) argue that non-$\mathrm{CO_2}$ greenhouse gases would be necessary. However, in the subsurface, under hydrothermal conditions, magnesium carbonates and serpentine can form independently of climate. It has

also been suggested that many phyllosilicates on Mars could form from water-rich magma-derived fluids in the subsurface (Meunier et al., 2012).

The geomorphology and chemistry of particular deposits can be examined to determine whether groundwater alteration of ashfall, sedimentary placement of eroded clays, or percolating rainfall through soils (pedogenesis) was the likely mechanism for forming the phyllosilicates. In Mawrth Vallis, alteration of volcanic ash is favored because smectites drape over topography (Bishop et al., 2008; McKeown et al., 2009). Subsequent alteration of the phyllosilicates is also possible. Some sulfates may be found alongside phyllosilicates because of sulfuric acid weathering of previously deposited phyllosilicates, which produces different phyllosilicates along with silica and Al or Fe sulfates (Altheide et al., 2010). However, it is doubtful that this can explain layers of sulfates that are sandwiched between phyllosilicate layers (Wray et al., 2009b).

Phyllosilicates in fan and delta deposits could be particularly convincing evidence of liquid water if they formed in the lakes where these features arose, thereby combining mineralogical with geomorphic evidence. CRISM has detected phyllosilicates in Eberswalde, Holden and Jezero Craters, which contain fan and delta deposits (Ehlmann et al., 2008b; Mustard et al., 2008). However, the phyllosilicates are the same type as those in surrounding sediment source regions and so are probably detrital, i.e. washed in.

17.3.2.4 Silica

Opaline silica also potentially sets environmental constraints. Its coexistence with jarosite on the Hesperian plateau surrounding Valles Marineris, indicates low pH (Milliken et al., 2008). The silica can derive directly from weathered basalts or from phyllosilicates. Low-temperature to hydrothermal conditions can produce silica (McAdam et al., 2008) but high concentrations of silica (such as up to ~91% detected by the Spirit Rover) generally favor a hydrothermal genesis (Squyres et al., 2008).

17.3.2.5 Chlorides

These have been spectroscopically detected in association with Fe–Mg smectites in Terra Sirenum, in the Noachian southern highlands (Osterloo et al., 2008). Chlorides are found in inter-crater plains and occasionally some crater floors. The chlorides are more recent and embay ancient phyllosilicate-rich knobs (Glotch et al., 2010). One hypothesis for formation of chlorides is groundwater discharge and evaporation, leaving chloride crusts (Jensen and Glotch, 2011).

17.3.2.6 Iron Oxides

High-albedo regions show absorption features characteristic of oxidized iron both from orbit (Bibring et al., 2006) and from landers (Bell et al., 2000), while orbital near-infrared spectra lack absorption signatures for hydration (Bibring et al., 2006). Mars' red surface (Morris et al., 2000) and the butterscotch sky color (Huck et al., 1977) are both due to ferric iron oxides in finely crystalline or poorly crystalline form on the surface and in airborne dust. How such iron oxides formed is debated, but

---

[4] Phyllosilicates are placed into mineral groups based on the layering. For example, the 1 : 1 group, exemplified by kaolinite ($\mathrm{Al_2Si_2O_5(OH)_4}$), comprises a layer of silicate tetrahedra linked by hydrogen bonds to OH groups on an octahedral aluminum layer. The 2 : 1 group, exemplified by smectite, has an octahedral sheet sandwiched between two tetrahedral sheets and linked to them via bonds.

deposition of atmospheric peroxides (accompanied by escape of hydrogen to space) should produce surface oxidation (Bullock et al., 1994; Zahnle et al., 2008), even in the Amazonian. When this oxidation process started is relevant for whether the early atmosphere was reducing or oxidizing.

17.3.2.7 Perchlorates

The Phoenix Lander detected ~0.6 wt.% ClO_4^-, perchlorate, in soil samples through wet chemistry, which was corroborated by the evolved oxygen from heated samples (Hecht et al., 2009). Spectral data indicate that perchlorate has been concentrated into patches, possibly in brines (Cull et al., 2010). Pyrolysis of soil on MSL has also released oxygen consistent with perchlorates (Glavin et al., 2013). Perchlorate is also tentatively detected on the north polar cap (Massé et al., 2010). Initially, Mg, Na and K perchlorates were favored forms at the Phoenix site (Kounaves et al., 2010; Marion et al., 2010) but reanalysis suggests the possible presence of Ca perchlorate (Kounaves et al., 2014). The eutectic temperatures of Mg and Ca perchlorates are very low, −57°C (Stillman and Grimm, 2011) and −77°C (Pestova et al., 2005), respectively, which means that liquids will exist on Mars above these temperatures wherever these perchlorates contact ice. Moreover, metastable solutions can super-cool tens of degrees below eutectic temperatures (Toner et al., 2013). The origin of the perchlorate may be similar to that on Earth, which is oxidation of chlorine volatiles in the air to perchloric acid, which dry-deposits (Catling et al., 2010), although presently unknown gas–solid reactions are required for Mars (Smith et al., 2014). This would have occurred in the past on Mars when chlorine gases were released by volcanism or impact volatilization.

17.3.2.8 Summary

Many alteration minerals have been detected in diverse areas on Mars. Phyllosilicates and magnesium carbonate are Noachian in age, while chlorides, layered sulfates, and opaline silica are found in Noachian and Hesperian units. The coexistence of Noachian phyllosilicates and carbonates suggests alkaline or weakly acidic conditions, while some sulfates, such as alunite or jarosite, imply acidic environments. The link between aqueous chemistry and climate is indirect, however. Climate implications depend on the extent to which rainfall was required for the aqueous alteration versus hydrothermal systems or low-temperature brines. Nonetheless, it is reasonably secure to conclude that groundwater systems, likely charged by rainfall, were necessary to form Noachian–Hesperian hematite concretions, and that hydrothermal waters were needed to form some Noachian phyllosilicate minerals such as serpentine. The current inventory of carbonate outcrops is tiny and taken at face value (i.e. assuming deeply buried carbonates are not significant in volume) would imply considerable escape of the atmosphere if the past inventory of CO_2 was large. In the present environment, a relatively large concentration of Mg or Ca perchlorate in the soil is permissive of brines at temperatures on Mars today. Consequently, low-temperature brines should exist today and such brines may have been important in the past.

17.3.3 Isotope Evidence

Strong evidence for atmospheric escape is preserved in the isotopes of Ar, H, Xe, and N. Modest evidence for escape is seen in the isotopes of C and O. Before examining the evidence, we briefly review the kinds of escape processes pertinent to Mars. (The reader is referred to Chapter 15 for details.) These provide a framework for interpreting the evidence. We then present the evidence and see how it fits the framework element by element, beginning with the noble gases.

Current escape rates are low. Escape can be thermal (Jeans escape) or non-thermal (e.g. sputtering, dissociative recombination). Low levels of escape are often highly fractionating, either because the lighter isotope extends higher above Mars and thus is more prone to escape, or because the lighter isotope is more likely to acquire escape velocity by chance (Jeans escape). Such escape can leave a strong signal in what is left behind, but because the escape rate is low, the signal can only be seen if the reservoir left behind is quite small. Large fractionations in Ar, H, and N have been attributed to such processes.

Although sputtering and dissociative recombination can cause escape from Mars today, these mechanisms would have been more effective on ancient Mars, because the young Sun would have been a stronger source of ionizing radiation and would have had a more intense solar wind (Luhmann et al., 1992). Some models have predicted ancient escape rates by sputtering that are several orders of magnitude higher than today (Luhmann et al., 1992). The exception to this is that, when young, Mars had a significant intrinsic magnetic field, which may have protected its atmosphere against sputtering.

Ancient escape processes to consider are impact erosion and hydrodynamic hydrogen escape. The roughly uniform hundredfold depletion of Mars' noble gases with respect to Earth suggests that Mars lost much of its atmosphere by a relatively efficient, non-fractionating process. Impact erosion – the expulsion of atmosphere by impacts – is a leading candidate (Melosh and Vickery, 1989; Zahnle, 1993, Brain and Jakosky, 1998). Impact erosion is expected to have been efficient on Mars because its escape velocity is small compared to typical impact velocities. Impact erosion discriminates strongly between volatiles in the atmosphere and those that are condensed on or under the surface. Thus impact erosion preferentially removes the noble gases and nitrogen, while favoring retention of water, carbonate rock, chlorine, and sulfur. Isotopic fractionations are possible if the condensed state favors one isotope over another. Elements that could have been fractionated in this way on Mars are C, O, and to some extent H.

Impact erosion is restricted to early in Mars' history when impact rates were high. We can set an upper bound on impact erosion by adopting the Melosh and Vickery (1989) "tangent plane" model for all impacts. In this model it takes roughly 500 craters bigger than 100 km diameter to erode a 1 bar atmosphere. Crater counts restrict this to times before 4 Ga (Hartmann, 2004). By the same model, it takes roughly 500 craters bigger than 30 km diameter to erode a 0.01 bar atmosphere. This corresponds to 3.5 Ga (Hartmann, 2004). If Mars had a thick (0.5–1 bar) atmosphere in the late Noachian or later, the atmosphere could not have been removed by impact erosion.

Hydrodynamic hydrogen escape is more speculative, but is probably required to explain fractionation in Xe. Hydrodynamic escape refers to escape driven by a pressure gradient. The gas flows into space; there is no clearly defined edge to the atmosphere. By definition, escape rates are much higher than today. If the escape rate is high enough, the hydrogen carries other gases with it and the gases left behind become mass-fractionated.

Hydrodynamic escape has two requirements. First, hydrogen must be a significant constituent of the atmosphere. This is likely shortly after Mars formed (Dreibus and Wänke, 1987), less likely at later times. Second, if hydrodynamic escape is to fractionate heavier gases like Ar and Xe, the escape flux must be high and this requires a solar EUV flux 10–50 times what it is today (Pepin, 1991). Given that the Sun's EUV luminosity declined inversely with time (Zahnle and Walker, 1982), hydrodynamic escape is restricted to earlier than ~4 Ga. It should be noted that hydrodynamic escape is not likely to be greatly hindered by a planetary magnetic field – the outflowing planetary wind is too strong.

17.3.3.1 Data

Data sources are: *in situ* measurements by Viking, Phoenix, and MSL; trapped gases in Martian meteorites; and telescopic observations. The *in situ* observations have not always been self-consistent nor always consistent with the gases found trapped in glasses of the youngest of meteorites, EETA 79001 in particular (Bogard et al., 2001). With one notable exception, the available MSL measurements resemble the meteorite data more closely than they resemble Viking, Phoenix, or telescopic data. As the meteorites are the chief source of historical information, we concentrate on them.

Nyquist et al. (2001) and Bogard et al. (2001) comprehensively reviewed Martian meteorites and their volatiles. Martian meteorites formed at different times. The shergottites come from surface flows that might sample the atmosphere at the time of their crystallization ages (~170 Ma for some, ~450 Ma for others; Nyquist et al., 2001). Others, the nakhlites and Chassigny in particular, are older, were more deeply buried, and may sample the atmosphere at 1.4 Ga. Collectively, the shergottites, nakhlites, and chassignites are referred to as the SNC meteorites. ALH 84001 (which is not classified as an SNC) crystallized 4.1 Ga (Lapen et al., 2010). If it truly samples the Martian atmosphere at that time, it is the most relevant to early Mars (Leshin et al., 1996; Marti and Mathew, 2000; Miura and Sugiura, 2000; Sugiura and Hoshino, 2000; Mathew and Marti, 2001; Greenwood et al., 2008, 2010). NWA 7034, "Black Beauty", is a 4.4 Ga impact melt breccia, but its trapped gases appear to be Amazonian in age (Cartwright et al., 2014). The meteorites also provide some glimpses into non-atmospheric (Mars interior) volatile reservoirs, although the latter are not always interpretable.

17.3.3.2 Noble Gases

First, it bears repeating that noble gases in the Martian atmosphere are a hundred times rarer than on Earth (Anders and Owen, 1977) and 10,000 times rarer than they are in carbonaceous chondrites or gas-rich enstatite chondrites (Pepin, 1991). Put another way, if Mars accreted 0.01% of its mass in

a volatile-rich late veneer (roughly equal to the sum total of all the visible impact basins on Mars, excluding the hemispheric dichotomy), the atmosphere would carry the isotopic signature of that late veneer in its noble gases (Zahnle, 1993). However, no such signature is seen. This implies that a conventional late veneer was not the source of current atmospheric gases.

We now address the isotopic evidence in noble gases. Argon and xenon show strong isotopic evidence for fractionating escape. By contrast, neon (lighter than Ar) and krypton (lighter than Xe) show little and none, respectively. The noble gases are therefore naturally grouped in pairs, with Ne complicating the interpretation of Ar, and Kr doing the same for Xe. Table 17.2 summarizes the discussion.

(a) Argon Argon has three stable isotopes. The heavy isotope ^{40}Ar would be exceedingly rare were it not for radioactive decay of ^{40}K (half-life 1.25 Gyr). If ^{40}Ar has not escaped, the amount of ^{40}Ar in the atmosphere implies that Mars is only 1% degassed (Dreibus and Wänke, 1987). This could be accomplished by degassing a kilometer-thick basaltic crust over all of Mars' history, or it could be accomplished by degassing Mars to the core during its first 20 Ma, or by any of many other scenarios.

The lighter isotopes ^{36}Ar and ^{38}Ar are non-radiogenic. The $^{36}Ar/^{38}Ar$ ratio on Earth is 5.32 and the chondritic ratio is 5.30 (Pepin, 1991); the solar wind ratio is 5.50 (Vogel et al., 2011). Against these, the Martian atmospheric ratio of 4.2±0.1 (Atreya et al., 2013) is strikingly heavy. The MSL measurement confirms what had been deduced from the SNC meteorites (Wiens et al., 1986; for a review see Bogard et al., 2001). The argon fractionation is unique to the Martian atmosphere and therefore requires substantial mass-fractionating escape of Ar from Mars.

The argon escape mechanism that can operate throughout Mars' history is sputtering (Jakosky et al., 1994; Hutchins and Jakosky, 1996; Hutchins et al., 1997). Broadly put, sputtering escape occurs when the solar wind picks off atoms that rise above the exobase. Because low-mass atoms rise highest, the process discriminates by mass. Jakosky et al. (1994) showed that sputtering implies the loss of at least 50–75% of ^{36}Ar to generate the observed $^{36}Ar/^{38}Ar$. Hutchins and Jakosky (1996) developed a more complete model that included radiogenic ^{40}Ar; they estimated loss of 85–95% of ^{36}Ar and loss of 70–88% of outgassed ^{40}Ar. Hutchins et al. (1997) revised the model to include shielding by an ancient magnetic field, which increased the uncertainties (e.g. 75–99% loss of ^{36}Ar).

Recent sputtering poses several problems. Krypton is not subject to sputtering escape, and thus to prevent the Ar/Kr ratio from shrinking too much the model implicitly requires that Ar be replenished preferentially. Neon is more susceptible to sputtering than Ar, and therefore it too needs to be replenished. Hutchins and Jakosky (1996) argue that recent volcanic degassing would fall one or two orders of magnitude short; the problems with neon will be addressed with that element.

It therefore appears that modern sputtering losses have been overestimated and that the $^{36}Ar/^{38}Ar$ fractionation may be an ancient feature of the atmosphere. This is consistent with sputtering expected to have been much more effective early in Mars' history (Luhmann et al., 1992). But an ancient origin also opens the door to hydrodynamic escape. In an H_2–CO_2 atmosphere, the maximum rate that hydrogen can flow through

Table 17.2. *Isotope data and their interpretation.*

Isotopic ratio	Mars atmosphere	Earth atmosphere	Interpretation
$^{36}Ar/^{38}Ar$	<3.9 (SNCs) 4.01 (MSL)	5.3	Mass-fractionating escape (sputtering, hydrodynamic escape)
$^{40}Ar/^{36}Ar$	3000 (Viking) 1800 (shergottites) 1800 (MSL)	300	Early escape of ^{36}Ar (hydrodynamic escape, impact erosion)
$^{20}Ne/^{22}Ne$	~10 (shergottites)	9.8	Possible escape
Kr isotopes	Solar-like and unfractionated	Solar-like and unfractionated	Noble gases acquired from a solar source; non-fractionating escape (impact erosion)
Xe isotopes (Figure 17.7)	Fractionated	Fractionated	Hydrodynamic escape
$^{129}Xe/^{132}Xe$	2.5	0.97	Early escape of ^{132}Xe (impact erosion)
$^{14}N/^{15}N$	170	272	Mass-fractionating non-thermal escape
D/H	Variable, (1–6.7)×SMOW[a]	SMOW	Mass-fractionating thermal escape
$\delta^{18}O$ in CO_2	48±5‰ (MSL)	0	Possible escape
$\delta^{13}C$ in CO_2	46±4‰ (MSL)	0	Possible escape

[a] Standard mean ocean water.

a CO_2 atmosphere that does not escape is the diffusion limit (Hunten and Donahue, 1976). What this means for Ar, which is slightly less massive than CO_2, is that Ar can escape with the hydrogen, but that escape is inefficient and more strongly mass-fractionating than sputtering (Zahnle et al., 1990). In this model, a $^{36}Ar/^{38}Ar$ of 4.2 requires only a third of the ^{36}Ar to escape, and none of the ^{40}Ar. Both sputtering and diffusion-limited hydrodynamic escape predict significant neon fractionation (e.g. Zahnle et al. (1990) predicted $^{20}Ne/^{22}Ne$ ~ 7), and thus both models require replenishing the atmosphere with a small amount of isotopically light neon.

Radiogenic Ar also sets useful constraints that are broadly consistent with Ar fractionation being ancient rather than recent. The $^{40}Ar/^{36}Ar$ ratio of ~1800±100 deduced from trapped gases in SNC meteorites (Bogard et al., 2001) and confirmed by MSL (1900±300; Mahaffy et al., 2014) is much higher than on Earth (294). There is evidence in trapped gases in SNCs that the $^{40}Ar/^{36}Ar$ ratio has not changed appreciably in the past 1.4 Ga (Bogard and Garrison, 2006; Cassata et al., 2012). This suggests that neither ^{40}Ar degassing nor ^{36}Ar sputtering losses have been important in the past 1.4 Ga, as either process would raise the $^{40}Ar/^{36}Ar$ ratio. For comparison, Hutchins and Jakosky (1996) predicted that the ^{40}Ar content in the atmosphere should have increased by a factor of 5 in the past 1.4 Ga.

A very ancient atmospheric $^{40}Ar/^{36}Ar$ ratio of 626±100 has been reported in a 4.16±0.04 Ga mineral in ALH 84001 (Cassata et al., 2012). This ratio is surprisingly high given that only 20% of Mars' ^{40}K had decayed and only a fraction of the ^{40}Ar could have reached the atmosphere. The implication is that the atmospheric inventory of ^{36}Ar was already small by Earth's standards. Cassata et al. (2012) developed a model around this datum to set an upper bound of 0.4 bar on atmospheric CO_2 at 4.16 Ga. Their argument is too model-dependent to stand as a true upper bound, but the overall sense that by 4.16 Ga Mars had lost more than 90% of the ^{36}Ar it began with seems inescapable.

(b) Neon Neon has three stable isotopes that to within (ample) error reveal little by themselves. The $^{20}Ne/^{22}Ne$ ratio in SNC impact glasses is ~10 (Bogard et al., 2001), quite comparable to Earth's. But when viewed against the backdrop of argon escape, the relatively modest fractionation of neon isotopes is telling. All proposed models to account for $^{36}Ar/^{38}Ar$ fractionation predict that the $^{20}Ne/^{22}Ne$ ratio should be lower than it is because Ne escapes more easily than Ar (e.g. Zahnle et al. (1990) predict $^{20}Ne/^{22}Ne$ ~ 7). Hence all models require that new neon be added to the atmosphere after argon escape (Hutchins and Jakosky, 1996). The source of fresh neon would constrain when Ar escape took place if the source could be quantified. However, sources with the required properties – a high Ne/Ar ratio and a high $^{20}Ne/^{22}Ne$ ratio – are solar. Very few meteorites have these properties. Pesyanoe, an enstatite chondrite, is an exception. If all post-Noachian impacts on Mars were Pesyanoe-like, it is possible for Ar fractionation to post-date the Noachian. Otherwise an exogenous source can work only if Ar fractionation is ancient. Volcanic sources are also problematic. The interior component in Chassigny is probably the best guide we have to the composition of Martian volcanic gases (Ott, 1988; Bogard et al., 2001). If the ^{36}Ar content of Chassigny is representative, and if the Ne/Ar content of the gases was solar, it would take Tharsis-like degassing to reset the atmosphere's neon isotopes. This is unlikely in the post-Noachian period.

Thus, three lines of evidence suggest that Ar fractionation was established early. The constant $^{40}Ar/^{36}Ar$ ratio in SNCs precludes significant Ar escape and significant Ar degassing after 1.4 Ga. The high $^{40}Ar/^{36}Ar$ ratio in ALH 84001 shows that most

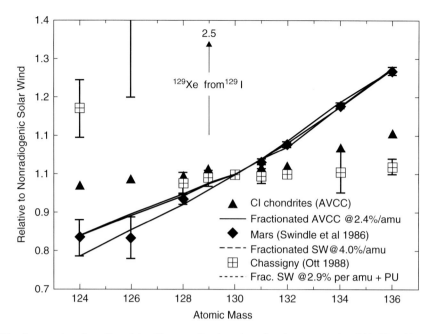

Figure 17.7. Xenon in Mars' atmosphere is well explained by mass fractionation of carbonaceous chondritic Xe with no fission products (Swindle et al., 1986), or it can be fit by mass fractionation of solar wind xenon with a small addition of fission Xe (Swindle and Jones, 1997; Mathew et al., 1998). The latter is preferred because Xe in Chassigny appears solar (Ott, 1988). Martian Kr (not shown) is identical to solar Kr. Martian ^{129}Xe plots at 2.5 (beyond the top of the diagram). These relationships superficially resemble what is seen on Earth, but differ in every detail, which implies that parallel processes have operated on different materials on the two planets.

^{36}Ar escape took place before 4.16 Ga. The relatively unfractionated Ne isotopes today require injecting new neon, so that Ar escape was complete before volcanic activity or impact bombardment subsided. Although any of these arguments can be questioned, they are independent, so each must be contradicted independently. Links between Ar and CO_2 are model-dependent, but at least one such model concludes that the atmosphere was no thicker than 0.4 bar at 3.8 Ga (Cassata et al., 2012).

(c) Krypton Krypton has six stable isotopes. Martian Kr is indistinguishable from solar Kr and clearly distinct from chondritic Kr and terrestrial Kr. Krypton has not been affected by fractionating escape. But Kr *has* escaped: its overall abundance is a little high but not grossly out of line with the abundances of Ar and Xe. This supports the hypothesis that the bulk of noble gas escape – including Ar – was by impact erosion, and not by a fractionating process like hydrodynamic escape. It also means that the total amounts of Ar and Xe lost during *fractionating* escape have been modest, no more than a factor 2 or 3.

(d) Xenon Xenon has nine stable isotopes, several of which are partially radiogenic. As shown in Figure 17.7, Mars' atmospheric Xe (trapped in SNC glasses) is strongly mass-fractionated with respect to solar or meteoritic Xe (Swindle et al., 1986; Wiens, 1988; Bogard et al., 2001). This observation suggests that some Xe was lost by a mass-fractionating process that did not affect Kr.

Atmospheric radiogenic xenon is very depleted. Best estimates (Swindle and Jones, 1997; Mathew et al., 1998) are that the atmosphere retains only ~0.1% of its possible radiogenic ^{129}Xe (from decay of ^{129}I, half-life 15.7 Myr) and ~0.3% of its possible radiogenic ^{136}Xe (from spontaneous fission of ^{244}Pu,

half-life 81 Myr). If Mars did accrete in fewer than several million years, as Dauphas and Pourmand (2011) suggest, both ^{129}I and ^{244}Pu were alive. Therefore, missing radiogenic xenon requires escape or a failure of the planet to degas, or both.

Martian internal xenon as seen in Chassigny differs strikingly from the atmosphere (Ott, 1988; Mathew et al., 1998; Bogard et al., 2001). Isotopically, it closely resembles solar xenon (Figure 17.7). Martian atmospheric xenon can be derived by mass-fractionating Chassigny xenon and adding radiogenic Xe from ^{129}I decay and fissiogenic xenon from ^{244}Pu decay (Swindle and Jones, 1997; Mathew et al., 1998). Chassigny is also rather Xe-rich. Degassing of 800 m of Chassigny-like basalt (~10^8 km^3) would double the xenon reservoir in the atmosphere (Zahnle, 1993). It takes less volcanic degassing to reset xenon's isotopes than neon's.

The ancient ALH 84001 contains several different xenons. The most interesting of these is something Mathew and Marti (2001) call "early atmosphere", which is radiogenically enriched (^{129}Xe/^{132}Xe = 2.16; 2.5 is the signature of the modern Martian atmosphere) and correlated with a rather Earth-like nitrogen (δ^{15}N = +7‰), but which is not as mass-fractionated as xenon in the atmosphere. If this really is a sample of the atmosphere 4.1 Ga, it means that most of the mass-fractionating Xe escape took place after 4.1 Ga.

Hydrodynamic hydrogen escape can fractionate xenon because a heavier isotope is dragged to space less easily than a lighter isotope (Sekiya et al., 1981; Hunten et al., 1987; Sasaki and Nakazawa, 1988; Pepin, 1991). This leads to the same problem with sputtering escape of argon and neon: if Xe is fractionated by hydrodynamic escape, then Kr should be as well. But it is not.

There are several ways to address this. One approach is to presume that the other noble gases are entirely lost, so that only fractionated Xe remains. Then fresh Ne, Ar, and Kr are added to the atmosphere, but fresh Xe is not. If this is done endogenously, Xe must remain inside Mars when Ne, Ar, and Kr degas (Pepin, 1991). Why Xe does not degas is unspecified. If the noble gases are replenished exogenously, a class of impactor is presumed in which Kr is much more abundant than Xe. This requirement might be met by some comets (Dauphas, 2003): experiments show that Kr is preferentially adsorbed by amorphous ice at ~50 K compared to Xe (Owen and Bar-Nun, 2001).

It is interesting to point out that the comet hypothesis does not actually require any Xe escape at all. The same story can be told if the mass-fractionated Xe were the result of gravitational settling inside porous planetesimals (Ozima and Nakazawa, 1980; Zahnle et al., 1990). This highly fractionated Xe is accompanied by a very low Kr/Xe ratio (also attributable to gravitational settling). Gravitational settling of atmospheric gases in deep snow is a well-known phenomenon in glaciology. If comets have very low Xe/Kr ratios and are the source of Kr, there is no need for hydrodynamic escape.

Another speculative possibility is that Xe escapes from planetary atmospheres by a different mechanism than Kr. This can be the case in hydrodynamic escape if Xe, the only noble gas more easily ionized than hydrogen, escapes as an ion in an ionized hydrogen wind. The strong Coulomb interaction between ions couples the Xe^+ to the flow. An ionized wind is possible along open (polar) magnetic field lines or in the absence of a magnetic field. In principle this could take place with relatively modest (but still significant) hydrogen escape rates and relatively recently, which makes it the only hypothesis that can accommodate fractionating Xe after 4.1 Ga.

17.3.3.3 Other Gases

(a) Nitrogen Nitrogen in the Martian atmosphere is isotopically heavy. Viking reported $^{14}N/^{15}N = 168 \pm 17$, a measurement confirmed by MSL, which reported 173 ± 11 (Wong et al., 2013; Earth is 272). In delta notation, $\delta^{15}N = 570 \pm 100‰$. SNC impact glasses also contain heavy nitrogen, but in no case is the nitrogen as heavy as atmospheric nitrogen. Several SNCs also carry isotopically light nitrogen ($\delta^{15}N = -30‰$; air is 0‰ by definition) that resembles nitrogen in enstatite chondrites. This appears to be an internal Martian reservoir and a plausible starting point for nitrogen evolution. The SNCs are therefore best interpreted as trapping a mixture of atmospheric and interior nitrogen (Bogard et al., 2001).

McElroy (1972) predicted that 99% of Mars' initial nitrogen would have been lost to space by dissociative recombination. This occurs when an N_2^+ ion (ionized by solar EUV radiation) recombines with an electron to form two high-velocity neutral atoms. The energy released is enough for the upward-moving atom to escape from Mars if it does not collide. Therefore, if dissociative recombination takes place above the exobase, a nitrogen atom is lost to space. Viking's discovery that Martian nitrogen is heavy was taken to confirm the theory (Nier and McElroy, 1977).

Nitrogen escape from Mars has been revisited many times since, with increasing sophistication (McElroy et al., 1977; Wallis, 1989; Fox and Hać, 1997a; Fox, 2007). Nitrogen can be lost by sputtering as well as by dissociative recombination. Both mechanisms favor ^{14}N escape because the $^{28}N_2/^{29}N_2$ ratio is higher at the exobase than in the lower atmosphere. Another effect is that, after dissociative recombination, the ^{14}N atoms have higher velocities than the ^{15}N atoms. Most calculations suggest that N escape is fast and rather too easily accounts for the ^{15}N enrichment in the remaining N_2. Indeed, the mechanism is often computed to be so efficient that it requires a substantial non-atmospheric nitrogen reservoir to damp the fractionation (McElroy et al., 1977; Bogard et al., 2001). In the limit, Wallis (1989) and Manning et al. (2009) suggested that the atmospheric $^{14}N/^{15}N$ ratio might be in steady state between a terrestrial-like source of N and escape. The range of elevated $^{14}N/^{15}N$ values in SNCs might then be taken as snapshots that chronicle the "steady-state" response of the atmosphere to rapid escape and stochastic resupply.

Nitrogen in ALH 84001 is paradoxical. Mathew and Marti (2001) report a correlation between $\delta^{15}N$ and $^{129}Xe/^{132}Xe$ that they use to define what they call the "early atmosphere". The nitrogen isotopic value corresponding to the highest $^{129}Xe/^{132}Xe$ (= 2.16) is +7‰. Mathew and Marti therefore suggest that, at 4 Ga, Martian N had evolved from an initial value of −30‰ to +7‰. However, Miura and Sugiura (2000) report that ALH 84001 contains the heaviest N found in a Martian meteorite, with $\delta^{15}N \approx 420‰$ ($^{14}N/^{15}N \approx 190$). Miura and Sugiura regard the heavy N as representative of the atmosphere at 4 Ga. There is thus a paradox: evidence has been reported that $^{14}N/^{15}N$ has evolved over the past 4 Ga and that it has not evolved over the past 4 Ga.

(b) Hydrogen Hydrogen is the most strongly fractionated element on Mars and, with Ar, the most certain to have escaped. The D/H ratio in water vapor measured by ground-based telescopic observations is 5–7 times higher than it is on Earth or in most meteorites ($5.5 \pm 2 \times SMOW$[5]; Owen et al., 1988; Owen, 1992; Krasnopolsky et al., 1997, 1998; Krasnopolsky, 2000; Villanueva et al., 2008, 2015; Novak et al., 2011). MSL has confirmed this (Webster et al., 2013), both for water vapor (5.95 ± 1.08) and for the surprisingly abundant adsorbed water in the Rocknest fines (6.88 ± 0.06). The other reported telescopic measurement is an inference that D/H = $2.4 \pm 1 \times SMOW$ in H_2 (Krasnopolsky, 2002), a measurement that is directly relevant to escape.

Heavy hydrogen is seen in SNCs (Watson et al., 1994; Leshin et al., 1996; Leshin, 2000; Greenwood et al., 2008, 2010). D/H ratios ranging from terrestrial to Martian water vapor are reported. The young (~170 Ma; Nyquist et al., 2001) SNC basalts Shergotty, Zagami, and Los Angeles have D/H ratios in apatites of 5.6×SMOW, 5.4×SMOW, and 4.5–5.1×SMOW, respectively (Greenwood et al., 2008). These are thought to represent atmospheric water. The extent to which the more Earth-like D/H ratios are attributable to terrestrial contamination versus indigenous water of Mars has been debated (Leshin et al., 1996; Boctor et al., 2003; Greenwood et al., 2008, 2010), but at least some of the low-D/H water is Martian. Leshin et al. (1996) report that a number of nakhlites (1.4 Ga) have D/H ~ 1.6×SMOW and a few shergottites (0.17 Ga) also have D/H ~ 2.2×SMOW in hydrous minerals. These are plausibly interpreted as sampling meteoric crustal water (Carr, 1996; Leshin et al., 1996). Bogard et al. (2001) list D/H = 1.09×SMOW for Martian interior water. Kurokawa et al. (2014) argue that Mars'

---

[5] Standard mean ocean water, i.e. Earth.

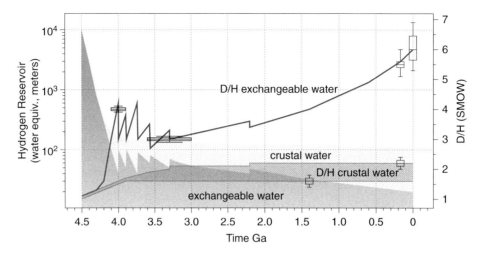

Figure 17.8. Time evolution of D/H (red line, right axis) and water reservoirs (blue shadings, left axis). Open box symbols are D/H SNC data with error bars representing ranges. Layered horizontal rectangles are D/H measurements from ALH 84001 (~3.9–4.1 Ga) and Gale clays (~3.0–3.5 Ga). Pale red hatched region represents D/H in the crustal reservoir. Jaggedness in some of the curves schematically illustrates how the system responds to catastrophic release of new water to the surface, such as outflow channels that release long-buried water from the deep crust. The general trend of D/H ratios growing more quickly before ~3 Ga than after is based on the assumption that H escape was faster then. A black and white version of this figure will appear in some formats. For the color version, please refer to the plate section.

original D/H ratio was 1.275×SMOW. It seems, therefore, that the starting point for D/H evolution on Mars was SMOW-like.

ALH 84001 also contains both heavy (4.0±0.1×SMOW; Greenwood et al., 2008) and light water (1.08×SMOW; Bogard et al., 2001; see also Boctor et al., 2003). As with the SNCs, the simple interpretation is that the rock records both internal undegassed water (low D/H) and environmental water (high D/H). This interpretation would imply that D/H fractionation and hydrogen escape were early processes that pre-date 4.1 Ga (Greenwood et al., 2008).

A recent *in situ* finding by MSL reports that D/H = 3.0±0.1×SMOW in water (hydroxide) that is chemically bound in clay minerals extracted from the mudstones of Yellowknife Bay (Mahaffy et al., 2014). Although the age of these minerals is uncertain, they appear to have been subjected to fluvial activity as late as 3.0–3.5 Ga (Grant et al., 2014; Grotzinger et al., 2014). It is likely that this surface water was exchanging with the atmosphere so that the D/H in these minerals would reflect the composition of the atmosphere at that time.

Interpretation of the D/H data is complicated by the fact that they vary so much (Villanueva et al., 2015), and that it is difficult to ascertain the source of the water (e.g. Boctor et al., 2003). The simplest approach to interpreting the data, illustrated in Figure 17.8, is through the use of Rayleigh distillation, which relates changes in fractionation to changes in the size of reservoirs (e.g. Yung et al., 1988; Carr, 1990; Yung and Kass, 1998; Kass and Yung, 1999; Villanueva et al., 2008, 2015). The only assumptions are (i) there is no resupply and (ii) the fractionation factor *F* that describes the relative efficiency of D escape to H escape is constant. A range of fractionation factors has been reported in the literature (*F* = 0.32, Yung et al., 1988; *F* = 0.016, Krasnopolsky et al., 1998). Since hydrogen appears to be escaping at close to the diffusion limit (Zahnle et al., 2008), we use *F* = 0.4, which is the ratio of D/H in H₂ (= 2.4, Krasnopolsky, 2002) to D/H in H₂O (= 6, see above references).

While the exact value of *F* in this range has an effect on reservoir sizes, the effect is modest and it does not change the basic conclusions about the need for significant escape over time.

In Rayleigh distillation, the remnant reservoir R_{now} is related to the initial reservoir R_{init} by

$$\left(\frac{R_{init}}{R_{now}} \right)^{1-F} = \frac{(D/H)_{now}}{(D/H)_{init}} \tag{17.1}$$

With *F* = 0.4, and a D/H now of 6×SMOW, only 5% of an initial SMOW water reservoir would remain. To convert relative reservoir sizes into absolute reservoir sizes requires another piece of information, either a known reservoir size or the hydrogen escape flux. If our best estimate of the current escape rate has been constant through time, Mars loses ~8 m of water in 4 Ga (Zahnle et al., 2008), leaving it with ~40 cm of water today. Thus, this model predicts that we are looking at Mars just before it dries up. These updated estimates differ only in details from what Yung et al. (1988) found when they first presented this argument. It is an inevitable consequence of starting D/H evolution from water initially at SMOW.

On the other hand, Rayleigh distillation works much better if evolution begins at ~3 Ga (a rough age for the Gale clays) with D/H = 3.0×SMOW. For D/H = 6.0 currently, *F* = 0.4, and today's H escape rate, the 6 m of water to escape over 3 Ga is two-thirds of an initial 9 m inventory, which leaves 3 m today. This model is also consistent with the younger SNCs, because it predicts that 170 million years ago D/H = 5.6×SMOW.

A feature of this simplest of possible models is that water inventories scale with the hydrogen escape rate. Carr and Head (2014) find that the total amount of near-surface unbound water on Mars today is ~50 m. If this represents the exchangeable reservoir, it implies that 150 m of exchangeable water was present when the Gale clays formed, and that 100 m of water have escaped since. This means that the average hydrogen escape rate for the past 3 Ga was 17 times bigger than it is today. The requirement for more H escape on a 3 Ga timescale is reasonable, although it needs to be emphasized that the rate of hydrogen escape is controlled by the oxygen sink (Hunten and Donahue, 1976; Zahnle et al., 2008). Escape and crustal oxidation are the main known sinks for oxygen. Since oxygen escape models cannot account for 100 m of water loss (Lammer et al., 2003), a sizable fraction of the excess oxygen must have been taken up by the crust. Of course, these numbers scale directly with the present reservoir size. A more likely scenario is that 20 m of water is presently exchangeable. This represents the size of the polar ice caps, which are thought to come and go with orbital variations (see Chapter 16). If only 20 m of the near-surface water is currently exchangeable, then 60 m of water was exchanging when the Gale clays formed, the average escape rate was only seven times the present value, and much less crustal oxidation is required.

One thing Rayleigh distillation models cannot generate is a lower D/H ratio at 3 Ga from the higher D/H ratio at 4.1 Ga in ALH 84001. If the D/H = 4 number truly represents atmospheric D/H at 4.1 Ga as Greenwood et al. (2008) claim, then a great deal of fresh isotopically light water must have been released to the surface after 4.1 Ga. The most attractive candidate is the liberation of old low-D/H crustal water by outflow events, which, as described by Carr (1996), are of the right order of magnitude. It is tempting to identify the old water with the D/H = 1.6 water in nakhlites and the D/H = 2.2 water in (the shallower) shergottites (Leshin et al., 1996). It is then reasonable to think that the lakes in Gale sample an unusual moment when an exceptionally large amount of old water had been recently mobilized. Exogenous water and juvenile volcanic water cannot be ruled out, but neither mechanism is quantitatively attractive. The amount of low-D/H exogenous water required is at the high end of what could be delivered after 4 Ga (cf. Chyba, 1990), while juvenile low-D/H volcanic water works only if the erupting matter is wet (>1% H_2O) or the volume of volcanism is in excess of the Tharsis construct.

To get to a D/H = 4 at 4.1 Ga requires significant early escape. Hydrodynamic escape, which best explains the xenon data, is weakly fractionating in a CO_2 atmosphere such that $F \approx 0.8$ (Zahnle et al., 1990). To get D/H = 4×SMOW at 4.1 Ga, Rayleigh distillation requires the loss of 99.999% of the initial reservoir, which Dreibus and Wänke (1987) estimate to be 150 km. Thus, 150 m would remain at 4.1 Ga. While the loss of this much water is possible because the process includes escape during accretion, it seems quite extreme. What is more likely is that hydrodynamic escape was responsible for raising D/H to something like the 1.6×SMOW seen in Nakhla during the loss of the first 90% of Mars' water, thereby setting the state of a crustal water reservoir that should still be considerable (>100 m) albeit currently inert. The need to achieve D/H = 4 at 4.1 Ga implies that hydrogen escape fluxes, now dominated by Jeans escape, were already low enough and the exchangeable reservoir small enough to be strongly mass-fractionating, at least episodically. This does not rule out subsequent episodes of hydrodynamic escape, but these would be associated with episodes of lower D/H. A generally similar story was explored by Carr (1996).

In summary, fitting the D/H data to a Rayleigh distillation model implies the loss of significant amounts of water over Mars' history. If the current exchangeable reservoir is ~20 m, and hydrogen has escaped at the diffusion limit ($F = 0.4$), then

the exchangeable reservoir at 3 Ga was ~60 m, the average H escape rate over the past 3 Ga was ~7 times higher than it is presently, and only a modest surface sink for oxygen is needed. An early epoch of hydrodynamic escape would have escaped many kilometers of water but fractionation of SMOW-like initial water to 4×SMOW at 4.1 Ga was likely accomplished after hydrodynamic escape ended and most of Mars' residual water was more or less permanently locked up in the crust. The subsequent lowering of D/H to 3×SMOW at 3 Ga could have been accomplished by mixing with low-D/H crustal waters brought to the surface by outflow channels.

(c) Oxygen Oxygen escape has been of interest since McElroy (1972) predicted that C and O could escape by dissociative recombination of CO^+. Oxygen escape by this and many other possible non-thermal mechanisms has been extensively studied (Hunten and Donahue, 1976; Liu and Donahue, 1976; McElroy et al., 1977; Luhmann et al., 1992; Fox, 1993; Jakosky et al., 1994; Fox and Haċ, 1997b, 2010; Luhmann, 1997; Lammer et al., 2003). Many of these studies were based on the expectation that O should escape at half the rate that H does (assuming no surface sinks), so that it is H_2O that on net escapes. The newer studies compute O escape rates that are much too small to balance H escape (Lammer et al., 2003). Whatever the details, in general the lighter isotopes escape more easily, so that, if escape has been important, mass-dependent fractionation is expected (Jakosky et al., 1994).

Oxygen in atmospheric CO_2 today is somewhat heavier than the terrestrial standard (SMOW), with $\delta^{18}O = 48\pm5‰$[6] from MSL (Mahaffy et al., 2013). This can be compared to ground-based telescopic observations ($\delta^{18}O = +18\pm18‰$; Krasnopolsky et al., 2007) and *in situ* measurements by the Phoenix Lander ($\delta^{18}O = +31\pm5.7‰$; Niles et al., 2010). H_2O in the Rocknest fines is quite heavy, with $\delta^{18}O = +84\pm10‰$ (Webster et al., 2013). Oxygen in water vapor is not reported (Webster et al., 2013). Oxygen in carbonate in SNCs and ALH 84001 is typically between +10‰ and +20‰ (Niles et al., 2010; Halevy and Eiler, 2011), with enough scatter to mask any obvious trends (Niles et al., 2010).

The fractionation in O is rather modest, and there is no sign that $\delta^{18}O$ has evolved from ALH 84001 and Shergotty in the carbonate isotopes. The fractionation is not outside the bounds of what geochemistry can do, and thus not so big that only O escape can explain it. The usual interpretation for a modest, slowly evolving fractionation is that either H_2O or CO_2 (or both) is exchanging with a much larger reservoir. An alternative explanation is that current and historic non-thermal O escape rates have been greatly overestimated.

Oxygen isotopes in meteorites also show multiple levels of strong, distinctive mass-independent fractionations (MIF) in silicates, water, carbonates, and sulfates in SNCs. Bulk Mars, as represented by the silicates, is distinct from Earth with $\Delta^{17}O \approx 0.3‰$[7]. Water and carbonates (in Lafayette and Nakhla) are

---

[6] The delta notation is defined as follows: $\delta R = ((R_{sample}/R_{standard}) - 1)\times1000‰$, where R is a ratio of two isotopes. For $\delta^{13}C$, R corresponds to $^{13}C/^{12}C$.

[7] Mass-independent fractionation is reported using the capital delta notation (Δ). It is the deviation of the measured isotope ratios from a theoretical curve based on mass-dependent fractionation.

more enriched in ^{17}O than the silicates with $0.5‰ < \Delta^{17}O < 1.0‰$ (Karlsson et al., 1992; Farquhar and Johnston, 2008), and sulfate in Nakhla is still more enriched with $\Delta^{17}O \approx 1.2‰$ (Farquhar and Johnston, 2008). A similar MIF is seen in carbonate in ALH 84001 (Farquhar and Johnston, 2008).

On Earth the chief source of ^{17}O enrichments is ozone. Most highly oxidized species in Earth's atmosphere (nitrates, sulfates, perchlorates, and peroxides) have at least some ^{17}O enrichment that can be traced back to ozone (Thiemens, 2006). Ozone is also an important photochemical species on Mars; it is plausible that the water and carbonate acquired their ^{17}O enrichment from atmospheric oxidants like hydrogen peroxide or ozone reacting with the soil (Farquhar and Thiemens, 2000).

Oxygen isotopes in Lafayette and Nakhla indicate that water was equilibrated with carbonate – i.e. that carbonate formed at temperate conditions with the aid of water – but neither is equilibrated with silicate or sulfate (Karlsson et al., 1992; Farquhar and Thiemens, 2000; Farquhar and Johnston, 2008). Thus, the SNCs appear to derive from at least three separate oxygen reservoirs (Carr, 1996). The silicates represent bulk Mars. The water and carbonate carry excess ^{17}O. This lack of interaction between the atmosphere and hydrosphere on the one hand and the rocky planet on the other is something that dates back at least as far as ALH 84001 (Farquhar and Johnston, 2008). It is an argument for an extremely dry planet (Carr, 1996). The sulfate carries an even greater excess of ^{17}O. Farquhar and Johnston (2008) point out that the lack of isotopic interchange between sulfates and water is characteristic of abiotic systems.

(d) Carbon Oxygen and carbon are usually considered together because they are tightly entangled in CO_2. Unfortunately, the reported $\delta^{13}C$ data vary widely. Ground-based telescopic observations find $\delta^{13}C = -22\pm20‰$ in atmospheric CO_2 today (Krasnopolsky et al., 2007). Viking measured $\delta^{13}C = 0\pm50‰$ (Bogard et al., 2001). The situation is exacerbated by differences in the composition of atmospheric CO_2 reported by Phoenix, $\delta^{13}C = -2.5\pm4.3‰$ (Niles et al., 2010), and MSL, $\delta^{13}C = 46\pm4‰$ (Mahaffy et al., 2013). Reported $\delta^{13}C$ values in Martian meteorite carbonates also span a huge range, from $-20‰$ to $+60‰$ (Niles et al., 2010). Carbonates in the 4.1 Ga ALH 84001 and 1.4 Ga Nakhla carry heavy carbon with $\delta^{13}C \approx +40‰$ (Romanek, et al., 1994; Jull et al., 1999; Halevy and Eiler, 2011), while the younger Zagami (170 Ma) has $\delta^{13}C \approx -20‰$ (Jull et al., 1997; all fractionations are with respect to the terrestrial PDB (Pee Dee belemnite standard carbonate). In all three cases the $\delta^{13}C$ was thought to reflect the composition of atmospheric CO_2 at the time.

As noted, McElroy (1972) predicted that carbon could escape from Mars following dissociative recombination of CO^+. Reports of heavy carbon in ALH 84001 and Nakhla were seen as confirming the hypothesis of massive fractionating CO_2 escape from early Mars (Jakosky and Jones, 1997). The recent MSL observation reinforces this interpretation. Niles et al. (2010) devised a model that also argued for early massive fractionating escape but, in order to accommodate the lighter telescopic and Phoenix data, replenished the atmosphere with isotopically light juvenile volcanic CO_2 after 1.4 Ga. The recent volcanism model appears to be contradicted by the constancy of the $^{40}Ar/^{36}Ar$ ratio over the same time span. More recently, Hu et al. (2015) have shown that photodissociation of CO and

subsequent sputtering escape can fractionate carbon to the MSL levels with only modest carbonate formation.

17.3.3.4 Summary

While the isotopic data are sometimes conflicting and difficult to interpret, they do support the hypothesis that much of the Martian atmosphere has escaped to space. H, N, Ar, and Xe are all isotopically heavy and are best explained by the preferential loss of light isotopes to space. The mechanism(s) of escape and its timing and duration remain uncertain, but they are all fractionating to some degree and have played some role in producing the observed patterns. The implication for early Mars is that the atmosphere was thicker than it is today. How much thicker is difficult to ascertain from the isotope data alone, though one estimate based puts an upper limit of ~400 mbar during the Noachian (Cassata et al., 2012). In the next section we review the origin and evolution of the Martian atmosphere and place some constraints on its mass and composition.

17.4 ORIGIN OF THE EARLY MARS ATMOSPHERE

17.4.1 Accretion and Core Formation

The origin of the Martian atmosphere is intimately linked to the materials from which it formed, and how they were assembled to construct the planet. The standard model of Solar System formation has planets forming from condensed materials within the solar nebula (e.g. Lin, 1986). Thus, the primitive atmosphere might consist of gases contained in the solar nebula. However, as mentioned above, noble gases on Mars are significantly depleted with respect to solar abundances (Owen, 1992), so a nebular origin of its atmosphere is not likely. Instead, the early Martian atmosphere must have formed by some process that extracted volatiles from the solid materials from which the planet formed, from volatiles contained in the asteroids and comets delivered after the planet formed, or some combination of both.

The materials from which the planet formed likely had compositions similar to the primitive chondritic meteorites, which constitute the bulk of known meteorites (see Scott and Krot (2005) for an overview of their origins and compositions). They are thought to be among the oldest objects in the Solar System and are rich in the silicate minerals olivine and pyroxene, but also contain significant volatile materials such as water (in the form of hydrated minerals) and carbon (as both inorganic and organic compounds).

In the carbonaceous chondrites, there are refractory inclusions of Ca and Al that are believed to be the first materials to condense in the solar nebula. These calcium aluminum inclusions (CAIs) have been precisely dated to 4.5672 Ga$\pm$0.6 Ma (Amelin et al., 2002) and thus provide an accurate time for when the Solar System began forming. One class of carbonaceous chondrites (CI) has a composition nearly identical to the Sun's photosphere, and therefore represents the unfractionated primitive material from which Solar System objects formed. The CI carbonaceous chondrites are often used to assess the degree of processing that Solar System objects have experienced.

Terrestrial planet formation models show three distinct phases of evolution beginning with the formation of kilometer-sized planetesimals, to lunar-sized planetary embryos, and finally to full-scale planets themselves (for a review see Lunine et al., 2011). Mars appears to have formed primarily through the accretion of planetesimals. Its small size implies that it escaped late-stage bombardment by lunar-sized objects, unlike the situation for Earth and Venus. This could have occurred if it accreted in a low-density region of the solar nebula (Chambers, 2001), suffered no collisions with planetary embryos by chance (Lunine et al., 2003), or migrated away from the embryo impact zone early in its history (Minton and Levison, 2011). Each of these scenarios implies that Mars would have a smaller inventory of volatiles compared to Earth, unless it formed locally from volatile-rich planetesimals (Drake and Righter, 2002).

As Mars grew, the energy released during accretion and the decay of short-lived radioisotopes (most notably ^{26}Al) eventually became large enough to melt the planet, causing it to differentiate into a metallic core and silicate mantle. From chemical analyses of the SNC meteorites, which provide inferences about the Martian mantle (McSween, 1994), we now know that accretion and differentiation were complete within 20 Ma after the CAIs first condensed (Lee and Halliday, 1997; Kleine et al., 2002). The main evidence for this comes from the measured excess of ^{182}W in the SNCs. The tungsten, which has an affinity for iron, is produced from the radioactive decay of ^{182}Hf, which has an affinity for oxygen[8]. The excess of ^{182}W in the mantle implies that the core must have formed before most of it was produced. Otherwise, it would have followed iron into the core if differentiation occurred after it was produced. The half-life of ^{182}Hf is 9 Ma. Using precise estimates of the mantle Hf/W ratio, Dauphas and Pourmand (2011) find that Mars reached half its present size in less than several million years. There also is evidence that the crust formed early as well. Theoretical predictions of rapid cooling from a possible magma ocean (Elkins-Tanton et al., 2005) and the breccia-like nature of the 4.4 Ga meteorite NWA 7533/7034 suggest that the bulk of the Martian crust formed within 100 Ma of Martian history (Humayun et al., 2013), Thus the core, mantle, and crust were largely in place very early in the planet's history.

17.4.2 Primary Atmosphere

17.4.2.1 Magma Ocean and Steam Atmosphere

The rapid establishment of these major geochemical reservoirs has several implications for the early atmosphere and climate system. The first is that the short accretion time may have led to the formation of a magma ocean. While there is no observational requirement for a magma ocean, some melting of the planet must have occurred to permit differentiation. Indeed, some have argued that core formation is not possible without complete melting (e.g. Terasaki et al., 2005). The energy released by accreting planetesimals and the decay of short-lived radioactive isotopes is more than enough to melt the entire

planet (see Elkins-Tanton et al., 2005). However, if a magma ocean did form, its energy must be retained against radiative loss to space. The most plausible way to do this is through the insulating effect of an optically thick atmosphere.

During accretion, volatiles will be released from the impacting planetesimals when their kinetic energy reaches a critical value. Water is the most common volatile in planetesimals of chondritic composition and it absorbs in many parts of the infrared spectrum (see Section 17.5.1). Thus a vapor atmosphere begins to develop. Whether it remains in the atmosphere or condenses on the surface depends on the accretion time since this determines the main energy input to the surface, and the infrared opacity of the atmosphere, since this determines the efficiency of heat retention. The accretion model of Matsui and Abe (1987) includes these processes. They found that impact devolatilization began on Mars when its radius grew to about 0.4 of its final value, and that the thermal blanketing effect of a steam atmosphere[9] becomes sufficient to melt the surface and maintain a magma ocean if the accretion time is less than 5 Ma. At the time their paper was published, these times were deemed improbably short. However, as we have seen, these short accretion times are consistent with the Hf/W data.

17.4.2.2 Hydrodynamic Escape

Whether or not Mars had a steam atmosphere and a magma ocean, it is likely that its primitive atmosphere was hydrogen-rich. Impact devolatilization and outgassing driven by high early heat flows would have created an atmosphere with abundant water, which would have been photodissociated by the higher EUV output of the young Sun and/or used to oxidize reduced minerals in the surface to produce a hydrogen-rich atmosphere. Such an atmosphere would have been prone to hydrodynamic escape whose outflow could drag off other volatiles of relevance to the climate system such as CO_2 and N_2. Hydrodynamic escape has been invoked to explain the noble gas isotopic patterns on Mars and Earth (see Section 17.3.3), as well as the high D/H ratio on Venus (Kasting and Pollack, 1983). Thus, the intensity and duration of an early epoch of hydrodynamic Mars will have a direct influence on the inventory of climatically significant volatiles that remain when it ends.

There is evidence that hydrodynamic escape did occur on early Mars. The hydrogen blow-off will preferentially remove the lighter elements and thus fractionate isotopic patterns. As discussed above, xenon is strongly fractionated and this is consistent with an early epoch of hydrodynamic escape. Furthermore, the high ^{129}Xe/^{132}Xe ratio implies that primordial ^{132}Xe was stripped off the planet by an epoch of hydrodynamic escape that ended within several half-lives (15.7 Myr) of ^{129}I (Zahnle, 1993). A longer period is possible if radiogenic contribution to the abundance of ^{136}Xe by fission of ^{244}Pu, which has a half-life of 82 Ma, is factored in (Jakosky and Jones, 1997).

---

[8] Elements with an affinity for iron are referred to as siderophiles, those with an affinity for oxygen are lithophiles, while calcophiles are paired with sulfur.

[9] Matsui and Abe (1987) make a distinction between the greenhouse effect and a thermal blanketing effect of an infrared optically thick atmosphere. The former refers to the case where the main heat source comes from the top of the atmosphere (i.e. the Sun), while the latter has the main heat source at the bottom (i.e. accretional heating).

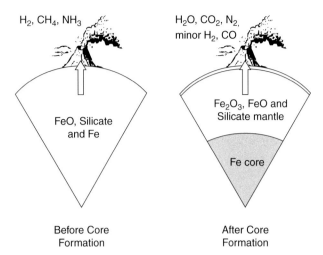

Figure 17.9. The oxidation state of outgassed materials transitions from reduced conditions (left) to mildly oxidizing conditions (right) after the core forms. From Catling and Claire (2005).

17.4.3 Secondary Atmosphere

There is no direct evidence for the mass and composition of the atmosphere after the original atmosphere escaped. However, it was likely derived from magmatic outgassing and its composition therefore would depend on the chemistry and oxidation state of the mantle. One line of thinking (Figure 17.9) is that, because the core formed early, iron quickly separated from the volatiles, leaving the mantle mildly oxidized (Beaty et al., 2005). In this case the key volatiles (H, C, and S) would have come out mainly as H_2O, CO_2, and SO_2, which are excellent greenhouse gases. However, the SNC meteorites suggest more reducing conditions (Wadhwa, 2008), where H_2, CO, CH_4, and H_2S would also be in the mix. These too are good greenhouse gases. We shall return to the question of the redox state of the mantle and its outgassing products in Section 17.5.2.

Historically speaking, CO_2 and H_2O were thought to be the main constituents of the early Mars atmosphere. Yet their evolution is difficult to assess, since we do not really know the degassing history of early Mars, and both are subject to removal from the atmosphere by escape out the top and chemical reactions at the surface. The best we can do is to provide an upper limit on how much was available, assess potential loss rates, and then speculate on various evolutionary scenarios.

17.4.3.1 Inventory of CO_2 and H_2O

Estimates for the original amount of CO_2 and water have been made on the basis of scaling arguments, geochemical data, and geological landforms. Venus, for example, has 90 bar of CO_2 in its atmosphere, while Earth has 60–90 bar of CO_2 locked up mostly in the rock reservoir as carbonates. Thus, Earth and Venus have similar amounts of CO_2. If we assume that Venus, Earth, and Mars accreted from materials with similar volatile contents, then scaling by mass and accounting for the difference in gravity would give Mars an initial inventory of ~15 bar of CO_2.

The C/^{84}Kr ratio also suggests a large initial inventory of CO_2. Krypton is not subject to the escape processes acting since the end of the late heavy bombardment and has therefore probably been retained in the atmosphere since then. The C/^{84}Kr ratio is ~4×10^7 for Venus and Earth, but is ~(4.4–6)×10^6 for Mars. This implies that Mars had 10 times the amount of CO_2 in its atmosphere at the end of the late heavy bombardment than it does today. Today, the mean annual surface pressure on Mars is about 6 mbar. If we assume this represents the bulk of the CO_2 presently available on the planet today, then at the end of the late heavy bombardment the minimum amount of CO_2 would have been 60 mbar. Furthermore, ^{84}Kr on Mars per unit mass is depleted by a factor of 100 with respect to Earth and Venus. If we assume that escape during and before the end of the late heavy bombardment accounts for the depletion, then it implies that Mars had an initial inventory of at least 6 bar of CO_2, which is 2.5 times less but roughly consistent with the scaling argument made in the paragraph above.

The initial inventory of water has been estimated from geochemical arguments. Dreibus and Wänke (1987) proposed a two-component model of accretion that began with 150 km of water but ended with only 130 m. Unlike the Earth, these two components, one volatile-poor and the other volatile-rich, accreted homogeneously, which allowed the water to react with iron, generating vast amounts of hydrogen that escaped to space, leaving the mantle dry. Of the 130 m left in the mantle, Dreibus and Wänke estimated that only 10 m was outgassed to the atmosphere. This amount is far less than needed to account for the Noachian fluvial features (Carr and Wänke, 1992), and smaller than the ~20 m in the polar caps (Carr and Head, 2014). As we have seen, early core formation would reduce the time that volatiles had to mix with iron, so that the Dreibus and Wänke (1987) assumption of complete mixing may be overstated.

The amount of water remaining after an early epoch of hydrodynamic escape and impact erosion has been estimated from the volume of water needed to carve various fluvial landforms on the surface, estimates of the volume of water present in possible ocean basins, and estimates of the volume of water outgassed from volcanoes. Carr (1986) estimated that 425–475 m was present from the volume of water needed to cut the large outflow channels and from geomorphic indicators of near-surface ground ice. As discussed in Section 17.3.1, the former presence of oceans on Mars has been suggested by a number of investigators and several shorelines have been tentatively identified (Lucchitta et al., 1986; Parker et al., 1989; Baker et al., 1991; Clifford and Parker, 2001). If these are true shorelines, then a near-surface inventory of ~1 km is implied. Finally, Baker et al. (1991) proposed an ocean model to explain Amazonian glacial features and post-Noachian valley networks. In this model, further elaborated upon in Baker et al. (2000) and Baker (2001), post-Noachian Mars was mostly cold and dry, but experienced episodes of warm and wet conditions brought about by ocean-forming events linked to magmatic activity in Tharsis that changed the climate system. They estimate that the volume of water involved in these events could have been as high as ~2.5 km.

The existence of former oceans on Mars is still being debated. Carr and Head (2003) argued against the shoreline interpretation, citing their large range in elevations, the fact that some sections have a clear volcanic rather than sedimentary origin, and that the volumes of water corresponding to the larger oceans are implausible since they are comparable to the size of the Earth's oceans. However, more recent developments favor the ocean hypothesis. The large-amplitude fluctuations in the elevations of the shorelines could be explained by true polar wander (Perron et al., 2007), and the enhanced potassium, thorium, and iron abundances within shoreline boundaries detected by the Gamma Ray Spectrometer (GRS) on Mars Odyssey are consistent with leaching and transport of these elements from the highlands to the lowlands (Dohm et al., 2009).

Whether or not oceans existed in the past, some ponding from the outflow flood events must have occurred. Thus to some extent the debate is more about the size of the ponded water, rather than the existence of an ocean *per se*. Certainly lakes and seas were likely in the past (see Section 17.3), and possibly even oceans. While most of the discussion has focused on post-Noachian times, it is relevant for our purposes in that it helps provide an estimate of how much water was available to the early Mars climate system. The volumes of water contained in the oceans proposed by the above investigators ranges from hundreds to thousands of meters global equivalent. As this represents only surface waters, it is a lower limit since it does not account for groundwater. Thus the inventory of water based on estimated ocean volumes is at least consistent with the geological estimates.

Thus the estimated initial inventory of CO_2 and water accreted by the planet is 6–15 bar of CO_2 and 150 km of water. How much of this initial inventory remained after the first 400 Ma is uncertain, but hundreds of meters to several kilometers of water must have been available during the Noachian to explain the fluvial landforms. This implies the loss of significant amounts of accreted water, which is consistent with the D/H data. For CO_2, much of its initial inventory must have been lost as well. The $^{40}Ar/^{36}Ar$ in ALH 84001 suggests a Noachian atmosphere of <400 mbar of CO_2 (Cassata et al., 2012), and secondary crater statistics put an upper limit of ~1–2 bar depending on surface properties (Kite et al., 2014). As discussed below, there are a variety of loss mechanisms that can significantly reduce the initial inventory of CO_2.

17.4.3.2 Outgassing History

The outgassing history of Mars has been estimated from the volume of volcanic lavas that have erupted onto the surface, their ages, and their presumed volatile compositions. Greeley and Schneid (1991) published estimates of the volume and ages of volcanic material since the onset of the geological record at about 3.8 Ga ago, and Craddock and Greeley (2009) used these data to estimate that ~0.8 bar of CO_2 has been outgassed to the atmosphere through the Amazonian. The synthesis of these data suggest that outgassing peaked during the early Hesperian ~3.3 Ga, and was minimal during the mid- to late Noachian. However, erosional modification of Noachian terrains makes it impossible to estimate the true volume of volcanic material for that epoch. Consequently, while this approach is useful for the

post-Noachian eras, it may greatly underestimate outgassing rates in the Noachian.

Phillips et al. (2001) estimated that 1.5 bar of CO_2 was outgassed during the Noachian. They based this estimate on the volume of magma that produced the Tharsis rise and assumed it had a composition similar to Hawaiian basalts. Follow-up work by Hirschmann and Withers (2008) and Grott et al. (2011) suggest much smaller outgassing totals. In particular, Grott et al. (2011) combined the solubility model of Hirschmann and Withers (2008) with a thermo-chemical evolution model to self-consistently calculate the amount of CO_2 dissolved in Martian magmas. Their calculations show that most of the outgassing occurs during the first few hundred million years and eventually ceased between 3.5–2 Ga depending on mantle oxidation state. They estimate that a total of ~1 bar of CO_2 was delivered to the atmosphere during this period, and that only ~250 mbar was outgassed between 4.1 and 3.7 Ga. These low outgassing volumes are contradicted by Wetzel et al. (2013), whose laboratory studies show that as much as 2.3 bar of carbon would have outgassed due to its greater solubility in the melt than previously assumed. Thus, there is a wide range of estimates for the outgassing volume.

However, none of these published estimates approach the 6–15 bar of CO_2 Mars is estimated to have begun with, nor can we account for this much CO_2 in the observable reservoirs. The reservoirs we can quantify include the atmosphere (~6 mbar, Haberle et al., 2008), the seasonal caps (~1 mbar, Kelly et al., 2006), the south polar residual cap (~0.1 mbar, Thomas et al., 2009), and the ice buried beneath it (5–6 mbar, Phillips et al., 2011). Together, these constitute less than 13 mbar of CO_2. The reservoirs whose volumes are difficult to quantify are the regolith and rock reservoirs. Zent and Quinn (1995) put an upper limit of ~40 mbar of CO_2 adsorbed in the regolith reservoir, though the presence of substantial volumes of near-surface ground ice at middle and high latitudes discovered later (e.g. Boynton et al., 2002) makes this seem generous. Carbonates have been detected in Martian dust and soil, the SNC meteorites, and regionally restricted outcrops, which could contain up to ~10 mbar of CO_2 (see Section 17.3.2, Ehlmann and Edwards, 2014). However, if the volume fractions of these detections (0.5–5%) are representative of the Martian crust, then 0.25–2.5 bar of CO_2 could be sequestered in a 1 km thick layer. Taken together, we can account for no more than several bars of CO_2, which implies that if the 6–15 bar estimate for the initial inventory is correct, then either outgassing was incomplete and the missing CO_2 is still in the magma, or outgassing was complete and substantial amounts of CO_2 have escaped to space.

17.4.3.3 Escape Mechanisms

Atmospheric escape is easier on Mars compared to Earth because of its smaller size. Also, the absence of an intrinsic magnetic field for most of Mars' history facilitates loss mechanisms not possible on the Earth. We have already mentioned that Mars may have experienced an early episode of hydrodynamic escape and that the bulk of its volatiles could have been carried off in the subsequent outflow. However, other escape processes will operate when hydrodynamic escape ends. For the pre-Noachian, the most important are thermal escape and impact erosion; non-thermal escape rates are slow by comparison, and sputtering would not have begun until the magnetic field shut down at about 4.1 Ga (Hutchins et al., 1997). However, after 4.1 Ga, sputtering can remove as much as 90% of the post-Noachian atmosphere (Jakosky and Jones, 1997), and non-thermal escape can account for the observed enrichment in the $^{15}N/^{14}N$ ratio (McElroy, 1972). A goal of the MAVEN mission now operating at Mars is to quantify present-day escape rates for these various mechanisms. (See Jakosky et al. (2015) for initial results, and Chapter 15 for details.)

Thermal escape (i.e. Jeans escape) occurs in the exosphere when molecules in the high-energy tail of the Maxwell distribution achieve velocities that exceed the escape velocities. Since molecular motions are proportional to temperature, the energy fluxes controlling those temperatures are important. For present-day Mars, thermal escape of molecules heavier than helium is negligible. However, on early Mars the higher EUV flux from the Sun could dramatically alter the temperature and altitude of the exobase. Using a one-dimensional thermosphere–ionosphere model that self-consistently calculates the ionization and excitation states of thermospheric constituents, Tian et al. (2009) found that the escape flux of carbon under the high EUV fluxes estimated by Ribas et al. (2005) ranged from 10^{12} molecule s^{-1} at 4.5 Ga to 10^{11} molecule s^{-1} at 4.1 Ga. Sustained fluxes of this magnitude would remove 1 bar of CO_2 in 1 Ma and 10 Ma, respectively. Though this model did not include H_2 and may therefore overestimate carbon escape (H_2 would take up some of the energy for escape), thermal escape of a thick CO_2 atmosphere during the period of high EUV fluxes on early Mars is likely to have been fast enough that even the most optimistic outgassing scenarios would have difficulty keeping pace with the loss to space. Under these circumstances, the only way to generate a thick CO_2 atmosphere before the end of the Noachian (3.7 Ga) is to outgas most of it after 4.1 Ga when the EUV fluxes declined enough for CO_2 to accumulate.

Impact erosion is another mechanism that makes it difficult to sustain a thick early atmosphere. Melosh and Vickery (1989) showed that large impactors could remove substantial atmospheric mass in the expanding vapor plume. If the impactor is large enough (~3 km for the current atmosphere) and fast enough (~14 km s^{-1}), it can remove the atmosphere above the tangent plane at the impact site. From the observed cratering record, they determined the flux of impactors capable of eroding the atmosphere and integrated it back in time to obtain estimates of the total amount of CO_2 that could be removed by this process. In this manner they estimated that the surface pressure on Mars 4.5 Ga was 60–100 times greater than it is today. Thus, ~350–600 mbar of CO_2 could have been eroded since the beginning of Martian history.

Brain and Jakosky (1998) elaborated on this model by focusing on the observed crater densities of the oldest terrains rather than the difficult-to-determine cratering rate as a function of time. They pointed out that it is the amount of erosion since the onset of the geological record during the late Noachian that is of interest rather than for all of the planet's history, since it was at that time that the observed fluvial features formed. Based on the observed range of crater densities, they estimate between about 50% and 90% of the late-Noachian atmosphere has been removed by impacts, which implies an atmosphere with surface pressures of ~12–60 mbar. However, when they factored in sputtering losses, which they took to be 90%, the combination could remove 95–99% of the late-Noachian atmosphere, leaving therefore 1–5% in the present atmosphere. The 120–600 mbar surface pressures implied by this calculation represent a lower limit, since CO_2 could have migrated into non-atmospheric reservoirs (see above). These higher early surface pressures are supported by the modeling work of Hu et al. (2015), who match the MSL $\delta^{13}C$ values through fractionation by photodissociation and sputtering modulated by carbonate formation.

17.4.3.4 Surface Sinks

As mentioned above, a potentially important surface sink for atmospheric CO_2 on Mars is conversion to carbonate rocks. On Earth, the long-term CO_2 content of the atmosphere is governed by the carbonate–silicate cycle schematically illustrated in Figure 17.10. CO_2 dissolves in rainwater producing a mildly acidic carbonic acid solution that leaches cations from silicate rocks on land surfaces and transports them into the ocean where they are eventually precipitated as carbonates onto the ocean floor by shell-forming organisms in the sea. Some of these carbonates are buried in sediments, subducted to great depths as the sea floor spreads, and eventually thermally decompose and release CO_2 gas back into the atmosphere through volcanoes. If oceans existed on early Mars that were maintained by an active hydrological cycle, carbonate precipitation should have occurred[10], and recycling through plate tectonics would not have been effective. Instead, hot spot volcanism (Pollack et al., 1987) or impact gardening (Carr, 1989) would have been the main recycling mechanisms. However, both would have declined at the end of the Noachian, at which point conversion of atmospheric CO_2 to carbonate would have been irreversible.

17.4.3.5 Summary

Mars formed quickly with accretion and core formation largely complete within the first several million years. Noble gas data rule out a primary atmosphere captured from the solar nebula. Instead, a steam atmosphere and possibly a magma ocean likely dominated the early environment. Much of this water was probably lost during a brief episode of hydrodynamic escape powered by enhanced EUV fluxes from the young Sun. A secondary atmosphere subsequently developed from volcanic outgassing. The mass and composition of this atmosphere are uncertain. Rapid core formation favors an atmosphere dominated by CO_2 and H_2O, though the meteorite data suggest more reduced gases (H_2, CO, CH_4). Estimates of their initial accreted volatile abundances range from 6 to 15 bar of CO_2 and up to 150 km for water. Although all indications are that most of the initial water escaped to space, enough (hundreds to a thousand meters) must have been present during the Noachian to form the observed fluvial features. And while there are a variety of loss mechanisms that can also limit the buildup of thick CO_2 atmospheres prior to 4.1 Ga, it is entirely plausible that, because outgassing rates

---

[10] Living organisms are not required to precipitate carbonates.

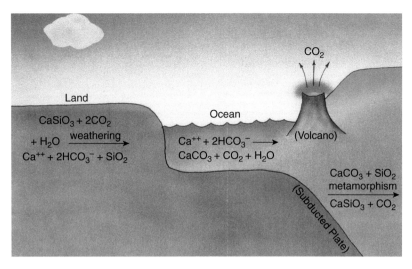

Figure 17.10. Earth's carbonate–silicate cycle. From Kasting and Catling (2003). See text for details. With permission from J.F. Kasting.

were higher, the Noachian atmosphere was thicker than it is at the present time. How much thicker is difficult to determine. Outgassing models put the number between 250 and 1500 mbar; a theoretical interpretation of the $^{40}Ar/^{36}Ar$ ratio in ALH 84001 suggests up to 400 mbar; secondary crater statistics can accommodate 1–2 bar depending on surface properties; and a reducing mantle could contribute up to 2.3 bar. Thus, published estimates of the size of a CO_2 Noachian atmosphere range from 0.25 to 2.3 bar.

17.5 THE CLIMATE OF EARLY MARS

The processes affecting the early Mars' climate are summarized in Figure 17.11. Some of these have already been discussed. In this section we discuss the potential mechanisms for warming early Mars. The principal challenge for greenhouse models is coping with the faint young Sun. Standard solar evolution models suggest that, when the Solar System formed at 4.56 Ga, the Sun was less luminous than it is today (Newman and Rood, 1977; Gough, 1981). The luminosity of Sun-like stars increases with time. The fusion of hydrogen into helium increases the mean molecular weight of the core. To maintain the balance between the pressure gradient force and gravity, the core contracts and warms. The increased densities and temperatures increase the rate of fusion and hence the star's luminosity increases with time. A good fit to this increase with time for standard solar models is given by (Gough, 1981)

$$L(t) = \frac{L(t_0)}{1 + \frac{2}{5}(1 - t/t_0)} \tag{17.2}$$

where L is luminosity, t is time, and t_0 is the age of the Solar System (4.56 Ga).

Figure 17.12 shows the solar luminosity and the planet's effective temperature, T_e, as functions of time. The latter is defined as the temperature at which a black body radiates away the total energy it absorbs. If A_p is the planetary albedo and S_o is the annual mean solar flux at Mars' orbit, then T_e is given by

$$T_e = \left[\frac{(1 - A_p)S_o}{4\sigma} \right]^{1/4} \tag{17.3}$$

where σ is the Stefan–Boltzmann constant (5.67×10^{-8} W m^{-2} K^{-4}). In Figure 17.12 we assume that $A_p = 0.25$ and is constant in time, and that S_o for Mars today is 590 W m^{-2}. Thus, today the planet's effective temperature is 210 K, but during the Noachian epoch (3.7–4.1 Ga), its effective temperature under these assumptions would have been 195 K. To raise global mean annual surface temperatures to the melting point of water thus requires the atmosphere to produce 78 K of greenhouse warming. Considering that the Earth's atmosphere provides only 33 K of greenhouse warming (Table 17.3), this presents a challenge. However, it is not impossible, as Venus' atmosphere produces over 500 K of warming (though with 90 bar of CO_2)!

Attempts to resolve the faint young Sun problem for Mars depend on what the required surface temperatures actually are. Unfortunately, the geological and mineralogical data are ambiguous on this point. Traditionally, a global and annual averaged surface temperature, T_s, of at least 273 K has been the assumed requirement. While this makes some sense, there is little discussion in the literature about the rationale. If, instead, the requirement is that mean annual temperatures at specific locations be above freezing, then this constraint could be relaxed. It could be further relaxed if these surface temperatures need only be above freezing seasonally or daily, and further yet if brines are permitted. Thus, it is not clear what requirements we are demanding from an early Mars greenhouse effect. What is clear, however, is that surface temperatures must be above freezing when liquid water is present.

A conceptual argument can be made that T_s must be close to or above 273 K if the valley networks, eroded terrains, and hydrated mineral deposits require a long-lived continuously warm and wet Noachian climate with an active hydrological cycle involving rainfall and runoff. If this is the case, then large bodies of open water comparable to seas or oceans must exist. Scattered lakes or ponds could not sustain such a climate system (Soto et al., 2010). These seas/oceans would exist in the northern lowlands, as has been suggested (see Section 17.3.1), which means that mean annual temperatures in most

Table 17.3. *Approximate surface and effective temperatures of the terrestrial planets.*

Planet	$S_o/4$ (W m^{-2})	A_p	T_{se} (K)	T_e (K)	$T_{se} - T_e$ (K)
Venus	657	0.75	740	232	508
Earth	342	0.30	288	255	33
Mars	148	0.25	215	210	5

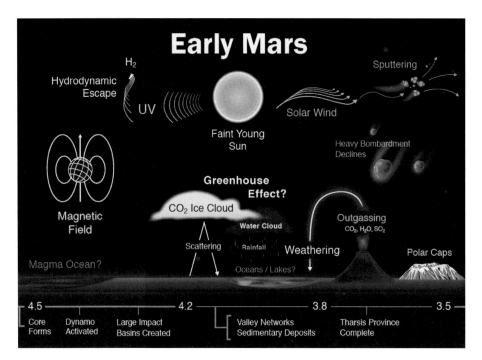

Figure 17.11. Cartoon summary of the processes acting on early Mars. Graphic courtesy of Christina Olivas. A black and white version of this figure will appear in some formats. For the color version, please refer to the plate section.

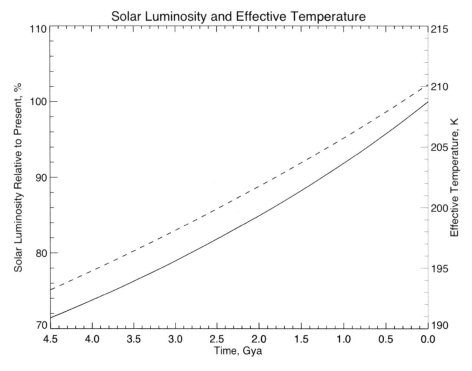

Figure 17.12. Solar luminosity (solid line, left axis) and effective temperature (dashed line, right axis) as functions of time.

of the northern hemisphere would have to be above freezing, otherwise a frozen surface would raise the albedo, lower surface temperatures, reduce evaporation rates, and slow the hydrological cycle. Similarly, mean annual temperatures in the southern hemisphere, where most of the valley networks are found, would also have to be above freezing to permit rainfall and runoff. Thus, if long-lived continuously warm and wet conditions are required, then the annual and globally averaged surface temperature T_s must be close to or above freezing. Somewhat lower temperatures are possible if the oceans become salty and/or only partially freeze over. How much lower is difficult to quantify. But if rainfall is truly required, mean annual global average surface temperatures cannot be too much below 273 K.

Much of the present debate about early Mars hinges on this issue. If long-lived warm wet conditions are required throughout the Noachian, then large bodies of open water are required, and the atmosphere must produce and sustain a greenhouse capable of raising global mean annual surface temperatures close to the melting point of water. Alternatively, the Noachian climate could have been mostly cold and dry with $T_s < 273$ K, but punctuated with transient periods of globally warm and wet conditions brought about by external forcings such as impacts or volcanism. Finally, it is possible that global mean annual surface temperatures were always less than 273 K and that warm wet conditions were only needed at certain places and certain times. Thus, after consideration of climate processes in Section 17.5.1, we divide this part of our review into three categories: long-lived greenhouse atmospheres (Section 17.5.2), transient greenhouse atmospheres (Section 17.5.3), and cold climates with locally wet conditions (Section 17.5.4). A fourth possibility, which we also consider, is that the luminosity of the early Sun did not follow the standard solar model and the Sun was actually bright enough to warm early Mars with only modest changes to the mass and composition of the atmosphere. This idea is more speculative, but since it has been explored in the literature we include it here for completeness (Section 17.5.5).

17.5.1 Climate Processes

17.5.1.1 Radiative Transfer

Because of its high temperature, the Sun radiates most of its energy at visible wavelengths, while planets, being much cooler, emit mainly in the thermal infrared. This provides a clean separation for assessing planetary energy budgets. Planets gain energy by absorbing the sunlight intercepted by their cross-sectional area, and lose energy by emitting in the infrared over their entire surface. In the absence of an atmosphere, T_e as defined by (17.3) is equivalent to the global and annual mean effective surface temperature[11] T_{se} and the total outgoing long-wave radiation (OLR) is σT_{se}^4 (Haberle, 2013). However, if we introduce to such a planet an atmosphere that is transparent at solar wavelengths but

Table 17.4. *Fundamental vibrational frequencies of common greenhouse gases (cm⁻¹).*

Gas	ν_1	ν_2	ν_3	ν_4
CO_2	1388[a]	667	2349	
H_2O	3657	1595	3756	
SO_2	1152	518	1362	
H_2S	2615	1183	2626	
CH_4	2917[a]	1534[a]	3019	1306
N_2O	2224	589	1285	
NH_3	3337	950	3444	1627

[a] These transitions do not change the dipole moment and are therefore infrared-inactive.

opaque in the infrared, the OLR will be reduced by an amount depending on the concentration, distribution, and absorption properties of those gases. Consequently, it will be absorbing more energy than it is emitting and the temperature of its atmosphere and surface will increase until a new balance is established. Though there are many subtle features that must be considered, this is the essence of the greenhouse effect. It explains why the surface temperature can be higher than the effective temperature, and it gives us a convenient measure of the greenhouse power of a given atmosphere (Haberle, 2013). Table 17.3 lists the difference between the effective temperature and effective surface temperature for Mars, the Earth, and Venus at the present time.

The main challenge is finding plausible greenhouse gases that can plug up the gaps in the OLR as a function of wavelength. The most common greenhouse gases that have been considered for early Mars include carbon dioxide (CO_2), water (H_2O), sulfur dioxide (SO_2), hydrogen sulfide (H_2S), methane (CH_4), nitrous oxide (N_2O), and ammonia (NH_3). In the infrared, these molecules absorb photons that mainly change their vibrational–rotational states. These changes occur at discrete frequencies that are listed in Table 17.4. Figure 17.13 shows their infrared absorption cross-sections as a function of wavenumber. The top panel is the Planck function for a surface temperature of 273 K. If an early Mars greenhouse produced global mean surface temperatures of 273 K, this plot shows the spectral regions in the infrared where greenhouse gases need to be absorbing. Ideally, the absorption should be strongest in the 20–1200 cm⁻¹ region where about 85% of the energy is radiated to the atmosphere by the ground.

The second panel shows the absorption cross-section for CO_2 and H_2O, gases that could have been present in the early Mars atmosphere in climatically significant amounts. CO_2 is a linear molecule and its bending modes at 667 cm⁻¹ are the most prominent absorption feature. Water vapor is a polar molecule and has a much richer set of absorption features throughout the infrared. Its vibrational–rotational modes allow it to contribute significant absorption near 200 cm⁻¹ and 1500 cm⁻¹. The combination of these two gases can provide considerable opacity throughout the infrared, except in the so-called "window" region between 800 and 1200 cm⁻¹. Since roughly 25% of the energy radiated by a 273 K surface is in the window, gases that can provide additional opacity in this

[11] T_{se} is the temperature at which a black body emits the equivalent amount of radiation emitted by the entire surface over a year.

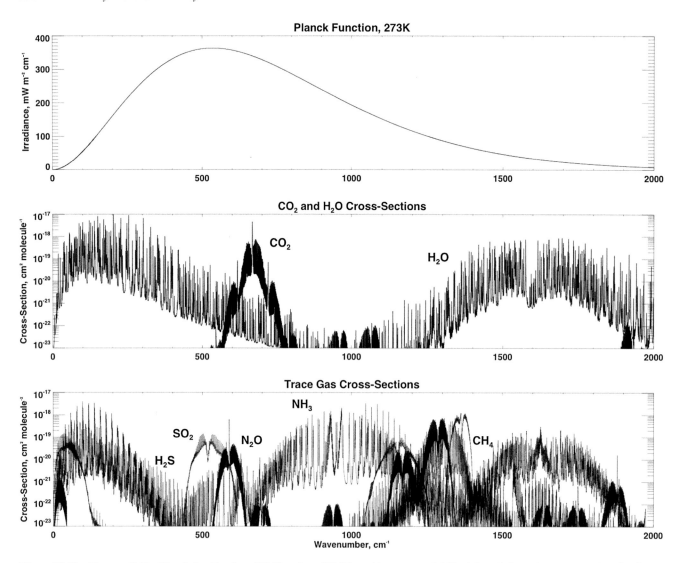

Figure 17.13. (Top panel) The Planck function for a 273 K surface. (Middle and bottom panels) The infrared absorption cross-sections for the gases listed in Table 17.4. A black and white version of this figure will appear in some formats. For the color version, please refer to the plate section.

region could significantly boost the greenhouse effect of a CO_2–H_2O atmosphere.

The bottom panel of Figure 17.13 shows absorption cross-sections for other potential greenhouse gases. By far, NH_3 has the greatest potential to plug up the window region, as it exhibits significant absorption from 700 to 1300 cm^{-1}. Other gases that can contribute opacity in the window are N_2O, SO_2, and H_2S. Outside the window, water vapor dominates the absorption spectra, though SO_2, N_2O, and CH_4 can contribute if present in sufficient quantities. Based on the Planck-weighted mean cross-sections listed in Table 17.5, on a per molecule basis water vapor is the most effective greenhouse

gas and H_2S is the least effective; NH_3, SO_2, and CO_2 are major contributors, while N_2O and CH_4 have only minor effects. Of course, the actual greenhouse potential of these greenhouse gases will depend on the thermal structure of the atmosphere, their abundance and vertical distribution, and their lifetimes (i.e. sources and sinks). More sophisticated calculations are therefore needed to determine their real greenhouse potential[12].

It is important to note that the ability of these gases to absorb at the listed frequencies is enhanced by a number of processes that broaden their lines. The most familiar are "pressure" and "Doppler" broadening. The former is due to collisions between molecules, while the latter is the result of molecular motion. Thus, the enhanced absorptions produced by these mechanisms depend on pressure and temperature. However, at high enough pressures, collisions can also *induce* absorption. This collision-induced absorption (CIA)

---

[12] For example, the Intergovernmental Panel on Climate Change (IPCC) has defined the greenhouse warming potential of a given gas as the ratio of its time-integrated forcing relative to a reference gas. Such a calculation is beyond the scope of this chapter.

Table 17.5. *Planck-weighted (273 K) cross-sections.*

Gas	Mean cross-section (10^{-20} cm^2 molecule^{-1})
H_2O	2.48
NH_3	2.06
SO_2	1.35
CO_2	0.99
N_2O	0.43
CH_4	0.14
H_2S	0.13

is complicated and not well understood, but generally results from temporary dipoles, forbidden transitions, or short-lived dimers. It should be emphasized that CIA is distinctly different from pressure broadening in that new lines appear, which absorb at different frequencies. For pure CO_2 atmospheres, CIA becomes important for surface pressures exceeding several hundred millibars with absorptions occurring in the 0–250 cm^{-1} and 1200–1500 cm^{-1} regions (Wordsworth et al., 2010). For H_2-rich atmospheres, CIA may also be important. Sagan (1977) first proposed the idea as a solution to the faint young Sun problem for both Earth and Mars, and it has been recently revived by Wordsworth and Pierrehumbert (2013) for the Earth, and by Ramirez et al. (2014) for Mars (see below).

In addition to the discrete line absorptions described above, gases often exhibit "continuum" absorption. Water vapor has this property. Its continuum absorption has been widely studied because of its relevance to the Earth's climate system. Laboratory measurements clearly show that there is more absorption in the window region (800–1200 cm^{-1}) than can be accounted for from nearby lines. The cause for this excess absorption is uncertain. Super-Lorentzian behavior (meaning that the line shapes differ from the shape commonly assumed and produce more absorption) in the far wings of the lines is the prevailing view, but the theoretical and observational basis for this is weak. The approach to including water vapor continuum absorption has therefore been empirical and is based on laboratory measurements and direct observations.

An extensive body of literature exists that describes the theoretical and experimental basis for these mechanisms and how they can be applied to radiative transfer calculations. Textbook discussions can be found in Petty (2006) and Pierrehumbert (2010), while more Mars-specific applications are presented in Halevy et al. (2009) and Mischna et al. (2012). The reader is referred to these works for further details.

17.5.1.2 Convection

In addition to greenhouse warming, convection in the lower atmosphere plays an important role in regulating surface temperatures. Convection sets the lapse rate, which is the rate of decrease of temperature with altitude, and therefore controls the thermal structure of the troposphere. Radiative processes

and surface heat exchange tend to destabilize the lower atmosphere and produce lapse rates that exceed the adiabatic lapse rate, the maximum lapse rate in a stable hydrostatic atmosphere. Convection acts to restore the adiabatic lapse rate by transporting heat from the lower atmosphere to the upper atmosphere through turbulent motions. Thus, convection cools the near-surface atmosphere and warms the upper atmosphere, with the amount of cooling or warming depending on the moisture content of the atmosphere. For the same initially unstable temperature profile, convection in dry atmospheres produces less surface cooling than convection in moist atmospheres because the latent heat released in rising air parcels containing enough water vapor to condense provides an additional heat source. Consequently, the moist adiabatic lapse rate is not as steep as the dry adiabatic lapse rate, i.e. temperatures fall off more slowly with height. This is also the case for dry atmospheres, where the main constituent can condense. In this instance, the lapse rate follows the frost-point temperature. This is the situation for the Martian atmosphere, which today is 95% CO_2, and for early Mars may have been an even a higher percentage.

17.5.1.3 Horizontal Heat Transport

Horizontal heat transport also plays a role in regulating surface temperatures, though in this case it is the spatial distribution of temperatures that is mostly affected rather than global mean temperatures. However, horizontal heat transport can affect global mean temperatures through feedbacks such as the ice–albedo feedback. For planets like Mars, where more energy is absorbed at low latitudes compared to high latitudes, horizontal wind systems will transport heat toward the poles, thereby lowering tropical temperatures and raising polar temperatures. The nature of the wind systems so produced and the strength of the transport they generate depends on a variety of factors, including the solar forcing, mass and composition of the atmosphere, and orbit properties (obliquity and rotation rate in particular).

For early Mars the most important factors for determining the degree to which transport can warm polar temperatures are the mass and composition of its atmosphere, since these will determine the partitioning of condensables between the atmosphere and polar caps. Mars' atmosphere today, for example, is too thin to transport enough heat into the polar regions to prevent CO_2 from condensing onto the surface (Leovy and Mintz (1969) were the first to show this). This weak transport is part of the reason that CO_2 ice at the south pole survives all year long. The existence of a permanent cap on Mars is critical since its heat balance will determine the mean annual surface pressure (see Chapter 12 for a review). However, as the mass of the atmosphere increases, more and more heat will be transported into the polar regions and the greenhouse effect will be strengthened. At some point, permanent polar caps will not be possible and all the CO_2 in the system will exist in the atmosphere.

This is a critical issue for early Mars since it means there is a minimum surface pressure, below which greenhouse warming and heat transport are insufficient to prevent the atmosphere from collapsing into a permanent polar cap. The value of that minimum surface pressure depends on the solar luminosity, the albedo of the cap, and the planet's orbit properties. Haberle et al. (1994), Kreslavsky and Head (2005), Manning et al. (2006), and Soto et al. (2011) have studied this issue. It is difficult to state with any certainty, since it depends on what other absorbers might be present, but based on these studies surface pressures on the order of at least several hundred millibars and possibly much higher are needed to stabilize the atmosphere of early Mars with present-day orbital conditions.

17.5.1.4 Climate Models

Climate models do not use the line-by-line approach to radiative transfer because of the complexity of the line distribution, which would require an overwhelming amount of computational time to find solutions from it. Consequently, alternative, more efficient, methods are used to calculate spectrally integrated radiative fluxes in the bands of interest, such as the 15 μm CO_2 band. There are a variety of approaches to radiative transfer in climate models, ranging from simple gray models, to band models, to exponential sums and correlated-k techniques that offer fast and reasonably accurate methods to compute temperature profiles.

Climate models also have a range of spatial resolution. One-dimensional globally averaged models were the main tool for the initial studies during the last several decades of the 20th century (e.g. Pollack, 1979). These models sought equilibrium solutions to a given radiative forcing and included convective adjustment in the lower atmosphere. Recent versions of these one-dimensional models use many spectral bands and include gaseous and aerosol scattering. With the increase in computing technology, most notably the speed and memory of modern machines, full three-dimensional general circulation models with quite sophisticated physics packages are now being utilized to study this problem (e.g. Wordsworth et al., 2013).

17.5.2 Long-Lived Greenhouse Atmospheres

17.5.2.1 CO_2–H_2O Atmospheres

Carbon dioxide and water vapor are the most extensively studied greenhouse gases for early Mars (see Haberle (1998) for a review of the earlier work). These gases are plentiful, relatively stable, and form the basis of greenhouse models of Earth, runaway Venus, and early Mars. As shown in Figure 17.13 they can provide significant infrared opacity if present in large enough quantities. The first modeling studies of these atmospheres indicated that between 5 and 10 bar of CO_2 was needed to raise temperatures to the melting point of water in the presence of the faint young Sun (Pollack, 1979; Pollack et al., 1987). The main problem with these atmospheres is that at some point the atmosphere saturates and CO_2 condenses. Kasting (1991) pointed out this problem and illustrated the consequences with a one-dimensional radiative–convective model. As surface pressures increase, CO_2 begins to condense in the atmosphere. The associated release of latent heat pins temperatures to the frost point and the lapse rate begins to follow a moist CO_2 adiabat instead of a dry adiabat. As a consequence, the upper atmosphere warms and the surface must cool to maintain energy balance. As surface pressures continue increasing, more and more of the atmosphere saturates and eventually the entire atmosphere follows the moist CO_2 adiabat. At this point,

further increases in CO_2 simply result in precipitation to the surface. For a solar luminosity appropriate to conditions 3.8 Ga (i.e. 75% of today's value), this ultimate limit was reached in Kasting's model at ~2 bar of CO_2 and the corresponding surface temperature was ~220 K.

Another setback for CO_2–H_2O atmospheres is the finding that CIA appears to be weaker than previously calculated (Figure 17.14). The standard parameterization for CIA comes from the work of Kasting et al. (1984) and this has been used in a number of studies of early Mars' greenhouse potential (Pollack et al., 1987; Kasting, 1991; Forget and Pierrehumbert, 1997; Mischna et al., 2000). However, this parameterization was found to overestimate CIA in pure CO_2 atmospheres (Halevy et al., 2009; Wordsworth et al., 2010). As shown in Figure 17.14, absorption due to the induced-dipole transitions in the 250–500 cm^{-1} range was greatly overestimated. Models using this parameterization have therefore overestimated the greenhouse effect. Wordsworth et al. (2010) showed that global mean surface temperatures may have been overestimated by as much as 20–30 K at high surface pressures with the original parameterization.

The Kasting (1991) model has since been updated (see Kopparapu et al., 2013). The most recent update includes new absorption coefficients for CO_2 and H_2O as well as the CIA parameterization suggested by Wordsworth et al. (2010). The model now predicts a maximum surface temperature of 230 K at 3.7 Ga for a pure CO_2–H_2O atmosphere (Ramirez et al., 2014). Figure 17.15 shows the latest results for four different values of the solar luminosity. For solar luminosities less than 80% of the present-day value, which corresponds in time to ~2.85 Ga, Ramirez et al.'s new model cannot raise surface temperatures to the melting point of water. The reduced temperatures at high pressures predicted by Kasting's latest model are due to the new CIA parameterization, which provides substantially less absorption than the old parameterization. This makes it even more difficult to warm early Mars with a pure CO_2–H_2O atmosphere.

However, the potential for CO_2 condensation complicates the picture. CO_2 clouds not only modify air temperatures through latent heat release, but also interact with the solar and infrared radiation fields in ways that depend on their optical properties. Kasting (1991) speculated that, because CO_2 ice clouds are much more transparent in the infrared than water ice clouds, they would have an additional net cooling effect since the clouds should be highly reflective in the visible. Thus, the clouds would not only reduce the greenhouse effect by decreasing the lapse rates, but also increase the planetary albedo and provide little infrared opacity to block upwelling surface emission. Implicit in this line of reasoning is the assumption that the cloud particles would be too small to interact with infrared radiation. However, one unique aspect of CO_2 ice clouds on Mars is that during condensation the flow of gas towards growing ice crystals will be hydrodynamic rather than diffusive (Rossow, 1978). This is likely to result in the rapid growth of particles large enough (10–100 μm) to efficiently scatter infrared radiation before they fall to the surface (Pierrehumbert and Erlick, 1998).

To assess this possibility, Forget and Pierrehumbert (1997) modified Kasting's (1991) model to include reflective CO_2 ice clouds and found that optically thick clouds containing 10

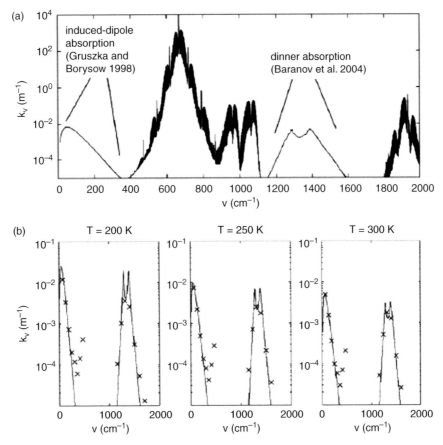

Figure 17.14. (a) Extinction for a 1 bar CO_2 atmosphere at 273 K as a function of wavenumber. (b) Comparison of the updated CIA extinction (solid lines) with the older parameterization of Kasting et al. (1984) (crosses) for several different temperatures. Note the much stronger absorption in the old parameterization in the 250–500 cm^{-1} range. From Wordsworth et al. (2010).

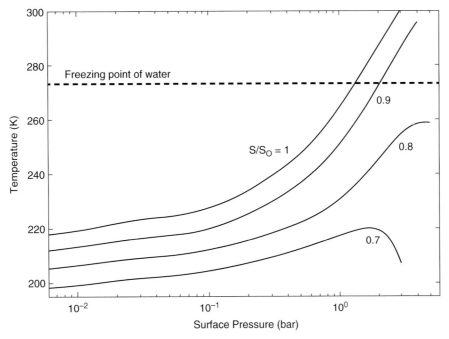

Figure 17.15. Global mean surface temperature as a function of surface pressure. The different lines correspond to different values of the solar luminosity S/S_0, where S_0 is the present value. From Ramirez et al. (2014).

μm particles in a 2 bar atmosphere could compensate for the increased planetary albedo and provide a net surface warming. Rather than absorbing upwelling radiation and emitting it to space at low temperatures, which is the conventional greenhouse effect, the clouds reflected the radiation back to the surface, creating a scattering greenhouse effect. So powerful was this effect in their model that surface temperatures of 273 K could be sustained in atmospheres with surface pressures as low as ~0.5 bar of CO_2.

The viability of the scattering greenhouse depends on the concentration and particle size of the ice particles. Poorly understood CO_2 microphysical processes control these. While models have begun simulating the behavior of CO_2 ice clouds (e.g. Forget et al., 1998, 2013; Mischna et al., 2000; Colaprete and Toon, 2003; Colaprete et al., 2003; Wordsworth et al., 2013), very few laboratory studies have been conducted to measure their nucleation and growth properties (Glandorf et al., 2002). Thus, there is uncertainty in the net radiative effect of CO_2 ice clouds. Furthermore, if the opacity and sizes do produce surface warming, then the atmosphere will also warm and this may limit cloud concentrations and surface warming. In their fully coupled self-consistent simulations, Colaprete and Toon (2003) obtained a modest 5–10 K surface warming from the scattering greenhouse effect because of this effect.

Spatial coverage is another issue for CO_2 ice clouds. Forget and Pierrehumbert (1997) assumed 100% cloud coverage, reasoning that clear skies might be more difficult to achieve when the major constituent of the atmosphere is condensing. However, they recognized the importance of fractional cloud cover, as simulations with 75% cloud coverage were 20–30 K colder than with complete coverage. This issue is best studied with full three-dimensional global circulation models, as dynamical processes will have a strong influence on cloud coverage. To date, only a few such studies have appeared in the peer-reviewed literature (Forget et al., 2013; Wordsworth et al., 2013). While it does appear that CO_2 ice clouds cover a major part of the planet in these models, their optical depths are not high enough to raise surface temperatures by more than 15 K even under the most favorable circumstances (Forget et al., 2013). Thus, CO_2 ice clouds do not provide as much warming as originally thought.

Thus, for a variety of reasons, CO_2–H_2O atmospheres are not as promising as they once were. To date, no model is able to generate a steady-state climate system with a pure CO_2–H_2O atmosphere capable of producing global mean annual surface temperatures at or above the melting point. However, radiatively active water ice clouds in such atmospheres could, in principle, enhance the greenhouse effect. Urata and Toon (2013) show this using an Earth global climate model modified for Mars conditions. Under the right conditions – large cloud particles, reduced precipitation, and complete cloud coverage in grid cells – mean surface temperatures near the melting point can be achieved. Segura et al. (2008) first recognized the potential for a cloud greenhouse effect on Mars in their studies of post-impact environments on early Mars. Also it can significantly raise surface temperatures at times of high obliquity (Haberle et al., 2012; Madeleine et al., 2013). However, the LMD (Laboratoire de Météorologie Dynamique) model of Wordsworth et al. (2013), which also includes radiatively active water ice clouds, does not find a significant role for a water ice cloud greenhouse. Thus, while the potential for a significant cloud greenhouse on early Mars is intriguing, it is model-dependent and therefore needs further study.

17.5.2.2 Supplemental Greenhouse Gases

The difficulty of raising surface temperatures to the melting point of water with pure CO_2–H_2O atmospheres has focused attention on finding additional trace gases to boost the greenhouse effect. Sulfur dioxide (SO_2), methane (CH_4), ammonia (NH_3), and nitrous oxide (N_2O) have all been mentioned in the literature, but only for SO_2 has there been significant published work on this topic.

Sulfur dioxide absorbs in parts of the spectrum not covered by CO_2 and H_2O (Figure 17.13). Its strongest absorption feature is due to the v_2 vibrational transitions near 7 μm. Postawko and Kuhn (1986) considered high levels of SO_2 (up to 1000 ppmv) in combination with a thick CO_2–H_2O atmosphere but were unable find solutions with global mean surface temperatures at 273 K. This negative result, and the fact the SO_2 is photochemically unstable, soluble in water, and readily oxidized to sulfate, did not motivate much further research at the time. One paper suggested that small amounts of SO_2 (0.1 ppmv) might stave off the CO_2 condensation problem (Yung et al., 1997). However, with the discovery of Noachian/Hesperian-aged sulfate deposits by the MER Rovers, and later mapped by OMEGA and CRISM, interest in SO_2 has been renewed, particularly since SO_2 could have acidified surface waters and inhibited carbonate formation (see Section 17.3.2).

While there are ways to form sulfates that do not involve the atmosphere, volcanic emission of reduced sulfur species (mainly SO_2 and H_2S) must certainly occur, and since Mars was much more volcanically active during the late Noachian and early Hesperian, an atmospheric delivery of sulfur to the surface is inevitable (e.g. Settle, 1979). The early Martian mantle is thought to be rich in sulfur (Wänke and Dreibus, 1994) such that outgassing during intense periods of volcanism could have emplaced enough SO_2 in the atmosphere to raise surface temperatures at least to levels where brines could exist (~250 K), or even temporarily well above the freezing point. The three-dimensional greenhouse model of Johnson et al. (2008) suggests this possibility. Their 500 mbar CO_2–H_2O atmosphere simulation, for example, gives global mean surface temperatures of 283 K and 315 K for SO_2 mixing ratios of 6.14 and 245 ppmv, respectively. Unfortunately, the warming estimated by this model appears to be overestimated (Mischna et al., 2013). And there are other reasons, discussed below, that warming global mean temperatures to the melting point is not likely to be correct for early Mars.

An SO_2 greenhouse strong enough to sustain liquid water on the surface in a cloud- and aerosol-free CO_2-dominated atmosphere requires high surface pressures and elevated SO_2 levels. The values required are model-dependent, but are in the ranges 0.5–4 bar of CO_2 and 1–100 ppmv of SO_2 (Halevy et al., 2007; Johnson et al., 2008; Tian et al., 2010). Whether SO_2 concentrations at this level are sustainable depends on the details of an early Martian sulfur cycle. Volcanic outgassing rates must be high enough to balance atmospheric losses. The main sinks are rainout, gas-phase reactions, and photolysis. Halevy et al. (2007) proposed a sulfur cycle that operated analogously to the CO_2 cycle on Earth. They focused principally on the late Noachian when Tharsis volcanism could have sustained sulfur outgassing rates twice those on present-day Earth. In their model, the generally reducing nature of the outgassed sulfur (as SO_2 and H_2S) would have exhausted the supply of oxidants and therefore limited the removal of SO_2 by gas-phase oxidation reactions. Oxidants on Mars are created when hydrogen escapes to space, and they reasoned that this was a relatively slow process compared to the faster supply of reductants from volcanoes. They further argued that Rayleigh scattering limited loss by photolysis in a thick CO_2 atmosphere. Rainout would not be significant until temperatures approached the melting point. Thus, with the main sinks of atmospheric SO_2 suppressed, its concentration could build until the greenhouse warming produced liquid water, thereby accelerating its removal and limiting further increases. This feedback on atmospheric SO_2 is similar to how the Earth's carbonate–silicate cycle regulates the long-term concentration of CO_2 in the atmosphere.

Some elements of this scenario are supported by the one-dimensional photochemical studies of Johnson et al. (2009) and Tian et al. (2010). Photolysis rates at low levels do significantly decline in thick CO_2 atmospheres (because of Rayleigh scattering) and this does stabilize SO_2 in the lower atmosphere. The abundance of oxidants also declines, which further limits the sink of SO_2 due to gas-phase reactions. However, these are not the main sinks for SO_2 in these models. Rainout is the main removal mechanism and this limits the lifetime of SO_2 to hundreds, perhaps thousands, of years. This is a much longer lifetime than that estimated by Wong et al. (2004) for the present Martian atmosphere (~0.5 year). Johnson et al. (2009) concluded that the enhanced lifetime of SO_2 on early Mars meant that volcanic outgassing events would be followed by transient periods of warm wet conditions that could help carve the observed fluvial features.

In practice, however, an SO_2 greenhouse still faces significant challenges. The main challenge is avoiding the production of sulfur-bearing aerosols that inevitably cool the planet (Tian et al., 2010). These were not considered by Halevy et al. (2007) and Johnson et al. (2008, 2009). On Earth, SO_2 is converted to sulfuric acid (H_2SO_4) by the oxidation of SO_2 to SO_3 (by reaction with OH), which then reacts with water to form H_2SO_4. The sulfuric acid then condenses to form sulfate aerosols, which are highly reflective at visible wavelengths. The cooling effect of sulfate aerosols is a well-known consequence of volcanic eruptions on Earth (e.g. Hansen et al., 2002), and the sulfuric acid clouds on Venus (e.g. Bullock and Grinspoon, 2001). On early Mars, OH might be limited by the supply of reductants from volcanic emissions as Halevy et al. (2007) suggest. However, there is an additional supply of oxidants on Mars that is not present on Earth. Carbon dioxide, the assumed main constituent of the early Mars atmosphere, photolyzes to produce CO and O. Thus, the oxidation of SO_2 to SO_3 can also proceed by reaction with the atomic oxygen so produced. Furthermore, even in highly reducing conditions, the photolysis of SO_2 (and H_2S) can produce elemental sulfur S_2 that can polymerize to make stable particles of S_8 (Kasting et al., 1989). These particles have optical properties that also cool (Tian et al., 2010). Thus, regardless

of the oxidation state of the atmosphere, elevated SO_2 levels will produce sulfur aerosols that offset the greenhouse warming.

There are other problems. The timescale for producing sulfur-bearing aerosols is on the order of months (Tian et al., 2010), which is much less than the timescale for removal of SO_2 by rainout. Thus, the duration of a transient greenhouse would be much less than envisioned by Johnson et al. (2009). And the injection of the minimum amount of SO_2 needed for a transient greenhouse (~10 ppmv) apparently requires volcanic eruptions much more powerful than those on Earth (Tian et al., 2010). Finally, there is still the high solubility of SO_2 in water to deal with. The rainout rates in the above models were parameterized and varied in sensitivity studies with results showing a lifetime of hundreds of years. However, a more recent study with a detailed cloud microphysical model indicates that in warm wet conditions when liquid water is present, the lifetime is closer to several months because precipitation rates are very high (40 cm yr^{-1}) under such conditions (McGouldrick et al., 2011). While it is possible that surface waters will saturate with SO_2 and the rainout could be balanced by return fluxes (Halevy et al., 2007), the resulting elevated SO_2 levels will inevitably lead to a cooling haze layer. Given all these issues, a sustainable powerful SO_2 greenhouse does not seem plausible for early Mars.

The few published papers on the viability of NH_3 and/or CH_4 as potential greenhouse gases are also not encouraging. Sagan and Mullen (1972) proposed that NH_3 in Earth's atmosphere at concentrations of 10–100 ppm could solve the faint young Sun paradox. Though it was subsequently shown to be photochemically unstable on very short timescales (Kuhn and Atreya, 1979; Kasting, 1982), the idea was revived by Sagan and Chyba (1997), who suggested that a UV-absorbing organic haze produced from the photolysis products of CH_4 could shield the ammonia from photolysis. Such a haze layer does exist in the atmosphere of Titan. However, the idea was discounted for Earth because the haze was also likely to strongly absorb at visible wavelengths, creating an offsetting anti-greenhouse effect (McKay et al., 1999; Pavlov et al., 2001). More recently, this picture has changed as the fractal nature of organic haze particles allows for the possibility of low visible opacities, thereby avoiding the anti-greenhouse effect, but very high UV opacities, thereby providing an effective shield (Wolf and Toon, 2010). So the possibility of an NH_3–CH_4 atmosphere for early Earth cannot be ruled out.

For early Mars, a CH_4–NH_3 atmosphere would face similar issues. These gases, though powerful infrared absorbers, need a strong source and an effective shield to survive. An NH_3 greenhouse would require a mixing ratio of 5×10^{-4} in a 4–5 bar CO_2 atmosphere to raise surface temperatures to 273 K (Kasting et al., 1992). Without shielding from ultraviolet radiation, maintaining this concentration requires a source strength several orders of magnitude greater than that estimated for early Mars (Brown and Kasting, 1993). If the shielding is provided by CH_4 photolysis products, as suggested for ancient Earth, then a reducing atmosphere (>10% CH_4) is required (Domagal-Goldman et al., 2008). Otherwise CH_4 will oxidize to CO_2 rather than polymerize to form hydrocarbons.

Alternatively, CH_4 itself was thought capable of providing the needed greenhouse effect. Initial calculations indicated that a several bar CO_2 atmosphere containing 1% CH_4 could have kept early Mars warm and wet (Kasting, 1997). Unfortunately, the discovery of a coding error in those calculations showed that CH_4 greenhouse warming was greatly overestimated (Haqq-Misra et al., 2008). This, combined with the downward revision of CIA and the fact that CH_4 absorbs solar radiation in the near-infrared, severely limits the greenhouse warming from CH_4 (Ramirez et al., 2014).

However, a reducing mantle could conceivably outgas H_2 as well as CH_4. Ramirez et al. (2014) use Kasting's updated model to show that CIA from the interaction of H_2 molecules with CO_2 could provide significant greenhouse warming on early Mars with temperatures reaching the melting point of water in atmospheres containing 1.3–4 bar of CO_2 and 5–20% H_2. The case for more reduced outgassing products is bolstered by recent laboratory experiments indicating that, for oxygen fugacities at or below iron–wüstite (IW) –0.55 $\log_{10}$ units, carbon in the mantle is stored as iron pentacarbonyl ($Fe(CO)_5$ rather than as carbonate (Wetzel et al., 2013). Upon degassing, the iron carbonyl disassociates to CH_4 and CO, which is then photochemically oxidized to CO_2. However, there is a wide range of oxygen fugacities in the Martian meteorites, ranging from just below IW all the way up to the quartz–fayalite–magnetite (QFM) buffer (Wadhwa, 2008). If the mantle during the Noachian was at the very low end of this range, then the outgassing products would have had a greater fraction of H_2, CH_4, and CO. The challenge then becomes simultaneously producing the large quantities of H_2 and CO_2 required. Batalha et al. (2015) find that reaching 5% H_2 levels is possible, but it requires active volatile recycling and either additional sources of H_2 (serpentinization), or an escape rate below the diffusion limit. If such conditions were met, thick hydrogen-rich CO_2 atmospheres could provide a solution to the faint young Sun problem for Mars[13].

17.5.2.3 Summary

A widely accepted long-lived greenhouse solution for early Mars has yet to emerge. State-of-the-art global climate models with pure CO_2–H_2O atmospheres are not capable of raising global mean annual surface temperatures above 250 K, and even in this case the conditions are either not realistic (e.g. the atmosphere is 100% saturated with respect to water vapor), or at the extreme end of the possibilities (e.g. surface pressure of 2 bar). Supplementing such atmospheres with trace greenhouse gases can boost the greenhouse warming of a CO_2–H_2O atmosphere, but all of the proposed gases have sustainability issues that have yet to be solved. SO_2, for example, is soluble and oxidizes to sulfate. Without shielding, NH_3 is photochemically unstable. And CH_4 requires highly reduced conditions to prevent oxidation. The most recent ideas, a cloud greenhouse and a hydrogen-rich atmosphere, show some promise, but the robustness of these solutions has yet been demonstrated. We do not know if clouds of the right optical depths will form at the right altitudes to maintain a warm stable climate system. And while a more reduced mantle for early Mars is possible, we do not know if hydrogen-rich atmospheres can be generated and sustained for long periods of time. We now review some ideas about transient greenhouse atmospheres.

17.5.3 Transient Greenhouse Atmospheres

17.5.3.1 Impact-Generated Climate Change

An alternative explanation for the late Noachian valley networks is that they were formed from the cumulative effects of temporary warm rainy climates following the impact of large asteroids onto the surface. To our knowledge this idea first appeared in print in a 1977 *Scientific American* article by Conway Leovy – see the last paragraph in Leovy (1977). It was later taken up by others (e.g. Matsui et al., 1988), but was first quantified by Segura et al. (2002), who demonstrated its feasibility with large impactors (>100 km). Follow-on studies focused on smaller impactors (Segura et al., 2008) and the possibility of multiple climate states (Segura et al., 2012). The basic idea is that the energy released by large impacts creates a globally disbursed ejecta blanket and a steam atmosphere that lasts for hundreds to thousands of years. As the atmosphere cools, water rains onto the surface and, when integrated over many impacts, the total volume of rainfall involved is enough to carve the valley networks we see today. This hypothesis has several attractive features. First, it does not require special conditions. Mars obviously experienced a high impact rate early in its history and the larger impactors can certainly affect the climate system. Second, the forcing function was mainly effective early in Mars' history when the valley networks formed. Finally, it gets around the problem of finding ways to generate and sustain an effective greenhouse on early Mars. This latter feature makes it particularly attractive.

However, there are several aspects of the impact hypothesis that have been questioned with respect to the observations. The first involves the relative timing of the large impacts and valley network formation. Jakosky and Mellon (2004) pointed out that the largest impact craters formed well before the epoch of valley network formation, which they took to be near the end of the Noachian around 3.8 Ga. Since then there have been attempts to improve the estimated ages of these events. Werner (2008) reevaluated the ages of 15 impact basins larger than 250 km and four smaller ones and found them to range in age from 3.7 to 4.1 Ga, with a mean crater age of ~3.9 Ga. Fassett and Head (2008) examined valley networks in the cratered uplands, which they estimated formed very near the Noachian–Hesperian boundary sometime between 3.7 and 3.9 Ga. This puts an average difference of no more than 200 Ma between the large impact era and valley network formation era. Of course, precise dating of specific events on early Mars is difficult, as it relies on visible crater populations that may not be representative, and models of the early Mars impact rate at Mars that may not be accurate. Consequently, the uncertainties can be large. However, on the basis of these new age estimates, it does appear that the interval of time between large impacts and valley network formation is narrower than envisioned by Jakosky and Mellon (2004).

Another criticism of the impact hypothesis has been its ability to produce the observed erosion rates. In their follow-on study of the impact hypothesis, Segura et al. (2008) estimated

---

[13] We note, however, that a hydrogen-rich atmosphere would also imply the absence of life, since both laboratory experiments (Kral et al., 1998) and calculations (Kasting et al., 2001) show that microbes tend to lower H_2 abundances to an energetic limit of ~10^{-5} bar.

that, depending on surface pressure, between about 1 and 50 m of material could have been removed by the post-impact rainfall resulting from impactors larger than 10 km in diameter. This is considerably less than the 50–2500 m they estimate would have been removed during the late Noachian if the erosion rates of Craddock and Maxwell (1993) lasted for 500 Ma. Their rainfall totals (~650 m) are also much less than the estimated hundreds to thousands of kilometers required from various models (Irwin et al., 2005; Howard, 2007; Barnhart et al., 2009; Hoke et al., 2011). These are large discrepancies to overcome and would appear to require that either the surface is much easier to erode than assumed – this seems unlikely given that they used an erodibility factor of $K \sim 0.12$, representative of a loamy sand surface – or their rainfall totals are significantly underestimated.

Barnhart et al. (2009) used a sophisticated landscape evolution model to determine the rainfall and surface characteristics of the Parana Valles drainage basin. A key observational constraint in their study was the lack of morphological evidence for breached craters in this basin area. This implies that water flowing through the valleys did not accumulate in craters to the point of overflowing the rims. They concluded that the Parana Valles system likely formed in a dry arid climate with sporadic but numerous (hundreds of thousands) flooding events of sufficient intensity to erode the surface, but not so intense as to breach the craters. They ruled out the impact hypothesis on the basis that they are fewer in number and would therefore require deluge-style flooding events that would have created numerous exit breaches on the crater rims. We note, however, that breached craters do exist on Mars and they could have been caused by short-lived events of heavy rainfall (see e.g. Fassett and Head, 2005).

17.5.3.2 Orbital Variations

Orbital variations are particularly large for Mars (see Chapter 16) and must have played some role in the early Mars climate system. Obliquity, eccentricity, and precession variations occur on 10^5–10^6 year timescales and, while they have a very small effect on the total annual insolation received, they do alter the latitudinal and seasonal distribution in ways that can be significant. However, this subject has received less attention for the Noachian era than it has for the Amazonian. Only a handful of studies for Noachian Mars have been reported in the literature. High-obliquity simulations with global circulation models show that dry, dense CO_2 atmospheres preferentially raise mean annual surface temperatures in the northern hemisphere due to increased insolation and an elevation effect, but these are not the regions where the valley networks and sedimentary units are located (Forget et al., 2013). These same simulations with a water cycle, however, do show that obliquity variations can shift the location of seasonal melting events between hemispheres, though the volume of meltwater generated is always small (1–2 mm yr^{-1}) and not necessarily in the right places (Wordsworth et al., 2013). More recently Mischna et al. (2013), also using a three-dimensional global climate model, studied the potential feedbacks between obliquity variations and volcanic emissions of SO_2 and found that, while such interactions do not raise global mean annual surface temperatures above freezing, they

can broaden the times and locations where surface temperatures are above freezing.

17.5.4 Cold Climates With Locally Wet Conditions

A relatively new line of thinking, motivated by the problems with greenhouse warming, holds that early Mars was both cold and wet. In this view the fluvial features could be the result of several different processes. One possibility is that water flows on the surface in response to seasonal warming (McKay, 2004). The dry valleys of Antarctica provide an example. There, mean annual temperatures are well below freezing (253 K), precipitation is minimal (1–2 cm yr^{-1}) and always in the form of snow, yet lakes and rivers exist with liquid water (e.g. Lake Vanda and the Onyx river). Another possibility is that water can erupt from confined subsurface aquifers and flow on the surface (Gaidos and Marion, 2003). A cold early Mars would produce a thick cryosphere that could force large volumes of deep warm primordial groundwater to the surface. More recently, Fairén (2010) has proposed yet another possibility in which the existence of widespread liquid water on the surface – even oceans – could be made possible by the presence of solutes from the weathering of typical Martian basalts. Solutes depress the freezing point and calculations suggest that liquids could be stable against evaporation at temperatures between 245 and 255 K (Fairén et al., 2009). A cold ocean on early Mars would help explain the apparent rare occurrence of phyllosilicates in exposed Noachian terrains of the northern hemisphere (Fairén et al., 2011), as well as glacial features such as fretted terrain and moraines near the hemispheric dichotomy boundary (Davila et al., 2013).

The most recent idea for forming the valley networks in a cold environment comes from general circulation modeling simulations of the early Mars hydrological cycle. Wordsworth et al. (2013) find that, for surface pressures higher than several hundred millibars, Mars becomes more Earth-like in the sense that high-altitude terrains will be colder than low-altitude terrains. This is due to the tighter coupling between the surface and atmosphere at higher pressures. This "adiabatic" effect favors snowfall in the southern highlands, which forms thick ice sheets that can later melt from impacts, volcanic activity, or basal melting. This "icy highlands" hypothesis is attractive in that it explains the predominance of valley networks in the southern hemisphere (though we do not know if they did form in the northern hemisphere but were subsequently buried), it does not require a thick atmosphere, and it makes use of plausible transient non-solar energy sources for melting. However, it is not clear if these energy sources can generate the meltwater volumes needed for the required erosion. In this view, early Mars is mostly a cold planet in which erosion and fluvial activity result from the episodic melting of surface ice deposits.

A common feature of these ideas is that they require higher surface pressures than Mars has today. For the icy highlands idea, high surface pressures are needed for the adiabatic effect. For the seasonal melting, groundwater eruption, and solutes in the water ideas, boiling is an important consideration. The mean annual surface pressure on Mars today is very close to the triple-point pressure of water. Below the triple point, water cannot exist as liquid. But even at those locations and times where pressures and temperatures exceed the triple point, liquid water on Mars today would be very close to the boiling point (Haberle et al., 2001). Boiling occurs when the vapor pressure of water is comparable to the surface pressure, and this leads to very high evaporation rates (Ingersoll, 1970; Kahn, 1985; McKay, 2004). Since the latent heat of vaporization is almost two orders of magnitude higher than the specific heat, evaporation will rapidly cool the water. Under these circumstances, water quickly freezes unless it is gaining energy from an external source. Thus, without higher surface pressures, liquid water will not last long enough to erode the surface to the levels observed. The exact value needed is situation-dependent, but it should be well above the triple point. McKay (2004) suggests a minimum value of about 100 mbar, which is a plausible value for the Noachian epoch.

17.5.5 A Brighter Early Sun?

If early in Mars' history the Sun was brighter than predicted by the standard solar model, it would make it easier to stabilize liquid water on the surface. This could be achieved if the Sun was more massive in its early main-sequence phase (Whitmire et al., 1995). Solar luminosity is proportional to the fourth power of solar mass $M_\odot$, and, since a planet's orbital distance r is inversely proportional to $M_\odot$, and the solar flux varies as r^{-2}, the flux at a planet scales as $F \sim M^6$. Hence, small changes in solar mass can result in large changes in the solar flux.

If the Sun was more massive when it joined the main sequence, then it must have lost the excess mass over time in order to arrive at its present mass. The amount of excess mass and its loss history have been the subjects of several observational (e.g. Morel et al., 1997; Wood et al., 2002, 2005) and theoretical (e.g. Willson et al., 1987; Boothroyd et al., 1991; Swenson and Faulkner, 1992; Guzik and Cox, 1995; Sackmann and Boothroyd, 2003; Minton and Malhotra, 2007; Turck-Chièze et al., 2011) studies. The main constraints on mass loss are the observed lithium depletion in the Sun's photosphere (Boothroyd et al., 1991), helioseismic observations that infer the speed of sound through its interior, and the fact that the Earth still has most of its water, which it would have lost if the Sun was too bright (Kasting, 1988). These considerations generally imply that the early Sun could not have been more than 10% more massive than it is today, and that most of its excess mass was lost early in the main-sequence phase (<1 Ga after birth).

Sackmann and Boothroyd (2003) have published perhaps the most definitive study on how early mass loss might have affected Mars and Earth. They considered three different mass-loss scenarios and tested how well their solar evolution models compared with helioseismic observations. The key results are shown in Figure 17.16. Their preferred scenario is for an initial mass of 1.07 $M_\odot$, which gives results that are slightly (but not significantly) more consistent with helioseismic observations than the standard solar model (SSM). In this case, the solar flux at Mars 4.5 Ga would have been ~5% higher than it is today. By 3.4 Ga it would have declined to a value ~16% lower than today's. Thus, throughout the Noachian, the solar flux would have been significantly higher than predicted by the SSM. Under these circumstances, the surface pressure requirement

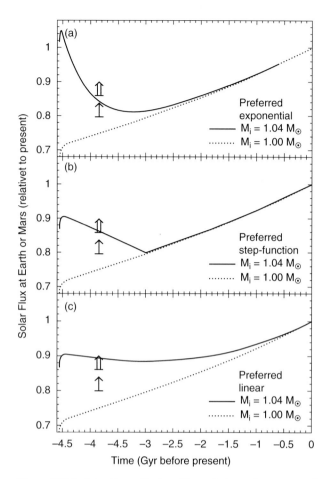

Figure 17.16. Solar flux at Earth or Mars relative to the present solar flux as a function of time. Solid and dashed lines refer to different initial solar masses (M_i) relative to the present mass ($M_\odot$). Each panel represents a different loss scenario, as indicated. Thick double arrows give the lower flux limit of Kasting (1991) for the presence of water on early Mars. Thin arrows give the same limit for a low surface albedo. From Sackmann and Boothroyd (2003).

for a sustained CO_2 greenhouse would be reduced in comparison with previous predictions (e.g. Kasting, 1991; Wordsworth et al., 2013).

Observational support for the early mass loss hypothesis is tenuous. To date, loss rates at the level favored by Sackmann and Boothroyd (2003) have not yet been confirmed or observed for stars similar to the Sun. At the present time, the solar wind removes ~3×10^{-14} $M_\odot$ yr^{-1}, which would amount to ~0.01% of its present mass if this loss rate were constant over time. Lunar and meteoritic data suggest the mean solar wind over the past 3 Ga was an order of magnitude higher than it is today (Kerridge et al., 1991). Thus a soft lower limit for the Sun's mass loss is ~0.1% of present mass. Of course, these data do not constrain mass loss rates during the first billion years, so observations of Sun-like analogs might help in this regard. Wood et al. (2002, 2005) have inferred mass loss rates from several nearby young Sun-like stars using observations of Ly α absorption that result from interactions of stellar winds with the interstellar medium. Their most recent results indicate loss rates of ~0.1% of initial mass for these stars, which is consistent with the lunar and meteoritic data for our Sun, but far less than the 7% mass loss favored by Sackmann and Boothroyd (2003). The removal timescale is also problematic, as it may be too short to keep the Earth from freezing after the first billion years or so (Minton and Malhotra, 2007). Thus, the size of the needed mass loss and its removal time are inconsistent with the available data. However, the observed mass loss rates are somewhat model-dependent, the sample size is very small, and exact solar analogs are difficult to find. Furthermore, the recent downward revision in the abundances of heavy elements in the Sun's interior (Asplund et al., 2009; Caffau et al., 2010) increases the discrepancy of SSMs with helioseismic data, which can be partly mitigated with even larger mass loss rates (10–30%) (Guzik and Mussack, 2010; Turck-Chièze et al., 2011). Thus, the early mass loss hypothesis cannot be completely ruled out.

17.6 DISCUSSION AND FUTURE PROSPECTS

In spite of the considerable research conducted since Mariner 9 images first revealed ancient fluvial systems, we still do not know for certain what environmental conditions prevailed during the late Noachian on Mars. There is compelling evidence and widespread agreement that water flowed on the surface, and that the atmosphere was thicker in those times. But the notion of a sustained warm and wet climate system has been difficult to justify.

The early optimism that the greenhouse effect of a CO_2–H_2O atmosphere could offset the diminished output of the young Sun by providing enough warming to raise global mean surface temperatures to the melting point has faded to skepticism as researchers have delved deeper into the details of such a climate system. Not only does it appear that the Noachian atmosphere was less massive than previously assumed (though this is somewhat model-dependent), it is also less capable of warming the surface because of limitations imposed by CO_2 condensation, a weaker CO_2 cloud scattering greenhouse effect, and greatly reduced collision-induced absorption. State-of-the-art global climate models with pure CO_2–H_2O atmospheres are unable to raise global mean surface temperatures above ~250 K for any surface pressure unless a strong cloud greenhouse can be generated.

Supplementing such atmospheres with additional greenhouse gases such as SO_2, H_2S, CH_4, NH_3, and H_2 could boost their greenhouse power if their concentrations can be maintained at the right levels. But this is difficult to do; each of these gases has sinks that limit their effectiveness. SO_2 converts to sulfate, H_2S and CH_4 are not very effective in the first place, NH_3 – the most powerful greenhouse gas – is photochemically unstable, and H_2 requires a reducing mantle and very large outgassing rates. This does not rule them out, but special circumstances that have yet to be demonstrated are required to sustain the right concentrations. More research is needed to assess the viability of supplemental greenhouse gases in solving the early Mars dilemma.

Similarly, more research is needed into the role of impact-generated climate change. The few studies done thus far show that 10 km impactors can alter the climate system and produce rainfall, but the rainfall and erosion they produce falls short of what is required. An interesting feature of these studies that offers a promising new direction for early Mars research is the potential for a cloud greenhouse effect. The one-dimensional model of Segura et al. (2008) showed that the (water) clouds warm the surface and that they have a significant role in maintaining liquid water on the surface. More recent separate studies confirm the potentially warming effect of clouds, though with diverse estimates of the quantitative effect. Whether such clouds, generated either by impacts or at times of high obliquity, play an important role in warming the early Mars climate system warrants further study.

Finally, the continued analysis of existing data and the future acquisition of new data will help refine our understanding of the true nature of the early Martian climate system. Image analysis that further quantifies the volume of eroded material and the timing and duration of precipitation will help distinguish between the need for continuous versus episodic climate change. A more precise determination of the mineralogy of the surface should help better define the environmental conditions that existed when they formed. And the continued refinement of isotopic data from the SNCs and MSL will be of great value in piecing together escape scenarios. This, along with measurements of the current escape rates to come from MAVEN, should go a long way towards estimating the mass and composition of the early atmosphere.

ACKNOWLEDGMENTS

We are grateful for helpful reviews from Brian Toon, Jim Kasting, and Bruce Jakoksy.

REFERENCES

Altheide, T., Chevrier, V. F., and Noe Dobrea, E. (2010) Mineralogical characterization of acid weathered phyllosilicates with implications for secondary Martian deposits, *Geochim. Cosmochim. Acta*, 74, 6232–6248.

Amelin, Y., Krot, A. N., Hutcheon, I. D., and Ulyanov, A. A. (2002) Lead isotopic ages of chondrules and calcium-aluminum-rich inclusions, *Science*, 297, 1678–1683.

Anders, E., and Owen, T. (1977) Mars and Earth – origin and abundance of volatiles, *Science*, 198, 453–465.

Ansan, V., and Mangold, N. (2013) 3D morphometry of valley networks on Mars from HRSC/MEX DEMs: implications for climatic evolution through time, *J. Geophys Res.*, 118, 1873–1894, doi:10.1002/jgre.20117

Asplund, M., Grevesse, N., Sauval, A. J., and Scott, P. (2009). The chemical composition of the Sun. *Annu. Rev. Astron. and Astrophys.*, 47, 481–582, doi:10.1146/annurev.astro.46.060407.145222.

Atreya, S. K., Trainer, M. G., Franz, H. B., et al. (2013) Primordial argon isotop fractionation in the atmosphre of Mars measured by the SAM instrument on Curiosity and implications for atmospheric loss, *Geophys. Res. Lett.*, 40, 1–5, doi:10.1002/2013GL057763.

Baker, V. R. (1982) *The Channels of Mars*, Texas University Press, Austin, TX.

Baker, V. R. (2001) Water and the Martian landscape, *Nature*, 412, 228–236.

Fassett, C. I., and Head, J. W. (2005) New evidence for fluvial sedimentary deposits on Mars: deltas formed in a crater lake in the Nili Fossae region, in *36th Lunar and Planet. Sci. Conference*, League City, TX, Abstract No. 1098.

Fassett, C. I., and Head, J. W. (2008) Valley network-fed, open-basin lakes on Mars: distribution and implications for Noachian surface and subsurface hydrology, *Icarus*, 198, 37–56, doi:10.1016/j.icarus.2008.06.016.

Fassett, C. I., and Head, J. W. (2010) Conditions on early Mars: scenarios, transitions and events. In *41st Lunar Planet Sci. Conf.*, abstract 1951, Houston, TX.

Fassett, C. I., and Head, J. W. (2011) Sequence and timing of conditions on early Mars, *Icarus*, 211, 1204–1214.

Forget, F., and Pierrehumbert, R. T. (1997) Warming early Mars with carbon dioxide clouds that scatter infrared radiation, *Science*, 278, 1273–1276.

Forget, F., Hourdin, F., and Talagrand, O. (1998) CO_2 snowfall on Mars: simulation with a general circulation model, *Icarus*, 131, 302–316, doi:10.1006/icar.1997.5874.

Forget, F., Wordsworth, R., Millour, E., et al. (2013) 3D modeling of the early Martian climate under a denser CO_2 atmosphere: temperatures and CO_2 ice clouds, *Icarus*, 222, 81–89, doi:10.106/j.icarus.2012.10.019.

Fox, J. L. (1993) On the escape of oxygen and hydrogen from Mars, *Geophys. Res. Lett.*, 20, 1847–1850.

Fox, J. L. (2007) Comment on the papers "Production of hot nitrogen atoms in the Martian thermosphere" by F. Bakalian and "Monte Carlo computations of the escape of atomic nitrogen from Mars" by F. Bakalian and R. E. Hartle, *Icarus*, 192, 296–301.

Fox, J. L., and Hać, A. B. (1997a) The $^{15}N/^{14}N$ isotope fractionation in dissociative recombination of N_2^+, *J. Geophys. Res.*, 102, 9191–9204.

Fox, J. L, and Hać, A. B. (1997b) Spectrum of hot O at the exobases of the terrestrial planets, *J. Geophys. Res.* 102, 24005–24011.

Fox, J. L., and Hać, A. B. (2010) Isotope fractionation in the photochemical escape of O from Mars, *Icarus*, 208, 176–191.

Frey, H. V. (2003) Buried impact basins and the earliest history of Mars, *6th International Conf. on Mars*, Pasadena, CA.

Gaidos, E., and Marion, G. (2003) Geological and geochemical legacy of a cold early Mars, *J. Geophys. Res.*, 198, doi:10.1029/2002JE002000.

Gaillard, F., and Scaillet, B. (2009) The sulfur content of volcanic gases on Mars, *Earth Planet. Sci. Lett.*, 279, 34–43.

Gendrin, A., Mangold, N., Bibring, J.-P. et al. (2005) Sulfates in Martian layered terrains: the OMEGA/Mars Express view, *Science*, 307, 1587–1591.

Glandorf, D. L., Colaprete, A., Tolbert, M. A., and Toon, O. B. (2002) CO_2 snow on Mars and early Earth: experimental constraints, *Icarus*, 160, 66–72, doi:10.1006/icar.2002.6953.

Glavin, D. P., Freissinet, C., Miller, K. E., et al. (2013) Evidence for perchlorates and the origin of chlorinated hydrocarbons detected by SAM at the Rocknest Aeolian deposit in Gale Crater, *J. Geophys. Res.*, 118, 1955–1973, doi:10.1002/jgre.20144.

Glotch, T. D., and Rogers, A. D. (2007) Evidence for aqueous deposition of hematite- and sulfate-rich light-toned layered deposits in Aureum and Iani Chaos, Mars, *J. Geophys. Res.*, 112, E06001, doi:10.1029/2006JE002863.

Glotch, T. D., Bandfield, J.L., Tomabene, L. L., et al. (2010) Distribution and formation of chlorides and phyllosilicates in Terra Sirenum, Mars, *Geophys. Res. Lett.*, 37, doi:10.1029/2010GL044557.

Golden, D. C., Ming, D. W., Morris, R. V., and Graff, T. G. et al. (2008) Hydrothermal synthesis of hematite spherules and jarosite: implications for diagenesis and hematite spherule formation in sulfate outcrops at Meridiani Planum, Mars, *American Mineralogist*, 93, 1201–1214.

Golombek, M. P., and Bridges, N. T. (2000) Erosion rates on Mars and implications for climate change: constraints from the Pathfinder landing site, *J. Geophys. Res.*, 105, 1841–1853.

Golombek, M. P., Grant, J. A., Crumpler, L., et al. (2006) Erosion rates at the Mars Exploration Rover landing sites and long-term climate change on Mars, *J. Geophys. Res.*, 111, doi:10.1029/2006JE002754.

Gooding, J. L. (1992) Soil mineralogy and chemistry on Mars – possible clues from salts and clays in SNC meteorites, *Icarus*, 99, 28–41.

Gough, D. O. (1981) Solar interior structure and luminosity variations, *Solar Physics*, 74, 21–34, doi:10.1007/BF00151270.

Grant, J. A., Wilson, S. A., Mangold, N., et al. (2014) The timing of alluvial activity in Gale Crater, Mars, *Geophys. Res. Lett.*, 41, 1142–1148, doi:10.1002/2013GL058909.

Greeley, R., and Schneid, B. D. (1991) Magma generation on Mars: amounts, rates, and comparisons with Earth, Moon, and Venus, *Science*, 254, 996–998.

Greenwood, J. P., Riciputi, L. R., McSween, H. Y. Jr., Taylor, L. A. (2000) Modified sulfur isotopic compositions of sulfides in the nakhlites and Chassigny, *Geochim. Cosmochim. Acta*, 64, 1121–1131.

Greenwood, J. P., Itoh, S., Sakamoto, N., Vicenzi, E. P., and Yurimoto, H. (2008) Hydrogen isotope evidence for loss of water from Mars through time, *Geophys. Res. Lett.*, 35, doi:10.1029/2007GL032721.

Greenwood, J. P., Itoh, S., Sakamoto, N., Vicenzi, E. P., and Yurimoto, H. (2010) D/H zoning in apatite of Martian meteorites QUE 94201 and Los Angeles: implications for water on Mars, *Amer. Met. Soc.*, 73, 5347.

Grott, M., Morschhauser, A., Breuer, D., and Hauber, E. (2011) Volcanic outgassing of CO_2 and H_2O on Mars, *Earth and Planet. Sci. Lett.*, 308, 391–400.

Grotzinger, J. P., Arvidson, R. E., Bell, J. F., et al. (2005) Stratigraphy and sedimentology of a dry to wet eolian depositional system, Burns formation, Meridiani Planum, Mars, *Earth and Planet. Sci. Lett.*, 240, 11–72, doi:10.1016/j.epsl.2005.09.039.

Grotzinger, J. P., Sumner, D. Y., Kah, L. C., et al. (2014) A habitable fluvio-lacustrine environment at Yellowknife Bay, Gale Crater, Mars, *Science*, 343, 6169, doi:10.1126/science.1242777.

Grotzinger, J. P., Gupta, S., Malin, M. C., et al. (2015) Deposition, exhumation, and paleoclimate of an ancient lake deposit, Gale Crater, Mars, *Science*, 350, 6257, doi:10.1126/science.aac7575.

Gruszka, M., and Borysow, A. (1998) Computer simulation of the far infrared collision induced absorption spectra of gaseous CO2, *Mol. Phys.*, 93, 1007–1016, doi:10.1080/002689798168709.

Gulick, V. C. (1998) Magmatic intrusions and a hydrothermal origin for fluvial valleys on Mars, *J. Geophys. Res.*, 103, 19365–19387.

Guzik, J. A., and Cox, A. N. (1995) Early solar mass loss, element diffusion, and solar oscillation frequencies, *ApJ.*, 448, 905–914.

Guzik, J. A., and Mussack, K. (2010) Exploring mass loss, low-Z accretion, and convective overshoot in solar models to mitigate the solar abundance problem, *ApJ*, 713, 1108–1119, doi:10.1088/0004-637X/713/2/1108.

Haberle, R. M. (1998) Early Mars climate models, *J. Geophys. Res.*, 103, 28467–28480, doi:10/1029/98JE01396.

Haberle, R. M. (2013) Estimating the power of Mars' greenhouse effect, *Icarus*, 223, 619–620.

Haberle, R. M., Tyler, D., McKay, C. P., and Davis, W. L. (1994) A model for the evolution of CO_2 on Mars, *Icarus*, 109, 102–120, doi:10.1006/icar.1994.1079.

Haberle, R. M., McKay, C. P., Schaeffer, J. et al. (2001) On the possibility of liquid water on present day Mars, *J. Geophys. Res.*, 106, 23317–23326, doi:10.1029/2000JE001360.

Haberle, R. M., Forget, F., Colaprete, A., et al. (2008) The effect of ground ice on the Martian seasonal CO_2 cycle, *Planet. and Space Sci.*, 56, 251–255.

Haberle, R. M., Kahre, M. A., Hollingsworth, J. L., et al. (2012) A cloud greenhouse effect on Mars: significant climate change in the recent past?, in *43rd Lunar and Planetary Science Conf.*, abstract 1665, Houston, TX.

Halevy, I., and Eiler, J. M. (2011) Carbonates in ALH 84001 formed in a short-lived hydrothermal system, in *42nd Lunar and Planet. Sci. Conf.*, abstract 2512, Houston, TX.

Halevy, I., Zuber, M. T., and Schrag, D. P. (2007) A sulfur dioxide climate feedback on early Mars, *Science*, 318, 1903–1907, doi:10.1126/science.1147039.

Halevy, I., Pierrehumbert, R. T., and Schrag, D. P. (2009) Radiative transfer in CO_2-rich paleoatmospheres, *J. Geophys. Res.*, 114, doi:10.1029/2009JD01915.

Hansen, J., Sato, M., Nazarenko, L., et al. (2002) Climate forcings in Goddard Institute for Space Studies SI2000 simulations, *J. Geophys. Res.*, 107, doi:10.1029/2001JD001143.

Haqq-Misra, J. D., Domagal-Goldman, S. D., Kasting, P. J., and Kasting, J. F. (2008) A revised, hazy methane greenhouse for the Archean Earth, *Astrobiology*, 8, 1127–1137.

Hartmann, W. K. (2004), Updating the crater count chronology system for Mars, in *35th Lunar and Planet. Sci. Conf.*, March 15–19, League City, TX, Abstract No. 1374.

Hartmann, W. K., and Neukum, G. (2001) Cratering chronology and the evolution of Mars, *Space Sci. Rev.*, 96, 165–194.

Hausrath, E. M., and Olsen, A. A. (2013) Using the chemical composition of carbonate rocks on Mars as a record of secondary interaction with liquid water, *Am. Mineral.*, 98, 897–906.

Head, J. W., and Marchant, D. R. (2014) The climate history of early Mars: insights from the Antarctic McMurdo Dry Valleys hydrologic system, *Antarctic Science*, 26, 774–800, doi:10.1017/S0954102014000686.

Head, J. W., Heisinger, H., Ivanov, M. A., et al. (1999) Possible ancient oceans on Mars: evidence from Mars Orbiter Laser Altimeter data, *Science*, 286, 2134–2137.

Hecht, M. H., Kounaves, S. P., Quinn, R. C., et al. (2009) Detection of perchlorate and the soluble chemistry of Martian soil: findings from the Phoenix Mars Lander, *Science*, 325, 64–67.

Hirschmann, M. M., and Withers, A. C. (2008) Ventilation of CO_2 from a reduced mantle and consequences for the early Martian greenhouse, *Earth and Planet. Sci. Lett.*, 270, 147–155.

Hoke, M. R. T., Hynek, B. M., and Tucker, G. E. (2011) Formation timescales of large Martian valley networks, *Earth and Planet. Sci. Lett.*, 312, 1–12.

Howard, A. D. (2007) Simulating the development of Martian highland landscapes through the interaction of impact cratering, fluvial erosion and variable hydrologic forcing, *Geomorphology*, 91, 332–363.

Howard, A. D., Moore, J. M., and Irwin, R. P. (2005) An intense terminal epoch of widespread fluvial activity on Mars: 1. Valley network incision and associated deposits, *J. Geophys. Res.*, 110, E12S14, doi:10.1029/2005JE002459.

Hu, R., Kass, D. M., Ehlmann, B. L., and Yung, G. E. (2015) Tracing the fate of carbon and the atmospheric evolution of Mars, *Nature*, 6, doi:10.1038/ncomms10003.

Huck, F., Jobson, D. J., Park, S. K., et al. (1977) Spectrophotometric and color estimates of the Viking Lander sites, *J. Geophys. Res.*, 82, 4401–4411.

Humayun, M., Nemchin, A., Zanda, B., et al. (2013) Origin and age of the earliest Martian crust from meteorite NWA 7533. *Nature*, 505, doi:10.1038/nature12764.

Hunten, D. M., and Donahue, T. M. (1976) Hydrogen loss from the terrestrial planets, *Ann. Rev. Earth and Planet. Sci.*, 4, 265–292, doi:10.1146/annurev.ea.04.050176.001405.

Hunten, D. M., Pepin, R. O., and Walker, J. C. G. (1987) Mass fractionation in hydrodynamic escape, *Icarus*, 69, 532–549.

Hutchins, K. S., and Jakosky, B. M. (1996) Evolution of Martian atmospheric argon: implications for sources of volatiles, *J. Geophys. Res.*, 101, 14933–14949.

Hutchins, K. S., Jakosky, B. M., and Luhmann, J. G. (1997) Impact of a paleomagnetic field on sputtering loss of Martian atmospheric argon and neon, *J. Geophys. Res.*, 102, 9183–9190.

Hynek, B. M., and Phillips, R. J. (2003) New data reveal mature, integrated drainage systems on Mars indicative of past precipitation, *Geology*, 31, 757–760.

Hynek, B. M., and Phillips, R. J. (2008) The stratigraphy of Meridiani Planum, Mars, and implications for the layered deposits' origin, *Earth Planet. Sci. Lett.*, 274, 214–220.

Hynek, B. M., Beach, M., and Hoke, M. R. T. (2010) Updated global map of Martian valley networks and implications for climate and hydrologic processes, *J. Geophys. Res.*, 115, doi:10.1029/2009JE003548.

Ingersoll, A. P. (1970) Mars: occurrence of liquid water, *Science*, 168, 972–973, doi:10.1126/science.168.3934.972.

Irwin, R. P., and Howard, A. D. (2002) Drainage basin evolution in Noachian Terra Cimmeria, Mars, *J. Geophys. Res.*, 107, doi:10.1029/2001JE001818.

Irwin, R. P., Maxwell, A. D., Howard, A. D., Craddock, R. A., and Moore, J. M. (2005) An intense terminal epoch of widespread fluvial activity of Mars: 2. Increased runoff and paleolake development, *J. Geophys. Res.*, 110, E12S15, doi:10.1029/2005JE002460.

Jakosky, B. M., and Jones, J. H. (1997) The history of Martian volatiles, *Rev. Geophys.*, 35, 1–16.

Jakosky, B. M., and Mellon, M. T. (2004) Water on Mars, *Physics Today*, 57, doi:10.1063/1.1752425.

Jakosky, B. M., Pepin, R. O., Johnson, R. E., and Fox, J. L. (1994) Mars atmospheric loss and isotopic fractionation by solar-wind-induced sputtering and photochemical escape, *Icarus*, 111, 271–288.

Jakosky, B. M., Grebowsky, J. M., Luhmann, J. G., and Brain, D. A. (2015) Initial results from the MAVEN mission, *Geophys. Res. Lett.*, 42, 8791–8802, doi:10.1002/2015GL065271

Jensen, H. B., and Glotch, T. D. (2011) Investigation of the near-infrared spectral character of putative Martian chloride deposits, *J. Geophys. Res.*, 116, doi:10.1029/2011JE003887.

Jerolmack, D. J., Mohrig, D., Zuber, M. T., and Byrne, S. (2004) A minimum time for the formation of Holden Northeast fan, Mars, *Geophys. Res. Lett.*, 31, doi:10.1029/2004GL021326.

Johnson, S. S., Mischna, M. A., Grove, T. L., and Zuber, M. T. (2008) Sulfur-induced greenhouse warming on early Mars, *J. Geophys. Res.*, 113, doi:10.1029/2007JE002962.

Johnson, S. S., Pavlov, A. A., and Mischna, M. A. (2009) Fate of SO_2 in the ancient Martian atmosphere: implications for transient greenhouse warming, *J. Geophys. Res.*, 114, doi:10.1029/2008JE003313.

Jull, A. J. T., Eastoe, C. J., and Cloudt, S. (1997) Isotopic composition of carbonates in the SNC meteorites, Allan Hills 84001 and Zagami, *J. Geophys. Res.*, 102 (E1), 1663–1670.

Jull, A. J. T., Beck, J. W., Burr, G. S., et al. (1999) Isotopic evidence for abiotic organic compounds in the Martian meteorite, Nakhla, *Meter. and Planet. Sci.*, 34, p.A60.

observations from the Mars Reconnaissance Orbiter, *J. Geophys. Res.*, 114, doi:10.1029/2009JE003342.

Murchie, S., Roach, L., Seelos, F., et al. (2009b) Evidence for the origin of layered deposits in Candor Chasma, Mars, from mineral composition and hydrologic modeling, *J. Geophys. Res.*, 114, doi:10.1029/2009JE003343.

Mustard, J. F., Murchie, S. L., Pelkey, S. M., et al. (2008) Hydrated silicate minerals on mars observed by the Mars Reconnaissance Orbiter CRISM instrument, *Nature*, 454, 305–309.

Mustard, J. F., Ehlmann, B. L., Murchie, S. L., et al. (2009) Composition, morphology, and stratigraphy of Noachian crust around the Isidis Basin, *J. Geophys. Res.*, 114, doi:10.1029/2009JE003349.

Newman, M. J., and Rood, R. T. (1977) Implications of solar evolution for the Earth's early atmosphere, *Science*, 198, 1035–1037.

Nier, A. O., and McElroy, M. B. (1977) Composition and structure of Mars' upper atmosphere – results from the neutral mass spectrometers on Viking 1 and 2, *J. Geophys. Res.*, 82, 4341–4350.

Niles, P. B., Boynton, W. V., Hoffman, J. H., et al. (2010) Stable isotope measurements of Martian atmospheric CO_2 at the Phoenix landing site, *Science*, 329, 1334–1337.

Niles, P. B., Catling, D. C., Berger, G., et al. (2013) Geochemistry of carbonates on Mars: implications for climate history and nature of aqueous environments, *Space Sci. Rev.*, 174, 301–328.

Novak, R. E., Mumma, M. J., and Villanueva, G. L. (2011) Measurement of the isotopic signatures of water on Mars: implications for studying methane, *Planet. Space Sci.*, 59, 163–168.

Nyquist, L. E., Bogard, D. D, Shih, C.-Y., et al. (2001) Ages and geologic histories of Martian meteorites, *Space Sci. Rev.*, 96, 105–154.

Osterloo, M. M., Hamilton, V. E., Bandfield, J. L., et al. (2008) Chloride-bearing materials in the southern highlands of Mars, *Science*, 319, 1651–1654.

Ott, U. (1988) Noble gases in SNC meteorites: Shergotty, Nakhla, Chassigny, *Geochim. Cosmochim. Acta*, 52, 1937–1948.

Owen, T. (1992) The composition and early history of the atmosphere of Mars, in *Mars*, Kieffer H. H., et al., editors, 818–834, University of Arizona Press, Tucson.

Owen, T., and Bar-Nun, A. (2001) From the interstellar medium to planetary atmospheres via comets, In *Collisional Processes in the Solar System*, ed. M. Ya. Marov and H. Rickman, Astrophysics and Space Science Library 261, 249–264, Kluwer.

Owen, T., Maillard, J.-P., de Bergh, C., and Lutz, B. L. (1988) Deuterium on Mars: the abundance of HDO and the value of D/H, *Science*, 240, 1767–1770.

Ozima, M., and Nakazawa, K. (1980) Origin of rare gases in the Earth, *Nature*, 284, 313–316.

Parker, T. J., Saunders, R. S., and Schneeberger, D. M. (1989) Transitional morphology in the west Deuteronilus Mensae region of Mars: implications for modification of the lowland/upland boundary, *Icarus*, 82, 111–145.

Parker, T. J., Gorsline, D. S., Saunders, R. S., et al. (1993) Coastal geomorphology of the Martian northern plains, *J. Geophys. Res.*, 98, 11061–11078.

Pavlov, A. A., Brown, L. L., and Kasting, J. F. (2001) UV shielding of NH_3 and O_2 by organic hazes in the Archean atmosphere, *J. Geophys. Res.*, 106, 23267–23288.

Pepin, R.O. (1991) On the origin and early evolution of terrestrial planet atmospheres and meteoritic volatiles, *Icarus*, 92, 2–79.

Perron, J. T., Mitrovica, J. X., Manga, M., et al. (2007) Evidence for an ancient Martian ocean in the topography of deformed shorelines, *Nature*, 447, 840843.

Pestova, O. N., Myund, L. A., Khripun, M. K., and Prigaro, A. V. (2005) Polythermal study of the systems $M(ClO_4)_2$-H_2O (M^{2+} = Mg^{2+}, Ca^{2+}, Sr^{2+}, Ba^{2+}), *Russian J. Appl. Chem.*, 78, 409–413.

Petty, G. W. (2006) *A First Course in Atmospheric Radiation*, 2nd edition, Sundog, Madison WI.

Phillips R. J., Zuber, M. T., Solomon, S, C., et al. (2001) Ancient geodynamics and global-scale hydrology on Mars, *Science*, 291, 2587–2591.

Phillips, R. J., Davis, B. J. Tanaka, K. L., et al. (2011) Massive CO_2 ice deposits sequestered in the south polar layered deposits of Mars, *Science*, 332, 838–841.

Pieri, D. C. (1980) Geomorphology of Martian valleys, Ph.D. dissertation, Cornell University.

Pierrehumbert, R. T. (2010) *Principles of Planetary Climate*, Cambridge University Press, Cambridge, UK.

Pierrehumbert, R. T., and Erlick, C. (1998) On the scattering greenhouse effect of CO_2 ice clouds, *J. Atmos. Sci.*, 55, 1897–1902.

Pollack, J. B. (1979) Climatic change on the terrestrial planets, *Icarus*, 37, 479–553.

Pollack, J. B., Kasting, J. F., Richardson, S. M., and Poliakoff, K. (1987) The case for a wet, warm climate on early Mars, *Icarus*, 71, 203–224.

Postawko, S. E., and Kuhn, W. R. (1986) Effect of the greenhouse gases (CO_2, H_2O, SO_2) on Martian paleoclimate, *J. Geophys. Res.*, 91, D431–D438.

Poulet, F., Bibring, J.-P., Mustard, J. F., (2005) Phyllosilicates on Mars and implications for early Martian climate, *Nature*, 438, 623–627.

Ramirez, R. M., Kopparapu, R. K., Zugger, M. E., et al. (2014) A CO_2–H_2 greenhouse for early Mars, *Nature Geosci.*, 7, 59–63, doi:10.1038/ngeo2000.

Ribas, I., Guinan, E. F., Güdel, M., and Audard, M. (2005) Evolution of the solar activitiy over time and effects on planetary atmospheres. I. High-energy irradiances (1–1700 Å) *ApJ.*, 622, 680–694.

Righter, K., Pando, K., and Danielson, L. R. (2009) Experimental evidence for sulfur-rich Martian magmas: implications for volcanism and surficial sulfur sources, *Earth Planet. Sci. Lett.*, 288, 235–243.

Roach, L. H., Mustard, J. F., Lane, M. D., et al. (2010a) Diagenetic haematite and sulfate assemblages in Valles Marineris, *Icarus*, 207, 659–674.

Roach, L. H., Mustard, J. F., Swayze, G., et al. (2010b) Hydrated mineral stratigraphy of Ius Chasma, Valles Marineris, *Icarus*, 206, 253–268.

Romanek, C. S., Grady, M. M, Wright, I. P., et al. (1994) Record of fluid-rock interactions on Mars from the meteorite ALH 84001, *Nature*, 372, 655–657.

Rossow, W. B. (1978) Cloud microphysics: analysis of the clouds of Earth, Venus, Mars, and Jupiter, *Icarus*, 36, 1–50, doi:10.1016/0019-1035(78)90072-6.

Sackmann, I.-J., and Boothroyd, A. I. (2003) Our Sun. V. A bright young Sun consistent with helioseismology and warm temperatures on ancient Earth and Mars, *ApJ*, 583, 1024–1039, doi:10.1086/345408.

Sagan, C. (1977) Reducing greenhouses and the temperature history of Earth and Mars, *Nature*, 269, 224–236.

Sagan, C., and Chyba, C. (1997) The early faint Sun paradox: organic shielding of ultraviolet-labile greenhouse gases, *Science*, 276, 1217–1221.

Sagan, C., and Mullen G. (1972) Earth and Mars: evolution of the atmospheres and surface temperatures, *Science*, 177, 52–56.

Sasaki, S., and Nakazawa, K. (1988) Origin and isotopic fractionation of terrestrial Xe: hydrodynamic fractionation during escape of the primordial H_2–He atmosphere, *Earth Planet. Sci. Lett.*, 89, 323–334.

Scott, E. R. D., and Krot, A. N. (2005) Chondrites and their components, In *Treatise on Geochemistry*, vol. 1, *Meteorites, Comets and Planets*. Elsevier, Amsterdam.

Sefton-Nash, E., and Catling, D. C. (2008) Hematitic concretions at Meridiani Planum, Mars: their growth timescale and possible relationship with iron sulfates, *Earth Planet. Sci. Lett.*, 269, 365–375.

Segura, T. L., Toon, O. B., Colaprete, A., and Zahnle, K. (2002) Environmental effects of large impacts on Mars, *Science*, 292, 1977–1980.

Segura, T. L. Toon, O. B., and Colaprete, A. (2008) Modeling the environmental effects of moderate-sized impacts on Mars, *J. Geophys. Res.*, 113, doi:10.1029/2008/JE003147.

Segura, T. L. McKay, C. P., and Toon, O. B. (2012) An impact-induced, stable, runaway climate on Mars, *Icarus*, 220, 144–148, doi:10.1016/j.icarus.2012.04.013.

Sekiya, M., Hayashi, C., and Kanazawa, K. (1981) Dissipation of the primordial terrestrial atmosphere due to irradiation of the solar far-UV during T Tauri stage, *Progress in Theoretical Physics*, 66, 1301–1316.

Settle, M. (1979) Formation and deposition of volcanic sulfate aerosols on Mars, *J. Geophys. Res.*, 84, 8343–8354.

Smith, M. L., Claire, M. W., Catling, D. C., and Zahnle, K. J. (2014) The formation of sulfate, nitrate and perchlorate salts in the Martian atmosphere, *Icarus*, 231, 51–64.

Soto, A., Richardson, M. I., and Newman, C. E. (2010) Global constraints on rainfall on ancient Mars: oceans, lakes, and valley networks, In *41st Lunar and Planet. Sci. Conf.*, Abstract 2395, Houston, TX.

Soto, A., Mischna, M. A., and Richardson, M. I. (2011) Ancient Mars and atmospheric collapse. In *Fourth International Workshop on the Mars Atmosphere: Modelling and Observations*, Paris.

Squyres, S. W., Knoll, A. H., Arvidson, R. E., et al. (2006) Two years at Meridiani Planum: results from the Opportunity Rover, *Science*, 313, 1403–1407.

Squyres, S. W., Arvidson, R. E., Ruff, S., et al. (2008) Detection of silica-rich deposits on Mars, *Science*, 320, 1063–1067.

Squyres, S. W., Knoll, A. H., Arvidson, R. E., et al. (2009) Exploration of Victoria Crater by the Mars Rover Opportunity, *Science*, 324, 1058–1061.

Stephens, S. K. (1995a) Carbonate formation on Mars: experiments and models, Ph.D. Thesis, California Institute of Technology, Pasadena, 276.

Stephens, S. K. (1995b) Carbonates on Mars: experimental results. In *26th Lunar Planet. Sci. Conf.*, 1355–1356.

Stepinski, T. F., and O'Hara, W. J. (2003) Vertical analysis of Martian drainage basins. In *35th Lunar Planet. Sci. Conf.*, abstract 1659.

Stillman, D. E., and Grimm, R. E. (2011) Dielectric signatures of adsorbed and salty liquid water at the Phoenix landing site, Mars, *J. Geophys. Res.*, 116, doi:10.1029/2011JE003838.

Sugiura, N., and Hoshino, H. (2000) Hydrogen-isotopic compositions in Allan Hills 84001 and the evolution of the Martian atmosphere, *Met. Planet. Sci.*, 35, 373.

Sutter, B., Boynton, W. V., Ming, D. W., et al. (2012) The detection of carbonate in the Martian soil at the Phoenix Landing site: a laboratory investigation and comparison with the Thermal and Evolved Gas Analyzer (TEGA) data, *Icarus*, 213, 290–296.

Swayze, G. A., Ehlmann, B. L., Milliken, R. E., et al. (2008) Discovery of the acid-sulfate mineral alunite in Terra Sirenum, Mars, using MRO CRISM: possible evidence for acid-saline lacustrine deposits? *AGU Fall Meeting*, abstract P44A-04, San Francisco, CA.

Swenson, F. J., and Faulkner, J. (1992) Lithium dilution through main-sequence mass loss, *ApJ*, 395, 654–674.

Swindle, T. D., and Jones, J. H. (1997) The xenon isotopic composition of the primordial Martian atmosphere: contributions from solar and fission components, *J. Geophys. Res.*, 102, 1671–1678.

Swindle, T. D., Caffee, M. W., and Hohenberg, C. M. (1986) Xenon and other noble gases in shergottites, *Geochim. Cosmochim. Acta*, 50, 1001–1015.

Terasaki, H. Frost, D. J., Rubie, D. V., and Langenhorst, F. (2005) The effect of oxygen and sulphur on the dihedral angle between Fe–O–S melt and silicate minerals at high pressure: implications for Martian core formation, *Earth and Planet. Sci., Lett.*, 232, 379–392.

Thiemens, M. H. (2006) History and applications of mass-independent isotope effects, *Ann. Rev. Earth and Planet. Sci.*, 34, 217–262.

Thomas, P. C., James, P. B., Calvin, W. M., et al. (2009) Residual south polar cap of Mars: stratigraphy, history, and implications of recent changes, *Icarus*, 203, 352–375.

Tian, F., Kasting, J. F., Solomon, S. C. (2009) Thermal escape of carbon from the early Martian atmosphere, *Geophys. Res. Lett.*, 36, doi:10.1029/2006GL036513.

Tian, F., Claire, M. W., Haqq-Misra, et al. (2010) Photochemical and climate consequences of sulfur outgassing on early Mars, *Earth Planet. Sci. Lett.*, 295, 412–418.

Toner, J. D., Catling, D. C., and Light, B. (2013) Experimental formation and persistence of metastable aqueous salt solutions on Mars, in *Present-day Habitability of Mars Conference*, UCLA.

Toon, O. B., Pollack, J. B., and Sagan, C. (1977) Physical properties of the particles composing the Martian dust storm of 1971–1972, *Icarus*, 30, 664–696.

Turk-Chièze, S., Piau, L., and Couvidat, S. (2011) The solar energetic balance revisited by young solar analogs, helioseismology, and neutrinos, *ApJ*, 731:L29, doi:10.1088/2041-8205/731/2/L29.

Urata, R. A., and Toon, O. B. (2013) Simulations of the Martian hydrologic cycle with a general circulation model: implications for the ancient Martian climate, *Icarus*, 226, doi/10.1016/j.icarus.2013.05.014

Villanueva, G. L., Mumma, M. J., Novak, R. E., et al. (2008) Mapping the D/H of water on Mars using high-resolution spectroscopy, *3rd International Workshop on Mars Atmosphere: Modeling and Observations*, Williamsburg, VA, 9101.

Villanueva, G. L., Mumma, M. J., Novak, R. E., et al. (2015) Strong water isotopic anomalies in the Martian atmosphere: probing current and ancient reservoirs, *Science*, 348, 218–221, doi:10.1126/science.aaa3630

Vogel, N., Heber, V. S., Baur, H., et al. (2011) Argon, krypton, and xenon in the bulk solar wind as collected by the Genesis mission, *Geochim. Cosmochim. Acta*, 75, 3057–3071.

Wadhwa, M. (2008) Redox conditions on small bodies, the Moon, and Mars, *Rev. Mineral. and Geochem.*, 68, 493–510, doi:10.2138/rmg.2008.68.1

Wallis, M. K. (1989) C, N, O isotope fractionation on Mars: implications for crustal H_2O and SNC meteorites, *Earth Planet. Sci. Lett.*, 93, 321–324.

Wang, A., Korotev, R. L., Jolliff, B. L., et al. (2006) Evidence of phyllosilicates in Wooly Patch, an altered rock encountered at West Spur, Columbia Hills, by the Spirit Rover in Gusev Crater, Mars, *J. Geophys., Res.*, 111, doi:10.1029/2005JE002516.

Wänke, H., and Dreibus, G. (1994) Chemistry and accretion history of Mars, *Phil. Trans. R. Soc. Lond.*, A. 349, 285–293.

Watson, L. L., Hutcheon, I. D., Epstein, S., and Stolper, E. M. (1994) Water on Mars: clues from deuterium/hydrogen and water contents of hydrous phases in SNC meteorites, *Science*, 265, 86–90.

Webster, C. R., Mahaffy, P. R., Glesch, G. J., et al. (2013) Isotope ratios of H, C and O in CO_2 and H_2O of the Martian atmosphere, *Science*, 341, 260–263, doi:10.1126/science.1237961.

Weitz, C. M., Anderson, R. C., Bell III, J. F., et al. (2006) Soil grain analyses at Meridiani Planum, Mars, *J. Geophys. Res.*, 111, doi:10.1029/2005JE002541.

INDEX

References to figures are indicated with f.; references to tables are indicated with t.

spherical harmonics method (SHDOM) 114, 116–117
Chapman function, 114
modified gamma distribution, 90, 126, 133, 136–138
HITRAN, 121, 124, 238
GEISA, 121
Raman scattering, 119
T-matrix method, 128
droxtal, 90, 135–136
local thermodynamic equilibrium (LTE) 142
non-local thermodynamic equilibrium (non-LTE or NLTE) 142, 157, 455–456
radiatively active clouds, 254, 279, 554
Rayleigh distillation, 540–541
reference atmospheres, 65–69
residual caps. *See* polar caps
Richardson number, 174, 175, 189, 205, 239
rock abundance, 240, 305, 503
Rossby Number, 205
Rossby radius of deformation, 16, 204, 205, 213, 221

saltation. *See* dust lifting by wind stress
seasonal ice cap, 57, 89, 96, 97, 344, 350f. 11.11
seasons. *See* areocentric longitude
slope winds. *See* winds
SNC meteorites, 24, 63, 536, 537, 543, 544, 545
sol. *See* Martian solar day
solar cycle, 142, 147, 434–439, 452–453
solar energetic particles (SEP) 147, 382, 475
solar radiation, 5, 91, 111, 134, 142, 148, 151, 158, 175, 187, 229, 235, 238, 241, 259, 377, 379, 433, 497, 510, 511, 556
solar storms, 475–476, 488
solar wind, 433, 464–471, 473–481
solstitial pause, 252
spacecraft instruments
Compact Reconnaissance Imaging Spectrometer for Mars (CRISM), 30, 58, 60, 76, 87, 99, 116, 139, 140, 155, 298, 303, 339, 341, 343, 348, 351, 381, 390, 391, 409, 416, 534, 554
Context Camera (CTX), 30, 306, 380
Gamma Ray Spectrometer (GRS), 28, 59, 339, 379, 382, 384, 388, 510, 512, 545
High Energy Neutron Detector (HEND), 382
High Resolution Imaging Science Experiment (HiRISE), 30, 306, 380, 394
High Resolution Stereo Camera (HRSC), 28, 76, 99, 306, 380
Infrared Interferometer Spectrometer (IRIS), 20, 26, 43, 44, 53, 64, 76, 130, 156, 243, 244, 281, 297, 339
Infrared Radiometer (IRR), 380
Infrared Spectroscopy (IRS), 297, 382
Infrared Thermal Mapper (IRTM), 22, 76, 232, 259, 339
Mars Advanced Radar for Subsurface and Ionospheric Sounding (MARSIS), 347, 349
Mars Atmospheric Water Detector (MAWD), 22, 76, 139, 338, 340, 341, 346, 354, 357, 364
Mars Climate Sounder (MCS), 30, 45, 51, 53, 65, 77, 82, 85, 86, 97, 98, 132, 140, 150, 210, 223, 232, 243, 254, 261, 282, 298, 303, 328, 380, 384, 385
Mars Color Imager (MARCI), 29, 30, 61, 77, 82, 111, 136, 155, 275, 298, 306, 380, 417
Mars Horizon Sensor Assembly (MHSA), 45
Mars Orbiter Camera (MOC), 25, 76, 82, 84, 98, 192, 275, 278, 298, 303, 306, 313, 351, 379, 380, 383, 390
Mars Orbiter Laser Altimeter (MOLA), 25
Mini-TES, 31–32, 47, 156, 180, 183, 304, 341

Neutron spectrometer, see Gamma Ray Spectrometer
Observatoire pour la Minéralogie, l'Eau, les Glaces et l'Activité (OMEGA), 28–29, 58, 60, 76, 96, 97, 99, 139, 141, 157, 185, 217, 298, 303, 314, 339, 341, 347, 348, 349, 351, 352, 378, 381, 390, 392, 393, 418, 427, 446, 554
Phoenix LIDAR, 33, 77, 78, 88–89, 183–184, 302
Planetary Fourier Spectrometer (PFS), 29, 45, 58, 62, 141, 147, 157, 298, 341, 346, 381, 384, 422
Radiation Assessment Detector (RADS), 147
Rover Environmental Monitoring Station (REMS), 33, 182, 216, 258, 307
Sample Analysis at Mars (SAM), 24, 33, 58, 62, 64, 65, 413, 419
Shallow Radar (SHARAD), 30, 347
Spectroscopy for the Investigation of the Characteristics of the Atmosphere of Mars (SPICAM), 29, 298, 303, 341, 346, 381, 383, 415, 417, 418, 419, 426, 427, 435, 447–449, 466
Thermal Emission Imaging System (THEMIS), 28, 45, 55, 76, 82, 99, 139, 140, 232, 298, 306, 422
Thermal Emission Spectrometer (TES), 25, 26, 29, 30, 44, 46, 51, 53, 55, 65, 76, 82, 87, 94, 98, 133, 139, 140, 154, 156, 191, 223, 233, 243, 252, 269, 272, 276, 298, 310, 312, 338, 339, 341, 343, 379, 380, 390, 392, 415
spacecraft missions, 230
Mariner 4, 20
Mariner 9, 20
Mars 96, 24
Mars Atmosphere and Volatile Evolution (MAVEN), 30–31
Mars Climate Orbiter (MCO), 9, 24, 29, 30
Mars Exploration Rovers (MER), 31–32
Mars Express (MEX), 28–29
Mars Global Surveyor (MGS), 25–28
Mars Observer, 1, 2, 9, 24, 30, 145
Mars Odyssey, 28
Mars Orbiter Mission (MOM), 2
Mars Pathfinder, 31
Mars Polar Lander, 48, 118, 122
Mars Reconnaissance Orbiter (MRO), 29–30
Mars Science Laboratory (MSL), 33
Nozomi, 24
Phobos 2, 23
Phoenix, 32–33
Viking, 20–23
standard mean ocean water (SMOW) 65, 539–541
stationary eddies
basic theory, 280–281
fundamental aspects, 279–280
modeling, 283–285
observations, 281–283
storm zones, 274–275
stream function, 252
sublimation, 12, 88, 92, 96, 343, 350
sulfates (gypsum, jarosite) 11, 354, 526, 533
superrotation, 249
surface frost, 32, 183, 344, 391
surface heat flux, 173, 186, 194, 324
surface pressure measurements
MSL, 33, 57, 231, 244, 307
from orbit, 20, 28, 185
Pathfinder, 57
Phoenix, 57
telescopic, 4, 20, 57, 60
Viking Landers, 6, 23, 57, 231, 244
surface properties, 234–235
surface roughness length, 173, 185, 239, 240, 316, 356, 503